数学教育研究手册

第三册

学生、教师和学习环境

人民教育出版社
·北京·

主编：蔡金法

策划：蔡金法　徐斌艳

译者：孙　伟　刘启蒙　陈　倩　于文华　姚一玲　张　波　张晋宇　吴颖康　韩继伟
　　　聂必凯　顾非石　贺真真

审稿：蔡金法

校对：李　俊　蔡金法　姚一玲　聂必凯　李旭辉　吴颖康

《数学教育研究手册》（Compendium for Research in Mathematics Education）一书的中英文本，经全美数学教师理事会（National Council of Teachers of Mathematics）授权，由人民教育出版社出版，在全世界范围内发行

Copyright©2017 by The National Council of Teachers of Mathematics, Inc

图书在版编目（CIP）数据

数学教育研究手册. 第三册，学生、教师和学习环境 /（美）蔡金法主编. —北京：人民教育出版社，2020.5
ISBN 978–7–107–34487–9

Ⅰ.①数… Ⅱ.①蔡… Ⅲ.①数学教学—手册 Ⅳ.①O1–62

中国版本图书馆 CIP 数据核字（2020）第 083100 号

数学教育研究手册　第三册　学生、教师和学习环境

出版发行	人民教育出版社
	（北京市海淀区中关村南大街 17 号院 1 号楼　邮编：100081）
网　　址	http://www.pep.com.cn
经　　销	全国新华书店
印　　刷	北京天宇星印刷厂
版　　次	2020 年 5 月第 1 版
印　　次	2020 年 6 月第 1 次印刷
开　　本	890 毫米 × 1240 毫米　1/16
印　　张	34.25
字　　数	1039 千字
印　　数	0 001~3 000 册
定　　价	90.00 元

版权所有·未经许可不得采用任何方式擅自复制或使用本产品任何部分·违者必究
如发现内容质量问题、印装质量问题，请与本社联系。电话：400-810-5788

谨将此丛书

献给

张奠宙先生

（1933—2018）

及其他

所有数学教育的前辈们

中文版序言

蔡金法

从正式接受全美数学教师理事会（NCTM）的邀请，策划并主编《数学教育研究手册》到英文版正式出版，历时五年六个月。然而，从策划翻译到交稿又花费了整整三年六个月，而且从目前的交稿到出版，又需要一些时日。这样从主编英文版到中文版的出版一共花了将近十年的时间。非常荣幸能够主持这么大的一个项目，也许这是一个人一辈子难得能花如此长时间完成的一件事了，所以我非常欣慰能在人民教育出版社出版这套《数学教育研究手册》的中英文对照版本。在手册即将付梓之际，有很多的感慨、很多的感恩，也有很多的期待，在这里写上几句，也希望读者在阅读的时候，能够明白翻译此著作的重要性。

翻译的初衷

翻译工作实际上是一个非常繁重的工作，这不仅涉及知识、内容，还需要背景。所以，准备这样一本《数学教育研究手册》需要很大的决心。那为什么还要做此事呢？具体来说，翻译这本手册主要有以下三方面的考虑。

第一，这本手册代表了目前最新的、最具综合性和前沿性的有关数学教育研究的过去、现在和将来。与以往的手册不同之处在于，这本手册中，我不仅要求每一章的作者们要把最前沿的研究发现总结出来，更更要的是要求每一章作者能够从历史的角度看待自己所写作的这一领域的有关内容，尤其是研究方法的使用是如何促使该领域得到突破性进展的。也就是说，要特别注意某一研究领域的历史进程，同时把应用于该领域中的研究方法变迁的历史进程详细地描述出来。同时在每一章中，我要求作者对今后五年，甚至十年的研究方向有清晰的阐述。毕竟阅读中文更为容易，所以翻译此著作就是为了把最前沿的思想介绍给读者。

第二，在过去几年中，我受邀指导了几位中国高校的博士生的论文写作，也参与了一些博士生的开题报告，发现他们大都会引用许多国外的文献，指出某某这么说，某某那么说。但基于我所了解的内容，我很快会发现，他们的不少引用都不那么精确。后来我才了解到，他们很多引用的都是二手资料。本来就存在一个翻译的问题，而翻译如果不准确的话，像这样传来传去，就会越传越不准确了。因此，这套中英文对照版手册一方面是希望把就目前来说这样一个相对最为前沿的内容给翻译出来，来帮助我们的学者了解数学教育研究方面的最新动态。

另一方面，我也希望他们能够了解和引用到最准确的内容。

第三，我国的数学教育已经受到世界各国的关注，我国的学者也在国际数学教育的舞台上越来越活跃，而且第十四届国际数学教育大会将于2021年在上海举办。这些都标志着我国数学教育的发展和地位。然而，我国数学教育的学者在国际上的声音和我国数学教育的受重视程度是不相称的。我国本土学者很少在主流的国际数学教育研究杂志上发表文章，且在大型国际会议上受邀做主题报告的比例也相对较小。即使受邀，往往只是可以很好地完成报告，但在问答阶段因不太能听懂英文而很难与国际学者有深入交流。总体上讲，我个人认为这与我们英文表达上的欠缺不无关系。然而，从研究本身而言，若是没有语言障碍，我国本土学者的研究完全可以在国际主流数学教育研究杂志上发表。因此，希望这套中英文对照版研究手册能够帮助我们的学者在英语交流上有所提高。

几点说明

我要就此中英文对照版本做几点说明，以协助读者更好地阅读。

首先，因为这套书是中英对照出版的，所以考虑到阅读的舒适性，将英文版的五部分内容划分为四册：第一、二部分为第一册；第三部分为第二册；第四部分为第三册；第五部分为第四册。同时，在单位的使用、引文的规范，以及格式上都尽量保持原貌，而没有改成国际单位或者完全遵循中文习惯。例如，英文用"mile"，在中文版中就保持了"英里"。为了便于阅读，我们把名字都统一翻译成了中文，但为了便于读者查找原文文献，在行文中的引文出处以及每章后面的参考文献都是按照原文列出的。

其次，建议读者在阅读每一章时，要有耐心，不仅了解研究的结论、研究方法、研究方向，而且要思考每一章都是用怎样一个构架写作的，这个构架的建立是非常值得学习的。

最后，由于语言等各方面的原因，原文作者很少引用中文发表的文章，建议读者在阅读每一章时可以有意识地做一些梳理，有哪些中文发表的研究可以纳入该章节中。若是有类似的研究，看其在方法上和结论上是否与该章节一致。这样的反思与挖掘会有助于我们的研究与世界接轨，也有助于我们的研究在英文数学教育研究的主流杂志上发表。

致 谢

出版这样的书籍，需要投入的心力以及各方面的支持都多到难以用语言形容，在此我想对大家的支持献上我深深的谢意和真诚的祝福。

感谢全美数学教师理事会（NCTM）愿意出版这样大型的专业性极强的著作，并给我们版权由人民教育出版社出版中英文对照版。

由衷感谢人民教育出版社愿意出版这一著作，特别要感谢李海东先生和王嵘女士非常快速地促成出版事宜。这其中也得到了杨刚、章建跃、王永春及周小川等人教社朋友的大力支持。他们在教材编写方面已经非常繁忙了，出版此书无疑会增加他们的负担。为此，我也要为人教社中学数学室在推进数学教育发展上的使命感深表敬意。特别感谢四位师妹，王嵘、张艳娇、王翠巧和宋莉莉担任责任编辑，她们一如既往地保持着人教社高质量出版的传统。在这长达近四年的工作中，王嵘师妹是我的联络人，我们彼此理解、支持，时不时给一句暖心的话，例如，"这类书籍的出版真的是靠对专业的一腔热情，很令人钦佩"。她的这句话也道出了我们合作出版此书的初衷和坚持的动力。

就翻译工作本身而言，我觉得它也是一个人才培养的过程。我也很希望能通过这样的工作，让我们参与翻译的学者有所收获。尽管如此，他们的贡献也同样是非常重要和伟大的，他们愿意参与到这个翻译工作中，是辛苦的，是不易的，所以要特别感谢所有参与翻译的学者及朋友。你们辛苦了！在这里要特别提到徐斌艳教授在前期所做的协调工作，如果没有她，我们也无法很快地组建这个翻译团队，谢谢徐老师！

为了确保翻译的质量，我们不仅协调了译者之间的互相校对，而且又花了大力气进行校对。老实说，校对工作比我原先想象的要挑战得多很多，有些章节甚至校对过7遍。在因为校对而担心不能按时交稿之际，我还得到了"天使"和"天兵天将"的协助。李俊教授就是我的天使，她协助我校对了近一半的译稿，她的帮助用雪中送炭来形容一点都不过分。李俊，一切的感谢都在不言中。"天兵天将"还包括聂必凯、李旭辉和吴颖康，他们每人协助我至少校对了三章。另一位"天使"是姚一玲，她当时在特拉华大学做博士后研究，不仅协助我在最后校对工作中的所有中文录入，还协助我一起商讨决定一些疑难词句的翻译，有时候一句话的精确翻译可能要花上几个小时。需要指出的是，以下老师们也协助我对个别章节做了前期的校对，在此一并致谢。他们是莫雅慈、张侨平、梁淑坤、丁美霞、贾随军、黄荣金、李业平、李小保、陈算荣、王闯及张玲。

为了确保中文版的质量，我还邀请以下老师对中文版进行了通读，他们是鲍建生、曹广福、曹一鸣、陈婷、代钦、董连春、顾泠沅、何小亚、胡典顺、金美月、孔企平、李建华、李士锜、刘坚、吕传汉、吕世虎、马云鹏、孙晓天、涂荣豹、汪晓勤、王光明、王尚志、夏小刚、叶立军、喻平、张红、张辉蓉、张维忠、张蜀青、章建跃、郑毓信、朱文芳。他们无需阅读英文版，只需从头至尾通读中文版的某一章，将那些拗口、表达不当处用批注的形式加以修改、重述，再将看不明白的标记出来，最后再由我来定稿。他们中的所有人都爽快地答应并认真地阅读给自己的章节。如此一来，本书质量得以进一步的提升。非常感谢他们能助力本书成为经典。也非常感谢他们对此著作给予的高度正面的评价，下面这样的反馈是很典型的："想不到你在做这么辛苦的工作，但是很值得做""内容很丰富，阅读的过程也学到许多东西，有些观点很有启发。可以为以后的研究参考。期待这本书的出版，将会对数学教育提供很有价值的参考。谢谢！"。

此著作被国家出版基金管理办公室确定为2018年国家出版基金项目。在此感谢国家出版基金管理办公室的大力支持！

在2019年岁末完成简短的序言，也是为2019年画上一个圆满且感恩的句号。期盼2020年中国数学教育再添辉煌，希望此著作可以为此做出点滴贡献。

目录

前言 2

21 语言多元化和数学：第二语言、双语和多语言学习者 2

22 种族与数学教育 44

23 数学教育研究中的身份 98

24 性别与数学的研究 120

25 数学参与的复杂性：动机、情感和社会互动 150

26 多模式的物质思维：数学教育中的具身理论 210

27 关注流派和构建任务：针对数学课堂话语文献的批判性分析 252

28 K-12数学教学核心实践的研究 332

29 数学专业发展研究 380

30 课程研究的若干重要问题：基于证据的发现和今后的研究方向 438

31 学习数学的技术 490

前言

为了知识积累与人才培养的研究手册

全美数学教师理事会（NCTM）将在2020年举办100周年庆祝大会。同时，《数学教育研究杂志》（JRME）也将庆祝其办刊50周年以来一直致力于发表高质量的研究以解决数学教育中的重大问题。此研究手册的出版则是另一个重要的里程碑时刻。多年来，NCTM出版了许多书籍和杂志，以数学教育实践者为目标读者居多。而且，NCTM非常强调知识的积累和人才的培养，它一直并将继续是数学教育每一领域的原创性研究的积极倡导者。事实上，包括JRME期刊在内，NCTM还出版了JRME的专刊系列以及这本《数学教育研究手册》。

根据字典的释义，"手册（compendium）"一词指的是"关于某一特定领域的简洁但详尽的内容的汇总集中"，这种汇总集中通常是一种"系统性汇集"（Hobson, 2004, 第84页）。该定义中有三个特别值得注意的关键词：系统性汇集、简洁、详尽。此研究手册就是对数学教育研究内容系统、简明而详尽的汇总。近100位作者利用自己的专业知识努力将该领域的知识提炼成可用的资源，为培养学生的数学学习提供了最佳、最关键性的证据。该研究手册代表了NCTM的另一目标，即努力给所有人提供最好的内容。

请读者注意，手册名称用"for"而非"of"是因为，该手册不仅仅是对已有研究文献的静态收集，而是将这些研究作为资源应用"于（for）"更高层次的数学教育。它是关于数学教育研究的汇总，更是用于研究数学教育的重要资源。这也反映了NCTM对数学教育领域促进学生数学学习的前沿研究的支持。

研究手册的结构

该研究手册包括五个主要部分。第一部分是基础，由六章组成，作者们主要考察了数学教育研究中的各种基础性内容，如研究的本质、研究与实践的结合，以及研究资金与政策的作用。此外，还有一些章节主要关注学习（如学习进程）、教学，以及运用不同的理论视角推动前沿性的研究等方面。

第二部分是方法，包括三章，主要关注定性与定量研究方法，以及设计研究方法。因为不同的研究问题会涉及选用不同的研究方法，因此这部分的几个章节主要用于帮助读者理解什么样的研究方法适用于什么样的研究问题。

第三部分是数学过程和内容。这部分包括了与数学有关的过程，如推理、数学建模，以及学生的理解。另外，这部分的研究还考察了数学教育研究的最新进展，涵盖了从早期的数到微积分及微积分后继内容领域。有两章讨论了代数的相关内容，涉及从小学到高中的相关代数概念和思维内容。除此之外，还有章节讨论了有关测量、几何、概率与统计，以及微积分的教与学。

第四部分是关于学生、教师和学习环境的。前六章涉及与多语言、种族、身份认同、性别、数学参与，以及具身认知等有关的研究。接下来三章考察了课堂对话、数学教学中的核心实践，以及教师专业学习的研究。最后两章讨论了能够潜在支持学生学习和教师教学的相关领域的研究：课程与技术。

Preface

A Research Compendium for Building Knowledge and Capacity

The National Council of Teachers of Mathematics (NCTM) will mark its 100th anniversary in 2020. At the same time, the *Journal for Research in Mathematics Education* (JRME) will celebrate its 50th year of fulfilling its charge to publish high-quality research dealing with significant problems in mathematics education. The publication of the *Compendium for Research in Mathematics Education* is an important milestone leading to these significant anniversaries. Over the years, NCTM has published many books and journals, primarily for practitioners, including research-based resources. In addition, NCTM has been, and continues to be, a strong advocate for original research in all aspects of mathematics education, placing great emphasis on building knowledge and capacity in the field. Indeed, alongside JRME, NCTM publishes the JRME monograph series and now this *Compendium for Research in Mathematics Education*.

According to the dictionary, the word "compendium" refers to "a collection of concise but detailed information about a particular subject" that has been "systematically gathered" (Hobson, 2004, p. 84). Three key terms in this definition deserve special attention: systematically gathered, concise, and detailed. This research compendium is precisely a systematically gathered, concise, and detailed volume about research in mathematics education. The nearly 100 authors who have contributed their expertise to its writing have striven to distill the knowledge of the field into a resource that provides the best, most critical evidence about what is known about fostering students' learning of mathematics. This research compendium represents another investment of NCTM in its effort to achieve excellence for all.

Please note that I deliberately chose to use "for" rather than "of" in the title, because the compendium is not simply a static collection but rather a resource to be used "for" advancing research in mathematics education. Although it is a compendium "of" research in mathematics education, it is to be used "for" research in mathematics education. This, too, reflects NCTM's charge to support cutting-edge research in mathematics education that can advance students' learning of mathematics.

Structure

The compendium consists of five major sections. The first section, Foundations, comprises six chapters in which the authors examine a variety of fundamental issues in mathematics education research, such as the nature of research, linking research and practice, and the role of funding and policy. Moreover, there are chapters focused on learning (e.g., trajectories), teaching, and pushing the frontiers of research by adopting different theoretical lenses.

The second section, Methods, contains three chapters focusing on qualitative and quantitative research methods, as well as design research. Because research questions circumscribe choices of research methods, the chapters in this section are designed to help readers understand what kinds of research questions each set of methods is able to appropriately address.

Section 3 is called Mathematical Processes and Content. This section covers research on mathematical processes, such as proof and mathematical modeling, as well as students' sense making. Also, this section examines the state-of-the-art in mathematics education research for mathematical content areas ranging from early number through postcalculus mathematics. Algebra is discussed in two chapters, which together span algebraic concepts and thinking in elementary school through high school. In addition, this section contains chapters that focus on teaching and learning measurement, geometry, probability and statistics, and calculus.

The title of Section 4 is Students, Teachers, and Learning Environments. The first six chapters of this section involve research related to language diversity, race, identity, gender, mathematical engagement, and embodiment. The next three chapters deal with research on teaching and teacher learning with a focus on classroom discourse, core practices in mathematics teaching,

最后一部分讨论了未来需要进一步研究的主题。这部分并不是对某个领域文献的系统性的回顾，而是讨论了某些研究领域所面临的挑战并提供了一些前瞻性的观点。有些章节讨论了与数学教育交叉的正在快速发展中的研究领域，如教育神经科学、天才生以及特殊学生的教育。鉴于对未来研究者能力发展的重要性，我们也纳入了一些数学教育领域博士生学习的研究。其他章节则讨论了日渐引起公众注意的领域，包括评估、社区大学的数学教育，以及非正式环境下的数学教育。

研究手册的特点

该研究手册提供了最全面和最新的调查研究，这些研究是数学教育中最优秀的、最新的，以及对关键性内容分析的研究。除了这些共有的特点之外，该研究手册还有三个独有的特点。

第一个特点是从**历史**发展的视角来看待数学教育的研究问题，尤其是文章中包含了关键性的里程碑式的研究主题。大部分章节都是基于过去已有的研究来追踪考察现在我们所知道的内容，并根据已有知识来展望未来（如下一步我们应该做什么）。

第二个特点是该研究手册非常强调**方法论**的重要性。尽管有三章对数学教育中一些关键的研究方法进行了详细的讨论，但几乎每一章都涉及了方法论问题及其在特定研究领域中的内涵，其中包括我们常用的方法、方法对知识产生的贡献，以及这些方法是如何发展的。

第三个特点是与NCTM出版刊物一致的理念，即努力为数学教育提供**国际化的**研究视角。尽管NCTM位于美国，但我一直努力纳入更多具有国际性视角的研究。首先，我邀请了很多国际上的作者来一起合作完成本手册。其次，尽管每一章的内容都主要关注的是美国的数学教育，但作者们都尽最大努力将国际范围上的研究发现、问题、视角，以及未来研究方向纳入自己的文章当中。

如何阅读手册的各章

在本书的每一章中，读者可以发现，作者们着重介绍了相应领域中最新且最为主要的研究发现。特别地，作者们对于数学教育每个领域中的研究现状及未来走向都给出了他们的见解。

当然，阅读本书也是极富挑战性的。它们讨论的是领域内长期以来较有难度的一些主题，综合分析的是庞大而复杂的研究体，并且为这些研究领域提出了较为深刻的见解。在审核和编撰本书的每一章时，我的头脑中始终萦绕着这样五组问题：

本章是如何组织起来的，为何以这种方式来组织，以及所选择的这种组织结构是否有助于读者领会作者关于这章的观点和内容？

概念框架是什么，本章是如何选择这一框架的，以及这个概念框架是否能让相关的研究主题或领域产生出新的见解？

作者是如何确定哪些研究需要包含在内（进行综合分析）的，以及哪些研究可以作为边缘内容（在本章中不直接提及）？

本章所述领域使用的研究方法这些年来是如何演变的？

本章所关注的特定研究领域中未来的研究方向可能有哪些，以及作者们是如何诠释这些未来方向的？

我发现这组问题是相当有用的，不仅仅是对于这本研究手册的主编，对每一位读者来说亦是如此。因此，我分享这些问题，希望读者可以通过它们对作者所要阐述的内容进行深层次的挖掘。

对未来的展望

现在，我终于可以将这本研究手册呈现给读者和教育研究者了，尤其是数学教育研究者。我非常荣幸能够被邀请主编如此重要的一本手册。我对这本高质量研究手册的问世感到非常满意。我希望，这不只是NCTM出版的第一本研究手册，可能在未来的10至15年还会有更成功的第二版、第三版……研究手册得以出版。这些手

and teacher professional learning. The section ends with two chapters about research in areas that have great potential to support students' learning and teachers' teaching: curriculum and technology.

The final section of the compendium is Futuristic Issues. Rather than a systematic review of literature in an area, each chapter in this section is a short essay discussing challenges and providing forward-looking views about the research in an area. Some chapters discuss burgeoning areas of research where mathematics education intersects with other fields, such as educational neuroscience and gifted and special education. Because of the importance of building capacity for the future of the field, we also include a chapter on doctoral studies in mathematics education. Other chapters in this section consider areas that are increasingly prominent in the research community public eye, including assessment, mathematics education in community college settings, and mathematics education in informal settings.

Features

This compendium provides the most comprehensive and up-to-date survey of the best research, new developments, and critical analyses of issues in mathematics education. In addition to these common features, this compendium has three unique features that cut across its chapters.

The first is the synthesis of research with an eye to the *historical* development of a research topic, in particular covering historical milestones of the research topic. In most chapters, the research in a given area is traced to examine what is now known in light of what has been done in the past, as well as to look into the future (i.e., what should we do next?) given what is now known.

The second unique feature in this compendium is that, in addition to incorporating a historical perspective, we have placed more emphasis on the *methodological* perspective. Although there are three chapters in the compendium specifically devoted to discussing key research methods in mathematics education, almost every chapter in the compendium addresses methodological issues and implications in a given area of research, including the methods typically used, the contribution of methodology to the generation of knowledge, and how methodology has advanced. In a number of chapters, the authors have also included detailed descriptions of the methods they used to select research articles to be included in their review of research and how they synthesized the research findings.

The third unique feature for this publication of NCTM is a concerted effort to provide an *international* perspective on research in mathematics education. Although NCTM is situated in the United States, I have striven to include international perspectives whenever possible. First, I have tried to accomplish this is by inviting international authors to coauthor a number of the chapters. Second, in each chapter, even those primarily focusing on the United States, the authors have worked very hard to situate findings, issues, perspectives, and future directions in an international context.

Reading a Compendium Chapter

In each chapter of this compendium, readers will find the most recent, major findings that the authors have highlighted from their area. In particular, the authors have provided insights on where the field is and where it is going in each area of study in mathematics education.

However, compendium chapters can be challenging to read. They tackle long-standing, difficult issues for the field; synthesize large, complex bodies of research; and provide deep insights into an area of research. It can be helpful to approach them with some orienting questions. When reviewing and editing each chapter in the compendium, I kept the following five sets of questions in mind:

How is the chapter structured, why is it structured that way, and how does the chosen structure help the reader grasp the authors' points and the content of the chapter?

What is the conceptual framework, how was it chosen for the chapter, and how does the conceptual framework allow for new insights into the topic or area of study?

How did the author or authors decide what needed to be included (synthesized) and what could be set to the margins (not mentioned explicitly in the chapter)?

How have the research methods evolved over the years for the specific research area reported in the chapter?

What are possible future directions for research in the specific research area in the chapter, and how do the authors elaborate on these future directions?

I found this set of questions extremely helpful, not just as the editor of this compendium, but also as a reader of each chapter. Therefore, I offer them as tools to help the reader dig deeply into what the authors are offering.

Looking to the Future

And now, finally, I am so pleased to present this compendium to the readers, to the educational research community in general, and to the mathematics education community in particular. I was honored to be invited to edit such a hefty volume. I have had a great sense of

册将会指导未来的数学教育研究学者，并赋予他们更艰巨的使命。因此，我觉得通过对编撰本研究手册的过程的思考来结束前言部分，将会对读者更有帮助。

过去5年中，我参与写作了三本研究手册或章节，包括本套研究手册中的一章（Lloyd, Cai, & Tarr, 2017）、《国际数学教育研究手册（第三版）》的一章（Cai & Howson, 2013），以及《数学教育心理学研究手册（第二版）》的一章（Santos & Cai, 2016）。作为这本手册的主编和其中一章的作者，我也反思了如何编撰一本研究手册或写作手册中的一章。从编撰这本手册的经历来看，我发现其实可以有不同的方式。一种是自下而上的方式，即系统地检索期刊或书籍，然后利用概念框架来分析这些文献。其中，课堂话语那一章（Herbel-Eisenmann, Meaney, Bishop, & Heyd-Metzuyanim, 2017，本套书）就做了这样的尝试。许多这一类型的章节更多都是基于数据来展开的。事实上，课堂话语这一章呈现了作者选择和编码相关文献的方法。另一种则是自上而下的方式，即在初始建立的概念框架基础上，有组织性地综合这些相关的研究发现。手册中关于课程研究那一章（Lloyd等，2017，本套书）的写作方式便是如此。当然，也有很多研究使用的方式是处于二者之间的。

那么，编撰一本手册的最佳方式是什么呢？尽管我并不想说每一章的写作只有唯一最佳的方式，但很明显的是，无论作者选择哪一种方式，他们的工作都不仅仅只是对已有文献的综述和总结。相反地，作者们必须要努力找到一种新的方式来建构这些文献（及整章）并给出这一领域的一个轮廓，以帮助读者理解该领域中有哪些内容，以及我们仍然不了解的内容有哪些。这些新的文章结构能够给予我们启发，去理解什么是研究的本质、方法论的发展，以及未来的研究方向。当然，这些新的结构也必须要基于合理的概念框架。

编撰这本研究手册是重要且关键的一步，它能够让我们总结数学教育到目前为止所取得的进展和积累的知识。这本手册不仅仅呈现了对目前数学教育研究领域关键问题的最佳理解，更是为了启发研究者如何才能更有效推进这些领域取得发展而作。在这个过程中，研究者可以对知识领域的状态进行反思，并阐释这些状态是如何建构自己的反思的，这些都将有助于发展自己推进该领域知识取得进展的能力。因此，我真诚地希望并相信，作为NCTM出版的第一本《数学教育研究手册》，它将能够很好地服务于每一位参与数学教育的人。

蔡金法
美国特拉华大学

（姚一玲译，杭州师范大学）

satisfaction editing such a high-quality research compendium and now have a big sense of relief having completed it. That said, this is, I hope, only the first in the research compendium series to be published by NCTM. Possibly every 10 to 15 years we will have a successor—a second, third, and fourth research compendium. Those volumes will be shepherded and authored by future mathematics education scholars. Looking toward this future, I thought it might be helpful to conclude this preface by providing some food for thought about the process of writing a compendium chapter.

Over the past 5 years, I have coauthored three research compendium or handbook chapters, including one in this compendium (Lloyd, Cai, & Tarr, 2017, this volume), one in the *Third International Handbook of Mathematics Education* (Cai & Howson, 2013) and one in the *Second Handbook of Research on the Psychology of Mathematics Education* (Santos & Cai, 2016). As both the editor of this compendium and a chapter coauthor, I have been reflecting on how one goes about writing a compendium or handbook chapter. From my experiences editing this compendium, I have observed that there are different ways of writing a compendium chapter. One way takes a bottom-up approach, which is to systematically search journals or books and then analyze related works using a conceptual framework. The compendium chapter on discourse (Herbel-Eisenmann, Meaney, Bishop, & Heyd-Metzuyanim, 2017, this volume) represents this type of effort. Chapters developed this way are more data-driven. Indeed, the discourse chapter includes a specific section on the methods the authors used to select and code the relevant articles. At the other end of the spectrum is the top-down approach, which involves first building a conceptual framework, followed by synthesizing the research findings based on the organization presented by the conceptual framework. The compendium chapter on curriculum studies (Lloyd et al., 2017, this volume) represents such an effort. There are also many approaches that fall somewhere between these two.

What is the best way to write a compendium chapter? Although I would not claim that there is a single best way to write such a chapter, it is clear that, no matter which method an author chooses, writing a research compendium chapter extends far beyond reviewing the literature to gather and summarize what has been done. Instead, authors must strive to find a novel way to structure the literature (and the chapter) to help readers understand and navigate the contours of what the field does and does not know for the specific area of research. That novel structure provides insights about the nature of research in a given area and can also serve to show methodological advances and future directions for research in that area. However, the novel structure must be based on a sound conceptual framework.

Assembling a research compendium like this is an important and critical step so the field of mathematics education can take stock of progress and the knowledge that has been accumulated thus far. This volume encodes not only the field's best understandings of the critical issues in mathematics education research today, but also its insights about how researchers can most effectively move forward as a field. This process, in which researchers both reflect on the state of knowledge and explain how they have come to structure their reflections, serves to build the field's capacity to continue to push the frontiers of its knowledge. It is, therefore, my sincere hope and belief that this first *Compendium for Research in Mathematics Education* will serve everyone involved in mathematics education well.

JINFA CAI
University of Delaware, United States

References

Cai, J., & Howson, A. G. (2013). Toward an international mathematics curriculum. In M. A. Clements, A. Bishop, C. Keitel, J. Kilpatrick, & F. K. S. Leung (Eds.), *Third international handbook of mathematics education* (pp. 949–978). New York, NY: Springer.

Herbel-Eisenmann, B., Meaney, T., Bishop, J., & Heyd-Metzuyanim, E. (2017). Highlighting heritages and building tasks: A critical analysis of mathematics classroom discourse literature. In J. Cai (Ed.), *Compendium for research in mathematics education* (pp. 722–765). Reston, VA: National Council of Teachers of Mathematics.

Hobson, A. (Ed.). (2004). *The Oxford dictionary of difficult words*. Oxford, United Kingdom: Oxford University Press.

Lloyd, G. M., Cai, J., & Tarr, J. E. (2017). Issues in curriculum studies: Evidence-based insights and future directions. In J. Cai (Ed.), *Compendium for research in mathematics education* (pp. 824–852). Reston, VA: National Council of Teachers of Mathematics.

Santos, L., & Cai, J. (2016). Curriculum and assessment. In A. Gutiérrez, G. Leder, & P. Boero (Eds.), *Second handbook of research on the psychology of mathematics education* (pp. 153–186). Rotterdam, The Netherlands: Sense.

21 语言多元化和数学：第二语言、双语和多语言学习者

理查德·巴韦尔
加拿大渥太华大学
朱迪特·N.莫谢科维奇
美国加利福尼亚大学圣克鲁斯分校
马穆克戈西·塞塔蒂·费肯
南非开普敦大学
译者：孙伟
美国陶森大学

很少有一个数学课堂中所有的学生和老师都只讲同一种语言，所有社会都会有一些语言多元化的特征。语言多元化不仅包括不同语言的存在，而且包括同一语言的不同变体存在，例如西班牙语中不同的方言以及不同地域所讲的英语。表示语言多元化的通用词汇包括多语言（指说话人在日常生活中使用多种语言）、双语（指说话人在日常生活中使用两种语言）、第二语言（指学生在校外主要讲一种或多种与教学语言不同的语言）。语言多元化可以是历史原因造成的多语言或双语社区、土著语言或移民人口的语言。由于多种形式的混合和交叉，语言多元化本身也变得更加多元（Blommaert, 2010；Blommaert & Rampton, 2011；Vertovec, 2007），这也使得诸如"双语者"和"本地语者"这样的词汇越来越受到质疑（例如，可以参见Blommaert & Rampton, 2011），的确，越仔细地观察语言的多元化，就会越难以在不同的场合中区分它们。本章所探讨的就是关于语言多元化与数学教与学研究。

范围与组织

我们对不同的语言多元化环境中所进行的数学的教与学研究作回顾整理，包括第二语言、双语和多语言课堂。虽然在20世纪70年代人们已经意识到这一问题的重要性，但在1980年以前，针对数学教育的这类研究还很少，因此本章中绝大部分讨论都是围绕着1980年之后的研究展开的。这些研究来自世界不同的地区，到目前为止，这一领域大多数的研究是在北美地区、欧洲、非洲南部的一些国家和澳大利亚进行的。

我们查阅了使用各种各样的词汇来表示语言多元化的研究。除了多语言、双语和第二语言这些词汇外，英语学校系统中经常使用的词汇还包括英语学习者（ELLs）、有限英语能力者（LEPs）、英语作为第二语言者（ESL）、英语作为附加语言者（EAL）。所有这些标签都隐含负面倾向，因为它们强调的是教学语言，而不是强调学生所讲的其他语言（见Barwell, 2003a）。

我们的讨论没有包括有关语言在数学思维、学习或教学中的角色研究，也没有包括探讨数学课堂话语的问题（见这套书中Herbel-Eisenmann, Meaney, Bishop, & Heyd-Metzuyanim, 2017一章），不过我们还是强调了这两个更具有普遍性的研究课题与数学教育有关语言多元化研究之间的联系。同样，我们不讨论发生在数学教室中的社会语言学研究，除非这些研究能够阐述语言多元化与数学教与学之间的关系（例如Burston, 1999；de Courcy & Burston, 2000）。然而，我们还是在涉及数学教育的环境时把语言多元化和社会语言学研究联系起来了。最后，我们没有讨论数学课堂中教学语言的不同方式，例如双语教育、浸入式教育、内容语言综合学习

21 | Language Diversity and Mathematics: Second Language, Bilingual, and Multilingual Learners

RICHARD BARWELL
University of Ottawa, Canada

JUDIT N. MOSCHKOVICH
University of California Santa Cruz, United States

MAMOKGETHI SETATI PHAKENG
University of Cape Town, South Africa

There are few mathematics classrooms in which all students and the teacher speak a single common language and no other. All societies feature some kind of language diversity. Language diversity encompasses not only the presence of different languages but also the presence of different varieties of the same language, such as dialects of Spanish or regional varieties of English. Commonly used terms that indicate some form of language diversity include *multilingualism*, in which speakers use several languages in their daily lives; *bilingualism*, in which speakers use two languages in their daily lives; and *second language*, in which students mainly speak a language or languages outside school that is different from the language of instruction. Language diversity can involve historically multilingual or bilingual communities, aboriginal languages, or languages of migrant populations. Language diversity is itself becoming more diverse, with multiple forms of mixing and intersecting languages (Blommaert, 2010; Blommaert & Rampton, 2011; Vertovec, 2007), such that terms like "bilingual" and "native speaker" are increasingly contested (see, for example, Blommaert & Rampton, 2011). Indeed, the more closely one looks at language diversity, the more difficult it becomes to make clear distinctions between different situations. This chapter is about research on language diversity and the learning and teaching of mathematics.

Scope and Organization

We review research on learning and teaching mathematics in various situations of language diversity, including second language, bilingual, and multilingual classrooms. Until around 1980, there was little research on this topic specific to mathematics education, although it was recognized as a significant issue in the 1970s. Most of the chapter is, therefore, devoted to work conducted since around 1980. We include research conducted in many parts of the world: To date, the majority of research in this area has been conducted in North America, Europe, southern Africa, and Australasia.

We consulted studies using a wide variety of terms to indicate language diversity. In addition to multilingual, bilingual, and second language, common labels used in English-language school systems include English language learners (ELLs), limited English proficient (LEPs), English as a second language (ESL), and English as an additional language (EAL). All these labels are implicitly deficit oriented, because they emphasize the language of instruction and deemphasize students' other languages (see Barwell, 2003a).

We do not include research on the role of language in general in mathematical thinking, learning, or teaching or questions about the nature of mathematics classroom discourse (see Herbel-Eisenmann, Meaney, Bishop, & Heyd-Metzuyanim, 2017, this volume). Nevertheless, we do highlight connections between these two more general research topics and language diversity in mathematics education where these connections are relevant. Similarly, we do not discuss sociolinguistic studies that happen to have been conducted in mathematics classrooms, unless these studies shed light on the relationship between language diversity and the learning and teaching of mathematics specifically (for example, Burston, 1999; de Courcy & Burston, 2000). We do, however, make links to sociolinguistics research on language diversity in educational settings where it has informed work in mathematics education. Finally, we do not review

（CLIL）以及基于内容的语言教学（对于这些不同的教学模式，见 Baker，2011；Dalton-Puffer，Nikula，& Smit，2010；Lotherington，2004；Lyster & Ballinger，2011；Teddick，Christian，& Fortune，2011）。

在讨论数学教育中语言多元化研究的历史背景后，本章主要内容则围绕研究人员在多元的语言环境下进行数学教与学研究中所广泛采用的三个理论观点来组织。

第一部分从认知主义的角度，总结了始于80年代的研究。这些研究寻求回答下面这些问题：在第二语言环境下学习数学在成就方面有什么区别？如何从认知过程的角度解释这些区别？这方面的研究通常是基于数学是要获取知识体的观点，并且主要涉及个体学习者的数学思维和知识的获取。对语言多元化的处理在本质上也是个人主义的，所以双语学习和第二语言学习通常被理解为个体学习者的认知活动，而不是复杂的社会文化能力或者更广泛的社会现象。

第二部分从语言交流的角度讨论自90年代开始的研究。这些研究所寻求回答的问题是：在语言多元化的情况下，数学课堂内到底发生了什么？这一研究基于数学是交流实践的观点，其中含有说、听、写、读，以及数学的符号抽象，强调的是语言的社会、文化和符号属性，语言学习，参与和学习数学。

最后一部分则从社会政治的角度讨论2000年以来的研究并寻求回答下面的问题：语言多元化的数学课堂对谁有利或不利？这样的有利或不利结果是怎样发生的？从这个角度来说，语言并不是中立的，而是政治和意识形态的载体。

这三个观点是按照它们在数学教育中语言多元化研究领域中出现的大致时间顺序来呈现的，这个顺序反映了数学教育研究作为一个整体所具有的广泛趋势。然而，不同观点之间并非泾渭分明，它们之间有很多重叠，同时，时间顺序也不像我们所说的那么清清楚楚。近期人们所持有的观点在多年前就已存在，而多年前的研究仍为目前的研究提供着依据。本文将这个研究划分为三个阶段多少有些人为的成分，但我们还是希望这一划分是有用的。

本章在最后对这些研究进行了一些反思。在考察了这一研究领域的现状及其对政策和实践的影响之后，本文对未来的研究给出了一些建议。

在我们论述之前，有必要对语言作一点说明。语言这个词可以有多种含义，它可以表示一个国家或地区使用的语言、课堂中使用的语言、家庭和社区使用的语言、数学家使用的语言、教科书或测试题中所使用的语言，等等。社会语言学告诉我们，语言及其使用具有社会性、文化性和历史性。我们使用语言来表示国家语言，如日语，也用它来指人类交流体系的一般现象。本文用数学语言来代表数学中使用的语言的特性，数学语言不仅是具有精确数学含义的词汇或技术语言，而且还指参与数学活动所必需的话语实践（使用语言和其他符号体系的方式）。有关数学语言特性的研究突出了其语言结构和组织的特征（Crowhurst，1994；Pimm，1987）、语法结构和符号形式（例如O'Halloran，2005），以及话语实践（Moschkovich，2007b）。

历史背景

在这一节里，我们简要总结了20世纪80年代之前促使语言多元化成为数学教育中一个研究重点的前期工作。同时，也讨论了导致对语言多元化不同理论理解的有关双语和多语言研究的历史发展。

数学教育和语言多元化：早期研究

1980年以前，数学教育研究很少关注数学课堂中所使用的语言。的确，对这类问题的兴趣似乎是由于数学课堂中语言的多样性所带来的感知挑战所引起的。例如，1972年在英国举行的第二届国际数学教育大会（ICME 2）期间，就明确了对数学学习和学习数学所使用的语言之

different approaches to language instruction in mathematics classrooms, such as bilingual education, immersion education, content-language integrated learning (CLIL) and content-based language teaching (for introductions to these various models, see Baker, 2011; Dalton-Puffer, Nikula, & Smit, 2010; Lotherington, 2004; Lyster & Ballinger, 2011; Teddick, Christian, & Fortune, 2011).

Following a discussion of the historical background of research on language diversity in mathematics education, the main part of the chapter is organized around three broad theoretical perspectives adopted by researchers to investigate learning and teaching mathematics in language-diverse situations.

The first section, on cognitivist perspectives, summarizes research that began in the 1980s. This work sought to answer the following questions: What difference does learning mathematics in a second language make in terms of achievement? How can these differences be explained in terms of cognitive processes? Research conducted from this perspective is generally based on a view of mathematics as a body of knowledge to be acquired and is largely concerned with the mathematical thinking or attainment of individual learners. The treatment of language diversity is also individualist in nature, so that bilingualism and second language learning are understood as cognitive aspects of individual learners rather than as complex sociocultural competencies or broader social phenomena.

The next section, on discursive perspectives, looks at research that, beginning in the 1990s, sought to answer the following question: What actually happens in mathematics classrooms in situations of language diversity? This work is based on a view of mathematics as a set of discourse practices that includes talking, listening, writing, reading, and symbolizing mathematics. Discursive perspectives emphasize the social, cultural, and semiotic nature of language, language learning, and participation in and learning of mathematics.

The final section, on sociopolitical perspectives, looks at research that, particularly since the 2000s, has sought to answer the following questions: Who gains and who loses in mathematics classrooms in situations of language diversity? How do such differences arise? From this perspective, language is not seen as neutral but as politically and ideologically loaded.

These three perspectives are presented in the approximately chronological order in which they have appeared in the area of research on language diversity in mathematics education, an order that reflects broader trends in mathematics education research as a whole. Nevertheless, the different perspectives are not neatly bounded; there is much overlap between them. Nor is the chronology as neat as our account may imply. Ideas that have taken hold more recently were apparent many years ago, while work from earlier periods still informs research today. Our characterization of this body of work into three phases is, then, a somewhat artificial organizing device, though, we hope, a useful one.

The chapter concludes with some reflection on this body of work. We consider the state of the art and its implications for policy and practice, and finally, we suggest what the future might hold.

Before we proceed, it is necessary to say something about language. The word *language* can have multiple senses: It can refer to the language of a nation or region, of classrooms, of home and community, of mathematicians, of textbooks or test items, and so on. Informed by sociolinguistics, we see language and its use as social, cultural, and historical in nature. We use *language* to refer either to national languages, like Japanese, or to the general phenomenon of human communication systems. We use *mathematical language* to refer to the specific nature of language used in mathematics. This term refers not only to vocabulary or technical words with precise mathematical meanings but also to the discursive practices (ways of using language and other sign systems) necessary for participation in mathematical activities. Research on the nature of mathematical language has highlighted particular features of syntax and organization (Crowhurst, 1994; Pimm, 1987), grammatical structure and forms of semiosis (e.g., O'Halloran, 2005), and discourse practices (Moschkovich, 2007b).

Historical Background

In this section, we briefly summarize preliminary work in mathematics education prior to the 1980s that prompted the emergence of language diversity as a research focus in mathematics education. We also look at historical developments in research on bilingualism and multilingualism that led to different theoretical understandings of language diversity.

Mathematics Education and Language Diversity: Antecedent Work

Before 1980, research in mathematics education paid little attention to the use of language in mathematics classrooms. Indeed, it seems likely that interest in such topics was prompted by perceived challenges arising from language diversity in mathematics classrooms. During the second International Congress on Mathemat-

间的关系进行基础性研究的必要性。此后，1974年由联合国教育、科学及文化组织（UNESCO）和国际数学教育委员会和海外教育发展中心（UNESCO, 1974）合作在肯尼亚内罗毕举办了一个"语言与数学教育相互作用"的国际研讨会。该研讨会强调了对语言和数学之间关系的研究匮乏，并得出结论：造成数学学习困难的原因部分来自于学习的语言。会议同时进一步肯定，所有语言都含有有益于学生数学概念学习的语言特征，因此所有语言都可以用于数学教学。

研讨会强调的议题之一是在所谓全球性语言，如英语或法语的环境中学习时，使用其他语言学习数学并不是很特别的。在很多国家，学生必须用不是母语的国家性语言来学习数学（例如，在坦桑尼亚或印度）。

内罗毕研讨会带来的结果之一是在数学教育杂志中出现了第一篇关于语言和数学教育的文章（Austin & Howson, 1979）。奥斯汀和豪森（1979）对语言多元化的问题给了相当大的关注，他们指出，并不像当时大家感受到的语言和数学教与学所遇到的挑战似乎只是发展中国家所面临的问题，在一些国家，挑战还可以出自非标准的多种地方语言（比如在牙买加）。奥斯汀和豪森承认双语是一个政治问题，因此语言的变化会导致政策的变化。

构筑双语：心理学和语言学的发展

虽然奥斯汀和豪森（1979）的论文是第一篇出现在数学教育杂志中关于语言与数学学习的文章，从心理学和语言学角度对双语对学习者的影响的争论已经持续了一段时间，且研究也证明了这种影响既有正面的也有负面的。

在语言学中对于双语和多语言这样的概念一直存有争论。虽然社会语言学家强调语言的社会性及其在不同情境中的使用，心理语言学家的立场则是来自于实验环境下个体的表现。从社会语言学的角度来看，心理语言学实验对人们如何使用语言所知甚少，如同哈库特和麦克劳林（1996）所给出的解释，"语言学知识不是存在于个人心理层面的，而是作为一个群体的集体语言规范"（第608页）。双语的定义是对两种语言有着母语般的熟练度（可能是最直观的定义），可以交替使用两种语言（例如De Avila & Duncan, 1981a），属于一个双语社群（Valdés-Fallis, 1978）。从心理语言学的角度来看，一个双语者是可以熟练使用两种或两种以上语言的任何个人，然而，从社会语言学的角度来看，双语者是指参与多语言社区活动的人，是"特定语言社区的产物，完成某些功能时使用一种语言，而完成另一些功能或在另一些情形时则使用另外一种语言"（Valdés-Fallis, 1978, 第4页）。其中第二个定义假定双语不仅是个人特性，而且涉及参与社区语言实践的社会和文化现象。

对双语的一个常见误解是认为双语者能够同样流利地使用两种语言，然而双语研究者们认为能够"母语般地使用两种或多种语言"是一个不现实的定义，不能反映大多数很少能够流利使用两种语言的双语者的特征（Grosjean, 1999），相反，格罗让提出，我们应该把词汇"单语"和"双语"当作个体使用每种语言的流利程度的标签转变为仅以是否会说或使用双语作为两个单独的标签。

1979年以前的多数研究都认为双语的使用对学习者的语言、认知和教育发展都有负面的影响。双语行为被认为是不自然的现象，研究人员认为使用双语的孩子们不能够像只学习一种语言那样完美地学好两种语言。还有一个流传更为广泛的观点，即认为与掌握一种语言相比，掌握两种语言所需的认知努力降低了儿童们学习其他可能和应该学习的事物的能力。备受尊敬的德国语言学家韦斯格伯认为双语并行会损害整个民族的智力（引自Saunders, 1988），而雷诺则认为双语并用会导致语言的混淆和困惑，以及精准思维与行为能力降低、智力下降、嗜睡增加、自律能力降低（引自Saunders, 1988）。由于双语现象的广泛存在，很难理解这些观点是如何通过严谨的研究而得出的。然而，直到最近，这些观点在学术领域都是很常见的，并且反映了有关语言、智力、种族和殖民主义的更广泛的以欧洲为中心的意识形态

ical Education (ICME 2), for example, held in the United Kingdom in 1972, the need for fundamental research on the relationship between the learning of mathematics and the language through which it is learned was identified. As a result, an international symposium on "interactions between linguistics and mathematical education" was held in Nairobi, Kenya, in 1974, sponsored by the United Nations Educational, Scientific and Cultural Organization (UNESCO), in cooperation with the International Commission on Mathematical Instruction and the Centre for Educational Development Overseas (UNESCO, 1974). The symposium highlighted the lack of research on the relationship between language and mathematics and concluded that difficulties in mathematics learning depend in part on the language of learning. It further affirmed that all languages include linguistic features that can benefit students' learning of mathematical concepts, and thus all languages can be used to learn and teach mathematics.

One of the issues that the symposium highlighted is that the problems of learning mathematics in an additional or foreign language are not peculiar to learning in a so-called global language such as English or French. In many countries, learners have to learn mathematics in a national language that is not their home language (e.g., in Tanzania or India).

The Nairobi symposium led to the first journal paper on language and mathematics education to appear in a mathematics education journal (Austin & Howson, 1979). Austin and Howson (1979) gave considerable attention to questions of language diversity and concluded that the challenge of language and mathematics learning and teaching is not, as perceived at the time, just an issue for developing countries. In some countries, the challenge is nonstandard local language varieties (e.g., in Jamaica). Austin and Howson acknowledged the fact that bilingualism is a political matter, and thus change in society may lead to policy change.

Framing Bilingualism: Developments in Psychology and Linguistics

Although Austin and Howson's (1979) paper was the first in a mathematics education journal, there had been extensive debates in psychology and linguistics about the effects of individual bilingualism on the learner for some time before, with research apparently demonstrating both positive and negative effects.

Notions such as bilingualism and multilingualism have a history of debate in linguistics. Although sociolinguists stress the social nature of language and its use in varying contexts, psycholinguists use an individual view of performance in experimental settings. From a sociolinguistic perspective, psycholinguistic experiments provide limited knowledge about how people use language, because, as Hakuta and McLaughlin (1996) explain, "linguistic knowledge is situated not in the individual psyche but in a group's collective linguistic norms" (p. 608). Definitions of bilingualism include native-like fluency in two languages (perhaps the most intuitive definition), alternating use of two languages (e.g., De Avila & Duncan, 1981a), and belonging to a bilingual community (Valdés-Fallis, 1978). From a psycholinguistic perspective, a bilingual person is any individual who is proficient in more than one language. In contrast, from a sociolinguistic perspective, a bilingual person is someone who participates in multiple language communities and is "the product of a specific linguistic community that uses one of its languages for certain functions and the other for other functions or situations" (Valdés-Fallis, 1978, p. 4). This second definition assumes that bilingualism is not only an individual characteristic but also a social and cultural phenomenon that involves participation in the language practices of communities.

A common misunderstanding of bilingualism is the assumption that bilingual people are equally fluent in their two languages. However, bilingualism scholars see "native-like control of two or more languages" as an unrealistic definition that does not reflect evidence that the majority of bilinguals are rarely equally fluent in both languages (Grosjean, 1999). Instead, Grosjean proposes a shift from using the terms "monolingual" and "bilingual" as labels for individuals or their fluency in each language to using them as labels for the endpoints on a continuum.

Most research before 1979 concluded that individual bilingualism had negative effects on learners' linguistic, cognitive, and educational development. Bilingualism was seen as unnatural, and researchers argued that bilingual children learn neither language as perfectly as they would if limited to one. There was also a widespread view that the cognitive effort required to master two languages instead of one diminishes a child's power to learn other things that might and ought to be learned. Weisgerber (cited in Saunders, 1988), a highly regarded German linguist, argued that bilingualism could impair the intelligence of a whole ethnic group, and Reynold (cited in Saunders, 1988) was concerned about the fact that bilingualism leads to language mixing and confusion and a reduction in the ability to think and act precisely, a decrease in intelligence, an increase in lethargy, and reduced self-discipline. It is hard to understand how such views could pass as

（例如，可以参见Makoni，2011）。

皮尔和兰伯特（1962）进行了第一批挑战这一意识形态的研究，证明双语能力能够成为一种优势。他们研究了双语能力对蒙特利尔法语沉浸式课程中的10岁儿童的智力功能的影响，在沉浸式课程中，一些或所有的科目都用"目标"语言教授，其目的是使学习者在他们的学校生涯中能够精通这种语言。皮尔和兰伯特发现，双语学习者并不是在"精神混乱"中受罪，而是从语言资源中获益。类似地，杨克-沃勒尔（1972）对南非荷兰语-英语双语儿童的研究和本-泽伊夫（1977）对希伯来语-英语双语儿童的研究都发现，双语儿童比单语儿童更早地意识到语言的任意性，即元语言意识。这个结果对研究双语儿童的认知能力有一定的意义。如同卡明斯（1981，第33页）所指出的，如果孩子要有效地使用一种语言作为思维的工具，那么具有从一个词的发音中分离出这个词的意思的能力是必要的。

斯温和卡明斯（1979）比较了正负两方面的研究并得出结论，正面的研究发现通常与沉浸式课程中多数语言群体相关，在这样的情况下，通晓两种语言就有很高的价值，第二语言对第一语言是一种免费的附加，而这些孩子父母的社会经济地位也相对较高。另一方面，负面的研究发现则是学生在当时的思维模式下被视为"淹没"而不是沉浸，这种情况也被称为"文化削减"的学校教育（Valenzuela，1999，2002），即学校的功能"不是以积累的方式促进双语、双语言和双文化主义，而是减少墨西哥裔美国儿童的文化、语言和自身基于社区的身份"（Valenzuela，2002，第236页）。在这种削减教育的情况下，学生被迫学习主流语言，并且不鼓励他们保留自己的第一语言（Nieto，1992）。斯温和卡明斯也认为虽然有多种因素影响儿童的智力发展，但双语是一个重要的可以产生积极影响的因素。

最新的心理语言学研究也报道了这样一种观点：除了元语言意识，双语儿童比起单语儿童在非语言的执行和控制方面也有优势，也就是说，他们发展了一种"有选择地处理信息和抑制误导信号的强化能力"（Bialystok，2001，第245页）。元语言优势在有关双语的研究中也有适量的报道（Bialystok，2001；Bialystok & Majumder，1998），执行控制方面的优势也在"新兴双语"的沉浸式课程中被发现。

尽管早期的双语教育研究没有考虑社会因素，但很明显双语教育对儿童的认知和智力发展不一定有特别的影响（消极的，积极的，或是中立的）这一说法已为人们所接受。相反，任何这样的影响需要在更为广泛的社会政治语言环境下来理解，比如学生是否是在增强或削弱的教育环境中学习，家庭语言在当地的地位，以及学生是否有机会用他们的母语来发展自己的读写能力。不考虑这些因素就无法对语言进行有效的比较。

认知主义者的观点

或许是很自然的现象，对数学教育中语言多元化感兴趣的研究人员所讨论的第一个广泛的问题应该是关于学生的表现：用第二语言学习数学在成就方面有什么不同？这样的研究主要是在第二语言学习者的框架下进行的，最初是在历史情况下或者当今移民到英语国家（如美国，英国和澳大利亚）的背景下考虑的，更具体的问题则包括：对教学语言的熟练掌握和数学成就之间是否有关系？双语能力和数学成就之间是否有关系？数学是否有特殊的语言特征会给第二语言或双语学习的人带来挑战？

与20世纪70年代和80年代数学教育中的许多研究一样，本节讨论的研究大多采用的是认知主义者的观点，反映了数学教育作为数学和教育心理学交叉领域的初始起点。以这类观点所做研究分析的对象通常是个体学习者，学习数学被理解为是对数学知识的获取和个人建构。个体学习者对数学知识的获取情况是通过考试或访谈来确定的。同时，语言作为认知术语也被理解为是一个被学习和存储在个人头脑中的系统。这类研究可分为三个

serious research, given the widespread occurrence of bilingualism. Nevertheless, such views were common in academic literature until relatively recently and presumably reflected broader Eurocentric ideologies about language, intelligence, race, and colonialism (see, for example, Makoni, 2011).

Peal and Lambert (1962) conducted one of the first studies to challenge this ideology, demonstrating that bilingualism can be an asset. They studied the effects of bilingualism on the intellectual functioning of 10-year-old children from French immersion programs in Montreal. In immersion programs, some or all subjects are taught in a "target" language with the goal of learners becoming proficient in this language over the course of their school career. Peal and Lambert found that instead of suffering from "mental confusion," bilingual learners were profiting from a language asset. Similarly, Ianco-Worrall's (1972) study of Afrikaans-English bilingual children and Ben-Zeev's (1977) study of Hebrew-English bilingual children both found that bilingual children realize sooner than monolingual children the arbitrary nature of language, known as metalinguistic awareness. This result had some implications for the bilingual children's cognitive abilities. As Cummins (1981, p. 33) argued, the ability to separate the meaning of a word from its sound is necessary if a child is to use language effectively as a tool for thinking.

Swain and Cummins (1979) compared positive and negative studies and concluded that the positive findings were usually associated with majority language groups in immersion programs. In such cases, there is a high value attached to knowing two languages. The second language is added at no cost to the first, and the parents are of relatively high socioeconomic status. Negative findings, on the other hand, were found with students who are, in the thinking of that time, seen as "submerged" rather than immersed. This situation has also been labeled "culturally subtractive" schooling (Valenzuela, 1999, 2002), where the function of schools "is not to promote bilingualism, biliteracy, and biculturalism in an additive fashion but rather to subtract Mexican American children's culture, language, and community-based identities" (Valenzuela, 2002, p. 236). In subtractive situations, students are forced to learn the majority language and are not encouraged to retain their first language (Nieto, 1992). Swain and Cummins also argued that although multiple factors influence children's intellectual development, bilingualism was a significant factor that could have a positive impact.

More recent research in psycholinguistics has reported that in addition to metalinguistic awareness, bilingual children show advantages over monolingual children in nonverbal executive control; that is, they develop an "enhanced ability to selectively attend to information and inhibit misleading cues" (Bialystok, 2001, p. 245). Metalinguistic advantages have been reported at modest levels of bilingualism (Bialystok, 2001; Bialystok & Majumder, 1998), and the advantage in executive control has been found also in so-called "emerging bilinguals" in immersion programs.

Although the early research on bilingualism did not foreground social factors, it is clear that there was an acceptance that bilingualism need not necessarily have any particular effect (negative, positive, or neutral) on the cognitive and intellectual development of children. Rather, any such effects need to be understood in relation to the wider sociopolitical language context, such as whether the student population is in an additive or subtractive educational setting, what the local status of the home language is, and whether there are opportunities for students to develop literacy in their home language. Comparisons across language settings cannot be usefully made without these considerations.

Cognitivist Perspectives

It is perhaps natural that the first broad question addressed by researchers interested in language diversity in mathematics education should be about student performance: What difference does learning mathematics in a second language make in terms of achievement? Such research was predominantly framed in terms of second language learners, initially in the context of historical or contemporary immigration to English-speaking countries like the United States, the United Kingdom, and Australia. More specific questions included the following: Is there a relationship between proficiency in the language of instruction and mathematics achievement? Is there a relationship between the degree of bilingualism and mathematics attainment? Are there specific linguistic features of mathematics that create challenges for second language or bilingual learners?

In common with much research in mathematics education in the 1970s and 1980s, the research discussed in this section tended to adopt a cognitivist perspective, reflecting the origins of mathematics education as a domain at the intersection of mathematics and educational psychology. The unit of analysis for research with this kind of perspective is generally the individual learner. Learning mathematics is understood to be the acquisition or individual construction of mathematical

广泛的主题：（1）语言能力和数学成就之间的关系，（2）数学文字题，（3）语言结构与数学认知。下面将对这三个主题进行详细讨论。

语言能力和数学成就之间的关系

研究人员一直试图确定教学语言掌握的熟练程度对数学学习效果的影响作用。语言能力和数学成就之间的关系并不那么直接，例如，根据科金和奇普曼（1988）对美国20世纪70年代的研究所做的回顾，有明确关系的证据是"不确凿的甚至是不一致的"（第25页）。研究这种关系的挑战之一是潜在的干预变量的数量，特别是在采用定量方法时。科金和奇普曼（1988）给出了下面几方面内容：语言技能、学习能力、教学质量、父母的帮助、各种动机、考试的语言以及使用的数学问题的类型。

然而很多使用了学生成就数据的研究都表明一些使用双语学习的学生在数学学习中表现不佳（例如，Cocking & Chipman，1988 所做的研究回顾；Howie，2001；Phillips & Birrell，1994；Secada，1992）。在美国，科金和齐普曼（1988）也得出结论"虽然没有证据表明少数族裔的学生在基本能力方面与讲英语的学生有不同，但在少数族裔学生和大多数学生群体之间的成就差异是显著的"（第25页）。塞科达（1992）（收录在由格劳斯（1992）主编的书中）则更为谨慎，他依据20世纪80年代和90年代的五项研究得出：

> 一个人对某种语言的精通程度与其数学成就表现之间存在关联。然而，尽管语言能力和数学成就之间的相关性是显著的，但这种相关性也只是在0.20到0.50之间，而且多数是处于这个范围的低段。因此这其中有很多差异需要给予解释（Secade，1992，第638页）。

事实上，最近对美国类似群体的研究表明，学生在学校取得的成就不仅与语言能力有关，而且与社会因素有关，社会因素在解释不同族裔的数学成就的差异方面起着重要作用（Oseguera，Conchas，& Mosqueda，2011）。同时，在原住民和非原住民的拉丁裔和讲英语者被分开安排居住的情况下，对拉丁裔的研究也发现数学表现的差异在加剧（Mosqueda，2010）。

另一个证据来自对沉浸式教育课程的研究。加拿大法语沉浸式课程中的学生在数学学习方面取得了和在采用普通英语授课的课程中的学生相似的分数（Lapkin，Hart，& Turnbull，2003；Swain & Lapkin，2005；Turnbull，Lapkin & Hart，2001）。事实上，在一些情况下，接受浸入式课程的学生的数学表现要优于采用普通英语授课的课程中的学生（Bournot-Trites & Reeder，2001）。

虽然这些发现都表明了一种共识，但它们也有一些重要的局限性。首先，与语言多元化的多种形式相比，这些研究专注于相对比较具体的情况。在美国和英国进行的研究关注的是来自少数族裔群体的，教学语言是英语的学生，而加拿大法语沉浸式课程代表的则是语言多元化的一个不寻常的现象。在每种情况下，重要的社会影响，如种族和社会阶层，都会与语言交叉（如同 Secada，1992 所指出的）。第二，有关数学测试题目中语言复杂性的研究表明，美国的英语学习者学生在数学成就上至少有一部分差异可能要归因于数学测试题目中语言的复杂性而不是被评估的内容（Abedi，2002；Abedi & Lord，2001；Shaftel，Belton-Kocher，Glasnapp，& Poggio，2006）。修改考试题以降低语言的复杂性能够导致这个群体中的一些学生成绩的提高（Abedi，2009；Abedi & Lord，2001）。

产生这些差异的一种可能的解释是算术计算的熟练性。例如，莫谢科维奇（2007c）总结了一组对单语和双语成年人进行算术计算时反应时间的比较研究（Marsh & Maki，1976；McLain & Huang，1982），研究人员的假设是，因为双语者用他们的第一语言做心算，所用时间也会长一些。然而总的来说，所有关于"对于双语者而言，对算术事实[①]的检索可能比单语者要慢一些"这一结论的

[①] 即表内加减法和表内乘除法。——译者注

knowledge. The acquisition of mathematical knowledge by individual learners is determined through tests or clinical interviews. Language, meanwhile, is also understood in cognitive terms, as a system that is learned and stored in the individual mind. Research in this strand can be grouped into three broad topics, which are discussed in more detail below: (1) the relationship between language proficiency and mathematics attainment, (2) mathematical word problems, and (3) language structure and mathematical cognition.

The Relationship Between Language Proficiency and Mathematics Attainment

Researchers have long sought to establish the extent to which proficiency in the language of instruction influences mathematics attainment. The relationship between language proficiency and mathematics attainment is not straightforward. According to Cocking and Chipman (1988), for example, in a review of research from the United States in the 1970s, evidence for a clear relationship is "by no means conclusive or even consistent" (p. 25). One of the challenges of investigating this relationship, particularly when adopting a quantitative methodological approach, is the number of potential intervening variables. Cocking and Chipman (1988) include the following: language skills, learning ability, quality of instruction, parental assistance, various aspects of motivation, the language of testing, and the type of mathematics problems used.

Nevertheless, numerous studies using achievement data have established that some bilingual students underperform in mathematics (e.g., studies reviewed in Cocking & Chipman, 1988; Howie, 2001; Phillips & Birrell, 1994; Secada, 1992). In the United States, Cocking and Chipman (1988) concluded that "although there is no evidence to suggest that the basic abilities of minorities are different from Anglo-speaking students, the achievement differences between minority and majority students are pronounced" (p. 25). Secada (1992), in his chapter in Grouws (1992), was more cautious, concluding on the basis of five studies conducted in the 1980s and early 1990s that there is

> a relationship between how proficient someone is in a language and performance on measures of mathematics achievement. However, though correlations between language proficiency and mathematics achievement are significant, they tend to be in the range of 0.20 to 0.50, with most being at the lower end of the scale. Hence there is much variance to be explained. (Secada, 1992, p. 638)

Indeed, recent research with similar populations in the United States has shown that school achievement in general is correlated not only with language proficiency but also with social factors that play a significant role in explaining variation in mathematics achievement of different ethnic groups (Oseguera, Conchas, & Mosqueda, 2011). Other research focused on Latinos/as, meanwhile, has found that disparities in mathematics performance were exacerbated by the track placement of native and nonnative Latina/o English speakers (Mosqueda, 2010).

An alternative source of evidence comes from research on immersion education programs. Students enrolled in French immersion programs in Canada achieve similar scores in mathematics as students enrolled in regular English programs (Lapkin, Hart, & Turnbull, 2003; Swain & Lapkin, 2005; Turnbull, Lapkin & Hart, 2001). Indeed, in some cases, students in immersion programs outperform students in regular English programs in mathematics (Bournot-Trites & Reeder, 2001).

Although these findings suggest a trend, they have some important limitations. First, when compared with the many forms of language diversity, they focus on relatively specific situations. Research conducted in the United States and United Kingdom has focused on students seen as being from minority backgrounds and as learners of English as the language of instruction. French immersion programs in Canada, meanwhile, represent a relatively unusual instance of language diversity. In each case, significant social effects, such as race and social class, intersect with those of language (as also noted by Secada, 1992). Second, work on the language complexity of mathematics test items suggests that at least some differences in the attainment of U.S. ELL students may be attributed to the linguistic complexity of mathematics test items rather than the content being assessed (Abedi, 2002; Abedi & Lord, 2001; Shaftel, Belton-Kocher, Glasnapp, & Poggio, 2006). Modifying test items to ameliorate language complexity has led to higher scores of some ELL students (Abedi, 2009; Abedi & Lord, 2001).

One possible explanation that has been proposed for these differences is fluency in arithmetic computation. For example, as summarized in Moschkovich (2007c), one set of studies compared response times between monolingual and bilingual adults carrying out arithmetic computation (Marsh & Maki, 1976; McLain & Huang, 1982); researchers hypothesized that since bilingual adults do mental arithmetic in their first language, they take longer to compute. Overall, however, the evidence is inconclusive and contradictory: "retrieval times for arithmetic facts *may* be slower for bilinguals than monolinguals" (Bialystok, 2001, p. 203). This possible difference

证据都是不确定而且相互矛盾的（Bialystok，2001，第203页）。然而这种可能的差别是非常小的（大约0.5s的差别），而且还只存在于成人之间，不是儿童之间，因此这些小的差异在年轻的学习者中可能并不存在。

事实上，如果在实验过程中双语成年人没有被要求切换语言的话，所报告的反应时间的微小差异也就消失了，如果要求双语成年人只使用一种语言，那么"首选语言优势"也就不存在了。这个结果似乎支持在课堂中允许双语学生在进行算术计算时选择自己的语言这样的课堂实践。一位研究人员总结了有关双语的使用对数学问题解决影响的心理语言学研究：

> 这些研究的结果呈现了一个复杂的画面，而且在某些情况下还相互矛盾。与数据一致的最明显的解释是，只要语言能力至少足以理解问题，那么双语的使用对数学问题解决就没有影响。甚至在某些情况下，使用较弱的语言解决问题时也没有阻碍（Bialystok，2001，第203页）。

第二类寻求解释数学成就差异的研究以卡明斯（2000）在双语教育方面的研究为起点。卡明斯（2000）的阈限假设提出，对于双语学生，对两种语言都具有高水平的能力会产生良好的整体学业表现，而对两种语言的掌握都比较低时会导致较差的整体学业表现。[1] 对一种语言具有高水平熟练程度的双语学生与单语学生的表现相当。

在澳大利亚和巴布亚新几内亚进行的一系列研究中，克拉克森（1991，1992，2007；Clarkson & Galbraith，1992）对双语学习者的数学表现和语言能力进行了比较。这些研究首先对学生的母语能力和英语能力进行了评估，随后，克拉克森将他的样本分为三组：（1）对两种语言都熟练掌握的学生，（2）对一种语言熟练掌握，但对另一种语言没有熟练掌握的学生（"一种语言占主导优势"），（3）对两种语言都没有熟练掌握的学生（例如Clarkson，2007）。分析比较了三组学生的数学表现，同时在一些情况下，还与单语学生样本的表现做了比较。这些研究的发现与阈限假设一致。克拉克森发现对两种语言都没有熟练掌握的学生的数学表现要显著低于单语学生，而对两种语言都熟练掌握的学生在数学表现方面一般要高于单语学生或对两种语言都没有熟练掌握的学生。克拉克森的研究被从事双语教育研究的人员作为重要的佐证所引用（见Baker，2001，第167页）。道（1983）在英国的研究以及最近在爱尔兰的研究项目（Ní Ríordáin & O'Donoghue，2009）也提供了进一步的佐证。

数学文字题

数学文字题在很多文献中都被明确地列为是有关第二语言或双语学习者特别关注的任务类别，文字题一直是很多研究的焦点，其中大多数研究都基于认知主义的观点。文字题在数学课中普遍存在，并且具有特殊的语言特征（Gerofsky，1996），这些语言特征对一些双语学生来说具有特别的挑战性（Barwell，2009）。多数这类课题的研究利用已经确定的问题类型，特别是由卡彭特和莫泽（1984）或赖利和格里诺（1988）所给出的类型，来关注个体对算术文字题的反应。

语言能力和解文字题的表现之间的确存在着显著关系，在尼日利亚（Adetula，1989，1990）、美国（例如，Mestre，1986，1988；Secada，1991）、菲律宾（Bernardo，1999）和爱尔兰共和国（例如，Ní Ríordáin & O'Donoghue，2011）所进行的研究中都显示出这种关系。然而，这种关系的特性会因研究的背景而有所不同。此外，由于研究的背景会涉及不同的因素及其不同的组合方式，因此解释这些发现也是困难的。

一个重要的因素是这类研究可以为参与者提供文字题所使用的不同语言。例如，在塞科达（1991）的研究中，为英语-西班牙语双语学生提供的是用英文或西班牙文给出的问题，塞科达发现整体上用英文回答的学生表现较强。阿德图拉（1989，1990）在尼日利亚所做的研

is, however, very small (of the order of 0.5 seconds), and has been documented clearly only with adults, not young children, so these small differences may not exist for young learners.

In fact, the slight reported difference in response time disappeared if bilingual adults were not asked to switch languages during an experimental session. If bilingual adults were required to use only one of their languages, the "preferred language advantage" was eliminated. This result seems to support practices allowing bilingual students to choose the language they use for arithmetic computation in the classroom. One researcher summarized the psycholinguistic research on the impact of bilingualism on mathematical problem solving:

> The results of these studies present a complex picture and appear in some instances to contradict each other. The most generous interpretation that is consistent with the data is that bilingualism has no effect on mathematical problem solving, providing that language proficiency is at least adequate for understanding the problem. Even solutions in the weaker language are unhampered under certain conditions. (Bialystok, 2001, p. 203)

A second line of research seeking to explain differences in mathematics achievement took as its starting point Cummins's (2000) work in bilingual education. Cummins's (2000) threshold hypothesis proposed that for bilingual students, high levels of linguistic proficiency in two languages would lead to stronger general academic performance and low levels of linguistic proficiency in both languages would lead to lower academic performance.[1] Bilingual students with high levels of proficiency in one language would perform comparably with monolingual students.

In a series of studies conducted in Australia and Papua New Guinea, Clarkson (1991, 1992, 2007; Clarkson & Galbraith, 1992) compared the mathematics performance of bilingual learners with their language proficiency. These studies first used instruments to evaluate students' proficiency in their home language, as well as in English. Clarkson then divided his samples into three groups: (1) students with high proficiency in both languages, (2) students with high proficiency in one language and not the other ("one dominant"), and (3) students with low proficiency in both languages (e.g., Clarkson, 2007). Analyses compared the mathematics performance of the three groups, as well as, in some cases, with the performance of a sample of monolingual students. The findings of these studies are consistent with the threshold hypothesis. Clarkson found the strongest effects for students with low proficiency in two languages, whose performance in mathematics was significantly lower than one-dominant students. Students with high proficiency in two languages tended to record higher performance in mathematics than one-dominant students or students with low proficiency in two languages. Clarkson's work has been cited as key evidence for the threshold hypothesis by researchers in bilingual education (see Baker, 2001, p. 167). Further evidence includes findings from Dawe's (1983) study in the United Kingdom, as well as a recent study conducted in Ireland (Ní Ríordáin & O'Donoghue, 2009).

Mathematical Word Problems

Mathematical word problems have been clearly identified in the literature as a task type of particular concern for second language or bilingual learners, and word problems have been the focus of numerous studies, the majority of which are based on a cognitivist perspective. Word problems are certainly widespread in mathematics classrooms and have particular linguistic features (Gerofsky, 1996) that may be particularly challenging for some bilingual students (Barwell, 2009). Much of the research on this topic focuses on individual responses to arithmetic word problems and draws on established typologies of such problems, particularly those of Carpenter and Moser (1984) or Riley and Greeno (1988).

There does seem to be a relationship between language proficiency and performance on word problems, demonstrated in research conducted in, among other places, Nigeria (Adetula, 1989, 1990), the United States (e.g., Mestre, 1986, 1988; Secada, 1991) the Philippines (Bernardo, 1999), and the Republic of Ireland (e.g., Ní Ríordáin & O'Donoghue, 2011). The nature of this relationship, however, varies according to the context of the research. Moreover, interpreting such findings is difficult, since several factors may be involved in different combinations according to the context of the research.

One important factor would appear to be the language in which the word problems are presented to the participants in the research. In Secada's (1991) study, for example, problems were presented to English-Spanish bilingual students in either English or Spanish. Secada found that overall performance was stronger in English. In Adetula's (1989, 1990) research in Nigeria, problems were presented in either English or Yoruba. In this context, English was only used at school, with Yoruba being the main language of society in the region in which the study was conducted. The students performed better on the problems presented in Yoruba. Secada (1991) con-

究中，由于英语只在学校使用，因此他们提供的问题是用英文或约鲁巴语呈现的，但是由于约鲁巴语是本研究所在区域社区中的主要语言，因此用约鲁巴语呈现问题的学生表现更好。在这一方面，塞科达（1991）考虑了他的结果与阿德图拉（1989）的结果不同的原因，提出在他的研究中，英语正在成为学生的主导语言，而约鲁巴语则是阿德图拉研究中学生的主导语言。这一解释在伯纳多（1999）在菲律宾所做的研究中也得以验证，在他的研究中，用母语呈现问题的学生表现好于用英文呈现问题的学生。梅斯特雷（1986）所做的针对美国大学中西班牙语-英语双语学生的研究中，发现学生的表现未受问题使用的语言影响，因为到了这个阶段，学生应该已经发展到了一个比较平衡的双语水平，并没有哪一种语言占据主导地位。因此，尽管问题使用的语言有所不同，但学生的表现取决于哪一种语言是主导语言。

人们对影响双语或多语言学习者在文字题表现上的大量其它影响因素也做了研究。这些因素包括文字题中的句法（Mestre & Gerace, 1986）、日常生活中的词汇和语境（Martiniello, 2008, 2010）和具有数学和非数学内涵的词汇（比如"较多"和"较少"，P. Jones, 1982），以及学生的阅读理解水平（Chamot, Dale, O'Malley, & Spanos, 1992; Mestre & Gerace, 1986）。使用不太复杂的语言来重新呈现文字题可以提高学生的表现，例如，伯纳多（1999）在他的研究中发现，对双语学生使用改写后的问题都取得了较好的成绩，这一结果与单语学生在文字题表现方面的研究结果是一致的（例如 De Corte, Verschaffel, & De Win, 1985），同时也在美国为英语学习者所做的改进考试题目的研究中得到验证（Abedi, 2009; Abedi & Lord, 2001）。阿德图拉（1989, 1990）发现，当问题是用约鲁巴语（学生更为熟悉的语言）而不是英语给出时，学生会使用更加复杂的解题策略。

总的来说，这些研究都表明文字题中语言的复杂性对于双语学生是一个特别的挑战，至少在某些情况是这样的，比如学生还处于熟练掌握教学语言能力之时。然而，语言复杂性不是在专业的数学词汇水平上面，而是在相关的背景知识（Martiniello & Wolf, 2012）、句子和段落的语言复杂程度以及语法的复杂程度上，比如被动语态、多个从句以及嵌套结构的使用（Cook & MacDonald, 2012）。更为重要的是，这些发现经常表明双语者所遇到的困难与文献记载中使用英语的人所遇到的困难很相似，例如，梅斯特雷（1986）发现参加研究的双语者与英语为母语的人一样，在将用自然语言给出的文字题"转换"为代数方程时都会出错（参见Mestre, Gerace, & Lochhead, 1982）。因此，很多关注双语学生和单语学生差异的研究可能漏掉了或者没有关注参与者在数学活动中那些重要的相似之处。

语言结构与数学认知

萨丕尔-沃尔夫假说（人们对世界的看法由他们的语言决定）和语言相对性（语言之间的差异导致人们对世界看法的不同）是语言和认知结构关系框架下的两种主要方式。一些研究试图在数学情境下来检验这一关系（例如 Miller, Kelly, & Zhou, 2005; Miura, Kim, Chang, & Okamoto, 1988; Towse & Saxton, 1998; Zepp, Monin, & Lei, 1987）。

缪拉等人（1988）的研究提到，有证据表明东亚几个国家（韩国、中国和日本）的语言中有规则的数字名称使得儿童比较容易发展数的概念，比如10进制结构。然而，陶斯和萨克斯顿（1998）则对这一说法提出质疑，并断定儿童对数的表征会受到实验条件的高度影响，比如实验者的引导（这对皮亚杰使用的任务也成立，见Donaldson, 1978），而语言对数的认知表征的影响也没有像早期研究中所提出的那么直接，例如，布利斯贝尔特、菲亚斯和诺埃尔（1998）就发现当要求参与者书写而不是说出答案时，算术计算时间方面的差异就消失了。

实证研究提供了很多理由来质疑把数学成就或成功归结于不同语言中早期数字名称的便利性的说法。计数或算术的能力是由不同部分构成的（Dowker, 1998），有些与语言无关，比如基数（Sarnecka & Carey, 2008）或者利用数轴做估算（Muldoon, Towse, Simms, Perra, & Menzies, 2013），这些能力遵循复杂的发展路径（Dowker, 1998; Muldoon 等, 2013; Sarnecka & Carey, 2008），即使有些研究结果表明计数体系对算术表现

sidered why his results differ from those of Adetula (1989) in this respect, proposing that the students in his study were becoming dominant in English, while those in Adetula's were probably dominant in Yoruba. This explanation would also account for Bernardo's (1999) findings in the Philippines, in which problems presented in English were solved less well than problems presented in students' first language, as well as Mestre's (1986) study of Spanish-English bilingual university students in the United States, where he found no difference in performance based on the language of the problem, since by this stage, the students have presumably developed a more balanced bilingualism and neither language is dominant. Hence, although the language of the problem does appear to make a difference, it appears to depend on which language is dominant.

A variety of other factors influencing bilingual or multilingual learners' performance on word problems have been examined. These factors include word problem syntax (Mestre & Gerace, 1986), the presence of everyday words or contexts (Martiniello, 2008, 2010) or words that have both mathematical and nonmathematical connotations (e.g., "more" and "less," P. Jones, 1982), and students' level of reading comprehension (Chamot, Dale, O'Malley, & Spanos, 1992; Mestre & Gerace, 1986). Rewording problems to use less complex language can improve performance. Bernardo (1999), for example, found that students performed better on reworded problems in both languages used in his study, a result that is consistent with other studies of word problem performance in monolingual students (e.g., De Corte, Verschaffel, & De Win, 1985), as well as research on modified test items for ELLs in the United States (Abedi, 2009; Abedi & Lord, 2001). Adetula (1989, 1990) found that students used more sophisticated strategies when the problems were presented in Yoruba (more familiar) than English.

Overall, these findings suggest that the linguistic complexity of word problems is a specific challenge for bilingual students, at least in some situations, such as when they are developing proficiency in the language of instruction. However, this linguistic complexity is not at the level of technical mathematical vocabulary but arises principally in relation to background knowledge (Martiniello & Wolf, 2012), linguistic complexity at the sentence and paragraph level, and grammatical complexity, such as the use of the passive voice without an agent, multiple subordinate clauses, and nested constructions (Cook & MacDonald, 2012). Importantly, these findings often show that bilingual learners face difficulties quite similar to those documented for native speakers. Mestre (1986), for example, found that bilingual participants, like native English speakers, "translated" word problems incorrectly from natural language to algebraic equations (see also Mestre, Gerace, & Lochhead, 1982). Hence, studies that have focused on differences between bilingual students and monolingual students may have missed or deemphasized important similarities in participants' mathematical activity.

Language Structure and Mathematical Cognition

The Sapir-Whorf hypothesis (that how people think about the world is determined by their language) and linguistic relativity (that differences among languages lead to differences in how people think about the world) are the two main ways that the relationship between language and cognitive structures has been framed. Several studies have sought to examine this relationship in the context of mathematics (e.g., Miller, Kelly, & Zhou, 2005; Miura, Kim, Chang, & Okamoto, 1988; Towse & Saxton, 1998; Zepp, Monin, & Lei, 1987).

Miura et al. (1988) claimed to have provided evidence that the regular structure of number names in several East Asian languages (Korean, Chinese, and Japanese) makes it easier for young children to develop number concepts such as the base-10 structure. Towse and Saxton (1998), however, critiqued this claim and concluded that children's representations of numbers can be heavily influenced by experimental conditions such as interviewer directions (also true for Piagetian tasks; see Donaldson, 1978) and that the influence of language on the cognitive representation of number was less direct than had been suggested in earlier studies. For example, Brysbaert, Fias, and Noel (1998) found that differences in arithmetic calculation times disappeared when participants were asked to type instead of say the answer.

Claims that mathematics achievement or success can be attributed causally to early facility with the names for numbers in different languages are suspect for many reasons documented through empirical work. Competence in counting or arithmetic is made up of several components (Dowker, 1998), some of them not linguistic, such as cardinality (Sarnecka & Carey, 2008) or using a number line for estimation (Muldoon, Towse, Simms, Perra, & Menzies, 2013). These competencies follow complex development paths (Dowker, 1998; Muldoon et al., 2013; Sarnecka & Carey, 2008). Even when findings suggest that a counting system can have some influence on arithmetic performance, the "effects tend to be limited to rather specific areas of arithmetic" (Dowker, Bala, & Lloyd, 2008, p. 536). Last, because studies have shown that cues provided by the interviewer can change those

有些影响，"其影响也是局限于相当特定的算术领域"（Dowker, Bala, & Lloyd, 2008，第536页）。最后，因为研究已经表明实验者提供的线索或结论可以改变这些影响（Alsawaie, 2004; Towse & Saxton, 1998），那么对于计数的语言结构中的劣势应该可以比较容易地通过教学来解决。

评论和批评

在一般性的学生发展研究以及针对文字题的特定研究中，一个非常明显的局限就是在调查中对语言多元化的本质没有给予足够的关注，学生被冠以"双语"或"第二语言学习者"或类似的称呼，好像这些称呼都有了明确的定义。事实上，双语是高度多元化的，在许多情况下，其他语言也可能存在。但是这些研究却将双语学习视为没有变化并且不依赖于任何情境的个体认知能力。

此外，许多早期的研究都使用数学活动的狭义概念并且侧重于算术计算的快速完成或者使用皮亚杰的诊断任务（例如，可以参见De Avila & Duncan, 1981a, 1981b）。后续的研究拓展了数学活动的观念，不仅考察算术计算，而且考察推理和问题解决等能力（例如Mestre & Gerace, 1986）、学生解决文字题的详细过程（例如Clarkson, 2007; Parvanehnezhad & Clarkson, 2008; Spanos, Rhodes, Dale, & Crandall, 1988）、学生用于解算术文字题的策略（Secada, 1991），以及学生对两位数数量概念的理解（Fuson, Smith, & Lo Cicero, 1997）。这样的进展并不出乎人们的预料，因为认知心理学和数学教育在数学活动的框架下都朝着相似的方向发展。

由于大多数研究都局限于算术计算和解文字题，结论不能推广到其他数学分支，如比例推理、几何、度量、概率或复杂的问题解决，算术计算反应时间的研究对于参与者在进行算术计算时所采用的策略几乎不能说明什么，同样地，侧重于将文字题的内容转换成代数方程的研究也没有反应出参与者的代数思维信息，因此，不可能利用这些研究来得出有关学生的高阶思维或概念理解方面的结论。

最后，更早期的认知研究还受到把语言简单化的影响，尽管不是所有的研究都是这样，例如，塞科达（1991）的研究使用了语言熟练度这一复杂的观念并考察了算术文字题的语义结构。人们还用语言评估量表（De Avila & Duncan, 1981b; Duncan & De Avila, 1986, 1987）、口述故事和口头数数等方法对语言能力进行了评估，这些工具不仅评估了语法、语音、词汇和语用，也包括与具体数学思维和研究课题紧密相关的语言任务。然而，大多数的认知主义研究通常使用简单的语言和双语理念，对语言结构和数学思维联系的研究就是一个典型例子。

交流的观点

20世纪90年代集中出现了很多关于数学课堂互动的研究，特别是一些在语言多元化的情境中的课堂互动。在这些研究中，研究人员采用了广泛的社会文化的理论导向，特别关注学生和教师的话语运用，持这类观点的研究所展开的分析通常是针对参与者的交流实践，数学学习被认为是学生在数学交流群体中的社会活动。对数学学习的研究通常是通过观察来进行的，通常包括对课堂互动的录音和详细的转录。这种方式也强调了利用社会为导向的观点来研究语言多元化的课堂互动，涉及的问题包括：学习者和教师利用什么样的资源来展示数学的意义？他们如何使用他们所知道的不同语言？教师是如何处理语言多元化的？数学学习如何与语言学习相结合？

本节的研究基本上是与数学教育中侧重交流方面有关的研究，包括调查数学课堂中参与数学讨论的互动（Lampert, 1990）、数学论证（Forman, 1996; Krummheuer, 1995），以及社会数学规范（Cobb, Wood, & Yackel, 1993）。从这一视角来看，学会数学交流不仅是学习词汇或理解含义，而且数学交流被视为使用社会的、语言的和物质的资源去参与数学实践（参阅下列研究：Kieran, Forman & Sfard, 2002; Seeger, Voigt & Waschescio, 1998; Sfard, 2008）。对于斯法德（2008）而言，学会数学思维即是学会与自己进行数学交流，同时通过与他人的互动而进行。数学交流（借

effects (Alsawaie, 2004; Towse & Saxton, 1998), it should be relatively easy to address any disadvantages related to language structure of counting numbers through instruction.

Commentary and Critique

A clear limitation of the attainment studies in general, and of the word problem studies in particular, is the lack of attention to the nature of the language diversity under investigation. Students are referred to as "bilingual" or "second language learners" or similar, as though these things are clearly defined states. In fact, bilingualism is highly diverse, and in many cases, other languages are also likely to be present. These studies treat bilingualism as an individual cognitive capacity that has no variation and is not in any way context dependent.

Furthermore, many early studies used narrow conceptions of mathematical activity and focused on quick performance on arithmetic computation or used Piagetian tasks (see De Avila & Duncan, 1981a, 1981b, for examples). Later studies developed a broader view of mathematical activity, not only examining arithmetic computation but also reasoning and problem solving (e.g., Mestre & Gerace, 1986), detailed protocols of students solving word problems (e.g., Clarkson, 2007; Parvanehnezhad & Clarkson, 2008; Spanos, Rhodes, Dale, & Crandall, 1988), the strategies children used to solve arithmetic word problems (Secada, 1991), and student conceptions of two-digit quantities (Fuson, Smith, & Lo Cicero, 1997). This progression is not surprising since cognitive psychology and mathematics education have moved in a similar direction in their framing of mathematical activity.

Because the majority of studies are restricted to arithmetic computation and solving word problems, conclusions do not generalize to other topics in mathematics, such as proportional reasoning, geometry, measurement, probability, or complex problem solving. Studies of response time during arithmetic computation reveal little about strategies participants use to carry out these computations. Studies that focused on translating word problems to algebraic equations reveal little about participants' algebraic thinking. It is not, therefore, possible to use these studies to reach conclusions about higher order mathematical thinking or conceptual understanding.

Finally, much early cognitivist work suffers from a simplistic view of language, although this was not the case for all studies. For example, Secada's (1991) study used a complex view of language proficiency and examined the semantic structure of arithmetic word problems. Language proficiency was assessed using the Language Assessment Scales (De Avila & Duncan, 1981b; Duncan & De Avila, 1986, 1987), oral storytelling, and verbal counting up and down. The instruments assessed syntax, phonetics, lexicon, and pragmatics and included language tasks that were closely related to the specific mathematical thinking and topic examined in the study. Most cognitivist work, however, has generally used a simplistic view of language and bilingualism exemplified by work on the relationship between language structure and mathematical thinking.

Discursive Perspectives

The 1990s saw the emergence of research that focused on mathematics classroom interaction in general and classroom interaction in situations of language diversity in particular. In this work, researchers adopted a broadly sociocultural theoretical orientation, with particular attention on students' and teachers' discursive practices. The unit of analysis for research with this kind of perspective is generally participants' discourse practices. Learning mathematics is understood in terms of students' socialization into a mathematical discourse community. Mathematics learning is researched through observation, often involving recording and detailed transcription of classroom interaction. This approach has also entailed a more socially oriented perspective on language diversity. Questions addressed have included the following: What resources do learners and teachers use to make mathematical meaning? How do they use the different languages they know? How do teachers deal with language diversity? How can mathematics learning be integrated with language learning?

Research in this section is related to discourse-focused research in mathematics education more generally, including research that examined mathematics classroom interaction in terms of participation in mathematical discussions (Lampert, 1990), mathematical argumentation (Forman, 1996; Krummheuer, 1995), and socio-mathematical norms (Cobb, Wood, & Yackel, 1993). From this perspective, learning to communicate mathematically involves more than learning vocabulary or understanding meanings. Instead, communicating mathematically is seen as using social, linguistic, and material resources to participate in mathematical practices (see studies in Kieran, Forman, & Sfard, 2002; Seeger, Voigt, & Waschescio, 1998; Sfard 2008). For Sfard (2008), learning to think mathematically is learning to communicate mathematically with oneself and, as such, arises through interaction with others. Mathematical discourse (drawing on Gee's, 1996, definition of discourses) includes ways of talking, acting,

鉴Gee，1996对交流的定义）包括说、做、互动、思考、确信、阅读和书写数学，以及关于数学的价值、信念和对情境的看法，词汇的含义被视为存在于社会和文化情境之中，而不是固定的。与认知观点不同，这一观点对学生必须纳入他们记忆词典的词汇并没有附加任何的预先设定的含义，相反，词汇的含义由参与者在互动中共同建构。近期的研究扩展了这些理念并借鉴了诸如系统功能语言学等更多的理论观点来记述更为详细的关于数学交流特性的研究，并突出视觉图像和符号系统的作用（例如O'Halloran，2005）。

这方面的研究探讨了三个相关的主题：（1）使数学有意义的资源，（2）语码转换，（3）教师和教学。下面对这三个主题进行详细的讨论。

使数学有意义的资源

本节所讨论的研究将一种语言（或多种语言）视为学习者和教师为使数学有意义所使用的一种资源（在某些情况下视为一种工具），这一立场与将语言多元化视为屏障或障碍形成对比（例如，可以参见Gorgorió & Planas，2001）。本节所讨论的文献作者都强调了在语言多元化情形下，学生使用的一些资源并探讨了学生和教师是如何调配这些资源的，这些资源包括不同的语言、语言类别、肢体语言、图表等，以及语码转换（下一节将做专门介绍），甚至还有参与不同形式的课堂互动的方式（Barwell，2012a）。

莫谢科维奇（1999，2002，2007a，2008，2009，2011b）记录了学生如何使用多种模式，包括肢体语言。她的分析显示了学生在学习情境下如何利用存在的资源，同时只有在把肢体语言和实物看作资源的情况下数学推理和意义建构的生成才会变得可见。她展示了学生如何使用肢体语言来表明他们的意思并用他们面前的具体实物，比如绘图来做参照，或者澄清所给出的说明（例如Moschkovich，2002，2008），这项研究还显示了学生如何将他们的第一语言作为资源来使用，例如在所参与的讨论中加入他们第一语言中的词汇（例如Moschkovich，1999，2002）。

这个多重含义的观点将重点从询问双语学生遇到什么样的困难转到探求教学如何能够支持双语学生参与讨论。教师可以通过利用多种解释、把方法建立在学生自己的推理之上、将学生的推理和数学概念联系起来等方式来支持数学讨论（Moschkovich，2011b），这些策略可以作为双语学生参与讨论的模式，同时还能把学习建立在学生推理之上，保持讨论与数学概念的联系（参见Khisty & Chval，2002；Razfar，2013）。

以下这些研究表明，英语学习者学生从小组讨论和解决问题的机会中获益（Adler，1997；Brenner，1998；Khisty，1995），特纳和同事进行了两项关于英语学习者学生在高认知需求的活动中互动情况的研究（Turner, Dominguez, Maldonado, & Empson, 2013；Turner & Celedon-Pattichis, 2011）。与其他研究一样，这些研究强调了多重符号资源（两种语言、多重表征）和模式（谈话、文本、指点、肢体语言）的重要性，类似地，普拉纳斯（2014）确认了在西班牙加泰罗尼亚的巴塞罗那的数学课堂中所使用的三种语言实践，其中一些学生有移民背景，这三种语言活动都与学生使用数学词汇的方式有关：第一，学生有时会谨慎地对待这些术语；第二，学生有时会发明替代的术语；第三，学生有时会使用直接翻译的术语。

在英国的民族志研究中，巴韦尔（例如2003b，2005b，2005c，2005d，2009）研究了小学高年级将英语作为附加语学习的学生理解算术文字题的不同方法。巴韦尔不是分析学生写出的解题过程，而是多次记录了20位学生以二人小组的方式书写和解文字题的过程。他提出了学生使用语言来建构和解释文字题的四个方面：（1）类别，（2）叙述经验，（3）数学结构，（4）语言形式。他的研究表明，即使是在英语学习的初期，学生也能识别文字题的类别特征并在解题过程中利用这些特征，利用他们自身的经验来理解文字题中的情境（比如他们自身购物的经历），也能够把文字题中的文字与其所具有的数学结构之间建立联系，学生关于文字题中具体语言特征的讨论也有助于他们对数学含义的理解。巴韦尔（2005b，2005d）的研究还显示了，学生在文字题方面的学习与其自身身份方面的构建有关（比如他们的社会阶

interacting, thinking, believing, reading, and writing mathematics, as well as mathematical values, beliefs, and points of view of a situation. Word meanings are seen as socially and culturally situated rather than fixed. In contrast to a cognitive perspective, this view does not attach a priori meanings to words that students must incorporate into a mental dictionary. Rather, meaning is coconstructed by participants in interaction. More recent work has extended these ideas and drawn on additional theoretical perspectives, such as systemic functional linguistics, to develop much more detailed accounts of the nature of mathematical discourse and highlight the role of visual images and the symbolic system (e.g., O'Halloran, 2005).

Research in this strand has addressed three related topics, which are discussed in more detail below: (1) resources for meaning making in mathematics, (2) code-switching, and (3) teachers and teaching.

Resources for Meaning Making in Mathematics

The work discussed in this section looks at language (and languages) as a resource (and in some cases a tool) which learners and teachers use to make mathematical meaning. This position is in contrast to a view of language diversity as a barrier or obstacle (see, for example, Gorgorió & Planas, 2001). The researchers whose work is discussed in this section highlighted some of the resources students use in situations of language diversity and explored how these resources are deployed by students and teachers. These resources include different languages, genres, gestures, diagrams, and so forth, as well as code-switching (to which the next section is devoted) and even ways of participating in different forms of classroom interaction (Barwell, 2012a).

Moschkovich (1999, 2002, 2007a, 2008, 2009, 2011b) has documented how students use multiple modes, including gestures. Her analyses show how students use resources available in the learning situation and how mathematical reasoning and sense making often become visible only if gestures and objects are seen as resources for meaning making. She has shown students using gestures to illustrate what they mean and to refer to the concrete objects in front of them, such as drawings, or to clarify descriptions (e.g., Moschkovich, 2002, 2008). This work also shows how students use their first language as a resource, for example interjecting words from their first language into their contributions during discussions (e.g., Moschkovich, 1999, 2002).

This perspective on multiple meanings shifts the emphasis from asking what difficulties bilingual students encounter to asking how instruction can support bilingual students in participating in discussions. Teachers can support mathematical discussion by using multiple interpretations, building on students' own reasoning, and connecting student reasoning to mathematical concepts (Moschkovich, 2011b). These strategies can serve as a model for engaging bilingual students in discussions that simultaneously build on student reasoning and keep the discussion connected to mathematical concepts (see also Khisty & Chval, 2002; Razfar, 2013).

Following work by other researchers showing that ELL students benefit from opportunities to discuss and work through problems in small groups (Adler, 1997; Brenner, 1998; Khisty, 1995), Turner and colleagues conducted two studies that examined ELLs' interactions in high cognitive demand tasks (Turner, Dominguez, Maldonado, & Empson, 2013; Turner & Celedon-Pattichis, 2011). As with other research, these studies underline the importance of multiple semiotic resources (two languages, multiple representations) and modes (talk, text, pointing, gesturing). Similarly, Planas (2014) identified three language practices used in mathematics classrooms in Barcelona, Catalonia-Spain, in which some students came from immigrant backgrounds. The three practices all related to the way mathematical vocabulary was treated by students: First, sometimes students treated such terms with caution; second, sometimes students invented alternative terms; and third, sometimes students used literal translations.

In ethnographic research in the United Kingdom, Barwell (e.g., 2003b, 2005b, 2005c, 2005d, 2009) examined ways in which upper elementary school students who were learners of English as an additional language made sense of arithmetic word problems. Rather than analyzing students' written solutions, Barwell recorded 20 students working in pairs on multiple occasions as they wrote and solved word problems. He identified four aspects of language that the students made use of in constructing and interpreting word problems: (1) genre, (2) narrative experience, (3) mathematical structure, and (4) linguistic form. His findings showed that even students in the beginning stages of learning English recognized features of the word problem genre and could use these features in their work. The students made use of narrative accounts of their experience to make sense of word problem scenarios (e.g., accounts of their experience of shopping), they made connections between the word problem text and its mathematical structure, and students' discussions about specific linguistic features of word problems contributed to their understanding of its mathematical meaning. Barwell (2005b, 2005d)

层或宗教信仰）。这些发现突出了话语资源的不同形式，比如之前没有考虑过的类别和表述。

最后，尽管数学教育研究人员通常只是关注数学学习的结果，但在很多课堂教学中，还是期望教师和学生能把数学的学习和语言的学习结合起来。而这一点又成为了应用语言学研究中持续增长的一个关注点（例如，可以参见 Barwell, 2005a; Dalton-Puffer, 2007; Davison & Williams, 2001; Mohan, 1986; Schleppergrell, 2007）。巴韦尔（2005c）在解小学数学文字题的背景下研究了这个问题，巴韦尔的分析强调了学生对语言形式的关注和对数学结构的关注之间的反射性相互作用，例如，一对学生在争论他们自己写出的文字题（"如果有5位儿童和20本书，你可以给每位儿童几本书？"）需要使用加法还是除法过程中，讨论了"给"这个词可能有的不同意义和时态。这样的讨论是语言学的，但也是数学的，因为它们有助于学生对文字题中数学结构的理解。

语码转换

在数学教育中关于语言多元化的初期研究阶段，对于语码转换[2]最常见的定义是指在发言或对话中使用两种或更多的语言：

> 语码转换是指一个人（或多或少是有意地）交替使用两种或更多的语言……语码转换是有目的的[且]语言之间的转换

有着重要的社会和权力方面的内容，就像方言之间的转换一样。（Baker, 1993, 第76—77页）

语码转换不同于整合，格罗让（1982）把整合描述为从另一种语言中借用一个词并将其整合到基础语言中。整合是世界各地许多多语言数学课堂中的一个常见现象，数学英语中的词汇被整合到当地的语言中（例如 Kazima, 2007; Moschkovich, 2007c; Setati, 2005）。在多数的数学教育研究中，整合并不属于语码转换。

从历史上看，语码转换的地位较低，有些人认为需要语码转换的人对每一种语言的掌握都不够好，以至于不能够只用一种语言交流。格罗让（1982, 第147页）指出，正是因为这样的一些态度，有些使用多种语言的人选择在一些情况下不进行语码的转换或者限制这样的转换以保证他们不会受到羞辱。在数学课堂中，学习者可能会在和其他学习者交流时转换语言，而不是在和教师的交流中进行转换（例如 Farrugia, 2009a, 2009b; Setati, 1998）。

关于数学课堂的研究表明了语码转换在数学教学过程中作为交流资源的价值（Adler, 2001; D.V. Jones, 2009; Khisty, 1995; Moschkovich, 1999, 2000, 2002, 2007a, 2007c, 2011b; Planas & Setati, 2009; Razfar, 2013; Setati, 1998, 2005; Setati & Adler, 2001; Setati & Barwell, 2006）。在大多数课堂中，语码转换的发生似乎是由认知和课堂管理因素引起的（Adendorff, 1993; Merritt, Cleghorn, Abagi, & Bunyi, 1992），有研究显示，教师使用语码转换来集中或重新获得学习者的注意力，或者澄清、优化及强调课堂学习材料。数学课堂中的语码转换只有一部分是受官方语言政策支配的（Farrugia, 2009a, 2009b; Setati, 2005），即使官方政策存在，教师也是根据有效交流的需要来随时选择要使用的语言（Adler, 2001; Farrugia, 2009a, 2009b; Setati, 1998, 2005）。学生之间的语码转换也被描述为，是在他们重复或者澄清自己的解释时出于礼貌的需要（Moschkovich, 2007a）或者是认知方面的需求（Clarkson, 2007）。一般来说，学生之间的语码转换可能会由情境中的多个方面所引起：环境、社会角色、主题、谈话对象或者身份标记（Torres, 1997; Zentella, 1997）。

塞塔蒂（2005）对南非多语言小学数学课堂中语言实践的分析表明，虽然语码转换受到重视，但英语仍是交流的主流语言，而且学习者和教师也倾向于使用英语来教授数学。塞塔蒂（2008）的分析还进一步表明，虽然学习者的家庭语言可以成为教学的资源并被用于进行概念沟通，但是教师使用语码转换主要是为了表现与学生的团结，英语通常仍然被用作权威语言和程序性话语（Setati, 2005）。

多明格斯（2011）在美国所做的研究中也发现了学生使用不同语言功能方面的不同，他考察了数学问题背景下（日常和学校）学生以二人小组方式解决问题过程中谈话的组织和协

also showed how students' work on word problems was related to students' construction of aspects of their identity (e.g., their social class or religious identities). These findings highlight forms of discursive resources, such as genre and narrative, not previously considered.

Finally, although researchers in mathematics education have only generally paid attention to mathematical outcomes, in many classrooms, teachers and learners are expected to combine mathematics learning with language learning. How to do this constitutes a growing focus of research in applied linguistics (see, for example, Barwell, 2005a; Dalton-Puffer, 2007; Davison & Williams, 2001; Mohan, 1986; Schleppergrell, 2007). Barwell (2005c) examined this issue in the context of solving elementary school arithmetic word problems. Barwell's analysis highlights the reflexive interplay between students' attention to linguistic form and their attention to mathematical structure. For example, one pair of students discussed the different possible meanings and tenses of "gave" as they debated whether a word problem they had written themselves ("if you have five children and twenty books how many can you gave the books") required addition or division. Such discussions are linguistic, but they are also mathematical, since they contribute to the students' understanding of the mathematical structure of their word problems.

Code-Switching

The most common definition of code-switching[2] used in initial work on this topic in mathematics education refers to the use of two or more linguistic varieties within the same utterance or conversation:

> Code-switching is when an individual (more or less deliberately) alternates between two or more languages.... Code-switches have purposes [and there] are important social and power aspects of switching between languages, as there are between dialects. (Baker, 1993, pp. 76–77)

Code-switching is distinguished from *integration*, which Grosjean (1982) describes as borrowing a word from another language and integrating it into the base language. Integration is a common feature in many multilingual mathematics classrooms all over the world, where words from mathematical English are integrated into the local languages (e.g., Kazima, 2007; Moschkovich, 2007c; Setati, 2005). In most research in mathematics education, integration is not regarded as code-switching.

Historically, code-switching has had an inferior status, with some believing that people who code-switch know neither language well enough to converse in either one alone. Grosjean (1982, p. 147) points out that because of these attitudes, some people who use multiple languages prefer not to code-switch or restrict their switching to situations in which they will not be stigmatized. In mathematics classrooms, learners may choose to switch only when interacting with other learners and not with the teacher (e.g., Farrugia, 2009a, 2009b; Setati, 1998).

Research in mathematics classrooms has demonstrated the value of code-switching as a resource for communication while teaching and learning mathematics by teachers and students (Adler, 2001; D. V. Jones, 2009; Khisty, 1995; Moschkovich, 1999, 2000, 2002, 2007a, 2007c, 2011b; Planas & Setati, 2009; Razfar, 2013; Setati, 1998; 2005; Setati & Adler 2001; Setati & Barwell, 2006). In most classrooms, the occurrence of code-switching seems to be motivated by cognitive or classroom management factors (Adendorff, 1993; Merritt, Cleghorn, Abagi, & Bunyi, 1992); teachers have been documented using code-switching to focus or regain learners' attention or to clarify, enhance, or reinforce lesson material. Code-switching in mathematics classrooms is only partially dictated by formal language policy (Farrugia, 2009a, 2009b; Setati, 2005). Even if official policy exists, teachers make individual moment-to-moment decisions about language choice that are mostly determined by the need to communicate effectively (Adler, 2001; Farrugia, 2009a, 2009b; Setati, 1998, 2005). Code-switching among students has also been described as motivated by politeness when they are repeating or clarifying an explanation (Moschkovich, 2007a) or due to cognitive demands (Clarkson, 2007). In general, code-switching among students may be motivated by multiple aspects of a situation: setting, social roles, topics, addressees, or markers of identity (Torres, 1997; Zentella, 1997).

Setati's (2005) analysis of language practices in multilingual primary school mathematics classrooms in South Africa showed that although code-switching was encouraged, English remained the dominant language of interaction, and learners and teachers preferred that mathematics be taught in English. Setati's (2008) analysis further shows that although the learners' home languages can be a resource for teaching and learning and were used for conceptual discourse, teachers used code-switching mainly to show solidarity with students. English was generally used as the language of authority and procedural discourse (Setati, 2005).

Domínguez (2011) also found a difference in the function of students' different languages in his examination of the organization and coordination of talk across mathematics problem contexts (everyday and school) and

调情况，他发现使用英语的讨论反映了在学校的交流模式，而使用西班牙语的讨论则反映了更为典型的家庭和社区的交流模式，学生更可能会用西班牙语而不是英语来分享他们的想法（无论是重复还是提出新的想法），与英语中更多的个体探索相比，学生在西班牙语环境中会更多地参与联合探索。

最近，塞塔蒂和她的同事们探索了在多语言课堂中数学教学的策略，探索的起因来自于一个研究的发现，即在南非，英语的社会价值意味者学习者倾向于用英语学习数学（见下一节关于社会政治观点的讨论）。塞塔蒂和她的同事们制定了在多语言课堂中教授数学时采取有意识、主动和有策略地使用学习者家庭语言的策略（例如 Molefe, 2009；Mpalami, 2007；Nkambule, 2009；Vorster, 2008），这一方法包括上课和给学生布置任务时用学生的家庭语言，但也用英语来给出任务，学生可以根据自己的需要参照任何一种语言，从而更好地获得数学知识。塞塔蒂（2012）认为这种方法更为可取，它使学生在口头和书面交流中主动地（而不是被动地）使用自己的家庭语言。

教师与教学

虽然从语言交流的角度来看，很多研究关注的是学生的意义建构，但也有一些研究对教师和他们在教学中制定的策略给予了更多的关注。阿德勒（1997，1998，1999，2001）在南非开展的数学课堂中语言多元化的研究是一个重要的贡献，她的研究凸显了在后续研究中被考虑的一些问题和想法，最值得注意的是阿德勒发现了数学教师在语言多元化的背景下所经历的"教学困惑"。

阿德勒所采用的方法的一个关键环节是视频录制教师的课堂教学，再和教师一起回放视频片段并讨论当时发生了什么。基于广泛的社会文化理论观点，阿德勒确定了三个经常出现的关键性困惑：（1）透明度，（2）语码转换，（3）调解。关于透明度的困惑涉及数学语言的可见性或缺乏可见性。在比较流畅的数学课堂交流中，语言本身会是透明的——能够被使用但不需要对其进行讨论，然而，在某些情况下，参与阿德勒研究的教师觉得应该关注数学语言的方方面面，同时，他们又不想打断学生的数学思维（Adler, 2001）。语码转换的困惑则与使用一种以上的语言做数学的实践有关。教师的困惑是如果学生使用他们的家庭语言，学生会较好地理解数学并能更流畅地做出数学解释，但是如果学生一直使用家庭语言，他们就可能不会达到熟练地使用英语做数学（Adler, 2001）。最后，调解的困惑包含了教师在学生科学概念发展中扮演的与调解角色有关的各种困惑（从维果茨基理论的角度看），阿德勒认为这一困惑与教师想要包括所有学习者的愿望和制定明确的数学概念的相关需求有关（Adler, 1998, 2001）。

一些研究人员研究了什么样的教学策略能够有效地支持学生的数学学习，特别是在第二语言环境中。在一项早期的研究中，基斯蒂（1995）比较了美国三个二年级班级和两个五年级班级的数学教学，教师都具有西班牙语-英语双语能力，所有班级都有一些讲西班牙语的学生。结果发现，教学效果好的教师都同时关注了数学语言和数学内容本身，这一发现意味着这些教师以自己特定的方式经历了阿德勒的困惑。

古铁雷斯（2002）同样研究了在美国一所学校中使用的策略，这些策略成功地鼓励了来自不同西班牙语-英语语言背景的学生（如以西班牙语为主的、双语的、以英语为主的）在高中修学高级数学课并取得了成功，这项研究中重要的策略包括西班牙语的使用和小组活动，以及明确地使用数学语言。根据古铁雷斯的发现，教师教学策略得以成功实现的一个关键因素是"教师对学生语言背景和数学需求有着敏锐的了解"（第1079页）。最后，特纳等人（2013）也表明了教师通过重新表达、提出具有挑战性的问题和其他有效举措对调控学生学习机会具有很重要的作用。

关于教师和教学的最后一个工作涉及教师谈论语言多元化所带来的他们自己之间的交流（例如 Chitera, 2009；Takeuchi & Esmonde, 2011）。例如在加拿大，竹内和埃斯蒙德（2011）分析了参与针对支持英语学习者学生学习的专业发展培训的教师之间的交流对话。他们

across languages that arose when students solved problems in pairs in the United States. Discussions in English reflected interaction patterns in school, and discussions in Spanish reflected interaction patterns more typical of the home and community. Students were more likely to share knowledge in Spanish than in English (both to reproduce and reinvent ideas), and students engaged in joint exploration of ideas in Spanish in contrast to more individual approaches in English.

More recently, Setati and her colleagues have explored strategies for teaching mathematics in multilingual classrooms, prompted by findings showing that the social value of English in South Africa meant that learners preferred to study mathematics in English (see the next section on sociopolitical perspectives). Setati and her colleagues developed strategies for the deliberate, proactive, and strategic use of the learners' home language when teaching mathematics in multilingual classrooms (e.g., Molefe, 2009; Mpalami, 2007; Nkambule, 2009; Vorster, 2008). This approach involved giving students tasks written in all the home languages present in the class, as well as in English. Students could refer to any language version whenever they needed to, thus enabling greater access to mathematical knowledge. Setati (2012) argues that this approach is preferable, since it involves the proactive (rather than reactive) use of the students' home languages in both oral *and* written communication.

Teachers and Teaching

Although much research from a discursive perspective focuses on students' meaning making, some studies have given more attention to teachers and the strategies they develop in their teaching. Adler's (1997, 1998, 1999, 2001) work, conducted in South Africa, has been a key contribution to research on language diversity in mathematics classrooms, highlighting a number of issues and ideas that have been taken up in subsequent research. Most notably, Adler identified what she terms "teaching dilemmas" experienced by mathematics teachers in contexts of language diversity.

A key part of Adler's methodology involved Adler video recording a teacher's lesson and then playing back extracts with the teacher and discussing what was happening. Based on a broadly sociocultural theoretical perspective, Adler identified three key dilemmas that seemed to arise quite frequently: (1) transparency, (2) code-switching, and (3) mediation. The dilemma of transparency concerns the visibility, or lack of visibility, of mathematical language. In smooth mathematics classroom interaction, language itself would be transparent—

used but not discussed. In some situations, however, the teachers in Adler's research felt that attention should be drawn to aspects of mathematical language; at the same time, they hesitated to interrupt students' mathematical thinking (Adler, 2001). The dilemma of code-switching arose for the teachers in relation to the practice of using more than one language to do mathematics. The dilemma for teachers is that their students may understand mathematics better and be more fluent in their own mathematical explanations if they use their home languages; if they do so all the time, however, they may not develop sufficient proficiency in doing mathematics in English (Adler, 2001). Finally, the dilemma of mediation encompassed a variety of dilemmas relating to the role of teachers in mediating the development of scientific concepts (in the Vygotskian sense). Adler sees this dilemma as relating to the desire to be inclusive of all learners and the associated need to make explicit key mathematical concepts, such as generalization (Adler, 1998, 2001).

Some researchers have examined what kinds of teaching strategies may be effective in supporting student achievement in mathematics, particularly in second language settings. In a key early study of this type, Khisty (1995) compared teaching in three second-grade and two fifth-grade mathematics classrooms in the United States. The teachers were all Spanish-English bilingual, and all the classes included some Spanish-speaking students. Teachers who were more effective paid more attention to the language of mathematics as well as to the mathematics itself, a finding that implies that these teachers navigated Adler's dilemmas in a particular way.

Gutiérrez (2002) similarly studied strategies, used in a school in the United States, that were successful in encouraging students from a variety of Spanish-English language backgrounds (e.g., Spanish dominant, bilingual, English dominant) to pursue and be successful in advanced mathematics courses in high school. Strategies that were significant in this work included the use of Spanish and group work and explicit use of mathematical language. According to Gutiérrez, a key underlying factor in the successful development of the teachers' specific strategies was "the teachers' keen understanding of students' linguistic backgrounds and mathematical needs" (p. 1079). Finally, Turner et al. (2013) showed the importance of the teacher in mediating students' opportunities to learn through revoicing, posing challenging questions, and other productive moves.

A final strand of work on teachers and teaching involves the discourses teachers themselves draw on to talk about language diversity (e.g., Chitera, 2009; Takeuchi & Esmonde, 2011). In Canada, for example,

注意到在培训介入开始时，教师有关学习数学的英语学习者学生的交流主要围绕着学生的学习障碍和阻碍，但教师参与了研究人员的几次讨论会并做了一项调查之后其对话内容有所转变，由此，竹内和埃斯蒙德追踪了一对教师在他们的调查过程中是如何把对话重点转移到英语学习者学生所带来的资源方面。

评论和批评

从交流角度开展的研究将语言多元化视为一种社会现象，并考察了数学课堂内学生理解数学意义的多种方法，以及教师所找到的支持学生学习的策略。这类研究表明，学生可以利用他们的家庭语言和经验、数学课堂类型的知识、图表、肢体语言，以及教师或同学给予的解释来成功地学习数学。因此，这项工作挑战了认为用双语、多语言和第二语言学习数学的人缺乏语言（教学中使用的语言）能力这一观点，指出了需要改进的是以缺失为核心的研究和政策，从这个角度来说，语言不是一把进入数学的钥匙（Barwell, 2005a），而是不同资源中的一种。

然而，这项研究一个很明显的局限是对学生课堂中意义建构的关注比较狭窄，同时对有效的教学策略也有着一些功利主义的关注。大量研究（包括本节讨论的有关认知观点的研究）表明，课堂互动不是数学学习结果的唯一因素，其他如社会经济状况、种族、学习的数学课程、家长的参与以及分流等因素都对学校数学成绩有影响，并早已得到承认（Mosqueda, 2010；Secada, 1992）。将这种较大规模的社会或学校因素与小型数学活动的细节联系起来仍然是一个需要解决的挑战，研究人员可以通过更多地关注数学课堂中语言使用的方式来解决这个问题，这些方式可能会使学习者处于不同的地位或拥有不同的权力，还可以关注课堂互动的方式，这些方式可能会导致语言资源不一样的使用。

第二个局限来自于研究对象的本地化、小规模等特点。鉴于在国家、地区、学校和课堂环境中语言多元化缺乏统一性，这类研究的发现是基于情境的，这种依赖于情境的研究既有优点也有不足。这些研究对背景的详细关注能够提高我们在语言多元化背景下深度理解数学课堂中所使用的语言，这些研究对实践的拓展是有价值的，他们可以提供详细的案例以促使教师反思和发展自身的实践。然而，这类研究的不足在于缺乏普遍性，这种普遍性可能来自对于理论发展的更多关注，特别是对学生或教师语言使用模式之间的更多关注，以及考虑他们所学习的语言环境的特征。

巴韦尔（2012b）最近试图通过综合已有研究来解决这个问题，他借鉴贝克汀（1981）的语言理论，展示了语言多元化和数学这一主题的文献中提到的许多不同背景下所出现的共同的问题，并突出强调了下面的问题：家庭和学校语言之间、正式和非正式的数学语言之间、语言政策和数学课堂实践之间，以及学习数学的语言和融入世界的语言之间的问题。例如，家庭和学校语言之间的矛盾在语码转换的研究中是显而易见的，但在语码转换没有发生的情况下也看得到。在前者中，这个问题可能会成为教师的困惑（Adler, 2001），而在后者中，缺乏语言转换的可能性会给学生带来挑战，关键在于尽管在给定的数学课堂环境下问题会以不同的形式出现，但这些问题也都是普遍存在的。的确，依据贝克汀的理论，巴韦尔（2012b，2014）认为这些问题是不可避免的。

社会政治的观点

近几十年来，数学教育研究中的政治问题越来越受到关注。许多研究都强调了数学的关键方面，以及从更为广泛的社会和政治背景中思考数学教育研究的重要性。一些词汇，如批判性数学教育（Skovsmose, 1994）、社会政治观点（例如Gutiérrez, 2013）、民族数

Takeuchi and Esmonde (2011) analyzed the discourse of teachers participating in a professional development intervention focused on supporting ELL students. They noted that at the start of the intervention, the teachers' discourse relating to ELL students in mathematics was couched in terms of barriers and obstacles. The teachers participated in discussion sessions with researchers and conducted an inquiry project. Takeuchi and Esmonde traced how the discourse of one pair of teachers shifted over the course of their inquiry project to focus more on the resources that ELL students bring.

Commentary and Critique

Research based on a discursive perspective sees language diversity as a social phenomenon and examines the diverse ways in which students make meaning in mathematics classrooms, as well as the strategies teachers have found to support students. This research has shown how students may draw on their home languages and experiences, knowledge of mathematics classroom genres, diagrams, gestures, and the meanings made available by their teacher or their peers to successfully learn mathematics. As a result, this work has challenged deficit-oriented research and policy in which bilingual, multilingual, or second language learners of mathematics were seen as lacking (in the language of instruction) and needing to be fixed. From this perspective, language is not a key giving access to mathematics (Barwell, 2005a), but a resource among several others.

A clear limitation of this research, however, is its rather narrow focus on students' classroom meaning making, coupled with a somewhat utilitarian concern for effective teaching strategies. A good deal of research, including work discussed in the section on cognitive perspectives, has shown that classroom interactions are not the only significant factor in mathematics learning outcomes. That factors such as socioeconomic status, race, access to mathematics courses, parental involvement, and tracking have an impact on school mathematics achievement has long been established (Mosqueda, 2010, Secada, 1992). Linking such larger scale societal or school factors to the minutiae of a small-group mathematics activity remains a challenge to be addressed. Researchers could address this issue by devoting more attention to the way language use in mathematics classrooms may position learners as having different status or power or to the way classroom interactions may result in differential use of language resources.

A second limitation arises from the localized, small-scale nature of much of the research. Given the diverse nature and lack of uniformity of language diversity across national, regional, school, and classroom contexts, the findings of these studies must be seen as highly situated. This situatedness is arguably both a strength and a weakness; the detailed attention to context in these studies adds depth to our understanding of how language is used in mathematics classrooms in situations of language diversity. Such studies are valuable for the development of practice in offering detailed examples with which to prompt teachers to reflect on and develop their own practice. The weakness in this work, however, is a lack of generality. Such generality is likely to come from more attention to theoretical development and, in particular, to a greater focus on the relationship between the patterns of language use by students or teachers and the consideration of the specific nature of the language context in which they are learning.

Barwell (2012b) has recently sought to tackle this issue in a synthesis of much of the work discussed in this section. Drawing on Bakhtin's (1981) theory of language, Barwell shows how a common set of tensions arise across the many different contexts referred to in the literature on language diversity and mathematics. Barwell (2012b) highlights the following tensions: between home and school languages, between formal and informal mathematical language, between language policy and mathematics classroom practice, and between a language for learning mathematics and a language for getting on in the world. For example, the tension between home and school languages is apparent in the research on code-switching, but also in contexts in which code-switching does not happen. In the former, the tension can appear as a dilemma for teachers (Adler, 2001); in the latter, however, the lack of the possibility of code-switching can present challenges for students. The point is that although the tension plays out differently in any given mathematics classroom context, the presence of such a tension is widespread. Indeed, based on Bakhtin's theory, Barwell (2012b, 2014) argues that such tensions are inevitable.

Sociopolitical Perspectives

In recent decades there has been an increasing concern with political issues in mathematics education research. Much research has emphasized critical aspects of mathematics and the importance of doing research that takes into account the broader social and political contexts of mathematics education. Terms such as critical mathematics education (Skovsmose, 1994), sociopolitical perspectives (e.g., Gutiérrez, 2013), ethnomathematics

学（D'Ambrosio，1991；Powell & Frankenstein，1997）、社会正义（例如Gutstein，2006），以及其他一些词汇都在研究中被提及，并且超越了主导多数数学教育研究的认知和教学法层面。勒曼（2000）将这一转向社会和政治关切的举措称为数学教育研究的社会转向，他使用社会转向来表示"把含义、思考和推理看作社会活动的产物开始出现在数学教育研究的理论之中"（第23页），如同他所解释的，社会转向超越了社会互动所提供的一种产生或刺激个人内部意义建构活动的火花，从而将思考和学习数学视为社会产生和组织的活动。这种思考也拓展到数学教与学中语言的作用，例如，泽温伯根（2000）用语言资本的概念来表示对于一些学生来说，学习数学既是学习特定的概念或方法，也像是学习学校数学交流的"破解代码"。这个想法与库珀和邓恩（2000）的研究结果是一致的，库珀和邓恩的研究在一定程度上显示了文字题的语言是如何按照社会阶层方式对学生进行分层的，在他们的研究中，中产阶级的孩子更有可能把阅读文字题作为课堂任务并以预期的方式进行回应，而低阶层的孩子对同样问题的阅读往往会导致与预期规范不一致的回应，尽管数学上也正确。这项研究显示了语言是如何在数学课堂中产生真正的政治影响，以及学生可以接触到与他们社会经济背景相关的强大的数学形态。在本节中，我们回顾采用社会视角的两个不同的研究方向：（1）基于语言政治概念化的研究，（2）利用民族数学理念的研究。

在多语言数学课堂中语言的政治作用

在20世纪初期，数学教育和语言多元化领域里的社会转向变得更为突出，特别是随着引用吉（1996）的研究的增加，研究人员意识到如果他们想要以一种一致和全面的方式来解释语言实践的话，他们在多语言的数学课堂中必须超越语言的认知和教学方法层面，考虑语言的政治作用（Setati，2005）。根据吉（1999）的观点，塞塔蒂认为语言是具有政治性的，语言的政治性不仅在于政策制定的宏观层面，也在于课堂互动的微观层面。在对话和决策中语言可以用来排除或包容不同的人，语言是界定一个人遵守群体价值观的一种方式（Zentella，1997）。因此，有关在多语言数学课堂里使用哪一种语言、如何使用以及使用目的就不仅仅是教学方法或认知方面的决定，也是政治性的决定——它们影响着社会物品如何被分配或应该如何分配（Setati，2005）。即使在数学课堂里，言语也从来不仅仅是言语，语言不仅是表达思想的一种媒介（一种文化工具），它还是教师用来制定（被认可的）一个特定的"谁"（身份）从事特定的"什么"（情景活动）的社会政治工具。

透过这一视角，塞塔蒂（2005，2008）发现在像南非这样的语言多元化的环境中，数学和英语都具有象征力量，尽管学习者的英语流利程度有限，但在教学中很难让数学超越程序化的交流，语言多元化也很难超越团结与支持的范畴。在这种情况下，塞塔蒂认为数学的成绩是由一系列复杂的相互关联的因素所决定的，双语和多语言学习者的较差表现不能单一地归结于学习者的语言能力，而不去考虑注入学校生活的更广范围的社会、文化和政治因素（Setati，2012）。例如，普拉纳斯和西维（2013）利用加泰罗尼亚的巴塞罗纳和亚利桑纳的图斯康的数据显示了语言作为资源教学法（来源于本章关于交流观念研究的讨论）的积极价值是如何被语言的政治维度所削弱的，例如，教师鼓励学生使用多种语言来做数学，但这一做法或许与他们对这些语言（低等）价值或地位的社会政治意识形态背道而驰。在数学教育研究和实践中忽略语言的政治作用就如同是假设社会中不存在权力关系。

一个新出现的研究方向开始研究在语言多元化的数学课堂中学生的身份感是如何产生的，这样的工作往往涉及边缘化和权力赋予的问题。在瑞典进行的一项研究中，诺伦（2011）展示了移民学生在数学和社会身份的话语建构中复杂的相互作用，她展示了一位知晓班级里几个学生的语言（阿拉伯语）的教师如何促进学生

(D'Ambrosio, 1991; Powell & Frankenstein, 1997), social justice (e.g., Gutstein, 2006), and others, have been used in research that moves beyond the cognitive and pedagogic perspectives that dominate the majority of research in mathematics education. Lerman (2000) has referred to this move toward social and political concerns as the social turn in mathematics education research. He uses the term *social turn* to signal "the emergence into the mathematics education research community of theories that see meaning, thinking, and reasoning as products of social activity" (p. 23). As he explains, the social turn goes beyond the idea that social interaction provides a spark that generates or stimulates an individual's internal meaning-making activity to see thinking and learning mathematics as socially produced and organized. This approach extends to the role of language in learning and teaching mathematics. Zevenbergen (2000), for example, referred to the notion of linguistic capital to show how for some students, learning mathematics is as much about "cracking the code" of school mathematics discourse as it is about learning particular concepts or methods. This idea is consistent with Cooper and Dunne's (2000) findings that show in some detail how the language of word problems serves to stratify students along the lines of social class. In their study, middle class children were more likely to read word problems as classroom tasks and respond in expected ways; working class children's reading of such problems often led to responses that were not consistent with expected norms, though no less mathematically correct. This work shows how language has real political effects in mathematics classrooms, with students' access to powerful forms of mathematics linked to their socioeconomic background. In this section, we review two distinct strands of research adopting a social perspective: (1) research based on a political conceptualization of language and (2) research drawing on ideas from ethnomathematics.

The Political Role of Language in Multilingual Mathematics Classrooms

The social turn became more prominent within the area of mathematics education and language diversity in the early 2000s. With the growing use of the work of Gee (1996) in particular, researchers saw that they must go beyond the cognitive and pedagogical aspects of language in multilingual mathematics classrooms and consider its political role if they were to explain language practices in a coherent and comprehensive way (Setati, 2005). Drawing on Gee (1999), Setati argued that language is political and that the political nature of language is not only at the macrolevel of policy making but also at the microlevel of classroom interaction. Language can be used to exclude or include people in conversation and decision making. Language is one way to define one's adherence to group values (Zentella, 1997). Therefore, decisions about which language to use in multilingual mathematics classrooms, how to use it, and for what purposes are not just pedagogical or cognitive but also political—they have implications for how social goods are or ought to be distributed (Setati, 2005). Even in a mathematics classroom, words are never just words; language is not just a vehicle to express ideas (a cultural tool) but also a sociopolitical tool that teachers use to enact (i.e., to be recognized as) a particular "who" (identities) engaged in a particular "what" (situated activity).

Using this lens, Setati (2005, 2008) found that in a language-diverse context like South Africa, where mathematics and English both have symbolic power, a practice is forged in which it is difficult to move mathematics beyond procedural discourse and language diversity beyond solidarity and support. This is despite learners having limited fluency in English. It is in this context that Setati has argued that mathematics achievement is determined by a complex set of interrelated factors. Poor performance of bilingual and multilingual learners cannot be attributed to the learners' language proficiencies in isolation from the wider social, cultural, and political factors that infuse schooling (Setati, 2012). Planas and Civil (2013), for example, drawing on data from Barcelona, Catalonia, and Tuscon, Arizona, show how the positive value of a language-as-resource pedagogy (derived from the research discussed in the section on discursive perspectives) can be undermined by the political dimension of language. For example, a teacher's encouragement of students to use multiple languages to do mathematics may run counter to sociopolitical ideologies about the (inferior) value or status of these languages. To ignore the political role of language in mathematics education research and practice would be to assume that power relationships do not exist in society.

An emerging line of research has begun to examine how students' identities are produced in mathematics classrooms in situations of language diversity. Such work is often concerned with issues of marginalization and empowerment. In research in Sweden, Norén (2011) shows the complex interactions in the discursive construction of immigrant students' mathematical and social identities. She shows how the presence of a teacher who shares the language of several students in her class (Arabic) promotes students' positive identities as mathematics learners and also disrupts prevailing normalizing

作为数学学习者的正面身份，同时也打乱了规范的主流瑞典语的交流。普拉纳斯（2011）在巴塞罗纳和巴韦尔（2005b）在英国所做的研究中也有类似的发现。

民族数学观念

民族数学的视角拓展了数学活动的种类，超出了在教科书中和学校里所学习的数学（Bishop，1986；D'Ambrosio，1991；Nunes, Schilemann & Carraher, 1993）。这一观点强调，数学活动在不同的环境下以不同的方式出现，所有的文化群体都会产生数学概念，西方的数学只是众多数学活动中的一种（Bishop，1986）。

关注语言多元化的民族数学研究考察了数学在不同语言下是如何表达的。例如，在新西兰记载的毛利数学就被作为早先数学表达发展的一个例子而被广泛引用（见Barton, Fairfall & Trinick, 1998；Trinick, Meaney & Fairhall, 2014）。这项工作被视为是构成毛利人中等教育和课程开发的大范围政治计划的一部分，并最终保留了毛利语（Barton 等，1998；Meaney, Trinick & Fairhall, 2011）。

民族数学研究也探讨了不同语言中出现的不同的数学，也就是说，它假定任何给定的语言的语法和结构会影响数学关系和概念的开发和建构方式。例如，在一些语言中，计数词语和其他数学概念是动词（Barton, 2009；Barton 等，1998；Lunney Borden, 2011, 2013），而这种语言结构可以用来提升教学（Lunney Borden, 2011, 2013）。同样，在巴西，门德斯（2007）与亚马逊人的教师合作，以理解语言在数学中的作用，并为教师专业发展做出贡献。尽管数学教与学应该被社会政治化这样的想法受到人们的质疑，但在民族数学研究用于边缘化人群的数学教与学的背景下，这样的民族数学研究就具有了社会政治性的方面（例如，研究人员关注边缘化、权力的失去和不平等的问题）（Barton, 2009；Wagner & Lunney Borden, 2012）。与之前提到的认知研究相比，采用民族数学方法的研究是以一种不同的方式来思考语言与思维的关系，民族数学方法更侧重于语言为谈论数学提供可能性，而不是认知结构。

评论与批评

从社会政治角度开展的研究促使人们关注在语言多元化的情况下数学教与学所面临的更广泛的挑战，这些挑战不能简化为理解个人认知或共享的交流过程的技术问题，并根据这些技术问题开发和应用精准的教学方法。本节所描述的研究强调了语言的选择、身份以及许多学习者的边缘化等问题的重要性，这些问题都是数学课堂中语言多元化直接导致的结果。这类研究开始记录了这些问题的影响，并显示了在某些情境下（尤其是在南非），无论是利用隐性的或明确的具有政治特性的语言理论还是民族数学的理论，它们是如何产生影响作用的。民族数学做法或许导致了更为彻底地制定政治行动的意识（特别参阅Lunney Borden，2011，2013）。

尽管如此，对这些问题的研究仍然不够，现有的研究仍有很大的发展空间，特别是今后的研究需要更仔细地考察和解释动态和多重身份的发展、社会分层和边缘化等问题。虽然有研究表明语言多元化与数学表现差异有关（如在某些情况下第二语言学习者的表现不佳，而在另一些情况下双语学习者的表现却有所提高），但这些差异只能通过求助于那些含有社会政治维度的理论以及详细考察产生这些差异的机制的研究来解释。

一般性讨论：本领域的现状

我们力争找出语言多元化和数学教育研究中的主要发现和趋势。在本节中，我们会指出这一研究领域的优点和不足，并给出今后的研究方向，我们围绕几个广泛的议题展开了讨论。

背景和范围

本章中所综述的研究明显地提升了我们对数学教育中语言多元化的理解。然而，这些进展是在特定的背景下取得的，这些背景可概括如下：

discourses of Swedishness. Similar findings can also be seen in Planas's (2011) study in Barcelona and Barwell's (2005b) study in the United Kingdom.

Ethnomathematical Perspectives

An ethnomathematical perspective expands the kinds of activities considered mathematical beyond the mathematics found in textbooks or learned in schools (Bishop, 1986; D'Ambrosio, 1991; Nunes, Schliemann, & Carraher, 1993). This perspective emphasizes that mathematical activity appears in different ways in different settings, all cultural groups generate mathematical ideas, and Western mathematics is only one type of mathematical activity among many (Bishop, 1986).

Ethnomathematics research with a focus on language diversity has considered how mathematics is expressed in different languages. For example, the development of a Maori mathematics register in New Zealand is widely cited as an example of proactive mathematics register development (see Barton, Fairhall & Trinick, 1998; Trinick, Meaney & Fairhall, 2014). This work can be seen as forming part of wider political projects to develop Maori-medium schooling and curriculum and ultimately to maintain the Maori language (Barton et al., 1998; Meaney, Trinick, & Fairhall, 2011).

Ethnomathematics research has also explored the different mathematics that arises in different languages. That is, it is hypothesized that the grammar and structure of any given language affects the way mathematical relationships and concepts are developed and construed. For example, in some languages, counting words and other mathematical ideas are verbs (Barton, 2009; Barton et al., 1998; Lunney Borden, 2011, 2013). Awareness of this kind of linguistic structure can then be used to enhance teaching (Lunney Borden, 2011, 2013). Similarly, in Brazil, Mendes (2007) worked with teachers from Amazonian peoples both to understand the role of language in mathematics and to contribute to teacher professional development. Such ethnomathematical studies have a sociopolitical dimension (for example, researchers pay attention to issues of marginalization, disempowerment, and inequity) when they are used in the context of the mathematics learning and teaching of marginalized peoples, although the idea that mathematics learning and teaching should be seen as sociopolitical has been challenged (Barton, 2009; Wagner & Lunney Borden, 2012). Studies that take an ethnomathematical approach are based on a different way of thinking about the relationship between language and thought than the cognitive studies mentioned earlier; the ethnomathematical approach is more focused on the possibilities offered by language to talk about mathematics than on cognitive structures.

Commentary and Critique

Research drawing on sociopolitical perspectives has drawn attention to the broader challenges of learning and teaching mathematics in situations of language diversity. These challenges cannot be reduced to a technical problem of understanding individual cognition or shared discursive processes, from which precise methods of teaching can be developed and applied. The research described in this section underlines the importance of issues of language choice, identity, and the marginalization of many learners—issues that arise as a direct result of language diversity in mathematics classrooms. This work has begun to document the effects of such issues and show how they play out in some contexts (notably in South Africa), whether drawing on implicit or explicit theories of the political nature of language or on ethnomathematics. This latter approach has perhaps led to a more thoroughly worked out sense of political action (see, in particular, Lunney Borden, 2011, 2013).

Nevertheless, these issues remain underresearched, and the work that does exist leaves much room for further development. In particular, research needs to more carefully examine and explain issues like the development of dynamic and multiple identities, social stratification, and marginalization. Although research shows that language diversity is associated with differential mathematics performance (e.g., underperformance of second language learners in some contexts, enhanced performance of bilingual learners in others), these differences can only be explained through recourse to theories that include a sociopolitical dimension and through research that looks in detail at the mechanisms that produce them.

General Discussion: The State of the Field

We have endeavored to capture the key findings and trends in research on language diversity and mathematics education. In this section, we highlight strengths and weaknesses in this area of research and suggest possible future directions for research in relation to these strengths and weaknesses. We have organized the discussion around several broad themes.

Contexts and Coverage

The research we have summarized in this chapter has clearly advanced understanding about language diversity

- 发达国家学校系统中的移民或语言少数族裔的以第二语言学习数学的学习者（如对美国拉丁裔学生的研究）；
- 非洲南部的多语言数学课堂；
- 教学语言环境是英语，无论是在发达国家（美国、英国、澳大利亚）还是在非洲南部；
- 小学阶段。

因此，以下几个方面需要更多的研究：

- 在许多亚洲国家和拉丁美洲发现的独特的多语言生态学；
- 范围更为广泛的教学语言，特别是不同于英语和西班牙语的不太国际化的语言；
- 从语言学的角度研究聋哑人的数学学习，其中很多人至少掌握两种语言，包括手语；
- 中学阶段、高等教育阶段、教师教育、教师学习和教师职业发展（对于最后这一领域，可参阅Chitera，2011；Essien，2014）。

在数学内容方面，数学文字题在语言多元化研究中占有明显的主导地位，这一主导地位可以通过文字题给第二语言学习者带来的特殊的语言挑战来解释（Barwell，2009），几类研究中的第二个方面是有关线性函数方面的内容（Moschkovich，2002，2011b；Setati，Molefe & Langa，2008）。不同的数学内容会涉及数学语言的不同方面，例如，证明需要用到逻辑关系，而概率则需要使用条件形式（见Kazima，2007），因此，未来需要对具体的课程领域进行更系统地研究，而且目前的研究还没有提出特别针对数学教学的、不同语言能力水平的、不同数学能力的或者不同年龄组的学生的详细而明确的指导意见。

目前这一领域需要有关第二语言学习者如何学习和阅读不同数学文本（教科书、文字题，等等）的研究。在设计这类研究时，研究人员对阅读教科书、阅读文字题和阅读其他类型的数学书面文章的区分是非常重要的。考察那些正在用第二语言学习和阅读的学生，区分他们是否有能力用第一语言进行学习和阅读。另外，还需要进行纵向的研究，通过一段时间跟踪学生的数学学习和实践以及他们的语言学习和实践，能更好地理解数学和语言之间的相互作用。

为了设计基于学生资源的教学，研究人员需要详细研究双语和多语言学生用于数学推理的资源，研究需要区分多种形式（书面的和口头的），接受能力和生成能力，听力理解和口语理解，口语言的理解和生成能力，以及书面文本的理解和生成能力。

最后，大多数研究项目的持续时间相对较短，即对学习者数学思维和知识的发展，或者学习者数学语言的发展过程所知甚少，因此，我们强调需要进行跟踪学习者数年的纵向研究来发现在较短时间研究中可能不太明显的现象。

理论导向

我们将这一领域的研究划分为三个阶段：（1）以认知为导向的研究，（2）以交流为导向的研究，（3）以社会政治为导向的研究。诚然，这种划分有些简单，例如，大多数社会政治方面的研究都借鉴了交流的观点，但都在课堂交流之外或是取代了课堂交流，突出了交流的意识形态和政治层面。同样地，一些研究人员把认知和交流方法结合起来（例如Moschkovich，2011a，2015；Turner等，2013）。莫谢科维奇（2011a）利用生态理论框架，研究了双语学习者的数学推理（以认知为焦点），该理论框架基于交流方法和人类学方法，融合了文化实践动态观念，并记录了日常思维中的资源。最近，莫谢科维奇（2015）还提出了一个数学学术素养的框架，其中包括数学活动的认知和交流方面的三个综合部分：（1）数学能力，（2）数学实践，（3）数学交流。特纳等人（2010）还结合社会文化和情境理论框架并利用以前认知研究中设计和使用过的任务，研究了学生在对这些任务所进行的数学讨论中是如何被定位的（参见Turner & Celedon-Pattichis，2011）。

in mathematics education. These advances are, however, in particular contexts, which can be characterized as follows:

- Second language learners of mathematics who are immigrants or language minorities in school systems in developed countries (such as research on Latinos/as in the United States)
- Multilingual mathematics classrooms in southern Africa
- Settings in which English is the language of instruction, whether in developed countries (United States, United Kingdom, Australia) or in southern Africa
- Elementary grades

There is therefore a need for more research—

- in the distinctive multilingual ecologies found in many Asian countries or in Latin America;
- with a wider range of languages of instruction, particularly languages that are less internationally widespread than English or Spanish;
- on linguistic aspects of learning mathematics for learners who are deaf, many of whom navigate at least two languages, one of which is a sign language; and
- in secondary grades, at tertiary level, and in teacher education, teacher learning, and teacher professional development (for this last area, see, for example, Chitera, 2011; Essien, 2014).

In terms of mathematical content, mathematical word problems are clearly dominant in the research on language diversity. This dominance may be explained by the particular linguistic challenges word problems can pose to second language learners in particular (Barwell, 2009). A second area featured in several studies is linear functions (Moschkovich, 2002, 2011b; Setati, Molefe & Langa, 2008). Different mathematical topics may involve different aspects of mathematical language. Proof, for example, requires use of logical relations, but probability requires the use of conditional forms (see Kazima, 2007). There is, therefore, a clear need for more systematic research on specific curriculum areas. Research has yet to produce detailed, explicit guidance that is specific to mathematics instruction and to students at different levels of language proficiency or mathematics proficiency or in different age groups.

Studies are needed that examine how second language learners learn to read different mathematical texts (textbooks, word problems, etc.). In designing this research, it will be important for researchers to differentiate between reading textbooks, reading word problems, and reading other genres in mathematical written discourse. When working with children who are learning to read in a second language, it will also be important to distinguish between children who are competent readers in a first language and those who are not. By extension, there is also a need for longitudinal studies that track students' mathematics learning and practices and students' language learning and practices over time to better understand how they interact.

To be able to design instruction that builds on student resources, research needs to examine in more detail the resources that bilingual and multilingual students use for mathematical reasoning. Studies will need to distinguish among multiple modalities (written and oral), receptive and productive skills, listening and oral comprehension, comprehension and production of oral contributions, and comprehension and production of written text.

Finally, most research projects are of relatively short duration, meaning that very little is known about the development of learners' mathematical thinking and knowing, or learners' languages in mathematics over time. We therefore highlight a need for longitudinal studies tracking learners over several years to uncover phenomena that may not be apparent in shorter studies.

Theoretical Orientations

We have organized research into three broad phases: (1) cognitively oriented research, (2) discursively oriented research, and (3) sociopolitically oriented research. Inevitably, this division is simplistic. Most of the sociopolitical research, for example, draws on discursive perspectives but highlights the ideological or political dimensions of discourse in addition to, or instead of, classroom discourse practices. Similarly, several researchers combine cognitive and discursive approaches (for example Moschkovich, 2011a, 2015; Turner et al., 2013). Moschkovich (2011a) has examined mathematical reasoning (a cognitive focus) among bilingual learners using an ecological theoretical framework that is based both on discursive approaches and on anthropological approaches that integrate a dynamic view of cultural practices and document the resources in everyday thinking. More recently, Moschkovich (2015) also proposed a framework for academic literacy in mathematics with three integrated components that include cognitive and discursive aspects of mathematical activity: (1) mathematical proficiency, (2) mathematical practices, and (3) mathematical discourse. Turner et al. (2010) also combined a sociocultural and situated theoretical framework to examine how students are positioned during

对于未来的理论工作，最紧迫的问题涉及将多语言主义的社会语言学和数学教育研究中语言多元化的最新发展结合起来，这些观点包括针对多语言主义的单一语言观点的评判。尽管事实上世界上的多数人在语言多元化的环境中生活和学习，但是，数学教育的研究仍然倾向于在单语主义的框架内运作并将其视为正常状态。今后的研究需要找出不同的方法从多语言的角度来看语言多元化，也就是说，研究不应建立在单一语言本质上是正常的、优越的、可取的假设之上，世界上大多数人都或多或少地使用着多种语言。同样地，今后的研究应该避免比较单语和双语学生。双语学生的数学学习需要依据他们自己的方法来描述和理解，而不仅仅是与单语学生进行比较，使用单语学生（或课堂）作为标准的比较给双语学习者强加了一个有缺陷的模式，如果研究确实要关注单语和双语学生的比较，则不能以单语主义为标准。

从双语或多语言角度来看，以第二语言学习数学可以是一个优势也可以是一个劣势，取决于具体情况。这些差异在一定程度上是政治性的。虽然已有研究已经提及甚至强调了这个问题，但是仍需要研究如何处理这个问题，特别是需要教师自己对这个问题进行研究。研究人员需要找到方法来回应以社会政治为导向的那些研究的结论，这些结论显示了政策和实践对不同群体学习者的不同影响。例如，新的理论发展应当探索语言多元化与种族、性别、社会阶层等的交叉问题。

最后，需要更多的理论方面的发展来综合来自不同情况下的研究发现，如语言、国家、环境和政治状况等。

方法问题

在语言多元化的情况下研究数学的教与学是具有挑战性的，因为会使用多种语言和潜在的语言变体，同时也由于双语、多语言或语言多样性的多种概念定义。

第一，研究人员需要清楚如何使用双语、多语言或语言学习者这样的标签来描述课堂和学习者，虽然这些术语通常被描述性地使用，但是它们常常掩盖了隐含的有关语言多元化的本质并且同等对待不同的群体（Barwell，2003a；Moschkovich，2010）。例如，同一个学生可以被描述为英语学习者或者是双语学习者，但是这两个标签对于语言多元化来说针对性是不同的。对于研究人员来说解决使用模糊混淆的标签的一个办法是对学生和其所在社区有更好的了解。

第二，理想的状态是研究人员不仅记录和报告学生使用每种语言的精炼程度，还要记录他们在不同的环境和任务中使用这种语言的历史、实践和经验（Moschkovich，2010），只要有可能，研究人员都应描述学生口头使用和书面使用每种语言的精炼程度。理想的状况是，研究不是一般性地评估学生的语言能力，而是特别针对学生对特定的数学内容进行书面和口头交流的能力，因为学生有不同的机会在非正式的或正式的教学环境中，在不同的数学内容中用不同的语言谈论和书写数学。

第三，研究人员需要认真注意转录和翻译活动（Barwell，2003a；Moschkovich，2010；Setati，2003）。转录的文本是理论的载体：关于如何表达课堂互动的选择反映了对语言的本质、多种语言和人际交流的假设（Ochs，1979；Poland，2002）。当参与者使用不止一种语言时，研究人员必须选择明确的方式来将课堂交流转录成文本并决定如何准确翻译这些内容。参与者用一种或多种语言所进行的实际表达和对此所做的翻译说明都需要包括在陈述和研究报告中。

对政策和课堂教学的影响

从认知角度进行的研究表明，学生能够在双语、多语言或者第二语言环境下成功地学习数学，在合适的条件下还能够超越单一语言学习者，然而，同样的研究还表明，在另外一些条件下，这些学生也可能会遇到学习困难。但从交流角度进行的研究表明，教师不应假设第

mathematical discussions with tasks previously designed and used in cognitive studies (see also Turner & Celedon-Pattichis, 2011).

For future theoretical work, the most pressing issues relate to integrating recent developments in the sociolinguistics of multilingualism and language diversity with research in mathematics education. These perspectives include critiques of monolingualist perspectives on multilingualism. Research in mathematics education still tends to operate within a framework of monolingualism as the normal state, although in fact most people in the world live and learn in situations of language diversity. Future research needs to find ways to look at language diversity from a multilanguage perspective. That is, research should not be based on the assumption that monolingualism is inherently normal, better, or desirable. Most people in the world draw on multiple languages to a greater or lesser extent. Similarly and relatedly, future research should move away from comparisons between monolingual and bilingual learners. Bilingual students learning mathematics need to be described and understood on their own terms and not just in comparison to monolingual students. Comparisons that use monolingual learners (or classrooms) as the norm impose a deficit model on bilingual learners. If studies do focus on comparing monolingual and bilingual learners, they should not assume that monolingualism is the norm.

Learning mathematics in a second language, bilingually, or multilingually can be an advantage or a disadvantage, depending on the circumstances. These differences are partly political in nature. Although this issue has been well highlighted and documented by research, there is a need for research that looks at how this issue can be tackled, particularly by teachers. Researchers need to find ways to respond to the findings of sociopolitically oriented research, which shows how policy and practice has different effects on various groups of learners. New theoretical developments should, for example, explore the intersection of language diversity with issues of race, gender, social class, and so forth.

Finally, there is a need for more theoretical development to draw together the sometimes disparate findings from research conducted in a range of situations, languages, nations, settings, and political circumstances.

Methodological Issues

Researching the learning and teaching of mathematics in situations of language diversity can be challenging due to the multiple languages and language varieties potentially in play, as well as multiple conceptualizations of bilingualism, multilingualism, or language diversity.

First, researchers need to be clear about how labels like bilingual, multilingual, or language learners are used to describe both classrooms and learners. Although such terms are often used descriptively, they often mask implicit assumptions about the nature of language diversity and tend to treat groups homogenously (Barwell, 2003a; Moschkovich, 2010). For example, the same student could be described as an English language learner or as a bilingual learner, but the two labels imply different orientations to language diversity. One solution to the use of vague and confusing labels is for researchers to know the students and the community well.

Second, researchers should ideally document and report not only students' proficiency in each language but also their histories, practices, and experiences with each language across a range of settings and tasks (Moschkovich, 2010). Researchers should describe both oral and written proficiencies in each language wherever possible. Ideally, studies should not assess language proficiency *in general* but specifically for communicating in writing and orally about a particular mathematical topic, since students have different opportunities to talk and write about mathematics in different languages, in informal or instructional settings, and within different mathematical topics.

Third, researchers need to pay careful attention to transcription and translation practices (Barwell, 2003a, Moschkovich, 2010; Setati, 2003). Transcripts are theory laden: Choices about how to represent classroom interactions reflect assumptions about the nature of language, multilingualism, and human interaction (Ochs, 1979; Poland, 2002). When participants use more than one language, researchers must choose clear ways to display transcripts and decide how the transcript will show translations. Both the actual utterances of the participants, in one or more languages, and any translations need to be included in presentations and publications reporting on the research.

Implications for Policy and Classroom Practice

Research using a cognitive perspective shows that students are successfully able to learn mathematics in a bilingual, multilingual, or second language setting and in the right conditions are able to outperform learners more oriented to a single language. The same research, however, also shows that in other conditions these same students may struggle. Research from a discursive per-

二语言、双语、多语言学习者有问题或者面临障碍或阻碍。从社会政治角度进行的研究则表明，学生在数学学习中成功与否的条件的本质是一个公平的问题，与社会阶层的深层分化紧密相关。

对于双语和多语言学生群体的教学政策有两个至关重要的问题。第一，目前在许多环境中使用的像"英语学习者"这样的标签是含糊不清的，有着不同的含义，不是基于客观标准，没有反映合理的分类，并且在不同环境下也不具有可比性或等同性。这些标签很可能反映或被用作人口统计标签的替代物，而没有准确描述正在学习第二语言（比如英语）的学生（例如，可以参见Gándara & Contreras, 2009）。第二，"语言能力"是一个复杂的构成，它反映的是在多种情境、模式和学科方面的熟练度。目前的语言能力测量方法可能无法准确地描述个人的实际语言能力。特别是由于语言能力的复杂性和目前使用的"英语学习者"这类标签的局限性，教学决策不应仅仅基于标签。

根据本章中所讨论的研究，我们提出以下建议，用来指导针对多语言学生的数学教学实践。针对这个学生群体的教学应该（a）不要只是教授词汇；（b）注重数学推理和实践；（c）利用多种资源——物体、绘图、图形、肢体语言等；（d）主动地将日常的谈话、家庭语言、校外的经验作为资源，而不是将这些作为数学交流的障碍（关于教学实践的指导，参见Moschkovich, 2013）。

针对学生学习英语的数学教学的研究表明，教师应当支持学生参与到数学讨论中去，即使这些学生正在学习英语，而不要等他们具备了英语能力之后再教数学（Khisty, 1995；Moschkovich, 2010）。学生在学习教学语言的同时是能够参与数学讨论的，教师可以通过使用不同的策略，如使用更正式的谈话方式重述学生的发言，让学生做出澄清，探索学生的思维等（Moschkovich, 1999）来支持学生畅游于非正式和正式的数学语言之间以及母语和教学语言之间（Clarkson, 2009）。

数学课堂的教学应该认识到并有策略地给予这些学生接触语言复杂性的机会。"语言"不仅是说话，还要考虑数学交流中三个符号体系的相互影响：（1）自然语言，（2）数学符号体系，（3）视觉显示。数学课堂中的语言是复杂的并且涉及多种形式（口头、书面、感受、表达，等等）、多种表征形式（实物、图片、文字、符号、表格、图像，等等）、不同类型的书面文本（教科书、文字题、学生的解释、教师的解释，等等）、不同类型的谈话（探究的、解释的，等等）以及展示的不同对象（对教师展示、对同学展示、由教师展示、由同学展示，等等）。教师应尽可能地与语言教育工作者（如双语教育的教师，第二语言的教师）合作，将自己数学教学方面的特长与语言教学方面的专业特长结合起来。

教学实践不要过分强调正确词汇和正式语言，因为这会限制教师和学生在课堂上可以用来学习理解数学的语言资源，相反，教学应该为学生积极地使用数学语言来交流和商讨数学情境中的含义提供机会。教学需要超越仅仅把数学语言解释为只针对数学的单词和短语的集合，相反，教学需要把数学语言这一复杂观念整合为不仅是特定的词汇——由熟悉的词汇引出的新词汇和新含义——而且是拓展的话语，包括语法、组织和特定的话语实践。

为了解决语言问题，教师专业发展项目应帮助教师看到，对语言的关注是一个听取学生数学思维并依据他们的思维制定今后教学的重要机会（Moschkovich, 1999, 2002, 2013），这些教师专业发展项目中的课程应注意支持教师学习，以提高学生表达所知的能力，并创造一个互相尊重的学习环境，为所有的学生提供学习用数学进行交流的机会（Moschkovich & Nelson-Barber, 2009）。

spective suggests that teachers should not assume that second language, bilingual, or multilingual learners have a problem or that they face barriers or obstacles. Research from a sociopolitical perspective suggests that the nature of the conditions through which students come to succeed or not in mathematics is a matter of equity, intimately bound up with deep-running social divisions in society.

Two particular issues are crucial to instructional policies for bilingual and multilingual student populations. First, labels like "English learner" as currently used in many settings are vague, have different meanings, are not based on objective criteria, do not reflect sound classifications, and are not comparable or equivalent across settings. Such labels are likely to reflect or be used as a proxy for demographic labels rather than accurately portray students who are learning a second language, such as English (see, for example, Gándara & Contreras, 2009). Second, "language proficiency" is a complex construct that can reflect proficiency in multiple contexts, modes, and academic disciplines. Current measures of language proficiency may not give an accurate picture of an individual's actual language competence. In particular, because of the complexity of language proficiency and the limitations of labels like "English learner" as currently implemented, instructional decisions should not be made solely based on the label.

Based on the research discussed in this chapter, we propose the following recommendations to guide instructional practices in mathematics for students who are learning the language of instruction. Instruction for this student population should (a) address much more than vocabulary; (b) focus on mathematical reasoning and practices; (c) draw on multiple resources—objects, drawings, graphs, and gestures; and (d) proactively treat everyday ways of talking, home languages, and experiences outside of school as resources rather than obstacles for mathematical communication (see also Moschkovich, 2013, for guidelines for teaching practices).

Research specific to mathematics instruction for students learning English shows that teachers should support students' participation in mathematics discussions even as the students are learning English rather than waiting until they are English proficient (Khisty, 1995; Moschkovich, 2010). Students who are learning the language of instruction can participate in mathematical discussions as they learn the language of instruction. Teachers can support these students' navigation between informal and formal mathematical language and between home languages and the language of instruction (Clarkson, 2009) through strategies such as revoicing student contributions using more formal ways of talking, asking for clarification, and probing a student's thinking (Moschkovich, 1999).

Instruction should recognize and strategically support these students' opportunities to engage with the linguistic complexity of language in mathematics classrooms. "Language" needs to expand beyond talk to consider the interaction of the three semiotic systems involved in mathematical discourse: (1) natural language, (2) mathematics symbol systems, and (3) visual displays. Language in mathematics classrooms is complex and involves multiple modes (oral, written, receptive, expressive, etc.), multiple representations (objects, pictures, words, symbols, tables, graphs, etc.), different types of written texts (textbooks, word problems, student explanations, teacher explanations, etc.), different types of talk (exploratory, expository, etc.), and different audiences for presentations (to the teacher, to peers, by the teacher, by peers, etc.). Teachers should, where possible and appropriate, collaborate with language educators (e.g., bilingual education teachers, second language teachers) to combine their expertise in mathematics teaching with professional expertise in language teaching.

Teaching practices need to move away from an overemphasis on correct vocabulary and formal language, which limits the linguistic resources teachers and students can use in the classroom to learn mathematics with understanding. Instead, instruction should provide opportunities for students to actively use mathematical language to communicate about and negotiate meaning for mathematical situations. Instruction needs to move beyond interpretations of the mathematical language as merely a set of words and phrases that are particular to mathematics. Instead, instruction needs to integrate a complex view of mathematical language as not only specialized vocabulary—new words and new meanings for familiar words—but also as extended discourse that includes syntax, organization, and specific discourse practices.

To address issues of language, teacher education and professional development should help teachers see a focus on language as an opportunity to hear students' mathematical thinking and to design further instruction based on that thinking (Moschkovich, 1999, 2002, 2013). Such programs will need to pay attention to supporting teachers in learning to improve student competence in articulating what students know, as well as to create respectful learning environments that provide opportunities for all students to learn to communicate mathematically (Moschkovich & Nelson-Barber, 2009).

教师专业发展应当支持教师以使他们做好准备来应对学习中教学语言方面和数学内容方面的问题，特别是，教师需要学会（a）如何听到并找出学生表述中的数学，即使这些数学内容是以意外浮现的、不完美的或者日常的语言所表达的；（b）什么时候和如何来支持学生从日常的交流方式转向更加数学化的交流方式；（c）什么时候和如何支持学生形成更加精准的数学交流方式。针对数学教学中学生学习教学语言的教师专业发展课程还应讨论文献中描述的常见问题，如学习内容和学习语言这样不同目标之间的问题以及在课堂上使用多种语言的问题。专业发展活动可以让教师对数学教学中的语言问题进行反思，讨论和与同事持续地交流互动。

注释

1. 在他原来提出的阈值假设中，卡明斯（1979）使用"半语"这个术语来描述那些"在两种语言中都不具有母语般能力"的儿童（第228页）。这一定义涉及一种猜测，即有些儿童对于他们所讲的一种或多种语言只有有限的或非母语般的能力（MacSwan, 2000）。近期以来，大多数学者，包括卡明斯（2000）都摒弃了半语的概念。因此，阈值假设应与半语这一术语的使用相分离。

2. 语码转换一词在社会语言学中越来越引起争议，特别是在那些挑战不同种族所使用的各类语言意识形态的研究人员中间（例如，可以参见Blackledge & Creese, 2010；Blommaert, 2010；Blommaert & Rampton, 2011；Makoni & Pennycook, 2007）。替代的术语包括语言转换（Cenoz & Gorter, 2015；Wei & Garcia, 2014）、交融语言（Blackledge & Creese, 2014）和编码协调（Young & Martinez, 2011）。有关这些观念如何应用于数学教育研究的讨论，见巴韦尔（2016）。

Professional development should support teachers so that they are better prepared to deal with the tensions of learning both the language of instruction and mathematical content. In particular, teachers need to learn (a) how to hear and uncover the mathematics in student contributions, even when these are expressed in emergent, imperfect, or everyday language; (b) when and how to support students in moving from everyday to more mathematical ways of communicating; and (c) when and how to support students in developing more mathematically precise ways of communicating. Teacher professional development programs focusing on mathematics instruction for students learning the language of instruction should also address common tensions described in the literature, such as that between content-learning and language-learning goals and multiple languages in the classroom. Professional development activities can involve teachers in reflection, discussion, and continuous interaction with colleagues about language issues in teaching mathematics.

Notes

1. In his original formulation of the threshold hypothesis, Cummins (1979) used the term "semilingualism" to describe children with "less than native-like skills in both languages" (p. 228). This definition involves the conjecture that some children have limited or nonnative ability in the language or languages they speak (MacSwan, 2000). More recently, most scholars have discarded the concept of semilingualism, including Cummins (2000). Therefore, the threshold hypothesis should be separated from the use of the term semilingualism.

2. The term code-switching has been increasingly contested in sociolinguistics, particularly by researchers who challenge the ideology of discrete languages spoken by distinct peoples (see, for example, Blackledge & Creese, 2010; Blommaert, 2010; Blommaert & Rampton, 2011; Makoni & Pennycook, 2007). Alternative terms include translanguaging (Cenoz & Gorter, 2015; Wei & Garcia, 2014), heteroglossia (Blackledge & Creese, 2014), and code-meshing (Young & Martinez, 2011). For a discussion of how some of these ideas apply to research in mathematics education, see Barwell (2016).

References

Abedi, J. (2002). Standardized achievement tests and English language learners: Psychometric issues. *Educational Assessment, 8*(3), 231–257.

Abedi, J. (2009). Validity of assessments for English language learning students in a national/international context. *Estudios sobre Educación, 16*, 167–183.

Abedi, J., & Lord, C. (2001). The language factor in mathematics tests. *Applied Measurement in Education, 14*(3), 219–234.

Adendorff, R. (1993). Codeswitching amongst Zulu-speaking teachers and their pupils *Language and Education, 7*(3), 141–162.

Adetula, L. O. (1989). Solutions of simple word problems by Nigerian children: Language and schooling factors. *Journal for Research in Mathematics Education, 20*(5), 489–497.

Adetula, L. O. (1990). Language factor: Does it affect children's performance on word problems. *Educational Studies in Mathematics, 21*(4), 351–365.

Adler, J. (1997). A participatory-inquiry approach and the mediation of mathematical knowledge in a multilingual classroom. *Educational Studies in Mathematics, 33*(3), 235–258.

Adler, J. (1998). A language of teaching dilemmas: Unlocking the complex multilingual secondary mathematics classroom. *For the Learning of Mathematics, 18*(1), 24–33.

Adler, J. (1999). The dilemma of transparency: Seeing and seeing through talk in the mathematics classroom. *Journal for Research in Mathematics Education, 30*(1), 47–64.

Adler, J. (2001). *Teaching mathematics in multilingual classrooms*. Dordrecht, The Netherlands: Kluwer Academic Press.

Alsawaie, O. N. (2004). Language influence on children's cognitive number representation. *School Science and Mathematics, 104*(3), 105–111.

Austin, J. L., & Howson, A. J. (1979). Language and mathematical education. *Educational Studies in Mathematics, 10*(2), 161–197.

Baker, C. (1993). *Foundations of bilingual education and bilingualism*. Clevedon, United Kingdom: Multilingual Matters.

Baker, C. (2001). *Foundations of bilingual education and bilingualism* (3rd ed.). Clevedon, United Kingdom: Multilingual Matters.

Baker, C. (2011). *Foundations of bilingual education and bilingualism* (5th ed.). Bristol, United Kingdom: Multilingual Matters.

Bakhtin, M. M. (1981). Discourse in the novel (C. Emerson & M. Holquist, Trans.). In M. Holquist (Ed.), *The dialogic imagination: Four essays by M. M. Bakhtin* (pp. 269–422). Austin, TX: University of Texas Press.

Barton, B. (2009). *The language of mathematics: Telling mathematical tales*. New York, NY: Springer.

Barton, B., Fairhall, U., & Trinick, T. (1998). Tikanga Reo Tatai: Issues in the development of a Maori mathematics register. *For the Learning of Mathematics, 18*(1), 3–9.

Barwell, R. (2003a). Linguistic discrimination: An issue for research in mathematics education. *For the Learning of Mathematics, 23*(2), 37–43.

Barwell, R. (2003b). Patterns of attention in the interaction of a primary school mathematics student with English as an additional language. *Educational Studies in Mathematics, 53*(1), 35–59.

Barwell, R. (2005a). Critical issues for language and content in mainstream classrooms: Introduction. *Linguistics and Education, 16*(2), 143–150.

Barwell, R. (2005b). Empowerment, EAL and the National Numeracy Strategy. *International Journal of Bilingual Education and Bilingualism, 8*(4), 313–327.

Barwell, R. (2005c). Integrating language and content: Issues from the mathematics classroom. *Linguistics and Education, 16*(2), 205–218.

Barwell, R. (2005d). Working on arithmetic word problems when English is an additional language. *British Educational Research Journal, 31*(3), 329–348.

Barwell, R. (2009). Mathematical word problems and bilingual learners in England. In R. Barwell (Ed.), *Multilingualism in mathematics classrooms: Global perspectives* (pp. 63–77). Bristol, United Kingdom: Multilingual Matters.

Barwell, R. (2012a). Discursive demands and equity in second language mathematics classrooms. In B. Herbel-Eisenmann, J. Choppin, D. Wagner, & D. Pimm (Eds.), *Equity in discourse for mathematics education: Theories, practices, and policies* (pp. 147–164). New York, NY: Springer.

Barwell, R. (2012b). Heteroglossia in multilingual mathematics classrooms. In H. Forgasz & F. Rivera (Eds.), *Towards equity in mathematics education: Gender, culture and diversity* (pp. 315–332). Heidelberg, Germany: Springer.

Barwell, R. (2014). Centripetal and centrifugal language forces in one elementary school second language mathematics classroom. *ZDM—The International Journal on Mathematics Education, 46*(6), 911–922.

Barwell, R. (2016). Mathematics education, language and superdiversity. In A. Halai & P. Clarkson (Eds.), *Teaching and learning mathematics in multilingual classrooms: Issues for policy, practice and teacher education* (pp. 25–39). Rotterdam, The Netherlands: Sense.

Ben-Zeev, S. (1977). The influence of bilingualism on cognitive strategy and cognitive development. *Child Development, 48*, 1009–1018.

Bernardo, A. B. I. (1999). Overcoming obstacles to understanding and solving word problems in mathematics. *Educational Psychology, 19*(2), 149–163.

Bialystok, E. (2001). *Bilingualism in development: Language, literacy and cognition*. Cambridge, United Kingdom: Cambridge University Press.

Bialystok, E., & Majumder, S. (1998). The relationship between bilingualism and the development of cognitive processes in problem solving. *Applied Psycholinguistics, 19*, 69–85.

Bishop, A. (1986). Mathematics education in its cultural context. *Educational Studies in Mathematics, 10*(2), 135–146.

Blackledge, A., & Creese, A. (2010). *Multilingualism: A critical perspective*. London, United Kingdom: Continuum.

Blackledge, A., & Creese, A. (Eds.). (2014). *Heteroglossia as practice and pedagogy*. Dordrecht, The Netherlands: Springer.

Blommaert, J. (2010). *The sociolinguistics of globalization*. Cambridge, United Kingdom: Cambridge University Press.

Blommaert, J., & Rampton, B. (2011). Language and super-diversity. *Diversities, 13*(2), 1–21.

Bournot-Trites, M., & Reeder, K. (2001). Interdependence revisited: Mathematics achievement in an intensified French immersion program. *The Canadian Modern Language Review/ La revue canadienne des langues vivantes, 58*(1), 27–43.

Brenner, M. (1998). Adding cognition to the formula for culturally relevant instruction in mathematics. *Anthropology & Education Quarterly, 29*(2), 214–244.

Brysbaert, M., Fias, W., & Noel, M. P. (1998). The Whorfian hypothesis and numerical cognition: Is "twenty-four" processed in the same way as "four-and-twenty"? *Cognition, 66*(1), 51–77.

Burston, M. (1999). Mathématiques en immersion partielle: Comment les enfants s'y prennent-ils pour résoudre un problème? *Le journal de l'immersion, 22*(1), 37–41.

Carpenter, T. P., & Moser, J. M. (1984). The acquisition of addition and subtraction concepts in grades one through three. *Journal for Research in Mathematics Education, 15*, 179–202.

Cenoz, J., & Gorter, D. (Eds.). (2015). *Multilingual education: Between language learning and translanguaging*. Cambridge, United Kingdom: Cambridge University Press.

Chamot, A., Dale, M., O'Malley, M., & Spanos, G. (1992). Learning and problem solving strategies of ESL students. *Bilingual Research Journal, 16*(3&4), 1–34.

Chitera, N. (2009). Code-switching in a college mathematics classroom. *International Journal of Multilingualism, 6*(4), 426–442.

Chitera, N. (2011). Language of learning and teaching in schools: An issue for research in mathematics teacher education? *Journal of Mathematics Teacher Education, 14*(3), 231–246.

Clarkson, P. C. (1991). Language comprehension errors: A further investigation. *Mathematics Education Research Journal, 3*(2), 24–33.

Clarkson, P. C. (1992). Language and mathematics: A comparison of bilingual and monolingual students of mathematics. *Educational Studies in Mathematics, 23*(4), 417–430.

Clarkson, P. C. (2007). Australian Vietnamese students learning mathematics: High ability bilinguals and their use of their languages. *Educational Studies in Mathematics, 64*(2), 191–215.

Clarkson, P. C. (2009). Mathematics teaching in Australian multilingual classrooms: Developing an approach to the use of classroom languages. In R. Barwell (Ed.), *Multilingualism in mathematics classrooms: Global perspectives* (pp. 145–160). Bristol, United Kingdom: Multilingual Matters.

Clarkson, P. C., & Galbraith, P. (1992). Bilingualism and mathematics learning: Another perspective. *Journal for Research in Mathematics Education, 23*(1), 34–44.

Cobb, P., Wood, T., & Yackel, E. (1993). Discourse, mathematical thinking, and classroom practice. In E. Forman, N. Minick, & C. A. Stone (Eds.), *Contexts for learning: Sociocultural dynamics in children's development* (pp. 91–119). Oxford, United Kingdom: Oxford University Press.

Cocking, R. R., & Chipman, S. (1988). Conceptual issues related to mathematics achievement of language minority children. In R. R. Cocking & J. Mestre (Eds.), *Linguistic and cultural influences on learning mathematics* (pp. 17–46). Hillsdale, NJ: Lawrence Erlbaum.

Cook, G., & MacDonald, R. (2012). Draft of "Can Do" descriptors for CCSS Standards for Mathematical Practice. Madison, WI: WIDA.

Cooper, B., & Dunne, M. (2000). *Assessing children's mathematical knowledge: Social class, sex and problem-solving*. Buckingham, United Kingdom: Open University Press.

Crowhurst, M. (1994). *Language and learning across the curriculum. Instructor's manual*. Scarborough, Canada: Allyn & Bacon.

Cummins, J. (1979). Linguistic interdependence and the educational development of bilingual children. *Review of Educational Research, 49*(2), 222–251.

Cummins, J. (1981). The role of primary language development in promoting educational success for language minority students. In California State Department of Education (Ed.),

Schooling and language minority students: A theoretical framework (pp. 3–49). Los Angeles, CA: Evaluation, Dissemination and Assessment Center, California State University.

Cummins, J. (2000). *Language, power, and pedagogy.* Clevedon, United Kingdom: Multilingual Matters.

Dalton-Puffer, C. (2007). *Discourse in content and language integrated learning (CLIL) classrooms.* Amsterdam, The Netherlands: John Benjamins.

Dalton-Puffer, C., Nikula, T., & Smit, U. (Eds.). (2010). *Language use and language learning in CLIL classrooms.* Amsterdam, The Netherlands: John Benjamins.

D'Ambrosio, U. (1991). Ethnomathematics and its place in the history and pedagogy of mathematics. In M. Harris (Ed.), *Schools, mathematics and work* (pp. 15–25). Bristol, PA: Falmer Press.

Davison, C., & Williams, A. (2001). Integrating language and content: Unresolved issues. In B. Mohan, C. Leung, & C. Davison (Eds.), *English as a second language in the mainstream* (pp. 51–70). Harlow, United Kingdom: Pearson Education.

Dawe, L. (1983). Bilingualism and mathematical reasoning in English as a second language. *Educational Studies in Mathematics, 14*(4), 325–353.

De Avila, E., & Duncan, S. (1981a). Bilingualism and the metaset. In R. Durán (Ed.), *Latino language and communicative behavior* (pp. 337–354). Norwood, NJ: Ablex.

De Avila, E., & Duncan, S. (1981b). *A convergent approach to oral language assessment: Theoretical and technical specification on the Language Assessment Scales (LAS) Form A* (Stock 621). San Rafael, CA: Linguametrics Group.

De Corte, E., Verschaffel, L., & De Win, L. (1985). Influence of rewording verbal problems on children's problem representations and solutions. *Journal of Educational Psychology, 77*(4), 460.

de Courcy, M., & Burston, M. (2000). Learning mathematics through French in Australia. *Language and Education, 14*(2), 75–95.

Domínguez, H. (2011). Using what matters to students in bilingual mathematics problems. *Educational Studies in Mathematics, 76*(3), 305–328.

Donaldson, M. (1978). *Children's minds.* London, United Kingdom: Fontana.

Dowker, A. (1998). Individual differences in normal arithmetical development. In C. Donlan (Ed.), *The development of mathematical skills* (pp. 275–302). Hove, United Kingdom: Psychology Press.

Dowker, A., Bala, S., & Lloyd, D. (2008). Linguistic influences on mathematical development: How important is the transparency of the counting system? *Philosophical Psychology, 21*(4), 523–538.

Duncan, S., & De Avila, E. (1986). *Pre-LAS user's manual* (Form A). San Rafael, CA: Linguametrics Group.

Duncan, S., & De Avila, E. (1987). *Pre-LAS Español user's manual* (Form A). San Rafael, CA: Linguametrics Group.

Essien, A. A. (2014). Examining opportunities for the development of interacting identities within pre-service teacher education mathematics classrooms. *Perspectives in Education, 32*(3), 62.

Farrugia, M. T. (2009a). Reflections on a medium of instruction policy for mathematics in Malta. In R. Barwell (Ed.), *Multilingualism in mathematics classrooms: Global perspectives* (pp. 97–112). Bristol, United Kingdom: Multilingual Matters.

Farrugia, M. T. (2009b). Registers for mathematics classrooms in Malta: Considering the options. *For the Learning of Mathematics, 29*(1), 20–25.

Forman, E. (1996). Learning mathematics as participation in classroom practice: Implications of sociocultural theory for educational reform. In L. Steffe, P. Nesher, P. Cobb, G. Goldin, & B. Greer (Eds.), *Theories of mathematical learning* (pp. 115–130). Mahwah, NJ: Lawrence Erlbaum.

Fuson, K., Smith, S., & Lo Cicero, A. (1997). Supporting Latino first graders' ten-structured thinking in urban classrooms. *Journal for Research in Mathematics Education, 28*(6), 738–766.

Gándara, P. C., & Contreras, F. (2009). *The Latino education crisis: The consequences of failed social policies.* Cambridge, MA: Harvard University Press.

Gee, J. P. (1996). *Sociolinguistics and literacies: Ideology in discourses.* London, United Kingdom: Falmer Press.

Gee, J. P. (1999). *An introduction to discourse analysis: Theory and method.* London, United Kingdom: Routledge.

Gerofsky, S. (1996). A linguistic and narrative view of word problems in mathematics education. *For the Learning of Mathematics, 16*(2), 36–45.

Gorgorió, N., & Planas, N. (2001). Teaching mathematics in multilingual classrooms. *Educational Studies in Mathematics, 47*(1), 7–33.

Grosjean, F. (1982). *Life with two languages: An introduction to bilingualism.* Cambridge, MA: Harvard University Press.

Grosjean, F. (1999). Individual bilingualism. In B. Spolsky (Ed.), *Concise encyclopedia of educational linguistics* (pp. 284–290). London, United Kingdom: Elsevier.

Grouws, D. A. (Ed.). (1992). *Handbook of research on mathematics teaching and learning.* New York, NY: MacMillan.

Gutiérrez, R. (2002). Beyond essentialism: The complexity of language in teaching mathematics to Latina/o students. *American Educational Research Journal, 39*(4), 1047–1088.

Gutiérrez, R. (2013). The sociopolitical turn in mathematics education. *Journal for Research in Mathematics Education, 44*(1), 37–68.

Gutstein, E. (2006). *Reading and writing the world with mathematics: Toward a pedagogy for social justice.* New York, NY: Routledge.

Hakuta, K., & McLaughlin, B. (1996). Bilingualism and second language learning: Seven tensions that define research. In D. Berliner & R. C. Calfee (Eds.), *Handbook of educational psychology* (pp. 603–621). New York, NY: Macmillan.

Herbel-Eisenmann, B., Meaney, T., Bishop, J., & Heyd-Metzuyanim, E. (2017). Highlighting heritages and building tasks: A critical analysis of mathematics classroom discourse literature. In J. Cai (Ed.), *Compendium for research in mathematics education* (pp. 722–765). Reston, VA: National Council of Teachers of Mathematics.

Howie, S. J. (2001). *Mathematics and science performance in grade 8 in South Africa 1998/1999: TIMSS-R 1999 South Africa.* Pretoria, South Africa: Human Sciences Research Council.

Ianco-Worrall, A. D. (1972). Bilingualism and cognitive development. *Child Development, 43*, 1390–1400.

Jones, D. V. (2009). Bilingual mathematics classrooms in Wales. In R. Barwell (Ed.), *Multilingualism in mathematics* classrooms: *Global perspectives* (pp. 113–127). Bristol, United Kingdom: Multilingual Matters.

Jones, P. (1982). Learning mathematics in a second language: A problem with more and less. *Educational Studies in Mathematics, 13*(3), 269–288.

Kazima, M. (2007). Malawian students' meanings for probability vocabulary. *Educational Studies in Mathematics, 64*(2), 169–189.

Khisty, L. (1995). Making inequality: Issues of language and meanings in mathematics teaching with Hispanic students. In W. G. Secada, E. Fennema, & L. B. Adajian (Eds.), *New directions for equity in mathematics education* (pp. 279–297). New York, NY: Cambridge University Press.

Khisty, L., & Chval, K. (2002). Pedagogic discourse and equity in mathematics: When teachers' talk matters. *Mathematics Education Research Journal, 14*(3), 154–168.

Kieran, C., Forman, E., & Sfard, A. (Eds.). (2002). *Learning discourse: Discursive approaches to research in mathematics education*. Dordrecht, The Netherlands: Kluwer.

Krummheuer, G. (1995). The ethnography of argumentation. In P. Cobb & H. Bauersfeld (Eds.), *The emergence of mathematical meaning: Interaction in classroom cultures* (pp. 229–269). Hillsdale, NJ, Lawrence Erlbaum.

Lampert, M. (1990). When the problem is not the question and the solution is not the answer: Mathematical knowing and teaching. *American Educational Research Journal, 27*(1), 29–64.

Lapkin, S., Hart, D., & Turnbull, M. (2003). Grade 6 French immersion students' performance on large-scale reading, writing, and mathematics tests: Building explanations. *Alberta Journal of Educational Research, 49*(1), 6–23.

Lerman, S. (2000). The social turn in mathematics education research. In J. Boaler (Ed.), *Multiple perspectives on mathematics teaching and learning* (pp. 19–44). Westport, CT: Ablex.

Lotherington, H. (2004). Bilingual education. In A. Davies & C. Elder (Eds.), *Handbook of applied linguistics* (pp. 697–718). Oxford, United Kingdom: Blackwell.

Lunney Borden, L. (2011). The "verbification" of mathematics: Using the grammatical structures of Mi'kmaq to support student learning. *For the Learning of Mathematics, 31*(3), 8–13.

Lunney Borden, L. (2013). What's the word for . . . ? Is there a word for . . . ? How understanding Mi'kmaw language can help support Mi'kmaw learners in mathematics. *Mathematics Education Research Journal, 25*(1), 5–22.

Lyster, R., & Ballinger, S. (Eds.). (2011). Content-based language teaching [Special issue]. *Language Teaching Research, 15*(3).

MacSwan, J. (2000). The threshold hypothesis, semilingualism, and other contributions to a deficit view of linguistic minorities. *Hispanic Journal of Behavioral Sciences, 22*(1), 3–45.

Makoni, S. B. (2011). Sociolinguistics, colonial and postcolonial: An integrationist perspective. *Language Sciences, 33*(4), 680–688.

Makoni, S. B., & Pennycook, A. (Eds.). (2007). *Disinventing and reconstituting languages*. Clevedon, United Kingdom: Multilingual Matters.

Marsh, L., & Maki, R. (1976). Efficiency of arithmetic operations in bilinguals as a function of language. *Memory and Cognition, 4*, 459–464.

Martiniello, M. (2008). Language and the performance of English language learners in math word problems. *Harvard Educational Review, 78*(2), 333–368.

Martiniello, M. (2010). Linguistic complexity in mathematics assessments and the performance of English language learners. TODOS *Research Monograph: Mathematics for All. Assessing English-Language Learners in Mathematics*. Vol. 2. Monograph 2: Linguistic complexity in mathematics assessments. Washington, DC: National Education Association.

Martiniello, M., & Wolf, M. K. (2012). Exploring ELLs' understanding of word problems in mathematics assessments: The role of text complexity and student background knowledge. In S. Celedón-Pattichis & N. Ramirez (Eds.), *Beyond good teaching: Advancing Mathematics Education for ELLs* (151–162). Reston, VA: National Council of Teachers of Mathematics.

McLain, L., & Huang, J. (1982). Speed of simple arithmetic in bilinguals. *Memory and Cognition, 10*, 591–596.

Meaney, T., Trinick, T., & Fairhall, U. (2011). *Collaborating to meet language challenges in Indigenous mathematics classrooms*. Dordrecht, The Netherlands: Springer.

Mendes, J. R. (2007). Numeracy and literacy in a bilingual context: Indigenous teachers education in Brazil. *Educational Studies in Mathematics, 64*(2), 217–230.

Merritt, M., Cleghorn, A., Abagi, J. O., & Bunyi, G. (1992). Socialising multilingualism: Determinants of codeswitching in Kenyan primary classrooms. *Journal of Multilingual & Multicultural Development, 13*(1–2), 103–121.

Mestre, J. (1986) Teaching problem-solving strategies to bilingual students: What do research results tell us? *International Journal of Mathematical Education in Science and Technology, 17*(4), 393–401.

Mestre, J. (1988). The role of language comprehension in mathematics and problems solving. In R. Cocking & J. Mestre (Eds.), *Linguistic and cultural influences on learning mathematics* (pp. 259–293). Hillsdale, NJ: Lawrence Erlbaum.

Mestre, J., & Gerace, W. (1986). A study of the algebra acquisition of Hispanic and Anglo ninth graders: Research findings relevant to teacher training and classroom practice. *NABE Journal, 10*, 137–167.

Mestre, J., Gerace, W., & Lochhead, J. (1982). The interdependence of language and translational math skills among bilingual Hispanic engineering students. *Journal of Research in Science Teaching, 19*(5), 399–410.

Miller, K., Kelly, M., & Zhou, X. (2005). Learning mathematics in China and the United States: Cross-cultural insights into the nature and course of mathematical development. In J. I. D. Campbell (Ed.), *Handbook of mathematical cognition* (pp. 163–178). Hove, United Kingdom: Psychology Press.

Miura, I. T., Kim, C. C., Chang, C.-M., & Okamoto, Y. (1988). Effects of language characteristics on children's cognitive representation of numbers: Cross-national comparisons. *Child Development, 59*, 1445–1450.

Mohan, B. A. (1986). *Language and content*. Reading, PA: Addison-Wesley.

Molefe, T. B. (2009). Using multiple languages to support mathematics proficiency in a grade 11 multilingual classroom of second language learners: An action research (Unpublished master's dissertation). University of the Witwatersrand, South Africa.

Moschkovich, J. N. (1999). Supporting the participation of English language learners in mathematical discussions. *For the Learning of Mathematics, 19*(1), 11–19.

Moschkovich, J. N. (2000). Learning mathematics in two languages: Moving from obstacles to resources. In W. Secada (Ed.), *Changing faces of mathematics: Perspectives on multiculturalism and gender equity* (pp. 85–93). Reston, VA: National Council of Teachers of Mathematics.

Moschkovich, J. N. (2002). A situated and sociocultural perspective on bilingual mathematics learners. *Mathematical Thinking and Learning, 4*(2&3), 189–212.

Moschkovich, J. N. (2007a). Bilingual mathematics learners: How views of language, bilingual learners, and mathematical communication impact instruction. In N. Nasir & P. Cobb (Eds.), *Diversity, equity, and access to mathematical ideas* (pp. 89–104). New York, NY: Teachers College Press.

Moschkovich, J. N. (2007b). Examining mathematical discourse practices. *For the Learning of Mathematics, 27*(1), 24–30.

Moschkovich, J. N. (2007c). Using two languages while learning mathematics. *Educational Studies in Mathematics, 64*(2), 121–144.

Moschkovich, J. N. (2008). "I went by twos, he went by one": Multiple interpretations of inscriptions as resources for mathematical discussions. *The Journal of the Learning Sciences, 17*(4), 551–587.

Moschkovich, J. (2009). How language and graphs support conversation in a bilingual mathematics classroom. In R. Barwell (Ed.), *Multilingualism in mathematics classrooms: Global perspectives* (pp. 78–96). Bristol, United Kingdom: Multilingual Matters.

Moschkovich, J. N. (2010). Language(s) and learning mathematics: Resources, challenges, and issues for research. In J. N. Moschkovich (Ed.), *Language and mathematics education: Multiple perspectives and directions for research* (pp. 1–28). Charlotte, NC: Information Age.

Moschkovich, J. N. (2011a). Ecological approaches to transnational research on mathematical reasoning. In R. Kitchen & M. Civil (Eds.), *Transnational and borderland studies in mathematics education* (pp. 1–22). New York, NY: Routledge.

Moschkovich, J. N. (2011b). Supporting mathematical reasoning and sense making for English learners. In M. Strutchens & J. Quander (Eds.), *Focus in high school mathematics: Fostering reasoning and sense making for all students* (pp. 17–36). Reston, VA: National Council of Teachers of Mathematics.

Moschkovich, J. N. (2013). Principles and guidelines for equitable mathematics teaching practices and materials for English language learners. *Journal of Urban Mathematics Education, 6*(1), 45–57.

Moschkovich, J. N. (2015). Academic literacy in mathematics for English learners. *Journal of Mathematical Behavior, 40,* 43–62.

Moschkovich, J., & Nelson-Barber, S. (2009). What mathematics teachers need to know about culture and language. In B. Greer, S. Mukhopadhyay, A. B. Powell, & S. Nelson-Barber (Eds.), *Culturally responsive mathematics education* (pp. 111–136). New York, NY: Routledge.

Mosqueda, E. (2010). Compounding inequalities: English proficiency and tracking and their relation to mathematics performance among Latina/o secondary school youth. *Journal of Urban Mathematics Education, 3*(1), 57–81.

Mpalami, N. (2007). *Teaching and learning linear programming in a grade 11 multilingual mathematics class* (Unpublished master's dissertation). University of the Witwatersrand, South Africa.

Muldoon, K., Towse, J., Simms, V., Perra, O., & Menzies, V. (2013). A longitudinal analysis of estimation, counting skills, and mathematical ability across the first school year. *Developmental Psychology, 49*(2), 250.

Nieto, S. (1992). We speak in many tongues: Language diversity and multicultural education. In C. Díaz (Ed.), *Multicultural education for the twenty-first century* (pp. 112–136). Washington, DC: National Education Association.

Ní Ríordáin, M., & O'Donoghue, J. (2009). The relationship between performance on mathematical word problems and language proficiency for students learning through the medium of Irish. *Educational Studies in Mathematics, 71*(1), 43–64.

Ní Ríordáin, M. N., & O'Donoghue, J. (2011). Tackling the transition—the English mathematics register and students learning through the medium of Irish. *Mathematics Education Research Journal, 23*(1), 43–65.

Nkambule, T. (2009). *Teaching and learning linear programming in a grade 11 multilingual mathematics class of English language learners: Exploring the deliberate use of learners home language* (Unpublished master's dissertation). University of the Witwatersrand, South Africa.

Norén, E. (2011). Students' mathematical identity formations in a Swedish multilingual mathematics classroom. *Nordic Studies in Mathematics Education, 16*(1–2), 95–113.

Nunes, T., Schliemann, A. D., & Carraher, D. W. (1993). *Street mathematics and school mathematics.* Cambridge, United Kingdom: Cambridge University Press.

Ochs, E. (1979). Transcription as theory. In E. Ochs & B. B. Schiffelin (Eds.), *Developmental pragmatics* (pp. 43–72). New York, NY: Academic Press.

O'Halloran, K. (2005). *Mathematical discourse: Language, symbolism and visual images.* London, United Kingdom: Continuum.

Oseguera, L., Conchas, G. Q., & Mosqueda, E. (2011). Beyond family and ethnic culture: Understanding the preconditions for the potential realization of social capital. *Youth & Society, 43*(3), 1136–1166.

Parvanehnezhad, Z., & Clarkson, P. (2008). Iranian bilingual students reported use of language switching when doing mathematics. *Mathematics Education Research Journal, 20*(1), 52–81.

Peal, E., & Lambert, W. E. (1962). The relation of bilingualism to intelligence. *Psychological Monographs: General and Applied, 76*(27), 1–23.

Phillips, C. J., & Birrell, H. V. (1994). Number learning of Asian pupils in English primary schools. *Educational Research, 36*(1), 51–62.

Pimm, D. (1987). *Speaking mathematically: Communication in mathematics classrooms.* London, United Kingdom: Routledge.

Planas, N. (2011). Language identities in students' writings about group work in their mathematics classroom. *Language and Education, 25*(2), 129–146.

Planas, N. (2014). One speaker, two languages: Learning opportunities in the mathematics classroom. *Educational Studies in Mathematics, 87*(1), 1–16.

Planas, N., & Civil, M. (2013). Language-as-resource and language-as-political: Tensions in the bilingual mathematics classroom. *Mathematics Education Research Journal, 25*(3), 361–378.

Planas, N., & Setati, M. (2009). Bilingual students using their languages in the learning of mathematics. *Mathematics Education Research Journal, 21*(3), 36–59.

Poland, B. (2002). Transcription quality. In J. Gubrium & J. Holstein (Eds.), *Handbook of interview research: Context and method* (pp. 629–649). Thousand Oaks, CA: Sage.

Powell, A. B., & Frankenstein, M. (Eds.). (1997). *Ethnomathematics: Challenging Eurocentrism in mathematics education*. Albany, NY: State University of New York Press.

Razfar, A. (2013). Multilingual mathematics: Learning through contested spaces of meaning making. *International Multilingual Research Journal, 7*(3), 175–196.

Riley, M. S., & Greeno, J. G. (1988). Developmental analysis of understanding language about quantities and of solving problems. *Cognition and Instruction, 5*(1), 49–101.

Sarnecka, B. W., & Carey, S. (2008). How counting represents number: What children must learn and when they learn it. *Cognition, 108*, 662–674.

Saunders, G. (1988). *Bilingual children: From birth to teens*. Clevedon, United Kingdom: Multilingual Matters.

Schleppegrell, M. (2007). The linguistic challenges of mathematics teaching and learning: A research review. *Reading & Writing Quarterly, 23*, 139–159.

Secada, W. (1991). Degree of bilingualism and arithmetic problem solving in Hispanic first graders. *Elementary School Journal, 92*(2), 213–231.

Secada, W. G. (1992). Race, ethnicity, social class, language and achievement in mathematics. In D. A. Grouws (Ed.), *Handbook of research on mathematics teaching and learning* (pp. 623–660). New York, NY: MacMillan.

Seeger, F., Voigt, J., & Waschescio, U. (1998). *The culture of the mathematics classroom*. Cambridge, United Kingdom: Cambridge University Press.

Setati, M. (1998). Code-switching and mathematical meaning in a senior primary class of second language learners. *For the Learning of Mathematics, 18*(1), 34–40.

Setati, M. (2003). "Re"-presenting qualitative data from multilingual mathematics classrooms. *ZDM—The International Journal on Mathematics Education, 35*(6), 294–300.

Setati, M. (2005). Teaching mathematics in a primary multilingual classroom. *Journal for Research in Mathematics Education, 36*(5), 447–466.

Setati, M. (2008). Access to mathematics versus access to the language of power: The struggle in multilingual mathematics classrooms. *South African Journal of Education, 28*, 103–116.

Setati, M. (2012). Mathematics in multilingual classrooms in South Africa: From understanding the problem to exploring possible solutions. In B. Herbel-Eisenmann, J. Choppin, D. Wagner, & D. Pimm (Eds.), *Equity in discourse for mathematics education: Theories, practices and policies* (pp. 125–145). Dordrecht, The Netherlands: Springer.

Setati, M., & Adler, J. (2001). Between languages and discourses: Code switching practices in primary classrooms in South Africa. *Educational Studies in Mathematics, 43*(3), 243–269.

Setati, M., & Barwell, R. (2006). Discursive practices in two multilingual mathematics classrooms: An international comparison. *African Journal of Research in Mathematics, Science and Technology Education, 10*(2), 27–38.

Setati, M., Molefe, T., & Langa, M. (2008). Using language as a transparent resource in the teaching and learning of mathematics in a grade 11 multilingual classroom. *Pythagoras, 67*, 14–25.

Sfard, A. (2008). *Thinking as communicating: Human development, the growth of discourses, and mathematizing*. Cambridge, United Kingdom: Cambridge University Press.

Shaftel, J., Belton-Kocher, E., Glasnapp, D., & Poggio, J. (2006). The impact of language characteristics in mathematics test items on the performance of English language learners and students with disabilities. *Educational Assessment, 11*(2), 105–126.

Skovsmose, O. (1994). *Towards a philosophy of critical mathematics education*. Dordrecht, The Netherlands: Kluwer.

Spanos, G., Rhodes, N. C., Dale, T. C., & Crandall, J. (1988). Linguistic features of mathematical problem solving: Insights and applications. In R. R. Cocking & J. Mestre (Eds.), *Linguistic and cultural influences on learning mathematics* (pp. 221–240). Hillsdale, NJ: Lawrence Erlbaum.

Swain, M., & Cummins, J. (1979). Bilingualism, cognitive functioning and education. *Language Teaching, 12*(1), 4–18.

Swain, M., & Lapkin, S. (2005). The evolving socio-political context of immersion education in Canada: Some implications for program development. *International Journal of Applied Linguistics, 15*(2), 169–186.

Takeuchi, M., & Esmonde, I. (2011). Professional development as discourse change: Teaching mathematics to English learners. *Pedagogies, 6*(4), 331–346.

Teddick, D. J., Christian, D., & Fortune, T. W. (Eds.). (2011). *Immersion education: Practices, policies, possibilities*. Bristol, United Kingdom: Multilingual Matters.

Torres, L. (1997). *Puerto Rican discourse: A sociolinguistic study of a New York suburb*. Mahwah, NJ: Lawrence Erlbaum.

Towse, J., & Saxton, M. (1998). Mathematics across national boundaries: Cultural and linguistic perspectives on numerical competence. In C. Donlan (Ed.), *The development of mathematical skills* (pp. 129–150). Hove, United Kingdom: Psychology Press.

Trinick, T., Meaney, T., & Fairhall, U. (2014). Teachers learning the registers of mathematics and mathematics education in another language: An exploratory study. *ZDM—The International Journal on Mathematics Education, 46*(6), 1–13.

Turnbull, M., Lapkin, S., & Hart, D. (2001). Grade 3 immersion students' performance in literacy and mathematics: Province-wide results from Ontario (1998–99). *The Canadian Modern Language Review/La revue canadienne des langues vivantes, 58*(1), 9–26.

Turner, E., & Celedon-Pattichis, S. (2011). Problem solving and mathematical discourse among Latino/a kindergarten students: An analysis of opportunities to learn. *Journal of Latinos in Education, 10*(2), 146–168.

Turner, E., Dominguez, H., Maldonado, L., & Empson, S. (2013). English language learners' participation in mathematical discussion: Shifting positionings and dynamic identities. *Journal for Research in Mathematics Education, 44*(1), 199–234.

United Nations Educational, Scientific and Cultural Organization. (1974). *Interactions between linguistics and mathematics education: Final report of the symposium sponsored by UNESCO, CEDO and ICMI.* Nairobi, Kenya: Author.

Valdés-Fallis, G. (1978). Code switching and the classroom teacher. *Language in education: Theory and practice.* Wellington, VA: Center for Applied Linguistics. (ERIC Document Reproduction Service No. ED153506)

Valenzuela, A. (1999). *Subtractive schooling: Issues of caring in education of US-Mexican youth.* Albany, NY: State University of New York Press.

Valenzuela, A. (2002). Reflections on the subtractive underpinnings of education research and policy. *Journal of Teacher Education, 53*(3), 235–231.

Vertovec, S. (2007). Super-diversity and its implications. *Ethnic and Racial Studies, 30*(6), 1024–1054.

Vorster, H. (2008). Investigating a scaffold to code-switching as strategy in multilingual classroom. *Pythagoras, 67,* 33–41.

Wagner, D., & Lunney Borden, L. (2012). Aiming for equity in ethnomathematics research. In B. Herbel-Eisenmann, J. Choppin, D. Wagner, & D. Pimm (Eds.), *Equity in discourse for mathematics education* (pp. 69–87). Dordrecht, The Netherlands: Springer.

Wei, L., & Garcia, O. (2014). *Translanguaging: Language, bilingualism and education.* Basingstoke, United Kingdom: Palgrave Macmillan.

Young, V. A., & Martinez, A. Y. (Eds.). (2011). *Code-meshing as world English: Pedagogy, policy, performance.* Urbana, IL: National Council of Teachers of English.

Zentella, A. C. (1997). *Growing up bilingual: Puerto Rican children in New York.* Malden, MA: Blackwell.

Zepp, R., Monin, J., & Lei, C. L. (1987). Common logical errors in English and Chinese. *Educational Studies in Mathematics, 18*(1), 1–17.

Zevenbergen, R. (2000). "Cracking the code" of mathematics classrooms: School success as a function of linguistic, social, and cultural background. In J. Boaler (Ed.), *Multiple perspectives on mathematics teaching and learning* (pp. 201–223). Westport, CT: Ablex.

22 种族与数学教育

丹尼·伯纳德·马丁
美国伊利诺伊大学芝加哥分校
西莉亚·卢梭·安德森
美国孟菲斯大学
尼罗尔·沙阿
美国密歇根州立大学
译者：刘启蒙
　　　北京师范大学中国基础教育质量监测协同创新中心

本章我们主要综述美国种族与数学教育研究的最新进展，并为未来的研究方向提供建议。我们将审视理论与实证研究中的相关发现，并讨论数学教育研究中与种族问题相关的一系列研究方法。我们将在实践和政策层面提出一些可能的建议，以便于数学教育研究者、数学教师和其他数学教育领导者都能够从中获得一些新的认识和思路。

在构思这篇综述的过程中，我们分析了已有的综述性研究，比如"种族，民族，社会阶层，语言与数学教育的成就"（Secada，1992）和"文化，种族，权力与数学教育"（数学教育中的多样化[DiME]，2007）。这些研究都关注了数学教育中的种族因素，并且反映了当时该领域的研究状况。但是，近期该领域取得了许多研究进展，新的学术视角以及实证研究方法使得我们不仅能够以欣赏的态度去回顾上述研究，而且能够以更加准确而扎实的理解对上述研究进行批判性分析。与此同时，我们参阅了引用率最高的几个数学教育研究期刊（如《数学教育研究学报》《数学思维与学习》《城市数学教育杂志》《数学教育研究》），参考了一些近期的与城市数学教育有关的著作（如《黑人儿童的数学才华：不止是数，走向新话语》，Leonard & Martin，2013；《黑人儿童生活中的数学教学、学习与解放》，Martin，2009c；《建立数学学习共同体：改善城市高中的学业结果》，Walker，2012），以及数学教育领域之外的一些期刊（如《美国教育研究学报》《教育研究综述》《师范学院档案》《城市教育》《高中杂志》《美国黑人男性教育杂志》）。

本章并不追求综述的详尽，而是重点选择一些研究和研究脉络，其中包含特定概念、研究方法或者研究范例。考虑到近10年间，明确关注种族问题的文章和书籍在数量上有明显上升，我们特别关注近10年的大量研究，并从中选取范例。近10年间的这种增长现象象征着数学教育中所谓的社会政治变革，虽然这种变革尚未完全体现出来（参见这套书的Stinson & Walshaw，2017一章），但是越来越多的文献表现出对数学教育中政治因素的不断关注，包括知识、权力和身份（例如Ernest, Greer, & Sriraman，2009；R. Gutiérrez，2013b；Gutstein，2008，2009；Martin，2009a，2013；Martin & Larnell，2013；Skovsmose & Valero，2001；Stinson，2011；Stinson & Bullock，2012；Valero & Zevenbergen，2004；Walshaw，2004，2013）。近期在《数学教育研究学报》特刊上（JRME；R. Gutiérrez，2013a）发表的一系列文章，帮助我们对这种变革有了更加清晰的认识和理解。此外，佐治亚州立大学于2008年创办与培育的《城市数学教育杂志》（JUME）期刊，开辟了一个重要的学术阵地，刊登并传播与种族问题直接相关的学术研究。

我们的目的是突出那些明确关注种族问题的研究和观点（如种族与数学学业表现、种族与学习机会、种族身份与数学身份、种族之间的轻微攻击、种族化的话语和

22 | Race and Mathematics Education

DANNY BERNARD MARTIN
University of Illinois at Chicago, United States

CELIA ROUSSEAU ANDERSON
University of Memphis, United States

NIRAL SHAH
Michigan State University, United States

In this chapter, we review extant research on race and mathematics education, primarily in the context of the United States, and provide recommendations for future research directions. In addition to examining theoretical and empirical findings, we discuss a range of methods that have been used to examine race in mathematics education. Where possible, we make recommendations for practice and policy so that emerging ideas and new understandings are accessible to teachers and other mathematics education leaders.

In constructing our review, we build on previous reviews such as "Race, Ethnicity, Social Class, Language, and Achievement in Mathematics Education" (Secada, 1992) and "Culture, Race, Power, and Mathematics Education" (Diversity in Mathematics Education [DiME], 2007). Although these chapters addressed race in ways that reflected the state of the field at those times, recent scholarly perspectives and empirical approaches allow us to look back on those chapters not only with appreciation but also with more refined and robust understanding that allows for critical analysis. In conducting this review, we consulted several of the most widely referenced mathematics education research journals (e.g., *Journal for Research in Mathematics Education; Mathematical Thinking and Learning; Journal of Urban Mathematics Education; Educational Studies in Mathematics*), a number of recent texts in urban mathematics education (e.g., *The Brilliance of Black Children in Mathematics: Beyond the Numbers and Toward New Discourse*, Leonard & Martin, 2013; *Mathematics Teaching, Learning, and Liberation in the Lives of Black Children*, Martin, 2009c; *Building Mathematics Learning Communities: Improving Outcomes in Urban High Schools*, Walker, 2012), and a number of journals outside mathematics education (e.g., *American Educational Research Journal; Review of Educational Research; Teachers College Record; Urban Education; High School Journal; Journal of African American Males in Education*).

Rather than attempt to provide an exhaustive review, we highlight select studies and lines of research that we believe serve as exemplars of particular concepts, research methods, and findings. We especially draw from a corpus of research produced within the last 10 years, given that there has been a noticeable rise in the number of articles and books explicitly addressing race during the decade. This increase is symbolic of, but not fully captured by, what has been called the *sociopolitical turn* in mathematics education (see also Stinson & Walshaw, 2017, this volume), a growing movement characterized by increased attention to the political aspects of mathematics education, including knowledge, power, and identity (e.g., Ernest, Greer, & Sriraman, 2009; R. Gutiérrez, 2013b; Gutstein, 2008, 2009; Martin, 2009a, 2013; Martin & Larnell, 2013; Skovsmose & Valero, 2001; Stinson, 2011; Stinson & Bullock, 2012; Valero & Zevenbergen, 2004; Walshaw, 2004, 2013). A series of articles published in a recent special issue of the *Journal for Research in Mathematics Education* (*JRME*; R. Gutiérrez, 2013a) helped to make this turn more explicit. Moreover, the founding and growth of the *Journal of Urban Mathematics Education* (*JUME*) at Georgia State University in 2008 opened up an important intellectual space where scholarship that explicitly addresses race is being disseminated.

Our choice is to highlight studies and perspectives that *explicitly* address race (e.g., race and mathematics achievement; race and opportunity to learn; racial and mathematics identities; racial microaggressions; racialized discourses and narratives about mathematics learners; racialized mathematics experiences). In our

关于数学学习者的叙述、种族化的数学学习经验）。在我们看来，将种族化的社会类属[1]与诸如数学能力、素养、动机和情感等概念相并列，常常有助于强化种族类别的社会意义，以及人们对不同种族人的印象（比如亚洲人擅长数学，黑人在数学方面有所欠缺，以及黑人智力劣势）。除了以上与种族问题直接相关的数学教育研究以外，我们没有分析与种族问题间接相关的研究。然而，我们承认，虽然许多研究并非重点关注种族，但是这些研究仍然可能会帮助研究者建构有关种族的认知（比如有助于对无肤色歧视问题的研究）。例如，当没有特意强调研究对象的种族身份时，读者也许会假定研究对象是白人，这就起到了把白人标准化的作用，并将白人群体定位为普遍主体。[2]事实上，我们认为每一项数学教育研究都会在某些方面与种族有关，并通过知识的生产和消费反映出一种或多种与种族和数学有关的意识形态。

人们普遍认为，关于种族政策、意义和结果的讨论通常较为困难，会令人不安，并且有时会因为敏感而需要回避。因此，本章的目的是不仅要吸引越来越多的数学教育者聚焦于种族及其相关话题，还要吸引数学教育以外的更广泛领域的学者，他们中的大部分都没有明确注意过这一话题（Berry，2005；Martin，2009a；Parks & Schmeichel，2012；Stinson，2011）。考虑到本章的教育作用，我们会慎重地界定重要概念，并提供相关例子，以便于读者理解。当然，种族并非是一个新的概念，只是在20多年前它才出现在数学教育领域中。

它不仅仅与教育均衡、教育多样化、城市数学教育、非白人学生的讨论有关。我们也相信，即使种族问题不是研究、政策或实践的一个明确的重点，但是讨论种族如何在社会意义和知识建构、政策、意识形态、经历、类别、身份和权力关系等方面表现出来还是很重要的，它一直渗透在数学教育中，也就是说，它的"缺席"引人注目（Apple，1999，第9页）。

我们在一开始就认识到种族这一概念及其物质实体并非孤立存在，而是与其他重要概念（如性别、社会阶层、语言背景）紧密相关。尽管如此，本章不会涉及所有相关的内容（如多样化、可获得性、学习机会、文化情境中的教育学、基于社会公正的数学教学）。虽然这些话题也非常重要，且经常会与种族问题紧密交织在一起（C. R. Anderson & Tate，2008；Lipman，2012；Weissglass，2002），但是本章遵循以种族为中心的原则并将种族作为首要思考的内容。此外，其他学者关注不平等问题并讨论改变不平等问题的方法，我们对此表示理解并认可，但是我们的种族中心观点并不会完全将注意力集中在表述数学学业成就结果的悬殊差异上（如种族的成绩差异），或参与度和学习资源的差异上（例如Lubienski，2008）。讨论这种差异是重要的，但我们并不希望将这些矛盾归因于某些学习者和社会群体的缺陷，以及另一些群体的优势。在适当的情况下，我们会指出当前研究是如何探讨这种差异对学生种族化经历的影响的。

为了能更好地报告有关种族和数学教育领域的研究现状，我们考虑了一系列的问题，这些问题是该领域学者正在关注的主题，也是新进入这一领域或对这一领域不熟悉的学者可能会认为与之相关的或具有重要意义的一些主题：

- 在不同类型的研究中，种族是如何在数学教育研究中被概念化（为理论）和（用什么方法加以）分析的？研究者是如何使用这些概念开展有关数学学习和学业表现研究的？
- 现有研究中揭示了哪些种族内和种族间学生数学学习和参与的种族特性？
- 学生交流和课堂实践中常见的种族话语和刻板印象体现在哪些方面？
- 作为一个知识生产领域，数学教育研究是如何讨论和产生与种族意识形态、意义、差异、层次和身份有关的研究的？

除了回顾当前的研究之外，上述四个问题还从教育的角度引导我们考虑其他可能会对这一领域有帮助的研究内容。例如，一些学者可能不清楚为什么在数学学习和教学中要把种族以及与种族有关的过程与实践作为重要的考虑因素。接下来有一节来专门说明这个问题。此外，当前正在从事种族研究或有意向从事这类研究的学者，作为种族知识的消费者以及生产者，对他们来说，至关重要的一点就是，能够建构严谨而清晰的概念并进行学术沟通。在本章的开头，我们给出了一个概念上的引子。在本章的结尾，

view, the juxtaposition of racialized social categories[1] with concepts like mathematics ability, competence, motivation, and disposition often helps reinforce societal meanings for the racial categories and beliefs about people thought to belong to those categories (e.g., Asians as good at math, Blackness as deficient, and Black intellectual inferiority). Beyond those studies that explicitly address race, we do not attempt to assign any intentionality or particular ideology to researchers whose work does not explicitly address race. However, we acknowledge that many studies may contribute to the construction of racial knowledge even when race is not the focus (e.g., contributing to color-blind ideology). For example, when there is no mention of participants' racial identities, it might be assumed that the participants are White, which has the effect of normalizing Whiteness and positioning the White subject as the universal subject.[2] In essence, we claim that every study in mathematics education is about race in some way and reflects—through knowledge production or consumption—one or more sets of ideologies about race and mathematics.

Recognizing that discussions focused on the politics, meaning, and consequences of race are often difficult, emotionally unsettling, and sometimes avoided, our goal for this chapter is that it appeal not only to the growing number of mathematics education scholars who focus on race and related topics but also to those in the broader mathematics education field, many of whom do not give explicit attention to this topic (Berry, 2005; Martin, 2009a; Parks & Schmeichel, 2012; Stinson, 2011). Given the educative role that the chapter can serve, we carefully define and provide examples of important concepts for those who are unfamiliar with the literature. However, we take care to note that race is not a novel consideration, only appearing in mathematics education during the last 20 years or so, nor is it only relevant in discussions of equity, diversity, urban mathematics education, and non-White students. We also believe it is important to discuss how race—manifested in social meanings and knowledge construction, policy, ideologies, experiences, categories and identities, and relations of power—has always permeated mathematics education, even when it has not been an explicit focus of research, policy, or practice; that is, conspicuous by its "absent presence" (Apple, 1999, p. 9).

We acknowledge at the outset that the concept of race and its material realities do not exist within a vacuum, divorced from other important considerations like gender, social class, and linguistic background. However, we avoid the chapter serving as a depository for all-things-equity (e.g., diversity, access, opportunity to learn, culturally relevant pedagogy, teaching mathematics for social justice).

Although these topics are important and often intersect with issues of race in important ways (C. R. Anderson & Tate, 2008; Lipman, 2012; Weissglass, 2002), there is value in centering race as a primary consideration. Finally, although we understand and appreciate the more common approach of drawing attention to inequities and pinpointing areas for change, our centering of race will not be focused entirely on recounting well-documented disparities in mathematics achievement outcomes (e.g., "racial achievement gaps"), participation, and resources (e.g., Lubienski, 2008). Such disparities are important, but we do not wish to suggest that these discrepancies reflect deficiencies in some learners and social groups and superiority in others. Where appropriate, we point out how the extant research explores how such differences reflect the consequences of students' racialized experiences.

In thinking through how we could best inform the field about extant research on race and mathematics education, we considered a set of questions that scholars who conduct research in this area have generally been pursuing and that scholars who are new to or unfamiliar with this work might find relevant and insightful:

- Across different levels of analysis, how has race been conceptualized (theory) and studied (methods) in mathematics education, and how have those conceptualizations been used to inform studies of learning and achievement?
- What does extant research reveal about the racialized nature of mathematics learning and participation for students within and across racial categories?
- What are the prevailing racial discourses and stereotypes that inform student interactions and classroom practices?
- How does mathematics education research, as a knowledge-producing domain, consume and produce racial ideologies, meanings, disparities, hierarchies, and identities?

From an educative perspective, these four questions also led us to consider what other content, beyond a review of extant research, might be helpful to the field. For example, it may be unclear to some scholars why race, and processes and practices related to race, are important considerations in mathematics learning and teaching. We offer a section devoted to addressing this question. In addition, it is important for those currently engaged in race-focused research, or who seek to be, to communicate with rigor and conceptual clarity, both as consumers and producers of racial knowledge. Early in the chapter, we offer a conceptual primer. At the end of

我们将会提供关于种族和数学教育研究的一个阅读指南。

基于对上述问题以及相关内容的思考，我们将这一章分为8个部分。在第一部分，我们以种族问题的最新研究进展为基础，概述了关注种族的合理性（DiME，2007；Martin，2009a；Parks & Schmeichel，2012；Stinson，2011）。我们对合理性的阐述是源于这样一种认知：尽管种族问题研究方面有一些进展，但种族问题依然是学术研究中被边缘化的主题，而且当阐述这一问题时，经常面临概念化不足的问题。在第二部分，我们提出一个概念导引，进而在一定程度上解决后一个问题，它包括了一些关键概念的定义及其关系，如种族、种族主义、种族化以及种族意识形态。我们还会区分种族、民族和文化等概念之间的差异，因为这些概念通常被看作是相同的，而后两个概念通常被认为是第一个概念的替代品。在第三部分，我们提供了5个宏观主题，以说明种族问题领域是如何对种族概念化的：（1）种族是一个范畴变量；（2）种族可以与文化概念互换使用；（3）种族的特性和种族化经历；（4）种族是一组相互交叉的话语体系和意识形态；（5）数学教育作为一种组织和结构，它的种族化特征。在第四部分，我们更为详尽地综述了在数学课堂研究中是如何讨论种族问题的，并重点介绍了研究所涉及的若干个子领域。在第五部分，我们综述的范围扩大到种族、社会和数学教育方面，特别关注了将种族、数学和社会化三者相关联的研究，以及种族化话语和刻板印象的相关研究。在第六部分，我们考察数学教育研究作为一个知识产出领域如何通过学术研究揭示其种族化特征。在第七部分，我们围绕如何推进关于种族和数学教育的知识议程（DiME，2007；Secada，1992）提出了一些建议，以促进这一领域的发展。最后，在第八部分，我们通过四个启发式问题提供一个阅读指南，有助于读者在阅读中批判性地分析我们工作中产生和表现出的种族知识和观点。我们注意到，在大多数明确指向种族的实证研究中，非裔美国学习者以及他们的学习机会和学业表现一直是研究焦点，部分原因在于种族学业表现差异话语体系的流行以及在种族对话和理论长期占据统治地位的黑-白二元论。在本章的最后，我们也将讨论如何超越黑-白二元论。

为什么要研究数学教育中的种族问题？

随着美国选出了第一位黑人总统，许多人主张将种族问题作为后种族主义（即社会超越了种族，种族不再是社会问题，以及种族主义已经成为过去；Bonilla-Silva，2010）和无肤色歧视意识形态（即把当代种族不平等解释为非种族动态的产物；Bonilla-Silva，2010）的一部分予以淡化。但是种族仍是一个重要的问题，它在美国社会中仍然是最具争议和两极分化的主题之一，并且它仍然具有强大的力量，在生活的各个领域构建着人类的交互、机会和身份。种族的意识形态、话语体系以及实践塑造和反映了人们关于智慧和能力的信念。已有证据表明，数学教育搭建了一个丰富情境，用来探索这些意识形态、话语体系和实践等的趋向和结果。例如，最近的批判性分析已经表明，数学教育研究是一个思想和物质的空间，能够促进种族意义、差异和等级的讨论和重新生成（DiME，2007；Martin，2009a，2013；Shah，2013；Stinson，2011）。从知识生产的角度，这些相关的分析表明了儿童数学行为的学术解释是如何为社会意识形态、话语体系以及关于种族、种族类别、才能和能力的信念提供信息的。此外，基于种族的社会意识形态、话语体系和来自不同社会群体的儿童的信念，也揭示了数学教育研究、政策与实践是如何以与儿童相关联的方式被概念化和配置的。

正如斯廷森（2011）所指出的那样，虽然"我们作为一个共同体，已经认识到种族、种族主义、白人至上主义在我们的社区、学校和班级的教育制度中所起到的作用"（第3页）可能是一个合理的假设，但现实却是只有在关注差异时才会涉及种族，主流数学教育中对种族的关注非常有限。基于鲁宾斯基和鲍恩（2000）早期的一个研究综述，斯廷森（2011）以及帕克斯和施迈歇尔（2012）指出了在数学教育和数学教育研究中解决种族问题时会遇到的若干障碍，包括：（1）种族和民族讨论被边缘化；（2）大众普遍而且不断地认为种族是一个独立变量；（3）没有机会对种族和民族作相关分析；（4）在公平性及其分析的讨论中，种族和民族问题被淡化。斯廷森分析了帕克斯和施迈歇尔（2011）以及鲁宾斯基和

the chapter, we offer a guide for reading research on race and mathematics education.

Based on the questions above and our additional considerations, we decided to organize the chapter into eight sections. In the first section, we briefly outline the rationales for our focus on race, which builds on the advances made in recent reviews on the status of race in extant research (DiME, 2007; Martin, 2009a; Parks & Schmeichel, 2012; Stinson, 2011). Our rationales are rooted in the recognition that, despite some measurable progress, race continues to be a marginalized topic of scholarly inquiry and that, when it has been addressed, it has often been subject to underconceptualization. In the second section, we partially address this latter issue by providing a conceptual primer that provides definitions and relationships among key concepts such as race, racism, racialization, and racial ideology. We also help readers distinguish between race, ethnicity, and culture, knowing that these are often conflated and that the latter two are often used as proxies for the first. In the third section, we provide five broad themes that capture how the field has conceptualized race: (1) race as a categorical variable; (2) race being used interchangeably with culture; (3) race in terms of racial identities and racialized experiences; (4) race as a set of intersecting discourses and ideologies; and (5) the racialized character of mathematics education as an institution and structure. In the fourth section, we provide a more detailed review of how race has been addressed in research focused on mathematics classrooms and highlight several areas of research. In the fifth section, we broaden the review to address race, society, and mathematics education, giving particular attention to studies focused on racial-mathematical socialization and studies focused on racialized discourses and stereotypes. In the sixth section we take a closer look at research that has attempted to make explicit the racialized character of mathematics education research as a knowledge-producing domain. In the seventh section, we offer some recommendations for moving the field forward with respect to advancing an intellectual agenda on race and mathematics education (DiME, 2007; Secada, 1992). Finally, in the eighth section, we provide a reading guide—in the form of four heuristic questions—for critically analyzing the racial knowledge and perspectives produced and represented in our work. We note that across most of the empirical studies that explicitly address race, African American learners and their opportunities and outcomes have been the focus. This is partly due to the prevalence of racial achievement gap discourse and the long-standing dominance of a Black-White binary in race conversations and theory. Later in this chapter, we argue for moving beyond this binary.

Why Study Race in Mathematics Education?

On the heels of the United States having elected its first Black president, many have argued for the minimization of attention to race as a part of postracial (i.e., that society has transcended race, that race no longer matters in society, and that racism is a thing of the past; Bonilla-Silva, 2010) and color-blind ideologies (i.e., explaining contemporary racial inequality as the product of non-racial dynamics; Bonilla-Silva, 2010). But race matters. It remains one of the most controversial and polarizing topics in American society, and it continues to be a powerful force that structures human interactions, opportunities, and identities in all areas of life. Racial ideologies, discourses, and practices shape and reflect beliefs about intelligence and competence. Mathematics education has proven to be a rich context to explore the convergence and consequences of these ideologies, discourses, and practices. Recent critical analyses have shown that mathematics education research, for example, is an ideological and material space that facilitates the consumption and reproduction of racial meanings, disparities, and hierarchies (DiME, 2007; Martin, 2009a, 2013; Shah, 2013; Stinson, 2011). With respect to knowledge production, these same analyses have demonstrated how scholarly interpretations of children's mathematical behaviors serve to inform societal ideologies, discourses, and beliefs about race, racial categories, abilities, and competence. Moreover, race-based societal ideologies, discourses, and beliefs about children from various social groups inform the ways that mathematics education research, policy, and practice are conceptualized and configured in relation to these children.

As noted by Stinson (2011), although it might be reasonable to assume that "as a community, we have grown to understand how race, racism, and white supremacy function in education policy and within our communities, schools, and classrooms" (p. 3), the reality is that beyond a focus on disparities, race has received limited attention in mainstream mathematics education. Stinson (2011) and Parks and Schmeichel (2012)—building on an earlier review by Lubienski and Bowen (2000)—noted several barriers to addressing race in mathematics education and mathematics education research, including (1) the marginalization of race and ethnicity discussions, (2) widespread reiterations of race as an independent variable, (3) the absence of race and ethnicity as a relevant site for analysis, and (4) the minimization of race and ethnicity in

鲍恩的综述，总结发现："数学教育文献中有关种族的研究实际上并没有增加（即在近三十年内，同行评议期刊中阐述数学教育和种族/民族问题的文章比例一直保持不变[3.7%—3.8%]）"（第4页）。具体而言，帕克斯和施迈歇尔（2012）指出，2008年至2011年间在《数学教育研究学报》上发表的46篇文章中，仅有5篇文章将种族作为一个类别变量进行讨论，5篇之中有3篇出现在最新一期的特刊中（R. Gutiérrez, 2013a）。他们还提出以下观点，类似马丁（2009a）在主流数学教育研究中提出的无肤色歧视取向的观点：

> 种族和民族屡次被描述为与理解社会交互和学习环境毫无关系的变量，逐渐地在话语中就会演变成为一种定论。这样所产生的后果是，人们越来越简单地表述涉及数学的人类互动，而不去讨论种族和民族可能产生的影响。对这一话题相关讨论的缺失，也会导致出现一种不良的学术环境，使得评论者可能会呼吁将种族和民族的讨论从有关数学教育文献中移除，因为他们看上去与数学教育并不相干（Parks & Schmeichel, 2012，第246—247页）。

DiME（2007）综述的作者同样指出，尽管他们努力研究课堂情境中以及"在一个更广泛、更加结构化的背景下"（第421页）的种族和权力，但几乎没有数学教育研究提供种族和权力的结构性分析。他们呼吁展开更多研究，因为他们发现"在这一领域中尚未形成关于种族、种族主义和权力的理论。换句话说，文献并没有充分阐述种族如何在数学教育领域中与有色人种或白人学生的经历产生关系"（第427页）。

回顾DiME（2007）给出的观点，我们注意到，尽管作者建议要在这方面进行更多的研究，但是他们并没有提供清晰的种族理论，以便于数学教育研究者可以在研究中使用。因此，在那一章中，一些重要的概念，如种族、种族主义、种族意识形态、种族结构、种族刻板印象和种族化等概念，并没有得以强调或者理论化。在教育领域之外有关种族的观点，比如社会学和批判性研究，可能会促进将种族作为一种社会的、政治的和法律的结构进行理解。此外，作者并没有给出可以有效考察种族话语、种族结构和种族化经历的一些特殊而且被证实有效的研究方法，而这些研究方法恰恰是可能会对数学教育者十分有益的。

明晰种族概念：一个概念性的导引

为使本章的思想易于理解，我们为贯穿本章的一些重要概念——种族、种族主义、种族化、种族意识形态——提供了一些简要的指向性定义。我们希望帮助读者理解这些词汇一般是如何被使用的，以及在某些情况下，当它们被用在数学教育中时，是如何或者应该是如何服务于讨论与争论的。我们的综述表明，那些真正专注于种族和数学教育的学者们经常借助于该领域之外的众多理论和观点，以便从概念和分析上明晰他们的研究。限于篇幅，我们简要地讨论这些领域和重要的相关概念。

我们首先要问：什么是种族？在科学领域有着这样的广泛共识，即种族没有生物学基础（Gould, 1996; Smedley & Smedley, 2005; Zuberi & Bonilla-Silva, 2008）。然而，数学教育领域外的研究文献揭示出，有很长一段历史时期，学者围绕种族是否是生物学上的（从肤色或体貌特征加以区别）或者社会建构的（Andreasen, 2000; Frankenberg, 1992; Smedley & Smedley, 2005）进行了丰富而激烈的争论。不过，本章不会对这些争论的细节进行回顾，就我们的目的而言，我们将本章中种族的讨论与那些将种族作为社会的、政治的和法律的构建以及将种族类别作为社会产物的理论和框架相结合（例如Bonilla-Silva, 2001; López, 1996）。社会学家迈克尔·欧米和霍华德·怀南特（1986）解释了种族的社会和政治架构特征，他们说："种族类别和种族意义是通过其所嵌入的特殊的社会关系和历史背景来具体表达的"（第60页）。这一观点表明，种族的含义不是固定的，甚至是有争议的。他们还指出："在微观层面上，种族是一个个体问题，随着人们的身份而形成……在宏观层面上，种族是一个社会结构形成的问题，包括政治、意识形态和经济"（第66—67页）。

在他们更宏观的理论框架下，欧米和怀南特注释道"种族是社会结构和文化表征（意义）的一种功能，也就是说，解释种族的意义就是从社会结构上框定它，即从社会结构认识种族维度"（1994，第55—56页）。这个意义

discussions of equity and analysis. Based on his analysis of reviews by Parks and Schmeichel (2011) and Lubienski and Bowen, Stinson concluded, "there has been virtually no increase (i.e., for nearly 3 decades, the percentage of peer-reviewed journal articles that address mathematics education and race/ethnicity has stayed constant [3.7%; 3.8%])" (p. 4). More specifically, Parks and Schmeichel (2012) noted, for example, that of the 46 articles published in the *Journal for Research in Mathematics Education* between 2008 and 2011, only 5 discussed race in ways that moved beyond race as a categorical variable. Of those 5, 3 appeared in the most recent special issue (R. Gutiérrez, 2013a). They further noted—similar to Martin's (2009a) claim of a *color-blind* orientation in mainstream mathematics education research:

> When again and again, race and ethnicity are portrayed as irrelevant to understanding social interactions and learning environments, this becomes more inscribed in the discourse. As a result, it becomes easier and easier to write about human interactions involving mathematics without discussing the ways that they may be impacted by discourses of race and ethnicity. Failing to make any reference to these topics also contributes to an intellectual environment in which reviewers may call for the removal of discussions of race and ethnicity from articles about mathematics education because these are seen as irrelevant. (Parks & Schmeichel, 2012, pp. 246–247)

The authors of the DiME (2007) review also noted that despite their efforts to examine race and power in the classroom context, as well as "within a broader, structural context" (p. 421), there existed very little mathematics education research that provided a structural analysis of race and power. They called for continued research based on their finding that "race, racism, and power remain undertheorized in the field. In other words, the literature does not sufficiently address how race interacts with the experiences of students of color or White students in mathematics education" (p. 427).

As we look back on the review provided by DiME (2007), we note that despite making recommendations for additional research, the authors provided no explicit theories of race that mathematics education researchers could use in their work. As a result, important concepts like race, racism, racial ideology, racial structure, racial stereotypes, and racialization were unaddressed or undertheorized in the chapter. Perspectives on race from fields outside education, such as sociology and critical studies, would have facilitated an understanding of race as a social, political, and legal construction. In addition, the authors did not highlight specific research methods that have proven useful in examining racial discourse, racial structure, and racialized experiences and that might be useful for mathematics educators.

Clarifying Race: A Conceptual Primer

Given our desire to make the ideas in this chapter accessible, we offer some brief orienting definitions for key concepts—race, racism, racialization, racial ideology—that appear throughout the chapter. We hope to support readers in understanding how these terms are generally used and, in some instances, how they are or should be subject to negotiation and contestation when used in mathematics education. Our review reveals that scholars who do focus on race and mathematics education often turn to various theories and perspectives outside the field to bring analytical and conceptual clarity to their work. Given the limitations of space, our presentation of these areas and key concepts is necessarily telegraphic.

We begin by asking: *What is race?* There is broad consensus within the scientific community that race has no biological basis (Gould, 1996; Smedley & Smedley, 2005; Zuberi & Bonilla-Silva, 2008). However, research literature outside mathematics education reveals a long history of rich and intense debates about whether race is biological (distinguished by skin color or physical features) or a social construction (Andreasen, 2000; Frankenberg, 1992; Smedley & Smedley, 2005). Reviewing those debates is beyond the scope of this chapter. For our purposes, we align the discussion of race in this chapter with theory and frameworks that speak to race as being socially, politically, and legally constructed and that view racial categories as social products (e.g., Bonilla-Silva, 2001; López, 1996). Sociologists Michael Omi and Howard Winant (1986) explain the socially and politically constructed nature of race by stating, "Racial categories and the meaning of race are given concrete expression by the specific social relations and historical context in which they are embedded" (p. 60). This latter point suggests that the meanings for race are not fixed and are contested. They also suggest that "at the microlevel, race is a matter of individuality, of the formation of identity.... At the macro-level, race is a matter of the formation of social structures: political, ideological, and economic" (pp. 66–67).

Within their broader theoretical framework, Omi and Winant note that "race is a function of social structure and cultural representation (meaning)—that is, to interpret the meaning of race is to frame it social structurally and to recognize the racial dimension in social structure is to interpret the meaning of race" (1994, pp. 55–56). Key to this meaning making is the process of *racialization*—"the extension of racial meaning to a previously racially

建构的关键是种族化的过程——"对过去种族中未分类的关系、社会实践和族群进行种族意义的扩展（1986，第64页）"。这一过程能够帮助解释以下这些社会方式，例如，非洲人中伊博人或约鲁巴人在奴隶制度产生种族意义，组成黑人，而各种欧洲移民，如爱尔兰人、意大利人（Ignative, 2009; Roediger, 2006）以及犹太人（Brodkin, 1998）成为白人一样。

在数学教育中，严谨处理种族问题时也需要澄清种族主义的含义。虽然种族主义在字面上通俗易懂，但我们常常把种族主义归结为个人信仰和偏见（Bonilla-Silva, 2001）。但实际上，种族主义存在于多个层面，如个人的、制度上的和结构化的，并且可以表现为日常经验或者体制所控制的环境。由于对种族主义的全面讨论超出了本章的范畴，我们建议读者参考博尼拉-席尔瓦（2001）的著作，他在文章中对种族主义的不同解释给出了令人信服的分析，包括马克思主义的观点、制度主义的观点、内部殖民主义的观点、种族形成的观点，以及种族主义是社会垃圾的观点等。其他一些有用的当代研究有《种族架构：种族和教育的多维理论》（Leonardo, 2013）、《沉默的种族主义：善意的白人如何延续种族的分裂》（Trepagnier, 2010）、《日常反种族主义：探寻学校中种族问题的现状》（Pollock, 2008）、《白人种族主义的日常用语》（J. H.Hill, 2009），以及《种族与种族主义理论》（Back and Solomos, 2000）。为了达到综述的目的，我们引用埃塞德（2002）对种族主义的定义：

> 种族主义是一种**结构**，由于种族的优势存在于该结构中，并通过制定和运用规章、法律、条例以及通过获得和分配资源而得到再现。最终，种族主义是一个**过程**，因为任何结构和意识形态只能存在于创造和确认它们的日常实践之中，这些实践能够适应并有助于改变社会、经济和政治条件。（第185页，强调为原文所加）

还需要指出的一个重要问题是：虽然有关种族的讨论通常会直接指向种族主义，但是也存在一些研究种族的方法，帮助我们在给定的背景下理解种族的意义和产生的后果，而不必把重点放在种族主义上。例如，研究种族意识形态——"行为者使用基于种族的框架去解释和证明或者挑战种族现状"（Bonilla-Silva, 2001，第63页）——可能会揭示信仰体系、实践和导向特定种族经验的话语，或有助于解释种族化结构的存在和功能。社会学家指出，并非所有的种族意识形态都是种族主义的或有害的，但他们有间接关系，需要进一步的研究。

虽然本章综述的许多研究并没有明确涉及种族意识形态，但是这些研究往往受到研究者的种族意识形态的影响，其研究结果往往反映了作者研究中隐含的种族意识形态。种族意识形态的例子包括种族优越、无肤色歧视、平等主义和白人至上主义等。比如，安斯利（1997）将白人至上定义为"一个政治、经济和文化系统，其中的白人控制着绝大部分权力和物质资源，很多人有意或无意地认为白人优越和白人应享有更多权力，并且白人占据主导地位、非白人处于从属地位的关系每天都会在大批机构和社会环境中不断上演"（第592页）。未来研究的一个有前景的方向就是将关于意识形态的研究延伸到数学教育领域，并且批判性地审视数学意识形态（Ernest, 2009）和种族意识形态的关系。

在总结这一段关于关键种族概念的简短内容时，我们相信，将种族概念与其他可能相混淆的一些重点或者相关概念区分开来是十分重要的。例如，社会学家和民族学家就指出过种族、文化和民族研究在概念和研究范式上的区别（例如Cornell & Hartmann, 1998; Omi & Winant, 1986, 1994）。又比如，文化族群往往不是由其生理结构或生理特征来区分的，而是由其信仰和价值观来区分的，包括他们的精神、宗教、地域、语言和生存方式等。类似地，斯梅德利和斯梅德利（2005）指出：

> 民族将拥有共同文化特征的人群和具有其他文化特征的人区分开来。拥有共同语言，居住在或者原来居住在相同的地域，有相同的宗教信仰、历史观、传统、价值观、信念、饮食习惯等的人群，会被外界和自身视为是一个民族群体（例如，可以参见Jones, 1997; Parrillo, 1997; A. Smedley, 1999b; Steinberg, 1989; Takaki, 1993）。（第17页）

斯梅德利和斯梅德利（2005）呼应斯坦伯格的观点，

unclassified relationship, social practice, or group" (1986, p. 64). This process helps explain, for example, the social practices by which Africans whose identities were Ibo or Yoruba came to be constituted as *Black* via the racial meanings growing out of slavery as well as how various European immigrants, such as Irish and Italians (Ignative, 2009; Roediger, 2006) and Jews (Brodkin, 1998) became *White*.

Rigorously addressing race in mathematics education also necessitates clarity on what is meant by *racism*. Although colloquial, everyday understanding often reduces racism to individual beliefs and prejudices (Bonilla-Silva, 2001), racism operates at many different levels—individual, institutional, and structural—and can unfold as everyday experience or through structural domination. Because a full discussion of racism is beyond the scope of this chapter, we refer readers to Bonilla-Silva (2001), who provides a cogent analysis of various critical frameworks used to interpret racism, including Marxist perspectives, institutionalist perspectives, internal colonialist perspectives, the racial formation perspective, and the racism as societal waste perspective. Other helpful contemporary texts include *Race Frameworks: A Multidimensional Theory of Racism and Education* (Leonardo, 2013); *Silent Racism: How Well-Meaning White People Perpetuate the Racial Divide* (Trepagnier, 2010); *Everyday Anti-Racism: Getting Real About Race in School* (Pollock, 2008); *The Everyday Language of White Racism* (J. H. Hill, 2009); and *Theories of Race and Racism* (Back and Solomos, 2000). For the purpose of this review, we draw on a definition of racism provided by Essed (2002):

> Racism is a *structure* because racial dominance exists in and is reproduced by the system through the formulation and applications of rules, laws, and regulations and through access to and the allocation of resources. Finally, racism is a *process* because structures and ideologies do not exist outside the everyday practices through which they are created and confirmed. These practices both adapt to and themselves contribute to changing social, economic, and political conditions in society. (p. 185, emphasis in original)

It is also important to note that although discussions of race often lead directly to discussions of racism, there are approaches to studying race that can lead to understanding the meanings and consequences of race in a given context without necessarily focusing on racism. For example, studies of *racial ideology*—"racially based frameworks used by actors to explain and justify or challenge the racial status quo" (Bonilla-Silva, 2001, p. 63)—may reveal belief systems, practices, and discourses that lead to particular kinds of racialized experiences or that help explain the presence and function of racialized structures. Sociologists note that not all racial ideologies are racist or injurious. However, they are consequential and need to be studied.

Although many of the studies reviewed in this chapter do not explicitly address racial ideology, these studies are often informed by the racial ideologies of the authors and the results often speak to the racial ideologies implicit in their studies. Examples of racial ideologies include racial uplift, color blindness, egalitarianism, and White supremacy. Ansley (1997), for example, defined White supremacy as "a political, economic, and cultural system in which Whites overwhelmingly control power and material resources, conscious and unconscious ideas of White superiority and entitlement are widespread, and relations of White dominance and non-White subordination are daily re-enacted across a broad array of institutions and social settings" (p. 592). Extending the study of ideologies to mathematics education and critically examining relationships between *mathematics* ideologies (Ernest, 2009) and *racial* ideologies may represent a promising direction for future research.

In concluding this brief section on key racial concepts, we believe it is important to distinguish race from other potential foci and related concepts that could be confounded with race. For example, sociologists and scholars of ethnic studies have pointed out the conceptual and paradigmatic differences between research on race, culture, and ethnicity (e.g., Cornell & Hartmann, 1998; Omi & Winant, 1986, 1994). Cultural groups, for example, are often distinguished not by their physical or biological features but by their beliefs and values, which include spirituality, religion, region, language, and livelihoods. Similarly, as Smedley and Smedley (2005) pointed out:

> Ethnicity refers to clusters of people who have common culture traits that they distinguish from those of other people. People who share a common language, geographic locale or place of origin, religion, sense of history, traditions, values, beliefs, food habits, and so forth, are perceived, and view themselves as constituting, an ethnic group (see, e.g., Jones, 1997; Parrillo, 1997; A. Smedley, 1999b; Steinberg, 1989; Takaki, 1993). (p. 17)

Smedley and Smedley (2005), echoing Steinberg, also distinguished between racism and ethnocentrism:

> In speaking of the differences in America between European immigrant minorities early in the 20th century and racial groups, [Steinberg] pointed out that immigrants were "disparaged for their cultural peculiarities," and they were discriminated against, but the message conveyed by the nation to them was, "You will become like us whether you

同样区分了种族主义和民族优越感：

> 在谈论20世纪早期移民美国的欧洲少数群体之间的差异以及种族群体时，斯坦伯格指出，移民的"文化特征遭到蔑视"，移民也受到歧视，但是这个国家传递给他们的信息是，"无论你是否愿意，你都将变成和我们一样的人。"被同化是必要的也是预料之中的。对于地位低下的种族群体，这个信息的意义是："无论你变得和我们有多像，你还是会被孤立"（Steinberg，1989，第42页）。民族被认为具有可塑性和传递性，但是种族传达出的概念是：差异是不可以被超越的。（第19页）

我们认为，弄清种族、民族和文化概念的差异不仅具有社会意义，而且对推动种族和数学教育方面的知识发展也有重要影响。许多研究都试图分析学生学业成就上的差异与种族和民族并无关联（例如Lubienski，2002；McGraw, Lubienski, & Strutchens, 2006；Strutchens & Silver, 2000）。通常，这些研究中既没有定义种族，也没有定义民族，在这些文章的标题和分析中会使用"种族/民族"或"种族和民族"，这种表达方式可能传达的信息是，在这些研究中种族和民族是一回事，这些研究也假定二者在更大的社会中是等同的（Grosfoguel，2004）。鉴于越来越多的学者坚持认为有必要理解种族化的经历以及形成这些经历的种族化结构，那么将二者加以区分在本综述中是极其重要的。当我们考虑不同群体的社会地位，以及如何基于种族和民族所被赋予的意义来推崇或诋毁这种社会地位时，将种族和民族进行区分，并指出他们当中重叠的部分（如种族化的民族，民族化的种族），可以帮助我们在更深层次上理解学业成就。我们注意到，上述三篇综述（Lubienski & Bowen, 2000；Parks & Schmeichel, 2012；Stinson, 2011）中均提到了种族和民族这两个概念，但没有区分这两个概念。

现有数学教育研究中的种族概念化方法

在上一节中，我们旨在对重要的概念提供一些概要。在这一节中，为了推进我们的讨论，我们将概述当前数学教育中的种族研究所涉及的主题。在一些案例中，这些主题关乎研究者如何将种族进行概念化的方式。在另一些案例中，这些主题反映了种族和其他概念类别之间的交集。本节中有意地对主题进行简略描述，因为我们将在研究综述时重新回顾它们。这里提供这些主题，仅仅是为了引导读者注意在整个讨论过程中可能会重新出现的观点。此外，我们注意到这些主题并不是分散的，通过综述我们也发现，一项研究或一条研究主线可能会涉及多个主题。

主题1：数学教育研究中的种族常常被当做一个类别变量

正如塞科达（1992）和DiME一章（2007）的作者所言，数学教育领域中的研究常常把种族作为一个变量，即作为一种区分学生群体的手段。例如，种族在定量研究中被视作一种标识符，特别是在那些聚焦于不同群体学生学业成就比较的研究中。这种比较学生学业成就的研究是数学教育领域中与种族相关研究的重要组成部分（例如Lubienski，2002，2008；Secada，1992；Strutchens & Silver, 2000；Tate，1997）。帕克斯和施麦歇尔（2012）分析了涉及种族/民族与数学教育的文献，结果发现，他们所综述的文章中有相当一部分（18%）发表在以实验设计为主的心理学期刊，而这些文章中种族很可能被用作定量分析的一个变量。

种族还会在课堂和教学研究项目中被用作标识符。在这些案例中，学生总体的种族构成被记录下来，但是并没有用于分析（Parks & Schmeichel, 2012）。例如早期数学教育改革项目，如认知指导教学（CGI）项目，所报告的研究涉及有色人种学生群体。在这些研究中，识别总体的种族构成是重要的，然而，种族主要是被当作一个标识符，而非分析框架中的一部分。比如，维拉森纽（1991）对一项以少数族裔学生为主的课堂中实施CGI项目的研究加以评论时，CGI项目的开发人员观察到"结论很重要，因为它们为CGI方法在弱势学生群体中的有效性提供了具体的证据"（Peterson, Fennema, & Carpenter, want to or not." Assimilation was necessary and expected. With the low-status racial groups, the message was, "No matter how much like us you are, you will remain apart" (Steinberg, 1989, p. 42). Ethnicity was recognized as plastic and transmissible, but race conveyed the notion of differences that could not be transcended. (p. 19)

We argue that conceptual differences between race, ethnicity, and culture are consequential not only in the larger society but also in furthering the intellectual agenda on race and mathematics education. A number of studies that have purported to analyze differences in achievement do not distinguish between race and ethnicity, for example (e.g., Lubienski, 2002; McGraw, Lubienski, & Strutchens, 2006; Strutchens & Silver, 2000). Quite often, neither race nor ethnicity is defined in these studies. The presentation of "race/ethnicity" versus "race and ethnicity" in the titles and analyses may signal that race and ethnicity mean the same thing in a given study and are assumed to be equivalent in the larger society (Grosfoguel, 2004). These latter points are particularly important for this review, given the increasing number of scholars who insist on the need to understand racialized experiences and racialized structures that shape those experiences. Distinguishing between race and ethnicity, as well as pointing out where they overlap (e.g., racialized ethnicities, ethnicized races), may also provide deeper understanding of achievement outcomes when we consider the social status of various groups and how that status is valued or denigrated based on the meaning assigned to race and ethnic categories. We note that all three reviews cited above (Lubienski & Bowen, 2000; Parks & Schmeichel, 2012; and Stinson, 2011) attended to both race and ethnicity without distinguishing the two concepts or related literature.

Extant Approaches to Conceptualizing Race in Mathematics Education Research

In the previous section, we sought to provide an outline of key concepts. In this section, we move the discussion forward by outlining broad themes that emerged from our current review of the research on race in mathematics education. In some cases, these themes relate to the ways in which researchers conceptualize race. In other cases, these thematic ideas reflect intersections between race and other conceptual categories. The description of the themes in this section is intentionally brief, since they will be revisited throughout our review of the research. We offer them here simply to orient the reader to ideas that will resurface throughout the discussion of the research. Moreover, we note that these themes are not discrete.

As will be demonstrated in the review of the research, a single study or line of research could reflect more than one of these themes.

Theme 1: Race in Mathematics Education Research Is Often Framed as a Categorical Variable

As noted by Secada (1992) and the authors of the DiME chapter (2007), research in mathematics education often treats race as a variable—as a means to identify student groups. For example, race has been employed as an identifier in quantitative studies, particularly those focused on comparison of achievement among student groups. Such studies comparing student achievement are a significant segment of the race-related research in mathematics education (e.g., Lubienski, 2002, 2008; Secada, 1992; Strutchens & Silver, 2000; Tate, 1997). Parks and Schmeichel (2012) conducted a review of the literature involving race/ethnicity and mathematics education. They found that a substantial number (18%) of the articles that they identified were located in psychology journals that focus on experimental design, where race was most likely used as a variable for quantitative analysis.

Race has also been used as an identifier in studies of classrooms and instructional programs. In these cases, the racial makeup of the student population is noted, but not used analytically (Parks & Schmeichel, 2012). For example, early mathematics reform programs, such as Cognitively Guided Instruction (CGI), reported studies involving populations of students of color. In these cases, the identification of the racial makeup of the population was significant. However, race was centralized as an identifier, not as part of an analytical framework. Commenting on a study by Villasenor (1991) about the implementation of CGI in predominantly minority classrooms, for example, CGI developers observed that the "results are important because they provide concrete evidence for the effectiveness of the CGI approach with a disadvantaged population of students" (Peterson, Fennema, & Carpenter, 1991, p. 78). Thus, race was a noted feature of the student population but was employed primarily as an identifier.

Theme 2: Culture and Race Are Often Used Interchangeably in Mathematics Education Research

Another theme in the study of race in mathematics education has focused on culture. This line of research generally reflects investigations of the everyday behavior of members of a particular racial group and the ways that this behavior influences teaching and learning. In many cases it is difficult, if not impossible, to disentangle the differences between race and culture in research,

1991，第78页）。因此，种族是学生群体的一个值得注意的特征，但是主要被用作一个标识符。

主题2：文化和种族在数学教育研究中经常被交换地使用

数学教育领域的种族研究聚焦的另外一个主题是文化。该研究路径通常对一个特定种族群体成员的日常行为进行调查，并分析这种日常行为对教与学产生影响的方式。在许多研究中，不可能或很难厘清种族和文化的区别，因为数学教育中许多关于种族的研究都将种族和文化混为一谈（DiME，2007；Nasir & Hand，2006）。在过去二十年中，数学教育种族研究者主要关注文化以及学生文化实践与数学教学过程之间的关系（González, Andrade, Civil, & Moll, 2001；Nasir, 2002；Nasir & de Royston, 2013；Taylor, 2009）。在接下来的几节中，我们将综述这众多文献中的一部分，分析数学教育背景下种族问题的含义。

在结束这一节之前，我们应该注意到DiME（2007）一章反映了这一特别的主题，该章对文化的讨论聚焦于"文化活动理论"（第407页），其中包括更详细地关注参与、身份、学习机会、课堂话语实践、拓展数学能力的概念，以及校内外数学知识的链接等问题，这些研究在很大程度上是基于社会文化的框架。作者承认，"如果不关注更广泛的权力和种族的问题，那么对文化的研究是不够分量的"（第406页），DiME一章中关于种族和权力的讨论包括：关注与文化相关的教育学（Ladson-Billings，1994；Tate，1995b），代数项目（Moses & Cobb，2001；Moses, Kamii, Swap, & Howard, 1989），以及基于社会公正的数学教学（Frankenstein，1997；Gutstein，2003，2005）。在本章的后续部分，我们会部分地回顾课堂中文化研究的一些方法，并且会聚焦于其他反映种族与文化交叉的研究。

主题3：学生的种族身份对其数学学习经历的影响

数学教育领域种族研究的第三个主题是身份问题。特别地，意识形态和实践不仅将种族上升为一种社会力量，更产生了人们需要在不同背景下进行沟通的种族身份感，这些种族身份与其他特定身份（如数学、性别、学术等）是同时存在的。随着种族与身份的研究不断取得进展，对教与学的社会文化方面感兴趣的学者和对数学教学实践中权力关系感兴趣的学者开始关注这一研究领域。研究者们特别关注的是，在学校环境中，权力关系如何在赋予一些学习者权力的同时，又排斥或剥夺另外一些学习者的权利（例如Berry，2008；English-Clarke, Slaughter-Defoe, & Martin, 2012；Esmonde, Brodie, Dookie, & Takeuchi, 2009；Gholson & Martin, 2014；Langer-Osuna, 2011；Martin, 2009b；Nasir & de Royston, 2013；Spencer, 2009；Stinson, 2013；Terry, 2011；在这套书Langer-Osuna & Esmonde, 2017一章中有更多关于身份研究的综合引文清单）。研究者考察学生对身份关注的心理发展，包括学习者如何关注自己、别人如何看待他们，以及他们如何将这些身份体现在行为和表现当中。通过这些考察，研究者试图从学习者自身的角度来理解社会地位、主体以及学习机会等问题。在本章的后续部分，我们回顾了那些关注于共同建构数学身份和种族身份的研究。这一领域的许多研究都将批判种族理论（CRT）作为重要的概念和方法论框架，本章的后面我们会对批判种族理论进行详细介绍。

主题4：种族作为形成数学教与学活动的一系列交互话语和意识形态

研究种族的学者开始关注数学教育中种族、话语和意识形态之间的交互关系（例如R. Gutiérrez, 2013b；Martin, 2009a, 2013；Nasir & Shah, 2011；Nasir, Snyder, Shah, & Ross, 2012；Stinson, 2008, 2013）。例如，研究者已经利用后结构主义框架去考虑学生在学校教育中的一般性经历，尤其是数学经历，是如何在更广泛的有关学生肤色的话语体系中形成的。学者们使用这种方法来追溯社会文化话语体系对学生数学种族化经历的影响。比如，依据纳西尔和沙哈（2011）的说法，"关于学生智力和数学能力的种族化叙述，在定位和识别学生的学习过程中起着重要作用"（第27页）。同样，研究者还探索了

because much of the research on race in mathematics education has treated race and culture interchangeably (DiME, 2007; Nasir & Hand, 2006). A primary focus of researchers on race in mathematics education over the past two decades has been culture and the relationship between students' cultural practices and the processes of mathematics teaching and learning (González, Andrade, Civil, & Moll, 2001; Nasir, 2002; Nasir & de Royston, 2013; Taylor, 2009). In the subsequent sections, we review some of this extensive literature with an eye toward implications for issues of race in the context of mathematics education.

Before leaving this section, however, we should note that the DiME (2007) chapter reflects this particular theme. The discussion of culture in that chapter focused on "theories of cultural activity" (p. 407), which included giving more detailed attention—largely through a sociocultural framing—to issues of participation, identity, opportunity to learn, classroom discourse practices, expanding conceptions of mathematical competence, and bridging in-school and out-of-school mathematics knowledge. The authors acknowledged "it is not enough to examine culture without attending to broader issues of power and race" (p. 406). The discussion of race and power in the DiME chapter included attention to culturally relevant pedagogy (Ladson-Billings, 1994; Tate, 1995b), the Algebra Project (Moses & Cobb, 2001; Moses, Kamii, Swap, & Howard, 1989), and teaching mathematics for social justice (Frankenstein, 1997; Gutstein, 2003, 2005). In subsequent sections of this chapter, we revisit some of these approaches to the study of culture in the classroom while also drawing attention to other research reflecting the intersection of culture and race.

Theme 3: The Study of Students' Racial Identities Sheds Light on Their Mathematical Experiences

A third theme in research on race in mathematics education is identity. Specifically, the ideologies and practices that give rise to race as a social force also give rise to racial identities that people must negotiate across various contexts. These racial identities are coconstructed along with other salient identities (e.g., mathematics, gender, academic). Not surprisingly, the study of race and identity has continued to evolve as an area of focus for scholars who are interested in the sociocultural aspects of learning and teaching and the ways in which power relations within mathematical practices, particularly in school settings, serve to empower some learners and marginalize and disempower others (e.g., Berry, 2008; English-Clarke, Slaughter-Defoe, & Martin, 2012; Esmonde, Brodie, Dookie, & Takeuchi, 2009; Gholson & Martin, 2014; Langer-Osuna, 2011; Martin, 2009b; Nasir & de Royston, 2013; Spencer, 2009; Stinson, 2013; Terry, 2011; see Langer-Osuna & Esmonde, 2017, this volume, for a more comprehensive list of identity references). By focusing on identity development—how learners come to see themselves, how they are seen by others, and how they instantiate these identities in behavior and performance—researchers have attempted to better understand, from the point of view of learners themselves, issues of social position, agency, and opportunity to learn. In subsequent sections of this chapter, we review studies that focus on the coconstruction of mathematics identity and racial identity. Much of the work in this area draws on critical race theory (CRT) as the key conceptual and methodological framework. We elaborate on CRT later in this chapter.

Theme 4: Race Operates as a Set of Intersecting Discourses and Ideologies That Shape Mathematics Teaching and Learning

Researchers studying race have begun to attend to the interaction between race, discourse, and ideology in mathematics education (e.g., R. Gutiérrez, 2013b; Martin, 2009a, 2013; Nasir & Shah, 2011; Nasir, Snyder, Shah, & Ross, 2012; Stinson, 2008, 2013). For example, researchers have leveraged poststructural frameworks to consider how students' experiences in schooling in general and mathematics in particular are shaped by larger discourses about students of color. Scholars using this approach have traced the influence of sociocultural discourses on students' racialized experiences in mathematics. According to Nasir and Shah (2011), for example, "racialized narratives about students' intellectual and mathematical abilities play a central role in processes of positioning and identification" (p. 27). In a related vein, researchers have also explored the influence of stereotypes on the mathematical experiences of students of color (e.g., McGee & Martin, 2011b; Nasir & Shah, 2011; Stinson, 2013). This research, although closely related to the work concerning student identity, offers insight into the impact of the larger racial narratives on mathematics teaching and learning. We outline additional examples of the role of racial discourse and ideology in subsequent sections of this chapter.

Theme 5: The Role of Race in Mathematics Education Extends to the Institutional and Structural Levels

A final theme that we identified as salient with regard to the role of race draws attention to the institution and

刻板印象对有色人种学生数学经历的影响（例如McGee & Martin, 2011b; Nasir & Shah, 2011; Stinson, 2013）。这方面研究虽然与学生身份有关的研究非常相近，但是它为我们提供了重要见解，便于我们从更广泛的种族话语叙述中分析其对数学教与学的影响作用。在本章的后面部分，我们将概述有关种族话语和意识形态作用的其他案例。

主题5：将数学教育中种族的作用拓展到制度和结构层面

本章中，关于种族作用的最后一个突出主题，提醒读者注意数学教育的制度和结构。这一视角让数学教育研究超越了课堂或学生的研究，并将数学教育作为一个整体进行考虑。虽然这个主题在前文涉及的文献中并不突出，但是我们注意到它在这一领域还是非常地突出的。这一视角也承认数学教育领域当中的种族化特征（R. Gutiérrez, 2013b; Martin, 2013）。此外，强调这个主题的原因还在于，本章所强调的理论观点是，在制度层面理解种族的作用时，我们需要超越数学教育传统范式界限的理论框架。

数学课堂中的种族

在本节中，我们将讨论在数学教育研究中的4条研究主线，这些主线涉及了不同程度的种族问题。4条主线是：（1）关于学业成就和学习机会的结构性障碍研究；（2）关于教学计划与学生文化实践的研究；（3）关于非裔美国数学学习者种族/种族化经历的研究；（4）关于数学教师种族身份和种族意识形态的研究。我们会讨论，前文所讨论的不同主题是如何反映在每一条研究主线研究者多样化的种族概念定义方式（如，作为标识符、作为文化、作为身份）中，以及他们所使用的研究方法中的。

学业成就和学习机会的结构性障碍

塞科达（1992）的研究综述聚焦在那些"按种族、社会阶层、民族和语言等将学业成就和不同的社会群体联系起来的定量研究文献"方面（第625页）。从历史角度看，数学教育中有关种族的研究有很大一部分关注学生的学业成就（Lubienski, 2002, 2008）。这些研究涉及不同种族群体间的比较，并倾向于将种族概念化为一种标识符（即一个类别变量）。如前文所述，我们在本章的目标并非聚焦于这一比较方向内的具体研究或者是这类研究的结果，而是会概述数学教育中的种族研究所带来的启示。

在过去的十年当中，对不同种族间以及其他因素间显现的成绩差异的持续研究已经引起了相当大的争论（见R. Gutiérrez, 2008; Lubienski & Gutiérrez, 2008; Martin, 2009a, 2009b）。关注成绩差异的研究，还包括将白人成绩标准化和将种族等级进行细化的发展趋势。此外，对成绩的强调和近期对学生经验的呼吁形成了鲜明的对比（DiME, 2007; Martin, 2007a）。虽然塞科达（1992）最终还是选取了那些关注种族群体之间差异的研究，但他也承认，将种族和民族仅仅狭隘地概念化为类别变量会让问题变得很复杂。这样的类别变量会预示社会政治的转向，以及近期对种族社会建构本质的研究，他明确地指出纯粹地用相对固定的人口统计学上的分类是不恰当的：

> 种族和民族性、社会阶层和母语这些分类各有其特定的含义，但我们应该注意，这些类别同时也具有社会的建构性和传统性……至少，我们应该注意到，不同群体中的成员对自己身份的解释不同于本群体之外个体的解释，而有关上述这种情况如何影响儿童的数学教育的研究才刚刚起步（第627页）。

塞科达（1992）还对当时种族定量研究的方向（即研究群体差异）提出了强烈的批评，特别是基于所属种族的群体给出因果解释和进行预测的做法：

> 就其本身而言，将群体作为结构变量来使用可能不是什么问题，但是仅从属于特定群体的角度而不谈别的来解释结果……用群体成员来作预测性的结论可能也是有问题的……如果要做这样的预测，且有这样的预测的话，那么应该是这样的：除非我们改变来自这些背景的学生所接受的教育经历，否

structure of mathematics education. This perspective moves mathematics education research beyond the study of classrooms or students to consider mathematics education as a whole. Although this theme is not as prominent in the existing literature as those outlined above, we note it here due to its salience for the field. This perspective acknowledges the racialized character of the domain itself (R. Gutiérrez, 2013b; Martin, 2013). Additionally, this theme is important to highlight because of the theoretical perspectives used in this chapter. Understanding the role of race at the institutional level requires theoretical frameworks beyond the traditional paradigmatic boundaries of mathematics education.

Race in Mathematics Classrooms

In this section, we discuss four lines of mathematics education research that to varying degrees engage issues of race: (1) research on achievement outcomes and structural barriers to opportunities to learn; (2) research on instructional programs and students' cultural practices; (3) research on the racial/racialized experiences of African American mathematics learners; and (4) research on mathematics teachers' racial identities and racial ideologies. Throughout our discussion we note how the various themes discussed in the previous section are reflected in the variety of ways researchers have conceptualized race (e.g., as identifier, as culture, as identity) within each line of research, as well as the related methodological approaches that have been employed.

Achievement Outcomes and Structural Barriers to Opportunity to Learn

Secada (1992) framed his review as an effort to focus on "examining the quantitative-research literature that links academic achievement to various social groups along lines of race, social class, ethnicity, and language" (p. 625). Historically, a significant segment of the race-related research in mathematics education has focused on student achievement (Lubienski, 2002, 2008). These types of studies involve comparisons between racial groups and tend to rely on conceptualizations of race as an identifier (i.e., a categorical variable). As previously noted, our goal in this chapter is not to focus on specific studies within this comparative vein or the results of this line of research. Rather, in this section, we seek to outline what this research tells us about race in mathematics education.

In the last decade, continued research on achievement gaps—racial and otherwise—has generated considerable debate (see R. Gutiérrez, 2008; Lubienski & Gutiérrez, 2008; Martin, 2009a, 2009b). Concerns about achievement-gap-focused research include its tendency to normalize White outcomes and to reify racial hierarchies. Further, the emphasis on *achievement* stands in contrast to more recent calls to focus on the *experiences* of students (DiME, 2007; Martin, 2007a). Although Secada (1992) ultimately drew on studies that focused on identifying disparities between racial groups, he did acknowledge the complexities associated with narrowly conceptualizing race and ethnicity only as categorical variables. Portending the sociopolitical turn and more recent studies of the socially constructed nature of race, he accurately argued against a strict focus on demographic categories as static:

> The categories of race and ethnicity, social class, and language may have conceptual cores, but we should remember that they are also socially constructed and maintained.... At the very least, we should note that members of diverse groups interpret their membership in ways that differ from the interpretation of individuals who are outside of those groups. Research into how this affects the mathematics education of diverse children is just beginning. (p. 627)

Secada (1992) also offered a strong critique of the direction that quantitative research on race (i.e., research on group disparities) had taken by that time, particularly in relation to causality and prediction based on racial group membership:

> In itself, this use of groups as structural variables might not be so problematic were findings to be interpreted in terms of specific group membership and nothing else.... Using membership in a group for making predictive claims is also problematic.... The prediction that should be made, if there is one, is that, unless we modify the educational experiences that students from these backgrounds receive, we will fail to educate them as fully as we believe they can be educated. A better way of thinking of these groups is that they help define the contexts in which students are educated and research gets conducted (p. 640)

Reflecting on the chapter by Secada (1992), we make two additional observations. First, his critical engagement with "student diversity" and critique of how social categories were conceptualized and addressed in prior research echoes more recent claims that mathematics education research facilitates the consumption and production of racial meanings (R. Gutiérrez, 2013b; Martin, 2009a, 2013). Research up to that point was continuing to imbue meaning to racial categories and, ultimately, to the concept of race itself. Second, we note how Secada's review was constrained by the dominance of the race-

则将无法像我们所认为的那样对他们进行充分的教育。一个更好的考虑这些群体的方式是，研究者能够帮助界定学生接受教育的情境以及开展研究的情境（第640页）。

反思塞科达（1992）的这一章，我们提出两项补充意见。首先，他对"学生多样性"的批判性评论，以及对已有研究中社会类别如何进行概念界定和处理的批评意见，均呼应了近期的一些主张，即数学教育研究促进了种族内涵的发展和建构（R. Gutiérrez, 2013b; Martin, 2009a, 2013）。此前的研究一直持续在为种族类别赋予意义，并最终为种族本身的概念赋予意义。其次，我们注意到塞科达综述的主要内容是，在有关美籍非裔、拉丁美洲和美国土著学生研究中的种族比较范式作为主导的情况下，已有研究受到了怎样的限制。很少有研究关注这些群体中学生数学发展的组内研究，这导致我们对很多内容都知之甚少，比如学生如何应对他们的种族化经历，如何构建他们作为特定社会群体成员的身份或是作为数学学习者的身份等方面。因此，20多年前学业成就比较的局限性已经凸显，虽然这些研究无疑是数学教育研究中的重要组成部分（Lubienski, 2008），但是聚焦于成就比较所产生的薄弱点，至少在一定程度上推动了其他研究种族方法的发展，本章将会概述这些方法。

需要明确的是，在呼吁人们关注研究种族成就差距的一些问题时，我们并非认为所有把种族作为一种标识符的研究都存在固有的问题。事实上，同样的理论取向也被用来确定结构性因素，这些因素造成了学生在广泛的、全系统范围内学习机会的种族不平等性。在DiME（2007）的一章中，作者注意到数学教育领域中很少有研究聚焦于学习机会问题，特别是课堂之外的结构条件问题。为此，作者借鉴了数学教育领域之外的其他开创性研究，这些研究记录了在学校教育的物质条件方面，白人学生与有色人种学生持续存在的种族差异，比如是否能够学习高等数学课程和高质量的数学课程（Darling-Hammond & Sykes, 2003; Oakes, Joseph, &Muir, 2003; Orfield, Frankenberg, & Lee, 2003）。类似的关于学校种族隔离、分流、课程设置以及对有色人种学生后继影响的评论也被其他学者所提及（见C. R. Anderson & Tate, 2008），我们并非试图要以一种综合的方式来拓展这些已有的综述。我们旨在突出与学习机会（OTL）相关的其他研究，并指出这些将种族作为标识符的研究可能有助于我们理解数学教育中平等的结构性障碍。

DiME（2007）帮助我们了解它所概括的因素，同时我们认为它已经开始关注讨论数学教育中的教师质量差异问题。例如，塔特（2008）探讨了教师质量在非裔美国男性学生数学学习机会中所起的作用。特别地，他在报告中指出数学教师至少在辅修科目上分布不均，不同学校类型（基于人口构成）以及同一学校内（基于分流水平）也存在差异。这些差异表明，那些非裔美国学生，特别是非裔美国男生的数学老师曾至少辅修过某一科目。类似地，希尔和鲁宾斯基（2007）使用数学教学知识（MKT）测试卷，对位于加利福尼亚州的一个大的教师样本进行了研究。他们发现，教师面向教学的数学知识和学校总体水平之间存在关系："那些招收了大批低收入和少数族裔学生的学校所聘用的教师的数学教学知识平均来说要略低于那些经费充裕的学校的教师"（第764页）。希尔和鲁宾斯基还注意到，教师通常在他们学生时代就读的学校附近教书（Loeb & Reininger, 2004），他们指出了这种现象造成的循环，"在数学知识方面不能够充分培育教师的学校，可能会在这些学生毕业若干年后返校任教时承受不良后果"（第765页）。

虽然这些研究只代表了很小一部分表明教师素质与种族差异问题有关的研究，但是我们认为，这些研究与本讨论十分相关，至少有两个原因。首先，这些研究反映了采用"种族作为变量"的研究方法在确定数学学习机会的结构性障碍方面能够发挥重要的作用。虽然在这些研究中，种族被作为一个变量或者静态的类别，但这类研究的结果对推进社会公正事业是极为重要的，因为这类研究让我们注意到种族结构不平等的问题。

强调这些特殊研究的第二个理由是为了将教师的优质教育作为一种跨代的传承资源（Tate, 2008）。对教师优质教育的这一理解潜在地展现了"非种族主义者的种族主义"的表现（Bonilla-Silva, 2010）。也就是说，教师质量的代际不平等模式不需要"种族主义者"个人的

comparative paradigm for studies involving African American, Latinx, and Indigenous students. Very little research focused on within-group analysis of mathematical development for students from these groups. As a consequence, little could be said about the various ways that students respond to their racialized experiences or construct their identities as members of particular social groups or as learners of mathematics, for example. Thus, the limitations of achievement comparisons were outlined over two decades ago. Although these studies are certainly a significant part of the research body within mathematics education (Lubienski, 2008), the voids created by a focus on comparisons of achievement motivate, at least in part, the other approaches to the study of race that are outlined here.

To be clear, in calling attention to some of the concerns with studying racial achievement gaps, we do not mean to suggest that *all* research that treats race as an identifier is inherently problematic. Indeed, this same theoretical orientation has also been used to identify structural factors that produce racial inequities in students' opportunities to learn on a broad, system-wide scale. In the DiME (2007) chapter, the authors noted that very little of the research within mathematics education had focused on issues related to opportunity to learn, particularly with regard to structural conditions beyond the classroom. For this reason, the authors drew from seminal research outside mathematics education that has documented persistent racial disparities between White students and students of color in terms of the material conditions of schooling, such as the availability of advanced mathematics courses and high-quality mathematics curricula (Darling-Hammond & Sykes, 2003; Oakes, Joseph, & Muir, 2003; Orfield, Frankenberg, & Lee, 2003). As similar reviews of school segregation, tracking, and course offerings and the subsequent impact on students of color have also been conducted elsewhere (see C. R. Anderson & Tate, 2008), we do not attempt to extend these previous reviews in a comprehensive manner. However, we do highlight some additional research that is relevant to opportunity to learn (OTL) and points to the role that research using race as an identifier might play in our understanding of structural barriers to equity in mathematics education.

In addition to the opportunity to learn factors outlined in the DiME chapter (2007), we also submit that important work has been done documenting disparities in teacher quality in mathematics education. For example, Tate (2008) has explored the role of teacher quality on African American males' opportunity to learn mathematics. Specifically, he reported the maldistribution of mathematics teachers with at least a minor in the subject, noting differences across school type (based on demographic composition) and within-school differences based on track level. These differences make it less likely that African American students, particularly male students, will have a mathematics teacher with at least a minor in the subject. Similarly, H. Hill and Lubienski (2007) studied a large sample of teachers in California. Using a measure of mathematics knowledge for teaching (MKT), they found a relationship between the teachers' mathematics knowledge for teaching and the school population: "Schools enrolling larger numbers of low-income and minority students employed teachers who had, on average, slightly less mathematical knowledge for teaching than their counterparts in more affluent schools" (p. 764). H. Hill and Lubienski also note previous research findings indicating that teachers often teach in schools near those they attended as a student (Loeb & Reininger, 2004). They point out the cycle that this creates. "Schools that fail to adequately prepare teachers, in terms of mathematical knowledge, may suffer the consequences when these students return to teach in later years" (p. 765).

Although these studies represent only a small number of those indicating differences in teacher quality related to race, we suggest that they are particularly relevant to this discussion for at least two reasons. First, these studies reflect the important role that research employing a "race as variable" approach can play in identifying structural barriers to opportunity to learn in mathematics. Although race was treated as a variable or static category in these cases, the results of this type of research are crucial for advancing the cause of social justice, as this research draws our attention to the ways that race structures inequality.

A second reason for highlighting these particular studies is to situate teacher quality as an intergenerational resource (Tate, 2008). This understanding of teacher quality potentially demonstrates the operation of "racism without racists" (Bonilla-Silva, 2010). That is, the intergenerational pattern of teacher inequality requires no involvement by "racist" individuals. The repeated "underfunding" of particular schools with regard to the resource of teacher quality is an intergenerational phenomenon. Thus, these studies indicate the ways that research on opportunity to learn related to race, including research based on a "race as identifier" perspective, can provide important insight into the structural operation of racism.

Instructional Programs and Students' Cultural Practices

Beyond studies of racial achievement gaps and structural barriers to students' opportunities to learn, the "race as variable" approach has also informed studies of mathematics teaching and learning. As previously noted, this can occur when the racial makeup of the student population is noted but not used analytically. For example, like CGI, the Quantitative Understanding: Amplifying

参与，某些学校在师资方面一再出现"资金不足"的现象就是一种代际现象。因此，这些研究表明了那些有关不同种族学习机会的研究，包括基于"种族作为一个标识符"观点的研究，能够帮助我们了解种族主义的结构运行规律。

教学计划与学生的文化实践

除了研究种族学业成就差异和学生学习机会的结构性障碍外，"种族作为变量"的方法还能够促进数学教与学的研究。如前文所述，当学生群体的种族构成被注意到但没有被分析使用时，这种情况就会发生，例如，CGI项目和量化理解：提高学生成绩和推理（QUASAR）项目，它们报告了涉及有色人种学生群体的研究。QUASAR项目的研究者通过使用种族分类，按种族来分解表现的方式，评估了教学干预带来的影响（Silver & Stein, 1996），这种方法有助于证明QUASAR项目在缩小种族学业表现差异方面是有效的。因此，关注种族很重要，但并非要严格地将其作为一个标识符。

当然，尽管种族的这种概念化在项目评估中继续发挥着关键作用，但是将种族作为标识符的比较分析也能够为创建改变种族不均衡的项目提供基础。例如，最初使用"种族作为变量"的方法时，特里斯曼（1992）发现加州大学伯克利分校的非裔美国大学生在微积分入门课程上的通过率要低于华裔美国大学生，但是与自始至终区分种族差异的研究不同，特里斯曼和他的团队使用种族志的方法去探索学生每周的学习习惯（即他们与学习相关的文化实践）。他们发现，虽然非裔美国学生和他们的同伴一样努力学习，但是他们更倾向于孤立的学习；与之相反，华裔美国学生则会合作完成作业并准备考试。这些发现促成了名为数学工作坊项目（MWP）的成功诞生，该项目专门设计用来增加教学时间和实施结构，以鼓励历史上处于边缘地位的有色人种的大一新生在数学入门课程上进行合作（Fullilove & Treisman, 1990）。从理论的观点来看，特里斯曼的研究展示了运用"种族即文化"的方法的具体过程。

与数学教学有关的研究，也考察了学生文化与课堂环境的规范与实践之间的互动关系。对文化相关教育学的研究（Ladson-Billings, 1997；Tate, 1995b）、文化特定教育学的研究（Leonard, 2008），以及文化不兼容的研究（Murrell, 1994）都聚焦于学生的文化与数学学习的相互作用方式上。例如，墨雷尔（1994）描述了一个案例，在这个案例中，一名教师对数学交流目的的解释不同于课堂上那些非裔美国男性学生对"数学谈话"的理解，墨雷尔将这种不兼容性归因于文化差异。类似地，塔特（1995a）也提供了一个白人职前教师与她的非裔美国学生之间文化不兼容的例子。最后，纳西尔（2002）的研究反映了学生在校外的文化经验与他们在校内的数学学习之间可以是脱节的，在纳西尔的研究中，这种脱节不能让非裔美国男性学生在课堂中表现出同等的数学能力，就像他们在其他情况（如在篮球场上）中所表现的那样。

这条研究主线上的一些研究者探索了如下问题：如何将学生的文化知识更加有效地转化到数学教学中去（Civil, 2006；Leonard, 2008；Leonard, Brooks, Barnes-Johnson, & Berry, 2010；Nasir, Hand, & Taylor, 2008；Tate, 1995b）。例如，研究者探索了在数学课堂上使用与文化相关的教学法。根据拉德森-比林斯（1994）的研究，与文化相关的教学法建立在三个关键组成部分上：学生学习、文化能力和社会政治意识。正如伦纳德等人（2010）所断言："有意义的与文化相关的教学法（CRP），包括在学习群体中使用那些吸引人的文化实践，以支持学生的话语交流并强化师生关系"（第263页）。拉德森-比林斯（1994，1997）在她对玛格丽特·罗西的叙述以及塔特（1995a）对桑德拉·曼森的叙述中都描述了一些针对非裔美国学生与其文化相关的数学教学的例子。类似地，葛斯丁（Gutstein, 2003, 2005）在一个墨西哥裔美国人的环境下描述了文化相关和种族意识的教学法。在这些案例中，研究人员记录了一种建立在学生的文化知识和实践之上的数学教学法，拓展学生学习并发展学生的社会政治意识。与记录文化不兼容的研究一样，这种研究反映了"种族即文化"主题的一种表现。

另一条考虑教学计划和学生文化实践的研究主线是，探索学生的数学"学习风格"是如何在不同的种族之间出现差异的。每当文化和认知被认为基本上不相关时，这些文化角度的学习风格的文献就会对贬低非标准文化差异的缺陷理论进行回应（参见Ginsburg & Russell, 1981）。为了将文化和认知联系起来，学者们会试图将有色人种学

Student Achievement and Reasoning (QUASAR) project reported studies involving populations of students of color. Researchers on the QUASAR project evaluated the impact of their pedagogical interventions by using racial categories to disaggregate performance by race (Silver & Stein, 1996). This approach facilitated their claims that QUASAR was effective in narrowing racial performance gaps. Thus, race was important to note but served strictly as an identifier.

Of course, though such conceptualizations of race continue to play a critical role in program evaluation, race-comparative analyses that treat race as an identifier can also serve as the basis for *creating* programs that redress disparities. For example, initially using a "race as variable" approach, Treisman (1992) found that African American undergraduates passed at lower rates than Chinese American undergraduates in introductory calculus at the University of California, Berkeley. But unlike research that begins and ends with the identification of racial disparities, Treisman and his team used ethnographic methods to explore the students' weekly study habits (i.e., their cultural practices related to learning). They found that although African American students worked just as hard as their peers, they tended to work in isolation; in contrast, the Chinese American students worked collaboratively on homework and to prepare for exams. These findings precipitated the creation of a successful program called the Mathematics Workshop Program (MWP), which was specially designed to increase instructional time and implement structures to encourage collaboration among historically marginalized freshmen of color taking introductory mathematics (Fullilove & Treisman, 1990). From a theoretical standpoint, Treisman's work exemplifies the use of a "race as culture" approach.

Research related to mathematics instruction also examines the interaction between students' culture and the norms and practices of the classroom environment. Research on *culturally relevant pedagogy* (Ladson-Billings, 1997; Tate, 1995b), *culturally specific pedagogy* (Leonard, 2008), and *cultural incompatibility* (Murrell, 1994) has focused on the ways that students' culture interacts with mathematics learning. For example, Murrell (1994) described a case in which the interpretation of the purpose of mathematical communication held by a teacher differed from the understanding of "math talk" held by the African American male students in the class. Murrell attributed this incompatibility to cultural differences. Similarly, Tate (1995a) offered an example of the cultural discontinuity between a White preservice teacher and her African American students. Finally, Nasir's (2002) research reflects the ways that students' cultural experiences outside school can be disconnected from their mathematics learning in school. According to Nasir, this disconnect prevents African American male students from demonstrating the same mathematical competence in the classroom that they organically demonstrate in other contexts, such as the basketball court.

Several authors within this line of research have explored how students' cultural knowledge can be translated more effectively into mathematics instruction (Civil, 2006; Leonard, 2008; Leonard, Brooks, Barnes-Johnson, & Berry, 2010; Nasir, Hand, & Taylor, 2008; Tate, 1995b). For example, researchers have explored the use of culturally relevant pedagogy in mathematics classrooms. According to Ladson-Billings (1994), culturally relevant pedagogy is built on three key components: student learning, cultural competence, and sociopolitical consciousness. As Leonard et al. (2010) assert, "meaningful CRP [culturally relevant pedagogy] includes the use of engaging cultural practices situated within learning communities that support student discourse and stronger teacher-student relationships" (p. 263). Examples of culturally relevant pedagogy for African American students in mathematics are described by Ladson-Billings (1994, 1997) in her account of Margaret Rossi and by Tate (1995a) in his depiction of Sandra Mason. Similarly, Gutstein (2003, 2005) describes culturally relevant and race-conscious pedagogy in a Mexican American context. In each of these cases, the researchers document a mathematics pedagogy that builds upon students' cultural knowledge and practices while extending student learning and developing sociopolitical consciousness. As with the research documenting cultural discontinuities, this work reflects a manifestation of the "race as culture" theme.

Another line of research that considers instructional programs and students' cultural practices is that which explores how students' mathematical "learning styles" diverge along racial lines. At a time when culture and cognition were seen as largely unrelated, the cultural learning styles literature offered a response to deficit models that devalued nonnormative cultural differences (see also Ginsburg & Russell, 1981). In connecting culture and cognition, scholars sought to recast students of color as intellectually sophisticated and cognitively capable. We address it here because the literature on "cultural learning styles" is an example of research in mathematics education that focuses on culture but addresses race head-on. And although it may no longer be a part of mainstream mathematics education research, the notion that different racial groups learn differently is still very much a part of the educational discourse among practitioners.

生的形象重塑为具有复杂智能和认知能力的群体。我们在这里提到它，是因为关于"文化学习风格"的文献是数学教育研究的一个例子，它聚焦于文化，但又直面种族问题，虽然它可能不再是主流数学教育研究的一部分，但是不同种族群体学习风格的不同仍然是教育实践中非常重要的教育共识。

以心理学研究为基础，假设不同种族群体之间存在基本的认知、相互作用和认识论上的差异（例如Hale-Benson，1986；Shade，1982），对文化学习风格的相关研究调查了文化倾向对学生数学问题解决的影响过程（Malloy & Jones，1998；Stiff & Harvey，1988）。特别地，研究者聚焦于非裔美国学生和白人学生在"学习风格"上的差异，例如，斯蒂夫和哈维（1988）将非裔美国学生描述为具有"场依赖型"特征的学习者，他们用"整体"的方式解决数学问题，与"场依赖型"相对的是白人学生的"面对问题情境，重视分析性思维和系统地解决问题的方法"（第196页）。其他学者（例如Malloy & Jones，1998）则致力于通过个案访谈的方法来证实这种主张，他们认为这种方式揭示了不同种族之间在数学思维上的明显差异。

对种族学习风格的研究可能存在问题，因为这种研究倾向于本质化和过度概括化（K. D. Gutiérrez & Rogoff，2003）。在认知功能和种族背景之间建立因果关系意味着人们如何思考和互相作用是他们种族属性确定的副产品，而不是文化和历史进程造成的结果。此外，主张非裔美国学生是"整体的"思考者，而白人是"分析的"思考者，意味着一个群体中所有的（或绝大多数）成员用同样的方式思考和行动，这种使种族同质化的主张不仅掩盖了组内存在的差异，还甚至假设对种族群体进行清晰的划分是可能的。"文化学习风格"的研究并没有给"种族"这个概念本身带来麻烦，它与前面提到的关注差异研究并没有本质上的区别，都将种族理解为一种与生俱来的固定类别。

相比之下，对学生校外数学实践的研究总体而言，避免了文化学习风格研究中这种棘手的结论（例如Civil，2006；Nasir，2000；Nasir & de Royston，2013；Taylor，2009）。尽管有时调查实践的研究常常都会离不开特定的种族群体（例如，非裔美国学生玩多米诺骨牌或打篮球），这个领域的研究者并没有坚称种族归属决定了人们在认知层面上参与某些文化实践的方式。从研究者的角度来看，任何关联都是参与特定文化实践历史产生的巧合的副产品（DiME，2007；K. D. Gutiérrez & Rogoff，2003），DiME（2007）一章中对很多相关的研究进行了回顾。

非裔美国数学学习者的种族化经历

另一条研究主线特别地聚焦于非裔美国学习者（青少年或是成人），研究数学身份、数学社会化、种族身份、种族社会化、主体和数学成功之间的相互关系（例如Aguirre，Mayfield-Ingram，& Martin，2013；Berry，2008；Ellington & Frederick，2010；English-Clarke 等，2012；Gholson & Martin，2014；Jackson，2009；Jett，2011；Larnell，2013；Leonard & Martin，2013；Martin，2009b，2012；McGee，2013a；McGee & Martin，2011a，2011b；Noble，2011；Nzuki，2010；Stinson，2013；Terry，2011；Thompson & Davis，2013；Walker，2012，2014）。这一研究有效地证明了数学学习、参与和为数学素养的努力可以被概念化为种族化经历形式（例如Matrin，2006，2012）；也就是说，随着社会和政治为种族建构意义，种族类别在塑造这些经验上也变得愈发突出。

这个领域中最早的一些研究来自于马丁（2000，2006，2007b），从这些研究理论的角度和分析模型的角度来看与现有的数学教育文献存在明显的不同，这些研究认为现有的文献更为关注以下几方面内容：（a）注重成功而不是失败；（b）留意第一手的学生经历资料，而不仅仅是学业成就；（c）以种族和数学特性的共同建构为中心；（d）在多水平的框架之下进行分析。研究中两个主要的理论构架是数学社会化和数学身份。马丁（2006）将数学社会化定义为一种经验范围，"个体和群体在学校、家庭、同伴群体和学习场所等各种情境中所拥有的促进、合法化或抑制有意义的数学参与的经历"（第206页）。他对数学身份的定义如下：

> 数学身份是指个体在其整体自我概念中形成的情感态度以及深层次的关于他们有效参与和进行数学学习、使用数学改变他们生活条件的信念和能力，

Grounded in psychological research that posits fundamental cognitive, interactional, and epistemological differences between racial groups (e.g., Hale-Benson, 1986; Shade, 1982;), research on cultural learning styles has investigated how cultural orientations shape students' mathematical problem solving (Malloy & Jones, 1998; Stiff & Harvey, 1988). In particular, researchers have focused on differences in the "learning styles" preferred by African Americans compared with Whites. For example, Stiff and Harvey (1988) characterized African American students as "field-dependent" learners who solve mathematical problems in "holistic" ways, as opposed to "field-independent" Whites who "value analytical thinking and systematic approaches to problematic situations" (p. 196). Other scholars (e.g., Malloy & Jones, 1998) have attempted to substantiate this theoretical claim through empirical studies involving clinical interviews, which they argue reveal clear differences in mathematical thinking across racial lines.

Research on racial learning styles can be problematic because of its tendency to essentialize and overgeneralize (K. D. Gutiérrez & Rogoff, 2003). Forging causal links between cognitive functioning and racial background implies that how people think and interact is a deterministic byproduct of their racial affiliation rather than an effect of cultural and historical processes. Further, claiming that African American learners are "holistic" thinkers and White learners are "analytical" thinkers implies that all (or even most) members of a particular group think and act in the same way. Not only does this kind of assertion homogenize racial groups in ways that mask within-group variation, but it also assumes that cleanly demarcating racial groups is even possible. In not troubling the idea of "race" itself, the "cultural learning styles" literature, then, does not differ substantially from the aforementioned gap-focused studies that understand race as a set of fixed categories into which people are born.

In contrast, research on students' out-of-school mathematical practices has by and large avoided the knotty conclusions of the cultural learning styles research (e.g., Civil, 2006; Nasir, 2000; Nasir & de Royston, 2013; Taylor, 2009). Despite sometimes investigating practices commonly associated with particular racial groups (e.g., African American children playing dominoes or basketball), scholars in this area have not argued that racial affiliation *determines* the ways in which people engage on a cognitive level with certain cultural practices. From the researchers' perspective, any correlations are coincidental by-products of histories of participation in particular cultural practices (DiME, 2007; K. D. Gutiérrez & Rogoff, 2003). Much of this body of work was reviewed in the DiME (2007) chapter.

Racialized Experiences of African American Mathematics Learners

Another line of research has focused specifically on African American learners—adolescents and adults—and attended to interrelationships among mathematics identity, mathematics socialization, racial identity, racial socialization, agency, and mathematics success (e.g., Aguirre, Mayfield-Ingram, & Martin, 2013; Berry, 2008; Ellington & Frederick, 2010; English-Clarke et al., 2012; Gholson & Martin, 2014; Jackson, 2009; Jett, 2011; Larnell, 2013; Leonard & Martin, 2013; Martin, 2009b, 2012; McGee, 2013a; McGee & Martin, 2011a, 2011b; Noble, 2011; Nzuki, 2010; Stinson, 2013; Terry, 2011; Thompson & Davis, 2013; Walker, 2012, 2014). An important contribution of this work is that it has effectively demonstrated that mathematics learning, participation, and the struggle for mathematics literacy can be conceptualized as *racialized forms of experience* (e.g., Martin, 2006, 2012); that is, as experiences where the socially and politically constructed meanings for race and racial categories become highly salient in shaping those experiences.

Some of the earliest research in this area was Martin's (2000, 2006, 2007b), which in terms of theoretical perspective and modes of analysis departed from the extant mathematics education literature in several significant ways: (a) focusing on success and not just failure; (b) attending to students' firsthand accounts of their experiences and not just achievement outcomes; (c) centering on the coconstruction of racial and mathematics identities; and (d) situating the analysis within a multilevel framework. The two primary theoretical constructs in this research were *mathematics socialization* and *mathematics identity*. Martin (2006) defined mathematics socialization as the range of "experiences that individuals and groups have within a variety of contexts such as school, family, peer groups, and the workplace and that facilitate, legitimize, or inhibit meaningful participation in mathematics" (p. 206). He defined mathematics identity as follows:

> Mathematics identity refers to the dispositions and deeply held beliefs that individuals develop, within their overall self-concept, about their ability to participate and perform effectively in mathematical contexts and to use mathematics to change the conditions of their lives. A mathematics identity encompasses a person's self-understanding of himself or herself in the context of doing mathematics. It also encompasses how others "construct" us in relation to mathematics. (p. 206)

数学身份包括一个人在做数学时的自我理解，它还包括其他人如何从数学角度"塑造"自己。（第206页）

关于种族、身份和数学学习，马丁（2000，2006，2007b，2009a，2009b，2012）提出了两个问题：（1）在数学学习和参与的背景下，黑人群体意味着什么？（2）在黑人学生的背景下，数学学习和实践意味着什么？这两个问题引出了更多的研究。根据威尔拉斯、马丁和凯恩（2012）的观点，这种聚焦于种族和身份的问题是重要的，因为数学身份与非裔美国学生的数学成功和失败紧密相连。在围绕身份问题进行研究的过程中，马丁认为，将非裔美国学生的数学学习经历置于恰当的情境水平下是非常必要的。在马丁看来，这种研究意味着要分析个体学习者、学校层面的因素、家长和其他社区成员的观点，以及社会历史因素，其中包括"基于历史上的歧视性政策和做法，这些政策和做法阻碍了非裔美国人在数学和其他社会领域中成为平等参与者的机会"（Martin，2000，第29页）。马丁（2012）认为，这种多水平的分析提供了一种方法，用于描述和理解"黑人学习数学"的现象特征。

例如，罗伯特·贝利（2008）和大卫·斯丁森（2006，2008，2013）分别针对数学上取得成功的非裔美国男性中学生和成人开展了一系列研究。批判种族理论（CRT）是批判研究的一个分支，它认为种族和种族歧视是社会组织结构的一个永久固定物（Bell, 1992），贝利和他的同事们（Berry, Thunder, & McClain, 2011）使用该理论来表明：他所访谈和观察到的当地中学男孩所经历的种族歧视，实际是更广泛的结构种族主义的一个实例，这种案例在美国还是十分普遍的。然而，当这些男孩在面对这种种族主义时，大量的支持和保护性因素为他们提供了缓冲。其中保护性因素包括积极的学前经历、家长作为监护人能够提供的学术资源以及学习机会、强烈的学业身份认同、喜欢数学，以及强大的宗教身份和体魄，以上这些帮助他们获得并保持成功。还有一点需要注意的是，大部分家长都把种族作为这些孩子学术和数学经历的一个方面。另外同样重要的是，一些男生讲述了他们对种族化经历产生的新理解。一个八年级的学生科德尔这样说：

> 我叫科德尔，我是纪念中学的一名八年级学生。我是家里唯一的孩子，我和妈妈一起住。我知道我妈妈作为一个单亲家长，有一份辛苦的工作，因此我需要比其他孩子承担更多的责任并学会独立。我的外祖母和姨妈们帮助我妈妈，鼓励我做出正确的选择并确保我走正道。我的外祖母和妈妈告诉我在学校要好好学习，监督我做好功课。我妈妈经常对我说，如果我计划上大学的话，我就要在学校做得更好。
>
> 数学是我喜欢的学科，因为在我看来它最简单。数学有意思且十分吸引人，因为在数学中你必须思考并不断地尝试，直到你得到正确的解答。我第一次迷上数学是在三年级，那时我们开始学习如何做乘法。我知道我很厉害，因为我比班上的其他学生更早就学会了乘法……
>
> 上四年级的时候，我开始遇到麻烦，因为我讨厌学校。我的老师教的知识我已经知道了，所以我开始在课堂上玩。我妈妈认为学校课程没有让我得到足够的学习机会，这是我遇到麻烦的原因。在与老师和校长谈过几次以后，我妈妈感觉我可以接受AG（学术天赋）项目的测试，可老师和校长不想让我参加这个测试，因为他们认为我没有天赋。我妈妈认为他们不想测试我的原因在于我是黑人，她继续坚持，最终老师和校长让我参加测试。我的成绩很好，在四年级中途进入到AG项目中。（Berry, 2008, 第473页）

斯丁森（2006，2008，2009，2013）和同事们（Stinson, Jett, & Williams, 2013）同样使用批判种族理论，并结合批判后现代和后结构，来探讨在高中曾取得成功的非裔美国成年男性的身份认同问题和行为问题。斯丁森研究中的参与者承认并面对了多种"缺陷论"，特别是与非裔美国人和非裔美国男性相关的。我们会在后面聚焦于数学教育中的种族话语和意识形态的研究，那时会详细探讨这一工作。

麦吉（2013a，2013b，2013c）和同事们（McGee & Martin, 2011a, 2011b; Terry & McGee, 2012）将数学

With respect to race, identity, and mathematics learning, Martin (2000, 2006, 2007b, 2009a, 2009b, 2012) has framed two questions that have led to additional research: (1) What does it mean to be Black in the context of mathematics learning and participation? (2) What does it mean to be a learner and doer of mathematics in the context of being Black? According to Varelas, Martin, and Kane (2012) such race- and identity-focused questions are critical because mathematics identities are strong correlates of mathematics success or failure for African American students. And in framing his work around questions of identity, Martin argued that it was imperative to situate African American students' mathematics learning experiences within the appropriate layers of context. For Martin, this meant analysis of the individual learner, school-level factors, perspectives of parents and other community members, and sociohistorical factors that include "the historically based discriminatory policies and practices that have prevented African Americans from becoming equal participants in mathematics and other areas in society" (Martin, 2000, p. 29). According to Martin (2012), these multilevel analyses offer one way to describe and understand the phenomenal realities that characterize "learning mathematics while Black."

Robert Berry (2008) and David Stinson (2006, 2008, 2013), for example, have developed lines of work focused on mathematically successful African American male middle school students and adults, respectively. Berry and his colleagues (Berry, Thunder, & McClain, 2011) used critical race theory (CRT), a branch of critical studies that views race and racism as permanent fixtures in the organizing structures of society (Bell, 1992), to show that the local racism experienced by the middle school boys he interviewed and observed was an instance of broader structural racism that remains pervasive in the United States. Yet, in the face of this racism, these boys were buffered by a number of supports and protective factors—including positive preschool experiences, parents as academic resources and guardians of opportunities, strong academic identities, liking mathematics, and strong religious and athletic identities—that allowed them to maintain their success. It is also important to note that most of the parents discussed race as a factor in these boys' academic and mathematics experiences. Equally important is that some of the boys narrated an emerging understanding of their racialized experiences. Cordell, an eighth grader, stated the following:

> My name is Cordell, and I am an eighth-grade student at Memorial Middle School. I am an only child, and I live with my mother. I know that my mother, being a single parent, has a tough job, so I have had to take on more responsibilities than other kids do and I have learned to be independent. My grandmother and aunts help my mother by encouraging me to make good decisions and make sure that I stay on the right track. My grandmother and mother talk to me about doing well in school and make sure I do my work. My mother is always saying I better do well in school if I plan on going to college.
>
> Math is my favorite subject because it is my easiest subject. Math is interesting and fun because in math you have to think and keep trying until you get it right. I was first drawn to math in the third grade when we started to learn how to multiply. I knew I was good because I learned to multiply earlier than the other kids in my class. . . .
>
> When I was in fourth grade, I started getting into trouble because I was bored with school. My teacher was teaching me things I already knew, so I would start playing around in class. My mother thought I was not being challenged enough and that is why I got into trouble. After a few conferences with the teacher and the principal, my mother felt that I should be tested for the AG [academically gifted] program. The teacher and principal did not want me tested because they felt I was not gifted. My mother thinks the reason they did not want to test me was because I am Black. She stayed on the teachers and principals until I was tested. I did well enough to be placed in the AG program midway through my fourth-grade year. (Berry, 2008, p. 473)

Stinson (2006, 2008, 2009, 2013) and colleagues (Stinson, Jett, & Williams, 2013), also used CRT, along with critical postmodern and poststructural theory, to examine identity and agency among African American adult males who had previously been successful in high school. Stinson's participants acknowledged and confronted various "discourses of deficiency" related to African Americans and African American males, in particular. We elaborate on this work later in our discussion of research that has focused on racial discourses and ideologies in mathematics education.

McGee (2013a, 2013b, 2013c) and colleagues (McGee & Martin, 2011a, 2011b; Terry & McGee, 2012) have extended the study of mathematics learning and participation as racialized forms of experience by focusing on resilience among African American students in mathematics and the range of protective supports and coping strategies used by these students. McGee's work also expands Claude Steele's (1997) seminal work on *stereotype threat* by demonstrating how individuals can manage and buffer themselves against racialized stereotypes typically thought to diminish academic performance. Undergirding this theory is the premise that African Americans are a stigmatized and devalued group in American society and, specifically, that this social status negatively influences their academic belief systems and behaviors (Chavous et al., 2003; Cokley et al., 2012;

学习和参与作为一种种族化的经验形式进行了拓展，将研究重点聚焦于非裔美国学生的数学毅力和获得的大量保护支持以及这些学生在面对种族问题时使用的应对策略。麦吉的研究还拓展了克劳德·斯蒂尔（1997）在刻板印象威胁方面所做的开创性工作，通过展示个体如何进行自我管理和自我缓冲，进而对抗通常被认为会降低学业成绩的种族化刻板印象。支持这一理论的前提是承认非裔美国人在美国社会中是一个被污蔑和贬低的群体，以及这种社会地位会对他们的学习信念系统和行为产生负面的影响（Chavous 等，2003；Cokley 等，2012；Steele，1997；Steele & Aronson，1995）。为了分析种族刻板印象在高成就、数学方面成功的黑人学生的生活中所扮演的角色，麦吉使用了生活故事的方法来区分面对种族刻板印象时脆弱的和更坚强的毅力状态。这种毅力概念的细微差异表明，在脆弱的状态下，学生对于实现数学成功的行为和动机受到刻板印象和外力的影响。如果学生保持这种脆弱的状态，那么他们的成功通常付出较高的情感代价，而那些沿着更加强健的轨迹发展的学生，其动力则来自于内在的、自我生成的数学成功。

在扩展刻板印象威胁研究的基础上，麦吉发展出了刻板印象管理的概念（McGee & Martin，2011b）。刻板印象管理的概念认为，获得高成就的黑人学生运用了一些策略来证实他们智力上和学术上的能力，以及向他们的老师、家庭、同伴，以及更大范围的机构证明他们的实力。作为日常生活的一部分，许多非裔美国学生每天都必须在一系列的情况下忍受种族歧视。因此，他们需要有一个适应的应对过程，这成为他们整个毅力故事的一部分。刻板印象管理，通过定性的反叙述记录下来，呈现出了这些学生每天忍受的种族化挑战和经历。尽管刻板印象威胁经常导致种族之间的排斥现象和黑人学生的低成就，但刻板印象管理的研究却发现，黑人学生受到激励后，有可能会获得更好的学业成绩和出众的数学学业表现。

研究非裔美国学习者的学者们多次使用批判种族理论，并在使用过程中举例论证了 R. 古铁雷斯（2013b）所描述的社会政治转向，特别是使用了以前从未在数学教育中应用过的理论框架方面来体现社会政治转向的内容。需要重点强调的一点是，以上研究表明，批判种族理论在数学教育研究领域中出现并与"种族即身份"研究的讨论相结合。迄今为止，绝大多数使用批判种族理论观点的研究都聚焦于这一主题。然而，批判种族理论的出现似乎促使社会文化理论家在分析社会活动的时候去考虑更加广泛的结构和过程（例如 Esmonde 等，2009；Nasir，2011；Nasir 等，2009；Nasir & de Royston, 2013）。因此，我们提供了批判种族理论的简要背景及其与数学教育中学生身份研究的联系。

拉德森-比林斯（1998）和其他研究者（例如 Dixson & Rousseau，2006；Ladson-Billings & Tate，1995；Solórzanao & Yosso，2002）强调，批判种族理论最早出现在法律研究中，自从20世纪70年代末期这个概念被引入法律研究后，批判种族理论就被拓展为各种不同版本，包括 LatCrit 和 FemCrit，它们分别关注拉丁裔和女性的交叉体验。在其最初的形式中，批判种族理论被作为：

> 不符合民权实证主义和自由主义法律话语的一种学问。这种学术传统反对美国缓慢的种族改革步伐。批判种族理论基于如下观念：种族主义在美国社会中是正常的。有时候它采用讲故事的方式，因此它有别于主流的法律知识。它批判自由主义，认为白人已经成为民权立法的主要受益者（Ladson-Billings，1998，第7页）。

迪克森和鲁索（2006，第33页）引用松田、劳伦斯、德尔加多和克伦肖（1993）的研究，概述了批判种族理论关于其产生以及对批判法律研究所做回应的以下六个统一的主题：

1. 批判种族理论认为种族主义普遍存在于美国人的生活中。
2. 批判种族理论对主流法律声明的中立、客观、无肤色歧视和英才教育表示怀疑。
3. 批判种族理论对非历史主义提出挑战，坚持对法律进行背景/历史的分析……批判理论家……采取了假设种族主义助长了当代群体优势和劣势的所有表现的立场。
4. 批判种族理论在分析法律和社会时，坚持对有色人种和我们起源社区的经验知识的认可。

Steele, 1997; Steele & Aronson, 1995). To explore the role of racial stereotypes in the lives of high-achieving, mathematically successful Black students, McGee has used life story methods to differentiate between fragile and more robust forms of resilience in the face of racial stereotypes. This nuancing of the resilience concept shows that, in the fragile form, student behavior and motivation to achieve in mathematics are driven by the influence of stereotypes and external forces. Success often comes with a high emotional price if students remain in this fragile form. Students who develop along the more robust trajectory are motivated by internal, self-generated mathematics success.

Extending the work on stereotype threat, McGee developed the concept of *stereotype management* (McGee & Martin, 2011b). The concept of stereotype management suggests that there are strategies that high-achieving Black students employ to substantiate their intellectual and academic value or credibility to their teachers, families, peers, and the larger educational community. As aspects of normal everyday experience, many African American students must endure everyday racism across a range of contexts. Thus, there is a need for adaptive coping processes, which become part of their overall stories of resilience. Stereotype management, documented through qualitative counternarratives, exposes the daily, racialized challenges and experiences these students endure. Whereas stereotype threat often leads to disengagement and lowered achievement, studies of stereotype management show how Black students are motivated to obtain higher grades and to excel in mathematics.

In their frequent use of CRT, researchers in this area of study on African American learners exemplify the *sociopolitical turn* described by R. Gutiérrez (2013b), particularly with regard to the use of theoretical frameworks that have not previously been applied in mathematics education. The emergence of CRT in mathematics education research is important to note in conjunction with a discussion of research on "race as identity." To date, the majority of research using a CRT perspective has focused on this theme. However, the emergence of CRT appears to have motivated sociocultural theorists to take broader structures and processes into account when analyzing social activity (e.g., Esmonde et al., 2009; Nasir, 2011; Nasir et al., 2009; Nasir & de Royston, 2013). For this reason, we offer a brief background on CRT and the connection to the study of student identity in mathematics education.

As noted by Ladson-Billings (1998) and others (e.g., Dixson & Rousseau, 2006; Ladson-Billings and Tate, 1995; Solórzanao & Yosso, 2002), CRT first emerged in legal studies. Since its introduction in legal studies in the late 1970s, CRT has expanded to include many variants, including LatCrit and FemCrit that focus on the intersectional experiences of Latinxs and women, respectively. In its original formulation, CRT served

> as a counterlegal scholarship to the positivist and liberal legal discourse of civil rights. This scholarly tradition argues against the slow pace of racial reform in the United States. Critical race theory begins with the notion that racism is normal in American society. It departs from mainstream legal scholarship by sometimes employing storytelling. It critiques liberalism and argues that Whites have been the primary beneficiaries of civil rights legislation (Ladson-Billings, 1998, p. 7)

Citing Matsuda, Lawrence, Delgado, and Crenshaw (1993), Dixson and Rousseau (2006, p. 33), outlined six unifying themes of critical race theory as it emerged from, and in response to, critical legal studies:

1. Critical race theory recognizes that racism is endemic to American life.
2. Critical race theory expresses skepticism toward dominant legal claims of neutrality, objectivity, colorblindness, and meritocracy.
3. Critical race theory challenges ahistoricism and insists on a contextual/historical analysis of the law.... Critical theorists ... adopt a stance that presumes racism has contributed to all contemporary manifestations of group advantage and disadvantage.
4. Critical race theory insists on recognition of the experiential knowledge of people of color and our communities of origin in analyzing law and society.
5. Critical race theory is interdisciplinary.
6. Critical race theory works toward the end of eliminating racial oppression as part of the broader goal of ending all forms of oppression. (Matsuda et al., 1993, p. 6)

CRT was first introduced into education by Gloria Ladson-Billings and William Tate (1995) and has been taken up by a number of researchers across the field of education because of its sharp focus on race and racism. The recent publication of the *Handbook of Critical Race Theory in Education* (Lynn & Dixson, 2013) attests to the increasing appeal of CRT for education researchers interested in issues of race. Although perhaps not as prominent in mathematics education research as in other education subdisciplines, a growing number of scholars have begun to apply CRT to the study of students of color in mathematics. Moreover, the introduction of CRT to mathematics education research is noteworthy, given the traditionally "raceless" nature of the mathematics education enterprise.

5. 批判种族理论是跨学科的。
6. 批判种族理论致力于消除种族压迫，作为结束所有形式的压迫这一更广泛目标的一部分。（Matsuda 等，1993，第6页）

格洛瑞亚·拉德森-比林斯和威廉姆·塔特（1995）首先将批判种族理论引入到教育领域，因其对种族和种族主义的尖锐关注而被教育领域的许多研究者采纳。最近出版的《教育中的批判种族理论手册》（Lynn & Dixson, 2013）证实了对种族问题感兴趣的研究者对于批判种族理论持续增加的诉求。虽然在数学教育研究中可能不像在其他教育子学科中那样突出，但越来越多的学者开始使用批判种族理论进行数学教育中有色人种学生的研究。而且，传统上数学教育事业具有"无种族"特征，因此将批判种族理论引入到数学教育研究中是值得注意的。

迄今为止，在数学教育研究中，批判种族理论的主要应用集中在松田等人（1993）所总结的第四个原则——有色人种的经验知识上。特别地，批判种族理论形成的"反面例子"方法被用以描述成功的非裔美国学生的经历（见 Berry, 2008; Jett, 2012; Martin, 2006; Terry, 2011）。这种对批判种族理论的使用与索洛萨诺和犹索（2002）所总结的批判种族方法论是一致的，该方法聚焦于构建、揭露和挑战绝大多数人的故事或者主流话语的反故事。主流话语是基于种族特权的历史和遗产，因此经常对"有色人种经历进行扭曲和压制"（第29页）。这样导致的结果是，"多数人虽然声称自己是中立和客观的，但是其隐含的假设是基于对有色人种的负面刻板印象作出的（Solorzano & Yosso, 2002, 第29页；Ikemoto, 1997）。另一方面，反面例子和批判种族方法论证实了"我们必须将学校内外的种族歧视、性别歧视、阶级歧视和同性恋歧视的经历和反应，视为有效、恰当和必要的数据形式"（第37页）。正如犹索（2006）所指出的，"反面例子并没有将注意力放在尝试说服人们相信种族主义的存在方面，而是试图从那些受种族主义伤害和迫害的人的角度来记录种族主义的持久性。此外，反面例子将人们的注意力引向那些勇敢地抵抗种族歧视、为建立一个社交和种族更加公平的社会而奋斗的人"（第10页）。

总之，近期有关数学教育中非裔美国人的研究与批判种族理论非常一致，因为批判种族理论赋予了被边缘化的有色人种经验知识以特殊的权利。批判种族理论家认为那些最能阐明种族压迫的重要后果的人是那些每天都在经历它的人（Ladson-Billings, 1998, 2012）。虽然并非所有研究非裔美国学生数学学习经历的研究都明确引用了批判种族理论，但是他们所选择的呈现数据的方式与反故事方法论是一致的。事实上，使用访谈的方法来阐明这类经验知识是十分有效的。

有关这一类研究还需要注意的一点是，几乎上述所有引用的研究都属于所谓的"成就论"（Stinson, 2006）。通常的成就差异研究往往强调非裔美国学生学业失败，与之不同的是，"成就论"研究有意让那些在数学上取得成功的非裔美国学生发声。选择聚焦于成功的学生是对那种将非裔美国学生描绘成失败和缺乏主动性言论的一种有意识的反驳。这种研究的一个缺陷是，他们大多数都聚焦于非裔美国男性，仅有少数研究（例如 Gholson & Martin, 2014; Y. A. Johnson, 2009; Moody, 2004; Strutchens & Westbrook, 2009）特别地关注于非裔美国女性。

对非裔美国数学学习者的研究，在很多方面推动了种族教育领域的发展。首先，它强调了种族身份在塑造学生对于自身在数学上取得成功的能力感知方面所起到的作用，并激发他们产生保持数学学习动机这一目标。在将种族问题以更加直接的方式融入到学习的概念化时，这一类研究帮助我们理解了学生的多重交叉身份和他们参与数学教育活动的关系。其次，这一研究同时聚焦了种族和种族主义，这两点也是参与研究的非裔美国学生在分流区别对待的做法中会经常经历到的。实际上，这一发现提供了定性的研究结果，补充了那些采用定量方法表明课堂不平等的研究（Oakes, 2005）。最后，从方法论的角度来看，该文献显示了访谈和叙述的力量，阐述了人们数学学习经历的故事。

数学教师的种族身份和种族意识形态

上一节所回顾的关于学生种族化经历的研究，至少与两类涉及教师的调查研究有关（即教师的种族身

Within mathematics education research, the primary application of CRT to date has focused on the fourth tenet outlined by Matsuda et al. (1993)—the experiential knowledge of people of color. Specifically, the CRT construct of "counterstory" has been used to depict the experiences of successful African American students (see Berry, 2008; Jett, 2012; Martin, 2006; Terry, 2011). This use of CRT is consistent with the critical race methodology outlined by Solórzano and Yosso (2002), which focuses on the construction of counterstories that expose and challenge majoritarian stories, or masternarratives. A masternarrative is based on histories and legacies of racial privilege and thereby often "distorts and silences the experiences of people of color" (p. 29). As a result, "majoritarian methods purport to be neutral and objective yet implicitly make assumptions according to negative stereotypes about people of color" (Solorzano & Yosso, 2002, p. 29; Ikemoto, 1997). On the other hand, counterstories and critical race methodology confirm "that we must look to experiences with and responses to racism, sexism, classism, and heterosexism in and out of schools as valid, appropriate, and necessary forms of data" (p. 37). And as Yosso (2006) notes, "counterstories do not focus on trying to convince people that racism exists. Instead, counterstories seek to document the persistence of racism from the perspectives of those injured and victimized by its legacy. Furthermore, counterstories bring attention to those who courageously resist racism and struggle toward a more socially and racially just society" (p. 10).

Overall, the recent literature on African Americans in mathematics education aligns well with CRT in that it privileges the experiential knowledge of marginalized people of color. Critical race theorists argue that those in the best position to articulate the material consequences of racial oppression are those who experience it daily (Ladson-Billings, 1998, 2012). And although not all researchers investigating the mathematics learning experiences of African American students have explicitly cited CRT in their work, the ways in which these scholars have chosen to present their data is consistent with the counterstory methodology. Indeed, the use of interviews has been particularly effective in shedding light on this type of experiential knowledge.

An additional point worth noting about this body of research is that nearly all the studies cited above belong to what has been called the "discourse of achievement" (Stinson, 2006). Unlike achievement gap studies that usually highlight the academic failures of African American students, research in this category has intentionally given voice to African American students who have achieved *success* in mathematics. The choice to focus on successful students represents a deliberate rebuttal to portrayals of African American students as failures and as lacking in agency. One limitation of this work is that most of the studies have tended to focus on African American males. Only a few studies (e.g., Gholson & Martin, 2014; Y. A. Johnson, 2009; Moody, 2004; Strutchens & Westbrook, 2009) have focused exclusively on African American females.

Research on African American mathematics learners has moved the field forward in a number of ways. First, it has highlighted the importance of racial identity in shaping students' perceptions of their capacity to succeed in mathematics, as well as the goals that fuel their desire to persist in learning mathematics. In bringing issues of race more directly into conceptualizations of learning, this body of work has broadened the understanding of how students' multiple, intersecting identities relate to their participation in mathematics education. Second, this research has put a spotlight on both race *and* racism, which the African American participants in these studies usually experienced through discriminatory tracking practices. In effect, this finding represents a qualitative complement to studies that have documented inequities in course taking using quantitative methods (Oakes, 2005). Finally, from a methodological standpoint, this literature has demonstrated the power of interviews and narratives to illuminate people's histories of lived experiences learning mathematics.

Mathematics Teachers' Racial Identities and Racial Ideologies

The research reviewed in the previous section on students' racialized experiences is also related to at least two other lines of inquiry involving teachers. One nascent line of research has begun to explore the influence of teachers' racial identities on their teaching practice.[3] For example, a 2013 special issue of *Teachers College Record* included articles based on the Mid-Atlantic Center for Mathematics Teaching and Learning (MACMTL) Algebra I Case Studies Project. The articles focus on two African American teachers identified as "well-respected, but not extraordinary" (Chazan, Brantlinger, Clark, & Edwards, 2013). This research on African American teachers included an exploration of teachers' perspectives on students, the ways that these perspectives shape the socialization practices experienced by students, and how these practices might shape the identity-development process for students (Clark, Badertscher, & Napp, 2013). Thus, this work examines how the practices of teachers can influence the racial and disciplinary identities of students.

份和个人身份）。其中一个新兴的研究主线是，开始探索教师的种族身份对其教学实践的影响。[3] 例如，2013年《师范学院记录》的一份特刊收录了基于大西洋中部数学教与学研究中心（MACMTL）代数I案例研究项目的文章，这些文章聚焦于两名被认为是"很受尊敬，但并非十分特别"的非裔美国教师（Chazan, Brantlinger, Clark, & Edwards, 2013）。具体内容包含了探索教师对学生的看法、这些看法如何形成学生所经历的社会化实践，以及这些实践如何塑造学生的身份发展过程（Clark, Badertscher, & Napp, 2013）。因此，该研究分析了教师的实践如何影响学生种族的和学科的身份。

然而，该研究还让人们深刻理解有关教师个人身份如何通过他们学生时期的种族化经历所形成，以及这些经验在建构他们的教学实践时所起到的作用。例如，在对麦迪森·摩根进行的案例研究中，研究者博奇、参赞和莫里斯（2013）注意到她的数学学习经验主要集中于技能和程序，其结果是，她试图帮助她的学生超越重复性的程序操作，找到数学上的意义并发展他们自己的问题解决策略。根据研究者所说，摩根的童年经历和这些经历对她的后续实践所产生的影响必须通过种族的视角来理解。他们断言，摩根的反应是基于她对自己学生时期经历的种族化的解释以及她的信念，即"她在教学中所追寻的意义可以为学生提供比她自己从教育系统中接受到的更好的数学理解，因为自己当初的非裔美国孩子的身份给她带来很差的服务"（第22页）。与以学生为研究对象的研究一样，对教师生活经历的关注揭示了教师的种族身份，也提供了对学生学习经历的深刻理解。

与教师有关的第二个研究方向是关注教师的种族意识形态。与聚焦于教师种族身份的研究一样，这一研究提供了对学生种族化经历的洞察。特别地，研究探讨了教师的种族意识形态在形成课程和教学决策中的作用。例如，福克纳、史蒂夫、马歇尔、尼特菲尔德和克罗斯兰（2014）在一项纵向研究中调查了学生进入八年级代数班的分班方式。作者分析了教师评估与之前的以数学成绩作为安排白人和黑人学生的预测指标的相对优势，他们发现教师评估对黑人学生的影响要大于白人学生。与此同时，过去的学业成绩作为安排的预测指标，其作用对黑人学生要小于白人学生。此外，尽管各个水平的黑人学生比同等成绩等级的白人学生进入代数的可能性更小，但这种影响对高成就的黑人学生更为严峻。作者认为：

> 黑人学生面对的是一种无法立足的障碍，因为他们的黑肤色（或者，正如我们在这里所说的，教师对这些黑人学生的一种内隐的反应），这是一种看不见的、可怕的、阻碍他们进入更高水平数学课程的障碍，无论他们表现如何（第306页）。

特别地，作者认为黑人学生受到了"负面的学术种族定性"（第307页）。在这个研究中，教师对黑人学生的负面感知让他们无法获得白人同学所享有的课程机会。

除了对课程决策的影响，研究者还探索了教师信念和意识形态在形成教学决策中的作用（Davis & Martin, 2008; Martin, 2007a）。例如，巴特和钱（2010）研究了教师对非裔美国学生数学能力不足的信念，并将这些意识形态的特征描述为与非裔美国学生更大范围的元叙述（例如，关于家长参与和天生数学能力的故事）有关系。如作者所言，这些数学上的元叙述的作用是将非裔美国学生的失败正常化。当老师内化了这些元叙述以后，他们就会坦然接受非裔美国学生的低成就，并且不会质疑他们自己在造成这一现象中所起到的作用。他们认为没有理由去质疑他们建立在这种缺陷信念基础上所做的教学决策。除了这一有关在职教师信念的研究，类似的关注教师把对非裔美国学生有缺陷的观点带入课堂的研究在关于职前教师的文献当中同样存在（Bartell 等, 2013）。

然而，教师的种族身份和种族意识形态在塑造数学教学和学习当中所扮演的角色并没有像其他与种族相关的研究一样受到很大重视，事实上，斯宾塞、帕克和斯特格达（2010）指出，当前的研究当中很少关注的一点是，教师的情感态度在构建学校数学教学以服务于低收入有色人种学生的过程中，起到何种作用。"在这些方面缺乏数学教与学的理论，是提醒我们这是数学教育研究中的一个巨大疏忽"（第217页）。我们自己对文献的回顾也证实了，这是在数学教育研究中比较缺乏的一个研究

Yet, this research also provides insight into how the teachers' own identities were shaped by their racialized experiences as students and the role of those experiences in constructing their teaching practice. For example, in their case study of Madison Morgan, researchers Birky, Chazan, and Morris (2013), note that her mathematics learning experiences had focused on skills and procedures. As a result of these experiences, she sought to help her students move beyond repeating procedures to find meaning in the mathematics and develop their own problem-solving strategies. According to the authors, Morgan's childhood experiences and their influence on her subsequent practice must be understood through the lens of race. They assert that Morgan's response was based on her racialized interpretations of her experiences as a student and her belief that "the meaning she strove for in her teaching would provide students with a better understanding of mathematics than she had received from an educational system that ill-served her as an African American child" (p. 22). As is the case in the research focused on students, this attention to the lived experiences of teachers sheds light on teachers' racial identities and provides insight into the learning experiences of students.

A second direction of research related to teachers has attended to teachers' racial ideologies. Like the research focused on teachers' racial identity, this work provides insight into students' racialized experiences. In particular, this work explores the role of teachers' racial ideologies in shaping both curricular and instructional decisions. For example, a longitudinal study by Faulkner, Stiff, Marshall, Nietfield, and Crossland (2014) examined patterns of student placement into eighth-grade algebra. The authors analyzed the relative strength of teacher evaluation and prior math performance as predictors of placement for White and Black students. They found that the impact of teacher evaluation was stronger for Black students than for White students. At the same time, the role of past achievement as a predictor of placement was weaker for Black students than for their White peers. Moreover, although Black students at all achievement levels were less likely than White peers at the same achievement levels to be placed into algebra, the effect was even more severe for high achieving Black students. The authors argue,

> Black students confront an untenable impediment in that their Blackness (or, as we suggest here, the teachers' implicit responses to these students' Blackness) serves as an invisible, albeit formidable, obstacle to gaining access to higher level mathematics courses, irrespective of their demonstrated performance. (p. 306)

In particular, the authors suggest that Black students are subject to "negative academic racial profiling" (p. 307). In this study, teachers' negative perceptions of Black students denied them access to curricular opportunities afforded their White peers.

In addition to the impact on curricular decisions, researchers have also explored the role of teacher beliefs and ideologies in shaping instructional decisions (Davis & Martin, 2008; Martin, 2007a). For example, Battey and Chan (2010) explored teachers' deficit beliefs regarding African American students in mathematics and characterized these ideologies as related to the larger metanarratives about African American students (e.g., stories regarding parental involvement, innate mathematical ability). According to the authors, the effect of these metanarratives in mathematics is to normalize the failure of African American students. When teachers internalize these metanarratives, they readily accept the low achievement of their African American students and do not question their own role in the construction of that achievement. They see no reason to interrogate the instructional decisions that are made on the basis of these deficit beliefs. In addition to this research on the beliefs of in-service teachers, similar attention to the deficit perspectives that teachers bring to the classroom about African American students can also be found in the literature on preservice teachers (Bartell et al., 2013).

Yet, the role of teacher racial identity and racial ideology in shaping mathematics teaching and learning has not received as much attention as some of the other lines of race-related research. In fact, Spencer, Park, and Santagata (2010) note that there is little extant research on the role of teacher dispositions in shaping mathematics instruction in schools serving low-income students of color. "The lack of theorizing about mathematics teaching and learning in these spaces strikes us as a gaping omission in mathematics education research" (p. 217). Our own review of the literature also confirms this as an underresearched area of inquiry in mathematics education research (Davis & Martin, 2008; Martin, 2007a).

Race, Society, and Mathematics Education

Whereas much research has focused on how issues of race intersect with mathematics teaching and learning at the classroom level, increasing attention is being paid to how racial phenomena at the community and societal level come to bear on mathematics education. In particular, we discuss research that is attempting to situate the classroom learning environment within a discursive context, specifically in terms of understanding the nature of

领域（Davis & Martin，2008；Martin，2007a）。

种族、社会和数学教育

虽然许多研究聚焦于种族如何与课堂层面的数学教与学交叉，但是越来越多的研究开始关注种族现象如何在社区和社会层面上对数学教育产生影响。特别地，我们讨论那些致力于将课堂学习环境置于一个话语环境中的研究，尤其是理解关于数学的种族化话语本质和这些话语如何在日常课堂活动层面塑造学生关于能力、同伴以及参与数学学习过程的感知的研究。

种族-数学的社会化研究

马丁早期的研究（2000）中有一部分是关于更好地理解儿童和成人数学社会化经历的。虽然我们在本章前面提供了一个更新的定义，但马丁最初将数学社会化定义为"个体和群体的数学身份在社会历史、社区、学校和个人内心情景下形成的过程和经历"（第19页）。为了研究数学社会化以及与之重叠的种族社会化经历，马丁使用了访谈的方法，要求非裔美国学生的父母说出他们在儿童时期学习数学的个人经历。这些被访谈对象是已经决定了未来要通过进入北加州的一所社区大学学习来重新接受数学教育的成年人。马丁运用自己的多层背景框架来解释访谈数据，发现参与者无法将他们的数学学习经历同他们非裔美国人的种族身份分离开，许多家长回忆起他们过去数学学习中的特别瞬间，并且从种族的视角来解释这些回忆，以32岁的非裔美国妈妈安布尔为例，谈到她在高中不让她读最高级的数学课程的经历时这样说道：

> 我已经在私立学校里完成了代数I，我应该继续学习三角学，然后是代数II和统计等知识。他们告诉我选三角学课程的人满了，代数II和统计也满了。所有孩子，这些课程里的所有人都是亚裔和白人。于是，他们让我修2门体育课。见鬼！为什么我需要修2门体育课？我上了2门体育课，一门舞蹈，一门保险学，这已经几乎占据了我一整天。所有上保险学课的学生都是黑人学生和西班牙裔学生，我为什么需要上保险学课？……我并没有打算成为一个小商人，它是为从事小型企业的人设计的。而其他所有想要成为一名医生或者律师的亚洲人和白人，都上了所有的科学课、代数I、代数II、统计课和三角学。（Martin，2006，第215页）

根据马丁的研究，像这种个人叙述让我们了解到一些非裔美国家长是如何让他们的孩子融入到学校的社交当中，向孩子传递数学工具性价值的特定信念以及在美国学习数学所涉及的种族化的制度壁垒。尽管基于种族的分流已经在文献中有着详尽的介绍（Oakes等，1990），但这里需要重点说明的一点是，这一现象由那些亲身经历过的人亲自证实了。用这种方式，马丁的工作阐明了将学生个体的身份，与学校内部、学生与家长之间以及社会上长期起作用的更为广泛的力量联系起来的重要性（Cobb，Gresalfi，& Hodge，2009）。

杰克逊（2009）以及英格利希-克拉克和同事们（2012）近期将种族数学社会化的研究进行了拓展，杰克逊对五年级学生的数学学习进行了为期14个月的人种志研究。她跟踪了来自于两个家庭、不同社区以及不同学校环境的一个男生和一个女生，发现这些学生的数学学习与他们的社会身份紧密相关，他们的社会化实践对更广泛的数学学习机会有影响。她的发现支持了这样的观点，即对学生数学学习的理解不能离开他们所处的背景、性别、种族、天分、年轻人的学术挑战等因素来单独研究。作为分析的一部分，她还质疑作为学生社会化媒介的教师和学校管理层，出于他们对这些来自低收入社区的非裔美国年轻人的看法，如何将学习数学"基本技能"的做法合法化，这种做法又如何危害了年轻人的数学发展。

英格利希-克拉克使用一种混合的方法来调查高中生接受到的与数学、种族、种族数学社会化有关的信息，以及年轻人感知和使用这种社会化信息的方法。虽然她主要关注非裔美国青年，但是她的研究包括了来自多个种族和民族的青年（白人、亚裔、拉丁裔）。她调查了来自美国东北部大都市地区三所高中的263名学生（168名非裔美国学生），并访谈了其中的39名学生（29名非裔美国

racialized discourses about mathematics and how those discourses shape students' sense making about ability, their classmates, and participation in the mathematics learning process at the level of everyday classroom activity.

Studies of Racial-Mathematical Socialization

One part of Martin's initial study (2000) consisted of trying to better understand the mathematics socialization experiences of both children and adults. Although we provided an updated definition earlier in this chapter, Martin originally defined *mathematics socialization* as "the processes and experiences by which individual and collective mathematics identities are shaped in sociohistorical, community, school, and intrapersonal contexts" (p. 19). To investigate mathematics socialization and the overlapping racial socialization experiences, Martin utilized interviews where he asked African American parents to reflect on their personal experiences learning mathematics as children. These were adults who had decided to reinvest in their mathematics education later in life by enrolling at a community college in Northern California. Using his multilevel contextual framework to interpret his interview data, Martin found that participants could not separate their mathematics learning experiences from their racial identities *as African Americans*. Many parents recalled specific moments in their mathematics education histories and interpreted them through a racial lens. Consider the case of Amber, a 32-year-old African American mother who spoke of being tracked out of the most advanced mathematics courses at her high school:

> Now in private school I already had Algebra I. I should be going onto Trig, and then Algebra II and Statistics, leaving out of there. They told me that trig was full, Algebra II was filled and Statistics was filled. All the kids, the people that was in there was Asian and White. So they gave us 2 PE (Physical Education) classes. Why the hell you need 2 PE? I had 2 PE, a dance and an insurance class. That was most of my day. And all the people who were in insurance were Black and Hispanics. Why do I need insurance? ... But I didn't plan on being a small business person in that kind of business. It was orientated for someone who was in small business. Everybody else who wanted to be doctors and lawyers who were the Asians and Whites, got all the science classes, all the Algebra, Algebra II, Statistics, Trigonometry. (Martin, 2006, p. 215)

According to Martin, personal narratives such as this provide insight into how some African American parents socialize their children into school and pass on particular beliefs about the instrumental value of mathematics, as well as perceptions of the racialized institutional barriers involved with learning mathematics in the United States. And although race-based tracking has been thoroughly documented in the literature (Oakes et al., 1990), what is important here is that evidence of the phenomenon is substantiated by the voices of those that experience it firsthand. In this way, Martin's work illustrates the importance of connecting an individual student's identity to the broader forces at play within a school, between students and parents, and in society over time (Cobb, Gresalfi, & Hodge, 2009).

Jackson (2009) and English-Clarke and colleagues (2012) have recently extended the study of racial-mathematical socialization. Jackson conducted an ethnographic study of fifth-grade children's mathematical learning over the course of 14 months. She followed the children, one boy and one girl, of two families across neighborhood, home, and school settings and found that these students' mathematical learning was tightly coupled with their social identification and their socialization into particular practices that have implications for wider mathematics learning opportunities. Her findings supported the argument that mathematical learning for students cannot be understood separately from their positioning across contexts and as gendered, racialized, "gifted," and "academically challenged" youth. As part of her analysis, she also questioned how the teachers and school administration, as agents of students' socializations, drew from their perceptions of who these youth were as African Americans from a low-income community to legitimate a "basic skills" approach to the learning of mathematics and how this approach served to undermine the mathematical development of the youth.

English-Clarke used a mixed-method approach to examine the mathematical, racial, and racial-mathematical socialization messages received by high school students and the ways that youth perceived and used this socialization. Although her primary focus was on African American youth, her study included youth from various racial and ethnic categories (White, Asian, Latinx). She surveyed 263 students (168 African American) and interviewed 39 of these students (29 African American) from three high schools in a northeastern U.S. metropolitan area. She found that a majority of youth of all races reported hearing messages or stories about math, which included mathematical experiences and self-perceptions, strategies used to cope with difficulty, messages about math's importance, or problem-solving strategies. About one-third of interviewed youth reported hearing racial-

学生)。她发现所有种族的绝大多数青少年报告他们听到过关于数学的讯息或故事,包括数学经历和自我的感知、克服困难的策略、关于数学重要性的讯息,以及问题解决的策略等。发现有大约三分之一接受访谈的学生表示听到过数学种族歧视的信息或者故事(与种族相关的数学故事或者与数学相关的种族故事),这些社会化信息大多描述了数学背景下的种族歧视,虽然也有一些提到了与数学相关的种族刻板印象,或者非裔美国学生在高水平数学上的缺陷。英格利希-克拉克指出,由于那些报告他们听到过种族数学社会化的年轻人对数学都有着积极感受,因此,种族数学社会化对他们而言所起到的可能是一种特殊的支持作用,而不仅仅是种族社会化的一个其他情境,因此,他们可能会对种族歧视的深远影响、他们生活中可能发生歧视的相关情境,以及他们在到达较高水平的数学层次时可能感知到的种族不平等,有更加深刻和复杂的理解。所以,即使非裔美国高中生还并没有亲历过种族主义和种族歧视,但这些因素也能够通过种族和种族数学社会化影响他们的信念、行动以及身份。

关于数学学习者和数学学习的种族化话语

尝试记录社会中更广泛的种族化话语与学生数学学习经历之间的关系,已经成为一个新兴的研究分支。例如,斯丁森(2008)鉴别出若干对与非裔美国男性学生数学教育轨迹相关的十分重要的社会历史话语。基于后结构理论范式,斯丁森指出了教育中两个突出的种族话语:(1)"缺陷的话语",它声称非裔美国人群体缺乏能够在学校取得成功的文化和遗传资源;(2)"拒绝的话语",它倾向于将非裔美国学生的成绩不佳解释为他们拒绝富有成效的学习行为,在白人主导的社会中,坚持自身种族身份来应对系统性的种族主义(参阅Ogbu, 1978)。依据斯丁森(2013)的研究,这些话语与数学教育中一种普遍的话语同时存在,即数学是一个主要由白人中产阶级男性取得成功的领域(即"白人男性数学神话")。

通过将学生的学业轨迹置于这种种族话语的情境下,斯丁森的后结构方法阐明了社会话语在构成数学学习者种族化经历和身份中的作用。斯丁森指出,事实上,参与者的数学身份"在脱离情境化的复杂情况下并不能够被完全理解,也即说明了非裔美国男性青年对围绕他们的某些广泛的、不公平的社会文化话语是如何应对的"(第87页)。为了说明话语如何作用于具有特定学术、数学和种族身份的学生,该研究构建和拓展了前面提到的种族、身份和数学学习之间的相互作用。当然,虽然话语能够调节身份的构建,但是它们并不能够决定学生如何识别他们是作为种族化的人还是作为学习者。事实上,除了记录种族化话语的存在,斯丁森和其他人还调查了学生在面对数学心理方面或者结构方面的障碍时,是如何锻炼他们的主体能力的(参见McGee & Martin, 2011b)。

已有文献中越来越多的研究开始关注与数学教育相关的种族刻板印象或说法,特别是在种族的说法如何形成学生每日的课堂交流方式方面(见Nasir等,2009;Nasir & Shah, 2011;Shah, 2013)。通过对许多涉及非裔美国男性学生数学学习的研究,纳西尔和沙哈(2011)发现小学、初中和高中学生能够意识到种族的说法,把非裔美国学生说成在数学以及在学校的学习方面能力都欠佳,这一结果与之前这一领域的研究是一致的。然而,他们还发现关于亚洲人的种族说法(如"亚洲人擅长数学")在他们研究的参与者中也十分突出。借鉴社会文化理论,纳西尔和沙哈将这些种族说法概念化为"文化产物",学生可以在学习情境中获得和使用这些文化产物。他们认为,数学学习和学生身份形成的种族化过程包含了多个种族群体的种族说法的同时展开。

建立在这种理论基础上,沙哈(2013)提出了"种族数学话语"的概念并将其作为一种概念化的手段,来说明种族作为给各个种族背景的数学学习者定位的一个语言、符号和实践系统,它是如何更广泛地发挥作用的。沙哈在一所种族多元化的高中对数学课堂和数学学习者进行了为期一年的人种志研究,发现学生在数学学习中对种族的意义构建是围绕着他们对种族数学说法的意识进行组织的,如"亚洲人擅长数学"。有意思的是,学生们并不是孤立地看待这些种族数学说法,而是通过引用数学外的种族话语把它们联系起来,比如,特定种族群体比其他种族更聪明这样的话语。除了将学生种族意识

mathematical messages or stories (stories or messages about math that had to do with race or vice versa); most of these socialization messages described racial discrimination in a mathematical setting, although others touched on mathematics-related racial stereotypes or the dearth of African Americans in high-level mathematics. English-Clarke noted that as all the youth that reported hearing racial-mathematical socialization had positive feelings about math, racial-mathematical socialization may serve as a special support for these youth rather than just an additional context for racial socialization; as a result, they may develop a deeper and more complex understanding of the far-reaching effects of discrimination, the youth-relevant contexts in which discrimination can occur, and the racial imbalances that they may perceive as they reach higher levels of mathematics. Thus, even if African American high school youth have not yet personally experienced racism and racial discrimination, these factors can affect youths' beliefs, actions, and identity by means of racial and racial-mathematical socialization.

Racialized Discourses About Mathematics Learners and Mathematics Learning

An emerging body of research has sought to document the relationship between broader racialized discourses in society and students' mathematics learning experiences. For example, Stinson (2008) identified several sociohistorical discourses consequential for the educational trajectories of African American male students in mathematics. Grounded in a poststructural theoretical paradigm, Stinson notes the prominence of two racial discourses in education: (1) the "discourse of deficiency," which asserts that African American communities lack the cultural and genetic resources to succeed in school, and (2) the "discourse of rejection," which tends to explain underperformance in terms of African American learners' rejecting academically productive behaviors in response to systemic racism while grappling with their own racial identities in a White-dominant society (cf. Ogbu, 1978). According to Stinson (2013), such discourses exist alongside a prevailing discourse in mathematics education that mathematics is a domain in which mainly White, middle-class males are able to succeed (i.e., the "White male math myth").

By situating students' academic trajectories within this racial discursive context, Stinson's poststructural approach illuminates the role of societal discourses in *constituting* the experiences and identities of mathematics learners as racialized. Stinson argues, in fact, that his participants' mathematics identities "cannot be fully understood in their complexities decontextualized from how these young men negotiated some of the broader unjust sociocultural discourses that surround African American male adolescents" (p. 87). In showing how discourses function to position students with particular academic, mathematical, and racial identities, this research builds on and extends the aforementioned body of work on the interaction between race, identity, and mathematics learning. Of course, although discourses mediate identity construction, they do not *determine* how students identify as racialized people or as learners. Indeed, beyond documenting the existence of racialized discourses, Stinson and others have investigated how students exercise agency in the face of psychological and structural barriers in mathematics (see also McGee & Martin, 2011b).

Racial stereotypes or narratives related to mathematics education have received increasing attention in the literature, particularly with respect to the ways that racial narratives shape students' everyday interactions in mathematics classrooms (see Nasir et al., 2009; Nasir & Shah, 2011; Shah, 2013). Through multiple studies involving African American male students in mathematics, Nasir and Shah (2011) found that elementary, middle, and high school students were aware of racial narratives positioning African American learners as less capable both in math and in school. This finding resonates with prior research in this area. However, they also found that racial narratives about Asians (e.g., "Asians are good at math") held salience for participants in their studies. Drawing on sociocultural theory, Nasir and Shah conceptualized these racial narratives as "cultural artifacts," which can become available to—and deployed by—students in learning situations. They argued that the process by which mathematics learning and student identity formation become racialized involves the concurrent deployment of racial narratives about multiple racial groups.

Building on this theoretical insight, Shah (2013) has offered the notion of "racial-mathematical discourse" as a way of conceptualizing how race functions more broadly as a system of language, signs, and practices that positions mathematics learners of all racial backgrounds. In a yearlong, ethnographic study of mathematics classrooms and mathematics learners at a racially diverse high school, Shah found that students' sense making about race in the context of mathematics learning was organized around their awareness of racial-mathematical narratives such as, "Asians are good at math." Interestingly, students did not perceive these racial-mathematical narratives in isolation. Instead, they connected them to

形态的内容和结构与数学学习相联系外，沙哈还记录了种族数学话语影响日常课堂实践的方法，他发现，信念系统不是静态的，学生在参与到特定的课堂交互活动中时，如参加一次考试或者参与全班讨论时，都会使用种族数学的说法。

从理论角度看，沙哈的研究与斯丁森（2008）的后结构理论相呼应，都通过话语强调数学学习者是种族主体。这项工作的另一贡献是，它洞察到超越黑白范式的数学教育种族动态模式。当然，它也确实揭示了美国种族数学话语中的亚洲人中心论，还呈现了这一话语如何暗示各个种族背景的学生群体，并按他们的相互关系将其定位。从方法论的角度来看，这项工作预示着一种转变，即使用观察的方法来调查种族在学生与学生、学生与教师之间日常实际的课堂的交互过程中是如何实时运作的。然而，迄今为止许多实证研究还在完全依赖回顾性叙事访谈的方法。

到目前为止，我们的文献回顾聚焦于这个领域学者所做的实证研究。因为知识的产生并非是一个中立的事业，下一步我们会将注意力转向数学知识生产领域的本质，也即数学教育种族化的特征是什么？

该领域中的种族：研究和政策

在2013年《数学教育研究学报》关于公平性的特刊上的一篇文章中，马丁（2013）提出了"种族和种族主义如何构成数学教育事业的本质？"以及"什么样的事业是数学教育？"的问题，用以呼应早些时候S. E. 安德森（1970）和鲍威尔（2002）的倡议，马丁呼吁调查这一领域的内部结构，以确定其发展和改革的意识形态基础，并研究围绕这些发展和改革产生的权力配置问题。他指出，这样的内部分析至少可以揭示两种可能性：

其一，数学教育作为一个探究领域，可以被认为不受种族争论、层次、阶级、意识形态等决定大多数其他社会背景特征的事物的影响。在这种假设下，会得出这个领域的现行做法和结构安排真正地在本质上是民主的结论，让该领域凌驾于种族争论、种族政治、种族歧视之上运转。其二，结构化的分析……使我相信数学教育事业与社会上其他种族化背景并无二致，该领域中权力和特权的分配并非仅仅是民主原则和实践的产物，种族的社会意义在构建这个领域的规范和关系时具有十分重要的意义（第322页）。

为了支持第二个观点，马丁（2013）为国家数学咨询委员会提供的一个分析思路，表明了数学教育可以作为一个"白人制度空间"的实例来构建，这是基于下面摩尔（2008）所描述的特点：

（a）白人在人数上占统治地位，并将有色人种排除在机构的职权之外；（b）发展一个白人构架，组织机构或者学科的逻辑；（c）课程模式的历史构建是基于白人是精英的思想；（d）知识的产生被认为是中立和公平的，与权力关系无关（第323页）。

数学教育的种族化特征同样明显地体现在组织环境当中。比如，最近对全美数学教师理事会（NCTM）成员的一项调查当中（26%的自我报告），人口统计数据显示92%是白人、4%是黑人和非裔美国人、2%是亚裔人、1%是西班牙裔人和拉美裔人，还有1%是多种族的。虽然同质的人口统计并不一定意味着同样的信仰、意识形态和经历，并且这种分类标签也并不能够反映种族复杂性的全部，但是它们确实提供了一些关于这些组织可以代表哪些声音和为哪些人的利益服务的见解。

对于数学教育作为一个"白人制度空间"的质疑还揭示了，数学教育研究将白人儿童作为所有儿童（R. Gutiérrez, 2013a；Martin, 2013）进行知识生产过程研究的历史趋势。如前文所提到的，数学教育研究中，对有色人种学生关注的重点是研究学业成就，更具体一点，在于"学业成就差异"。然而，如马丁所言，这种关注同样反映了制度空间的主导作用，关注成就的差异让研究者和其他人"假定并接受关于非裔美国人、拉丁美洲人和原住民儿童及其在数学能力层次上的位置都低人一等的观念"（第316页）。依据马丁（2009a）的研

racialized discourses outside mathematics by citing, for example, narratives about certain racial groups being more intelligent than others. In addition to mapping the content and structure of students' racial ideologies related to mathematics learning, Shah also documented the ways in which racial-mathematical discourse mediates everyday classroom practices. That is, rather than a static belief system, Shah found that students deploy racial-mathematical narratives as they engage in typical classroom interactions, such as taking a test and participating in whole class discussions.

From a theoretical standpoint, Shah's research resonates with Stinson's (2008) poststructural emphasis on the constitution of mathematics learners as racial subjects through discourse. An additional contribution of this work is that it offers insight into the dynamics of race in mathematics education beyond the Black-White paradigm. Indeed, it sheds light on the centrality of Asians to racial-mathematical discourse in the United States and also reveals how the discourse implicates and positions students of all racial backgrounds in relation to each other. From a methodological standpoint, the work signals a shift toward the use of observational methods to investigate how race operates in real time during everyday student-student and teacher-student interactions in classrooms. In contrast, much empirical research to date has relied exclusively on retrospective narrative interview methods.

Up to this point in our review, we have focused on empirical studies produced by scholars in the field. Because knowledge production is not a neutral enterprise, we next turn our attention to the nature of the domain where mathematics knowledge is produced. What is the racialized character of mathematics education?

Race in the Field: Research and Policy

In an article appearing in the 2013 equity special issue of the *Journal for Research in Mathematics Education*, Martin (2013) raised the questions, "How do race and racism structure the very nature of the mathematics education enterprise?" and "What kind of project is mathematics education?" Echoing earlier calls by S. E. Anderson (1970) and Powell (2002), Martin called for examining the internal structure of the domain to characterize the ideological bases of its movements and reforms, as well as examining the configurations of power that surround those movements and reforms. He noted that such an internal analysis would reveal at least two possibilities:

> First, mathematics education, as a domain of inquiry, could be considered immune to the racial contestation, stratification, hierarchies, and ideologies that characterize most other societal contexts. Under this assumption, it follows that the prevailing practices and structural arrangements in the field are truly democratic in nature, allowing the domain to function above the fray of racial formation, racial politics, and racial subordination. Second, a structural analysis ... lead[s] me to believe that the enterprise of mathematics education is not unlike other racialized contexts in society; configurations of power and privilege in the domain are not simply the result of democratic principles and practices. Societal meanings for race are highly salient in structuring norms and relations in the domain. (p. 322)

In supporting this second assertion, Martin (2013) presented an analysis of the National Mathematics Advisory Panel to show that mathematics education could be framed as an instantiation of "White institutional space," based on the following characteristics adapted from Moore (2008):

> (a) numerical domination by Whites and the exclusion of people of color from positions of power in institutional contexts, (b) the development of a White frame that organizes the logic of the institution or discipline, (c) the historical construction of curricular models based upon the thinking of White elites, and (d) the assertion of knowledge production as neutral and impartial, unconnected to power relations. (p. 323)

The racialized character of mathematics education is also evident in key organizational contexts. For example, in a recent survey (26% self-reporting) of National Council of Teachers of Mathematics (NCTM) membership, the demographics were reported as 92% White, 4% Black/African American, 2% Asian, 1% Hispanic/Latin American, and 1% multiracial. Although homogeneous demographics do not necessarily imply homogenous beliefs, ideologies, and experiences, and categorical labels do not reflect all the complexities of race, they do offer some insight into which voices may be represented and whose interests may be served by such organizations.

Additional interrogation of mathematics education as a White institutional space may reveal the historical tendency of mathematics education research to study White children as all children (R. Gutiérrez, 2013a; Martin, 2013) in the process of knowledge production. As previously noted, a significant portion of the attention paid to students of color in mathematics education research has focused on the study of achievement and, more specifically, on the "achievement gap." However, according to Martin, this focus also reflects the dominance of the institutional space insofar as the focus on the achievement gap positions researchers and others "to assume and

究,"主流数学教育研究很少阐明参与模式和机会模式是如何在历史上发展的,也很少去挑战学生和他们的能力在种族背景下的建构方式"(第322页)。类似地,马丁(2013)指出,即使是那些从批判的视角来进行数学教育研究的学者,也常常不去探究种族、种族主义和种族化的作用。事实上,马丁断言,数学教育的批判性调查倾向于反映"种族和种族主义独特的背景以及所展现出的概念性缺陷"(第328页)。

马丁(2008,2011,2013)对数学教育制度结构特征的描述还建议,如果要想更全面地理解种族在制度空间中的作用,就需要考察种族在数学教育中塑造和被更广泛的种族动态所塑造的方式。认识到数学教育在历史上一直被作为一种特殊的社会、民主工程,学者们注意到数学教育已经开始服务于许多市场导向的、民主和民族主义的项目当中(参见Gustein,2008,2009;Nielsen,2003),马丁(2013)询问数学教育是否也是一项种族的项目,它又是如何服务于在社会层面运作的各种种族项目的(例如,新保守主义、极右主义、自由主义、新自由主义的项目)。社会学家奥米和怀南特(1994)将这样的项目定义为:

> 同时对种族动态做出解释、表征和说明,并且致力于沿着特定的种族路线对资源进行重组和重新分配。种族项目将种族在一个特定话语背景下的意义与基于这个意义的两种社会结构的种族组织方式联系起来(第56页)。

在我们的综述中,发现近期只有四篇文章(Berry, Ellis, & Hughes, 2014; Berry, Pinter, & McClain, 2013; Martin, 2011, 2013)尝试为数学教育改革中那些反对种族背景并有影响的种族项目做实记录。这些研究认为,在美国历史上特定历史时期所流行的种族意义和种族意识形态为理解种族在这些时期是如何影响数学教育,以及数学教育又如何服务于更大的种族议程提供了一个背景。例如,20世纪50年代末至60年代初的新数学改革发生在美国的民权运动时期,这是一个种族隔离合法,并且秉承白人至上主义意识形态的时期。数学教育改革在这一时期呼吁要教育最聪明的学生。总的来说,这些学生不包括那些非白人群体或者女性(Berry等,2013,2014;Martin,2011,2013)。最近,有一些证据表明,一种无肤色歧视的种族意识形态开始在最近一些备受瞩目的政策实例中占据主导地位,例如美国国家数学咨询委员会和美国州际共同核心数学标准的倡议(Gutstein,2010;Martin,2008)。虽然在一个更广阔的种族背景下,充分描述这一历史转变超出了本章的范围,但它们与数学教育的关系还是值得以一种简要的形式来说明,并且值得进行进一步研究的。在提出上述观点时,我们的目标是强调在数学教育中考虑种族问题并不能够脱离大的社会环境去在一个真空的社会或政治环境中讨论。

对未来研究的展望

上述所回顾的研究提供了研究的一个横断面,这些研究推动了过去十余年种族和数学教育的学术进程。然而,正如贝利(2005)、马丁(2009a)、帕克斯和施麦歇尔(2012)以及斯丁森(2011)所言,主流数学教育对种族问题的研究仍然不足。在展示现存的工作如何解决教学、学习、课程和评估问题时,我们希望在更广的研究领域内的研究者,开始去探索种族如何同他们自己的工作相关,或者他们的工作如何能够为种族知识的发展和种族意义的构建做出贡献。数学教育当中社会政治的转变(R. Gutiérrez,2013b)表明种族、权利和身份问题必须继续得到关注。下面,我们概括出一份数学教育领域种族未来研究的规划。虽然我们的建议不应该被当作是一种规定,但我们相信有几个可行的领域需要进一步研究。

在种族研究中利用新的理论观点

数学教育领域中的种族研究,已经开始超越这个领域的传统界限,并开始使用新的理论观点。我们的文献综述表明,目前仍然有额外空间去利用这些"新"的关于种族的理论观点。我们所谓的使用"新"理论并不排除利用当前的和历史上的有关种族的理论。虽然我们指出了很多当前的观点,但是其中许多观点是基于早期研

accept beliefs about the inferiority of African American, Latino, and Indigenous children and their locations on the hierarchy of mathematics ability" (p. 316). According to Martin (2009a), "mainstream mathematics education research has done little to shed light on how patterns of participation and opportunity have developed historically and done little to challenge the ways that students and their abilities have been constructed against the backdrop of race" (p. 322). Similarly, Martin (2013) notes that even those researchers who approach the study of mathematics education through a critical lens often fail to interrogate the role of race, racism, and racialization. In fact, Martin asserts that critical examinations of mathematics education tend to reflect "peculiar backgroundings and conceptually flawed foregroundings of race and racism" (p. 328).

Martin's (2008, 2011, 2013) characterizations of the institutional structure of mathematics education also suggests that a fuller understanding of the role of race in the institutional space requires examination of the ways that race in mathematics education shapes and is shaped by broader racial dynamics. Acknowledging that mathematics education has historically been framed as a particular kind of social, democratic project and that, more recently, scholars have noted mathematics education being put in service to various market-oriented, democratic and nationalist projects (also see Gustein, 2008, 2009; Nielsen, 2003), Martin (2013) asked whether mathematics education is also a *racial project* and how it has been put in service to various racial projects (e.g., neoconservative, far right, liberal, neoliberal) that operate at the societal level. Sociologists Omi and Winant (1994) define such a project as

> simultaneously an interpretation, representation, or explanation of racial dynamics, and an effort to reorganize and redistribute resources along particular racial lines. Racial projects connect what race means in a particular discursive practice and the ways in which both social structures are racially organized, based upon that meaning. (p. 56)

In our review, we found only four recent papers (Berry, Ellis, & Hughes, 2014; Berry, Pinter, & McClain, 2013; Martin, 2011, 2013) that attempted to historicize mathematics education reform against the backdrop of race and prevailing racial projects. These papers suggest that the prevailing meanings for and ideologies of race at particular historical moments in U.S. history provide a context for understanding how race has shaped mathematics education during these moments and how mathematics education has served larger racial agendas. For example, the *new math* reforms of the late 1950s and early 1960s took place during the U.S. civil rights movement, an era of legalized racial segregation and White supremacist ideology. Mathematics education reforms calling for educating the brightest students, by and large, did not include those who were non-White or female (Berry et al., 2013, 2014; Martin, 2011, 2013). More recently, there is some evidence that a prevailing color-blind racial ideology has been prominent in some recent high-profile policy instances such as the National Mathematics Advisory Panel and the Common Core State Standards initiative (Gutstein, 2010; Martin, 2008). Although a full accounting of these historical shifts in the larger racial context is beyond the scope of this chapter, their relationships to mathematics education are worth noting in abbreviated form and worthy of further study. In making the points above, our goal is to stress that considerations of race in mathematics education do not exist in a social or political vacuum isolated from the realities of race in the larger society.

An Agenda for Future Research

The studies reviewed above provide a cross-section of research that has advanced the intellectual agenda on race and mathematics education over the last decade or so. Yet, as noted by Berry (2005), Martin (2009a), Parks and Schmeichel (2012), and Stinson (2011), race still remains understudied in mainstream mathematics education. In showing how extant work addresses issues of teaching, learning, curriculum, and assessment, we hope researchers in the broader community may come to see how race could be relevant to their own work or how their work may contribute to the production of racial knowledge and racial meaning. The sociopolitical turn in mathematics education (R. Gutiérrez, 2013b) suggests that issues of race, power, and identity must continue to receive attention. Below, we outline an agenda for future research on race in mathematics education. Although our suggestions should not be taken as prescriptive, we believe there are several viable areas in need of further study.

Leveraging New Theoretical Perspectives on Race

The study of race in mathematics education has begun to engage theoretical perspectives beyond the traditional paradigmatic boundaries of the field. Our review suggests that there is additional space to leverage these "new" theoretical perspectives on race. Our use of "new" does not preclude drawing from both contemporary and historical perspectives on race. Although we point to

究的（例如Cox，1948；DuBois，1903，1940；Myrdal，1944）。这些观点可以服务于很多目的：（1）帮助将数学教育建构为一种特殊的物质和意识空间，在这里生产和讨论种族的意义；（2）从概念上理解种族意义和种族类别含义的改变，是如何在一个更大的系统中以一种对数学教育而言重要的方式展开的；（3）使人们认识到种族与其他社会标志和意义系统的交叉方式；（4）将注意力集中在可以超越个体层面对种族和种族主义进行概念化的方式；（5）聚焦于种族化的主体和角色的经历，以及他们对种族化经历的解释和反应。

当前理论杠杆的一个潜在来源是批判种族理论。正如前文所述，批判种族理论至今为止主要通过反故事方式去理解有色人种学生在数学上的经历。虽然使用这种方式有助于我们理解数学中的学生身份，但它仅仅是批判种族理论中能够运用到数学教育研究中的一种概念方式。利益趋同和白人即财富的概念同样是由批判种族理论在法律研究中的概念建构而来的，并且它们已经被用于教育研究当中（Dixson & Rousseau，2006；Ladson-Billings & Tate，1995）。这两种概念提供了一个框架，来分析能够对有色人种学生的教育经历产生影响的条件和事件，并且提供了一个关键词汇来命名和强调那些可能看不见的结构（Crenshaw, Gotanda, Peller, & Thomas，1995）。数学教育研究者，特别是那些对种族作为一种结构和制度力量的作用感兴趣的学者，可能会从批判种族理论文献中探索这些以及其他结构而获益。福克纳等（2014）指出白人即财富的作用是为了理解他们关于学生安置决策的研究结果。类似地，批判种族理论利益趋同的概念在阐释种族和数学改革努力之间的微妙交叉关系上可能特别有用（Martin，2011）。利益趋同表明"当精英群体通常为了在世界上露面或者国际竞争的需要，使得他们需要非裔美国人（和其他少数群体）有所突破时，黑人（和其他少数群体）的利益就与白人的自身利益一致了"（Delgado，2002，第371页）。我们引用这些例子来强调，截止到目前为止，批判种族理论作为一个理论还未被开发出潜力。利益趋同和白人即财富也仅仅是批判种族理论中可能对学者寻求理解种族在数学教育中所起作用的两个概念（Battey，2013）。

批判种族理论中的反故事方法，通常会与种族微攻击方面的研究相结合。种族微攻击是精神病学家查尔斯·皮尔斯和他的同事们在对美国黑人的研究当中使用过的一个术语（Pierce, Carew, Pierce-Gonzalez, & Willis，1978）。皮尔斯将微攻击定义为"微小的、惊人的、通常是自动化的、非语言的交流，是一种'贬低'（Pierce等，1978，第66页）。"随后在1995年，皮尔斯指出：

> 可能对种族主义和性别歧视的受害者最严重的攻击机制就是这种微攻击，这种攻击是微小的、看似无伤大雅的、潜意识的，或者无意识的降级，或者说是贬低，通常是动作的，但能够用口头和动作表示出来。就其本身而言，一次微攻击可能看上去是无害的，但是在生活当中不断积累的微攻击所形成的负担从理论上说可导致死亡率降低、发病率增加和失去信心（第281页）。

休及其同事（例如Sue，2010；Sue等，2007）把微攻击的研究拓展到其他社会群体中。批判种族学者（例如Solórzano, Ceja, &Yosso，2000；Yosso, Smith, Ceja, & Solórzano，2009）研究了种族微攻击与拉丁裔和非裔美国学生的校园氛围之间的关系，其他学者则聚焦于特定的种族群体如非裔美国男性（Harper，2009；Henfield，2011）。在数学教育中，种族微攻击可以帮助我们研究学生如何应对负面的、基于种族的刻板印象和与能力有关的信息。这种研究已经开始出现了（McGee & Martin，2011a，2011b；Nasir & Shah，2011；Shah，2013）。

菲洛米娜·埃斯德（2002）的研究也为探索学生在特定情境和不同情境下种族数学经历提供了可能性。埃斯德利用批判种族和性别研究将日常种族主义作为她论点的一部分，她认为种族主义不仅仅是一种结构和意识形态，更是需要在日常生活的经历中被明晰的现象。埃斯德（2002）做出了一个重要的区分，认为"日常种族主义是种族主义，但是并非所有的种族主义都是日常种族主义"（第177页）。依据埃斯德的说法，日常种族主义将结构性的力量与日常生活中的常规经历结合在一起，这些经历构成了种族主义研究中的重要数据。她将日常

a number of contemporary perspectives, many of them rest on earlier work by scholars (e.g., Cox, 1948; DuBois, 1903, 1940; Myrdal, 1944). Such perspectives can serve several purposes: (1) helping frame mathematics education as a particular kind of material and ideological space where racial meanings are produced and consumed, (2) offering conceptual understanding for how changes in the meanings of race and racial categories in the larger system unfold in ways that might be consequential to mathematics education, (3) bringing awareness to the ways that race intersects with other social markers and systems of meanings, (4) focusing attention on the ways that race and racism can be conceptualized beyond the level of individuals, and (5) drawing attention to the experiences of racialized subjects and actors and their interpretations and responses to racialized experiences.

One potential contemporary source of theoretical leverage comes from critical race theory. As noted in the review, CRT has, to date, been engaged primarily via counterstory as a means to understand the experiences of students of color in mathematics. Although this use of counterstory has made an important contribution to our understanding of student identity in mathematics, it is but one of the constructs from CRT that could be applied to the study of mathematics education. The concepts of *interest convergence* and *Whiteness as property* are constructs from CRT in legal studies that have been previously applied to the study of education (Dixson & Rousseau, 2006; Ladson-Billings & Tate, 1995). They provide a framework for analyzing conditions and events that affect the educational experiences of students of color. And they offer a critical vocabulary to name and highlight structures that otherwise might remain invisible (Crenshaw, Gotanda, Peller, & Thomas, 1995). Mathematics education researchers, particularly those interested in the role of race as a structural and institutional force, might benefit from exploring these and other constructs from the CRT literature. Faulkner et al. (2014) pointed to the role of Whiteness as property to understand the results of their study regarding student placement decisions. Similarly, the CRT concept of interest convergence could be particularly useful in illuminating subtle intersections between race and mathematics reform efforts (Martin, 2011). Interest convergence suggests that "gains for blacks [and other minority groups] coincide with White self-interest and materialize at times when elite groups need a breakthrough for African Americans [and other minority groups], usually for the sake of world appearances or the imperatives of international competition" (Delgado, 2002, p. 371). We cite these examples to highlight the, as of yet, underexplored potential of CRT as a theoretical perspective. Interest convergence and Whiteness as property are but two of the concepts from CRT that could be useful for scholars seeking to understand the role of race in mathematics education (Battey, 2013).

The use of counterstories in CRT has been often been coupled with research on *racial microaggressions,* a term used by psychiatrist Charles Pierce and colleagues in their research on Black Americans (Pierce, Carew, Pierce-Gonzalez, & Willis, 1978). Pierce defined microaggressions as "subtle, stunning, often automatic, and nonverbal exchanges which are 'put downs' " (Pierce et al., 1978, p. 66). Later, in 1995, Pierce stated:

> Probably the most grievous of offensive mechanisms spewed at victims of racism and sexism are microaggressions. These are subtle, innocuous, preconscious, or unconscious degradations, and putdowns, often kinetic but capable of being verbal and/or kinetic. In and of itself a microaggression may seem harmless, but the cumulative burden of a lifetime of microaggressions can theoretically contribute to diminished mortality, augmented morbidity, and flattened confidence. (p. 281)

Sue and colleagues (e.g., Sue, 2010; Sue et al., 2007) have extended work on microaggressions to other social groups. Critical race scholars (e.g., Solórzano, Ceja, & Yosso, 2000; Yosso, Smith, Ceja, & Solorzano, 2009) have studied racial microaggressions in relation to campus climate for Latinx and African American students, and other scholars have focused on particular populations such as African American males (Harper, 2009; Henfield, 2011). The importance of racial microaggressions for mathematics education lies in the study of how students respond to negative, race-based stereotypes and messages about competence and ability. Such research is beginning to take place (McGee & Martin, 2011a, 2011b; Nasir & Shah, 2011; Shah, 2013).

The work of Philomena Essed (2002) also offers possibilities for exploring students' racial-mathematical experiences within and across specific contexts. Essed draws on critical race and gender studies to consider *everyday racism* as part of her argument that racism is more than structure and ideology and that the everyday lived experiences of racism need to be made explicit. Essed (2002) makes an important distinction in suggesting that "everyday racism is racism, but not all racism is everyday racism" (p. 177). According to Essed, everyday racism links structural forces to routine experiences in daily life. These experiences constitute important data in the study of racism. She defines everyday racism as

种族主义定义为：

> 一个过程，在这个过程中，（a）社会化的种族主义概念被赋予具体意义，使得实际做法立即变得确定且可控；（b）具有种族主义含义的实际做法本身变得熟悉与重复；（c）潜在的种族和民族关系是通过这些日常情况下的例行做法来实现和强化的（第190页）。

在数学教育中探索种族和结构问题的另一种可能性是基于种族化社会制度的概念的（Bonilla-Silva，2001），这一概念是指"社会当中的经济、政治、社会和意识形态水平在一定程度上由种族类别或种族角色安排构建"（2001，第37页）。这一概念阐述了本综述当中所强调的几个观点：（1）"种族化社会制度中的角色会以受益者（占据主导地位的种族中的成员）或者从属者（被主导的种族当中的成员）的形式参与到种族关系中"（第11页）；（2）"在所有的种族化社会制度中，角色在种族类别里的位置会涉及某种形式的等级制度，这种等级制度会在种族之间产生明确的社会关系"（第37页）；（3）"处于种族地位的角色占据这些地位不是因为他们来自X或Y种族，而是因为X和Y是被社会定义了的种族。因为种族是社会建构的，所以种族在种族结构中的意义和地位总处于竞争状态"（第40—41页）。

最后，我们认为一个有关种族的文化研究框架（Leonardo，2013），可能有助于理解种族意义通过话语构建的过程，以及种族群体媒体形象中呈现的过程。基于斯图亚特·霍尔（1997）和米歇尔·福柯（1972）等后结构主义理论家的研究，文化研究框架强调了权力如何通过这种话语表征来循环和使用，这一框架强调将语言作为一种基本的媒介，它为数学教育研究者提供了一种新的方式去考虑种族化的数学身份是如何构建和制定的。此外，对话语的关注有助于研究种族如何在数学课堂上调节社会互动。

超越非裔美国学习者的种族化研究

我们注意到，在过去十年当中，许多关于种族与数学教育的研究都明确地关注非裔美国学习者的种族化经历，或者学业成就的种族比较，我们认为应该有更多的研究去关注社会建构的其他种族群体的数学学习经历，包括白人和多民族的学生。例如，数学教育领域缺乏那种致力于质疑或者打破"模范少数群体"神话及其后果的研究，因为这种神话已经被运用到各种亚裔美国群体了（例如Chou & Feagin，2008；Lee，1996；Li & Wang，2008）。我们不仅主张研究要涵盖这些群体的学习者——就像许多研究，即使不是大多数，涵盖了白人学生一样——而且这些研究要聚焦于这些群体经历的种族化本质。这种研究将清晰地界定种族、种族身份，以及种族化经历在这些研究中的意义，这些意义在分析中何等重要。例如，在聚焦身份问题时，研究者可以依据马丁形成的那些问题（2007b，2009a）来构建问题模型：在数学学习和参与的背景下，作为（"白人""亚洲人""拉美人""本地土著"等）意味着什么？在这些不同种族（"白人""亚洲人""拉美人""本地土著"等）背景下，作为一个数学实践者和学习者意味着什么？这些问题强调了种族以及种族身份的意义是社会协商结果的思想，它们在特定的背景下可能会有特别的含义，而学习者表面的、数学之外的经历中的某些方面，可能也要在分析中加以考虑（Oppland-Cordell & Martin，2015）。

从理论上讲，呼吁将研究拓展到社会构建的其他种族群体之中，是扎根于种族是关系这样的概念化过程中的，这样，对某一种族群体数学经历的全面理解取决于从与其他种族群体的关系结构当中来理解这个种族的关系结构。从白人研究中获得的理论观点，也许有助于理解白人学生作为数学学习背景中种族角色的关系结构（见Leonardo，2009）。重要的是，关系的视角不同于种族比较的方法，后者常常会导致种族层次和白人标准。

教师作为种族数学社会化中的种族角色和代理

这里还需要更多的研究去关注教师的种族身份、意

a process in which (a) socialized racist notions are integrated into meanings that make practices immediately definable and manageable, (b) practices with racist implications become in themselves familiar and repetitive, and (c) underlying racial and ethnic relations are actualized and reinforced through these routine practices in everyday situations. (p. 190)

Another possibility for exploring race and structure in mathematics education is anchored by the concept of *racialized social systems* (Bonilla-Silva, 2001), which refers to "societies in which economic, political, social, and ideological levels are partially structured by the placement of actors in racial categories or races" (2001, p. 37). This concept elaborates several points highlighted in this review: (1) "actors in racialized social systems participate in race relations as either beneficiaries (members of the dominant race) or subordinates (members of the dominated race or races)" (p. 11); (2) "in all racialized social systems the placement of actors in racial categories involves some form of hierarchy that produces definite social relations among the races" (p. 37); (3) "actors in racial positions do not occupy those positions because they are of X or Y race, but because X or Y has been socially defined as race. Because races are socially constructed, both the meaning and the position assigned to races in the racial structure are always contested" (pp. 40–41).

Finally, we suggest that a cultural studies framework on race (Leonardo, 2013) may be useful in understanding how racial meanings are constructed through discourse and how racial groups are represented in media imagery. Based on the work of poststructural theorists such as Stuart Hall (1997) and Michel Foucault (1972), for example, the cultural studies framework emphasizes how power circulates and is exercised through discursive representations. With its focus on language as a constitutive medium, this perspective offers mathematics education researchers new ways of thinking about how racialized mathematical identities are constructed and enacted. Further, the focus on discourse can facilitate studies of how race mediates social interactions in mathematics classrooms.

Beyond the Racialization of African American Learners

Noting that much of the research on race and mathematics education produced within the last 10 years has focused explicitly on the racialized experiences of African American learners or on race-comparative achievement outcomes, we suggest that more research is needed on the mathematics learning experiences of other socially constructed racial groups, including Whites and multiracial students. For example, there is a paucity of research in mathematics education that has attempted to raise questions about or disrupt the "model minority" myth, and its consequences, as it has been applied to various Asian American groups (e.g., Chou & Feagin, 2008; Lee, 1996; Li & Wang, 2008). We are not merely suggesting that studies include learners from these groups—as many, if not most, studies include White students—but that these studies focus on the racialized nature of these experiences. Such studies would clearly define what race, racial identity, and racialized experiences mean in these studies and how these meanings are consequential in the analyses. For example, in focusing on issues of identity, questions could be modeled on those formulated by Martin (2007b, 2009a): What does it mean to be ("White," "Asian," "Latinx," "Indigenous," etc.) in the context of mathematics learning and participation? What does it mean to be a doer and learner of mathematics in the context of being ("White," "Asian," "Latinx," "Indigenous," etc.)? Such questions reinforce the idea that the meanings for race and racial identity are socially negotiated; that they may have particular meanings in a given context; and that certain aspects of learners' phenomenal, extra-mathematical experiences may also need to be accounted for in the analysis (Oppland-Cordell & Martin, 2015).

Theoretically, this call for expanding research on other socially constructed racial groups is grounded in a conceptualization of race as *relational*, such that a comprehensive understanding of the experiences of one racial group in mathematics depends on understanding the positionality of that group with respect to the positionality of other groups. Theoretical insights from Whiteness studies may be useful in illuminating the positionality of White students as racial actors in mathematics learning contexts (see Leonardo, 2009). Importantly, a relational perspective is not the same as a race-comparative approach, the latter of which often results in racial hierarchy and the normalization of Whiteness.

Teachers as Racial Actors and Agents of Racial-Mathematical Socialization

Additional research is needed on teachers' racial identities, ideologies, and sense making about issues of race in mathematics education. Recent studies, such as those produced by Faulkner et al. (2014), Leavitt and Washington (2013), and the Mid-Atlantic Center for Mathematics Teaching and Learning (MACMTL) Algebra I Case Studies Project (Birky et al., 2013; Chazan et al.,

识形态和关于数学教育中种族问题的意义构建。近期的研究中，如福克纳等（2014）、莱维特和华盛顿（2013），以及大西洋中部数学教与学研究中心（MACMTL）代数 I 案例研究项目（Birky 等，2013；Chazan 等，2013；Clark 等，2013；W. Johnson, Nyamekye, Chazan, & Rosenthal, 2013），在《师范院校实录》近期的一期特刊中，都强调了这一发展方向代表着未来的研究趋势。根据美国师资力量的人口统计学资料，该领域可以借鉴白人研究并从中获得启发，就像已经在数学教育领域之外的研究中所做的那样，它将告诉我们如何将教师种族意识形态概念化，并理解这些意识形态造成的后果，包括它们如何调节教师的教学决策。然而，正如我们强调需要拓展学生种族经验方面的研究那样，我们可以看到从数学教育角度研究所有教师种族意识形态所蕴含的极大价值。还需要更多的研究关注于非传统空间（例如种族同质空间）中的教学干预，这样可以重新安排课堂活动以提供更平等的种族学习环境。

拓展调查的方法

聚焦于叙事的访谈方法已经产生了大量有关学生学习经历的回顾性记录，由于这种记录是从实际数学行为和表现中暂时被剥离出来的，因此需要更多的研究聚焦于种族化实时发生的过程。这表明，需要采用一些其他能够捕捉到在日常教学活动中外显的和内隐的种族和种族主义表现的方法（如，使用视频分析技术、内隐联想测量方法）。从理论上来讲，通过话语来了解种族的方法可能是有用的，因为它关注的是作为种族和种族主义的实施和发展场所的语言和社会互动。

一个相关的理论观点是，将广泛的社会结构现象与本地日常的课堂活动相联系会带来一定的挑战，未来的研究应该努力尝试使用一些方法去捕捉与种族相关的宏观力量是如何在中观和微观层面上得到体现或协调的（Martin, 2013；Shah, 2013），像参与式行动研究和设计研究（例如 Gutstein, 2005；Terry, 2011）这些方法在这方面可能会有所帮助。在参与式行动研究中，"调查和行动的共同体不断发展，并解决对作为共同研究者参与其中的人具有重要意义的一般问题和重大问题"（Reason & Bradbury, 2008，第1页）。例如，特里（2011）曾与非裔美国高中男生合作，来调查监狱中的非裔美国男性是否多于大学中的非裔美国男性，在他的研究中，学生们能够使用多种统计分析方法来挑战这种说法。类似地，葛斯丁（2007）与非裔美国和拉丁裔的高中生合作，使用统计的和离散动态系统来调查抵押贷款歧视和邻居搬家、2004年总统选举中的统计异常、艾滋病流行，以及数学性别歧视。这些主题中的每一个都与种族有交叉，对这些主题的学习和教学发生在一个处于高度种族化的城市和社区背景之下。特里和葛斯丁的研究表明，方法论的转变可能需要关注交叉部分（种族、阶级、性别、地理位置、语言等）的理论观点，以领会课堂上社会交互的复杂性（参见 Esmonde & Langer-Osuna, 2013）。

从研究到政策与实践

虽然事实并非总是如此，但所有数学教育改革和政策方面的努力都是在当前的种族背景下进行的，从这个意义上说，他们有可能受到更大环境的影响或是对更大的环境作出贡献。正如前文所述，新的数学改革发生在民权时代，同样，美国国家数学咨询委员会所做的工作也是在2001年9·11事件后所形成的加强民族主义的背景下开始的。近期，美国州际共同核心数学标准（http://www.corestandards.org）的改革也是在美国是否处在后种族背景下的争论之中展开的。由于许多改革和政策的努力常常从公平、权利和机会方面进行表述，因此，它们代表着社会和政治的工程。我们认为，需要更多的研究去关注特定的改革进程如何促进社会广大的种族秩序，并反过来受其影响。事实上，这种研究可能代表了这一领域中一条新的研究思路，它需要致力于数学教育、种族和政策研究交叉领域的研究人员。

在本节的最后，我们在教师实践和专业发展的背景下，讨论从种族和数学教育的研究中得到的一些启发。一个近期所做的研究尝试了来自于 DEBT-M 的项目（通过思考数学和有关数学的思考来进行设计以达到均衡的目的；http://debt-m.mspnet.org），这个项目由匹兹堡公立学

2013; Clark et al., 2013; W. Johnson, Nyamekye, Chazan, & Rosenthal, 2013) highlighted in a recent special issue of *Teachers College Record*, represent promising directions. Given the demographics of the U.S. teaching force, the field would benefit from mining Whiteness studies—as has been done outside of mathematics education—for ways of conceptualizing teachers' racial ideologies and understanding the consequences of those ideologies, including how they mediate teachers' pedagogical decisions. However, just as we noted the need to expand research on students' racialized experiences, we see great value in studying all teachers' racial ideologies with respect to mathematics education. More research is also needed on pedagogical interventions in nontraditional spaces (e.g., racially homogeneous spaces) that can reconfigure classroom activity to produce more racially equitable learning environments.

Expanding Methods of Inquiry

Narrative-focused interview methods have produced many rich retrospective accounts of students' learning experiences. Because such accounts are temporally removed from actual mathematical behavior and performance, more research is needed on how processes of racialization occur in real time. This points to a need for employing other methodologies (e.g., using video analysis techniques, implicit association measures) that are able to capture both explicit and implicit manifestations of race and racism in the context of regular classroom activity. Theoretically, approaches that attend to race via discourse may be useful, to the extent that they attend to language and social interaction as sites for the enactment and reproduction of race and racism.

A related theoretical point is the challenge of connecting broad social-structural phenomena to local everyday classroom activity. Future research efforts could use methods that capture how macrolevel forces related to race are manifested or negotiated at mesolevels and microlevels (Martin, 2013; Shah, 2013). Methods like participatory action research and design research (e.g., Gutstein, 2005; Terry, 2011) might be useful in this regard. Within participatory action research, "communities of inquiry and action evolve and address questions and issues that are significant for those who participate as co-researchers" (Reason & Bradbury, 2008, p. 1). Terry (2011), for example, has worked with African American males in high school to investigate the societal narrative that there are more African American men in prison than college. The students in his study were able to use various statistical analyses to challenge this narrative. Similarly, Gutstein (2007) has worked with African American and Latinx high school students to use statistics and discrete dynamical systems to investigate mortgage discrimination and neighborhood displacement, statistical anomalies in the 2004 presidential election, the HIV-AIDS epidemic, and the mathematics of sexism. Each of these topics overlapped with race and the learning and teaching of these topics unfolded in a highly racialized city and community context. The work of Terry and Gutstein suggests that methodological shifts may require theoretical perspectives that attend to intersectionality (race, class, gender, geographic location, language, etc.) to fully appreciate the complexity of social interactions in classrooms (also see Esmonde & Langer-Osuna, 2013).

From Research to Policy and Practice

Although not always acknowledged, all mathematics education reform and policy efforts are situated in the prevailing racial context. In this sense, they have the potential to be impacted by or contribute to that larger context. As noted earlier, the new math reforms were situated in the civil rights era. Similarly, the efforts of the National Mathematics Advisory Panel were situated in the heightened nationalism following the events of September 11, 2001. More recently, the reforms of the Common Core State Standards initiative (http://www.corestandards.org) are unfolding in the contestation of whether the United States is, indeed, in a postracial context. Because reforms and policy efforts are often couched in terms of equity, access, and opportunity, they represent social and political projects. In our view, more research is needed on how specific reforms contribute to and are affected by the larger racial order in society. In fact, this may represent a new line of research for the field, calling for researchers who work at the intersection of mathematics education, race, and policy studies.

We end this subsection by suggesting that insights from research on race and mathematics education be brought to bear in the context of teacher practice and professional development. One recent attempt is the DEBT-M project (Designing for Equity by Thinking in and About Mathematics; http://debt-m.mspnet.org), a collaboration between the Pittsburgh Public Schools, Carnegie Mellon University, the University of Pittsburgh, and the Education Development Center. This 5-year project, which began in 2015, involves an intensive program of professional development for teachers centered on inquiry both in and about mathematics. However, the project will also focus on helping teachers develop sociohistorical and critical perspectives of mathematics and

校、卡耐基梅隆大学、匹兹堡大学以及教育发展中心共同完成，开始于2015年的这个5年计划项目包含了密集的教师专业发展课程，这些课程以数学内部的和对数学的探究为中心。然而，这个项目还将聚焦于帮助教师发展对数学和数学教育的社会历史和批判的观点，这些观点涉及数学经历如何塑造学习者的信念和身份，以及数学教育如何让许多学习者渐渐疏远。该项目的一个基本元素是聚焦种族，包括正在进行的与教师的讨论、对学生进行的调查及访谈。这种对种族和数学的共同关注是一个有希望的进步。

国际热点

我们有理由将我们对种族和数学教育的综述拓展到全球和国际的背景下。虽然伴随而来的社会条件、种族意义、结构和层次在这些国际化的背景下，其重要性并不亚于在美国，但是它们在这些背景之间（以及背景之中）确实表现出不同的效果。关于全球背景下种族的讨论也超出了一章可以描述的范围，在这里，我们给美国和其他国际学者的一个建议是，充分考虑这些背景之中种族条件和意义的相同性和差异性，以及它们各自对数学教育的影响（例如Valoyez-Chavez, & Martin, 待出版）。例如，在美国种族隔离（Berry等，2013）、南非种族隔离（Khuzwayo, 2005）、以色列对埃塞俄比亚犹太人的歧视（Mulat & Arcavi, 2009）和澳大利亚白人政策（Matthews, Watego, Cooper, & Baturo, 2005；Meaney & Evans, 2013；Windschuttle, 2004）背景中，数学改革的性质和目的存在着怎样的异同，在这种社会政策颁布之后，改革措施导致了怎样的趋同或者分化？

开展种族和数学教育研究的指导性问题

在文献综述过程中，我们注意到近期的批判性分析认为数学教育研究是一种意识形态和物质空间，例如，以种族意义、差异和阶层的消费和再生产为特征的这样一种意识形态和物质空间。数学教育工作者利用现有的种族知识去指导他们的工作，而他们的工作反过来对现有的种族知识进行补充。本章中用于呈现材料的整个框架在不同层面处理了这些问题——学习者的个人经历、课堂和学校、社区和社会。所有涉及的研究都以种族为视角，同时为读者提供关于种族的知识。虽然我们能够指出特定研究和观点中与种族相关而重要的内容，我们还是相信，十分有必要让所有数学教育工作者都能够理解在这个领域的研究中"种族是如何起作用的"。为了达到这个目的，我们提供了一组启发式的问题作为阅读指南，供那些将继续研究本章所讨论主题的研究者们，和那些虽然不从事这种研究，但是可能会希望考虑他们的工作如何能够通过"种族的缺失"（Apple, 1999）来对种族知识作出贡献的研究者使用。

本研究中是如何定义种族的？ 正如我们和其他研究者所指出的那样，种族在数学教育研究中最突出的特征是将种族作为一个分类变量来分析学生学业成就的数据。当种族作为一个分类变量时，它在特定研究中可能与生物学和现象学有关。最近，学者们开始借鉴社会学和民族研究的文献，例如，把种族作为社会政治的一个架构来考察，其意义是基于谈判和争论的。我们鼓励读者提出以何种方式在研究中定义种族的问题，并理解这些定义的来源。此外，读者应该去思考种族是否被用作更具体的概念和过程的总称，如种族化、种族意识形态、种族主义和种族身份等。当使用这些概念时，作者如何给出定义？以及用了什么来源来定义它们？

在这一研究中，谁或者什么被种族化了？为什么？ 在这里，我们鼓励读者去探求种族意义是否被拓展到特定的个体、群体、关系、结构和过程中了。此外，为什么种族意义要以这种方式来拓展？这些都向读者明确地描述了吗？提供了哪些理由来支撑？种族意义的延伸在理论上合理吗？种族化过程符合特定的种族意识形态吗（比如无肤色歧视）？同样重要的是，如果在这个研究中没有明确地提到种族问题，那么研究的假设、问题、方法以及结果能够汇聚在一起建构特定的种族意义吗？例如，在一个白人儿童的案例研究中，案例分析是以一种有助于将白人孩子构建为普通孩子的方式进行的吗？是否将白人的身份作为标准，而认为"非白人"身份有问题？同样，在聚焦非裔美国学习者的种族和数学身份的

mathematics education relating to how mathematical experiences shape the beliefs and identities of learners, and how mathematics education alienates many learners. A fundamental element of the project is its focus on race, including ongoing discussions with teachers and surveys and interviews with students. The joint focus on race and mathematics is a promising advance.

International Focus

An argument could be made for extending our review of race and mathematics education to global, international contexts. Although no less consequential than in the United States, the accompanying social conditions, racial meanings, configurations, and hierarchies play out differently across (and within) these contexts. The discussion of race in global contexts is also beyond the scope of a single chapter. One recommendation that we make here is for U.S. and international scholars to consider similarities and differences in racial conditions and meanings across such contexts and the implications for mathematics education in each (e.g., Valoyez-Chavez, & Martin, in press). For example, what similarities and differences exist in the nature and purpose of mathematics reforms in the contexts of U.S. racial segregation (Berry et al., 2013), South African apartheid (Khuzwayo, 2005), discrimination against Ethiopian Jews in Israel (Mulat & Arcavi, 2009), and the White Australia policy (Matthews, Watego, Cooper, & Baturo, 2005; Meaney & Evans, 2013; Windschuttle, 2004) and how have those reforms converged or diverged in the aftermath of such social policies?

Guiding Questions for Conducting Research on Race and Mathematics Education

In the course of this review, we noted that recent critical analyses have argued that mathematics education research, for example, is an ideological and material space that can be characterized by the consumption and reproduction of racial meanings, disparities, and hierarchies. Mathematics educators draw on extant racial knowledge to inform their work, and their work, in turn, contributes to extant racial knowledge. Our overall framework for presenting material in this chapter addressed these issues at many different levels—individual experiences of learners, classrooms and schools, and community and society. All the studies reviewed take up a perspective on race while providing readers with knowledge about race. Although we were able to highlight the relevance and salience of race in particular studies and perspectives, we believe it is important that all mathematics educators be able to understand how "race works" in the research that is produced in the field. To that end, we offer a reading guide—a set of heuristic questions—that can be used by those who continue to conduct research on the topics addressed in this chapter and by those who do not but who may wish to consider how their work contributes to racial knowledge via the "absent presence of race" (Apple, 1999).

How is race defined in this study? As we and other scholars have noted, the predominant characterization of race in mathematics education research has been as a categorical variable used to disaggregate achievement data. When used as a categorical variable, it may appear that race in a particular study is linked to biology and phenotype. Only recently have scholars begun to draw on sociological and ethnic studies literatures, for example, to examine race as a sociopolitical construction whose meaning is subject to negotiation and contestation. We encourage readers to raise questions about the ways race is defined in studies and to understand the sources of those definitions. Moreover, readers should ask whether race is being used as an umbrella term for more specific ideas and processes such as racialization, racial ideology, racism, and racial identity. When these concepts are invoked, how are they defined and what sources are used to define them?

Who or what is being racialized in this study and why? Here we encourage readers to ask whether racial meaning is being extended to particular individuals, groups, relationships, structures, or processes. Moreover, why is racial meaning being extended in the ways it is? Are these clearly articulated for readers? What justifications are provided? Is the extension of racial meaning theoretically justified? Does the racialization process align with particular racial ideologies (e.g., color blindness)? Just as important, if race is not being explicitly addressed in this study, do the assumptions, research questions, methods, and findings converge to construct particular racial meanings? For example, in a case study of a White child, does the analysis proceed in a way that contributes to a construction of the White child as the universal child? Is it the case that White identity is normalized while "non-White" identities are made problematic? Similarly, in studies focused on racial and mathematical identities of African American learners, do the analyses contribute to essentialized constructions of African American learner identities?

What claims are being made about race in this study, and how is the inclusion of race consequential for mathematics learning and participation? Readers should raise these questions to better understand how race

研究中，这样的分析是否有助于非裔美国学习者身份的本质建构？

在这一研究中，作者关于种族提出了怎样的主张？种族的引入是如何影响数学学习和参与的？ 读者应该提出这些问题，以便能更好地理解种族如何作为一个视角来认识我们所关注的现象。还有，包含种族或排除种族对研究的框架、数据的分析和解释会产生怎样的影响？主张、数据和结论一致吗？这种框架和分析是否是由种族的多种因素决定的？哪些其他相关变量，连同种族一起，可能让研究者对数据产生更加深入的了解？提出种族是否是重要的这样的问题还会引出其他的问题，如在特定的分析中是否应该注意种族问题并进行分析？例如，有人可能会假设，由于某一特定研究中存在或者关注亚裔学生，所以研究者必须提供一个对种族背景的认识，但我们认为，只有在研究者看来种族（或种族身份）对分析很重要的情况下，种族背景才是必须的。相反，读者也应该对没有包含种族分析，但很可能应该包含种族分析的案例加以调整，在这些情况下，没有能够明确地分析解决种族问题可能会造成对研究结果的质疑（见 Parks & Schmeichel, 2012）。结合他们自身的种族知识，评论者和读者能够指出并在一定程度上明确种族问题的局限性和优势。

这一研究如何促进该领域和社会中普遍存在的种族知识？ 虽然这个问题对那些不熟悉文献的人而言可能十分困难，但是关注种族和数学教育的人一定能够回答它。在这里，我们要希望读者超越特定研究的特殊性，在更多的研究文献以及更广泛的种族和数学教育的话语体系下去考虑如下问题：一个特定的研究作为一个数据点如何服务于特定的种族说法和话语？它是否有助于使有关种族等级或种族缺陷的信念变得具体化？这一研究是否加入到反叙事当中？

这些只是读者在做研究或阅读和解释他人研究时可能要考虑的几个问题，这些问题提醒我们，无论我们的工作是否明确地涉及种族问题，它们都处于具有种族意义和理解的本土以及更广泛的全球背景下，我们都应该努力确保那些意义和理解能够让种族研究富有成效。

注释

1. 在本章中，我们使用白人、亚裔美国人、黑人、非裔美国人、拉丁裔，和本地土著这样的种族类别，我们这样做是因为现存的文献就是这样使用的。在讨论特定的文章时，我们会使用原作者使用的词汇。此外，诸如"黑人"和"非裔美国人"这样的词汇包括了这一群体成员的多种自我认同方式。我们承认所有这些词汇的社会政治和权力含义。

2. 研究者（Carbado & Harris, 2008）在其他教育情境下也注意到了这种默认假设。

3. 对教师在数学教育中种族身份的研究，继承了这个领域中由拉德森-比林斯（1994）、班克斯（2006）、麦金太尔（1997）和其他研究者所做的重要研究，我们承认这一研究的历史渊源，但是对这些研究的综述超出了本章的范围。

is being used as a lens to understand the phenomena of concern. Moreover, how does the inclusion or exclusion of race seem to affect the framing of the study or the analysis and interpretation of the data? Do the claims, data, and conclusions cohere? Do the framing and analysis seem to be overdetermined by race? What other relevant variables, in conjunction with race, might lead to more robust understanding of the data? Raising questions about whether race is consequential also leads to the question of whether race should even be noted or pursued in a particular analysis. For example, one might assume that because of the presence of or focus on Asian students in a particular study, the researcher must provide an understanding of the racial context. We would argue that this may only be necessary if, in the view of the researcher, race (or racial identity) is consequential for the analysis. Conversely, readers should also be attuned to cases in which race is not included in the analysis but likely should have been. In these cases, the failure to explicitly address race analytically can raise questions about the results of the study (see Parks & Schmeichel, 2012). Reviewers and readers, informed with their own racial knowledge, can then point out the limitations and advantages of being more or less explicit about race.

How does this study contribute to prevailing racial knowledge in the field and in society? Although this question may be more difficult for those who are not familiar with the literature, those who focus on race and mathematics education can certainly contend with it. Here, we ask readers to go beyond the particularities of a given study to consider the larger corpus of studies and the larger discourse on race and mathematics education. How does a particular study serve as a data point in particular racial narratives and discourses? Does it help to reify beliefs about racial hierarchies or racial deficiencies? Does the study add to a counternarrative?

These are but a few of the considerations that readers might take up in formulating research or reading and interpreting research. These questions serve to remind us that our work is situated in both local and more global contexts of racial meanings and understandings, whether that work addresses race explicitly or not. We should all work to ensure that those meanings and understandings contribute to making race work productively.

Notes

1. In this chapter, we use the racial categories White, Asian American, Black, African American, Latinx, and Indigenous, as they have been used in the extant literature. In discussion referring to particular articles, we use the terms of the authors. In addition, terms like "Black" and "African American" encompass the diverse ways that groups' members self-identify. We acknowledge the sociopolitical and power implications of all these terms.

2. Scholars (Carbado & Harris, 2008) have noted this default assumption in other educational contexts.

3. Research on teachers' racial identities in mathematics education follows important research in this area by Ladson-Billings (1994), Banks (2006), McIntyre (1997), and others. A review of that literature is beyond the scope of this chapter, but we acknowledge the historical lineage of this work.

References

Aguirre, J., Mayfield-Ingram, K., & Martin, D. (2013). *The impact of identity in K–8 mathematics learning and teaching: Rethinking equity-based practices.* Reston, VA: National Council of Teachers of Mathematics.

Anderson, C. R., & Tate, W. F. (2008). Still separate, still unequal: Democratic access to mathematics in U.S. schools. In L. D. English & M. G. Bartolini Bussi (Eds.), *Handbook of international research in mathematics education* (2nd ed., pp. 299–319). New York, NY: Routledge.

Anderson, S. E. (1970). Mathematics and the struggle for Black liberation. *The Black Scholar, 2*(1), 20–27.

Andreasen, R. O. (2000). Race: Biological reality or social construct. *Philosophy of Science, 67*, S653–S666.

Ansley, F. L. (1997). "White supremacy (and what we should do about it)," in R. Delgado and J. Stefancic (Eds.), *Critical White studies: Looking behind the mirror* (pp. 592–595). Philadelphia, PA: Temple University Press.

Apple, M. W. (1999). The absent presence of race in educational reform. *Race, Ethnicity and Education, 2*(1), 9–16.

Back, L., & Solomos, J. (2000). *Theories of race and racism: A reader.* London, United Kingdom: Routledge.

Banks, J. A. (2006). *Race, culture, and education: The selected works of James A. Banks.* New York, NY: Routledge.

Bartell, T., Foote, M., Drake, C., McDuffie, A., Turner, E., Aguirre, J. (2013). Teachers of Black children: (Re)orienting thinking in an elementary mathematics methods course. In D. B. Martin & J. Leonard (Eds.), *The brilliance of Black children in mathematics* (pp. 341–365). Charlotte, NC: Information Age.

Battey, D. (2013). Access to mathematics: A possessive investment in whiteness. *Curriculum Inquiry, 43*(3), 332–359.

Battey, D., & Chan, A. (2010). Building community and relationships that support critical conversations on race: The case of cognitively guided instruction. In M. Foote (Ed.), *Mathematics teaching and learning in K–12: Equity and professional development* (pp. 41–58). New York, NY: Palgrave Macmillan.

Bell, D. (1992). *Faces at the bottom of the well: The permanence of racism.* New York, NY: Basic Books.

Berry, R. Q., III. (2005). E "racing" myths about race and mathematics education through a critical race theory analysis. In S. Hughes (Ed.), *What we still don't know about teaching race: How to talk about it in the classroom* (pp. 13–34). Lewiston, NY: The Edwin Mellon Press.

Berry, R. Q., III. (2008). Access to upper-level mathematics: The stories of African American middle school boys who are successful with school mathematics. *Journal for Research in Mathematics Education, 39*(5), 464–488.

Berry, R. Q., III, Ellis, M., & Hughes, S. (2014). Examining a history of failed reforms and recent stories of success: Mathematics education and Black learners of mathematics in the United States. *Race Ethnicity and Education, 17*(4), 540–568.

Berry, R. Q., III, Pinter, H, & McClain, O. L. (2013). A critical review of American K–12 mathematics education, 1954–present: Implications for the experiences and achievement of Black children. In D. B. Martin & J. Leonard (Eds.), *Beyond the numbers and toward new discourse: The brilliance of black children in mathematics*. Charlotte, NC: Information Age.

Berry, R. Q., III, Thunder, K., & McClain, O. L. (2011). Counter narratives: Examining the mathematics and racial identities of Black boys who are successful with school mathematics. *Journal of African American Males in Education 2*(1), 10–23.

Birky, G., Chazan, D., & Morris, K. (2013). In search of coherence and meaning: Madison Morgan's experiences and motivations as an African American learner and teacher. *Teachers College Record, 115*(2), 1–42.

Bonilla-Silva, E. (2001). *White supremacy and racism in the post-civil rights era*. Boulder, CO: Lynne Reinner.

Bonilla-Silva, E. (2010). *Racism without racists: Color-blind racism & racial inequality in contemporary America*. Lanham, MD: Rowman & Littlefield.

Brodkin, K. (1998). *How Jews became white folks and what that says about race in America*. New Brunswick, NJ: Rutgers University Press.

Carbado, D. W., & Harris, C. I. (2008). The new racial preferences. *California Law Review*, 1139–1214.

Chavous, T. M., Bernat, D. H., Schmeelk-Cone, K., Caldwell, C. H., Kohn-Wood, L., & Zimmerman, M. A. (2003). Racial identity and academic attainment among African American adolescents. *Child development, 74*(4), 1076–1090.

Chazan, D., Brantlinger, A., Clark, L., & Edwards, A. (2013). What mathematics education might learn from the work of well-respected African American mathematics teachers in urban schools. *Teachers College Record. 115*(2), 1–36.

Chou, R. S., & Feagin, J. R. (2008). *The myth of the model minority: Asian Americans facing racism*. Boulder, CO: Paradigm.

Civil, M. (2006). Building on community knowledge: An avenue to equity in mathematics education. In N. Nasir & P. Cobb (Eds.), *Improving access to mathematics: Diversity and equity in the classroom* (pp. 105–117). New York, NY: Teachers College Press.

Clark, L., Badertscher, E., & Napp, C. (2013). African American mathematics teachers as agents in their African American students' mathematics identity formation. *Teachers College Record, 115*(2), 1–36.

Cobb, P., Gresalfi, M., & Hodge, L. (2009). An interpretive scheme for analyzing the identities that students develop in mathematics classrooms. *Journal for Research in Mathematics Education, 40*(1), 40–68.

Cokley, K., McClain, S., Jones, M., & Johnson, S. (2012). A preliminary investigation of academic disidentification, racial identity, and academic achievement among African American adolescents. *The High School Journal, 95*(2), 54–68.

Cornell, S., & Hartmann, D. (1998). *Ethnicity and race: Making identities in a changing world*. London, United Kingdom: Pine Forge Press.

Cox, O. (1948). *Race, caste and class*. New York, NY: Monthly Review Press.

Crenshaw, K., Gotanda, N., Peller, G., & Thomas, K. (Eds.). (1995). *Critical race theory: The key writings that formed the movement*. New York, NY: New Press.

Darling-Hammond, L., & Sykes, G. (2003). Wanted: A national teacher supply policy for education: The right way to meet the "highly qualified teacher" challenge. *Education Policy Analysis Archives, 11*(33). Retrieved from http://epaa.asu.edu/epaa/v11n33/

Davis, J., & Martin, D. (2008). Racism, assessment, and instructional practices: Implications for mathematics teachers of African American students. *Journal of Urban Mathematics Education, 1*(1), 10–34.

Delgado, R. (2002). Explaining the rise and fall of African American fortunes: Interest convergence and civil rights gains. *Harvard Civil Rights-Civil Liberties Law Review, 37*, 369–387.

Diversity in Mathematics Education (DiME) Center for Learning and Teaching. (2007). Culture, race, power, and mathematics education. In F. K. Lester Jr. (Ed.), *Second handbook of research on mathematics teaching and learning* (pp. 405–433). Charlotte, NC: Information Age; Reston, VA: National Council of Teachers of Mathematics.

Dixson, A., & Rousseau, C. (2006). And we are still not saved: Critical race theory in education ten years later. In A. Dixson & C. Rousseau (Eds.), *Critical race theory in education: All God's children got a song* (pp. 31–54). New York, NY: Routledge.

DuBois, W. E. B. (1903). *The souls of Black folk*. New York, NY: Penguin Press.

DuBois, W. E. B. (1940). *Dusk of dawn: An essay toward an autobiography of a race concept*. New Brunswick, NJ: Transaction.

Ellington, R. M., & Frederick, R. (2010). Black high achieving undergraduate mathematics majors discuss success and persistence in mathematics. *Negro Educational Review, 61*, 61–84.

English-Clarke, T., Slaughter-Defoe, D., & Martin, D. (2012). What does race have to do with math? Relationships between racial-mathematical socialization, mathematical identity, and racial identity. In D. Slaughter-Defoe (Ed.), *Racial stereotyping and child development* (pp. 55–79), Contributions to Human Development book series. Basel, Switzerland: Karger.

Ernest, P. (2009). Mathematics education ideologies and globalization. In P. Ernest, B. Greer, & B. Sriraman (Eds.), *Critical issues in mathematics education* (pp. 67–110). Charlotte, NC: Information Age.

Ernest, P., Greer, B., & Sriraman, B. (Eds.). (2009). *Critical issues in mathematics education* (Vol. 6). Charlotte, NC: Information Age.

Esmonde, I., Brodie, K., Dookie, L., & Takeuchi, M. (2009). Social identities and opportunities to learn: Student perspectives on group work in an urban mathematics classroom. *Journal of Urban Mathematics Education, 2*(2), 18–45.

Esmonde, I., & Langer-Osuna, J. M. (2013). Power in numbers: Student participation in mathematical discussions in heterogeneous spaces. *Journal for Research in Mathematics Education, 44*(1), 288–315.

Essed, P. (2002). Everyday racism: A new approach to the study of racism. In P. Essed & D. Goldberg (Eds.), *Race critical theories* (pp. 176–194). Malden, MA: Blackwell.

Faulkner, V., Stiff, L., Marshall, P., Nietfield, J., & Crossland, C. (2014). Race and teacher evaluations as predictors of algebra placement. *Journal for Research in Mathematics Education, 45*(3), 288–311.

Foucault, M. (1972). *The archaeology of knowledge.* London, United Kingdom: Tavistock Publications.

Frankenberg, R. (1992). *The social construction of whiteness: White women, race matters.* London, United Kingdom: Routledge.

Frankenstein, M. (1997). In addition to the mathematics: Including equity issues in the curriculum. In J. Trentacosta & M. Kenney (Eds.), *Multicultural and gender equity in the mathematics classroom* (pp. 10–22). Reston, VA: National Council of Teachers of Mathematics.

Fullilove, R. E., & Treisman, P. U. (1990). Mathematics achievement among African American undergraduates at the University of California, Berkeley: An evaluation of the mathematics workshop program. *Journal of Negro Education, 59*(3), 463–478.

Gholson, M., & Martin, D. B. (2014). Smart girls, Black girls mean girls, and bullies: At the intersection of identities and the mediating role of young girls' social network in mathematics communities of practice. *Journal of Education, 194*(1), 19–33.

Ginsburg, H. P., & Russell, R. L. (1981). Social class and racial influences on early mathematical thinking. *Monographs of the Society for Research in Child Development, 46*(6), 1–69.

González, N., Andrade, R., Civil, M., & Moll, L. (2001). Bridging funds of distributed knowledge: Creating zones of practices in mathematics. *Journal of Education for Students Placed at Risk, 6*(1–2), 115–132.

Gould, S. J. (1996). *The mismeasure of man.* New York, NY: WW Norton & Company.

Grosfoguel, R. (2004). Race and ethnicity or racialized ethnicities? Identities within global coloniality. *Ethnicities, 4*(3), 315–336.

Gutiérrez, K. D., & Rogoff, B. (2003). Cultural ways of learning: Individual traits or repertoires of practice. *Educational Researcher, 32*(5), 19–25.

Gutiérrez, R. (2008). A "gap-gazing" fetish in mathematics education? Problematizing research on the achievement gap. *Journal for Research in Mathematics Education, 39*(4), 357–364.

Gutiérrez, R. (Ed.). (2013a). *JRME* Equity Special Issue. *Journal for Research in Mathematics Education, 44*(1).

Gutiérrez, R. (2013b). The sociopolitical turn in mathematics education. *Journal for Research in Mathematics Education, 44*(1), 37–68.

Gutstein, E. (2003). Teaching and learning mathematics for social justice in an urban, Latino school. *Journal for Research in Mathematics Education, 34*(1), 37–73.

Gutstein, E. (2005). *Reading and writing the world with mathematics.* London, United Kingdom: Routledge Falmer.

Gutstein, E. (2007). "And that's just how it starts": Teaching mathematics and developing student agency. *The Teachers College Record, 109*(2), 420–448.

Gutstein, E. (2008). The political context of the National Mathematics Advisory Panel. *The Montana Mathematics Enthusiast, 5*(1), 415–422.

Gutstein, E. (2009). The politics of mathematics education in the United States: Dominant and counter agendas. In B. Greer, S. Mukhopadhyay, S. Nelson-Barber, & A. Powell (Eds.), *Culturally responsive mathematics education* (pp. 137–164). New York, NY: Routledge.

Gutstein, E. (2010). The Common Core State Standards Initiative—A critical response. *Journal of Urban Mathematics Education, 3*(1), 9–18.

Hale-Benson, J. (1986). Black children: Their roots. *Culture, and learning styles* (Rev. ed.). Baltimore, MD: Johns Hopkins University Press.

Hall, S. (1997). Introduction. In S. Hall (Ed.), *Representation: Cultural representations and signifying practices.* Thousand Oaks, CA: Sage.

Harper, S. R. (2009). Niggers no more: A critical race counter-narrative on Black male student achievement at predominantly White colleges and universities. *International Journal of Qualitative Studies in Education, 22*(6), 697–712.

Henfield, M. S. (2011). Black male adolescents navigating microaggressions in a traditionally White middle school: A qualitative study. *Journal of Multicultural Counseling and Development, 39*(3), 141.

Hill, H., & Lubienski, S. (2007). Teachers' mathematics knowledge for teaching and school context: A study of California teachers. *Educational Policy, 21*(5), 747–768.

Hill, J. H. (2009). *The everyday language of White racism.* Malden, MA: Wiley-Blackwell.

Ignatiev, N. (2009). *How the Irish became White.* London, United Kingdom: Routledge.

Ikemoto, L. (1997). Furthering the inquiry: Race, class, and culture in the forced medical treatment of pregnant women. In A. Wing (Ed.), *Critical race feminism: A reader* (pp. 136–143). New York: New York University Press.

Jackson, K. (2009). The social construction of youth and mathematics: The case of a fifth-grade classroom. In D. B. Martin (Ed.), *Mathematics teaching, learning, and liberation in the lives of Black children* (pp. 175–199). New York, NY: Routledge.

Jett, C. C. (2011). "I once was lost, but now am found": The mathematics journey of an African American male mathematics doctoral student. *Journal of Black Studies, 42*(7), 1125–1147.

Jett, C. (2012). Critical race theory interwoven with mathematics education research. *Journal of Urban Mathematics Education, 5*(1), 21–30.

Johnson, W., Nyamekye, F., Chazan, D., & Rosenthal, W. (2013). Teaching with speeches: Using the mathematics classroom to prepare students for life. *Teachers College Record, 115*(2), 1–26.

Johnson, Y. A. (2009). "Come home, then": Two eighth-grade Black female students' reflections on their mathematics experiences. In D. B. Martin (Ed.), *Mathematics teaching, learning, and liberation in the lives of Black children,* (pp. 289–203). New York, NY: Routledge.

Khuzwayo, B. (2005). A history of mathematics education research in South Africa: The apartheid years. In R. Vithal, J. Adler, & C. Keitel (Eds.), *Researching mathematics education* in South Africa: Perspectives, practices and possibilities, (pp. 234–286). Cape Town, South Africa: The Human Sciences Research Council.

Ladson-Billings, G. (1994). *The dreamkeepers: Successful teachers of African American children*. San Francisco, CA: Jossey-Bass.

Ladson-Billings, G. (1997). It doesn't add up: African American students' mathematics achievement. *Journal for Research in Mathematics Education, 28*(6), 697–708.

Ladson-Billings, G. (1998). Just what is critical race theory and what's it doing in a nice field like education? *International Journal of Qualitative Studies in Education, 11*(1), 7–24.

Ladson-Billings, G. (2012). Through a glass darkly: The persistence of race in education research & scholarship. *Educational Researcher, 41*(4), 115–120.

Ladson-Billings, G., & Tate, W. (1995). Toward a critical race theory of education. *Teachers College Record, 97*(1), 47–68.

Langer-Osuna, J. M. (2011). How Brianna became bossy and Kofi came out smart: Understanding the trajectories of identity and engagement for two group leaders in a project-based mathematics classroom. *Canadian Journal of Science, Mathematics and Technology Education, 11*(3), 207–225.

Langer-Osuna, J. M., & Esmonde, I. (2017). Identity in research on mathematics education. In J. Cai (Ed.), *Compendium for research in mathematics education* (pp. 637–648). Reston, VA: National Council of Teachers of Mathematics.

Larnell, G. V. (2013). On "New Waves" in mathematics education: Identity, power, and the mathematics learning experiences of all children. *New Waves—Educational Research and Development, 16*(1), 146–156.

Leavitt, D. R., & Washington, E. N. (2013). "I teach like you are all gifted": Leading lowest track students to become confident mathematics learners. In A. Cohan & A. Honigsfeld (Eds.), *Breaking the mold of education: Innovative and successful practices for student engagement, empowerment, and motivation* (pp. 171–178). Lanhan, MD: Rowman and Littlefield.

Lee, S. J. (1996). *Unraveling the "model minority" stereotype: Listening to Asian American youth*. New York, NY: Teachers College Press.

Leonard, J. (2008). *Culturally specific pedagogy in the mathematics classroom*. New York, NY: Routledge.

Leonard, J., Brooks, W., Barnes-Johnson, J., & Berry, R. (2010). The nuances and complexities of teaching mathematics for cultural relevance and social justice. *Journal of Teacher Education, 61*, 261–271

Leonard, J., & Martin, D. B. (Eds.). (2013). *The brilliance of Black children in mathematics: Beyond the numbers and toward new discourse*. Charlotte, NC: Information Age.

Leonardo, Z. (2009) *Race, whiteness, and education*, New York, NY: Routledge

Leonardo, Z. (2013). *Race frameworks: A multidimensional theory of racism and education*. New York, NY: Teachers College Press.

Li, G., & Wang, L. (2008). *Model minority myth revisited: An interdisciplinary approach to demystifying Asian American educational experiences*. Charlotte, NC: Information Age.

Lipman, P. (2012). Neoliberal urbanism, race, and equity in mathematics education. *Journal of Urban Mathematics Education, 5*(2), 6–17.

Loeb, S., & Reininger, M. (2004). *Public policy and teacher labor markets: What we know and why it matters*. East Lansing: Education Policy Center at Michigan State University.

López, I. F. H. (1996). *White by law: The legal construction of race*. New York: New York University Press.

Lubienski, S. (2002). A closer look at black-white mathematics gaps: Intersections of race and SES in NAEP achievement and instructional practices data. *Journal of Negro Education, 71*(4), 269–287.

Lubienski, S. (2008). On "gap gazing" in mathematics education: The need for gaps analyses. *Journal for Research in Mathematics Education, 39*(4), 350–356.

Lubienski, S. T., & Bowen, A. (2000). Who's counting? A survey of mathematics education research 1982–1998. *Journal for Research in Mathematics Education, 31*(5), 626–633.

Lubienski, S. T., & Gutiérrez, R. (2008). Bridging the gaps in perspectives on equity in mathematics education. *Journal for Research in Mathematics Education, 39*, 365–371.

Lynn, M., & Dixson, A. (Eds.). (2013). *Handbook of critical race theory in education*. New York, NY: Routledge.

Malloy, C. E., & Jones, M. G. (1998). An investigation of African American students' mathematical problem solving. *Journal for Research in Mathematics Education, 29*(2), 143–163.

Martin, D. (2000). *Mathematics success and failure among African American youth: The roles of sociohistorical context, community forces, school influence, and individual agency*. Mahwah, NJ: Lawrence Erlbaum Associates.

Martin, D. (2006). Mathematics learning and participation as racialized forms of experience: African American parents speak on the struggle for mathematics literacy. *Mathematical Thinking and Learning, 8*(3), 197–229.

Martin, D. (2007a). Beyond missionaries or cannibals: Who should teach mathematics to African American children? *The High School Journal, 91*(1), 6–28.

Martin, D. (2007b). Mathematics learning and participation in African American context: The co-construction of identity in two intersecting realms of experience. In N. Nasir & P. Cobb (Eds.), *Diversity, equity, and access to mathematical ideas* (pp. 146–158). New York, NY: Teachers College Press.

Martin, D. (2008). E(race)ing race from a national conversation on mathematics teaching and learning: The National Mathematics Advisory Panel as White institutional space. *The Montana Math Enthusiast, 5*(2&3), 387–398.

Martin, D. (2009a). Researching race in mathematics education. *Teachers College Record, 111*(2), 295–338.

Martin, D. B. (2009b). Liberating the production of knowledge about African American children and mathematics. In D. Martin (Ed.), *Mathematics teaching, learning, and liberation in African American contexts* (pp. 3–36). New York, NY: Routledge.

Martin, D. B. (Ed.). (2009c). *Mathematics teaching, learning, and liberation in the lives of Black children*. New York, NY: Routledge.

Martin, D. (2011). What does quality mean in the context of White institutional space? In B. Atweh, M. Graven, W. Secada, & P. Valero (Eds.), *Quality and equity agendas in mathematics education* (pp. 437–450). New York, NY: Springer Publishing.

Martin, D. (2012). Learning mathematics while Black. *The Journal of Educational Foundations, 26*(1–2), 47–66.

Martin, D. (2013). Race, racial projects, and mathematics education. *Journal for Research in Mathematics Education, 44*(1), 316–333.

Martin, D. B., & Larnell, G. (2013). Urban mathematics education. In R. Milner & K. Lomotey (Eds.), *Handbook of urban education* (pp. 373–393). London, United Kingdom: Routledge.

Matsuda, M., Lawrence, C., Delgado, R., & Crenshaw, K. (Eds.). (1993). *Words that wound: Critical race theory, assaultive speech, and the first amendment.* Boulder, CO: Westview.

Matthews, C., Watego, W., Cooper, T. J., & Baturo, A. R. (2005). Does mathematics education in Australia devalue Indigenous culture? Indigenous perspectives and non-Indigenous reflections. In P. Clarkson, A. Downton, D. Gronn, M. Horne, A. McDonough, R. Pierce, & A. Roche (Eds.), *Building connections: Research, theory and practice—Proceedings 28th annual conference of Mathematics Education Research Group of Australasia, Melbourne, Vol. 2* (pp. 513–520). Sydney, Australia: MERGA.

McGee, E. (2013a). Young, Black, mathematically gifted, and stereotyped. *The High School Journal, 96*(3), 253–263.

McGee, E. O. (2013b). Threatened and placed at risk: High achieving African American males in urban high schools. *The Urban Review, 45*(4), 448–471.

McGee, E. O. (2013c). High-achieving Black students, biculturalism, and out-of-school STEM learning experiences: Exploring some unintended consequences. *Journal of Urban Mathematics Education, 6*(2), 20–41.

McGee, E., & Martin, D. (2011a). From the hood to being hooded: A case study of a Black male PhD. *Journal of African American Males in Education, 2*(1), 46–65.

McGee, E., & Martin, D. (2011b). You would not believe what I have to go through to prove my intellectual value! Stereotype management among successful Black college mathematics and engineering students. *American Educational Research Journal, 48*(6), 1347–1389.

McGraw, R., Lubienski, S. T., & Strutchens, M. E. (2006). A closer look at gender in NAEP mathematics achievement and affect data: Intersections with achievement, race/ethnicity, and socioeconomic status. *Journal for Research in Mathematics Education, 37*(2), 129–150.

McIntyre, A. (1997). *Making meaning of Whiteness: Exploring racial identity with White teachers.* New York, NY: SUNY Press.

Meaney, T., & Evans, D. (2013). What is the responsibility of mathematics education to the Indigenous students that it serves? *Educational Studies in Mathematics, 82*(3), 481–496.

Moody, V. R. (2004). Sociocultural orientations and the mathematical success of African American students. *The Journal of Educational Research, 97*(3), 135–146.

Moore, W. (2008). *Reproducing racism: White space, elite law schools, and racial inequality.* New York, NY: Rowman & Littlefield.

Moses, R., & Cobb, C. E. (2001). *Radical equations.* Boston, MA: Beacon Press.

Moses, R. P., Kamii, M., Swap, S. M., & Howard, J. (1989). The Algebra Project: Organizing in the spirit of Ella. *Harvard Educational Review, 59*(4), 423–444.

Mulat, T., & Arcavi, A. (2009). Success in mathematics within a challenged minority: The case of students of Ethiopian origin in Israel (SEO). *Educational Studies in Mathematics, 72*(1), 77–92.

Murrell, P. (1994). In search of responsive teaching for African American males: An investigation of students' experiences of middle school mathematics curriculum. *The Journal of Negro Education, 63*(4), 556–569.

Myrdal, G. (1944). *An American dilemma: The Negro problem and American democracy.* New York, NY: Harper.

Nasir, N. I. S. (2000). "Points ain't everything": Emergent goals and average and percent understandings in the play of basketball among African American students. *Anthropology & Education Quarterly, 31*(3), 283–305.

Nasir, N. S. (2002). Identity, goals, and learning: Mathematics in cultural practice. *Mathematical Thinking and Learning, 4*(2&3), 211–245.

Nasir, N. S. (2011). *Racialized identities: Race and achievement among African American youth.* Stanford, CA: Stanford University Press.

Nasir, N. S., Atukpawu, G., O'Connor, K., Davis, M., Wischnia, S., & Tsang, J. (2009). Wrestling with the legacy of stereotypes: Being African American in math class. In D. B. Martin (Ed.), *Mathematics teaching, learning and liberation in the lives of Black children* (pp. 231–248). New York, NY: Routledge.

Nasir, N., & de Royston, M. M. (2013). Power, identity, and mathematical practices outside and inside school. *Journal for Research in Mathematics Education, 44*(1), 264–287.

Nasir, N. S., & Hand, V. (2006). Exploring sociocultural perspectives on race, culture, and learning. *Review of Educational Research, 76*(4), 449–475.

Nasir, N. S., Hand, V., & Taylor, E. (2008). Culture and mathematics in school: Boundaries between "cultural" and "domain" knowledge in the mathematics classroom and beyond. *Review of Research in Education, 32,* 187–240.

Nasir, N. S., & Shah, N. (2011). On defense: African American males making sense of racialized narratives in mathematics education. *Journal of African American Males in Education, 2*(1), 24–45.

Nasir, N. S., Snyder, C. R., Shah, N., & Ross, K. M. (2012). Racial storylines and implications for learning. *Human Development, 55,* 285–301.

Nielsen, R. H. (2003). How to do educational research in university mathematics? *The Mathematics Educator, 13,* 33–40.

Noble, R. (2011). Mathematics self-efficacy and African American male students: An examination of two models of success. *Journal of African American Males in Education, 2*(2), 188–213.

Nzuki, F. M. (2010). Exploring the nexus of African American students' identity and mathematics achievement. *Journal of Urban Mathematics Education, 3*(2), 77–115.

Oakes, J. (2005). *Keeping track: How schools structure inequality.* New Haven, CT: Yale University Press.

Oakes, J., Joseph, R., & Muir, K. (2003). Access and achievement in mathematics and science: Inequalities that endure and change. In J. A. Banks & C. A. Banks, (Eds.), *Handbook on research in multicultural education* (2nd ed., pp. 69–90). San Francisco, CA: Jossey-Bass.

Oakes, J. (with Ormseth, T., Bell, R., & Camp, P.). (1990). *Multiplying inequalities: The effects of race, social class, and* tracking on opportunities to learn mathematics and science. Santa Monica, CA: RAND.

Ogbu J. (1978). *Minority education and caste: The American system in cross-cultural perspective.* New York, NY: Academic Press.

Omi, M., & Winant, H. (1986). *Racial formation in the United States: From the 1960s to the 1980s.* New York, NY: Routledge.

Omi, M., & Winant, H. (1994). *Racial formation in the United States.* New York, NY: Routledge.

Oppland-Cordell, S., & Martin, D. B. (2015). Identity, power, and shifting participation in a mathematics workshop: Latin@ students' negotiation of self and success. *Mathematics Education Research Journal, 27*(1), 21–49.

Orfield, G., Frankenberg, E., & Lee, C. (2003). The resurgence of school segregation. *Educational Leadership, 60*(4), 16–20.

Parks, A. N., & Schmeichel, M. (2011, April). *Theorizing of race and ethnicity in mathematics education literature.* Paper presented at the annual meeting of the American Educational Research Association, New Orleans, LA.

Parks, A. N., & Schmeichel, M. (2012). Obstacles to addressing race and ethnicity in the mathematics education literature. *Journal for Research in Mathematics Education, 43*(3), 238–252.

Peterson, P., Fennema, E., Carpenter, T. (1991). Using children's mathematical knowledge. In B. Means & M. Knapp (Eds.), *Teaching advanced skills to educationally disadvantaged students* (pp. 103–128). Washington, DC: U.S. Department of Education.

Pierce, C. M. (1995). Stress analogs of racism and sexism. In C. V. Willie, P. P. Rieker, B. M. Kramer, & B. S. Brown (Eds.), *Mental health, racism, and sexism* (pp. 277–293). Pittsburgh, PA: University of Pittsburgh Press.

Pierce, C., Carew, J., Pierce-Gonzalez, D., & Willis, D. (1978). An experiment in racism: TV commercials. In C. Pierce (Ed.), *Television and education* (pp. 62–88). Beverly Hills, CA: Sage.

Pollock, M. (2008). *Everyday antiracism: Getting real about race in school.* New York, NY: The New Press.

Powell, A. (2002, April). Ethnomathematics and the challenges of racism in mathematics education. In P. Valero & O. Skovsmose (Eds.), *Proceedings of the Third International Mathematics Education and Society Conference* (pp. 17–30). Copenhagen, Denmark: Centre for Research in Learning Mathematics.

Reason, P., and Bradbury, H. (Eds.). (2008). *The Sage handbook of action research: Participative inquiry and practice.* Thousand Oaks, CA: Sage.

Roediger, D. R. (2006). *Working toward Whiteness: How America's immigrants became White: The strange journey from Ellis Island to the suburbs.* New York, NY: Basic Books.

Secada, W. (1992). Race, ethnicity, social class, language, and achievement in mathematics. In D. Grouws (Ed.), *Handbook of research in mathematics teaching and learning* (pp. 623–660). New York, NY: Macmillan.

Shade, B. J. (1982). Afro-American cognitive style: A variable in school success? *Review of Educational Research, 52*(2), 219–244.

Shah, N. (2013). *Racial discourse in mathematics and its impact on student learning, identity, and participation* (Unpublished doctoral dissertation). University of California, Berkeley.

Silver, E. A., & Stein, M. K. (1996). The Quasar Project: The "revolution of the possible" in mathematics instructional reform in urban middle schools. *Urban Education, 30*(4), 476–521.

Skovsmose, O. & Valero, P. (2001). Breaking political neutrality: The critical engagement of mathematics education with democracy. In B. Atweh, H. Forgasz, & B. Nebres (Eds.), *Sociocultural research on mathematics education: An international perspective* (pp. 37–55). Mahwah, NJ: Lawrence Erlbaum.

Smedley, A., & Smedley, B. D. (2005). Race as biology is fiction, racism as a social problem is real: Anthropological and historical perspectives on the social construction of race. *American Psychologist, 60*(1), 16.

Solorzano, D., Ceja, M., & Yosso, T. (2000). Critical race theory, racial microaggressions and campus racial climate: The experiences of African American college students. *Journal of Negro Education, 69,* 60–73.

Solorzano, D. G., & Yosso, T. J. (2002). Critical race methodology: Counter-storytelling as an analytical framework for education research. *Qualitative Inquiry, 8*(1), 23–44.

Spencer, J. A. (2009). Identity at the crossroads: Understanding the practices and forces that shape African American success and struggle in mathematics. In D. B. Martin (Ed.), *Mathematics teaching, learning and liberation in the lives of Black children* (pp. 200–230). New York, NY: Routledge.

Spencer, J. A., Park, J., & Santagata, R. (2010). Keeping the mathematics on the table in urban mathematics professional development: A model that integrates dispositions toward students. In M. Foote (Ed.), *Mathematics teaching and learning in K-12: Equity and professional development* (pp. 199–218). New York, NY: Palgrave Macmillan.

Steele, C. M. (1997). A threat in the air: How stereotypes shape intellectual identity and performance. *American Psychologist, 52*(6), 613.

Steele, C. M., & Aronson, J. (1995). Stereotype threat and the intellectual test performance of African Americans. *Journal of Personality and Social Psychology, 69*(5), 797.

Stiff, L. V., & Harvey, W. B. (1988). On the education of Black children in mathematics. *Journal of Black Studies, 19*(2), 190–203.

Stinson, D. W. (2006). African American male adolescents, schooling (and mathematics): Deficiency, rejection, and achievement. *Review of Educational Research, 76*(4), 477–506.

Stinson, D. (2008). Negotiating sociocultural discourses: The counter-storytelling of academically (and mathematically) successful African American male students. *American Educational Research Journal, 45*(4), 975–1010.

Stinson, D. W. (2009). Negotiating Sociocultural Discourses: The Counter-Storytelling of Academically and Mathematically Successful African American Male Students. In D. B. Martin (Ed.), *Mathematics teaching, learning, and liberation in the lives of Black children* (pp. 265–288). New York, NY: Routledge.

Stinson, D. (2011). "Race" in mathematics education: Are we a nation of cowards? *Journal of Urban Mathematics Education, 4*(1), 1–16.

Stinson, D. W. (2013). Negotiating the "White male math myth": African American male students and success in school mathematics. *Journal for Research in Mathematics Education, 44*(1), 69–99.

Stinson, D. W., & Bullock, E. C. (2012). Critical postmodern theory in mathematics education research: A praxis of uncertainty. *Educational Studies in Mathematics, 80*(1–2), 41–55.

Stinson, D. W., Jett, C. C., & Williams, B. A. (2013). Counterstories from mathematically successful African American male students: Implications for mathematics teachers and teacher educators. In J. Leonard & D. B. Martin (Eds.), *The brilliance of Black children in mathematics: Beyond the numbers and toward new discourse* (pp. 221–245). Charlotte, NC: Information Age.

Stinson, D. W., & Walshaw, M. (2017). Exploring different theoretical frontiers for different (and uncertain) possibilities in mathematics education research. In J. Cai (Ed.), *Compendium for research in mathematics education* (pp. 128–155). Reston, VA: National Council of Teachers of Mathematics.

Strutchens, M. E., & Silver, E. A. (2000). NAEP findings regarding race/ethnicity: Students' performance, school experiences, and attitudes and beliefs. In E. A. Silver & P. A. Kenney (Eds.), *Results from the seventh mathematics assessment of the National Assessment of Educational Progress* (pp. 45–72). Reston, VA: National Council of Teachers of Mathematics.

Strutchens, M. E., & Westbrook, S. K. (2009). Opportunities to learn geometry: Listening to the voices of three African American high school students. In D. B. Martin (Ed.), *Mathematics teaching, learning, and liberation in the lives of Black children* (pp. 249–264). New York, NY: Routledge.

Sue, D. W. (2010). *Microaggressions in everyday life: Race, gender, and sexual orientation.* New York, NY: John Wiley & Sons.

Sue, D. W., Capodilupo, C. M., Torino, G. C., Bucceri, J. M., Holder, A., Nadal, K. L., & Esquilin, M. (2007). Racial microaggressions in everyday life: Implications for clinical practice. *American Psychologist, 62*(4), 271–286.

Tate, W. (1995a). Mathematics communication: Creating opportunities to learn. *Teaching Children Mathematics, 1*(6), 344–349, 369.

Tate, W. (1995b). Returning to the root: A culturally relevant approach to mathematics pedagogy. *Theory Into Practice, 34*(3), 166–173.

Tate, W. F. (1997). Race-ethnicity, SES, gender, and language proficiency trends in mathematics achievement: An update. *Journal for Research in Mathematics Education, 28*(6), 652–680.

Tate, W. F. (2008). The political economy of teacher quality in school mathematics: African American males, opportunity structures, politics and method. *American Behavioral Scientist, 51*(7), 953–971.

Taylor, E. V. (2009). The purchasing practice of low-income students: The relationship to mathematical development. *The Journal of the Learning Sciences, 18*(3), 370–415.

Terry, L. (2011). Mathematical counterstory and African American male students: Urban mathematics education from a critical race theory perspective. *Journal of Urban Mathematics Education, 4*(1), 23–49.

Terry, C. L., Sr., & McGee, E. O. (2012). "I've come too far, I've worked too hard": Reinforcement of support structures among Black male mathematics students. *Journal of Mathematics Education at Teachers College, 3*(2), 73–85.

Thompson, L., & Davis, J. (2013). The meaning high-achieving African-American males in an urban high school ascribe to mathematics. *The Urban Review, 45*(4), 490–517.

Treisman, U. (1992). Studying students studying calculus: A look at the lives of minority mathematics students in college. *The College Mathematics Journal, 23*(5), 362–372.

Trepagnier, B. (2010). *Silent racism: How well-meaning White people perpetuate the racial divide.* Boulder, CO: Paradigm.

Valero, P., & Zevenbergen, R. (Eds.). (2004). *Researching the socio-political dimensions of mathematics education: Issues of power in theory and methodology.* Mathematics Education Library, Vol. 35. New York, NY: Kluwer Academic.

Valoyes-Chavez, L., & Martin, D. B. (in press). Exploring racism inside and outside mathematics classrooms in two different contexts: Colombia and the United States. *Intercultural Education.*

Varelas, M., Martin, D. B., & Kane, J. (2012). Content learning and identity construction (CLIC): An interpretive framework to strengthen African American students' mathematics and science learning in urban elementary schools. *Human Development, 55* (5–6), 319–339.

Villasenor, A. (1991). *Teaching the first grade curriculum from a problem-solving perspective* (Unpublished doctoral dissertation). University of Wisconsin, Milwaukee.

Walker, E. N. (2012). *Building mathematics learning communities: Improving outcomes in urban high schools.* New York, NY: Teachers College Press.

Walker, E. N. (2014). *Beyond Banneker: Black mathematicians and the paths to excellence.* Albany, NY: SUNY Press.

Walshaw, M. (Ed.). (2004). *Mathematics education within the postmodern.* Charlotte, NC: Information Age.

Walshaw, M. (2013). Post-structuralism and ethical practical action: Issues of identity and power. *Journal for Research in Mathematics Education, 44*(1), 100–118.

Weissglass, J. (2002). In Focus... Inequity in mathematics education: Questions for educators. *The Mathematics Educator, 12*(2), 34–39.

Windschuttle, K. (2004). *The White Australia policy.* Sydney, Australia: Macleay Press.

Yosso, T. J. (2006). *Critical race counterstories along the Chicana/Chicano educational pipeline.* New York, NY: Routledge.

Yosso, T. J., Smith, W. A., Ceja, M., & Solórzano, D. G. (2009). Critical race theory, racial microaggressions, and campus racial climate for Latina/o undergraduates. *Harvard Educational Review, 79*(4), 659–691.

Zuberi, T., & Bonilla-Silva, E. (Eds.). (2008). *White logic, White methods: Racism and methodology.* Lanham, MD: Rowman & Littlefield.

23 数学教育研究中的身份

珍妮弗·M.兰格-奥苏纳
美国斯坦福大学
因迪哥·埃斯蒙德
加拿大多伦多大学
译者：陈倩
　　　四川师范大学数学科学学院

身份是人们用来捕捉他们凭直觉对自己和他人的了解的一个概念。每个人都有自己的独特之处，可被称为自我概念或个性。然而，人们也有与其所属社会群体的其他成员共享的东西：比如女孩子、白人或富人。他们与这些社会群体的关系依赖于历史时间和地理位置，这些因素确定了这些群体的内涵以及群体成员对群体的体验。身份是一个被用来描述这些复杂现象的概念。在这一章，我们将区分个人的身份与成员的身份，前者是个人在与某一社会情境的关联中形成的，比如数学课堂上"挣扎的学生"或篮球队中的"明星队员"，而后者则是以种族、性别和语言相关的社会成员关系为基础的。

在北美的学术界，关于数学教学中身份的研究是相当新的。可以说，从大约15年前博勒和格里诺（2000）、马丁（2000）以及纳西尔（2002）的研究开始，数学教育中的身份研究主要是从社会文化的视角来看，认为身份和学习从根本上是相互关联的，而且产生于社会实践（Lave & Wenger, 1991；Wenger, 1999）。这类研究帮助我们阐明了学习者通过在课堂内外参与各种文化实践或社团实践，从而形成与数学相关的身份的种种方式。社会文化的视角将身份从根本上与学习联系起来，因为学习被概念化为某种特定的人的变成过程，而这一过程与数学活动有关，包括活动涉及的各种技能、知识库和社会实践。博勒（2002b）进一步思考了数学身份是如何变

得性别化的，而纳西尔（2002）和马丁（2000）关注了种族化的数学身份。因此，从一开始，研究者们就已经关注于理解数学中的个人身份（如，"有能力的""能力强的""正常的"；Black, Mendick, Rodd, &Solomon, 2009，第5页）是如何与更广的成员身份相互联系的。

我们回顾文献的目的是要梳理当前数学教育研究中对身份的各种不同定义，并探索背后的理论框架。由于一些原因，这篇文献综述并不能做到十分详尽。首先，也是最简单的原因是，身份研究在数学教育中是一个新生的研究领域，一张理论框架图将比一份详细的研究成果集更能推动该领域的发展。此外，由于对身份概念的定义和表述方式缺乏连贯性，很难将数学教育中关于身份研究的成果综合起来。这种不连贯性的增加，是因为发表的成果常常没有定义身份，或者只是以模糊或非操作化的方式来定义身份。

其次，由于我们最感兴趣的是身份与数学思维、教学和学习之间的关系，所以我们关注的是那些假定身份是出现在社会情境中并随着时间的推移可以变化的定义和理论。身份一词的某些用法是针对一个人的内部而言，类似于自我概念或内在的自我意识。尽管关注的是个人，这些身份的观念也捕捉了通过参与诸如课堂的社会情境而形成的东西。然而，这种与社会情境的根本关联在身份发展的理论中通常是模糊的。因此，在我们关注的理论中，虽然焦点是放在个人身上，但是身份发展从根本

23 | Identity in Research on Mathematics Education

JENNIFER M. LANGER-OSUNA
Stanford University, United States

INDIGO ESMONDE
University of Toronto, Canada

Identity is a construct that people use to capture something they intuitively grasp about themselves and other humans. There is something unique about each individual, something that could be called a self-concept or a personality. And yet, there is something people share with other members of their social groups: girl-ness or white-ness or wealth-ness. Their relationships to these social groups depend on historical time and geographic place, which shape what these groups mean and how group members experience them. Identity is a construct that has been used to describe these complex phenomena. In this chapter, we make a distinction between *individual* identities—those that an individual develops in relation to a social context, such as "struggling student" in the mathematics classroom or "star player" on the basketball team—and *membership* identities—those based on social membership related to race, gender, and language.

Studies of identity in mathematics teaching and learning are fairly new in North American scholarship. Arguably starting about a decade and a half ago with Boaler and Greeno (2000), Martin (2000), and Nasir (2002), identity research in mathematics education has largely been approached from a sociocultural perspective, which frames identity and learning as fundamentally linked to one another and arising from participation in social practices (Lave & Wenger, 1991; Wenger, 1999). This research has helped illuminate the ways in which learners develop mathematics-linked identities through their participation in varied cultural practices or communities of practice, both in and out of the classroom. This perspective frames identity as fundamentally linked to learning because learning is conceptualized as the *process of becoming* a certain kind of person in relation to mathematical activity, including its varied skills, knowledge base, and social practices. Boaler (2002b) went on to consider how mathematical identities become gendered, whereas Nasir (2002) and Martin (2000) focused on racialized mathematical identities. Thus, from the start researchers have been concerned about understanding how individual identities in mathematics (e.g., "able," "competent," "normal"; Black, Mendick, Rodd, & Solomon, 2009, p. 5) are related to broader membership identities.

Our goal for this review is to tease apart the different definitions of identity currently at play in mathematics education research and explore the underlying theoretical frameworks. This is not an exhaustive review for several reasons. First and most simply, the study of identity is a nascent area of research in mathematics education, and a map of the theoretical terrain will advance the field more than would a detailed collection of findings. Further, it is difficult to synthesize the findings of research on identity in mathematics education due to the lack of coherence in how the construct is defined and framed. This incoherence is increased because published works frequently do not define identity or define identity in a vague or unoperationalized way.

Second, because we are most interested in the relationship between identity and mathematics thinking, teaching, and learning, we focus on definitions and theories that posit identity as occurring in social contexts and as changeable over time. Some uses of the term *identity* refer to something that is interior to a person, like a self-concept or inner sense of self. Although focused on the individual, these conceptions of identity capture something that nevertheless develops through participation in social contexts, like classrooms. However, this fundamental connection to social context is often obscured in theories of identity development. For this reason, we focus on

上是与身份形成的环境相联系的。

本章的结构如下：首先，我们将讨论关于个人身份发展的研究，特别地，将关注身份研究的四个理论方法：（1）话语法（后结构主义法），（2）定位法，（3）叙事法，（4）心理分析法。然后，我们会讨论有关个人身份与成员身份（如，作为种族或性别群体成员的身份）关系的研究成果。针对身份的每一种理论方法，我们先给出定义、理论基础和典型的方法论，再思考每一种方法对于我们如何概念化和组织数学教与学空间带来的启示。最后，随着身份这一新生研究领域的成长和更多地进入到数学教育的主流对话中，我们也对该研究领域提出了一些建议。

身份研究的四个理论方法

我们先较为广泛地讨论个人身份研究的理论方法。尽管我们将这些方法刻画为四种不同的理论框架，但是这些框架之间存在着很多关联和牵扯。必须承认的是，我们的讨论掩盖了这个领域的某些复杂性，但是我们发现对四种不同方法的划分是对当前研究进行概念化的一种十分有用的方式。我们先借鉴福柯（1982）的研究成果来论述话语法（后结构主义法），然后主要引用拉弗和温格（1991），霍兰、拉奇科特、斯金纳和凯恩（1998），以及斯法德和普鲁萨克（2005）的研究成果来讨论各种社会文化理论，并且关注已经出现的身份的两种变化形式：即定位性身份和叙事性身份。[1] 最后，我们讨论了心理分析理论，虽然它是北美数学教育研究中很少用到的一种方法，但它提供了独特的见解。表23.1总结了这些方法以及它们对数学身份和数学教育的启示。

数学自我的话语产生

身份的话语法或后结构主义理论描述的是"个人是复杂多样且支离破碎的主体，该主体受制于——但非决定于——构成这个人的各种社会文化话语（Butler，1990/1999）"（Stinson，2013，第77页）。布莱克等人（2009）描述了后结构主义方法是如何构建数学教育研究的：

> 我们不问"x是真的吗？"，而问"什么使得x可能发生？"和"它的发生有什么影响？"比如，我们不问"一些人在数学上比其他人更有天分，这是真的吗？"，而问"是什么使得人们可能会认为一些人在数学上比其他人更有天分？"和"这种看法对我们思考数学、自我、他人及其相互联系的方式有什么影响？"类似地，重新表述我们对选择进行分析的问题，我们必须从"为什么某人（不）选择继续学习数学？"转移到"是什么使我们可能认为某人（不）选择做数学？"和"这种看法有什么影响？"……这是一种焦点的转换，突出的不是个人和他们的"选择"及"能力"，而是人们通过各种关系模式、物质性等被组合成（不）选择数学和（没）有能力的方式（第72—73页）。

在方法论上，采用话语法或后结构主义理论的研究通常依赖于诸如访谈等各种叙事，再将它们与"更广阔的社会文化情境"（Stinson，2013，第87页）联系起来分析。这些情境包括生活经验和在当地环境（如学校）的参与，以及围绕数学、教师和学生的更广泛的交流（例如Stinson，2013；Walshaw，2013）。

比如，沃尔肖（2013）在对参加教学实习的职前教师进行分析时，"突出了学习教学的政治性和策略性特征"（第79页）。政策制定者、政府以及学校对教学的管理提供了一种语言，通过使用（或者不用）这种语言，教师可以建构自己的身份认同。这些规章制度通过权力的组织使某些看待和从事教学的方式成为可能。权力在每天的课堂生活中运作，通过各种互动来塑造教学身份，而这些互动是受政策和课程现实引发的讨论所影响的。

这些研究阐明了更广的权力结构如何产生以及限制教师和学生身份的各种可能性。在结构与能动性的相互作用下，话语法或后结构主义的研究工作揭示了结构如何塑造着人们发挥能动性的这一场所。在下一部分，我们将关注社会文化方法，这些方法也考虑了更广的权力结构，但突出的是个体能动性在操纵权力关系中的作用。

theories in which identity development is fundamentally bound to the contexts in which the identity is enacted even when the focus remains on the individual.

The organization of the chapter is as follows. We begin with research that considers individual identity development. In particular, we focus on four theoretical approaches to identity: (1) discursive (poststructural), (2) positional, (3) narrative, and (4) psychoanalytic. We then discuss related research that examines the relationships between individual identities and membership identities (e.g., identities as members of racialized or gendered groups). For each approach to identity, we provide definitions, theoretical underpinnings, and typical methodologies. We then consider the implications of each approach for how we conceptualize and organize mathematical teaching and learning spaces. We close the chapter by considering recommendations for this nascent area of research as it grows and increasingly enters mainstream conversations about mathematics education.

Four Theoretical Approaches to the Study of Identity

We begin by considering the breadth of theoretical approaches to the study of individual identities. Although we characterize these approaches as four distinct theoretical frameworks, there are many connections and tensions between them. We acknowledge that our discussion masks some of the complexity of the field, but we find that the division into four distinct approaches is a useful way to conceptualize the ongoing research. We start with what have been termed discursive or poststructural theories of identity, drawing on the work of Foucault (1982). We then discuss sociocultural theories—drawing primarily on the work of Lave and Wenger (1991); Holland, Lachicotte, Skinner, and Cain (1998); and Sfard and Prusak (2005)—and focus on two variations on identity that have come to the fore: positional identities and narrative identities.[1] We end with psychoanalytic theory, an approach that is infrequently used in North American mathematics education research but nevertheless offers unique insights. Table 23.1 offers a summary of these approaches and their implications for both mathematical identities and mathematics education.

Discursive Production of Mathematical Selves

Discursive or poststructural theories of identity describe "the person as a multiplicitous and fragmented subject who is subjugated—but not determined—by the sociocultural discourses that constitute the person (Butler, 1990/1999)" (Stinson, 2013, p. 77). Black et al. (2009) offer a characterization of how a poststructural approach frames inquiries into mathematics education:

> We do not ask "Is x true?," but rather "What makes x possible?" and "What are its effects?" For example, we do not ask "Is it true that some people have greater natural ability in mathematics than others?," but rather "What makes it possible to think about some people having greater natural ability in mathematics than others?" and "What effects does this idea have" on the ways we think of mathematics, ourselves, others, and the relations among these? Similarly, reformulating our analytic questions around choice, we must move from, "Why did someone choose (not to) continue studying mathematics?" to "What makes it possible for us to think of someone as making a choice (not) to do mathematics?" and "What effects does this have?" ... This is a switch of focus that foregrounds not the individual and their "choices" and "abilities," but the ways that people are assembled as (not) choosing mathematics and as un/able through patterns of relationships, materiality, and so on. (pp. 72–73)

Methodologically, research that draws from discursive or poststructural theories typically relies on narratives, such as interviews, which are then analyzed in relation to the "broader sociocultural context" (Stinson, 2013, p. 87). These broader contexts include life experiences and participation in local environments (e.g., schools), as well as broader discourses around mathematics, teachers, and students (e.g., Stinson, 2013; Walshaw, 2013).

For example, Walshaw (2013) "foregrounded the political and strategic nature of learning to teach" (p. 79) in an analysis of preservice teachers engaged in practicum teaching. The regulation of teaching by policy makers, governments, and schools provides a language through which (and against which) teachers construct identities. These regulations make some ways of seeing and engaging in teaching possible through the organization of power. Power operates in the routines of everyday classroom life, shaping teaching identities through interactions influenced by the discourses made available to them through policy and curricular realities.

This research illuminates how broader structures of power enable and constrain possibilities for teacher and student identities. In the interplay between structure and agency, discursive or poststructural work foregrounds how structures shape the arenas in which people exercise agency. In the next section, we focus on sociocultural approaches, which also consider broader structures of power, but foreground the role of individual agency in navigating relationships of power.

表23.1 身份研究的理论方法及其性质以及对数学教育的启示

	后结构主义理论	定位理论	叙事理论	心理分析理论
身份的性质	广泛的话语使某些主体地位得以存在,并限制了人们运用能动性的方式。人们联系这些话语来进行身份建构。	个人在互动中,特别是通过谈话,建构与他人相关的主体地位,从而对特定社会背景下的人做出判断。相对于后结构主义理论而言,定位理论强调能动性。	个人和他们周围的人在理解各种社会环境之间的经历时创造关于他们自己的故事。	个人被强烈的无意识的恐惧和欲望力量所控制,并在处理这些内在力量时身份得以发展。
对数学身份的启示	数学的身份确认从根本上是由诸如教育政策、资金分配、课程选择等更广的权力结构所塑造的。	数学的身份是通过人们就数学空间中的自我和他人做出各种判断而形成的。	当人们理解他们的数学体验,形成成功和失败、归属或远离的故事时,数学的身份得以发展。	当人们处理他们依附于数学体验的焦虑,恐惧,欲望以及渴望时,数学身份得以发展。
对数学教育研究的启示	研究强调了思考更广的权力结构在形成个人如何认同(从而参与)数学教与学的可能性中,组织作用的重要性。	研究强调了课堂(及相关)谈话对如何塑造数学能力感和归属感的重要性。	研究强调了特定类型(积极或消极)的体验是如何形成数学教育以及其他领域(例如STEM方向)的轨迹的。	研究强调了个人与数学教与学之间所形成的情感联系,以及(不)参与数学如何成为处理这些焦虑和渴望的方式。

数学关联世界中的地位身份

社会文化理论或情境理论将身份与个人在社会实践中的参与形式联系起来(Lave & Wenger, 1991)。在关注身份的数学教育研究中,伴随着定位理论,主要是通过戴维斯和哈勒(1999)及霍兰等人(1998)的研究工作,这类研究已经越来越多。霍兰等人认为:

> 我们所说的方言,在演讲中采用的正式程度,我们做的事情,我们去的地方,我们表达的感情,以及我们穿戴的服装都可以视为相对于我们互动对象而言的社会类别和特权地位的宣称和识别的指标。(第127页)

分析师的工作是要考虑如何用大量的数据——谈话、行动及具体空间——来检验正在被建构或在互动中显现的地位身份,并辨别权力和特权是如何被分配、实施及接纳的(Davies & Harré, 1999)。

定位法认为身份是通过社会互动来建构的,社会互动表明了一个人相对于一个社会情境中的他人来说是谁(Davies & Harré, 1999)。当人们彼此互动时,他们用行动来定位自己应该展现出的某些特定的品质(如友好、聪明或者权威)或扮演某个特定的角色(如教师、学生或团队领导者)。人们也用行动定位他人,接受或拒绝关于他们自己或他人的定位行为。虽然个体总是能动的,但是个人没有能力完全确定他们自己的身份。事实上,人们是在某个特定的团体中通过一个持续的定位行为过程来互动地协商他们的身份(Engle, Langer-Osuna, & McKinney de Royston, 2014)。这些行为来自于当地规范和活动结构,它们在一定程度上决定了什么是可能的。

身份的定位理论与基于福柯研究的话语法或后结构主义方法有很多共同之处。事实上,霍兰等人(1998)将福柯对权力和话语的关注整合于他们描绘的世界和身份的理论框架之中。[2]我们这里所做的区别是双重的:(1)个体能动性在社会文化方法中的相对前置;(2)更密切地关注局部描绘的世界和身份的建构(社会文化),

TABLE 23.1. Theoretical Approaches to Identity and Their Natures and Implications for Mathematics Education

	Poststructural theory	Positioning theory	Narrative theory	Psychoanalytic theory
Nature of identity	Broad discourses make certain subject positions available and constrain the ways in which people exercise agency. People construct identifications in relation to these discourses.	Individuals construct subject positions in relationship with others during interactions, in particular through talk, that make claims about who people are within particular social contexts. Relative to poststructural theories, positioning theories emphasize agency.	Individuals and the people around them create stories about themselves as they make sense of their experiences across social settings.	Individuals are governed by strong unconscious forces of fear and desire and develop identities as they cope with these inner forces.
Implication for mathematical identities	Mathematical identifications are fundamentally shaped by broader power structures, such as education policy, funding allocation, curricular options, and so on.	Mathematical identities develop through the claims people make about themselves and others in mathematical spaces.	Mathematical identities develop as people make sense of their experiences with mathematics and develop stories of success and failure and belonging or distance.	Mathematical identities develop as people cope with their anxieties, fears, desires, and longings attached to mathematical experiences.
Implication for mathematics education research	Research highlights the importance of considering the organizing role of broader power structures in shaping possibilities for how individuals identify with (and thus engage in) mathematics teaching and learning.	Research highlights the importance of how classroom (and related) talk shapes identities of competence and belonging in mathematics.	Research highlights how particular kinds of (positive or negative) experiences shape trajectories of mathematics education and beyond (i.e., the STEM pipeline).	Research highlights the emotional relationships individuals develop with mathematics teaching and learning and how (dis)engagement with mathematics are ways of coping with these anxieties and desires.

Positional Identities in Mathematics-Linked Worlds

Sociocultural or situative perspectives link identities to individuals' forms of participation in social practices (Lave & Wenger, 1991). In mathematics education research focused on identity, this line of work has been increasingly augmented with positioning theory, primarily through the work of Davies and Harré (1999) and Holland et al. (1998). Holland et al. argue that

> the dialect we speak, the degree of formality we adopt in our speech, the deeds we do, the places we go, the emotions we express, and the clothes we wear are treated as indicators of claims to and identification with social categories and positions of privilege relative to those with whom we are interacting. (p. 127)

The analyst's job is to consider this vast array of data—talk, action, and embodied space—to examine the positional identities being constructed or made salient in interaction and to discern how power and privilege are distributed, enacted, and taken up (Davies & Harré, 1999).

Positional approaches define identity as constructed through social interactions that make claims about who a person is in relation to others in a social context (Davies & Harré, 1999). As people interact with one another, they act to position themselves as displaying certain qualities (e.g., friendly, smart, authoritative) or as having a particular role (e.g., teacher, student, group leader). People also act to position others and accept or reject acts of positioning about themselves or others. Although always agentive, individuals do not have the capacity to solely determine their own identity. Rather, people interactionally negotiate their positions within a particular community through an ongoing process of positional acts (Engle, Langer-Osuna, & McKinney de Royston, 2014). These acts draw from local norms and activity structures that partially determine what is possible.

Positional theories of identity have much in common with discursive or poststructural approaches that build on the work of Foucault. Indeed, Holland et al. (1998) integrated Foucault's attention to power and discourse into their theoretical frame of figured worlds and identities.[2] The distinction we make here is twofold: (1) the relative foregrounding of individual agency in sociocultural

而非广泛的社会话语及权力和控制的循环（话语的后结构主义）。

在方法论上，定位身份的研究各不相同，取决于它们是更多地关注所谓的微身份（"一个人在某一时刻的定位"；Wood，2013，第780页）还是厚身份，厚身份是微身份随时间累积而建构的一种看似稳定的身份，即一个学生越来越可能被一种特定的方式所定位（如，麻烦制造者；Wortham，2006；多米诺骨牌的玩家；Nasir，2005；或一场辩论的赢家；Engle等，2014）。微身份的研究考虑的是，特定的定位行为如何限制了课堂活动中的学习机会。一些微身份会赋予数学能力，比如被定位为一名数学的解释者（Wood，2013）、专家或促进者（Esmonde，2009）。其它微身份赋予的则是数学无能，比如被定位为一名卑微的佣工（某个接受来自同龄人的社会指令的人；Wood，2013）或新手（Esmonde，2009）。这些微身份能迅速地转变，提供一个关于互动如何能就地产生或限制学习机会的即时理解。关于厚身份的研究聚焦于一系列随时间推移的定位行为，注意到它们是如何建立在过去的互动基础之上的，从而强调了各种数学学习者的产生（Langer-Osuna，2011，2015；Nasir，2005；Wortham，2006）。比如，纳西尔（2005）关注了非洲裔美国学生的数学思维，他们的数学思维在校外的情境中更强，他们在校外环境中比在数学课堂上体会到更强的归属感。能力和归属感的认同带来了更加重要的参与和更强的推理表现。

定位法认为教师和学生是课堂的共同参与者，他们的地位身份是共同演变的。这类研究讲述了很多关于教师（以及其他学生）的定位行为是如何使学生获得某一些定位身份，而非其它身份的。通过让学生参与各种数学意义建构的活动，体现其能力，教师们能使积极的[3]定位行为更可能发生在更多的学生身上（Cobb, Gresalfi, & Hodge, 2009）。支持学生发挥概念能动性和智力权威的课堂规范能使学生将自己看成是富有成效的数学实干家（Boaler, 2002a；Boaler & Greeno, 2000；Cobb等, 2009；Gresalfi & Cobb, 2006；Gresalfi, Martin, Hand, & Greeno, 2009；Langer-Osuna, 2015）。博勒和斯特普尔斯（2008）认为，被定位为能动者的学生往往更喜欢数学，坚持更久，学习更多。这一发现与兰格-奥苏纳（2015）关于一个数学课堂的研究相呼应，该课堂给予学生很大的自主权，对学生参与、身份认同和学习等方面产生了积极的影响。

定位分析的视角研究了解释特定数学关联身份发展的话语机制。特别地，此类工作阐明了学生是如何接纳、改变和抵制特定数学空间中提供的各种参与、学习和身份确认的机会的，它也阐明了能使更大范围的学生建构积极的数学身份，更深入地参与到学习活动中，并更经常地利用学习机会的数学学习空间种类。因此，这类工作使得营造公平并富有成效的数学学习空间的设计研究成为可能。

然而，这种方法也存在一些局限性。定位分析通常包含对一个具体互动案例（可能仅仅持续几分钟）作深入研究，去记录正在发挥作用的话语机制。这种深入分析的劳动密集型特征使得很难在参与者和学习空间之间看到模式。身份研究的这一理论方法适用于关注学习过程的细节，而不适用于对学习者群组进行概括。由于这类研究涉及所考虑的许多案例，一个元综述将有助于获得这一领域中对跨群体和跨情境的与过程相关的模式的认识。然而，这需要研究者让每个案例中使用的理论假设和方法技巧更加透明。接下来我们将要讨论的叙事法能使这些模式的检验成为可能，但对特定教学技巧或策略的影响只能提供较少的信息。

叙事性身份：来自数学教师和学生的故事

当人们反思和理解他们在数学空间中的体验时，他们叙述他们自己关于数学的故事。斯法德和普鲁萨克（2005）认为，身份应该被定义为我们所叙述的关于自己和他人的故事。他们说，"冗长的思考引导我们将身份与关于人的故事等同起来。不，没错：我们不是说在故事中发现身份的表达，我们是说它们本身就是故事"（第14页，原文中强调）。

approaches and (2) the closer focus on the construction of local figured worlds and identities (sociocultural) rather than on broad societal discourses and the circulation of power and control (discursive poststructural).

Methodologically, studies of positional identity differ according to whether they focus more on what have been called *micro-identities* ("the position of a person in a moment of time"; Wood, 2013, p. 780) or on *thickened* identities, in which micro-identities accumulate over time to construct a seemingly stable identity, that is, when a student becomes more and more likely to be positioned in a particular way (e.g., as a trouble-maker; Wortham, 2006; as a dominoes player; Nasir, 2005; or as the winner of a debate; Engle et al., 2014). Studies of micro-identities consider how particular acts of positioning constrain opportunities for learning during classroom activities. Some micro-identities afford mathematical competence, such as being positioned as a mathematical explainer (Wood, 2013), expert, or facilitator (Esmonde, 2009). Other micro-identities afford mathematical incapability, such as being positioned as a menial worker (someone who accepts social directives from peers; Wood, 2013) or novice (Esmonde, 2009). These micro-identities can shift rapidly, offering a moment-to-moment understanding of how interactions enable or constrain opportunities for learning in situ. Studies of thickened identities focus on a series of acts of positioning over time, noting how they build upon past interactions in ways that highlight the production of kinds of mathematics learners (Langer-Osuna, 2011, 2015; Nasir, 2005; Wortham, 2006). Nasir (2005), for instance, focused on African American students' mathematical thinking, which was more robust in out-of-school contexts where they experienced greater belonging than in the mathematics classroom. Identities of competence and belonging beget more central participation and robust reasoning.

Positional approaches consider teachers and students as coparticipants in the classroom, and their positional identities coevolve. The research has much to say about how positioning acts by teachers (and other students) make some positional identities available for students but not others. Teachers can make positive[3] acts of positioning more likely to occur for more students by framing a range of student participation in mathematical sense making as enactments of competence (Cobb, Gresalfi, & Hodge, 2009). Classroom norms that support students in taking up conceptual agency and intellectual authority enable students to see themselves as productive doers of mathematics (Boaler, 2002a; Boaler & Greeno, 2000; Cobb et al., 2009; Gresalfi & Cobb, 2006; Gresalfi, Martin, Hand, & Greeno, 2009; Langer-Osuna, 2015). Boaler and Staples (2008) argue that students who are positioned as agentive tend to like mathematics better, persevere longer, and learn more. This finding is echoed by Langer-Osuna's (2015) study of a mathematics classroom that afforded students a great deal of autonomy, resulting in positive consequences for engagement, identity, and learning.

Positional perspectives study the discursive mechanisms that explain the development of particular mathematics-linked identities. In particular, this work illuminates how students take up, alter, and resist the opportunities for engagement, learning, and identification that are offered in particular mathematical spaces. It also illuminates the kinds of mathematical learning spaces that enable a greater range of students to construct positive mathematics identities, engage more deeply in learning activities, and take up learning opportunities more often. In doing so, this work enables possibilities for design-based research into equitable and productive mathematics learning spaces.

However, there are also limitations to this approach. Positioning analysis typically includes a deep dive into a specific case of interaction, which may last only a few minutes, to document the discursive mechanisms at play. The labor-intensive nature of this type of deep analysis makes it difficult to see patterns across participants and learning spaces. This theoretical approach to identity lends itself to focusing on the details of the learning process rather than generalizing across groups of learners. As this work builds in number of cases considered, a metasynthesis would help the field gain a sense of process-related patterns across populations and contexts; however, this requires researchers to be far more transparent about the theoretical assumptions and methodological techniques used in each case. A narrative approach, which we discuss next, enables examinations of these patterns, but offers less data about the impact of particular techniques or strategies for teaching and learning.

Narrative Identities: Stories From Students and Teachers of Mathematics

As people reflect on and make sense of their experiences in mathematical spaces, they tell stories about their mathematical selves. Sfard and Prusak (2005) argue that identity should be defined *as* the stories we tell about ourselves and about other people. They state, "Lengthy deliberations led us to the decision to *equate identities with stories about persons*. No, no mistake here: We did not say that identities *were finding their expression* in stories—we said they *were* stories" (p. 14, emphasis in original).

身份即故事，或者说叙事性身份就是故事。与许多身份理论相反，这些理论假定身份是一个实际对象（如一棵树或一只脚，一个圆或一条线），而研究者的任务就是通过各种陈述（如故事）去接近它。当身份被定义为故事或"关于人的故事集"时（Sfard & Prusak, 2005, 第16页），一些使身份特别难以研究的有问题的假设就被解决了。作为故事的身份是人造的，并非是内部的和不可知的，它由社会所塑造，因时间、情境和叙述者的变化而变化，研究者可以通过诸如访谈、观察或收集书面传记等实证方法来获得。

在斯法德和普鲁萨克的表述中，身份是故事，这些故事来源于我们在构成自己生活的许多社会实践中的定位或被别人定位的经历，这似乎是合理的。换句话说，叙事性身份可以被认为是定位性身份的具体化（Wenger, 1999）。我们如何与他人一起构建、叙述以及理解这些经历，塑造着我们如何理解我们自己，并且当新框架出现时我们可能会随时间的推移而改变我们的认识。虽然斯法德和普鲁萨克（2005）认为他们的方法明显不同于温格的方法，因为温格关注的是经历而非故事，但我们相信他们都认为各种情境的参与影响着我们叙述的故事，而且那些故事有能力塑造未来参与的种种可能性。事实上，温格认为，我们对自己的参与形式赋予意义；我们塑造故事，而故事也塑造我们。

叙事性身份并不是一个单一的故事，而是一个由不同叙述者讲述给不同听众的故事集（Sfard & Prusak, 2005）。这一定义反映了身份的社会性，一个人永远无法完全控制他的身份。然而，采用叙事性身份的实证研究倾向于关注讲述自我的故事，从而隐含地构建了一个类似于自我概念的身份叙事。方法论上，叙事性身份的研究会很好地模仿定位性身份的文献，后者不仅强调一个人如何定位自己，而且强调一个人如何定位别人和被别人所定位。比如，出自学习者自己以及他们的父母和教师的关于学习者的故事集，可能会为一个给定学生作为一个数学学生的身份提供更完整的可能性。

关于学生叙事性身份的研究阐明了归属的重要性。我们常常听人说，他们"不是一个擅长数学的人"，将数学与他们自我概念的某个部分联系起来（Boaler, 2002a）。学生关于数学的故事往往关注他们是如何产生一种归属感或排斥感的（Bartholomew, Darragh, Ell, & Saunders, 2011；Rodd & Bartholomew, 2006；Solomon, 2007）。比如说，罗德和巴塞洛缪（2006）发现，学生与数学的关系不能与他们的社会和感情生活相脱离。事实上，罗德和巴塞洛缪的研究表明，一个学生的归属感与身份问题的关联比与数学成绩的关联更多（Solomon, Lawson, & Croft, 2011）。

关于教师叙事性身份的研究阐明了教师过去作为数学学习者的个人历史是如何与教师的数学教学方法相关联的（Adams, 2013；Akkoç, Yeîildere-Dmreb, & Ali Balkanlıo°luc, 2014；Teixeira & Cyrino, 2014；Williams, 2011）。这些身份，以教师叙述自己所成为的人为基础，是与教师在教室里的行为相关的（Battey & Franke, 2008）。许多关于教师身份的研究都聚焦于分析教师在特定的专业发展经历故事中的转折点（Battey & Franke, 2008；Wager & Foote, 2013）。通过检查专业发展前后所写的故事来寻找实践转变的可能性。

叙事性身份研究不仅强调了能让学生体会学校数学归属感的数学经验的重要性，而且强调了教师实施有效教学法的种种可行途径。然而，对于教师和学生身份实际上是怎样相互塑造的，我们仍然不清楚，而且，故事讲述是概括性的，这种概括对讲故事的人有意义，但不能捕捉到有助于创建这些故事的课堂互动细节。这种层次的细节，以及教师和学生身份之间的关系，在定位法中能被更好地捕捉到，因为定位法检视了局部互动中身份的建构以及关系权力在组织此类互动中的作用。但是，与定位法不同，叙事法往往利用访谈研究或自传，能够在一个研究中分析更多人的故事，从而分析经历和故事中的模式。总之，这些方法可以为研究者提供对某个特定课堂、学校或社区中的归属感模式的一个认识，又能让研究者深入分析一个案例，阐明能导致（非）归属感的互动。

身份的心理分析理论

在北美的数学教育文献中，身份的社会文化方法是最常见的，此外还有一些后结构主义传统的方法。数学

Identity-as-story, or narrative identity, is framed in opposition to many theories of identity that assume that identity is a real object (like a tree or a foot, a circle or a line) and that the researcher's job is to approximate it through representations (like stories). When identities are defined instead *as* stories, or "collection of stories told about persons" (Sfard & Prusak, 2005, p. 16), some problematic assumptions that make identities particularly difficult to research are resolved. Rather than being internal and unknowable, identities as stories are human made; shaped by society; changeable across time, context, and narrator; and accessible to researchers through empirical methods such as interviews, observations, or collecting written biographies.

In Sfard and Prusak's formulation that identities *are* stories, it is plausible that these stories arise from our experiences positioning and being positioned by others across the many social practices that make up our lives. In other words, narrative identities can be thought of as the reification (Wenger, 1999) of positional identities. How we frame, narrate, and make meaning of these experiences with others shape how we understand ourselves and may change over time as new frames become available. Although Sfard and Prusak (2005) argue that their approach differs significantly from Wenger's, in that Wenger focused on experience as opposed to stories, we believe that both argue that participation in various contexts influences the stories we tell, and those stories have power to shape future possibilities for participation. Indeed, Wenger argues that we make meaning of our forms of participation; we shape stories, and stories shape us.

Narrative identity is not a single story, but a collection of stories, told by different narrators and to different audiences (Sfard & Prusak, 2005). This definition reflects the social nature of identity; one is never entirely in control of one's identity. However, empirical research that takes up narrative identity tends to focus on stories told about the self, thus implicitly framing an identity narrative as akin to a self-concept. Methodologically, research on narrative identity would do well to imitate the positioning literature, which emphasizes not only how one positions oneself, but how one positions and is positioned by others. For instance, collections of stories about learners by the learners themselves, as well as their parents and teachers, may offer a more complete sense of the possibilities available for a given student's identity as a mathematics student.

Studies of students' narrative identities illuminate the centrality of belonging. It is quite common to hear people assert that they are "not a math person," linking mathematics to some part of their self-concept (Boaler, 2002a). Students' stories about mathematics tend to focus on how they develop a sense of belonging or exclusion (Bartholomew, Darragh, Ell, & Saunders, 2011; Rodd & Bartholomew, 2006; Solomon, 2007). For example, Rodd and Bartholomew (2006) found that students' relationship to mathematics could not be separated from their social and emotional lives. Indeed, Rodd and Bartholomew's study showed that a student's sense of belonging was tied more to issues of identity than to mathematics achievement (Solomon, Lawson, & Croft, 2011).

Studies of teachers' narrative identities illuminate how past, personal histories as mathematics learners relate to teachers' approaches to mathematics instruction (Adams, 2013; Akkoç, YeÎildere-Ðmreb, & Ali Balkanlıo°luc, 2014; Teixeira & Cyrino, 2014; Williams, 2011). These identities, based on who teachers narrate themselves to *be*, are connected to what teachers *do* in the classroom (Battey & Franke, 2008). Many studies of teacher identity focus analytically on shifts in teachers' stories after particular professional development experiences (Battey & Franke, 2008; Wager & Foote, 2013). Stories written before and after professional development are examined to locate possibilities for shifts in practice.

Narrative identity research highlights both the importance of mathematical experiences that allow students to feel a sense of belonging to school mathematics and possible avenues for teachers to implement effective pedagogies. However, it remains unclear *how* teacher and student identities actually shape one another. Further, stories are told at a broad grain size, one that is meaningful to the person telling the story, but which cannot capture the details of classroom interaction that contribute to the creation of these stories. This level of detail, and the relations between teacher and student identities, is better captured in positional approaches, which examine the construction of identity in local interactions and the role of relational power in organizing such interactions. However, unlike positional approaches, narrative approaches tend to draw on interview studies or autobiographies and can analyze the stories of many more individuals within one study, enabling analysis of patterns of experiences and stories. Together, these approaches might offer a researcher both a sense of patterns of belonging within a particular classroom, school, or community, as well as a deep dive into a sample case that illuminates the interactions that lead to a sense of (un)belonging.

Psychoanalytic Theories of Identity

In the North American mathematics education literature, sociocultural approaches to identity are by far the most common, with some representation of poststructural traditions. Mathematics education research, with

教育研究植根于学习的认知方法，有注重理性的倾向。身份研究的社会文化方法常常也有这个倾向，主要关注与内容学习相关的身份认同，而且关注学生和教师的理性和有意决策。后结构主义方法考虑的是日常形成行为与更广的权力体系之间的关系。相比之下，心理分析方法，尤其是在英国的用法，更仔细地考查了身份是如何与情感、焦虑和潜意识的愿望相联系的。[4]毕比（2010）指出，心理分析理论能提供"不同的隐喻并关注困难的部分：恐惧和焦虑、幻想和愿望、爱和恨，不够理性和我们的激情和潜意识中奇怪的逻辑"（第3页）。数学教育研究者从心理分析的理论和方法中提取概念，并将它们应用于对叙事或课堂互动的理解。这些研究中采用的概念包括防卫和防卫机制、心理现实、潜意识以及"镜像"（Bibby，2010；Black等，2009）。

理解数学身份的心理分析方法与社会文化理论有一些连接点。以心理现实这一心理分析概念为例，心理分析学家使用这个词来描述我们对我们周围世界的解释，以及有意识和无意识的反应（Bibby，2010）。我们的心理现实既是内部的，也是外部的，它被我们对世界的体验所塑造，也塑造着我们对世界的体验。毕比引述弗罗施和巴雷特瑟（如Bibby，2010所引）的话说，心理现实是"主体生活的地方"（第8页，原文中强调）。这表明心理现实可能类似于一个假定的世界（Holland等，1998）。不同之处在于心理现实更关注个人，强调个人如何解释周围的世界，而假定的世界是被广泛分享的关于我们生活经验的故事情节。

另外一个连接点是心理分析的镜像概念，它与定位理论相呼应。毕比（2010）引用拉坎的观点，拉坎认为我们将自己与他人的互动当作一面镜子：他人对待我们的方式告诉我们他们看到的是谁，他们认为我们是谁。在课堂上，教师的评价和与同学的互动就像一面镜子，学生们不得不将其融入他们的自我意识："我们通过观察他人提供的镜像功能来认识和了解我们自己"（Bibby，2010，第32页）。从方法论上看，镜像分析看起来与定位分析极为相像，可能更敏锐地关注对课堂互动的情感反应：被反映为好的或聪明的快乐，被反映为无知的羞耻和耻辱。这使得身份发展的观点成为超越了此时此地的数学课堂，思考的是人类的憧憬、愿望和耻辱。

比如，布莱克等人（2009）对三位作者的数学自传进行了心理分析，他们的研究提出，尽管这三位作者有很好的数学成绩，但是在学校中塑造"数学能力"导致了他们的焦虑。这一焦虑使女性建构起一系列防卫：用一种与数学的新关系所产生的新自我取代了那个旧自我、将数学作为一种与他人建立关系的方式（在其中一个案例中，是与她的父亲）、用其它相关领域的成功进行补偿。这些分析注重带动联系数学的"幻想"之间的相互作用，以此来"为我们提供支持我们自我意识的方式"（第14页）。

个人身份与社会身份之间的关系

我们这里关注的是数学教育身份研究中的一部分，该部分明确地将个人身份发展与诸如种族、性别和语言的社会成员身份连接起来，这一连接通常是与社会身份的一种社会政治立场相联系的。研究者们认为，如果避而不谈社会政治结构，那么学校数学将优先考虑父权规范（Esmonde，2011）、男子气概（Langer-Osuna，2011；Mendick，2006；Solomon，2007）、白人（Russell，2012；Varelas, Martin, & Kane，2012）、财富（Esmonde，2014）、主要的语言（Moschkovitch，2010），及其各种交集（Esmonde & Sengupta-Irving，2015）。比如，一个课堂案例研究发现，同伴在定位一个男生展现权威时会持积极态度，而同样的同伴会拒绝一个女生展现其权威，最终导致这个女生在群体中被边缘化（Langer-Osuna，2011）。这一分析说明，权力身份并不是对一个给定课堂中所有的学生都是同样可得的方式，难于获得强有力的数学身份，这既不是教师的明确选择，也不一定是学生的有意识选择。但是，有权力的学生和非裔美籍女生的交叉身份会制造紧张的局面，这种局面塑造了她的权力表现——在这个例子中，她会被其他人解读为一个发号施令的小组领导者。

在本章的这一节，我们对使数学相关个人身份与诸如种族、性别和语言等社会身份相协调的研究理论基础进行回顾。当我们描绘协调个人身份与社会身份之间

roots in cognitive approaches to learning, has a tendency to focus on the rational. Sociocultural approaches to identity often share this tendency, concerned primarily with identity as it relates to content learning and focused on rational and intentional decisions on the part of both students and teachers. Poststructural approaches consider the relation of everyday acts of becoming to broader systems of power. By contrast, psychoanalytic approaches, used particularly in the United Kingdom, look more closely at how identities relate to emotions, anxieties, and subconscious desires.[4] Bibby (2010) states that psychoanalytic theories can provide "different sets of metaphors and give attention to the difficult bits: the fears and anxieties, the fantasies and desires, the loves and hates, the less than rational and the strange logics of our passions and our unconsciouses" (p. 3). Researchers in mathematics education draw concepts from psychoanalytic theory and method and apply them to an understanding of narratives or classroom interactions. Concepts utilized in this work include defenses and defense mechanisms, psychic reality, the unconscious, and "mirroring" (Bibby, 2010; Black et al., 2009).

Psychoanalytic approaches to understanding mathematics identity have some points of connection with sociocultural theories. For instance, consider the psychoanalytic concept of psychic reality. Psychoanalysts use the term to describe our interpretation of, and conscious and unconscious responses to, the world around us (Bibby, 2010). Our psychic reality is both internal and external. It is shaped by, and shapes, our experiences with the world. Bibby quotes Frosh and Baraitser (as cited in Bibby, 2010) to say that psychic reality is "what the subject *lives in*" (p. 8, emphasis in original). This suggests the psychic reality may be akin to a figured world (Holland et al., 1998). The difference is that psychic reality is more individually focused, with an emphasis on how individual people interpret the world around them, whereas figured worlds are broadly shared storylines about our lived experiences.

Another point of connection is the psychoanalytic concept of mirror, which echoes positioning theory. Bibby (2010) draws from Lacan, who argues that we use our interactions with others as a kind of mirror: the way others treat us shows us who they see, who they think we are. In the classroom, teacher assessments and interactions with peers function as a mirror that students have to integrate into their sense of self: "We come to recognise and make sense of ourselves by looking to others for the mirror-functions they provide" (Bibby, 2010, p. 32). Methodologically, analyses of mirror can look strikingly similar to positioning analyses, with perhaps a sharper focus on emotional responses to classroom interactions: the pleasure of being mirrored as good or clever, the shame and humiliation of being mirrored as ignorant. This enables a view of identity development that goes beyond the here and now of the mathematics classroom to consider human longing, desire, and shame.

For example, Black et al. (2009) present a psychoanalysis of three of the authors' mathematical autobiographies. They point out how the construction of "mathematical ability" in school led these three authors to anxiety, despite their high mathematical achievement. This anxiety led the women to construct a series of defenses: replacing the old self with a new self that held a new relationship to mathematics; using mathematics as a way to build relationship with others (in one case, with her father); and compensating with success in other, related fields. These analyses focus on the drives for connection that interact with "phantasies" of mathematics to "provide us with ways to support our sense of self" (p. 14).

Relationships Between Individual and Social Identities

We focus here on a subsection of mathematics education research on identity that explicitly connects individual identity development with social membership identities such as race, gender, and language. This connection is typically framed in relation to a sociopolitical stance on social identities. These researchers argue that if sociopolitical structures are left unaddressed, school mathematics privileges heteronormativity (Esmonde, 2011), masculinity (Langer-Osuna, 2011; Mendick, 2006; Solomon, 2007), whiteness (Russell, 2012; Varelas, Martin, & Kane, 2012), wealth (Esmonde, 2014), the dominant language (Moschkovitch, 2010), and its intersections (Esmonde & Sengupta-Irving, 2015). For example, one classroom case study found that peers positioned displays of authority by a boy positively, whereas the same peers rejected displays of authority by a girl, ultimately leading to her marginalization in the group (Langer-Osuna, 2011). This analysis illuminates ways that identities of power were not equally available to all students in a given classroom. This lack of access to powerful mathematics identities was not an explicit choice by the teacher and may not have been a conscious choice by the students; rather, the intersecting identities of powerful student and African American girl created tensions that served to shape how her performance of power—in this case, her role as group leader who issued directives—was interpreted and responded to by others.

In this section, we review the theoretical basis for research that coordinates mathematics-linked individual

的身份发展的研究理论空间时，我们只考虑将社会成员关系理论化为可变的、受他人影响的以及多重的那些研究。因此，我们不会关注使用"身份"这个词来指代社会成员关系，但却不将身份转变理论化的那些研究。比如，社会心理学中关于刻板印象威胁的大量研究考虑的是一个人的种族、性别和与数学的关系是如何影响实验测试成绩的（例如，可以参见 Gresky, Ten Eyck, Lord, & McIntyre, 2005; Schweinle & Mims, 2009; Spencer, Steele, & Quinn, 1999; Steele, Spencer, & Aronson, 2002; Tagler, 2012; Tine & Gotlieb, 2013; Tomasetto, Alparone & Cadinu, 2011）。尽管这类研究对社会成员关系在教育环境中的重要性给出了非常深刻的见解，但它们不能解释这些社会成员关系与个人身份发展是如何被共同建构起来的，而且，很多与数学学习相关的研究都受到了种族和性别的二元对立观点的阻碍。这些研究通常将性别划分为非此即彼，将种族分类为一套选择框，而不是与人们拥有相互关系的社会建构。比如，我们中的两位（兰格-奥苏纳和埃斯蒙德），尽管在出生时都被标记为女孩，但却有着非常不同的性别身份（特别是与女人这个种类的关系）。同样地，我们其中一个人（兰格-奥苏纳）的种族身份因具体的地理位置而发生变化，当地对白人的定义有时候明确地接纳或拒绝她，有时候又模棱两可、含糊不清。将社会身份看作是简单二元体的研究并不能捕捉到我们社会成员关系的复杂性以及这些关系是如何随时间和地点而改变的。考虑到这一复杂性的分析方法和理论框架能够在我们跨团体应用这项成果时提供重要的见解。

关于社会身份的数学教育研究中经常引用的另一个理论框架是批判种族理论（Martin, Anderson, & Shah, 2017，这套书; Stinson & Walshaw, 2017，这套书）。批判种族理论并不是一个种族身份的理论，但是它在美国是一个种族和种族主义的理论，是围绕种族的社会结构如何塑造体制结构的理论（Ladson-Billings & Tate, 1995）。同样，个人与种族关系的复杂性，以及这些关系如何随时间或地点而变化——身份理论中的关键问题——通常没有被明确地考虑。因此，刻板印象威胁理论或批判种族理论可以被纳入到身份研究当中，以帮助突出社会成员关系的某些方面，但这些理论本身并不是身份理论。

因此，在这一节，我们讨论研究者如何使身份发展的理论和各种社会政治框架或见解相互协调，以此作为思考个人身份和成员身份如何得以共建的一种方式。为了阐明这是如何实现的，我们提供两个例子：一个来自批判种族理论，另一个来自女性主义的酷儿理论（例如 Esmonde, Takeuchi, & Dookie, 2012）。在此后的几节，我们阐明哪些身份理论被使用以及它们是如何从社会政治的角度被理解的。我们再次注意到，为了清晰起见，以上的回顾区分了四种理论视角——后结构主义/话语的、定位的、叙事的和心理分析的。事实上，许多学者在他们的研究工作中采用了不止一种方法。在接下来的讨论中，我们会关注在什么时候使用多种方法，以及谨慎的概念协调能提供什么重要的见解。

种族身份与数学的理论化

从叙事的角度讲，聚焦身份的研究强调了数学归属过程按种族组织的方式。有时，会赋予学习者不参与数学这样的身份，正如马丁（2006）与非裔美国成年人的访谈中所例证的那样。当这些成年人回想他们在K-12中小学的学习经历时，他们讲述了很多无论他们的数学兴趣和成绩如何，都被忽视、批评，或者被推入低层次数学班级的例子。一些学习者将这些故事内化，并开始相信自己在数学上不够聪明；其他学习者认为他们被对待的方式存在种族歧视，但不能改变别人讲述的关于他们的故事。贝里（2008）讲述了多个在数学上成功的非裔美国男孩的故事，这些故事突出了种族化体验可以充当障碍，也可能是支持系统。

瓦尔拉斯等人（2012）的研究通过关注学生作为数学实干者的学科身份、作为非裔美国人的种族身份以及作为学术任务和课堂实践参与者的学术身份这三者之间的交集，他们提出了内容学习和身份建构框架（CLIC），以此作为一种检视非裔美国人数学学习身份的方式。内容学习和身份建构框架使关于身份的叙事性和定位性视角都与批判的社会政治观点相协调。也就是说，身份被认为是一个故事集（Sfard & Prusak, 2005） identities with social identities such as race, gender, and language. As we mapped the theoretical space of research that coordinates identity development across individual and social identities, we considered only research that theorizes these social memberships as changeable, influenced by others, and multiple. We therefore do not focus on bodies of research that use the word identity to refer to social memberships but that do not theorize identity shifts. For example, the vast body of research on stereotype threat in social psychology considers how a person's race, gender, and affiliation with mathematics influence achievement on experimental tests (see, e.g., Gresky, Ten Eyck, Lord, & McIntyre, 2005; Schweinle & Mims, 2009; Spencer, Steele, & Quinn, 1999; Steele, Spencer, & Aronson, 2002; Tagler, 2012; Tine & Gotlieb, 2013; Tomasetto, Alparone & Cadinu, 2011). Although this research offers critically important insights into how social memberships matter in educational environments, it cannot explain how these social memberships are coconstructed with individual identity development. Further, much of the research related to mathematics learning is also hampered by an overwhelmingly binary view of race and gender. Research typically categorizes gender as either/or, and race as a set of checkboxes, rather than social constructs to which people have relationships. The two of us (Langer-Osuna and Esmonde), for example, have very different gender identities (and specifically, relationships to the category woman), despite both being designated girls at birth. Likewise, the racial identity of one of us (Langer-Osuna) shifts in relation to particular geographic locations, where local definitions of whiteness sometimes explicitly include or reject her and sometimes remain ambiguous and tenuous. The complexity of our social memberships and how they change over time and location is not captured in research that considers social identities to be simple binaries. Analytic methods and theoretical frameworks that take this complexity into account enable insights that matter when applying this work across various communities.

Another theoretical frame that is frequently drawn into mathematics education research concerned with social identities is critical race theory (Martin, Anderson, & Shah, 2017, this volume; Stinson & Walshaw, 2017, this volume). Rather than a theory of racial identity, however, critical race theory is a theory of race and racism in the United States and how institutional structures are shaped by social constructs around race (Ladson-Billings & Tate, 1995). Again, the complexity of individual relationships to race and how these relationships change over time or place—key concerns in theories of identity—are typically not explicitly considered. Thus, stereotype threat theory or critical race theory could be *drawn into* a study of identity to help highlight some aspects of social membership, but they are not themselves theories of identity.

Therefore, in this section we discuss how researchers coordinate theories of identity development with a variety of sociopolitical frames or insights as a way to consider how individual and membership identities are coconstructed. To illustrate how this is done, we offer two examples: one that draws from critical race theory and one that draws from feminist queer theory (e.g., Esmonde, Takeuchi, & Dookie, 2012). In the sections that follow, we elucidate which theories of identities are used and how they are understood in light of sociopolitical perspectives. We note again that the review above drew distinctions across the four theoretical perspectives—poststructural/discursive, positional, narrative, and psychoanalytic—for the purposes of clarity. Indeed, many scholars draw upon more than one approach in their work. In the discussion below, we note when multiple approaches are used and the insights careful conceptual coordination can offer.

Theorizing Racial Identity and Mathematics

From a narrative perspective, research focused on identity highlights ways in which the process of belonging to mathematics can organized by race. At times, identities of nonparticipation in mathematics are chosen for learners, as exemplified in Martin's (2006) interviews with African American adults. When these adults reflected on their experiences as learners in K–12 schools, they reported many instances of being ignored, criticized, or pushed into low-track mathematics classes despite interest and achievement in mathematics. Some learners internalized these stories and began to believe that they were not smart enough for mathematics; others recognized the racism in the way they were treated but were not able to change the stories others told about them. Berry (2008) offers stories of mathematically successful African American boys that highlight racialized experiences that served as possible obstacles, as well as support systems.

Varelas et al. (2012) offer the content learning and identity construction framework (CLIC) as a way to examine African American mathematics learning identities by focusing on the intersection between disciplinary identities as doers of mathematics, racial identities as African Americans, and academic identities as participants in academic tasks and classroom practices. The CLIC framework coordinates both narrative and positional perspectives on identity with critical, sociopolitical insights.

和在假定世界中的各种表现（Holland等，1998）。来自批判种族理论中的概念再被用来构建在一个国家（美国）中的身份结构，在这个国家，种族化的社会等级制度被嵌入到每个层次的社会结构中（Martin，2013）。这种现象的协调使（1）学科身份（如，关于数学的身份），（2）种族身份，（3）学术身份（即作为学校学生的身份）这三种身份之间的分析相互联系起来，所有这三种身份都被证明是以各种方式相互影响的。这个框架强调了斯法德和普鲁萨克所称的"指定的身份"以及霍兰等人（1998）所称的"假定的身份"。当其他人对我们将要或应该取得的身份有种种期望时，指定的或假定的身份就产生了，种族的刻板印象以有害的方式影响着这些期望。

基于用内容学习和身份建构框架得到的分析，瓦尔拉斯等人（2012）提出，教育者应该鼓励身份之间的各种积极连接，这些连接可以通过支持以下学生行为来实现：(a)反思他们在学科实践中的参与，(b)想象并呈现他们自己在校内外及以后的生活中做数学，(c)思考他们自己与专业及学科从业人员之间的相似与不同。内容学习和身份建构框架关注种族身份，但不局限于种族身份，能被交叉地用于同时考虑不止一种成员身份（如，种族和性别；Gholson & Martin，2014）。

性别与数学的理论化

关于性别与数学的研究通常接受了性别的二元观，甚至是女性的不足观（Esmonde，2011；Lubienski & Ganley，2017，这套书）。然而，有少数研究采用了叙事的、后结构主义的和心理分析的理论去探究性别与学校数学之间的关系。比如，罗德和巴塞洛缪（2006）研究了年轻的女性数学本科生，发现这些女性陷入了一个身份难题，在构造的身份中她们是无形的（被男性，学生和教授都忽视），同时也是特别的（十分显眼且被视为独特的）。这个难题有助于解释所罗门等人（2011）所描述的现象，即许多年轻女性形成了关于数学的"脆弱的身份"；当这些女性被他人评判时，这些身份容易瞬间发生变化。

为了理解性别与数学关联的各种方式，门迪克（2006）诉诸于女性主义的酷儿理论以获得她对数学身份的理解，"其中的关键因素是，性别是我们做的事，不是我们是谁，男性主义与女性主义是相对对方而构建的，而数学被牢牢地固定在这个鸿沟的男性那边"（第1页）。门迪克与青少年数学学习者的访谈揭示了几种数学观，数学是有序的、逻辑的、结构化的、权威的和可靠的，在西方，这些都是与男人和男子气概相关的特点。因此，学好数学就像要展现出男子气概一样，这是一个对不同性别的人来说明显不同的任务，也就是说，展现男子气概不是中立的或简单的任务，女性在数学课堂（Langer-Osuna，2011）及之外的地方（Brescoll，2011）展现出男子气概可能会被解读为是不适当的。

门迪克（2006）提出要"做出社会性的选择，支持性别越界以及开放数学"（第2页）。首先，作出社会性的选择（从事数学与否）意味着停止将选择看作是个人身份的反映；相反，选择是在某个社会情境中做出的。选择从事数学也严重依赖于数学能力的建构，这与男性作为理性的、逻辑的和权威的指定身份一致。第二，支持性别越界远远不止是将女性刻画为本质上不同于男性的各种方法，这些方法要求女性更加自信（即有男子气概），而没有认识到人们看待女孩和男孩对自信的展现是不同的（Langer-Osuna，2011）。相反，门迪克认为"我们需要让更广泛的人群获得更广泛的（性别）主体性"（2006，第112页）。最后，开放数学意味着承认数学除了是精确的、逻辑的和权威的以外，数学也是创造性的、直觉的和合作性的。

未来研究的启示和方向

学数学从根本上说是关于身份的工作。正如温格（1999）所说，"学习是一个身份形成的过程；反过来，身份形成也是一个学习的过程"（第143页）。如果学生们的数学故事、身份表现和情感世界是基于能使个人身份、成员身份和数学身份更有力地相结合的体验，那么他们的数学故事、身份表现和情感世界会是什么样子？支持教师设计这种课堂的能力和学生的课堂驾驭能力，可以说是大多数与数学相关的身份研究的核心。因此，我们

That is, identity is conceived as a set of stories (Sfard & Prusak, 2005) and as performances within figured worlds (Holland et al., 1998). Concepts from critical race theory are then used to frame the construction of identity in a country (the United States) where the racialized social hierarchy is embedded within social structures at every level (Martin, 2013). This coordination of perspectives enables analytic interconnections between three types of identity: (1) disciplinary identity (e.g., identity with respect to mathematics), (2) racialized identity, and (3) academic identity (i.e., identity as a school student). All three identities are shown to influence one another in various ways. The framework emphasizes what Sfard and Prusak call "designated identities" and what Holland et al. (1998) call "figured identities." Designated or figured identities arise when other people have expectations about the identities we will or should achieve. Racial stereotypes influence these expectations in pernicious ways.

Drawing from their analyses using CLIC, Varelas et al. (2012) argue that educators can encourage positive connections across identities by supporting the following student actions: (a) reflecting on their engagement in disciplinary practices, (b) imagining and presenting themselves doing mathematics in and out of school as well as later in life, and (c) considering similarities and differences between themselves and professional and disciplinary practitioners. The CLIC framework focuses on, but is not limited to, racialized identities and can be used intersectionally to consider more than one membership identity at once (e.g., race and gender; Gholson & Martin, 2014).

Theorizing Gender and Mathematics

Research on gender and mathematics has often taken a binary view of gender, and even a deficit view of girls (Esmonde, 2011; Lubienski & Ganley, 2017, this volume). However, there is a small body of research that takes up narrative, poststructural, and psychoanalytic theories to investigate the relationship between gender and school mathematics. For example, Rodd and Bartholomew (2006) studied young women undergraduate mathematics students and found that the women were in an identity conundrum, constructing identities in which they were simultaneously invisible (unnoticed by the men, both students and professors) and special (highly visible and seen as unique). This conundrum contributes to the phenomenon that Solomon et al. (2011) describe, in which many young women develop "fragile identities" with respect to mathematics; these identities are apt to change at a moment's notice when the women are judged by others.

To understand the myriad ways gender relates to mathematics, Mendick (2006) turns to feminist queer theory to inform her understanding of mathematical identity, "key elements of [which] are that gender is something we do, not something we are, and that masculinity and femininity are constructed in opposition to each other, with maths firmly fixed on the masculine side of the divide" (p. 1). Mendick's interviews with adolescent mathematics learners revealed conceptions of mathematics as ordered, logical, structured, authoritative, and reliable. These are all qualities that have, in the West, been associated with men and masculinity. Therefore, to do mathematics well is akin to performing masculinity, a task that is distinctly different for people of different genders. That is, performing masculinity is no neutral or simple project. Women's performances of masculinity can be interpreted as inappropriate in mathematics classrooms (Langer-Osuna, 2011) and beyond (Brescoll, 2011).

Mendick (2006) suggests "making choices social, supporting gender transgression and opening up mathematics" (p. 2). First, making choices (to pursue mathematics or not) social means to stop considering choices as reflections of individual identity. Instead, choices are made in a social context. The choice to pursue mathematics is also heavily reliant on constructions of mathematical ability, which align with designated identities of masculinity as rational, logical, and authoritative. Second, supporting gender transgression goes well beyond approaches that characterize girls as essentially different from boys. These approaches ask girls to be more assertive (i.e., masculine), without recognizing that a girl's display of assertiveness is taken up differently than a boy's (Langer-Osuna, 2011). Instead, Mendick argues that "we need to make a wider range of [gender] subjectivities available to a wider range of people" (2006, p. 112). Finally, opening up mathematics means to acknowledge that in addition to being precise, logical, and authoritative, mathematics is also creative, intuitive, and collaborative.

Implications and Directions for Future Research

Learning mathematics is fundamentally about identity work. As stated by Wenger (1999), "Learning is a process of identity formation and, conversely, identity formation is a process of learning" (p. 143). What would students' mathematical stories, performances of identity, and emotional landscapes be like if they were based on experiences that more robustly enabled intersecting individual, membership, and mathematical identities? Supporting teachers' capacities to design such classrooms, and students' capacity to navigate them, is arguably at the heart of much of the research on mathematics-linked identities. We thus conclude with a discussion on the implications of this body of research for the *identity work*

最后讨论这类研究对学生和教师参加的身份工作的启示，以及新的和需要研究的可能方向。以下部分将按这样的顺序来组织：首先，我们讨论这类研究对与数学教师和学生相关的各种不同身份工作的启示。其次，我们讨论对数学课堂上教师和学生身份工作中权力作用的启示。最后，我们讨论未来研究的方向。

教师与学生身份的工作

关于如何针对学生和教师，对身份工作进行有区别的概念化，关注学生或教师在数学中身份的研究揭示出了一些深刻的见解。教师在发展他们身份的过程中的能动性体现在围绕着一个元意识，选择他们想要成为的教师类型。比如，在叙事法中，教师将专业发展经历的特定方面引入到他们作为数学教师的故事中时，教师被认为是具有能动性的。通常来说，这项工作的目的是，支持教师在通过强调平等性的专业发展而增强的社会政治意识与他们作为教师的意识之间建立联系。相反，学生很少被要求通过反思性的故事写作来为自己选择新的身份。其实，学生的能动性是根据他们在智力上参与数学的过程中，所体验到的自由度来建立的，身份工作的责任不在于学生深思熟虑地和有意识地选择他们想成为何种学习者，而在于学生在和合作者的互动中感到他们自己是有能力的数学思考者和问题解决者。因此，积极的学生身份工作的一个前提条件就是，要有支持创造真正包容性课堂的教师身份工作（Clark, Badertscher, & Napp, 2013; De Freitas 等, 2012）。

互动的或定位的方法很少被用于理解教师身份，尽管教师在许多关注学生的定位分析中是关键人物，但这类研究迄今为止往往并没有思考课堂互动如何帮助教师身份的发展（一个例外，见 Walshaw, 2008）。对课堂中教师定位性身份进行分析可以帮助我们了解在特定学校、地区或政策环境中，教师工作环境是如何影响教师在学校数学中的实践以及他们的身份的。此外，学生对特定教学方法的接纳或抵制是教师身份塑造的关键因素。学生认为：使学习成功的教师，是能干的教师；使学生集体挣扎或抵制某种活动的教师，是无能的教师。这些定位行为可能会影响教师如何采用新的教学实践。考虑到人们希望教师能让学生参与从认知角度看比较复杂的数学思维形式，这一点似乎特别重要，而学生们，至少刚开始，可能会抵制这些复杂的数学思维形式，通过设计，学生们会努力克服的。

身份工作中权力的作用

这类身份研究的另一个相关启示是，学生和教师的身份工作是围绕着权力关系组织的，而那些权力关系与数学的归属感是相关联的。学生讲述的关于数学的故事强调了归属数学的挣扎，能精准地指出一个人生命过程中与数学形成了长期关系的重要事件。定位的视角，尤其是对课堂定位的分析，能对权力在学习者身份的即时和更长期的建构中所起的作用给出深刻的见解。在数学课堂上，权力关系包括谁掌握数学的权威，谁被定位为有能力的人（Dunleavy, 2015; Gresalfi & Cobb, 2006; Gresalfi 等, 2009）。比如，格雷沙菲等人（2009）认为，能力不是个人特征，而是从参与课堂活动体系中涌现出来的。课堂活动向学生提供多种可能的方式参与数学活动，同时让学生对自己和同学的想法负责，这可以以数学权威和能力形式来定位学生（Gresalfi 等, 2009）。比如，在博尔纳和汉弗莱斯（2005）的研究中，一位初中数学教师让学生解释常见错误背后的逻辑，了解学生是否理解同伴的想法，以及以学生的名字命名不同的问题解决方法。所有这些行为都是为了将包括犯错者在内的各种学生定位为数学意义建构的共同的、富有成效的贡献者。同样地，学生有机会创作、论证和争论对数学的看法，也能定位他们的数学权威。这些主体地位，随着时间的推移，支持着包含归属感和智力力量在内的数学关联身份的建构。

尽管这类研究已经开始展现数学课堂是如何被设计成培养强有力的定位身份的，但是我们仍然需要研究，来阐明当来自刻板印象群体的学生在说话、扮演领导角色、提供帮助等取得这些数学权力的定位时，所出现的不同后果。如何设计课堂情境以产生积极数学身份工作的方式来构建社会身份的方方面面呢？一个例子就是，

Teacher and Student Identity Work

Research focused on either student or teacher identity in mathematics reveals insights as to how identity work is conceptualized differently for students and teachers. Teachers' agency in developing their identities is framed around a meta-awareness of choosing the kind of teacher they want to be. For instance, in narrative approaches, teachers are framed as agentive in bringing particular aspects of the professional development experience into their stories as mathematics teachers. Often, this work is meant to support connections between teachers' increased sociopolitical awareness through equity-focused professional development and their sense of themselves as teachers. In contrast, rarely are students asked to choose new identities for themselves through reflective story writing. Rather, students' agency is framed in terms of the degrees of freedom they experience in being able to intellectually engage with mathematics. The onus of the identity work lies not so much in students choosing what kind of learner they want to be reflectively and consciously, but rather in students experiencing themselves as competent mathematical thinkers and problem solvers in local interactions with collaborating others. Thus, a precursor to positive student identity work is the kind of teacher identity work that supports the creation of genuinely inclusive classrooms (Clark, Badertscher, & Napp, 2013; de Freitas et al., 2012).

Interactional or positional approaches are used far less often to understand teacher identities. Although teachers are key players in many positioning analyses that focus on students, the body of work to date does not tend to consider how classroom interactions contribute to teachers' identity development (for an exception, see Walshaw, 2008). Analyses of teacher positional identities in the classroom could shed new light on how the context of teachers' work within particular schools, districts, or policy climates shape teachers' practices in school mathematics, and so their identities. Further, students' uptake or resistance to particular instructional approaches are critical shapers of teachers' identities. Where students seem to succeed, teachers are positioned as competent; where students collectively struggle or resist particular kinds of activity, teachers are positioned as incompetent. These acts of positioning may shape how teachers take up new instructional practices. This seems particularly important given that teachers are expected to engage students in cognitively complex forms of mathematical thinking, which students, at least initially, may resist and which, by design, students struggle through.

The Role of Power in Identity Work

A related implication of this body of work is that students' and teachers' identity work is organized around relationships of power and those relationships of power are connected to a sense of belonging to mathematics. Stories students tell about mathematics highlight the struggle to belong and can pinpoint consequential events in a person's life course that shaped a long-lasting relationship to mathematics. Positional perspectives, and in particular, analyses of classroom positioning, enable insight into the workings of power in the moment-to-moment and longer term construction of identities as learners. In mathematics classrooms, relationships of power include who holds the mathematical authority, and who is positioned as competent (Dunleavy, 2015; Gresalfi & Cobb, 2006; Gresalfi et al., 2009). For example, Gresalfi et al. (2009) argue that competence is not an individual attribute but rather emerges out of participation in classroom activity systems. Classroom activities that offer multiple possible ways to engage in mathematics while holding students accountable for making sense of their own and peers' ideas can serve to position students with mathematical authority and forms of competence (Gresalfi et al., 2009). For instance, in Boaler and Humphreys (2005), a middle school mathematics teacher asks students to explain the logic behind common mistakes, inquires as to whether students understand a peer's mathematical ideas, and names students as authors of different problem-solving methods. All these moves serve to position a range of students, including those who made errors, as contributors to collective and productive mathematical sense making. Likewise, students' opportunities to author, justify, and debate mathematical ideas position them with mathematical authority. These subject positions, when experienced over time, support the construction of mathematics-linked identities that include belonging and a sense of intellectual power.

Although this work has begun to show how mathematics classrooms might be designed to foster powerful positional identities, we also need studies that illuminate the differential consequences when students from stereotyped groups speak out, take up leadership roles, offer

在美国的很多地方，西班牙语通常被认为是一门不如英语的学术语言，因为它不是主要的教学语言。特纳、多明格斯、马尔多纳多和恩普森（2013）展示了在一个数学学习情境中教师和学生的定位性身份工作，在这个情境中，不仅数学活动设计得使更多学生能具有更大的数学权威，而且西班牙语也被明确认定为是与英语地位平等的数学工作语言。学生们彼此依赖，共同创作和争论对数学的想法，为彼此翻译，创造了一个充满活力的西班牙-英语的双语空间，以这两种语言聚焦数学工作，这使得常常沉默的说西班牙语的学生能贡献于集体的工作，也使只说英语的学生能感受到他们说西班牙语的同伴们的数学能力并能回应他们的想法。这个学习情境是为所有的学习者提供坚实的学习和强有力的数学身份而设计的。

前行：未来研究的方向

未来有很多的可能性。这部分给出了一些我们认为会在前行中有帮助的建议。首先，数学教育中身份的研究者必须更加清楚地表达在他们的研究中使用了哪些理论框架和分析方法。在这个领域中，对研究理论和方法的更加清楚的交流，能使我们辨认出哪些理论和方法最适合回答某些特定种类的问题，也能对特定的实践领域有所启示。其次，与数学相关的个人身份和成员身份的发展都需要获得进一步的理论化和研究。这些理论应该反映出社会政治观点的更大层面的整合。

在方法论上，使个人身份发展和成员身份发展相互协调仍然是一个挑战。叙事法更适合于捕捉特定群体之间共同的经历，也适合于描述学生和教师是如何看待他们的数学社会经历的。话语法和定位法能揭示运作机制——学生和教师如何在局部的和结构性的力量作用于他们时就地协商他们的身份。这些方法有可能很好地协同工作。此外，我们还需要实现进一步的理论化和方法上的进步，以此来引领这一研究领域的发展，理解并设计出针对成为数学教师和数学学习者的过程的学习情境。

注释

1. 布莱克等人（2009）将这四种方法假定为三种，把身份的定位法和叙事法合成一个单一的社会文化种类，并与话语法和心理分析法形成对比。我们决定区分这两种社会文化方法，因为他们借鉴了不同的方法，而且就身份、身份如何发展和身份为何重要提出了不同的主张。

2. 描绘的世界是社会建构和历史情境中的故事情节，既影响个人的可能行为，也影响他人如何解释这些行为。描绘的身份是特定故事情节中提供的可用的主体身份，比如说，在学校教育这个假定的世界中，教师和学生描绘的身份是可用的。个人的行为，比如，发布一道命令，是通过可得的故事情节来解释的。教师向学生发布命令被认为是合适的，而学生向教师发布命令被认为是行为犯规。假定的世界的社会建构和历史性可以通过故事情节自身中的局部差异和暂时转变来证明，比如，今天的课堂可能把一个学生向同伴发布命令解释为学生领导的合作性工作的恰当元素，标志着学校教育这个假定的世界中更广泛的历史变化。然而，这些解释在局部环境之间是变化的，特别是先进的学校和更加独裁的"无借口"学校在他们对标志学生权威的学生行为的解释上形成了鲜明对比。

3. 我们通过把一些定位行为和身份表示为"积极的"，把其它的表示为"消极的"，来理解这一领域的研究工作。尽管严格地说，我们不能评估一个定位行为或一个身份的价值，但是文献通常把将学生描述为有能力的或有价值的定位行为称作是积极的，把包含对数学价值的欣赏的身份称作是积极的。

4. 正如瓦莱罗（2009）所指出的，数学教育中的心理分析研究似乎很少使用"身份"这个词，而更喜欢使用"关系"这个词。尽管我们承认在这个理论以及本章中讨论的其他理论中，身份一词并不是最合适的，但是为了简便，我们在这个部分通篇使用身份这个词。

help, and otherwise claim these positions of mathematical power. How could classroom contexts be designed to frame aspects of social identities in ways that enable positive mathematics identity work? As an illustrative example, in many parts of the United States, the Spanish language is often framed as a less academic language than English by virtue of not being the dominant language of instruction. Turner, Dominguez, Maldonado, and Empson (2013) illuminate the positional identity work of teachers and students in a mathematics-learning context where not only were mathematical activities designed in ways that enabled greater mathematical authority among students, but also the Spanish language was explicitly framed as a language of mathematical work on a par with English. Students relied on one another both to author and debate mathematical ideas and to translate for one another, creating a dynamic Spanish-English bilingual space. Mathematical work was centered on both languages. This enabled typically silenced Spanish speakers to contribute to the collective work and monolingual English speakers to experience their Spanish-speaking peers' mathematical competence and respond to their ideas. The learning context was designed for both robust learning and powerful mathematical identities for all learners.

Moving Forward: Directions for Future Research

There are great possibilities ahead. We close with some suggestions that we believe will be helpful in moving forward. First, researchers studying identity in mathematics education must more clearly communicate which theoretical frameworks and analytic methods are used in their work. Clearer communication about the theories and methods we bring to this work will enable our field to recognize which ones are best suited to answer particular kinds of questions and have implications for particular areas of practice. Second, the development of mathematics-linked individual and membership identities both need to be further theorized and studied. These theories should reflect greater integration of sociopolitical insights.

Methodologically, coordinating individual and membership identity development remains a challenge. Narrative approaches are better suited to capture common experiences across particular groups of people, as well as how students and teachers make meaning of their social experiences with mathematics. Discursive and positional approaches can shed light on mechanisms at play—how students and teachers negotiate their identities in situ as local and structural forces act upon them. These approaches have the potential to work well together. Additionally, we see a need for further theorizing and methodological advances to guide our field in understanding and designing learning contexts that centrally take into account the process of becoming mathematics teachers and learners.

Notes

1. Black et al. (2009) posit these four approaches as three, folding positional and narrative approaches to identity into a single sociocultural category and contrasting with discursive and psychoanalytic. We have decided to distinguish between the two sociocultural approaches because they draw from different methodologies and make different claims about identity, how it develops, and why it is important.

2. Figured worlds are the socially constructed and historically situated storylines that shape both possible actions by individuals and how those actions are interpreted by others. Figured identities are the available subject positions offered within particular storylines. For example, in the figured world of schooling, the figured identities of teacher and student are available. The actions of individuals, say, issuing an order, are interpreted through the available storyline. A teacher issuing an order to students is interpreted as appropriate, whereas a student issuing an order to teachers is interpreted as a breach of conduct. The socially constructed and historical nature of figured worlds is evidenced through local differences and temporal shifts in the storylines themselves. For example, today's classrooms might interpret a student issuing an order to peers as an appropriate element to student-led collaborative work, signaling broader, historical shifts in the figured world of schooling. Yet, these interpretations vary across local settings, where particularly progressive schools and more authoritarian "no excuses" schools contrast sharply on their interpretations of student actions that signal student authority.

3. We follow research in this area by denoting some acts of positioning and identities as "positive" and others as "negative." Although, strictly speaking, we cannot assess the value of an act of positioning or an identity, the literature generally refers to acts of positioning that frame a student as competent or worthy as positive and identities that include an appreciation for the value of mathematics as positive.

4. As Valero (2009) points out, psychoanalytic studies in mathematics education seem to rarely use the word "identity," preferring instead "relationship." For ease, we use the term identity throughout this section, although recognizing that in this theory, more so than the others discussed in this chapter, it is not the preferred term.

References

Adams, G. (2013). Women teachers' experiences of learning mathematics. *Research in Mathematics Education, 15*(1), 87–88.

Akkoç, H., Yeŝildere-Ðmreb, S., & Ali Balkanlıo°luc, M. (2014). Examining professional identity through story telling. *Research in Mathematics Education, 16*(2), 204–205.

Bartholomew, H., Darragh, L., Ell, F., & Saunders, J. (2011). "I'm a natural and I do it for love!": Exploring students' accounts of studying mathematics. *International Journal of Mathematical Education in Science and Technology, 42*(7), 915–924.

Battey, D., & Franke, M. L. (2008). Transforming identities: Understanding teachers across professional development and classroom practice. *Teacher Education Quarterly, 35*(3), 127–149.

Berry, R. Q., III. (2008). Access to upper-level mathematics: The stories of successful African American middle school boys. *Journal for Research in Mathematics Education 39*(5), 464–488.

Bibby, T. (2010). *Education—An "impossible profession"?: Psychoanalytic explorations of learning and classrooms.* New York, NY: Routledge.

Black, L., Mendick, H., Rodd, M., & Solomon, Y. (2009). Pain, pleasure and power: Selecting and assessing defended subjects. In L. Black, H. Mendick, & Y. Solomon (Eds.), *Mathematical relationships in education: Identities and participation* (pp. 19–30) London, United Kingdom: Routledge.

Boaler, J. (2002a). *Experiencing school mathematics: Traditional and reform approaches to teaching and their impact on student learning.* New York, NY: Routledge.

Boaler, J. (2002b). Paying the price for "sugar and spice": Shifting the analytical lens in equity research. *Mathematical Thinking and Learning, 4*(2–3), 127–144.

Boaler, J., & Greeno, J. G. (2000). Identity, agency, and knowing in mathematics worlds. In J. Boaler (Ed.), *Multiple perspectives on mathematics teaching and learning* (pp. 171–200). Westport, CT: Ablex.

Boaler, J., & Humphreys, C. (2005). *Connecting mathematical ideas: Middle school video cases to support teaching and learning* (Vol. 1). Portsmouth, NH: Heinemann Educational Books.

Boaler, J., & Staples, M. (2008). Creating mathematical futures through an equitable teaching approach: The case of Railside School. *Teachers College Record, 110*(3), 608–645.

Brescoll, V. (2011). Who takes the floor and why: Gender, power, and colubility in organizations. *Administrative Science Quarterly, 56*(4), 622–641.

Clark, L., Badertscher, E., & Napp, C. (2013). African American mathematics teachers as agents in their African American students' mathematics identity formation. *Teachers College Record, 115*(2), 1–36.

Cobb, P., Gresalfi, M., & Hodge, L. L. (2009). An interpretive scheme for analyzing the identities that students develop in mathematics classrooms. *Journal for Research in Mathematics Education, 40*(1), 40–68.

Davies, B., & Harré, R. (1999). Positioning and personhood. In R. Harré & L. van Langenhove (Eds.), *Positioning theory: Moral contexts of intentional action* (pp. 32–52). Malden, MA: Blackwell.

de Freitas, E., Wagner, D., Esmonde, I., Knipping, C., Lunney Borden, L., & Reid, D. (2012). Discursive authority and sociocultural positioning in the mathematics classroom: New directions for teacher professional development. *Canadian Journal of Science, Mathematics and Technology Education, 12*(2), 137–159.

Dunleavy, T. K. (2015). Delegating mathematical authority as a means to strive toward equity. *Journal of Urban Mathematics Education, 8*(1), 62–82.

Engle, R. A., Langer-Osuna, J., & McKinney de Royston, M. (2014). Towards a model of influence in persuasive discussions: Negotiating quality, authority, and access within a student-led argument. *Journal of the Learning Sciences. 23*(2), 245–268.

Esmonde, I. (2009). Mathematics learning in groups: Analyzing equity in two cooperative activity structures. *The Journal of the Learning Sciences, 18*(2), 247–284.

Esmonde, I. (2011). Snips and snails and puppy dog tails: Genderism and mathematics education. *For the Learning of Mathematics, 31*(2), 27–31.

Esmonde, I. (2014). "Nobody's rich and nobody's poor ... It sounds good, but it's actually not": Affluent students learning mathematics and social justice. *Journal of the Learning Sciences, 23*(3), 348–391.

Esmonde, I., & Sengupta-Irving, T. (2015, April). Leveraging feminist theory to disrupt gendered mathematics teaching and learning: Classroom perspectives. Paper presented at the Research Conference of the National Council of Teachers of Mathematics, Boston, MA.

Esmonde, I., Takeuchi, M., & Dookie, L. (2012). Integrating insights from critical race and queer theories with cultural-historical learning theory. In *Proceedings of the 10th International Conference of the Learning Sciences (ICLS 2012)—Volume 2* (pp. 491–492). Sydney, Australia: International Society of the Learning Sciences.

Foucault, M. (1982). The subject and power. *Critical inquiry, 8*(4), 777–795.

Frosh, S., & Baraitser, L. (2008). Psychoanalysis and psychosocial studies. *Psychoanalysis, Culture & Society, 13*(4), 346–65.

Gholson, M., & Martin, D. B. (2014). Smart girls, black girls, mean girls, and bullies: At the intersection of identities and the mediating role of young girls' social network in mathematical communities of practice. *Journal of Education, 194*(1), 19–34.

Gresalfi, M. S., & Cobb, P. (2006). Cultivating students' discipline-specific dispositions as a critical goal for pedagogy and equity. *Pedagogies, 1*(1), 49–57.

Gresalfi, M., Martin, T., Hand, V., & Greeno, J. (2009). Constructing competence: An analysis of student participation in the activity systems of mathematics classrooms. *Educational Studies in Mathematics, 70*(1), 49–70.

Gresky, D. M., Ten Eyck, L. L., Lord, C. G., & McIntyre, R. B. (2005). Effects of salient multiple identities on women's performance under mathematics stereotype threat. *Sex Roles, 53*(9–10), 703–716.

Holland, D., Lachicotte, D., Jr., Skinner, D. D., & Cain, C. (1998). *Identity and agency in cultural worlds.* Cambridge, MA: Harvard University Press.

Ladson-Billings, G., & Tate, W., IV. (1995). Toward a critical race theory of education. *Teachers College Record, 97*(1), 47–68.

Langer-Osuna, J. (2011). How Brianna became bossy and Kofi came out smart: Understanding the trajectories of identity and engagement for two group leaders in a project-based mathematics classroom. *Canadian Journal of Science, Mathematics and Technology Education, 11*(3), 207–225.

Langer-Osuna, J. (2015). From getting "fired" to becoming a collaborator: A case on student autonomy and the co-construction of identity and engagement in a project-based mathematics classroom. *Journal of the Learning Sciences, 24*(1), 53–92.

Lave, J., & Wenger, E. (1991). *Situated learning: Legitimate peripheral participation*. Cambridge, MA: Cambridge University Press.

Lubienski, S. T., & Ganley, C. M. (2017). Research on gender and mathematics. In J. Cai (Ed.), *Compendium for research in mathematics education* (pp. 649–666). Reston, VA: National Council of Teachers of Mathematics.

Martin, D. B. (2000). *Mathematics success and failure among African-American youth: The roles of sociohistorical context, community forces, school influence, and individual agency*. Mahwah, NJ: Lawrence Erlbaum.

Martin, D. B. (2006). Mathematics learning and participation as racialized forms of experience: African American parents speak on the struggle for mathematics literacy. *Mathematical Thinking and Learning, 8*(3), 197–229.

Martin, D. B. (2013). Race, racial projects, and mathematics education. *Journal for Research in Mathematics Education, 44*(1), 316–333.

Martin, D. B., Anderson, C. R., & Shah, N. (2017). Race and mathematics education. In J. Cai (Ed.), *Compendium for research in mathematics education* (pp. 607–636). Reston, VA: National Council of Teachers of Mathematics.

Mendick, H. (2006). *Masculinities in mathematics*. Maidenhead, United Kingdom: Open University Press.

Moschkovich, J. N. (2010). *Language and mathematics education: Multiple perspectives and directions for research*. Charlotte, NC: Information Age.

Nasir, N. I. S. (2002). Identity, goals, and learning: Mathematics in cultural practice. *Mathematical Thinking and Learning, 4*(2–3), 213–247.

Nasir, N. I. S. (2005). Individual cognitive structuring and the sociocultural context: Strategy shifts in the game of dominoes. *The Journal of the Learning Sciences, 14*(1), 5–34.

Rodd, M., & Bartholomew, H. (2006). Invisible and special: Young women's experiences as undergraduate mathematics students. *Gender and Education, 18*(1), 35–50.

Russell, N. M. (2012). Classroom discourse: A means to positively influence mathematics achievement for African American students. *Curriculum and Teaching Dialogue, 14*(1–2), 35.

Schweinle, A., & Mims, G. A. (2009). Mathematics self-efficacy: Stereotype threat versus resilience. *Social Psychology of Education, 12*(4), 501–514.

Sfard, A., & Prusak, A. (2005). Telling identities: In search of an analytic tool for investigating learning as a culturally shaped activity. *Educational Researcher, 34*(4), 14–22.

Solomon, Y. (2007). Not belonging? What makes a functional learner identity in undergraduate mathematics? *Studies in Higher Education, 32*(1), 79–96.

Solomon, Y., Lawson, D., & Croft, T. (2011). Dealing with "fragile identities": Resistance and refiguring in women mathematics students. *Gender and Education, 23*(5), 565–583.

Spencer, S. J., Steele, C. M., & Quinn, D. M. (1999). Stereotype threat and women's math performance. *Journal of Experimental Social Psychology, 35*(1), 4–28.

Steele, C. M., Spencer, S. J., & Aronson, J. (2002). Contending with group image: The psychology of stereotype and social identity threat. *Advances in Experimental Social Psychology, 34*, 379–440.

Stinson, D. W. (2013). Negotiating the "white male math myth": African American male students and success in school mathematics. *Journal for Research in Mathematics Education, 44*(1), 69–99.

Stinson, D. W., & Walshaw, M. (2017). Exploring different theoretical frontiers for different (and uncertain) possibilities in mathematics education research. In J. Cai (Ed.), *Compendium for research in mathematics education* (pp. 128–155). Reston, VA: National Council of Teachers of Mathematics.

Tagler, M. J. (2012). Choking under the pressure of a positive stereotype: Gender identification and self-consciousness moderate men's math test performance. *The Journal of Social Psychology, 152*(4), 401–416.

Teixeira, B. R., & Cyrino, M. C. D. C. T. (2014). O estágio de observação e o desenvolvimento da identidade profissional docente de professores de matemática em formação inicial [Classes observations and the development of preservice mathematics teachers' professional identity]. *Educação Matemática Pesquisa. Revista do Programa de Estudos Pós-Graduados em Educação Matemática, 16*(2), 599–622.

Tine, M., & Gotlieb, R. (2013). Gender-, race-, and income-based stereotype threat: The effects of multiple stigmatized aspects of identity on math performance and working memory function. *Social Psychology of Education, 16*(3), 353–376.

Tomasetto, C., Alparone, F. R., & Cadinu, M. (2011). Girls' math performance under stereotype threat: The moderating role of mothers' gender stereotypes. *Developmental Psychology, 47*(4), 943–949.

Turner, E., Dominguez, H., Maldonado, L., & Empson, S. (2013). English learners' participation in mathematical discussion: Shifting positionings and dynamic identities. *Journal for Research in Mathematics Education, 44*(1), 199–234.

Valero, P. (2009). What has power got to do with mathematics education. In P. Ernest, B. Geer, & B. Sriraman (Eds.), *Critical issues in mathematics education*, (pp. 237–254). Charlotte, NC: Information Age.

Varelas, M., Martin, D. B., & Kane, J. M. (2012). Content learning and identity construction: A framework to strengthen African American students' mathematics and science learning in urban elementary schools. *Human Development, 55*(5), 319–339.

Wager, A. A., & Foote, M. Q. (2013). Locating praxis for equity in mathematics lessons from and for professional development. *Journal of Teacher Education, 64*(1), 22–34.

Walshaw, M. (2008). Developing theory to explain learning to teach. In T. Brown (Ed.), *The psychology of mathematics education: A psychoanalytic displacement* (pp. 119–138). Rotterdam, The Netherlands: Sense.

Walshaw, M. (2013). Explorations into pedagogy within mathematics classrooms: Insights from contemporary inquiries. *Curriculum Inquiry, 43*(1), 71–94.

Wenger, E. (1999). *Communities of practice: Learning, meaning, and identity*. Cambridge, MA: Cambridge University Press.

Williams, J. (2011). Teachers telling tales: The narrative mediation of professional identity. *Research in Mathematics Education, 13*(2), 131–142.

Wood, M. B. (2013). Mathematical micro-identities: Moment-to-moment positioning and learning in a fourth-grade classroom. *Journal for Research in Mathematics Education, 44*(5), 775–808.

Wortham, S. E. F. (2006). *Learning identity: The joint emergence of social identification and academic learning*. Cambridge, MA: Cambridge University Press.

24 性别与数学的研究

萨拉·特勒·鲁宾斯基
美国伊利诺伊大学厄巴纳-香槟分校
科琳·M.甘利
美国佛罗里达州立大学
译者：于文华
　　　山东师范大学数学与统计学院

莱德于1992年对性别和数学的相关研究进行了综述，发现数学教育研究团体对学业成绩、态度和高中数学课程参与度方面的性别差异给予了高度关注，当时的数学教育期刊有9%—10%的内容与性别有关（Leder, 1992；Lubienski & Bowen, 2000）。[1]然而，在过去的25年里，相关研究结论已经发生了巨大的转变：修读高中高级数学课程的差距已经大幅缩减（Dalton, Ingels, Downing, & Bozick, 2007），女性在数学成就和科学、技术、工程、数学（STEM）等领域的参与性上获得了进展（Hyde, Lindberg, Linn, Ellis, & Williams, 2008；经济合作与发展组织[OECD]，2015；Perez-Felkner, McDonald, & Schneider, 2014）。关注点已经转向男生相对较差的阅读能力与女生更可能从高中和大学毕业这一事实上，对女生在数学方面的担忧在进一步减少（例如Sommers, 2001）。

虽然在数学教育领域中性别研究似乎有些过时，[2]但是这方面的研究在其他领域仍很普遍。的确，有些期刊专门致力于性别这一主题，特别是用心理学或社会学的视角（例如，《性别角色》《女性心理学季刊》），有关性别和数学的文章经常出现在这些期刊中。虽然对性别的研究渗透了许多学科，但在许多工业化国家，男性在数学上表现出的优势的独特性表明，对这个主题的研究不仅可以揭示数学教育中的问题，也可以使人们对性别问题有更广泛的了解。这一章的作者分别是数学教育研究者和心理学研究者。我们致力于揭示这两个领域取得的进展，并推动对数学教育中常见的、令人困惑的性别问题重新给予更多关注。

本章概述：为什么要研究性别与数学教育？

显然，在男孩和女孩之间相对差异性而言更多的是相似性，当人们在大众媒体上呼吁关注性别差异时，是冒着使陈旧的模式化的见解长存的风险（Boaler & Sengupta-Irving, 2006；O'Connor & Joffe, 2014）。尽管存在这些相似性和风险，而且对美国和许多其他工业化国家的男孩相对较低的阅读能力和大学入学率一直存在合理的担忧（例如Loveless, 2015；OECD, 2015），但是，数学教育学者依然有重要的理由，特别要继续致力于男性优势差距方面的研究。

虽然从整体数学学生的成绩上来看，与男女生性别相关的差距很小，并且各国之间的结果并不一致，但是男性在自信心以及最高水平数学的学业成绩方面具有较大优势（OECD, 2015）。在美国，虽然有些州测试有让人鼓舞的证据，但最新的全国性的证据表明，尽管男孩和女孩在幼儿园入学时有相似的数学水平，但是从小学三年级开始，他们的学业成绩和自信心方面出现了令人惊讶的差距（Cimpian, Lubienski, Timmer, Makowski,

24 | Research on Gender and Mathematics

SARAH THEULE LUBIENSKI
University of Illinois at Urbana-Champaign, United States

COLLEEN M. GANLEY
Florida State University, United States

In 1992, Leder synthesized existing research on gender and mathematics at a time when the mathematics education research community gave substantial attention to sex differences in performance, attitudes, and participation in advanced high school mathematics courses. At that time, 9–10% of the contents of mathematics education journals pertained to sex or gender (Leder, 1992; Lubienski & Bowen, 2000).[1] However, in the past 25 years, concerns about gender in education have shifted dramatically. Disparities in high school advanced-mathematics course taking have narrowed substantially (Dalton, Ingels, Downing, & Bozick, 2007), and women have gained ground in mathematics achievement and participation in some science, technology, engineering, and mathematics (STEM) fields (Hyde, Lindberg, Linn, Ellis, & Williams, 2008; Organisation for Economic Co-Operation and Development [OECD], 2015; Perez-Felkner, McDonald, & Schneider, 2014). Attention has turned to boys' relatively poor reading performance and the fact that girls are more likely to graduate from high school and college, further diminishing concerns about girls in mathematics (e.g., Sommers, 2001).

Although gender research seems to have gone somewhat out of vogue in the mathematics education community,[2] such work is still quite prevalent in other fields. Indeed, there are entire journals devoted to the topic of gender, typically with a psychological or sociological lens (e.g., *Sex Roles*, *Psychology of Women Quarterly*), and articles about gender and mathematics often appear in these journals. Although the study of gender permeates many disciplines, the uniqueness of males' advantage in mathematics within many industrialized countries suggests that research on this topic not only can inform mathematics education but also could inform broader understandings of gender. The authors of this chapter are respectively, a mathematics education researcher and a psychology researcher. Our goals are to highlight developments that have occurred in both fields and to push for renewed and more focused attention to persistent, puzzling gender issues specific to mathematics education.

Chapter Overview: Why Study Gender and Mathematics Education?

Clearly, there are more similarities than differences between boys and girls, and when those in the popular media call attention to gender differences, they run the risk of perpetuating stereotypes (Boaler & Sengupta-Irving, 2006; O'Connor & Joffe, 2014). Despite such similarities and risks, and despite ongoing, valid concerns about boys' relatively low reading performance and college attendance in the United States and many other industrialized countries (e.g., Loveless, 2015; OECD, 2015), there are important reasons for mathematics education scholars, in particular, to continue to attend to disparities favoring males.

Although sex-related disparities in overall mathematics achievement are small and inconsistent across countries, substantial disparities in confidence and in performance at the highest mathematics levels persistently favor males (OECD, 2015). Within the United States, despite some encouraging evidence from state tests, recent national evidence indicates that although boys and girls begin kindergarten with similar mathematics proficiency, surprising disparities in achievement and confidence develop by grade 3 (Cimpian,

& Miller, 2016; Lubienski, Robinson, Crane, & Ganley, 2013; Penner & Paret, 2008; Wiest, 2011)。形成对比的是,女孩在幼儿园时期表现出的阅读优势在小学期间已经缩减(Robinson & Lubienski, 2011)。

而且,虽然女性比男性更有可能进入大学,但是她们进入大学后不太可能从事与数学相关的职业(Perez-Felkner等,2014)。这与由社会经济地位(SES)或种族/民族产生的差距不同,对后者来说,数学相关职业的追求上的差异与大学入学准备及出勤率不同有关(Riegle-Crumb & King, 2010)。在从事与数学相关的主要职业上的差距仍然很大,这几乎对所有经济合作与发展组织成员的女性和劳动力素质有严重的影响(OECD, 2015)。例如,根据美国人口普查的数据,女大学生在毕业10年后全职工作的收入仅为同等男性工资的69%,这样的收入差距很大程度上归因于一个事实,那就是更多的男性从事与数学相关的职业,因为在这些领域的女性和男性的收入都高于其他领域的同行(Dey & Hill, 2007)。

目前的证据清楚地表明,性别和数学教育有一些特殊且持久的东西需要我们关注。本章,我们聚焦于那些在数学中男性占优势的领域。我们考察国际和美国背景下的研究,并在可能时,包括有关美国人口中性别与人种、种族、社会经济地位交叉的文献。我们从进化理论观点讨论出发,之后我们把数学学业成绩和参与的模式记录下来。然后我们讨论与这些模式有关的因素,最后以未来研究与干预的方向结束。

理论观点的发展

在过去的几十年里,女权主义观点影响了学者们研究性别和数学的方法。传统意义上,"自由女权主义者"专注于帮助女性在数学方面和男性一样成功,而"激进女权主义者"则挑战了主张同化主义的人、"改变女孩"的倾向,转而关注如何让社会和数学领域改变方式,以停止给予男性特权和他们喜欢的认知方式(Lacampagne, Campbell, Herzig, Damarin, & Vogt, 2007; Leder, 2010)。后现代女权主义的观点(包括酷儿理论)已经对社会的性别建构给予了更大的关注,这对传统"男性"和"女性"对立划分的观点是一个挑战(Esmonde, 2011)。

在过去,关于男性和女性差异原因的理论一直存在先天与后天两方面的对立倾向。由于数学教育研究者致力于改善教学,因此该领域更倾向于关注"后天",包括女孩的数学经验和由此产生的态度。然而,数学教育领域之外,一些学者认为,数学中的性别差异是由于生物因素产生的(Benbow & Stanley, 1980; Geary, 1996; Kimura, 1992; Pinker, 2002)。最近的理论认为,性别差异的存在是由于遗传和环境因素之间复杂的相互作用产生的(Halpern, 2004; Halpern, Wai, & Saw, 2005; Nuttall, Casey, & Pezaris, 2005)。

当前的观点鼓励我们聚焦于个体力量、价值观和选择,而更少地强调将天性和培育作为产生性别差异的决定性因素来分析。更具体地说,与其企图让女生觉得数学重要或改变数学学习经历来让数学成绩平衡,学者们现在呼吁我们考虑女生的优势和价值观,以及考虑她们可能不适合高强度的数学工作(Eccles & Wang, 2015; Vale & Bartholomew, 2008)。特别是,埃克尔斯(2009)的期望价值理论强调在职业选择时女性的价值观和对成功的期望的作用。

虽然我们应该尊重女孩的价值观和选择,应该质疑为什么社会不在价值上对女性主导的职业与男性主导的职业一视同仁(Noddings, 1998),但是从更直接的意义上说,如上所述,目前在数学相关领域中女性任职人数不足的现象严重影响了劳动力和薪酬公平性。即使女性选择去从事利润较低、地位较低的职业,而不是那些需要更多数学知识的职业,但问题是是否这些选择受到了教育机会、社会化或歧视的制约。

因此,仍然有很多根本问题需要探讨,本章以下的内容正是做的此项工作。从数学成绩中的性别差异开始,到这些模式表面下的潜在的原因。在反应这个领域的不同的情况(Leder, 2010)时,我们从与上文梳理的理论角度有关的研究中得到结论,并主要聚焦于这10年完成的工作。我们专注于环境因素而不是生物因素,因为大量的研究指出了社会因素的重要性,因此关注数学教育共同体能够很好地解决的可塑因素是有意义的。

Lubienski, Timmer, Makowski, & Miller, 2016; Lubienski, Robinson, Crane, & Ganley, 2013; Penner & Paret, 2008; Wiest, 2011). In contrast, the gap favoring girls in reading is present in kindergarten but narrows during elementary school (Robinson & Lubienski, 2011).

Moreover, although women are more likely than men to attend college, they are less likely to pursue mathematics-related careers once there (Perez-Felkner et al., 2014). This is different than gaps by socioeconomic status (SES) or race/ethnicity, where disparities in the pursuit of mathematics-related careers are linked to differences in college preparation and attendance (Riegle-Crumb & King, 2010). Gaps in key mathematics-related careers remain large, with serious implications for women and for workforce quality across almost all OECD countries (OECD, 2015). For example, according to U.S. census data, women college graduates employed full-time still earn only 69% of comparable men's salaries 10 years after graduation. Much of this income disparity is attributable to the fact that more men pursue mathematics-related careers, as both women and men in those fields earn more than their counterparts in other fields (Dey & Hill, 2007).

Current evidence clearly suggests that there is something special and persistent about gender and mathematics education that deserves our attention. In this chapter we focus on areas in which we see disparities favoring males in mathematics. We examine research in both international and U.S. contexts and include literature on the intersections of gender with race, ethnicity, and SES within U.S. populations, when available. We begin with a discussion of evolving theoretical perspectives, after which we take stock of patterns in mathematics performance and participation. We then discuss factors related to these patterns, and we close with directions for future research and interventions.

Evolving Theoretical Perspectives

Feminist perspectives have influenced scholars' approaches to studying gender and mathematics over the past several decades. Traditionally, "liberal feminists" have focused on helping women become as successful in mathematics as men, whereas "radical feminists" have challenged assimilationist, "change the girl" tendencies and focused instead on ways in which society and the field of mathematics should change to stop privileging men and their preferred ways of knowing (Lacampagne, Campbell, Herzig, Damarin, & Vogt, 2007; Leder, 2010). More recently, postmodern feminist perspectives (including queer theory) have given greater attention to the social construction of gender, challenging the traditional view of "male" and "female" as dichotomous categories (Esmonde, 2011).

In the past, theories about the causes of differences between males and females have tended to fall along nature versus nurture lines. Given mathematics education researchers' focus on improving instruction, the field has tended to focus on "nurture," including girls' experiences with mathematics and resulting attitudes. However, outside of mathematics education, some scholars have suggested that sex differences in mathematics are due to biological factors (Benbow & Stanley, 1980; Geary, 1996; Kimura, 1992; Pinker, 2002). More recent theories posit that sex differences exist due to a complex interplay between genetic and environmental factors (Halpern, 2004; Halpern, Wai, & Saw, 2005; Nuttall, Casey, & Pezaris, 2005).

Current perspectives encourage us to focus on individual agency, values, and choices, with less emphasis on parsing the role of nature or nurture as deterministic factors that create sex differences. More specifically, in contrast to attempts to equalize mathematics outcomes by convincing girls that mathematics is important or by changing mathematics learning experiences, scholars now urge us to consider girls' strengths and values and how they might or might not fit with mathematics-intensive occupations (Eccles & Wang, 2015; Vale & Bartholomew, 2008). Specifically, Eccles's (2009) expectancy-value theory emphasizes the role of women's values and expectations for success when making occupational choices.

Although we should respect girls' values and choices and should also question why society does not value female-dominated occupations to the same degree as those that are male-dominated (Noddings, 1998), in the more immediate sense, there are severe workforce and pay-equity consequences to the current underrepresentation of women in mathematics-related fields, as argued above. Even if women are *choosing* to pursue less lucrative, lower-status careers over those that are more mathematics-intensive, *the question remains whether those choices are constrained by educational opportunities, socialization, or discrimination.*

Hence, there remain many fundamental questions to be explored, and the remainder of this chapter does so, beginning with sex differences in mathematics outcomes before moving to potential factors underlying those patterns. Reflecting the varied state of the field (Leder, 2010), we draw from studies aligned with the array of theoretical perspectives outlined above, focusing primarily on work completed in the past decade. We focus on

数学学业成绩

国际趋势

在过去的20年中,两个大型国际研究,国际数学与科学趋势研究(TIMSS)和国际学生评估项目(PISA),使得考察不同国家和文化中与性别相关的差异成为可能。2011年的TIMSS数据呈现出混合模式,与八年级相比,四年级的男生更明显地表现出在数学方面的优势(Provasnik等,2012)。与TIMSS相比,PISA中的性别差异一致地倾向于男孩优势,这可能是因为PISA更注重数学素养[3],而TIMSS与学校课程紧密相连(Else-Quest, Hyde, & Linn, 2010)。来自PISA的数据显示,在经济合作与发展组织成员中,男孩在数学上的平均表现比女孩好,相当于多接受了5个月的教育(OECD,2015)。例如,2012年的PISA数据显示,65个参与国中,38个国家男性得分显著超过女性,22个国家没有显著的性别差异,5个国家(约旦、卡塔尔、泰国、马来西亚、冰岛;OECD,2014)女性得分超过男性。然而,在表现最好的学生中,没有一个国家的女孩超过男孩(OECD,2015),这与澳大利亚和美国发现在成绩分布的顶端,男性优势最为明显的研究结果相一致(Forgasz & Hill, 2013;Robinson & Lubienski, 2011)。一些研究还发现,数学学业成绩中男性优势差距较小的国家更加强调性别平等,例如,以女性外出工作和担任政治职务的比例为衡量标准(Else-Quest 等, 2010;González de San Román & De la Rica Goiricelaya, 2012;Guiso, Monte, Sapienza, & Zingales, 2008)。

总的来说,与数学信念的差异相比,PISA数学学业成绩的性别差异较小(OECD, 2013, 2014)。经济合作与发展组织(2014)的报告总结了下面这些国际层面的结果:

> 即使有时女孩在数学方面和男孩一样好,但是她们表现出毅力更少,学习数学的动力更小,对自己数学技能的信任更少,对数学的焦虑程度也更高。女孩在数学上的平均表现不如男孩,而同时男孩的性别差异优势在最高学业成就的学生之中表现得甚至更明显,当这些年轻女性进入劳动力市场时,这些发现将意蕴深远……(第4页)[4]

美国趋势

像TIMSS和PISA这样的国际评估,都没有考察到女孩和男孩之间的差距第一次出现在何时以及差距是如何随着学生在学校的发展而改变的。美国的两个数据集,美国教育进展评估(NAEP)和美国早期儿童纵向研究-幼儿园(ECLS-K)项目,使在小学、初中、高中范围内对美国性别差异的深入考查成为可能。

最初,美国数学学业成绩的性别差异似乎是在初中和高中期间形成的(Hyde, Fennema, & Lamon, 1990)。然而,在17岁孩子的NAEP数学成绩中,男性优势从1973年约为0.3个标准差下降到最近评估中的0.1个标准差(Perie, Moran, & Lutkus, 2005)。相反,1973年美国教育进展评估表明,9岁和13岁孩子的数学学业成绩差异女孩优势很小但很显著,但到20世纪90年代初,这一差异逆转为男孩具有优势,并且在长期趋势评估和主要的美国教育进展评估中都维持着微小却显著的(约0.1个标准差)男孩优势(McGraw, Lubienski, & Strutchens, 2006;Perie 等, 2005)。这些随时间而发生的变化表明,性别差异不应该被视为固定的或不可避免的。事实上,在过去的20年里,随着男孩和女孩的主要美国教育进展评估数学成绩的提高,在任何给定时间节点上,女孩的分数至少与几年前男孩的分数一样高(McGraw等, 2006;最新的美国教育进展评估结果可以在http://nces.ed.gov/nationsreportcard/nde查阅)。

过去的20年中,ECLS-K(1988和2011)的支持者允许研究者更仔细地考察中小学的模式,这些数据表明,从小学一年级到三年级,男生的优势差距增长到约0.25个标准差(Cimpian 等, 2016;Fryer & Levitt, 2010;Husain & Millimet, 2009;Penner & Paret, 2008)。事实上,ECLS-K的男女差异的增长至少与在小学低年级与种族和社会经济地位相关的差异的增长持平(Fryer & Levitt, 2010;Reardon & Robinson, 2008)。但是在5年级之后,ECLS-K中数学学业成绩的男性优势逐渐减少(Robinson & Lubienski, 2011)。

不同种族/民族群体中的ECLS-K数据进一步表明,拉

environmental—not biological—factors because ample research points to the importance of social factors, and it makes sense to focus on malleable factors that the mathematics education community is well-positioned to address.

Mathematics Performance

International Trends

In the past 20 years, two large international studies, the Trends in International Mathematics and Science Study (TIMSS) and the Program for International Student Assessment (PISA), have made it possible to examine sex-related differences across a number of countries and cultures. The 2011 TIMSS data reveal a mix of patterns, with mathematics disparities favoring males occurring more often at grade 4 than at grade 8 (Provasnik et al., 2012). Gaps more consistently favor boys on PISA than on TIMSS, which could be due to PISA's focus on mathematical literacy[3] in contrast with TIMSS's tighter connection with school curricula (Else-Quest, Hyde, & Linn, 2010). Data from PISA reveal that boys, on average, outperform girls in mathematics by the equivalent of 5 months of schooling across OECD countries (OECD, 2015). More specifically, the 2012 PISA data reveal that males significantly outscored females in 38 of the 65 participating countries, no significant sex differences were found in 22 countries, and females outscored males in 5 countries (Jordan, Qatar, Thailand, Malaysia, and Iceland; OECD, 2014). However, in no country did girls outperform boys among the highest performers (OECD, 2015), consistent with studies from Australia and the United States indicating that mathematics gaps favoring males are largest at the top of the distribution (Forgasz & Hill, 2013; Robinson & Lubienski, 2011). Some research finds that mathematics performance gaps favoring males are smaller when countries have more gender equality, as measured, for example, by the percentage of women working outside the home and holding political office (Else-Quest et al., 2010; González de San Román & De la Rica Goiricelaya, 2012; Guiso, Monte, Sapienza, & Zingales, 2008).

Overall, PISA mathematics performance disparities by sex are small compared with disparities in mathematics-related beliefs (OECD, 2013, 2014). The OECD (2014) report summarizes these international patterns as follows:

> Even when girls perform as well as boys in mathematics, they report less perseverance, less motivation to learn mathematics, less belief in their own mathematics skills, and higher levels of anxiety about mathematics. While the average girl underperforms in mathematics compared to the average boy, the gender gap in favour of boys is even wider among the highest-achieving students. These findings have serious implications...when these young women enter the labour market. (p. 4)[4]

United States Trends

No international assessment, such as TIMSS or PISA, detects when gaps between girls and boys first appear and how the gaps change as students progress through school. Two datasets within the United States, the National Assessment of Educational Progress (NAEP) and the Early Childhood Longitudinal Study-Kindergarten (ECLS-K), allow an in-depth examination of U.S. disparities by sex across elementary, middle, and high school.

Initially, sex differences in U.S. mathematics performance seemed to develop during middle or high school (Hyde, Fennema, & Lamon, 1990). However, the gap favoring males in 17-year-olds' NAEP mathematics scores declined from roughly .3 standard deviations in 1973 to .1 standard deviations in more recent assessments (Perie, Moran, & Lutkus, 2005). Conversely, in 1973 there were small but significant NAEP mathematics performance gaps *favoring girls* for both 9- and 13-year-olds, but by the early 1990s, the gap reversed to favor boys and has remained small but significantly (roughly .1 standard deviations) in favor of boys on both the Long Term Trend and Main NAEP (McGraw, Lubienski, & Strutchens, 2006; Perie et al., 2005). These changes over time suggest that disparities by sex should not be viewed as fixed or inevitable. In fact, as Main NAEP mathematics scores have risen for both boys and girls over the past 2 decades, at any given point in time girls scored at least as high as boys did just a few years earlier (McGraw et al., 2006; the most recent NAEP results can be viewed at http://nces.ed.gov/nationsreportcard/nde).

Over the past decade, two cohorts of the ECLS-K dataset (1998 and 2011) have allowed researchers to more carefully examine patterns in elementary and middle school, and these data indicate that gaps favoring boys appear by grade 1 and grow to about .25 standard deviations by grade 3 (Cimpian et al., 2016; Fryer & Levitt, 2010; Husain & Millimet, 2009; Penner & Paret, 2008). In fact, ECLS-K male-female gaps grow at least as much as race- and SES-related gaps do during early elementary grades (Fryer & Levitt, 2010; Reardon & Robinson, 2008). After grade 5, however, the gaps favoring males in ECLS-K mathematics performance diminish slightly (Robinson & Lubienski, 2011).

丁裔中男女孩差异最小（0.1个标准差或更小；Lubienski 等，2013），与之前非裔美国女孩和男孩之间没有数学差距的研究结果不同（例如 McGraw 等，2006），ECLS-K 数据显示，三、五、八年级的非裔美国男孩更有优势（标准差为0.1—0.4）。

测试成绩与等级

在美国和其他地方数学学业成绩呈现出的模式，可能会与那些由老师评定的成绩和州测试的成绩没有观察到男生优势的研究报告（Catsambis，1994；Hyde 等，2008；Pomerantz, Altermatt, & Saxon, 2002）有冲突。然而，当教学评估内容严格与学校教的内容一致时，女生的表现往往比较好（Downey & Vogt Yuan, 2005；Kimball, 1989），在最具挑战性的题目上男生更具优势，这与在成绩最好的那部分学生中，男女学生差距最大的研究结论相一致（McGraw 等，2006；Robinson & Lubienski, 2011）。事实上，虽然林德伯格、海德、彼得森和林（2010）对242项研究作的元分析得出总体上男性具有数学优势的差异是可以忽略不计的（$d=0.05$），但是他们发现在高度选择的样本中存在显著差异（$d=0.4$）。在美国，州数学测试有相对较少的挑战性问题（Hyde 等，2008），旨在判断学生是否达到特定的内容标准，而不是精确考察学业成绩的分布。这些因素有助于解释为什么在美国的数学评估中男性优势更明显，因为包括SAT和ECLS-K评估在内的美国数学评估很少与课程内容相联系，而是旨在考察学业成绩的分布（College Board, 2013；Robinson & Lubienski, 2011）。

数学各内容领域呈现的模式

不同内容领域中或不同数学问题类型中的性别差异是不一致的。PISA在2012年的研究表明，尽管在四个内容领域的量表上，都是男生占优势，但空间与图形领域的量表与其他三个领域的量表相比，男女差异更大——空间和图形部分是15分，而变化和关系、数量、不确定性这三个领域在9—11分之间（OECD, 2014）。

美国的相关研究与这些国际层面的结果相呼应。例如，美国教育进展评估结果显示男女学生在代数方面没有性别差异，但在测量领域方面的差异特别明显，尤其是涉及测量量表解释的题目（McGraw 等，2006）。另外的证据表明男生在估计和测量物体时有极强的表现，尤其是当这些物体肉眼不可见时（Vasilyeva, Casey, Dearing, & Ganley, 2009）。另一方面，当测量问题涉及公式运用时女孩比男孩的表现好，这可能又一次表明女孩和男孩对学校教学内容依赖的差异性（Vasilyeva 等，2009）。这些模式可能与空间思维能力（下面会详细讨论）和问题解决方面的性别差异有关。

问题解决

越来越多的研究已经确定了女孩和男孩在解决问题方法上的差异。例如，在一项研究中，德国男生在需要"逆向推理"的题目中得分特别高，而女生在通过计算或绘图才能解决的题目中做得比较好（Winkelmann, van den Heuvel-Panhuizen, & Robitzsch, 2008）。类似地，解决比例推理问题时，研究发现美国女生使用程序性或加法策略的频率几乎是男生的两倍（Che, Wiegert, & Threlkeld, 2012）。这些发现可能与解决问题时使用言语还是空间思维能力的性别差异有关（例如 Klein, Adi-Japha, & Hakak-Benizri, 2010）。这方面的证据也与女生往往使用老师教的熟悉的方法，而不是自己发明新策略来解决数学问题这一研究成果相一致（Carr & Jessup, 1997；Fennema, Carpenter, Jacobs, Franke, & Levi, 1998；Gallagher & De Lisi, 1994；Gallagher 等，2000；Goodchild & Grevholm, 2009；Zhu, 2007）。这些研究与一个更普遍的模式相吻合，即女生在课堂上往往表现出教师所期望的行为（Downey & Vogt Yuan, 2005；Kimball, 1989；Rathbun, West, & Germino-Hausken, 2004），这也许可以解释为什么女孩通常在学校教的内容上表现得更好。很可能循规蹈矩的"好女孩"的行为有短期的益处，但是女生在数学自信心和表现方面却有着长期的劣势，因为她们以后遇到越来越复杂的数学任务时会需要更灵活的问题解决的方法（Goodchild & Grevholm, 2009）。

大学专业与职业选择

如上所述，相对于男性，女性较

A further examination of ECLS-K data within different racial/ethnic groups revealed that the differences between boys and girls were smallest among Latino/a students (.1 SDs or less; Lubienski et al., 2013). Unlike prior research indicating no mathematics gap between African American girls and boys (e.g., McGraw et al., 2006), the ECLS-K gaps tended to favor African American boys at grades 3, 5, and 8 (.1–.4 SDs).

Test Performance Versus Grades

Patterns in mathematics performance in the United States and elsewhere might appear to conflict with reports of others who have observed no disparities favoring boys in teacher-assigned grades and in performance on state tests (Catsambis, 1994; Hyde et al., 2008; Pomerantz, Altermatt, & Saxon, 2002). However, girls tend to perform better when mathematics assessments are aligned closely with school-taught material (Downey & Vogt Yuan, 2005; Kimball, 1989), and gaps favor boys on the most challenging items, consistent with the male-female gap being largest at the upper end of the achievement distribution (McGraw et al., 2006; Robinson & Lubienski, 2011). In fact, although Lindberg, Hyde, Petersen, and Linn (2010) concluded from their meta-analysis of 242 studies that the overall mathematics gap favoring males is negligible ($d = .05$), they found a substantial difference ($d = .4$) among highly selective samples. Within the United States, state mathematics tests have relatively few challenging problems (Hyde et al., 2008) and are designed to determine whether students meet particular content standards, as opposed to precisely measuring performance across the full distribution. These factors help explain why a male advantage is more evident on U.S. mathematics assessments that are less tied to curricular content and designed to measure performance across the distribution, including the SAT and ECLS-K assessments (College Board, 2013; Robinson & Lubienski, 2011).

Patterns by Mathematical Strand

Differences by sex are not consistent across content areas or types of mathematics problems. Although gaps consistently favor males across its four strands, PISA's 2012 Space and Shape subscale has a significantly larger disparity than that of the other mathematics content subscales—15 points versus 9–11 points in the areas of Change and Relationships, Quantity, and Uncertainty (OECD, 2014).

Evidence from the United States mirrors these international patterns. NAEP results show no sex differences in algebra, for example, but disparities tend to be particularly pronounced in the area of measurement, especially on items involving the interpretation of measurement scales (McGraw et al., 2006). Other evidence points to boys' particularly strong performance on estimation or measurement of objects, especially when some of the object is not visible (Vasilyeva, Casey, Dearing, & Ganley, 2009). On the other hand, girls perform better than boys when measurement problems involve formula use—which might, again, suggest differences in girls' and boys' reliance on school-taught material (Vasilyeva et al., 2009). These patterns could be related to gender differences in spatial skills (discussed in more detail below) and problem solving.

Problem Solving

A growing number of studies have identified differences in girls' and boys' problem-solving approaches. For example, in one study, German boys scored particularly well on items requiring "backwards reasoning," whereas girls did better on items that could be solved by counting or drawing (Winkelmann, van den Heuvel-Panhuizen, & Robitzsch, 2008). Similarly, when solving proportional reasoning problems, U.S. girls were found to use procedural or additive strategies almost twice as often as boys (Che, Wiegert, & Threlkeld, 2012). Such findings could relate to sex differences in the use of verbal versus spatial skills when solving problems (e.g., Klein, Adi-Japha, & Hakak-Benizri, 2010). This evidence is also consistent with studies suggesting that girls tend to use familiar, teacher-given rules to solve mathematics problems instead of inventing novel strategies (Carr & Jessup, 1997; Fennema, Carpenter, Jacobs, Franke, & Levi, 1998; Gallagher & De Lisi, 1994; Gallagher et al., 2000; Goodchild & Grevholm, 2009; Zhu, 2007). These studies align with a more general pattern showing that girls tend to exhibit teacher-expected behavior in the classroom (Downey & Vogt Yuan, 2005; Kimball, 1989; Rathbun, West, & Germino-Hausken, 2004), which may explain why girls often perform better on school-taught material. It could be that compliant, "good girl" behavior has short-term benefits but long-term drawbacks in terms of girls' mathematical confidence and performance as they later encounter increasingly complex mathematics tasks that require more flexible problem-solving approaches (Goodchild & Grevholm, 2009).

College Major and Career Choices

As noted above, women are still vastly underrepresented in some mathematics-intensive fields, including engineering and computer science (National Science Foundation [NSF], 2011; OECD, 2015). Within the United

少从事数学密集型职业，包括工程和计算机科学（美国国家科学基金会[NSF]，2011；OECD，2015）。在美国，现在女性广泛出现在生物和医学科学领域内，但工程学士学位中女性仅占16%，计算机科学学士学位中女性仅占18%（Snyder & Dillow，2011）。劳动力上的差异更加明显，持有学士学位与博士学位的科学家和工程师中女性依次仅占11%和6%（NSF，2011）。[5]

尽管在学业成绩分布上，顶端男性的优势导致了职业道路上的差异（Lubinski & Benbow，2006），但是数学成绩与所修课程的差异却远不能解释大学专业选择的差异。例如，里格尔-克拉姆和金（2010）发现，即使考虑到高中预备课程和偏向数学的态度，美国男性主修物理科学或工程领域的几率仍然大致是女性的两倍，尽管高中预备课程可以解释为何黑人和西班牙裔的男性在科学、技术、工程、数学领域的低代表率，然而，无论是数学预备课程还是数学态度（2002年的教育纵向研究[ELS]中测量到的）[6]都不能完全解释性别差异，因此，我们在寻找女性比例偏低的原因时，必须超越大学预备课程的范畴，来考察女性最终选择从事其他领域的复杂原因（Perez-Felkner 等，2014；Riegle-Crumb, King, Grodsky, & Muller，2012；Tyson, Lee, Borman, & Hanson，2007）。

与数学成绩和职业道路差异相关的因素

根据到目前为止所回顾的证据，我们发现一个在数学学业成绩方面证据充分但细小的有利于男性的差异，而在大学专业和职业选择上的差异则更大。下一个问题自然就是："为什么我们会发现这些差异？"正如上面提到的，多年来研究人员已经从多方面来探求这个问题，但这里我们只研究与数学教育特别相关的六个领域：（1）社会陈旧的传统观念；（2）教师的作用与教学；（3）高等数学的状况；（4）学生的态度、信念和价值观；（5）语言技能；（6）空间思维能力。我们强调似乎准备提供新见解的研究领域，并指出那些可能不像曾经那样富有成效的其他领域。

关于性别与数学的社会传统观念

对男性主导数学这一传统观念的调查研究已经产生了不一致的结果，差异在于研究的国家和对传统观念的测量方式。隐式测量的研究（即内隐的联想测验）表明，一些国家的儿童都知道性别与数学传统观念的存在（新加坡：Cvencek, Kapur, & Meltzoff，2015；美国：Cvencek, Meltzoff, & Greenwald，2011；德国：Steffens, Jelenec, & Noack，2010）。然而性别传统观念的外显（自我报告）测量研究结果是不一致的，一些研究发现了数学是男性主导的这种传统观念的证据（Cvencek 等，2015；Cvencek 等，2011；Steffens 等，2010），但是其他研究表明在一些国家没有传统观念（瑞典：Brandell, Leder, & Nyström，2007；澳大利亚和美国：Forgasz, Leder, & Kloosterman，2004；以色列：Forgasz & Mittelberg，2008）。[7]总之，这方面的研究表明，孩子们可能知道数学的性别传统观念，但是有时他们并不太相信那些传统观念。

研究人员发现，表现出数学内隐性别传统观念的女性对数学有较低的自我概念，大概是因为她们将传统观念应用到了自己身上（Cvencek 等，2015；Nosek, Banaji, & Greenwald，2002；Steffens 等，2010）。此外，对成年女性的研究表明，那些含蓄地认同传统观念的女性不太可能在数学和科学领域中有很好的表现，亦或是继续在数学和科学领域发展（Cvencek 等，2015；Nosek & Smyth，2011）。因此，性别传统观念会对女性的数学结局产生负面影响。

研究人员认为，性别传统观念可能影响女性的机制是一种传统观念威胁。传统观念威胁是指在数学测试之前被提醒在传统观念中女性存在数学弱势的女性，比那些没有被提醒的女性表现差的这一现象。许多研究已经发现了传统观念威胁论影响成年女性（例如 Spencer, Steele, & Quinn，1999）和年轻女孩（例如 Ambady, Shih, Kim, & Pittinsky，2001；Tomasetto, Alparone, & Cadinu，2011）的证据。然而，研究人员也开始质疑这些结果的稳健性，这表明文献中有不一致的结果以及发表偏倚的证据（Flore & Wicherts，2015；Ganley 等，2013；Stoet & Geary，2012）。此外，在现实环境中对传统观念威胁的研

States, women are now well-represented in the biological and medical sciences but earn only 16% of bachelor's degrees in engineering and 18% in computer science (Snyder & Dillow, 2011). Differences in the workforce are even more pronounced, with women representing 11% of bachelor-degree-holding and 6% of doctorate-degree-holding scientists and engineers (NSF, 2011).[5]

Although disparities favoring males at the top of the performance distribution contribute to differences in career paths (Lubinski & Benbow, 2006), gaps in mathematics achievement and course taking leave much of the disparities in college major choices unexplained. For example, Riegle-Crumb and King (2010) found that, even after accounting for high school preparation and attitudes toward mathematics, the odds of majoring in a physical science or engineering field were roughly double for U.S. males versus females, and that whereas high school preparation accounted for the underrepresentation of Black and Hispanic males in STEM, neither mathematics preparation nor attitudes (as measured in the Education Longitudinal Study of 2002 [ELS])[6] could fully explain sex differences. Hence, our search for explanations of the underrepresentation of women must look beyond college preparation and examine the complex reasons why women ultimately choose to pursue other fields (Perez-Felkner et al., 2014; Riegle-Crumb, King, Grodsky, & Muller, 2012; Tyson, Lee, Borman, & Hanson, 2007).

Factors Related to Disparities in Mathematics Performance and Career Paths

According to the evidence reviewed thus far, we see a fairly well-documented, but small, difference in mathematics performance favoring males and a larger difference in some college major and career choices. The next natural question is "Why do we see these differences?" As outlined above, researchers have approached this question in many ways over the years, but here we examine six areas particularly relevant to mathematics education: (1) societal stereotypes; (2) the role of teachers and instruction; (3) climate in postsecondary mathematics; (4) student attitudes, beliefs, and values; (5) verbal skills; and (6) spatial skills. We highlight areas of research that seem to be poised to offer new insights and point to other areas that may not be as fruitful as they seemed in the past.

Societal Stereotypes About Gender and Math

Research examining the stereotyping of mathematics as a male domain has yielded inconsistent results, with variation depending on the country studied and the way that stereotyping is measured. Research using implicit measures (i.e., the Implicit Association Test) indicates that children in a number of countries are aware that stereotypes about gender and mathematics exist (Singapore: Cvencek, Kapur, & Meltzoff, 2015; United States: Cvencek, Meltzoff, & Greenwald, 2011; Germany: Steffens, Jelenec, & Noack, 2010). However research with explicit (self-report) measures of gender stereotypes is mixed, with some studies finding evidence of stereotyping of mathematics as a male domain (Cvencek et al., 2015; Cvencek et al., 2011; Steffens et al., 2010), whereas other work suggests no stereotyping in some countries (Sweden: Brandell, Leder, & Nyström, 2007; Australia and the United States: Forgasz, Leder, & Kloosterman, 2004; Israel: Forgasz & Mittelberg, 2008).[7] Altogether, the work in this area suggests that children may be aware of gender stereotypes about mathematics but are sometimes less likely to indicate that they believe those stereotypes.

Researchers have found that females who show implicit gender stereotyping about mathematics have lower self-concepts in mathematics, presumably because they apply the stereotype to themselves (Cvencek et al., 2015; Nosek, Banaji, & Greenwald, 2002; Steffens et al., 2010). In addition, research with adult women has shown that those who implicitly endorse the stereotype are less likely to perform well or continue in mathematics and science (Cvencek et al., 2015; Nosek & Smyth, 2011). Hence, gender stereotyping may have negative effects on females' mathematics outcomes.

Researchers have suggested that one mechanism through which gender stereotypes might affect females is stereotype threat. Stereotype threat is a phenomenon by which females who are reminded of the stereotype about women and mathematics before a mathematics test perform worse than those with no reminder. A number of studies have found evidence for stereotype threat effects in adult women (e.g., Spencer, Steele, & Quinn, 1999) and young girls (e.g., Ambady, Shih, Kim, & Pittinsky, 2001; Tomasetto, Alparone, & Cadinu, 2011). However, researchers have begun to question the robustness of these findings, suggesting that there are inconsistent results and evidence of publication bias in the literature (Flore & Wicherts, 2015; Ganley et al., 2013; Stoet & Geary, 2012). In addition, studies of stereotype threat in real-world settings have not found an effect (Stricker & Ward, 2004; Wei, 2012). Therefore, based on the current literature, stereotype threat is unlikely to be a main factor explaining sex differences in mathematics performance, but it is possible that more research examining particular groups of girls who might be especially susceptible could illuminate important nuances in the role of stereotype threat in girls' mathematics performance.

究并没有发现其影响效果（Stricker & Ward, 2004; Wei, 2012）。因此，在现有文献的基础上，传统观念威胁不太可能是解释数学学业成绩性别差异的主要因素，但可能的是，对易受影响的这一特殊女孩群体的研究可以阐明，传统观念威胁对女孩数学学业成绩的重要的且细微的作用。

教师和教学的作用

虽然传统观念威胁不大可能是性别传统观念影响女孩数学成绩的主要机制，但研究的确表明了传统观念的重要影响。教师、家长和同龄人可以通过多种方式向学生传达传统观念，但是对孩子的这些社会影响中，课堂里的教师是数学教育界中影响最大的，因此，我们主要关注教师。

教师对性别与数学的传统观念。研究发现，教师在课堂中只提"男孩们"和"女孩们"群体（例如，当问候学生或让他们排队时），这会使学生产生更顽固的传统观念（Hilliard & Liben, 2010）。虽然这项研究并不针对数学，但它突出了学生能够从教师那里接收到的微妙暗示。

一些研究指出，教师关于性别的传统观念会影响他们对学生数学能力的看法。鲁宾逊和鲁宾斯基（2011）发现，在数学（而非阅读）上，只有当美国小学数学老师意识到女生比男生更努力并且比男生表现更好时，他们才把女生与那些成绩相当的男生一视同仁。里格尔-克拉姆和汉弗莱斯（2012）在研究美国高中老师时也发现了类似的模式，但这种模式仅存在于白人学生中。ECLS-K 的证据表明，教师对性别的看法可能促进了早期关于男生具有数学成绩差距优势的这种观念（Robinson-Cimpian, Lubienski, Ganley, & Copur-Gencturk, 2014）。

尽管人们很容易把教师对乖女孩（循规蹈矩、努力）行为的看法当成一种错误的传统观念，但是研究确实表明，女孩在课堂上更勤奋并做了更多取悦教师的行为（Downey & Vogt Yuan, 2005; Kimball, 1989; Rathbun 等, 2004）。看起来教师正逐渐意识到这些问题，但最后结论是女孩需要比男生更努力才能在数学上获得成功。正如上文所提到的，女孩取悦教师的行为很可能是性别社会化的结果，有可能与之后的数学问题解决方法的差异相联系，即女生往往比男生更遵循老师所教授的方法。在瑞典和德国的中学教学中的研究表明，教师和学生认同男生和女生在数学课堂上所使用的方法有差异（Kaiser, Hoffstall, & Orschulik, 2012; Sumpter, 2015），问题是这种认同在多大程度上是由传统观念或是实际观察所形成的。

教师数学焦虑。研究还探讨了教师自身的数学焦虑对女生成绩的影响。小学低年级教师中绝大多数（98%）是女性（Robinson-Cimpian 等, 2014）。一般来说，女性比男性有更高的数学焦虑（Hembree, 1990; Miller & Bichsel, 2004），并且教育专业人员比其他大多数专业人员有更高的数学焦虑（Hembree, 1990）。贝洛克、冈德森、拉米雷斯以及莱文（2010）发现，一年级和二年级女教师的数学焦虑会对女生关于谁擅长数学的信念产生负面影响，从而对女生的数学学业成绩造成消极影响。因此，即使女教师力图平等地对待男生和女生，但是她们对自己作为数学学习者的认知仍然可能会潜移默化地传达给学生。重要的是，数学知识越强，数学焦虑越少，并且已经证明注重提高教师数学知识的课程可以减少数学焦虑（Battista, 1986; Gresham, 2007; Ma, 1999; Rayner, Pitsolantis, & Osana, 2009）。需要更多的研究来探讨究竟教师的数学焦虑或者对女生的低估，是如何与影响学生的课堂实践相联系的。

教学方法。过去的几十年里，关注性别平等的学者关注了数学教学的各个方面。例如，有学者主张将"女性的认知方式"（Belenky, Clinchy, Goldberger, & Tarule, 1986）与数学教学改革相融合，促进更多与"注重联系的认知"相结合的合作教学（Becker, 1995; Boaler & Sengupta-Irving, 2006; Jacobs, 1994）。然而最近海德和林德伯格（2007）的结论是，女孩更喜欢合作教学的证据很少，而且即使教学偏好因性别而异，但是我们并不清楚是否应该迎合他们。韦尔和巴塞缪（2008）同意这一观点，认为为女孩创造一个"安全"的环境可以抑制女孩的数学独立性，并指出在单一性别的课堂中，教师的教学往往是基于他们对男孩和女孩偏好预设的传统观念。这些研究结果与那些综述了单一性别学校教育研究的人的想法一致，得出的结论是针对女生的教学方式的优势并不明显（Bishop & Forgasz, 2007; Pahlke, Hyde, & Allison, 2014）。基于更多目前研究的新型教

Role of Teachers and Instruction

Although it is unlikely that stereotype threat is the primary mechanism through which gender stereotypes affect girls' mathematics outcomes, research does suggest that stereotypes matter. Teachers, parents, and peers can communicate gender stereotypes to students in a number of ways, but of these social influences on children, the classroom teacher is the one that the mathematics education community can affect most, and we therefore focus on teachers.

Teacher stereotypes about gender and mathematics. Research has found that teachers merely mentioning "boys" and "girls" as groups in the classroom (e.g., when greeting children or lining them up) produces more rigid stereotyping among students (Hilliard & Liben, 2010). Although not specific to mathematics, this study highlights the subtle cues that students can receive from teachers.

Several studies point to ways in which teachers' gendered stereotypes might affect their perceptions of students' mathematical abilities. Robinson and Lubienski (2011) found that, within mathematics (but not in reading), U.S. elementary teachers rate girls on par with similarly achieving boys only if they perceive those girls as working harder and behaving better than those boys. Similar patterns have been noted by Riegle-Crumb and Humphries (2012), who found a parallel pattern in their study of U.S. high school teachers, but the pattern held only among White students. Evidence from ECLS-K indicates that teachers' gendered perceptions may contribute to the early development of mathematics performance gaps favoring boys (Robinson-Cimpian, Lubienski, Ganley, & Copur-Gencturk, 2014).

Although it might be tempting to dismiss teachers' notions of good-girl (compliant, hard-working) behavior as a faulty stereotype, research does suggest that girls are more diligent in the classroom and engage in more behavior meant to please the teacher (Downey & Vogt Yuan, 2005; Kimball, 1989; Rathbun et al., 2004). It appears that teachers are picking up on this, but then concluding that girls need to work harder than boys do to succeed in mathematics. Girls' teacher-pleasing behavior is likely a consequence of gender socialization and, as noted above, is likely linked to later differences in mathematical problem-solving approaches, with girls following teacher-given rules more often than boys. Studies in Swedish and German secondary classrooms suggest that teachers and students perceive such differences in boys' and girls' approaches in mathematics classrooms (Kaiser, Hoffstall, & Orschulik, 2012; Sumpter, 2015), raising the question of the extent to which such perceptions are shaped by stereotypes versus observation.

Teacher mathematics anxiety. Studies have also explored the impact of teachers' own anxiety about mathematics on girls' performance. The vast majority (98%) of early elementary teachers are women (Robinson-Cimpian et al., 2014). In general, women have higher mathematics anxiety than men (Hembree, 1990; Miller & Bichsel, 2004), and education majors have higher mathematics anxiety than most other majors (Hembree, 1990). Beilock, Gunderson, Ramirez, and Levine (2010) found that first- and second-grade female teachers' mathematics anxiety exerts a negative influence on girls' beliefs about who is good at mathematics and ultimately on girls' mathematics achievement. Hence, even if they strive to treat boys and girls equitably, female teachers' own perceptions of themselves as mathematics learners might be subtly conveyed to students. Importantly, greater mathematics knowledge corresponds with less mathematics anxiety, and courses that focus on enhancing teachers' mathematics knowledge have been shown to reduce mathematics anxiety (Battista, 1986; Gresham, 2007; Ma, 1999; Rayner, Pitsolantis, & Osana, 2009). More research is needed to examine exactly how teachers' mathematics anxiety or underrating of girls may relate to classroom practices that affect students.

Teaching methods. Scholars concerned about gender equity have focused on various aspects of mathematics instruction over the past several decades. For example, scholars have argued for merging "women's ways of knowing" (Belenky, Clinchy, Goldberger, & Tarule, 1986) with the reform of mathematics teaching, promoting more collaborative classrooms that align with "connected knowing" (Becker, 1995; Boaler & Sengupta-Irving, 2006; Jacobs, 1994). However, more recently, Hyde and Lindberg (2007) concluded that evidence of girls preferring more collaborative instruction is thin, and even if instructional preferences do differ by sex, it is not clear that we should cater to them. Vale and Barthomew (2008) concur, arguing that creating a "safe" environment for girls could inhibit girls' mathematical independence and noting that teachers in single-sex settings too often base instruction on stereotypes of girls' and boys' preferences. These findings align with those of others who have reviewed studies on single-sex schooling and concluded that benefits are not clearly evident (Bishop & Forgasz, 2007; Pahlke, Hyde, & Allison, 2014). It is certainly possible that new pedagogies based on more current research could produce benefits in some single-sex settings, but further research is needed.

学方法在一些单一性别的环境中产生效益当然是可能的，但还需要进一步的研究。

学者们还基于性别平等研究了教学工具和教学材料。比如，在过去15年里的研究发现，如果在数学教学中使用技术，更可能的是，对男孩比女孩更有利（Ursini & Sánchez, 2008; Vale & Bartholomew, 2008; Vale & Leder, 2004）。其他学者认为教科书中对男性和女性的描述存在偏见，比起那些至少在课程材料中的公然偏见被处理掉的国家，这个问题现在似乎在发展中国家更突出（例如Weldeana, 2014）。研究人员也一直在思考，特定任务是否会对女孩或男孩更具激励性，他们发现女孩往往在具有传统的女性背景特点（例如，购物、布娃娃）的题目中表现特别好，而男孩往往在传统的男性环境的题目中表现更好（例如，建筑、汽车; Norton, 2006; Zohar & Gershikov, 2008）。虽然这些环境偏好影响数学评估公平性的建立，但他们对混合性别课堂教学的影响却不那么明确。

师生互动。师生互动中的性别公平性研究产生了些许不一致的发现，这方面的工作主要出现在莱德1992年的综述中，但自那以后的研究已经不那么有说服力。以前的研究表明，男孩自愿回答问题的次数往往比女孩多，因此他们也收到了不同学科教师给予的不成比例的反馈（Altermatt, Jovanovic, & Perry, 1998; Duffy, Warren, & Walsh, 2001; Sadker & Sadker, 1994）。然而，在师生交往中性别差异似乎比以前认为的更微妙，一些研究发现教师对待女孩和男孩没有差异（例如Harrop & Swinson, 2011），另外一些研究则认为教师与男孩的互动更负面（Jones & Dindia, 2004）。有可能是过去几十年里的文化变化导致了课堂互动性别偏见减少。然而，由于教师本身的数学焦虑和对女孩能力的看法影响了女孩的数学表现和态度，有害信息会以某种方式传达给儿童（尤其是女孩）。为了更好地理解这种情况究竟是如何发生的，需要对课堂互动进行更细致入微的测量。

总之，最近的证据表明，合作学习环境、单一性别的课堂、平等的师生互动可能不会像曾经希望的那样可以促进性别平等。相反，最近的证据指出，鼓励女孩独立、灵活的问题解决技能，挑战教师的传统观念以及减少教师的数学焦虑就有重要作用。然而，需要进一步的研究来指导实施这些想法并评估他们对性别平等的影响。

高等数学中的氛围

一些研究表明，大学数学课程的氛围可以阻止女性的参与。例如，赫齐格（2004）强调女性和有色人种学生在数学博士项目中感到被疏远，感觉此领域太具竞争力，奉行个人主义，与有意义的背景毫无关联。阿伦和马丁（2006）通过访谈工程本科专业的女大学生，认为确实存在"寒冷氛围"，包括教员公然的打击。然而，这些问题在他们的调查数据中似乎不是很普遍，最近其他证据提出了寒冷氛围对女性影响的问题，例如，洛德等人（2009）发现，美国妇女，包括非裔美国人和西班牙裔女性至少有可能与男性一样坚持在本科STEM专业学习。里格尔-克拉姆、金和莫顿（2014）发现"学术融合"的程度（例如，参与研究小组、与教授进行社会互动）强烈预示着女性对本科STEM专业的坚持，但对男性来说只有微弱的预测作用，这也表明氛围对女性的重要性。然而，他们也发现，女性坚持学习STEM专业的比例和资质相当的男性一样多，高学术融合的女性坚持的比例甚至比同样融合的男性还要高。里格尔-克拉姆等人提出，大众媒体对寒冷氛围的关注可能会不必要地阻止女性选择STEM专业。然而，与赫齐格（2004）不同的是，里格尔-克拉姆等人的研究集中于STEM专业的大学生，而不是数学专业的学生或研究生。另外，近年来，女性在数学和其他STEM项目中的学习环境可能发生了变化，学习环境问题正变得不太重要了。

总的来说，最近的研究对寒冷氛围实际上阻碍了女性继续她们的STEM学习的程度提出了质疑。大多数证据表明，男子和女子在大学专业上的差异主要源于早年，这也说明进入大学前的经验和决定的重要性（Maltese & Tai, 2010; Tai, Liu, Maltese, & Fan, 2006）。不过，如上所述，考虑到女性在与数学相关的职业中所占比例比大学专业中所占比例更低，女性有可能是坚持熬过了大学阶段这种具有寒冷气氛的数学学习环境，但这种环境吓退了一些女性在大学毕业后继续攻读与数学相关的研究生学位和从事与数学相关的职业。

Scholars have also examined instructional tools and materials with an eye toward gender equity. For example, in the past 15 years, studies have tended to find that technology use in mathematics instruction, if anything, is more favorable for boys than girls (Ursini & Sánchez, 2008; Vale & Bartholomew, 2008; Vale & Leder, 2004). Other scholars have considered bias in the portrayal of males and females in textbooks, an issue that now seems more pertinent for developing countries (e.g., Weldeana, 2014) than countries in which at least blatant bias in curriculum materials has been addressed. Researchers have also continued to ask whether particular tasks are more motivating for girls or boys, finding that girls tend to perform particularly well on items with stereotypically female contexts (e.g., shopping, dolls) whereas boys do better on items with stereotypically male contexts (e.g., construction, cars; Norton, 2006; Zohar & Gershikov, 2008). Although these contextual preferences have implications for creating fair mathematics assessments, their implications for instruction in mixed-sex classes are less clear.

Teacher-student interactions. The research on gender equity in teacher-student interactions has produced somewhat inconsistent findings. This area of work was featured more in Leder's 1992 review, but research since then has been less convincing. Previous research suggested that boys volunteer to answer questions more often than girls and that they receive a disproportionate amount of teacher feedback across subjects (Altermatt, Jovanovic, & Perry, 1998; Duffy, Warren, & Walsh, 2001; Sadker & Sadker, 1994). However, sex differences in teacher-student interactions appear to be more subtle than previously thought, with some studies finding no difference in the treatment of girls and boys (e.g., Harrop & Swinson, 2011) and others suggesting that teachers initiate more negative interactions with boys (Jones & Dindia, 2004). It is possible that cultural changes in the past few decades have led to classroom interactions being less gender biased. However, given that teachers' own mathematics anxiety and perceptions of girls' proficiency affects girls' mathematics performance and attitudes, detrimental messages are somehow communicated to children (especially girls). More nuanced measures of classroom interactions are needed to better understand how exactly this might occur.

Overall, recent evidence suggests that cooperative learning environments, single-sex classrooms, and equalizing teacher-student interactions may not be as promising for promoting gender equity as once hoped. Instead, recent evidence points toward the potential importance of encouraging independent, flexible problem-solving skills among girls, as well as challenging teacher stereotypes and decreasing teachers' mathematics anxiety. However, further research is needed to guide implementation of these ideas and assess their impact on gender equity.

Climate in Postsecondary Mathematics

Some research has suggested that the climate of university mathematics programs can deter women's participation. For example, Herzig (2004) highlighted ways in which females and students of color in mathematics doctoral programs can feel alienated, experiencing the field as competitive, individualistic, and disconnected from meaningful contexts. Allan and Madden's (2006) interviews with undergraduate women in engineering identified substantial evidence of a "chilly climate," including overt faculty discouragement. However, these issues appeared less prevalent in their survey data, and other recent evidence raises questions about the impact of chilly climate on women. For example, Lord et al. (2009) found that U.S. women, including African American and Hispanic women, were at least as likely as their male peers to persist in STEM undergraduate programs. Riegle-Crumb, King, and Morton (2014) found that the degree of "academic integration" (e.g., study group participation, social interaction with professors) was a strong predictor of women's persistence in a STEM major, but only a weak predictor for men, pointing to the importance of climate for women. However, they also found that women persist in STEM majors as much as equally qualified males, and women with high academic integration persist at even higher rates than similarly integrated males. Riegle-Crumb et al. raise the concern that a focus on chilly climate in the popular media could unnecessarily deter women from choosing STEM majors. However, Riegle-Crumb et al.'s study focused on STEM undergraduate majors, not mathematics majors or graduate students, in contrast with Herzig (2004). Additionally, it is possible that the climate for women in mathematics and other STEM programs has changed in recent years and is becoming a less critical issue.

Overall, recent research raises questions about the extent to which a chilly climate actually deters women from staying in their STEM programs. Most evidence suggests that differences in men's and women's college majors are rooted primarily in earlier years, pointing to the importance of precollege experiences and decisions (Maltese & Tai, 2010; Tai, Liu, Maltese, & Fan, 2006). Still, as noted above, given that women's underrepresentation in mathematics-related careers is more

态度、信念和价值观

大量的研究表明，学生对数学的态度和信念对他们以后取得的成就和职业选择非常重要。正如前面引用的经济合作与发展组织的观点所指出的，数学态度和数学信念的性别差异大于数学成绩的差异，这表明，至少关注这些因素是很重要的（Louis & Mistele，2012；Lubienski 等，2013；OECD，2014）。在这里，我们讨论一些已经受到最多的研究所关注的态度，特别是数学的自信心和焦虑、对成功的恐惧、固定心态与成长心态、数学兴趣以及职业价值观。

数学的自信心与焦虑。 不论是在美国还是国际上，男孩对自身的数学能力似乎比女孩表现得更有自信（例如 Bench，Lench，Liew，Miner，& Flores，2015；Else-Quest 等，2010；Lubienski 等，2013；OECD，2015）。在一些研究中，这些差异早在一年级就显现出来了（Fredricks & Eccles，2002）。另有研究发现，女孩对数学往往比男孩更容易焦虑（Ganley & McGraw，2016；Herbert & Stipek，2005；Hyde，Fennema，Ryan，Frost，& Hopp，1990）。然而，一项对德国学生的研究发现，从数学课堂或考试中的焦虑感的实时报告中并没有发现性别差异，但是在一次比较传统的问卷调查中显示女孩的数学焦虑感更严重些（Goetz，Bieg，Lüdtke，Pekrun，& Hall，2013）。还需要更多的研究来对利用自然的数据收集方法捕捉学生的瞬时"状态"，以及用传统方法进行"特质"问卷调查所得的结果进行比较。

值得注意的是，许多国家的研究也发现，信心和焦虑是预测数学学业成绩强有力的因素，部分原因是由于缺乏信心或高度焦虑会导致学生逃避数学，让他们不太可能学好数学（例如 Ashcraft & Krause，2007；Ganley & Vasilyeva，2014；Krinzinger，Kaufmann，& Willmes，2009；Ma，1999；Marsh，Trautwein，Ludtke，Köller，& Baumert，2005；Z. Wang，Osterlind，& Bergin，2012）。虽然与其他因素相比，过去的成绩更能预测信心，但是研究确实表明，数学态度（焦虑和自信）和数学成绩之间存在互惠关系，这表明较高的焦虑和较低的自信导致学业成绩较差，从而进一步增加焦虑并降低自信（Ganley & Lubienski，2016；Ma & Xu，2004；Marsh 等，2005；Pinxten，Marsh，De Fraine，Van Den Noorgate，& Van Damme，2014）。

一些研究发现，过高的焦虑水平可能通过对工作记忆资源的负面影响（Beilock，2010；Ganley & Vasilyeva，2014；Maloney，Schaeffer，& Beilock，2013）或自我感知空间能力的负面影响（Maloney，Waechter，Risko，& Fugelsang，2012）降低女孩的数学成绩。重要的是，焦虑和自信不仅与数学学业成绩有关，而且与学生选择与数学相关领域的专业有关，而对专业选择的焦虑和需要的自信远在对数学学业成绩的焦虑和自信之上（Goldman & Penner，2014；M. T. Wang，Eccles，& Kenny，2013；Watt 等，2012）。

对成功的恐惧。 数学中"对成功的恐惧"的概念代表着态度研究的另一个领域。这一理论正好和数学焦虑（害怕失败）理论相反，这一理论认为，一些学生尤其是女孩已经被社会化了，所以他们害怕在数学方面取得成功，害怕这会降低他们的社会地位，害怕因在数学上成功而被视为"怪物"或"书呆子"这样的传统观念。对德国、加拿大和以色列的数学高成就学生的一项研究发现，中学女生比男生更容易受到被视为书呆子的压力的影响（Boehnke，2008）。在这一领域的实验研究表明，当计算机科学与怪物这一传统观念相联系时，越来越多的女性被阻止进入这一领域（Cheryan，Plaut，Davies，& Steele，2009）。因此，也许是与数学成功相关的传统观念导致女孩远离这些领域。

固定心态与成长心态。 学生对自己数学能力的本质的看法会影响他们的成功。研究表明，有些人认为数学能力是可以改变和发展的（一种"成长"的心态——"如果我努力，我就能更好地学习数学"），而另外一些人有更"稳固"的心态（"不管我做什么，我的数学能力都不会改变"）。虽然，没有多少直接关于这个课题的研究，但是有一些研究表明，女孩具有成长心态的可能性较小（Dweck，2007；Nix，Perez-Felkner，& Thomas，2015）。此外，还有丰富的相关研究表明，具有成长心态的学生在学校的表现更好（Dweck，1999；Stipek & Gralinski，1996）。这背后隐含的意义是，如果学生不相信他们能够进步，那么他们就几乎没有动力努力学习（Dweck，1999）。

severe than in college majors, it is possible that women choose to persist through a chilly climate in college but that a chilly climate deters some women from pursuing mathematics-related graduate degrees and careers after completing college.

Attitudes, Beliefs, and Values

A large body of research has shown the importance of students' attitudes and beliefs about mathematics in shaping later achievement and career choices. As noted in the OECD quote above, sex differences in attitudes and beliefs about mathematics are larger than differences in mathematics achievement, suggesting it is at least as important to focus on these factors (Louis & Mistele, 2012; Lubienski et al., 2013; OECD, 2014). Here we discuss some of the attitudes that have gotten the most research attention, specifically mathematics confidence and anxiety, fear of success, fixed versus growth mindsets, mathematics interest, and occupational values.

Mathematics confidence and anxiety. Boys tend to express greater confidence in their mathematical abilities than do girls, both in the United States and internationally (e.g., Bench, Lench, Liew, Miner, & Flores, 2015; Else-Quest et al., 2010; Lubienski et al., 2013; OECD, 2015). These differences show up as early as first grade in some studies (Fredricks & Eccles, 2002). Additional research finds that girls tend to be more anxious about mathematics than boys (Ganley & McGraw, 2016; Herbert & Stipek, 2005; Hyde, Fennema, Ryan, Frost, & Hopp, 1990). However, in one study of German students, no sex differences were found in real-time reports of anxiety felt during mathematics classes or exams, but girls reported higher general levels of mathematics anxiety on a more traditional survey (Goetz, Bieg, Lüdtke, Pekrun, & Hall, 2013). Additional research is needed to compare results from naturalistic data collection methods capturing students' moment-to-moment "states" with traditionally administered "trait" surveys.

Notably, research in many countries also finds that confidence and anxiety are strong predictors of mathematics achievement, partially because a lack of confidence or high anxiety can cause students to avoid mathematics, making them less likely to learn it well (e.g., Ashcraft & Krause, 2007; Ganley & Vasilyeva, 2014; Krinzinger, Kaufmann, & Willmes, 2009; Ma, 1999; Marsh, Trautwein, Ludtke, Köller, & Baumert, 2005; Z. Wang, Osterlind, & Bergin, 2012). Although past achievement is a stronger predictor of confidence than the other way around, research does indicate the existence of reciprocal relations between mathematics attitudes (anxiety and confidence) and mathematics achievement, suggesting that higher anxiety and lower confidence lead to poorer performance, which then further increases anxiety and decreases confidence (Ganley & Lubienski, 2016; Ma & Xu, 2004; Marsh et al., 2005; Pinxten, Marsh, De Fraine, Van Den Noorgate, & Van Damme, 2014).

Some research has found that heightened levels of anxiety may diminish girls' mathematics performance through its negative impact on working memory resources (Beilock, 2010; Ganley & Vasilyeva, 2014; Maloney, Schaeffer, & Beilock, 2013) or on self-perceived spatial abilities (Maloney, Waechter, Risko, & Fugelsang, 2012). Importantly, anxiety and confidence are related not only to mathematics achievement but also to students' choices to major in a mathematics-related field, above and beyond mathematics performance (Goldman & Penner, 2014; M. T. Wang, Eccles, & Kenny, 2013; Watt et al., 2012).

Fear of success. The concept of a "fear of success" in mathematics represents another area of attitudes research. This theory, which could be considered the opposite of mathematics anxiety (a fear of failure), posits that some students, especially girls, have been socialized to fear mathematical success as it could decrease their social standing, given "geek" or "nerd" stereotypes of mathematically successful students. In one study of high-achieving mathematics students in Germany, Canada, and Israel, middle-school girls were more affected than were boys by the pressure to avoid being perceived as a nerd (Boehnke, 2008). Experimental research in this area has shown that when computer science is associated with the geek stereotype, more women are deterred from entering the field (Cheryan, Plaut, Davies, & Steele, 2009). Thus, perhaps the stereotypes associated with mathematical success steer girls away from these fields.

Fixed or growth mindsets. Students' perceptions about the nature of mathematics ability can influence their success. Research shows that some people tend to view mathematics ability as something that can be changed and developed (a "growth" mindset—"If I work at it, I can be better at mathematics") whereas others have a more "fixed" mindset ("No matter what I do, my mathematics ability will not change"). Some research suggests that girls are less likely to have a growth mindset (Dweck, 2007; Nix, Perez-Felkner, & Thomas, 2015), though there is little research directly on this topic. In addition, there is a wealth of correlational research suggesting that students with a growth mindset per-

实验研究还表明，那些被告知数学智力可以随着时间的推移而提高的学生更有可能坚持完成数学任务并且表现良好（Blackwell, Trzesniewski, & Dweck, 2007）。尽管其他研究发现成长心态干预对男孩和女孩同样有帮助（Paunesku 等，2015），但是还有一些研究表明，成长心态对女孩的学业成绩尤为重要（Dweck, 2007; Good, Aronson, & Inzlicht, 2003）。

数学兴趣。研究普遍表明，男孩对数学的兴趣比女孩更浓厚（Frenzel, Pekrun, & Goetz, 2007; Köller, Baumert, & Schnabel, 2001; Watt, 2004）。然而，有些研究发现兴趣的性别差异比信心的性别差异要小（Ganley & Lubienski, 2016），有些研究人员则没有发现差异（Fredricks & Eccles, 2002; Hyde, Fennema, Ryan 等，1990），或发现女孩对数学的兴趣比男孩更浓厚（Hemmings, Grootenboer, & Kay, 2011）。最近对TIMSS和PISA的数据分析表明，在相对富裕的国家，男性和女性在数学的兴趣上存在更大的差距。在这些国家，经济需求可能不太能够推动职业决策，从而给予女孩更多的自由去选择与她们自我认知一致的职业（Charles, Harr, Cech, & Hendley, 2014; Else-Quest 等，2010）。

关于数学兴趣、学业成绩、大学专业与职业选择之间的关系的研究结果是多样的（Ganley & Lubienski, 2016; Hemmings 等，2011; Köller 等，2001; Marsh 等，2005; Pinxten 等，2014; Wigfield & Eccles, 2002）。总体而言，数学兴趣的性别差异小，而且对学业成绩和大学专业选择有微小且不一致的影响，从而不能解释我们所看到的职业差异。即使女孩喜欢数学并且对数学感兴趣，她们也不一定有信心去追求它，或者她们可能不认为数学密集型的职业与她们的价值观（下一节的主题）是一致的。

职业价值观。埃克尔斯（1986, 2009）认为，女性不应该被视为"回避"数学，而我们应该试着去理解她们选择特定研究领域的原因。最近大规模的研究表明，女性看重的职业，包括帮助人们和承担家庭责任，而男性往往优先考虑做一些工作、赚钱和地位高的职业。与期望-价值理论相一致，埃克尔斯和王（2015）发现这些职业价值观以及数学自信心的差异，很大程度上解释了男性和女性在十二年级职业选择上的差异。同样，萨法尔（2013）发现，男性优先考虑未来的收益潜力比数学学业成绩更能解释专业选择的性别差异。因此，这一部分的工作值得进一步关注。在职业价值观上，种族、阶级和性别的交叉作用也需要研究。

语言能力

与期望价值理论一致的是，学生进入特定领域的决定是基于他们对可行选择的看法，而女性强大的语言表达能力可能使非STEM领域职业选择相对更具有吸引力。一些研究表明，如果一个女性数学很强，那么她比同等数学成绩的男性更有可能选择非STEM领域职业，因为她可能具有更强的语言能力（Correll, 2001; Riegle-Crumb 等，2012; M. T. Wang 等，2013）。因此，具有较强数学技能的女性往往比男性同龄人有更多的选择，平均而言，男性的语言表达能力较弱。

对职业价值观的研究以及数学相关领域女性人数比例偏低表明，解决职业差距的一个关键要素是帮助女性看到这些领域与她们自己的身份认同和价值观相关。具体来说，如果女性更了解数学相关领域（例如，工程或计算机科学）能在哪些方面提供灵活性，看到它能够提供帮助社会的机会以及锻炼语言表达能力，那么这些领域会更加吸引女性的关注。

空间思维

心理学的研究表明，除了学生的态度，空间思维也与数学成绩密切相关，它在促进性别平等方面的作用可能比数学教育学者所认为的更为重要。空间思维是指对空间中物体进行心理表征和操作的能力，研究发现，一些特定的空间思维能力，如女性对物体及其位置的记忆更有优势，而大多数空间思维能力则显示出不同程度的男性优势，这个幅度最大有0.6个标准差（Halpern 等，2007; Linn & Petersen, 1985; Voyer, 1996; Voyer, Voyer, & Bryden, 1995; Weiss, Kemmler, Deisenhammer, Fleischhacker, & Delazer, 2003）。如前所述，空间思维能力的性别差异可以解释为什么在涉及空间思维的数学领域

form better in school (Dweck, 1999; Stipek & Gralinski, 1996). The idea behind this is that if students do not believe they can improve, they have little motivation to work hard (Dweck, 1999). Experimental studies have also shown that students who are told that mathematics intelligence can be improved over time are more likely to persist in mathematics tasks and perform well in mathematics (Blackwell, Trzesniewski, & Dweck, 2007), and some work suggests that a growth mindset is particularly important for girls' achievement (Dweck, 2007; Good, Aronson, & Inzlicht, 2003), though other work found that a growth mindset intervention was similarly helpful for both boys and girls (Paunesku et al., 2015).

Mathematics interest. Research generally indicates that boys are more interested in mathematics than girls (Frenzel, Pekrun, & Goetz, 2007; Köller, Baumert, & Schnabel, 2001; Watt, 2004). Still, differences in interest are smaller than differences in confidence (Ganley & Lubienski, 2016), with some researchers finding no difference (Fredricks & Eccles, 2002; Hyde, Fennema, Ryan, et al., 1990) or that girls are more interested than boys in mathematics (Hemmings, Grootenboer, & Kay, 2011). Recent analyses of TIMSS and PISA data indicate that male-female gaps in mathematics interest are actually larger in wealthier countries. In these countries economic need may be less likely to drive career decisions, giving girls more freedom to choose careers aligned with their conceptions of themselves (Charles, Harr, Cech, & Hendley, 2014; Else-Quest et al., 2010).

Research on the relationship between mathematics interest, achievement, college majors, and career choices is mixed (Ganley & Lubienski, 2016; Hemmings et al., 2011; Köller et al., 2001; Marsh et al., 2005; Pinxten et al., 2014; Wigfield & Eccles, 2002). Overall, sex differences in mathematics interest are small and have such small and inconsistent effects on achievement and college major choices that they cannot account for the occupational differences we see. Even if girls like mathematics and are interested in it, they do not necessarily have the confidence to pursue it, or they may not perceive mathematics-intensive occupations as aligning with their values—the subject of the next section.

Occupational values. Eccles (1986, 2009) has argued that women shouldn't be viewed as "avoiding" mathematics but instead as choosing particular areas of study for reasons that we should try to understand. Recent large-scale studies indicate that women value careers that involve helping people and that accommodate family responsibilities, whereas men tend to prioritize working with things, making money, and high-status tasks. Aligned with expectancy-value theory, Eccles and Wang (2015) found that these differences in occupational values, along with mathematical confidence, explained much of the disparity in male and female 12th graders' career choices. Similarly, Zafar (2013) found that males' prioritization of future earnings potential explained the gender gap in choice of major more than did mathematics performance. Hence, this line of work merits further attention. Studies that investigate intersections of race, class, and gender in occupational values are also needed.

Verbal Skills

Consistent with expectancy value theory, a student's decision to enter a particular field is based on their perceptions of viable options, and women's strong verbal skills may make non-STEM options relatively more attractive. Several studies indicate that if a woman is very strong in mathematics, she is more likely to consider non-STEM career options than males with equivalent mathematics scores because she likely has stronger verbal abilities (Correll, 2001; Riegle-Crumb et al., 2012; M. T. Wang et al., 2013). Thus, women with strong mathematics skills tend to have more choices than their male peers, who, on average, have weaker verbal skills.

The research on occupational values, taken together with girls' underrepresentation in mathematics-related fields, suggests that one critical element of addressing occupational disparities is helping women see how those fields relate to their own identities and values. Specifically, if women were more aware of the ways in which mathematics-related fields such as engineering or computer science offer flexibility, provide opportunities to help society, and can draw upon verbal skills, perhaps they would be more attracted to those fields.

Spatial Thinking

In addition to student attitudes, psychology research reveals that spatial thinking is strongly related to mathematics outcomes, and it may be more important for promoting gender equity than mathematics education scholars have assumed. Spatial thinking involves the ability to mentally represent and manipulate objects in space. Research finds that some particular spatial skills, such as memory for objects and their locations, show a female advantage whereas most show a male advantage of varying magnitudes, with some of these differences being as large as .6 standard deviations (Halpern et al., 2007; Linn & Petersen, 1985; Voyer, 1996; Voyer, Voyer, & Bryden, 1995; Weiss, Kemmler, Deisenhammer, Fleischhacker, & Delazer, 2003). As

（例如测量、空间和形状）上存在较大的性别差距。

研究表明，空间思维能力可以预测数学学业成绩和数学相关的职业选择（Casey, Nuttall, Pezaris, & Benbow, 1995；Ganley & Vasilyeva, 2011；Wai, Lubinski, & Benbow, 2009）。实际上，韦等人（2009）发现空间思维能力是STEM专业和职业选择的最佳预测因素。研究人员对350 000多人进行了抽样调查，他们考察了高中学生的空间思维能力与11年后参加数学密集型职业生涯之间的关系。结果表明，大约有一半在STEM领域获得博士学位的被试者在高中的空间思维能力测试中都居于前4%。

一些研究特别发现，空间思维能力有助于解释数学表现中的性别差异（Casey 等，1995；Halpern 等，2007；Mix & Cheng, 2012）。另外，拉斯基等人（2013）发现一年级女生的空间思维能力能够预测他们为完成简单的算术问题而使用更高层次的分解和检索策略（而不是数数）的能力。这些结果与前面提到的有关更高层次策略对解决数学问题的重要性的工作有关。看来，空间思维能力的弱势以及遵循教师给定的规则，可能导致女孩使用标准的算法策略。

空间思维能力具有高度的可塑性（Uttal 等，2013），在这样的研究背景下解释空间思维能力的性别差异是至关重要的。以色列的一项研究发现，教学生如何进行空间想象可以消除空间思维能力的性别差异（Tzuriel & Egozi, 2010）。此外，一些实验研究发现，一门空间思维能力课程可以带来许多好处，包括提高数学测试的成绩（Cheng & Mix, 2014）、提高数学课程的成绩（Sorby, Casey, Veurink, & Dulaney, 2013）以及女性在工程专业不断坚持的毅力（Sorby, 2009）。然而，最近的一项研究提出了空间思维能力的干预措施是否可以提高数学学业成绩的问题（Hawes, Moss, Caswell, & Poliszczuk, 2015）。空间思维能力训练对女生数学成绩、大学专业选择与毅力的影响方面还需要做更多的纵向研究。

讨论

在本章中，我们看到，尽管在过去的几十年中性别平等有一些进步，但在许多国家的数学评估，特别是与学校所教内容联系较少的评估中，男性仍然保持微小却持续的优势。此外，在大学主修专业的选择上仍然存在大量的、无法解释的性别差异，而在STEM的职工总数上甚至存在更大的差异。数学自信心与焦虑的性别差异比数学学业成绩中的性别差异更大，空间思维能力的性别差异也是如此，每一种差异似乎都有助于解释数学学业成绩的差异。教师的数学焦虑和期望，以及女孩、男孩的价值观和对各种职业成功期望的差异，似乎也会影响在大学专业和职业选择方面的差异。尽管如此，女性和男性之间的差异会随着时间的推移而有所减小，并且因国家而异，这也说明，这些差距可以通过仔细的研究、干预和社会变革而有所改善。

未来的研究方向

既然在查明女孩数学自信心、焦虑、问题解决的方法、空间思维能力和职业选择的持续差异方面已经做了大量的工作，接下来的研究应该集中在进一步阐明这些模式的原因以及有针对性的干预方式上。例如，为了理解为什么女性认为自己的数学能力不如调查数据表明的那样，我们需要对女孩形成这些看法的过程进行纵向研究，正如已经完成的关于成功黑人男性的数学身份认同的发展以及关注客观和微观影响做过的细致研究一样（Berry, 2008；Stinson, 2008），我们需要做更多的工作来研究女性的数学身份认同是如何从幼儿时期发展到大学专业选择、职业决策的，这些研究可以阐明女孩看待自己与数学关系的方式，包括吸引或排斥她们的领域方面（例如，与数学相关的氛围或传统观念）。

数学教育研究人员尤其有必要回到性别与数学教育的研究，因为他们在课程、教学和数学本质方面有特殊的专长。例如，我们需要研究男生和女生从低年级开始，在不同的教学环境中自我认知和问题解决方法是如何发展的。高度结构化的数学教学环境更有可能培养好女孩行为吗？这样的行为最终会限制女孩问题解决的方法吗？此外，女孩通常被认为比一般的男孩更勤奋，那么为什么这种勤奋的看法似乎会降低教师和学生对女孩数学能力的评估，但没有降低对女孩读写能力的评估

noted, sex differences in spatial skills could explain why there are larger gaps in mathematical areas that involve spatial thinking (e.g., measurement, space, and shape).

Research shows that spatial skills are a predictor of mathematics performance and mathematics-related career choices (Casey, Nuttall, Pezaris, & Benbow, 1995; Ganley & Vasilyeva, 2011; Wai, Lubinski, & Benbow, 2009). In fact, Wai et al. (2009) found that spatial skills are one of the best predictors of STEM major and career choices; with a sample of over 350,000 participants, they examined the relation between spatial skills in high school and participation in mathematics-intensive careers 11 years later. The results showed that about half of the participants who received PhDs in STEM fields were in the top 4% on spatial ability tests in high school.

Several studies have specifically found that spatial skills help explain sex differences in mathematics performance (Casey et al., 1995; Halpern et al., 2007; Mix & Cheng, 2012). In addition, Laski et al. (2013) found that first-grade girls' spatial skills predicted their use of higher-level decomposition and retrieval strategies (as opposed to counting) for completing simple arithmetic problems. These results connect to the work mentioned earlier regarding the importance of higher level strategies for problem solving in mathematics. It appears that both spatial skills and girls' tendency to follow teacher-given rules might contribute to girls' greater use of standard, algorithmic strategies.

It is critical to interpret sex differences in spatial skills in the context of research suggesting that spatial skills are highly malleable (Uttal et al., 2013). One study in Israel found that teaching students how to think spatially eliminated the sex difference in spatial skills (Tzuriel & Egozi, 2010). In addition, some experimental studies have found that a course in spatial skills can produce benefits, including enhanced mathematics test performance (Cheng & Mix, 2014), improved grades in a mathematics course (Sorby, Casey, Veurink, & Dulaney, 2013), and women's increased persistence in an engineering major (Sorby, 2009). However, one recent study has raised questions about whether spatial skills interventions can improve mathematics performance (Hawes, Moss, Caswell, & Poliszczuk, 2015). More longitudinal work should be done to test the transfer of spatial skills training to girls' mathematics performance, college major choices, and persistence.

Discussion

In this chapter, we saw that, despite advances in gender equity over the past decades, small but persistent differences favoring males tend to remain on mathematics assessments in many countries, particularly assessments that are less tied to school-taught material. Moreover, there remain substantial, unexplained sex differences in college major choices and even larger disparities in the STEM workforce. Sex differences in mathematical confidence and anxiety are larger than achievement gaps, as are sex differences in spatial skills, and each appears to contribute to sex differences in mathematics outcomes. Teachers' mathematics anxiety and expectations, as well as differences in girls' and boys' values and expectations for success in various occupations also appear to contribute to disparities in college majors and careers. Still, the fact that disparities between females and males have decreased over time and vary by country suggest that these disparities can be ameliorated with careful research, interventions, and societal change.

Future Directions for Researchers

Now that substantial work has been done to pinpoint persistent disparities in girls' mathematical confidence, anxiety, problem-solving approaches, spatial skills, and occupational choices, the time seems right for more focused studies that will further illuminate the reasons for these patterns and that can point the way toward targeted interventions. For example, to understand why women perceive themselves as less mathematically able than achievement data would indicate, we need longitudinal studies on girls as they develop these perceptions. Much like the careful work that has been done on the development of mathematical identities of successful black males with attention to microlevel and macrolevel influences (Berry, 2008; Stinson, 2008), we need more research on the ways girls' mathematical identities develop from early childhood through college major and career decisions. Such studies could illuminate the ways girls view themselves in relation to mathematics, including aspects of the field that attract or repel them (e.g., climate or stereotypes associated with mathematics).

There is a need for mathematics education researchers, in particular, to return to conducting research on gender and mathematics education, as they have particular expertise in curriculum, instruction, and the nature of mathematics. For example, we need current studies of how boys' and girls' self-perceptions and problem-solving approaches develop in various instructional contexts, beginning in early grades. Are highly structured mathematics instructional environments more likely to foster good-girl behaviors, and do such behaviors ulti-

（Robinson-Cimpian等，2014）？是因为数学被认为是一个人要么"擅长"要么"不擅长"的事（例如，是"学数学的料"或者不是"学数学的料"），从而女孩在数学上的努力就表明他们不是"天生"擅长数学的吗？

数学教育工作者也需要研究在数学课堂中强调空间思维和大胆解决问题的方法，这些方法可能最终关系到数学学业成绩和后期职业选择中的性别差异。例如，一些实验和准实验工作表明，加强空间思维能力会影响数学测试或课程表现，但在K-12课堂上考察空间思维能力的养成还有许多工作要做（Cheng & Mix，2014；Sorby，2009；Sorby等，2013）。数学教育研究人员应该很好地与其他领域（例如，认知心理学）的研究人员共同合作，努力在数学教学中创设包括对空间思维能力和问题解决的关注的最合适的方式。

此外，考察教师的数学焦虑和他们对男孩、女孩的数学能力的看法，可以帮助我们更好地了解这些因素是如何可能转化为学生学业成绩上的性别差异的以及是如何影响学生的。例如，有数学焦虑的教师在课堂上表现出哪些类型的行为会让女孩们留意？关于谁擅长数学或谁不擅长数学，有哪些微妙的信息会传递给学生？

最后，由于一些性别问题在不同群体中的表现不同，对种族/民族、社会经济地位以及性别的交互作用的研究是很有必要的。在这方面的工作可能会受益于新的州纵向数据系统，因为该系统可以以新的方式促进对相对较小的子群的研究。这些研究可以帮助我们确定对特定学生群体（例如美国土著女性）进行深层次跟踪调查的最迫切的需求。

潜在的干预措施

虽然潜在的性别干预的优化设计与其影响的研究很有必要，但是本章所评论的研究为实践提供了一些初步的启示。第一，有证据表明，在数学成绩和情感上的性别差异在早期就已经发展，干预措施应从小学阶段开始并利用女孩在学校的日常数学经验，这与大多数干预措施针对大龄学生，并通过课外项目实施的事实形成了鲜明的对比（American Association of University Women，2004）。

第二，如上所述，数学教育工作者应该更加重视培养女孩的空间思维能力和问题解决的方法，并将这些纳入日常学校课程。教师应该放弃在数学课堂上奖励好女孩的行为，而是应该鼓励女孩冒险，采用新的解决问题的方法并对数学问题进行推理。如前所述，数学教育研究人员需要开发和检测支撑学生学习的资源，以帮助教师实施教学手段提高女孩的空间思维能力和问题解决能力。

第三，研究指出了通过改变女孩对数学的看法和态度来提高女孩成绩这种干预措施的潜力。一些研究表明，干预措施可以减轻数学焦虑的影响（Ramirez & Beilock，2011）。此外，帮助女孩发展成长心态可能会培养更有益的数学信念，使女孩远离她们的数学能力是固定的想法，错误地认为她们"不擅长数学"（Dweck，2007；Good等，2003）。

最后，考虑到女孩的职业价值观，以及他们相对较强的语言能力和随之而来更广泛的职业选择（Correll，2001；Eccles，1986；M. T. Wang等，2013），女孩需要机会去了解，在与数学相关的职业中，强大的沟通技巧可以是一种宝贵财富，这些职业涉及与人合作和帮助社会的方式。更普遍的是，关于数学密集领域的信息应强调这些职业怎样与女孩的自我认知相一致（Charles等，2014），应该让女孩做出更明智的选择。值得注意的是，在大学数学和物理科学方面的性别差异相对较小，但在计算机科学和工程学方面的性别差异较大。这可能是因为高中女生熟悉数学和物理专业的教学职业选择，她们更容易设想自己从事在这些职业生涯中，而不是在工程和计算机相关的职业生涯中。这也可能与女性由于察觉到家庭需求缺乏灵活性而回避STEM职业有关（Eccles & Wang，2015）——这是STEM雇主（包括大学）在试图吸引和留住女性时必须要解决的问题。

美国教育科学研究所委托的一份以研究为基础的关于

mately constrain girls' problem-solving approaches? Additionally, girls are viewed as more diligent than boys in general, so why does this perception of diligence seem to diminish teachers' and students' assessment of girls' abilities in mathematics but not in literacy (Robinson-Cimpian et al., 2014)? Is this because mathematics is viewed as something one is either "good at" or not (e.g., being a "mathematics person" or not), and therefore girls' efforts in mathematics signal that they are not "naturally" good at mathematics?

Mathematics educators are also needed to study ways in which emphasizing spatial thinking and bold problem-solving approaches in mathematics classrooms might ultimately relate to sex differences in mathematics performance and later career choices. For example, some experimental and quasi-experimental work has shown that enhanced spatial skills affect mathematics test or course performance, but more work is needed that examines the infusion of spatial skills into K–12 classrooms (Cheng & Mix, 2014; Sorby, 2009; Sorby et al., 2013). Mathematics education researchers are well positioned to work with researchers in other fields (e.g., cognitive psychology) to try to develop the most appropriate ways to include attention to spatial skills and problem solving in mathematics instruction.

Additionally, research examining teachers' mathematics anxiety and their perceptions of boys' and girls' mathematical abilities can help us better understand how these factors might translate into sex differences in performance and affect among students. For example, what types of behaviors do mathematics-anxious teachers show in the classroom that girls might pick up on? What subtle messages might be conveyed to students regarding who is and is not good at mathematics?

Lastly, additional studies examining intersections of race/ethnicity, SES, and gender are needed, given that some gender issues play out differently across groups. Work in this area may benefit from new state longitudinal data systems, which can facilitate examinations of relatively small subgroups in new ways. Such studies can help us identify the most pressing need for in-depth, follow-up studies on specific student groups, such as Native American females.

Potential Interventions

Although research is clearly needed on the optimal design and impact of potential gender interventions, the research reviewed in this chapter offers some tentative implications for practice. First, given evidence suggesting that sex differences in mathematics achievement and affect develop early, interventions should begin in primary grades and should leverage girls' day-to-day mathematics experiences in school. This stands in stark contrast to the fact that most interventions target older students and are implemented via extracurricular programs (American Association of University Women, 2004).

Second, as mentioned above, mathematics educators should give greater attention to enhancing girls' spatial skills and problem-solving approaches, infusing these into the regular school curriculum. Teachers should move away from rewarding good-girl behavior in the mathematics classroom and instead encourage girls to take risks, employ novel problem-solving approaches, and reason about mathematics problems. As noted previously, mathematics education researchers need to develop and test scaffolded resources that can help teachers implement spatial skills and problem solving in mathematics classrooms.

Third, research points to the potential of interventions to improve girls' performance through addressing their perceptions and attitudes about mathematics. Some research has shown that interventions can alleviate the effects of mathematics anxiety (Ramirez & Beilock, 2011). Additionally, helping girls develop a growth mindset may foster more productive mathematics beliefs, moving girls away from the idea that their mathematical abilities are fixed and that mistakes reveal they "aren't good at math" (Dweck, 2007; Good et al., 2003).

Finally, given girls' occupational values, as well as their relatively strong verbal skills and consequent broader career options (Correll, 2001; Eccles, 1986; M. T. Wang et al., 2013), girls need opportunities to learn about ways in which strong communication skills can be a valuable asset in mathematics-related careers, as well as the ways in which these careers can involve working with people and helping society. More generally, information about mathematics-intensive fields should emphasize how these careers align with girls' self-perceptions (Charles et al., 2014) and should allow girls to make more informed choices. It is worth noting that sex differences are relatively small in undergraduate mathematics and physical sciences but large in computer science and engineering. This could be because high school girls are familiar with teaching as a career option for mathematics and physics majors and can envision themselves in those careers more than in engineering and computer-related occupations. This could also relate to females' avoidance of STEM professions due to perceived lack of flexibility for family needs (Eccles &

如何提高女孩在数学和科学方面的参与度的报告突出了上述许多要点，包括培养成长心态和教授空间思维能力的必要性（Halpern等，2007）。然而，教师是否准备实施这些建议，这些建议应如何实施（例如，在单一性别还是男女混合的背景下），或者说如果实施这些建议会产生什么样的效果，目前还不清楚。

精心设计的干预措施和随后对其影响的研究对解决这些问题是非常必要的。此外，教师教育工作者可以通过专注于提高小学教师的内容知识，减少他们的数学焦虑，克服教师关于男孩和女孩的数学能力，预期课堂行为以及教师对这些行为的解释等方面的传统观念来做出自己的贡献。

结论

虽然女孩和男孩在数学学业成绩方面的差距不大，但在分布的上层差距是巨大的，在数学自信心和职业选择上的性别差异也是巨大的。虽然女孩的选择应该被重视，但数学教育研究者更应该研究不公平的教育机会约束这些选择的方式。本章指出了影响女孩数学学业成绩和进行选择的因素，因此干预和进一步研究的时机已经成熟，包括研究女孩解决问题的方法、空间思维能力、女孩和教师对数学的自我认知。

数学相关的职业中女性比例低对女性和数学相关领域都造成了巨大的损失。尽管在过去的几十年里，世界各地的性别平等得到了显著的提高，但仍然存在着令人费解的、持续的差异，这些差异是数学独有的，值得我们重新关注。

注释

1. 莱德在她1992年的那一章中指出，术语"性"和"性别"使用混乱。在本章中，我们按照目前对这些术语的思考（Damarin & Erchick，2010；Glasser & Smith，2008），用术语"性"来指离散的、生物学的范畴，而"性别"包括自我认同和社会文化角色。

2. 例如，与前几十年报道的10%的文章相比，我们在统计《数学教育研究学报》上自2000年以来发表的文章数量时发现，只有5%的文章在某种程度上（通常是直接的）涉及性或性别。

3. 经济合作与发展组织（2014）将"数学素养"定义为"个人在各种情境下表述、使用和解释数学的能力"（第37页）。

4. 经济合作与发展组织关于女生"表现不佳"的这句话说明了进行比较的危险，如男性被描绘成衡量女性的标准（Boaler，2002；Walkerdine，1998）。另一方面，这句话也表明应期望女性的表现与男性相当，从而说明这样的比较是一把双刃剑。

5. 在这里，科学家和工程师包括计算机和信息科学家、数学家、物理科学家和工程师。

6. 虽然教育纵向研究测量了数学的自信心和情感，那些自信心的问题主要集中在学校的任务上（例如，数学作业），而不是典型的那些具有很大性别差异的更一般的问题，例如"我擅长数学"（Lubienski等，2013）。

7. 隐式测量与显式测量所捕捉的内容一直存在争议。一种解释是，隐式测量比显式测量更能显示真正的传统观念，因为隐式测量不受社会期望偏差的影响（Greenwald, McGhee, & Schwartz，1998）。另一种可能的解释是，内隐联想测验能捕捉自动联想，这主要是由于意识到文化的传统观念，但不一定赞同它们（Arkes & Tetlock，2004）。有趣的是，大多数研究表明，隐式和显式的度量是不相关的（Cvencek等，2015；Greenwald, Poehlman, Uhlmann, & Banaji，2009），这表明它们可能在测量不同的结构体。

Wang, 2015)—something that STEM employers (including universities) must address in their attempts to attract and retain women.

The Institute of Education Sciences commissioned a research-based report about ways to enhance girls' participation in mathematics and science, highlighting many of the above points, including the need to foster a growth mindset and teach spatial skills (Halpern et al., 2007). However, it is not clear if teachers are prepared to implement the recommendations, how the recommendations should be implemented (e.g., in single-sex settings or mixed), or what the effects would be if they were implemented.

Carefully designed interventions and subsequent research on their impact are needed to address these questions. Additionally, teacher educators can contribute to these efforts by focusing on enhancing elementary teachers' content knowledge; reducing their mathematics anxiety; and confronting teachers' stereotypes about girls' and boys' mathematical abilities and expected classroom behaviors, as well as teachers' interpretations of those behaviors.

Conclusion

Although disparities in girls' and boys' mathematics performance tend to be small, achievement disparities at the top of the distribution are substantial, as are sex differences in mathematical confidence and occupational choices. Although girls' choices should be valued, mathematics education researchers should examine ways in which those choices may be constrained by inequitable educational opportunities. This chapter points toward factors that influence girls' mathematics achievement and choices and are therefore ripe for intervention and further study, including girls' problem-solving approaches, spatial skills, and girls' and teachers' perceptions of themselves in relation to mathematics.

Women's underrepresentation in mathematics-related careers has substantial costs for both women and mathematics-oriented fields. Despite remarkable gains in gender equity around the world over the past decades, there remain puzzling, persistent disparities that are unique to mathematics and worthy of our renewed attention.

Notes

1. In her 1992 chapter, Leder noted the emergence of confusion about the terms "sex" and "gender." In this chapter we use the term "sex" to refer to discrete, biological categories and "gender" to encompass self-identity and sociocultural roles, in accordance with current thinking about these terms (Damarin & Erchick, 2010; Glasser & Smith, 2008).

2. For example, in contrast to the 10% of articles reported in earlier decades, our count of articles in the *Journal for Research in Mathematics Education* since 2000 revealed that only 5% pertained to sex or gender in some way (often peripherally).

3. OECD (2014) defines "mathematical literacy" as "an individual's capacity to formulate, employ, and interpret mathematics in a variety of contexts" (p. 37).

4. This OECD quote about girls' "underperformance" illustrates a danger of making comparisons such as these—that is, males are portrayed as the standard by which females are to be measured (Boaler, 2002; Walkerdine, 1998). On the other hand, this quote also conveys that females should be expected to perform on par with males, thereby illustrating the double-edged sword of such comparisons.

5. Here, scientists and engineers comprise computer and information scientists, mathematicians, physical scientists, and engineers.

6. Although ELS measured both mathematics confidence and affect, the confidence questions focused on school tasks (e.g., mathematics assignments) as opposed to more general items that typically have larger sex differences, such as "I am good at math" (Lubienski et al., 2013).

7. What is captured by implicit measures compared to explicit measures has been debated. One interpretation is that implicit measures show true stereotype endorsement more than explicit measures because they are free of the effects of social desirability bias (Greenwald, McGhee, & Schwartz, 1998). The other possible explanation is that the implicit association test is capturing automatic associations, which are largely due to being aware of cultural stereotypes, but not necessarily endorsing them (Arkes & Tetlock, 2004). Interestingly, most research shows that implicit and explicit measures are not correlated (Cvencek et al., 2015; Greenwald, Poehlman, Uhlmann, & Banaji, 2009), indicating that they are likely measuring different constructs.

References

Allan, E. J., & Madden, M. (2006). Chilly classrooms for female undergraduate students: A question of method? *The Journal of Higher Education, 77*(4), 684–711.

Altermatt, E. R., Jovanovic, J., & Perry, M. (1998). Bias or responsivity? Sex and achievement-level effects on teachers' classroom questioning practices. *Journal of Educational Psychology, 90*(3), 516–527.

Ambady, N., Shih, M., Kim, A., & Pittinsky, T. L. (2001). Stereotype susceptibility in children: Effects of identity activation on quantitative performance. *Psychological Science, 12*, 385–390.

American Association of University Women. (2004). *Under the microscope: A decade of gender equity projects in the sciences.*

Retrieved from http://www.aauw.org/learn/research/upload/underthemicroscope.pdf

Arkes, H. R., & Tetlock, P. E. (2004). Attributions of implicit prejudice, or "would Jesse Jackson fail the Implicit Association Test?" *Psychological Inquiry, 15*(4), 257–278.

Ashcraft, M. H., & Krause, J. A. (2007). Working memory, math performance, and math anxiety. *Psychonomic Bulletin & Review, 14*(2), 243–248.

Battista, M. T. (1986). The relationship of mathematics anxiety and mathematical knowledge to the learning of mathematical pedagogy by preservice elementary teachers. *School Science and Mathematics, 86*(1), 10–19.

Becker, J. R. (1995). Women's ways of knowing in mathematics. In P. Rogers & G. Kaiser (Eds.), *Equity in mathematics education: Influences of feminism and culture* (pp. 163–174). London, England: Falmer Press.

Beilock, S. L. (2010). *Choke: What the secrets of the brain reveal about getting it right when you have to.* New York, NY: Simon and Schuster.

Beilock, S. L., Gunderson, E. A., Ramirez, G., & Levine, S. C. (2010). Female teachers' mathematics anxiety affects girls' mathematics achievement. *Proceedings of the National Academy of Sciences, USA, 107*(5), 1060–1063.

Belenky, M., Clinchy, B., Goldberger, N., & Tarule, J. M. (1986). *Women's ways of knowing: The development of self, voice, and mind.* New York, NY: Basic Books.

Benbow, C. P., & Stanley, J. C. (1980). Sex differences in mathematical ability: Fact or artifact? *Science, 210*(4475), 1262–1264.

Bench, S. W., Lench, H. C., Liew, J., Miner, K., & Flores, S. A. (2015). Gender gaps in overestimation of math performance. *Sex Roles, 72*(11) 536–546.

Berry, R. Q. (2008). Access to upper-level mathematics: The stories of successful African American middle school boys. *Journal for Research in Mathematics Education, 39,* 464–488.

Bishop, A. J., & Forgasz, H. J. (2007). Issues in access and equity in mathematics education. In F. K. Lester Jr. (Ed.), *Second handbook of research on mathematics teaching and learning* (pp. 1145–1167). Reston, VA: National Council of Teachers of Mathematics.

Blackwell, L., Trzesniewski, K., & Dweck, C. S. (2007). Implicit theories of intelligence predict achievement across an adolescent transition: A longitudinal study and an intervention. *Child Development, 78,* 246–263.

Boaler, J. (2002). Paying the price for "sugar and spice": Shifting the analytical lens in equity research. *Mathematical Thinking and Learning, 4,* 127–144.

Boaler, J., & Sengupta-Irving, T. (2006). Nature, neglect and nuance: Changing accounts of sex, gender and mathematics. In C. Skelton, B. Frances, & L. Smulyan (Eds.), *The SAGE handbook of gender and education* (pp. 205–220). London, England: SAGE.

Boehnke, K. (2008). Peer pressure: A cause of scholastic underachievement? A cross-cultural study of mathematical achievement among German, Canadian, and Israeli middle school students. *Social Psychology of Education, 11*(2), 149–160.

Brandell, G., Leder, G., & Nyström, P. (2007). Gender and mathematics: Recent development from a Swedish perspective. *ZDM—The International Journal on Mathematics Education, 39*(3), 235–250.

Carr, M., & Jessup, D. L. (1997). Gender differences in first-grade mathematics strategy use: Social and metacognitive influences. *Journal of Educational Psychology, 89,* 318–328.

Casey, M. B., Nuttall, R., Pezaris, E., & Benbow, C. (1995). The influence of spatial ability on gender differences in mathematics college entrance test scores across diverse samples. *Developmental Psychology, 31*(4), 697–705.

Catsambis, S. (1994). The path to math: Gender and racial-ethnic differences in mathematics participation from middle school to high school. *Sociology of Education, 67*(3), 199–215.

Charles, M., Harr, B., Cech, E., & Hendley, A. (2014). Who likes math where? Gender differences in eighth-graders' attitudes around the world. *International Studies in Sociology of Education, 24*(1), 85–112.

Che, M., Wiegert, E., & Threlkeld, K. (2012). Problem solving strategies of girls and boys in single-sex mathematics classrooms. *Educational Studies in Mathematics, 79*(2), 311–326.

Cheng, Y. L., & Mix, K. S. (2014). Spatial training improves children's mathematics ability. *Journal of Cognition and Development, 15*(1), 2–11.

Cheryan, S., Plaut, V. C., Davies, P. G., & Steele, C. M. (2009). Ambient belonging: How stereotypical cues impact gender participation in computer science. *Journal of Personality and Social Psychology, 97*(6), 1045–1060.

Cimpian, J. R., Lubienski, S. T., Timmer, J. D., Makowski, M. B., & Miller, E. K. (2016). Have gender gaps in math closed? Achievement, teacher perceptions, and learning behaviors across two ECLS-K cohorts. *AERA Open, 2*(4), 1–19. doi:10.1177/2332858416673617

College Board. (2013). *2013 college-bound seniors: Total group profile report.* New York, NY: Author.

Correll, S. J. (2001). Gender and the career choice process: The role of biased self-assessments. *American Journal of Sociology, 101,* 1691–1730.

Cvencek, D., Kapur, M., & Meltzoff, A. N. (2015). Math achievement, stereotypes, and math self-concepts among elementary-school students in Singapore. *Learning and Instruction, 39,* 1–10.

Cvencek, D., Meltzoff, A. N., & Greenwald, A. G. (2011). Math-gender stereotypes in elementary school children. *Child Development, 82,* 766–779.

Dalton, B., Ingels, S. J., Downing, J., & Bozick, R. (2007). *Advanced mathematics and science coursetaking in the spring high school senior classes of 1982, 1992, and 2004* (NCES 2007-312). Washington, DC: U.S. Department of Education, Institute of Education Sciences, National Center for Education Statistics.

Damarin, S., & Erchick, D. B. (2010). Toward clarifying the meanings of "gender" in mathematics education research. *Journal for Research in Mathematics Education, 41*(4), 310–323.

Dey, J. G., & Hill, C. (2007). *Beyond the pay gap.* Washington, DC: American Association of University Women Educational Foundation.

Downey, D. B., & Vogt Yuan, A. S. (2005). Sex differences in school performance during high school: Puzzling patterns and possible explanations. *The Sociological Quarterly, 46*(2), 299–321.

Duffy, J., Warren, K., & Walsh, M. (2001). Classroom interactions: Gender of teacher, gender of student, and classroom subject. *Sex Roles, 45*(9–10), 579–593.

Dweck, C. S. (1999). *Self-theories: Their role in motivation, personality, and development.* Philadelphia, PA: Psychology Press.

Dweck, C. S. (2007). Is math a gift? Beliefs that put females at risk. In S. J. Ceci & W. Williams (Eds.), *Why aren't more women in science? Top researchers debate the evidence* (pp. 47–55). Washington, DC: American Psychological Association.

Eccles, J. S. (1986). Gender-roles and women's achievement. *Educational Researcher, 15*(6), 15–19.

Eccles, J. S. (2009). Who am I and what am I going to do with my life? Personal and collective identities as motivators of action. *Educational Psychologist, 44*(2), 78–89.

Eccles, J. S., & Wang, M. (2016). What motivates females and males to pursue careers in mathematics and science? *International Journal of Behavioral Development 40*(2). doi:10.1177/0165025415616201

Else-Quest, N. M., Hyde, J. S., & Linn, M. C. (2010). Cross-national patterns of gender differences in mathematics: A meta-analysis. *Psychological Bulletin, 136,* 103–127.

Esmonde, I. (2011). Snips and snails and puppy dogs' tails: Genderism and mathematics education. *For the Learning of Mathematics, 31*(2), 27–31.

Fennema, E., Carpenter, T. P., Jacobs, V. R., Franke, M. L., & Levi, L. W. (1998). A longitudinal study of gender differences in young children's mathematical thinking. *Educational Researcher, 27*(5), 6–11.

Flore, P. C., & Wicherts, J. M. (2015). Does stereotype threat influence performance of girls in stereotyped domains? A meta-analysis. *Journal of School Psychology, 53*(1), 25–44.

Forgasz, H. J., & Hill, J. C. (2013). Factors implicated in high mathematics achievement. *International Journal of Science and Mathematics Education, 11*(2), 481–499.

Forgasz, H. J., Leder, G. C., & Kloosterman, P. (2004). New perspectives on the gender stereotyping of mathematics. *Mathematical Thinking and Learning, 6*(4), 389–420.

Forgasz, H. J., & Mittelberg, D. (2008). Israeli Jewish and Arab students' gendering of mathematics. *ZDM—The International Journal on Mathematics Education, 40*(4), 545–558.

Fredricks, J. A., & Eccles, J. S. (2002). Children's competence and value beliefs from childhood through adolescence: Growth trajectories in two male-sex-typed domains. *Developmental Psychology, 38*(4), 519–533.

Frenzel, A. C., Pekrun, R., & Goetz, T. (2007). Girls and mathematics—a "hopeless" issue? A control-value approach to gender differences in emotions towards mathematics. *European Journal of Psychology of Education, 22,* 497–514.

Fryer, R. G., & Levitt, S. D. (2010). An empirical analysis of the gender gap in mathematics. *American Economic Journal: Applied Economics, 2*(2), 210–240.

Gallagher, A. M., & De Lisi, R. (1994). Gender differences in Scholastic Aptitude Test—Mathematics problem solving among high-ability students. *Journal of Educational Psychology, 86,* 204–211.

Gallagher, A. M., De Lisi, R., Holst, P. C., McGillicuddy-DeLisi, A. V., Morely, M., and Cahalan, C. (2000). Gender differences in advanced mathematical problem solving. *Journal of Experimental Child Psychology, 75,* 165–190.

Ganley, C. M., & Lubienski, S. T. (2016). Mathematics confidence, interest, and performance: Gender patterns and reciprocal relations. *Learning and Individual Differences, 47*(April), 182–193.

Ganley, C. M., & McGraw, A. L. (2016). The development and validation of a revised version of the Math Anxiety Scale for Young Children. *Frontiers in Psychology, 7,* 1181. doi:10.3389/fpsyg.2016.01181

Ganley, C. M., Mingle, L. A., Ryan, A., Ryan, K., Vasilyeva, M., & Perry, M. (2013). An examination of stereotype threat effects on girls' mathematics performance. *Developmental Psychology, 49*(10), 1886–1897.

Ganley, C. M., & Vasilyeva, M. (2011). Sex differences in the relation between math performance, spatial skills, and attitudes. *Journal of Applied Developmental Psychology, 32*(4), 235–242.

Ganley, C. M., & Vasilyeva, M. (2014). The role of anxiety and working memory in gender differences in mathematics. *Journal of Educational Psychology, 106*(1), 105–120.

Geary, D. C. (1996). Sexual selection and sex differences in mathematical abilities. *Behavioral and Brain Sciences, 19*(2), 229–247.

Glasser, H. M., & Smith, J. P. (2008). On the vague meaning of "gender" in education research: The problem, its sources, and recommendations for practice. *Educational Researcher, 37*(6), 343–350.

Goetz, T., Bieg, M., Lüdtke, O., Pekrun, R., & Hall, N. C. (2013). Do girls really experience more anxiety in mathematics? *Psychological Science, 24*(10), 2079–2087.

Goldman, A. D., & Penner, A. M. (2014). Exploring international gender differences in mathematics self-concept. *International Journal of Adolescence and Youth, 21*(4). doi:10.1080/02673843.2013.847850

González de San Román, A., & De la Rica Goiricelaya, S. (2012). Gender gaps in PISA test scores: The impact of social norms and the mother's transmission of role attitudes. (IZA Discussion Paper No. 6338). Retrieved from http://ftp.iza.org/dp6338.pdf

Good, C., Aronson, J., & Inzlicht, M. (2003). Improving adolescents' standardized test performance: An intervention to reduce the effects of stereotype threat. *Journal of Applied Developmental Psychology, 24,* 645–662.

Goodchild, S., & Grevholm, B. (2009). An exploratory study of mathematics test results: What is the gender effect? *International Journal of Science and Mathematics Education, 7*(1), 161–182.

Greenwald, A. G., McGhee, D. E., & Schwartz, J. L. (1998). Measuring individual differences in implicit cognition: the implicit association test. *Journal of Personality and Social Psychology, 74*(6), 1464–1480.

Greenwald, A. G., Poehlman, T. A., Uhlmann, E. L., & Banaji, M. R. (2009). Understanding and using the Implicit Association Test: III. Meta-analysis of predictive validity. *Journal of Personality and Social Psychology, 97*(1), 17.

Gresham, G. (2007). A study of mathematics anxiety in pre-service teachers. *Early Childhood Education Journal, 35,* 181–188.

Guiso, L., Monte, F., Sapienza, P., & Zingales, L. (2008). Diversity: Culture, gender, and math. *Science, 320*(5880), 1164–1165.

Halpern, D. F. (2004). A cognitive-process taxonomy for sex differences in cognitive abilities. *Current Directions in Psychological Science, 13*(4), 135–139.

Halpern, D., Aronson, J., Reimer, N., Simpkins, S., Star, J., & Wentzel, K. (2007). *Encouraging girls in math and science* (NCER 2007-2003). Washington, D.C.: National Center for Education Research, Institute of Education Sciences, Department of Education.

Halpern, D. F., Wai, J., & Saw, A. (2005). *A psychobiosocial model: Why females are sometimes greater than and sometimes less than males in math achievement.* New York, NY: Cambridge University Press.

Harrop, A., & Swinson, J. (2011). Comparison of teacher talk directed to boys and girls and its relationship to their behaviour in secondary and primary schools. *Educational Studies, 37*(1), 115-125.

Hawes, Z., Moss, J., Caswell, B., & Poliszczuk, D. (2015). Effects of mental rotation training on children's spatial and mathematics performance: A randomized controlled study. *Trends in Neuroscience and Education, 4*(3), 60-68.

Hembree, R. (1990). The nature, effects, and relief of mathematics anxiety. *Journal for Research in Mathematics Education, 21*, 33-46.

Hemmings, B., Grootenboer, P., & Kay, R. (2011). Predicting mathematics achievement: The influence of prior achievement and attitudes. *International Journal of Science and Mathematics Education, 9*(3), 691-705.

Herbert, J., & Stipek, D. (2005). The emergence of gender differences in children's perceptions of their academic competence. *Applied Developmental Psychology, 26*, 276-294.

Herzig, A. H. (2004). "Slaughtering this beautiful math": Graduate women choosing and leaving mathematics. *Gender and Education, 16*(3), 379-395.

Hilliard, L. J., & Liben, L. S. (2010). Differing levels of gender salience in preschool classrooms: Effects on children's gender attitudes and intergroup bias. *Child Development, 81*(6), 1787-1798.

Husain, M., & Millimet, D. L. (2009). The mythical "boy crisis"? *Economics of Education Review, 28*(1), 38-48.

Hyde, J. S., Fennema, E., & Lamon, S. J. (1990). Gender differences in mathematics performance: A meta-analysis. *Psychological Bulletin, 107*, 139-155.

Hyde, J. S., Fennema, E., Ryan, M., Frost, L. A., & Hopp, C. (1990). Gender comparisons of mathematics attitudes and affect: A meta-analysis. *Psychology of Women Quarterly, 14*, 299-324.

Hyde, J. S., & Lindberg, S. M. (2007). Facts and assumptions about the nature of gender differences and the implications for gender equity. In S. S. Klein (Ed.), *Handbook for achieving gender equity through education* (2nd ed., pp. 19-32). Mahwah, NJ: Erlbaum.

Hyde, J. S., Lindberg, S. M., Linn, M. C., Ellis, A. B., & Williams, C. C. (2008). Gender similarities characterize math performance. *Science, 321*(5888), 494-495.

Jacobs, J. E. (1994). Feminist pedagogy and mathematics. *ZDM—The International Journal on Mathematics Education, 26*(1), 12-17.

Jones, S. M., & Dindia, K. (2004). A meta-analytic perspective on sex equity in the classroom. *Review of Educational Research, 74*(4), 443-471.

Kaiser, G., Hoffstall, M., & Orschulik, A. B. (2012). Gender role stereotypes in the perception of mathematics: An empirical study with secondary students in Germany. In H. Forgasz & F. Rivera (Eds.), *Towards equity in mathematics education: Gender, culture, and diversity* (pp. 115-140). Berlin, Germany: Springer.

Kimball, M. M. (1989) A new perspective on women's math achievement. *Psychological Bulletin, 105*, 198-214.

Kimura, D. (1992). Sex differences in the brain. *Scientific American, 267*, 139-155.

Klein, P. S., Adi-Japha, E., & Hakak-Benizri, S. (2010). Mathematical thinking of kindergarten boys and girls: Similar achievement, different contributing processes. *Educational Studies in Mathematics, 73*(3), 233-246.

Köller, O., Baumert, J., & Schnabel, K. (2001). Does interest matter? The relationship between academic interest and achievement in mathematics. *Journal for Research in Mathematics Education, 32*, 448-470.

Krinzinger, H., Kaufmann, L., & Willmes, K. (2009). Math anxiety and math ability in early primary school years. *Journal of Psychoeducational Assessment, 27*(3), 206-225.

Lacampagne, C. B., Campbell, P. B., Herzig, A. H., Damarin, S., & Vogt, C. M. (2007). Gender equity in mathematics. In S. S. Klein (Ed.), *Handbook for achieving gender equity through education* (pp. 235-253). New York, NY: Routledge.

Laski, E. V., Casey, B. M., Yu, Q., Dulaney, A., Heyman, M., & Dearing, E. (2013). Spatial skills as a predictor of first grade girls' use of higher level arithmetic strategies. *Learning and Individual Differences, 23*, 123-130.

Leder, G. C. (1992). Mathematics and gender: Changing perspectives. In D. A. Grouws (Ed.), *Handbook of research on mathematics teaching and learning: A project of the National Council of Teachers of Mathematics* (pp. 597-622). New York, NY: Macmillan.

Leder, G. C. (2010). Commentary 1 on feminist pedagogy and mathematics. In B. Sriraman & L. English (Eds.), *Theories of mathematics education: Seeking new frontiers* (pp. 447-454). New York, NY: Springer.

Lindberg, S. M., Hyde, J. S., Petersen, J. L., & Linn, M. C. (2010). New trends in gender and mathematics performance: A meta-analysis. *Psychological Bulletin, 136*, 1123-1135.

Linn, M. C., & Petersen, A. C. (1985). Emergence and characterization of sex differences in spatial ability: A meta-analysis. *Child Development, 56*, 1479-1498.

Lord, S. M., Camacho, M. M., Layton, R. A., Long, R. A., Ohland, M. W., & Wasburn, M. H. (2009). Who's persisting in engineering? A comparative analysis of female and male Asian, Black, Hispanic, Native American, and White students. *Journal of Women and Minorities in Science and Engineering, 15*(2), 167-190.

Louis, R. A., & Mistele, J. M. (2012). The differences in scores and self-efficacy by student gender in mathematics and science. *International Journal of Science and Mathematics Education, 10*(5), 1163-1190

Loveless, T. (2015). *The 2015 Brown Center report on American education: How well are American students learning?* Retrieved from http://www.brookings.edu/~/media/Research/Files/Reports/2015/03/BCR/2015-Brown-Center-Report_FINAL.pdf?la=en

Lubienski, S. T., & Bowen, A. (2000). Who's counting? A survey of mathematics education research 1982-1998. *Journal for Research in Mathematics Education, 31*(5), 626-633.

Lubienski, S. T., Robinson, J. P., Crane, C. C., & Ganley, C. M. (2013). Girls' and boys' mathematics achievement, affect, and experiences: Findings from ECLS-K. *Journal for Research in Mathematics Education, 44*(4), 634–645.

Lubinski, D., & Benbow, C. P. (2006). Study of mathematically precocious youth after 35 years: Uncovering antecedents for the development of math-science expertise. *Perspectives on Psychological Science, 1,* 316–345.

Ma, X. (1999). A meta-analysis of the relationship between anxiety toward mathematics and achievement in mathematics. *Journal for Research in Mathematics Education, 30*(5), 520–540.

Ma, X., & Xu, J. (2004). The causal ordering of mathematics anxiety and mathematics achievement: A longitudinal panel analysis. *Journal of Adolescence, 27*(2), 165–179.

Maloney, E. A., Schaeffer, M. W., & Beilock, S. L. (2013). Mathematics anxiety and stereotype threat: Shared mechanisms, negative consequences and promising interventions. *Research in Mathematics Education, 15*(2), 115–128.

Maloney, E. A., Waechter, S., Risko, E. F., & Fugelsang, J. A. (2012). Reducing the sex difference in math anxiety: The role of spatial processing ability. *Learning and Individual Differences, 22*(3), 380–384.

Maltese, A. V., & Tai, R. H. (2010). Eyeballs in the fridge: Sources of early interest in science. *International Journal of Science Education, 32*(5), 669–685.

Marsh, H. W., Trautwein, U., Ludtke, O., Köller, O., & Baumert, J. (2005). Academic self-concept, interest, grades, and standardized test scores: Reciprocal effects models of causal ordering. *Child Development, 76*(2), 397–416.

McGraw, R., Lubienski, S. T., & Strutchens, M. E. (2006). A closer look at gender in NAEP mathematics achievement and affect data: Intersection with achievement, race/ethnicity, and socioeconomic status. *Journal for Research in Mathematics Education, 37*(2), 129–150.

Miller, H., & Bichsel, J. (2004). Anxiety, working memory, gender, and math performance. *Personality and Individual Differences, 37,* 591–606.

Mix, K. S., & Cheng, Y. L. (2012). Space and math: The development and educational implications. In J. Benson (Ed.), *Advances in child development and behavior* (pp. 179–243). New York, NY: Elsevier.

National Science Foundation. (2011). *SESTAT: Scientists and Engineers Statistical Data System.* Retrieved from http://www.nsf.gov/statistics/sestat/

Nix, S., Perez-Felkner, L., & Thomas, K. (2015). Perceived mathematical ability under challenge: A longitudinal perspective on sex segregation among STEM degree fields. *Frontiers in Psychology, 6*(530). doi: 10.3389/fpsyg.2015.00530.

Noddings, N. (1998). Perspectives from feminist psychology. *Educational Researcher, 27*(5), 17–18.

Norton, S. (2006). Pedagogies for the engagement of girls in the learning of proportional reasoning through technology practice. *Mathematics Education Research Journal, 18*(3), 69–99.

Nosek, B. A., Banaji, M. R., & Greenwald, A. G. (2002). Math = male, me = female, therefore math not π me. *Journal of Personality and Social Psychology, 83,* 44–59.

Nosek, B. A., & Smyth, F. L. (2011). Implicit social cognitions predict sex differences in math engagement and achievement. *American Educational Research Journal, 48,* 1125–1156.

Nuttall, R. L., Casey, M. B., & Pezaris, E. (2005). Spatial ability as a mediator of gender differences on mathematics tests: A biological-environmental framework. New York, NY: Cambridge University Press.

O'Connor, C., & Joffe, H. (2014). Gender on the brain: A case study of science communication in the new media environment. *PLOS One, 9*(10). doi:10.1371/journal.pone.0110830

Organisation for Economic Co-Operation and Development. (2013). *PISA 2012 results: Ready to learn: Students' engagement, drive and self-beliefs* (Vol. III). Retrieved from http://www.oecd.org/pisa/keyfindings/pisa-2012-results-volume-III.pdf

Organisation for Economic Co-Operation and Development. (2014). *PISA 2012 results: What students know and can do—Student performance in mathematics, reading and science* (Vol. I, rev. ed.). Retrieved from http://www.oecd.org/pisa/keyfindings/pisa-2012-results-volume-I.pdf

Organisation for Economic Co-Operation and Development. (2015). *The ABC of gender equality in education: Aptitude, behaviour, confidence.* doi:10.1787/9789264229945-en

Pahlke, E., Hyde, J. S., & Allison, C. M. (2014). The effects of single-sex compared with coeducational schooling on students' performance and attitudes: A meta-analysis. *Psychological Bulletin, 140*(4), 1042.

Paunesku, D., Walton, G. M., Romero, C., Smith, E. N., Yeager, D. S., & Dweck, C. S. (2015). Mind-set interventions are a scalable treatment for academic underachievement. *Psychological Science, 26*(6). doi:10.1177/0956797615571017

Penner, A. M., & Paret, M. (2008). Gender differences in mathematics achievement: Exploring the early grades and the extremes. *Social Science Research, 37*(1), 239–253.

Perez-Felkner, L., McDonald, S.-K., & Schneider, B. L. (2014). What happens to high-achieving females after high school? Gender and persistence on the postsecondary STEM pipeline. In I. Schoon & J. S. Eccles (Eds.), *Gender differences in aspirations and attainment* (pp. 285–320). Cambridge, United Kingdom: Cambridge University Press.

Perie, M., Moran, R., & Lutkus, A. (2005). *NAEP 2004 trends in academic progress: Three decades of student performance in reading and mathematics* (NCES 2005–464). Washington, DC: U.S. Department of Education, Institute of Education Sciences, National Center for Education Statistics.

Pinker, S. (2002). *The blank slate: The modern denial of human nature.* New York, NY: Penguin.

Pinxten, M., Marsh, H. W., De Fraine, B., Van Den Noortgate, W., & Van Damme, J. (2014). Enjoying mathematics or feeling competent in mathematics? Reciprocal effects on mathematics achievement and perceived math effort expenditure. *British Journal of Educational Psychology, 84*(1), 152–174.

Pomerantz, E. M., Altermatt, E. R., & Saxon, J. L. (2002). Making the grade but feeling distressed: Gender differences in academic performance. *Journal of Educational Psychology, 94,* 396–404.

Provasnik, S., Kastberg, D., Ferraro, D., Lemanski, N., Roey, S., and Jenkins, F. (2012). *Highlights From TIMSS 2011: Mathematics and science achievement of U.S. fourth- and eighth-grade students in an international context* (NCES 2013–009 Revised). Washington, DC: U.S. Department of Education, Institute of Education Sciences, National Center for Education Statistics.

Ramirez, G., & Beilock, S. L. (2011). Writing about testing worries boosts exam performance in the classroom. *Science, 331*(6014), 211–213.

Rathbun, A. H., West, J., & Germino-Hausken, E. (2004). *From kindergarten through third grade: Children's beginning school experiences* (NCES 2004–007). Washington, DC: U.S. Department of Education, Institute of Education Sciences, National Center for Education Statistics.

Rayner, V., Pitsolantis, N., & Osana, H. (2009). Mathematics anxiety in preservice teachers: Its relationship to their conceptual and procedural knowledge of fractions. *Mathematics Educational Research Journal, 21*(3), 60–85.

Reardon, S. F., & Robinson, J. P. (2008). Patterns and trends in racial/ethnic and socioeconomic academic achievement gaps. In H. F. Ladd & E. B. Fiske (Eds.), *Handbook of research in education finance and policy* (pp. 499–518). New York, NY: Routledge.

Riegle-Crumb, C., & Humphries, M. (2012). Exploring bias in math teachers' perceptions of students' ability by gender and race/ethnicity. *Gender & Society, 26*(2), 290–322.

Riegle-Crumb, C., & King, B. (2010). Questioning a white male advantage in STEM: Examining disparities in college major by gender and race/ethnicity. *Educational Researcher, 39*(9), 656–664.

Riegle-Crumb, C., King, B., Grodsky, E., & Muller, C. (2012). The more things change, the more they stay the same? Prior achievement fails to explain gender inequality in entry into STEM college majors over time. *American Educational Research Journal, 49*(6), 1048–1073.

Riegle-Crumb, C., King, B., & Morton, K., (2014, April). *Staying the course in STEM: How academic achievement and integration shape persistence by gender and race/ethnicity.* Paper presented at the annual meeting of the American Educational Research Association, Philadelphia, PA.

Robinson, J. P., & Lubienski, S. T. (2011). The development of gender achievement gaps in mathematics and reading during elementary and middle school: Examining direct cognitive assessments and teacher ratings. *American Educational Research Journal, 48*(2), 268–302.

Robinson-Cimpian, J. P., Lubienski, S. T., Ganley, C. M., & Copur-Gencturk, Y. (2014). Teachers' gender-stereotypical ratings of mathematics proficiency may exacerbate early gender achievement gaps. *Developmental Psychology, 50*(4), 1262–1281.

Sadker, M., & Sadker, D. (1994). *Failing at fairness: How America's schools cheat girls.* New York, NY: Macmillan.

Snyder, T. D., & Dillow, S. A. (2011). *Digest of education statistics 2010* (NCES 2011–015). Washington, DC: U.S. Department of Education, Institute of Education Sciences, National Center for Education Statistics.

Sommers, C. H. (2001). *The war against boys: How misguided feminism is harming our young men.* New York, NY: Touchstone Books.

Sorby, S. A. (2009). A course in spatial visualization and its impact on the retention of female engineering students. *Journal of Women and Minorities in Science and Engineering, 7*(2), 153–172.

Sorby, S., Casey, B., Veurink, N., & Dulaney, A. (2013). The role of spatial training in improving spatial and calculus performance in engineering students. *Learning and Individual Differences, 26*, 20–29.

Spencer, S. J., Steele, C. M., & Quinn, D. M. (1999). Stereotype threat and women's math performance. *Journal of Experimental Social Psychology, 35*, 4–28.

Steffens, M. C., Jelenec, P., & Noack, P. (2010). On the leaky pipeline: Comparing implicit math-gender stereotypes and math withdrawal in female and male children and adolescents. *Journal of Educational Psychology, 102*, 947–963.

Stinson, D. W. (2008). Negotiating sociocultural discourses: The counter-storytelling of academically (and mathematically) successful African American male students. *American Educational Research Journal, 45*(4), 975–1010.

Stipek, D., & Gralinski, J. H. (1996). Children's beliefs about intelligence and school performance. *Journal of Educational Psychology, 88*, 397–407.

Stoet, G., & Geary, D. C. (2012). Can stereotype threat explain the gender gap in mathematics performance and achievement? *Review of General Psychology, 16*(1), 93–102.

Stricker, L. J., & Ward, W. C. (2004). Stereotype threat, inquiring about test takers' ethnicity and gender, and standardized test performance. *Journal of Applied Social Psychology, 34*, 665–693.

Sumpter, L. (2015). Investigating upper secondary school teachers' conceptions: Is mathematical reasoning considered gendered? *International Journal of Science and Mathematics Education*, 1–16.

Tai, R. H., Liu, C. Q., Maltese, A. V., & Fan, X. (2006). Planning early for careers in science. *Science, 312*, 1143–1144.

Tomasetto, C., Alparone, F. R., & Cadinu, M. (2011). Girls' math performance under stereotype threat: The moderating role of mothers' gender stereotypes. *Developmental Psychology, 47*(4), 943–949.

Tyson, W., Lee, R., Borman, K. M., & Hanson, M. A. (2007). Science, technology, engineering, and mathematics (STEM) pathways: High school science and math coursework and postsecondary degree attainment. *Journal of Education for Students Placed at Risk, 12*(3), 243–270.

Tzuriel, D., & Egozi, G. (2010). Gender differences in spatial ability of young children: The effects of training and processing strategies. *Child Development, 81*(5), 1417–1430.

Ursini, S., & Sánchez, G. (2008). Gender, technology and attitude towards mathematics: A comparative longitudinal study with Mexican students. *ZDM—The International Journal on Mathematics Education, 40*(4), 559–577.

Uttal, D. H., Meadow, N. G., Tipton, E., Hand, L. L., Alden, A. R., Warren, C., & Newcombe, N. S. (2013). The malleability of spatial skills: A meta-analysis of training studies. *Psychological Bulletin, 139*(2), 352–402.

Vale, C., & Bartholomew, H. (2008). Gender and mathematics: Theoretical frameworks and findings. In H. Forgasz et al. (Eds.), *Mathematics education research in Australasia: 2004-2007* (pp. 271–290). Rotterdam, The Netherlands: Sense.

Vale, C. M., & Leder, G. C. (2004). Student views of computer-based mathematics in the middle years: Does gender make a difference? *Educational Studies in Mathematics, 56*(2-3), 287–312.

Vasilyeva, M., Casey, B. M., Dearing, E., & Ganley, C. M. (2009). Measurement skills in low-income elementary school students: Exploring the nature of gender differences. *Cognition and Instruction, 27*(4), 401–428.

Voyer, D. (1996). The relation between mathematical achievement and gender differences in spatial abilities: A suppression effect. *Journal of Educational Psychology, 88*, 563–571.

Voyer, D., Voyer, S., & Bryden, M. P. (1995). Magnitude of sex differences in spatial abilities: A meta-analysis and consideration of critical variables. *Psychological Bulletin, 117*(2), 250–270.

Wai, J., Lubinski, D., & Benbow, C. P. (2009). Spatial ability for STEM domains: Aligning over 50 years of cumulative psychological knowledge solidifies its importance. *Journal of Educational Psychology, 101*(4), 817–835.

Walkerdine, V. (1998). *Counting girls out: Girls and mathematics* (2nd ed.). London, England: Falmer Press.

Wang, M. T., Eccles, J. S., & Kenny, S. (2013). Not lack of ability but more choice: Individual and gender differences in choice of careers in science, technology, engineering, and mathematics. *Psychological Science, 24*(5), 770–775.

Wang, Z., Osterlind, S. J., & Bergin, D. A. (2012). Building mathematics achievement models in four countries using TIMSS 2003. *International Journal of Science and Mathematics Education, 10*(5), 1215–1242.

Watt, H. M. G. (2004). Development of adolescents' self-perceptions, values, and task perceptions according to gender and domain in 7th through 11th grade Australian students. *Child Development, 75*, 1556–1574.

Watt, H. M. G., Shapka, J. D., Morris, Z. A., Durik, A. M., Keating, D. P., & Eccles, J. S. (2012). Gendered motivational processes affecting high school mathematics participation, educational aspirations, and career plans: A comparison of samples from Australia, Canada, and the United States. *Developmental Psychology, 48*(6), 1594–1611.

Wei, T. E. (2012). Sticks, stones, words, and broken bones: New field and lab evidence on stereotype threat. *Educational Evaluation and Policy Analysis, 34*, 465–488.

Weiss, E. M., Kemmler, G., Deisenhammer, E. A., Fleischhacker, W. W., & Delazer, M. (2003). Sex differences in cognitive functions. *Personality and Individual Differences, 35*(4), 863–875.

Weldeana, H. N. (2015). Gender positions and high school students' attainment in local geometry. *International Journal of Science and Mathematics Education, 13*(6), 1331–1354.

Wiest, L. R. (2011). Females in mathematics: Still on the road to parity. In B. Atweh, M. Graven, W. Secada, & P. Valero (Eds.), *Mapping equity and quality in mathematics education* (pp. 325–339). Dordrecht, The Netherlands: Springer Netherlands.

Wigfield, A., & Eccles, J. S. (2002). The development of competence beliefs, expectancies for success, and achievement values from childhood through adolescence. In A. Wigfield & J. S. Eccles (Eds.), *Development of achievement motivation* (pp. 173–195). San Diego, CA: Academic Press.

Winkelmann, H., van den Heuvel-Panhuizen, M., & Robitzsch, A. (2008). Gender differences in the mathematics achievements of German primary school students: Results from a German large-scale study. *ZDM—The International Journal on Mathematics Education, 40*(4), 601–616.

Zafar, B. (2013). College major choice and the gender gap. *The Journal of Human Resources, 48*(3), 545–595.

Zhu, Z. (2007). Gender differences in mathematical problem solving patterns: A review of literature. *International Education Journal, 8*(2), 187–203.

Zohar, A., & Gershikov, A. (2008). Gender and performance in mathematical tasks: Does the context make a difference? *International Journal of Science and Mathematics Education. 6*(4), 677–693.

25 数学参与的复杂性：动机、情感和社会互动

詹姆斯·米德尔顿
美国亚利桑那州立大学
阿曼达·詹森
美国特拉华大学
杰拉尔德·A.戈尔丁
美国罗格斯大学
译者：姚一玲
杭州师范大学教育学院

本研究手册大部分内容是关于人们所学习的数学、数学学习的质量以及促进有效教学的条件。所有的研究都表明了一个数学学习的事实，那就是学生必须参与到数学中。参与数学活动能够用一些有意思的形式所表现，如在学习者与学习内容之间建立连接、帮助学习者在数学探究过程中愿意去实践、形成概念和技能，并建立与数学的亲密感。本章，我们将会用"参与"一词来指代个体与其当下所处环境之间的即时性的关系，其中个体所处的环境是指任务、内部特征以及与个体互动的其他要素。参与能够在活动中表现出来，这些活动包括可观察到的行为以及与注意、努力、认知及情感等相关的心智活动。

参与或许是产生有质量的数学学习的一种有效行为，这也是我们作为教育者所希望的。但是，从数学角度来看参与也有可能会完全无效，例如，某个学生只想着少花力气去强化其在社会交往中的地位或维持自尊。我们将有效参与视为有效学习的关键，但是总体上我们也认为参与会根据具体课堂中的约束条件以及学生们广泛的社交目标、个人目标和情感喜好作出调整。

本章将会从三个主要观点贯穿全文。第一个观点是学习与参与学习发生过程完全是不可分的，这里的学习不仅是针对概念和技能的，还包括对社会规范、学校数学实践以及与数学活动有关的情感、信念和价值观的学习。第二个观点是学生课堂参与的复杂性。动机、情感和社交互动三者形成了一个动态的系统，影响着学生的参与行为，并且能够对学生行为中的认知、情感和社交适应性提供反馈，因此学生个体或群体的参与可以自然地产生，依赖于环境并且对原始条件具有敏感性。虽然如此，我们仍然可以合理地对其进行预测。第三个观点是可塑因素的重要性，即教师、课程设计者或者教育部门领导者可以鼓励学生产生更深入更有效的参与，最终，能适应当下所处的学习环境。这些观点基本上涵盖了从心理学角度对学校参与的讨论（例如 Shernoff, 2013），但是，本章我们将会从数学学科的角度来讨论这些问题。

本章的主要内容是讨论动机、情感和社会互动问题。动机是指个体通过参与满足自己的需要、渴望或目的——也就是参与的原因。动机可能是短暂的，也可能是长期和持久的。人们或许能够意识到自己的动机，但至少会有一部分刺激是来自于无意识的需求。我们认为情感是与情绪及其相关内涵最接近的一个领域，它包括影响参与的情绪及其它相关因素。情感维度包括态度、信念、价值观、情绪、情绪管理及相关概念。另外，由于数学课堂是包括学生、教师、学习内容及参与活动的一个社会空间，所以社会互动是一个基础性的维度。社会互动不总是外显的，在学生与教师尝试揣测他人意图以及在数学活动中思考如何回应时也会发生隐性互动。

在本章中，我们主张用综合的理论视角考察参与。

25 | The Complexities of Mathematical Engagement: Motivation, Affect, and Social Interactions

JAMES MIDDLETON
Arizona State University, United States

AMANDA JANSEN
University of Delaware, United States

GERALD A. GOLDIN
Rutgers University, United States

Research about the mathematics a person learns, the quality of the learning, and the conditions promoting effective teaching and learning constitute most of this compendium. Each discussion is predicated on the fact that to learn mathematics, students must *engage* with mathematics. Engagement evokes interesting metaphors, such as forming a connection between the learner and what is to be learned and developing intimacy with and commitment to the practices, concepts, and skills of mathematical inquiry. In this chapter, we ultimately take engagement to refer to the in-the-moment relationship between someone and her immediate environment, including the tasks, internal states, and others with whom she interacts. Engagement manifests itself in activity, including both observable behavior and mental activity involving attention, effort, cognition, and emotion.

Engagement may be productive, leading to qualities of mathematics learning we desire as educators. But engagement may also be mathematically counterproductive; for example, if a student is primarily focused on minimizing school effort to enhance social status or to preserve a sense of self-worth. We regard productive engagement as essential to effective learning, but we see engagement generally as adaptive—to local classroom constraints and to students' wider social and personal goals and emotional priorities.

Three main ideas recur throughout this chapter. The first is the *fundamental inseparability* of learning from the engagement through which learning takes place—involving not only concepts and skills but also the social norms and practices of school mathematics and the emotions, beliefs, and values associated with mathematical activity. The second idea is the *complexity* of students' engagement in the classroom. Motivation, emotion, and social interaction form a dynamic system, influencing engaged behavior and providing feedback about the cognitive, emotional, and social fit of that behavior. The engagement of a student, or group, is thus in a sense emergent, context dependent, and sensitive to initial conditions, but still reasonably predictable over the longer term. The third idea is the importance of *malleable factors* that a teacher, curriculum designer, or educational leader can influence to encourage deeper, more productive engagement—ultimately, through adaptation of in-the-moment conditions of learning. These ideas generally characterize discussions of the psychology of school engagement (e.g., Shernoff, 2013); here we focus on them in relation to mathematics.

Major sections of this chapter address motivation, affect, and social interactions. By *motivation*, we mean the need, desire, or purpose the person is trying to fulfill by engaging—that is, the reasons why someone engages. Motivation may be transient, or it may be longer term and enduring. People may be consciously aware of their motivations or motivated at least partially by unconscious needs. We take *affect* to refer to the domain most involving emotions and their meanings, inclusive of the many associated ways in which emotions interact with other factors to influence engagement. The affective domain thus includes attitudes, beliefs, and values, as well as emotions, the regulation of emotion, and related constructs. We regard *social interactions* as fundamental because mathematics classrooms are inherently social spaces embracing students, the teacher, the subject of learning, and the engaged activity. Social interactions are not always overt; tacit interactions occur as students and teachers attempt to anticipate the meanings of others and plan responses during mathematical activity.

虽然本章每一部分都在讨论与参与相关的不同因素，但是，在参与过程中或学习者经历中，这些因素表现出来的特征并不像在研究中经常表现出来的特征所推断出来的那么明显，也就是说，情感、认知、动机和社会互动是互相交错、相互影响的，并且情境因素高度影响着学习者的参与。另外，虽然学生在各学科上会表现出非常多的参与行为模式，但是我们认为数学参与具有学科特征，即反映出关于数学的特征，并不一定适用于其他学科。

在本章的每一节，我们都对近20至25年来学生在数学课堂上的参与、动机、情感和社交互动方面的研究进行了综述。麦克劳德（1992）总结了过去40年关于数学教育中情感的研究，J.A.米德尔顿和斯班尼厄斯（1999）也回顾了先前关于动机的研究。因此，我们尽可能从这些研究综述截止的时间开始，但也会涉及部分与本文内容特别相关的之前的研究。麦克劳德在他的文章结尾提出，直至当时，关于情感的研究尚未对数学教育产生特别的影响。目前来看，我们认为这个说法已经不正确了。麦克劳德还指出，数学课程和教学的研究很少关注情感及其相关结构如何影响学生概念学习及学习的过程，这一现状至今依然存在，因此，希望本章对这方面的研究能有所帮助。

为了能从大量的研究中清晰地提炼出关键性内容，我们主要关注一些重要的观点和视角，概述一些具有代表性的文献和具体研究结果，同时引证一些相关的研究。这一章，我们先讨论数学参与的意义和内涵，之后我们将从方法论的角度分析在这些研究中用到的主要研究方法。接下来的三个部分，我们将分别讨论动机及其要素、情感维度和结构、以及社会互动和教育实践。最后一部分，着重用综合的视角讨论参与的各个重要维度并提供一些未来研究方向。

数学参与的内涵解析

定义

当个体将认知和情感投入到活动中时，参与就自然而然发生了。个体的参与通常是指向具体对象的，这个对象有可能是活动本身或者活动当中涉及的学术、社交或其它环境因素。因此，参与是表现在情感、认知或行为上，具有多维度的结构。例如，阿普尔顿、克里斯坦森和弗朗（2008）考察了学校心理学文献中19个参与的定义，所有的定义都将参与视为是对某些活动的参与，其中包括行为上的和认知上的参与。其中，有11个定义指出参与可以通过个体在情感或情绪上的表现而体现出来；有4个定义通过描述归属感或关系来体现参与的社会维度；有6个定义则提到了个体的投入、价值观和追求能够促使参与的发生。

从参与对象来看，这19个定义中有12个定义指出参与对象都与学校或学习有关，部分定义认为这些对象可以涉及社会交往关系或者非学术类的活动，有些定义则未具体指出参与的对象。因此，在参与的定义中，对象有可能会被重点关注也有可能会被忽略掉。

只有两个定义是将参与置于情境中来界定的。一个来自于2003年弗朗等人在《加州学校心理学》杂志上发表的文章，他们指出情感、认知和行为的"子类"是发生在"学生内部、同伴团体、课堂和学校的广泛情境中的"（引自Appleton等，2008，第371页，重点标注如原文）。另一个定义是在1991年由康奈尔和韦尔伯恩提到的，"当心理需求在如家庭、学校和工作等文化组织内部得以满足的时候"，参与就发生了（The Minnesota Symposia on Child Psychology 第23卷；引自Appleton等，2008，第371页；重点标注如原文）。

我们认为，参与的本质内容包括学生个体的投入、目标或欲望以及参与的对象，这些都是从学生视角（即参与者）来讨论的。有了参与的欲望和对象并不意味着要减少对学生数学学习参与效果的关注。此外，数学参与存在于广泛的数学课程、国家和文化的传统、国家和州一层的教育政策、学校环境、教师的教学实践、同伴文化、学习者的家庭文化和期望、学生的个人经历中，等等。

即时性参与和长期性参与

即时和长期参与之间的相互作用是贯穿本章内容的核心。即时数学参与包括欲望、想法、感受及在特殊场合中的交互作用，它通常会涉及特殊的任务和人。即时参与既是个体动机、情感和认知共同作用结果，也是

In this chapter we advocate taking an *integrative* theoretical perspective on engagement. While chapter sections focus on different factors, these are not as distinct during engagement or in the experiences of learners as might be inferred from their frequent characterization in research. That is, we see affect, cognition, motivation, and social interactions as intertwined, mutually interacting, and highly situational influences on engagement. Moreover, although students may display broad behavior patterns across academic subjects, we regard mathematical engagement as *domain specific*—involving features particular to mathematics—and not necessarily generalizable to other subjects.

In each section of this chapter, we review an illustrative subset of the peer-reviewed research on engagement, motivation, affect, and social interactions in mathematics classrooms over the past 20 to 25 years. McLeod (1992) summarized the preceding 40 years of research on affect in relation to mathematics education. J.A. Middleton and Spanias (1999) likewise reviewed prior research on motivation. We endeavor to begin where these sources leave off, but we also include earlier studies that seem particularly relevant. McLeod concludes his chapter by noting that research on affect up to that time had not been particularly influential in mathematics education. In our view, this is no longer true. McLeod also suggests that studies in mathematics curriculum and instruction gave insufficient attention to how affect and related constructs influence students' learning of concepts and procedures. We think this is still the case and hope the present chapter inspires improvement in this situation.

To distill a large volume of research coherently we focus here on major ideas and perspectives, summarizing a few representative references and research findings in more detail and citing related work. The next section discusses meanings and interpretations of mathematical engagement. We then explore methodological issues, describing some exemplars of the methods primarily utilized in the research. In the three sections that follow, we focus respectively on motivation and its subconstructs, on affective variables and structures, and on social interactions and educational practice. In the final section, we highlight integrative aspects of the critical dimensions of engagement and propose some directions for future research.

Interpretations of Mathematical Engagement

Definitions

Engagement occurs when someone participates in an activity with some cognitive or affective investment. A person's engagement is typically directed toward an *object*, which may involve the activity itself or its academic, social, or other context. It may be expressed affectively, cognitively, or behaviorally. As such, engagement is a multidimensional construct. Appleton, Christenson, and Furlong (2008), for example, examined 19 definitions from the school psychology literature. All definitions describe engagement as *participation* in some activity, either *behavioral* or *cognitive* participation. In 11 of these definitions, engagement manifests itself in *affective* or *emotional expression;* 4 incorporate the social dimension of engagement by describing *belonging* or *relatedness* as a manifestation. The idea that personal *investments, values,* and *desires* drive engagement is reflected in 6 definitions.

Among the 19 definitions, 12 describe the object of engagement as related to school or learning. Other definitions acknowledge that this object could involve social relationships or nonacademic activities. Some definitions do not specify the object of engagement. Thus, the object of engagement may take a foreground or background position in the definition.

Only two definitions situate engagement in contexts. First, Furlong et al., in their 2003 article in *The California School Psychologist* indicate that affective, cognitive, and behavioral "subtypes" occur "within *student, peer group, classroom,* and *school wide contexts*" (as cited in Appleton et al., 2008, p. 371, emphasis in original). Second, in 1991 Connell and Wellborn wrote that engagement occurs as "*psychological needs ... are met* within cultural enterprises such as family, school, and work" (Vol. 23 of *The Minnesota Symposia on Child Psychology;* as cited in Appleton et al., 2008, p. 371; emphasis in original).

In our perspective, essential aspects of engagement include students' personal investments, goals or desires, and the objects of engagement. These are in the eye of the student (i.e., the one engaged). The presence of a range of desires or objects of engagement does not necessarily detract from the effectiveness of engagement in learning mathematics. Furthermore, mathematical engagement is situated in the wider discipline of mathematics, traditions of countries and cultures, national and state level educational policies, school environments, teachers' instructional practices, peer cultures, learners' home cultures and expectations, students' personal histories, and so forth.

In-the-Moment vs. Longer Term Engagement

The interplay between the here and now and the longer term is central throughout this chapter. The phenomenon of *in-the-moment* engagement with mathematics includes the desires, thoughts, feelings, and interactions occurring on a particular occasion, usually involving a

个人和社交同时作用的结果。个体中那些更有持久性的个性特质，如性格、取向、态度和信念、价值观、能力和关系，在某种程度上都会影响即时性参与的本质，而且这些特质也会在当下的情境中以特定的方式表现出来。慢慢地，新的记忆和调整后的情感以及认知取向、关系等通过持续的数学参与经验慢慢形成。

每一个影响数学参与的重要因素都具有相应的状态（即时的）和特质（长期的、持续的）。动机包括对情境的兴趣、任务目标、内生效用（状态）以及个人的兴趣、目标取向和外生效用（特质）。情感包括即时情绪和局部情感（状态）和情绪取向、态度以及全局情感（特质）。情感结构（长期的）是能够在当下作为一种预期的结构或模式帮助个体参与到当下的目标中，并提供情绪反馈。课堂数学教学（社交状态）在社会数学规范（类似于社交系统的特质）中表现出一种自然而然的稳定性（Cobb & Yackel，1996）。简而言之，即时参与是一个基础性概念，可以用来描述个体在数学参与上短暂和不稳定的经验如何转变为更为持久的理解和能力。

行为、情感和认知维度

弗雷德里克斯、布卢门菲尔德和帕里斯（2004）认为参与是一种元结构，融入了学习者对学习的承诺或投入。他们将参与分为三种方式：行为、情绪和认知。行为参与是指可观测性的事件：学生个体在其他人面前表现的参与模式，包括遵循课堂已有参与规范的程度、对学校任务作出的努力、以及他们与同伴之间的合作或捣乱行为。情绪参与包括各种感觉，如好奇心和高兴，以及与这些感觉相反的不满情绪，如厌倦、挫折或焦虑。认知参与是指学生的自我调节策略，即学生采取调节注意力、调整自己的努力，并将信息整合成有意义的记忆的一些方法。

任何一种参与方式要么是一种状态（即时性的），要么是一种特质（长期性的）。每一种方式都视学校学习为一种参与对象，每一种参与方式也都提出了一些可用于区分学生是否参与的方法，以便于从行为参与、情感参与或认知参与这三个方面对学生在特定时刻的参与情况进行分类。

我们可否将数学参与也划分为这三个维度或成分呢？若可以，就情绪参与而言，提供鼓励和成功经验的方式是否足以满足培养个体积极情感，避免消极情感的要求呢？教师是否应该像许多研究所建议的那样，主要依赖于内在奖励而非外在奖励方式？教师在学生所熟悉的、现实的情境中运用数学，是否有利于学生的参与？

接下来的内容表明上述问题的答案或许是"情况可能远比这些问题复杂"。然而，系统地关注情感、行为和认知方面的变量是必要且有价值的，因为这些变量能够说明参与的复杂性，以便于我们通过周密的教学设计、教师准备、专业发展及政策，更深入、更有效地吸引学生。

数学参与的复杂性

上世纪90年代以前，大部分与数学教育的动机、情感和参与有关的研究都在关注鉴定和测量个体的特质或取向，即通过大规模问卷和相关测量调查学校数学教育成就，从而确定学生长期性的特征，而且这种研究趋势一直持续到今天。由于麦克劳德（1989）、德贝利斯和戈尔丁（1997）、汉诺拉（2002）和埃文斯（2002）等人的倡导，一些研究者越来越关注处于具体情境中的即时性事件的发生，而且，尤其会采用定性而非定量的方法研究个体的渴望、情绪、信念、社交互动和参与。但除了偶尔的例外，这两种趋势之间的交互作用似乎非常小，有时二者看上去还存在一种相互矛盾的范式，与其把这些研究方向综合起来，倒不如论述数学参与的内在复杂性。

我们也注意到有很多研究强调参与的程度和强度。当然，这里所说的参与程度依赖于个体对参与对象的看法——有些人虽然没有参与到数学课堂当中，但与同伴

particular task and particular people. In-the-moment engagement is simultaneously motivational, affective, and cognitive; it is simultaneously personal and social. More enduring individual characteristics—traits, orientations, attitudes and beliefs, values, competencies, and relationships—partially influence the nature of this engagement, but these manifest themselves in unique ways in immediate situations. New memories and modified affective and cognitive orientations, relationships, and so forth, are in turn built up through successive engaged experiences with mathematics over time.

Each important factor influencing mathematical engagement has *state* (in-the-moment) and *trait* (longer term, enduring) counterparts. Motivation involves situational interest, task goals, and endogenous utility (states) as well as personal interests, goal orientations, and exogenous utility (traits). Affect includes immediate emotions and local affect (states) as well as emotional orientations, attitudes, and global affect (traits). Affective structures (longer term) may function in the moment as anticipatory structures or schemas that help individuals engage in immediate goals and provide emotional feedback. Classroom mathematical practices (social states) exhibit emergent stability in sociomathematical norms, an analog of traits in social systems (Cobb & Yackel, 1996). In short, *in-the-moment engagement* is a fundamental construct in describing how transient, fluctuating experiences in the mathematical here and now become integrated into more enduring understandings and competencies.

Behavioral, Affective, and Cognitive Dimensions

Fredericks, Blumenfeld, and Paris (2004) characterize engagement as a *meta construct*, incorporating the commitment to, or investment in, studies by the learner. Their literature review is organized according to three ways in which engagement may be manifested: behavioral, emotional, and cognitive. Behavioral engagement involves observable events: the patterns of participation students display with others, including the degree to which they conform to the norms of participation established in the class, the effort they put forth in school work, and the cooperative or disruptive behavior they exhibit with their peers. Emotional engagement involves feelings such as curiosity and joy as opposed to disaffection associated with feelings such as boredom, frustration, or anxiety. Cognitive engagement involves self-regulatory strategies students may employ to direct their attention, adjust their effort, and consolidate information into meaningful memories.

Each of these ways engagement may be manifested lends itself either to in-the-moment or longer term interpretation. Each is defined by taking school studies as an object of engagement. Each also suggests ways to distinguish engagement from disengagement, so that students can in principle be characterized as more or less engaged behaviorally, affectively, or cognitively at any given moment.

Can we meaningfully separate mathematical engagement into these three dimensions or components? If so, with respect to emotional engagement is it sufficient or even desirable to aim toward fostering positive feelings and avoiding negative ones, for example, by offering encouragement and providing success experiences? Should teachers rely heavily on intrinsic reward as opposed to reward extrinsic to mathematics, as suggested by much of the research? Is it beneficial to engagement for teachers to situate mathematics in familiar, real-life contexts with relevant applications?

The balance of this chapter indicates that the answer to all these questions may be, "The situation is far more complex than that." Nevertheless, systematic attention to affective, behavioral, and cognitive variables is necessary and valuable in navigating the complexities of engagement so that we can engage students more deeply and productively through careful instructional design, teacher preparation and professional development, and policy.

The Complexity of Mathematical Engagement

Before the 1990s, most of the research pertaining to motivation, affect, and engagement in mathematics education focused on the identification and measurement of traits or orientations, that is, longer term characteristics operationalized through large-scale questionnaires and surveys and correlated with measures of success in school mathematics. This strand of research continues to the present day. Due to advocacy by McLeod (1989), DeBellis and Goldin (1997), Hannula (2002), Evans (2002), and others, some researchers increasingly focused on context-dependent, in-the-moment occurrences—typically operationalized using qualitative rather than quantitative methods—to study desires, emotions, beliefs, social interactions, and engagement. But with only occasional exceptions, interaction between these two strands seems to have been minimal. Sometimes they are seen as competing paradigms. We argue instead for integrating these research directions to address the complexity inherent in mathematical engagement.

We also note the frequent research emphasis on the degree or intensity of engagement. This degree depends, of course, on what one considers to be the object of engagement—someone disengaged with classroom mathematics may be highly engaged with peer interactions or

或其他对象却有很高的互动。当讨论即时性参与时，或许需要定义一个非常具体的对象，例如，一个特殊的数学任务或一些其它的内容。而不同时刻的对象可能都会不一样，这也是参与复杂性的一种体现。当某种参与是由于个体的个性特质所导致时，我们通常用更一般的方式来定义对象，例如，数学参与或学校参与的程度（参考Bodovski & Farkas，2007）。

我们认为，即时性参与包含这三个维度之间连续的相互作用：认知、情感、行为。有一些质性研究结果表明：在做数学的过程中，学生对学习的认知、问题解决和社交互动作出反应时，情绪会迅速改变，而且，行为也从明显的状态和行动变为注意力分散和走神（例如Alston 等，2007；Op't Eynde，De Corte，& Verschaffel，2007）。因此，学生参与的效果（如，他们当前的目标、认知策略、伴随的情绪、有效或无效的行为、社会背景和交互作用）、每一维度的参与程度以及参与的潜力和意愿，随时都会变化。

构建解决复杂性的模式

考虑到即时性参与的变化和复杂性，分析个体和个体在群体中所表现出的模式很重要。戈尔丁、爱泼斯坦、肖尔和沃纳（2011）用参与结构一词来描述一个"行为／情感／社会的集群"，描述了数学课堂上的学生的参与模式。他们认为，这一心理结构是由不同要素相互作用而组成的，而这每一个要素都在已有文献中被作为单独的变量或特征加以讨论。这些变量或特征包括（a）短期目标或积极的欲望；（b）行为模式，包括社交互动特征；（c）情感路径；（d）公开的情感表达；（e）情绪感觉的意义编码；（f）元情感；（g）自我对话；（h）信念和价值观之间的相互作用；（i）其它的个性特质和取向之间的相互作用；（j）与问题解决策略和探索过程之间的相互作用。

戈尔丁等人所讨论的九个参与成分是：（1）完成任务，即完成分配给自己任务的积极欲望；（2）看到自己有多聪明，即表现自己数学能力的欲望；（3）检查，即一种获得来自于内部或外部回报的欲望；（4）我真的参与其中了，即为了自己而获得或保持参与的欲望，类似于奇凯岑特米哈伊（1990）深入讨论过的动态现象；（5）不要不尊重我，即被挑战之后挽回颜面的欲望；（6）远离麻烦，即避免冲突、不赞成甚至是丢脸的欲望；（7）这不公平，即纠正不公平的欲望；（8）让我来教你，即帮助其他同学的欲望；（9）伪参与，即没有实质性数学参与的表面参与欲望。弗纳、曼瑟威和邵蒂（2013）在研究多文化教师群体从各自不同的传统文化出发设计几何图形的过程时，增加了一个成分：（10）承认我的文化，即让其他人承认自己民族的数学遗产的积极欲望。莱克和纳迪（2014）采用质性研究方法证实了数学教师具有这样的参与成分。这些研究结果为我们研究即时性参与方面提供了从参与对象、效果和程度方面确定参与模式的一种方法，而且为这些模式的建立提供了可能的研究基础。

即时性参与的可塑性

我们认为，课堂参与是非常关键的，因为不像学生的长期特质、取向等，课堂参与会随时受到教师教学目标、教学质量和教学程度的影响。尽管学生的特质与其数学学业成功紧密相关，但这些特质不大会在短期内得到改变，可是会因任务设计、教师期望和社交互动规范而产生变化。

总之，在数学学习和学生的长期发展中，即时性参与是一个复杂的、依赖于情境的、具有可塑性的综合体。关注数学课堂参与的复杂性可以加强教师教育，还有助于更好地设计课程、工具技术以及动机性策略。

数学参与的研究方法

工具和数据收集技术

研究者们基于本人对参与的界定和研究范式，采用了不同的工具和技术对学生的数学参与进行了调查研究。传统的研究主要是用自我报告（例如Kong, Wong, other objects. When the discussion is on in-the-moment engagement, it becomes possible to define an object very specifically—for example, a particular mathematical task or some other focus. That the object may change from moment to moment is part of the complexity. When engagement is considered trait-like, its object is usually defined in a more general way—for example, the degree of mathematical engagement or school engagement (as in Bodovski & Farkas, 2007).

We suggest that in-the-moment engagement involves the incessant interaction of all three dimensions: cognitive, affective, and behavioral. Qualitative research suggests how emotions while doing mathematics change rapidly in response to cognitive aspects of learning, problem solving and social interactions, while behaviors range from observable statements and overt actions to moments of apparent inattention and digression (e.g., Alston et al., 2007; Op 't Eynde, De Corte, & Verschaffel, 2007). Thus, qualities of students' engagement (e.g., their immediate goals, cognitive strategies, attendant emotions, productive or possibly unproductive behaviors, social context, and interactions), the intensity of their engagement in each dimension, and their underlying capabilities and willingness to engage vary greatly from situation to situation.

Constructs Addressing Complexity

Given the variability as well as the complexity of in-the-moment engagement, it becomes important to discern identifiable patterns in individuals and across groups of individuals. Goldin, Epstein, Schorr, and Warner (2011) use the term *engagement structure* to describe a "behavioral/affective/social constellation" characterizing a pattern of engagement that occurs in students in mathematics classrooms. They suggest that such a psychological structure consists of mutually interacting strands, each of which has been discussed in the literature as a separate variable or characteristic. These are (*a*) the immediate goal or *motivating desire;* (*b*) patterns of *behavior,* including characteristic *social interactions;* (*c*) *affective pathways of emotions* experienced; (*d*) overt *expressions* of affect; (*e*) *meanings* encoded by emotional feelings; (*f*) *meta-affect;* (*g*) *self-talk;* (*h*) interactions with *beliefs and values;* (*i*) interactions with other *personality traits and orientations;* and (*j*) interactions with *problem-solving strategies and heuristic processes.*

The nine engagement structures proposed and discussed by Goldin et al. are (1) Get the Job Done, where the motivating desire is to complete a task as assigned; (2) Look How Smart I Am, where the desire is to show off mathematical ability; (3) Check This Out, where the desire is for either an intrinsic or extrinsic payoff; (4) I'm Really Into This, where the desire is to achieve and maintain the engaged experience for its own sake, as in the phenomenon of *flow* discussed extensively by Csikszentmihalyi (1990); (5) Don't Disrespect Me, where the desire is to save face after being challenged; (6) Stay Out of Trouble, where the desire is to avoid conflict, disapproval, or possible humiliation; (7) It's Not Fair, where the desire is to rectify an inequity; (8) Let Me Teach You, where the desire is to help another student; and (9) Pseudo-Engagement, where the desire is to seem engaged without actually doing the mathematics. Verner, Massarwe, and Bshouty (2013), in studying a multicultural group of teachers working with geometric designs drawn from different traditions, propose an additional engagement structure, (10) Acknowledge My Culture, where the motivating desire is for recognition of one's own ethnomathematical heritage by others. Lake and Nardi (2014) report qualitative evidence for such engagement structures in mathematics teachers. The in-the-moment perspective taken in this line of research allows identification of patterns in the object, quality, and intensity of engagement and the positing of plausible reasons for such patterns.

Malleability of in-the-Moment Engagement

Classroom engagement is, in our view, pivotal because unlike students' longer term traits, orientations, and so forth, it is open to immediate influence by the teacher with respect to its objects, qualities, and intensity. While certain traits indeed correlate with academic success in mathematics, these traits may be unlikely to change in a short period of time, whereas the immediate state constraints of task design, teacher expectations, and norms of social interaction are malleable by the teacher.

In short, in-the-moment engagement is a complex, context-dependent, malleable composite central to mathematical learning and to longer term student development. Attention to the complexity of classroom engagement with mathematics could enhance teacher education and enable better designed curriculum, tools and technologies, and motivational strategies.

Methods for the Study of Mathematical Engagement

Instruments and Data Collection Techniques

Researchers of mathematical engagement employ a wide range of instruments and techniques depending on their conceptions of engagement or one of its facets and

& Lam，2003；Plenty & Heubeck，2013）或访谈法调查研究对象（例如 Jansen，2006；J. A. Middleton，1995）。经验取样法（例如 Schiefele & Csikszentmihalyi，1995；Uekawa, Borman, & Lee，2007）为利用电子手段收集即时性调查数据提供了方法。

观察法关注的对象有个体、小组或班级（例如 Esmonde，2009；Gresalfi，Martin，Hand，& Greeno，2009；Webb 等，2014），数据类型有活动录像、现场笔记、出声思维或访谈。另一种收集信息的有效方式是，通过访谈或传统的调查方法来了解教师和父母对学生典型参与的报告（例如 Bodovski & Farkas，2007）。

最后，研究者经常使用多种研究方法和技巧，例如，结合研究者的观察与学生的自我报告（例如 Jansen，2008；Webel，2013）或者教师的报告与学生的报告（例如 J. A. Middleton，1995）。下面，我们对这些研究方法进行简要的举例说明，并着重谈这些方法对构建数学参与的内涵所带来的启示。

调查法

大部分关于参与动机和情感的研究采用了调查方法，自我报告是最主要的调查方法，而且，调查的变量有数学自我效能感、情感反应、情境兴趣、目标取向和数学焦虑。强迫选择和李克特式量表的索引通常设计有子量表来评估参与的各种子结构。

例如，A.J. 马丁、鲍勃、韦和安德森（2015）用强迫选择李克特量表，从9个维度调查了澳大利亚1600多名来自44个学校5—8年级共200个班级学生的参与情况。这9个指标中5个是认知维度（计划、学习管理、坚持、自我妨碍，以及不参与），3个是行为参与维度（课堂参与、努力以及完成家庭作业），1个是情感参与维度（愉快）。同时，他们收集了性别、个体、班级、学校层面的人口统计学资料以及家庭对数学价值的看法。该研究对每个学生进行了两次调查，两次调查的时间间隔为一年。作者对比了5—6年级学生（进入中学前）和7—8年级学生（进入中学后）从各科整合的小学班级环境过渡到各科独立的学习环境时的数学参与的变化。他们发现，学生从小学过渡到中学后，其认知参与、行为参与和情感参与均显著下降。

这些研究认为个体的综合特质决定数学参与，并且试图寻找这些特质和其他与参与或成就相关的变量之间的关系。为此，研究者使用了开放式问卷，如：通常要求学生反思自己的经历，而且，问卷题目有高度结构性提示语，如："这两个数学主题，哪一个更有趣并解释为什么"（J.A. Middleton，1995），还有许多其它开放式问题，如"你能想起来你在数学课堂上被提问而且你回答错误的时候吗？你当时的反应和感受是什么？"（Jansen，2009）。

舍夫力和特米哈衣（1995）为了从根本上提高调查法的有效性，采用经验取样法来说明学生在做数学活动期间的即时想法和感受。尽管，该研究只例证了一种研究范式，但它展示了如何用技术手段大规模收集学生即时性敏感反应数据的方法，比如，学生的传呼机（手机等）在随机选定的时间会响起并要求他们回答当下的感受。对活动即时性的评价使该研究方法比回顾性调查法更具有效度，但是，也有一些不足，比如，利用传呼机和提问的方式会打断学生即时性参与活动进程。

舍夫力和特米哈衣以来自于芝加哥和伊利诺伊高中一、二年级的108名数学天才生为研究对象，调查了他们的兴趣和成就动机（利用纸笔测验方法进行评价）、数学能力（利用PSAT分数进行评价）、数学经验的特性（利用经验取样法进行评价）以及数学成就（用学科成绩来评价）之间的关系。研究结果显示，学生经验质量的综合得分与其同年的学业成绩之间存在显著的正相关性，然而，学生的每一种体验维度（情感、效能、专注水平、内部动机、自尊、重视程度和感知技能）与其学业成绩之间却没有显著的相关性，但是，学生的能力和兴趣能够较强地预示其成绩。这些结论更加确信学生的参与是一个复杂的结构，它涉及多种情感、动机和认知变量之间的相互作用。经验取样法

depending on their paradigmatic approaches. Conventional surveys use mainly written self-report instruments (e.g., Kong, Wong, & Lam, 2003; Plenty & Heubeck, 2013). Researchers also interview participants verbally (e.g., Jansen, 2006; J. A. Middleton, 1995). Experience sampling methods (e.g., Schiefele & Csikszentmihalyi, 1995; Uekawa, Borman, & Lee, 2007) provide ways of gathering in-the-moment survey data electronically.

Observational methods may focus on individuals, small groups, or classrooms (e.g., Esmonde, 2009; Gresalfi, Martin, Hand, & Greeno, 2009; Webb et al., 2014) and include videos of activity, observer field notes, and interview data based on thinking aloud or retrospective protocols. Another useful way to gather information is to seek teacher or parent reports of students' typical engagement (e.g., Bodovski & Farkas, 2007) through either interview or traditional survey methods.

Finally, researchers frequently use multiple methods, combining techniques such as researcher observations and student self-reports (e.g., Jansen, 2008; Webel, 2013) or teacher reports and student reports (e.g., J. A. Middleton, 1995). We illustrate each of these approaches briefly, with attention to their implications for how mathematical engagement is conceived.

Survey Methods

The vast majority of studies addressing motivational and affective aspects of engagement use some variation of survey methods. Self-report measures predominate in this literature, focusing on variables such as mathematical self-efficacy, affective response, situational interest, goal orientation, and math anxiety. Forced-choice and Likert-scale indices are typically designed with subscales to assess various subconstructs of engagement.

A. J. Martin, Bobis, Way, and Anderson (2015), for example, surveyed over 1,600 Australian students from 200 classrooms across 44 schools in grades 5–8 regarding nine indices of engagement, using a forced-choice Likert scale. Five of their indices were adapted specifically to cognitive engagement in mathematics (planning, study management, persistence, self-handicapping, and disengagement), three focused on behavioral engagement (class participation, effort, and homework completion), and one focused on affective engagement (enjoyment). They collected gender and demographic information at the individual, class, and school levels as well as individual information on home valuation of mathematics. The survey was administered twice to each student, with a one-year interval between administrations. The authors compared students in grades 5–6 (before the move to middle school) with those in grades 7–8 (after the move) to examine change in mathematical engagement associated with the transition from self-contained elementary classrooms to a more departmentalized school environment. They found this transition to be strongly associated with a drop in cognitive, behavioral, and affective engagement.

Such studies operationalize mathematical engagement as a composite trait and look for relationships between this trait and other engagement- or achievement-related variables. To study more nuanced aspects of students' engagement with enhanced sensitivity to individual differences, researchers use open-ended questionnaires, which typically ask students to reflect on their experiences. Items range from highly structured prompts, such as, "For these two mathematics topics, indicate which one is more fun, and state why you think so" (J. A. Middleton, 1995), to more open-ended questions, such as, "Can you think about a time when you were called on in math class and your response was incorrect? What was your reaction and how did you feel about this situation?" (Jansen, 2009).

Schiefele & Csikszentmihalyi (1995) radically enhanced survey research with their use of experience sampling methods to uncover students' in-the-moment thoughts and feelings during mathematics related activity. This study exemplifies a survey paradigm, but shows how technology can enable large-scale data collection with in-the-moment sensitivity. The student's pager (cell phone, etc.) buzzes at randomly selected times, and the student is directed to answer questions about experiences at the moment of the signal. Temporal proximity to the activity being assessed may lend this method greater validity than retrospective surveys, but with the limitation that the paging and the questions interrupt in-the-moment engaged activity.

In their study, Schiefele and Csikszentmihalyi investigated relationships between interest and achievement motivation (assessed using paper-and-pencil measures), mathematical ability (assessed using incoming PSAT scores), quality of mathematical experience (assessed using experience sampling), and mathematics achievement (assessed by course grades). Data were reported for a sample of 108 high school freshmen and sophomores in Chicago, Illinois, identified as being mathematically talented. The composite score of quality of experience positively and significantly correlated with grades obtained that year. Individual dimensions of quality of experience (affect, potency, level of concentration, intrinsic motivation, self-esteem, importance, and perceived skill), however, did not significantly correlate with grades. Rather, ability and interest were shown to be stronger predictors of grades. Such results support the perspective that

保证了研究人员能够很好地调查学生的即时性数学参与，同时，为情境因素促进学生的数学兴趣这一观点提供了充分且重要的证据。

莱德和弗盖兹（2002）也采用经验取样法比较了两名学生（来自于较大研究样本中的20个样本）的日常活动。尽管可以预料的是，两名学生在参与数学活动时的方法是不同的，但是，经验取样法为研究者可以直接比较两名学生在数学课堂或数学学习过程中的经历提供了方便。有一位名叫博伊德的学生，他似乎较缺乏数学动机，专注水平也较低，成功的体验也比较少，然而，从报告中可以发现他没有学习的原因是为了读一本名叫《法国数学家》的书，该书是关于埃瓦里斯特·伽罗瓦的。这件事情表明，捕捉学生即时参与的方法如何有助于阐明他们的潜在动机、参与的目标以及它们在数学、社会或学业方面获得的成效。我们相信博伊德的参与是富有成效的，而且，他自己也认为是这样，尽管，比起其他的同学，博伊德并没有比他的同学更能确定自己阅读关于伽罗瓦的书的效果。

J.A.米德尔顿和斯班尼厄斯（1999）指出书面自我报告方法始终存在的一个问题是，学生是否或在多大程度上意识到了自己的参与，并能用动机、情感和社会性的词汇把它表达出来。不过，自我报告有一个明显的好处，即研究者可以站在学生的视角观察课堂，但是，自我报告法的可靠性和有效性受到挑战是不可避免的，特别在小样本调查中更为突出。为此，更多直接的方法，如，访谈法和观察法，可作为自我报告数据的验证手段来使用。

直接观察法和访谈法

为了直接而合理地获取学生即时性参与经历的数据，研究者必须采用观察法。观察法也常常结合访谈法进行，以便于获得学生和教师对观察到的行为的观点。在一个小组合作目标发展案例研究中，韦贝尔（2013）研究了位于美国中大西洋地区的一所职业高中的数学课堂。他在为期12周以上的23节90分钟的课堂上观察了高中学生的参与情况，并在观察前后访谈了8名学生。访谈中，他首先给学生看了自己参与行为的录像，然后问他们行为背后的原因。韦贝尔发现，在组层面上，由于存在既定的规范，学生的参与行为都较为稳定，可是在个人层面上，由于学生的个人目标与既定规范和小组目标之间存在冲突或不一致性，使得他们的参与行为存在很大的变异性。能否保持自我价值目标、成就目标以及其它个人目标与表现行为有很大的关系，而这些行为反过来取决于这些个人目标与小组目标的一致性。韦贝尔强调即便是在同一小组，学生不同的个人目标会影响他们的行为。这种有情境的个别化的参与模式是无法用纸笔的自我报告来发现的。没有记录学生的真实反应行为以及提供给他们进行反思的情况下，即便是经验取样法可能也无法揭示个人目标与小组目标之间的细微差别（参见Helme & Clarke, 2001）。

构建凯利方格是另外一种描述学生参与模式的访谈/调查方法。J.A.米德尔顿（1995）观察了6所美国中学课堂，并就在课堂中主要使用的数学活动访谈了6名数学教师。该研究选择了教师认为在班级中数学动机水平低的3名学生和水平高的3名学生，并提供了他们在上个月参与过的"最重要的"10个数学活动的成对列表。学生从每一对活动中选择"最有趣"的活动并说明理由，并将学生提供的理由概念化为学生在活动中与其内部动机相关的个体认知结构。然后，将个体认知结构填入凯利方格行中，再将10个数学活动填入列中，要求每一位学生给出1至10的评价，用来表示每个认知结构反映每一个活动的程度。研究者用聚类分析方法对每一个学生的凯利方格的交叉间距进行建模，并推断出一种可能的心理组织结构。研究中，每一名学生的认知结构和评级都是独特的。学生们的数学老师也同样需要填写自己的凯利方格，然后再根据自己的判断填写一个关于自己学生的凯利方格。结果表明，学生的内部动机集中体现在他们对自己所参与活动的个人兴趣、活动的挑战性以及个人对活动的控制方面。教师也能较好地预测出学生的动机与自己的模

engagement is complex, involving the interaction of several affective, motivational, and cognitive variables. Experience sampling enables the authors to tie their measures to the students' in-the-moment mathematical engagement, providing substantial evidence of the importance of situational factors in promoting mathematical interest.

Leder and Forgasz (2002) also used experience sampling methods to compare the daily activities of 2 students (of 20 sampled in a larger study). Although it is to be expected that any two students will differ in their approaches to mathematical activity, experience sampling enabled direct comparison of the two students' experiences in mathematics class and when studying mathematics. One student, Boyd, seemingly lacked motivation in mathematics, reporting less concentration while studying and less frequent feelings of success; however, these reports were tied to his election not to study in order to read a book, *The French Mathematician*, about Évariste Galois. This anecdote illustrates how a method that captures the in-the-moment engagement of students can help illuminate their underlying motives, their objects of engagement, and the extent to which these are mathematically, socially, or academically productive. We believe Boyd's engagement to have been productive, and he too believed it so; while Boyd felt less sure of his productivity while reading about Galois than did his classmate.

A consistent issue with written self-report methods is whether or to what degree students are aware of their own engagement and are able to express it in motivational, affective, and social terms; J. A. Middleton and Spanias (1999) provide a critique. A clear benefit of self-reporting is gaining access to the world of the classroom through students' eyes, but both the reliability of self-report methods and their validity, particularly at small sample sizes, have been challenged. More direct methods, such as interviews and observations, can serve as means of validating self-report survey data.

Direct Observational and Interview Methods

To obtain reasonably direct access to students' engagement experiences as they occur, researchers must use observational methods. These are often coupled with interviews to gain students' and teachers' personal insights into the observed behavior. In his case study of goal development in small group work, Webel (2013) studied a mathematics class in a vocationally focused high school in the mid-Atlantic region of the United States. He observed high school students' engagement during 23 90-minute lessons over 12 weeks and interviewed eight students before and after his observations. In the interviews, he showed students their video-recorded engagement behaviors and asked them to report on the reasons behind their behaviors. In general Webel finds that, at the group level, engagement behaviors can be highly stable due to established norms, but individually, there is high variability depending on the personal goals of students and the degree to which those goals are in conflict or in common with the norms and goals set by the group. Goals for preserving self-worth, goals for achievement, and other personal goals are associated with different manifested behaviors depending on their congruence with group goals. Webel's methods highlight that even within the same group, students' different individual goals guide their behavior. Such situational, individual patterns of engagement are not detectable using paper-and-pencil self-reports. Even experience sampling methods might not uncover subtle distinctions between personal and group goals without the actual behaviors having been recorded and presented for students to reflect on (see also Helme & Clarke, 2001).

Constructing a *repertory grid* is another interview/survey method for describing students' engagement patterns. J. A. Middleton (1995) observed six U.S. middle school classrooms and interviewed six mathematics teachers regarding the important mathematics activities in which their classes engaged. Three children from each class identified by the teacher as low in mathematical motivation and three identified as highly motivated were provided with a list of pairs of the ten "most important" mathematics activities in which they had engaged in the past month. For each pair, the students were asked to choose the "most fun" activity and state the reason. The reasons offered were conceptualized as the student's *personal constructs* related to intrinsic motivation in the given activities. Personal constructs were then arranged in rows on a grid, with the 10 mathematics activities arranged in columns; each student was asked to rate, on a scale of 1 to 10, the degree to which each construct described each activity. Cluster analysis was used to model the interconstruct distances for each child's grid, allowing the researcher to infer a probable mental organization of the constructs. Each set of constructs and ratings was unique to each child in the study. The children's mathematics teachers likewise filled out their own repertory grids and then filled out their nominated students' grids as they envisioned the students would. Results suggest that students' intrinsic motivation centers on their personal interests, the challenge of the activities, and the degree of personal control they are

式有相似的关系。

访谈法与其他自我报告一样存在数据的有效性问题。韦贝尔（2013）和J.A.米德尔顿（1995）的研究表明，将访谈中的提示性语言与学生参与的真实活动联系起来是一种有效研究即时性参与的方法。此外，观察学习环境对了解学生参与的条件和交叉验证学生的自我报告数据都非常重要（参见Walter & Hart, 2009）。

一些近期的关于参与的研究意在观察作为一种群体现象它是如何进行的。例如，格里索菲等人（2009）比较了两所美国学校的6年级和8年级中学代数项目课程的课堂，从同一学年的一月份到五月份共观察了10—12次，并用两台摄像机分别记录了一个学生小组和整体课堂活动。作者认为，学生应负的责任内容、负责任的对象以及个人的能动性都是通过开放的数学任务和课堂标准形成的，这里，课堂标准指的是，讲道理、成功地完成任务、论证、确定提出的观点和想法归属于谁以及谁可以批判他人。总之，这些因素决定了一个班级的竞争力，而且，两个班在这些方面的差异表明参与的能力或成效不存在普遍性，它来自于特定的课堂规范和约定。因此，由谁、由什么界定成功的数学参与，目标和能力如何得到支持以及对成功或失败的潜在情感反应都是数学课堂当中潜在的可塑因素。

多种方法

直到最近，该领域的研究者才意识到，与参与行为相关的情感的复杂度和社交诱因的微妙性都需要多种且常常是混合的观察方法。

例如，孔等人（2003）用了一个三层结构的方法，利用两周时间观察了8名中国学生，访谈了22名中国学生。根据研究结果，研究者建构了一个用于测量学生数学参与的认知和行为的自我报告工具。这个工具就是用表层认知参与表示记忆和练习，用深度认知参与表示理解问题、概括所学内容以及将新知识和已有学习方法建立联系，行为参与则用专注度、努力程度和花费的时间来衡量。他们通过观察和访谈确立的工具，选取了将近300名5年级学生进行了两次试测，并将工具完善之后又调查了大约550名学生。调查结果显示，参与的行为、认知和情感三者之间存在统计学意义上的显著相关性。与大多数从动机和情感方面进行的参与研究一样，该研究也使用了纸笔测试方法。然而，该研究工具并不具有普遍性，因为它是在对学习过程的真实观察和对研究对象的深度访谈基础上而设计的针对某个特定样本的调查工具。

乌卡瓦等人（2007）研究了在数学课堂中课堂活动是如何影响学生的参与、感知和沟通的。他们推测学生的感知和沟通是教师教学策略和学生参与之间的中介因素。在课堂结构与社交系统中，民族和种族的相互作用也是作者感兴趣的。该研究的数据来自于三年来在美国四个城市收集到的大量数据。作者从这四个城市各选择了两所高中，并从中分别选取了两位教师，然后从每个教师的两个教学班中招募了约10名学生，形成了一个由320名学生组成的研究样本。该研究采用经验取样法记录了学生5天的经历，旁听他们的课，平均每个学生约有7次观察记录。课堂观察与经验取样是同时进行的，并以10名学生为一组展开后续的小组访谈。因为损耗及其它因素，作者最后对1936个观察记录结果进行了分析。

多层线性模型结果显示，学生经历的大部分数学学习是讲授式教学方法，学生的参与水平与其种族之间有很高的相关性。而且，在小组学习中，亚裔学生的参与更有效，相反，拉丁裔学生参与效果较差，黑人学生在所有类型的环境中都表现出很高的参与性。这种差异也同样存在于不同社区的各民族之间（如在艾尔帕索和得克萨斯的拉丁裔学生与在迈阿密和佛罗里达的拉丁裔学生），研究表明

able to exercise in the activities. Teachers are better able to predict the motivational responses of students with response patterns similar to their own.

Interview methods raise the same validity issues as other self-report data. The work of Webel (2013) and of J.A. Middleton (1995) suggests that tying interview prompts to the actual activities in which students engage is a promising method for studying engagement-in-the moment. Moreover, observation of the learning environment is important for understanding the conditions of engagement and for cross-validating student self-reports (see also Walter & Hart, 2009).

Some more recent research on engagement focuses on observation of how engagement unfolds as a group phenomenon. Gresalfi et al. (2009), for example, compared two middle school classrooms in the United States using the Algebra Project Curriculum: one sixth-grade classroom and one eighth-grade classroom, each observed 10–12 times from January to May of the same school year. Two video cameras were used, each capturing a group of students as well as a view of the overall action in the classroom. The authors concluded that what students are accountable for, to whom students are accountable, and what personal agency students are afforded are shaped by the openness of the mathematical task and classroom norms: norms for sense making, for successful completion of the task, for argumentation, for who is recognized as the author of ideas, and for who is permitted to critique whom. Together, these factors create a working definition of competence for a class. The fact that the two classes differed on these factors suggests that competence or productive engagement is not defined universally but emerges from the norms and negotiations of the class. Thus, who and what define successful mathematical engagement, how goals and abilities are supported, and the potential affective responses to success or nonsuccess are all potential malleable factors in the mathematics classroom.

Multiple Methods

Only relatively recently has the field recognized that the complexity of affect in relation to engaged behavior and the subtlety of social cues spurring the interest of students require multiple and often mixed observational methods.

For example, Kong et al. (2003) used a three-tiered approach: They observed eight Chinese students for 2 weeks, following up with interviews of 22 Chinese students. From the results, the researchers constructed a self-report instrument to measure cognitive and behavioral engagement in mathematics: They operationalized surface cognitive engagement as memorizing and practicing and deeper cognitive engagement as understanding the question, summarizing what was learned, and connecting new knowledge with old ways of learning. Behavioral engagement was measured with respect to attentiveness, diligence, and time spent. The instrument, constructed from observations and interviews and pretested twice on approximately 300 fifth-grade students, was revised and administered to about 550 students, revealing statistically significant correlations among the behavioral, cognitive, and affective measures of engagement. Like most of the research on engagement with respect to motivation and affect, this study used paper-and-pencil measures. It is atypical, however, because it began with actual observation of the learning process and probed more deeply with interviews to develop a survey measure tailored to the sample population.

Uekawa et al. (2007) studied how classroom activities affect student engagement, student perceptions, and conversations during mathematics class. They theorized that student perceptions and conversations are intervening factors between teachers' pedagogical strategies and student engagement. The interaction of race and ethnicity with classroom structures and social systems was also of interest to the authors. Data for this study came from a larger body of data collected over three years in four urban sites in the United States. The authors selected two high schools from each of the four sites, and from each school they then selected two teachers. Approximately 10 students were recruited from each of two classes taught by each teacher, yielding a potential sample of 320 students. Experience sampling methods were used to record students' experiences, with the authors following the same students for 5 days, attending their classes, and obtaining an average of around seven observations per student. Classroom observations were conducted simultaneously with the experience sampling, and a subset of 10 students participated in follow-up focus group interviews. Due to attrition and other factors, the authors analyzed results from 1,936 observations.

Hierarchical linear modeling revealed that the students received lecture-based instruction for most of their mathematics experiences. The level of engagement was found to be highly related to ethnicity. Asians showed more productive engagement when working in groups, whereas Latinos showed less productive engagement working in groups. Blacks appeared relatively highly engaged in all circumstances. Differences are also apparent for ethnic groups in different communities (e.g., Latinos in El Paso, Texas, compared with Latinos in Miami, Florida),

当地文化对建立数学课堂的参与范式有重要作用。作者很谨慎地指出，这些案例中的文化也代表了相应社区的学校文化。在他们的研究中，一些学校似乎在延续一些不公平的做法，这些做法会影响不同的参与模式。

研究方法和参与的可塑性

正如乌卡瓦等人（2007）所强调的，心理和社会结构的可塑性与稳定性之间存在着矛盾。因为许多关于参与的研究者都假设动机和情感是个体的个性特质，所以关于参与的研究设计通常不会考虑到可塑性，相反，研究者会假设参与具有稳定性——总的来说，他们会尝试获取学生常见的参与活动或是数学课堂和课程内容强加在学生想法和信念上的一般压力。研究可塑性，尤其把数学参与看做一种综合特质时，有必要进行持续一段时间的研究，因为，这样可以了解到学生的参与是否有变化及变化原因。

例如，在普伦蒂和休贝克（2013）对澳大利亚7—10年级的519名学生的调查中，他们考察了那些在学生参与和动机上连续两年内有所变化的和稳定不变的因素。他们采用多群组和多场景的设计形式以收集数据并分析群体层面和个人层面的横向数据和纵向数据。他们的研究是为了改进过去中学生在校学习中参与和动机的调查研究（见A.J. Martin, 2011），使其更加突出数学特征——他们在研究中会把题目中"学校"一词用"数学"来代替，并在"考试"一词的前面加上"数学"二字。动机和参与这两个术语几乎可以互相替代，并用于刻画学生经历的11个方面（包括一般的经历和数学经历）：（1）自我效能；（2）价值判断；（3）学习目标取向；（4）计划；（5）任务管理；（6）坚持；（7）焦虑；（8）避免失败；（9）对不确定性的控制；（10）自我设限；（11）不参与。研究发现，学生的数学动机低于一般学业动机，而且所有学生对数学的重视和数学任务管理能力都随时间逐渐降低，然而群组的趋势却稳定。该研究在方法论上的意义在于，它对不同群组学生的动机进行了较长时间的研究，并且揭示了动机具有很强的情境性和个体性等特点。尽管，学生个人的动机在群组内部会有一定波动，有些学生升高而有些学生会有所降低，然而整体群组层面的动机具有稳定性。作者表示，他们的研究结果表明关注高中阶段学生数学动机的变化是合理的。

讨论

无论采用哪种研究方法，关键是以谁（研究者、教师或学生）的视角作为优先来研究参与。我们并不认为，存在某个所谓的"正确的"参与研究的视角，多种视角才能让我们更全面地研究参与。在研究中，教师和学生对同一事件或情境的看法通常是不一样的。如果忽略学生的看法就会遗漏用经验取样法所观察到的个体差异，忽略教师和群体的看法则会遗漏与参与有关的关键性环境、情境和社会等要素。

很多大规模定量研究采用自我报告调查工具考察数学参与和学业成绩之间的关系，通过项目分析、因素分析及其它心理测量工具发现一些稳定的潜在变量并设计能够被验证的关于这些变量之间交互作用的假设。从方法论的角度看，研究者更倾向于研究特质类的结构及与其最相关和最重要的因素，因为这样的研究能够用较低的开支调查大样本。然而，难以测量的那些特质或许是研究学生数学参与的重要方面，尤其是研究者关注即时性现象的时候。例如，元情感（将在情感视角的部分讨论）很重要，但目前社会科学领域中缺乏相应的工具来研究它（Goldin, 2002；Schlöglmann, 2006）。

目前，参与领域的研究趋势是用动机、情感和归属感在其它多个变量中的交互作用来研究，并用验证性因素分析、结构方程模型和潜在增长曲线分析（见Ahmed, van der Werf, Kuyper, & Minnaert, 2013；Marsh 等, 2013；Pinxten, Marsh, De Fraine, Van Den Noortgate, & Van Damme, 2014；Trautwein, Lüdtke, Köller, & Baumert, 2006）对这种交互作用进行建模分析。观察法倾向于使用一种更具描述性和叙事性的方式关注参与的

indicating that local culture is important in establishing norms of engagement in mathematics classes. The authors are careful to point out that the culture in these cases is the culture of schooling in the respective communities. A number of schools in their study appear to be perpetuating inequitable practices that affect the different engagement patterns uncovered.

Methods of Research and the Malleability of Engagement

As the Uekawa et al. (2007) study highlights, there is a tension between malleability and the stability of psychological and social constructs. Because many researchers on engagement assume motivational and affective factors to be trait-like, the designs of studies of engagement do not always allow for malleability; instead, researchers assume stability—attempting to capture, on balance, a students' habitual engagement activity or the general press of the mathematics classroom and curriculum on the student's thoughts and beliefs. Studies that do address malleability, particularly where mathematical engagement is regarded as a composite trait, necessarily take place over a substantial period of time, attempting to measure whether changes in engagement occur and what conditions help explain the changes.

For example, in their study of 519 students in grades 7–10 in Australia, Plenty and Heubeck (2013) examined factors that changed and those that remained stable in students' engagement and motivation over two consecutive years. They used a multicohort and multioccasion design to provide both cross-sectional data and change over time data for group level and individual analyses. Their purpose was to modify an existing survey for studying engagement and motivation for school learning among secondary students (see A. J. Martin, 2011) to be more mathematics specific—changing the items so that references to "school" are replaced with "math" and putting "math" before words like "exams." The terms motivation and engagement were used almost interchangeably to capture a multidimensional set of 11 elements in students' experience (both generally and mathematically): (1) self-efficacy, (2) valuing, (3) learning goal orientation, (4) planning, (5) task management, (6) persistence, (7) anxiety, (8) failure avoidance, (9) uncertain control, (10) self-handicapping, and (11) disengagement. The students' mathematics-specific motivation was found to be lower than their general academic motivation. Valuing of mathematics and task management in mathematics decreased over time for all students. However, group trends generally remained stable. The methodological importance of this study is that studying students' motivation over time across different cohorts reveals the highly situational and individual nature of motivation. Individual students' motivations fluctuated—within a group, some students' ratings for a scale increased while others' decreased, leading to the appearance of group-level stability. The authors claim their study suggests the validity of focusing on changes in students' mathematics motivation during high school.

Discussion

Across various approaches, it is important to identify whose perspectives on engagement are privileged—those of the researchers, teachers, or students. It is not our stance that there is one "true" perspective on engagement; multiple perspectives provide a more holistic picture. When studies report both teachers' and students' perspectives on the same event or situation, the perspectives often differ. Disregarding students' voices misses the individual differences observed using experience sampling methods, whereas disregarding teacher and group perspectives misses essential contextual, situational, and social components of engagement.

Many large-scale quantitative studies use self-report survey instruments to examine relationships between mathematical engagement and performance, often using item analyses, factor analyses, and other psychometric tools to uncover latent variables that are robust and to develop verifiable hypotheses about the interactions among these variables. The methodology tends to direct researchers toward trait-like constructs and relationships as most relevant and important, due to their adaptability to large-scale measurement at relatively low cost. Nevertheless, constructs that are more challenging to measure may be the more important ones for the study of students' mathematical engagement, particularly when researchers focus on in-the-moment phenomena. For example, meta-affect (discussed in the section on affective perspectives) is important, but social science currently lacks instrumentation to study it (Goldin, 2002; Schlöglmann, 2006).

The field has moved toward describing engagement in a way that treats motivation, affect, and belongingness as interactions among many variables. Confirmatory factor analyses, structural equations modeling, and latent growth curve analyses (e.g., Ahmed, van der Werf, Kuyper, & Minnaert, 2013; Marsh et al., 2013; Pinxten, Marsh, De Fraine, Van Den Noortgate, & Van Damme, 2014; Trautwein, Lüdtke, Köller, & Baumert, 2006) are analytic methods for modeling such interactions.

复杂性。

当然，方法的使用要根据具体的研究问题。我们更倾向于认真界定参与这一概念来指导调查，无论将其解释为即时性的还是长期性的。无论是一个还是多个参与目标都应该是明确的，并且所定义的那些关键的可观测变量无论在何种情况下都是可操作的。研究者要认识到参与的复杂性，而且，针对不同研究对象（如个体、小组、班级）采用多种测量方法，分析框架也应该考虑到各种关键变量之间的交互作用并进行长期的研究。最后，可测量工具的简便性不应该成为将数学参与视作最重要或基础的理论视角的主要驱动力。

动机以及它与参与和数学学习的关系

学生是否参与数学任务有很多原因。激励和强迫都能够起到即时的和长时间的推动作用（Bolles, 1972）。刺激强化、目标、兴趣和自我知觉的交互作用会影响学生的数学定向——他们选择做什么、和谁一起做以及要达到什么目的。综合起来，我们将这些参与的原因称作数学动机（更详细的讨论见J.A. Middleton & Jansen, 2011）。在本节中，我们将讨论对数学动机有重要贡献的因素，即自我调节、目标、兴趣、实用性。动机的情感维度会在下一节讨论，社会维度会在这之后进行讨论。

被广泛运用的心理学概念，如自我效能和目标取向，它们在数学学科的学习情境中具有不同于其他学科，哪怕是在与数学非常相近的科学学科的形式和功能。例如，马什等人（2013）利用2007年的国际数学与科学趋势研究（TIMSS）的评价，考察了四个阿拉伯国家以及美国、澳大利亚、英格兰和苏格兰学生数学和科学的自我效能感、自我概念和课程作业的情况，并发现存在明显的文化差异。例如，阿拉伯国家学生的成绩水平都低于西方国家的学生，但在动机方面却高于西方国家的学生。同时，不同国家学生的学业成绩与其动机之间的关系在个体层面上具有相同的模式，即在数学（或科学）上有较高自我效能感、价值判断和自我概念的学生也同样有较好的修课情况和较高的数学（或科学）成绩。然而，学生的数学动机和科学动机仅有中度的相关性。

下面我们来讨论有关数学动机文献中三个关键的概念：自我调节、目标、兴趣。

自我调节

自我调节一般是指学生在计划和适应个人目标得以实现时的想法、情绪和行为（K. Duckworth, Akerman, MacGregor, Salter, & Vorhaus, 2009；Zimmerman, 2000）。它能够预测数学参与的动机因素，比如：学生设定学习目标、管理注意力、有策略地组织自己的观点、选择和使用可用的资源、监督自己的表现、管理时间以及对自己能力的信念（Schunk & Ertmer, 2000）。这些因素都具有可塑性，而且还能够被用于制定有效的任务设计、指导和示范调节策略（Boekaerts, 2006；Boekaerts & Corno, 2005；Diamond, Barnett, Thomas, & Munro, 2007；Perry & VandeKamp, 2000）。通过自我管理更能适应学习环境的学生更有可能获得学业上的成功（Duncan 等, 2007；McClelland, Acock, & Morrison, 2006；Yen, Konold, & McDermott, 2004）。

最近，自我调节的拓展概念——长期执着地追求有价值的目标——被称为毅力，受到了研究者的关注（A.L. Duckworth, Peterson, Matthews, & Kelly, 2007）。毅力的意思是尽管在追求目标过程中有失败、逆境和停滞，但人们还是会迎接挑战，数年来保持努力和兴趣。在拓展的自我调节的已有研究中将毅力视为个体的一种具体特质，具有这种特质的人要比其他人的毅力更强，在讨论一个人成功和失败的原因时，毅力也常常被提及。

然而，这种毅力的概念或许能够用于描述一个人坚持不懈的状态，却不能用来解释一个人坚持的原因。有效的自我调节是一种坚忍不拔的毅力，但是毅力还没有被视为一种在人类活动中所表现出的一般的个性特征。将学生在数学学科上所用的坚持的自我调节推广到其它学科或所有他们所追求的长期目标上，将是一种质的飞跃（J.A. Middleton, Tallman, Hatfield, & Davis, 2014）。

Observational studies tend to focus on the complexity of engagement in a more descriptive, narrative fashion.

Methodological approaches should, of course, be tailored to the questions being asked. We favor careful definitions of engagement, whether interpreted as in the moment or not, to guide investigations. The object or objects of engagement should be made clear, and the key observable variables defined operationally wherever possible. The complexity of engagement should not be downplayed, and multiple measures and methods at multiple levels (e.g., individual, group, class) should be built into longer term research programs. The framework of analysis should allow for interactions among the key variables. Finally, ease of reliable measurement should not be the main driver of theoretical perspectives on which aspects of mathematical engagement are most important or fundamental.

Motivation and Its Relation to Engagement and Mathematics Learning

There are myriad reasons people engage or disengage with mathematical tasks. Incentives and coercives pull and push behavior, both in the moment and long term (Bolles, 1972). Reinforcers, goals, interests, and self-perceptions interact to occasion students' mathematical directions—what they choose to do, with whom, and to what ends. Collectively, we call these reasons for engagement *mathematical motivations* (see J. A. Middleton & Jansen, 2011, for a more detailed discussion). In this section we discuss some of the most important contributors to mathematical motivations that have received research attention: self-regulation, goals, interests, and utility. The affective dimension of motivation is addressed in the section that follows, and its social dimension in the section after that.

Widely used psychological constructs, such as self-efficacy and goal orientation, have forms and functions in the context of mathematics learning that differ in other subject contexts, even the closely related subject of science. Marsh et al. (2013), for example, used the 2007 Trends in International Mathematics and Science Study (TIMSS) assessment to examine the relationships among self-efficacy, self-concept, and coursework taken in mathematics versus science across four Arab countries, the United States, Australia, England, and Scotland. They found clear cultural differences: For example, Arab students scored consistently lower across achievement measures, but higher across motivation measures, than Western students. Meanwhile, the relationship between subject matter performance and motivation variables shows the same patterns at the individual level, with higher ratings of self-efficacy, value, and self-concept for mathematics (or science) associated with higher course taking and performance in mathematics (or science). Students' ratings of mathematics motivation constructs, however, correlate only moderately with their corresponding ratings for science.

Let us next consider three critical constructs in the mathematical motivation literature: self-regulation, goals, and interests.

Self-Regulation

Self-regulation refers broadly to the thoughts, emotions, and behaviors of students as they plan for and adapt to the attainment of personal goals (K. Duckworth, Akerman, MacGregor, Salter, & Vorhaus, 2009; Zimmerman, 2000). It encompasses facets of motivation that predict engagement in mathematics. Students set goals for learning, manage their attention, strategically organize ideas, select and use available resources, monitor performance, manage time, and hold efficacious beliefs about their abilities (Schunk & Ertmer, 2000). These facets appear to be malleable and responsive to effective task design, guidance, and modeling strategies for regulation (Boekaerts, 2006; Boekaerts & Corno, 2005; Diamond, Barnett, Thomas, & Munro, 2007; Perry & VandeKamp, 2000). Learners more able to adapt to learning environments through self-regulation are more likely to succeed academically (Duncan et al., 2007; McClelland, Acock, & Morrison, 2006; Yen, Konold, & McDermott, 2004).

Recently, the notion of extended self-regulation—perseverance over time in pursuit of a passionately valued goal—has been termed *grit* and has gained research attention (A. L. Duckworth, Peterson, Matthews, & Kelly, 2007). Grit means working strenuously to meet challenges and maintaining effort and interest over years despite failure, adversity, and plateaus in progress. The literature on extended self-regulation treats grit as a trait that crystallizes in individuals, with some having more grit than others; it is discussed as a general cause of successes and failures.

However, this conceptualization of grit may be more a description of perseverance rather than an explanation of why one persists. Productive self-regulation is gritty, but grit has not yet been established as a general personality characteristic that crosses the wide panoply of human activity. To generalize a students' tendency to self-regulate in gritty ways in mathematics to other subject matter, or to all long-term goals to which the learner aspires, is a huge leap (J. A. Middleton, Tallman, Hatfield, & Davis, 2014).

不过，研究毅力有助于明确那些与长期的、自我管理的数学参与有关的因素：目标、兴趣、努力。目标会指导行为，兴趣及相关的情绪会调节自我调节策略，让学生即便是在遇到失败的时候仍然能够坚持参与，培养学生在数学任务上的毅力。努力既依赖于学生的数学兴趣，也取决于任务对个人未来目标的感知效用（J.A. Middleton，2013）。

目标

当学习者有自主权通过选择任务制定个人目标时，当他们学习一门有挑战性活动的课程并有机会展开合作时，以及当教师能够提供相关信息和选择机会以最大程度地降低压力和要求时，都是培养数学自我调节的最佳时机（Boekaerts & Corno，2005；Deci & Ryan，1985；Fredricks 等，2004）。在参与过程的早期，学生形成了一定的目标认同，这个目标和他们对未来可能实现的目标的评价直接相关。认同不仅与个体对自我效能的认知有关，还与自由想象有关，这种自由想象是个体对未来的积极想象，它能对除策略性思维之外的目标认同产生影响（Oettingen，1999）。对未来结果的积极想象、有效地计划实施策略、运用恰当的技能以及监控元认知发展都是把目标价值转变为有效的学习结果的关键手段（Gollwitzer，1999；Sheeran, Webb, & Gollwitzer，2005）。那些突出对未来追求的个人目标可以是短期的，如考试获得高分，也可以是长期的，如从事某个特殊职业。个人目标可包含多个预期结果，如完善对某个概念的学习，或追求某种情感状态，如成就感（Pintrich，2000）。

韦贝尔（2013）利用以上所讨论的方法描述了，社会、个人、个体和数学目标是如何随着学生动机的发展而相互影响的。他描述了学生在小组工作中的角色（如专家角色）、行为类型（如不愿意帮助他人）以及对特定行为产生原因的回顾（如，因为每个人都要对自己的理解负责）。对某些学生来说，他们的非数学目标，如只要得到答案或实现人际交往，比数学目标更重要；某个学生觉得数学太难而将数学学习目标转变为避免其他同学认为自己没有能力学习数学。而后，他的数学参与就变成了社交参与。

有些学生会受到个人数学目标的激励，想在某个策略或学习内容的正确性方面得到认可。这些目标会产生非常不同的行为模式：想要得到外界认可的亲社会行为和想要学习某一概念但又感觉得不到他人帮助的反社会性行为。韦贝尔（2013）的研究显示，个体与群体目标一致还是冲突，与数学内容保持一致还是冲突，这些都依赖于学生个人的目标取向。该研究表明，学生的即时性参与是一种动态的、协商的过程，而且这个过程得以进行的动机是来自于个体的偏好、群体规范、个人需要和与数学目标相冲突（或有助于数学目标实现）的压力。戈尔丁等人（2011）有关参与结构中的激发动机的研究，沃尔特和哈特（2009）的情境动机理论中关于智能数学动机和个人社会动机的研究都对学生的不同类型目标和即时变化性有详细的讨论。

J.A. 米德尔顿等人（2014）认为数学目标在四个维度上表现出差异：（1）目标明确性；（2）目标与学习者当前状态的接近程度；（3）目标焦点（如，在学习、社会比较或逃避任务方面）；（4）目标倾向（即接近或逃避预期的最终状态）。

目标明确性。目标越明确，学生实现目标的可能性就越大。教师可以鼓励学生根据具体的数学任务设定明确的目标。但关键是，学生要创建一个怎么实现目标以及如何评价结果的计划（Ford，1992；Harackiewicz & Sansone，1991；Latham & Locke，1991）。

目标的接近度。学生对即时性需要和长远性需要的关注会导致他们的目标有所不同。近期目标能够在短期内实现，而远期目标可能需要数周、数月、甚至是数年才能实现。如果是个体自己提出的远期目标，通常是其对未来职业身份的期望（Husman & Lens，1999）。能够设想未来目标并阐明实现目标的计划的学生往往会付出更多的努力（Hester，2012），更好地管理时间（Harber, Zimbardo, & Boyd，2003），更有效地处理信息（Horstmanshof & Zimitat，2007），并取得比同龄人更高的学业成绩（Zimbardo & Boyd，1999）。

然而，目前的研究表明学生设定的目标越临近（如，越接近一次数学考试的时间）就越有可能（例如，在学

Nevertheless, research on grit has been helpful in identifying correlates of long-term, self-regulated mathematical engagement: *goals, interest,* and *effort*. Goals direct behavior. Interest and associated emotional states mediate self-regulation strategies, keeping students engaged even in the face of failure and fostering perseverance in mathematical tasks. Effort depends on both a student's interest in mathematics and the task's perceived utility in relation to future personal goals (J. A. Middleton, 2013).

Goals

Optimal conditions for developing mathematical self-regulation occur when learners have autonomy to set personally meaningful goals through choice of tasks, when they learn from a curriculum with challenging activities and opportunities to collaborate, and when their instructor provides pertinent information and opportunities for choice while minimizing pressures and demands (Boekaerts & Corno, 2005; Deci & Ryan, 1985; Fredricks et al., 2004). Relatively early in the process of engagement, students develop degrees of goal commitment directly related to their assessment of probable goal attainment at the end. Commitment is associated not only with perceptions of efficacy but also with free fantasy, where positive imaginings of the future can influence commitment to a goal apart from strategic thinking (Oettingen, 1999). Imagining positive future endpoints, effectively planning an implementation strategy, recruiting appropriate skills, and monitoring progress metacognitively are critical for translating the value of the goal into a productive learning outcome (Gollwitzer, 1999; Sheeran, Webb, & Gollwitzer, 2005). Personal goals projecting a future desired state can be near term, such as obtaining a high score on a quiz, or long term, such as entering a particular career. Personal goals can involve multiple intended outcomes, such as the improved learning of a concept or a desired affective state such as a sense of achievement (Pintrich, 2000).

Webel (2013), using methods discussed above, describes how social, personal, individual, and mathematical goals interact as students' motivations develop. He describes students' roles in group work (e.g., the role of expert); the kind of behavior exhibited (e.g., reluctant to help others); and retrospective reasons for the particular behavior (e.g., because each person is responsible for their own understanding). For some students, nonmathematical goals, such as just getting answers or socializing, can trump mathematical goals: one participant, finding the mathematics too difficult, changed his goal from learning mathematics to preventing other students from perceiving him as mathematically incompetent. He subsequently abandoned mathematical engagement to socialize.

Some students are motivated by personal mathematical goals, wanting to be convinced of the correctness of a strategy or to learn the content. These goals lead to very different behavior patterns: prosocial for the student wanting to be convinced and antisocial for the student who wants to learn a concept but feels the other students can't help. Webel's (2013) study documents that individual and group goals can be consistent or in conflict with each other and consistent or in conflict with mathematics, depending on the student's orientation. This study illustrates that, in the moment, a student's engagement is a dynamic, negotiated process that derives motivation from individual predilections, group norms, personal needs, and the various other presses that can conflict with (or contribute to) mathematical goals. The variety of students' goals, and the moment-to-moment changes in them, are elaborated further in the discussion of motivating desires associated with Goldin et al.'s (2011) engagement structures and the intellectual-mathematical and personal-social motivations identified in Walter and Hart's (2009) contextual motivation theory.

J. A. Middleton et al. (2014) describe mathematical goals as varying along four dimensions: (1) goal specificity; (2) goal proximity to the learner's current state; (3) goal focus (e.g., on learning, social comparison, or work avoidance); and (4) goal tendency (i.e., approach versus avoidance of an anticipated end state).

Goal specificity. The more specific a goal is, the more likely it is for a student to achieve it. Teachers can encourage students to set specific goals for a particular mathematical task. However, it is more critical that students create a plan for how they will accomplish the goal and evaluate their outcomes (Ford, 1992; Harackiewicz & Sansone, 1991; Latham & Locke, 1991).

Goal proximity. Goals differ in how much they address immediate needs vs. needs further in the future. A *proximal goal* can be reached in a short period of time. A *distal goal* may be weeks, months, or even years removed. Distal goals, if developed by the individual, often focus on a projected identity the learner wishes to take on in the future (Husman & Lens, 1999). Students who are able to envision future goals and articulate plans to reach them tend to expend more effort (Hester, 2012), manage time better (Harber, Zimbardo, & Boyd, 2003), process information more efficiently (Horstmanshof & Zimitat, 2007), and achieve higher school grades (Zimbardo & Boyd, 1999) than do their peers.

However, recent research shows that the more proximal the goal (e.g., the closer the time for a mathematics

习过程中）采取一些自我调节的策略，如认知练习、组织、详细阐述、批判性思维和元认知控制、时间管理、付出努力、寻求帮助及同伴学习。学生运用这些策略的强度取决于他们是否能够看到当前活动是如何影响到自己实现未来目标的（Zhang, Karabenick, Maruno, & Lauermann, 2011）。为了实现远期目标，学生需要设计一个方案，一系列能够通向实现远期目标的近期目标和实现这些近期目标的策略（Gollwitzer & Oettingen, 2011），这样的目标结构不仅能够让学生理解例如代数学习不仅对今后数学学习有帮助，而且对他们将来能够成为什么样的人有用。

目标焦点。在大多数关于什么是目标取向，及其与动机、成就和自我效能感相关的文献中，目标被分为两类。学习目标（通常称为精熟目标）主要关注在理解方面，而自我目标（通常称为表现性目标）则关注与其他人不同的个体对价值的感知（Covington, 2000）。学习目标包括想象拥有新知识的未来状态，并相信努力的应用将产生新的数学技能和理解的预期结果（例如 Morrone, Harkness, D'Ambrosio, & Caulfield, 2004）。相反，自我目标指的是个体对自我价值、表现出有能力或避免表现出落后于同伴的状态的一种想象。

已有文献高度一致地报告称，学习目标（与自我目标相反）是一种取向，与更高的数学自我效能感、较高的毅力并倾向于挑战数学任务、对学习的自我调节以及对所学内容的积极情感之间都存在相关性（例如 Kaplan & Maehr, 2007；Harackiewicz, Barron, Pintrich, Elliot, & Thrash, 2002；Urdan, & Midgley, 2003）。因此，学习目标会更容易因学习环境改变而改变。但另一种更强调即时性的视角认为，两种目标都是高度依赖于情境的（Goldin 等，2011）。例如，数学考试和竞赛的流行就体现出了自我目标的功效。试图表现出有能力是因为数学在社会中具有特殊的地位而产生的一种适应社会压力的方式。避免错误作为一种学习目标，可能适用于对数学程序性内容的熟练掌握，而它对于最佳概念学习则不合适。

目标倾向。另一个维度将目标分为趋近与逃避两种倾向（Harackiewicz 等，2002；Pintrich, 2000）。具有趋近倾向的学生会积极寻找学习新概念和技能的机会（学习方式），或者寻找展示自己卓越能力的机会（自我方式）。具有逃避倾向的学生会试图去避免学习不理解的内容（逃避学习）或跟不如自己的同学作比较（自我逃避）。

罗、帕里斯、霍根和罗（2011）利用典型的动机研究方法考察了1700名新加坡高中学生数学学习目标的材料。他们进行了各种调查，研究了学生的学习水平，自我方式和自我逃避目标与各种自我调节策略和对数学情感之间的关系。他们利用潜在聚类分析，对学生在这些变量上的回答进行了分类，在此基础上，研究人员认为学生会同时具备学习目标和自我目标，并且也会同时具备逃避倾向和趋近倾向，只是表现程度上有所不同。他们给出了四个不同的学生类型：（1）以成功为导向的学生（具有较高的学习目标以及较高的自我方式和自我逃避目标的学生），这类学生有学习的欲望，而且会将自己的同伴视为竞争对象，并对学习不好的同伴抱有恶意；（2）中等程度的学生（具有中等程度的学习目标和低水平的自我取向），这类学生并不关心偶尔的表现不佳，他们想要学习数学，但并没有感到特别兴奋；（3）散漫的学生（所有学习目标和自我目标都处于中等水平）；（4）主动学习的学生（具有较高的学习目标和自我主动学习的倾向目标，但具有较低的自我逃避目标），这类学生想学习并且让别人看起来自己有能力，从不担心自己看起来是否能力不强。通过比较具有不同自我调节策略的学生群体，罗等人发现，具有自我目标并没有什么问题，但强加于自我目标上的相关价值却是有问题的。那些以成功为导向，追求比自己较高学习目标还要高的自我目标的学生在完成数学任务时会不够努力。这一结论也得到了其他研究的证实，即当一个学生的能力低于他的同伴的时候，达不成目标会导致放弃（自我逃避；Bounoua 等，2012）。然而，中等程度学习取向的学生会表现出更高程度的努力，这也与另一个研究结果一致，即主要强调自我目标会削弱学生在学习任务上付出的努力，尤其是，自我逃避对学生学习结果产生的不利影响要多于具有主动学习倾向的学生（例如 Jõgi, Kikas, Lerkkanen, & Mägi, 2015；M. J. Middleton & Midgley, 1997）。

另一个基本的目标取向是回避任务（Nicholls,

test), the more likely self-regulatory strategies—such as cognitive rehearsal, organization, elaboration, critical thinking and metacognitive control, time management, effort put forth, help seeking, and peer learning—are to be employed (e.g., during studying). These strategies all increase in intensity as a function of the student's ability to see how current activity contributes to a distal goal (Zhang, Karabenick, Maruno, & Lauermann, 2011). To achieve distal goals, students need to devise a *plan*, a sequence of proximal goals that result in achieving the distal outcome coupled with strategies to reach those proximal goals (Gollwitzer & Oettingen, 2011). Such a goal structure enables students to understand why algebra, for example, matters not just for later mathematical study but for whom they want to become.

Goal focus. In the vast literature on what is termed *goal orientation*—and its association with motivation, performance, and self-efficacy—goals are classified into two types. *Learning goals* (often called *mastery goals*) focus on understanding, whereas *ego goals* (often called *performance goals*) focus on a person's perceived worth in comparison with others (Covington, 2000). Learning goals involve imagining a future state of having new knowledge and believing that the application of effort will yield the desired outcomes of new mathematical skills and understandings (e.g., Morrone, Harkness, D'Ambrosio, & Caulfield, 2004). Ego goals, in contrast, involve imagining a future state of self-worth, appearing competent, or avoiding the appearance of inferiority in relation to peers.

The literature is fairly consistent in reporting that learning goals—as opposed to ego goals, regarded as orientations—correlate with higher mathematical self-efficacy, greater persistence in and preference for challenging mathematical tasks, self-regulated learning, and positive affect toward the subject matter (e.g., Kaplan & Maehr, 2007; Harackiewicz, Barron, Pintrich, Elliot, & Thrash, 2002; Urdan, & Midgley, 2003). Thus learning goals are more adaptive to learning environments. However an alternative, more in-the-moment perspective is to regard both kinds of goals as adaptive in highly situation-dependent ways (Goldin et al., 2011). The popularity of mathematics contests and competitions, for example, suggests the efficacy of ego goals. Seeking to appear competent is an adaptation to social pressures surrounding the special status of mathematics in society. Avoiding mistakes, a kind of learning goal, may be adaptive for mastering mathematical procedures, while maladaptive for optimal conceptual learning.

Goal tendencies. An additional dimension along which goals are classified is *approach* vs. *avoidance* tendencies (Harackiewicz et al., 2002; Pintrich, 2000). With an approach orientation, a student actively seeks opportunities to learn new concepts and skills (*learning approach*) or seeks opportunities to show off superior competence (*ego approach*). With an avoidance orientation, the learner tries to avoid misunderstanding (*learning avoidance*) or looking inferior (*ego avoidance*).

Luo, Paris, Hogan, and Luo (2011), with methods typical of motivational studies, focused on the goal profiles of 1,700 Singaporean high school mathematics students. They administered a variety of surveys, examining how levels of learning, ego approach, and ego avoidance goals relate to various self-regulatory strategies and students' affect with respect to mathematics. They used latent cluster analysis to sort students into groups based on profiles of responses across these variables. This allowed the authors to show that students have both learning and ego goals and both avoidance and approach tendencies, but to different degrees. They reported four different profiles: (1) success oriented students (those with high learning goals as well as high ego-approach and ego-avoidance goals), who possess a desire to learn as well as to look competent in relation to peers and are ill-disposed to looking incompetent; (2) moderate mastery students (those with moderate levels of learning goals and low ego orientation), who do not particularly care about looking bad occasionally—they want to learn mathematics, but are not especially excited about it; (3) diffuse students (those with moderate levels of all learning and ego goals); and (4) approach students (those with high learning goals and high ego-approach goals, but low ego-avoidance goals), who want to learn and to look competent, without worrying about looking incompetent. Comparing the clusters of students across self-regulatory strategies, the authors found that holding ego goals is not particularly problematic, but the relative value placed on ego goals is. The success oriented students with ego goals even higher than their high learning goals report a low level of effort on mathematics tasks. This is echoed by research indicating that when a student's ability is not superior to peers, having goals thwarted tends to result in giving up (ego-avoidance; Bounoua et al., 2012). The moderate learning-oriented students, however, reported high levels of effort. This is consistent with other research finding that a primary emphasis on ego goals detracts from the effort students will expend on academic tasks. In particular, ego avoidance seems to affect students' learning outcomes in a maladaptive way more than ego approach (e.g., Jõgi, Kikas, Lerkkanen, & Mägi, 2015; M. J. Middleton & Midgley, 1997).

An additional basic goal orientation—*work-avoidance* (Nicholls, Patashnik, & Nolen, 1985), not classified as

Patashnik, & Nolen, 1985), 它既不是学习取向, 也不是自我取向, 而是被描述为避免数学参与的一种目标。由于在数学任务上的多次失败, 有些回避任务的学习者会倾向于完成最低限度的数学任务, 或者尽可能地逃避这些任务。希尔沃宁、托尔瓦宁、奥诺拉和努尔米（2012）针对儿童回避任务的情况, 调查了225名小学低段的芬兰学生及他们的老师和家长, 他们将这些数据与学生的基本运算表现情况进行了比较。逃避数学任务的倾向性与较差的学业成绩之间存在高度相关。潜在增长曲线分析结果显示, 虽然研究者未必知道哪一个或哪些是诱因, 但是任务回避目标与学业成绩目标是同时发展的。最初的数学能力与回避任务并不相关, 但随着学生学业成绩逐渐落后于他们的同伴, 回避任务就会越来越多。

兴趣

兴趣是指学生倾向于探究的活动或主题, 他们能够从中发现乐趣, 而且这些内容是比较符合自己的性格倾向的。一项强有力的证据表明, 数学兴趣是积极的数学课堂情感体验最重要的预测指标之一（Schiefele & Csikszentmihalyi, 1995）。如何培养学生的兴趣是非常重要的, 这与学生在数学学习过程中的外部奖励和惩罚以及与他们设定的近期目标和远期目标有关。一些研究表明, 随着时间的推移, 学生在具体情境下所产生的即时性兴趣会导致数学兴趣作为特质的发展（见Hidi & Renninger, 2006所作的文献综述）。

米切尔（1993）在一个经典模型中提出, 即时性兴趣的产生需要通过任务中的某些关键要素来捕捉, 如：新颖性和社交刺激, 之后再通过个人对学习意义的认识以及活动参与的深度来保持这些兴趣。对情境性兴趣的捕捉和保持并不是任务本身的特性, 而是学生的个人兴趣与任务本身赋予的认知、情感和社会约束之间的一种协商。对350名大学预备班的高中生所做的一项研究发现, 让学生产生兴趣变量的活动包括有智力游戏、小组合作和电脑, 维持兴趣变量包括丰富的意义, 这是学生认为能够维持他们兴趣的数学课程的特点之一。对调查数据进行量化分析研究发现, 个体的兴趣独立于对数学情境的兴趣, 情境兴趣不同于长期兴趣, 但与长期兴趣有关, 而学术兴趣则与具体的内容有关（参见Hidi & Renninger, 2006）。

在学生接受正规的学校教育过程中, 他们的数学兴趣逐渐降低, 而且对数学越来越感到厌倦（Eccles等, 1993; Gottfried, Marcoulides, Gottfried, Oliver, & Guerin, 2007; Kloosterman & Gorman, 1990）。例如, 弗伦泽尔、戈茨、佩克伦和瓦特（2010）对来自德国的基本处于三个能力水平上的42所学校83个班级（每个学校两个班级）, 5—9年级的3000多名学生进行了数学兴趣纵向发展研究。他们用含有6个问题的兴趣量表调查了学生对自己同学的数学价值以及数学老师的教学热情的认识, 以及学生父母的数学价值。潜在增长曲线模型结果显示, 学生的数学兴趣在刚开始的时候呈现迅速下降的趋势, 之后, 从8年级到9年级的时候变得平稳。大约从7年级起, 女生的数学兴趣开始低于男生, 而且女生数学兴趣的下降速度更大。处于最低能力水平的学生开始时表现出很低的数学兴趣, 但是, 他们的增长曲线比高能力水平学生的下降速度要小。这些结果与已有的关于美国和澳大利亚青少年的纵向研究结果一致（例如Mangu, Lee, Middleton, & Nelson, 2015; Watt, 2004）, 即学生的数学兴趣从一年级开始下降, 一直持续到中学（Fredricks & Eccles, 2002）。

在美国, 初中生对数学的兴趣直接影响着他们对高中数学课程学习的兴趣与成绩, 也决定着他们是否进一步学习数学课程。尽管K-12阶段学生的数学兴趣几乎不存在性别差异, 但从初中开始, 女生就会表现出比男生较低的数学兴趣, 较少参与课外的数学活动以及具有较低的数学自我概念（Simpkins, Davis-Kean, & Eccles, 2006）。高中阶段, 学生对数学的兴趣和对数学重要性的认识能够很好地预测他们选修数学课程的意愿以及所选数学课程的数量（Meece, Wigfield, & Eccles, 1990; Thorndike-Christ, 1991）。

情境兴趣。无论个体的数学兴趣处于怎样的水平, 他们在具体任务上的兴趣表现都会有很大差别。当学生的已有经验或长期兴趣没有产生影响时, 学生会在参与之前就对任务的潜在趣味性产生一种特定的观点, 而且, 还会

either learning or ego oriented—is descriptive of goals serving the avoidance of mathematical engagement. Due to repeated failures in mathematics tasks, some work avoiders tend to do the minimum to complete their mathematical work or try to escape it if possible. Hirvonen, Tolvanen, Aunola, and Nurmi (2012) surveyed 225 Finnish early elementary children, their teachers, and their parents regarding work avoidance in children; they compared these data with performance in basic arithmetic. A tendency to avoid mathematics tasks is strongly associated with poorer performance. Latent growth curve analysis shows work avoidance and performance developing together, though researchers don't necessarily know the originating factor or factors. Initial mathematics ability is uncorrelated with work avoidance, but the latter grows as performance drops relative to peers.

Interests

Interests refer to activities or topics that students tend to seek out and find enjoyable and toward which they orient their identities. There is compelling evidence that mathematical interests are among the strongest predictors of positive affective experiences in mathematics class (Schiefele & Csikszentmihalyi, 1995). How interests develop in relation to the extrinsic rewards and punishments that learners must negotiate in mathematics, and the proximal and distal goals they hold, is of great importance. Some literature suggests that students' in-the-moment situational interest leads over time to the development of mathematical interest as a trait (see a review by Hidi & Renninger, 2006).

Mitchell (1993) posits in a now classic model that interest in-the-moment needs to be *caught* by some key facets of the task, such as novelty and social stimulation, and then *held* through personal meaningfulness of the learning and depth of active participation. The catch and hold facets of situational interests are not properties of the task per se but occur as a negotiation between the student's individual interests and the cognitive, affective, and social constraints imposed by the task. In a study of 350 high school students in college preparatory classes, catch variables included puzzles, group work, and computers. Hold variables included meaningfulness, one of the characteristics of mathematics lessons cited by students as desirable in keeping them interested. Quantitative analysis of survey data showed personal interest to be clearly separable from situational interest in mathematics; situational interest to be distinct from, but associated with, longer term interests; and academic interests to be content specific (see also Hidi & Renninger, 2006).

Mathematics interest begins to taper off and boredom to increase as students progress through formal schooling (Eccles et al., 1993; Gottfried, Marcoulides, Gottfried, Oliver, & Guerin, 2007; Kloosterman & Gorman, 1990). Frenzel, Goetz, Pekrun, and Watt (2010), for example, studied the longitudinal development of mathematical interest in over 3,000 adolescents, grades 5–9, in Germany, where the secondary education system is strictly ability tracked. A total of 83 classrooms from 42 schools (2 classes per school) were sampled, roughly equally from the three ability tracks. Students were assessed on a six-item interest scale asking their perceptions of their classmates' value of mathematics and their mathematics teachers' enthusiasm. Students' parents also rated the value with which the family held mathematics. Latent growth curve modeling showed that mathematical interest decreased rapidly at first, then plateaued through eighth and into ninth grade. Girls' mathematical interest became lower than that of boys starting at about grade 7, and its rate of change was greater in the negative direction. Students in the lowest track started with the lowest mathematical interest, but their growth curve showed a shallower decline than the upper ability tracks. These results are consistent with longitudinal research on U.S. and Australian adolescents (e.g., Mangu, Lee, Middleton, & Nelson, 2015; Watt, 2004), where the decline starts in first grade and continues through secondary school (Fredricks & Eccles, 2002).

In the United States, interest in mathematics in middle school students directly influences their later interest and achievement in high school and their subsequent decisions to take further mathematics courses. While there are relatively few gender differences in mathematics achievement in the K–12 years, beginning in the middle grades girls still tend to show less interest in mathematics, less participation in mathematical extracurricular activities, and lower mathematical self-concepts than boys (Simpkins, Davis-Kean, & Eccles, 2006). Students' intentions to take elective mathematics courses in high school, and the number of courses they take, is predicted significantly by their interest in mathematics and their view of its importance (Meece, Wigfield, & Eccles, 1990; Thorndike-Christ, 1991).

Situational interest. Regardless of their levels of personal interest in mathematics, the interest students show in any particular task can vary widely. When prior experience or long-term interests provide no guidance, learners develop an ad hoc argument for the potential interestingness of a task before engagement and an evaluation of the activity subsequent to engagement in relation to their long-term interests (Ainley, 2006; Ainley & Hidi, 2014).

对后续所参与的与自己长期兴趣有关的活动做出自己的评估（Ainley, 2006；Ainley & Hidi, 2014）。我们通常称这种即时性的监控为情境兴趣（Renniner, Hidi, & Krapp, 2014）。

情境兴趣比那些只是匹配个人兴趣的任务更微妙。最佳挑战水平能够唤起并维持情境兴趣，而且这种优化是个人和任务交互所独有的。例如，有着主动学习倾向性目标的学生倾向于探寻高挑战性的任务，而有着自我逃避目标的学生在遇到高挑战性任务时则可能会感到焦虑。一个人对任务的个人控制程度，即情境效能，也会影响情境兴趣（Mangu等，2015；J. A. Middleton, 1995, 2013；J. A. Middleton, Leavy, & Leader, 2012）。

近期的研究结果表明，低年级学生倾向于用自己的情绪反应来定义兴趣（如，有趣、刺激），而中学生则会在对任务要求、对自己能力的主观认识，以及他们对任务的活动和方法的自主选择能力方面，显现出越来越复杂的推理能力（Frenzel, Pekrun, Dicke, & Goetz, 2012；Hidi & Renninger, 2006）。当学生有理由认为任务及其结果具有价值的时候，他们会使用一些策略，比如改变解决问题的方法，使无趣的任务变得更加有趣（Sansone & Thoman, 2005）。优化挑战的教学策略包括：允许学生从具有不同挑战水平的可用任务中进行选择，提供不同的任务情境和解决方法，以及为他们提供战胜困难的工具。

持续不断地参与到任务中能够让学生产生情境兴趣，这种情境兴趣能够帮助学生建立一个基本的数学学习经验，从而将这种经验转变为一种个体的兴趣（Hidi & Renninger, 2006；Krapp, 2002；J. A. Middleton & Toluk, 1999）。学生的个人兴趣为重视数学任务提供了理由，他们越快速解决任务，越表现出深层次的认知过程，也越喜欢更有挑战性的任务，未来更愿意选修数学课程以及将数学作为自己的事业。此外，对数学有兴趣的学生更易于享受其中并且对个人成就有更高的满足感（Ainley, Hidi, & Berndorff, 2002；Betz & Hackett, 2006；Deci & Ryan, 2002；Eccles & Wigfield, 2002；Efklides & Petkaki, 2005；Hackett & Betz, 1989；Köller, Baumert, & Schnabel, 2001；Laukenmann等，2003；Linnenbrink & Pintrich, 2004；Tulis & Ainley, 2011）。

实用性的作用

学生越觉得数学概念和技能对他们有用，而且有利于自己远期目标的实现，他们表现出的情境性兴趣就越大（J. A. Middleton, 2013）。研究者将这种现象称为感知手段（Husman, Derryberry, Michael Crowson, & Lomax, 2004；Malka & Covington, 2005）。例如，如果一个学生认为数学能够提供技能使自己实现成为一个海洋生物学家的长期目标，那么他就更有可能把数学本身视为个人兴趣，并把数学任务视为有趣的情境。乔伊纳德和罗伊（2008）对一组样本进行了纵向研究，他们调查了加拿大1130名讲法语的7至11年级的学生，该调查历时3年，且每年开展两次（例如，开始时是9年级的学生最后一次参加调查是在他们11年级快结束的时候）。研究使用的是芬尼马-谢尔曼的数学态度量表的翻译版本（Fennema & Sherman, 1976），评价了包括实用价值在内的一些有关动机的因素。乔伊纳德和罗伊利用分层模型调查了3年中学生在这一变量以及其它动机的变量上发生的变化，发现学生对于数学实用价值的信念在7至9年级有显著下降，这一下降趋势在9年级、10年级一直到11年级变得更为明显。虽然回避任务目标与实用价值之间表现出显著的负相关关系，但是学生们的学习方式目标导向和自我方式目标导向与实用价值之间存在中度相关关系。与实用价值类似，在中学阶段，学习方式目标也趋于减少，但自我方式目标和回避任务目标相对比较稳定。对于那些随时间推移而逐渐减少的变量来说，大部分是在学年中减少的，即学生对数学实用价值的感知和他们的学习方式目标会在一学年中急剧下降，而在假期又会有轻微的回升，到下一学年又会下降，如此反复。

就个人兴趣而言，研究广泛地表明，从初中开始经过高中一直到大学，学生对数学工具性的评价有一个急剧的下降（Chouinard & Roy, 2008；Hackett & Betz, 1989；Kessels & Hannover, 2007）。当学生把对数学实用性的认知转变为具体的实施计划时，那些有着远期目标的学生就会将职业愿景与未来学习的课程结合起来（Lens,

The term often used for such in-the-moment monitoring is *situational interest* (Renninger, Hidi, & Krapp, 2014).

Situational interest is subtler than a mere matching of task to individual interest. An optimal level of challenge evokes and maintains situational interest, and this optimization is unique to the individual and task interaction. Students with learning-approach goals, for example, tend to seek out high challenge; those with ego-avoidance goals may feel anxious when faced with too high a challenge. The degree of personal control a person feels over the task—that is, *situational efficacy*—also contributes to situational interest (Mangu et al., 2015; J. A. Middleton, 1995, 2013; J. A. Middleton, Leavy, & Leader, 2012).

Recent research suggests that younger students tend to define interest based on their emotional responses (e.g., it is fun, exciting), whereas secondary students show increasingly sophisticated reasoning about the task requirements, their subjective sense of competence, and their free choice of activities and approaches to the task (Frenzel, Pekrun, Dicke, & Goetz, 2012; Hidi & Renninger, 2006). Students use strategies such as varying their approach to the problem to make uninteresting tasks more interesting when they have reasons to value the task and its outcomes (Sansone & Thoman, 2005). Instructional strategies to optimize challenge include allowing students to choose from available tasks that differ in their level of challenge, offering a variety of task contexts and solution methods, and providing tools to overcome learners' difficulties.

Continued engagement with tasks that evoke situational interest builds a base of experiences for developing mathematics as an individual interest (Hidi & Renninger, 2006; Krapp, 2002; J. A. Middleton & Toluk, 1999). The student's individual interest provides reason to value mathematical tasks—the student takes to them more readily, shows deeper cognitive processing, prefers the more challenging tasks, and tends to orient toward future course taking and mathematical careers. Moreover, students with individual mathematical interest tend to enjoy it more and feel a greater satisfaction of personal accomplishment (Ainley, Hidi, & Berndorff, 2002; Betz & Hackett, 2006; Deci & Ryan, 2002; Eccles & Wigfield, 2002; Efklides & Petkaki, 2005; Hackett & Betz, 1989; Köller, Baumert, & Schnabel, 2001; Laukenmann et al., 2003; Linnenbrink & Pintrich, 2004; Tulis & Ainley, 2011).

The Role of Utility

The more that students feel that mathematics concepts and skills are useful to them, contributing to their distal goals, the greater situational interest they display (J. A. Middleton, 2013). Researchers call this *perceived instrumentality* (Husman, Derryberry, Michael Crowson, & Lomax, 2004; Malka & Covington, 2005). A student who views mathematics as providing skills to reach his long-term goal of becoming a marine biologist, for instance, is more likely to view mathematics itself as an individual interest and mathematical tasks as interesting situationally. Chouinard and Roy (2008) employed an overlapping cohort longitudinal design, surveying 1,130 French-speaking Canadian students, in grades 7 through 11, twice per year for 3 years (e.g., 9th graders at the beginning are last surveyed as finishing 11th graders). The study assesses several motivational factors including *utility value*, measured using a translated version of the Fennema-Sherman Mathematics Attitude Scales (Fennema & Sherman, 1976). Chouinard and Roy used multilevel modeling to measure the change in this and other motivational variables across 3 years, finding that students' beliefs in the utility value of mathematics decreased significantly between seventh and ninth grades, and the decrease was greater as students moved through grades 9, 10 and 11. Learning-approach and ego-approach goal orientations showed moderate to strong correlations with utility value, while work-avoidance goals showed a significant negative correlation with it. In parallel with utility value, learning-approach goals tended to decrease during the secondary school years, while ego-approach goals and work-avoidance goals stayed relatively stable. For the variables that decreased over time, most of the decrease occurred within-year: that is, students' perceptions of the utility value of mathematics, and their learning-approach goals, tended to drop sharply over the academic year, rebound slightly during the vacation months, and drop again over the next academic year in a repeating cycle.

As with individual interest, research broadly shows a sharp decline in students' assessment of the instrumentality of mathematics, beginning in the middle grades and continuing through high school into college (Chouinard & Roy, 2008; Hackett & Betz, 1989; Kessels & Hannover, 2007). The distal goals students develop relate to occupational aspirations and future courses as they translate their perceptions of mathematics utility into action plans (Lens, Paixão, & Herrera, 2009; Nuttin & Lens, 1985).

In-the-moment perceived instrumentality is influenced by the fact that some tasks provide immediate, useful knowledge whereas other tasks are only way stations on the path to reaching a goal. *Endogenous instrumentality* is the belief that a task provides important learning or skill development that will help reach the long-term goal. *Exogenous instrumentality*, in contrast, is the belief that a task is just a hurdle to be overcome (Hilpert et al., 2012).

Paixão, & Herrera, 2009; Nuttin & Lens, 1985)。

即时性感知数学的工具性会受到当下所接触到的任务的影响，有的任务能够提供直接、有用的知识，而有些任务仅仅是实现目标过程中的某些中间点。内源性工具是一种信念，这种信念使学习者相信某个任务能够为他们提供重要的知识学习和技能发展以帮助他们实现长远的目标。相反，外源性工具是将任务作为一个需要克服的障碍的一种信念（Hilpert 等，2012）。对数学实用性的内生观点，有助于促进学习者更深入、更专注于学习策略，而且能够增强他们的兴趣、学业表现和毅力（Hilpert 等，2012；Simons, Dewitte, & Lens, 2000；Simons, Vansteenkiste, Lens, & Lacante, 2004）。

其它的复杂性内容

动机、情感和社交这些与学生参与有关的因素似乎并没有像文献中描述地那么明晰。通常，学生同时具有学习的和自我的目标、个性的和标准的目标、与成功和失败相关的情绪，以及与他人的友谊和其它关系。所有的这些目标在一起会在具体情境中促进或阻碍学生的数学参与，而且这些情况会随着时间的变化而改变。

动机因素，如学习目标、兴趣和内在感知的工具性，除了在有效数学学习的测量方面是一致的之外，它们二者还存在中度的相关关系。然而，这些变量在解释数学成绩时的变异性很小（约为6%）（Mangu 等，2015；J. A. Middleton, 2013），相较于其它一些重要的中介变量（例如，课程、规范与实践及数学认同感），动机因素对学生数学成绩的影响作用似乎更少、更间接。

情感视角

麦克劳德（1989，1992，1994）有影响力的文章鼓舞研究者利用质性研究方法在复杂的数学教与学及问题解决过程中研究情感。他的研究将我们的注意力引到对即时性数学参与情感特征的研究，我们基于麦克劳德的作为数学参与主要子维度的情绪、态度和信念开始研究。他认为，情绪是改变最快且与认知最不直接相关的维度。

信念是最稳定而且是三个当中与认知联系最紧密的维度，而态度是介于二者之间的维度。此外，德贝利斯和戈尔丁（2006）区分了价值观的范畴，包括伦理、道德和对什么是重要的或正确的感知。他们指出，这四个维度当中的任何一个维度都可以被解释为暂时性的或相对稳定的学生的特征。这些维度还可以与学生的动机及社会环境中其他人的情绪、态度、信念和价值观动态地相互作用。参与过程的可塑性或许取决于情感的这些方面，它们在行为参与和认知参与方面起着推动和反馈作用。

状态与特征的区别

麦克劳德（1992）用情绪一词专指情感状态；态度和信念所包含的特质。在这个解释中，广义上的数学焦虑感是一种态度，它不同于学生所感受到的情绪上的焦虑。态度被认为是在特定情境下对某种特定行为模式或特定的情绪感受的倾向（即特质）。但有一些研究认为，态度是一种状态，例如：在某天或某个时刻学生对数学课堂的心情和做法。类似地，信念不仅仅被认为是一种长期的个体对数学以及自己与数学关系（即特质）的坚定信念，还被描述为学生在进行数学学习活动中（如，对自己目前是否具备某一能力的坚定看法）产生的一种短暂的信念（即相信的状态）。这样的信念可能会因为任务的要求或是同伴、老师给出的支持等而很快改变。总之，关于情感的文献中，价值被认为具有很高的稳定性，具有类似于特质一样的结构。然而，这个术语在动机的期望价值理论（Wigfield & Eccles, 2000；Wigfield, Tonks, & Klauda, 2009）中的使用有很大不同，该理论认为，所选取的特定任务对学生而言具有的价值或意义都是不断发生变化的。

除了数学焦虑感，特质变量还包括数学的自我概念、自我效能感、成就价值、我们在动机部分所讨论的目标取向和其它各种态度和信念。然而，其中的一些变量则在某一时刻才表现出来，从而给研究结构的恰当解释和操作带来了困惑。为了减少这种困惑，有人建议使用特殊情感一词（Goldin, 2000；Gómez-Chacón, 2000），它不仅被用于描述情感状态，还被用于表述个体的情绪

The endogenous perspective on utility promotes deeper and more focused learning strategies and enhanced interest, performance, and persistence (Hilpert et al., 2012; Simons, Dewitte, & Lens, 2000; Simons, Vansteenkiste, Lens, & Lacante, 2004).

Other Complexities

Motivational, affective, and social factors are not nearly as distinct in the engagement of learners as much of the literature seems to suggest. Students simultaneously hold learning and ego goals, personal and normative goals, emotions associated with success and failure, and friendships and relationships with others. Together, these facilitate or hinder mathematical engagement in context-dependent ways, and the patterns exhibit change over time.

Motivational factors such as learning goals, interest, and endogenous perceived instrumentality correlate modestly but consistently with measures of effective mathematical learning. These variables, however, account for relatively little variance (on the order of 6%) in measures of mathematics achievement (Mangu et al., 2015; J. A. Middleton, 2013). They seem to be less important as direct influences than as mediators of other important variables (e.g., curriculum, norms and practices, and mathematical identity).

Affective Perspectives

The influential articles of McLeod (1989, 1992, 1994) helped inspire the qualitative study of affect during complex episodes of mathematics teaching, learning, and problem solving. Directing our focus toward affective characteristics of in-the-moment mathematical engagement, we begin with McLeod's characterization of *emotions*, *attitudes*, and *beliefs* as major subdomains. He takes emotions as the most rapidly changing of these and the least directly linked with cognition. Beliefs are seen as the most stable and most highly cognitive of the three, and attitudes are viewed as intermediate in both respects. DeBellis and Goldin (2006) additionally distinguish the domain of *values*, including ethics, morals, and a sense of what is important or right. They suggest that each of these four kinds of constructs can be interpreted as transitory or as relatively stable characteristics of a student. They can also interact dynamically with a student's motivations and the emotions, attitudes, beliefs, and values of others in the social environment. The malleability of engagement processes may well depend on these aspects of affect as both drivers of, and feedback from, behavioral and cognitive engagement.

The State vs. Trait Distinction

McLeod (1992) uses the term emotion to refer exclusively to the state of affect; attitude and belief encompass traits. Generalized *math anxiety*, in this interpretation, is an attitude, to be distinguished from when a student is experiencing the emotion of anxiety. Attitude is interpreted as a propensity toward certain patterns of behavior or certain sets of emotional feelings in a defined context (i.e., a trait), but in some research it may refer to states, such as a student's mood and approach to mathematics class on a particular day or at a particular moment of the day. Similarly, belief has been used to describe not only long-term, deeply held convictions about mathematics or oneself in relation to mathematics (i.e., traits) but also the momentary convictions (i.e., states of belief) that occur as mathematics is being done (e.g., the conviction that a particular idea is within or not within one's present ability to grasp). Such beliefs may change quickly depending on the requirements of the task, available support from peers or the teacher, and so on. In general, values in the affect literature are seen as highly stable, trait-like structures. However, the term is used rather differently in the expectancy-value theory of motivation (Wigfield & Eccles, 2000; Wigfield, Tonks, & Klauda, 2009), where particular tasks are taken to have an associated value or worth to the student that is variable from moment to moment.

In addition to math anxiety, trait variables include mathematics self-concept and self-efficacy, achievement values, the goal orientations reviewed in our discussion of motivation, and various other attitudes and beliefs. However, some of these variables have been found to have in-the-moment manifestations, leading to some confusion regarding appropriate interpretation and operationalization of constructs for research. To alleviate such confusion, the term *local affect* has been suggested (Goldin, 2000; Gómez-Chacón, 2000) to be descriptive not only of affective states, but also of the *in-the-moment interactions* between emotions and cognition, the social environment, and the individual's trait-like characteristics. The term *global affect*, in contrast, refers not only to affective traits, but also to longer term structures that incorporate affect, cognition, and social factors—constructs such as mathematical self-efficacy, identity, or integrity, and the engagement structures discussed earlier.

Dettmers et al. (2011) examined data from approximately 3,500 German students in the Program for International Student Assessment (PISA) 2003 study. Students were surveyed in grade 9 and again in grade 10 on their expectations of success in mathematics and valuation of mathematics, achievement emotions in mathemat-

与认知、社交环境与个性特征之间的即时相互作用。相反，一般情感一词不仅指情感特征，还指包含个体情感、认知和社交因素的长期结构，如数学自我效能感、认同感或归属感，以及我们前文讨论过的参与结构。

德特默等人（2011）分析了2003年国际学生评估项目（PISA）研究中约3500名德国学生的数据。该研究从学生对数学成功的期望、对数学价值的看法、数学成就感以及在家庭作业上花费的精力方面调查了9年级学生，并在10年级的时候又进行了一次调查。和其他大样本数据的二次分析一样（例如 J. A. Middleton, 2013），研究者利用两个或三个问题对学生的情感进行了评估，这些问题的信度在0.6至0.7之间。德特默等人用多层模型发现，情感变量、所报告的努力程度与学生的数学成绩之间存在显著相关性，只是其相关系数较低，通常在0至0.3左右范围内。但是，这些变量整体上对处于个人层面上学生的努力程度和成绩水平有25%的解释作用，在班级层面上则波动更大。

这样的研究结果通常表明，对数学的积极情绪与其他积极情绪以及成功之间都存在相关关系（Laurent 等，1999；Pekrun, Goetz, Frenzel, Barchfeld, & Perry, 2011）。好奇心和高兴与自我调节学习策略的使用相关（Ahmed 等，2013）。那些为自己的学习负责并且在努力中取得成功的人往往会对他们所努力的目标产生积极的情绪反应。此外，情绪可能会驱动个体对数学进行更深入、更有效或更精细的认知处理。例如，对家庭作业的认知水平会导致高中生产生与问题相关的积极的或消极的情绪。这些情绪会反映出个体所要付出努力的程度：积极情绪与更多的努力相关，而消极情绪与较少的努力相关（Dettmers 等，2011）。自豪和高兴等情绪与使用灵活的策略（Fredrickson, 2001；Pekrun, 2006）和详细描述、组织和归类所获得的信息都有关系（Isen, 2004）。生气和厌烦等消极情绪同样也会产生很大的作用，尤其是当学生在数学任务中失败时（Tulis & Ainley, 2011）。

从这些研究结论中可以推断，积极情绪能够用来表示在数学上的情感参与，而消极情绪则用于表示个体在数学参与上的一种不满情绪。然而，我们认为这是毫无根据的。虽然数学上的成功通常与较低的焦虑感和挫折感相关，但是，这种关系是非线性的，而且很大程度上取决于情境。弗伦泽尔、佩克伦和戈茨（2007）对德国5年级到10年级的共1623名学生进行了调查，考察学生对课堂环境的认知对其数学情感体验的影响。多层模型分析结果显示，学业成绩较高的班级学生的焦虑感会提高，比起学业成绩较低的班级学生，这种焦虑感表现出更低的数学快乐感和更高的愤怒感。相反，个人数学成绩越高的学生，对数学的焦虑感、愤怒感越低，快乐感越高。这些结果显示学习文化具有重要的作用：高学业成绩的班级可能会有更多的社会性比较，对失败有更多的负担，为成功付出了更高的代价。尽管个人的成就能够使其对数学产生更多的快乐感和较少的焦虑感（例如 Tulis & Ainley, 2011），但是在社交的约束下，这种感觉有可能消退也有可能增进。

所以，这些研究都有自身的局限性。在一些严格控制变量的大型研究中，与数学成绩有关的个体的特质通常对其学业表现仅有10%或更少比例的影响作用（J. A. Middleton, 2013）。积极情绪之间的相互关系并没有为研究者提供这种关系背后的因果关系的机制，最为重要的是，我们不知道教师可以做哪些事情来影响学生的即时性参与，并通过对即时性参与的影响来产生长期的影响作用。此外，如果不是太大的、太令人失望的或太严重的挫折感，就能够被视为一种动力帮助学生重新努力学习数学（在学生重视数学上取得的成功的前提下），并在之后能使学生对成功产生强烈的满足感。积极的情绪，例如满意，通常很容易产生，但当学习者所面对的任务具有较高的价值的时候，瞬间的挫折、急躁或气愤也是有效参与的重要指标，这就需要制定自我调节策略去提高个体对任务的控制和能动性（Pekrun, Frenzel, Goetz, & Perry, 2007）。

因此，数学成功的特质之间的关系并不能为教学策略优化方面提供明确的建议。由于学生的特质各不相同，而且特质的变化需要相当长的时间，所以，如果将特质的变化视为数学参与的一个重要前提的话，那么我们将要面临巨大的挑战。

麦克劳德（1992）呼吁对即时的数学活动的精细情感分析，这促使了更多人使用质性方法研究数学情境

ics, and effort expended in homework. As in other large-scale secondary analyses (e.g., J. A. Middleton, 2013), affective variables were assessed using two- or three-item measures, with reported reliabilities in the .6 to .7 range. Using multilevel modeling, Dettmers et al. found significant correlations between affective variables, reported effort expended, and mathematics achievement. Correlations were small, typically ranging from near 0 to around .30. Together, however, these variables accounted for about 25% of the variability in effort and achievement at the individual level and much more variation at the class level.

Such survey research shows generally that positive emotions about mathematics correlate with other positive emotions and with success (Laurent et al., 1999; Pekrun, Goetz, Frenzel, Barchfeld, & Perry, 2011). Curiosity and enjoyment are associated with the use of self-regulated learning strategies (Ahmed et al., 2013). People who take charge of their learning, and are successful in their efforts, tend to develop positive emotional responses that focus on the object of their efforts. Moreover, emotions may drive deeper and more efficient or elaborate cognitive processing of mathematics in the individual. For example, the perceived quality of homework assignments leads high school students to develop positive or negative emotions associated with the problems. These emotions provide feedback regarding the degree of effort to expend: positive emotions associated with greater effort, negative emotions with less (Dettmers et al., 2011). Emotions such as pride and enjoyment are associated with the use of flexible strategies (Fredrickson, 2001; Pekrun, 2006) and with elaborating, organizing, and categorizing incoming information (Isen, 2004). Negative emotions such as anger and boredom also play a strong role, especially when students experience failure in mathematical tasks (Tulis & Ainley, 2011).

Such findings have led to the inference that positive emotion is a useful surrogate for affective mathematical engagement and negative emotion for disaffection. We take this to be unwarranted. While success in mathematics is generally correlated with lower feelings of anxiety and frustration, the association is nonlinear and highly context dependent. Frenzel, Pekrun, and Goetz (2007) surveyed 1,623 German students in grades 5 through 10 regarding the impact of perceived classroom environment on their emotional experiences in mathematics. Multilevel modeling found that for higher achieving classrooms, anxiety tended to *increase*. This anxiety was manifest in lower enjoyment of mathematics and higher expression of anger in mathematics than in lower achieving classrooms. Conversely, the more a student achieves in mathematics individually, the *lower* her anxiety and anger and the *higher* her enjoyment. These results suggest the key role played by the learning culture: high achieving classrooms may involve more social comparison, greater onus associated with failure, and higher stakes for success. Although individual achievement evokes greater enjoyment and less anxiety in mathematics (e.g., Tulis & Ainley, 2011), this occurs within social constraints that may negate or enhance such feelings.

Thus, such research has inherent limitations. In very large studies with tight controls, the observed correlations of traits with mathematical performance generally account for around 10% or less of the variance in performance measures (J. A. Middleton, 2013). Correlations among positive emotions do not inform researchers about the causal mechanisms behind the correlations or, most important, what teachers can do to influence students' engagement in the short term and, through in-the-moment influence, in the long term. Moreover, frustration, if it is not too great, or disappointment, if it is not too severe, can serve as an impetus for renewed effort in mathematics (when success in the mathematics is valued) and a stronger subsequent experience of satisfaction in success. Positive emotions such as satisfaction are generally good to engender, but states of momentary frustration, impatience, or anger can also be important indices of productive engagement when the value of the task is high—cues to institute self-regulation strategies that increase personal control and agency over the task (Pekrun, Frenzel, Goetz, & Perry, 2007).

The correlation of certain traits with mathematical success, therefore, does not clearly suggest optimal teaching strategies. Since traits differ from student to student and are changeable only over substantial time periods, the challenges are steep if we regard trait change as an important prerequisite to mathematical engagement.

McLeod's (1992) call for fine-grained, affective analyses of in-the-moment mathematical activity led to greater use of qualitative methods in the study of emotions in mathematical contexts (DeBellis & Goldin, 2006; Goldin, 2000; Gómez-Chacón, 2000; Hannula, 2006; Malmivuori, 2006; Op 't Eynde, De Corte, & Verschaffel, 2006, 2007; Zan, Brown, Evans, & Hannula, 2006) and in the study of mathematical beliefs (Di Martino & Zan, 2011; Hannula, 2002, 2006; Leder, Pehkonen, & Törner 2002). Philipp (2007) describes this trend in detail as it pertains to mathematics teachers' beliefs and, more generally, their affect.

To illustrate, in a collection of papers published as a special issue of *Educational Studies in Mathematics*, each author sought to interpret the recorded protocol of a

中的情绪（DeBellis & Goldin, 2006；Goldin, 2000；Gómez-Chacón, 2000；Hannula, 2006；Malmivuori, 2006；Op't Eynde, De Corte, & Verschaffel, 2006, 2007；Zan, Brown, Evans, & Hannula, 2006）和数学信念中的情绪（Di Martino & Zan, 2011；Hannula, 2002, 2006；Leder, Pehkonen, & Törner, 2002）。菲利普（2007）详细描述了这种趋势，因为它与数学教师的信念有关，更一般地说，与教师的情感有关。

为了说明这一点，在《数学教育研究》特刊发表的一系列论文中，每个作者都在试图对深度参与到数学问题解决中的一名学生（"弗兰克"）的观察记录进行解释（Op't Eynde & Hannula, 2006）。其中，观察记录的数据有该学生解决问题的手稿，从课堂录像观察到的他的面部表情和身体动作，以及基于录像刺激的回忆性访谈内容中获得的观点。基于这些数据，研究者们对弗兰克的情绪发展顺序进行了讨论，戈尔丁（2000）称其为"情感路径"。有证据表明，弗兰克的状态是由快乐/自信变为担心，然后释怀。之后从担心/恐慌变为失望和生气，然后是快乐，接着是生气，再是快乐。到他问题解决快要结束的时候，弗兰克的状态又再次变为紧张，最终变为开心。研究者认为这一复杂的情感经历，即在消极和积极情感之间的这种交替变化，对弗兰克参与任务来说是非常关键的。消极情绪对他的参与起到建设性的作用，有助于提高他的参与度。

迪·马蒂诺和赞（2011）分析了1—13年级的共1662名意大利学生的作文，这些文章的主题为"我与数学：到目前为止我与数学的关系"。迪·马蒂诺和赞分析了学生的叙述，首先将那些使用消极情感的语言谈到自己对数学的负面情绪（如"不喜欢"等）的作文挑选出来，之后，他们考察了学生描述的情感状态与这种情感产生的原因（如，"因为有太多的规则"）之间的联系。他们将这些联系分为两类：（1）情感倾向与数学观之间的联系，（2）情感倾向与能力知觉之间的关系。学生的情绪反映了他们对数学作为某个领域的构想：他们将数学描述为枯燥的，不提供个人情感或个人表现的空间，令人困惑的，充斥着需要记忆的规则和公式。情感和自我能力知觉之间的联系与认知评价研究有关：成功与失败的原因（内部的还是外部的），成功与失败的稳定性（稳定的还是不稳定的）以及可控性（可控的还是不可控的）。然而，表现具有学科特殊性，例如，学生将记忆、了解和运用规则看作是数学成功的稳定定义能力，相信数学是枯燥的。有些学生似乎是将成功的标准完全外化了，服从教师对他们的界定。此外，个别学生对数学的信念还涉及很多情绪方面的陈述。迪·马蒂诺和赞将自我概念和自我效能感视为影响学生趋近或逃避态度的关键情感结构内容，并且推断出这些情感因素中非常重要的情感组成部分。

情感复杂结构的方方面面

为了理解情感在数学学习和参与中的作用，研究者需要一种方式去综合地考察相关变量之间的关系，即基于教育学和社会心理学中关于情感的相关理论但又充分考虑到数学学科的特性。在这里我们将着重强调一些重要的方面，戈尔丁（2014）提供了一个更详细的列表。

情绪能够表现出目标的结果和目标的取向，因为目标有所进展会产生积极的情绪，如缓解、快乐或自豪，而缺乏这种进展则会产生消极的情绪（Anderman & Wolters, 2006；Linnenbrink & Pintrich, 2000, 2002；Schunk, 2001）。情绪也与学生的成绩直接相关（Pekrun等, 2007）。

成就情绪的控制价值理论（Pekrun, 2006；Pekrun & Perry, 2014）明确了这类情绪的起因和影响作用以及它们之间的相互关系，该理论同时指出，这些情绪都是针对具体领域的。学生关于主观控制自己的学习活动及结果的信念（如，他们期待通过努力而获得成功）与他们对参与的主观判断（如，对成功认知的重要性）一起，组成了情感反应的预期结构。当学生期望在一个重要的任务上获得成功的时候，他们就会感受到预期的情绪，如希望和开心。当成功的价值或失败的代价很高，而期望又很低或不确定的时候，学生就会产生焦虑情绪。对成功或失败的回顾性分析会导致回顾性情感的产生，如自豪（在高成功价值的任务上获得成功）、失望（在高成功价值的任务上失败）、厌倦（在低成功价值的任务上成

student ("Frank") deeply engaged in solving a mathematical problem (Op 't Eynde & Hannula, 2006). The available data include a transcription of his problem solving, observations of his facial expressions and body movements derived from a video recording of his classroom work, and comments during a video-based stimulated recall interview. Based on these data, the contributors discuss Frank's inferred sequence of emotions, a kind of "affective pathway" in the sense of Goldin (2000). The evidence suggests his state changes from happiness/confidence to worry, followed by relief. This is followed by states of worry/panic moving into frustration and anger, then happiness, then anger again followed by happiness. Toward the end of his problem-solving session, Frank's state again becomes nervous, but ends as happy. This complex affective experience, in which emotions alternate between the negative and the positive, is interpreted by contributors as essential to Frank's engagement with the task. Negative emotions play a constructive role and contribute to enhancing his engagement.

Di Martino and Zan (2011) analyzed essays of 1,662 Italian students in grades 1–13 on the topic "Me and mathematics: my relationship with maths up to now." Di Martino and Zan analyzed students' narratives by first identifying essays that made reference to a negative emotional disposition toward mathematics through the use of language indicating negative affect ("dislike," etc.). They then examined the linkage between students' statements of affect and their reasons for feeling it (e.g., "because there are a lot of rules"). They categorized the linkages into two types: (1) connections between emotional dispositions and vision of mathematics and (2) connections between emotional dispositions and perceived competence. Students' emotions reflect their conceptualization of mathematics as a domain: They describe it as being dry, as not leaving room for feelings or personal expression, as confusing, and as made up of rules and formulas to be remembered. The links between emotions and perceived competence connect with research on cognitive appraisal: the locus of success or failure (internal vs. external), stability of success or failure (stable vs. unstable), and controllability (controllable vs. uncontrollable). The manifestations, however, are subject-specific: for example, students take remembering, knowing, and applying rules as stable defining competencies for mathematical success in believing mathematics to be dry. Some students, however, seem to externalize success criteria entirely, deferring to the teacher to define them. Moreover, individual students' beliefs about mathematics reference many emotional statements. Di Martino and Zan identified self-concept and self-efficacy as key affective structures influencing students' approach or avoidance, inferring a strong emotional component in these structures.

Aspects of the Complex Architecture of Affect

To understand the role of affect in mathematical learning and engagement, researchers need a way to consider interactions among such relevant variables coherently—that is, a theoretical view of the affective domain using what has been learned from educational and social psychology, but taking adequate account of features domain-specific to mathematics. Here we highlight some important aspects; Goldin (2014) provides a more detailed account.

Emotion serves a *representational* function as a consequence of goals and goal orientations, with positive emotions such as relief, joy, or pride resulting from appraisal of progress toward a goal and negative emotions resulting from lack of progress (Anderman & Wolters, 2006; Linnenbrink & Pintrich, 2000, 2002; Schunk, 2001). Emotions are also directly linked to achievement outcomes (Pekrun et al., 2007).

The control-value theory of achievement emotions (Pekrun, 2006; Pekrun & Perry, 2014) identifies some of the antecedents and effects of such emotions, with interactions among these. It points also to the domain-specificity of such emotions. Students' beliefs about the *subjective control* they hold over achievement activities and their outcomes (e.g., their expectation that application of effort will lead to success) coupled with their *subjective value* of engagement (e.g., the perceived importance of being successful) work in concert as anticipatory structures for affective responses to action. Students experience anticipatory emotions such as hope or joy when they expect success in a valued task. Anxiety occurs when the value of success or the cost of failure is high but expectations are low or uncertain. Retrospective appraisal of successes or failures results in retrospective affect such as pride (success in a high-value task) or disappointment (failure in a high-value task), boredom (success in a low-value task), or apathy (failure in a low-value task). Both anticipatory emotions and retrospective emotions are tied to situational interest through enjoyment, frustration, and other in-the-moment affective appraisals of students' engagement.

Taking a full representational perspective, emotional states (or affective states generally) are not only responses to experienced situations, but encode information that interacts with cognitive representation in the individual's ongoing pattern of responses (Zajonc, 1980). Such affect has anticipatory and evaluative functions. Examples in

功）或冷漠（在低成功价值的任务上失败）。无论是预期的情绪还是回顾性情绪都与情境兴趣相关，与学生在参与过程中体验到的快乐、沮丧和其它即时性情感密切相关。

从一个完整的表征视角来看，情绪状态（或者通常也叫情感状态）不仅是对经历过的情境的反应，还是对信息的一种编码，这些信息与个体认知表征的反应模式之间存在相互作用的关系（Zajonc, 1980）。而且，这些情感具有预测和评价的作用。研究者已经讨论过数学教育中的相关案例，比如，各种情绪方式对问题信息、成功或失败的策略、学生与问题的关系或学生解决问题过程中与他人的关系进行的编码，这些情绪方式包括好奇心、困惑、挫折或满意度等（DeBellis & Goldin, 2006；Goldin, 2000）。这些在情感、认知和反应之间存在的持续不断且复杂的相互关系使得我们从认知参与和行为参与中很难区分出情感因素。

还有一个重要的情感因素被称为是元情感（DeBellis & Goldin, 1997, 2006；Goldin, 2002, 2014；Gómez-Chacón, 2000），它非常类似于元认知的概念（Flavell, 1976），元情感包括元心情和元情绪能力（利用自我调节策略应对消极情绪、监督和控制情感），通常出现在情商研究（Fitness & Curtis, 2005）和数学教育研究（Op't Eynde 等, 2007）中。然而，元情感也包括了关于情感的情感和关于认知的情感，反过来说，也就是对情感的认知。因此，我们要重点强调个体对自己所体验到的情绪的想法是会改变的，甚至可以完全转变，例如，当个体竭尽全力爬山的时候他所经历的痛苦可以转变为快乐，而如果是通过不诚信的方式获得奖励，那么快乐就会变为羞耻。元情感的概念对理解这一情况是很重要的，即在数学活动中，对新概念所产生的混淆或困惑如何用兴趣和好奇心来代表说明，或者当获得正确的数学解决方法时所产生的快乐，却又因为解决方法并不是因为真正的理解而是用死记硬背的方式来解决的，从而如何变的不安和内疚。总之，个体以混合或相互转化的形式，多水平上体验积极的或消极的情绪。有些体验可能是有意识的，也有可能是无意识的。作为情感结构的一个特征，元情感或许可以帮助研究者理解在通往成功的道路上，占主导的消极情绪是如何增强学生在成绩上的自豪感和满意度以及之后是如何提高他们的参与度的。

元情感对于我们理解短暂且温和的情绪在处理数学信息上的作用也十分重要。德贝利斯和戈尔丁（2006）介绍了一个名叫"朗达"的学生，他们一开始对朗达展开了一个基于任务的访谈，但是朗达在摄影机、摄制组和研究者面前感到不舒服或不安。随后，在解决问题过程中，她对精确地将一个圆分割成三份所遇到的困难也感到不安。作者认为，两种"负面情绪状态"之间最大的区别在于元情感——朗达对不安情绪的来源和控制的解释。第二次的时候，"她让自己在一个假设的情境中感受自己的情绪。不同于访谈刚开始时的不安，这次的情绪是她能够控制并且完全能够对情绪处之泰然。她不仅变得更加温和，而且自己创设了元情感是积极且安全的情境"（第141页，强调语气为原文所有）。

数学的情感结构

情感体系中的一个重要方面是情感结构概念，它类似于认知结构或图式的概念。情感结构在特定情境中是有效的，而且，可以帮助研究者理解参与，就好比认知结构、策略和试探方式帮助研究者理解数学学习和问题解决中的认知维度一样。错综复杂的要素可以包括情绪、态度、信念和价值观以及特有的目标、行为、认知、社交互动等。与数学参与相关的要素有数学自我认同感和自我效能感，通常也被描述为信念或理论（Dweck, 2000），但也涉及情绪、价值观、行为模式等，还有数学亲密感和数学诚信（DeBellis & Goldin, 2006）以及前面讨论过的参与结构。

数学亲密感（DeBellis & Goldin, 1999, 2006）是一种可以通过行为观察到的特征，例如，用某种姿势保护自己的工作，不愿意与别人分享自己的工作，深呼吸以及闭上眼睛，非常慢地、平静地或兴奋地讲话。数学亲密的体验包括：敏感、紧张和兴奋、喜爱和热情、变得"特别"、审美满足和自豪等感觉。数学亲密感也许与数学有关，也许与个体所在意的其他对象有关系（父母、老师和同伴），如：个体如何从别人的角度看待自己以及个体和社会是如何评价数学的。因此，就像其它情感结构一样，亲密

mathematics education that have been discussed include ways emotions such as curiosity, bewilderment, frustration, or satisfaction encode information about a problem, the success or lack of success of a strategy, the student in relation to the problem, or the student solving the problem in relation to others (DeBellis & Goldin, 2006; Goldin, 2000). Such ongoing, complex interactions between affect, cognition, and overt responses cause difficulty in distinguishing affective from cognitive or behavioral engagement.

Another important aspect of affect has been termed meta-affect (DeBellis & Goldin, 1997, 2006; Goldin, 2002, 2014; Gómez-Chacón, 2000). Closely analogous to the notion of metacognition (Flavell, 1976), meta-affect embraces the ideas of meta-mood and meta-emotional competencies (coping with negative emotions, monitoring and controlling affect through self-regulation), frequently considered in emotional-intelligence research (Fitness & Curtis, 2005), as well as in mathematics education (Op 't Eynde et al., 2007). However, meta-affect also includes affect about affect and affect about cognition that is, in turn, about affect. Thus we highlight the idea that the emotion someone experiences about an emotion can modify, or even wholly transform, the experience—for example, pain can be felt joyfully, as in straining one's limits to climb a high mountain, and pleasure can be felt shamefully, as when a reward is achieved through a breach of integrity. The idea of meta-affect is important to understanding how, during mathematical activity, bewilderment or confusion with a novel concept might be interpreted with interest and curiosity, or alternatively, how the pleasure of obtaining a mathematically correct solution might be experienced with unease or guilt that the solution method was not understood but followed by rote. In short, emotions of positive or negative valence can be experienced in a mixed or transformed way and at many levels. Some of this experience may be preconscious or unconscious. Meta-affect, as a feature of the architecture of affect, may help researchers understand how major, negative emotions along a path of achievement can heighten pride and satisfaction in the achievement and strengthen subsequent engagement.

Meta-affect may also be important in understanding the role that momentary, mild emotions play in encoding mathematical information. DeBellis and Goldin (2006) describe a student, "Londa," who initially in a task-based interview is inferred to feel discomfort or unease at the presence of video cameras, camera-crew, and researchers. Later, during the problem solving, Londa is inferred to feel discomfort with the difficulty of cutting a circle into thirds accurately. The authors suggest that an important difference in the two "negative emotional states" lies in the meta-affect—Londa's interpretation of the source and controllability of the discomfort. In the second instance, "She *allows* herself to feel the emotion, in a hypothetical context. Unlike her discomfort at the beginning of the interview, this is an emotion she controls and feels entirely *comfortable* about feeling. It is not merely milder. She has shaped the context herself; the meta-affect is positive and safe" (p. 141, emphasis in original).

Structures of Affect Pertaining to Mathematics

An important feature of the architecture of affect is the notion of *affective structure*, analogous to the idea of cognitive structure or schema. Affective structures become active in particular contexts and may help researchers understand engagement, analogous to the way that cognitive structures, strategies, and heuristics assist researchers in understanding cognitive aspects of mathematical learning and problem solving. Intertwined components may include emotions, attitudes, beliefs, and values, as well as characteristic goals, behaviors, cognitions, social interactions, and so forth. Examples relevant to mathematical engagement include mathematical self-identity and self-efficacy, often described as beliefs or theories (Dweck, 2000) but also entailing emotions, values, behavior patterns, and so forth; mathematical intimacy and mathematical integrity (DeBellis & Goldin, 2006); and the engagement structures discussed earlier.

Mathematical intimacy (DeBellis & Goldin, 1999, 2006) is characterized by observable behavior such as physical gestures protective of the persons' work; reluctance to share the work; deep breathing and closing of the eyes; or exceptionally slow, quiet, or excited speech. The intimate experience of mathematics may include feelings of vulnerability, thrill and excitement, affection and warmth, being "special," aesthetic satisfaction, and pride. It may be tied not only to mathematics, but also to relationships with others about whom one cares (a parent, a teacher, or peers), how one sees oneself as perceived by others, and how mathematics is valued personally and socially. Thus, like other affective structures, it is both personal and social, and may occur outside of school as well as in school contexts. For instance, Else-Quest, Hyde, and Hejmadi (2008) describe the emotions of mothers and their fifth-grade children (age 11) working on mathematics homework problems. They find the emotions of mother and child to be highly correlated (suggesting an intimate social relationship) and associate humor, pride, and positive interest with better test performance and tension with lower performance.

感既具有个体性也具有社会性，而且，可能发生在校内也可能发生在校外。例如，埃尔斯-奎斯特、海德和荷马迪（2008）研究了母亲的情绪和她们读5年级的孩子（11岁）一起解决数学作业问题时的情况，发现母亲和孩子的情绪有非常高的相关性（一种亲密的社会关系），学生的高学业成绩与其幽默、自豪和积极兴趣的情绪相关，学生的低学业成绩则与紧张不安相关。

数学诚信（DeBellis & Goldin, 1999, 2006；Vinner, 1997）是学习者在对问题有充分理解，问题得到解决，成绩是真实获得的，或得到应得的认可之后对自己表现的想法。即便学习者成功且完整地解决了某个数学问题，它要求学习者承认自己理解的不充分性。多年前，斯肯普（1976）就已经对数学的工具性理解和关系性理解进行了区分。事实上，如果没有关系性理解，学习者也有可能表现很好，但如果学习者意识到了这个问题，那么缺乏关系性理解就会让学习者感到不安并且在数学诚信方面产生问题。另外，加入社会维度之后，学习者要考虑的事情是如何在他人（老师或同伴）面前表现自己，也就是承认或坚持自己对数学的某些内容并没有理解。

有相当多的研究关注数学认同感、心态和自我效能感——人们是如何界定或相信自己与数学和数学能力的关系的。学生的自我效能信念是预测动机的指标（例如 Dweck, 2000；Heyd-Metzuyanim & Sfard, 2012；Zimmerman, 2000）。这种数学信念体系不仅仅是认知结构，它还与情感体验的各种特征交织在一起，这些情感体验从羞愧和耻辱到开心、自豪和满足。情感体验和元情感对产生稳定的信念结构有重要的作用（Goldin, 2002），但根据伍德（2013）对4年级学生的分析结果，数学认同感可以在不同时刻有不同的诠释。

在有关认知情感、认同感和自我效能感相互作用的研究中，数学焦虑感恐怕是最常被研究的一种情感特质。许多成人的痛苦经历中有与学校数学有关的，并且这些共同的焦虑感体验有碍于他们的数学参与（在其他学科上都没有如此让人心力交瘁的经历）。对学生来说，数学焦虑感与数学成绩之间存在负相关性（Hembree, 1990；Ma, 1999；Ma & Xu, 2004），而且，这些关系最多达到中等程度。李（2009）利用PISA2003提供的数据，选取了到目前为止最大样本的数据研究了学生的数学焦虑感和成绩之间的关系。作者指出，数学焦虑感与成绩之间的相关系数从印度尼西亚的$r=-0.12$到丹麦的$r=-0.51$，整体均值的相关系数为$r=-0.39$（在显著水平$p<0.01$下均为显著）。但在较小样本的研究中，结果没有这么明显。例如，迪瓦恩、福西特、苏奇和多克（2012）选取了英国的学生，在控制了他们的考试焦虑水平之后，发现仅有女生的数学焦虑感与数学成绩之间存在负相关关系。不同总体有不同的发现，这取决于研究者是否将数学焦虑分为不同的特定维度（如，将学生的考试焦虑与对自己数学能力的焦虑感合并在了一起；Bessant, 1995），或者仅仅只有一个维度（如，学生在面对任何数学问题时都会感到焦虑；Beasley, Long, & Natali, 2001）。他们之间的因果关系仍然不清楚（Sherman & Wither, 2003）。数学焦虑感的产生缘由、与一般考试焦虑之间的关系、引起和导致数学焦虑感产生的参与模式（Zeidner, 2014）以及在学校教育中它的发展过程，这些都需要进一步研究。

海蒂-梅祖雅尼（2015）详细描述了一个名叫依蒂特的以色列女生的数学焦虑感的发展过程，这个女生在7年级的时候数学成绩很好，但到了9年级，她的数学成绩却不及格。研究者试图采用"对话角度"去发现情绪、认知和社会维度之间的关系。通过这一分析视角，依蒂特的参与被描述为"礼节性的"，并且，她的数学焦虑感的发展机制被解释为一种"基于礼仪对话的一种恶性循环"（第504页）。他们采用精细分析法分析了学生的对话及其父母和教师的背景，该研究是理解数学焦虑感产生复杂原因的重要参考。

很多关于长期的情感或情感特征对学生动机和参与的表现的影响研究也是很重要的，借助一些模型，研究人员已经把这些联系表示出来了（例如Pekrun, 2006；Wigfield & Eccles, 2000；并参见本章对动机的讨论）。

已有研究将参与的结构描述为一种理论方法，这种理论方法综合了情感和动机、社交互动及学生数学课堂即时性参与模式的其他维度（Epstein等，2007；Goldin等，2011；Schorr, Epstein, Warner, & Arias, 2010a, 2010b）。当运用某一种参与结构的时候，需要注意到这些因素之间的相互关系。其中，七个是状态变量（欲望

Mathematical integrity (DeBellis & Goldin, 1999, 2006; Vinner, 1997) pertains to a learner's stance with regard to when sufficient understanding has been attained, when a problem has been solved, when achievement is genuine, or when approval and recognition are deserved. It requires the learner's acknowledgement of insufficiency in his own understanding, even after he may have implemented a procedure successfully to solve a mathematical problem. Many years ago Skemp (1976) distinguished instrumental from relational understanding in mathematics. The fact that it is possible to perform well without necessarily acquiring relational understanding may, if recognized by the learner, create discomfort and raise integrity issues. Again the social dimension enters, as what is at issue may be how the learner presents herself to others—to the teacher or to her peers—in admitting or asserting that something does not make sense mathematically.

A considerable literature is devoted to mathematical identity, mindset, and self-efficacy—how people define themselves and what they believe about themselves in relation to mathematics and in relation to their mathematical ability. Self-efficacy beliefs of students are a predictor of motivation (e.g., Dweck, 2000; Heyd-Metzuyanim & Sfard, 2012; Zimmerman, 2000). Such systems of mathematical belief are not only cognitive but interwoven with characteristic emotional feelings that may range from shame and humiliation to pleasure, pride, and satisfaction. Emotional feelings and meta-affect can contribute importantly to the stability of the belief structures (Goldin, 2002); but mathematical identity can also have a moment-by-moment interpretation, as described by Wood (2013) in her analysis of a student in grade 4.

An example of cognitive-emotional interaction in connection with identity and self-efficacy occurs in perhaps the most-examined affective trait, math anxiety. Many adults report histories of painful experience in connection with school mathematics and commonly experience anxiety impeding their engagement with mathematics (without such debilitating experience in other subjects). In students, math anxiety appears to be correlated negatively with achievement measures (Hembree, 1990; Ma, 1999; Ma & Xu, 2004), though these correlations are modest at best. Lee (2009), using data provided by the 2003 PISA, performed the largest scale study to date on the relationship between math anxiety and achievement. The author reports correlations of math anxiety with performance scores ranging from $r = -.12$ in Indonesia to $r = -.51$ in Denmark, with a mean correlation of $r = -.39$ (all significant to $p < .01$). In smaller-scale studies, results can be less clear. Devine, Fawcett, Szűcs, and Dowker (2012), for example, determined that among British students, after controlling for test anxiety, there was a negative correlation between math anxiety and mathematics performance for girls only. Different findings for different populations are reported depending on whether math anxiety is conceptualized as several distinct factors (such as students' test anxiety together with anxiety about their mathematical ability; Bessant, 1995) or just one factor (e.g., students' feelings of anxiety when confronted with any aspect of mathematics; Beasley, Long, & Natali, 2001). The causal direction of the relationship remains unclear (Sherman & Wither, 2003). The origins of math anxiety, its relation to general test anxiety, the engagement patterns that occasion it and result from it (Zeidner, 2014), and its progression during schooling all require further research.

Heyd-Metzuyanim (2015) describes in detail the development of math anxiety in the case of a female Israeli student, Idit, who was a high-achiever in seventh grade but by ninth grade was failing mathematics. The author contrasts several possible theoretical lenses through which Idit's math anxiety may be viewed and adopts a "communicational lens" as a way to capture interactions among the emotional, cognitive, and social dimensions. Through this lens, Idit's engagement is described as "ritual," and the mechanism for her development of math anxiety is interpreted as a "vicious cycle that thrives on the basis of a ritualistic discourse" (p. 504). This study is rare in its effort to understand the complex origins of math anxiety through the fine-grained analysis of discourse, seen against the background narratives of the student's parents and teacher.

Many other longer term emotional feelings or affective traits associated with achievement are also important in characterizing students' motivations and engagement, and such associations have been described through carefully developed models (e.g., Pekrun, 2006; Wigfield & Eccles, 2000; see also our discussion of motivation in this chapter).

The engagement structures described previously are intended as a theoretical approach toward integrating the description of affect with that of motivation, social interactions, and other dimensions in characterizing patterns of students' in-the-moment classroom engagement with mathematics (Epstein et al., 2007; Goldin et al., 2011; Schorr, Epstein, Warner, & Arias, 2010a, 2010b). Note that among the components that interact when an engagement structure is active, seven refer to state variables (motivating desire; behavior, including social interactions; emotional states; expressions of emotion; meanings encoded by emotions; meta-affect; and self-talk). Three refer to interactions with the person's traits or competencies (beliefs and values, personality traits and

动机；行为，包括社交互动；情绪状态；情绪表达；情绪所表示的含义；元情感；自我对话），三个是个体特质或能力之间的相互影响变量（信念与价值观，个性特质与取向，问题解决策略与探索过程）。例如，肖尔等人（2010a）利用"不要不尊重我"的结构和其它参与结构，通过观察课堂录像的方式理解学生对数学思想进行批判和愤怒地辩护的相互作用。

环境压力

长期以来，社会心理学家一直用压力这个概念，这里，压力是指社会环境会强加一种情境约束作用使其影响个体实现心理上的追求（Murray，2008）。在数学课堂上，压力有可能来自于老师或其他同学的即时性情境要求（如，教师的某个要求，其他同学的挑战，或寻求帮助）。它也有可能包括父母的内在期望、与课堂社会文化准则一致的期望，等等。在小组合作中，学生的角色可能是同学与教师默许的或明确期望。例如，数学亲密感、诚信、认同感或自我效能感等情感结构都含有社会期望和环境压力，它们会对这些结构如何作用于学生的即时性参与产生影响（Epstein 等，2007；Goldin 等，2011；Schorr 等，2010b）。

数学领域的情感维度

我们对这一内容所描述的很多观点看上去都可以应用在学校的其他课程中，其实，有很多研究观点是针对数学所特有的问题的，例如"社会学中的焦虑"和"人文学科的焦虑"并不像数学学科有相应的焦虑感研究，不过也有一小部分关于"外语阅读焦虑"和"科学焦虑"的研究，而且数量正在增加（依次参见 Saito, Garza, & Horwitz, 1999 与 Bryant 等，2013）。因此，应该考虑数学教育中有别于其它学科的情感因素。

长期以来，问题解决都是学校数学活动的中心（Koiche，2014）。但是，数学问题不同于大部分常规程序，即学习者会进入一个困境，不知道下一步该怎么办的一种经历（Schoenfeld，1985）。事实上，困境是促使学生学习问题解决的主要方法。在解决一个富有内涵的复杂问题时，个体所遇到的每一个困境都与其从情绪角度对困境作出的评价以及在解决（或无法解决）困境时的情绪反应有关。这些情绪包括遇到障碍或挫折时的好奇心和高涨的兴趣以及依赖于学生对努力后结果的厌烦和愤怒或安慰和开心（D'Mello & Graesser，2014；Goldin，2000；Op't Eynde 等，2007）。如何处理这些情绪，即学生的元情感，将有助于促进或阻碍学生的参与方式，因为，这些具体的情感体验会影响学生的策略选择。

数学学科的其它方面的特征（见 Goldin，2014），包括过程性知识和概念性理解频繁的脱节；学校课程的分层结构，即引入新内容时往往假设学生已经掌握了必要的概念和技能；以及与逐渐增多的抽象概念相关的一系列认识或认知方面的障碍。

社会上普遍承认的一些信念（Handal，2003；Leder 等，2002；Maasz & Schlöglmann，2009）会将数学成功归因于天生的能力或天赋，把在数学上成功的学生形容为没有吸引力的"书呆子"或宣扬一种刻板印象，即数学是男生、白人或亚洲人的领域。这些信念会影响从学生的个人想法和班级社会关系方面评价和解释个体在数学上的成功（Yeager & Dweck，2012）。

总之，即时参与的情感复杂性反应了一个事实：学生的独特个性、长期性的兴趣、目标和对实用性的信念都体现在他们的情绪上。学生对数学的自我调节策略、价值观、态度和信念等包括在内，具有不同的思维习惯。另外，他们还具有与数学有关的一般性的学业自我概念结构、自我效能感结构和个性特质。在一些特殊场合中，这些特质结构会对评价学生的参与机会产生影响，这里，参与机会是指学生正在经历和形成各种追求、目标和预期情绪的机会。其中，有些特质可能与数学和数学学习有关，有些与教师有关，有些与个体、同伴或其他对自己而言重要的人有关，有些他们对自己状态或能力的认知有关，还有些是与远离当下课堂场景的想象力或幻想能力有关。而且，下一节我们要讨论的这些追求、目标和情绪会很大地受到环境压力的影响，即在当下的社会环境中感知来自他人和社会文化所赋予的更广泛的期望，他们与活动或任务相关的状态都会随时改变。

orientations, problem-solving strategies, and heuristic processes). In one example, Schorr et al. (2010a) used the Don't Disrespect Me structure and other engagement structures to understand the interactions of students criticizing and angrily defending mathematical ideas during a video-recorded classroom episode.

Environmental Press

Social psychologists have long made use of the concept of *press*—that is, the idea that the social environment imposes situational constraints affecting the individual's pursuit of the fulfillment of psychological needs (Murray, 2008). In a mathematics classroom, press may include in-the-moment situational demands stemming from the teacher or other students (e.g., a demand by the teacher, a challenge from another student, or a request for help). It may also include the internalized expectations of parents, expectations of conformity to sociocultural classroom norms, and so forth. In group work, assumption of a specific role may be tacitly or explicitly expected of a student by her peers or by the teacher. Affective structures such as mathematical intimacy, integrity, identity, or self-efficacy incorporate such social expectations, and the environmental press influences how such structures affect student engagement in the moment (Epstein et al., 2007; Goldin et al., 2011; Schorr et al., 2010b).

Aspects of Affect Specific to the Mathematical Domain

Many of the ideas described to this point might seem applicable across the school curriculum. Yet much of the research points to issues specific to mathematics. Math anxiety, for example, does not have a counterpart of "social studies anxiety" or "art anxiety," but there is a small but growing body of research on "foreign language reading anxiety" and "science anxiety" (see Saito, Garza, & Horwitz, 1999, and Bryant et al., 2013, respectively). It is worthwhile, therefore, to consider the affective features that distinguish mathematics from other disciplines.

Problem solving has long been central to mathematical activity in schools (Koichu, 2014). Yet mathematical problems other than the most routine involve *impasse*—the experience of not knowing what to do next (Schoenfeld, 1985). Impasse-driven learning, in fact, is a major approach to the study of problem solving. Each impasse encountered in the solution of a meaningfully complex problem has associated with it some emotional appraisal on encountering the impasse and an emotional response stemming from resolution (or absence of resolution) of the impasse. These emotions may include curiosity and heightened interest when encountering the roadblock or frustration, annoyance and anger, or relief and happiness depending on the outcome of the students' efforts (e.g., D'Mello & Graesser, 2014; Goldin, 2000; Op 't Eynde et al., 2007). How such emotions are handled, that is, the student's meta-affect, contributes to or impedes patterns of engagement, since specific emotional feelings also influence strategic choices.

Other discipline-specific features of mathematics include (see Goldin, 2014) the frequent disconnect between procedural knowledge and conceptual understanding; the hierarchical organization of the school curriculum, where introduction of a new topic typically presupposes prior mastery of prerequisite concepts and skills; and the sequence of epistemological or cognitive obstacles associated with increasingly abstract concepts.

Some socially prevalent beliefs (Handal, 2003; Leder et al., 2002; Maasz & Schlöglmann, 2009) that attribute mathematical success to inherent ability or genius. Some beliefs depict mathematically successful students as unattractive "nerds," or stereotype mathematics as a domain for boys, Whites, or Asians. Such beliefs influence how success in mathematics is valued and interpreted in the personal thoughts of the student and in classroom social relationships (Yeager & Dweck, 2012).

To sum up, the affective complexity of in-the-moment engagement reflects the fact that students have distinct personal, longer term interests, goals, and utility beliefs encoded in their emotions. Students bring with them different habits of mind, including self-regulatory strategies, values, attitudes, and beliefs about mathematics. Moreover, they have mathematics-related as well as general academic self-concept structures, self-efficacy structures, and personality traits. On any particular occasion, these trait-like structures contribute to the appraisal of the engagement opportunities students are experiencing, generating a variety of desires, goals, and anticipatory emotions. Some of these may pertain to mathematics and its learning, some to the teacher, some to one or more peers or to others important in their lives, some to the perception of their status or ability, and some to flights of imagination or fancy far removed from the immediate classroom scene. Moreover, as discussed in the next section, these desires, goals, and emotions are influenced heavily by the environmental press—the perceived expectations of others in the immediate social environment and the broader social expectations of their social and cultural milieu. Their activity- or task-related states change from moment to moment.

As cognitive appraisal theory suggests, while doing mathematics, students experience emotional feelings that range from curiosity to frustration, from anxiety

认知评价理论认为，在数学学习过程中，学生的情感体验伴随着可能的一些生理或心理反应，从好奇心到挫折感、从焦虑或害怕到开心和满意以及从自豪到尴尬或痛苦的羞耻感。学生会感知自己的感觉、感觉的适当性、与自己相关的感觉以及其他人的感觉。他们会采用各种复杂的策略，包括问题解决策略和逃避管理这些情感体验的技巧，并为他们实现积极的自我追求而提供行为指导。学生在以下行为中都用到了这些策略，如：写作、举手、大喊、看天花板、与其他同学争论、把头枕在胳膊或桌子上。在活动开始和进行之中，复杂的社交互动便会发生，比如：提供帮助、挑战、对同伴表示出赞成和不赞成、教师的教学和干预等。

然而，以上提到的这些方面还很少在数学课堂上被系统地研究过。为了充分研究数学情感结构对数学活动的评估和评价作用，研究者应该同时关注暂时的和长期维持的特质类情绪、情境兴趣和个人兴趣以及数学社交活动和社交规范。

作为社交互动和社交机会的参与

我们从以下四个主题来讨论数学参与的社交维度：（1）学校的数学学习是一种需要学生去感知别人希望他们如何参与的社交努力；（2）关系与归属是数学参与经历中的维度；（3）学校和课堂社交情境为学生的数学参与提供了机会；（4）在课堂之外更大的社会文化背景中，学生的数学参与也会发生。前两个主题指的是，学生的参与和学习是分不开的，而后两个主题则提出了参与的可塑性和复杂性。如果参与既具有情境性又具有可塑性，那么探究参与的复杂性之一就表现在如何可以将其放在某个情境中。

作为社交努力的数学学习

学生数学参与的机会受他们对课堂社交结构感知的影响。帕特里克、赖安和卡普兰（2007）把弗雷德里克等人（2004）对参与的分类（行为的、认知的和情感的）拓展到包括社交维度：与任务有关的互动。它包括学生在全班讨论时提出的观点和方法、解释他们的想法及理由、在小组合作活动中与其他人讨论其它的方法以及在独自完成课堂作业过程中与他人分享想法或提供非正式的帮助。帕特里克等人调查了602名以白人和中产阶级家庭为主的学生，他们分别来自于6所伊利诺伊州小学5年级的31个数学教学班级，他们考察了学生对课堂环境（包括教师支持、促进相互尊重、促进与任务相关的互动以及学生的支持）的感知与他们参与（包括自我调节和与任务相关的互动）的关系。他们还调查了学生的学习目标及其对学业和社会效能的感觉。该研究从人与人之间与任务相关的互动角度描述了参与的特性：在学生回答问题、解释内容以及与同学分享自己数学观点方面的互动程度。作者通过结构方程模型分析发现，在对个体差异进行控制后，社交环境变量（如，来自老师的情感支持、老师鼓励学生参与到与任务相关的互动过程中，以及来自同伴的学业帮助）与参与变量（愿意参与、使用自我调节策略、与他人互动）有直接相关关系，而且参与变量又直接与成绩相关。他们还发现，动机变量（学习目标、学业和社会效能感）是这些变量中的首要调节变量。

韦伯等人（2014）利用偏相关和逻辑回归对6个南加州的3年级和4年级课堂进行了研究，其结果也同样表明了，如果学生越能参与到同学的数学思考中，那么相应地就越能取得较高的数学成绩。当学生能够加入到其他同学所建议的策略时，不管是提出不同意见提出其它方法，或是加入了进一步的细节，又或者是利用之前提到过的想法提出另一种替代方法（比如指出这个替代方法比之前的好或者与之前的不同），他就会有较高程度的参与。这些互动方式的形成，部分是依赖于教师对学生互动的认可、促进和跟进而得以形成的。帕特里克等人（2007）和韦伯等人（2014）的研究结果都表明，教师和同伴有助于学生形成参与行为，而且学生的动机对社交规范有动态的影响作用。

学生在数学课堂上的角色是由课堂文化决定的，其中，课堂文化包括学生如何认识他们的参与机会。尽

or fear to pleasure and satisfaction, and from pride to embarrassment or painful humiliation, accompanied possibly by physical as well as psychological responses. Students also have feelings about their feelings, about the appropriateness of their feelings, about themselves in relation to their feelings, and about the feelings of others. They employ diverse and complex strategies, including strategies for problem solving and techniques of avoidance to manage these feelings, and orient their behavior in a direction they perceive as helping to fulfill a motivating desire. Students evidence this in behavior such as writing, raising their hands, calling out, looking up at the ceiling, arguing with other students, or laying their heads on their arms or their desks. Complex social interactions occur as the activity commences and proceeds, including offers of help, challenges, expressions of approval or disapproval from peers, teacher instructions and interventions, and so forth.

As yet very few of these aspects have been studied systematically in mathematics classes. To study adequately the appraisal and evaluative roles affective structures play in mathematical activity, researchers should consider both state and trait emotions, both situational interest and individual interest, and both mathematical social practices and social norms.

Engagement as Social Interaction and Social Opportunity

We address the social dimension of engagement with mathematics through four main themes: (1) learning mathematics in school is a social endeavor involving students' perceived expectations for how to engage; (2) relatedness and belonging are among the dimensions of the experience of mathematical engagement; (3) school and classroom social contexts shape students' opportunities to engage mathematically; and (4) mathematical engagement occurs in a larger sociocultural context beyond the classroom. The first two themes express the idea that engagement is inextricably intertwined with learning. The second two provide insight into the malleability and complexity of engagement. If engagement is both situated and malleable, then one of the complexities to explore is just how situated it can be.

Learning Mathematics as a Social Endeavor

Students' opportunities for mathematical engagement are mediated by their perceptions about classroom social structures. Patrick, Ryan, and Kaplan (2007) extend the characterization by Fredricks et al. (2004) of engagement (as behavioral, cognitive, and emotional) to include a social dimension: *task-related interaction*. This includes students suggesting ideas and approaches during whole-class discussions, explaining their thoughts or reasoning, discussing alternatives with others during small group activities, and sharing ideas or informally giving help during individual seatwork. Patrick et al. surveyed 602 predominantly White, middle-class fifth-grade students from 31 mathematics classes in six Illinois elementary schools, examining relationships between the perceived classroom environment (including teacher support, promotion of mutual respect, promotion of task-related interaction, and student support) and students' engagement (involving both self-regulation and task-related interaction). They also surveyed students' learning goals and their sense of academic and social efficacy. This research characterizes engagement as task-related interactions among people: the extent to which students answered questions, explained content, and shared ideas about mathematics with their classmates. The authors used structural equation modeling, finding after controlling for individual differences that social environment variables (e.g., emotional support from the teacher, encouragement by the teacher to engage in task-related interactions, and academic support from peers) are directly related to engagement variables (tendency to engage, use of self-regulatory strategies, interactions with each other), and the latter are directly related to achievement. Motivational variables (learning goals, academic and social efficacy) are found to mediate the first of these direct relations.

Webb et al.'s (2014) study of six third- and fourth-grade classrooms in Southern California, using partial correlation and logistic regression, similarly demonstrates that higher levels of students' engagement with their peers' mathematical thinking relate directly to stronger mathematics performance. High-level engagement occurred when students added to another student's suggested strategy, either by disagreeing and suggesting an alternative, adding further detail, or providing an alternative approach that explicitly referenced the idea previously posed (e.g., by stating that the alternative approach is better than or different from the original). These ways of interacting were shaped, in part, by teachers promoting them as appropriate and following up on students' interactions. Patrick et al. (2007) and Webb et al. (2014) both show that teachers and peers help shape individuals' engagement behaviors and that individual students' motivations interact dynamically with social norms.

Students' roles in the mathematics classroom are shaped by the classroom culture, including how students

管，积极的自我效能感以及较高的学习目标与高的数学成绩具有相关关系，自我满足与低的学业成绩也具有相关性（Rolland，2012），但是，我们想强调的是即时性动机具有很高的变异性，而且，课堂文化对即时性动机的影响作用具有高度复杂性。学生对具体课堂活动目的的认识是影响他们参与性的主要因素。例如，詹森（2009）研究美国小学职前教师对学生参与全班数学知识内容讨论的认识，发现全班讨论的目的是分享形成过程中的思考。就像巴恩斯（1992）所描述的那样，他们对学生的"草稿"进行讨论，因此这些学生在他们思考过程中就会更愿意分享自己的思路，即便是他们不确定自己的思路。相反，那些认为讨论的目的是为了分享已经完善和完整的思路（讨论"定稿"）的学生，会不愿意参与，除非他们已经确定自己的思路是正确的。课堂上与任务有关的互动会随学生对教学目标认识的改变而改变；对目标的认识不仅依赖于课堂文化，还取决于学生的个人目标。

合作以及共享解决策略等行为方式受多种因素的影响而得到改变（如，追求个人目标；协商集体目标；以及在个人和集体的即时性目标中对数学关注的程度；Webel，2013）。詹森（2006，2008）研究了美国某中产阶层乡村学校的7年级两个班级的15名学生，她发现学生们并不愿意参与到批判同伴解决办法的这种可能富有成效的对话当中，学生这种不参与行为与其如何认为怎样的课堂讨论是适当的信念是一致的。虽然，有的时候学生是按照教师认可的标准进行参与，但是，学生可能有自己的参与理由。利文森、提罗什和撒米尔（2009）对以色列二年级的5个班学生的研究中发现，一个学生遵循老师要求给出理由时，他的目的可能只是对数学内容给出一长串的解释，而不是对数学的意义进行交流。这个案例体现了另一种参与的复杂性，因为，学生们会根据自己对参与的社交目标的认识来评价、比较和调整自己的行为。

支持数学学习的人际关系和归属感

青少年的归属感和社交支持感都与其学业动机相关联（Goodenow，1993）。近来研究表明，归属感和自我效能感及自信心等构成的社交系统对学生在数学活动中表现出的动机和毅力起着非常关键的作用。与学校学习体验相关的这些感觉提高学生的兴趣和欣赏能力，并且，还提高学生的自我调节和有效的元认知能力（Lawson & Lawson，2013）。在数学课堂上至少迎合一部分的社交需要将能促进学生有效参与。

学生的参与程度取决于他们感到舒适的社交风险水平。有些行事谨慎的人看似有充分的参与，但实际上却是在规避风险，就像前面参与结构中提到过的不惹麻烦和伪参与模式（Goldin 等，2011）。而且，其他人则在还不确定这个想法是否正确的情况下，尝试冒险并且分享自己的想法。詹森（2008）的研究显示，那些害怕在课堂讨论中出错的学生，却愿意大声回答程序性的问题，因为，他们感觉回答这类问题要比回答概念性问题的风险更低。还有一些学生则愿意冒险与同学交流自己的想法。考斯科（2012）观察了美国东南部某高中10节课，研究了学生在对话过程中用含糊的表达来获得他人对自己观点的明确反应。研究发现，学生越是含糊其辞，学生回答问题的自主权（感受到可以掌控某种行为的开始和持续状态）越高，他们也越愿意展示自己的想法，即便这些方法可能是不正确的。

教师的支持与初中生的学校学习兴趣有正相关性（Wentzel，1998）。这个年纪的学生能够将教师的教学选择（如数学任务）视为是对自己努力的关心（或不关心），甚至是教师在不明显或无意识状态下的关心（Jansen & Bartell，2013）。赫齐格（2004）研究了来自美国高校数学专业6名女博士生，发现数学精英即便在缺乏老师支持下，也会觉得自己不适合于自己的研究项目组，而且，与指导教师的关系疏远或不好也会影响坚持还是离开自己博士研究项目的决定。

同伴的关系也影响学生的参与。詹森（2006，2008）发现，7年级学生的社交目标，如：帮助同伴，与其学业目标保持一致性。然而，并不是所有的同伴都想要给予或得到学业上的帮助，而且，同伴文化可能有助于学生追求学业成就，也可能毫无帮助作用（Lawson & Lawson，2013）。弗兰克等人（2008）基于青少年健康与

perceive their opportunities to engage. While positive self-efficacy and higher learning goals are associated with higher mathematical performance, and ego goals are associated with lower academic achievement (Rolland, 2012), we stress that in-the-moment motivating desires are highly variable, and the influences of classroom culture far more complex. Students' perception of the purposes of particular classroom activities is a major influence on the nature of their engagement. Jansen (2009), for example, studied prospective U.S. elementary teachers' perspectives on participating in whole-class discussions in mathematics content courses. She found that students who perceived that the purpose of whole-class discussions was to share in-progress thinking, or exploratory "rough draft" talk as described by Barnes (1992), were more willing to share their thinking, even if they were not certain they were correct. In contrast, students who perceived the purpose of discussion to be sharing complete and finished thinking ("final draft" talk) tended to hesitate to participate unless they were certain they were correct. The task-related interactions in a class change as the perceived goals of instruction change; the perceived goals depend not only on classroom culture, but also on students' individual goals.

Behaviors such as cooperation and sharing solutions vary along multiple dimensions (e.g., the pursuit of individual goals; the negotiation of collective goals; and the extent of the mathematical focus of in-the-moment goals, both individual and collective; Webel, 2013). In a study of 15 seventh-grade students in two classrooms from a U.S. middle-class rural school, Jansen (2006, 2008) found that students disengaged from potentially productive discourse, that is, from critiquing peers' solutions, in ways consonant with their beliefs about what counts as appropriate during class discussions. When students do participate according to teacher-approved norms, they may do so for their own reasons. In a study of two grade 5 classrooms in Israel, Levenson, Tirosh, and Tsamir (2009) described a student conforming to the norm of providing elaborated justification but stating that the purpose was to provide a longer explanation rather than to communicate mathematical meaning. Such cases illustrate another complexity of engagement, as students assess, compare, and adjust their behavior to what they perceive the social purpose of their engagement to be.

Relatedness and Belonging as Supports for Mathematics Learning

Feelings of belonging and a sense of social support have long been associated with adolescent academic motivation (Goodenow, 1993). More recently, the importance of belonging and the extent to which the social system is perceived as supportive of a student's self-efficacy and confidence have been shown to be especially critical for motivation and perseverance in mathematics activity. Such feelings connected to the experience of learning in school enhance interest and appreciation and increase self-regulation and effective metacognition (Lawson & Lawson, 2013). Meeting at least some social needs in the mathematics classroom can enable students' productive engagement.

Engagement varies depending on the level of social risk students feel comfortable taking. Some play it safe by posing as if participating fully but actually avoiding risk—patterns captured in the engagement structures Stay Out of Trouble and Pseudo-Engagement (Goldin et al., 2011). Others take intellectual risks and share their thinking even when unsure. In Jansen's (2008) study, students who seemed threatened by potentially being incorrect during whole class discussions nevertheless participated aloud in response to questions about procedures, which may have felt less risky to them than questions about concepts. Other students take risks that encourage interaction of peers with their thinking. Observing 10 periods of a high school class in the southeastern United States, Kosko (2012) studied which students hedged their statements during discourse so as to implicitly invite responses to their ideas. More hedging was done by students reporting higher autonomy (perceived control over initiating and maintaining particular behaviors), who are more comfortable with revealing their thinking even if it may be incorrect.

Support from teachers is positively associated with middle grade students' interest in school (Wentzel, 1998). Students of this age perceive teachers' instructional choices, such as mathematical tasks, as efforts to care for them (or as evidence of lack of care), even when teachers are not overtly or consciously enacting care (Jansen & Bartell, 2013). Herzig (2004), in a study of six female mathematics PhD students in a U.S. university, found that even the mathematically elite perceived that they did not fit in their graduate program's department when teacher support was felt to be lacking. Limited or negative relationships with faculty mentors affected these students' decisions to persist in or leave their doctoral programs.

Relationships with peers also influence students' engagement. Jansen (2006, 2008) found that seventh-grade students' *social goals,* such as helping peers, are held concurrently with their *academic goals.* However, not all peers want to give or receive academic help, and peer cultures may or may not be supportive of the pursuit

学业成绩研究中的课堂记录数据，采用一种新的网络算法，确定了78所美国高中学生（约15 000名）在当地所处的水平。他们发现，那些选择高层次数学课程的女生受到选择同样课程的女同学的影响。

总之，教师和同伴都能对学生的恰当行为提供解释理由。学生在选择如何参与数学学习的时候会留心观察权威数据和同伴群体，并且经由社交暗示影响自己的动机和情感参与。

社交情境形成参与机会

正如格雷索菲（2009，第361页）所说，"要理解学生为什么会消极或积极地参与到数学学习中，就需要考察他们当下和未来所参与的情境"。数学课堂中的一些广为流传的情境模式可能会降低学生的参与度。学生从小学过渡到初中阶段的那些研究表明，在学校和课堂上，改变学生之间以及学生与老师之间的关系，学生一般都表现出消极参与或不愿意参与（例如 Eccles 等，1993；Roeser, Eccles, & Sameroff, 2000；Wigfield, Eccles, & Roderiguez, 1998）。在这个强调自我的时代，个体在社会交往中的互相比较也会随之增多，教师控制多了，学生的选择就少了，这些都与个体强调自我的需要相冲突。服从导向的参与机会和以严厉的权威为基础的学校做法已经被认为是孤立的、疏远的或压抑的（Lawson & Lawson, 2013）。当教师与学生的互动越来越有距离感的时候，就无法满足培养青少年与手握权利的大人之间的紧密关系的需要了。

广义社会文化情境中的数学参与

学生与学校和学习内容之间的常规互动方式早在家庭文化背景中就已经建立了。而且，在学校学习过程中，一些来自校外的影响作用会被高估，而其它的社会文化因素的影响作用则会被低估。研究表明，学生的参与受经济水平、父母教育水平和参与、民族、种族和社区特征，以及个体特质的影响；参与力则受家庭文化与学校实践方式之间一致性程度的影响。

卢宾斯基（2000）的研究指出，美国中西部的某个课堂使用的是以问题为中心的课程，社会经济地位（SES）低下的学生喜欢老师提供更多的指导，但学校提供的课程和教学却与学生的需求不一致。这个班级中社会经济地位低下的学生偏爱于用情境化的方式学数学，这种情境化的方式能让他们积极参与其中。尽管已有研究证明了基于问题的教学方式是成功的，但对教师而言如何把重要的数学思想置于活动之中存在一定困难，这并不是说基于问题的课程和教学不适用于社会经济地位低下的学生（Silver & Stein, 1996）。然而，要理解学生对学校数学教学方法的不同体验，就必须要关注他们的校外经历。

贝里（2008）以8名成功的非洲裔美国初中男生为研究对象，考察了他们是如何形成积极的数学认同感的。他的案例重点描述了在家庭成员支持的背景下，学生是如何参与到一些具体情境中的，如：大学预科学习计划、宗教团体以及运动项目中，如何以一种人文的恰当方式形成各种特殊身份认同感。这一研究冲击了非洲裔美国男生低期望以及因为成人的信念学生拒绝参与机会这一刻板印象（Aronson 等，1999；Leonard & Martin, 2013；D.Martin, 2000）。伦纳德和马丁（2013）提供了反驳论述，通过深入的历史分析，提倡数学教学要考虑到非洲裔美国学生在精神、历史及文化方面的认同感。这一研究让我们了解到一些影响这些学生数学认同感的变量，以帮助他们在自己家庭环境和学校所提供环境不一致的情况下的学习。

下面我们要讨论有助于形成学生有效数学参与的社交因素。首先，从任务层面开始，讨论即时性参与是如何在任务要求中形成的。然后，我们继续讨论课堂互动及教师的支持和期望。最后，我们要讨论学校情境和更宽泛的社会文化情境。

数学任务。有效参与包括同伴合作，因此，任务就需要以培养学生之间互动的方式来设计（和靠近）。比如，设计可以用多种方法解决的问题使学生想出有价值的想法（Boaler & Staples, 2008），而用仅有一种解决方法的问题来奖励第一个完成任务，并且希望成为小组中提出数学想法最多的学生。正如前文提到的，越多的开

of academic success (Lawson & Lawson, 2013). Frank et al. (2008), using course transcript data from the Adolescent Health and Academic Achievement Study, employed a new network algorithm to identify local positions in 78 U.S. high schools (about 15,000 students). They found that high school girls' choices to take higher level mathematics courses were influenced by their female peers also doing so.

In short, both teachers and peers provide cues for appropriate behavior. Students look to authority figures and to their peer group in choosing how to engage in mathematics, and they filter their own motivation and affective engagement through social cues.

Social Contexts Shaping Opportunities to Engage

As Gresalfi (2009, p. 361) writes, "understanding why students are passively or actively engaging in mathematics requires examining the contexts in which they have been, and will be, engaging." Some widespread contextual patterns in mathematics classes contribute to diminished student engagement. Studies of the transition from elementary school to middle school indicate that schools and classrooms become generally less engaging or welcoming to students due to changing relationships among students and between students and the teacher (e.g., Eccles et al., 1993; Roeser, Eccles, & Sameroff, 2000; Wigfield, Eccles, & Roderiguez, 1998). At this time of heightened self-focus for adolescents, practices that involve social comparison often receive more emphasis, teacher control increases, and students' choices decrease, conflicting with adolescents' increasing need for autonomy. Compliance-oriented engagement opportunities or rigid, authority-based school practices have been identified as isolating, alienating, or oppressive (Lawson & Lawson, 2013). When teachers become more distant in their interactions, adolescents' need to foster stronger relationships with adults in authority is unmet.

Mathematical Engagement Situated in Wider Sociocultural Contexts

Students' normative ways of interacting with school and academic content become established early in their home cultures. In school, some ways of acting outside of school are valued, and others devalued. Variation in student engagement is influenced by economic class, parental education and involvement, race, ethnicity, and community characteristics, as well as individual traits; the productivity of engagement is influenced by the alignment of home culture and school practices.

Lubienski (2000) reports that in a midwestern U.S. classroom using a problem-centered curriculum, lower socioeconomic status (SES) students prefer more teacher direction, but the curriculum and pedagogy does not align with these preferences. The lower SES students in this class tend to approach mathematical ideas in a contextualized manner, which leads to their active involvement but poses challenges in locating the significant mathematical ideas in the activities. It does not follow that problem-based curriculum and instruction are inappropriate for lower SES students; prior research documents successes with problem-based instruction (Silver & Stein, 1996). However, understanding variations in how students *experience* approaches to school mathematics must include attention to their experiences outside of school.

Likewise, Berry (2008) studied eight successful African American middle school boys to examine how positive mathematical identities develop. His cases highlighted how, with support from family members, students engaged in contexts such as precollege programs, religious communities, and athletic programs, developing various efficacious identities in a culturally appropriate manner. This work counters the stereotype of lower expectations of African American boys and the pattern of Black students denied access to opportunities due to adults' beliefs (Aronson et al., 1999; Leonard & Martin, 2013; D. Martin, 2000). Leonard and Martin (2013) offer further detailed counternarratives and extensive historical perspective, advocating instruction in mathematics that takes account of African American students' spiritual as well as historical and cultural identities. Such work leads to an understanding of malleable factors for students whose home environments and school environments may not be well-aligned.

We next turn to a discussion of social factors with the potential to foster productive engagement among students of mathematics. We begin at the task level, discussing how in-the-moment engagement is shaped by a task's requirements. We then move to classroom interactions and teacher support and expectations. Finally we discuss school contexts and larger sociocultural contexts.

Mathematical tasks. As productive engagement includes peer collaboration, tasks need to be designed (and approached) in a way that fosters interactions among students. Problems that can be solved in multiple ways allow different students to contribute relevant and valuable ideas (Boaler & Staples, 2008), whereas tasks yielding to only one solution method reward students who are first to complete the task and who subsequently may do most of the mathematical thinking for their group. As noted earlier, more open tasks also tend to evoke more

放式问题会让学生对数学产生越多的乐趣和兴趣。即便是刚开始对开放式问题有抵触的学生也会变得喜欢挑战（Boaler，1998；Boaler & Staples，2008）。

埃斯蒙德的研究（Esmonde，2009；Langer-Osuna & Esmonde，2017，这套书）表明，研究者不仅要考虑到学生是否参与，还应该关注学生是如何参与以及任务要求所能提供给学生公平参与的程度。在2009年的研究中，她选择了来自加州某高中同一数学教师教授的三个班学生。当学生们一起完成某个小组测验时，他们更有可能参与到"不对等的帮助"中，如听从某个能人；但是当他们一起准备一个演讲的时候，他们更有可能参与到与同伴的合作互动中。后一种任务设计可以促进学生公平地分享观点，而前一种任务设计只能帮助需要帮助的学生获得在小组中分享的知识。沙利文等人（2014）研究了来自澳大利亚的34个小学课堂和15个中学教师的课堂所使用的挑战性的任务。他们发现高挑战性任务对提高全体学生富有成效的参与来说是最佳的方式，但是课程设计者需要建构有用的操作性提示（如，减少步骤数、简化复杂的数字、或使用多种表述方法）来确保那些感到过于困难的学生也能够很好地参与。这首先取决于教师是否将所教数学定位在学生学的水平，使任务使用的数学原理很清晰，并且提出合理的活动建议（如，"有的时候，解决乘法和除法问题就是在寻找其中的模式，第一个任务关注的焦点在于识别哪些数字经过乘法运算其答案总以0结尾"）。

教师能够从挑战性、时间分配和学生回答完整性等方面评估学生的学习情况。例如，霍恩（2007）提供了一个关于加州某教师的案例，该教师描述了与完成任务非常"迅速"的学生一起工作的挑战。该教师的同事们对快速完成任务是否应该得到很高的分数产生质疑。他们认为，这类学生可能会因为没有参与到不同推理方法或没有从多个视角思考问题甚至有可能不够仔细，从而错过了一些学习的机会。当然，该教师已经有了一些有效的方式来评估学生的学习和参与，因为她并没有在学生内部讨论学生参与的困境，而是给学生提供了更多具有挑战性的任务。她的同事们认为她应该重新思考与"快速"完成任务相关的学生状态，而且用多种方法来培养学生的数学能力。对学生的期望可以在教师与学生的谈话中暗示出来并影响学生，教师通过重新认识学生，能够为更多学生提供有效学习和参与的空间。

社会支持和课堂互动。教师的社会性情绪支持对学生的学业成绩、自我效能感、学习兴趣、亲社会行为和目标具有积极的作用（Rolland，2012）。哈肯伯格（2005，2010）对4名6年级学生展开了一项为期8个月的教学实验，研究表明学生们愿意承受在面对挫折和认知障碍时的情感耗竭，如果在这些时刻他们得到了教师相应的支持并且能够克服或享受挑战。这些结果提醒我们要去辨别数学关怀，即教师不仅要对学生掌握的数学知识，还要对他们的情感状态给予反馈。教师、课程内容和学生之间的互动关系被视为是数学关怀的表现。巴蒂（2013）对4年级的一个课堂进行了一个案例研究，该班级几乎所有学生都是拉丁裔和非洲裔美国学生。他将教师互动分为四个维度，每个维度都可以或支持或阻碍学生的参与：（1）强调学生的行为，（2）构建学生的能力，（3）认可学生的贡献，（4）留意学生的文化和语言。

教师对学生的期望：能力的支持。教师对哪些或哪类学生是能胜任（或不能胜任）数学学习的会表现出相应的期望。教师期望的良好数学参与是：例如，能快速得到正确答案或理解同学的解法。这样的期望给学生提供了不同的参与机会，前者，学生能很轻易地参与到算法计算过程中，后者，则利于学生互相交流想法（Gresalfi，2009；Jansen，2012）。

无论学生是否相信他们会越来越有能力学习数学，无论数学能力被认为是可塑的或是不可改变的，这些都可以用对学生参与和成绩结果有影响的心态来诠释（Dweck & Leggett，1988）；这些心态可以通过教学来影响。布莱克韦尔、特兹纽斯基和德韦克（2007）对正在过渡到7年级的来自纽约州的373名公立学校学生进行了研究，他们在学年之初对

enjoyment and interest in mathematics. Even students who initially resist open tasks come to appreciate the challenge (Boaler, 1998; Boaler & Staples, 2008).

Esmonde's work (Esmonde, 2009; Langer-Osuna & Esmonde, 2017, this volume) indicates that researchers must consider not only whether but how students engage and the degree to which task requirements support equitable engagement. In her 2009 study, students from three classrooms in a California high school, all taught by the same mathematics teacher, were more likely to engage in "asymmetric help," such as deferring to an expert, when working together on a group quiz but more likely to engage in collaborative interactions when working together on a presentation. The latter task design promotes equitable sharing, whereas the former only positions the students who ask for help to obtain knowledge to share with the group. Sullivan et al. (2014) studied challenging tasks introduced in 34 primary classrooms and the classrooms of 15 secondary teachers in Australia. They found that high challenge is optimal for increasing overall student productive engagement, but curriculum designers need to build in enabling prompts (for example, reducing the number of steps, simplifying the complexity of the numbers, or varying the forms of representation) to ensure the productive engagement of students for whom the challenge is above optimal. This is predicated on teachers first situating the mathematics with the students, making the mathematical rationale for the task clear, and suggesting appropriate activity (e.g., "Sometimes solving multiplication and division problems is about finding patterns. In the first task, the focus is on identifying numbers that when multiplied have an answer that ends in 0").

There is power in teachers being aware of their assumptions about students in relation to task requirements such as challenge, time allotted, and thoroughness of response. Horn (2007), for instance, presented the case of a teacher in California who described the challenge of working with "fast" students who quickly complete tasks. The teacher's colleagues questioned whether completing the task quickly should be given high status. They noted that such a student might miss some opportunities to learn by not engaging in various ways of reasoning, not seeing the task from multiple perspectives, or perhaps not working carefully. The teacher already had some productive ways to think about students' learning and engagement because she had not essentialized the engagement dilemma within the student, focusing instead on finding more challenging tasks. Her colleagues challenged her to reconceptualize aspects of status associated with "fast" work completion and to work on multiple ways of developing competence in mathematics. Expectations of students can be communicated implicitly by how teachers frame students in their talk, and reframing how teachers think about students can open spaces for more students to learn and engage productively.

Social support and classroom interactions. Teachers' socioemotional support relates positively to students' academic achievement, self-efficacy, interest in learning, and prosocial behaviors and goals (Rolland, 2012). Hackenberg's (2005, 2010) work with four sixth-grade students in an eight-month teaching experiment shows that students are willing to bear periods of emotional depletion during times of frustration or cognitive roadblocks if these moments are balanced with teacher support and a sense that students can overcome or enjoy the challenge. These findings led to identifying *mathematical caring relations*, ways in which the teacher responds not only to what students appear to know about mathematics, but also to students' emotions. Relational interactions among teacher, content, and students are then seen as acts of mathematical care. Battey (2013), in a case study of a fourth-grade classroom composed almost entirely of Latino and African American students, identifies four dimensions of teacher interaction, each of which can be enacted in ways that support or constrain students' engagement: (1) addressing behavior, (2) framing students' ability, (3) acknowledging students' contributions, and (4) attending to students' culture and language.

Teachers' expectations of students: supporting competence. Teachers communicate expectations about which students or kinds of students are mathematically capable (or not capable). What counts as engaging competently with mathematics is established by teachers' expectations: for example, obtaining a correct answer quickly or understanding the solutions of peers. Such expectations afford different opportunities for engagement—students can engage more readily with computational algorithms in the former case and more readily with each other's thinking in the latter (Gresalfi, 2009; Jansen, 2012).

Whether or not students believe they can become more competent and whether capabilities in mathematics are understood to be malleable or fixed have been described as mindsets with consequences for engagement and achievement (Dweck & Leggett, 1988); such mindsets can be influenced through instruction. In their study of 373 public school students in New York City transitioning to seventh grade, Blackwell, Trzesniewski, and Dweck (2007) assessed students' mindsets at the beginning of the year, along with other motivation-relevant variables. Although students with "fixed" and "growth" mindsets entered seventh grade with equal prior math achievement, after one semester the mathematics grades of

学生的心态及与动机相关的其它变量进行了评估。尽管有"固定型"和"成长型"心态的学生在进入7年级时候他们的数学成绩相同，但在一学期之后，"成长型"心态的学生数学成绩更好。但是，心态并不是固定不变的个性特质。在该研究中，有一个小组的学生被告知，当他们在用大脑的时候，大脑会变得更强大，而且学习能够在大脑中形成新的连接，这些连接可以用在他们的学校学习中。接受这种干预的学生数学成绩都有所提高。教师可以帮助学生用一种成长型的心态对待数学能力，即通过认识、命名以及标记发挥数学能力的种种办法来提高学生参与状态。作为一种复杂教学的原则（Cohen, 1994），这些状态上的提升变化能够让更多学生认识到，当他们和同伴表现出自己的能力时，会让更多学生被认为同样有能力，并且能够提供给更多学生机会去感受自己具有数学能力，进而产生更富有成效的参与和更深层次的理解。当学生参与到有助于他们探究多种解决方案的开放性问题中时，教师可以强调"擅长数学"的表现是，提出好的问题、对任务有很好的解释，以及思考其他同学的方法（Boaler & Staples, 2008），这种方式为我们提供了一个不同于仅仅关注快速得出正确答案的能力的视角。

超越课堂：从学校层面支持学生的参与。学生数学参与的发展不仅仅只限于在一节课上的实践。学校课程设置进程结构的方式、提供给学生发展和转变的机会和学生是否及如何利用这些机会等，都会促进或限制学生在数学上对自己的认同感（Horn, 2008）。对某一个教师来说，仅在课堂上为一批学生创造参与机会也是不够的，除非在课前和课后也提供这样的机会。詹森、赫柏林-艾森曼和史密斯（2012）的研究表明，在关于贝瑟尼的案例研究中，学生们是通过对比从初中过渡到高中后的差异来建立自己的数学认同感的。学生更喜欢并采用与某情境相符的参与方式，而且会坚持这种方式，即便是下一堂课的情境与之前的不一致。正如教师与学生的互动模式一样（例如Battey, 2013），这种认同感可能能够产生富有成效的数学参与，也可能产生适得其反的效果。

教师对学生动机和参与的理解。影响学生参与的很多可塑性的任务和课堂因素都依赖于教师的知识以及教师对动机和参与的理解。克拉克等人（2014）在对来自美国的259位小学高年级教师和184位初中教师的数据进行的二次分析中发现，前者并没有像后者那样相信，教师应该给学生提供有困难的任务，这也是学生学习数学的一部分。不过，那些有着很强数学知识的小学高年级教师比有较弱数学知识的教师更相信，在提供干预之前学生应该自己先努力。J.A.米德尔顿（1995）发现，他所研究的5位威斯康星州初中学校的数学教师不了解学生的动机的本质及其与数学参与的关系。教师的信念会影响他们的教学，也就是说，那些相信实用性可以激发学生动机的老师更倾向于选择真实生活情境中的问题。

虽然我们不清楚帮助数学教师思考将动机应用于他们教学实践的大样本研究的结果会怎样，但有案例研究提供了一点证据。例如，来自美国中西部某一所初中的6位教师在参与了为期一年的专业发展实践中了解了动机的原理，使得他们比之前更少责备自己的学生缺乏动机或参与，并且愿意承担通过教学影响参与的责任（Turner, Warzon, & Christensen, 2010）。通过具体的教学实践，教师关于激励学生的知识也得到了提高。在开展一个旨在指导学生在具体情境中探究数学概念的教学之后，教师转变了自己对于什么能够刺激学生参与到更具挑战性任务的信念（J.A. Middleton, 1999）。事实上，目睹学生良好的数学参与会对教师的信念有很大的影响作用。

但是，教师对学生参与的解释技能存在一定差异。韦杰（2014）研究了威斯康星州13位小学数学教师，发现教师所注意到的学生参与模式与学生实际之间存在很大差别。初中数学教师中，有色人种教师要比白人教师更能意识到学生的情绪（Clark等，2014）。

教师是课堂参与过程中社交活动的核心，因此，教师对数学参与的复杂性理解，包括它的社会维度，对于培养学生的即时性参与起着非常重要的作用。

讨论与展望

本章我们通过案例说明研究动机、情感和社交互动的关键是这些因素产生最多作用的特定时刻——数学

students with "growth" mindsets were higher. But mindsets are not permanent personality traits. In the same study, a group of these students were taught that brains grow stronger when they are used and that learning leads to forming new connections in the brain and how this can be applied to their schoolwork. Mathematics performance improved among the students who received this intervention. Teachers can help students view mathematical competence through a "growth" lens by explicitly raising the status of various ways to engage—recognizing, naming, and labeling ways of enacting mathematical competence. Such status-raising moves, a principle of complex instruction (Cohen, 1994), allow more students to recognize when they and their peers are enacting competence, position more students as competent, and provide more students with opportunities to *feel* competent, leading in turn to more productive engagement and deeper understanding. When students engage with open-ended tasks in a setting that promotes exploring multiple solution methods, teachers can highlight that "being good at mathematics" involves asking good questions, explaining work well, and thinking about others' solutions (Boaler & Staples, 2008), offering a view of competence different from getting the right answer quickly.

Beyond one classroom: supporting engagement at the school level. The development of students' mathematical engagement goes beyond their single-classroom experiences. The ways in which course tracking structures are created within schools, opportunities students have to move through them or to shift from one to another, and whether and how students take these opportunities, support or constrain students' identities with respect to mathematics (Horn, 2008). It is not enough for one teacher to create engagement opportunities for a range of students, unless classes before and after also provide such opportunities. Jansen, Herbel-Eisenmann, and Smith (2012) show, in their case of Bethany, that students develop mathematical identities by contrasting such discontinuities as they transition from middle school to high school. Students prefer and adopt ways of engaging more aligned with one context than another and may assert those preferences even if the next classroom setting is not aligned with it. As with teachers' patterns of interactions with students (e.g., Battey, 2013), such identity framing can be mathematically productive or counterproductive.

Teachers' understanding of students' motivation and engagement. Much of the malleability of task and classroom factors influencing engagement is dependent upon teachers' knowledge and understanding of motivation and engagement. Clark et al. (2014), in a secondary analysis of data from 259 upper elementary and 184 middle grades teachers in the United States, reported that the former are less likely than the latter to believe teachers should allow students to engage in struggle as a part of learning mathematics. However, upper elementary teachers with stronger mathematical knowledge are more likely than less knowledgeable teachers to believe students should struggle prior to intervention. J.A. Middleton (1995) found that the five Wisconsin middle school mathematics teachers he studied were uninformed about principles of students' motivation and how it relates to mathematical engagement. Teachers' beliefs influence their instruction; that is, those who believe utility motivates students tend to choose tasks set in real-life contexts.

We do not know of a large-scale study of the effects of helping mathematics teachers consider motivational implications in their practice, but there is case study evidence. Six middle grades teachers in a school in the midwestern United States, introduced to principles of motivation through a year-long professional development experience, were less likely than before to blame their students for a lack of motivation or engagement and take more responsibility for influencing engagement through teaching (Turner, Warzon, & Christensen, 2010). Teachers' knowledge of what motivates students also develops by engaging students in particular practices. After using a mathematics curriculum targeting students' exploration of concepts in context, teachers shifted their beliefs about what motivates students to include more challenging tasks (J.A. Middleton, 1999). Seeing students actually engage productively with mathematics greatly influenced teachers' beliefs.

But, teachers vary in their skill at interpreting their students' engagement. Wager (2014), in a study of 13 Wisconsin elementary mathematics teachers' noticing of students' participation, reported great variation in which teachers noticed student participation patterns and which students did or did not participate. Middle grades mathematics teachers of color reported greater awareness of student dispositions than White teachers (Clark et al., 2014).

Teachers' understandings of mathematical engagement in its complexity, including its social dimension, are essential to fostering students' in-the-moment engagement, as the teacher is central to the social affordances of engagement in the classroom.

Discussion and Future Directions

We make a case throughout this chapter that the key focus for the study of motivation, affect, and social interaction should be the very moment when these factors

参与发生的时刻。虽然，通过已有的经验形成了学生的动机和情感结构，并且学生会在社会规范情境中表现他们的信念和态度，但是一般来说过去的经验会错失教育机会。作为教育工作者，我们只能从我们对学生的了解出发，力求从身边既有的事物中创建富有成效的即时性参与。随着时间的推移，我们将会看到我们的努力是否会在提高学生的数学毅力、自我调节能力、学习、亲社会行为和积极且强烈的情感等方面取得成果。

本章讨论的一个主题是状态和特质在参与研究中的区别。我们分析了一系列属于特质变量的研究，如学生对数学的动机取向、个性情感、态度及树立起的信念以及课堂规范。无论样本大小，研究都发现这些变量之间及其与学生数学学业成绩（通常用考试成绩或学业成绩）之间存在明显的中度相关关系。其它方法关注的是学生数学参与的状态或很快会变化的特征。这些包括学生的即时性目标或欲望动机、连续的情感状态、数学活动过程中情感与认知的互动关系以及课堂社交互动关系。这些研究都一致地说明了，在参与活动过程中动态的互动结构下，学生数学参与的复杂性和变异性。

研究操作性定义和特质变量的最佳方法是综合使用收集数据调查法和统计分析方法。采用这些工具便于进行大样本研究，可以得到一般性的实证研究结果——这些结果都是有效的，且都是定量数据。另外，质性的观察方法更多地被用于研究动态的即时性参与。这种方法对小样本的案例研究非常好，目的是对具体情境中的特定学生进行精细分析。该方法需要很大的人力支持。经验取样法为从大样本中获得即时性数据提供了很好的方法，但是这种方法以中断体验参与的过程为代价。

特质变量是一种较为稳定的变量，不易受教师和学校体系的影响。然而，这些变量虽然与成功的数学学习之间是中度相关关系，但是也能说明这些变量具有可塑性，如鼓励学习趋向型取向和内在动机、提高数学的乐趣并降低数学焦虑感、形成一个可以鼓励学生分享想法并尊重其他同学的想法的课堂文化。另一方面，即时性参与更容易受到教师的影响。它包括学生的长期特质和能力的发展过程。

研究者可能会忽视当下发生的事情，而更多的去关注更容易测量的特质变量。之后，他们可能会试图去发现一个控制变量的组合，以便达到可测量的数学教育的改善的目的。在这一方面，有很多优秀的研究都指出了多个相互作用的变量。但是，我们认为这个方法已接近其实用性的极限。变量之间的相关性没必要联系到我们最看重的学习的测量，特别是这些相关性对大部分的变异都不能给出各自或总体性的解释。

另外一种选择是研究即时性数学参与的复杂性。当然，这一工作的困难在于该现象具有交互影响作用，它是动态的、自然发生的且多维度的，而且，在大样本研究中对资金和人力也有较高的要求。虽然如此，我们仍然认为这类研究是非常重要的。关键问题包括：数学课堂参与中，哪些特征和结构具有稳定性？稳定性发生的条件是什么？在什么条件下，参与的各方面是可塑的？如何塑造？

比如，我们前面提到的动态结构（Csikszentmihalyi，1990）及和参与结构"我真的很投入"（Goldin 等，2011）有关的内容。动态的学习过程是即时性参与的一种方式。然而，结构又具有稳定性，因为它是可以被定义的和在不同情况下很多人都会遇到的一种现象。它会在恰当的条件下发生，而且具有长远的影响。因此，在被问到参与的哪些特征或结构是稳定的时，通常指的是在功能或结构上是稳定的，而不是暂时的稳定性。

暂时的稳定性会对个体的耐力或特质有意义。功能上的稳定性指的是，某一个结构会在某个时刻重复发生，而无需持续很长时间。结构上的稳定性是指，无论是即时性的还是长期性的（或二者都有），该结构都会重复发生且特征具有复杂性。在数学参与经常发生的情况下区别功能性稳定和结构性稳定，能够让研究者、教

have the most opportunity for expression—that moment when mathematical engagement occurs. Although antecedent experiences shape students' motivational and affective structures, and students express their beliefs and attitudes in the context of social norms, past experience generally constitutes missed educational opportunities. As educators, we can only take what we know about students and seek to create productive engagement in the moment at hand. Over time, we can see whether our efforts have borne fruit through increased mathematical perseverance, self-regulation, learning, prosocial behavior, and positive and powerful affect.

One of the themes of this chapter has been the state vs. trait distinction in research on engagement. We have seen how one set of approaches focuses on trait-like variables such as students' motivational orientations toward mathematics, trait emotions, attitudes and well-established beliefs, and classroom norms. In large-scale as well as smaller-scale studies, correlations at modest but noticeable levels are consistently found among these variables, as well as between them and measures of mathematics learning—most often, test performance or school grades. Another set of approaches focuses on state-like or rapidly changing features of mathematical engagement. These include students' in-the-moment goals or motivating desires, sequences of emotional states, interactions of affect with cognition during mathematical activity, and classroom social interactions. Such studies consistently describe the complexity and variability of student engagement with mathematics, with the relevant constructs interacting dynamically during engaged activity.

Survey methods of gathering data, together with statistical analyses, serve best for the operational definition and study of trait-like variables. Such instruments are well-suited to large-scale studies aiming toward generalizable empirical results—they are efficient, and findings are quantitative. Qualitative observational methods, on the other hand, are more frequently employed to study the dynamics of in-the-moment engagement. They serve well for smaller-scale case studies, aiming toward fine-grained analyses of specific students in specific situations. The methods are labor-intensive. Experience sampling provides ways to obtain in-the-moment data on larger scales, but at the expense of interrupting the experience of engagement.

Trait-like variables, being temporally stable, are not easily susceptible to modification by the teacher or school system. Nevertheless the correlation of certain variables with measures of mathematical success, albeit modest in size, has lent support to the idea of their malleability—encouraging learning goal orientation and intrinsic motivation, increasing mathematical enjoyment and reducing math anxiety, developing a classroom culture that encourages sharing ideas and valuing other students' ideas. In-the-moment engagement, on the other hand, is more easily susceptible to immediate influence by the teacher. It embraces the *processes* whereby students' longer term traits and capabilities develop.

Researchers may disregard what happens in the moment and focus on more easily-measurable trait variables. They may then try to find a combination of controllable variables to achieve measurable improvement in mathematics education. Some excellent work of this kind points to structures of multiple interacting variables. But we think this approach is close to the limit of its usefulness. The correlations found between and across variables do not necessarily involve the measures of learning we most value and typically do not explain a large fraction of the variance, separately or collectively.

The alternative is to study the complexity of mathematical engagement in the moment. A deterrent to doing this work, of course, is that the phenomenon is interactive, dynamical, emergent, and multidimensional and that its study on a larger scale is expensive and labor intensive. Nevertheless, we see such study as vital. Key questions include, what features or constructs of engagement in mathematics classrooms are stable and under what conditions? Under what conditions are aspects of engagement malleable and how?

As an example, we mentioned the construct of flow (Csikszentmihalyi, 1990) and the related engagement structure I'm Really Into This (Goldin et al., 2011). Flow is a manner of engaging in the moment. Yet the construct has stability, in that it is a phenomenon that can be defined and occurs with many people in many situations. It can be induced under appropriate conditions, and it has longer term consequences. Thus, in asking what features or constructs of engagement are stable, we mean functional or structural stability, not temporal stability.

Temporal stability makes sense directly for enduring or trait-like characteristics. Functional stability refers to a construct that occurs repeatedly in the moment but does not necessarily last for a long period of time. Structural stability captures the conditions under which the construct recurs and the complexity of its features, either in the moment or long term (or both). Identifying functionally and structurally stable recurrent phenomena of mathematical engagement allows researchers, teachers, and students to see them, explore them, discover the conditions of their occurrence, and potentially support their recurrence. Tools ranging from enhanced experience sampling methods to concurrent multiple-video

师和学生观察、探究和发现二者发生的条件以及潜在的支持因素。从增强经验取样方法到同时使用多个录像设备收集数据等工具都能够用于研究即时性数学参与，但是这类研究涉及的不仅仅是数据收集工具。还需要大量的进一步的理论研究，并辅以精细观察和分析法，来区别稳定的、复杂的和与情境有关的结构。我们需要那种能有助于我们从现象中找到有效且有用的特质的研究，包括有助于培养学生丰富而有效的数学参与结构的那些条件。

我们非常需要知道怎样的课堂环境、课程和课堂社会规范有助于促进学生有效的即时性参与。参与的可塑性意味着，教育者可以通过改变学习条件来鼓励学生进行他们所希望的参与。教育工作者对参与有一定了解后，一些研究结果也能帮助他们判断影响不同学生在不同情境下参与的最有效的方式。

未来研究应该提供视角去思考如何同时且动态交互地考虑动机、情感和社会互动这些要素，这有助于教师寻找干预和支持学生的情感参与以及认知参与的机会。由于参与的复杂性，帮助教师参与学生活动并不像确定有效干预方式并且将其教给教师那么简单。因此，一条有效的研究主线应该是：（1）认识功能性稳定结构，（2）帮助教师理解这些结构，（3）确定引起这些结构产生的条件，（4）为了促进这些结构的建立，给教师介绍这些结构所发生的条件，（5）调整激励的条件，使其适用于不同学生和不同情境。

我们还需要研究教师如何提高对参与现象的理解。我们选择哪些参与结构可以帮助教师有更清醒的认识？哪种结构有助于教师努力去注意并看到课堂中参与所起的作用？例如，基于特纳等人（2010）的研究，我们有必要问，哪些方式能够为教师提供机会，让他们从有关动机的研究中学到更多？进一步，我们提到一个关键问题：哪些方式可以为教师提供机会，去理解和应用他们从数学动机研究中所学到的相关内容？

未来研究还应该考察，教师怎样应用新的认识去应对和解释学生的参与，包括识别挑战和提供指引。当教师要应对学生的参与时，他们可能对分析每个学生的心理状态感到有一定的压力。然而，学生如何参与到当下情境中的模式是存在的；参与结构的概念是描述它们的其中一种方式（Goldin等，2011）。用于描述学生典型的相互作用方式的框架能够为教师提供不同的方法去看待和解释学生的数学参与。那么，我们怎样帮助教育实践者发现相应的干预机会呢？教师对他们所观察到的学生参与状况又应有怎样的反应呢？

最后还有一件事需要我们思考。当儿童学习勾股定理的时候，她学的不仅是直角三角形的几何知识，还有数学术语、直角三角形的直角边与斜边的代数关系，以及一些（可能的）非正式和正式的证明。她也知道，研究代数模型令人沮丧，但很有趣，而且有潜在的回报。她了解到约瑟知道很多平方根而且可以帮助她们小组进行估算。她学到了工程师可以用这个定理去决定合力的大小。她学到了，当自己感到沮丧的时候尝试一种新的方法通常是有用的，并且教师会帮助她，给予她关心。她了解到了自己的数学能力并经历其发展。她学到了在具体数学情境中要与其他人互动。这些经历最终都会变成她自己的数学。她也知道没有其它的方式能够产生这种感觉。于是对即时性参与的研究不仅丰富了我们对动机、情感和社交互动的知识，而且揭示了数学知识的本质和它在学生生活中所起的作用。

captures can support study of in-the-moment mathematical engagement, but such study involves more than data collection tools. Considerable further theoretical work is needed, supported by fine-grained observation and analysis, to identify stable, complex, context-dependent structures. We need research that leads us to find valid, useful distinctions in these phenomena, including the conditions that promote the development of productive, generative engagement structures in mathematics.

We point to the critical need for information that will help us orient classroom environments, curriculum, and classroom social norms toward productive, in-the-moment engagement in students. The malleability of engagement suggests that educators can change the conditions of learning to encourage students to engage as educators would want them to. Educators have some control over engagement, and research findings can aid in making informed judgments about the most effective ways to influence engagement with different students, in different contexts.

Future research should provide insight into how motivation, affect, and social interactions can be taken into account simultaneously and interactively. It should help teachers recognize opportunities to intervene and support students' engagement affectively as well as cognitively. Due to the complexity of engagement, supporting teachers in engaging students is not as simple as identifying productive interventions and teaching those interventions to teachers. Rather, a fruitful line of inquiry would be to (1) identify functionally stable constructs, (2) help teachers understand these constructs, (3) identify structural conditions that occasion these constructs, (4) help teachers introduce these structural conditions in order to promote these constructs, and (5) modify motivational conditions to adapt to individual differences among students and varying contexts.

Studies are needed about how teachers develop understandings of engagement phenomena. How do we pick which engagement constructs to help teachers become more aware of? Which constructs support teachers' efforts to notice and see engagement at play in the classroom? Building on prior work by Turner et al. (2010), for example, we must ask, in what ways can teachers be given opportunities to learn more about research on motivation? Going further, we show in this chapter that it is critical to ask, in what ways can teachers be given opportunities to understand and implement their learning of research on motivation *in mathematics*.

Future studies should examine how teachers apply newly developed awareness to attend to and interpret students' engagement, including identifying challenges and navigating them. As teachers attend to engagement, they may feel pressure to analyze each child psychologically. However, patterns exist in how children engage in the moment; the engagement structures construct is one approach to describing them (Goldin et al., 2011). A framework for describing how students *typically* interact can provide teachers with ways of seeing and interpreting students' mathematical engagement. How can we then help practitioners recognize opportunities to intervene? How do teachers respond to what they notice about their students' engagement?

Here is one last thing to think about. When a child learns the Pythagorean theorem, she not only learns about the geometry of right triangles, mathematical terminology, an algebraic relationship between the side lengths and the length of the hypotenuse, and (perhaps) some method of informal or formal proof. She also learns that examining algebraic patterns is frustrating but fun and potentially rewarding. She learns that Jose knows a lot about square roots and can help their group make estimates. She learns that engineers can use the theorem to help them determine resultant forces. She learns that when she feels frustrated, trying a new strategy often helps, and that the teacher provides care through her helping behavior. She learns about her own mathematical competence and experiences its development. She learns about relating to others in a mathematical context. Such realizations *become* mathematics for her. She knows no other way to feel. The study of in-the-moment engagement thus promises to shed light on more than motivation, affect, and social interactions. It promises to illuminate the very nature of mathematical knowledge and its function in the life of the student.

References

Ahmed, W., van der Werf, G., Kuyper, H., & Minnaert, A. (2013). Emotions, self-regulated learning, and achievement in mathematics: A growth curve analysis. *Journal of Educational Psychology, 105*(1), 150–161.

Ainley, M. (2006). Connecting with learning: Motivation, affect and cognition in interest processes. *Educational Psychology Review, 18*(4), 391–405.

Ainley, M., & Hidi, S. (2014). Interest and enjoyment. In R. Pekrun & L. Linnenbrink-Garcia (Eds.), *International handbook of emotions in education* (pp. 205–227). New York, NY: Routledge.

Ainley, M., Hidi, S., & Berndorff, D. (2002). Interest, learning, and the psychological processes that mediate their relationship. *Journal of Educational Psychology, 94*(3), 545–561.

Alston, A., Goldin, G. A., Jones, J., McCulloch, A., Rossman, C., & Schmeelk, S. (2007). The complexity of affect in an urban mathematics classroom. In T. Lamberg & L. R. Wiest (Eds.), *Exploring mathematics education in context: Proceedings of the 29th annual meeting of PME-NA, Lake Tahoe, NV* (pp. 326–333). Reno: University of Nevada.

Anderman, E. M., & Wolters, C. A. (2006). Goals, values, and affect: Influences on student motivation. In P. Alexander

& P. Winne (Eds.), *Handbook of educational psychology* (pp. 369–389). Mahwah, NJ: Erlbaum.

Appleton, J. J., Christenson, S. L., & Furlong, M. J. (2008). Student engagement with school: Critical conceptual and methodological issues of the construct. *Psychology in the Schools, 45*(5), 369–386.

Aronson, J., Lustina, M. J., Good, C., Keough, K., Steele, C. M., & Brown, J. (1999). When White men can't do math: Necessary and sufficient factors in stereotype threat. *Journal of Experimental Social Psychology, 35*, 29–46.

Barnes, D. (1992). *From communication to curriculum*. Portsmouth, NH: Heinemann.

Battey, D. (2013). "Good" mathematics teaching for students of color and those in poverty: The importance of relational interactions within instruction. *Educational Studies in Mathematics, 82*(1), 125–144.

Beasley, T. M., Long, J. D., & Natali, M. (2001). A confirmatory factor analysis of the Mathematics Anxiety Scale for children. *Measurement and Evaluation in Counseling and Development, 34*, 14–26.

Berry, R. Q., III. (2008). Access to upper-level mathematics: The stories of successful African American middle school boys. *Journal for Research in Mathematics Education, 39*(5), 464–488.

Bessant, K. C. (1995). Factors associated with types of mathematics anxiety in college students. *Journal for Research in Mathematics Education, 26*(4), 327–345.

Betz, N. E., & Hackett, G. (2006). Career self-efficacy theory: Back to the future. *Journal of Career Assessment, 14*(1), 3–11.

Blackwell, L. S., Trzesniewski, K. H., & Dweck, C. S. (2007). Implicit theories of intelligence predict achievement across an adolescent transition: A longitudinal study and an intervention. *Child Development, 78*(1), 246–263.

Boaler, J. (1998). Open and closed mathematics: Student experiences and understandings. *Journal for Research in Mathematics Education, 29*(1), 41–62.

Boaler, J., & Staples, M. (2008). Creating mathematical futures through an equitable teaching approach: The case of Railside School. *Teachers College Record, 110*(3), 608–645.

Bodovski, K., & Farkas, G. (2007). Do instructional practices contribute to inequality in achievement? The case of mathematics instruction in kindergarten. *Journal of Early Childhood Research, 5*(3), 301–322.

Boekaerts, M. (2006). Self-regulation and effort investment. In K. A. Renninger, I. E. Sigel, W. Damon, & R. M. Lerner (Eds.), *Handbook of child psychology* (6th ed., pp. 345–377). New York, NY: John Wiley and Sons.

Boekaerts, M., & Corno, L. (2005) Self-regulation in the classroom: A perspective on assessment and intervention. *Applied Psychology, 54*, 267–299.

Bolles, R. C. (1972). Reinforcement, expectancy, and learning. *Psychological Review, 79*(5), 394–409. doi:10.1037/h0033120

Bounoua, L., Cury, F., Regner, I., Huguet, P., Barron, K. E., & Elliot, A. J. (2012). Motivated use of information about others: Linking the 2 ¥ 2 achievement goal model to social comparison propensities and processes. *British Journal of Social Psychology, 51*(4), 626–641.

Bryant, F. B., Kastrup, H., Udo, M., Hislop, N., Shefner, R., & Mallow, J. (2013). Science anxiety, science attitudes, and constructivism: A binational study. *Journal of Science Education and Technology, 22*(4), 432–448.

Chouinard, R., & Roy, N. (2008). Changes in high-school students' competence beliefs, utility value and achievement goals in mathematics. *British Journal of Educational Psychology, 78*(1), 31–50.

Clark, L. M., DePiper, J. N., Frank, T. J., Nishio, M., Campbell, P. F., Smith, T. M., . . . Choi, Y. (2014). Teacher characteristics associated with mathematics teachers' beliefs and awareness of their students' mathematical dispositions. *Journal for Research in Mathematics Education, 45*(2), 246–284.

Cobb, P., & Yackel, E. (1996). Constructivist, emergent, and sociocultural perspectives in the context of developmental research. *Educational Psychologist, 31*(3–4), 175–190.

Cohen, E. G. (1994). Restructuring the classroom: Conditions for productive small groups. *Review of Educational Research, 64*(1), 1–35.

Covington, M. V. (2000). Goal theory, motivation, and school achievement: An integrative review. *Annual Review of Psychology, 51*(1), 171–200.

Csikszentmihalyi, M. (1990). *Flow: The psychology of optimal performance*. New York, NY: Cambridge University Press.

DeBellis, V. A., & Goldin, G. A. (1997). The affective domain in mathematical problem solving. In E. Pehkonen (Ed.), *Proceedings of the 21st Conference of the International Group for the Psychology of Mathematics Education (PME), Lahti, Finland* (Vol. 2, pp. 209–216). Helsinki, Finland: University of Helsinki Department of Teacher Education.

DeBellis, V. A., & Goldin, G. A. (1999). Aspects of affect: Mathematical intimacy, mathematical integrity. In O. Zaslavsky (Ed.), *Proceedings of the 23ird Conference of the International Group for the Psychology of Mathematics Education (PME), Haifa, Israel* (Vol. 2, pp. 249–256). Haifa, Israel: Technion Printing Center.

DeBellis, V. A., & Goldin, G. A. (2006). Affect and meta-affect in mathematical problem solving: A representational perspective. *Educational Studies in Mathematics, 63*, 131–147.

Deci, E. L., & Ryan, R. M. (1985). The general causality orientations scale: Self-determination in personality. *Journal of Research in Personality, 19*(2), 109–134.

Deci, E. L., & Ryan, R. M. (Eds.). (2002). *Handbook of self-determination research*. Rochester, NY: University of Rochester Press.

Dettmers, S., Trautwein, U., Lüdtke, O., Goetz, T., Frenzel, A. C., & Pekrun, R. (2011). Students' emotions during homework in mathematics: Testing a theoretical model of antecedents and achievement outcomes. *Contemporary Educational Psychology, 36*(1), 25–35.

Devine, A., Fawcett, K., Szűcs, D., & Dowker, A. (2012). Gender differences in mathematics anxiety and the relation to mathematics performance while controlling for test anxiety. *Behavioral and Brain Functions, 8*, 33.

Diamond, A., Barnett, W. S., Thomas, J., & Munro, S. (2007). Preschool program improves cognitive control. *Science, 318*(5855), 1387–1388.

Di Martino, P., & Zan, R. (2011). Attitude towards mathematics: A bridge between beliefs and emotions. *ZDM—The International Journal on Mathematics Education, 43*(4), 471–482.

D'Mello, S. D., & Graesser, A. C. (2014). Confusion. In R. Pekrun & L. Linnenbrink-Garcia (Eds.), *Handbook of emotions in education* (pp. 289–310). New York, NY: Taylor & Francis.

Duckworth, A. L., Peterson, C., Matthews, M. D., & Kelly, D. R. (2007). Grit: Perseverance and passion for long-term goals. *Journal of Personality and Social Psychology, 92*(6), 1087–1101.

Duckworth, K., Akerman, R., MacGregor, A., Salter, E., & Vorhaus, J. (2009). *Self-regulated learning: A literature review.* [*Wider Benefits of Learning Research Report No. 33*]. London, United Kingdom: Centre for Research on the Wider Benefits of Learning, Institute of Education, University of London.

Duncan, G. J., Dowsett, C. J., Claessens, A., Magnuson, K., Huston, A. C., Klebanov, P., . . . Japel, C. (2007). School readiness and later achievement. *Developmental Psychology, 43*(6), 1428–1446.

Dweck, C. S. (2000). *Self-theories: Their role in motivation, personality, and development.* Philadelphia, PA: Taylor & Francis.

Dweck, C. S., & Leggett, E. L. (1988). A social-cognitive approach to motivation and personality. *Psychological Review, 95*(2), 256–273.

Eccles, J. S., & Wigfield, A. (2002). Motivational beliefs, values, and goals. *Annual Review of Psychology, 53*(1), 109–132.

Eccles, J. S., Wigfield, A., Midgley, C., Reuman, D., MacIver, D., Feldlaufer, H. (1993). Negative effects of traditional middle schools on students' motivation. *Elementary School Journal, 93*(5), 553–574.

Efklides, A., & Petkaki, C. (2005). Effects of mood on students' metacognitive experiences. *Learning and Instruction, 15*(5), 415–431.

Else-Quest, N. M., Hyde, J. S., & Hejmadi, A. (2008). Mother and child emotions during mathematics homework. *Mathematical Thinking and Learning, 10*, 5–35.

Epstein, Y., Schorr, R. Y., Goldin, G. A., Warner, L., Arias, C., Sanchez, L., . . . Cain, T. R. (2007). Studying the affective/social dimension of an inner-city mathematics class. In T. Lamberg & L. Wiest (Eds.), *Proceedings of the 29th Annual Conference of the North American Chapter of the International Group for the Psychology of Mathematics Education* (pp. 649–656). Stateline: University of Nevada, Reno.

Esmonde, I. (2009). Mathematics learning in groups: Analyzing equity in two cooperative activity structures. *Journal of the Learning Sciences, 18*, 247–284.

Evans, J. (2002). *Adults' mathematical thinking and emotions: A study of numerate practices.* London, United Kingdom: Routledge.

Fennema, E., & Sherman, J. A. (1976). Fennema-Sherman Mathematics Attitude Scales: Instruments designed to measure attitudes toward the learning of mathematics by females and males. *Journal for Research in Mathematics Education, 7*, 324–326.

Fitness, J., & Curtis, M. (2005). Emotional intelligence and the Trait Meta-Mood Scale: Relationships with empathy, attributional complexity, self-control, and responses to interpersonal conflict. *E-Journal of Applied Psychology: Social Section, 1*, 50–62.

Flavell, J. H. (1976). Metacognitive aspects of problem solving. In L. B. Resnick (Ed.), *The nature of intelligence* (pp. 231–236). Hillsdale, NJ: Erlbaum.

Ford, M. E. (1992). *Motivating humans: Goals, emotions, and personal agency beliefs.* Thousand Oaks, CA: Sage.

Frank, K. A., Muller, C., Schiller, K. S., Riegle-Crumb, C., Mueller, A. S., Crosnoe, R., & Pearson, J. (2008). The social dynamics of mathematics coursetaking in high school. *American Journal of Sociology, 113*(6), 1645–1696.

Fredricks, J. A., Blumenfeld, P. C., & Paris, A. H. (2004). School engagement: Potential of the concept, state of the evidence. *Review of Educational Research, 74*(1), 59–109.

Fredricks, J. A., & Eccles, J. S. (2002). Children's competence and value beliefs from childhood through adolescence: Growth trajectories in two male-sex-typed domains. *Developmental Psychology, 38*(4), 519–533.

Fredrickson, B. L. (2001). The role of positive emotions in positive psychology: The broaden-and-build theory of positive emotions. *American Psychologist, 56*(3), 218–226.

Frenzel, A. C., Goetz, T., Pekrun, R., & Watt, H. M. (2010). Development of mathematics interest in adolescence: Influences of gender, family, and school context. *Journal of Research on Adolescence, 20*(2), 507–537.

Frenzel, A. C., Pekrun, R., Dicke, A. L., & Goetz, T. (2012). Beyond quantitative decline: Conceptual shifts in adolescents' development of interest in mathematics. *Developmental Psychology, 48*(4), 1069–1082.

Frenzel, A. C., Pekrun, R., & Goetz, T. (2007). Perceived learning environment and students' emotional experiences: A multilevel analysis of mathematics classrooms. *Learning and Instruction, 17*, 478–493.

Goldin, G. A. (2000). Affective pathways and representation in mathematical problem solving. *Mathematical Thinking and Learning, 2*, 209–219.

Goldin, G. A. (2002). Affect, meta-affect, and mathematical belief structures. In G. Leder, E. Pehkonen, & G. Torner (Eds.), *Beliefs: A hidden variable in mathematics education?* (pp. 59–72). Dordrecht, The Netherlands: Kluwer.

Goldin, G. A. (2014). Perspectives on emotion in mathematical engagement, learning, and problem solving. In R. Pekrun & L. Linnenbrink-Garcia (Eds.), *Handbook of emotions in education* (pp. 391–414). New York, NY: Taylor & Francis.

Goldin, G. A., Epstein, Y. M., Schorr, R. Y., & Warner, L. B. (2011). Beliefs and engagement structures: Behind the affective dimension of mathematical learning. *ZDM—The International Journal on Mathematics Education, 43*, 547–556.

Gollwitzer, P. M. (1999). Implementation intentions: Strong effects of simple plans. *American Psychologist, 54*(7), 493–503.

Gollwitzer, P. M., & Oettingen, G. (2011). Planning promotes goal striving. *Handbook of Self-Regulation: Research, Theory, and Applications, 2*, 162–185.

Gómez-Chacón, I. M. (2000). Affective influences in the knowledge of mathematics. *Educational Studies in Mathematics, 43*, 149–168.

Goodenow, C. (1993). Classroom belonging among early adolescent students' relationships to motivation and achievement. *The Journal of Early Adolescence, 13*(1), 21–43.

Gottfried, A. E., Marcoulides, G. A., Gottfried, A. W., Oliver, P. H., & Guerin, D. W. (2007). Multivariate latent change modeling of developmental decline in academic intrinsic math motivation and achievement: Childhood through adolescence. *International Journal of Behavioral Development, 31*(4), 317–327.

Gresalfi, M. S. (2009). Taking up opportunities to learn: Constructing dispositions in mathematics classrooms. *Journal of the Learning Sciences, 18*(3), 327–369.

Gresalfi, M., Martin, T., Hand, V., & Greeno, J. (2009). Constructing competence: An analysis of student participation in the activity systems of mathematics classrooms. *Educational Studies in Mathematics, 70*(1), 49–70.

Hackenberg, A. (2005). A model of mathematical learning and caring relations. *For the Learning of Mathematics, 25*(1), 45–51.

Hackenberg, A. J. (2010). Mathematical caring relations in action. *Journal for Research in Mathematics Education, 41*(3), 236–273.

Hackett, G., & Betz, N. E. (1989). An exploration of the mathematics self-efficacy/mathematics performance correspondence. *Journal for Research in Mathematics Education, 20*(3), 261–273.

Handal, B. (2003). Teachers' mathematical beliefs: A review. *The Mathematics Educator, 13*, 47–57.

Hannula, M. S. (2002). Attitude towards mathematics: Emotions, expectations and values. *Educational Studies in Mathematics, 49*, 25–46.

Hannula, M. S. (2006). Motivation in mathematics: Goals reflected in emotions. *Educational Studies in Mathematics, 63*, 165–178.

Harackiewicz, J. M., Barron, K. E., Pintrich, P. R., Elliot, A. J., & Thrash, T. M. (2002). Revision of achievement goal theory: Necessary and illuminating. *Journal of Educational Psychology, 94*(3), 638–645. doi:10.1037//0022-0663.94.3.638

Harackiewicz, J. M., & Sansone, C. (1991). Goals and intrinsic motivation: You can get there from here. *Advances in Motivation and Achievement, 7*, 21–49.

Harber, K. D., Zimbardo, P. G., & Boyd, J. N. (2003). Participant self-selection biases as a function of individual differences in time perspective. *Basic and Applied Social Psychology, 25*(3), 255–264.

Helme, S., & Clarke, D. (2001). Identifying cognitive engagement in the mathematics classroom. *Mathematics Education Research Journal 13*(2), 133–153.

Hembree, R. (1990). The nature, effects, and relief of mathematics anxiety. *Journal for Research in Mathematics Education, 21*, 33–46.

Herzig, A. H. (2004). "Slaughtering this beautiful math": Graduate women choosing and leaving mathematics. *Gender and Education, 16*(3), 379–395.

Hester, A. (2012, June). *The effect of personal goals on student motivation and achievement.* Action Research Projects presented at annual research forum, Studies in Teaching: 2012 Research Digest. Winston-Salem, NC.

Heyd-Metzuyanim, E. (2015). Vicious cycles of identifying and mathematizing: A case study of the development of mathematical failure. *Journal of the Learning Sciences 24*(4), 504–549.

Heyd-Metzuyanim, E., & Sfard, A. (2012). Identity struggles in the mathematics classroom: On learning mathematics as an interplay of mathematizing and identifying. *International Journal of Educational Research, 51–52*, 128–145.

Hidi, S., & Renninger, K. A. (2006). The four-phase model of interest development. *Educational Psychologist, 41*(2), 111–127.

Hilpert, J. C., Husman, J., Stump, G. S., Kim, W., Chung, W. T., & Duggan, M. A. (2012). Examining students' future time perspective: Pathways to knowledge building. *Japanese Psychological Research, 54*(3), 229–240.

Hirvonen, R., Tolvanen, A., Aunola, K., & Nurmi, J. E. (2012). The developmental dynamics of task-avoidant behavior and math performance in kindergarten and elementary school. *Learning and Individual Differences, 22*(6), 715–723.

Horn, I. S. (2007). Fast kids, slow kids, lazy kids: Framing the mismatch problem in mathematics teachers' conversations. *Journal of the Learning Sciences, 16*(1), 37–79.

Horn, I. S. (2008). Turnaround students in high school mathematics: Constructing identities of competence through mathematical worlds. *Mathematical Thinking and Learning, 10*(3), 201–239.

Horstmanshof, L., & Zimitat, C. (2007). Future time orientation predicts academic engagement among first-year university students. *British Journal of Educational Psychology, 77*(3), 703–718.

Husman, J., Derryberry, W., Michael Crowson, H., & Lomax, R. (2004). Instrumentality, task value, and intrinsic motivation: Making sense of their independent interdependence. *Contemporary Educational Psychology, 29*(1), 63–76.

Husman, J., & Lens, W. (1999). The role of the future in student motivation. *Educational Psychologist, 34*(2), 113–125.

Isen, A. M. (2004). Positive affect facilitates thinking and problem solving. In A. S. R. Manstead, N. Frijda, & A. Fischer (Eds.), *Feelings and emotions: The Amsterdam symposium* (pp. 263–281). Cambridge, United Kingdom: Cambridge University Press.

Jansen, A. (2006). Seventh graders' motivations for participating in two discussion-oriented mathematics classrooms. *The Elementary School Journal, 106*(5), 409–428.

Jansen, A. (2008). An investigation of relationships between seventh grade students' beliefs and their participation during mathematics discussions in two classrooms. *Mathematical Thinking and Learning, 10*(1), 68–100.

Jansen, A. (2009). Prospective elementary teachers' motivation to participate in whole-class discussions during mathematics content courses for teachers. *Educational Studies in Mathematics, 71*(2), 145–160.

Jansen, A. (2012). Developing productive dispositions during small-group work in two sixth-grade mathematics classrooms: Teachers' facilitation efforts and students' self-reported benefits. *Middle Grades Research Journal, 7*(1), 37–56.

Jansen, A., & Bartell, T. (2013). Caring mathematics instruction: Middle school students' and teachers' perspectives. *Middle Grades Research Journal 8*(1), 33–50.

Jansen, A., Herbel-Eisenmann, B., & Smith, J. P. (2012). Detecting students' experiences of discontinuities between middle school and high school mathematics programs: Learning during boundary crossing. *Mathematical Thinking and Learning: An International Journal, 14*(4), 285–309.

Jõgi, A. L., Kikas, E., Lerkkanen, M. K., & Mägi, K. (2015). Cross-lagged relations between math-related interest, performance goals and skills in groups of children with different general abilities. *Learning and Individual Differences, 39*, 105–113.

Kaplan, A., & Maehr, M. L. (2007). The contributions and prospects of goal orientation theory. *Educational Psychology Review, 19*(2), 141–184.

Kessels, U., & Hannover, B. (2007). How the image of math and science affects the development of academic interests. In M. Prenzel (Ed.), *Studies on the educational quality of schools. The final report on the DFG priority programme* (pp. 283–297). Münster, Germany: Waxmann Verlag GmbH.

Kloosterman, P., & Gorman, J. (1990). Building motivation in the elementary mathematics classroom. *School Science and Mathematics, 90*(5), 375–382.

Koichu, B. (2014). Reflections on problem solving. In M. Fried & T. Dreyfus (Eds.), *Mathematics & mathematics education: Searching for common ground* (pp. 113–135). New York, NY: Springer.

Köller, O., Baumert, J., & Schnabel, K. (2001). Does interest matter? The relationship between academic interest and achievement in mathematics. *Journal for Research in Mathematics Education, 32*(5), 448–470.

Kong, Q. P., Wong, N. Y., & Lam, C. C. (2003). Student engagement in mathematics: Development of instrument and validation of construct. *Mathematics Education Research Journal, 15*(1), 4–21.

Kosko, K. W. (2012). Geometry students' hedged statements and their self-regulation of mathematics. *The Journal of Mathematical Behavior, 31*(4), 489–499.

Krapp, A. (2002). Structural and dynamic aspects of interest development: Theoretical considerations from an ontogenetic perspective. *Learning and Instruction, 12*(4), 383–409.

Lake, E., & Nardi, E. (2014). Looking for Goldin: Can adopting student engagement structures reveal engagement structures for teachers? The case of Adam. In P. Liljedahl, C. Nicol, S. Oesterle, & D. Allan (Eds.), *Proceedings of the 38th Conference of the International Group for the Psychology of Mathematics Education* (Vol. 4, pp. 49–56). Vancouver, Canada: PME.

Langer-Osuna, J. M., & Esmonde, I. (2017). Identity in research on mathematics education. In J. Cai (Ed.), *Compendium for research in mathematics education* (pp. 637–648). Reston, VA: National Council of Teachers of Mathematics.

Latham, G. P., & Locke, E. A. (1991). Self-regulation through goal setting. *Organizational Behavior and Human Decision Processes, 50*(2), 212–247. doi:10.1016/0749-5978(91)90021-K

Laukenmann, M., Bleicher, M., Fuß, S., Gläser-Zikuda, M., Mayring, P., & von Rhöneck, C. (2003). An investigation of the influence of emotional factors on learning in physics instruction. *International Journal of Science Education, 25*(4), 489–507.

Laurent, J., Catanzaro, S. J., Joiner, T. E., Jr., Rudolph, K. D., Potter, K. I., & Lambert, S. (1999). A measure of positive and negative affect for children: Scale development and preliminary validation. *Psychological Assessment, 11*, 326–338.

Lawson, M. A., & Lawson, H. A. (2013). New conceptual frameworks for student engagement research, policy, and practice. *Review of Educational Research, 83*(3), 432–479.

Leder, G. C., & Forgasz, H. J. (2002). Measuring mathematical beliefs and their impact on the learning of mathematics: A new approach. In G. C. Leder, E. Pehkonen, & G. Törner (Eds.), *Beliefs: A hidden variable in mathematics education?* (pp. 95–113). Dordrecht, The Netherlands: Kluwer.

Leder, G. C., Pehkonen, E., & Törner, G. (Eds.). (2002). *Beliefs: A hidden variable in mathematics education?* Dordrecht, The Netherlands: Kluwer.

Lee, J. (2009). Universals and specifics of math self-concept, math self-efficacy, and math anxiety across 41 PISA 2003 participating countries. *Learning and Individual Differences, 19*, 355–365.

Lens, W., Paixão, M. P., & Herrera, D. (2009). Instrumental motivation is extrinsic motivation: So what??? *Psychologica, 50*, 21–40.

Leonard, J., & Martin, D. B. (Eds.). (2013). *The brilliance of Black children in mathematics: Beyond the numbers and toward new discourse*. Charlotte, NC: Information Age.

Levenson, E., Tirosh, D., & Tsamir, P. (2009). Students' perceived sociomathematical norms: The missing paradigm. *The Journal of Mathematical Behavior, 28*(2–3), 171–187.

Linnenbrink, E. A., & Pintrich, P. R. (2000). Multiple pathways to learning and achievement: The role of goal orientation in fostering adaptive motivation, affect, and cognition. In C. Sansone & J. M. Harackiewicz (Eds.), *Intrinsic and extrinsic motivation: The search for optimal motivation and performance* (pp. 195–227). San Diego, CA: Academic Press.

Linnenbrink, E. A., & Pintrich, P. R. (2002). Achievement goal theory and affect: An asymmetrical bidirectional model. *Educational Psychologist, 37*, 69–78.

Linnenbrink, E. A., & Pintrich, P. R. (2004). Role of affect in cognitive processing in academic contexts. In D. Y. Dai & R. J. Sternberg (Eds.), *Motivation, emotion, and cognition: Integrative perspectives on intellectual functioning and development* (pp. 57–87). New York, NY: Taylor & Francis.

Lubienski, S. T. (2000). Problem solving as a means toward mathematics for all: An exploratory look through a class lens. *Journal for Research in Mathematics Education, 31*(4), 454–482.

Luo, W., Paris, S. G., Hogan, D., & Luo, Z. (2011). Do performance goals promote learning? A pattern analysis of Singapore students' achievement goals. *Contemporary Educational Psychology, 36*(2), 165–176.

Ma, X. (1999). A meta-analysis of the relationship between anxiety toward mathematics and achievement in mathematics. *Journal for Research in Mathematics Education, 30*, 520–554.

Ma, X., & Xu, J. (2004). The causal ordering of mathematics anxiety and mathematics achievement: A longitudinal panel analysis. *Journal of Adolescence, 27*, 165–179.

Maasz, J., & Schlöglmann, W. (Eds.). (2009). *Beliefs and attitudes in mathematics education: New research results*. Rotterdam, The Netherlands: Sense.

Malka, A., & Covington, M. V. (2005). Perceiving school performance as instrumental to future goal attainment: Effects on graded performance. *Contemporary Educational Psychology, 30*(1), 60–80.

Malmivuori, M. L. (2006). Affect and self-regulation. *Educational Studies in Mathematics, 63*, 149–164.

Mangu, D., Lee, A., Middleton, J. A., & Nelson, J. K. (2015). Motivational factors predicting STEM and engineering career intentions for high school students. In *2015 IEEE Frontiers in Education conference proceedings* (pp. 2285–2291). El Paso, TX: IEEE.

Marsh, H. W., Abduljabbar, A. S., Abu-Hilal, M. M., Morin, A. J., Abdelfattah, F., Leung, K. C., . . . Parker, P. (2013). Factorial, convergent, and discriminant validity of TIMSS math and science motivation measures: A comparison of Arab and Anglo-Saxon countries. *Journal of Educational Psychology, 105*(1), 108–128.

Martin, A. J. (2011). Holding back and holding behind: Grade retention and students' non-academic and academic outcomes. *British Educational Research Journal, 37*(5), 739–763.

Martin, A. J., Bobis, J., Way, J., & Anderson, J. (2015). Exploring the Ups and Downs of Mathematics Engagement in the Middle Years of School. *Journal of Early Adolescence, 35*(2), 199–244.

Martin, D. (2000). *Mathematics success and failure among African American youth: The roles of sociohistorical context, community forces, school influence, and individual agency*. Mahwah, NJ: Lawrence Erlbaum Associates.

McClelland, M. M., Acock, A. C., & Morrison, F. J. (2006). The impact of kindergarten learning-related skills on academic trajectories at the end of elementary school. *Early Childhood Research Quarterly, 21*(4), 471–490.

McLeod, D. B. (1989). Beliefs, attitudes and emotions: New views of affect in mathematics education. In D. McLeod & V. Adams (Eds.), *Affect and mathematical problem solving: A new perspective* (pp. 245–258). New York, NY: Springer.

McLeod, D. B. (1992). Research on affect in mathematics education: A reconceptualization. In D. Grouws (Ed.), *Handbook of research on mathematics teaching and learning: A project of the National Council of Teachers of Mathematics* (pp. 575–596). New York, NY: Macmillan.

McLeod, D. B. (1994). Research on affect and mathematics learning. *Journal for Research in Mathematics Education, 25*, 637–647.

Meece, J. L., Wigfield, A., & Eccles, J. S. (1990). Predictors of math anxiety and its influence on young adolescents' course enrollment intentions and performance in mathematics. *Journal of Educational Psychology, 82*(1), 60–70.

Middleton, J. A. (1995). A study of intrinsic motivation in the mathematics classroom: A personal constructs approach. *Journal for Research in Mathematics Education, 26*(3), 254–279.

Middleton, J. A. (1999). Curricular influences on the motivational beliefs and practice of two middle school mathematics teachers: A follow-up study. *Journal for Research in Mathematics Education, 30*(3), 349–358.

Middleton, J. A. (2013). More than motivation: The combined effects of critical motivational variables on middle school mathematics achievement. *Middle Grades Research Journal, 8*(1), 77–95.

Middleton, J. A., & Jansen, A. (2011). *Motivation matters and interest counts: Fostering engagement in mathematics*. Reston, VA: National Council of Teachers of Mathematics.

Middleton, J. A., Leavy, A. M., & Leader, L. (2012). A path analysis of the relationship among critical motivational variables and achievement in reform-oriented mathematics curriculum. *Research in Middle Level Education, 36*(8), 1–10. Retrieved from http://www.amle.org/portals/0/pdf/publications/RMLE/rmle_v0136_n08.pdf

Middleton, J. A., & Spanias, P. (1999). Motivation for achievement in mathematics: Findings, generalizations, and criticisms of the recent research. *Journal for Research in Mathematics Education, 30*(1), 65–88.

Middleton, J. A., Tallman, M., Hatfield, N., & Davis, O. (2014). Taking the *severe* out of perseverance: Strategies for building mathematical determination. In N. Alpert (Ed.), *The collected papers*. Chicago, IL: Spencer Foundation. Retrieved from http://www.spencer.org/collected-papers-april-2015

Middleton, J. A., & Toluk, Z. (1999). First steps in the development of an adaptive, decision-making theory of motivation. *Educational Psychologist, 34*(2), 99–112.

Middleton, M. J., & Midgley, C. (1997). Avoiding the demonstration of lack of ability: An underexplored aspect of goal theory. *Journal of Educational Psychology, 89*(4), 710–718.

Mitchell, M. (1993). Situational interest: Its multifaceted structure in the secondary school mathematics classroom. *Journal of Educational Psychology, 85*(3), 424–436.

Morrone, A. S., Harkness, S. S., D'Ambrosio, B., & Caulfield, R. (2004). Patterns of instructional discourse that promote the perception of mastery goals in a social constructivist mathematics course. *Educational Studies in Mathematics, 56*(1), 19–38.

Murray, H. A. (2008). *Explorations in personality* (70th anniversary ed.). New York, NY: Oxford University Press.

Nicholls, J. G., Patashnick, M., & Nolen, S. B. (1985). Adolescents' theories of education. *Journal of Educational Psychology, 77*(6), 683–692.

Nuttin, J., & Lens, W. (1985). *Future time perspective and motivation: Theory and research method*. Hillsdale, NJ: Erlbaum.

Oettingen, G. (1999). Free fantasies about the future and the emergence of developmental goals. In J. Brandstädter & R. M. Lerner (Eds.), *Action & self-development* (pp. 315–342). London, United Kingdom: Sage.

Op 't Eynde, P., De Corte, E., & Verschaffel, L. (2006). Accepting emotional complexity: A socioconstructivist perspective on the role of emotions in the mathematics classroom. *Educational Studies in Mathematics, 63*, 193–207.

Op 't Eynde, P., De Corte, E., & Verschaffel, L. (2007). Students' emotions: A key component of self-regulated learning? In P. Schutz & R. Pekrun (Eds.), *Emotion in education* (pp. 185–204). Burlington, MA: Academic Press.

Op 't Eynde, P., & Hannula, M. S. (2006). The case study of Frank. *Educational Studies in Mathematics, 63*(2), 123–129.

Patrick, H., Ryan, A. M., & Kaplan, A. (2007). Early adolescents' perceptions of the classroom social environment, motivational beliefs, and engagement. *Journal of Educational Psychology, 99*(1), 83–98.

Pekrun, R. (2006). The control-value theory of achievement emotions: Assumptions, corollaries, and implications for educational research and practice. *Educational Psychology Review, 18*, 315–341.

Pekrun, R., Frenzel, A., Goetz, T., & Perry, R. P. (2007). The control–value theory of achievement emotions: An integrative approach to emotions in education. In P. A. Schutz & R. Pekrun (Eds.), *Emotion in education* (pp. 13–36). San Diego, CA: Academic Press.

Pekrun, R., Goetz, T., Frenzel, A. C., Barchfeld, P., & Perry, R. P. (2011). Measuring emotions in students' learning and performance: The Achievement Emotions Questionnaire (AEQ). *Contemporary Educational Psychology, 36*, 36–48.

Pekrun, R., & Perry, R. P. (2014). Control-value theory of achievement emotions. In R. Pekrun & L. Linnenbrink-Garcia (Eds.), *International handbook of emotions in education* (pp. 120–141). New York, NY: Routledge.

Perry, N. E., & VandeKamp, K. J. (2000). Creating classroom contexts that support young children's development of self-regulated learning. *International Journal of Educational Research, 33*(7), 821–843.

Philipp, R. A. (2007). Mathematics teachers. Beliefs and affect. In F. K. Lester Jr. (Ed.), *Second handbook of research on mathematics teaching and learning* (pp. 257–315). Charlotte, NC: Information Age; Reston, VA: National Council of Teachers of Mathematics.

Pintrich, P. R. (2000). Multiple goals, multiple pathways: The role of goal orientation in learning and achievement. *Journal of Educational Psychology, 92*(3), 544–555.

Pinxten, M., Marsh, H. W., De Fraine, B., Van Den Noortgate, W., & Van Damme, J. (2014). Enjoying mathematics or feeling competent in mathematics? Reciprocal effects on mathematics achievement and perceived math effort expenditure. *British Journal of Educational Psychology, 84*(1), 152–174.

Plenty, S., & Heubeck, B. G. (2013). A multidimensional analysis of changes in mathematics motivation and engagement during high school. *Educational Psychology, 33*(1), 14–30.

Renninger, A., Hidi, S., & Krapp, A. (Eds.). (2014). *The role of interest in learning and development.* New York, NY: Psychology Press.

Roeser, R. W., Eccles, J. S., & Sameroff, A. J. (2000). School as a context of early-adolescents' academic and socio-emotional development: A summary of research findings. *Elementary School Journal, 100*(5), 443–471.

Rolland, R. G. (2012). Synthesizing the evidence on classroom goal structures in middle and secondary schools: A meta-analysis and narrative review. *Review of Educational Research, 82*(4), 396–435.

Saito, Y., Garza, T. J., & Horwitz, E. K. (1999). Foreign language reading anxiety. *The Modern Language Journal, 83,* 202–218. doi:10.1111/0026-7902.00016

Sansone, C., & Thoman, D. B. (2005). Interest as the missing motivator in self-regulation. *European Psychologist, 10*(3), 175–186.

Schiefele, U., & Csikszentmihalyi, M. (1995). Motivation and ability as factors in mathematics experience and achievement. *Journal for Research in Mathematics Education, 26*(2), 163–181.

Schlöglmann, W. (2006). Meta-affect and strategies in mathematics learning. In M. Bosch (Ed.), *Proceedings of the fourth congress of the European Society for Research in Mathematics Education.* Sant Feliu de Guixols, Spain: FUNDEMI IQS, Universitat Ramon Llull.

Schoenfeld, A. H. (1985). *Mathematical problem solving.* New York, NY: Academic Press.

Schorr, R. Y., Epstein, Y. M., Warner, L. B. & Arias, C. C. (2010a). Don't disrespect me: Affect in an urban math class. In R. Lesh, P. L. Galbraith, C. R. Haines, & A. Hurford (Eds.), *Modeling students' mathematical modeling competencies: ICTMA 13* (pp. 313–325). New York, NY: Springer-Verlag.

Schorr, R. Y., Epstein, Y. M., Warner, L. B., & Arias, C. C. (2010b). Mathematical truth and social consequences: The intersection of affect and cognition in a middle school classroom. *Mediterranean Journal for Research in Mathematics Education, 9,* 107–134.

Schunk, D. H. (2001). Social-cognitive theory and self-regulated learning. In B. Zimmerman & D. Schunk (Eds.), *Self-regulated learning and academic achievement: Theoretical perspectives* (2nd ed., pp. 125–151). Mahwah, NJ: Erlbaum.

Schunk, D. H., & Ertmer, P. A. (2000). Self-regulation and academic learning: Self-efficacy enhancing interventions. In M. Boekaerts, P. Pintrich, & M. Zeidner (Eds), *Handbook of self-regulation* (pp. 631–649). San Diego, CA: Academic Press.

Sheeran, P., Webb, T. L., & Gollwitzer, P. M. (2005). The interplay between goal intentions and implementation intentions. *Personality and Social Psychology Bulletin, 31*(1), 87–98.

Sherman, B. F., & Wither, D. P. (2003). Mathematics anxiety and mathematics achievement. *Mathematics Education Research Journal, 15*(2), 138–115.

Shernoff, D. J. (2013). *Optimal learning environments to promote student engagement.* New York, NY: Springer.

Silver, E. A., & Stein, M. K. (1996). The Quasar project: The "Revolution of the Possible" in mathematics instructional reform in urban middle schools. *Urban Education, 30*(4), 476–521.

Simons, J., Dewitte, S., & Lens, W. (2000). Wanting to have vs. wanting to be: The effect of perceived instrumentality on goal orientation. *British Journal of Psychology, 91*(3), 335–352.

Simons, J., Vansteenkiste, M., Lens, W., & Lacante, M. (2004). Placing motivation and future time perspective theory in a temporal perspective. *Educational Psychology Review, 16*(2), 121–139.

Simpkins, S. D., Davis-Kean, P. E., & Eccles, J. S. (2006). Math and science motivation: A longitudinal examination of the links between choices and beliefs. *Developmental Psychology, 42*(1), 70–83.

Skemp, R. R. (1976). Relational understanding and instrumental understanding. *Mathematics Teaching, 77,* 20–26.

Sullivan, P., Askew, M., Cheeseman, J., Clarke, D., Mornane, A., Roche, A., & Walker, N. (2014). Supporting teachers in structuring mathematics lessons involving challenging tasks. *Journal of Mathematics Teacher Education, 18*(2), 123–140.

Thorndike-Christ, T. (1991). *Attitudes toward mathematics: Relationships to mathematics achievement, gender, mathematics course-taking plans, and career interests.* Washington, DC: ERIC.

Trautwein, U., Lüdtke, O., Köller, O., & Baumert, J. (2006). Self-esteem, academic self-concept, and achievement: How the learning environment moderates the dynamics of self-concept. *Journal of Personality and Social Psychology, 90*(2), 334–349.

Tulis, M., & Ainley, M. (2011). Interest, enjoyment and pride after failure experiences? Influences on students' state-emotions after success and failure during learning in mathematics. *Educational Psychology, 31*(7), 779–807.

Turner, J. C., Warzon, K. B., & Christensen, A. (2010). Motivating mathematics learning: Changes in teachers' practices and beliefs during a nine-month collaboration. *American Educational Research Journal, 48*(3), 718–762.

Uekawa, K., Borman, K., & Lee, R. (2007). Student engagement in U.S. urban high school mathematics and science classrooms: Findings on social organization, race, and ethnicity. *The Urban Review, 39*(1), 1–43.

Urdan, T., & Midgley, C. (2003). Changes in the perceived classroom goal structure and pattern of adaptive learning during early adolescence. *Contemporary Educational Psychology, 28*(4), 524–551.

Verner, I., Massarwe, K., & Bshouty, D. (2013). Constructs of engagement emerging in an ethnomathematically-based teacher education course. *The Journal of Mathematical Behavior, 32*(3), 494–507.

Vinner, S. (1997). From intuition to inhibition—mathematics, education, and other endangered species. In E. Pehkonen (Ed.), *Proceedings of the 21st Conference of the International*

Group for the Psychology of Mathematics Education (*PME*), *Lahti, Finland* (Vol. 1, pp. 63–78). Helsinki, Finland: University of Helsinki Department of Teacher Education.

Wager, A. A. (2014). Noticing children's participation: Insights into teacher positionality toward equitable mathematics pedagogy. *Journal for Research in Mathematics Education, 45*(3), 312–350.

Walter, J. G., & Hart, J. (2009). Understanding the complexities of student motivations in mathematics learning. *The Journal of Mathematical Behavior, 28*(2), 162–170.

Watt, H. M. (2004). Development of adolescents' self-perceptions, values, and task perceptions according to gender and domain in 7th-through-11th-grade Australian students. *Child Development, 75*(5), 1556–1574.

Webb, N. M., Franke, M. L., Ing, M., Wong, J., Fernandez, C. H., Shin, N., & Turrou, A. C. (2014). Engaging with others' mathematical ideas: Interrelationships among student participation, teachers' instructional practices, and learning. *International Journal of Educational Research, 63*, 79–93.

Webel, C. (2013). High school students' goals for working together in mathematics class: Mediating the practical rationality of studenting. *Mathematical Thinking and Learning, 15*(1), 24–57.

Wentzel, K. R. (1998). Social relationships and motivation in middle school: The role of parents, teachers, and peers. *Journal of Educational Psychology, 90*(2), 202–209.

Wigfield, A., & Eccles, J. S. (2000). Expectancy-value theory of achievement motivation. *Contemporary Educational Psychology, 25*(1), 68–81.

Wigfield, A., Eccles, J. S., & Rodriguez, D. (1998). The development of children's motivation in school contexts. *Review of Research in Education, 23*, 73–118.

Wigfield, A., Tonks, S., & Klauda, S. L. (2009). Expectancy-value theory. In K. R. Wentzel & A. Wigfield (Eds.), *Handbook of motivation at school* (pp. 55–75). New York, NY: Routledge.

Wood, M. B. (2013). Mathematical micro-identities: Moment-to-moment positioning and learning in a fourth-grade classroom. *Journal for Research in Mathematics Education, 44*(5), 775–808.

Yeager, D. S., & Dweck, C. S. (2012). Mindsets that promote resilience: When students believe that personal characteristics can be developed. *Educational Psychologist, 47*(4), 302–314.

Yen, C., Konold, T. R., & McDermott, P. A. (2004). Does learning behavior augment cognitive ability as an indicator of academic achievement? *Journal of School Psychology, 42*, 157–169.

Zajonc, R. B. (1980). Feeling and thinking: Preferences need no inferences. *American Psychologist, 35*, 151–175.

Zan, R., Brown, L., Evans, J., & Hannula, M. S. (2006). Affect in mathematics education: An introduction. *Educational Studies in Mathematics, 63*(2), 113–121.

Zeidner, M. (2014). Anxiety in education. In R. Pekrun & L. Linnenbrink-Garcia (Eds.), *Handbook of emotions in education* (pp. 265–288). New York, NY: Taylor & Francis.

Zhang, L., Karabenick, S. A., Maruno, S. I., & Lauermann, F. (2011). Academic delay of gratification and children's study time allocation as a function of proximity to consequential academic goals. *Learning and Instruction, 21*(1), 77–94.

Zimbardo, P. G., & Boyd, J. N. (1999). Putting time in perspective: A valid, reliable individual-differences metric. *Journal of Personality and Social Psychology, 77*(6), 1271–1288.

Zimmerman, B. J. (2000). Self-efficacy: An essential motive to learn. *Contemporary Educational Psychology, 25*(1), 82–91.

26 多模式的物质思维：数学教育中的具身理论

路易斯·雷德福
加拿大劳伦蒂安大学教育学院
费迪南多·阿扎雷洛
意大利都灵大学
劳瑞·爱德华兹
美国加利福尼亚圣玛丽学院
克里斯蒂娜·萨贝纳
意大利都灵大学
译者：张波
扬州大学数学科学学院

在一个五年级的课堂上，学生们正投身于探究正棱柱的面数。教师提出了如下问题："如果已知这个正棱柱的名称，我们能否推演出这个正棱柱的面数呢？"班上一位叫吉姆的学生总结了他的发现，对大家说道：

吉姆：是的，如果我们知道正棱柱的名字那么我们就可以推演出它的面数，因为，如果我们以正六棱柱作为例子，它的底面是个六边形……[他摸着右手上正六面体塑料模型的底面。见图26.1，图片1]。

教师：太棒了。

吉姆：……正如棱柱的名字。所以，每条边有一个面[摸着其中的一条边，见图26.1，图片2]，所以，一个正六边形有6条边[用两手触摸正六棱柱的几个面，见图26.1，图片3]。所以，一共有6个侧面，并且，如果我们算上两个底面的话[用他的右手食指做了个画圈的手势，见图26.1，图片4]，它是8个面。

在这段来自渥太华的课堂教学片断中，有几个要素和当代数学教育的讨论有关，这些要素涉及对几何实物教具作用的清晰认识、吉姆的手围绕这个塑料教具所做的触觉运动和在触摸的同时所表现出的语言活动，以及吉姆在意义建构过程中的感知和想象。对这些要素的关注都指向一个观点，即在教学过程中产生的数学意义是多模式的，更广泛地，对这些要素的关注来自于对人类认知的新观念，特别是对身体、语言和物质文化作用的新理解。和传统方法不一样，这些概念强调了数学思维以及教与学中的符号和具身的认知作用。在这些新观念中，手势、体态、肢体动作、人工制品以及广泛意义上的各类符号都被认为是在调查学生如何学、教师如何教时需要考虑的大量资源（例如Arzarello, 2006; Bautista & Roth, 2012; Borba & Villareal, 2006; Edwards, Radford, & Arzarello, 2009; Forest & Mercier, 2012; Radford & D'Amore, 2006; Radford, Schubring, & Seeger, 2008）。这些可以感知到的物质资源并不被认为是单纯的教与学的附带现象，而是作为学生及教师数学思维的核心要素被加以概念化。

然而，对于人类关于这个世界的触觉-动感的肢体经验以及人们与人工制品、物质文化之间的交互作用在人类思考、逐渐形成认识的道路上到底扮演了何种角色，

26 The Multimodal Material Mind: Embodiment in Mathematics Education

LUIS RADFORD
Faculté d'Éducation, Université Laurentienne, Canada

FERDINANDO ARZARELLO
Università di Torino, Italy

LAURIE EDWARDS
Saint Mary's College of California, United States

CRISTINA SABENA
Università di Torino, Italy

In a grade 5 class, the students are engaged in investigating the number of faces of a regular prism. The teacher has asked the following question: "If I know the name of a prism, can I deduce the number of its faces?" Addressing the class, Jim, one of the students, summarizes his findings and says:

Jim: Uh, yes, we can deduce the number of faces if we know the name of the prism because, if we take the example of a hexagonal prism, its bases are hexagons as the name says... [*He touches one of the bases of a plastic hexagonal prism that he is holding in his right hand; see Figure 26.1, Picture 1*].

Teacher: Excellent.

Jim: ... of the prism. So each edge [*touching one of the edges; see Figure 26.1, Picture 2*] has a face so a hexagon has 6 edges [*touching with his hands several faces of the hexagon; see Figure 26.1, Picture 3*]. So there are 6 lateral faces, and if we count the bases [*making a round gesture with the index of his right hand; see Figure 26.1, Picture 4*], it's 8 faces.

There are several elements in this short passage, which comes from a classroom in Ottawa, that have become relevant in contemporary discussions of mathematics education. These elements deal with a clear understanding of the roles played by the material geometric artifact, the tactile movement of Jim's hands around the physical geometric artifact and the linguistic activity that Jim deploys while touching it, and Jim's perception and imagination in the course of his embodied meaning-making process. Attention to these elements points to the idea that mathematical meanings that arise in teaching and learning are multimodal. More broadly, attention to these elements comes from new conceptions about human cognition, marked in particular by new understandings of the role of the body, language, and material culture. Distinct from traditional approaches, these conceptions highlight the cognitive role of semiotics and embodiment in mathematics thinking, teaching, and learning. Within these new conceptions, gestures, body posture, kinesthetic actions, artifacts, and signs in general are considered a fruitful array of resources to be taken into account when investigating how students learn and how teachers teach (e.g., Arzarello, 2006; Bautista & Roth, 2012; Borba & Villareal, 2006; Edwards, Radford, & Arzarello, 2009; Forest & Mercier, 2012; Radford & D'Amore, 2006; Radford, Schubring, & Seeger, 2008). These sensible and material resources are not considered mere epiphenomena of teaching and learning: They are conceptualized as central elements of the students' and teachers' mathematical thinking.

There are, however, a variety of interpretations of the role that humans' tactile-kinesthetic bodily experience of the world and their interaction with artifacts and material culture play in the way humans think and come

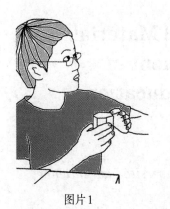   

图片1　　　　　　图片2　　　　　　图片3　　　　　　图片4

图26.1　吉姆触摸正六棱柱塑料模型并做手势

这里有多种解释。这些解释取决于不同的认知观念，比如，有些途径受认知语言学的启发（例如Fauconnier & Turner，2002；Friedrich，1970），强调语言的隐喻方面以及具体心理空间的综合构成（例如，可以参见Edwards，2009；Lakoff & Núñez，2000；Yoon，Thomas，& Dreyfus，2011）。受现象学研究的启发，其他的途径有些强调思想"丰富"的本质（Bautista & Roth，2012；Roth，2010；Thom & Roth，2011），还有一些则强调认知的物质性和它的文化历史维度（de Freitas & Sinclair，2013；Malafouris，2012；Radford，2013）。总之，这些途径主张意义和认知深深扎根于实际的、具身的存在物，并试图回答意义是如何产生的，以及思想如何与行动、情绪、感知相联系等问题。

本章对数学教育中的概念化和对具身的使用进行了批判性讨论。在接下来的两节，我们将在一般意义下讨论人类认知的概念以及具身和多模式的作用。再接下来的一节中，我们回顾与数学教学中具身有关的几种理论，包括受皮亚杰学派影响的理论、以符号学为导向的理论、行动主义、现象学途径和唯物主义途径。之后的两节，我们分别在文化历史理论和认知语言学方面对具身理论进行了更为细致的研究。最后，本章讨论了一些尚未解决的问题以及新的可能的探索方向。

人类思维

在前面一节提到的五年级学生们探索正棱柱的例子中，师生已经在之前的两节课中讨论了三维空间物体间的一些关键差异，已经区分过圆形物体和棱柱。在第三次课中，教师将讨论转移到棱柱的面数上来，这个问题不针对某一具体的棱柱，是在一般意义下被提出来的。和班上其他同学一样，吉姆关注了一个特殊的棱柱，这里，他挑选了六棱柱。他确切地知道那个问题不是关于这个特殊的棱柱，他的回答也试图包括其他棱柱。我们如何来解释吉姆的认识过程呢？

理性主义认识论

笛卡尔学派的理性主义认识论提出，为认知某物，人类思维把想要认识的物体分为某些部分。人类思维被认为是分析性操作，的确，我们就是通过对事物各部分的分析来最终认识事物的。对某个或某些部分的研究（在这个例子中，被考虑的不是一般意义下的棱柱，而是某个特殊的棱柱）使得吉姆能够认识到整个事物的属性，即一般意义下的棱柱的性质。17世纪笛卡尔派逻辑学家安托尼·阿尔诺（1861）把这一认识过程称为抽象：

> 人类思维范围的局限性使得我们无法完全理解稍微复杂的事物，除非我们把它们分解成一些部分来考虑，或者通过我们能感受到的若干阶段来认识它们。这一过程可被一般化地称为抽象。（第45页）

在本文情境中，吉姆通过特殊来认识一般。被教师关于一般棱柱的问

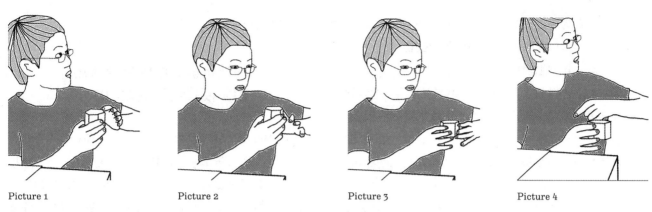

Picture 1　　Picture 2　　Picture 3　　Picture 4

FIGURE 26.1. *Jim touching the plastic hexagon and making gestures.*

to know. These interpretations depend on conceptions of cognition. For instance, some approaches inspired by cognitive linguistics (e.g., Fauconnier & Turner, 2002; Friedrich, 1970) emphasize the metaphoric dimension of language and the integrative constitution of embodied mental spaces (see, e.g., Edwards, 2009; Lakoff & Núñez, 2000; Yoon, Thomas, & Dreyfus, 2011). Other approaches, inspired by research in phenomenology, emphasize the "fleshy" nature of thought (Bautista & Roth, 2012; Roth, 2010; Thom & Roth, 2011), and yet others stress the materiality of cognition and its cultural-historical dimensions (de Freitas & Sinclair, 2013; Malafouris, 2012; Radford, 2013). All in all, these approaches claim that meaning and cognition are deeply rooted in physical, embodied existence and try to offer an answer to the questions of how meaning arises, and how thought is related to action, emotion, and perception.

This chapter provides a critical discussion of conceptualizations and applications of embodiment in mathematics education. In the next two sections, we discuss conceptions of human cognition and the role of embodiment and multimodality in general. This is followed by a section where we review several theories of embodiment in mathematics teaching and learning, including Piagetian inspired theory, semiotic oriented theory, enactivism, and phenomenological and materialist approaches. In the subsequent two sections, we delve in more detail into embodiment as featured in cultural-historical theory and cognitive linguistics, respectively. The chapter ends with a discussion of some open problems and possible new lines of inquiry.

The Human Mind

In the grade 5 example in the previous section, the students are exploring regular prisms. In the two previous lessons, the teacher and the students had discussed some key differences among three-dimensional solids; they had also distinguished between round solids and prisms. In this third lesson, the teacher shifts the investigation to the prisms' number of faces. The question is not asked for a specific prism. The question is asked *in general*. Like the other students in the class, Jim focuses on a specific prism; in his case, he chooses the hexagonal prism. He does know that the question is not about this particular prism. His answer attempts to cover other prisms as well. How can we interpret Jim's process of knowing?

Rationalist Epistemologies

Rationalist epistemologies of the Cartesian type argue that to cognize something, the human mind separates the thing to be known into parts. The human mind is supposed to operate analytically. It is, indeed, through the analysis of the thing's parts that the thing is finally known. The investigation of one or some of these parts (in this case, the consideration not of a prism in general, but a particular prism) would allow Jim to know properties of the whole thing to be known—the prism *in general*. The 17th century Cartesian logician Antoine Arnauld (1861) calls this knowing process *abstraction*:

> The limited extent of our mind renders us incapable of comprehending perfectly things which are a little complex, in any other way than by considering them in their parts, and, as it were, through the phases which they are capable of receiving. This is what may be termed, generally, knowing by means of abstraction. (p. 45)

Within this context, Jim targets something general through something particular. Prompted by the teacher's

题所驱动，吉姆通过一个特殊的棱柱——六棱柱，甚至可以说，通过六棱柱的一个特殊例子，也就是他手里拿的那个，来讨论一般棱柱。

事实上，吉姆的认识过程可追溯到亚里士多德哲学关于抽象的观念——一个通过忽略属性来定义的观念。为了思考一般的六棱柱，吉姆手中那个六棱柱的塑料属性，它的颜色、重量以及其他很多属性都被忽略掉了，考虑一般的六棱柱显然就不能考虑六棱柱的具体属性。在这一意义上的认识，人类思维被设想为安装了必要的识别程序来忽略事物的某些属性而保留事物的其他属性。另外，知识目标被设想为可以用分解和分析的方式来处理。

经验主义认识论

经验主义认识论为吉姆的认识过程提供了另一种解释。激进的经验主义认识论——比如，大卫·休谟在《人性论》（1739/1965）中清晰阐明的——数学对象的一般性质不是物体自身的性质，这些性质是个人基于感知的（或者说感觉到的，感觉上的）可能性的范围而赋予物体的。沿着这一思路，正棱柱的面数不是棱柱本身所具有的东西，也就是说，棱柱作为一个柏拉图多面体是独立于我们的感官的。因此，不如说，关于面数的结论是吉姆对在教室中、生活中的各种棱柱感知的结果。这是对感知经验的印象，休谟称之为想法，通过触摸、感受、抓握、在棱柱上移动他的手，吉姆形成了想法。把某个想法和其他想法联系起来，吉姆形成了越来越复杂的想法。在《人类理解研究》这本书中，休谟（1748/1921）提到：

> 尽管我们的思维看起来拥有这种无限的自由，但仔细一看我们就会发现，思维其实被限制在非常狭窄的范围内，所有思维的创造性只不过是我们对感知和经验给予我们的物质进行组合、改变、增加或减少的能力。（第16页）

在经验主义解释中，棱柱的面数是源于吉姆对于这个世界的经验的联想——休谟（1748/1921）称之为思维的"习惯"（第43页）。

理性认识论中的具身认知和多模式具身认知的作用

在理性主义的阵营中，文化产品和有知觉的身体并非知识的来源。笛卡尔（1641/1982）在《沉思录》中指出，对事物的理解不是经由感性经验而是仅通过理性获得的。事物"能被感知并非是由于它们能被看到、能被触摸，而是由于能被思维正确地理解"（我们翻译件的第26页）。同样地，另一理性主义者，戈特弗里德·威廉·莱布尼兹主张"必然真理，比如在纯数学中发现的，尤其是在算术和几何中，必须有不依赖于实例来证明的原理，因此也不依赖于感官的证据"（1705/1949，第44页）。理性主义的教育学在数学教学的过程中几乎不会涉及感官经验。

经验认识论中的具身

相比之下，对经验主义阵营而言，有知觉的身体才是知识的来源。但由于人类的感觉有其局限性，因此身体的认知作用经常表现出对我们所能知道的事物的限制。休谟（1739/1965）非常好地阐述了这一观点，他指出几何的基本原则（基于此推断出的那些命题被宣称为普适的和精确的），依赖于对感觉和想象的仔细检查和"宽松判断"（第70—71页）。这就是为什么休谟（1739/1965）认为几何是不精确的科学。事实上，几何的基本原则的确是通过"事物的一般特征"提炼获得的，而当我们考察这些事物容易改变的、令人惊叹的精密部分时，这些特征从未给我们提供保证（Hume, 1739/1965，第71页）。因为我们不能超越人类能力去感知非常小的角、直线或其他几何对象，因为我们无法超越人类知觉极限的极小值，"我们没有标准……来保证几何命题的正确性"（1739/1965，第71页）。不管我们身体感知的极限是否会带来认知的极限，一个经验主义的教育者仍会依赖并鼓励知觉经验，因为在经验主义认识论中，除了我们的身体和感官，我们没有其他来源可以用来学习以获得思想。

question about prisms in general, he talks about the prism in general through a specific prism—the hexagonal prism, or rather, through a specific instance of the hexagonal prism, namely the one he holds in his hands.

This account of Jim's process of knowing goes back, in fact, to the Aristotelian concept of abstraction—a concept defined by the omission of attributes. The plastic nature of the hexagonal prism that Jim holds in his hand, its color, weight, and many other attributes, are omitted in order to think of the *hexagonal prism in general,* which in turn is apparently considered without its specific hexagonal property to think of the *prism in general.* In this account of knowing, the human mind is conceived of as equipped with the required discriminatory procedures that allow it to discard some attributes and to keep others. Additionally, objects of knowledge are conceived of as amenable to decomposition and analysis.

Empirical Epistemologies

Empirical epistemologies provide a different account of Jim's process of knowing. Thus, within a radical empirical epistemology—for instance, the one articulated by David Hume in *A Treatise of Human Nature* (1739/1965)— general properties of mathematical objects are not properties of the objects *per se.* They are properties that the individual, within the realm of her sensorial (also called sensible or sensuous) possibilities, bestows on objects. In this line of thought, the number of faces of a regular prism is not something that pertains to the prism *as such,* that is, to the prism as a kind of Platonic object independent of our senses. That number is rather the result of the sensorial experience that Jim makes of the various prisms he encounters in the classroom and in life. It is the impression of the sensorial experiences that Hume calls *ideas.* So, by touching, perceiving, holding, and moving his hands on the prism, Jim forms ideas. And in associating one idea with others, Jim forms more and more complex ideas. In a passage of *An Enquiry Concerning Human Understanding,* Hume (1748/1921) notes that although

> our thought seems to possess this unbounded liberty, we shall find, upon a nearer examination, that it is really confined within very narrow limits, and that all this creative power of the mind amounts to no more than the faculty of compounding, transposing, augmenting, or diminishing the materials afforded us by the senses and experience. (p. 16)

In this account, the number of faces of a prism is an *association of ideas* that originates in Jim's empirical experience of the world—what Hume (1748/1921) calls a "habit" of thought (p. 43).

The Epistemological Role of Embodiment and Multimodality Embodiment in Rationalist Epistemologies

For the rationalist camp, cultural artifacts and the sentient body are not a source of knowledge. In his *Meditations,* Descartes (1641/1982) argues that things are grasped not through sensuous experiences but by the intellect alone: things "are not perceived because they are seen and touched, but only because they are correctly comprehended by the mind" (p. 26; our translation). In the same vein, another rationalist—Gottfried Wilhelm Leibniz—contended that "necessary truths such as found in pure mathematics, and particularly in arithmetic and in geometry, must have principles whose proof does not depend upon examples, nor consequently upon the testimony of the senses" (1705/1949, p. 44). A rationalist pedagogy would make little room for a sensuous experience in mathematics teaching and learning.

Embodiment in Empiricist Epistemologies

For the empiricist camp, by contrast, the sentient body is *the* source of knowledge. But because there is a limit to what humans can sense, the epistemic role of the body appears often as a constraint to what can be known. Hume (1739/1965) illustrated this point very well, arguing that the first principles of geometry, from which propositions are derived with alleged universality and exactness, rest, on closer examination, on "loose judgments of the senses and imagination" (pp. 70–71). This is why Hume (1739/1965) considered geometry an inexact science. Geometry's first principles are indeed still drawn from "the general appearance of the objects; and that appearance can never afford us any security, when we examine the prodigious minuteness of which nature is susceptible" (Hume, 1739/1965, p. 71). Because we cannot perceive minute angles, lines, and other geometric objects beyond human capacities, because we cannot transcend the threshold of human *perceptual minima,* "we have no standard ... as to assure us of the truth of [geometric] proposition[s]" (1739/1965, p. 71). Regardless of the limits of what can be known as a result of the sensorial limits of our body, an empiricist pedagogy would nonetheless rely on, and encourage, sensorial experiences, since in this epistemological account we do not have any other source other than our body and our senses from which to learn and form ideas.

康德认识论中的具身

前面的讨论突出了经验主义和理性主义的巨大差异。经验主义者认为没有感知在前，什么都不会进入思维，而理性主义者认为没有理性在前，什么都不会进入感知。康德试图做一件特别困难的事情——为经验主义和理性主义提供一个理论上连贯的中间点，即一种经验理性的认知理论。这一课题被收录在出版于1781年的《纯粹理性批判》这本书中（Kant, 1781/2003）。

1770年，康德区别出了两类不同的知识：经验知识和理性知识。康德认为经验知识包含了所有经由我们的身体和感觉而获得的知识，而理性知识则包含了所有不能由感知获得而仅能依靠理性或思维获得的知识。换句话说，经验知识是感知的结果，而理性知识则源于在感知之上的事物的表示，而不是这些事物自身自然的样子。康德是在为"感性知识和理性知识是两码事"这一观点辩护，他甚至进一步宣称不存在感性（也叫做感觉）知识和理性知识之间的连续性，必须分离这两个领域的知识。他提出了一种方法论上的"戒律"：注意不要让感性认识的原则超越其界限，进而影响理性概念（Kant, 1770/1894）。对于前面给出的吉姆的课堂片断，1770年的康德可能会说，通过感觉和对这个棱柱的触摸活动，吉姆获得了一个感知概念（从经验中获得的概念），而不是一个先验概念（在康德的术语中，独立于任何经验和感知印象的概念）。康德可能会警告我们要遵循他的规则，也要避免混淆吉姆所建立的经验性概念和棱柱的纯粹概念。

在《纯粹理性批判》一书中，经验知识和理性知识看似不再是两个分离的领域，而是与人类认知相联系的元素。在1781年给出上述课堂片断的话，康德可能会说，通过人类能被物质事物所影响的这种非常特殊的能力，吉姆感知到了物质的棱柱。这物质的棱柱对吉姆来说并不那么直接，而是一种被动的或接受的形式，这种认知模式是经由视觉、听觉、触觉、味觉、嗅觉等人类感官所形成的感觉。对实物的接受性接触给吉姆带来的影响就是康德所谓的感知。感知，换句话来说，是受到影响而产生的主观行为，是由物质的事物对感官作用而产生的，它可以是一种特殊的颜色、声音、热度，等等。然而，感觉和感知本身并不能导致棱柱的概念，它们并不能形成关于任何客体的知识（Kant, 1781/2003, 第73页, A28/B44），它们是在吉姆身体里面发生的改变，不是使物体成为一个棱柱的特质。那么，如果感觉经验有所欠缺的话，吉姆能形成棱柱的概念吗？关于这点，康德引入了一个至关重要的概念：直觉，这一概念可以被粗略地理解为一种被动的表示形式，事物（比如说棱柱）通过这个形式呈现在吉姆面前。尽管吉姆对棱柱及其性质的认识在精细化，但是问题仍然存在：如果感觉经验有所欠缺的话，吉姆能形成棱柱的概念吗？这里是康德的回答：棱柱的概念并不是吉姆仔细观察实物教具，从而读取其性质加以辨别出来的，对棱柱的被动表示（即实物教具）使得吉姆"可以说出这一概念所暗示的必须要的部分"（Kant, 1781/2003, 第19页, Bxii）。换句话说，如果吉姆能够这样认出棱柱，不是由于他的身体活动（这还是主观的，不能超越他个人的经验背景），也不是由于吉姆通过他的感官感受到这一实物被动性的表示，而是由于吉姆运用了在任何经验之前他就具有的棱柱的概念。事实上，在康德眼中，棱柱的概念和其他所有的数学概念本身就携带着它们自有的概念性质，这些性质是具备普遍性和逻辑必要性的，它们并不依赖于吉姆或其他个体，因此，他们不能从经验中获得。由于康德认为数学概念并非源自经验，因此康德称它们是先验的，"我们需要通过先验知识来理解，先验知识不是独立于这个或那个经验的知识，而是绝对独立于一切经验的知识"（1781/2003, 第43页, B3）。

总结一下，在康德的理论中，具身和多模式具身的作用是基于对两类知识的区分：经验知识和理性知识。尽管在他1770年的学位论文中，这两种知识被设想为完全分离的和不同的，但在1781年出版的《纯粹理性批判》一书中，它们被设想为互相协作的（Kant, 1781/2003, 第92页, A50/B74）。因此，具身和多模式具身成为中心并在认识论中获取了更重要的地位，正如康德在《纯粹理性批判》一书中的一段著名的话中所言，没有感知那我们就不能获得任何理性知识，没有理性知识，我们就不能思考任何感知的客体（Kant, 1781/2003,

Embodiment in Kant's Epistemology

The previous discussion highlights the tremendous differences between the empiricist and the rationalist camps. The empiricists claim that nothing can be in the mind if it has not been in the senses first; the rationalists claim that nothing can be in the senses if it has not been in the mind first. Kant tried something daunting: to offer a theoretically coherent middle point between empiricism and rationalism—an empiricist-rationalist theory of knowing. This project is contained in *Critique of Pure Reason*, published in 1781 (Kant, 1781/2003).

In 1770 Kant was engaged in making a distinction between two kinds of knowledge: sensible and intelligible knowledge. Kant suggested that *sensible* knowledge involves all that can be known through our body and our senses, whereas *intelligible* knowledge comprises all that cannot be known by the senses but by the intellect or the mind only. In other words, sensible knowledge is what results from sensations, whereas intelligible knowledge is what results from representations of things that cannot by their own nature come before the senses. Kant was defending the idea that sensible and intelligible knowledge are two separate things. He went even further and claimed that there is no continuity between sensible (also called sensuous) and intelligible knowledge. These two realms of knowledge should be kept apart. He stated a methodological "precept": to take care not to allow the principles of sensuous cognition to transgress their limits and affect the intellectual concepts (Kant, 1770/1894). Presented with Jim's classroom episode discussed previously, in 1770 Kant would have probably said that through the perceptual and tactile activity with the prism Jim came up with a *sensible concept* (i.e., a concept derived from experience), not with an *a priori concept* (in Kant's terminology, a concept independent of all experience and all impression of the senses). Kant would have warned us to follow his precept and to avoid confounding Jim's constructed experimental concept with the intelligible concept of the prism.

In *Critique of Pure Reason*, the sensible and the intelligible appear no longer as two separate realms but as related elements of human cognition. Presented with the same classroom episode in 1781, Kant would probably have said that Jim senses the physical prism through a very specific human capacity for being affected by material things. The physical prism appears to Jim not as such, directly, but as a kind of passive or receptive form of encountering the object mediated by the human modes of *sensibility*: sight, hearing, touch, taste, smell, and so forth. The effect on Jim that results from the receptive encounter of the object is what Kant calls *sensation*. Sensation, in other words, is the subjective act of being affected that results from the action of physical things on sensibility. It can be a specific color, sound, heat, and so forth. However, sensibility and sensation cannot lead by themselves to the concept of prism. They cannot yield knowledge of any object (Kant, 1781/2003, p. 73, A28/B44). They are *alterations in Jim's body*, not the *qualities of the object* that make it a prism. How, then, can Jim come up with the concept of prism if sensuous experience is not enough? Kant introduces a crucial concept in this regard: the concept of *intuition*, which can roughly be translated as a passive form of representation through which an object (e.g., the prism) appears to Jim. Despite this refinement in the account of Jim's coming to know the prism and its properties, the question remains: How can Jim come up with the concept of prism if sensuous experience is not enough? Here is Kant's answer: The concept of prism does not arise out of what Jim can possibly discern through inspection of the physical artifact where he could read off its properties. The passive representation of the prism (i.e., the physical artifact) allows him "to bring out what was necessarily implied in the concept." (Kant, 1781/2003, p. 19, Bxii). In other words, if Jim was able to recognize the prism as such, it was not as a result of his embodied activity (which remains subjective and incapable of transcending the situatedness of his own experience), nor was it because of the manner in which the physical object appeared to Jim through his senses in its passive representation. The reason is that Jim mobilized a concept of prism that *he already had before any possible experience*. Indeed, in Kant's eyes, the concept of prism and all mathematical concepts carry with them their own conceptual properties. These properties are universal and logically necessary. They do not depend on Jim or another individual. As a result, they cannot be derived from experience. Because for Kant mathematical concepts are not derived from experience, Kant called them *a priori:* "We shall understand by *a priori* knowledge, not knowledge independent of this or that experience, but knowledge absolutely independent of all experience" (1781/2003, p. 43, B3).

To sum up, the role of embodiment and multimodality in Kant's theory of knowing was based on the distinction between two kinds of knowledge: the sensible and the intelligible. Although in the 1770 *Dissertation* these kinds of knowledge were conceived of as separate and different, in 1781 in *Critique of Pure Reason* they were conceived of as cooperating with each other (Kant, 1781/2003, p. 92, A50/B74). As a result, embodiment and

第93页；A51/75改述过）。

尽管在康德认识论中对感知的讨论使得具身和多模式的具身凸显出了重要的作用，但它们的作用仍被限定为，感知是为理性提供启动它的原始材料。在康德看来，经验材料之所以成为可思考的材料，仅仅是因为理性选择了它们并赋予了它们概念性的内容，因为理性知识并不是一般经验的内容。这就是为什么吉姆，在他的数学经验中，不可能获得棱柱在数学意义上成为棱柱的普适的和必要的性质，一般意义下的必要性是不能从明显的"必要性"中得到的，明显的必要性来自于经验，而经验总是处于时空中的。如果吉姆认识到该棱柱是具有一般数学意义的一个理想对象，那么这并非是出于经验，在康德看来，人类思维的组织结构，连同其先验知识和理论原则的储备，给吉姆提供了一个"可称之为可能经验的理论图式"（Kant, 1781/2003, 第258页, A236—237/B295—296）。

但是，对康德来说图式到底是什么呢？图式是一种类推的过程，把"几个字母组合成一个图案"（Kant, 1781/2003, 第183页, A 142/B 181），它体现了在经验执行的过程中理性和感知之间的链接，一方面，图式是理性的；另一方面，图式又是感知的，但不可以将图式和图像相混淆：

> 比如，如果我把五个点一个接一个地标出来……，这就是5这个数的图像。反之，如果我只是思考一个一般的数，无论是5或者100，那么这种思维与其说是一个图像的本身，不如说是按照一定的概念把一个数目（比如说1000）表现在某个图像中的方法。像1000这样的一个数字，很难对其图像进行计数，也很难将它与该概念加以比较。这种为概念提供它的图像的一般化的想象过程的表象，我把它叫做这个概念的图式。（Kant, 1781/2003, 第182页, A 140/B 179）

在说图式是一种理论或者说一种一般化的过程时，康德的意思是它可以一遍又一遍地重复实施，实际上，图式需要一个迭代的指导思想，以连接知识和行动（Radford, 2005）。

皮亚杰认识论中的具身

尽管皮亚杰认为自己是一个优秀的康德主义者，但他对康德的先验论热情不高，皮亚杰赞同康德所说的理性的目的在于认识经验，但他不同意康德所说的理性是先验存在的。28岁的皮亚杰认为理性来自于经验："经验和理性并非是我们可以隔离开的两个术语，理性管理经验，经验改建理性"（Piaget, 1924, 第587页）。但是理性是如何从经验中产生的呢？

如前所述，在康德的认识论中，在感知和理性之间始终存在一个鸿沟，因此后者不能被认为是前者的一般化和抽象，康德也不需要抽象化这一概念。皮亚杰则不同，他真的需要一个能说清楚理论如何从经验中产生的概念。他求助于前面提到的康德的图式概念，出于自身的需要，皮亚杰通过强调抽象这一活动调整了图式概念，皮亚杰（1970）辩称对于"抽象"，这里有两种可能性：

> 第一种抽象是，当我们作用于一个对象时，我们的知识来源于对象本身，这是一般经验主义的观点，而且它在大部分实验或经验性知识的情况下是有效的。但还有另外一种可能性：当我们操作某个对象时，我们也可以对我们的动作进行思考，或者你也可以说是对操作进行思考。……按照这一假设，抽象并非来自于我们所操作的物体，而是来自于动作本身。对我来说，这就是逻辑和数学抽象的基础。（第16页）

皮亚杰尤为关注第二种抽象，称其为反省抽象。从被操作的对象中解放出来，反省抽象互相之间可以协调，比如叠加地、暂时性地和循序地来产生图式，皮亚杰把图式解释为和数学一样具有"平行的逻辑结构"（Piaget, 1970, 第18页）：

> 任何给定的图式本身并不具有逻辑成分，但是一些图式可以和其他图式相协调，这意味着动作的一般协调性。这种来自于动作的逻辑上的协调性本质上就是逻辑数学结构的出发点。（第42页）

multimodality came to the forefront and gained a more central epistemological role. As Kant asserts in a famous passage in *Critique of Pure Reason*, without sensibility no intelligible object would be given to us. Without the intellect, no sensible object would be thought (Kant, 1781/2003, p. 93; A51/75 rephrased).

Although embodiment and multimodality came to play a more prominent role in Kant's epistemological account of the senses, their contribution remained confined to providing the intellect with the raw material for it to be set into motion. In Kant's view, the empirical data becomes thinkable only because the intellect picks it up and endows it with conceptual content. For intelligible knowledge is *not* the content of generalized experience. This is why Jim, in the course of his mathematical experience, cannot possibly derive the universal and necessary properties of the prism that make it a prism in the mathematical sense. Universal necessity is unattainable from the apparent "necessity" that arises from experience, which remains always situated in space and time. If Jim comes to recognize the prism as an ideal object with its universal mathematical properties, it is not because of experience. In Kant's account, the architectonic constitution of the human mind, with its arsenal of a priori knowledge and pure principles, provides Jim with "nothing but what may be called the pure schema of possible experience" (Kant, 1781/2003, p. 258, A236–237/B295–296).

But what is a schema for Kant exactly? The schema is a kind of analogical procedure—a "monogram" (Kant, 1781/2003, p. 183, A 142/B 181)—that *unveils* the link between the intellectual and the sensual in the course of its empirical execution. In one respect, the schema must be intellectual; in another, it must be sensible. But the schema does not have to be confounded with an image:

> If five points be set alongside one another, thus, ... I have an image of the number five. But if, on the other hand, I think only a number in general, whether it be five or a hundred, this thought is rather the representation of a method whereby a multiplicity, for instance a thousand, may be represented in an image in conformity with a certain concept, than the image itself. For with such a number as a thousand the image can hardly be surveyed and compared with the concept. This representation of a universal procedure of imagination in providing an image for a concept, I entitle the schema of this concept. (Kant, 1781/2003, p. 182, A140/B179)

In saying that the schema is a method or universal procedure, Kant meant that its execution can be repeated again and again. The schema entails, in fact, a principle of iteration that links, thereby, knowledge and action (Radford, 2005).

Embodiment in Piaget's Epistemology

Considering himself a good Kantian, Piaget was not thrilled by Kant's *apriorism*. Piaget agreed with Kant that the object of reason is to inform experience. However, he did not accept Kant's idea that reason is something given a priori. The 28-year-old Piaget considered reason as constituted *in* experience: "Experience and reason are not two terms that we can isolate: Reason regulates experience and experience adapts reason." (Piaget, 1924, p. 587). But how could reason *emerge* from experience?

As we previously said, in Kant's theory of knowing, there always was an unbridgeable gap between the sensible and the intelligible, with the result that the latter cannot be thought of as a generalization or an abstraction of the former. Consequently, Kant did not need a concept of abstraction. Piaget, by contrast, was in need of one that would account for the emergence of reason *in* experience. He resorted to the aforementioned concept of *schema*, a concept that Piaget adapted to his needs by putting an emphasis on abstracting actions. Piaget (1970) argued that in abstraction there are two possibilities:

> The first is that, when we act upon an object, our knowledge is derived from the object itself. This is the point of view of empiricism in general, and it is valid in the case of experimental or empirical knowledge for the most part. But there is a second possibility: when we are acting upon an object, we can also take into account the action itself, or operation if you will.... In this hypothesis the abstraction is drawn not from the object that is acted upon, but from the action itself. It seems to me that this is the basis of logical and mathematical abstractions. (p. 16)

Piaget was eyeing the second kind of abstraction, which he termed *reflective abstraction*. Emancipated from its objects, reflective abstractions can be coordinated among themselves, for instance additively, temporally, and sequentially, giving rise to schemas that he interpreted as having "parallels in logical structures" (Piaget, 1970, p. 18) of a mathematical kind:

> Any given scheme in itself does not have a logical component, but schemes can be coordinated with one another, thus implying the general coordination of actions. These coordinations form a logic of actions that are the point of departure for the logical mathematical structures. (p. 42)

对皮亚杰而言，让吉姆认识到棱柱这一数学对象的是这么多年以来他在这门课中所遇到的大量的图式之间的逻辑协调性：学龄前感知和身体运动触觉活动是形象图式的来源，接着产生大小守恒的图式，大小守恒的图式大约在9岁或10岁的时候获得，并且以无中心的面对面的对象之间的关系为特征，比较的图式来源于外部形象和内部形象的关系，等等（例如，可以参见Piaget，1973）。那些各种各样的图式协调的最终产物就是射影几何关系和欧几里得几何关系，从这些关系中，吉姆看到的棱柱就是：一个具有一般性质的数学意义上的棱柱。

尽管康德把工作分为感性的和理性的，皮亚杰却主张感性知识和理性知识之间的发展关系。在皮亚杰的认识论中，感觉运动行为产生了一种实际知识（以尚不完整的逻辑-数学结构为特征），进而实际知识扩展为概念性知识，从而，随着"符号功能"的出现，动作和手势成为概念性的表征，符号功能是指"用信号、符号和其他物体来表示某个对象的能力"（Piaget，1970，第45页）。但是，这里要注意的关键点是，对皮亚杰来说，个人的多模式的和具身的活动逐渐消退，智力活动被逻辑-数学结构所支配，客体和教具也是如此。反省抽象是一种摆脱了客体和教具的抽象。它使得行为变成运算，运算变成符号。但是，据称这种人类思维的结构化本质使得皮亚杰可以在高级思维阶段的成因分析中撇开教具、手势、感知和所有具身的活动。皮亚杰写道：

> 反省抽象，来源于主体行为的原始概念，使主体行为转化为运算，这些运算迟早会使用符号展开而不再对最初操作的物体有任何关注，这些物体从一开始就成为在任何情况下的"任何东西"。（Beth & Piaget，1966，第238页）

小结一下，皮亚杰强调动作和手势的认识论作用。然而，对运算结构的强调给操作内容的主题化、慎重考虑的符号系统以及孩子们使用的文化产品仅留下了很小的余地。因此，尽管在研究中皮亚杰巧妙地使用了一系列令人敬佩的特别器具（如积木、液体容器、火车、汽车、各种重量和形状的物体），但操作对象在图式中的地位其实并不重要，就像刚才所引用的话那样，这些物体从一开始就可以是"任何东西"。威利龙和拉巴德尔评论道，相对于皮亚杰学派的主体，客体基本上是非历史的、非社会的，"它的主要属性是由物理规律构成的……在典型的皮亚杰学派实验中引入教具主要是因为它们能方便地凸显出真理的不变性质"（Verillon & Rabardel，1995，第80页）。

皮亚杰所依赖的结构主义（尽管是动态的结构）引出了在其认识论中无法解决的矛盾——我们可以说，这些矛盾的大小与康德在其认识论中引入那个先验论而带来的矛盾大小是相当的。对于康德而言，这种矛盾表现为感性和理性，对皮亚杰来说则表现为结构和对象。吉姆之所以能够跨越结构和对象之间的鸿沟，即吉姆所具有的关于具体棱柱的情境化的经验到具有普遍的、必备性质的数学对象之间的鸿沟，是因为所谓的认知结构和数学结构的"平行性"。不过，对上述两种认识论而言，共同的部分是发达心智的活动最终往往在很大程度上局限于抽象的心理活动（Radford，2005）。

在下一节中，我们将概述数学教育中关于具身的一些观点，并考察这些观点是如何解决上面我们所描述的矛盾的。

数学教育中的具身

皮亚杰流派的馈赠

皮亚杰的认识论对数学教育产生了重要的影响，还影响了具身这一概念，这种影响尤为细致地体现在被称为"过程-对象"的理论中，这一理论认为，思想是从学习者的行为发展到知识结构运算的。这里有两个例子，一是APOS理论（Dubinsky，2002；Dubinsky & McDonald，2001），二是"数学的三个世界"（Tall，2013）。APOS是指活动、过程、对象、图式，而"数学

In Piaget's account, what allows Jim to recognize the prism as a mathematical object is the logical coordination of a series of schemas that he has constructed over the course of many years: figural schemas resulting from perceptual and tactile sensorimotor activities in the preschool years, followed by schemas of invariability of magnitudes, which are acquired when a child is around 9 or 10 years old and characterized by relations of de-centration vis-à-vis the objects, schemas of comparison of forms, going from intrafigural to interfigural relations, and so forth (see, e.g., Piaget, 1973). The logical coordination of those varied schemas culminates in the generation of projective and Euclidean geometric relations out of which the prism appears to Jim as it is: a mathematical prism with its universal properties.

Although Kant kept a separation of labor between the sensible and the intelligible, Piaget pleaded for a *developmental* relationship between the sensible and the intelligible. In Piaget's epistemology, sensorimotor actions give rise to a practical intelligence (characterized by an as-yet incomplete logical-mathematical structure) that expands itself into conceptual knowledge. Thus, with the arrival of the "semiotic function"—a function that has "the ability to represent something by a sign or a symbol or another object" (Piaget, 1970, p. 45)—actions and gestures become conceptual representations. Yet, the crucial point to note here is that, for Piaget, the individual's multimodal and embodied activity fades away. Intelligence becomes governed by logical-mathematical structures. The same is true of objects and artifacts. Reflective abstraction is abstraction emancipated from objects and artifacts. It converts actions into operations and operations into signs. But the allegedly structural nature of human thinking authorizes Piaget to remove artifacts, gestures, perception, and all embodied activity from the genetic analysis of the higher stages of intelligence. Piaget wrote:

> Reflective abstraction, which derives from the first concepts from the subject's actions, transforms the latter into operations, and these operations can sooner or later be carried out symbolically without any further attention being paid to the objects which were in any case "any whatever" from the start. (Beth & Piaget 1966, p. 238)

To sum up, Piaget emphasized the epistemological role of action and gesture. However, the emphasis on the operations' structure left little room for a thematization of the content of the operations and for a serious consideration of the semiotic systems and the cultural artifacts that children use. Thus, although in his experiments Piaget cleverly introduced a formidable series of ad hoc objects (blocks, fluid containers, trains, cars, objects of different weight and form), the object that the hand holds in the schema is unimportant. It may be "any whatever" from the start, as he says in the last quotation. Verillon and Rabardel comment that the object submitted to the Piagetian subject is fundamentally nonhistorical and nonsocial: "its main property is that it is structured by physical laws.... The introduction of artifacts in classic Piagetian experiments is mainly due to their convenience for highlighting the invariant properties of reality" (Verillon & Rabardel, 1995, p. 80).

Piaget's recourse to structuralism (even if it was a dynamic one) introduced irresoluble tensions in his epistemology—tensions that are proportional, we may say, to the ones Kant introduced in his by having recourse to apriorism. In Kant the tension appears between the sensible and the intelligible; in Piaget the tension appears between structure and object. What allows Jim to overcome the gap between his situated experience of the physical prism and the mathematical one with all its universal and necessary properties is the alleged "parallelism" between cognitive and mathematical structures. In both epistemologies, nevertheless, the common denominator is that the activity of the developed mind tends, in the end, to be largely confined to abstract mental activity (Radford, 2005).

In the following section we provide an overview of some perspectives on embodiment in mathematics education and examine how these approaches deal with the tensions identified above.

Embodiment in Mathematics Education

The Piagetian Legacy

Piaget's epistemology has had a significant impact on mathematics education and has also influenced the conception of embodiment. The influence is particularly patent in the so-called "process-object" theories; that is, theories that conceive of thinking as moving from the learner's actions to operation knowledge structures. Two examples are APOS theory (Dubinsky, 2002; Dubinsky & McDonald, 2001) and the "three worlds of mathematics" (Tall, 2013). APOS stands for actions, processes, objects, and schemas. The "three worlds of mathematics" refers to—

1. *conceptual embodiment*, which builds on perception and action to develop mental images that "become

的三个世界"则是指——

1. 概念具身，这是以知觉和行为为基础进而发展为心理表象并成为"完美的心理实体"（Tall，2013，第16页）。举个例子来说，"数轴来自具体世界中用铅笔和尺所画的实体线，然后，发展到'完美的'柏拉图作图，线只有长度没有宽度"（Tall，2008，第14页）；

2. 运算符号，这是指从物理操作"发展"为或多或少灵活的数学过程；

3. 公理形式化，这是指"使用在集合论定义公理系统下建构形式的知识"（Tall，2013，第16页）。

APOS与"三个数学世界"观点的区别之一为，APOS理论侧重于图式组织与起源的研究（Arnon等，2014），而"三个数学世界"的方法强调了符号的作用，并根据学习者的注意力是否集中在对象、过程或符号上来研究过程的符号压缩（Gray & Tall，1994；Tall等，2001）。

"三个数学世界"方法包括了具身的具体思路以及数学思想（de Lima & Tall，2008；Tall，2004，2008，2013；Tall & Mejia-Ramos，2010；Watson & Tall，2002）。因此，沿着前面小节中讨论过的，融合了经验主义和理性主义的传统哲学道路，韬尔（2013）认为"数学思维起源于人的感觉运动、感知和行为，通过语言和符号化得到发展"（第11页）。在"数学的三个世界"中具身这一术语的意思是"使通俗意义上的'给个具体实例'与抽象的思想相一致起来"（Tall，2004，第32页）。

注意，思想以一种抽象的、非具体化的形式存在，可以用心理意象（或者其他形式的表征）"表述"或"得到体现"，这是与二元论相一致的，即把思想的王国与物质的、感知的王国区分开来。数学哲学家大卫·博斯托克（2009，第232页）称为"概念主义"的这一理论的承诺，使方法论意义上的调查方式有了明确的含义。比如，不需要对符号、数学定义或证明之类的实践从具身的、文化的和概念上的来源进行明确的分析，然而，在"数学的三个世界"的方法中，要从毫无疑问的数学和数学家的世界来分析这些思考和做数学的元素。以实数为例，对照上述三个层次的数学世界的含义来分析是这样的：具身分析，用手指沿着数轴追踪"连续运动"来说明；符号分析，和"$\sqrt{2}=1.4142\cdots\cdots$"一起出现；形式分析，定义为一个完全有序域。韬尔（2008）对这种等级体系的说明如下：

从物理上讲，数轴可被手指追踪，随着手指从1移到2，感觉上似乎经过了1到2之间的所有的点。但是当这一过程用小数来表示时，每个10进制小数都是一个不同的点（除了反复出现9这种困难情况），因此似乎不可能想象在有限时间内跑过所有的介于1和2之间的点……形式上，实数**R**是满足完备性公理的有序域。这就卷入了一个完全不同的世界，在这里，加法不再由整数或者小数加法的算法来定义，取而代之的是简单地宣称，对任何一个实数对a、b，必有第三个实数称为a、b的和，并被表示为$a+b$（第14—15页）。

小结一下，APOS理论为研究图式的产生提供了一个精妙的视角，同时，"数学的三个世界"方法为研究越来越多的符号转化和压缩提供了一个有力的研究框架，这种符号化源于具身水平，并向灵活使用符号和记法方式发展。然而，具身仍然保持在一个一般范畴，具身动作的命运就是被灵活的符号运算所取代。

多模式

在其他具身方法中，多种多样的被学生和教师采用的呈现模式来到我们面前，特别地，即使在抽象思维中，也普遍地更加认可具身的本质特征作用，这些方法可被命名为多模式。"多模式"这一术语是从外部研究领域借用到数学教育中来的，范围从神经系统学（例如，可以参见Gallese & Lakoff，2005）到通讯研究（Kress，2001，2010）。正如爱德华兹和罗布迪所指出的（2014，第7页），它们"在各研究领域中的含义并不互相排斥，而是相互交叉，相得益彰"。在数学教育中，多模式这一术语常被用来强调一系列不同的认知的、物质的、感知的（比如知觉、听觉、触觉等）形态或资源的相关

perfect mental entities" (Tall, 2013, p. 16). For instance, "the number line develops in the embodied world from a physical line drawn with pencil and ruler to a 'perfect' platonic construction that has length but no thickness" (Tall, 2008, p. 14);

2. *operational symbolism*, which "grows out" of physical action into more or less flexible mathematical procedures; and
3. *axiomatic formalism*, which "builds formal knowledge in axiomatic systems specified by set-theoretic definition" (Tall, 2013, p. 16).

One of the differences between APOS and the "three worlds of mathematics" perspective is the following. APOS theory focuses on the investigation of schema organization and genesis (Arnon et al., 2014); the "three worlds of mathematics" approach emphasizes the role of symbols and investigates the symbolic compression of processes according to whether the learner's attention is focused on objects, procedures, or symbols (Gray & Tall, 1994; Tall et al., 2001).

The "three worlds of mathematics" approach includes specific ideas about embodiment and mathematical thinking (de Lima & Tall, 2008; Tall, 2004, 2008, 2013; Tall & Mejia-Ramos, 2010; Watson & Tall, 2002). Thus, following a blend of the empiricist and rationalist philosophical traditions discussed in previous sections, Tall (2013) acknowledges that "mathematical thinking begins in human sensorimotor perception and action and is developed through language and symbolism" (p. 11). The meaning of the term embodiment in the "three worlds of mathematics" approach is explained as something that is "consistent with the colloquial notion of 'giving a body' to an abstract idea" (Tall, 2004, p. 32).

Notice that the notion that ideas exist in an abstract, nonembodied form that can be "expressed" or "receive a body"—in mental imagery (or another type of representation)—is consistent with dualistic theories that separate the realm of ideas from the realm of the material and the sensible. This theoretical commitment, that philosopher of mathematics David Bostock (2009, p. 232) calls "conceptualism," has a clear implication for the manner in which methodological investigations are conducted. For instance, there is no need for an explicit analysis of the embodied cultural and conceptual sources of such things as symbols, mathematical definitions, or practices like proof. Instead, in the "three worlds of mathematics" approach these elements of thinking about and doing mathematics are analyzed from a taken-as-given world of mathematics and mathematicians. Real numbers, for example, are analyzed by contrasting their meaning and use within the three hierarchical worlds mentioned above: *embodied*, illustrated by a finger tracing "continuous motion" along a number line; *symbolic*, accompanied by "$\sqrt{2} = 1.4142\ldots$"; and *formal*, with a definition of a complete ordered field. Tall (2008) explains the hierarchical relationship as follows:

> Physically the number line can be traced with a finger and, as the finger passes from 1 to 2, it feels as if it goes through all the points in between. But when this is represented as decimals, each decimal expansion is a different point (except for the difficult case of recurring nines) and so it does not seem possible to imagine running through *all* the points between 1 and 2 in a finite time.... Formally, the real numbers $\mathbb{R}$ is an ordered field satisfying the completeness axiom. This involves entering a completely different world where addition is no longer defined by the algorithms of counting or decimal addition, instead it is simply asserted that for each pair of real numbers a, b, there is a third real number call[ed] the sum of a and b and denoted by $a+b$. (pp. 14–15)

To sum up, APOS theory offers a refined perspective to investigate the genesis of schemas, whereas the "three world of mathematics" approach offers a powerful framework to study the increasing transformation and compression of symbolism starting from an embodied level and moving toward flexible ways of using symbols and notations. Embodiment, nonetheless, remains a general category; the fate of embodied actions is to be superseded by flexible actions with symbols.

Multimodality

In other approaches to embodiment, the variety of embodied modalities to which students and teachers resort comes to the forefront. In particular, there is generally a stronger commitment to the essential role of the body even in abstract thought. These approaches may be termed *multimodal*. The term "multimodality" entered the mathematics education field after being borrowed from external research domains, ranging from neuroscience (for example, see Gallese & Lakoff, 2005) to communication studies (Kress, 2001, 2010). As Edwards and Robutti note (2014, p. 7), the "meanings used in these different fields of study are not mutually exclusive but intersect and complement each other." In mathematics education, the term multimodality is often used to underline both the relevance and mutual coexistence of a range of different cognitive, physical, and sensuous (e.g., perceptual, aural, tactile) modalities or resources

性和共存性，它们在教学过程中，更广义地说，在数学意义的产生中发挥作用，"这些资源或形态既包括口头的与书面的符号交流，也包括绘图、手势、操作实物教具和电子教具，以及各种肢体运动"（Radford, Edwards, & Arzarello, 2009，第91—92页）。

亚伯兰罕森（2014）给出了一个多模式综合法的例子，用他自己的术语则叫"具身设计"。这一过程包含实际操作任务和计算环境的设计以容纳"主动的多模式感知的交互"（Hutto, Kirchhoff, & Abrahamson, 2015，第375页）。比如，数学意象训练器（MIT）可让学习者通过肌肉运动和视觉感知参与到比例概念的学习中。只有当学生的一只手离桌子的距离是另一只手的两倍，并且在上下移动手的时候保持这个比例时，他才能改变电脑屏幕的颜色。因此，对比例概念的介绍完全体现在"非符号感知运动图式"中（Abrahamson, 2014，第1页）。通过使用语言、手势以及最终的书面文字，学习者在协调他个人幼稚的、具身的、实施的经验与更为形式的数学结构时得到了帮助。

这里该如何理解具身呢？学生的手势在这里起到了什么作用呢？具身表现为身体的一种能力，它因为有助于在学习过程中创建数学结构而具有构建功能（Alibali & Nathan, 2012；McNeill, 2000, 2005）。

符号束

另一个多模式综合法的实例来自于阿扎雷洛和他的合作者。受维果茨基的工作和神经科学研究的指引，他们强调了在教学情境中学生符号化活动的多样化的重要性。这里，重点不在于像我们之前提到过的受皮亚杰影响的过程对象理论那样强调图式，重点在于符号的演变（Arzarello, 2006）。与维果茨基早期关于符号的观点一致，阿扎雷洛认为符号是思维的物质中介，正如工具是劳动的物质中介一样。在这一语境下，师生所使用的手势和其他具身资源成为了符号，即使它们不像语言、代数学和笛卡尔的图形符号那样，通过明确的语法或句法规则呈现相对正式的结果规则。对阿扎雷洛而言，多模式综合法源于各种符号集合之间的联系（比如，口头语言的集合、手势语言的集合、代数符号语言的集合），基于它们自身（正式或非正式）的本质发生作用、产生变化进而构成一个"符号束"。一个符号束由"i）一堆各种各样的符号集，ii）各种符号集之间的联系所构成的集合"精确地组成（Arzarello, 2006，第281页）。

正如我们看到的那样，符号束把各种符号资源看成一个统一的范式，这可让我们通过对课堂参与者使用的各种符号的演进来描述学习。阿扎雷洛、保拉、罗布蒂和萨贝纳（2009）解释道：

通常，一个符号束是由一个学生或一群学生在解决一个问题和/或讨论一个数学题目时所产生的符号组成的，可能教师也参与了这一产生过程，因此，这一符号束也可能包含了教师所使用的符号。（第100页）

由于我们考虑的是符号束最一般的意义，因此符号束不仅包含了符号的经典体系（Ernest, 2006）或表示的显示（Duval, 2006）这些特例，也包含了手势和具体的符号。使用符号束，可以做两类有关联的分析：（1）共时分析，它关注某一个时刻各种符号的相互关系；（2）历时分析，它关注符号的演进（以及符号之间相互关系的演进）。共时分析可从符号的观点为师生的数学活动拍个"照片"，而历时分析可对师生数学活动拍摄一种基于多元符号的"电影"。

用共时观点考察到的现象的一个例子就是论文最开始描述的吉姆活动中的手势–语言间的关系。在这个简要的例子中，手势和所说的语言不能被分开考虑，因为它们互相补充了对方的意义（McNeill, 2000）。如上所述，比如，我们可以在文字记录的第三行看到手势和口头语言是同时的，感知方面（触摸、凝视）和语言高度混合在一起表述了一个论断：它们共生在符号束中。

playing a role in teaching-learning processes and, more broadly, in the production of mathematical meanings: "These resources or modalities include both oral and written symbolic communication as well as drawing, gesture, the manipulation of physical and electronic artifacts, and various kinds of bodily motion" (Radford, Edwards, & Arzarello, 2009, pp. 91–92).

An example of a multimodal approach is provided by the work of Abrahamson (2014), in what he terms "embodied design." This process involves the creation of physical tasks and computational environments that allow "proactive multimodal sensorimotor interaction" (Hutto, Kirchhoff, & Abrahamson, 2015, p. 375). For example, the Mathematical Imagery Trainer (MIT) allows the learner to engage kinesthetically and perceptually with the idea of proportionality. The student can change the color of a computer screen only when he holds one of his hands twice as far from the table as his other and maintains this ratio while moving his hands up or down. Thus, the introduction to proportionality is fully embodied in "non-symbolic perceptuomotor schemas" (Abrahamson, 2014, p. 1). Through the use of language, gesture, and, eventually, written inscriptions, the learner is assisted in reconciling his naïve, embodied, enacted experiences with more formal mathematical constructions.

How is embodiment understood here? What is the role of the students' gestures? Embodiment appears as a faculty of the body that has a constructive function in that it helps the creation of mathematical constructs in the course of learning (Alibali & Nathan, 2012; McNeill, 2000, 2005).

Semiotic Bundles

Another example of the multimodal approach has been developed by Arzarello and collaborators. Drawing on Vygotsky's work and neuroscience research, they stress the importance of the multimodal character of the students' semiotic activity in teaching and learning contexts. Here, the emphasis is not on schemas, as is the case of the Piagetian-influenced process-object theories mentioned before, but on the *evolution of signs* (Arzarello, 2006). In accordance with Vygotsky's early concept of sign, Arzarello refers to signs as mediating entities of thinking, much as tools are conceived of as mediating entities of labor. Within this context, gestures and other embodied resources to which students and teachers resort become signs, even if they do not present relatively formal rules of production as do language and algebraic and Cartesian graphic symbolism, through explicit grammatical or syntactic rules. In Arzarello's approach multimodality occurs through relationships between sets of signs (e.g., the set of speech language, the set of gestures, the set of algebraic symbols), produced and transformed according to their (formal or informal) nature and constituting a "semiotic bundle." A semiotic bundle is precisely formed by "i) A collection of semiotic sets. ii) A set of relationships between the signs" (Arzarello, 2006, p. 281).

As we can see, the semiotic bundle considers the semiotic resources in a unifying manner, allowing for the description of learning through the evolution of signs as they are produced in the classroom by all participants. Arzarello, Paola, Robutti, and Sabena (2009) explain,

> Typically, a semiotic bundle is made of the signs that are produced by a student or by a group of students while solving a problem and/or discussing a mathematical question. Possibly the teacher too participates to this production and so the semiotic bundle may include also the signs produced by the teacher. (p. 100)

Due to the very general nature of the sign that is considered, the semiotic bundle includes the classical systems of signs (Ernest, 2006) or registers of representation (Duval, 2006) as particular cases, but also gestures and embodied signs. Using the semiotic bundle, two kinds of interrelated analysis can be done: (1) a *synchronic analysis*, which focuses on the relationships between different signs in a certain moment, and (2) a *diachronic analysis*, which focuses on the evolution of signs (and the evolutions of the relationships between signs). Synchronic analysis allows for taking a kind of "picture" of the students' and teacher's mathematical activity from a semiotic point of view; diachronic analysis allows for obtaining a sort of multimodal semiotic "movie" of such an activity.

An example of a phenomenon detected with the synchronic view is the gesture-speech relationship in Jim's activity as described in the initial vignette. In this brief example, gestures and words cannot be considered separate because the meaning of one set completes the meaning of the other one (McNeill, 2000). As pointed out above, we can see for instance in line 3 of the transcript that gestures are co-timed with speech, and the sensuous aspects (touching, gazes) are deeply intermingled with speech to jointly express a reasoning: they co-live in the semiotic bundle.

The diachronic analysis is at the heart of the analysis carried out within the semiotic bundle perspective

历时分析在符号束内部实施的分析中占据了中心地位,因为它可以让研究者判断在学生活动中,有没有以及如何发生意义的演变。比如,随着和教师以及同伴的讨论,吉姆在一张纸上画了一个五边形底面的棱柱来表示一般棱柱,并说道[图26.2,图片1]:

> 吉姆:一条边,嗯,呃,每条边[他的手指在他所画的图的边上滑动]是一个侧面[图26.2,图片2],所以它有五个侧面。

所以吉姆创作了一个书面图解,根据这个图解,我们辨认出与他交互的教具(的一个变形),他所使用的手势(的一个变形),以及他曾做的推理(的一个变形)。这种变化不仅包括边数上的变化(五边形而不是六边形),还包括与他互动的符号(是一幅画而不是那个塑料教具):符号束发生了演进,这种变化可能暗示了吉姆推理中的一般化方法,而这种一般化又体现在符号束的关系和演变中。

观察学生的符号演进过程,教师可以得到关于学生理解的线索,因此,活动的多模式特征可以帮助她决定,到底要不要为了帮助学生而进行干预。在文献中提到的一个教学现象叫做"符号游戏"(Arzarello等,2009),当教师协调学生所使用的某些符号集(通常是模仿某一手势),将之与其他符号集匹配(通常是言语或书面的数学符号)以建立个人的和大家都认可的数学含义之间的联系时,这一游戏就发生了。因此,在文化意义上的符号含义共享的使用过程中,符号游戏成了一个重要的策略。

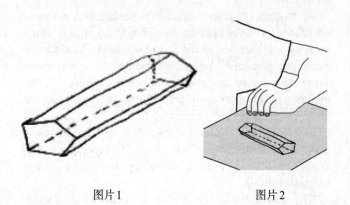

图片1　　　　　图片2
图26.2　历时分析的例子:图示以及稍后吉姆使用的手势

行动主义

正如马图拉纳和瓦雷拉(1992)的著作所描述的,行动主义的起源是人类的生物学根源和梅洛-庞蒂(1945/1962)的现象学理念。行动主义和许多具身理论在批判笛卡尔二元论观点方面是一致的,如客体和主体、理性和感性、思维和身体以及认知和现实,它试图为我们前面概述的理性主义和主观主义认识论的传统提供一种折中的观点。对一个行动主义者而言,世界并不具备那种独立于人类认知系统之外的先验的性质,同样地,认知系统也不能被认为是投射其自身的一个世界——一个其真实性是"系统内部规则反映"的世界(Varela, Thompson, & Rosch, 1991, 第172页)。行动主义者反对客观理性主义,他们认为认知范畴是经验的。这就是为什么行动主义者认为认知的功能并不是去反映这个世界:"认知并不反映世界,而是创造一个世界"(Reid & Mgombelo, 2015,第176页)。同时,行动主义者反对主观主义的观点,认为人类的认知范畴从属于他们所共享的生物和文化世界,与客观主义和主观主义观点不同,行动主义者认为世界就在那里,每个个体"相互有别"(第172页)。瓦雷拉、汤普森和罗施说,"我们的意图是通过研究认知,不把认知当作恢复或投射,而是作为具体的行动,从而完全绕过内部与外部这种逻辑地理关系"(第72页)。他们这样来解释具身行动的含义:

> 通过使用"具身的"这个术语,我们想要强调两点:首先,认知依赖于经验的种类,这些经验出自于具有各种感觉运动能力的身体;其次,这些个体的感觉运动能力本身根植于一个更广泛的生物的、心理的和文化的情境中。通过使用"活动"这个术语,我们再次强调感知和运动过程,感知和活动在有活力的认知中是完全不能分割的,事实上,两者不仅仅是偶然地联系在一起的个体,它们也一起演进。(Varela等,1991,第172—173页)

在这种观点下,认知和感知被构架为直接通过在认知主体和环境之间相互作用而发生的一个活动过程,感

because it allows the researcher to determine whether and how an evolution of meanings occurred during the students' activity. For instance, going on with the discussion with the teacher and his mates, Jim draws on a sheet of paper a representation of a prism with a pentagonal base and says (Figure 26.2, Picture 1):

Jim: An edge, uh, um, every edge [*He slips his finger on the edge drawn on the sheet.*] is a side face [*Figure 26.2, Picture 2*] so it has five lateral faces.

So Jim produces a written diagram, in which we recognize (a variation of) the artifact he had interacted with, performs (a variation of) the gesture he had produced, and produces (a variation of) the reasoning he had done before. The variation consists both in the number of edges (5 for a pentagon and no longer 6 for a hexagon) and in the signs with which he interacts (a drawing and not the plastic artifact): the semiotic bundle has evolved. The changes are a possible hint of the generality with which Jim is reasoning, which, in turn, is embodied in the semiotic bundle relationships and evolution.

Looking at the evolution of the students' signs, the teacher can gain clues with respect to the students' understanding: the multimodal aspects of the activity can therefore help her decide whether or not to intervene in order to support the students. A didactic phenomenon reported in the literature is the so-called "semiotic game" (Arzarello et al., 2009), which happens when the teacher attunes to a certain semiotic set employed by the students (typically imitating a certain gesture) and couples it with another set (typically speech words or written mathematical symbols) to build a connection between personal and shared mathematical meanings. Therefore, semiotic games constitute an important strategy in the process of the appropriation of the culturally shared meaning of signs.

Enactivism

The origins of enactivism are in the biological roots of humans as described in the work of Maturana and Varela (1992) and in the phenomenological ideas of Merleau-Ponty (1945/1962). Enactivism shares with many embodied theories a critique of the Cartesian dualistic view of object and subject, the intellectual and the sensual, mind and body, and cognition and reality. It attempts to offer a middle view of the rationalist and subjectivist epistemological traditions that we outlined before. From an enactivist viewpoint, the world does not have pregiven properties that would exist independently from the human cognitive system. Nor can the cognitive system be thought of as projecting its own world—a world whose reality would be "a reflection of internal laws of the system" (Varela, Thompson, & Rosch, 1991, p. 172). Enactivists object to the objectivist rationalist view by arguing that cognitive categories are *experiential*. This is why, enactivists argue, the function of cognition is not to represent the world: "cognition does not represent a world, it creates one" (Reid & Mgombelo, 2015, p. 176). At the same time, enactivists object to the subjectivist view by arguing that human cognitive categories belong to their shared biological and cultural world. Against the objectivist and subjectivist views they claim that the world out there and the individual "specify each other" (p. 172). "Our intention," Varela, Thompson, and Rosch say, "is to bypass entirely this logical geography of inner versus outer by studying cognition not as recovery or projection but as *embodied action*" (p. 72). They explain the idea of embodied action as follows:

> By using the term *embodied* we mean to highlight two points: first, that cognition depends upon the kinds of experience that come from having a body with various sensorimotor capacities, and second, that these individual sensorimotor capacities are themselves embedded in a more encompassing biological, psychological, and cultural context. By using the term *action* we mean to emphasize once again that sensory and motor processes, perception and action, are fundamentally inseparable in lived cognition. Indeed, the two are not merely contingently linked in individuals; they have also evolved together. (Varela et al., 1991, pp. 172–173)

Within this perspective, knowing and perception are framed as active processes occurring directly through the interaction between the cognizing subject and the

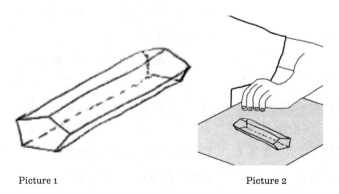

Picture 1   Picture 2

FIGURE 26.2. *Example of diachronic analysis: Diagram and gesture produced later by Jim.*

知由感知者的结构所决定，这些结构在操作上被认为是封闭和自发的。因此，学习的概念被认为是和环境交互作用而导致的适应和重构的过程，学习者和环境形成了一个复杂的动态系统（一种结构耦合）。戴维斯、萨马拉和基伦在1996年撰写的名为《认知、共同出现、课程》的论文中提到"环境中的学习者"并强调在行动主义者看来，"情境不仅仅是容纳学习者的场所，学生实际上就是情境的一部分"（引自 Reid & Mgombelo, 2015, 第177页）。

现象学途径

借鉴实验和发展心理学、认知科学和神经科学，与行动主义一样，内米洛夫斯基及其同事提出了关于数学思维和学习的一种非二元式的具身的观点。他们方法中的一个特别之处就是其现象学取向以及赋予想象力和知觉运动整合在学习者经验中的重要作用（Nemirovsky & Ferrara, 2009；Nemirovsky, Kelton, & Rhodehamel, 2013；Nemirovsky, Rasmussen, Sweeney, & Wawro, 2012）。知觉运动整合是在使用工具或身体动作中，将感知和运动深度交织构成的，特别地，在胡塞尔（1991）工作的基础上，内米洛夫斯基等人（2013）主张"(a) 数学思维由不同程度的显性和隐性表达的身体活动构成，(b) 数学学习由学习者在数学实践中身体活动的演变构成"（第376页）。这种方法不仅要克服身心二元论观点，而且要克服一些关于工具在心理学和数学教育中的作用的辩证观点（例如 Verillon & Rabardel, 1995, 以及工具性方法或者 Vygotsky, 1978, 早期的中介理论）。在内米洛夫斯基看来，辩证法往往过于倾向于认知结构而非身体经验，选择的方法是基于微人种志的研究，这种方法会极为细致地分析在很短的时间段内（以秒或分来计）活动的多模式方面，重视特定情境的多个方面。尽管感觉-运动整合包括社会文化因素，但是文化和物质层面仍然停留在背景之中，还没有被充分地整合到这一新兴理论框架所提供的整体图景中（类似的评论，见Stevens, 2012）。

唯物主义现象学

罗斯和他的同事们也采纳了现象学方法（Bautista & Roth, 2012；Hwang & Roth, 2011；Roth, 2012；Thom & Roth, 2011），他们称之为"唯物主义现象学"。他们借鉴了梅洛-庞蒂（1945/1962）、曼恩·德·比朗（1859）以及法国新生代现象学家的工作，比如马里恩（2002）和南希（2008）。他们的出发点是对理性主义具身观点的批判，以及对激进具身材料现象学的发展。事实上，他们不同意关于具身的一些观点，如在学生概念发展仍缺乏词汇时，认为手势和其他身体资源提供了理论上的内容，关键问题在于，这种理性主义对具身的理解，至少在某种程度上，假定学生已经具备了所需的概念模型和意向，即便学生还不能用一致连贯的语言方式来表达它们（类似的评论，见 Sheets-Johnstone, 2009, 第213—216页）。相反，"唯物主义现象学方法从先于思维和意向的原始形式的经验开始，再将认识理论化"（Roth, 2010, 第9页）。经验的原始形式的根源不是身体，是肉体：

> 我建议把肉体而不是身体作为一切知识的根源：知识是化身，非肉体的。因为数学知识是非肉体的，它也是被具身了的。基于肉体，我们才能找到触觉（触摸）、接触和可能性，因此也就有了知识的基础，这样身体的感觉就成为了理智的主要部分（Roth, 2010, 第9页）。

尽管在其他现象学方法中（如那些借鉴胡塞尔的工作），知觉被认为处于感觉的首位，在罗斯及其同事的激进的具身物质现象学方法中，触觉被认为是主要的感觉。围绕触觉，来自眼睛和其他感官的感觉才得以协调，"尤其是来自双手的触觉"（Roth, 2010, 第11页）。

罗斯讨论了一个二年级学生克里斯的例子，他比较了一个立方体模型和一个披萨盒，老师问，让披萨盒成为一个立方体应该怎样操作？在这一问题的推动下，克里斯用手摸着披萨盒的两条边移动，当再次指向这两条边

environment. Perception is determined by the structures of the perceiver, which are to be considered operationally closed and autonomous. Consequently, learning is conceptualized as a process of adaptation and restructuration caused by interactions within the environment, with learner and environment forming a complex dynamic system (a *structural coupling*). In their 1996 article "Cognition, Co-Emergence, Curriculum," Davis, Sumara, and Kieren speak of "the learner-in-her/his environment" and underline that in enactivist views, "context is not merely a place which contains the student; the student literally is part of the context" (quoted in Reid & Mgombelo, 2015, p. 177).

Phenomenological Approaches

Drawing on experimental and developmental psychology, cognitive science, and neuroscience, Nemirovsky and colleagues propose, like enactivism, a *nondualistic* embodied perspective on mathematical thinking and learning. One of the particularities of their approach is its phenomenological orientation and the important role ascribed to imagination and perceptuomotor integration in the learner's experience (Nemirovsky & Ferrara, 2009; Nemirovsky, Kelton, & Rhodehamel, 2013; Nemirovsky, Rasmussen, Sweeney, & Wawro, 2012). Perceptuomotor integration consists of a deep intertwining of perceptual and motor aspects of tool use or body movements. In particular, Nemirovsky et al. (2013) take a phenomenological perspective based on Husserl's work (1991) to advocate that "(*a*) mathematical thinking is constituted by bodily activity at varying degrees of overt and covert expression, and (*b*) mathematical learning consists of transformations in learners' lived bodily engagement in mathematical practices" (p. 376). This approach seeks to overcome not only dualistic views on mind and body, but also some dialectical perspectives on the role of tools in psychology and mathematics education (e.g., Verillon & Rabardel, 1995, and the instrumental approach or Vygotsky's, 1978, early theory of mediation). From Nemirovsky's perspective, dialectical approaches are too often inclined to privilege mental structures over bodily experiences. The chosen methodology is based on microethnographic studies, in which the multimodal aspects of the activities are analyzed in great detail over short periods of time (the order of seconds or minutes), taking into account multiple aspects of the situated context. Although perceptuomotor integration includes sociocultural factors, the cultural and material plane remains in the background and is not yet fully integrated in the overall picture provided by this emerging framework (for a similar remark, see Stevens, 2012).

A Materialist Phenomenology

Roth and colleagues also adopt a phenomenological approach (Bautista & Roth, 2012; Hwang & Roth, 2011; Roth, 2012; Thom & Roth, 2011) that they term "materialist phenomenology." They draw on the work of Merleau-Ponty (1945/1962), Maine de Biran (1859), and a new generation of French phenomenologists, such as Marion (2002) and Nancy (2008). Their departure point is a critique of rationalist versions of embodiment and the development of a radical embodied material phenomenology. Indeed, they disagree with some views of embodiment in which gestures and other bodily resources are understood as supplying with theoretical content the words that the students may still be lacking in their conceptual development. The question at stake is that such rationalist understanding of embodiment presupposes, at least to some extent, that the student already has the required conceptualizations and intentions, without being able to express them in a coherent verbal manner (for a similar critique, see Sheets-Johnstone, 2009, pp. 213–216). Instead, "A materialist phenomenological approach theorizes knowing beginning with primitive forms of experiences that precede mind and intention" (Roth, 2010, p. 9). At the root of the primitive forms of experience is not the body; it is the flesh:

> I suggest taking the flesh rather than the body as the ground of all knowing: knowledge as incarnated, enfleshed. Because it is enfleshed, mathematical knowledge also is embodied. It is the flesh where we find tact (touch), contact, and contingency, and therefore, the ground of knowledge so that the sense of the body comes to be the body of sense. (Roth, 2010, p. 9)

Although in other phenomenological approaches (e.g., those drawing on Husserl's work) the primacy of senses is attributed to perception, in the radical embodied material phenomenology that Roth and colleagues propose, touch is the main sense. It is around touch that the sensations from the eyes and the other organs are coordinated, "especially with touch from the hands" (Roth, 2010, p. 11).

Roth discusses the example of Chris, a grade 2 student, comparing a material cube and a pizza box. The teacher asks what the pizza box would have to have to make it a cube. Prompted by the question, Chris moves

的时候，他说出了"正方形"这个词。罗斯（2010）解释道：

> 随着眼睛和手的运动和协调，世界开始从触摸中呈现出它的样子。克里斯现在的经验是基于他手眼的协调，所以他会看着这个披萨盒并沿着它的一条边移动手，然后是另一条边。这种经验来自于手眼协调的实现，以及自己具备移动手眼这一能力的具体实现。（第11页）

罗斯所倡导的激进的具身物质现象学和前面所描述的行动主义现象学之间有一定的契合点，但它们并不一致。核心的区别在于内在的观点，唯物主义认识论认为：生物体所喜爱的那种原始的被动性的存在，为他们提供了影响事物同时被事物影响的可能性。在披萨盒这个例子中，这种内在特性经由克里斯的以下前概念和前意识的行为表明自己的存在：克里斯的手沿着披萨盒的边移动，后来又沿一个立方体教具移动。"是肉体，利用其触觉、接触（摸和被摸）和可能性，成为了一切感知、意义建构的努力以及由此而来的知识的基础。"（Roth，2010，第13页）。

但激进的具身物质现象学与行动主义之间还存在更多的不一样。罗斯（2010）的表述如下：

> 瓦雷拉等（1991）提出在"主体、社会和文化的交互中"寻找知识（第179页）。在这里要明确表达的是不存在什么交互界面：思维存在于社会和文化中，同样，社会和文化也存在于思维中。类似的观点可见于活动理论，从L.S.维果茨基到A.N.莱昂特耶夫，直到今天。马图拉那和瓦雷拉（1980）认为所谓社会就是"耦合的人类系统"（第118页），然而，我们这里的观点恰恰与之相反，具体的人类是社会的产物，而**不是**前面所说的耦合，或如活动理论中所说的，因为有社会才有思维。（第16页；重点为原著所加）

包容性唯物主义

德·福雷塔斯和辛克莱（2014）汲取了拜拉德（2007）、夏特雷（2000）和德勒兹（1968/1994）等人的工作，提出了他们称之为包容性唯物主义的方法。注意到具身理论仍经常将焦点放在学习者个体身上并设想数学概念是一个被动的存在，他们主张对主体进行概念重建，以延拓传统概念。他们建议假设主体是一个集合物，包含了"人类和非人类的组合成分"（2014，第25页），即一个由有机物质、概念、工具、符号、图表和物品等多种成分组成的一个集合物（2014，第225页）。

这种唯物主义本体论的立场为我们打开了一个讨论主体的空间，它不仅仅是人类皮肤下的东西，它也为讨论数学的主体和人们在数学活动中所使用的工具主体提供了空间。"我们所倡导的这一新的唯物主义"，他们说，"目的在于接受把数学的'主体'看成一个集合体，包含了做数学的人的身体，也包含了她的工具/符号/图表的身体"（de Freitas & Sinclair，2013，第454页）。

德·福雷塔斯和辛克莱所倡导的这一观念使得讨论的主题从作为个人活动主要特征之一的意向性，转移到个体的能动性及能动性的媒介这个领域。正如他们所指出的，"此处我们的目的在于减少关注人的意向，更多地去关注人各部分的能动性。我们想对一些本体论的原则提出质疑，这些原则支撑着人类身体是其自身参与的主要管理者这些特定概念"（2014，第19页）。

尽管人们通常认为人类具有能动性，天生具有智能是人类的属性，但是在包容性唯物主义中，这些属性并非仅限于人类。在这一观点下，把物质看作是能动体进行讨论就变得有意义了，因此，关于前面提到的罗斯（2010）对立方体的分析，德·福雷塔斯和辛克莱主张，"立方体的问题和数学概念的问题也都是能动的个体"（2014，第24页）。

包容性唯物主义延拓的不仅是身体这个概念，还延拓了智能中介这个概念，智能中介的概念必须被重新考虑，因为包容性唯物主义

> 质疑了集合体的任何一个部分都是活动、意图或意志的来源这一前提。这种质疑将意味着需扭转诸如学生智能中介观念，以及为改进或支持学生的智能中介而进行的宣传或干预。我们将需要重新考

his hand along two edges of the pizza box. He utters the word "square" while pointing to the two sides again. Roth (2010) explains:

> With the movements and coordination of movements of eyes and hands, the world begins to emerge from touch. Chris's present experience is based on the coordination of hands with eyes, so that seeing the pizza box and moving the hand along one edge, then another edge, is but a realization of the coordination of hands and eyes and the concrete realization of the ability of moving them. (p. 11)

There are certain converging points between the radical embodied materialist phenomenology that Roth advocates and the enactivist approach previously described, but they do not coincide. A central difference is the immanent viewpoint that the materialist epistemology adopts: the existence of an original passivity that living organisms enjoy and which provides them with the possibility of affecting something and to be simultaneously affected—in the pizza box example, this immanence expresses itself in the preconceptual and preintentional movement of Chris's hand along the edge of the pizza box and, later on, a material cube: "It is the flesh, with its capacity of tact (i.e., sense of touch), contact (i.e., touched and being touched), and contingency that is the ground of all senses, sense-making efforts, and, therefore, knowledge" (Roth, 2010, p. 13).

But there is more to the difference between the radical embodied materialist phenomenology and enactivism. Roth (2010) expresses it as follows:

> Varela et al. (1991) propose to look for knowledge at the "interface between body, society, and culture" (p. 179). In the position articulated here there is no interface: mind *is* in society and culture as much as society and culture are in the mind. Similar positions can be found in activity theory from L. S. Vygotsky via A. N. Leont'ev to the present day. Maturana and Varela (1980) take societies to be "systems of coupled human beings" (p. 118), whereas the position here is the converse: the specifically *human* being is a result of society rather than preceding coupling or, in activity theoretic terms, there is mind because there is society. (p. 16; emphasis in original)

Inclusive Materialism

De Freitas and Sinclair (2014) draw on the work of Barad (2007), Châtelet (2000), and Deleuze (1968/1994) to articulate an approach that they term *inclusive materialism*. Noting that theories of embodiment often remain focused on the individual learner and conceive of mathematical concepts as passive entities, they argue for a reconceptualization of the body that stretches conventional concepts. They suggest conceiving of the body as an assemblage "of human and non-human components" (2014, p. 25)—a heterogeneous assemblage of organic matter, concepts, tools, signs, diagrams, and objects (2014, p. 225).

This materialist ontological stance opens a space to talk of the human body as something that is more than what goes under the skin. It also makes room to talk of the *body of mathematics* and the *body of the tools* one uses in mathematical activity. "The new materialism we propose," they say, "aims to embrace the 'body' of mathematics as that which forms an assemblage with the body of the mathematician, as well as the body of her tools/symbols/diagrams" (de Freitas & Sinclair, 2013, p. 454).

The idea of the body that de Freitas and Sinclair propose moves the discussion away from intentionality as one of the chief characteristics of individuals' actions and focuses on the field of agents and agency. As they note, "our aim here is to focus less on human intention and more on distributed agency. We want to problematize some of the ontological tenets underpinning particular conceptions of the human body as the principal administrator of its own participation" (2014, p. 19).

Although to be an agent and to be endowed with agency have usually been considered attributes of humans, in inclusive materialism, these attributes are not restricted to humans only. In this perspective, it makes sense to talk about matter as agentic entities. Thus, referring to Roth's (2010) analysis of the cube discussed above, de Freitas and Sinclair contend that "the matter of the cube and the matter of the mathematical concepts are also agents" (2014, p. 24).

Inclusive materialism stretches not only the concept of the body but also the concept of agency. The concept of agency must be rethought, because inclusive materialism

> problematizes the premise that any one part of the assemblage is the source of action, intention or will. Such problematizing will mean revising notions such as student agency, as well as advocacy or interventions for improving or supporting student agency. We will need to reconceive agency as operating within the relations of an ever-changing assemblage. (de Freitas & Sinclair, 2014, p. 33)

And this is precisely what Roth's (2010) analysis of the cube discussed above does not address adequately. Indeed, de Freitas and Sinclair (2014) argue that Roth's analysis fails to notice that the cube is not an inert object but rather an animated entity in "intra-action" with the

虑智能中介，使其在不断变化的集合体的关系中发挥作用（de Freitas & Sinclair, 2014, 第33页）。

这正是罗斯（2010）在以上对立方体的分析中讨论不够充分的地方。事实上，德·福雷塔斯和辛克莱（2014）认为，罗斯的分析没有注意到立方体不是一个惰性的物体，它是在和学生的"内部作用"中，以及在数学概念的形成过程中的一个活跃的实体：

> 尽管罗斯关于立方体的例子阐明了身体在学习中的作用，但他的分析没能公平地对待立方体或是数学的重要性。也就是说，这一分析未能处理好立方体之所以成为立方体的方法，通过与孩子的接触，在变成立方体的过程中，它改变着自己的边。在这种（人与物）的接触过程中，罗斯把非人类的物体看成被动的、惰性的。……此外，立方体的数学概念保持不变，不受接触者的干扰，似乎它确实是一个非物质的和不变的概念，恰巧在这个特定的情况下表现出来了。（第23—24页）

总之，在现有关于数学的物质方面的文献中，德·福雷塔斯和辛克莱没有看到的观点就是"数学概念如何用一种有效的、智能中介的方式进入物质"（第40页）。

他们将数学概念的"有说服力的、有生气的、变化的、活泼的和物质的"（de Freitas & Sinclair, 2014, 第226页）本质定位于一个被称为虚拟的概念范畴，为了理解虚拟的意义和虚拟性，我们需要回到德勒兹的虚拟概念。德勒兹（1968/1994, 第209页）主张"必须严格地把虚拟定义为真实对象的一部分，仿佛这个对象本身就有一部分陷入在虚拟中，好像进入了一个客观的维度中"，这就是为什么"每个物体都是双重的，而不是说它的两个部分彼此相似，一个是虚拟图像，另一个是真实图像"（第209页）。包容性唯物主义把这个观点延展到了数学对象，所以，"数学不能脱离'可感知的物体'，而正是物体的虚拟维度激活了数学概念。因此数学实体是具有虚拟和真实两个维度的物质对象"（de Freitas & Sinclair, 2014, 第201—202页；强调为原著所加）。

因此，一个对象（数学的或其他的）是一个双重对象，由一个真实的图像和一个虚拟的图像组成，正是从它虚拟的一面，我们发现了概念的可变性。从这种后人本主义观点考虑，传统数学教学未能关注这种虚拟性，它关注的是逻辑。现在，根据这个观点，可以唤起或调用虚拟。虚拟性是这样一种东西，可被"激发""恢复""释放"和"想象"，但也会被"宰杀"（de Freitas & Sinclair, 2014, 第213页）。手势、图表和数学符号被认为是在"调用一个动态的挖掘过程，在可感知的物质中想象虚拟"（2014, 第67页）。

在本章的其余部分，我们将讨论数学教育中另外两种关于具身的方法。第一种方法来自文化-历史的文化理论以及它的辩证唯物主义哲学，第二种方法来自认知语言学。

辩证唯物主义

和前面所讨论的一些具身方法类似，辩证唯物主义者（Ilyenkov, 1982; Lefebvre, 2009）强调身体、物质和物质对象在认知和形成中的作用。然而，实物（比如，Ruth, 2010中的立方体的例子，或者我们在前面引入中提到的六棱柱）并不被认为是智能中介，但是它们也不只是我们手摸到的、耳朵听到的或者眼睛看到的东西，它们被认为是人类劳动积淀的承载物。换言之，它们承载着人类智能和人类生产的特殊历史形式，用某种确切的方式，影响着我们了解这个世界的方式。因此，从辩证唯物主义的观点来看，罗斯（2010）例子中克里斯手持的那个立方体和吉姆手里拿的那个棱柱（见引言中的图26.1）并非是中立的概念，这些物体承载着在文化发展的进程中产生和提炼出来的历史智慧，它们给学生提供了潜在的几何概念类别，以此来使他们对世界作出分类和理解。吉姆手中那个中国制造的塑料六棱柱已然体现了认识世界的一种方式，它是一种文化产物，已经嵌入到一个具有大规模学校教育和特定的知识传递形式的社会所拥有的特定的工业历史形式中，这些知识与古希腊、中世纪或前哥伦布时代的玛雅文化有本质区别。

student and the mathematical concept in a process of becoming:

> While Roth's cube example sheds light on the role of the body in learning, the analysis fails to do justice to the materiality of either the cube or the mathematics; that is to say, it fails to reckon with the way in which the cube is itself becoming-cube through its encounter with the child, shifting its own boundaries in this process of becoming. Roth treats the nonhuman material in this encounter as passive and inert.... Moreover, the mathematical concept of cube remains untouched and untroubled by the encounter, as though it were indeed an immaterial and inflexible concept that happens to be somehow manifest in this particular instance. (pp. 23–24)

In general, what de Freitas and Sinclair do not see in the current literature on the material aspects of mathematics is "how mathematical concepts partake of the material in operative, agential ways" (p. 40).

They locate the "forceful, animate, mobile, alive and material" (de Freitas & Sinclair, 2014, p. 226) nature of mathematical concepts in a conceptual category called the virtual. To understand the meaning of the virtual and virtuality, we need to go back to Deleuze's concept of the virtual. Deleuze (1968/1994, p. 209) contended that "the virtual must be defined as strictly a part of the real object—as though the object had one part of itself in the virtual into which it is plunged as though into an objective dimension." This is why "Every object is double without it being the case that the two halves resemble one another, one being a virtual image and the other an actual image" (p. 209). Inclusive materialism expands this idea to mathematical objects as well. As a result, "mathematics cannot be divorced from 'sensible matter,' and it is the virtual dimension of this matter that animates the mathematical concept. Mathematical entities are thus material objects with *virtual* and *actual* dimensions" (de Freitas & Sinclair, 2014, pp. 201–202; emphasis in original).

An object (mathematical or other) is hence a double object, made up of an actual image and a virtual one, and it is in the virtual one that we find the mobility of the concept. Considered from this posthumanist account, traditional teaching of mathematics does not attend to the virtual; it focuses on the logical. Now, the virtual, as considered in this approach, can be summoned or invoked. The virtual is something that can be "provoked," "recovered," "unleashed," and "conjured," but also "massacred" (de Freitas & Sinclair, 2014, p. 213). Gestures, diagrams, and mathematical notations are considered as "invoking a dynamic process of excavation that conjures the virtual in sensible matter" (2014, p. 67).

In the rest of the chapter we discuss two other approaches to embodiment in mathematics education. The first one comes from cultural-historical cultural theory and its dialectical materialist philosophy and the second one from cognitive linguistics.

Dialectical Materialism

Like some of the embodiment approaches discussed previously, dialectical materialism (Ilyenkov, 1982; Lefebvre, 2009) emphasizes the role of the body, matter, and the material world in knowing and becoming. Yet, material objects (e.g., the cube in Roth's, 2010, example mentioned above or the hexagonal prism alluded to in our introduction) are not conceived of as agentic entities. But neither are they considered the mere stuff that we touch with our hands, hear with our ears, or perceive with our eyes. They are considered bearers of sedimented human labor. That is, they are bearers of human intelligence and specific historical forms of human production that affect, in a definite way, the manner in which we come to know about the world. Thus, from the viewpoint of dialectical materialism, the cube that Chris holds in his hand in Roth's (2010) example and the hexagon that Jim holds in his (see Figure 26.1 in the introduction) are not conceptually neutral. These objects are bearers of a historical intelligence that has been produced and refined in the course of cultural development, providing the students with *potential* conceptual geometric categories through which they sort out and make sense of the world. The made-in-China plastic hexagon that Jim holds in his hand already intimates a way of seeing the world. It is a cultural artifact already embedded in a particular historical form of industry in a society of massive schooling and specific conveyed forms of knowing that are substantially different from those of Ancient Greece, the Middle Ages, or the pre-Columbian Maya cultures, for instance.

Dialectical materialism offers a conception of knowledge and the knowing subject as cultural, historical entities entangled in, and emerging from, material human activity (Leont'ev, 1978; Mikhailov, 1980). Within this perspective, the human subject is not a mere body. The human subject is the unique *individuation* of an ensemble of culturally and historically constituted ethical, social, political, and economic *relations*. As unique individuation of societal relations, the human subject is rather an entity in perpetual becoming—an unfolding and endless social, cultural, historical, material, and ideal (i.e., nonmaterial) *project* of life. The human subject is something always resisting the identity with itself, $I \neq I$.

辩证唯物主义提出了这样一种概念，即知识和认识主体是文化的、历史的实体，它们与人类物质活动纠缠在一起，但也从中涌现出来（Leont'ev, 1978; Mikhailov, 1980）。从这个角度看，人类主体并非是纯粹的肉体，人类主体是由文化和历史构成的道德、社会、政治、经济关系组合成的独特的个体。作为一个社会联系中的独特个体，人类主体是一个不断发展的实体，这个变化是一个不断伸展的、无穷无尽的社会、文化、历史、物质和理想的（即非物质的）生命计划。人类主体总是有点在抵制自己的同一性，我不等于我。

雷德福（2009b，2013，2014b）曾经探索过人类主体的这个话题以重新审视具身，并以新的视角思考认知、感觉、感知和物质。认知被定义为同时具有概念的、具身的和物质的特性，因此，认知不是通过"概念论的镜头"来看待的，也就是说，认知是一些发生在头脑中的想法（Bostock 2009; Stevens, 2012）。认知、身体、感觉、感知和物质被认为是各种交缠在一起的历史本性。这种理论方法被雷德福（2014b）称为"感觉认知，"

> 基于对感觉、感知、物质和概念范畴的特定历史的理解，从这个理论的角度来看，我们的认知领域只能被理解为一种文化和历史构成的感知形式，即创造性地反应、行动、感觉、想象、转化和理解这个世界。（第350页）

因此，人类的感官并不仅仅是我们生物系统进化发育装备的一部分，比如，感觉，它不再被认为是由一个沉思的笛卡尔主体（de Freitas, 2016）所操控的知觉综合，而被认为是"人类行为或实践的高度进化、特殊的模式，……仅从生物和生理来描述其特征……是不够的"（Wartofsky, 1979, 第189页）。采用这种感性认知方法，我们生来就有的生物定向-调节反应经历着文化转型，我们的感知感觉器官被转化为经历史沉淀而成的复杂的感觉形式（如，看、摸、听、尝的方式），通向人类发展的特殊形式（Radford, 2014b）。感觉的文化转型和它们在认知中的作用只能在情境中被理解为"把个人嵌入在这个世界的特定地区，也就是说，凭借这个联结的集合体，个体能够与其他个体和世界共存"（Fischbach, 2014, 第8页；经由我们翻译）。在辩证唯物论中，个体在社会中的这种动态的不断伸展和变化的嵌入过程叫做活动——材料联结活动。

为了说明这些观点，雷德福（2014a）讨论了一个例子，它发生在由7到8岁的学生组成的一个二年级普通班。在这个例子中，学生们按图26.3所示的顺序来解题。

学生们被邀请来画出第5项、第6项。在随后的问题中，这些学生又被邀请找出较远的一些项中的正方形个数，比如第12项和第25项。

雷德福指出，数学家往往倾向于认为这一序列的项是由两行组成。然后，他们审视这些行，找出项数和这一行以及那一行的方块数之间的函数线索，数学家很快就发现下面一行的方块数和项数是相等的，上面一行的方块数则比项数多1，他们总结出一般公式是 $y = n + n + 1$，也就是 $y = 2n + 1$。或者他们会注意到递推关系 $T_{n+1} = T_n + 2$（一个等差数列，将重复的加法转化为乘法）。所有这些发生得非常快，借用托夫斯基（1968，第420页）的说法，似乎两行方块以及递推关系就盯着我们的"脸"。然而，如图26.4所示，对年幼的学生来说，事情并非如此，图26.4呈现了两种典型的回答，由卡洛斯和詹姆士这两

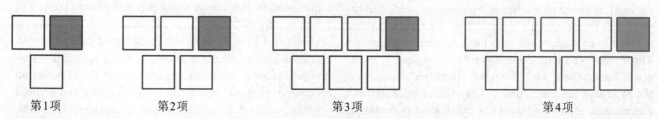

图26.3　代数课上二年级学生研究的序列的前几项。摘自 "The Progressive Development of Early Embodied Algebraic Thinking" by L. Radford, 2014, Mathematics Education Research Journal, 26(2), p.262。

Radford (2009b, 2013, 2014b) has explored this line of theorizing the human subject to revisit embodiment and to consider cognition, sense, sensation, and matter in a new light. Cognition is conceptualized simultaneously as conceptual, embodied, *and* material. As a result, cognition is not seen through "conceptualist lenses"—that is, as something about *ideas* occurring *in* the head (Bostock 2009; Stevens, 2012). Cognition, the body, sense, sensation, and matter are considered kinds of a historical nature *intertwined* with each other. This theoretical approach, which Radford (2014b) terms "sensuous cognition,"

> rests on a specific historical understanding of sense, sensation, materiality, and the conceptual realm. Within this theoretical perspective, our cognitive domain can only be understood as a culturally and historically constituted sentient form of creatively responding, acting, feeling, imagining, transforming, and making sense of the world. (p. 350)

As a result, the human senses are not conceived of as merely part of our phylogenetic evolved biological equipment. Perception, for instance, instead of being considered a sensorial synthesis carried out by a contemplative Cartesian subject (de Freitas, 2016) is considered a "highly evolved and specific mode of human action or praxis... [whose] characterization as only biological or physiological... is inadequate" (Wartofsky, 1979, p. 189). Within the sensuous cognition approach, the biological orienting-adjusting reactions with which we are born undergo cultural transformation. Our sensorial-perceptual organs are converted into historically constituted complex forms of sensing (e.g., ways of seeing, touching, hearing, and tasting), leading to specific forms of human development (Radford, 2014b). The cultural transformation of the senses and their role in knowing can only be understood in the context of the "insertion of the individuals in that specific region of the world that is society, that is to say the set of connections by which the individuals come to exist with one another and with the world" (Fischbach, 2014, p. 8; our translation). In dialectic materialism, the name of this dynamic continuously unfolding and changing process of insertion of individuals in society is *activity*—material joint-activity.

To illustrate these ideas Radford (2014a) discusses an example that comes from a regular grade 2 class of 7- to 8-year-old students. In this example, the students worked on the sequence shown in Figure 26.3.

The students were invited to draw Terms 5 and 6. In subsequent questions they were invited to find out the number of squares in remote terms, such as Terms 12 and 25.

Radford notes that mathematicians often tend to *see* the terms of the sequence as made up of two rows. Then they scan the rows for *functional* clues between the number of the term and the number of squares on one row and the other. Mathematicians quickly realize that there are as many squares on the bottom row as the number of the term, and that there is one more square on the top row than the number of the term. They conclude that the general formula is $y = n + n + 1$, that is, $y = 2n + 1$. Or they notice the recursive relationship $T_{n+1} = T_n + 2$ (an arithmetic sequence where the repeated addition is transformed into a multiplication). All this happens so quickly that it seems that the two rows and the recursive relationship stare us "in the face," to borrow an expression from Wartofsky (1968, p. 420). Yet, as Figure 26.4 intimates, things do not necessarily go that way for young students. Figure 26.4 shows two paradigmatic answers provided by two students: Carlos and James.

Figure 26.4, middle, shows in the interior of the squares the points left by the hand-pen counting device. These points are traces of Carlos's counting, which was also supported by uttered number names and a sustained perceptual activity. Against conceptualist trends, Radford (2009b) argues that "[T]hinking [is] not occur[ring] solely *in* the head but also *in* and *through* a sophisticated semiotic coordination of speech, body, gestures, symbols and tools" (p. 111).

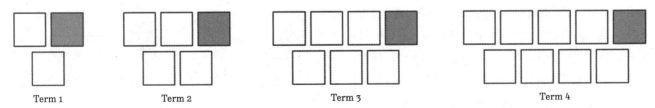

FIGURE 26.3. *The first terms of a sequence that grade 2 students investigated in an algebra lesson. From "The Progressive Development of Early Embodied Algebraic Thinking" by L. Radford, 2014,* Mathematics Education Research Journal, *26(2), p. 262.*

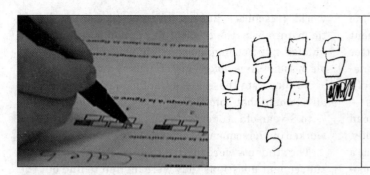

图26.4 左边,卡洛斯大声数数,按顺序指向第三项的上排的方块。中间,卡洛斯画的第5项。右边,詹姆士画的第5项。摘自"The Progressive Development of Early Embodied Algebraic Thinking" by L. Radford, 2014, Mathematics Education Research Journal, 26(2), p.263。

位学生给出。

图26.4的中间,在方块内部的点是手写计数装置留下的点,这些点是卡洛斯计数留下的痕迹,这也由清楚的数字名称以及持续的感知活动所支持。不同于概念论的倾向,雷德福(2009b)认为"思考不仅发生在大脑中,也在通过对语言、身体、手势、符号和工具进行复杂的符号学的协调中发生"(第111页)。

运用感性认知的方法,如图26.4所示的学生的回答并不意味着假定学生在序列的空间结构方面不寻求帮助。正如图26.4左边所示,卡洛斯一丝不苟地指着上面那行的方块,一个接一个,有条不紊,并且,当他数完后,他开始数下一行的方块。然而,这些项的空间结构仅仅表现为有助于执行连续计数的过程,主要活动集中于数量表征。

当然,连续计数的方法没有任何问题,只是它可能会显示出该方法在想象和研究较后面的项时有很大的局限性,正如这一课所显示的结果那样。学生看到了这些项,但他们未必会把这些项看成由两行组成,要能看出这些项是由两行组成已经需要一种理论的眼光,一种代数的观察方式,学生的眼睛还没有从一个普通人的感知器官转变为理论家的眼睛(Radford, 2010)。这就是为什么在感觉认知方法中,从文化和历史上构成的认知形式来认识是与我们感觉的转换密切相关的。我们运用并通过我们的感觉来进行实践和理论上的思考。这就是为什么在感觉认知方法中,以文化和历史的认知形式来认识是与我们感觉的转换同步的,我们借由并通过我们的感官(实际地和理论地)思考,这就是为什么在感觉认知方法中,感知、触感、手势、声音、动作和实物不是思维的媒介,它们就是思维的一部分。

雷德福(2010)记录了上述二年级学生感觉转换的一个关键时刻。教师与一个三人学生小组一起工作,在学生们以类似于图26.5所示的方式画出第5项和第6项之后,教师让他们参与探究这些行所显现出来的模式(见图26.5)。

老师说,"我们可以只看下面这行的正方形",同时,为了形象化地强调注意的对象和意图,老师做了三个连续划过的手势,每个手势都是从第一项的下面一行滑到第四项的下面一行。图26.5,左边,显示了第一次滑动的开始,老师继续道:"只看下面这些,不是上面那些。在第一项[她用她的两个食指指着第一项的下面,见图26.5,右边],那里有多少[正方形]?"其中一位学生指着回答:"一个"。老师和学生通过手势和语言,继续有节奏地探究第2、3、4项的下面这行,同时,通过手势和语言,也探索了无法被感知的第5、6、7、8项。然后,

图26.5 左边,教师指着下面的行。右边,学生和老师一起计数。摘自"Perceiving With the Eyes and With the Hands" by L. Radford, 2013, RIPEM, 3(1), p.64。

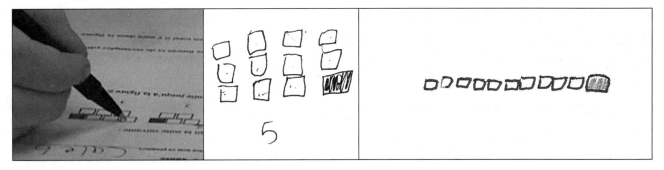

FIGURE 26.4. *Left, Carlos, counting aloud, points sequentially to the squares in the top row of Term 3. Middle, Carlos's drawing of Term 5. Right: James's drawing of Term 5. From "The Progressive Development of Early Embodied Algebraic Thinking" by L. Radford, 2014, Mathematics Education Research Journal, 26(2), p. 263.*

In the sensuous cognition approach, the students' answers shown in Figure 26.4 are not assumed to mean that the students do not seek help in the *spatial configuration* of the terms of the sequence. As shown in Figure 26.4, left, Carlos meticulously points to the squares in the top row, one after the other in an orderly manner, and, once he finishes counting them, he starts counting the squares in the bottom row. However, the spatial configuration of the terms appears as an aid to perform a consecutive counting process only. The leading activity is centered on *numerosity*.

There is nothing wrong, of course, with the consecutive counting approach, except perhaps that it may reveal itself as very limiting to imagine and investigate remote terms, as it turned out to be in the case of this lesson. The students *see* the terms, but they do not necessarily see them as made up of two rows. To see the terms as made up of two rows already requires a kind of *theoretical seeing*, an algebraic way of perceiving. The eye has yet to be transformed from an organ of quotidian perception into a theoretician (Radford, 2010). This is why in the sensuous cognition approach, coming to know in a culturally and historically constituted form of knowing goes hand in hand with a transformation of our senses. We think (practically and theoretically) with and through our senses. This is why, within the sensuous cognition approach, perception, tactility, gestures, sounds, movement, and material objects do not mediate thinking. They are part of it.

Radford (2010) documents a key moment in the transformation of the senses of the grade 2 students mentioned above. Working with a small group of three students, and after the students had drawn Terms 5 and 6 in ways similar to those shown in Figure 26.5, the teacher engaged the students in an exploration of the patterns in which the rows come to the fore (see Figure 26.5).

The teacher says, "We will just look at the squares that are on the bottom." At the same time, to visually emphasize the object of attention and intention, the teacher makes three consecutive sliding gestures, each one going from the bottom row of Term 1 to the bottom row of Term 4. Figure 26.5, left, shows the beginning of the first sliding gesture. The teacher continues: "Only the ones on the bottom. Not the ones that are on the top. In Term 1 [she points with her two index fingers to the bottom row of Term 1; see Figure 26.5, right] how many [squares] are there?" Pointing, one of the students answers: "one." The teacher and the students continue rhythmically exploring the bottom row of Terms 2, 3, and 4, and also, through gestures and words, the nonperceptually accessible Terms 5, 6, 7, and 8. Then, they turn to the top row. Imagining the nonperceptually accessible terms is a fully sensuous process out of which an algebraic sense of the relations between the number of the terms and the number of squares in their bottom and top rows starts to emerge. Through the complex coordination of gestures, words, figures, and rhythm,

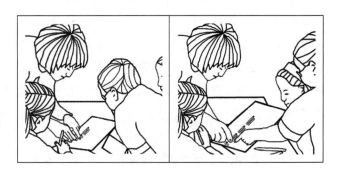

FIGURE 26.5. *Left, the teacher pointing to the bottom rows. Right, the students and the teacher counting together. From "Perceiving With the Eyes and With the Hands" by L. Radford, 2013, RIPEM, 3(1), p. 64.*

他们回到上面这行，想象那些无法被感知的项是一个完全的感知过程，经过这个过程，项数和其上、下两行正方形的数目之间关系的代数感觉开始显现。通过手势、语言、图片和节奏之间复杂的协作，学生们开始注意到一个文化和历史构成理论观下的观察和使用手势的方式，学生开始辨认出一种新的感知方法，一种代数的数形结合结构开始从中显现出来，并能够将之运用到学生感知领域之外的序列的其他项。

采用感觉认知理论的方法，教室中物质的部分和感知的部分联合活动，由此，数学的思维形式（在这个例子中，就是关于序列的代数思维形式）在理性意识上逐渐显现，这个被称为符号节点（Radford, 2009a; Radford, Demers, Guzmán, & Cerulli, 2003）。符号节点并不是一组信号，它是联合活动的一部分，通常包含对各种知觉和符号寄存器的复杂协调，为了关注某些东西（比如，一个数学结构或正在研究的一个数学概念），学生和教师要调动这些寄存器。在前面的例子中，当学生和教师一起计数的时候，符号节点包括了活动单上的符号，教师一系列的手势，教师和学生同时说出的语言，教师和学生协调的感知，教师和学生的身体位置，以及节奏，它集手势、感知、语言和符号于一体，是一个综合符号。

第二天，对代数结构意识的觉醒引导学生们得出，第12项的方块数是"12加12，加1"。符号节点的结构已经发生了变化：尽管节奏仍然出现在话语的韵律流中，但它表现为一种较短的、更为直接的方式。另外，空间指示词，比如"上面"和"底部"已经消失了，同样地，手指的手势也消失了。雷德福（2008）称这一现象为符号收缩，它是在相关的和无关的、需要说和不需要说之间做出选择的结果，这导致了"表达的收缩"，这是"更深层次的意识水平和智力水平"的物质表征（2008，第90页）。

符号节点的概念与感觉认知论中的思维理念是一致的，思维确实被认为是由物质和观念的成分组成的，包括（内部和外部的）语言、想象的感觉形式、手势、触感和我们对文化产品的实际动作。现在，把思维想象成一个感知和物质的过程，这就要求助于肉体和物质文化，但并不意味着思维是物品的集合，相反，思维是物质和观念成分的动态统一体（Rieber & Carton, 1987）。思维是处于运动和伸展中的东西——这种运动包括多种物质、语言、符号、手势、触觉、知觉、身体、美学和感情色彩与定位。

在感觉认知理论中，对符号节点及其符号收缩的研究是理解教学过程的关键。从方法论的角度来看，问题是要理解在课堂活动中，各种符号感觉模式、符号化的信号（语言、书面符号、图表等）以及文化产品是如何相互关联、协调以及纳入到一个新的感觉动态统一体中的（Radford, 2012）。

在下一节中，我们将转而讨论一个不同的具身概念，这一概念来自于意义生成的现代方法——认知语言学——它对数学教育中的具身概念产生了重大的影响。

认知语言学

认知语言学的原理是建立在具身认知理论基础上的，与行动主义一样，这一理论认为作为出生并成长于特定物理（或文化）世界的生物有机体而存在的共享经验为人类的语言、思想和意义提供了基础（Gibbs, 2006; Johnson, 1987, 2007; Lakoff & Johnson, 1999; Varela 等, 1991）。更具体地说，认知语言学的支持者认为，语言元素与其指代物之间通常不是正式和随意的联系，而是在世界上的行为、语言、思想和意义之间紧密相联的关系（Fauconnier, 1997; Fauconnier & Lakoff, 2009; Fauconnier & Sweetser, 1996; Fauconnier & Turner, 2002; Lakoff, 1987; Lakoff & Johnson, 1980, 1999）。比如说，回到开篇那个小故事，我们的五年级学生，吉姆，采用口头语言和手势的混合来论证他的观点"如果我们知道棱柱的名称，我们就可以推断出它的面数"。根据认知语言学的框架，不管是吉姆的语言还是他的肢体动作都和他对情境的考虑方式有关。当他提到棱柱的"面"时，他所使用的这个术语并非在数学内部独有，相反地，因为与人脸的联系，它具有了我们在社会世界中呈现和反应的某些平面特征。这样，以一非独有的方式，我们的物质形态成为了给数学实体命名的来源。

the students start noticing a culturally and historically constituted theoretical way of seeing and gesturing. The students start discerning a new way of perceiving out of which an algebraic numerical-spatial structure becomes *apparent* and can be now applied to other terms of the sequence that are not in the students' perceptual field.

Within the sensuous cognition theoretical approach, the segment of the material and sensuous classroom joint activity out of which a mathematical form of thinking (in this example, an algebraic form of thinking about sequences) progressively appears in sensible consciousness is called a *semiotic node* (Radford, 2009a; Radford, Demers, Guzmán, & Cerulli, 2003). A semiotic node is not a set of signs. It is a *segment of joint activity* that usually includes a complex coordination of various sensorial and semiotic registers that the students and teachers mobilize in order to notice something (e.g., a mathematical structure or a mathematical concept at work). In the previous example, when the students are counting along with the teacher, the semiotic node includes the signs on the activity sheet, the teacher's sequence of gestures, the words that the teacher and the students pronounce simultaneously, the coordinated perception of the teacher and the students, the corporeal position of the students and the teacher, and rhythm as an encompassing sign that links gestures, perception, speech, and symbols.

The next day, the emerging awareness of the algebraic structure leads the students to suggest that the number of squares in Term 12 is "12 plus 12, plus 1." The structure of the semiotic node has been transformed: although rhythm still appears there in the prosodic flow of the utterance, it appears in a shorter and more direct manner. Also, the spatial deictic terms, such as "top" and "bottom," have disappeared, as have the pointing gestures. Radford (2008) calls this phenomenon a *semiotic contraction*, which results from making a choice between what counts as relevant and irrelevant, what needs to be said and not said, leading "to a contraction of expression" that is a material token of a "deeper level of consciousness and intelligibility" (2008, p. 90).

The concept of semiotic node is consistent with the idea of thinking featured in the sensuous cognition approach. Thinking is indeed considered to be made up of material and ideational components including (inner and outer) speech, sensuous forms of imagination, gestures, tactility, and actual actions with cultural artifacts. Now, conceiving of thinking as a sensuous, material process that resorts to the body and material culture does not mean that thinking is a collection of items. Rather, thinking is a dynamic *unity* of material and ideal components (Rieber & Carton, 1987). Thinking is something moving and unfolding—a movement of multiple corporeal, linguistic, symbolic, gestural, tactile, perceptual, physical, aesthetic, and emotional tonalities and positions.

Within the sensuous cognition approach, the investigation of semiotic nodes and their semiotic contraction is a crucial point in understanding teaching-learning processes. From a methodological viewpoint, the problem is to understand how, in classroom activity, the diverse sensorial modalities, the semiotic signs (linguistic, written symbols, diagrams, etc.), and cultural artifacts are *related*, *coordinated*, and *subsumed* into a new sensuous dynamic *unity* (Radford, 2012).

In the next section, we turn to a different conception of embodiment that comes from a contemporary approach to meaning making—cognitive linguistics—that has had a significant impact on conceptions of embodiment within mathematics education.

Cognitive Linguistics

The discipline of cognitive linguistics is based on the theory of embodied cognition, which, like enactivism, holds that the shared experience of existing as biological organisms who are born and grow up in a specific physical (and cultural) world provides the foundation for human language, thought, and meaning (Gibbs, 2006; Johnson, 1987, 2007; Lakoff & Johnson, 1999; Varela et al., 1991). More specifically, proponents of cognitive linguistics suggest that the relationship between elements of language and their referents is, in general, not formal and arbitrary, but rather that the relationship is a close linkage between action in the world, language, thought, and meaning (Fauconnier, 1997; Fauconnier & Lakoff, 2009; Fauconnier & Sweetser, 1996; Fauconnier & Turner, 2002; Lakoff, 1987; Lakoff & Johnson, 1980, 1999). For example, if we look back at the opening vignette, our fifth-grade student, Jim, uses a combination of spoken words and gestures to justify his claim that "we can deduce the number of faces if we know the name of the prism." According to a cognitive linguistic framework, neither Jim's language nor his bodily actions are unrelated to the way he thinks about the situation. When he refers to the "faces" of the prism, he is using a term that did not develop arbitrarily within the mathematical community, but rather because of its association with the human face, a somewhat planar feature that we present and respond to in the social world. Thus, our physical form serves as a source for naming a mathematical entity, in a nonarbitrary way.

同样地,吉姆所做的画圈的动作,在物理上环绕了棱柱的所有的面,与认知语言学中所称的包容意象图式有关(Johnson, 1987; Lakoff, 1987; Lakoff & Núñez, 2000; Talmy, 2000)。意像图式是"反复出现的、稳定的感觉运动体验模式……保持知觉整体的拓扑结构……具有能够引发有约束的推断的内部结构"(Johnson, 2007,第144页)。这种包容意象图式来自孩子放满或清空某个容器的物理经验,这种经验建立了"里面""外面"和"边"或"边界"的概念。意象图式使得吉姆可以把棱柱的面当做一个可被计数的集合中的元素,他的画圈的手势表明这个集合的边界。包容意象图式为许多后面的理解提供了基础,不管是数学的内部还是外部,包括集合元素、函数的定义域和值域以及有界区域(Lakoff & Núñez, 2000; Núñez, 2000)。

意象图式有助于解释这样一个事实,即许多数学表达式还有一些符号能唤起空间和空间关系,即使主题不是几何的时候也是如此(如,"极限""域""映射到"⇔)。当我们从认知语言学的角度讨论"配平"一个方程,或"支持"某个观点时,由于我们都有平衡和支持身体(以及搭积木、骑车,等等)的共同经验,这些词是可以被理解的。

意象图式在认知语言学中起到了概念隐喻这个强大机制的源领域的作用。概念隐喻是两个概念域之间的无意识的映射,其中,第一个概念域的推断性结构被映射到第二个概念域上(Johnson, 1987; Lakoff, 1987, 1992; Lakoff & Johnson, 1980, 1999)。举个一般的意象结构的例子,起源-路径-目标,就是基于我们从一个地方(起点),沿着一条给定的路径,到达另一个地方(目标)的旅行实际经验(Johnson, 1987)。这个意象图式可在多个数学领域中发现,从使用数轴进行加法(Lakoff & Núñez, 2000),到函数和图像(Bazzini, 2001; Ferrara, 2003, 2014; Font, Bolite, & Acevedo, 2010),到连续性(Núñez, Edwards, & Matos, 1999),它甚至可以为理解一个和空间运动没有明显关系的概念的证明提供源领域。图26.6阐明了从"起源-路径-目标"这一图式的内在结构到数学证明的外显形式的隐喻映射(Edwards, 2010)。

如图26.7所示,这个隐喻的实证支持可以在一个博士生谈论数学证明的语言和手势中找到(Edwards, 2010):

> 学生:因为你开始运算,我从**点a**开始,**到点b结束,会有某条路径//它会经由哪里呢?** 我能演示下**我怎么到那里的吗?**(粗体表示说话与手势协调)
> (Edwards, 2010,第333页)

概念隐喻的源领域可以提取自物理世界中的经验(在这种情况下,被称为基础性隐喻),或者可以从现

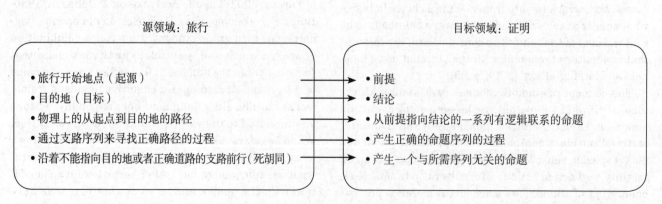

图26.6 概念隐喻"证明就是旅行。"改编自"Doctoral Students, Embodied Discourse and Proof" by L. D. Edwards, in M. M. F. Pinto and T. F. Kawasaki (Eds.), *Proceedings of the 34th Conference of the International Group for the Psychology of Mathematics Education* (Vol.2), 2010, Belo Horizonte, Brazil: PME。

Similarly, the circular gesture Jim uses to physically encompass all the faces of the prism is related to what cognitive linguistics calls the *image schema* of containment (Johnson, 1987; Lakoff, 1987; Lakoff & Núñez, 2000; Talmy, 2000). Image schemas are "recurrent, stable patterns of sensorimotor experience . . . [that] preserve the topological structure of the perceptual whole . . . having internal structures that give rise to constrained inferences" (Johnson, 2007, p. 144). The containment image schema arises from the child's physical experience of filling and emptying containers, experience which builds the notions of "inside," "outside," and "edge" or "boundary." This image schema allows Jim to think about the faces of the prism as members of a collection that can be counted, and his circular gesture indicates the boundary of the collection. The containment image schema provides the foundation for many later understandings, both within and outside mathematics, including set membership, the domain and range of a function, and bounded regions (Lakoff & Núñez, 2000; Núñez, 2000).

Image schemas can help account for the fact that many mathematical expressions and some symbols evoke space and spatial relationships, even when the subject is not geometry (e.g., "limit," "field," "onto," ⇔). When we talk about "balancing" an equation, or "supporting" an argument, from the perspective of cognitive linguistics, this is comprehensible because of our shared experience of balancing and supporting our bodies (as well as building blocks, bicycles, and so on).

An image schema can serve as the source domain for a powerful mechanism in cognitive linguistics, *conceptual metaphor*. Conceptual metaphors are unconscious mappings between two conceptual domains, in which the inferential structure of the first domain is mapped onto the second (Johnson, 1987; Lakoff, 1987, 1992; Lakoff & Johnson, 1980, 1999). As an example, a common image structure, source-path-goal, is based on our physical experience of traveling from one location (the source) to another (the goal), along a given path (Johnson, 1987). This image schema can be found in multiple areas of mathematics, from addition using the number line (Lakoff & Núñez, 2000) to functions and graphing (Bazzini, 2001; Ferrara, 2003, 2014; Font, Bolite, & Acevedo, 2010) to continuity (Núñez, Edwards, & Matos, 1999). It can even provide the source domain for understanding proof, a concept with no obvious relationship to movement through space. Figure 26.6 illustrates the metaphorical mapping from the internal structure of the source-path-goal schema to the explicit form of a mathematical proof (Edwards, 2010).

Empirical support for this metaphor can be found in both the speech and gestures of a doctoral student talking about mathematical proof, shown in Figure 26.7 (Edwards, 2010):

Student: 'cause you start figuring out, I'm starting at **point a and ending up at point b**. There's gonna be **some road//where does it go through**? And can I show that **I can get through there**? (bold indicates speech coordinated with gestures). (Edwards, 2010, p. 333)

The source domain for a conceptual metaphor can be drawn from experience in the physical world (in which case it is called a grounding metaphor), or it can be drawn from an existing conceptualization (creating a linking

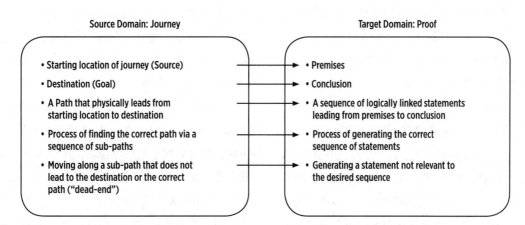

FIGURE 26.6. *Conceptual metaphor "a proof is a journey." Adapted from "Doctoral Students, Embodied Discourse and Proof" by L. D. Edwards, in M. M. F. Pinto and T. F. Kawasaki (Eds.),* Proceedings of the 34th Conference of the International Group for the Psychology of Mathematics Education *(Vol. 2), 2010, Belo Horizonte, Brazil: PME.*

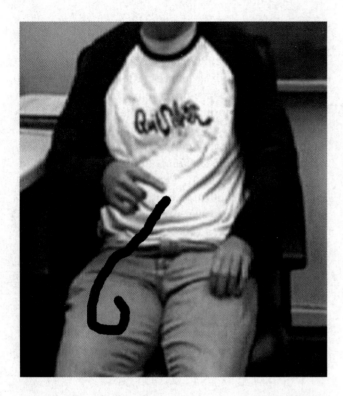

图26.7 手势暗示起源-路径-目标的意象图式。改编自"Doctoral Students, Embodied Discourse and Proof" by L.D. Edwards, in M. M. F. Pinto and T. F. Kawasaki (Eds.), Proceedings of the 34th Conference of the International Group for the Psychology of Mathematics Education (Vol. 2), 2010, Belo Horizonte, Brazil: PME。

存的概念模型中提取（创造一个链接隐喻，这可以产生更为抽象的概念链接，比如，数学中的子域，Lakoff & Núñez, 2000; Núñez, 2008）。在对大学生极限隐喻的分析中，可以发现一个在更复杂的数学层次上的概念隐喻的例子。在微积分导引课上，就极限概念，厄特曼（2009）采用访谈和书面作业的方法对学生提问。他辨认出他的学生们一贯使用的5组隐喻："(a)维度上的消失，(b)近似和误差分析，(c)逼近于空间中的第一个点，(d)一个小的物理刻度，超过它就没有东西存在了，(e)把无穷大看成一个数"（第396页）。比如，当学生被问到$\sin x$的泰勒级数时，他们会在陈述中使用"接近程度"或者"物理上接近"的隐喻，如"更为接近的话，多项式将会缠绕（增加了强调语气）在原始函数的周围"和"多项式将会越来越松散地分布（增加了强调语气）在曲线的

周围"（第417页）。

以上例子阐述了概念隐喻理论在分析具体的数学主题和概念时的使用，在数学交流之下还有一个更为基础的隐喻，这个隐喻可能导致了关于数学本体论地位的长期争论。当我们谈论数学实体时，就好像它们具有物理存在一样，也就是说，它们像是客观对象一样，这就是在使用隐喻（Font, Godino, Planas, & Acevedo, 2009; Lakoff & Núñez, 2000）。当人们讨论"配平"一个方程或者问"20中有几个5"的时候，这种隐喻就非常明显。拉考夫和努涅兹（2000）清楚地阐述了这种语言使用，他们称之为本体隐喻，它采用物理对象作为源领域来概念化数学对象，然而，这一现象在早先就被皮姆（1987）和斯法德（1994）记录下来了。斯法德（1994）指出，"我们使用'存在'一词来指代抽象的对象（比如在那些存在性定理中），这一事实以最有说服力的方式反映了抽象观念世界的隐喻特征"（第48页）。正如丰特及其同事定义的那样（Font等, 2009），客观隐喻具有极大优势，它允许人们在从事数学工作时，用处理客观物体那样的方式来处理符号和抽象观念，"移动"和"变换"它们，从而从根本上降低了认知负荷，如果每个数学符号都要以它的逻辑的、数学的定义为根据，那么认知负荷是必然的。

概念集成的观念差不多和概念隐喻同时出现在认知语言学中，然而，当概念隐喻包含着从一个源领域恰好对应到一个目标领域的单向映射时，概念集成可能包含了多重输入空间（Fauconnier & Lakoff, 2009）。概念集成，又名概念整合，"连接着输入空间，有选择地投射到一个整合的空间，并发展新兴结构"（Fauconnier & Turner, 2002，第89页）。

概念整合经常用一种类似于关联隐喻（Lakoff & Núñez, 2000）的方式建立在现存的整合上。比如，复数的映射概念有赖于数轴整合以及笛卡尔平面坐标系整合的存在，每个整合自身也是整合（Fauconnier & Turner, 2002; Lakoff & Núñez, 2000）。"复数"整合的第一个输入空间包含具有向量运算的有向坐标平面，对于第二个输入空间，这种整合吸收了正实数、负实数以及它们相

FIGURE 26.7. *Gesture indicating the source-path-goal image schema. Adapted from "Doctoral Students, Embodied Discourse and Proof" by L. D. Edwards, in M. M. F. Pinto and T. F. Kawasaki (Eds.),* Proceedings of the 34th Conference of the International Group for the Psychology of Mathematics Education *(Vol. 2), 2010, Belo Horizonte, Brazil: PME.*

metaphor, which can yield more abstract concepts—connecting, for example, subdomains within mathematics; Lakoff & Núñez, 2000; Núñez, 2008). An example of conceptual metaphor at a more sophisticated level of mathematics can be found in an analysis of undergraduate students' metaphors for limits. Oehrtman (2009) used interviews and written assignments to ask students in an introductory calculus class about limits. He identified five clusters of metaphors for limits used consistently by his students: "(a) a collapse in dimension, (b) approximation and error analyses, (c) proximity in a space of point-locations, (d) a small physical scale beyond which nothing exists, and (e) the treatment of infinity as a number" (p. 396). For example, students who were asked about the Taylor series for sin x used the "proximity" or "physical closeness" metaphor in statements such as, "the closer the polynomials will *wrap themselves* [emphasis added] around the original function" and "the polynomial becomes more and more *loosely fitted* [emphasis added] around the curve" (p. 417).

The examples above illustrate the use of the theory of conceptual metaphor to analyze specific mathematical topics and concepts; however, there is a more fundamental metaphor underlying mathematical discourse, one which may have contributed to perennial arguments over the ontological status of mathematics. This is the metaphor used when talking about mathematical entities as if they had a physical existence, that is, as objects (Font, Godino, Planas, & Acevedo, 2009; Lakoff & Núñez, 2000). This metaphor is evident when people talk about "manipulating" an equation or ask "how many fives in twenty?" Lakoff and Núñez (2000) spelled out this use of language in what they call the ontological metaphor, which takes physical objects as the source domain for conceptualizing mathematical entities; however, this phenomenon was noted earlier by Pimm (1987) and Sfard (1994). Sfard (1994) noted, "The fact that we use the word 'existence' with reference to abstract objects (as in existence theorems) reflects in the most persuasive way the metaphorical nature of the world of abstract ideas" (p. 48). The objectual metaphor, as it is termed by Font and colleagues (Font et al., 2009), offers great advantages—it allows someone carrying out mathematical work to treat symbols as well as abstract ideas as objects, "moving" and "transforming" them, thus radically reducing the cognitive load that would be required if every mathematical sign had to be grounded in its logical mathematical definition.

The construct of conceptual integration emerged within cognitive linguistics at about the same time as conceptual metaphor; however, where metaphor comprises a unidirectional mapping from exactly one source domain to exactly one target domain, conceptual integration can involve multiple input spaces (Fauconnier & Lakoff, 2009). Conceptual integration, also known as conceptual blending, "connects input spaces, projects selectively to a blended space, and develops emergent structure" (Fauconnier & Turner, 2002, p. 89).

Conceptual integration often builds on existing blends, in a manner similar to linking metaphors (Lakoff & Núñez, 2000). For example, the conceptual mapping for the complex numbers relies on the existence of the blend for the number line as well as the blend for the Cartesian coordinate plane, each of which are blends themselves (Fauconnier & Turner, 2002; Lakoff & Núñez, 2000). The first input space for the "complex numbers" blend consists of the oriented coordinate plane with vector arithmetic. For the second input space, the blend draws on positive and negative real numbers and their associated operations and properties. The blended space yields the complex numbers in the complex plane, in which each element is simultaneously a number and a vector, an emergent quality that

关的运算和性质，整合空间促生了复平面上的复数，这里每个元素同时是数和向量，这个新兴的性质无论在哪个输入空间都是不具备的，整合创造了其他新的突生的结构，比如，具有正负数的坐标轴的混合促生了由显示在 x 轴上的实数部分以及在 y 轴上的虚数部分所组成的复数。另外，"运转这个整合"（通过"精细化"机制来得出它的逻辑蕴含）使得加法和乘法运算重新被定义，以便在这个新空间中可以保持一致性和连贯性（Fauconnier & Turner, 2002）。福柯尼耶和特纳（2002）指出，这种整合的通用空间（也就是，两个空间共有的一些元素）是由两种具有一套特定性质的运算所组成的，这些性质有结合律、交换律、单位元、负元以及一个运算对另一运算的分配律，这种运算和性质的结合逐渐被认识并命名，作为一个数学实体自身，叫做交换环。

另一个关于概念整合的例子引自赞迪厄、罗和克那普（2014）的工作，他们分析了一组学生一起合作形成一个证明时运用的逻辑框架。学生们的任务是做出一个证明来表明第一个条件语句蕴含着第二个条件语句，具体来说就是"欧几里得的第五公设（EFP）蕴含了普莱费尔的平行公设（PPP）或者普莱费尔的平行公设蕴含了欧几里得的第五公设"（第213页）。针对他们证明的逻辑框架，研究者记录下两种不同的概念整合："一个简单的证明框架"（对这个特殊任务来说是不充分的）以及一个"条件隐含条件的证明框架"（第214页）。研究者也分析了学生使用与每个公设有关的视觉信息的方式，并提出概念整合的机制能让学生融合这些信息来找到他们证明所需的关键思想。他们还提出，概念整合并非总是指向正确的思路，并提供了一个三个学生合作的例子，他们"用了一种丢失隐含结构的方法来压缩欧几里得的第五公设的前提和结论"（第228页）。

尹等人（2011）的研究为如何将手势分析融入认知语言学又提供了一个例子。他们研究了他们称之为"虚拟数学结构"的内容——这一结构是经由对手势、言语和其他相关符号系统的多模式使用，在数学手势空间中通过感觉认知所创造的（第893页），也就是说，学生或者教师可以利用身体，尤其是手和手臂所提供的情境支持，经由手势和语言的联合来构建数学意义（McNeill, 2000, 2005），举个例子，某学生会绷直手掌做出（变化的）角度来表示一条不定积分曲线变化的倾斜度。这是可能的，因为物理世界的元素（这里是手和手臂）可被利用作为概念整合的输入空间，一种被称为真实空间的特殊类型的输入（Liddell, 1998）。（学生理解的）数学领域是第二个输入空间，在整合过程中，学生平放的或具有某个角度的手掌和手指被映射到切线的倾斜度，手的运动和位置沿着积分曲线而移动（Yoon 等, 2011）。经由概念整合，物理动作成为学生构建数学内容理解的一种重要来源。

认知语言学领域为理解数学提供了一个有力的理论框架和一套富有成效的工具，它同样适用于儿童早期的数感构建，也适用于数学家对于抽象架构的详细阐述。认知语言学的理论在概念上是连贯一致的，通过多种方法得到实证支持，它和认知科学（包括神经科学）研究的进展相联系（例如 Fields, 2013；Guhe 等, 2011；Winter, Marghetis, & Matlock, 2015）。具身认知的原则之一就是认知连续性原则（Johnson, 2007），在此原则下，数学在本体论上与认知及行为的其他领域并没有什么不同，反而，让人类在几千年来能够生存并繁荣兴旺的认知机制同样支持数学思维以及其他概念领域的发展。

展望未来：新问题、矛盾和问题

在这一章，我们综述了数学教育中的具身。出发点是一个现代具身理论关于意义和认知的一条普遍的宣言：意义和认识深深地扎根于实际的、物质的和具体的存在。比如，希茨-约翰斯通（2009）辩称，由于我们的生物构成，我们天生就具备一系列的原型的肉体运动形式和联系，他们组成了我们找到自己的途径来进入这个世界的基础（参见 Roth, 2012；Seitz, 2000）。基于这一关于意义和认知的具身本质的普遍宣言，具身理论尝试为意义如何产生，思想如何和行动、情绪及感觉相联系等问题提供答案（Edwards, 2011）。由于不是只有一种方法来对身体在认知中的作用进行理论化，我们对发现当代有种种关于具身认知的观点并不惊讶（综述可参看 Wilson, 2002；讨论可参看 Radford, 2013）。它们的分歧尤其存

was not present in either of the input spaces. The blend creates other new, emergent structures; for example, the blend of the coordinate axes with the positive and negative numbers yields a complex number made up of a real part, displayed on the x-axis, and a complex part on the y-axis. In addition, "running the blend" (working out its entailments through a mechanism called "elaboration") allows addition and multiplication to be redefined to work consistently and coherently in the new space (Fauconnier & Turner, 2002). Fauconnier and Turner (2002) point out that the generic space for this blend (that is, the elements that the two spaces share in common) is made up of two operations with a specific set of properties, namely, associativity, commutativity, identities, inverses, and distributivity of one operation over the other. This combination of operations and properties has come to be seen, and labeled, as a mathematical entity in itself, the commutative ring.

Another example of conceptual integration is drawn from the work of Zandieh, Roh, and Knapp (2014), who analyzed the logical frameworks that a group of students used in working together to create a proof. The task set for the students was to create a proof showing that one conditional statement implied a second conditional statement, specifically, that "either Euclid's Fifth Postulate (EFP) implies Playfair's Parallel Postulate (PPP) or PPP implies EFP" (p. 213). The researchers noted two different conceptual blends for the logical framing of their proof: "a simple proving frame" (which was inadequate for this particular task) and a "conditional implies conditional proving frame" (p. 214). The researchers also analyzed the ways in which the students utilized the visual information associated with each postulate and proposed that the mechanism of conceptual integration allowed the students to merge this information to find the key idea needed for their proof. They also note that conceptual blending does not always lead to correct thinking, offering an example in which three students "condense the premise and conclusion of EFP in a way that loses the implication structure" (p. 228).

The research of Yoon et al. (2011) provides another example of how the analysis of gesture has been integrated into cognitive linguistics. They have investigated what they refer to as "virtual mathematical constructs— constructs that are created via sensuous cognition in a mathematical gesture space through the multimodal use of gestures, speech and other linked semiotic systems" (p. 893). That is, they point out that a student or teacher can utilize the affordances offered by the body, specifically the hand and arm, to establish mathematical meanings through linked gesture and speech (McNeill, 2000, 2005), giving the example of a student using a straight hand held at a (varying) angle to represent the changing gradient of an antiderivative graph. This is possible because elements of the physical world (in this case the hand and arm) can be recruited to serve as one input space for a conceptual blend, a particular type of input called *real space* (Liddell, 1998). The (student's understanding of the) mathematical domain is the second input space. In the blend, the student's flat or angled palm and fingers are mapped to the gradient of the tangent line, and the hand's motion and location to movement along the antiderivative graph (Yoon et al., 2011). Via conceptual integration, physical action becomes an important resource for the students in constructing an understanding of the mathematical content.

The field of cognitive linguistics offers a powerful theoretical framework and a set of productive tools for understanding mathematics that can be applied equally well to a child's earliest construction of number sense or a mathematician's elaboration of abstract structures. The theory of cognitive linguistics is conceptually coherent, supported empirically via multiple methodologies, and connects with other advances in cognitive science, including neurological research (e.g., Fields, 2013; Guhe et al., 2011; Winter, Marghetis, & Matlock, 2015). One of the principles of embodied cognition is that of cognitive continuity (Johnson, 2007), under which mathematics is not ontologically different from other realms of cognition and action. Instead, the cognitive mechanisms that have allowed humans to survive and thrive over millennia have also supported the development of mathematical thought and other conceptual domains.

Looking to the Future: New Problems, Tensions, and Questions

In this chapter we presented an overview of embodiment in mathematics education. The starting point is a general claim that contemporary embodiment theories make about meaning and cognition: that meaning and cognition are deeply rooted in a physical, material, embodied existence. For instance, Sheets-Johnstone (2009) argues that, as a result of our biological makeup, we are naturally equipped with a range of archetypal corporeal-kinetic forms and relations that constitute the basis on which we make our ways into the world (see also Roth, 2012; Seitz, 2000). Based on this general claim about the embodied nature of meaning and cognition, embodied theories attempt to provide answers to questions of how meaning arises and of how thought is related to action, emotion, and perception (Edwards, 2011). Since there is not just one way in which to theorize the cognitive

在于对感觉、物质、认知和身体本身的看法上。这些分歧可以追溯到早就存在的哲学问题，即身体、感觉和心灵之间的关系问题。本章的第一部分确实简要阐述了主要认识论下的西方传统中的具身，也就是理性主义和经验主义的传统。第一部分试图说明西方思想在理解身体认识论问题上的痛苦挣扎，以更好地理解当代具身理论出现的历史背景。

当代具身理论所提出的理论观点为以新的方式来设想教学开创了新的可能。在本章的第二部分，我们提及了一些已纳入数学教育并获得发展的途径。尽管具有局限性，我们的综述还是展示了多种途径以及它们之间的一些分歧（比如，韬尔和杜宾斯基的"过程-对象"理论、罗斯的激进现象学理论、涅米罗夫斯基及其合作者胡塞尔式的现象学理论、德·福雷塔斯和辛克莱的包容性唯物主义理论、阿扎雷洛和合作者维果茨基式的符号学理论）。在本章剩下的部分，我们更详细地展开了另外两种方法，一个是"感觉认知"方法，这个方法根植于文化历史活动理论，并受到新黑格尔辩证唯物主义的启发，另一个是"认知语言学"方法，它受到拉考夫（1987，1992）和努涅斯（2000）等人工作的启发。这些途径强调认知的具身本质，并为我们提供了机会，让我们看到在我们考虑身体、手势、感觉、语言和教具时，在方式方法上有着不同的主题和不同的概念。由于这些主题的和概念性的差异，相应的研究问题和方法也有所不同。

数学教育中的具身仍然是一个新兴的正在发展中的研究领域，在概念和方法论层面还有大量的工作要做。从"感觉认知"理论的角度来看，我们需要更好地去理解学生认知的发展（包括思维、意志和情感），不是把学生认知发展看成一个严格的心智方面的事件，而是同时有概念、具身和物质的有形现象。这就是说，我们需要把认知作为一种现象来考察，这种现象从黑格尔的辩证观出发，通过更广泛的语言、身体、教具和符号活动，把有知觉的主体和思想的文化形式结合起来。比如，我们需要更好地理解（内部和外部）语言、想象的感觉形式、手势、触觉与操作文化产品（包括数学符号）之间的社会和个人活动限制的辩证关系。

从认知语言学的观点来说，一个将身体体验与数学知识和实践联系起来的综合框架的轮廓正在形成，这个框架为我们提供了明确的工具来为分析数学中的活动和从语言到符号到意象到手势这些广泛的表达形态（Edwards & Robutti, 2014）。然而，仍然存在很多问题，特别是"事后"分析数学思维是不够的，事实上，我们需要更多地学习如何在数学教学中更好地利用多种模式。此外，将实际经验与数学思维联系起来的生理和神经机制的研究相对来说还未开启。举个例子，对通过具体的动手操作来学习算术和用其他方法或用死记硬背的方式来学习算术进行比较，考察其在神经学上的相关性是很有趣的，认知语言学理论应该能预测算术概念隐喻在结果上的差异，如果能够发现这些差异，那么它们是否也会反映在神经结构或功能上？

当前数学教育中对具身的日益增长的兴趣为这一领域继续吸引更多的研究者带来了希望，这些研究者将会延续、拓展和想象出新的理论和实践途径来改进数学教与学。

role of the body, it is not surprising to find a variety of contemporary perspectives on embodied cognition (for an overview see Wilson, 2002; for a discussion see Radford, 2013). Their differences rest, among others, on conceptions about the senses, matter, cognition, and the body itself. The differences can be tracked back to the long-standing philosophical problem concerning the relationship of the body, the senses, and the mind. The first part of the chapter dealt indeed with a brief account of embodiment in the main epistemological Western traditions, namely the rationalist and the empiricist traditions. The first part was an attempt to show the legendary painful struggle of Western thought to understand the epistemological question of the body to better understand the historical background from which contemporary embodiment theories emerge.

The theoretical perspectives that contemporary embodiment theories bring forward open up new possibilities to envision teaching and learning in new ways. In the second part of the chapter we referred to some approaches that have been incorporated and developed within mathematics education. Our overview, although limited, nonetheless shows the variety of approaches and some of their differences (e.g., Tall's and Dubinsky's "process-object" theories, Roth's radical phenomenological approach, Nemirovsky and collaborators' Husserlian phenomenological approach, de Freitas and Sinclair's inclusive materialist approach, Arzarello and collaborators' Vygotskian semiotic approach). In the rest of the chapter we developed with more detail two other approaches—the "sensuous cognition" approach embedded in cultural historical activity theory and inspired by a neo-Hegelian dialectic materialism and the cognitive linguistics approach inspired by the work of Lakoff (1987, 1992) and Núñez (2000), among others. These approaches emphasize the embodied nature of cognition and provide us with an opportunity to see thematic and conceptual differences in the manner in which the body, gestures, the senses, language, and artifacts are considered. As a result of these thematic and conceptual differences, the corresponding research questions and methodologies vary.

Embodiment in mathematics education is still an emerging and developing research area. There is considerable work to be done at the conceptual and methodological levels. From the viewpoint of the "sensuous cognition" approach, we need to better understand the development of students' cognition (which includes thinking, volition, and emotion) not as a strict mental event but rather as a tangible phenomenon that is simultaneously conceptual, embodied, *and* material. That is, we need to examine cognition as a phenomenon that arises from and brings together, in the Hegelian dialectical sense, the sentient subjects and cultural forms of thought through language, body, artifacts, and semiotic activity more generally. We need, for instance, to better understand the social-and-individual activity-bound dialectic relationship between (inner and outer) speech, sensuous forms of imagination, gestures, tactility, and actions with cultural artifacts (including mathematical symbolism).

From the perspective of cognitive linguistics, the outline of a comprehensive framework linking bodily experience with mathematical knowledge and practices is emerging. This framework offers specific tools for the analysis of the wide range of modalities found in mathematical actions and expression, from speech to symbols to images to gestures (Edwards & Robutti, 2014). Yet many questions remain. In particular, it is not enough to analyze mathematical ideas "after the fact"; instead, we need to learn more about how best to employ various modalities in teaching mathematics. In addition, the physiological and neural mechanisms involved in linking physical experience to mathematical thought remain relatively unexplored. For example, it would be interesting to compare the neurological correlates of learning arithmetic via specific hands-on manipulatives versus others, or versus rote memorization. Cognitive linguistic theory would predict differences in the resulting conceptual metaphors for arithmetic; if found, would such differences also be reflected in neurological structure or function?

The current increasing interest in embodiment in mathematics education offers hope that this research field will continue to attract more researchers who will continue, expand, and envision new theoretical and practical paths to improving the teaching and learning of mathematics.

References

Abrahamson, D. (2014). Building educational activities for understanding: An elaboration on the embodied-design framework and its epistemic grounds. *International Journal of Child-Computer Interaction, 2*(1), 1–16.

Alibali, M. W., & Nathan, M. J. (2012). Embodiment in mathematics teaching and learning: Evidence from learners' and teachers' gestures. *Journal of the Learning Sciences, 21*(2), 247–286.

Arnauld, A. (1861). *The Port-Royal logic*. Edinburgh, Scotland: James Gordon.

Arnon, I., Cottrill, J., Dubinsky, E., Oktaç, A., Fuentes, S., Trigueros, M., & Weller, K. (2014). *APOS theory*. New York, NY: Springer.

Arzarello, F. (2006). Semiosis as a multimodal process. In L. Radford & B. D'Amore (Guest Eds.), *Revista Latinoamericana de Investigación en Matemática Educativa, Special Issue on Semiotics, Culture, and Mathematical Thinking* (pp. 267–299).

Arzarello, F., Paola, D., Robutti, O., & Sabena, C. (2009). Gestures as semiotic resources in the mathematics classroom. *Educational Studies in Mathematics, 70*(2), 97–109.

Barad, K. (2007). *Meeting the universe halfway.* Durham, NC: Duke University Press.

Bautista, A., & Roth, W.-M. (2012). Conceptualizing sound as a form of incarnate mathematical consciousness. *Educational Studies in Mathematics, 79*(1), 41–59.

Bazzini, L. (2001). From grounding metaphors to technological devices: A call for legitimacy in school mathematics. *Educational Studies in Mathematics, 47*(3), 259–271.

Beth, E. W., & Piaget, J. (1966). *Mathematical epistemology and psychology.* Dordrecht, The Netherlands: D. Reidel.

Borba, M., & Villareal, M. (2006). *Humans-with-media and the reorganization of mathematical thinking.* New York, NY: Springer.

Bostock, D. (2009). *Philosophy of mathematics.* Malden, MA: Wiley-Blackwell.

Châtelet, G. (2000). *Figuring space: Philosophy, mathematics and physics.* Dordrecht, The Netherlands: Kluwer.

de Freitas, E. (2016). Material encounters and media events: What kind of mathematics can a body do? *Educational Studies in Mathematics, 91,* 185–202.

de Freitas, E., & Sinclair, N. (2013). New materialist ontologies in mathematics education: The body in/of mathematics. *Educational Studies in Mathematics, 83,* 453–470.

de Freitas, E., & Sinclair, N. (2014). *Mathematics and the body.* Cambridge, England: Cambridge University Press.

de Lima, R. N., & Tall, D. (2008). Procedural embodiment and magic in linear equations. *Educational Studies in Mathematics, 67*(1), 3–18.

Deleuze, G. (1994). *Difference and repetition* (P. Patton, Trans.). New York, NY: Columbia University Press. (Original work published 1968)

Descartes, R. (1982). *Méditations [Meditations].* Paris, France: Librairie philosophique Vrin. (Original work published 1641)

Dubinsky, E. (2002). Reflective abstraction in advanced mathematical thinking. In D. Tall (Ed.), *Advanced mathematical thinking* (pp. 95–123). New York, NY: Kluwer.

Dubinsky, E., & McDonald, M. (2001). APOS: A constructivist theory of learning in undergraduate mathematics education research. In I. Arnon, J. Cottrill, E. Dubinsky, A Oktaç, S. R. Fuentes, M. Trigueros, & K. Weller (Eds.), *The teaching and learning of mathematics at university level: An ICMI study* (pp. 275–282). Dordrecht, The Netherlands: Kluwer.

Duval, R. (2006). A cognitive analysis of problems of comprehension in a learning of mathematics. *Educational Studies in Mathematics, 61,* 103–131.

Edwards, L. D. (2009). Gestures and conceptual integration in mathematical talk. *Educational Studies in Mathematics, 70*(2), 127–141. doi:10.1007/s10649-008-9124-6.

Edwards, L. D. (2010). Doctoral students, embodied discourse and proof. In M. M. F. Pinto & T. F. Kawasaki (Eds.), *Proceedings of the 34th Conference of the International Group for the Psychology of Mathematics Education* (Vol. 2, pp. 329–336). Belo Horizonte, Brazil: PME.

Edwards, L. D. (2011). Embodied cognitive science and mathematics. In B. Ubuz (Ed.), *Proceedings of the 35th Conference of the International Group for the Psychology of Mathematics Education* (Vol. 2, pp. 297–304). Ankara, Turkey: PME.

Edwards, L., Radford, L., & Arzarello, F. (2009). Gestures and multimodality in the teaching and learning of mathematics. *Educational Studies in Mathematics, 70*(2), 91–215.

Edwards, L. D., & Robutti, O. (2014). Embodiment, modalities and mathematical affordances. In L. D. Edwards, F. Ferrara, & D. Moore-Russo (Eds.), *Emerging perspectives on gesture and embodiment in mathematics* (pp. 1–23). Charlotte, NC: Information Age.

Ernest, P. (2006). A semiotic perspective of mathematical activity. *Educational Studies in Mathematics, 61,* 67–101.

Fauconnier, G. (1997). *Mappings in thought and language.* Cambridge, England: Cambridge University Press.

Fauconnier, G., & Lakoff, G. (2009). On metaphor and blending. *Cognitive Semiotics, 5*(1–2), 393–399.

Fauconnier, G., & Sweetser, E. (1996). *Spaces, worlds, and grammar.* Chicago, IL: University of Chicago Press.

Fauconnier, G., & Turner, M. (2002). *The way we think: Conceptual blending and the mind's hidden complexities.* New York, NY: Basic Books.

Ferrara, F. (2003). Bridging perception and theory: What role can metaphors and imagery play? In I. Schwank (Ed.), *Proceedings of the Third Conference of the European Society in Mathematics Education.* Bellaria, Italy: University of Pisa. Retrieved from http://www.dm.unipi.it/~didattica/CERME3/proceedings/Groups/TG1/TG1_ferrara_cerme3.pdf

Ferrara, F. (2014). How multimodality works in mathematical activity: Young children graphing motion. *International Journal of Science and Mathematics Education, 12*(4), 917–939.

Fields, C. (2013). Metaphorical motion in mathematical reasoning: Further evidence for pre-motor implementation of structure mapping in abstract domains. *Cognitive Processing, 14*(3), 217–229.

Fischbach, F. (2014). *La production des hommes [The production of men].* Paris, France: Vrin.

Font, V., Bolite, J., & Acevedo, J. (2010). Metaphors in mathematics classrooms: Analyzing the dynamic process of teaching and learning of graph functions. *Educational Studies in Mathematics, 75,* 131–152.

Font, V., Godino, J., Planas, N., & Acevedo, J. (2009). The existence of mathematical objects in the classroom discourse. In V. Durand-Guerrier, S. Soury-Lavergne, & F. Arzarello (Eds.), *Proceedings of the Sixth Congress of the European Society for Research in Mathematics Education* (pp. 984–995). Lyon, France: INRP. Retrieved from http://www.inrp.fr/editions/editions-electroniques/cerme6/

Forest, D., & Mercier, A. (2012). Classroom video data and resources for teaching: Some thoughts on teacher education. In G. Gueudet, B. Pepin, & L. Trouche (Eds.), *From text to "lived" resources* (pp. 215–230). Dordrecht, The Netherlands: Springer.

Friedrich, P. (1970). Shape in grammar. *Language, 46*(2), 379–407.

Gallese, V., & Lakoff, G. (2005). The brain's concepts: The role of the sensory-motor system in conceptual knowledge. *Cognitive Neuropsychology, 22*(3/4), 455–479.

Gibbs, R. W. (2006). *Embodiment and cognitive science.* Cambridge, England: Cambridge University Press.

Gray, E., & Tall, D. (1994). Duality, ambiguity and flexibility: A proceptual view of simple arithmetic. *Journal for Research in Mathematics Education, 25*(2), 116–140.

Guhe, M., Pease, A., Smaill, A., Martinez, M., Schmidt, M., Gust, H., . . . Krumnack, U. (2011). A computational account of conceptual blending in basic mathematics. *Cognitive Systems Research, 12*(3/4), 249–265. doi:10.1016/j.cogsys.2011.01.004

Hume, D. (1921). *An enquiry concerning human understanding and selections from a treatise of human nature.* Chicago, IL: Open Court. (Original works published 1748 and 1739, respectively)

Hume, D. (1965). *A treatise of human nature.* (L. A. Selby-Bigge, Ed.). Oxford, England: Oxford University Press. (Original work published 1739)

Husserl, E. (1991). *Collected works: Vol. IV. On the phenomenology of the consciousness of internal time* (R. Bernet, Ed., & J. B. Brough, Trans.). Dordrecht, The Netherlands: Kluwer.

Hutto, D. D., Kirchhoff, M. D., & Abrahamson, D. (2015). The enactive roots of STEM: Rethinking educational design in mathematics. *Educational Psychology Review, 27*(3), 371–389.

Hwang, S., & Roth, W.-M. (2011). *Scientific & mathematical bodies. The interface of culture and mind.* Rotterdam, The Netherlands: Sense Publishers.

Ilyenkov, E. V. (1982). *The dialectic of the abstract and the concrete in Marx's capital.* Moscow, Soviet Union: Progress.

Johnson, M. (1987). *The body in the mind.* Chicago, IL: University of Chicago Press.

Johnson, M. (2007). *The meaning of the body: Aesthetics of human understanding.* Chicago, IL: University of Chicago Press.

Kant, I. (1894). *Inaugural dissertation.* New York, NY: Columbia College. (Original work published 1770)

Kant, I. (2003). *Critique of pure reason* (N. K. Smith, Trans.). New York, NY: St. Martin's Press. (Original work published 1781)

Kress, G. (2001). *Multimodal discourse: The modes and media of contemporary communication.* London, England: Arnold.

Kress, G. (2010). *Multimodality. A social semiotic approach to contemporary communication.* London, England: Routledge.

Lakoff, G. (1987). *Women, fire, and dangerous things.* Chicago, IL: University of Chicago Press.

Lakoff, G. (1992). The contemporary theory of metaphor. In A. Ortony (Ed.), *Metaphor and thought* (2nd ed., pp. 203–204). Cambridge, England: Cambridge University Press.

Lakoff, G., & Johnson, M. (1980). *Metaphors we live by.* Chicago, IL: University of Chicago Press.

Lakoff, G., & Johnson, M. (1999). *Philosophy in the flesh.* New York, NY: Basic Books.

Lakoff, G., & Núñez, R. (2000). *Where mathematics comes from: How the embodied mind brings mathematics into being.* New York, NY: Basic Books.

Lefebvre, H. (2009). *Dialectical materialism.* Minneapolis, MN: University of Minnesota Press.

Leibniz [or Leibnitz], G. W. (1949). *New essays concerning human understanding.* La Salle, IL: The Open Court. (Original work published 1705)

Leont'ev, A. N. (1978). *Activity, consciousness, and personality.* Englewood Cliffs, NJ: Prentice-Hall.

Liddell, S. K. (1998). Grounded blends, gestures and conceptual shifts. *Cognitive Linguistics, 9*(3), 283–314.

Maine de Biran, P. (1859). *Oeuvres inédites, tomes I et II* [*Unpublished works, I and II*]. Paris, France: Dezobry & Magdeleine.

Malafouris, L. (2012). Prosthetic gestures: How the tool shapes the mind. *The Behavioral and Brain Sciences, 35*(4), 230–1.

Marion, J.-L. (2002). *Being given. Toward a phenomenology of givenness.* Stanford, CA: Stanford University Press.

Maturana, H. R., & Varela, F. J. (1992). *The tree of knowledge: the biological roots of human understanding.* Boston, MA: Shambhala.

McNeill, D. (2000). *Language and gesture.* Cambridge, England: Cambridge University Press.

McNeill, D. (2005). *Gesture and thought.* Chicago, IL: University of Chicago Press.

Merleau-Ponty, M. (1962). *Phenomenology of perception* (C. Smith, Trans.). London, England: Routledge & Kegan Paul. (Original work published 1945)

Mikhailov, F. T. (1980). *The riddle of the self.* Moscow, Soviet Union: Progress.

Nancy, J.-L. (2008). *Corpus.* New York, NY: Fordham University Press.

Nemirovsky, R., & Ferrara, F. (2009). Mathematical imagination and embodied cognition. *Educational Studies in Mathematics, 70*, 159–174.

Nemirovsky, R., Kelton, M. L., & Rhodehamel, B. (2013). Playing mathematical instruments: Emerging perceptuomotor integration with an interactive mathematics exhibit. *Journal for Research in Mathematics Education, 44*(2), 372–415.

Nemirovsky, R., Rasmussen, C., Sweeney, G., & Wawro, M. (2012). When the classroom floor becomes the complex plane: Addition and multiplication as ways of bodily navigation. *Journal of the Learning Sciences, 21*, 287–323.

Núñez, R. E. (2000). Mathematical idea analysis: What embodied cognitive science can say about the human nature of mathematics. In T. Nakahara & M. Koyama (Eds.), *Proceedings of the 24th International Conference for the Psychology of Mathematics Education* (Vol. 1, pp. 3–22). Hiroshima, Japan: PME.

Núñez, R. E. (2008). Mathematics, the ultimate challenge to embodiment: Truth and the grounding of axiomatic systems. In P. Calvo & A. Gomila (Eds.), *Handbook of cognitive science: An embodied approach* (pp. 333–353). Philadelphia, PA: Elsevier.

Núñez, R. E., Edwards, L. D., & Matos, J. F. (1999). Embodied cognition as grounding for situatedness and context in mathematics education. *Educational Studies in Mathematics, 39*(1–3), 45–65.

Oehrtman, M. (2009). Collapsing dimensions, physical limitation, and other student metaphors for limit. *Journal for Research in Mathematics Education, 40*(4), 396–426.

Piaget, J. (1924). L'expérience humaine et la causalité physique [Human experience and physical causality]. *Journal de Psychologie Normale et Pathologique, 21*, 586–607.

Piaget, J. (1970). *Genetic epistemology.* New York, NY: W. W. Norton.

Piaget, J. (1973). *Introduction à l'épistémologie génétique* (Vol. 1) [Introduction to genetic epistemology (Vol. 1)]. Paris, France: Presses Universitaires de France.

Pimm, D. (1987). *Speaking mathematically: Communication in mathematics classrooms*. London, United Kingdom: Routledge and Kegan Paul.

Radford, L. (2005). The semiotics of the schema. Kant, Piaget, and the calculator. In M. H. G. Hoffmann, J. Lenhard, & F. Seeger (Eds.), *Activity and sign. Grounding mathematics education* (pp. 137–152). New York, NY: Springer.

Radford, L. (2008). Iconicity and contraction: A semiotic investigation of forms of algebraic generalizations of patterns in different contexts. *ZDM—The International Journal on Mathematics Education, 40*(1), 83–96.

Radford, L. (2009a). "No! He starts walking backwards!": Interpreting motion graphs and the question of space, place and distance. *ZDM—The International Journal on Mathematics Education, 41*, 467–480.

Radford, L. (2009b). Why do gestures matter? Sensuous cognition and the palpability of mathematical meanings. *Educational Studies in Mathematics, 70*(2), 111–126.

Radford, L. (2010). The eye as a theoretician: Seeing structures in generalizing activities. *For the Learning of Mathematics, 30*(2), 2–7.

Radford, L. (2012). On the development of early algebraic thinking. *PNA, 6*(4), 117–133.

Radford, L. (2013). Sensuous cognition. In D. Martinovic, V. Freiman, & Z. Karadag (Eds.), *Visual mathematics and cyberlearning* (pp. 141–162). New York, NY: Springer.

Radford, L. (2014a). The progressive development of early embodied algebraic thinking. *Mathematics Education Research Journal, 26*(2), 257–277.

Radford, L. (2014b). Towards an embodied, cultural, and material conception of mathematics cognition. *ZDM—The International Journal on Mathematics Education, 46*, 349–361.

Radford, L., & D'Amore, B. (2006). Semiotics, culture, and mathematical thinking. *Special issue of Revista latinoamericana de investigación en matemática educativa*. Retrieved from http://luisradford.ca

Radford, L., Demers, S., Guzmán, J., & Cerulli, M. (2003). Calculators, graphs, gestures, and the production of meaning. In N. A. Pateman, B. Dougherty, & J. Zilliox (Eds.), *Proceedings of the 27th Conference of the International Group for the Psychology of Mathematics Education (PME27-PMENA25)* (Vol. 4, pp. 55–62). Hilo, HI: University of Hawaii.

Radford, L., Edwards, L., & Arzarello, F. (2009). Introduction: Beyond words. *Educational Studies in Mathematics, 70*(2), 91–95.

Radford, L., Schubring, G., & Seeger, F. (2008). *Semiotics in mathematics education: Epistemology, history, classroom, and culture*. Rotterdam, The Netherlands: Sense.

Reid, D., & Mgombelo, J. (2015). Survey of key concepts in enactivist theory and methodology. *ZDM—The International Journal on Mathematics Education, 47*, 171–183.

Rieber, R. W., & Carton, A. S. (Eds.). (1987). *The collected works of L. S. Vygotsky* (Vol. 1). New York, NY: Plenum.

Roth, W.-M. (2010). Incarnation: Radicalizing the embodiment of mathematics. *For the Learning of Mathematics, 30*(2), 8–17.

Roth, W.-M. (2012). Tracking the origins of signs in mathematical activity: A material phenomenological approach. In M. Bockarova, M. Danesi, & R. Núñez (Eds.), *Cognitive science and interdisciplinary approaches to mathematical cognition* (pp. 209–247). Munich, Germany: Lincom Europa.

Seitz, J. A. (2000). The bodily basis of thought. *New Ideas in Psychology, 18*, 23–40.

Sfard, A. (1994). Reification as the birth of metaphor. *For the Learning of Mathematics, 14*(1), 44–55.

Sheets-Johnstone, M. (2009). *The corporeal turn*. Exeter, Devon, United Kingdom: imprint-academic.com.

Stevens, R. (2012). The missing bodies of mathematical thinking and learning have been found. *The Journal of the Learning Sciences, 21*, 337–346.

Tall, D. (2004). Building theories: The three worlds of mathematics. *For the Learning of Mathematics, 24*(1), 29–32.

Tall, D. (2008). The transition to formal thinking in mathematics. *Mathematics Education Research Journal, 20*(2), 5–24.

Tall, D. (2013). *How humans learn to think mathematically*. Cambridge, England: Cambridge University Press.

Tall, D., Gray, E., Bin Ali, M., Crowley, L., DeMarois, P., McGowen, M., . . . Yusof, Y. (2001). Symbols and the bifurcation between procedural and conceptual thinking. *Canadian Journal of Science, Mathematics, and Technology Education, 1*(1), 81–104.

Tall, D., & Mejia-Ramos, J. P. (2010). The long-term cognitive development of reasoning and proof. In G. Hanna, H. N. Jahnke, & H. Pulte (Eds.), *Explanation and proof in mathematics* (pp. 137–149). New York, NY: Springer.

Talmy, L. (2000). *Toward a cognitive semantics. Volume I: Concept structuring systems*. Cambridge, MA: MIT Press.

Thom, J., & Roth, W. (2011). Radical embodiment and semiotics: Towards a theory of mathematics in the flesh. *Educational Studies in Mathematics, 77*(2–3), 267–284.

Varela, F., Thompson, E., & Rosch, E. (1991). *The embodied mind: Cognitive science and human experience*. Cambridge, MA: MIT Press.

Verillon, P., & Rabardel, P. (1995). Cognition and artifacts: A contribution to the study of thought in relation to instrumented activity. *European Journal of Psychology of Education, 10*, 77–101.

Vygotsky, L. S. (1978). *Mind in society*. Cambridge, MA: Harvard University Press.

Wartofsky, M. (1968). *Conceptual foundations of scientific thought*. New York, NY: Macmillan.

Wartofsky, M. (1979). *Models, representation and the scientific understanding*. Dordrecht, The Netherlands: D. Reidel.

Watson, A., & Tall, D. (2002). Embodied action, effect, and symbol in mathematical growth. In A. Cockburn & E. Nardi (Eds.), *Proceedings of the 26th Conference of the International Group for the Psychology of Mathematics Education, Norwich, UK* (Vol. 4, pp. 369–376). Norwich, England: PME.

Wilson, M. (2002). Six views of embodied cognition. *Psychonomic Bulletin & Review, 9*(4), 625–636.

Winter, B., Marghetis, T., & Matlock, T. (2015). Of magnitudes and metaphors: Explaining cognitive interactions between space, time, and number. *Cortex, 64*, 209–224.

Yoon, C., Thomas, M., & Dreyfus, T. (2011). Grounded blends and mathematical gesture spaces: Developing mathematical understandings via gestures. *Educational Studies in Mathematics, 78*, 371–393.

Zandieh, M., Roh, K. H., & Knapp, J. (2014). Conceptual blending: Student reasoning when proving "conditional implies conditional" statements. *Journal of Mathematical Behavior, 33*, 209–229.

27 关注流派和构建任务：针对数学课堂话语文献的批判性分析

贝思·赫贝尔-艾森曼
美国密歇根州立大学
塔姆辛·米尼
挪威卑尔根大学学院
杰西卡·皮尔森·毕肖普
美国得克萨斯州立大学
艾纳特·海德-梅朱亚尼姆
以色列理工学院
译者：张晋宇
华东师范大学数学科学学院

联合国教育、科学及文化组织（UNESCO；国际数育教学委员会和海外教育发展中心提供协作）于1974年9月发起的"语言学与数学教育的互动"研讨会（见Jacobsen, 1975）催生了聚焦于数学教育中的语言和交流这一数学教育研究的新领域。该领域的研究工作在20世纪70年代后期和80年代中期取得了突出进展，相关出版物包括奥斯汀和豪森（1979）与道（1983）在《数学教育研究》中发表的文章，奎瓦斯（1984）及卡拉尔、卡拉尔和谢里曼（1987）在《数学教育研究学报》中发表的文章以及大卫·皮姆（1987）的高引用著作《数学式地说话》。本章将回顾自1985年（该研究领域的决定性时刻）起发表的经同行评议的数学课堂话语研究文章。

关于数学课堂话语的专项研究在过去的30年中得到了长足的发展。事实上，以"数学课堂话语"为关键词进行Google学术搜索，可以获得25万个匹配结果，其中前10页的大部分文献都引用了过去20年间发表的文章。尽管此前已经有人专门针对数学课堂话语的文献进行了评述（例如Ellerton & Clarkson, 1996; Forman, 2003; Lampert & Cobb, 2003; Ryve, 2011; Walshaw & Anthony, 2008; Webb, 1991），本文的独特之处在于跨越大量同行评审期刊包含了近30年的文献。举例来说，沃肖和安东尼（2008）的文章仅关注教师的作用，韦伯（1991）则特别关注对小组的研究，而本文包括了所有与教师、学生、教科书和数学课堂中的书面文本相关的文章。此外，尽管里维（2011）并不特别针对这一研究领域的某个特定方面（如教师的作用），但是与本文相比，他的文章还是限定在较小的期刊范围内。本综述的另一个独特之处是，我们利用了前述大量文献的理论流派来组织和批判性地分析其研究重点。

本研究受到美国国家科学基金会（批准号#0918117，赫贝尔-艾森曼，PI；西里洛&斯蒂尔，co-PIs）的部分资助。本文所述观点、发现和结论或建议源自作者，并不代表美国国家科学基金会的看法。特此感谢熙珠·舒赫为本文收集文章，感谢利萨·霍金斯、马克·麦卡锡、科里·麦肯齐、埃米·雷和杰米·韦内特协助编写本章汇总表格。此外还要感谢安娜玛丽·康纳、扎因·戴维斯、托尼·埃辛、珍妮弗·兰格-奥苏纳、大卫·皮姆和大卫·瓦格纳提供的意见和建议。最后我们还要感谢蔡金法和三位匿名审稿人的见解和反馈。

27 Highlighting Heritages and Building Tasks: A Critical Analysis of Mathematics Classroom Discourse Literature

BETH HERBEL-EISENMANN
Michigan State University, United States

TAMSIN MEANEY
Bergen University College, Norway

JESSICA PIERSON BISHOP
Texas State University, United States

EINAT HEYD-METZUYANIM
The Technion—Israel Institute of Technology, Israel

In September 1974, a symposium sponsored by the United Nations Educational, Scientific and Cultural Organization (UNESCO; in cooperation with the International Commission on Mathematics Instruction and the Centre for Educational Development Overseas) called Interactions Between Linguistics and Mathematical Education (see Jacobsen, 1975) contributed to the launch of a new line of research in mathematics education focused on language and communication in mathematics education. The work gained prominence in the late 1970s and mid-1980s with the publication of Austin and Howson's (1979) and Dawe's (1983) articles in *Educational Studies in Mathematics;* Cuevas's (1984) and Carraher, Carraher, & Schliemann's (1987) articles in *Journal for Research in Mathematics Education;* and David Pimm's (1987) highly cited book *Speaking Mathematically*. In this chapter, we review peer-reviewed articles on mathematics classroom discourse published since 1985, a defining time in this line of work.

Research focused on mathematics classroom discourse has grown dramatically over the past three decades.

Indeed, entering "mathematics classroom discourse" into Google Scholar results in 250,000 hits, with most of the first 10 pages of references including citations from the past two decades. Although previous literature reviews have focused on mathematics classroom discours (e.g., Ellerton & Clarkson, 1996; Forman, 2003; Lampert & Cobb, 2003; Ryve, 2011; Walshaw & Anthony, 2008; Webb, 1991), ours is distinctive in its inclusion of three decades of literature across a large set of peer-reviewed journals. Walshaw and Anthony (2008), for example, focused only on the teacher's role and Webb (1991) focused particularly on research on small groups, whereas we included all articles related to teachers, students, and textbooks and written texts in mathematics classrooms. And, although Ryve (2011) was less specifically focused on a particular area of this research, such as the teacher's role, he narrowed his search to a smaller set of journals than we have here. Another distinctive aspect of this review is that we have used the theoretical heritages of this broad set of literature to organize and critically analyze its foci.

This research was supported, in part, by the National Science Foundation (Grant No. #0918117, Herbel-Eisenmann, PI; Cirillo & Steele, co-PIs). Opinions, findings, and conclusions or recommendations expressed here are the authors' and do not necessarily reflect the views of NSF. We thank Heejoo Suh for her support gathering articles and Lisa Hawkins, Mark McCarthy, Cori McKenzie, Amy Ray, and Jamie Wernet for their assistance with creating summary tables for this chapter. We also thank AnnaMarie Conner, Zain Davis, Tony Essien, Jennifer Langer-Osuna, David Pimm, and David Wagner for their comments and suggestions. Finally, we appreciated the insights and feedback from Jinfa Cai and three anonymous reviewers.

在这一章中，我们以具体文章使用的理论流派来组织我们的综述，这种方法使我们能够阐明与数学课堂话语相关的各种观点，以及这些流派对我们大家理解数学课堂话语的不同方面所产生的影响。因此，我们不像一些文献综述那样综合已有的发现，而是仔细分析这些文献是如何注意数学课堂话语的。为了使得对这些流派的讨论更加集中，我们使用吉（2011a；2011b）的那些构建任务。下一节将介绍这种方法的背景。

文献综述方法

包括与排除杂志进行综述的标准

在准备这一章的写作时，我们改变了"话语"的工作定义，使我们的讨论有一个出发点。例如，在我们使用的所有定义中，在特殊性、注意程度和对更广泛的社会和意识形态影响的考虑方面都存在差异，并不是所有人都在我们的工作中始终使用"话语"这个词，比如有的选择使用一些相关的术语，像数学语义（Halliday，1978）。因此，我们在本章开始时不定义"话语"，而是使用一系列我们认为在"话语"属下的概念：口语和书面语、语言、互动、手势、符号系统、表征、交流类型、文本、体裁、语义、社会定位的实践、话语实践、思想的实践，等等。使用一套与话语相关的观点而不是一个具体的定义似乎是恰当的，因为从我们自己看到的差异就表明，在我们正在回顾的大量工作中，观点的多样性可能会更大。为了描述国际数学教育界数学课堂话语研究的现状，我们还确定了同行评审的期刊，它们代表了广泛的视角和研究传统，与世界不同地区都有联系。在一名博士生的帮助下，我们对17个发表这类工作的期刊的电子目录进行了关键词（即话语、语言、互动、语义、交流和交谈）搜索。我们查看了其中许多文章的参考文献，并从另外7个期刊中又找到了一些文章（期刊列表见附录A）。除了少数例外，我们还包括了在2013年8月之前发表的或以DOI文件形式发布的论文。几乎所有这些文章都是用英语发表的，可能还有其他用我们不熟悉的语言写的因而我们不能识别的文章。通过搜索一共找到了475篇文章，为了确保搜索尽可能地广泛，我们征求了同事的意见，又获得了9篇文章供评论。

我们没有纳入关于课堂话语的教师专业发展或教师教育的文章（例如Chamberlin，2005；Crespo，2006；de Freitas等，2012；Herbel-Eisenmann，Drake & Cirillo，2009；Males, Otten & Herbel-Eisenmann，2010；Staples & Truxaw，2010），也没有纳入以研究者和实践者为共同读者对象的、与我们的关注点有关的数学教育的许多优秀专著和编著（列表参见附录B）。我们没有在综述中纳入上述内容的原因是受篇幅限制无法对其进行真正有意义的探讨。

就我们所知，本章与本书中相关主题的其他章节之间有着密切的联系：手势、具身和符号中介（Radford, Arzarello, Edwards, & Sabena，2017）、证明（Stylianides, Stylianides, & Weber，2017），双语和多语言学习者（Barwell, Moschkovich, & Setati Phakeng，2017）、种族（Martin, Anderson, & Shah，2017）、身份（Langer-Osuna & Esmonde，2017）和学习的情感维度（Middleton, Jansen, & Goldin，2017）。由于与上述其他章节内容有本质上的重叠，我们决定在本文中排除这些领域（除了使用与话语相关的构念来理解或研究身份认同的研究，原因是很多作者都在使用与话语相关的新观点）。另一个例外是我们纳入了关于双语和多语言学习者的一小部分研究，我们认为其中的一些开创性工作能说服数学教育研究人员更为一般地进入与话语相关的课题。

分析过程

上述文章被分配至作者团队。在判断哪些文章与本综述相关时，我们团队同意如上所述的标准，而且还确保这些文章都明确地聚焦在话语上（而不是只提及与话语有关的概念却没有将其作为文章的核心部分）。该过程将文章数量减少到266篇。之后我们决定按照文章的学术流派对其进行分组，下文将详细说明学术流派的含义。

按学术流派分组。 在我们阅读这些文章时，可以明

In this chapter, we organize our review using the theoretical heritages employed by specific articles. This approach allows us to illustrate the variety of ideas that have been connected to mathematics classroom discourse and the influence that these heritages have had on our community's understanding of different aspects of mathematics classroom discourse. Thus, rather than synthesizing the findings as some literature reviews do, we instead do a careful analysis of how this literature attends to mathematics classroom discourse. To provide some cohesion to how these heritages are discussed, we use Gee's (2011a; 2011b) building tasks. Background to this approach is included in the following section.

Review Methods

Journals, Inclusion, and Exclusion Criteria for the Review

In preparation for writing this chapter, we exchanged our working definitions of "discourse" to have a starting point for our discussions. Across our definitions there were variations in, for example, specificity, level of attention, and consideration of broader social and ideological influences. Not all of us have used the word "discourse" throughout our work, choosing, for example, to write about related terms like the mathematics register (Halliday, 1978). Thus, rather than begin this chapter by defining discourse, we use a range of ideas that we include under the umbrella of discourse: spoken and written words, language, interactions, gestures, semiotic systems, representations, types of communication, texts, genres, registers, socially situated practices, discursive practices, ideological practices, and so on. The use of a set of ideas related to discourse rather than a specific definition seemed appropriate because our own differences highlighted that the diversity of views was likely to be even greater across the enormous body of work that we were reviewing. In an attempt to depict the current state of mathematics classroom discourse research across mathematics education's international community, we also identified peer-reviewed journals that represent a wide range of perspectives and research traditions and are associated with various areas of the world. With the support of a doctoral student, we searched for key terms (i.e., discourse, language, interaction, register, communication, and talk) using electronic tables of contents for 17 journals that publish such work. We looked at the references from many of these articles and found additional pieces from 7 more journals (see Appendix A for the list of journals). With a few exceptions, we included articles that were published or available as a DOI document up until August of 2013. Almost all these articles were published in English, suggesting that there may be other articles in languages we are not fluent speakers of and therefore could not identify. Our search yielded some 475 articles. To ensure that our search had been as inclusive as possible, we solicited feedback from colleagues. This yielded an additional 9 articles for review.

We did not include articles on professional development or teacher education focused on classroom discourse (e.g., Chamberlin, 2005; Crespo, 2006; de Freitas et al., 2012; Herbel-Eisenmann, Drake, & Cirillo, 2009; Males, Otten, & Herbel-Eisenmann, 2010; Staples & Truxaw, 2010). We also did not include the many excellent books and edited books in mathematics education, for both research and practitioner audiences, that pertain to our focus (see Appendix B for a list). We decided not to include these in the review because authentic treatment was not possible given our page constraints.

We are aware that our chapter is closely connected to other chapters in this compendium with related subjects: gesture, embodiment, and semiotic mediation (Radford, Arzarello, Edwards, & Sabena, 2017), proof (Stylianides, Stylianides, & Weber, 2017), bilingual and multilingual learners (Barwell, Moschkovich, & Setati Phakeng, 2017), race (Martin, Anderson, & Shah, 2017), identity (Langer-Osuna & Esmonde, 2017), and affective dimensions of learning (Middleton, Jansen, & Goldin, 2017). Because there would be substantial overlap with the content of these other chapters, we decided to exclude those areas (except for research that used discourse-related constructs to understand or study identity, because many authors are applying novel discourse-related perspectives). The one other exception was to incorporate a small set of research on bilingual and multilingual learners, which, as seminal pieces, we believe may have convinced mathematics education researchers more generally to pursue discourse-related work.

Analytic Process

The set of articles was distributed across the author team. In determining which articles were relevant for our review, we agreed upon criteria across the team as articulated above; we also made sure the articles had a clear focus on discourse (rather than, for example, mentioning a discourse-related idea but not taking it up as a central part of the article). This process reduced the set to 266 articles. We then decided to group the readings based on the article's intellectual heritage, a notion that we elaborate below.

显发现这些文章在理论的、概念的和方法的取向上存在很大的差异,这一点里维(2011)也曾提及。我们将文章分为两组:(1)利用特定的话语学、语言学、社会语言学或与话语相关的学术流派的文章;(2)没有或不重点关注这些学术流派类型的文章。受篇幅限制,我们排除了第二组的46篇文章,原因是这些文章倾向于关注谈话或提问(并较多借鉴其他数学教育文献等)或使用扎根理论而不是诸如系统功能语言学、批判话语分析、话语心理学等话语分析框架。还有一些文章在其文献综述或理论/概念框架中借鉴或列出了许多流派,使我们难以辨别其主要的流派而被排除。因此,本文的分析结果仅汇报了我们对约220篇文章的综合和批判性评论。在本章中,我们特意使用"文章"而非"作者",原因是不同作者的工作可能在不止一个部分出现,这取决于在一篇文章中他们的哪个框架或思想最为突出。

我们基于作者对其工作的定位所涉及的一个或多个传统对这些文章进行了分类,但决定如何区分这些文章并不容易,经过细致地分析和讨论,我们决定将这些文章分为四种工作流派:(1)文化、社会和话语心理学;(2)社会语言学和话语研究;(3)"概念重建";(4)"线索"。

第一组文章从社会文化、话语心理学、定位以及哲学取向等理论中借鉴了像哈里和万．兰根霍夫、哈钦斯、莱夫和威戈、维果斯基和维特根斯坦等学者的观点。这一传统下的文章都以各种心理学视角为其出发点,其中部分作者在其文章中较大幅度地修改了这些传统。

第二组文章从系统功能语言学、符号互动主义、互动社会语言学、语用学、言语行为理论、修辞与交际以及语言人类学等理论中借鉴了像布鲁默、戈夫曼、格赖斯、哈利迪、图尔敏、萨克斯和瑟尔等学者的观点,社会语言学和话语研究是这些文章的出发点。

我们对第三组"概念重建"的名称选择是基于课程研究领域的一项运动。正如格鲁梅特(1989)所解释的,概念重建者希望"批判保守派、自由派和激进派推崇的个人和团体的意识形态,并推动这些意识形态的深思和改造"(第13页)。这一特殊的课程理论家群体倾向于借鉴批判主义、后结构主义、精神分析学和现象学传统,以及后来出现的女权主义理论。尽管上述传统之间存在差异,但在概念重建者所关注的问题上仍然存在相似性:即抛开那些"打断我们对工作想当然的理解,并再次提出在实践活动中被忽略的基本问题"的干预措施(Pinar & Grumet, 1982, 第54页)。该组文章借鉴了上述课程理论家(以及解释学)遵循的相同传统以及像伯恩斯坦、德里达、费尔克拉夫、福柯、吉、葛兰西、万·迪克和齐泽克等学者的工作。

最后,还有一小部分文献,在我们综述的35年间的工作中仅发表不超过三篇,且其借鉴的理论流派无法归入我们确定的任何类别。这些文章借鉴了交互写作方法、心理语言学和语言学理论,并引用像罗森布拉特和拉考夫等学者的观点。

鉴于前面两大流派的文章数量众多,我们决定将这两个流派进一步细分为子流派以便分析。举例来说,文化、社会和话语心理学流派的文章可以归入如下三个不同的子流派:(1)主要借鉴维果斯基的文章,(2)主要借鉴斯法德的文章,(3)主要借鉴哈里的文章。同样,社会语言学和话语研究流派的文章可以进一步分类为借鉴哈利迪的、借鉴图尔敏的和借鉴布鲁默的(例如保罗·科布及其同事的工作)、借鉴格赖斯的或借鉴爱德华兹的文章(尤其是理查德·巴维尔的工作)。此外我们还将使用交际社会语言学的一系列文章也归入这一流派。我们的目标是能够描述各个子流派中的作者研究数学课堂话语系统的前因后果。对于包含多于一个子流派的文章,我们只关注其借鉴程度最突出的子流派并将该文章归入其中。

阐述分析框架。在一组研究生的协助下,我们创建了涵盖所有文章的汇总表以方便描述每个小组的趋势。汇总表提供的信息包括研究问题、所研究的教室和学校类型(例如小学、中学、大学、城市、郊区)、作者关注的话语相关的构念类型(例如提

Grouping by intellectual heritages. As we read the articles, it became apparent that there was great variation in the range of theoretical, conceptual, and methodological approaches, an observation also made by Rvye (2011). We separated the articles into two groups: (1) articles that drew on particular discursive, linguistics, sociolinguistics, or discourse-related intellectual heritages and (2) articles that did not or were not centrally focused on these kinds of intellectual heritages. Due to page constraints, we do not include the 46 articles in the latter group because they tended to focus on talk or questioning (and drew on other mathematics education literature, for example) or used grounded theory instead of discursive analytic frames such as systemic functional linguistics, critical discourse analysis, discursive psychology, and so on. There were also some articles that drew on or listed many traditions in their literature reviews or theoretical/conceptual frameworks, making it difficult to discern what the primary heritage was. Thus, the findings from our analyses report only on our synthesis and critical review of about 220 articles. Throughout this chapter, we use "articles" (rather than authors) purposefully because different authors' work may be included in more than one section, depending on which of their frameworks or ideas were most prevalent in a particular article.

The articles were segmented, based on the tradition or traditions in which the authors seemed to locate their work. Deciding where to put articles was not always easy. After careful analysis and discussions, we settled on categorizing the articles into four heritages: (1) cultural, social, and discursive psychology; (2) sociolinguistics and discourse studies; (3) "reconceptualists"; and (4) "threads" of work.

The articles included in the first group drew on scholars like Harré and van Langenhove, Hutchins, Lave and Wenger, Vygotsky, and Wittgenstein, from theories like sociocultural, discursive psychology, and positioning, and on philosophical orientations. All the articles in this tradition took as their starting points various psychological perspectives. Some of these authors strongly modified these traditions in their work.

The second group of articles drew from scholars like Blumer, Goffman, Grice, Halliday, Toulmin, Sacks, and Searle and from theories like systemic functional linguistics, symbolic interactionism, interactional sociolinguistics, pragmatics, speech act theory, rhetoric and communications, and linguistic anthropology. Sociolinguistics and discourse studies were beginning points of this work.

Our choice of the name for the third group, "reconceptualist," was based on a movement in curriculum research. As Grumet (1989) explained, reconceptualists wanted "to critique the ideologies of the individual and the group that were being promoted by conservative, liberal, and radical groups and to bring them forward for deliberation and transformation" (p. 13). This particular group of curriculum theorists tended to draw on critical, poststructuralist, psychoanalytic, and phenomenological traditions, and later, feminist theories. Although differences among these traditions exist, there are still similarities in what reconceptualists were concerned with: setting aside interventions to "interrupt our taken-for-granted understandings of our work and ask again the basic questions practice activity silences" (Pinar & Grumet, 1982, p. 54). This group of articles drew on the same traditions as these curriculum theorists (and also on hermeneutics) and from the work of people like Bernstein, Derrida, Fairclough, Foucault, Gee, Gramsci, van Dijk, and Žižek.

Finally, there was a small set of literature in which three or fewer articles were published in the 35 years of work that we were reviewing that drew on a particular heritage not otherwise included in our list. These articles drew from transactional writing approaches, psycholinguistics, and linguistics and cited people like Rosenblatt and Lakoff.

Because the first two of the heritages had many articles in them, we decided to further subdivide these two heritages into subheritages for analysis. For example, in the cultural, social, and discursive psychology heritage, we were able to group articles into three different subheritages: (1) articles that drew primarily on Vygotsky, (2) articles that drew primarily on Sfard, and (3) articles that drew primarily on Harré. Similarly, in the sociolinguistics and discourse studies heritage, we decided on subheritages for articles that drew on Halliday, on Toulmin, on Blumer (Paul Cobb and colleagues, for example), on Grice, or on Edwards (Richard Barwell's work, in particular). We also grouped a set of articles together that used interactional sociolinguistics and included it in this heritage. Our goal was to be able to describe what authors in each of these subheritages foregrounded and backgrounded in relationship to their investigations of mathematics classroom discourse. For articles that included more than one of these subheritages, we attended to the subheritage that seemed to be drawn on most extensively and included the article in that subheritage.

Articulating analytic frameworks. With some assistance from a group of graduate students, we created summary tables for all the articles, which allowed us to describe trends in each grouping. These tables provided information about research questions, what kinds of classrooms and schools authors were researching (e.g., elementary, secondary, university, urban, suburban), and the types of

问、谈话、互动模式和话语；参见社会文化那一节中对话语的定义）以及作者如何定义（或没有定义）那些构念的。这些汇总表还使我们能够综合各组文章的发现并描述各小组的趋势。

之后，我们根据吉（2011a，2011b）的"构建任务"理念对每组文章进行了批判性分析。吉（2011b）认为，我们使用语法和词汇来"构建结构及其伴随的意义"，以便能"用语言做事情"，如思考或行动（第87页）。他描述了如下七项构建任务：

- 意义：通过语言强调了什么或什么变重要了；
- 活动：用语言做了什么或完成了什么；
- 身份：接受或建立了什么身份；
- 关系：语言如何反映或建立人与人之间或人与其他实体之间的关系；
- 政治：如何分配社会资源，什么是有价值的或被视为好的；
- 联系：哪些关键想法被认为是相关的；
- 符号系统和知识：一个团体使用和改变的了解方式与沟通方式，这些了解和沟通方式是该团体所看重的（更多细节见Gee，2011b，第88—90页）。

这些构建任务具有自反性，原因是它们既构造也反映了某种情境或背景。并不是所有构建任务都在任何单一的话语分析中被用到，尽管这些任务是相互关联的。鉴于此，我们的目标是在不同的学术流派下确定这些构建任务是如何被解释及其前因后果。除了对构建任务作详细描述以外，吉还提出了一系列需要考虑的问题。比如，有时为了使问题更加贴近数学教育，我们会对这些问题进行调整。经过初步分析，我们决定以两种方式使用此框架。首先，我们根据作者明确关注的内容分析每篇文章，之后我们把这篇文章当作一篇文本进行分析，换句话说，我们专注于作者在这篇文章中的话语工作。举例来说，某篇文章所述活动可能关注学生在做数学时如何使用不确定性用语（一级分析），同时，采用实证研究来回答指明的研究问题（二级分析）。表27.1提供了用来指导分析的问题示例，我们在每个部分的汇总表中总结了每个流派和子流派的主题或趋势（"线索型"流派除外）。

为建立该分析的共同基础，我们每两人一组针对同一个分类进行分析工作，从主要引用系统功能语言学（SFL）和概念重建流派的文章开始。每组作者分析了该流派下的4篇文章，讨论他们的分析结果，之后将剩余的

表27.1 用于构建任务分析的问题示例

意义	活动	身份	关系	政治	联系	符号系统和知识
作者要表达的观点是什么？ 文章的目的或要点是什么？	作者关注什么类型的活动（例如具有参与批判性话语的意识）？	作者是否明确地关注身份问题以及以何种方式关注此类问题？ 作者阐述了什么样的自身定位？	作者如何处理学生、教师和机构（数学、学校或更宽泛的机构）之间的关系？	作者是否明确提及权力？ 作者是否明确质疑霸权行为？	作者关注哪些关键概念或理念以及这些概念或理念之间是如何关联的（例如，发声与能动性，信念与实践）？	知道数学和做数学意味着什么？
作者认为什么是重要的？ 有哪些关键想法被反复提及？	作者在研究和撰写过程中做了什么事情（例如说服某人相信某事；回答研究问题）？	研究者的身份/定位/角色是什么？ 关于其身份/定位/角色，研究者的定位声明告诉了我们什么？	作者提倡什么样的关系？	作者以何种方式关注社会资源的分配？ 作者看重什么？	作者提倡什么样的关系（例如语言分析和数学内容之间的关系）？	语言使用（符号系统）和知识之间的关系有哪些认识论和哲学假设？ 研究的目的是什么？

discourse-related constructs to which authors attended (e.g., questioning, talk, interaction patterns, and Discourse; see the sociocultural section for a definition of Discourse) and how they defined (or did not define) those constructs. These summary tables also allowed us to synthesize findings of sets of articles and to describe trends in each grouping.

We then critically analyzed each set of articles, based on Gee's (2011a, 2011b) "building tasks." Gee (2011b) contended that we use grammar and words to "build structures and their accompanying meaning" in order to "do things with language," such as think or perform actions (p. 87). He described the following seven building tasks:

- *Significance:* what is emphasized or made important through language;
- *Activities:* what is being done or accomplished with language;
- *Identities:* what identities are taken on or built;
- *Relationships:* how language reflects or builds relationships between people or between people and other entities;
- *Politics:* how social goods are distributed and what is valued or seen as desirable;
- *Connections:* what key ideas are seen as related; and
- *Sign Systems and Knowledge:* ways of knowing and communicating that are used and changed by a community who values those ways of knowing and communicating (see Gee, 2011b, pp. 88–90 for more detail).

These building tasks are reflexive in that they both construct and reflect situations or contexts. Not all building tasks are addressed in any single discourse analysis, though these tasks are linked and interrelated. Given that, our goal was to identify how building tasks were construed and foregrounded or backgrounded within different intellectual heritages. In addition to lengthy descriptions of these building tasks, Gee offered a set of questions to consider. Sometimes we modified the questions because, for example, we needed to make the questions more specific to mathematics education. After an initial analysis, we decided to use this framework in two ways. First, we analyzed each article in terms of what the author or authors explicitly focused on, and then we analyzed the article as a piece of text—in other words, we focused on the discursive work the authors were doing with the article. For example, the Activity of an article might focus on how students used the language of uncertainty when doing mathematics (level one analysis) while, at the same time, engaging in empirical research to answer the stated research question (level two analysis). Table 27.1 provides examples of questions that we used to guide our analysis. We summarize the major themes or trends for each heritage and subheritage (except threads) in a summary table in each section.

To develop some common ground for this analysis, we worked in pairs on one grouping, beginning with articles that drew primarily on systemic functional linguistics (SFL) and the reconceptualist heritage. Each pair of

TABLE 27.1. Examples of Questions Used for Building Tasks Analyses

Significance	Activities	Identities	Relationships	Politics	Connections	Sign systems and knowledge
What is the author trying to make a case for? What is the goal or point of the paper?	What kind of activities does the author focus on (e.g., engaging in critical language awareness)?	Does the author explicitly attend to identities and in what ways? What does the author articulate about the author's positionality?	How does the author address relationships between and among students, teachers, and institutions (mathematics, school, broader institutions)?	Does it address power explicitly? Does the author explicitly question hegemonic practices?	What key concepts or ideas is the author focusing on and how are they related to each other (e.g., voice-agency, beliefs-practices)?	What does it mean to know and do mathematics?
What does the author assume to be important? What key ideas get repeated?	What activity is the author engaging in as he or she does research and writes (e.g., convincing people of X; answering a research question)?	What are the identities/positionings/roles that the researcher takes on? What does the researcher's positionality statement tell us about the researcher's identities/positioning/roles?	What kind of relationships does the author promote?	In what ways does the author attend to the distribution of social goods? What does the author seem to value?	What kind of connections does the author promote (e.g., the relationship between linguistic analysis and mathematics content)?	What are the epistemological and philosophical assumptions about the relationship between language use (sign systems) and knowledge? What is the purpose of the research?

文章分给本章各位作者进行独立分析。我们发现上述过程非常耗时，但同时，这个过程使得我们能较为一致地运用我们的分析框架。我们在完成了对这两组文章的分析之后，就我们将关注的核心主题、这些主题与作者所基于的主要内容以及他们是如何处理每一项构建任务的达成了共识。

当我们从还在进行的分析中讨论那些重要的主题时，我们就认定，哪怕只阅读了这些文章的一部分，其核心主题已经是非常明显的。在某些情况下，为了提高分析的效率，我们通过分析该领域中具有重大影响的工作（基于我们的经验和Google学术搜索中该文章的引用次数）来缩短研究过程，选择剩余的三分之一的文章进行了仔细的阅读和分析，并根据这些文章生成了主题。之后我们阅读了其余的文章，审视这些主题是否在这些文章中得到呼应，注意与其他文章之间的差异，并寻找差异点，以反映该类别范围内的多样性。因此，本章的参考文献清单仅包括我们仔细分析的文章和与我们确定的核心主题有偏差的文章（给本章作者贝思·赫贝尔－艾森曼发邮件到bhe@msu.edu可以获得参考文献的完整列表）。在新的流派中，我们指定那些被其他数学教育学者引用的文章作为"种子"文章，但我们认识到，还没有足够长的时间来确定这些文章是否真的有重大影响。如出现无法确定的情况，我们会询问从事该研究领域的同事来帮助我们做决定。

在本章的其余部分，首先呈现了我们从数学课堂话语文献分析中析出的主题。我们通过关注每一个流派和子流派如何处理这些构建任务对相关研究结果进行了批判性分析。在表述每一构建任务类别时，我们都将这些术语大写（中文版用黑体），以表示吾对这些词语的特殊用法。之后，我们会审视各小节的内容并提出方法论上的观点和问题。

对具有话语流派的
文章作综合性与批判性分析

我们现有四种流派：（1）文化、社会和话语心理学；（2）社会语言学和话语研究；（3）概念重建；（4）线索。对每个类别，我们首先给出背景信息和源于一般分析的概述，该概述对这些流派/子流派进行了一般的、宽泛的描述，并解释这些流派/子流派如何处理话语。然后我们罗列了与构建任务相关的趋势汇总表，紧接着是源于每个流派/子流派的构建任务分析（即涉及了哪些构建任务和用什么方式）的概述。

社会、文化和话语心理学类别的子流派

在本节中，我们将关注三类子流派：（1）社会文化理论，（2）话语心理学/沟通，（3）定位理论，这些文章的作者从某些形式的心理学开始他们的工作。

社会文化理论。归入该子流派的学者将学习的社会文化理论作为研究话语的主要理论框架。社会文化理论强调知识的社会起源并将文化融入更广泛的公共和历史实践。沃兹奇（1991）将社会文化方法描述为"人类心理过程的记录，该记录认识到这些过程与其文化、历史和制度背景之间的基本关系"（第6页）。该取向下的学习与建构主义及更传统的聚焦于个体认知的心理学取向完全不同。思维和学习的关键要素包括参与社会实践、语言和其他中介工具的作用，以及情境特征。

对总体分析的概述。该子流派共纳入43篇文章，其中25篇文章经过了细致分析，其余文章只是用于对在这里汇报的趋势的一致性进行检验。虽然该子流派的作者往往采用不同方法研究话语取向，但是他们都使用了一般意义下的社会文化理论，假设学习是通过协调的社会活动发生的，并且是由工具的使用（特别是语言）和学习所嵌入的情境的特征所协调的。

社会文化理论这一子流派下的文章借鉴了各种理论，包括学习和认知以及实践群体的情境视角（J.S. Brown, Collins, & Duguid, 1989; Lave & Wenger, 1991）；社会传播的认知（Hutchins, 1995）；对话性和言语类型（Bakhtin, 1981, 1996）；内化，最近发展区（ZPD）和中介活动（Vygotsky, 1978, 1986）的构念。一般来说，该子流派下的研究人员寻找各种方式来理解话语、语言

authors analyzed four articles within their heritage, discussed their analyses, and then distributed to individual authors the remaining articles for independent analysis. We found the process to be very time intensive, but at the same time, it allowed us to develop a strong common application of our analytic frameworks. Once we completed the analyses of these two groupings, we came to a consensus about the core themes we would focus on in relationship to what the authors foregrounded and how they handled each of the building tasks.

As we discussed the main themes from our ongoing analyses, we decided that the core themes became apparent even after reading a subset of the articles. In some cases, to be more efficient in our analyses, we truncated our process by analyzing the pieces that were seminal in the field (based on our experience and the number of times articles had been cited in Google Scholar), selecting one-third of the remaining articles to read and analyze carefully, and then generating themes based on this subset of articles. We then read the remaining articles to see whether the themes resonated with these articles, noted differences in other articles, and looked for discrepant points so as to capture the range of diversity in the group. Thus, our reference list in this chapter only includes the articles we carefully analyzed and articles that showed deviation from the core themes we identified. (For a full list of references email chapter author Beth Herbel-Eisenmann at bhe@msu.edu.) In heritages that were new, we designated articles that were cited by other mathematics education scholars as our "seminal" pieces but recognized that not enough time had elapsed to know for sure if these would become seminal. When unsure, we also asked colleagues who drew on that line of research to inform our decisions.

In the remainder of the chapter, we first present the themes that emerged from our analysis of literature addressing mathematics classroom discourse. We offer a critical analysis of this work by attending to *how* these building tasks are handled by each of the heritages and subheritages. Whenever we denote a building task category, we capitalize the terms to indicate Gee's specialized use of these words. We then look across these subsections to raise methodological issues and questions.

Synthesis and Critical Analysis of Articles With Discursive Heritages

We now offer the four heritages: (1) cultural, social, and discursive psychology; (2) sociolinguistics and discourse studies; (3) reconceptualists, and (4) threads. In each case, we first provide background information and an overview from our general analysis that shares some broader descriptions of the heritage/subheritage and explains how the heritage/subheritage addressed discourse. We then include a summary table of trends associated with the building tasks, followed by an overview from the building tasks analyses (i.e., which building tasks are addressed and in what ways) for each heritage/subheritage.

Social, Cultural, and Discursive Psychology Subheritages

In this section, we focus on three subheritages in which authors began their work with some form of psychology: (1) sociocultural theories, (2) discursive psychology/commognition, and (3) work on positioning.

Sociocultural theories. Scholars who work within this subheritage use sociocultural theories of learning as their primary theoretical framework to study discourse. Sociocultural theories emphasize the social origins of knowledge and enculturation into broader communal and historical practices. Wertsch (1991) describes a sociocultural approach as "an account of human mental processes that recognizes the essential relationship between these processes and their cultural, historical, and institutional settings" (p. 6). This approach to learning is in contrast to constructivist and more traditional psychological approaches that focus on individual cognition. Key aspects of thinking and learning include participation in social practices, the role of language and other mediating tools, and contextual features.

Overview from the general analysis. We included 43 articles in this subheritage, of which 25 articles were carefully analyzed and the remaining were checked for congruence with the trends reported here. The authors in this subheritage often took different approaches to discourse but were unified by their use of general sociocultural theories, positing that learning occurs through coordinated social activity and is mediated by both the use of tools (language in particular) and the characteristics of the contexts in which learning is situated.

Articles within the sociocultural theory subheritage drew from a variety of theories, including situated perspectives on learning and cognition and communities of practice (J. S. Brown, Collins, & Duguid, 1989; Lave & Wenger, 1991); socially distributed cognition (Hutchins, 1995); dialogicality and speech genres (Bakhtin, 1981, 1996); and the constructs of internalization, zone of proximal development (ZPD), and mediating activity (Vygotsky, 1978, 1986). Broadly speaking, researchers within this subheritage looked for ways to understand

和数学讨论在学习中的作用。总的来说，该子流派下的文章可以被归为以下研究类型中的一种：（a）将话语视为一种学习机制的学习研究或认知研究，（b）调查在建立符合改革目标的课堂集体时话语的作用或演变的研究，（c）以数学话语本身作为研究对象的研究。（有些文章可归入多个类别。）

大多数文章是关于数学学习的研究。在这些文章中，话语在学习中的作用被视为文章的主要焦点或作为补充文章主要焦点的关键特征。举例来说，有些作者认为学生的学习与特定的微观动作、话语的模式或形式有关，如复述、篇章衔接和提问模式）（例如 A. Anderson, Anderson & Shapiro, 2004；Bill, Leer, Reams, & Resnick, 1992；Enyedy, 2003；Goos, 2004；Goos, Galbraith, & Renshaw, 2002；Mercer, 2008；Waywood, 1994）；其他作者则对探索性谈话或对现实世界的交流等更广泛的话语构造进行了研究（Mercer, 2008；Mercer & Sams, 2006；Price, 2000）。还有学者对学习的基本机制进行探索，如最近发展区、对话式交谈和维果斯基的文化发展遗传规律（例如 Enyedy, 2003；Goos, 2004；Goos 等, 2002；Hussain, Monaghan, & Threlfall, 2013；Lau, Singh, & Hwa, 2009；Zack & Graves, 2001）。最后，还有少数学者从理论角度出发通过拓展和批判现有理论并将新的构念融入社会文化理论来研究话语与学习之间的关系（例如 Hoyles, 1985；Lerman, 2001；Mercer, 2008；van Oers, 2001）。

第二大类的文章并没有把认知或学习本身作为研究的目标，相反，这些学者关注探究式或改革式数学集体的发展（即促进学生问题解决、数学讨论和学生学习的责任，例如，可以参见 Goos, 2004；Hufferd-Ackles, Fuson, & Sherin, 2004；和 McCrone, 2005）。他们研究了融入课堂的话语类型。其中许多研究描述了特定类型的话语或理想的教学形式发生的情景并记录了这些环境如何随时间的推移而发展（Goos, 2004；Hufferd-Ackles 等, 2004；Hussain 等, 2013；Lau 等, 2009；McCrone, 2005；Moschkovich, 2008；Olson & Knott, 2013；Truxaw & DeFranco, 2007）。

有少量研究文章属于第三类：以数学话语本身为主要研究对象（即没有将话语作为研究某种其他现象的工具）。这些文章或者考虑数学话语的具体特征，例如主体间性、谈话的时间维度、对话性、日常用语和科学术语的交织性（Bill 等, 1992；Mercer, 2008；Nathan, Eilam, & Kim, 2007；Truxaw & DeFranco, 2007, 2008），或者研究教师或其他成年人如何支持学习者进行特定数学语言实践（Adler, 1999；A. Anderson 等, 2004；van Oers, 2001）。阿德勒（1999）的文章因其对数学语言教学的明确关注而对该子流派作出了独特的贡献，尤其是阿德勒强调了学生获得数学语义（参见系统功能语言学子流派部分以获得更多信息）和课堂实践的重要性，这些都能帮助学生进入更大规模的数学集体实践。

尽管在更广泛的社会文化流派下有许多文章引用了维果斯基和巴赫金的著作，但有些文章更加清晰和系统地将维果斯基和巴赫金所启发的构念融入到自己研究的问题、分析和讨论中，这些作者使用了诸如符号中介、最近发展区、对话性、科学的和自发的或日常的概念等构念作为构建或指导他们分析学习话语的关键要素。举例来说，戈斯和她的同事们以维果斯基的最近发展区为他们框架的基础，调查了探究式数学集体的发展和学生的元认知活动（Goos, 2004；Goos 等, 2002）。戈斯（2004）的框架描述了通过提高数学课堂讨论的参与水平推动学习者进入自身最近发展区的关键活动和实践，这一提法支持通过内化过程将社会现象转化为心理现象。扎克和格拉夫（2001）也使用了维果斯基的最近发展区，以及巴赫金的对话性和挪用性（他性和自为）的思想，作为他们纵向考察小组中"数学认知话语构建"的关键结构（第241页）。

该子流派下的研究发生于澳大利亚、巴林、加拿大、英国、马来西亚、荷兰、西班牙和美国。这些研究的背景主要是公立学校的课堂，只有一项研究涉及非正式的校外背景（A. Anderson 等, 2004）。其分析的对象几乎都是数学课录像的转录文字，关注其中的口头交流。维伍德（1994）对中学生数学日记的分析和里扎特（2006）对课本的分析是仅有的关注

the role of discourse, language, and mathematical discussions in *learning*. In general, articles within this subheritage could be categorized as one of the following kinds of studies: (a) studies of learning or cognition that treated discourse as a mechanism for learning, (b) studies that investigated the role or evolution of discourse over time in establishing classroom communities consistent with reform goals, or (c) studies focusing on mathematics discourse itself as an object of study. (Some articles fell into multiple categories.)

The majority of articles were studies of mathematics learning. In these articles, the role of discourse in learning was either the primary focus of the paper or was a critical feature that supplemented the primary research focus. For example, some authors considered student learning in relationship to particular microlevel moves, patterns, or forms of discourse such as revoicing, textual cohesion, and questioning patterns (e.g., A. Anderson, Anderson, & Shapiro, 2004; Bill, Leer, Reams, & Resnick, 1992; Enyedy, 2003; Goos, 2004; Goos, Galbraith, & Renshaw, 2002; Mercer, 2008; Waywood, 1994); other authors studied broader discursive constructs like exploratory talk or real-world scripts (Mercer, 2008; Mercer & Sams, 2006; Price, 2000). Some scholars explored foundational mechanisms of learning such as ZPD, dialogic talk, and Vygotsky's genetic law of cultural development (e.g., Enyedy, 2003; Goos, 2004; Goos et al., 2002; Hussain, Monaghan, & Threlfall, 2013; Lau, Singh, & Hwa, 2009; Zack & Graves, 2001). And, finally, a smaller number of scholars theoretically investigated the relationship between discourse and learning by extending and critiquing existing theories and integrating new constructs into sociocultural theory (e.g., Hoyles, 1985; Lerman, 2001; Mercer, 2008; van Oers, 2001).

The second broad category of articles did not take cognition or learning per se as their goal. Instead, these scholars focused on the development of inquiry-based or reform-based mathematics communities (i.e., promoting student problem solving, mathematical discussions, and student responsibility for learning—see, for example, Goos, 2004; Hufferd-Ackles, Fuson, & Sherin, 2004; and McCrone, 2005). They studied the types of discourse that were integral within these classrooms. Many of these studies described contexts in which certain kinds of discourse and/or desirable forms of teaching occurred and chronicled how these environments developed over time (Goos, 2004; Hufferd-Ackles et al., 2004; Hussain et al., 2013; Lau et al., 2009; McCrone, 2005; Moschkovich, 2008; Olson & Knott, 2013; Truxaw & DeFranco, 2007).

A small number of studies fell into the third category: articles with a focus on mathematics discourse itself as the primary object of study (as opposed to discourse as a tool with which to study some other phenomenon). These articles either considered specific features of mathematics discourse, such as intersubjectivity, the temporal dimension of talk, dialogicality and the interweaving of everyday and scientific terminology (Bill et al., 1992; Mercer, 2008; Nathan, Eilam, & Kim, 2007; Truxaw & DeFranco, 2007, 2008) or investigated how teachers or other adults supported learners to take up the specific language practices of mathematics (Adler, 1999; A. Anderson et al., 2004; van Oers, 2001). Adler's (1999) article provided a distinctive contribution to this subheritage, as she had an explicit focus on mathematics language instruction. In particular, Adler highlighted the importance of student access to the mathematics register (see the systemic functional linguistics subheritage for more information) and to classroom practices that broker students' entry into the larger mathematics community's practices.

Although many manuscripts in the broader sociocultural heritage set cited Vygotsky and Bakhtin, some articles more clearly and systematically integrated Vygotskian- or Bakhtinian-inspired constructs into research questions, analysis, and discussion. Authors used constructs like semiotic mediation, ZPD, dialogicality, and scientific and spontaneous or everyday concepts as key aspects of learning that framed or informed their analysis of discourse. For example, Goos and her colleagues used Vygotsky's ZPD as the basis of their frameworks to investigate the development of inquiry-based mathematics communities and students' metacognitive activity (Goos, 2004; Goos et al., 2002). Goos's (2004) framework described key activities and practices wherein learners are drawn forward into their ZPD through increasing levels of participation in mathematical discussion. This drawing forward supported the transformation of social phenomena into psychological phenomena through a process of internalization. Zack and Graves (2001), too, used Vygotsky's ZPD, along with the ideas of dialogism and appropriation (otherness and own-ness) from Bakhtin, as key constructs in their study to investigate the "discursive construction of mathematical knowing" (p. 241) over time in small groups.

Studies within this subheritage were situated in Australia, Bahrain, Canada, England, Malaysia, the Netherlands, Spain, and the United States. The contexts for these studies were primarily public school classrooms, with only one involving an informal, out-of-school context (A. Anderson et al., 2004). The objects of analysis were almost exclusively transcripts from video recordings of mathematics lessons focusing on spoken discourse. Waywood's (1994) analysis of secondary students'

书面文本的研究。然而，南森等人（2007）、莫谢科维奇（2008）和卡尔森（2009）使用了多模式分析，不仅融入口头话语，还纳入了图形表示、说明文字和手势。该子流派下研究的年级涵盖了从幼儿园到大学课堂。总体而言，大多数研究是细致的、观察性的、样本量小的定性案例研究，唯一例外的是莫瑟和萨姆斯（2006）涉及400名学生和14名小学教师的实验研究。这些作者表明，如果给予学生关于如何使用谈话更有效地合作和解决问题的明确指导，那么他们会比那些经历过典型教学形式的学生学得更多。

大多数作者在未定义的情况下使用"话语"这一术语并将其视为讨论、数学对话、互动、对话、参与或叙述等词语的同义词。有些作者则给出了话语/话语实践的定义；例如：

- "话语"被理解为包括语言的所有形式，包括手势、符号、人为构造、模仿等（Lerman，2001，第88页）。
- 话语实践是指"由课堂外更大社区组织的支持学生参与"的言语行为（Enyedy，2003，第365页）。
- 数学话语是"在学习环境中，使用正式或非正式数学语言交流数学思想和信息"（McCrone，2005，第112页）。

虽然许多作者没有给出话语的定义，但他们在分析过程中定义和使用了相关的话语构建，例如篇章衔接（Mercer，2008）、演讲类型（van Oers，2001）、探索性谈话（Mercer，2008；Mercer & Sams，2006）、对话式交谈（Mercer & Sams，2006；Truxaw & DeFranco，2007）、明确的数学语言教学（Adler，1999）、数学话语实践（Moschkovich，2008）和交互性陈述（Goos等，2002）。使用话语（大写D）的研究人员几乎都会提及吉（1996）的工作，吉对该术语的定义如下：

> 得到社会普遍接受的一种联系，用于表达思考、感觉、信任、重视和行动的语言、其他符号表达和"人为构造"的方式，它们都可以用来表明自己是社会性群体或"社交网络"的一员或表明（某人正在扮演）一个有社会意义的角色（第131页）。

构建任务分析得出的概论。在表27.2中我们汇总了该子流派下基于吉的定义的相关文章的任务构造的方式。

大多数源自社会文化理论的文章均认可学习和话语在社会建构数学认知和理解中的重要作用，特别是教师在协调学生在讨论中的思维和形成重视质疑、解释、问题解决和分担学习责任的课堂集体方面所发挥的作用。在大多数情况下，话语被视为学习的中介工具或机制

表27.2 与社会文化理论相关的构建任务趋势

意义	活动	身份	关系	政治	联系	符号系统和知识
强调学习数学和作为学习机制的话语和讨论的重要作用。数学课堂集体的形成以及参与者在其中合作进行推理，还有教师在这些集体中的作用也是主要的焦点。	通过讨论，突出了学生参与数学讨论和其他文化上被接受的数学活动、产生具有某些特征的数学话语，以及教师在促进和调整学生思维中的行为和作用。	基本未提及身份。	始终强调师生和生生关系。在某些文章中，学习者与集体和文化的关系被列为影响数学思维和学习的重要背景特征。	一小部分文章论述了教师权威、地位、参与可能性以及数学领域的主要社会实践等问题。	该子流派下最普遍的联系是数学话语与学习之间的联系。	数学被视为一种文化活动，具有制度化的、历史上被接受的实践，以及在特定课堂中的局部实践。学习数学就是参与数学实践。某些文章将数学描述为狭义的话语活动，从这个角度看，学习数学就是熟练掌握其话语。

mathematics journals and Rezat's (2006) analysis of textbooks were the only studies focused on written texts. Nathan et al. (2007), Moschkovich (2008), and Carlsen (2009), however, did use multimodal analyses and incorporated not only spoken discourse but also graphical representations, inscriptions, and gestures. The grade levels studied in this subheritage ranged from preschool to university classrooms. Overall, most studies were fine-grained, observational, qualitative case studies with small sample sizes. The exception was an experimental study by Mercer and Sams (2006) that involved 400 students and 14 primary grades teachers. These authors showed that if students were given explicit instruction on how to use talk to collaborate and engage in problem solving more effectively, they learned more than students who experienced more typical forms of instruction.

The majority of the authors used the term "discourse" without defining it and treated it as synonymous with terms like discussion, mathematics talk, interaction, dialogue, participation, or narrative. Some authors did define discourse/discourse practices; for example:

- "'Discourse' is to be taken to include all forms of language, including gesture, signs, artefacts, mimicking, and so on" (Lerman, 2001, p. 88).
- Discourse practices are speech acts that "are organized by and support students' participation in the larger communities outside of the classroom" (Enyedy, 2003, p. 365).
- Mathematical discourse is "the exchange of mathematical thoughts and information, in a learning environment, using either formal or informal mathematical language" (McCrone, 2005, p. 112).

Though many authors did not define discourse, they defined and used related discursive constructs in their analyses, such as textual cohesion (Mercer, 2008), speech genre (van Oers, 2001), exploratory talk (Mercer, 2008; Mercer & Sams, 2006), dialogic talk (Mercer & Sams, 2006; Truxaw & DeFranco, 2007), explicit mathematics language instruction (Adler, 1999), mathematical Discourse practices (Moschkovich, 2008), and transactive statements (Goos et al., 2002). Researchers who use Discourse (with a capital D) almost always refer to the work of Gee (1996), who defined it as,

a socially accepted association among ways of using language, other symbolic expressions, and "artifacts," of thinking, feeling, believing, valuing and acting that can be used to identify oneself as a member of a socially meaningful group or "social network," or to signal (that one is playing) a socially meaningful role. (p. 131)

Overview from the building tasks analyses. In Table 27.2 we summarize the ways in which articles in this subheritage addressed Gee's building tasks.

The majority of articles drawing from sociocultural theory give Significance to learning and the role that discourse plays in the social construction of mathematical

TABLE 27.2. Building Task Trends Related to Sociocultural Theory

Significance	Activities	Identities	Relationships	Politics	Connections	Sign systems and knowledge
Learning mathematics and the role of discourse and discussions as mechanisms for learning are highlighted as significant. The development of mathematics classroom communities, wherein participants reason collaboratively, and the teacher's role in those communities is also a main focus.	Student participation in mathematics discussions and other culturally accepted mathematical activities, the production of mathematical discourse with certain characteristics, and teacher actions and roles in facilitating and mediating student thinking through discussion are highlighted.	Identities are not generally addressed.	Teacher-student and student-student relationships are consistently emphasized. In some articles, learners' relationships to their communities and cultures are included as important contextual features influencing mathematical thinking and learning.	A small subset of articles addressed issues like teacher authority, status, access to participation, and dominant social practices within mathematics.	The most prevalent connection across this subheritage is the one between mathematics discourse and learning.	Mathematics is viewed as a cultural activity with institutionalized and historically accepted practices as well as local practices within a given classroom. Learning mathematics is participating in mathematical practices. Some papers described mathematics more narrowly as a discursive activity and, from this viewpoint, to learn mathematics is to become fluent in its discourse.

(亦强调联系构建任务),它发生于学生参与诸如数学讨论等实践活动之时。该子流派主要涉及的活动包括学生参与数学讨论和共同生成某些类型的数学话语(例如,探索性谈话、恰当使用数学术语的能力)。有些文章重点关注了教师的行为和动作,包括支持学生参与这些话语实践和创造一种允许学生交流彼此想法的环境等活动。大部分文章通过对师生或生生之间互动的详细分析来讨论他们之间的关系,主要聚焦在那些具有数学学习可能性的互动。

该流派的研究很少探索身份,对研究者自身的身份或定位的明确探讨也很少。扎克和格拉夫斯(2001)属于例外情况,他们通过探索在互动过程中学生所处的不同位置、角色、倾向和立场,明确讨论了身份问题。同样地,莱尔曼(2001)的理论性文章将定位和声音作为研究学习的关键工具,并将其与身份联系起来。

与身份构建任务类似,大多数作者没有将政治置于突出位置。然而,关注教师权威是一种讨论权力和政治的隐性方式。举例来说,戈斯(2004)、霍福德-阿克斯等人(2004)、侯赛因等人(2013)、麦克隆(2005)及扎克和格拉夫斯(2001)主张并记录了课堂,在其课堂中,学生在确定解答有效性和正确性方面其权威得以发展,其他作者则提出,当学生合作学习时会关注到权力动态和社会关系的重要性(Civil & Planas, 2004; Hoyles, 1985)。有少量文章则通过对主流社会实践或群体的认识或质疑(例如 Civil & Planas, 2004; Lerman, 2001; van Oers, 2001)来探讨政治。

该子流派下的文章中,很少给出数学的明确定义。因此,我们根据其文章中的数学任务和活动的意义推断了作者对数学的立场以及他们认为了解数学和做数学意味着什么(即符号系统与构建知识之任务)。不出所料的是,这些文章都认为数学学习就是参与数学实践。举例来说,万·奥尔(2001)提出,"数学是一种源自一个共同体的社会文化实践的文化活动"(第66页)。数学被认为是在特定背景下的,但同时,是一种制度化的实践,与思维的历史形式和世界上受重视的工具有关,换句话说,数学是在当地定义的,并以特定课堂内的规范化的文化实践为基础,该课堂支持社会化和文化融入到更广泛的数学社区。

部分作者笼统地描述了数学活动或探讨了多种数学实践(Adler, 1999; Black, 2004; Civil & Planas, 2004; Enyedy, 2003; Lerman, 2001; Moschkovich, 2008; van Oers, 2001),但大多数作者会指定学生应该参与的特定数学任务,这些数学任务包括概括、猜想、说服和论证、社会情景的数学化、问题解决、交流、下定义、意义建构和参与话语实践(Adler, 1999; Civil & Planas, 2004; Goos, 2004; Goos 等, 2002; Hoyles, 1985; HufferdAckles 等, 2004; Hussain 等, 2013; Lau 等, 2009; McCrone, 2005; Price, 2000; Truxaw & DeFranco, 2007)。部分论文将数学本身描述为一种话语活动(Adler, 1999; Lerman, 2001; Moschkovich, 2008; van Oers, 2001),并提出学习数学不仅仅是获取技能和概念,"还需要积极参与具体话语的重建"(Civil & Planas, 2004,第7页)。

我们发现研究的构念和话语分析的方法论取向在该子流派下存在很大的差异。研究主题包括了从中学生书面数学文本的制作,到给学龄前儿童阅读故事书时提出的数学问题,到各种数学课堂话语的分析和描述。上述差异性表明,社会文化理论提供了一个足够宽泛的基础用于研究与话语相关的多种主题,但与其他子流派不同,除了话语和学习之间的一般联系之外,该类别的研究之间几乎没有其他共同点,尤其是我们发现该组研究几乎没有方法论上的关联性。此外,社会文化理论如何被用于分析话语并不总是那么明确,举例来说,我们经常怀疑,哪些话语特征可以作为学习的证据以及为什么或者什么是分析的单位。该子流派下的研究可以通过更明确

knowing and understanding, particularly the role teachers play in mediating student thinking in discussions and developing classroom communities that value questioning, explaining, problem solving, and shared responsibility for learning. In most cases, discourse is viewed as a mediating tool, or mechanism, for learning (addressing the Connections building task as well), which occurs when students participate in practices such as mathematical discussions. The Activities foregrounded in this subheritage included students' participation in mathematical discussion and joint production of certain types of mathematical discourse (e.g., exploratory talk, an ability to appropriately use mathematical terms). Some articles focused on teacher moves and actions as Activities that supported students in engaging in these discursive practices and in creating an environment that allowed students to come into contact with one another's ideas. The majority of articles addressed Relationships through a detailed analysis of either teacher-student or student-student interactions, with a primary focus on those interactions with potential for learning mathematics.

Few studies across this heritage explored Identity, and there was very little explicit discussion of the researchers' own identity or positionality. An exception is Zack and Graves (2001), who explicitly addressed identity by exploring the different positions, roles, dispositions, and stances students took up during interactions. Similarly, Lerman's (2001) theoretical piece identified positioning and voice, which he tied to identity, as key tools to study learning.

Similar to the Identity building task, the majority of authors did not foreground Politics. Attention to teacher authority, however, was one way that power and politics were implicitly addressed. For example, Goos (2004), Huffered-Ackles et al. (2004), Hussain et al. (2013), McCrone (2005), and Zack and Graves (2001) argued for and documented classrooms in which students developed authority in determining the validity and correctness of solutions. Other authors argued for the importance of attending to power dynamics and social relationships when students work together (Civil & Planas, 2004; Hoyles, 1985). A few articles addressed Politics by recognizing or questioning dominant social practices or groups (e.g., Civil & Planas, 2004; Lerman, 2001; van Oers, 2001).

Across the articles in this subheritage, few explicit definitions of mathematics were provided. Thus, we inferred the authors' stances toward mathematics and what they thought it meant to know and do mathematics (i.e., the Sign Systems and Knowledge building task) based on the mathematical tasks and activities given significance in their writings. Unsurprisingly, a common theme across these articles was that to learn mathematics was to *participate* in mathematical practices. For instance, van Oers (2001) claimed, "Mathematics is a cultural activity that emerges out of sociocultural practices of a community" (p. 66). Mathematics was considered to be context specific yet, at the same time, an institutionalized practice related to historical forms of thinking and tools valued in the world. In other words, mathematics was locally defined and based on normative cultural practices within a given classroom that supported socialization and cultural appropriation into the broader mathematical community.

Some authors described mathematical activity in general terms or discussed a plurality of mathematical practices (Adler, 1999; Black, 2004; Civil & Planas, 2004; Enyedy, 2003; Lerman, 2001; Moschkovich, 2008; van Oers, 2001), but most authors specified particular mathematical tasks in which students should be engaged; these included generalizing, conjecturing, convincing and arguing, mathematizing social contexts, problem solving, communicating, defining, sense making, and engaging in discursive practices (Adler, 1999; Civil & Planas, 2004; Goos, 2004; Goos et al., 2002; Hoyles, 1985; Hufferd-Ackles et al., 2004; Hussain et al., 2013; Lau et al., 2009; McCrone, 2005; Price, 2000; Truxaw & DeFranco, 2007). A few papers described mathematics itself as a discursive activity (Adler, 1999; Lerman, 2001; Moschkovich, 2008; van Oers, 2001), claiming that learning mathematics involved more than acquiring skills and concepts: "There also needs to be an active participation in the reconstruction of a specific discourse" (Civil & Planas, 2004, p. 7).

Within this subheritage we noticed substantial variation in the constructs studied as well as the methodological approaches to the analysis of discourse. Topics ranged from students' production of written mathematical texts in secondary schools, to mathematical questions posed during storybook reading with preschool children, to analyses and descriptions of discourse in a variety of mathematics classrooms. This variability suggests that sociocultural theories provide a broad enough basis for studying multiple discourse-related topics, but unlike other subheritages there is little cohesion across the studies beyond the general connection of discourse and learning. In particular, we saw little methodological cohesion within this group. Moreover, it was not always clear how sociocultural theory was being applied in the analysis of discourse. For example, we often wondered what discursive characteristics were being taken as evidence of learning and why, or what the unit of analysis was. Research within this subheritage could be strengthened

地确认和解释方法论工具及相应的分析单位和这些工具如何与社会文化理论的特定方面产生联系得以加强,我们并不是说要统一,而是说要更加地透明和清晰。

话语心理学/交流认知。话语心理学的这条线索,从这里开始被称为交流认知,在很大程度上是受哈里和吉勒特的著作《话语思维》(1994)的启发,这是一种消除思维与身体之间的笛卡尔式区分方法,强调自我的话语起源。交流认知借鉴了维果斯基的工作,但利用了维特根斯坦(1953)的后期哲学对其观点进行了修正。该研究方向的一个主要观点是,思考基本上是一个公共的、社会文化和历史上产生的过程,它变得内在化和个性化,但本质上仍然是一种社会现象,而安娜·斯法德则是该领域的创始人。

交流认知将话语研究定义为对人类认知的研究,而并不仅仅是对人际交往的研究,人类思维是内心交流的一种形式。"交流认知"(Sfard,2007)这一术语的产生是有目的的,意图将"交流"和"认知"的概念融合成一种。

综合性分析的概述。该子流派包括24篇文章,其中6篇发表在由斯法德主编的《国际教育研究杂志》2012年特刊中。该组文章中的大部分包括了对数学课堂中(Güçler,2013;Kieran,2001;Nachlieli & Tabach,2012;Sfard,2000b,2001,2002,2007;Sfard & Kieran,2001;Sinclair & Moss,2012;Wood & Kalinec,2012;Xu & Clarke,2013)与课堂以外情境下(Ben-Yehuda,Lavy,Linchevski,& Sfard,2005;Caspi & Sfard,2012;Heyd-Metzuyanim,2013;Heyd-Metzuyanim & Sfard,2012;Ryve,Nilsson,& Pettersson,2013;Sfard & Lavie,2005)生生和师生交互的非常具体的审视。应用交流认知的数学领域包括早期数字推理(Sfard & Lavie,2005)、几何(Sfard,2007;Sinclair & Moss,2012)、代数(Caspi & Sfard,2012;Kieran,2001;Nachlieli & Tabach,2012;Sfard & Kieran,2001)、无穷大和极限的概念,以及高等教育中的线性代数(Güçler,2013;Kim,Ferrini-Mundy & Sfard,2013,Ryve,2004,2006)。

这些文章中大部分都是理论性的(Sfard,2000a,2000b,2001,2002,2007;Sfard & Kieran,2001;Sfard & Lavie,2005;Ben-Yehuda等,2005),尽管它们之中几乎所有的文章都用了很大篇幅分析实证数据。有些文章是实证调查,将框架应用于新的领域和研究焦点(Caspi & Sfard,2012;Güçler,2013;Heyd-Metzuyanim & Sfard,2012;Kim 等,2012;Nachlieli & Tabach,2012;Sinclair & Moss,2012;Wood & Kalinec,2012;Xu & Clarke,2013)。即使这些文章依然将很大的篇幅用于理论,但因为读者被假定为不熟悉该框架,所以需要对其词汇做详细的解释。该组中的一小部分文章将前述框架与其他理论或方法相结合(Jankvist,2011;Ryve,2004,2006;Ryve等,2013)。举例来说,简韦斯特(2011)运用了"元话语规则"的概念和数学教育中"以历史为目标"的观点,来调查学生对数学历史发展的元特征问题的讨论是如何立足于一个数学单元的学科内容的。在该子流派下,作者一贯地将话语称为"聚集部分人而排除其他人的不同类型的交流"(Sfard,2007,第573页)。斯法德将话语定义为"数学的",如果"它以数学词汇为特征,比如说那些与数量和形状有关的词汇"(第573页)。在许多这类文章中,作者强调话语包括非言语和言语的交流。里维等人(2013)重点探讨了在构建有效数学交流时视觉中介物及其与技术性术语之间的联系。海德-梅朱亚尼姆和斯法德(2012)审视了数学讨论中的情绪表达和语调,推导出由四位七年级学生所传达的"身份识别"信息(或构建学生身份的信息)。

构建任务分析得出的概论。在表27.3中,我们汇总了交流认知子流派下的文章在处理吉构建任务的一些常见方法。

挖掘话语发展的某些机制具有重要意义,其中一种这样的机制被称为"交流认知冲突",在这种冲突中,话语者遵循不同的元规则(定义活动模式的那些规则),并在不知情的情况下以不同的方式使用相同的关键词,从而形成沟通障碍(Sfard,2007)。因此,研究所关注的活动往往涉及诸如词汇使用的改变或对某些沟通线索的模式化回应等数学交流的细节。就活动而言,斯法德(2007)将元层级的学习定义为涉及话语元规则变化的学习,而非只涉及将已知规则应用于熟悉的数学对象的对象级学习。该子流派的许多作者都建立在元层级学习的

by more clearly identifying and explaining the methodological tools and corresponding units of analysis and how these tools are connected to particular aspects of sociocultural theory. We are not suggesting uniformity, but more transparency and clarity.

Discursive psychology/commognition. This line of discursive psychology, called commognition from here on, was largely inspired by Harré and Gillett's book *The Discursive Mind* (1994), an approach that offered to dissolve the Cartesian distinction between mind and body and emphasized the discursive origins of the self. Commognition draws upon the writings of Vygotsky but modifies his views using the later philosophy of Wittgenstein (1953). A major claim in this line of work is that thinking is basically a public, socioculturally and historically produced process that becomes internalized and individualized, but essentially remains a social phenomenon. Anna Sfard is the founding scholar of this line of work.

Commognition defines the study of discourse as the study of human cognition, not just that of interpersonal communication: human thinking is a form of intrapersonal communication. The coining of the term "commognition" (Sfard, 2007) was purposeful, fusing the ideas of "communication" and "cognition" into one phenomenon.

Overview from the general analysis. This subheritage included 24 articles, 6 of which were published in a 2012 special issue of *International Journal of Educational Research* edited by Sfard. Most of this set of articles included very detailed examinations of student-student and student-teacher interactions in mathematics classrooms (Güçler, 2013; Kieran, 2001; Nachlieli & Tabach, 2012; Sfard, 2000b, 2001, 2002, 2007; Sfard & Kieran, 2001; Sinclair & Moss, 2012; Wood & Kalinec, 2012; Xu & Clarke, 2013) and in out-of-class settings (Ben-Yehuda, Lavy, Linchevski, & Sfard, 2005; Caspi & Sfard, 2012; Heyd-Metzuyanim, 2013; Heyd-Metzuyanim & Sfard, 2012; Ryve, Nilsson, & Pettersson, 2013; Sfard & Lavie, 2005). The mathematical domains to which commognition has been applied span early numerical reasoning (Sfard & Lavie, 2005), geometry (Sfard, 2007; Sinclair & Moss, 2012), algebra (Caspi & Sfard, 2012; Kieran, 2001; Nachlieli & Tabach, 2012; Sfard & Kieran, 2001), conceptions of infinity and limits, and linear algebra in postsecondary education (Güçler, 2013; Kim, Ferrini-Mundy, & Sfard, 2013, Ryve, 2004, 2006).

Many of these articles were theoretical (Sfard, 2000a, 2000b, 2001, 2002, 2007; Sfard & Kieran, 2001; Sfard & Lavie, 2005; Ben-Yehuda et al. 2005), though almost all of them devoted a major portion to examining empirical data. Some were also empirical investigations, applying the framework to new domains and foci of research (Caspi & Sfard, 2012; Güçler, 2013; Heyd-Metzuyanim & Sfard, 2012; Kim et al., 2012; Nachlieli & Tabach, 2012; Sinclair & Moss, 2012; Wood & Kalinec, 2012; Xu & Clarke, 2013). Even these, however, still devoted a considerable part of each paper to theory, since the framework was presumed unfamiliar to readers and, hence, required extensive introduction of its vocabulary. A small number of the articles in this set have combined this framework with other theories or methodologies (Jankvist, 2011; Ryve, 2004, 2006; Ryve et al., 2013). For instance, Jankvist (2011) used the idea of "meta-discursive rules," together with ideas about "using history as a goal" in mathematics education, to examine how students' discussions of metaperspectival issues of the historical development of mathematics may be anchored in the subject matter of a mathematical unit. Within this subheritage, authors consistently referred to discourse as "the different types of communication that bring some people together while excluding some others" (Sfard, 2007, p. 573). Sfard defined a discourse as "mathematical" if "it features mathematical words, such as those related to quantities and shapes" (p. 573). In many of these writings, authors stressed that discourse included nonverbal as well as verbal communication. Ryve et al. (2013) concentrated on visual mediators and the links between them and technical terms in constructing effective mathematical communication. Heyd-Metzuyanim and Sfard (2012) examined emotional expressions and tone of voice in a mathematical discussion to derive the "identifying" messages (or messages that constructed students' identities) that were communicated by four seventh-grade students.

Overview from the building tasks analyses. In Table 27.3 we summarize the ways in which articles in the commognitive subheritage generally addressed Gee's building tasks.

Significance was mostly given to unearthing certain mechanisms of discursive development. One such mechanism has been conceptualized as a "commognitive conflict," in which discursants follow different metarules (rules that define patterns in the activity), unknowingly using the same key words in different ways and hence form a breach in communication (Sfard, 2007). Thus, the Activities that have gained attention often involve minute details of mathematical communication such as change in word-use or patterned responses to certain communicational cues. With regard to the Activities, Sfard (2007) defined *metalevel* learning as learning that involves change in the metarules of the discourse, as opposed to *object-level* learning that only involves the application of already-known rules onto familiar mathematical objects. Many of the authors in this subheritage have built on the idea of metalevel learning. For

思想上。举例来说，尔森和布鲁姆（2012）研究了在大学阶段学习数学史对理解数学元话语规则的用处。其他研究者发现，即使教师有意引导元层级的学习，但是儿童的学习还是往往停留在对象级上（Nachlieli & Tabach, 2012；Sinclair & Moss, 2012）。

表27.3 与交流认知有关的任务建构趋势

意义	活动	身份	关系	政治	联系	符号系统和知识
重要的是（1）数学对象的话语构造和数学活动的元规则（规范），（2）理论概念的准确和一致的定义以及（3）理论和方法论的哲学基础。	大多数研究集中于谈话和交互的细节。理论的和分析的论证与通过实证数据的理论说明相结合。	尽管交流认知流派声称提供统一的理论框架用数学认知来考查身份，但只有一小部分的文章关注了身份。	主要研究的是学生与数学活动之间的关系。	该流派没有明确提及这个方面的工作。	在数学历史发展和个人学习过程之间建立关联。在哲学理念（以维特根斯坦为主）和学数学及做数学的理论之间也建立了联系。	重点关注数学的符号系统及其在教与学的交互中的建构方式。该框架强调，符号系统并非思考的窗口，而是形成思维的窗口。

总的来说，该子流派最重要的构造任务之一就是符号系统与知识。交流认知理论不承认符号与意义（或"知识"）的二元论，并将符号（或交流的手段）视为形成思想的"材料"。因此，交流认知理论强调数学话语的发展及其构成要素的概念化。举例来说，出现在很多文章中的一个重要概念是对象化。根据斯法德和拉维（2005）的观点，对象化是"成年人话语"和"儿童话语"之间的显著差异之一，成年人喜欢用数字，就好像它们代表了外部有形实体，而儿童话语则没有这个特性。举例来说，本-耶胡达等人（2005）就报告了两位数学成绩相近，但都比较差的十一年级学生在数学概念对象化程度上存在的显著差异。他们声称，尽管对象化程度较高的女孩认为自己的学习成绩较差，但实际上，与用更严格的句法格式进行数学计算的男同学相比，她更有可能取得进步。牛顿（2012）的研究表明，关联数学项目（CMP；Lappan, Fey, Fitzgerald, Friel & Phillips, 2006）课程的文本课程材料（更具体地，CMP 2中"与分数相乘"单元）几乎在每个方面都不同于在六年级课堂所实施的课程，包括对象现实化（有理数）的相对重要性以及这些对象具有的不同性质的重要性。

该子流派中突出的关系主要涉及学生及其参与的数学活动。除研究身份的工作（如下所述）以外，其他文章对人际关系的关注较少。在学生和数学两者关系方面，有几篇文章提及数学交流的有效性，是通过"焦点分析"（Sfard, 2000b）和"职业关注"（Sfard & Kieran, 2001）来调查参与者在数学对话中的参与程度。有时，这种调查可能会产生有争议的发现，比如斯法德和基兰（2001）就对分组学习的效果提出了质疑，是因为他们的焦点分析表明，配对学习时其中一人会不加思考地跟随另一人，前者关心的是人际关系而非数学本身。里维（2004, 2006）在上述方法的基础上研究了工科学生对概念图的讨论，得出的结论是学生进行了有效交流，而且他们的讨论在数学上是富有成效的。他还批判性地考察了焦点分析和职业关注的方法论，其结论是此类方法适合于说明话语失效的原因而不是说明在数学上交流如何富有成效，他提出用"意图分析"（源自言语行为理论；Searle, 1979）补充该方法论框架，从而为分析对话者的意图提供更为明确的工具。

TABLE 27.3. Building Task Trends Related to Commognition

Significance	Activities	Identities	Relationships	Politics	Connections	Sign systems and knowledge
Significance is given to (1) the discursive construction of mathematical objects and metarules (norms) of mathematical activities, (2) accurate and consistent definition of theoretical concepts, and (3) philosophical underpinnings of theory and methodology.	Most of the studies focus on minute details of talk and interaction. Both theoretical and analytical arguments are combined with illustrations of the theory through empirical data.	Although commognition claims to offer a unified theoretical framework for examining identity with mathematical cognition, only a small subset of this group devotes attention to identity.	The main relationships explored are between students and the mathematical activity.	Politics are not explicitly addressed in this line of work.	Connections are made between historical developments of mathematics and individual learning processes. Connections are also made between philosophical ideas (mainly Wittgenstein) and theories of learning and doing mathematics.	A major emphasis is put on the system of mathematical signs and the way these are constructed in teaching-learning interactions. The framework emphasizes that sign systems are not a window to thinking but rather form it.

instance, Kjeldsen and Blomhøj (2012) examined the utility of studying the history of mathematics (at the college level) for understanding the metadiscursive rules of mathematics. Other researchers have found that children's learning often remains at the object level, despite teachers' attempts to move to the metalevel (Nachlieli & Tabach, 2012; Sinclair & Moss, 2012).

In general, one of the most important building tasks in this subheritage is that of Sign Systems and Knowledge. The commognitive theory rejects a dualism of sign and meaning (or "knowledge") and treats the signs (or means of communication) as the "stuff" from which thought is made. Therefore, commognitive theory puts emphasis on conceptualizing the development of the mathematical discourse and the components of which it is made. For instance, an important concept that appeared in many articles is that of objectification. Objectification, according to Sfard and Lavie (2005), is one of the main salient differences between "grown-ups' discourse" and "children's discourse." Whereas grown-ups use number words as if signifying externally given tangible entities, young children's discourse is devoid of this property. Ben-Yehuda et al. (2005), for example, showed how two 11th graders with similar low achievements in mathematics varied considerably in their degree of objectifying mathematical concepts. They claimed that, despite the fact that the girl with the more objectified language thought of herself as lower achieving, she was actually the one with more potential for improvement than her male peer who performed mathematical calculations in a strict syntactic form. Newton (2012) showed that the written *Connected Mathematics Project* (CMP; Lappan, Fey, Fitzgerald, Friel, & Phillips, 2006) curriculum materials (specifically, the unit "Multiplying With Fractions" from CMP 2) differed in almost every aspect from the enacted curriculum in a sixth-grade classroom, including the relative prominence of the objects' realizations (rational numbers) and the importance attributed to the different properties of these objects.

The Relationships foregrounded in this subheritage mostly involved students and the mathematics with which they were engaged. Less attention was given to interpersonal relationships, apart from the works dealing with identity (as described below). Within the relationship between students and mathematics, several papers dealt with the effectiveness of mathematical communication. This was done using "focal analysis" (Sfard, 2000b) and "preoccupational analysis" (Sfard & Kieran, 2001) to examine participants' engagement in mathematical conversations. Sometimes, this examination led to provocative findings. For instance, Sfard and Kieran (2001) questioned the usefulness of learning in groups since their focal analysis showed how one of the members of a learning dyad was merely following thoughtlessly the other member, being concerned with interpersonal matters rather than the mathematics itself. Ryve (2004, 2006) built on these methods to examine engineering students' discussions of concept maps, concluding that the students communicated efficiently and their discussions could be characterized as mathematically productive. He also critically examined the methodology of focal and preoccupational analysis, concluding that it was suited for showing why discourses were not effective rather than how communication was mathematically productive. He proposed complementing the methodological framework with "intentional analysis" (taken from speech act theory; Searle, 1979) to provide a more explicit tool for analyzing interlocutors' intentions.

在该子流派的文献中，另一种将学生-数学关系概念化的方法是通过"仪式性"与"探索性"参与的概念，这些概念起源于斯法德和拉维（2005）对幼儿数字谈话学习的研究。根据该研究的结果，斯法德假设学习的过程就是从仪式性参与（新成员遵循集体规则以便与掌握话语的人建立联系）发展至探索性参与（行动的目的是为自己创建数学叙述）。本-耶胡达等人（2005）以及海德-梅朱亚尼姆（2013）的研究表明，存在学习困难的学生通常为仪式性地参与，"被困"在学习某个数学内容的"仪式"阶段，而他们的同伴已经进入了更具探索性的参与阶段。

该子流派的数篇文章突出了我们在更大范围文献回顾中发现的具有独特性的联系。例如，数学话语的历史发展（称为系统发生发展）与儿童话语的个体发育发展之间通过数学学习产生联系。卡斯皮和斯法德（2012）在基于代数发展历史形成他们的代数话语前身理论时就显性化了这种联系，然后继续在尚未接触学校形式代数的学生中寻找类似的发展过程。

该子流派下的一小部分文章明确提及了身份（Heyd-Metzuyanim, 2013; Heyd Metzuyanim & Sfard, 2012; Wood & Kalinec, 2012）。交流认知框架将数学学习和身份建构都定义为话语活动，根据海德-梅朱亚尼姆和斯法德（2012）的观点，应该使交流认知框架成为一个在数学学习中的理论化身份建构的潜在平台。海德-梅朱亚尼姆和斯法德（2012）在比较与数学交流相关的信息时，他们发现识别活动（或"身份认同"）可能会对学生之间的有效数学沟通造成明显障碍。

该子流派的大部分文章都没有将政治或权力问题明确作为其研究的重点。然而，有些作者在其结论中指出了数学专业知识被视为一种社会福利的方式，举例来说，本-耶胡达等人（2005）讨论了两位有困难生活经历的、无法获得有意义的数学教学的女生的数学失败的社会建构。类似的，海德-梅朱亚尼姆（2013）质疑家长和教师将数学失败看作仅仅涉及学生"思维"或大脑的个体现象的倾向。不同于斯法德拒绝认知主义（例如 Sfard, 2000b, 2007; Sfard & Lavie, 2005），其他使用交流认知的作者（例如 Jankvist, 2011; Ryve, 2004; Xu & Clarke, 2013）对习得主义或认知观念的批评声较弱，尽管其中一些人也质疑某些以"个人主义"看待学习的观点。

交流认知框架看上去有助于从社会文化角度和互动环境下审视所谓的"数学内容"。该子流派的文章所使用的主要概念都是保持一致的，这要么是因为斯法德所使用的关键概念较为明确，要么是因为斯法德本人是文章的作者之一。该子流派的早期文章相对密集，其方法可能难以应用，然而，最近发表的文章（2012年及以后）为作为学习理论和方法论工具集的交流认知框架的应用提供了例子。

定位。使用定位理论的作者倾向于将他们的研究建立在至少两项数学教育领域之外的主要工作上：即罗姆·哈里及其同事的定位理论（例如 Harré & Moghaddam, 2003; Harré & Slocum, 2003; Harré & van Langenhove, 1999）和多罗西·霍兰德及其同事的自我认同和他人的身份认同（例如 Holland & Eisenhart, 1990; Holland, Lachiotte, Skinner, & Cain, 1998）。尽管大多数提及定位的数学教育研究者同时引用这两项工作（K. Anderson, 2009; Bishop, 2012; Gresalfi, 2009; Hand, 2012; Herbel-Eisenmann & Wagner, 2010; Herbel-Eisenmann, Wagner, Johnson, Suh, & Figueras, 2015; Kotsopoulos, 2014; Langer-Osuna, 2011; Turner, Gutiérrez, & Sutton, 2011; Wagner & Herbel-Eisenmann, 2009），但还是有少数人仅引用了霍兰德及其同事（Bell & Pape, 2012; Esmonde & Langer-Osuna, 2013; Langer-Osuna, 2014）或哈里及其同事（Esmonde, 2009b）的工作。以"倾向"为研究重点的格里赛非和科布（2006）以及亨特和安东尼（2011）的文章则属于例外，尽管有些作者会同时使用倾向与定位，但瓦格纳和赫贝尔-艾森曼（2009）提出倾向始终为名词并指示"明显稳定的身份"（第9页），而定位可以而且应该用作动词，因为它反映了当时的行为和可变角色之间的调整。

哈里和万·兰根霍夫（1999）认为，位置

Another way in which the student-mathematics relationship has been conceptualized within this subheritage is through the concepts of "ritual" vs. "explorative" participation. These concepts originated in Sfard and Lavie's (2005) examination of young children's learning of numerical talk. Based on that research, Sfard hypothesized that learning progresses from ritual participation, in which new members follow the community's rules for the sake of connecting with those who have mastered the discourse, to explorative participation, in which the actions are geared toward creating mathematical narratives for their own sake. Ben-Yehuda et al. (2005), as well as Heyd-Metzuyanim (2013), have shown that students with difficulties often participate ritually, getting "stuck" in the ritual phase of learning certain mathematical content while their peers have advanced to more explorative forms of participation.

Several works in this subheritage highlight Connections that we found to be unique within our broader set of articles. For instance, a connection was made between the historical development of mathematical discourses (termed phylogenetic development) and the ontogenetic development of the child's discourse through mathematical learning. Caspi and Sfard (2012) made this connection very explicit when they formed their theory of the antecedents of algebraic discourse, based on the history of algebra, and then moved on to search for a similar course of development in students who have not yet been introduced to formal algebra in school.

Identities were explicitly mentioned in a small subset of this subheritage (Heyd-Metzuyanim, 2013; Heyd-Metzuyanim & Sfard, 2012; Wood & Kalinec, 2012). The commognitive framework defines both mathematical learning and identity construction as discursive activities, making it, according to Heyd-Metzuyanim and Sfard (2012), a platform with potential for theorizing identity construction within mathematical learning. When comparing and aligning the identifying messages next to the mathematical ("mathematizing") statements, Heyd-Metzuyanim and Sfard (2012) concluded that the identifying activity (or "identity struggles") could create a significant barrier to effective mathematical communication between students.

Most of the articles in this subheritage did not explicitly deal with Politics or issues of power as the focus of their study. In their conclusions, however, some authors made claims about the ways in which mathematical expertise was treated as a social good. For instance, Ben-Yehuda et al. (2005) discussed the social construction of mathematical failure in relation to two girls with difficult life histories who were prevented from access to meaningful mathematics instruction. Similarly, Heyd-Metzuyanim (2013) questioned the tendency of parents and teachers to see mathematical failure as an individual phenomenon, located solely in the student's "mind" or brain. Unlike Sfard's rejection of cognitivism (e.g., Sfard, 2000b, 2007; Sfard & Lavie, 2005), other authors who used commognition (e.g., Jankvist, 2011; Ryve, 2004; Xu & Clarke, 2013) were less explicit about critiquing acquisitionist or cognitive ideas, though some of them questioned certain "individualist" ways of viewing learning.

The commognitive framework seems to be useful for examining what is often referred to as "mathematical content" within a sociocultural perspective and within interactional settings. The articles in this subheritage are consistent in the application of the major concepts used, either because Sfard is inconsistent about clear definitions of her key concepts or because Sfard herself was a coauthor of the article. The early writings of this subheritage were relatively dense, and their methodology may be difficult to apply. Recently published articles (2012 and onward), however, provide examples for the application of the commognitive framework as both a theory of learning and a methodological tool set.

Positioning. Authors using positioning tended to ground their research in at least two primary sets of work from outside of mathematics education: Rom Harré and colleagues' positioning theory (e.g., Harré & Moghaddam, 2003; Harré & Slocum, 2003; Harré & van Langenhove, 1999) and Dorothy Holland and colleagues' theory of self and identity (e.g., Holland & Eisenhart, 1990; Holland, Lachiotte, Skinner, & Cain, 1998). Although most mathematics education researchers who write about positioning cited both of these sets of work (K. Anderson, 2009; Bishop, 2012; Gresalfi, 2009; Hand, 2012; Herbel-Eisenmann & Wagner, 2010; Herbel-Eisenmann, Wagner, Johnson, Suh, & Figueras, 2015; Kotsopoulos, 2014; Langer-Osuna, 2011; Turner, Gutiérrez, & Sutton, 2011; Wagner & Herbel-Eisenmann, 2009), a few cited only Holland and colleagues (Bell & Pape, 2012; Esmonde & Langer-Osuna, 2013; Langer-Osuna, 2014) or only Harré and colleagues (Esmonde, 2009b). The two exceptions were Gresalfi and Cobb (2006) and Hunter and Anthony (2011), who focused primarily on "dispositions." Although some authors used disposition and position/positioning together, Wagner and Herbel-Eisenmann (2009) argued that disposition is always a noun and indexes "apparently stable identities" (p. 9), whereas position can and should be used as a verb because it indexes action and the negotiation of flexible roles in the moment.

是以不同方式构成的、复杂的、通用的个人属性的集合，通过将该集合所维持的某些权利、职能和义务分配给个人，从而对人际间、群体间甚至个人内心行动的可能性产生影响。（第1页）

该子流派的另一个中心思想是霍兰德及其同事提出的"假定世界"概念，其广义的定义为"源自社会的文化建构活动"（Holland等，1998，第40—41页），这些作者提出，假定世界是产生个人和社会身份的四个场所之一。在定位理论中，假定世界的描述与故事情节有相似之处，霍兰德等人（1998）认为哈里是"在考虑自我本质与他人的关系及假定世界和故事情节之间的联系方面的领导者"（Wagner & Herbel-Eisenmann，2009，第3页）。尽管许多数学教育研究人员同时借鉴了这两项工作，但目前为止还没有人对其理论的互通性进行细致的研究。

综合性分析得出的概述。 定位理论直到近期才得到应用。在本节回顾的16篇文章中，有7篇经过仔细分析，它们大部分发于2009年及以后的发表高潮期。该子流派下的部分作者并没有使用"话语"一词，而是关注了其他的术语，例如互动（Gresalfi & Cobb，2006；Hand，2012；Hunter & Anthony，2011；Langer-Osuna，2014），互动与对话（Kotsopoulos，2014；Wagner & Herbel-Eisenmann，2009），使用的语言（Bell & Pape，2012），参与结构（Langer-Osuna，2011）或定位行为（Turner等，2011）。话语被描述为"展开的戏剧"（Wagner & Herbel-Eisenmann，2009），且被定义为"参与者在使用语言进行交流、互动和行动时的口头和书面文字、符号系统、表征和手势"（Bishop，2012，第44页）。有些作者关注如何使用话语来表达意义（Bell & Pape，2012）、话语如何在身份构建方面扮演"关键角色"，以及"塑造和向他人传达自身身份的主要方法"（Bishop，2012，第43—44页）。在一些研究中，话语和假定世界被认为是相似的，因为它们"都捕捉了社会组织和构建的思维、互动、解释和识别方式"（Gresalfi，2009，第332页；Esmonde & Langer-Osuna，2013，第291页）。有些文章还使用了"话语过程"一词，因为它是定位内涵的一部分："话语过程中，自我作为可观察的和主观连贯的参与者而被定位于谈话中"（Davies & Harré，1990，第48页）。

除了K.安德森（2009）和库索保罗斯（2014）以外，该子流派大多数研究都以中学数学课堂为背景，许多文章都重点关注小组互动，目的是验证合作学习是一种更好的学习方式这一普遍假设（例如K. Anderson，2009；Esmonde & Langer-Osuna，2013；Kotsopolous，2014；Langer-Osuna，2011）。这些文章的作者非常注意背景，这是因为背景对说明定位或假定世界如何起作用有贡献，举例来说，埃斯蒙德（2009b）调查了两项特定的课堂活动：小组测验和汇报准备。她在讨论部分专门花了一节篇幅来说明读者应该或不应该假设什么是更普遍适用的，因为她的分析和发现是在特定背景下进行的。

构建任务分析得出的概论。 表27.4汇总了该子流派与吉构建任务相关的趋势。多数情况下，作者将定位概念作为理解能动性、权威和学生身份发展问题的一种方式。因此，该子流派的作者特别突出政治与身份构建任务，他们发现因为受公平性的影响，主张拓展数学的观念并特别注意背景具有重要意义。

由于在假定世界和故事情节上强调定位，该子流派下许多文章都将联系视为关键主题。举例来说，K.安德森（2009）通过微观层面、中观层面和宏观层面的细致分析来理解个体的失败行为是如何形成和"凝固"的，以至于某些学生被视为"失败者"。赫贝尔-艾森曼和瓦格纳（2010）研究了反复出现的四字词组（称为"词块"）并将其与更广泛的权威问题联系起来。埃斯蒙德和兰格-奥苏纳（2013）将小组交互中的特殊定位与两性浪漫关系的假定世界联系起来。然而，这些跨界联系需要更频繁地发生，并且可以更精准地确定与测量的关系（Herbel-Eisenmann等，2015）。

对政治的关注表现在每篇文章都使用了"权力"或"授权"这两个词，这些作者关注在小组和全班讨论中、在制度和院校以及社会中的各种权力关系。许多文章引入了知识资金、文化相关实践、民族数学、社会和文化资本以及包含校外情境的其他构念，并主张在审视数学教与学时需要考虑这些概念。

According to Harré and van Langenhove (1999), a position

> is a complex cluster of generic personal attributes, structured in various ways, which impinges on the possibilities of interpersonal, intergroup and even intrapersonal action through some assignment of such rights, duties and obligations to an individual as are sustained by the cluster. (p. 1)

Holland and colleagues' idea of "figured worlds," broadly defined as "socially produced, culturally constructed activities" (Holland, et al., 1998, pp. 40–41), is another central idea in this subheritage. These authors propose that figured worlds is one of four sites in which personal and social identities are produced. There is a resemblance between the description of figured worlds and storylines in positioning theory, and Holland et al. (1998) credited Harré as "leading the way in considering the nature of self in relation to others and noted connections between figured worlds and storylines" (Wagner & Herbel-Eisenmann, 2009, p. 3). Although many mathematics education researchers drew on both of these sets of work, careful examination of the theoretical commensurability has not yet been done.

Overview from the general analysis. Positioning has been used only recently. Of the 16 articles reviewed in this section, 7 of which were carefully analyzed, most came in a surge of work published in 2009 and later. Some of the authors in this subheritage did not use the term "discourse" and instead focused on other terms, for example, interaction (Gresalfi & Cobb, 2006; Hand, 2012; Hunter & Anthony, 2011; Langer-Osuna, 2014), interaction and conversation (Kotsopoulos, 2014; Wagner & Herbel-Eisenmann, 2009), language-in-use (Bell & Pape, 2012), participation structures (Langer-Osuna, 2011), or positioning acts (Turner et al., 2011). Discourse was described as "unfolding dramas" (Wagner & Herbel-Eisenmann, 2009) and defined as "the spoken and written words, semiotic systems, representations, and gestures of participants as they use language to communicate, interact, and act" (Bishop, 2012, p. 44). There was attention to how discourse was used for meaning making (Bell & Pape, 2012), how it played a "critical role in enacting identities," and the "primary method of shaping and conveying one's identity to others" (Bishop, 2012, pp. 43–44). Discourse and figured worlds were seen as similar in some of this work, because they "both capture socially organized and constructed ways of thinking, interacting, interpreting, and identifying" (Gresalfi, 2009, p. 332; Esmonde & Langer-Osuna, 2013, p. 291). Articles also used "discursive processes," because it is part of the definition of positioning: "the discursive process whereby selves are located in conversations as observably and subjectively coherent participants" (Davies & Harré, 1990, p. 48).

Most of the research contexts in this subheritage were secondary mathematics classrooms, with K. Anderson (2009) and Kotsopolous (2014) the exceptions. Many articles focused on small-group interactions to investigate the pervasive assumption that cooperative learning was always good (e.g., K. Anderson, 2009; Esmonde & Langer-Osuna, 2013; Kotsopolous, 2014; Langer-Osuna, 2011). Articles attended seriously to context because of its contribution to how positionings or figured worlds played out. Esmonde (2009b), for example, examined two specific activities in the classroom: the group quiz and presentation preparation. She devoted a section of her discussion to stating what a reader should or should not assume as more generally applicable, because her analysis and findings were located in a particular context.

Overview from the building tasks analyses. Table 27.4 summarizes the trends in this subheritage related to Gee's building tasks. Most often authors employed the idea of positioning as a way to understand issues of agency, authority, and students' developing identities. As such, authors in this subheritage foregrounded the Political and Identity building tasks and found Significance in arguing for broadening the view of mathematics and attention to context because of the ramifications for equity.

A key theme that appeared across many of these articles was Connecting, because of the emphasis on positioning alongside figured worlds and storylines. For example, K. Anderson (2009) carefully laid out analyses at the microlevel, mesolevel, and macrolevel to understand how individual acts of failing build up and "stick," resulting in some students being seen as "failures." Herbel-Eisenmann and Wagner (2010) examined recurring four-word groupings called "lexical bundles" and linked them to broader issues of authority. Esmonde and Langer-Osuna (2013) linked particular positionings in small-group interactions to figured worlds of gender and romance. These connections across scales, however, need to occur more often and can be located more precisely in relationship to scale (Herbel-Eisenmann et al., 2015).

A focus on Politics appeared through the use of the words power or empowerment in every article. The authors were concerned with power relations in interactions in small groups and whole-class discussions, in systems and institutions, and in society. Many articles

表27.4 与定位有关的构建任务趋势

意义	活动	身份	关系	政治	联系	符号系统和知识
赋予背景、试图找出交流和更广泛的话语之间的联系（特别是在身份类别的关系上）以及拓展数学观以意义。	大多数文章关注定位的行为、描述背景的不同层次、调查小组活动，并将定位与故事情节或假定世界相联系。	身份是这项工作的中心，聚焦于互动过程中角色调整的灵活性。在有些文章中，多重身份（例如，性别、种族）被认为对交互有重要作用。	几乎所有文章都关注教师与学生或学生与学生之间的关系。大多数作者认为学生和数学之间的关系对公平性有影响。	几乎每篇文章都提到权力或授权。许多文章对现有观点提出质疑（例如，小组合作始终有效的假设）。	很多文章尝试将不同层次的话语联系起来（例如，关于故事情节或假定世界的定位）。有些文章将身份与交互相联系。	传统的"学校"数学经常受到质疑，文章呼吁拓展数学学习的概念以包括学习与身份发展之间的关系。许多文章都提到知识资金或类似概念。

有四篇文章是概念性或理论性的，主要关注以下活动：阐述定位理论的核心思想，分析定位在数学教育中的应用方式，并提出方法论上的建议（Herbel-Eisenmann 等，2015；Wagner & Herbel-Eisenmann，2009），提出内容概念应拓展并需要更多关注学生的发展倾向（Gresalfi & Cobb，2006）和建立教师如何力争让学生在数学课堂内外"占据自己空间"的理论（Hand，2012）。

大部分此类文章的核心目标是提倡拓展学校数学的范围（符号系统和知识）并由此实现数学的"再神化"（Wagner & Herbel-Eisenmann，2009），通过考虑"一门学科的观念、价值观和参与方式，……在一个特定课堂上的体现，以及学生对这门学科的认同程度"（Gresalfi & Cobb，2006，第50页）。有些文章认为上述观点可以支持教师开展公平的数学教学（Hand，2012）。在使用定位理论的一项实证研究中，作者提出应拓展数学的范围，即在其中纳入对学生身份发展的关注以及学生认同该学科的方式。这些论点的一个基本核心假设是，对"数学"采取更广泛的视角可能会对那些历史上被学校数学边缘化的学生产生影响，并有助于使数学教育更为公平。

这类工作也表明，小组活动这种活动形式可能是有问题的。有些文章显示了一位学生是如何被定位为能力较差、问题较多或落后于合作伙伴或小组中的其他学生（K. Anderson，2009；Bishop，2012；Kotsopolous；2014）。他们提出了这样的问题，随着时间的推移，重复的定位可能会如何"加重"或积累，从而使学生的"类型"在不同情境下变得更稳定，并与身份相联系。其他文章指出，随着时间的推移，某些此类反复定位行为可能会对不同学生产生不同的影响，例如，兰格–奥苏纳（2011）展示了一位女学生如何从一位领导者转变为被视作"专横"的人，而处于同一组且前期参与度不高的一位男学生却成为领导者和数学贡献者。兰格–奥苏纳（2014）追踪了一位学年初期数学参与度较低的学生在整个学年的假定世界和定位状态，并发现该学生如何"在一个学年中发展成一名更具自我实现能力的学习者，他能够认清实物和表征资源并利用它们提高自己的产出能力"（第66页）。

埃斯蒙德和兰格–奥苏纳（2013）也揭示了学生在小组互动中如何出现多重假定世界，包括一个数学的假定世界和一个充满友谊与浪漫的假定世界。他们认为上述假定世界被种族化和性别化，是由学生积极构建和竞争形成的。他们展示了这些假定世界如何让一位非洲裔美国学生在课堂话语中把自己定位为强势之人，从而她能够参与更高水平的数学实践。

考虑到聚焦身份，特别是公平性，这一研究主线非

TABLE 27.4. Building Task Trends Related to Positioning

Significance	Activities	Identities	Relationships	Politics	Connections	Sign systems and knowledge
Significance is given to context, to trying to identify connections between interactions and broader discourses at play (especially in relationship to identity categories), and to broadening views of mathematics.	Most articles focus on acts of positioning, describe various levels of context, investigate small-group work, and connect positioning to storylines or figured worlds.	Identities are central to this work, with a focus on how flexible roles are negotiated as people interact with one another. In some of the work, multiple identities (e.g., gender, race) are recognized as important to the interactions.	Almost all articles focus on the relationship between teachers and students or students and students. Most argue that the relationship between students and mathematics has implications for equity.	Almost every article mentions power or empowerment. Many articles call into question status quo views (e.g., the assumption that small-group work is always good).	Many articles try to connect various levels of discourse, (e.g., positions to storylines or figured worlds). Some connect interactions to identities.	Conventional "school" mathematics is often questioned and articles call for a broader view on mathematical learning to include the relationship between learning and identity development. Many articles mention funds of knowledge or similar constructs.

brought in the ideas of funds of knowledge, culturally relevant practices, ethnomathematics, social and cultural capital, and other constructs that include out-of-school contexts, arguing that these need to be considered when examining mathematics teaching and learning.

Four articles were conceptual or theoretical and focused on the following Activities: articulating core ideas in positioning theory, analyzing how positioning is used in mathematics education, and making methodological suggestions (Herbel-Eisenmann et al., 2015; Wagner & Herbel-Eisenmann, 2009); arguing that conceptions of content should be broadened with increased focus on students' developing dispositions (Gresalfi & Cobb, 2006); and theorizing how teachers aim for students to "take up their space" in and beyond the mathematics classroom (Hand, 2012).

A central goal in most of these articles was to argue for a broad version of school mathematics (Sign Systems and Knowledge) that "re-mythologizes" (Wagner & Herbel-Eisenmann, 2009) mathematics, by considering "ideas about, values of, and ways of participating with a discipline...as it is realized in a particular classroom and the extent to which students come to identify with the discipline" (Gresalfi & Cobb, 2006, p. 50). Some articles argued that such views could support teachers to engage in equitable mathematics instruction (Hand, 2012). In the empirical work that used positioning, authors advocated for an extended version of mathematics that included attention to students' developing identities and the ways in which students identified with the discipline. A core underlying assumption of these arguments was that adopting a broader perspective on "mathematics" could have implications for students who have been historically marginalized by school mathematics and could contribute to making mathematics education more equitable.

This set of work also showed how the Activity of small-group work might be problematic. Some articles showed how one student was positioned as less capable, problematic, or inferior to other students in partner or small-group work (K. Anderson, 2009; Bishop, 2012; Kotsopolous, 2014). They raised issues related to how repeated positionings might "thicken" or accrue over time, becoming more stable "kinds" of students across contexts and linked to Identities. Other articles showed that some of these repeated positionings might have different impacts on different students over time. Langer-Osuna (2011), for example, showed how a female student shifted from being a leader to being seen as "bossy," while a male student in the same group, who was less engaged early on, became a leader and mathematical contributor. Tracing the figured worlds and positionings of one student who was not very mathematically engaged early in the school year, Langer-Osuna (2014) showed how he "developed into a more actualized learner across the academic year, one who identified physical and symbolic resources and leveraged them for productivity" (p. 66).

Esmonde and Langer-Osuna (2013) also demonstrated how multiple figured worlds surfaced in students' interactions in small groups, including a figured world of mathematics and a figured world of friendship and romance. They argued that these figured worlds were racialized and gendered and were actively constructed and contested by the students. They illustrated how these figured worlds allowed one African American student to position herself as powerful in the classroom discourse

常强大，令人惊讶的是只有一篇文章有明确的作者定位声明（见 Esmonde & Langer-Osuna，2013）。正如富特和巴特尔（2011）提出的有力看法那样，一个人的身份在很大程度上会影响他看待、解释和体验研究过程的方式。

社会语言学与话语研究流派

在本节中，我们将重点关注六个子流派，其作者均以某种形式的社会语言学或话语研究为出发点开始他们的课堂话语工作，具体包括：系统功能语言学、图尔敏的论证模型、浮现观（基于符号互动论[1]）、互动社会语言学、格赖斯门派和话语心理学（主要借鉴德里克·爱德华兹的工作）。

系统功能语言学。迈克尔·哈利迪自20世纪60年代开始开发的系统功能语言学（SFL），是用来描述书面和口头文本如何与它们所服务的交流功能相联系，从而支持要交换的意义的。哈利迪认为，文本是在特定的情境中产生的，而情境又被文化背景所包围（Halliday & Hasan，1989），文本会受到"场"（或正在发生的事情）、"说话者"（或参与者以及他们之间的关系）和"风格"（或使用的语言形式）的影响，上述几点又分别与语言的"观念""人际"和"文本"这些元功能有关。

哈利迪（例如 Halliday & Hasan，1989）设想，在产生文本时使用元功能的特定组合可能有助于形成特定的情境背景。当在特定的情况下为特定的目的而频繁使用特定的组合时，就形成了语言的语义。该组文章经常提及语义的概念。施莱配格里尔（2007）描述了为什么数学语义的概念对那些研究课堂互动的人具有吸引力："数学语义这一概念能帮助我们理解语言以不同于其他学科的方式构建数学知识的方式"（第140页）。

其他语言学家（例如马丁、克雷斯和万·利文）在类型和多模态方面发展了系统功能语言学。与哈利迪最初的语义概念相比，数学教育界吸收这些思想要慢一些，这可能是因为哈利迪是1974年参加内罗毕峰会的语言学家之一，并随后在他的著作《语言的社会符号作用》（Halliday，1978）中发表了与数学语义相关的一章内容。

综合分析得出的概论。像其他流派一样，借鉴系统功能语言学及其相关理论的一系列文章之间亦各有不同。我们回顾了29篇涉及系统功能语言学的数学课堂话语研究文章。虽然研究者，比如皮姆（1987），已经借鉴了哈利迪（1978）的部分观点，特别是数学语义的观点，但大多数文章的发表时间均在我们检索的30年间的后20年，只有两篇借鉴系统功能语言学相关理论的数学教育文章发表于1995年以前，分别是克劳福德（1990）及马克思和莫斯雷（1990）。

尽管系统功能语言学和相关理论在过去20年间有了更多的应用，但是这套完整理论似乎为我们提供了更多的启发，而不只是一种理论或方法框架。部分可能的原因是，哈利迪对语法的描述是复杂的，它提供了对语法所履行的沟通功能的深刻见解，且包含许多不同的组成部分，它的主要目的是描述语言，特别是英语，是如何运作的，因此数学教育研究人员在进行研究时不太可能利用到其所有组成部分。

构建任务分析得出的概论。表27.5汇总了系统功能语言学和相关理论如何与吉构建任务明显地相关联。

在我们的数据集中，数学教育研究者利用该子流派的方式各不相同，从在文献综述部分包含某一个讨论（例如 Han & Ginsburg，2001；Staples & Truxaw，2012）到成为影响方法论决定的主要理论框架（例如 Mesa & Chang，2010）。摩根（1996，2005）、赫贝尔–艾森曼及其同事（Herbel-Eisenmann，2007；Herbel-Eisenmann & Wagner，2007；Herbel-Eisenmann, Wagner, & Cortes，2010）和左科沃及其同事（Shreyar, Zolkower, & Pérez，2010；Zolkower & de Freitas，2012；Zolkower & Sheyar，2007）在他们的研究中较一致地使用了系统功能语言学。可能由于该子流派在数学教育领域中没有标准化的使用，许多文章的作者都使用实证的例子来说明系统功能语言学作为其部分活动的理论或方法论取向的优势。举例来说，赫贝尔–艾森曼和奥腾（2011）用实证的例子说明了识别词汇链在描绘两个中学几何课堂教学实录全貌时的有效性。

so that she could engage in high-level mathematical practices.

Given the strong thread of work that focused on identity, and equity in particular, it was surprising that only one article had an explicit author positionality statement (see Esmonde & Langer-Osuna, 2013). As Foote and Bartell (2011) convincingly argued, one's identity very much affects the ways in which one sees, interprets, and experiences the research process.

Sociolinguistic and Discourse Studies Heritage

In this section, we focus on six subheritages in which authors began their work on classroom discourse with some form of sociolinguistics or discourse studies: systemic functional linguistics, Toulmin's model of argumentation, an emergent perspective (based on symbolic interactionism[1]), interactional sociolinguistics, a Gricean umbrella, and discursive psychology (drawn primarily from the work of Derek Edwards).

Systemic functional linguistics. Systemic functional linguistics (SFL) has been developed by Michael Halliday since the 1960s to describe how texts, written and spoken, are connected to the communicative functions they serve and thus support meanings to be exchanged. Halliday suggested that texts are produced within a specific context of situation, which is surrounded by context of culture (Halliday & Hasan, 1989). A text will be affected by the "field" (or what is happening), the "tenor" (or who is taking part and the relationship between them), and the "mode" (or the form of language that is used). These are connected, respectively, to "ideational," "interpersonal," and "textual" metafunctions of language.

Halliday (e.g., Halliday & Hasan, 1989) envisaged that the use of particular combinations of metafunctions in producing a text could contribute to a specific context of situation coming into existence. When particular combinations are used frequently in specific circumstances and for a particular purpose, then a register of the language is developed. The idea of register was mentioned frequently in this set of articles. Schleppegrell (2007) described why the idea of mathematics register is appealing to those who research classroom interactions: "This notion of a mathematical register helps us understand the ways that language constructs mathematical knowledge in different ways than it constructs other academic subjects" (p. 140).

Other linguists (e.g., Martin, Kress, and van Leeuwen) have developed SFL in regard to genres and multimodality. These have had a slower uptake into mathematics education than Halliday's original ideas about register. This may be because Halliday was one of the linguists who attended the Nairobi summit in 1974 and subsequently published a chapter in his book *Language as Social Semiotic* (Halliday, 1978), which was about the mathematics register.

Overview from the general analysis. Like other heritages, the set of articles that drew on systemic functional linguistics and related theories was varied. We examined 29 articles that were related to mathematics classroom discourse and connected to SFL. Although researchers such as Pimm (1987) had used some of Halliday's (1978) ideas, particularly that of the mathematics register, the majority of the articles were written in the last 20 years of the 30-year period of our search. Only two articles drawing on SFL-related theories within mathematics education were published prior to 1995—One by Crawford (1990) and one by Marks and Mousley (1990).

Although there has been more use of SFL and related theories in the last 20 years, the full theory seems more to have provided inspiration than to have established itself as a major theoretical or methodological framework. It may be that part of the reason for this is that Halliday's description of grammar as providing insights into the communication functions that they fulfill is complex, with many different components. Its primary purpose has been to describe how language, particularly English, operates, and so it would be unlikely that any mathematics education researcher could make use of all of its components when conducting research.

Overview from the building tasks analyses. Table 27.5 provides a summary of how SFL and related theories were evident in relationship to Gee's building tasks.

In our data set, the ways in which the mathematics education researchers drew on this subheritage varied from including a discussion in the literature review (e.g., Han & Ginsburg, 2001; Staples & Truxaw, 2012) to being a main theoretical framework that informed methodological decisions (e.g., Mesa & Chang, 2010). Morgan (1996, 2005), Herbel-Eisenmann and colleagues (Herbel-Eisenmann, 2007; Herbel-Eisenmann & Wagner, 2007; Herbel-Eisenmann, Wagner, & Cortes, 2010), and Zolkower and colleagues (Shreyar, Zolkower, & Pérez, 2010; Zolkower & de Freitas, 2012; Zolkower & Sheyar, 2007) used SFL more consistently in their research. Perhaps because there did not seem to be a standard use of this subheritage in mathematics education, many authors of the articles used empirical examples to illustrate the benefits of SFL as a theoretical or methodological approach as part of their Activities. For example, Herbel-Eisenmann and Otten (2011) used empirical examples to illustrate the usefulness of identifying lexical chains

表27.5　与系统功能语言学有关的构建任务趋势

意义	活动	身份	关系	政治	联系	符号系统和知识
重要性经常与学生（通常是成绩不好的学生）如何需要支持相联系。	部分文章讨论了数学课堂的语境特征与其中使用或需要的语言之间的关系。其他文章则关注如何将系统功能语言学或相关理论用作方法论。	一般情况下，作者不关注参与者或自己的身份。有些文章着眼于作者如何使用语言来突出他们的角色或定位。	几乎所有文章都关注教师和学生之间的关系。部分文章专注于数学（通过教科书或教师发起的互动）和学生/新手之间的关系。	部分文章试图确定数学是如何通过语言构建的以及一些新手怎么会错过理解这一构建过程的。	每篇文章对联系都有不同的侧重点，但都讨论了联系。	数学意义是通过文本的生成和解释来发展的。不同的情境会影响学生/新手如何看待数学。

其他作者则使用实证材料来说明自己希望表达的观点。20世纪90年代，马克思和莫斯雷（1990）以及所罗门和奥尼尔（1998）都使用了20世纪80年代从系统功能语言学发展而来的类型理论来探讨课堂写作类型。他们的研究目的类似，即强调所有学生都应有机会接触适当的数学类型。他们还提出，无法就数学类型获得明确指导可能是导致数学表现不佳的原因之一（意义）。有人担心，如无法获得这些数学类型的具体指导，有些学生将无法在数学上取得成功。因此，这些文章都会涉及社会资源的分配（政治）。

与探讨数学类型的文章一样，该子流派的大多数文章亦明确提及了政治。与数学课堂相关的重要社会资源包括获取并了解数学语义（Morgan, 2006）、数学语言选择意识（Chapman, 1995, 1997）、能动性（Morgan, 2005）或选择、权力关系（Herbel-Eisenmann, 2007）、在各种文本中学生的定位（包括书面和口头文本；Herbel-Eisenmann & Wagner, 2007; Herbel Eisenmann 等, 2010; Veel, 1999）以及分工（谁来控制或选择任务分配；González & DeJarnette, 2012）。

总体而言，实证研究中使用的数据来自教室或课堂相关材料，如教科书或评估报告。证明是梅希亚-拉莫斯和英格利斯（2011）调查数学语言的一个领域，他们考虑了"证明"这个词的各种日常内涵如何影响着学生所用语言的概念。有两篇文章因使用了计算机软件，从而对大量的数据进行了考察（Herbel-Eisenmann 等, 2010; Monaghan, 1999）。在莫纳亨的研究中，对源于一组资源材料中的1 418项数据进行了分析，以确定"对角线"一词在不同年级的用法。在赫贝尔-艾森曼等的文章中，数据集合由来自148份课堂观察的记录组成，该研究聚焦于对词块的识别和分析。

总的来说，该子流派的研究能清楚地解释分析工具和分析细节（活动）。尽管如此，其中有大约三分之一的稿件缺乏有关工具或分析过程的细节，这些稿件的发表时间基本都在2000年之前，例如，阿特维、布雷彻和库伯（1998）就哈利迪的三种语言元功能给出了结果，但没有提供取得这些结果的细节。

在有些文章中，数学教育研究人员并未按照提出系统功能语言学或相关构念的语言学家的意图，以与之相匹配的方式使用这些构念，而是将这些理念开发成自己的分析工具（活动）。举例来说，奥哈罗兰（2000）将系统功能语言学拓展为符号数学和视觉表示的词汇语法，然而，哈利迪（Webster, 2003）对上述系统功能语言学的拓展方式表示不满意，因为数学符号可以通过各种方式转化为文字描述。西格尔、博拉西和房兹（1998）从系统功能语言学对功能的关注中获得启示，提出了"阅读在数学探究循环中可以发挥的30种不同功能"（第410页），然而，这些作者与这里所讨论的流派的联系并不明显。查普曼（1995, 1997）的工作也有类似问题，她认为自己的工作属于社会符号学，但其与系统功能语言学的联系更为隐性而非显性。

TABLE 27.5. Building Task Trends Related to Systemic Functional Linguistics

Significance	Activities	Identities	Relationships	Politics	Connections	Sign systems and knowledge
Significance is often tied to how students (generally low-achieving students) need support.	Some articles address the relationship between contextual features of mathematics classrooms and the language used or needed in them. Others focus on how SFL or related theories could be used as a methodology.	Generally, authors do not attend to the identities, either of participants or of themselves. Some articles look at how writers use language to highlight their role or position.	Almost all articles focus on the relationship between teachers and students. Some focus on the relationship between mathematics (which could be through textbooks or teacher-initiated interactions) and students/novices.	Some articles try to determine how mathematics is constructed through language and how some novices might miss out on understanding how this construction process occurs.	Each article has a different focus on connections, but they are present.	Mathematical meanings are developed through the production and interpretation of texts. Different contexts affect how students/novices come to view mathematics.

to map the mathematics in transcripts from two middle school geometry classes.

Other authors used empirical material to illustrate points that they wanted to make. In the 1990s, both Marks and Mousley (1990) and Solomon and O'Neill (1998) used genre theory, which had developed from SFL in the 1980s, to consider the kinds of writing done in classrooms. They had a similar purpose in highlighting the need for all students to be given access to appropriate mathematical genres and suggested that restricted access to explicit teaching of mathematical genres could provide some reasons for mathematical underachievement (Significance). There was concern that without specific instruction in these genres, some students would be unable to achieve in mathematics. These articles, therefore, were concerned with the distribution of social goods (Politics).

As was the case with the articles that discussed mathematical genres, Politics were evident in the majority of articles in this subheritage. Critical social goods related to the mathematics classroom included, for example, access to and knowledge of the mathematics register (Morgan, 2006), awareness of mathematical language choices (Chapman, 1995, 1997), agency (Morgan, 2005) or choice, authority relations (Herbel-Eisenmann, 2007), students' positioning in a variety of texts (both written and spoken; Herbel-Eisenmann & Wagner, 2007; Herbel-Eisenmann et al., 2010; Veel, 1999), and division of labor (who controlled or selected task components; González & DeJarnette, 2012).

On the whole, the data used in empirical research came from classrooms or classroom-related materials, such as textbooks or assessments. Proof was one area of mathematical language investigated by Mejía-Ramos and Inglis (2011), who considered how everyday connotations of the word "prove" affected students' conceptions of the language that they used. In two of the articles, large amounts of data were examined using computer software (Herbel-Eisenmann et al., 2010; Monaghan, 1999). In Monaghan's case, 1,418 items from a set of resource materials were examined to locate the use of the term "diagonal" for different year levels. In the article by Herbel-Eisenmann et al., the data sets consisted of classroom transcripts from 148 classroom observations and the research focused on identifying and analyzing lexical bundles.

In general, research using this subheritage clearly explained both the analytical tools and the details of the analysis (Activities). Nevertheless, about a third of the manuscripts lacked details about the tools or analytic process. Often these were papers written before 2000. For example, Atweh, Bleicher, and Cooper (1998) provided their results in regard to Halliday's three metafunctions of language but did not provide details about how they arrived at these results.

In some articles, rather than using SFL or related constructs in a way considered compatible with the intentions of the linguists who proposed them, mathematics education researchers developed the ideas into analytical tools of their own (Activities). For example, O'Halloran (2000) extended SFL to consider the lexicogrammar of symbolic mathematics and visual displays. Halliday (Webster, 2003), however, expressed some unease at extending SFL in this way because mathematical symbols can be turned into word descriptions in a variety of ways. Siegel, Borasi, and Fonzi (1998) took inspiration from SFL's focus on functions to propose "30 different functions that reading could play in a mathematics inquiry cycle" (p. 410). These authors' link to this heritage, however, is somewhat tenuous. A similar

有部分文章在系统功能语言学和相关理论与其他理论框架之间建立了联系。举例来说，韦尔（1999）用系统功能语言学衍生的见解与伯恩斯坦的社会学联系起来，还有一些情况是该流派与批判性话语分析结合作为方法论（Herbel-Eisenmann & Wagner, 2007, 2010; Morgan, 2005），这些作者所借鉴的部分相关理论框架表明，这些文章可以被放置在"概念重建"流派类别下，但鉴于它们以更一致和彻底的方式将系统功能语言学作为主要的理论和方法，我们选择将它们归入本流派。

综观所有数据集，伯顿和摩根（2000）的研究是使用系统功能语言学分析身份的少数例子之一。通过对53篇数学研究论文中的概念和人际元功能的分析，伯顿和摩根（2000）确定了数学家写作时在灵活性方面具有的局限。该分析的部分篇幅着重探讨了数学家如何建立他们作为各自领域内的权威这一身份。白汉姆（1996）在研究成人数学班时，受系统功能语言学的启发，用它探讨了参与者的身份构建过程。赫贝尔-艾森曼（2007）也研究了某种数学文本的人际要素，关注的是来自某数学课程的一个单元，她的聚焦点不是身份，而是学生与教科书编写者的解释权威之间的关系定位。尽管系统功能语言学可以用来研究参与者如何构建身份，但是调查课堂话语中身份问题的研究者通常会借鉴其他文献，而系统功能语言学和相关理论则支持通过能动性等构念来调查参与者之间的关系，如，冈萨雷斯和德贾内特（2012）就比较了学生和教师在高中几何课中的能动性。

也许因为对参与者身份的关注有限，很少有关于研究人员身份及其对当前研究影响的讨论。有趣的是，明确提及研究人员身份的两篇论文（即Mesa & Chang, 2010; Roberts, 1998）均与成年人相关。罗伯茨（1998）讲述了她作为澳大利亚北部偏远社区担任语言学教师的角色，而梅萨和昌（2010）则探讨了他们与参与他们研究的数学本科生不同的身份，上述两类角色的确影响了所开展的研究。考虑到很少有文章涉及中小学以后的情况，他们对身份关注的优势还是很有意思的。赫贝尔-艾森曼（2007）及其同事（Herbel-Eisenmann & Otten, 2011; Herbel-Eisenmann & Wagner, 2007; Herbel-Eisenmann等, 2010）为工作提供了思想基础，这可以看作是他们所从事的研究所具有的价值。

该子流派似乎被主要用于理解情境特征如何影响人们所使用的语言，这有助于课堂中的数学阐述，系统功能语言学和相关理论对此具有极为重要的作用。然而，由概述可以得出的意外结果之一就是，在试图更好理解数学课堂话语中其他相关语言学理论的使用是有限的，此类理论包括由克雷斯和万·利文开发的多模态理论，可用于教科书分析以及马丁和同事关于类型的工作。诸如克劳福德（1990）等研究将技术等课堂资源视为中性的或在某些情况下有天然的积极性，也许使用其他相关理论可以发展对什么有助于课堂话语及它如何影响学生可能获得的数学理解类型的更细致的理解。如果该子流派下各种研究的目的是探讨数学成就的社会益处，那么就需要一种更为细致的方法。

图尔敏论证模型。受维特根斯坦的影响，图尔敏在20世纪50年代开发了自己的论证模型，用以确定非形式论证的结构和组成部分（Toulmin, 1958/2003）。尽管图尔敏的工作主要关注伦理和哲学，但美国修辞学和传播学的学者热情地接受了他的论证框架（O'Grady, 2010）。他确立了论证的六个组成部分，虽然并非所有这些组成部分都会出现在每一种非形式的论证中。英格利斯、梅佳-拉莫斯和辛普森（2007）概述了这些组成部分：

> 结论（C）是辩论者希望说服听众的陈述。数据（D）是论证所依据的基础，即观点的相关证据。证明（W）通过诸如规则、定义或者类比来论证论据与结论之间的关系是成立的。逻辑支援（B）支持证明，提供进一步的

statement could be made about the work of Chapman (1995, 1997), who described her work as being social semiotics, but the link to SFL was more implicit than explicit.

In some articles, Connections were made between SFL and related theories to other theoretical frameworks. For example, Veel (1999) linked the understandings derived from using SFL to Bernstein's sociology. In other cases, this heritage was combined with critical discourse analysis for the methodology (Herbel-Eisenmann & Wagner, 2007, 2010; Morgan, 2005). Some of the related theoretical frameworks these authors drew on suggest that the articles could have been placed in the reconceptualist heritage, but we chose to include them here because they more consistently and thoroughly drew on SFL as a primary theory and method.

Across the whole data set, Burton and Morgan's (2000) research was one of the few examples of using SFL to analyze Identity. Burton and Morgan (2000), through their analysis of the ideational and interpersonal metafunctions in 53 mathematics research papers, were able to identify the limits of the flexibility that mathematicians had in their writing. Part of this analysis focused on how the mathematicians constructed their identities as authorities in their field. Baynham (1996), examining adult numeracy classes, used SFL as inspiration for discussing the identity work done by participants. Although Herbel-Eisenmann (2007) also considered the interpersonal aspect of a type of mathematical text, in this case in a unit from a mathematics curriculum, identity was not the focus. Rather, she focused on the positioning of the student in relationship to the authority construed by the textbook writers. Although SFL could be used to consider how participants' identities were constructed, researchers investigating issues of identity in classroom discourse usually drew upon other literature. SFL and related theories, instead, supported investigations of the Relationships between participants, using constructs such as agency. For example, González and DeJarnette (2012) compared student and teacher agency in review geometry lessons in a high school.

Perhaps because of the limited focus on participants' identity, there was also little discussion of the identity of the researchers and how this affected the research being undertaken. Interestingly, the two papers in which the researchers were explicit about their own identities (i.e., Mesa & Chang, 2010; Roberts, 1998) dealt with adults. Roberts (1998) described her role as a teacher linguist in a remote community in Northern Australia, whereas Mesa and Chang (2010) discussed their identities as being different from those of the undergraduate mathematics students who were the participants in their study. Both sets of roles did affect the research they undertook. Given that there were very few articles in the data set that dealt with postschool contexts, it is interesting to note their predominance in relationship to focusing on Identity. Herbel-Eisenmann (2007) and colleagues (Herbel-Eisenmann & Otten, 2011; Herbel-Eisenmann & Wagner, 2007; Herbel-Eisenmann et al., 2010) provided the ideological underpinnings for the work, which could be considered as the values that they attached to undertaking their research.

This subheritage seems to have been used mostly for trying to understand how contextual features affect the language being used, which then contributes to the mathematics being construed within classrooms. SFL and related theories seem particularly valuable for this. One of the surprising outcomes from doing this overview, however, is the limited use of other related linguistic theories in trying to better understand mathematics classroom discourse. These theories include multimodalities developed by Kress and van Leeven, which could be used within textbook analysis, and the work of Martin and colleagues on genres. Research such as that by Crawford (1990) present classroom resources such as technology as being neutral or in some cases naturally positive. Perhaps using other related theories could further develop a nuanced understanding about what contributes to classroom discourse and how this affects the kinds of mathematical understandings it is possible for students to gain. A more nuanced approach seems important if the aim for the kinds of studies undertaken with this subheritage is to discuss the social good of mathematics achievement.

Toulmin's model of argumentation. Influenced by Wittgenstein, Toulmin developed his model of argumentation in the 1950s to identify the structure and components of informal arguments (Toulmin, 1958/2003). Although Toulmin's work was primarily focused on ethics and philosophy, scholars in rhetoric and communication in the United States enthusiastically took up his framework for argumentation (O'Grady, 2010). He identified six components of arguments, although not all of them would be present in every kind of informal argument. Inglis, Meija-Ramos, and Simpson (2007) summarized descriptions of the components:

> The conclusion (C) is the statement of which the arguer wishes to convince their audience. The data (D) is the foundations on which the argument is based, the relevant evidence for the claim. The warrant (W) justifies the connection

证据。模态限定词（Q）通过表达置信程度来限定结论。反驳（R）通过陈述它不成立的条件来驳斥结论。（第4页）

英格利斯等人（2007）指出，图尔敏将数学证明归为基于逻辑的，因而在结构上不同于图尔敏模型可以表达的非形式化论证。同样讨论图尔敏的工作，克鲁姆霍伊尔（2000）将逻辑演绎论证视为"分析性的"，而"实质性的"论证往往通过归纳或溯因的视角（Pedemonte, 2007）逐渐支持某一个陈述。在数学教育领域，大多数使用图尔敏模型的研究调查了学生使用的非形式化的、"实质性"的论证，这些论证导致形式化证明的发展，要么作为直接的结果，要么作为学生在之后学年要完成的事情。

一般分析得出的概论。我们对该子流派下的26篇文章中的16篇做了详细分析，对其他文章则进行了主题一致性检查。大多数文章认为克鲁姆霍伊尔（1995）是最早使用图尔敏模型的数学教育家。在经过详细分析的文章中，参与者的年龄组和他们论证的复杂程度似乎会影响识别出的组成论证部分的数量。关于中小学生论证的探讨往往不包括模态限定词或反驳（例如，可以参见 Forman, Larreamendy-Joerns, Stein, & Brown, 1988；Krummheuer, 2007）。然而，其他研究人员对成年人、大学数学专业学生、教师教育专业学生或教师进行了研究，呈现了在考虑论证时他们使用模态限定词或反驳的有力案例（例如，可以参见 Giannakoulias, Mastorides, Potari & Zachariades, 2010；Inglis 等, 2007）。在这些案例中，论证所处的情境似乎有助于考虑模态限定词和反驳在论证形成过程中的作用。康纳、辛格尔特里、史密斯、瓦格纳和弗朗西斯科（2014）对一位职前中学教师支持学生发展论证的分析中识别出了所有的论证组成成分，包括反驳和限定词。然而，在该职前教师的提问或其他帮助学生发展论证的支持行为中，没有发现明显的逻辑支援成分。

与本章讨论的一些其他话语流派不同，图尔敏的论证模型通常被用作研究一个实证问题或提出一个主张的方法论。在许多文章中，还使用了其他理论，有时是在其它方法中结合了图尔敏模型来描述论证，例如，克鲁姆霍伊尔（2000）同时使用了埃里克森（1982）的学术任务结构，而米尼（2007）则使用了哈桑（Halliday & Hasan, 1989）的文本结构。其他人则从使用图尔敏模型的分析中建立理论，例如课堂中的数学实践（Stephan & Rasmussen, 2002）以及数学理论建设（Walter & Barros, 2011）。纳迪、笔札和扎卡－瑞尔德斯（2011）则采用了弗里曼（2005）对图尔敏证明的解读来更好地理解教师如何评估学生的反应，并确定采取适当的教学方法。然而，这些综合理论的做法似乎并没有为学生论证提供不同的见解，相反，它们似乎只提供了详细的说明。康纳等人（2014）就图尔敏的论证理论与其他分析集体论证的方法进行了广泛的讨论，并指出了他们认为使用这种方法的优点和局限性。

几乎所有的文章都考虑了学习者的论证以及他们是如何与数学学习相联系的。纳迪等人（2011）的研究则不是这样，他们调查了教师针对学生的错误回答所提供的论证——"从他们的话语中，我们看到了他们提出的用来支持或避免某些教学行为的论证"（第159页）。这项工作与康纳等人（2014）关于职前教师如何支持学生发展他们论证的研究有相似之处。

构建任务分析得出的概论。表27.6汇总了在借鉴图尔敏论证模型的文章中出现的吉的构建活动。一般来说，研究的意义在于论证的重要性。一些研究者专注于论证与证明之间的关系（Inglis 等, 2007；Pedemonte, 2007, 2008），而其他研究则关注学习者对其主张的证明是如何有助于其数学学习的（Forman 等, 1998；Giannakoulias 等, 2010；Walter & Johnson, 2007）。洋科（2002）的研究重点是教师在支持课堂讨论形成论证中所起的作用，其他人关注的是集体推理（Krummheuer, 2007；Stephan & Rasmussen, 2002）。康纳等人（2014）关注的是教师在集体论证中的作用。

between data and conclusion by, for example, appealing to a rule, a definition or by making an analogy. The warrant is supported by the backing (B) which presents further evidence. The modal qualifier (Q) qualifies the conclusion by expressing degrees of confidence; and the rebuttal (R) potentially refutes the conclusion by stating the conditions under which it would not hold. (p. 4)

Inglis et al. (2007) stated that Toulmin classified mathematical proof as being based on logic and therefore different in structure to the informal arguments that Toulmin's model could describe. Also discussing Toulmin's work, Krummheuer (2000) described logical deductive arguments as "analytic," whereas "substantial" arguments gradually support a statement, most commonly through inductive or abductive perspectives (Pedemonte, 2007). In mathematics education, most research that has used Toulmin's model investigated the informal, "substantial" arguments used by students, which could lead to the development of formal proofs, either as an immediate outcome or as something to be done in students' later school years.

Overview from the general analysis. Of the 26 articles identified for this subheritage, 16 were analyzed in detail while the others were checked for consistency with the themes reported here. Most articles identify Krummheuer (1995) as the first mathematics educator to use Toulmin's model. In the articles analyzed in detail, the age group of the participants and the sophistication of their arguments seemed to affect the number of components that were identified. Discussions of elementary and middle school students' arguments tended not to include the modal qualifier and rebuttal components (see, for example, Forman, Larreamendy-Joerns, Stein, & Brown, 1998; Krummheuer, 2007). Other researchers working with adults, university mathematics students, teacher education students, or teachers, however, present strong cases for the inclusion of modal qualifiers or rebuttals in considering arguments (see, for example, Giannakoulias, Mastorides, Potari, & Zachariades, 2010; Inglis et al., 2007). In these cases, the context in which the argument is operating seems to contribute to the need to also consider the role that modal qualifiers and rebuttals form in these arguments. Conner, Singletary, Smith, Wagner, and Francisco's (2014) analysis of a preservice secondary teacher's support for students' developing arguments identified all the components, including rebuttals and qualifiers. In the preservice teacher's questions or other supportive actions that were used to help students develop arguments, however, there were no explicit ones related to backings.

Unlike some of the other discourse heritages discussed in this chapter, Toulmin's model of argumentation is typically used as a methodology for either researching an empirical question or putting forward a proposition. In many of the articles, other theories are also used. Sometimes Toulmin's model was combined with other methodologies to describe arguments. For example, Krummheuer (2000) also used Erickson's (1982) academic task structure, whereas Meaney (2007) used Hasan's (Halliday & Hasan, 1989) text structure. Others built theories that started from an analysis that used Toulmin's model, such as classroom mathematics practices (Stephan & Rasmussen, 2002) and mathematical theory building (Walter & Barros, 2011). Nardi, Biza, and Zachariades (2011) used an elaboration by Freeman (2005) of Toulmin's warrants to better understand how teachers evaluated students' responses and identified appropriate pedagogical approaches to adopt. These combined theories, however, do not seem to have provided different insights into student argumentation; rather, they seem to provide an elaboration. Conner et al. (2014) provided an extensive discussion of how Toulmin's argumentation compared to other methods for analyzing collective argumentation and highlighted what they saw as the advantages and limitations of using this methodology.

Almost all the articles consider learners' arguments and how they connected to mathematics learning. Nardi et al.'s (2011) research was different in that they investigated arguments teachers put forward to respond to students' incorrect responses to questions—"From their utterances, we elicit what we see as the arguments they put forward to support or steer away from certain pedagogical actions" (p. 159). This work has similarities with Conner et al.'s (2014) research on how a preservice teacher supported students to develop their arguments.

Overview from the building tasks analyses. Table 27.6 summarizes Gee's building activities for the articles that draw on Toulmin's model of argumentation. Generally, the Significance of the research is on the importance of arguments. Some researchers focus on the relationship between arguments and proof (Inglis et al., 2007; Pedemonte, 2007, 2008), whereas others focus on how learners' justification of their claims could contribute to their learning of mathematics (Forman et al., 1998; Giannakoulias, et al., 2010; Walter & Johnson, 2007). The teachers' role in supporting the classroom discussion that led to argument development was the focus for Yackel (2002), whereas the collective reasoning was the focus for others (Krummheuer, 2007; Stephan & Rasmussen, 2002). Conner et al. (2014) focused on the teacher's role in collective argumentation.

表27.6　与图尔敏论证模式相关的构建任务趋势

意义	活动	身份	关系	政治	联系	符号系统和知识
理解论证的结构有助于深入理解论证与证明之间的关系以及集体推理的作用。	大多数文章着眼于课堂讨论，这些讨论有助于论证及其所依据的原则。	大多数情况下，身份不是这些文章的重点。研究人员偶尔会反思自己的角色。	人与人（教师-学生，学生-学生，教师-教师）之间的关系应支持对论证的质疑，从而使数学或教学原则变得透明。	这些文章没有明确讨论权力。	大多数文章涉及教师与课堂讨论、论证和证明之间联系的某些方面。	大多数文章认为数学论证是学数学和做数学内涵的的一部分。

图尔敏的模型很少用于考虑与身份或政治有关的活动，除非是以间接的方式。尽管大多数文章并没有明确探讨身份，但洋科（2002）和韦伯、马赫、鲍威尔和李（2008）的研究却与之不同。在韦伯等人（2008）的例子中，有可能涉及了身份与政治，因为该研究是在校外环境中进行的纵向研究。纳迪等人（2011）也讨论了教师的信念及其对课堂讨论的影响。福曼等人（1998）以类似的方式探讨了学生作为给出数学主张、质疑和理由的人可以发挥的作用。

论及学生论证对其学数学和做数学的重要性的一些文章中，作者含蓄地讨论了权力问题。米尼（2007）探讨了问题背景如何导致一些学生在其解答中没能表现出他们的数学能力，克鲁姆霍伊尔（2000）也强调了叙述性课堂结构如何在有效论证结构方面误导了学生。

尽管已有20年的应用历史，图尔敏模型的用途依然相当有限，其中最主要的就是突出展现学习者的数学思维与其论证之间的关系。有些研究以不同的方式使用图尔敏模型，例如理解教师的推理（Nardi等，2011）和教师对发展学生论证的支持（Conner等，2014），这表明图尔敏模型能为课堂互动中其他参与者的推理提供见解。

令人惊讶的是，论证模型中的证明被视为是最重要的组成部分，即使只确认了几个组成部分，证明在理解推理过程方面也起到了实质性的作用，看来大部分人均认为证明的透明化可能有助于从元层级探讨数学或教学法的原则。

在演绎、溯因和归纳论证以及它们与证明发展的关系方面，已经确定了不同种类的论证（Inglis等，2007；Pedemonte，2007，2008）。英格利斯等人（2007）提出要全面应用图尔敏模型的问题，并严厉批评了没有这样做的其他研究者。因此，似乎需要更多的研究来探讨如何使用模型的不同版本来回答不同的研究问题及其对研究结果产生的影响。

交际社会语言学。我们仔细分析了18篇宽泛界定为属于或借鉴了"交际社会语言学"的文章，这些文章包括利用对话分析的工作、语言人类学和社会学家欧文·戈夫曼的著作。这些文章引用的社会语言学内容非常多样化，没有一种理论或一个作者在所有文章中均被引用。尽管如此，可以看出约翰·古姆普雷兹、弗雷德里克·埃里克森、休·米恩和凯瑟琳·奥康纳与萨拉·麦克的著作在这个领域有较大的影响力。该领域的工作开始于20世纪90年代早期，其中朗佩（1990）及奥康纳和麦克（1993）影响较大。

综合分析得出的概论。该类别的许多工作具有双重目标。一方面引入一个方法论工具或一个理论概念（Brilliant-Mills，1994；Forman & Ansell，2002；Forrester & Pike，1998；Jurow，2005；Lobato, Ellis, & Munoz，2003；O'Connor & Michaels，1993），另一方面列举一些被认为是富有成效的且大多具有"改革"意向的教学实践。其他工作主要集中在解释一种分析的方法（Staats，2008），展示某一教学实践（Lampert，1990）或应用社

TABLE 27.6. Building Task Trends Related to Toulmin's Model of Argumentation

Significance	Activities	Identities	Relationships	Politics	Connections	Sign systems and knowledge
Understanding the structure of arguments allows for insights into relationships between arguments and proofs and the role of collective reasoning.	Most of the articles looked at classroom discussions that contribute to arguments and the principles on which those arguments are based.	Mostly, identities are not the focus of the articles. Occasionally the researchers reflect on their own role.	Relationships between people (teachers-students, students-students, teachers-teachers) should support challenging of arguments so mathematical or pedagogical principles become transparent.	Power is not discussed explicitly in any of the articles.	Most of the articles deal with some aspect of the connection between teachers and classroom discussion and argumentation and proof.	Most of the articles consider mathematical argumentation to be part of what it means to learn and do mathematics.

Toulmin's model was rarely used for considering activities to do with Identity or Politics, except in indirect ways. Although most of the articles do not discuss identities explicitly, Yackel (2002) and Weber, Maher, Powell, and Lee (2008) do. In the case of Weber et al. (2008) it may have been because the research was undertaken longitudinally in an out-of-school setting. Nardi et al. (2011) also addressed teachers beliefs and how these beliefs affected classroom discussions. In a similar manner, Forman et al. (1998) discussed the roles of students as authors of mathematical claims, challenges, and justifications.

Power issues were discussed implicitly in some articles in relationship to the importance of students' argumentation in learning and doing mathematics. Meaney (2007) discussed how the contexts of problems contributed to some students not showing their mathematical literacy competencies in their solutions. Krummheuer (2000) also highlighted how the structure of narrative classrooms can mislead the students in regard to valuable argumentation structures.

Although researchers have used Toulmin's model for 20 years, its use has been fairly limited, predominantly to highlight the relationship between learners' mathematical thinking and their argumentation. Some research has used Toulmin's model in different ways, such as understanding teachers' reasoning (Nardi et al., 2011) and teachers' support for developing students' arguments (Conner et al., 2014), suggesting that it can provide insights into the reasoning of other participants in classroom interactions.

Surprisingly, the component of the argumentation model that was treated as most Significant was the warrant. Even when only a restricted set of components were identified, warrants took on a substantial role in providing insights into reasoning. It seems to be agreed that making warrants transparent is likely to contribute to metalevel discussions about mathematical or pedagogical principles.

Different kinds of warrants have been identified in relationship to deductive, abductive, and inductive arguments and how they relate to the development of proofs (Inglis et al., 2007; Pedemonte, 2007, 2008). Inglis et al. (2007) raised the issue of using all aspects of Toulmin's model and were heavily critical of other researchers who did not. Therefore, it seems that more research is needed about how the different versions of the model can be used to answer different research questions and the impact that this has on the outcomes of the research.

Interactional sociolinguistics. We carefully analyzed 18 articles that we have broadly identified as belonging to or building on "interactional sociolinguistics." These include works drawing on conversation analysis, linguistic anthropology, and the writings of the sociologist Ervin Goffman. The papers include references to very diverse writings in the domain of sociolinguistics and there is no one theory or author that is referenced by all of them. Still, writings that have been influential on this group are John Gumprez, Frederick Erickson, Hugh Mehan, and the work of Catherine O'Connor and Sarah Michaels. Work in this domain started in the early 1990s with the influential works of Lampert (1990) and O'Connor and Michaels (1993).

Overview from the general analysis. Many of the works in this strand have a dual goal. One is introducing a methodological tool or a theoretical concept (Brilliant-Mills, 1994; Forman & Ansell, 2002; Forrester & Pike, 1998; Jurow, 2005; Lobato, Ellis, & Munoz, 2003; O'Connor & Michaels, 1993), the other is exemplifying some sort of teaching practice that is assumed to be productive, mostly in line with "reform" ideas. Other works focus mainly on illustrating a method of analysis (Staats, 2008), showcasing a certain instructional practice (Lampert,

会语言学的观点解决特定教学问题（Brodie，2007）。

该组文章提出的方法非常多样，范围从研究动词等特殊词语（O'Connor & Michaels，1993），甚至语调和韵律（Forrester & Pike，1998；Staats，2008），到对课堂活动的更宽泛地描述，具体包括事件的时间分配和对教师、学生行为的一般描述（Jurow，2005；Lampert，1990）。然而，所有这些工作均着眼于课堂对话，课堂对话如果不是唯一的分析对象，至少也是主要分析对象之一。这些研究的背景主要是公立学校的课堂，但是年龄段涵盖从一年级（Empson，2003）到小学高年级（Forrester & Pike，1998；Lampert，1990）、初中（Brilliant-Mills，1994；Forman & Ansell，2002；Jurow，2005；Lobato，2012）、高中（Brodie，2007），甚至有些大学课堂（Staats，2008）。

构建任务分析得出的概论。表27.7提供了我们根据交际社会语言学领域如何关注构建任务所确定的趋势汇总。

这组文章中有数篇对某些分析方法工具的具体启示意义重大。例如，布里连特－米尔斯（1994）提出一种通过社会语言学和人种学的视角来研究数学课堂的方法。该方法并不预先设置数学的定义，相反，她的目标是通过发生在特定课堂中的社会和话语实践来审视在那个特定的班级里什么才算是数学。福雷斯特和派克（1998）的文章也重点关注了方法论，他们使用对话分析，揭示在一个六年级课堂里，围绕测量和估计的教与学过程中隐含着的各种思想和概念。借鉴语言人类学，斯塔兹（2008）建议使用"诗意结构"，它包含了学生谈话中的重复和对仗用语，斯塔兹展示了诗意结构如何成为学生用来表达归纳和演绎论证以及在协作讨论中建构数学思想的形式。

该组的其他研究则将意义赋予实证发现（Empson，2003；Forman & Ansell，2002；Jurow，2005；Lobato等，2003），这些实证研究使用了来自交际社会语言学的方法或概念，要么作为更广泛的定性描述的出发点（通常基于扎根理论方法），要么作为描述课堂活动的一种工具。举例来说，洛巴托（2012）使用古德温（1994）"聚焦交互"的思想来调查注意（通过聚焦框架来理解）是否可以用来合理解释在直线斜率学习中发生的迁移。朱罗（2005）将"假定世界"理论（Holland等，1998，详见定位子流派中的说明）与社会语言学术语如"参与框架"和"立足点"相结合，来研究一个模拟现实项目是如何吸引一组八年级学生的，她还用"分层"的思想来描述在这个模拟项目中，所提出的不同假定世界的活动框架是如何分层的。

在活动方面，该组中的许多文章均关注教师谈话以及该谈话能为学生提供的社会框架（或参与者框架）（例如Lampert，1990；O'Connor & Michaels，1993）。受到关注的其他活动是教室里使用的板书等，例如图表、图

表27.7 交际社会语言学有关的构建任务趋势

意义	活动	身份	关系	政治	联系	符号系统和知识
文章使用理论视角，这些理论视角强调语言作为一种工具来处理社交互动、特定教学法（通常"以学生为中心"）和学生参与以及课堂实践和科学实践之间的联系。	所研究的活动大多为学生间的谈话或为学生提供某些"参与结构"的教师谈话。重点是通过特定语言互动构建的社会结构。	身份不是重点。某些文章在传递学生定位或将在课堂上形成的参与框架与学生的身份/定位联系起来时会提到。	文章主要通过教师交谈中传递给学生角色或位置信息的方式来研究师生关系。	没有明确涉及政治。	文章结合数学家所做的数学或"现实世界"中的数学以及课堂上的数学学习，探讨了社会实践和话语实践。	文章大多关注做数学与学数学的社会/互动方面，并将语义（或内容）方面后置。很多论文要么没有定义他们所说的做数学，要么将其泛称为数学家的实践。

1990), or applying ideas from sociolinguistics to particular problems of instruction (Brodie, 2007).

The methods put forward by these papers are quite diverse. They range from examining specific words, such as verbs (O'Connor & Michaels, 1993), and even intonation and prosody (Forrester & Pike, 1998; Staats, 2008), to a much broader description of classroom activity, including the time allocated for events and general descriptions of teacher and students actions (Jurow, 2005; Lampert, 1990). All these works, however, look at classroom talk, if not as the sole object of analysis, at least as one of the major ones. The contexts examined in these studies are mostly public school classrooms, but the age levels vary from first grade (Empson, 2003) to higher elementary grades (Forrester & Pike, 1998; Lampert, 1990), middle school (Brilliant-Mills, 1994; Forman & Ansell, 2002; Jurow, 2005; Lobato, 2012), high school (Brodie, 2007), and even some university level classrooms (Staats, 2008).

Overview from the building tasks analyses. Table 27.7 provides a summary of the trends we identified in terms of how the interactional sociolinguistics group attended to the building tasks.

Several works in this set give Significance to the specific affordances of certain methodological tools of analysis. For instance, Brilliant-Mills (1994) put forward a methodology of looking at a mathematics classroom through a sociolinguistic and ethnographical lens that makes no preliminary assumptions regarding what mathematics is. Rather, she aimed to examine, through the social and discursive practices taking place in a particular classroom, what "counts as mathematics" in that particular class. Another article with a main focus on methodology was that of Forrester and Pike (1998), who used conversation analysis to uncover the implicit ideas and concepts that surround the teaching and learning of measurement and estimation in a sixth-grade classroom. Drawing on linguistic anthropology, Staats (2008) suggested using "poetic structure," which included repetition and parallelism in students' talk. Staats showed how poetic structure can be the form that students use to express both inductive and deductive arguments and to build mathematical ideas in collaborative discussion.

Other studies in this strand gave Significance to empirical findings (Empson, 2003; Forman & Ansell, 2002; Jurow, 2005; Lobato et al., 2003) using methods or concepts taken from interactional sociolinguistics either as a point of departure for a broader qualitative description (often done by means of grounded theory methodology) or as one of a set of tools for describing the activity in the classroom. For instance, Lobato (2012) used Goodwin's (1994) concept of "focusing interactions" to investigate whether noticing (as conceived via the focusing framework) can be used to offer a plausible explanatory account of the occurrence of transfer in the learning of linear slopes. Jurow (2005) combined the theory of "figured worlds" (Holland et al., 1998, see positioning subheritage for more information), with terms taken from sociolinguistics such as "participation frameworks" and "footings," to examine the ways in which a simulated real-world project was engaging for a group of eighth-grade students. She also used the notion of "lamination" to describe how activity frames that proposed different figured worlds were layered onto one another in this simulated project.

In terms of Activities, many of the articles in this group focused on teachers' talk and the social frameworks (or participant frameworks) that talk afforded students (e.g., Lampert, 1990; O'Connor & Michaels, 1993).

TABLE 27.7. Building Task Trends Related to Interactional Sociolinguistics

Significance	Activities	Identities	Relationships	Politics	Connections	Sign systems and knowledge
Articles use theoretical lenses that put emphasis on language as a tool for handling social interaction, specific pedagogies (often "student centered") and student engagement, and connections between classroom practice and scientific practice.	Activities discussed are mostly students' talk or teachers' talk that affords certain "participation structures" for students. Emphasis is put on the social structure that is constructed through particular linguistic interactions.	Identities are not foregrounded. A few papers mention in passing student positioning or link participation frameworks formed in the classroom with students' identities/positions.	Articles mostly attend to teacher-student relationships through the roles or positions that teachers assign to students with teachers' talk.	Politics is not dealt with explicitly.	Articles discuss social and discursive practices associated with doing mathematics by mathematicians or "in the real world" and learning mathematics in the classroom.	Articles mostly foreground the social/interactional aspect of doing and learning mathematics and backgrounds the semantic (or content) aspect. Many papers either do not define what they mean by doing mathematics or refer to it generally as the practice of mathematicians.

画和其他形式的视觉中介物（Forman & Ansell, 2002；Jurow, 2005）。关系也受到较多关注，以师生之间形成的关系以及学生与数学之间的关系为主（Empson, 2003；Forman & Ansell, 2002；Lampert, 1990；O'Connor & Michaels, 1993）。政治和身份的受关注程度最低，只有恩普森（2003）、朗佩（1990）及福曼和安思尔（2002）明确提及学生所处的定位或角色，但即使这些文章中亦没有把权力和公平问题放在首位。

该子流派所含的文章提出了很多有趣的和方法论上的工具。总的来说，这些作者大都提出了一个衍生自社会语言学的有用概念或提出了一个更广泛的方法论观点，因此，该组文章对于思考与话语相关的结构、形式和一般方法都是有用的，但这些想法并没有和更广泛的学习理论联系起来。

浮现的观点。浮现的观点借鉴了布鲁默的符号互动论并在20世纪90年代早期经一系列讨论而获得高度关注，参与讨论的是一群德国数学教育家，包括海因里希·鲍尔斯菲尔德、耶格·沃伊特和格兹·克鲁姆霍尔，和一群美国数学教育家，包括保罗·科布、埃纳·洋科和特里·伍德。可以说，该理论以前所未有的方式引起了对数学课堂话语的关注，其中一些研究成为本章引用次数最多的内容（例如Yackel & Cobb, 1996，经Google学术搜索，有1500次引用）。

浮现观点认为知识创造是在课堂互动后浮现出来的，因此个体学习被认为是通过参与对数学活动的共同讨论和反思而发生的，这些作者描述了学习的心理学观点与社会学观点之间的反射性关系（见Cobb & Yackel, 1996，第177页）。举例来说，在探讨七年级课堂中关于统计推理的课堂干预时，科布（1999）指出："学习就是参与共同实践并为其发展做出贡献。被排斥的学生不仅被剥夺了学习机会，也丧失了获得数学成长的机会"（第35页）。

综合分析得出的概论。我们仔细分析了本类别的15篇文章，其中9篇是由与该理论的美国创始研究者有关系的研究小组撰写的，因此，发现几乎所有文章都使用了社会数学规范这一概念也就不足为奇了，是洋科和科布（1996）提出了该构念。

考虑到该组论文的焦点主要是在小学，只有一个例外是对七年级学生的一个干预项目，大多数文章都考虑了低年龄学生的数学发展。文章的另一条主线起源于洋科和拉斯马森之间的合作，出现了一批关注社会数学规范的文章并利用浮现的观点分析本科生数学课堂话语，此类研究尤其侧重于微分方程的教和学（例如Ju & Kwon, 2007；Rasmussen & Kwon, 2007；Rasmussen & Marrongelle, 2006；Yackel, Rasmussen, & King, 2000）。

由于同一群研究者在该领域开展研究，因此，尽管学生年龄不同，大多来自1、2、5和7年级，以及课题不同，包括位值、算术和统计推理，但得到了类似的研究结果。例如，麦克莱恩（2002）是这样总结她自己的教学实践研究的：

> 最初的课堂规范并不支持我的努力。只有通过持续的协商以后，才有可能促进我最初设想的那种讨论。在新规范的支持下，只有当我把注意力转向学生活动后，这些任务、工具和板书才成为支持我教学活动的工具。（第246页）

尽管撰写的时间在十年后，但该陈述与科布、洋科和伍德（1992）的早期文章中的发现有很多共同之处：

> 在整篇文章中，我们关注的焦点在个体儿童的解释和他们的数学知识的共享方面进行来回转换（参见Cobb, 1989）。在这个过程中，我们反复说明了反思性、循环性和相互依存性是指导我们试图去理解课堂情境的关键概念。（第119页）

上述两个概括都关注如何解释学生的学习和研究者与教师需要了解学生当前的理解情况并明确自己希望达成的目标，从共同发展的视角，而不是个人心理学的视角来考虑儿童的贡献。

这里所读文章的重点是教师和学生如何互动以及由

Additional activities that received attention were the use of inscriptions in the classroom, such as graphs, drawings, and other forms of visual mediation (Forman & Ansell, 2002; Jurow, 2005). Much attention was also given to Relationships, mainly those formed between the teacher and the students and those formed between the students and the mathematics (Empson, 2003; Forman & Ansell, 2002; Lampert, 1990; O'Connor & Michaels, 1993). The least attention was given to Politics and Identities. Only Empson (2003), Lampert (1990), and Forman and Ansell (2002) explicitly mentioned positioning or roles taken up by students, and even in their works, issues of power and equity were not at the fore.

There were many interesting and methodological tools to be found in the papers included in this subheritage. As a group, the authors mostly put forward either a useful concept taken from sociolinguistics or a more general methodological perspective. Thus, this group of work is useful for thinking about structures, forms, and general methods associated with discourse, but these ideas have not been linked to broader theories of learning.

Emergent perspective. The emergent perspective drew on Blumer's symbolic interactionism and came to prominence in the early 1990s, following discussions between a group of German mathematics educators, including Heinrich Bauersfeld, Jörg Voigt, and Götz Krummheuer, and a group of mathematics educators in the United States, including Paul Cobb, Erna Yackel, and Terry Wood. Arguably, this particular perspective brought attention to mathematics classroom discourse in unprecedented ways, with some of this research being the most highly cited in this chapter (e.g., Yackel & Cobb, 1996, has over 1,500 Google Scholar citations).

An emergent perspective situates knowledge creation as emerging out of classroom interactions; consequently, individual learning is considered to occur through participation in communal discussions and reflections about mathematical activities. The authors describe a reflexive relationship between psychological views and social views of learning (see Cobb & Yackel, 1996, p. 177). In discussing a classroom intervention on statistical reasoning in a seventh-grade classroom, for example, Cobb (1999) asserted, "to learn is to participate in and contribute to the evolutions of communal practices. Students who are excluded are deprived not merely of learning opportunities but of the very possibility of growing mathematically" (p. 35).

Overview from the general analysis. We carefully analyzed 15 articles for this chapter, 9 of which were written by the research group connected to the original researchers in the United States. Consequently, it is not surprising to find that almost all of the articles make use of the idea of sociomathematical norms. Yackel and Cobb (1996) introduced this construct.

Given that this group's focus was generally in elementary school, with the exception of an intervention project with year 7 students, most of the work has considered younger students' development of mathematics. Another line of work that began with a collaboration between Yackel and Rasmussen resulted in a set of articles focused on sociomathematical norms and drawing on the emergent perspective to analyze undergraduate mathematics classroom discourse. This work has focused, in particular, on the teaching and learning of differential equations (e.g., Ju & Kwon, 2007; Rasmussen & Kwon, 2007; Rasmussen & Marrongelle, 2006; Yackel, Rasmussen, & King, 2000).

Because the same group of researchers did much of this research, although with different ages of students, mostly from grades 1, 2, 5, and 7, and with different topics, including place value, arithmetic, and statistical reasoning, similar sets of results appeared. For example, this is how McClain (2002) summarized research on her own teaching practices:

> The norms as they were initially constituted in the classroom did not support my efforts. Only through a process of constant renegotiation did it become possible to facilitate the type of discussions of that I initially had envisioned. Only after I shifted my focus to the students' activity as supported by the emerging norms did the tasks, tools, and inscription function as means of supporting my agenda. (p. 246)

Although written a decade later, this statement had much in common with the findings of an earlier article by Cobb, Yackel, and Wood (1992):

> Throughout this paper, we have switched back and forth from a focus on individual children's interpretations to a focus on the taken-as-shared aspects of their mathematical knowing (cf. Cobb, 1989). In doing so, we have repeatedly illustrated that reflexivity, circularity, and mutual interdependence are key notions that guide our attempts to make sense of classroom situations. (p. 119)

The focus in both summaries concerned how to interpret students' learning and the need for researchers and teachers to make sense of students' current understandings, as well as be clear about what they wanted to achieve. Individual children's contributions were

这种互动所产生的共同理解。正如奥康纳（2001）所建议的，由于教师必须关注各个不同方面，所以课堂讨论是混乱的，但这并不意味着教师不能利用这些课堂讨论来支持学生的学习。对于部分学生来说，他们并不清楚正在实施的社会数学规范，除非教师通过互动强调其重要性。

尽管大多数项目都包含大量数据，但具体的分析往往局限于一两节课中发生的情况，提出的观点是关于正在形成的或实施的社会数学规范以及它们如何有助于后续学习。例如，洋科和科布（1996）在观察后指出，"就方法论而言，通过识别社会互动模式中的规律可推断出一般社会规范和社会数学规范"（第460页）。然而，他们没有解释或举例说明这些规范看上去是怎样的以及如何识别它们。

有时，一些统计信息会被使用以揭示数据分析的模式（例如 Bowers, Cobb & McClain, 1999；Cobb 等，1991；Kazemi & Stipek, 2001；Ju & Kwon, 2007）。通常上述方法是为了确定特定的实例，并方便随后进行定性程度更高的探讨。相对地，利文森、提罗什和撒米尔（2009）则使用统计数据来确定学生对教室中恰当的数学解释包括哪些要素的感知结果，这些统计数据与来自两位教师关于同一组解释的反馈、对他们自己班级的观察以及对学生和教师的访谈结合起来。从该数据组合中，利文森等人通过展示学生并不总是持有与他们的教师一样的社会数学规范，从而得以对混乱课堂的讨论做出了贡献。

构建任务分析得出的概论。表27.8总结了不同构建任务如何在该组文章中进行表述的。如前一节所述，这类研究的意义在于解释和促进社交互动与学习之间的关系。从下面对洋科和科布（1996）早期文章的摘录和很多其他文章都能明显看出这一点：

> 本文的目的是提出一种解释课堂生活的方法，旨在说明学生如何发展特定的数学信念和价值观，从而说明他们如何成为数学上的智力自主者，即他们如何形成一种数学倾向。（第458页）

因而，研究人员调查的活动普遍是以小组讨论和全班讨论形式开展的就不足为奇了。利文森等人（2009）的研究是一个例外，他们在以色列的两个五年级班级中以问卷调查和对学生和教师面谈的方式来确定在哪些情况下学生和教师对数学解释的社会数学规范的认识会相互一致或发生偏差。

从表27.8可以看出，借鉴浮现观点的文章侧重于

表27.8 浮现观点相关构建任务趋势

意义	活动	身份	关系	政治	联系	符号系统和知识
文章讨论了社会交往和个人学习之间的反射性关系。	研究人员调查了小组讨论和全班讨论。	很少提及身份。研究人员有时会在干预中担任老师/讲师的角色。	重点是学生和老师之间以及学生之间的关系。	对不重视学生相互交谈的教学会受到一定的批评（通常为隐含的）。	小组协作等教学实践被认为会影响学习。由于与学习自主性之间的关系，许多文章中认可社会数学规范的重要性。	很多文章都没有给出数学的确切定义。然而，有些文章将数学与问题解决和探究相联系，将其视为学习的重要组成部分。也有作者重视认识方式的多元化，认为需要提供解释。

considered from a communal development perspective rather than from an individual psychological perspective.

The focus for the articles read here was on how teachers and students interacted and the shared understandings that arose from those interactions. As O'Connor (2001) suggested, classroom discussions are messy because of the number of different aspects that teachers must attend to, but this does not mean that teachers cannot make use of those discussions to support students' learning. For some students, the sociomathematical norms that came into operation may not be clear unless the teacher, through her interactions, highlighted their importance.

Although most projects included a large amount of data, the analyses tend to be limited to what happened in one or two lessons, with points being made about the sociomathematical norms being developed or enacted and how they contributed to subsequent learning. For example, Yackel and Cobb (1996) observed, "methodologically, both general social norms and sociomathematical norms are inferred by identifying regularities in patterns of social interaction" (p. 460). What regularities looked like and how they were identified, however, were not explained or exemplified.

On occasion, some statistics were included to show patterns in the data analysis (for example, Bowers, Cobb, & McClain, 1999; Cobb et al., 1991; Kazemi & Stipek, 2001; Ju & Kwon, 2007). Usually this was done to identify specific instances, which were then discussed more qualitatively. In contrast, Levenson, Tirosh, and Tsamir (2009) used statistics to determine the outcomes of students' perceptions of what constituted appropriate mathematical explanations in their classrooms. These statistics were combined with feedback from two teachers on the same set of explanations, observations in their own classrooms, and interviews with students and the teachers. From this combination of data, Levenson et al. were able to contribute to the discussion of messy classrooms by showing that students did not always hold the same sociomathematical norms as their teachers.

Overview from the building tasks analyses. Table 27.8 summarizes how the different building tasks are presented in this set of articles. As noted in the previous section, the Significance of this research is to illustrate and promote the relationship between social interactions and learning. This can be seen in the following quote from Yackel and Cobb's (1996) early article but which is also evident in many of the other articles:

> The purpose of this paper is to set forth a way of interpreting classroom life that aims to account for how students develop specific mathematical beliefs and values and, consequently, how they become intellectually autonomous in mathematics, that is, how they come to develop a mathematical disposition. (p. 458)

Consequently, it is not surprising that the Activities that researchers investigated were generally small-group and whole-class discussions. The exception to this was the research undertaken by Levenson et al. (2009), in which they used questionnaires and conducted interviews with students and teachers in two fifth-grade classes in Israel to identify when students and teacher sociomathematical norms about mathematical explanations were mismatched or in alignment.

From Table 27.8, it can be seen that articles that draw on an emergent perspective focused on Relationships and

TABLE 27.8. Building Task Trends Related to the Emergent Perspective

Significance	Activities	Identities	Relationships	Politics	Connections	Sign systems and knowledge
Articles discussed reflexivity between social interactions and individual learning.	Researchers investigated small group and whole-class discussions.	Very little is mentioned about identities. Researchers sometimes take on the teacher's/lecturer's role in the intervention.	The focus is on the relationship between students and the teacher and between students.	There is some criticism (often implicit) of teaching where children talking to each other is not valued.	Instructional practices, such as group work, are considered as affecting learning. Sociomathematical norms are viewed as important in many articles because of their connection to intellectual autonomy.	Mathematics is not specifically defined in much of this work. Some articles, however, relate mathematics to problem solving and inquiry and see this as an important component of learning it. There is also a valuing of more than one way of knowing and the need to provide explanations.

关系和联系。大多数研究者强调师生之间的关系（甚至大学研究人员全程参与任务设计并和教师一起进行教学后的研讨会）和持续发展的社会数学规范与数学学习之间的联系。然而，一般来说，教师是坚持遵守社会数学规范很有必要的人（Kazemi & Stipek, 2001），学生的数学解释往往被作为调查前述联系的工具，原因是话语形式会影响到教师和学生在全小组讨论中的发展可能（O'Connor, 2001）。

该组文章起源于建构主义者对儿童如何学习数学的讨论，由于学习与学生所参与的社会交往相联系，因此，关于应该学习的数学内容的认识也发生了变化（符号系统与知识）。在其他研究中，科布及其同事区分了学校数学和探究性数学（Cobb, Wood, Yackel, & McNeal, 1992）。许多文章都认为通过个人重复练习习得的程序性数学与通过讨论和探索习得的概念性数学之间的区别是很重要的。

很少有作者关注身份和政治的构建任务。具和匡（2007）指出了其中一位作者是微分方程课的讲师，但文中很少再有关于该身份的其他论述，因为该文章的焦点是通过学生在全班课堂讨论时所给出的解释中的代词用法，来调查学生的定位如何随着时间的推移而变化（相关研究将在下一节关于语用学的部分讨论）。相对地，麦克莱恩（2002）的研究则集中于她以教师身份为促进学生建构数学论证而进行的工作，这是我们回顾的15篇文章中唯一一篇将身份放在首位的文章。在其余的文章中，教师的身份是一个重点，教师被假定或提升作为一名促进者（例如，可以参见 Bowers 等, 1999；Kazemi & Stipek, 2001；O'Connor, 2001）并由此与构建任务的关系产生联系。考虑到大多数围绕社会数学规范的文章均提倡教师作为与学生交互的促进者，很多研究人员担当起教育顾问的角色来推动不同类型的课堂互动的需求也就不足为奇了。

政治往往被隐性地处理，原因是学生对数学解释的运用似乎是有利于支持他们与社会数学规范的协商，从而有助于他们概念性地学习数学。科布（1999）明确指出过一个假设，没有机会参与上述社会数学规范协商活动的学生可能无法开展概念性的数学学习。科布（1999）是唯一明确探讨多样性和公平性的文章，尽管他所在团体几乎所有的研究都是在社会经济地位较低地区的学校进行的，就像奥康纳（2001）等人的研究一样。然而，有一些关于研究环境的讨论，迫使研究人员进行定量研究，"考虑到我们与教师之间合作所处的更广泛社会政治环境的本质，我们发现必须对参与和没有参与项目学习的学生的算术成绩、信念和个人目标进行定量比较"（Cobb, Yackel & Wood, 1992, 第100页），他们在1991年发表的论文可能就源于该压力。

自20世纪90年代初期浮现观点首次被提出，其理论和方法论取向始终保持一致。利文森等人（2009）是该类别下唯一试图在方法论方面将其拓展至课堂观察以外的作者，然而，他们经过研究发现，使用社会数学规范的观念是复杂的，是不容易克服的。他们的文章提出了很多问题，包括学生和教师是否有必要共同遵守一套社会数学规范，学生在课堂互动时，遵守教师选择的准则是否具有充分理由。虽然该观点强调了师生互动方式对儿童发展概念性数学理解的重要性，但可能需要更细致的理论才能更好地解析这种关系的微妙之处。

格赖斯门派。我们将该子流派称为格赖斯门派的原因是作者大量借鉴了 H. P. 格赖斯（1975）的著作（及其后续拓展性工作）中的会话准则和含义以及奥斯汀和赛尔等其他相关语言哲学家的工作。该子流派的一个重要思想是情境在理解数学话语涵义方面的作用，也就是说，讲话者（或作者）要表达的意思是由实际说出的词汇与背景信息、情境知识和其他参与者共同遵守的谈话规则（通常称为会话含义）组成的，比如，发言者有时会故意

Connections. Most researchers highlighted the relationship between teachers and students (even though university researchers were integrally involved in designing tasks and debriefing the teaching episodes with teachers) and the connections between the continuous development of sociomathematical norms and mathematical learning. Generally, however, the teacher is the one who enforces the necessity of following the sociomathematical norms (Kazemi & Stipek, 2001). Students' mathematical explanations were often the vehicle for investigating this connection, because discourse formats affect teachers' and students' possibilities within whole-group discussions (O'Connor, 2001).

This set of articles originated in the constructivist discussions about how children learn mathematics. Thus, as learning was connected to the social interactions that students engaged in, perceptions of what the mathematics was that should be learned also changed (Sign Systems and Knowledge). In other work, Cobb and colleagues distinguish between school mathematics and inquiry mathematics (Cobb, Wood, Yackel, & McNeal, 1992). The distinction between procedural mathematics, which could be learned by repetitively completing exercises individually, and mathematics that was learned conceptually through discussion and exploration, was important in many of the articles.

Little attention was paid to the building tasks of Identity and Politics. Although Ju and Kwon (2007) acknowledged that one of the authors was a lecturer in the differential equations class, little was made of this because their focus was on how student positions changed over time through investigating their use of pronouns in explanations given in whole-class discussions (related research is discussed in the next section on pragmatics). In contrast, McClain's (2002) research focused on what she did as the teacher to promote students' construction of mathematical arguments. This was the only article of the 15 reviewed in which Identity was at the fore. In the remaining articles, the Identity of the teacher is a focus in that it is assumed or promoted as being a facilitator (see, for example, Bowers et al., 1999; Kazemi & Stipek, 2001; O'Connor, 2001) and is thus related to the building task of Relationships. It is perhaps not surprising that many researchers took on the role of educational advisor promoting the need for different kinds of classroom interaction, given that most work around sociomathematical norms promotes the teacher as a facilitator of interactions with students.

Politics was generally dealt with implicitly, in that it appeared that students' use of mathematical explanations was considered beneficial in supporting their negotiation of sociomathematical norms that would contribute to them learning mathematics conceptually. There was a presumption, stated explicitly by Cobb (1999) above, that students who did not have opportunities to be involved in these kinds of negotiations of sociomathematical norms may have been impeded from learning conceptual mathematics. Cobb (1999) was the only article that explicitly discussed diversity and equity, even though almost all research by his group, as was also the case for others such as O'Connor (2001), was undertaken in schools in low socioeconomic areas. There was some discussion, however, about the research environment that pressured researchers to conduct quantitative studies, "given the nature of the wider socio-political setting in which we collaborate with teachers, we found it essential to conduct a quantitative comparison of project and non-project students' arithmetical achievement, beliefs, and personal goals" (Cobb, Yackel & Wood, 1992, p. 100). It was perhaps because of this pressure that they wrote their 1991 paper.

Since the early 1990s, when the emergent perspective was first proposed, its theoretical and methodological perspective has remained consistent. Levenson et al. (2009) were the only ones in this group who tried to extend it methodologically, going beyond classroom observation. Their research, however, identified complications in using perceptions of sociomathematical norms, which are not easy to overcome. Their paper raised many questions about whether it is important that students and teachers share the same set of sociomathematical norms or whether it is sufficient that students enact the teacher's preferred norms while interacting in the classroom. Although this perspective highlighted the importance of the ways that teachers and students interacted in order for children to develop conceptual mathematical understandings, it may be that a more nuanced theory is needed to better understand the subtleties of this relationship.

A Gricean umbrella. We describe this subheritage as the Gricean umbrella because the authors drew heavily from H. P. Grice's (1975) work (and later extensions of his work) on conversational maxims and implicature as well as other related philosophers of language such as Austin and Searle. An important idea within this subheritage is the role of context in understanding the meaning of mathematical utterances. That is, a speaker's (or writer's) intended meaning is a combination of the actual words that were spoken *in addition to* background information, contextual knowledge, and rules of conversation that other participants are assumed to share (this is often described as conversational implicature). For example,

不说出隐含的意思（如使用像"它"这样的代名词而不是像"等边三角形"这样的名词短语），原因是他们假定听众能够作出适当的推论来补充信息。

综合分析得出的概论。 经过我们仔细分析的这七篇文章通过分析语言特征来阐述意义的形成和说话者的意向，这些语言特征包括：(a)学生在学习数学时所生成的模糊限制语和非确定用语的使用和功能（例如，使用可能、应该、大约、左右等修饰语，见Meaney, 2006；Rowland, 1995)；(b)在讨论和推广数学思想时，对诸如"它"和"你"等代词和非特定指示词的使用（Gerofsky, 1996；Rowland, 1992, 1999)；(c)在数学谈话中，说话人如何通过话语保全他人的面子（Bills, 2000)；(d)使用手势来支持理解（Alibali等, 2013）。该类别的几位作者认为他们的工作借鉴了语用学这一语言学分支。

这些文章的分析对象主要是一对一或小组式的访谈，大多数作者分析的是在非正式学习环境中与学生进行数学对话时的口头话语（Bills, 2000；Rowland, 1995, 1999）或作为正式评估的一部分（Meaney, 2006），戈洛夫斯基对以往和现在的数学教科书中数学文字题的分析是对书面文本唯一的分析研究，而阿力巴里及其同事（2013）对中学数学课的分析是唯一一项涉及数学课堂教学分析的研究。大多数文章都关注小学生（Meaney, 2006；Rowland, 1995, 1999）或初中学生（Alibali等，2013；Meaney, 2006）。总的来说，作者没有定义话语或话语分析，相反，他们对自己研究中所调查的特定话语构念进行了界定。

构建任务分析得出的概论。 表27.9汇总了该子流派所含文章对吉的构建任务的处理方式。

该组文章将意义放在师生互动时的特定语言特征上（例如，代词的使用和不确定的用语），目的是描述学生如何使用语言来投入和做数学，以及来管理学习环境下的情感维度（例如维持脸面、权威、礼貌）。这些文章突出的对应活动是借助于这些语言特征的产生及其分析以理解它们是如何在情境中起作用的。该组文章均涉及的一个联系是话语和语境之间的联系：通过考虑教室、文化知识和其他背景信息等更广泛的背景特征，可以理解数学情境中的说话和发言的方式。特殊的语言特征也与更广泛的数学语义（例如如何使用不确定性用语或模糊限定词来进行猜测和估计）以及当地数学课堂的语义（例如自主性和社会距离如何反映在语言使用中）相联系。

综观整个数据集，大多数作者都明确讨论了关系并重点关注师生关系（有时还将与学生面谈的研究者认定为"教师"）。例如，罗兰（1999）认为，通过研究代词的使用方式可以揭示师生关系：教师使用"你"这个词的时机大多是指代学生，这可以传达一种权力（反过来，

表27.9 有关格赖斯门派的构建任务趋势

意义	活动	身份	关系	政治	联系	符号系统和知识
重点关注师生互动中的语言特征，例如代词使用、不确定性用语、模糊限定语、面子威胁行为和礼貌策略。	文章强调某些语言特征在情境中的产生、分析及作用方式，并与更广泛数学语义之间的关联。	没有处理身份问题。	文章侧重于师生关系，具体是指代词、模糊限定语、礼貌策略等的使用揭示了这些关系。	政治并非该组文章的重点，但有些文章考虑到了师生之间的权力失衡，并将面子和权威视为社会益处。	由于意义是基于本地（和文化）的知识和实践推断出来的，互动和话语是与语境相联系的。本组文章重点关注特定语言实践如何与更广泛的数学语义和本地课堂数学语义之间建立联系的。	学习数学的一部分是学习如何以及何时以与更广泛的数学集体相一致的方式表达不确定性和一般性。

speakers sometimes leave information out or unsaid (using a pronoun like "it" instead of a noun phrase like "equilateral triangle"), because they assume the listener will make appropriate inferences to fill in that information.

Overview from the general analysis. This set of seven articles that we carefully analyzed addressed meaning making and speaker intentionality through an analysis of linguistic features such as (*a*) the use and function of student-generated hedges and uncertainty devices when engaging in mathematics (e.g., the use of modifiers such as might, should, about, around—see Meaney, 2006; Rowland, 1995); (*b*) the use of pronouns and nonspecific referents such as "it" and "you" when discussing and generalizing about mathematical ideas (Gerofsky, 1996; Rowland, 1992, 1999); (*c*) how speaker utterances can allow others to save face during mathematical conversations (Bills, 2000); and (*d*) the use of gestures to support understanding (Alibali et al., 2013). Several authors in this group described their work as drawing from pragmatics, a subfield of linguistics.

The objects of analyses in these articles were primarily one-on-one or small-group interviews; the majority of authors analyzed spoken discourse occurring in mathematical conversations with students in informal learning environments (Bills, 2000; Rowland, 1995, 1999) or as part of a formal assessment (Meaney, 2006). Gerofsky's analysis of mathematics word problems in historical and current mathematics textbooks was the only analysis of written texts, and Alibali and colleagues' (2013) analysis of middle grades mathematics lessons was the only study that involved an analysis of classroom mathematics instruction. Most of the articles focused on students in the elementary grades (Meaney, 2006; Rowland, 1995, 1999) or middle grades (Alibali et al., 2013; Meaney, 2006). In general, authors did not define discourse or discourse analysis; instead, they defined the specific discourse constructs under investigation in their studies.

Overview from the building tasks analyses. Table 27.9 summarizes the ways in which articles in this subheritage generally addressed Gee's building tasks.

This group of articles gave Significance to specific linguistic features of teacher-student interactions (e.g., pronoun use and uncertainty devices) to describe how students used language to engage with and do mathematics and to manage affective dimensions of their learning environments (e.g., maintaining face, authority, politeness). The corresponding Activities these articles foregrounded were the production and analysis of these linguistic features to understand how they functioned *in context*. One of the Connections made consistently across this set of papers is the connection between an utterance and its context: ways of talking and speaking in mathematical situations are understood by considering broader contextual features of classrooms, cultural knowledge, and other background information. Particular linguistic features are also connected to the broader mathematics register (e.g., how uncertain language or hedging are used to make conjectures and estimates) as well as local mathematics classroom registers (e.g., how autonomy and social distance are reflected in language use).

Across the data set, the majority of authors explicitly discussed Relationships with a focus on teacher-student relationships (and sometimes took themselves as researchers to be the "teacher" as they interviewed students). For example, Rowland (1999) argued the teacher-student relationships were revealed in the ways in which pronouns were used: Teachers' use of the word "you"

TABLE 27.9. Building Task Trends Related to a Gricean Umbrella

Significance	Activities	Identities	Relationships	Politics	Connections	Sign systems and knowledge
Linguistic features in teacher-student interactions such as use of pronouns, uncertainty devices, hedging, face threatening acts, and politeness strategies are given significance.	Articles foreground production of, analysis of, and ways that certain linguistic features function in context and are related to the broader mathematics register.	Identities are not addressed.	Articles focus on teacher-student relationships; specifically, what the use of pronouns, hedges, politeness strategies, and so forth, reveals about these relationships.	Politics is not an explicit point of emphasis for this set of articles, although a few articles consider power imbalances between teachers and students and treat face and authority as social goods.	Interactions and utterances are connected to their contexts as meaning is inferred based on local (and cultural) knowledge and practices. These articles focus on how particular linguistic practices are connected to the larger mathematics register and to local classroom mathematics registers.	Part of learning mathematics is learning how and when to express uncertainty and generality in ways consistent with the broader mathematics community.

学生很少使用"你"来代指教师）。米尼（2006）则借鉴应用语言学家肯·海兰的工作，来解释产生不确定性用语的原因在本质上是如何相互关联的，例如，不确定性用语可以认可发言者与倾听者之间的关系，具体通过隐藏（或限制）发言者对所作陈述有效性的个人责任（例如在"我只是做了……"中使用"只是"这样的基于发言者的模糊限定语）或通过承认倾听者的个人责任或权威去评估信息的有效性并作出判断（例如"我认为"这样的基于倾听者的模糊限定语，它传递的是其他选择也可能存在）。

身份构建任务并非该子流派下一个明确的焦点，因为没有文章关注身份的话语构建。同样，大多数文章亦未涉及政治。然而，有两篇文章关注了师生之间的权力失衡问题（Bills，2000；Rowland，1999）。比尔斯将发言者和听众的相对权力视为判断面子威胁行为严重性的一种特征，罗兰则展示了教师如何通过使用"我们"和"你们"来表达权威。该类别的作者所关注的社会益处非常广泛，但通常是暗示而非明示，其中包括面子、表达不确定性的社区或文化特定方式、概括数学思想并将其视为对象并支持学生发展这种理解的能力，以及解决文字题的能力。

总的来说，了解、做和理解数学（符号系统与知识）与学习和做数学的具体方面有关。该子流派中了解和做数学的一部分是了解何时以及如何以特定的方式使用语言和认识数学表达的特定形式或类型。例如，米尼（2006）和罗兰（1995）关于我们在谈话时如何通过使用模糊限定语、推测、估计和近似的语言来表达不确定性的多或少的研究，承认了不确定性在数学思想中的重要性。事实上，罗兰（1995）将数学描述为"与惰性化知识体系相对立的人类工作领域"（第327页），其中不确定性是学习数学时意料之中的、公认的和明确的。米尼（2006）调整了罗兰的说法，他认为：

> 不确定性在数学课堂中的作用尚不明确。在某些情况下，例如给出估值，学生被期望给出模糊的答案。在其他情况下，例如呈现结果，教师会期望学生强调真实大小，这时使用不确定用语就不恰当了。（第387页）

那么，学习数学的一部分是学习何时以及如何表达不确定性。与确定性和不确定性概念相关的是数学中的一般性和一般化的作用，老师和学生通常可以通过从第一人称转至第二人称或使用非特定语言，来思考和谈论还没有人命名的数学对象来实现和表达这一点（Rowland，1995，1999）。该子流派中，有一些与关键的数学思想（如学生应该掌握的不确定性、推测和一般性）相关的使用数学语言的特定方式。

话语心理学。话语心理学理论源自德里克·爱德华兹（1997）的《话语与认知》和其他相关著作（例如Edwards，1993；Edwards & Potter，1992），其热衷的是"'心灵是什么样子的'是如何由话语构成的"（与前文所述斯法德话语心理学系统的出发点相反；巴维尔，个人沟通，2015年1月31日）。巴维尔（2003）解释称该版本话语心理学的基础源自加芬克尔（1967）关于人们如何理解他们所参与的社会情境的观点，以及萨克斯（Jefferson，1992）关于此类谈话的社会组织形式的研究。他认为，由于互动涉及人们的社会、文化和语言背景，这些想法可以通过为话语分析提供全面的方法和通过更好地解释教室的多元文化背景，来发展社会文化工作。

综合分析得出的概论。该子流派下经过仔细分析的文章包括理查德·巴维尔[2]（2003，2005，2009，2013a，2013b）撰写的五篇数学教育文章。巴维尔的工作重点在于参与者自身如何处理诸如"知道"或其他典型的心理学主题等问题，巴维尔（2013b）将该子流派看作是"一种对人类认知理论和方法论的观点，目的是理解本地产生的方法，参与者借助这些方法，处理彼此在交互中的心理过程"（第599页）。他将话语定义为一种社会实践和互动，即"发生于并构成特定社会情境的活动模式"（2003，第202页）。

这五篇文章中，有一篇解释和说明了话语心理学的关键概念并阐述了其在数学教育领域中的发展可能和挑战（Barwell，2003），另一篇对面向教学的数学知识进行了批判性分析（Barwell，2013b），另两篇是实证分析（一篇是数学家的谈话[Barwell，2013a]，另一篇则用

most often occurred as a form of address to students, which could convey power (conversely, students rarely use "you" as a form of address to teachers). Meaney (2006), drawing on the work of applied linguist Ken Hyland, explained how reasons for making uncertain utterances could be relational in nature. For example, uncertainty devices can acknowledge the relationship between speaker and listener by concealing (or limiting) the speaker's personal responsibility for the validity of a statement made (a speaker-based hedge such as the use of "just" as in "I just did...") or by recognizing the listener's personal responsibility or authority to assess the validity of information and to make judgments (listener-based hedge such as "I think," which communicates that other alternatives potentially exist).

The Identity building task was not an explicit focal point within this subheritage, as none of these articles focused on the discursive construction of identity. Similarly, most papers also did not address Politics. However, two of the papers did attend to power imbalances between teachers and students (Bills, 2000; Rowland, 1999). Bills identified the relative power of speakers and hearers as one feature determining the seriousness of a face-threatening act, and Rowland demonstrated how teachers' use of "we" and "you" was a way of asserting authority. The social goods at stake for authors in this group were wide-ranging but usually implied instead of being explicitly identified; they included face, community- or culturally specific ways of expressing uncertainty, the ability to generalize and see mathematical ideas as objects and support students to develop this understanding, and the ability to solve word problems.

By and large, coming to know, do, and understand mathematics (Sign Systems and Knowledge) was related to specific aspects of learning and doing mathematics. Part of knowing and doing mathematics in this subheritage was knowing when and how to use language in particular ways and recognizing specific forms or genres of mathematical expression. For example, Meaney's (2006) and Rowland's (1995) work on how we use language to express more and less certainty in our talk through the use of hedges, conjectures, and estimations and approximations acknowledged the importance of uncertainty in mathematical thinking. In fact, Rowland (1995) described mathematics as a "field of human endeavor as opposed to an inert body of knowledge" (p. 327), wherein uncertainty is expected, acknowledged, and explicit in learning mathematics. Meaney (2006) tempered Rowland's claim, asserting:

> The role of uncertainty in mathematics classrooms is not clear. In some situations such as giving estimations, students are expected to give fuzzy answers. In other situations, such as when presenting results, teachers would expect students to strengthen the truth value and using uncertainty devices would be inappropriate. (p. 387)

Part of learning mathematics, then, is learning when and how to express uncertainty. Related to the notion of certainty and uncertainty was the role of generality and generalizations in mathematics. This was often achieved and expressed linguistically by teachers and students shifting from first to second person or by using nonspecific language to think and talk about mathematical objects that no one had yet named (Rowland, 1995, 1999). Within this subheritage, there are specific ways of using language within mathematics related to key mathematical ideas such as uncertainty, conjecturing, and generality that students should acquire.

Discursive psychology. This line of discursive psychology began with Derek Edward's (1997) *Discourse and Cognition* and other related work (e.g., Edwards, 1993; Edwards & Potter, 1992), which was interested in "how 'what the mind is like' is discursively constructed" (in contrast to the starting point of Sfard's line of discursive psychology, as stated earlier; Barwell, personal communication, January 31, 2015). Barwell (2003) explained the basis of this version of discursive psychology as deriving from Garfinkel's (1967) idea about how people make sense of social situations they participate in and Sacks' (Jefferson, 1992) focus on the social organization of such talk. He argued that these ideas can develop sociocultural work by providing thorough approaches to discourse analysis and by helping to account for the multicultural context of classrooms since interactions relate to people's social, cultural, and linguistic backgrounds.

Overview from the general analysis. The articles that were carefully analyzed for this subheritage include five mathematics education articles authored by Richard Barwell[2] (2003, 2005, 2009, 2013a, 2013b). The focus of Barwell's work has been on how participants *themselves* dealt with issues like "knowing" or other typical psychological topics. Barwell (2013b) saw this subheritage as being both "a theoretical and methodological perspective on human cognition that seeks to understand the locally produced methods through which participants deal with each other's mental processes in interaction" (p. 599). He defined discourse as a social practice and interaction as "patterns of activity, which take place in and constitute particular social situations" (2003, p. 202).

Of the five articles, one article explains and illustrates key concepts from discursive psychology and

实例突出歧义及其在意义产生过程中的重要性[Barwell, 2005]），最后一篇使用话语心理学揭示研究者对于数学思维的解释如何被理解为一种话语构造（Barwell, 2009）。这几篇文章本身给出了关于各种想法（例如数学知识、学术语言和日常语言）和研究过程的一般性想法，而不局限于某个特定年级或具体的数学内容。

在该组文章中，巴维尔使用了以下术语和短语：话语实践、话语分析、话语活动和话语作为一种社会实践。他解释到，互动"可以被看作产生自特定社会情境并构成特定社会情境的活动模式"（Barwell, 2003, 第202页），且在该观点下话语本身就被视为研究焦点。其目标是探究话语如何解释它们所构成的活动，而不是概述话语的某个抽象版本。说到数学，巴维尔（2005）指出：

> 因此，数学是通过话语活动构建的，也就是说，通过使用口头、书面、符号的互动方式，包括使用手势和其他非语言的互动方法。这种话语活动由话语实践、使用语言资源做数学或学数学、教数学的方法所构成。（第119—120页）。

因此在分析数学课堂话语时，目标是理解学生如何通过互动构建数学，该目标假设了一种相对主义的数学观。

构建任务分析得出的概论。 表27.10 汇总了我们在该子流派中确定的主要趋势。

该组文章的主要活动是介绍、说明和使用话语心理学的主要思想，分析来自不同情境的数据并对其他文章中的理念或解释进行批判性分析。鉴于上述研究焦点，巴维尔所述的"联系"主要发生在话语心理学的理念之间，以及与数学教育其他方面的工作之间。巴维尔发现该观点下的"意义"提供了以下的可能性：它有助于研究数学"思想"如何在互动中构建，以及这些"思想"又带来了什么、它基于参与者做了什么和说了什么，而非他们的言辞或行为"可能传达的意思或想法"（Barwell, 2003, 第206页）。他所主张的关系可被看作表明几乎所有事情都可以通过人与人之间的交互协调，其中包括人们"知道"什么或"意思"是什么。他还解释称，该观点面临的一个挑战在于，符号系统与知识基于明确的相对主义认识论立场，他将这种观点描述为"反认知、反现实和反结构主义"（Barwell, 2013a, 第210页）。

该领域的另一个共同活动是对数学教育中一些常用的理念和理论提出批评。例如，巴维尔（2005）研究了数学课堂的模糊用语，阐明所对照的两种模型之间的差异：即认为意义固定且语言基本没有疑问的"形式模型"和认为意义具有主观性和不稳定性的"话语模型"。他后来质疑"日常"语言和"学术"语言的二分法，并研究数学家如何在互动过程中构建和利用这种区别（Barwell, 2013a）。他认为学习数学要求学习者对学术语言和日常语言都要越来越熟练。

表27.10 话语心理学相关构建任务趋势

意义	活动	身份	关系	政治	联系	符号系统和知识
众多文章均以批评各种理念和方法为核心。	文章批评了数学教育中经常使用的理念和理论，并说明了这种观点在阐述不同研究结果或解释时的有用性。	没有真正地涉及身份。	关系并非重点，关注点集中于重新分析其他文章的内容，由此利用了巴维尔的话语心理学和其他作者文章之间的关系。	除了对数学教育中其他理念和理论的批判之外，没有明确针对政治的讨论。文章重视将认知理念视为社会资源的情境化观点，而不是研究者先入为主的理念/框架。	文章将民族方法学和会话分析的理念与数学相联系，例如，他们挑战了以前关于日常语言和学术语言的思维方式。	文章明确采用相对认识观，关注人们的言行，而非参与者的意思或想法。

articulates some of the possibilities and challenges of this perspective for mathematics education (Barwell, 2003), one offers a critique of mathematical knowledge for teaching (Barwell, 2013b), two others are empirical analyses (one of mathematicians' talk [Barwell, 2013a] and the other on using examples to highlight ambiguity and its importance in meaning making [Barwell, 2005]), and the final piece uses discursive psychology to show how researchers' interpretations of mathematical thinking can be understood as a discursive construction (Barwell, 2009). The articles themselves make more general points about various ideas (e.g., mathematical knowledge, academic and everyday language) and research processes rather than being grounded in a particular grade level or specific mathematical content.

Across the set of articles, Barwell uses the following terms and phrases: discursive practices, discursive analyses, discursive activity, and discourse as a social practice. He explains that interaction "can be seen as patterns of activity which take place in and constitute particular social situations" (Barwell, 2003, p. 202) and that that discourse itself is the focus of the inquiry in this viewpoint. The goal is to investigate how discourses construe the activities they constitute, not to outline some abstracted version of a discourse. In relationship to mathematics, Barwell (2005) states,

> Mathematics, therefore, is constructed through discursive activity, i.e., through the use of spoken, written, symbolic interaction, including the use of gestures and other nonlinguistic aspects of interaction. This discursive activity is made up of discursive practices, ways of using linguistic resources to do mathematics or to do the teaching and learning of mathematics. (pp. 119–120)

In analyses of mathematics classroom discourse, then, the goal is to understand how students construct mathematics through their interactions, which assumes a relativist view on mathematics.

Overview from the building tasks analyses. Table 27.10 summarizes the key trends we identified in this subheritage.

A primary Activity of these articles was to introduce, illustrate, and use key ideas from discursive psychology to analyze data from various contexts and to critique ideas or interpretations in other articles. Because of these foci, Barwell's Connections occurred mainly between and among the ideas from discursive psychology and in relationship to other lines of work in mathematics education. Barwell found Significant the possibilities afforded by this perspective: it helps to examine how mathematical "thoughts" are constructed in interaction and, in turn, what these "thoughts" bring about; and it is grounded in what participants do and say rather than in what their words or behavior "might tell us about what they mean or think" (Barwell, 2003, p. 206). The Relationships he advocated for may be seen as indicating that almost everything is interactionally negotiated between people, including what people "know" or "mean." He also explained that a challenge of this view is that Sign Systems and Knowledge are based on an explicitly relativist epistemological stance. He described this view as "anti-cognitive, anti-realist, and anti-structuralist" (Barwell, 2013a, p. 210).

Another common Activity in this line of work was to offer critiques of some commonly adopted ideas and theories in mathematics education. Barwell (2005), for example, examined ambiguity in mathematics classrooms to illuminate the differences between two contrasting

TABLE 27.10. Building Task Trends Related to Discursive Psychology

Significance	Activities	Identities	Relationships	Politics	Connections	Sign systems and knowledge
Critiquing various ideas and methods were central to many of the articles.	Articles critique commonly adopted ideas and theories in mathematics education and illustrate the usefulness of this perspective for illuminating different findings or interpretations.	Identities was not really addressed.	Relationships was not a central focus. Rather, articles focused on reanalyzing transcripts from other articles, thus employing a relationship between Barwell's discursive psychology and other authors' articles.	Politics was not discussed explicitly except through critique of other ideas and theories in mathematics education. Articles valued situated views of cognitive ideas as social goods, rather than researchers' preconceived ideas/frameworks.	Articles connected ideas from ethnomethodology and conversation analysis to work in mathematics. They challenged, for example, previous ways of thinking about everyday versus academic language.	Articles used an explicitly relativistic view of knowing. A focus on what people do and say rather than what participants mean or think.

巴维尔（2013b）还对面向教学的数学知识的现有学识进行了批评，该知识始于教学内容知识的理念（Shulman，1987）。具体而言，他研究了这些研究人员可能对"知识"意义的认知，因为这方面工作还没有认识论层面的讨论，他认为，该理念观点基于对知识的表征和分类观点，并不代表话语心理学发掘出的数学课堂的知识话语。

有趣的是，该方面研究没有明确触及身份和政治问题，包括研究员本人都没有明确指明自己的位置。考虑到巴维尔（2005）曾指出，"交互的社会性质，例如身份的构建和维护、权力关系和联系，可以被看作是互动组织的主要基础"（第120页），因此很奇怪的是这些建构任务并没有更加明显。该论断的一个例外是2009年的一篇文章，在该文章中，他用了两个研究解释来表明：

> 期刊文章并非研究人员思想的"窗口"，实际上也不是研究参与者思想的"窗口"，它们是精心构建的文本，其目的之一是对一项研究提出一种合理的解释，包括对所发生的事情和参与者数学思维的各种解释。（第256页）

在这篇文章中，巴维尔（2009）重新诠释了格雷和韬尔（1994）以及斯法德（2001）的观点，并提出研究本身是一个话语过程，研究人员通过话语过程构建了不同版本的数学思维，并利用各种理论和概念观点，对所发生的事情进行合理的解释。尽管他通过重新解释两个例子来说明自己的观点，但他的观点与研究中更普遍构建的解释相关联。

概念重建流派

正如"方法评论"部分所述，我们将这些不同的观点归为一组的原因是他们借鉴了相似的观点，与被称为概念重建主义者的课程理论家一样。在仔细分析这21篇文章后我们发现，和概念重建主义者的课程理论家一样，利用这些传统的数学教育研究人员也对数学课堂和课本的现状提出了有意义的批评或补充。通常这些作者会提及隐性的政治，并批评数学课堂或数学教育研究领域的主流或流行实践。

综合分析得出的概论。大约有一半的文章本质上是理论性的，有些严格使用了理论来进行论证或说服读者如何改进数学教育中的普及性理论（T. Brown，1994；Evans，1999；Skovsmose，2012），其他的文章也在进行这类理论构建，但同时还运用实证的例子说明和阐述其理念（Barbosa，2006；T. Brown & Heywood，2011；Dowling，1996；Esmonde，2009a；Evans，Morgan，& Tsatsaroni，2006；Moschkovich，2007；Straehler-Pohl & Gellert，2013）。有差不多数量的文章是实证性研究，会产生一些研究结果（Esmonde，Brodie，Dookie，& Takeuchi，2009；Hoadley，2007；LeRoux，2008；Mordant，1993；Planas & Gorgorio，2004；Setati，2005，2008；Wagner，2007，2008；Wagner & Herbel-Eisenmann，2008），其中有些文章还明确指出其引用了数学教育领域之外的方法论（Hoadley，2007；LeRoux，2008；Wagner & Herbel-Eisenmann，2008）。有一篇文章则汇总了作者参与的一系列研究（Hoadley & Ensor，2009）。

在审视数学教育研究和实践的各个方面时，这些作者用后结构主义理论扩充了更为传统的心理学理论（T. Brown，1994；Evans，1999）；将社会文化理论与批判种族理论、刻板印象威胁以及社会地位观点相结合（Esmonde等，2009）；用吉的"文化模型"理念去理解与权力相关的规则转换（Setati，2005，2008）；使用后结构主义、系统功能语言学和人类学理论来实施"分类"（Bernstein，1996；Straehler-Pohl & Gellert，2013）；对学生表达中的一个关键词（"只"）进行定量分析，并通过批判性话语分析来理解该词的潜在影响（Wagner & Herbel-Eisenmann，2008）以及通过汇集社会学、符号学和心理分析学的观点采用一种批判性的方法来避免诸如在情绪研究中出现的个人/社会和认知/情感的二元论（Evans等，2006）。这些文章质疑将学校几何过早地变为依赖代数的一种练习的认识论基础（T. Brown & Heywood，2011），并认为数学重新定义了形式世界和操作世界，还强加了"符号霸权"（Skovsmose，2012）。数学文本被视为"学校数学的更广泛实践"的表征形式

models: the "formal model," which sees meaning as fixed and language as relatively unproblematic, and the "discursive model," which sees meaning as subjectively situated and in a state of flux. He later questioned the dichotomy of "everyday" and "academic" language, examining how mathematicians constructed and utilized this distinction in their interactions (Barwell, 2013a). He argued that learning mathematics required learners to gain increasing facility with both academic and everyday language.

Barwell (2013b) also critiqued current scholarship on mathematics knowledge for teaching, beginning with the idea of pedagogical content knowledge (Shulman, 1987). In particular, he examined what these researchers might mean by "knowledge," because there were no epistemological discussions in this work. He argued that the view was based on a representational and categorical view of knowledge and did not represent the discourses of knowledge in mathematics classrooms that discursive psychology unearthed.

Interestingly, Identity and Politics were often not explicitly addressed in this work, including the researcher himself not making his own positionality apparent. The fact that these building tasks were not more apparent was surprising given that Barwell (2005) stated that the "social nature of interaction, such as the construction and maintenance of identities, power relations and relationships, can be seen as the primary basis for the organisation of interaction" (p. 120). An exception to this claim can be found in a 2009 article in which he used two research accounts to show that

> journal articles are not "windows" on the minds of researchers or indeed on the minds of the participants in the research. They are carefully constructed texts, one of the purposes of which is to propose a plausible account of a piece of research, including various accounts of what happened and of participants' mathematical thinking. (p. 256)

In this article, Barwell (2009) reinterpreted aspects of Gray and Tall (1994) and Sfard (2001) and showed how research itself is a discursive process by which researchers construct versions of mathematical thinking, drawing on various theoretical and conceptual viewpoints to suggest reasonable accounts of what happened. Although he illustrated his point through reinterpreting two examples, his point relates to accounts constructed in research more generally.

Reconceptualist Heritage

As stated in the Review Methods section, we grouped this diverse set of perspectives into one group because they drew on similar perspectives as curriculum theorists who were dubbed reconceptualists. As we carefully analyzed these 21 articles, we found that, like reconceptualist curriculum theorists, mathematics education researchers who drew on these traditions also made Significant critiquing or augmenting status quo views on mathematics classrooms and texts. Oftentimes, these authors addressed implicit Politics and critiqued mainstream or hegemonic practices in mathematics classrooms or in mathematics education research.

Overview of the general analysis. About half of these articles were theoretical in nature; some strictly used theories to make arguments or persuade readers about how to improve popular theories in mathematics education (T. Brown, 1994; Evans, 1999; Skovsmose, 2012) and other articles engaged in this kind of theory building but also used empirical examples to illustrate and bring the ideas to light (Barbosa, 2006; T. Brown & Heywood, 2011; Dowling, 1996; Esmonde, 2009a; Evans, Morgan, & Tsatsaroni, 2006; Moschkovich, 2007; Straehler-Pohl & Gellert, 2013). A similar number of articles were empirical research with findings (Esmonde, Brodie, Dookie, & Takeuchi, 2009; Hoadley, 2007; LeRoux, 2008; Mordant, 1993; Planas & Gorgorio, 2004; Setati, 2005, 2008; Wagner, 2007, 2008; Wagner & Herbel-Eisenmann, 2008) and some of these articles also explicitly stated that they were introducing a methodology from outside mathematics education (Hoadley, 2007; LeRoux, 2008; Wagner & Herbel-Eisenmann, 2008). One piece provided a summary of a line of research in which the authors had engaged (Hoadley & Ensor, 2009).

In examining the aspects of mathematics education research and practice, these authors augmented more conventional psychological theories with poststructuralist theory (T. Brown, 1994; Evans, 1999); combined sociocultural theory with critical race theory, stereotype threat, and sociological views of status (Esmonde et al., 2009); brought Gee's idea of "cultural models" to understand code-switching as it relates to power (Setati, 2005, 2008); used poststructuralist, systemic functional linguistics and anthropological theories to operationalize "classification" (Bernstein, 1996, in Straehler-Pohl & Gellert, 2013); quantitatively analyzed a key word students expressed concern about ("just") along with a critical discourse analysis to understand the potential impact of that word (Wagner & Herbel-Eisenmann, 2008); and adopted a critical approach by bringing together sociological, semiotic, and psychoanalytic perspectives to avoid dualisms like individual/social and cognitive/affect in the study of emotions (Evans et al., 2006). These articles challenged the epistemological basis of school

(Dowling，1996，第389页），它使不同的社会阶层得以延续。研究数学建模实践（Barbosa，2006）和数学解释（Esmonde，2009a）的常用方法分别被批评为，不考虑学生如何成为世界上有批判眼光的数学用户和只专注于讲话的内容而忽略其所处背景的性质。最后，对教育学进行了仔细考察，研究其如何再现社会阶层（Hoadley，2007；Hoadley & Ensor，2009）、未明确注意提高学生的语言意识（Wagner，2007）和学生在使用语言表达意义时精确的数学语言所起作用的普遍看法（也就是说，即使学生未使用精确的数学语言，但他们仍然可以富有成效地讨论数学意义；Moschkovich，2007）。

可以从上文的一些描述中看到，这些文章中分析的对象范围较大，研究的背景也有很大差异，包括小学（Hoadley，2007；Hoadley & Ensor，2009；Moschkovich，2007；Setati，2005，2008）、初中（大约6—8年级或12—14岁左右的学生；Barbosa，2006；Evans等，2006；Straehler-Pohl & Gellert，2013）、高中（Dowling，1996；Esmonde，2009a；Esmonde等，2009；Planas & Gorgorio，2004；Wagner，2007，2008；Wagner & Herbel-Eisenmann，2008）、针对师范生的大学课程（Barbosa，2006；T. Brown & Heywood，2011）、大学数学教科书（LeRoux，2008）以及数学课堂以外的事物（Evans，1999）。值得注意的是，几乎所有研究人员（莫谢科维奇和赫贝尔-艾森曼除外）都在美国境外生活或工作，且只有一篇文章（Esmonde等，2009）发表在美国的期刊上。

关于书面文本的实证研究结果表明：(a)数学教科书包含许多不同类型的嵌入式文本，在面对改革呼吁下，这些文本表现出趋向于概念性数学的微小变化，但仍然有偏向于数学语义和学校数学文字题类型的强烈倾向（LeRoux，2008）；(b)因为教科书只给出了没有意义的行动清单，数学课本中用于引入课题的语言可能导致学生出现思路"冻结"（Mordant，1993）。在数学课堂交互方面的研究结果表明：(a)小团体工作的成功受到互动风格、数学理解水平、友谊以及诸如种族和性别的社会身份的影响（Esmonde等，2009）；(b)学生批判性的语言意识可解释他们自我定位的一系列方式，提供给教师关于学生如何主动参与的见解（Wagner，2007），并给予学生在数学课堂中实现新的可能（Wagner，2008）；(c)学生在数学课堂中觉得困难的一些词语（例如"每次只做一步"中的"只"）普遍存在，并可能导致自言自语（Wagner & Herbel-Eisenmann，2008）；(d)与教师对学生的期望相联系的（主要是隐性的）教学决策和行动，能使来自工人阶级或贫困家庭的学生以及移民学生被持续地边缘化（Hoadley，2007；Planas & Gorgorio，2004）。此外，赛塔蒂（2005，2008）的文章明确说明多语言环境下如何考虑权力问题："为了在多语言数学课堂中充分描述和解释语言的使用，我们需要超越教学法和认知方面，去考虑语言的政治作用"（Setati，2005，第464页）。

关于这些文章如何处理"话语"一词，有些文章使用了该术语，以及像叙述、反驳和语言等相关术语，但没有对其进行定义（例如T. Brown & Heywood，2011；Esmonde等，2009；Hoadley，2007；Hoadley & Ensor，2009；Mordant，1993）。其他文章则给出了话语的定义，例如，它指的是——

- "所有类型的语言，包括符号、手势、人工制品、模仿，等等"（Barbosa，2006，第297页，跟随Lerman，2001）；
- "在社会实践情境中发生的并影响个人和社会意义建构的一系列行动和互动"（Planas & Gorgorio，2004，第21页）；
- "与社会生活的其他非语言形式相联系的口头或书面语言的使用，话语并不相互排斥或固定，而是可重叠、可争议和可变化的"（LeRoux，2008，第313页，跟随Fairclough，1992）；
- "组织和规范特定社会和制度实践的一个符号系统"（Evans等，2006，第210页）。

geometry as a practice that too quickly relies on algebra (T. Brown & Heywood, 2011) and argued that mathematics formats the formal and mechanical world and imposes "symbolic power" (Skovsmose, 2012). Mathematical texts were seen as representations of a "wider practice of school mathematics" (Dowling, 1996, p. 389) that perpetuated various social strata. The common ways of researching the practices of mathematical modeling (Barbosa, 2006) and mathematical explanations (Esmonde, 2009a) were critiqued for not considering how students might become critical users of mathematics in the world and for focusing exclusively on content of the talk rather than the situated nature of it, respectively. Finally, pedagogy was scrutinized for how it reproduces social class (Hoadley, 2007; Hoadley & Ensor, 2009), for not attending explicitly to raising students' language awareness (Wagner, 2007), and for commonly held ideas about the role of precise mathematical language when students use language to make meaning (i.e., that students can talk about mathematical meanings quite productively, even if they do not use precise mathematical vocabulary; Moschkovich, 2007).

The range of objects of analyses in these articles were quite varied, as can been seen in some of the descriptions above. There was also great variation in the contexts that were studied, including elementary grades (Hoadley, 2007; Hoadley & Ensor, 2009; Moschkovich, 2007; Setati, 2005, 2008), middle grades (about grades 6–8 or students who are around 12–14 years old; Barbosa, 2006; Evans et al., 2006; Straehler-Pohl & Gellert, 2013), high school (Dowling, 1996; Esmonde, 2009a; Esmonde et al., 2009; Planas & Gorgorio, 2004; Wagner, 2007, 2008; Wagner & Herbel-Eisenmann, 2008), university-level courses for prospective teachers (Barbosa, 2006; T. Brown & Heywood, 2011), and university mathematics textbooks (LeRoux, 2008), as well as some focus on examples from outside mathematics classrooms (Evans, 1999). It is interesting to note that almost all these researchers (except Moschkovich and Herbel-Eisenmann) live or work outside the United States and that only one article (Esmonde et al., 2009) was published in a U.S. journal.

The findings in the empirical research about written texts reported that (a) mathematics textbooks include many different kinds of embedded texts and these texts, in the face of encouragement to reform, showed small changes toward conceptual mathematics, but still strong inclination toward the mathematics register and the school math word problem genre (LeRoux, 2008); and (b) the language used to introduce topics in mathematics textbooks could contribute to how students "freeze" at the ideas because the textbooks only gave lists of actions that have no meaning (Mordant, 1993). In relationship to interactions in the mathematics classroom, research findings indicated that (a) success of small group work was influenced by interactional style, levels of mathematical understanding, friendships, and social identities such as race and gender (Esmonde et al., 2009); (b) students' critical language awareness can illuminate a range of ways they position themselves, offer insights to teachers about how students engage in the dance of agency (Wagner, 2007), and can afford students new possibilities in their mathematics classrooms (Wagner, 2008); (c) some words that students find troublesome in mathematics classrooms (i.e., "just" as in "just do it one step at a time") were pervasive and can contribute to monoglossic discourse (Wagner & Herbel-Eisenmann, 2008); and (d) teachers' (mainly implicit) pedagogical decisions and actions related to their expectations for students can perpetuate the marginalization of students who are working class or poor and students who are immigrants (Hoadley, 2007; Planas & Gorgorio, 2004). Additionally, Setati's (2005, 2008) articles clearly illustrated how issues of power must be considered in multilingual contexts: "to fully describe and explain the use of language(s) in multilingual mathematics classrooms we need to go beyond the pedagogic and cognitive aspects and consider the political role of language" (Setati, 2005, p. 464).

In the case of how these articles addressed "discourse," some used this term, and related terms like narrative, counternarrative, and language, without defining them (e.g., T. Brown & Heywood, 2011; Esmonde et al., 2009; Hoadley, 2007; Hoadley & Ensor, 2009; Mordant, 1993). Others provide definitions of discourse, for example, as referring to—

- "all types of language, including signs, gestures, artefacts, mimics, and so on" (Barbosa, 2006, p. 297, following Lerman, 2001);
- "a set of actions and interactions that take place in a context of social practices and affect the construction of both personal and social meanings" (Planas & Gorgorio, 2004, p. 21);
- "spoken or written language use which is linked to other nonlanguage forms of social life—discourses are not mutually exclusive or fixed, but can overlap, be contested, and can change" (LeRoux, 2008, p. 313, following Fairclough, 1992); and
- "a system of signs that organizes and regulates specific social and institutional practices" (Evans et al., 2006, p. 210).

赛塔蒂（2005，2008）和莫谢科维奇（2007）都借鉴了吉对话语的定义（关于话语的更多信息参见社会文化子流派部分）。

许多作者又继而探讨了话语的作用。例如，埃文斯等人（2006）指出话语——

> 为参与者提供了资源去构建意义和身份、体验情感和解释行为。话语能具体说明什么样的对象和概念是重要的以及在实践中参与者可以有哪些定位——可以采纳的各种角色，以及这些角色采取行动的可能性和与其他参与者的关系。话语还提供评估标准，……构成权力社会关系的基础［并且］控制了参与者的定位如何发生变化，即个人如何从可用的特定话语定位中作出选择。（第210页）

许多作者描述了这些不同层次的话语（例如，互动、实践和更大范围的关系或权力的社会要素；Esmonde等，2009；Evans等，2006；Hoadley，2007；LeRoux，2008；Straehler-Pohl & Gellert，2013），而有些作者指出，话语和意义的形成、认知或思考是不可分割的（例如Barbosa，2006）。继齐泽克之后，斯科维斯莫斯（2012）关注语言和某些话语的"暴力"层面。

借鉴伯恩斯坦批判社会学的学者撰写了关于"描述语言"的文章，并考虑了实践描述对了解学生不同定位方式的益处（Evans等，2006）以及这些定位如何与教育学中的社会阶层（Hoadley，2007；Hoadley & Ensor，2009）和为分流后学习不同课程学生所写的数学文本（Dowling，1996）中的社会阶层的再现相联系。也有人关注伯恩斯坦（1999）关于框架和分类的想法，这些与水平话语（"可能是口头的、局部的、情境依赖和特定的、缄默的、多层次的、前后矛盾但不局限于情境的，关键特征在于分段组织"；Straehler-Pohl & Gellert，2013，第317页）和垂直话语（"采用连贯、明确和系统的原则性结构形式、分层组织，类似于科学，或者采取类似于社会科学和人文科学那样的一系列专业语言的形式"；Straehler-Pohl & Gellert，2013，第317页）有关。那些借鉴索绪尔（T. Brown，1994；Evans，1999）的作者着重于意指和能指的理念：词或图像及其相关的概念。这两个理念共同构成"符号"。T. 布朗（1994）指出，索绪尔的工作中还有一个隐含的要素："指称对象或对象本身"。最后，只有瓦格纳（2007，2008）明确聚焦于吸引学生关注语言。他运用摩根（1998）的"批判性语言意识"概念与学生一起练习和探索数学语言的各个方面，并借此考察数学课堂话语中个体的"能动性表现"（Pickering，1995）。

构建任务分析得出的概论。表27.11汇总了我们在阅读本流派所有文章时发现的关键趋势。

正如上文强调的研究结果所示，这些文章通常关注社会资源不公平分配或政治方面的问题。数学教育的政治性出现在这些作者所使用和强调的许多构念中：批判意识以及与数学和世界的批判关系（Barbosa，2006）、数学活动的"神秘感"（源自巴特）（T. Brown，1994）或数学的"符号暴力"或"符号力量"（Skovsmose，2012）、人们对语言的看法和与权力相关的语言观念中固有的

表27.11 概念重建相关构建任务趋势

意义	活动	身份	关系	政治	联系	符号系统和知识
数学及其教与学都受到了批评。语境处于作者的解释和论证的中心。	在每篇文章中批评都被视为重要的活动。	部分文章考虑身份与作为一种社会实践的数学之间的关系，有些则将中心身份视为更广泛的社会类别。	师生之间、同学之间、数学与学生之间、教科书与学生之间的关系都被考虑在内。有些作者还考虑了身份类别和学生之间的关系。	政治在这里是核心的问题。	所有文章均根据其基础理论建立各种理念之间/内部的联系。鉴于该流派兼收并蓄的性质，有太多联系因而没有在这里指出。	数学被宽泛地定义为一种政治、历史和社会实践，许多文章将学校数学与学术数学、现实数学、工作生活数学等区分开来。

Both Setati (2005, 2008) and Moschkovich (2007) drew on Gee's definition of big-D discourse (see sociocultural subheritage for more information about Discourse).

Many of these authors also went on to address what discourse does. For example, Evans et al. (2006) asserted that discourse

> provides resources for participants to construct meanings and identities, experience emotions, and account for actions. Discourses specify what objects and concepts are significant and what *positions* are available to participants in the practice—the various roles that may be adopted, together and with their possibilities for action and relationships with other participants. They also provide standards of evaluation... [which] form the basis of social relations of power [and] regulate how the positionings of participants come about—how individuals come to *take up* particular discursive positions from those available. (p. 210)

Many authors described these various levels of discourse (e.g., interaction, practices, and larger relations or social aspects of power; Esmonde et al., 2009; Evans et al., 2006; Hoadley, 2007; LeRoux, 2008; Straehler-Pohl & Gellert, 2013), and some authors stated that discourse and meaning making, cognition, or thinking were inseparable (e.g., Barbosa, 2006). Following Žižek, Skovsmose (2012) focused on language and "violent" aspects of certain discourses.

The scholars who drew from Bernstein's critical sociology wrote about the "language of description" and considered the ways in which descriptions of practice helped to see how students were positioned differently (Evans et al., 2006) and how these positionings related to the reproduction of social class both in pedagogy (Hoadley, 2007; Hoadley & Ensor, 2009) and in mathematics texts written for different tracks or streams of students (Dowling, 1996). There was also attention to Bernstein's (1999) ideas of framing and classification, which related to horizontal discourse ("likely to be oral, local, context dependent and specific, tacit, multi-layered, and contradictory across but not within contexts, crucial feature is that it is segmentally organized"; Straehler-Pohl & Gellert, 2013, p. 317) and vertical discourse ("takes the form of a coherent, explicit, and systematically principled structure, hierarchically organized, as in the sciences, or it takes the form of a series of specialized languages as in the social sciences and humanities"; Straehler-Pohl & Gellert, 2013, p. 317). Those authors who draw on Saussure (T. Brown, 1994; Evans, 1999) focused on the ideas of signified and signifier: the word or image and the associated concept. These two ideas, together, form a "sign." T. Brown (1994) pointed out that there was an additional implied element in Saussure's work: "the referent or the object itself." Finally, only Wagner (2007, 2008) focused on drawing students' attention to language explicitly. He applied Morgan's (1998) idea of "critical language awareness" to practice and explored aspects of mathematical language with students, which led him to consider the "dance of agency" (Pickering, 1995) in mathematics classroom discourse.

Overview from the building tasks analyses. In Table 27.11, we summarize the key trends we found in reading the set of articles in this heritage.

As the findings highlighted above suggest, these articles often had a focus on the inequitable distribution of social goods or Political issues. The Politics of mathematics education appeared in the many constructs these authors used and highlighted: critical consciousness and a critical relationship with mathematics and the world (Barbosa, 2006); "myths" (following Barthes) about mathematical activity (T. Brown, 1994) or the "symbolic violence" or "symbolic power" of mathematics

TABLE 27.11. Building Tasks Trends Related to Reconceptualists

Significance	Activities	Identities	Relationships	Politics	Connections	Sign systems and knowledge
Mathematics and the teaching and learning of mathematics were all critiqued. Context is central to authors' interpretations and arguments.	Critique is treated as a significant activity in every article.	Some articles considered identity in relationship to considering mathematics as a social practice and some included making central identity a broader social category.	Relationships between teacher and students, students and students, mathematics and students, and textbooks and students were all considered. Some articles also considered relationships between identity categories and students.	Politics is front and center in this perspective.	All articles make connections between/among various ideas, based on the theories from which they drew. Because this was an eclectic heritage, there are too many to name here.	Mathematics was treated broadly as a political, historical, and social practice, and many articles distinguished school mathematics from academic mathematics, real-life mathematics, work-life mathematics, and so on.

"文化模式"（Setati，2005，2008），"独白式"话语（Wagner & Herbel-Eisenmann，2008），与学生边缘化相关的机会、层级、地位和其他问题（例如Dowling，1996；Esmonde，2009a；Esmonde等，2009；Straehler-Pohl & Gellert，2013；Wagner，2007）。文章有时也会涉及国家的政治历史，例如南非（Hoadley，2007；Hoadley & Ensor，2009；LeRoux，2008；Setati，2005，2008），探索数学特定领域的历史（T. Brown & Heywood，2011；Skovsmose，2012）或学校本身的历史和承诺（Esmonde等，2009）。即使当特定的理论得到加强时，作者有时也会将这些理论与"政治和教育承诺"联系起来，并确定这些理论背后的价值观和假设（Evans，1999）。

尽管之前的论述中已经暗示，该流派将数学（符号系统和知识）处理地非常宽泛，并把学校数学与学术数学、现实数学、工作生活数学等区分开来。有些作者没有定义什么是数学，而是指出其可能具有的"多种形式"，例如："制作预算、计算工资、进行投资、阅读地图、完成设计、解决学校数学练习题、解决工程问题，以及不忘做数学研究"（Skovsmose，2012，第121页；另见Evans，1999；Evans等，2006）。数学被描述为一个"深奥的领域"（Dowling，1996）或"专业知识"（Hoadley，2007；Hoadley & Ensor，2009）。一位作者甚至说，数学家作为"完全不具代表性的少数派为数学本身的缘故而构建了数学"（Mordant，1993，第23页）。因此，这些文章通常不关注传统主题或大概念（例如，函数、有理数、整数运算），也不关注以理想形式存在的脱离人类的柏拉图式的理念，相反，数学被视为可能导致将学校和社会中的学生分层或边缘化的一种政治性、历史性、局部性的话语活动或实践。

有些文章将身份视为分析的核心，例如关注种族、阶级、性别和语言背景（例如Dowling，1996；Esmonde等，2009；Hoadley，2007；Hoadley & Ensor，2009；Planas & Gorgorio，2004；Setati，2005）。在本领域中，通过数学课堂中的交互对学生身份的塑造和被塑造有一个假设和明确处理，特别地，埃斯蒙德等人（2009）的文章涵盖了该流派中最彻底的身份论述。该文不仅使身份成为分析和研究结果的核心，还详细描述了作者的定位，后者在数学课堂话语（以及更一般的数学教育研究）领域的许多工作中是缺失的。

线索

本节的标题为线索，原因是每一小节要么将新的理论框架与其他流派的框架进行了整合使这一工作在本质上有所不同，要么是对一个领域的首次尝试，但是研究人员会在该领域继续前行的。由于极其多样化的性质，我们没有使用构造任务来寻找趋势，因为这组文献并不基于特定的流派。经仔细分析，我们将这一领域的14篇论文分为四个小节，反映着不同的观点。其中每一小节有2—5篇论文，这也表明了使用相关观点的研究人员很少。

第一小节着重于隐喻并由五篇论文组成。其中一篇侧重于对11岁学生的访谈（Abrahamson，Gutiérrez，& Baddorf，2012），而其余的文章则分析了高年级学生的课堂互动过程中的数据（Carreira，2001；Font，Bolite，& Acevedo，2010；Font，Godino，Planas，& Acevedo，2010）或工作场所的成年人以及高中生与老师之间的讨论（Williams & Wake，2007）。目前还不清楚为什么关于隐喻的研究主要关注较年长的参与者。

对于所有的文章来说，当一个抽象的数学概念以诸如物理属性等其他事物的形式出现时，隐喻就产生了。正如丰特、戈迪诺等人（2010）所指出的，"对象隐喻在教师的话语中是始终存在的，原因就是数学实体被呈现为'具有属性的对象'，可以物理地表示出来（在黑板上、用教具、用手势等）"（第15页）。从这个角度来看，存在一个源域和一个目标域，将源域投射至目标域。

尽管所有五篇文章均提及拉考夫的工作，尤其是他与努涅兹的著作（Lakoff & Núñez，1997，2000），但很多文章的观点或灵感却都来自于其他资料。威廉姆斯和韦克（2007）将拉考夫和努涅兹（2000）对隐喻的

(Skovsmose, 2012); the "cultural models" embedded in people's views on language and seeing language in relationship to power (Setati, 2005, 2008); "monoglossic" discourse (Wagner & Herbel-Eisenmann, 2008); and access, hierarchies, status, and other issues associated with the marginalization of students (e.g., Dowling, 1996; Esmonde, 2009a; Esmonde et al., 2009; Straehler-Pohl & Gellert, 2013; Wagner, 2007). Articles also sometimes situated the work in the political history of the country, such as South Africa (Hoadley, 2007; Hoadley & Ensor, 2009; LeRoux, 2008; Setati, 2005, 2008); the history of the particular area of mathematics being explored (T. Brown & Heywood, 2011; Skovsmose, 2012); or the history and commitments of the school itself (Esmonde et al., 2009). Even when particular theories were being augmented, authors sometimes located these theories in relationship to "political and pedagogical commitments" and identified values and assumptions underlying those theories (Evans, 1999).

Although it is implicit in the points already made, this heritage treated mathematics (Sign System and Knowledge) very broadly and distinguished school mathematics from academic mathematics, real-life mathematics, work life mathematics, and so on. Some authors did not define what mathematics was, but rather pointed out the "many forms" it could take, for example: "constructing a budget, calculating a salary, making an investment, reading a map, completing a design, solving school mathematics exercises, solving an engineering problem, not forgetting doing mathematical research" (Skovsmose, 2012, p. 121; see also Evans, 1999; Evans et al., 2006). Mathematics was described as an "esoteric domain" (Dowling, 1996) or "specialized knowledge" (Hoadley, 2007; Hoadley & Ensor, 2009). One author went so far as to say that mathematicians constructed mathematics as an "utterly unrepresentative minority who attune to mathematics for its own sake" (Mordant, 1993, p. 23). Thus, there was typically not a focus on traditional topics or big ideas (e.g., function, rational number, whole number operations), nor was there a primary focus on Platonic ideas that existed somehow in ideal forms separate from human beings. Rather, mathematics was treated as a political, historical, situated, discursive activity or practice that could contribute to the stratification or marginalization of students in schools and society.

Some articles made Identity central to the analyses. There was a focus on, for example, race, class, gender, and language background (e.g., Dowling, 1996; Esmonde et al., 2009; Hoadley, 2007; Hoadley & Ensor, 2009; Planas & Gorgorio, 2004; Setati, 2005). In this work, there is an assumption and explicit treatment of students' identities shaping and being shaped by the interactions in mathematics classrooms. In particular, the article by Esmonde et al. (2009) included the most thorough treatment of identity in this heritage. This article not only made identity a central part of the analysis and findings but it also detailed the authors' positionality, something that is lacking in much of this work on mathematics classroom discourse (as well as mathematics education research, more generally).

Threads

This section was labeled threads because each subsection either had integrated new theoretical frameworks with frameworks from other heritages in a way that made this work substantively different or was an initial foray into an area but from which the researchers moved on. Due to its extremely diverse nature, we did not look for trends using the building tasks because this set of papers was not based on a particular heritage. The 14 papers carefully analyzed for this section were subdivided into four smaller subsections, reflecting different perspectives. Each of these had 2–5 papers within them, indicating the small numbers of researchers using these perspectives.

The first subsection focused on metaphors and consisted of five papers. Of these, one focused on interviews with 11-year-old students (Abrahamson, Gutiérrez, & Baddorf, 2012), whilst the remaining articles analyzed data from interactions in classrooms with older students (Carreira, 2001; Font, Bolite, & Acevedo, 2010; Font, Godino, Planas, & Acevedo, 2010) or adults in the workplace and senior secondary students' discussions about these with their teacher (Williams & Wake, 2007). It is not clear why this work on metaphors focused mostly on older participants.

For all the articles, metaphors occurred when something such as an abstract mathematical concept took on the form of something else, such as physical attributes. As Font, Godino et al. (2010) claimed, "the object metaphor is always present in the teacher's discourse because here the mathematical entities are presented as 'objects with properties' that can be physically represented (on the board, with manipulatives, with gestures, etc.)" (p. 15). From this view there is a source and a target domain, with the source domain being projected onto the target domain.

Although all five papers mentioned the work of Lakoff, particularly his work done with Núñez (Lakoff & Núñez, 1997, 2000), many drew inspiration from additional sources. Williams and Wake (2007) combined Lakoff and

描述与托夫斯基（1979）的工作相结合，他们对源域和目标域的关系提出质疑，建议将其称为"两个领域的意义递归交互"（第353页；字体突出为原文所加），因而，不是将源域的特征映射到目标域，而是存在一种双向过程，因为参与者就目标域和源域达成了一种共同理解。他们的研究使用了在工作场所记录的音频数据，当时一名员工正在向一名带班参观的老师解释一个公式，之后，老师和学生们就该公式进行了讨论，表明数轴在促进两组互动的共同理解方面发挥了重要作用。

所有文章都强调了隐喻对数学交互学习的贡献。亚伯拉罕森等人（2012）指出："教师和学生利用隐喻这一话语手段来创造或修改数学学习环境中的共同观点、意义和工具使用方法"（第56页）。丰特及其同事（Font, Bolite, & Acevedo, 2010；Font, Godino等, 2010）借鉴一位联合作者阿塞维多（2008）的工作来考察关于绘图教学方面隐喻和学习之间的关系。

和威廉姆斯与韦克（2007）一样，卡雷拉（2001）认为建模暗示了隐喻的使用。她借鉴C.S.皮尔斯（1978；Hartshorne & Weiss, 1931）的观点将概念性隐喻定义为：

> 使我们能够通过更加熟悉或接近我们日常经验的领域来理解另一个领域的一种机制。拉考夫（1993）认为这种对应关系是从源域到目标域的真实映射或投影。（Carreira, 2001，第264—265页）

由此，可以认为学习是"一个无尽的解释链，每一个解释总是提供一个特定阶段的认知发展"（Carreira, 2001，第264页）。尽管在互动中使用隐喻可能会带来误导学生的歧义，但通过讨论和论证，学生能够扩展他们的隐喻，使得自己的理念得以发展。

第二小节包括四篇属于心理语言学的文章，每篇都有不同的视角。利恩、克莱门兹和德尔·坎波（1990）评估了澳大利亚和巴布亚新几内亚5—15岁学生对文字题的理解，这些文字题使用了较小数值的数以聚焦文字题的语言成分。他们认为，由于信息处理差异而导致的不同类型的文字题的感知困难，更可能与学生对极端化的比较术语的误解有关。利恩等人（1990）指出，心理语言学的研究文献表明，一对正反词中的正面词（比如多）比反面词（比如少）更容易习得。他们指出，可用于各种情境的术语（比如大小）比只适用于一个属性的术语（比如高矮）更容易习得。最后，他们还提出，当孩子没有词汇可以用来描述一个情境时，他们会使用来自类似语义领域的词语。但是，社会文化研究可能会让心理语言学研究受到影响，例如，沃克迪内（1988）关于母子互动的研究表明，没有一个孩子会提到更少这个词，这表明对于孩子来说，更少并不是一个难以掌握的概念，而只是一个更不熟悉的概念。

罗博蒂（2012）与布罗姆和斯坦因布林（1994）都没有将其研究明确纳入心理语言学，然而，由于他们关注了数学语言认知学习，从而我们将其纳入本小节论述。布罗姆和斯坦因布林（1994）使用奥格登和理查兹（1923）的观点来讨论由物体、符号和概念组成的三角形的含义。布罗姆和斯坦因布林（1994）能够根据对符号、关系和对象的关注对两位教师的谈话进行分类，从而考虑他们是如何讨论和发展一个概念的。他们的结果表明，经验越丰富的教师能够在符号和对象的讨论之间作出更柔和的转换，这表明其关联水平更为清晰。

在本节的四篇论文中，罗博蒂（2012）的文章是最近发表的。尽管她的理论框架提及了社会文化理论，但是她使用了心理语言学家布隆卡特（1985）的理念，与布罗姆和斯坦因布林（1994）一样，她的分析也集中于"学生话语中图形领域（绘图）和理论领域（理论对象）之间的多重转变"（Robotti, 2012，第434页）。她的研究结果表明，分析学生对自然语言的使用为他们了解学生的认知发展提供了线索。

本小节的最后一篇文章是关于教科书如何在文字题之前呈现符号题的（Nathan, Long, & Alibali, 2002）。在这篇文章中，作者调查了10年间代数和预代数课程的教科书的修辞结构，"修辞结构是指在不同类型的段落中，如比较和对比文本，发现的层级组织的差异，作用

Núñez's (2000) description of metaphors with the work of Wartofsky (1979). They queried the source to target domain relationship, suggesting that it was a "*recursive interaction* of meanings in two domains" (p. 353; italics in original). Thus, rather than a mapping of features from the source onto the target domain, there was a two-way process as a joint understanding was developed by participants about both the target and the source domains. Their research used data audio recorded in a workplace situation in which an employee was explaining a formula to a teacher who was there with her class. Later, the teacher and the students discussed the workplace formula, showing how the number line took an important role in facilitating joint understanding in both sets of interactions.

Metaphors' contribution to learning in mathematics interactions was highlighted in all papers. Abrahamson et al. (2012) stated "instructors and students utilize metaphor as a discursive means of striving to create or amend shared views, meanings, and uses of artifacts in mathematics learning environments" (p. 56). Font and colleagues (Font, Bolite, & Acevedo, 2010; Font, Godino, et al., 2010) drew upon work of one joint author, Acevedo (2008), to consider the relationship between metaphors and learning in regard to teaching graphing.

Like Williams and Wake (2007), Carreira (2001) considered that modeling implied the use of metaphors. She used C. S. Peirce's (1978; Hartshorne & Weiss, 1931) work to define a conceptual metaphor as

> a mechanism that allows us to understand one domain in terms of another, usually more familiar or closer to our daily experiences. In the words of Lakoff (1993), this correspondence is a real mapping or projection from an origin domain onto a target domain. (Carreira, 2001, pp. 264–265)

Consequently, learning came to be seen as "an endless chain of interpretants, with each interpretation always supplying a cognitive development of a given stage" (Carreira, 2001, p. 264). Although the use of metaphors within the interactions allowed for ambiguities that could mislead students, students' ideas developed as they extended their metaphors through discussion and argumentation.

The second subsection included four papers located within psycholinguistics, each of which took a different perspective. Lean, Clements, and Del Campo (1990) assessed students aged 5–15 in Australia and Papua New Guinea on comprehension of word problems using small numbers to focus on the linguistic components of the word problems. They suggested that the perceived difficulties with different types of word problems that had been attributed to information processing differences were more likely to be related to students' misunderstandings of polarized comparative terms. Lean et al. (1990) stated that the research literature from psycholinguistics suggested that positive members of pairs, such as *more*, are learned before negative members such as *less*. They pointed out that terms that can be used in a variety of situations (e.g., *big-small*) are learned before terms that can only be used for one attribute, such as *tall-short*. Finally, they suggested that when children did not have the words to describe a situation, they would use a term from a similar semantic field. Sociocultural research, however, may have over-shadowed pycholinguistic research. For example, Walkerdine's (1988) research on mother-child interactions showed that no child asked for less, suggesting that less was not a harder concept for children to grasp, just a less familiar one.

Both Robotti (2012) and Bromme and Steinbring (1994) did not explicitly situate their research studies within psycholinguistics. Their focus on cognitive learning of mathematical language, however, resulted in their inclusion in this subsection. Bromme and Steinbring (1994) used Ogden and Richards' (1923) work to discuss the meaning of triangle as being made up of object, symbol, and concept. Bromme and Steinbring (1994) were able to categorize the talk of two teachers by their foci on symbol, relation, and object to consider how a concept was discussed and developed. Their results showed that the more experienced teacher was able to make softer transitions between the discussions of the symbol and object, suggesting that the relational level was more in focus.

Of the four papers in this section, Robotti's (2012) paper was the most recent one. Although her theoretical framework mentioned sociocultural theories, she used the ideas of Bronckart (1985), a psycholinguist. Like Bromme and Steinbring (1994), her analysis focused on "the multiple shifts between the graphical domain (drawing) and the theoretical domain (theoretical object) in the discourse produced by students" (Robotti, 2012, p. 434). Her findings indicated that analysis of pairs of students' use of natural language provided indications of their cognitive development.

The final paper in this subsection is on how textbooks presented symbolic problems before verbal ones (Nathan, Long, & Alibali, 2002). In this paper, the authors investigated the rhetorical structure of textbooks in algebra and prealgebra courses over a 10-year period. "*Rhetorical structure* refers to differences in hierarchical organization found in different types of passages, such as compare and contrast texts, that help link to readers' common schemas

是联系读者的共同图式并引导其预期"（第2页）。他们发现，与预代数教科书相比，代数教科书更多地是将符号的代数练习置于文字题之前，这表明文字题对于学生来说比符号题更难掌握。这个发现与教师关于代数学习的信念研究一致，但与已知的学生解答代数题的发现相反。

第三小节的三篇文章侧重于阅读策略，其中两篇属于博拉西和西格尔及其同事的一整套研究，他们调查了如何利用关于阅读的知识，使阅读成为一个基于学生先前知识和兴趣的主动过程。（他们的另一篇文章已在系统功能语言学部分进行了讨论）。在博拉西、西格尔、房兹和史密斯（1998）及博拉西和西格尔（1990）的文章中都使用了罗森布拉特的交互阅读理论。这一理论指出，阅读不仅仅能够"解码这些文本中使用的数学符号和语言，以便他们可以提取文本中包含的信息并理解概念或解决问题"（Borasi等，1998，第277页）。博拉西及其同事先与一组美国教师合作，之后又与其中一些教师的8至11年级学生合作，呈现了不同形式的数学写作，并让这些学生参与阅读活动，以支持他们与先前知识之间建立联系。

本小节的另一篇文章来自谢泼德、塞尔登和塞尔登（2012），他们采取了与博拉西等人（1998）不同的方法，关注学生如何理解数学文本。他们使用了建设性响应阅读（CRR），该方式整合了一系列不同的阅读理论，包括罗森布拉特的理论（Pressley & Afflerbach, 1995）。研究发现，尽管他们调查的这些大学生在进入课程时已经具备良好的阅读成绩和数学成绩，也有许多建设性响应阅读策略，但他们还是需要努力地去理解那些必须阅读的教科书段落。作者的结论是，阅读数学课本需要一套不同于其他类型阅读的策略，学生需要支持来学习这些。

线索部分的最后一小节侧重于写作。两篇文章（Shepard, 1993; Shield & Galbraith, 1998）都对学生的数学写作进行了分类。在希尔德和加尔布雷斯（1998）的文章中，为了考虑如何通过写作来提高学生的数学学习，他们还对学生的教科书进行了分类。谢泼德（1993）这篇文章讨论了布里顿、伯吉斯、马丁、麦克劳德和罗森（1973）的交易性信息写作类别与概念发展的学习阶段之间可能存在的联系。相对地，希尔德和加尔布雷斯（1998）使用八年级学生的说明性写作和他们的教科书来开发不同类型写作的编码方案。他们注意到学生的写作方式和教科书中使用的写作风格之间有许多相似之处。

讨论

在本章中，我们将数学教育话语领域的文献按理论流派进行划分，然后对每个流派和子流派所含文章进行了三项分析。第一项分析使得我们能看清研究的进展趋势，该研究在哪里进行、和谁开展、分析的类型、焦点以及结果。另外两项分析借鉴吉（2011a, 2011b）的构建任务对各流派和子流派中突出的构建任务以及相关处理方式的性质进行了批判性的评论。总的来说，我们注意到这方面的研究大多——

- 侧重于二至十年级的课堂，很少关注更高年级的高中或大学数学或校外教与学的背景（如成人教育），且几乎不涉及幼儿教育；
- 调查口语或互动，而不是文本的读或写或现有书面文本（例如特别是教科书或白板上的书面文字）或课堂的非语言方面；[3]
- 侧重于对较小数据集的详细分析，很少使用混合方法或纯粹定量分析。

某些子流派已经在该领域内获得人气（例如，交流认知、定位、系统功能语言学、社会文化），同时在有些领域内的出版物数量已经减少，特别是在过去的十年里（例如在格赖斯门派、交际社会语言学和线索型这些领域）。这使得我们思考在数学教和学领域内可能丧失了哪些观点，因为这些视角提供了其他子流派所不具备的有趣发现。举例来说，这些领域已经揭示了许多事情，如在数学话语中一些特定词语是如何使用的（例如，"你"作为允许推广的代名词或模糊限定语在数学中的作用），礼貌如何形成数学交互，学生如何投入文本阅读，等等。我们还注意到，在更大的范围内，其中部分流派和子流派几乎全部发表在美国以外的期刊上（例如互动社会语言学、概念重建），而其他的类别又几乎全部出自美国的期刊

and guide their expectations" (p. 2). They found that in algebra, more so than prealgebra, textbooks had situated symbolic algebra exercises before word problems, suggesting that word problems would be harder for students to master than symbolic problems. This was in alignment with research on teachers' beliefs about algebraic learning, but in opposition to what was known about students' completion of algebraic problems.

The third subsection of three articles focused on reading strategies, two of which belonged to a body of work by Borasi and Siegel and colleagues and investigated how to utilize understandings about reading as an active process drawing on students' prior knowledge and interest. (Another of their articles was discussed in the SFL section.) In articles by Borasi, Siegel, Fonzi, and Smith (1998) and Borasi and Siegel (1990), Rosenblatt's transactional reading theory is used. This theory stated that reading is more than being able to "decode the mathematical symbols and language used in such texts so they can extract information contained in the text and understand the concept or solve the problem" (Borasi et al., 1998, p. 277). Borasi and colleagues worked first with a group of U.S. teachers and then with some of these teachers' year 8 to year 11 classes to present different forms of writing about mathematics and engaged students in reading activities, which supported them in making connections to their prior knowledge.

The other article in this subsection is by Shepherd, Selden, and Selden (2012), who took a different approach from that of Borasi et al. (1998) by being interested in how students made sense of their mathematics texts. They used constructively responsive reading (CRR), which was synthesized from a range of different reading theories, including that of Rosenblatt (Pressley & Afflerbach, 1995). They found that, although their university students had good reading and mathematics grades on entering the courses and had many of the CRR strategies, they struggled to make sense of the textbook passages they had to read. The authors concluded that reading mathematics textbooks required a different set of strategies than those used for other kinds of reading and that students needed support to learn these.

The final subsection in Threads focuses on writing. Both articles (Shepard, 1993; Shield & Galbraith, 1998) classified students' mathematical writing. In the case of Shield and Galbraith (1998), students' textbooks were classified in order to consider how students' mathematical learning could be improved through writing. Shepard's (1993) article was a discussion of the possible relationship between the transactional-informative writing categories of Britton, Burgess, Martin, McLeod, and Rosen (1973) and learning phases for conceptual development. In contrast, Shield and Galbraith (1998) use the expository writing of year 8 students and their textbooks to develop a coding scheme for the different types of writing. They noted a number of parallels between the ways that the students wrote and the writing styles used in their textbooks.

Discussion

In this chapter, we divided the literature on discourse in mathematics education into theoretical heritages and then did three analyses of articles included in each heritage and subheritage. One analysis allowed us to see trends in where this research was done, with whom, the kinds of analyses and foci, as well as the findings. The other two analyses drew on Gee's (2011a, 2011b) building tasks to critically review not only which building tasks the heritages and subheritages highlighted but also the nature of how these were handled. Overall, we notice that this body of research mostly—

- focused on classrooms between second and tenth grade, with very few focused on higher level high school or university-level mathematics or out-of-school teaching and learning contexts (say, with adults) and almost nothing from early childhood;
- examined spoken language or interactions, rather than the reading or writing of texts or existing written texts (especially textbooks or written words on a whiteboard, for example) or nonverbal aspects of classrooms;[3] and
- engaged in detailed analyses of a rather small set of data, with few mixed-method or solely quantitative analyses.

Some of these subheritages have been gaining popularity in the field (e.g., commognition, positioning, SFL, sociocultural) and, in some areas, the number of publications has diminished, especially over the last decade (e.g., under a Gricean umbrella, interactional sociolinguistics, and threads). This makes us wonder what we may be losing in terms of insights into mathematics teaching and learning, as these latter perspectives offer interesting findings that the other subheritages do not. For example, these areas have shed light on things like how particular words are used in mathematics discourse (e.g., "you" as a pronoun that allows people to generalize or the role of hedges in doing mathematics), how politeness shapes mathematical interactions, how students engage in reading texts, and so on. We also noticed, more generally, that some of these heritages and subheritages appear

（例如浮现）。

在本章的其余部分，我们首先提出一般的方法论问题和我们认为本领域需要考虑的问题。在讨论的第二部分，我们比较并对比了各子流派中的一些要素。特别是，突出展示了流派和子流派在如何处理、前后因问题，吉所提出的构建任务上的一些相似和不同之处。

方法论问题和疑问

在我们的分析中，一个明显的方法论问题是不同作者阐述和使用理论的方式是截然不同的。正如我们删除46篇文章所依据的那样，有些研究根本不借助话语理论或方法，即使在使用了话语理论和方法的文章中，我们注意到理论的阐释和使用也存在很大区别。有些作者只是顺便提及"数学语义"或"最近发展区"或"社会文化"等术语，其主要目的是将他们的研究定位于与数学学习的社会层面相关的领域。其他作者可能已经在他们文章的一节中引用并阐明了这些理念，但很少有作者会在整篇文章中清楚地阐述和使用这些概念和理论。我们认为，这些"质量不一"的表达和使用表明迄今为止还没有人，例如期刊论文评审人，对该领域的理论和概念的严谨性提出要求。

一个相关的方法论问题是所用理论和概念缺乏可操作性。只有部分文章的作者详细描述了其编码框架（例如，可以参见Conner等，2014），许多文章缺乏细节使得他人很难复制其研究和方法，这还使得比较不同的发现变的困难，原因是这些理论的使用可能互相并不一致。因此，构建一套强大的工具和概念来研究数学课堂话语将较为困难，我们发现该问题涉及之前提到的另一个更大问题：大量的论文提供方法论（或更为精确地说是方法论工具），这些文章提出了很多关于工具的建议，但只有其中一小部分建议被一致的采纳。因此，该领域包含许多方法论的"萌芽"及有用的建议，但很少有"较大分支研究"真正采取这些建议，并在其基础上接连发表论文。造成该现状的原因很多，可能是因为该领域还很年轻，或者是因为有一个"自然选择"的过程，也可能是由于政治原因——提出新东西总会比借鉴前人更受欢迎（或在学术生涯和出版过程中受到高度重视），或者，正如我们已经指出的，其原因可能与缺乏足够的可操作性有关，这种可操作性有助于研究者基于原创工作进行拓展，尤其是当框架复杂而全面时。

在理论的阐述、使用和操作方面，我们观察到一个潜在的时间因素在起作用，也就是说，与后期的研究相比，早期的数学课堂话语研究倾向于较少借鉴理论，这可能表明随着时间的推移，数学课堂话语研究的理论和方法越来越明确，这一领域的研究变得更为成熟且越来越被数学教育所接受。然而，我们也观察到，在某些情况下，大量的理论工作是在某流派下早期发表的文章中完成的，而后来的文章大多引用了这些工作，而没有进一步拓展或发展这些思想（例如，浮现的观点）。我们希望强调这些观点可以加速用于数学课堂话语研究的理论和概念在阐述、使用和可操作性方面的明确程度，我们也希望它将引发关于理论和方法论在该研究领域中的作用及其为当前研究贡献的价值的更广泛的讨论。

作为一个相关的观点，我们强调更深入的理论工具对该领域尤为有用。例如，在纵览众多知识流派和子流派时，我们注意到"定位"一词在不止一个领域出现（例如格赖斯门派、互动社会语言学、概念主义）。在某些领域，该术语被定义了，但在其他领域，它被使用但未给出定义，比如，恩普森（2003）、埃斯蒙德（2009a）、罗兰（1999）、奥康纳和麦克（1993）都详细介绍了与定位有关的过程（并将定位用作动词），但没有明确定义"定位"。此外，当研究人员已经使用了定位并引用定位理论和假定世界时，他们没有讨论这些理论如何结合或分离，所以尽管文章使用了同一词汇，但它也可能有着不同的意义。而本文回顾的大部分研究均很少探讨理论或概念的可通性，我们认为更深入的理论工作可以帮助数学教育研究人员追溯其研究的本源，让相关

almost exclusively in journals outside the United States (e.g., interactional sociolinguistics, reconceptualists), whereas others appear almost exclusively in U.S.-based journals (e.g., emergent).

In the remainder of this discussion, we first raise general methodological issues and questions that we think need to be considered by the field. In the second part of the discussion, we compare and contrast a few elements across the various subheritages. In particular, we highlight some of the similarities and differences in how the heritages and subheritages handle, foreground, and background the building tasks articulated by Gee.

Methodological Issues and Questions

A methodological issue that became apparent in our analysis was the very different ways in which authors articulated and used theory. As highlighted by our decision to remove 46 articles, some of this research did not draw on discursive theories or methods at all. Even within the articles that did draw on discursive theories and methods, we noticed a broad range in terms of the articulation and use of theory. Some authors only mentioned terms like "mathematics register" or "zone of proximal development" or "sociocultural" in passing, mainly to locate their research in an area concerned with social aspects of learning mathematics. Other authors may have referred to and unpacked the ideas in one section of their article, but fewer authors clearly articulated and used the concepts and theories consistently throughout their articles. We suggest that these observations of "spotty" articulation and use indicate that the field has so far not been required, for example by journal reviewers, to be theoretically and conceptually rigorous.

A related methodological issue was the lack of operationalization of the theories and concepts being used. Only in some articles did the authors describe in detail their coding frameworks (see, for example, Conner et al., 2014). The lack of detail in many of the articles makes it difficult for the replication of studies and methods. It can also make it difficult to compare various findings because the applications of these theories may not be consistent or consonant with others' use. Building a robust set of tools and concepts with which to study mathematics classroom discourse, then, would be difficult. We see this issue being related to a larger issue that we mentioned earlier: the tremendous number of papers offering a methodology (or more accurately, a methodological tool). These articles have offered many suggestions for tools yet only a fraction of them have been taken up in any consistent manner. Thus, the field consists of many little "sprouts" of methodologies with useful suggestions but very few "thick branches" that actually take these suggestions and build upon them one paper after the other. There could be many reasons for this state of the field. It may be because this field is still young or because there is a process of "natural selection." It may also be due to politics—it can be more popular (or highly valued in the academic tenure and publication process) to suggest something new than to build on what someone else has done. Or, as we have pointed out, the reason could be associated with the lack of sufficient operationalization that enables researchers to build on original works, especially when the frameworks are complex and comprehensive.

In relationship to the articulation, use, and operationalization of theory, we observed a potential time factor at play. That is, earlier research on mathematics classroom discourse tended to draw on theories less than later pieces. This might indicate that the work on mathematics classroom discourse is getting more explicit about its discourse theories and methods over time as this area of research becomes more mature and accepted within mathematics education. We also observed, however, that in some cases substantial theoretical work was done in articles published early in the heritage, and later pieces mostly referenced that work without extending or developing the ideas further (e.g., emergent). We hope that highlighting these points may increase the rate at which theories and concepts for studying mathematics classroom discourse become more clearly articulated, used, and operationalized. We also hope that it will spark a more general discussion about the role of theories and methodologies in research of this kind and the value that they add to the research being undertaken.

As a related point, we emphasize that deeper theoretical work would be especially useful to the field. For example, when we looked across many of these heritages and subheritages, we noticed that the words "position" or "positions" appears in more than one area (e.g., the Gricean umbrella, interactional sociolinguistics, reconceptualists). In some of these areas, this term is defined, and in others, it is used but not necessarily defined. For example, Empson (2003), Esmonde (2009a), Rowland (1999), and O'Connor and Michaels (1993) all detail processes related to positioning (and use position as a verb) but do not explicitly define position or positioning. Additionally, when researchers have used position and cite both positioning theory and the figured worlds work, there is no discussion of how these theories go together or do not. So even when articles use the same word, it may not mean the same thing, yet there was little discussion of theoretical or conceptual commensurability in much of the

观点随着时间的推移而逐渐明晰化，同时还能帮助本领域更好地理解理论上可行的方式以整合各种理论和概念。

最后，我们还需要强调的事实是，虽然我们在方法论的某些方面取得了进展，但参与研究的研究人员的定位是一个始终被忽略的方法论领域。在我们回顾的几乎所有文章中，很少谈及谁是研究者，研究者已有的什么经验有助于他们读取数据，前述主张的反例包括，例如埃斯蒙德等人（2009年）、埃斯蒙德和兰格-奥苏纳（2013）、海德-梅朱亚尼姆（2013）、梅萨和昌（2010），以及罗伯特（1998）。尽管在其他内容领域有研究者如何考虑其定位的清晰实例（例如Milner，2007；Peshkin，1988，2001），但在数学教育研究和本研究领域却很少有人做到这一点。我们呼吁期刊编辑和审稿人注意这个问题，更具体地就是在研究中要求有这样的说明。鉴于很多理论认为语境是解释交流的核心并且认识到个人在各种话语实践中的社会化塑造了他对个人经历（包括研究）的解释，这是一个需要关注的领域。在关注多样性、公平性和权力问题的工作中，对定位的关注尤为重要（Foote & Bartell，2011），正如我们回顾的很多文章所做的那样。

对流派分析的反思。本章关注数学课堂话语，因此我们首先考虑"话语"一词在这一领域中是如何被使用的问题。之后我们讨论了关键理念之间的对比并就吉的构建任务在各种流派和子流派如何被强调（或不强调）提供了几点意见。应该指出的是，我们并不是建议在基于不同流派和子流派的研究中需要明确呈现所有吉的构建任务，相反，更有意思的是找出某一特定流派或子流派在使用吉的构建任务时，哪个任务更为突出。

"话语"这个概念和术语本身在本研究中起着一系列的作用。该发现引导我们探讨这样的研究，即话语是理解一些其他现象（例如数学理解、信念）的透镜，而话语则是探究的对象。例如，在许多社会文化研究中，话语被用来研究数学的学习，相对照地，在格赖斯门派下，话语则是研究的对象，因为这些作者试图理解在数学教与学中使用语言的具体方式。

我们承认，并不是所有的作者使用"话语"一词（他们在使用时也没有给出其定义，在Ryve，2011中就已经讨论过这个发现），且不同流派下的关键构念差异很大。这可能是由于每种流派对"话语"在其关键思想和概念的集合或结构中有不同的定位。在表27.12中，我们列出了数学教育文献的作者借鉴相关流派和子流派作品时突出的一些关键思想，我们希望强调的一点是，关于从这些流派理论中引入何种思想来进行数学教育研究的决定并不是包罗万象的。

通过对以上问题的关注，我们可以确定一些相关的思想，并在此基础上强调在这些思想之间建立更多联系，从而比较/对比数学课堂内外的各种话语[4]的潜在价值。在这里我们举出一个特例，并说明还有可供进一步探索的例子。本章有许多研究区分"数学"话语和"非数学"话语，例如，社会文化理论侧重于日常和科学术语，交流认知理论对识别的信息和数学化语句之间进行区分。其他个别文章区分了，如非形式化的和形式化的数学语言（Setati & Adler，2000）、数学的和非数学的话语（Setati，2005）以及话语实践和数学实践（Enyedy，2003）。因此，有不同的想法和概念来尝试区分不同类型的话语。然而，也有几篇文章主张我们需要理解和借鉴数学和日常话语这两者来进行数学的交流。米尼（2005）使用系统功能语言学分析了两位数学家通过电子邮件所讨论的内容，巴维尔（2013a）使用话语心理学分析了三位数学家在没有剧本的广播直播节目中的讨论。这两位作者都认为，在审视数学教育情境时有必要考虑数学话语和日常话语。然而，在本领域中很少有理论或概念通过区分（或不区分）这些话语类型来讨论我们的得与失。

在本领域中优先考虑的关系类型几乎都是师生或生

research reviewed here. We think that deeper theoretical work could help mathematics education researchers trace the genealogy of their work to make transparent how ideas change over time, and it could help the field better understand theoretically justifiable ways to put various theories and concepts together.

Finally, we emphasize the fact that, although we have made progress in some aspects of methodology, one methodological area that has been virtually ignored is the positionality of the researchers involved in the work. In almost all the articles we reviewed, there were few statements about who the researcher was, what experiences the researchers had that informed their reading of the data, and so on. Some counterexamples to this claim include, for example, Esmonde et al. (2009), Esmonde and Langer-Osuna (2013), Heyd-Metzuyanim (2013), Mesa and Chang (2010), and Roberts (1998). Although there are clear examples in other content areas of how researchers can consider their positionality (e.g., Milner, 2007; Peshkin, 1988, 2001), little of this is done in mathematics education research and in this body of work, in particular. We appeal to journal editors and reviewers, more specifically, to require such statements in research. Given that many of these theories deem context as central to the interpretation of communication and recognize that one's socialization in various discourse practices shapes one's interpretations of one's experiences (including research), this is an area in need of attention. Attention to positionality is especially important in work that attends to issues of diversity, equity, and power (Foote & Bartell, 2011), as many of the articles in our set did.

Reflections on heritage analyses. This chapter focuses on mathematics classroom discourse, and consequently we first consider the question of how "discourse" is employed in this work. We then discuss some comparisons of key ideas and make several points about how Gee's building tasks were highlighted (or not) in various heritages and subheritages. It should be noted that we are not suggesting that all of Gee's building tasks should be clearly present in the research based on the different heritages and subheritages. Rather, it is interesting to see which of Gee's building tasks are more prominent when a particular heritage or subheritage is used.

The idea of "discourse" and the term itself plays a range of roles in this research. This observation has led us to talk about research in which *discourse is a lens with which to understand some other phenomenon* (e.g., mathematical understandings, beliefs) versus *discourse as the object of inquiry*. For example, in much of the sociocultural work, discourse is used to investigate the learning of mathematics. In contrast, under a Gricean umbrella, discourse is the object of inquiry, as authors try to understand specific ways of using language within the teaching and learning of mathematics.

We acknowledge that not all authors even use the term "discourse" (nor define it when they do use it, a finding previously discussed in Ryve, 2011), and that the key constructs offered by these various heritages vary immensely. This may be a result of each of the underlying heritages having a different place for "discourse" in the collection or structure of their key ideas and concepts. In Table 27.12, we include some of the key ideas authors in mathematics education highlight as they draw on these heritages and subheritages. We emphasize the point that the decisions about which ideas to bring to mathematics education research from these heritage theories is not all encompassing.

By drawing attention to these, we can identify some related ideas that appear and, from this, highlight the potential value in making more Connections between and among ideas to compare/contrast various d/Discourses[4] in the mathematics classroom and beyond. We highlight one particular example here and suggest that there are others that would be fruitful for further exploration. There is a range of work in this chapter that distinguishes "mathematical" discourses from "nonmathematical" discourses. For example, sociocultural theories have some focus on everyday and scientific terminology and commognition differentiates between identifying messages and mathematizing statements. Other individual articles distinguished, for instance, informal language and formal mathematical language (Setati & Adler, 2000), mathematical and nonmathematical Discourses (Setati, 2005), and discourse practices and mathematical practices (Enyedy, 2003). Thus, there are varying ideas and conceptualizations that try to separate these types of discourse. A couple of pieces, however, contend that we need to comprehend and draw on *both* mathematical and everyday discourse to communicate about mathematics. Meaney (2005) used SFL to analyze the email discussion of two mathematicians, and Barwell (2013a) used discursive psychology to analyze a discussion amongst three mathematicians in the context of a live, unscripted radio show. Both of these authors argue that it is necessary to consider mathematical *and* everyday discourses in our examinations of mathematics education contexts. Across this line of work, however, there has been little theoretical and conceptual consideration of what we gain and lose by separating out these kinds of discourse (or not).

The kinds of Relationships that took precedence in this work were almost all relationships between teachers

表27.12　数学教育研究所用关键构念示例

流派和子流派	数学教育研究所用关键构念示例
社会文化理论	最近发展区、内化、中介活动、文化发展的一般规律、对话的话语、言语类型、复述、标记动作、篇章衔接、提问模式、探索性谈话、现实脚本、主体间性、谈话的时间维度、日常和科学术语、讨论、数学交谈、互动、对话、参与和叙述
话语心理学	交流认知、交流认知冲突、客体化和具体化、焦点分析、职业分析、目的性分析、识别信息、数学化语句、仪式性和探索性参与（斯法德） 社会实践、话语实践、社会行动、修辞（巴维尔）
定位	位置/定位、言语行为、故事情节、假定世界、倾向、互动、对话、使用中的语言、参与结构、话语过程
系统功能语言学	数学语义、人际的、概念的、文本的元功能、类型、文本
图尔敏论证	论证（分析和实质论证等）、结论、数据、正当理由、主张、逻辑支援、模态限定词、反驳
浮现观点	社会数学规范、社会规范、课堂数学实践
格赖斯门派	模糊限定语、不确定性用语、类型、代词、面子、手势、礼貌理论、言语和言语行为力量
交际社会语言学	复述、参与者框架、模拟、诗歌结构、重复、并行性、重点互动、立足点
概念重建	交互风格、社会身份、文本、数学文字题类型、批判性语言意识、自言自语、批判意识、虚构、暴力/权力、反驳、描述性语言、水平话语、垂直话语、表征/符号/标识
线索型	隐喻、概念隐喻、数学语言、对象、符号与概念、修辞结构、交互阅读理论、建设性响应阅读、写作结构

生关系，学生和数学之间的关系则较少受到关注。鉴于一系列研究表明，许多学生对数学感到不满且对数学抱有可能会起反作用的信念，因此，我们可以从进一步考虑学生与数学之间的关系中获益良多。此外，在大学层面或校外的教与学环境中对人与数学之间关系的关注程度更低。

作为一个关于身份的更广泛的陈述，我们注意到对这一概念的关注主要出现在定位和概念重建子流派中，在较小程度上，也出现在交流认知子流派中。公平问题在很多这类研究中都特别重要，因而，在定位和概念重建的研究中往往强调身份以及政治的构建任务并不奇怪，例如，在几乎所有的定位子流派下的文章中，作者都使用"权力"和"授权"两个词来描述学生在小组中的互动或教师与学生之间的互动。概念重建流派的作者提供了一组更广泛的政治导向的术语，并不仅仅是在互动的关系上，还在它对理论和数学本身的批判上。另一个将政治放到重要位置的流派是借鉴了系统功能语言学的研究，在这些研究中数学语义的流畅性被视为每个学生都需要获得的重要社会资源，在将系统功能语言学与批判性话语分析和定位理论等其他观点结合使用并利用哈利迪与伯恩斯坦的联系时，这种关注尤其明显。

我们回顾的所有文章都聚焦在数学和数学教育，而且对符号系统和知识构建任务的审视表明这些作者对于"数学"的理解和他们明确阐述其认识论或哲学立场的程度上存在着有趣的偏差范围。例如，有些作者没有阐述他们对"数学"的观点或标准，但因为他们写的是关于学校数学的想法，所以我们假定他们的观点源自柏拉图主义、形式主义或相关思想。我们也想知道，这种表达的缺失是否表明这些观点在该领域已经成为惯例，因为虽然他们并没有提及但已经这么做了。随着在建构主义学习理论的基础上发展出不同的数学教学方法，浮现观点逐渐凸显了出来，然而，除最初的文章以外，这一点

TABLE 27.12. Sampling of Key Constructs Used in Mathematics Education Research

Heritages and subheritages	Examples of key constructs used in mathematics education research
Sociocultural theory	Zone of proximal development, internalization, mediating activity, genetic law of cultural development, dialogic discourse, speech genre, revoicing, labeling moves, textual cohesion, questioning patterns, exploratory talk, real-world scripts, intersubjectivity, the temporal dimension of talk, everyday and scientific terminology, discussion, math talk, interaction, dialogue, participation, and narratives
Discursive psychology	Commognition, commognitive conflict, objectification and reification, focal analysis, preoccupational analysis, intentional analysis, identifying messages, mathematizing statements, ritual and explorative participation (Sfard)
	Social practice, discursive practices, social action, rhetoric (Barwell)
Positioning	Position/positioning, speech act, storyline, figured world, disposition, interaction, conversation, language in use, participation structures, discursive processes
Systemic functional linguistics	Mathematics register, interpersonal, ideational, textual metafunctions, genre, text
Toulmin's argumentation	Argument (analytic, substantial, etc.), conclusion, data, warrant, claim, backing, modal qualifiers, rebuttal
Emergent	Sociomathematical norms, social norms, classroom mathematical practices
A Gricean umbrella	Hedges, uncertainty devices, genre, pronouns, face, gestures, politeness theory, locutionary and illocutionary force
Interactional sociolinguistics	Revoicing, participant frameworks, animation, poetic structure, repetition, parallelism, focusing interactions, footings
Reconceptualists	Interactional style, social identities, texts, math word problem genre, critical language awareness, monoglossic discourse, critical consciousness, myth, symbolic violence/power, counternarrative, language of description, horizontal discourse, vertical discourse, signified/signifier/sign
Threads	Metaphor, conceptual metaphor, mathematical language, object, symbol and concept, rhetorical structure, transactional reading theory, constructively responsive reading, writing structures

and students or students with other students. Much less attention has been given to the relationships between students and mathematics. Given the range of research that shows that many students are disaffected from mathematics and have beliefs about mathematics that are potentially counterproductive, we have much to gain from considering further the relationships between students and mathematics. Additionally, there was even less attention at the university level or teaching and learning environments outside school about the relationships between and among people and mathematics.

As a broader statement about Identity, we note that the majority of attention to this idea has appeared in the positioning and reconceptualist subheritages and, to a lesser extent, occurred in the work within the commognition subheritage. In many of these pieces, equity issues are particularly central. Therefore, it is not surprising that the Identity and Politics building tasks often get highlighted together in positioning and reconceptualist work. In almost all the articles in the positioning subheritage, for example, authors use the word "power" and "empowered" in relationship to interactions that involve students in small groups or interactions between teachers and students. The authors in the reconceptualist heritage offer an even broader set of politically oriented terms, not just in relationship to interactions but also in its critique of theories and of mathematics itself. The other primary heritage that had articles that foregrounded Politics was the work that drew on SFL where fluency in the mathematics register was seen as an important social good to which every student needed access. This focus was particularly apparent when SFL was used in combination with other perspectives, like critical discourse analysis and positioning theory, and made use of Halliday's connections to Bernstein.

All the articles we reviewed focus on mathematics and mathematics education, and examining the Signs Systems and Knowledge building task illuminated an interesting range of variation in what the authors took "mathematics" to be and the extent to which they explicitly articulated their epistemological or philosophical stance. There were authors, for example, who did not articulate their perspectives or criteria for "mathematics," but because they wrote about school mathematics ideas, we assumed their views were grounded in Platonist, formalist, or related views. We also wonder whether such a lack of articulation pointed toward these views being taken as the status quo in this research because they went without saying. The emergent perspective came into prominence with the move toward building on constructivist learning theory to develop different approaches to the teaching of mathematics. After initial articles, however, this was

很少再被明确提及，而是在诸如问题解决和探究的文献中推断得出。有些子流派的确明确表明了他们的认识论立场，话语心理学研究对了解数学和做数学的含义采取了明确的相对主义观点，许多其他子流派，包括大部分社会文化研究和一些概念重建、定位和系统功能语言学的文章，都指出数学是一种社会实践。后三组中的许多研究的关注点不仅包括"内容"，还包括存在方式、人际关系、涉及的权力动态以及与所描述的集体或实践相关的历史。浮现观点和社会文化流派的一些作者似乎依赖于准经验主义的数学观，因为前者借鉴了理查兹（1991）的工作并对"学校数学"和"探究数学"（Cobb, Wood等，1992）进行了区分，而后者引用了拉卡托斯的观点。在大多数情况下，概念重构主义流派在各种流派中对学校数学现状的狭隘观点是最具批判性的。

关注活动的作者让我们看到，一些子流派似乎在早期作品中证明并解释了他们的想法，但随着时间的推移，这些想法虽然被使用，但不再予以证明或解释。然而，其他子流派则使用了大量篇幅来证明其方法的合理性。我们想知道，前一种情况是否是在该领域中，思想变成"常识"的一个标志（Edwards & Mercer, 1987），它是否指出了思想或理论的现状，如果它突出了作者对新的或较少采用的观点或其他东西不信任的状态。例如在引入并解释了"社会数学规范"之后，许多不同的作者都使用了这个思想，但几乎没有阐明这样一个事实，即它是从一个特定的理论框架（布鲁默的符号互动论）发展出来的，并被修改为用于浮现子流派（它也不总是包括在文章的基础之中）。这让我们想知道，那些不是原始研究团队成员的研究人员是否是以与原始研究团队相同的方式使用了这个思想。我们也思考失去基础理论是否会导致后续工作把最初提出的想法"撵出去"；一个脱离其基础理论的概念可能会阻止理论在进一步的情境中得到发展。在后一种情况下，文章使用了大量篇幅来论证他们的方法（如系统功能语言学和交流认知），并花费了大量笔墨来解释和证明这些方法，这种理由可能表明该领域对这些方法不太熟悉，或者可能表明作者认为这些观点还未能被接受，我们把这些理论表达中的扩展、缩小和停滞看作可能预示着某些理论框架的生命周期。然而，我们警告不要随波逐流地拾起某些流行的概念，而丢弃其广阔的理论网络。

除了提供一个像文献综述这样的基础论证，我们注意到另一个频繁的活动是依赖于政策文件，如全美数学教师联合会的标准（2000），以证明研究在美国和其他地方研究的合理性。本文回顾的著作中，许多作者也强调了特定的学习观（比如探究性学习）。许多论文依赖于这样的政策文件，但几乎没有一篇论文试图确定基于讨论或丰富的对话教学的有用性在更大范围内是否确实有效。我们强调审读少量文本的重要性，就像我们做这个文献回顾时一样，因为这样的关注有助于理解数学课堂话语的细微差别以及为什么某些学生小组可能不如其他小组成功。不过，在其他内容领域进行的其他类型的研究也可以提供信息，比如，尼斯特兰德、吴、盖莫兰、泽塞和隆（2003）使用事件历史分析法来考察高中英语和社会研究课堂中的话语过程。他们分析了来自25所初中和高中200多个教室的数百份观察数据，关注话语是如何随着时间的推移而展开的，并使用了巴赫金对独白和对话式话语的区分。他们的研究结果表明，"真正的教师提问、理解和学生提问就像是对话式的投标，其中学生提问显示出特别大的作用"（第137页，字体突出为原文所加）。这项研究在理论上和方法上都很严谨，能够揭示各种话语行为对后续话语模式的影响。如此大规模的、证实或否定我们假设的工作还未在我们回顾的文献中尝试过，也许是因为我们所回顾的文献中的大部分文章除了关注话语的特定特征以外，还关注为什么这些特征似乎有助于特定的结果，如测试结果或协商的意义。

此外，尽管学校数学依然充当筛选的工具，但是明确质疑学校数学现状的文章却很少。我们知道政策文件对于研究的价值，把文章中提及探究性学习看做是反映了数学教育写作的主流方式。也就是说，如果你知道哪些想法是该领域内可以接受的，那么这些想法就更有可能被视为是有充分理由或解释的。然而，我们想知道的是，这样的举动是否限制了这一工作，特别是当许多流派和子流派用于数学教育之外的话语研究，卓有成效地指出当前社会科学和主流教育政策思潮具有局限性的时候。

rarely articulated explicitly but instead could be inferred from references to, for example, problem solving and inquiry. Some subheritages did make their epistemological stances explicit. The discursive psychology work took an explicit relativist view on what it means to know and do mathematics. Many other subheritages, including much of the sociocultural work and some of the articles in the reconceptualist, positioning, and SFL work stated that mathematics was a social practice. Much of the work in these latter three groups included attention to the "content" but also to ways of being, the interpersonal, power dynamics involved, and history associated with the communities or the practice described. The emergent perspective and some authors in the sociocultural heritage seemed to rely on a quasi-empiricist view of mathematics, as they drew on the work of Richards (1991) and distinguished between "school math" and "inquiry math" (Cobb, Wood et al., 1992) in the former and some authors, in the latter, cited Lakatos. For the most part, reconceptualists were the most critical of any of our heritages of narrow views of status quo school mathematics.

Attending to the Activities of the authors allowed us to see that some subheritages seemed to justify and explain their ideas in early pieces but, over time, the ideas were used but no longer justified or explained. Other subheritages, however, used a lot of space justifying their approaches. We wonder whether the former case was an indicator of ideas becoming "common knowledge" (Edwards & Mercer, 1987) in the field, whether this pointed toward status quo ideas or theories, if it highlighted a state of distrust by authors of newer or less used perspectives, or something else. For example, after "sociomathematical norms" was introduced and explained, the idea was used by many different authors but with little articulation of the fact that this had grown out of a particular theoretical framework (Blumer's symbolic interactionism) and was modified for use in the emergent subheritage (which was also not always included in the grounding of the articles). This made us wonder whether researchers who were not part of the original research team used the idea in the same way as the original research team. We also ponder whether losing grounding theories might result in successive works "flattening out" the ideas originally presented; a concept detached of its grounding theory may keep the theory from evolving when further contexts are explored. In the latter case in which articles used a lot of space justifying their approaches (e.g., SFL and commognition), articles spent a great deal of text explaining and justifying the approaches. This justification may have been an indicator that the field was less familiar with these approaches

or it may have indicated that the authors felt these views have not yet been readily accepted. We see these expansions, shrinkages, and stagnations in the articulation of theory as possibly indicative of the life cycles of certain theoretical frameworks. We warn, however, against "trends" that pick up certain popular concepts and leave behind their broader theoretical network.

Another frequent Activity we noticed was a reliance on policy documents like the National Council of Teachers of Mathematics Standards (2000) to justify research in the United States and elsewhere, in addition to providing a rationale based on, for example, a review of the literature. There was an emphasis on particular views of learning (like inquiry-based), too, by many of the authors in the work reviewed here. Many papers relied on policy documents like these but hardly any of them attempted to determine if the claims for the usefulness of discussion-based or rich discursive instruction were in fact valid on a larger scale. We emphasize the importance of doing the kinds of close reads of smaller sets of texts, as much of this literature we reviewed does, because such a focus can help the field understand the nuances of mathematics classroom discourse and why some groups of students may not be as successful as others. Yet, additional kinds of work being done in other content areas could be informative, too. Nystrand, Wu, Gamoran, Zeiser, and Long (2003), for example, used an *event-history* analysis to investigate discourse processes in high school English and social studies classrooms. They analyzed data from hundreds of observations of more than 200 classrooms in 25 middle and high schools, attending to how the discourse unfolded over time and using Bakhtin's distinctions between monologic and dialogic discourse. Their results showed that "authentic teacher questions, uptake, and student questions function as *dialogic bids* with student questions showing an especially large effect" (p. 137, italics in original). This work was theoretically and methodologically rigorous and was able to show the impact of various discourse moves on subsequent discourse patterns. Such large-scale work to confirm or disconfirm our assumptions has not been attempted in the literature we have reviewed, perhaps because much of the work in the set of articles we have reviewed has focused not just on specific features of discourse but on *why* the features seem to have contributed to specific outcomes, such as test results or negotiated meaning.

Additionally, the number of articles in which authors called the status quo school mathematics explicitly into question was few, even though school mathematics has continued to be a filtering device. We see the use of policy documents as justification for research and references

注释

1. 我们认识到，在开始用布鲁默的符号互动论拓展其工作之前，科布及其同事的开创性工作并未使用话语框架。因此，由于布鲁默对社交互动的关注，我们在这里包含了浮现观点。

2. 巴维尔还撰写过其他文章，但其中很多文章关注英语作为附加语言者（EAL）的学生。因为有一章专门讨论了这个领域，所以本章并未对其进行讨论。

3. 然而，我们认识到，这句话受到以下事实的影响，即我们没有包括那些涉及符号学、手势和其他相关理论的文章，因为这些观点包含在这套书的另一章中。

4. 关于话语的定义，请参见社会文化子流派一节。

to inquiry-based learning in articles as reflecting mainstream ways of writing in mathematics education; that is, if you know which ideas are acceptable domains in the field, it is more likely to be seen as sufficient for rationale or explication. We wonder, however, whether such moves constrain this line of work, especially when many of the heritages and subheritages used for studying discourse *outside* mathematics education have been productive in pointing to the limitations of current ideas in the social sciences and in mainstream educational policy.

Notes

1. We recognize that Cobb and colleagues' original work did not draw on discursive frameworks until they began to augment that work with Blumer's symbolic interactionism. Thus, we include the emergent perspective here because of Blumer's focus on social interaction.

2. Barwell has authored additional articles, but many of them focused on students for whom English was an additional language (EAL). Because there is a chapter focused on this area, we did not include discussion of these articles here.

3. We recognize, however, that this statement is impacted by the fact that we did not include articles that drew on semiotics nor gestures and other related theories because these perspectives are included in another chapter in this volume.

4. See the section on the sociocultural subheritage for a definition of Discourse.

References

Abrahamson, D., Gutiérrez, J. F., & Baddorf, A. K. (2012). Try to see it my way: The discursive function of idiosyncratic mathematical metaphor. *Mathematical Thinking and Learning, 14*(1), 55–80.

Acevedo, J. I. (2008). Fenómenos relacionados con el uso de metáforas en el discurso del profesor. El caso de las gráficas de funciones [Phenomena related with the use of metaphors in teachers' discourse] (Unpublished doctoral dissertation). University of Barcelona, Barcelona, Spain.

Adler, J. (1999). The dilemma of transparency: Seeing and seeing through talk in the mathematics classroom. *Journal for Research in Mathematics Education, 30*(1), 47–64.

Alibali, M. W., Nathan, M. J., Church, R. B., Wolfgam, M. S., Kim, S., & Knuth, E. J. (2013). Teachers' gestures and speech in mathematics lessons: Forging common ground by resolving trouble spots. *ZDM—The International Journal on Mathematics Education, 45*, 425–440.

Anderson, A., Anderson, J., & Shapiro, J. (2004). Mathematical discourse in shared storybook reading. *Journal for Research in Mathematics Education, 35*(1), 5–33.

Anderson, K. (2009). Applying positioning theory to the analysis of classroom interactions: Mediating micro-identities, macro-kinds, and ideologies of knowing. *Linguistics and Education, 20*, 291–310.

Atweh, B., Bleicher, R. E., & Cooper, T. J. (1998). The construction of the social context of mathematics classrooms: A sociolinguistic analysis. *Journal for Research in Mathematics Education, 29*(1), 63–82.

Austin, J. L., & Howson, A. G. (1979). Language and mathematical education. *Educational Studies in Mathematics, 10*(2), 161–197.

Bakhtin, M. M. (1981). *The dialogic imagination*. Austin: University of Texas Press.

Bakhtin, M. M. (1996). *Speech genres & other late essays*. Austin: University of Texas Press.

Barbosa, J. C. (2006). Mathematical modeling in classroom: A socio-critical and discursive perspective. *ZDM—The International Journal on Mathematics Education, 38*(3), 293–301.

Barwell, R. (2003). Discursive psychology and mathematics education: Possibilities and challenges. *ZDM—The International Journal on Mathematics, 35*(5), 201–207.

Barwell, R. (2005). Ambiguity in the mathematics classroom. *Language and Education, 19*(2), 117–125.

Barwell, R. (2009). Researchers' descriptions and the construction of mathematical thinking. *Educational Studies in Mathematics, 72*(2), 255–269.

Barwell, R. (2013a). The academic and the everyday in mathematicians' talk: The case of the hyper-bagel. *Language and Education, 27*(3), 207–222.

Barwell, R. (2013b). Discursive psychology as an alternative perspective on mathematics teacher knowledge. *ZDM—The International Journal on Mathematics Education, 45*(4), 595–606.

Barwell, R., Moschkovich, J. N., & Setati Phakeng, M. (2017). Language diversity and mathematics: Second language, bilingual, and multilingual learners. In J. Cai (Ed.), *Compendium for research in mathematics education* (pp. 583–606). Reston, VA: National Council of Teachers of Mathematics.

Baynham, M. (1996). Humour as an interpersonal resource in adult numeracy classes. *Language and Education, 10*(2–3), 187–200.

Bell, C. V., & Pape, S. J. (2012). Scaffolding students' opportunities to learn mathematics through social interactions. *Mathematics Education Research Journal, 24*, 423–445.

Ben-Yehuda, M., Lavy, I., Linchevski, L., & Sfard, A. (2005). Doing wrong with words: What bars students' access to arithmetical discourses. *Journal for Research in Mathematics Education, 36*(3), 176–247.

Bernstein, B. (1996). *Pedagogy, symbolic control and identity: Theory, research, critique*. London, United Kingdom: Taylor & Francis.

Bernstein, B. (1999). Vertical and horizontal discourse: An essay. *British Journal of Sociology of Education, 20*(2), 157–173.

Bill, V. L., Leer, M. N., Reams, L. E., & Resnick, L. B. (1992). From cupcakes to equations: The structure of discourse in a primary mathematics classroom. *Verbum, 1&2*, 63–85.

Bills, L. (2000). Politeness in teacher-student dialogue in mathematics: A socio-linguistic analysis. *For the Learning of Mathematics, 20*(2), 40–47.

Bishop, J. P. (2012). "She's always been the smart one. I've always been the dumb one": Identities in the mathematics classroom. *Journal for Research in Mathematics Education, 43*, 34–74.

Black, L. (2004). Teacher-pupil talk in whole-class discussions and processes of social positioning within the primary school classroom. *Language and Education, 18*(5), 347–360.

Borasi, R., & Siegel, M. (1990). Reading to learn mathematics: New connections, new questions, new challenges. *For the Learning of Mathematics, 10*(3), 9–16.

Borasi, R., Siegel, M., Fonzi, J., & Smith, C. F. (1998). Using transactional reading strategies to support sense-making and discussion in mathematics classrooms: An exploratory study. *Journal for Research in Mathematics Education, 29*(3), 275–305.

Bowers, J., Cobb, P., & McClain, K. (1999). The evolution of mathematical practices: A case study. *Cognition and Instruction, 17*(1), 25–66.

Brilliant-Mills, H. (1994). Becoming a mathematician: Building a situated definition of mathematics. *Linguistics and Education, 5,* 301–334.

Britton, J., Burgess, T., Martin, N., McLeod, A., & Rosen, H. (1973). *The development of writing abilities (11–18).* London, United Kingdom: Macmillan.

Brodie, K. (2007). Dialogue in mathematics classrooms: Beyond question-and-answer methods. *Pythagoras, 66,* 3–13.

Bromme, R., & Steinbring, H. (1994). Interactive development of subject matter in the mathematics classroom. *Educational Studies in Mathematics, 27*(3), 217–248.

Bronckart, J. P. (1985). *Le fonctionnement des discours.* [The operation of the speeches]. Paris, France: Delachaux et Niestlé.

Brown, J. S., Collins, A., & Duguid, P. (1989). Situated cognition and the culture of learning. *Educational Researcher, 18*(1), 32–42.

Brown, T. (1994). Creating and knowing mathematics through language and experience. *Educational Studies in Mathematics, 27,* 79–100.

Brown, T., & Heywood, D. (2011). Geometry, subjectivity and the seduction of language: The regulation of spatial perception. *Educational Studies in Mathematics, 77,* 351–367.

Burton, L., & Morgan, C. (2000). Mathematicians writing. *Journal for Research in Mathematics Education, 31*(4), 429–453.

Carlsen, M. (2009). Reasoning with paper and pencil: The role of inscriptions in student learning of geometry. *Mathematics Education Research Journal, 21*(1), 54–84.

Carraher, T., Carraher, D., & Schliemann, A. (1987). Written and oral mathematics. *Journal for Research in Mathematics Education, 18*(2), 83–97.

Carreira, S. (2001). Where there's a model, there's a metaphor: Metaphorical thinking in students' understanding of a mathematical model. *Mathematical Thinking and Learning, 3*(4), 261–287.

Caspi, S., & Sfard, A. (2012). Spontaneous meta-arithmetic as a first step toward school algebra. *International Journal of Educational Research, 51–52,* 45–65.

Chamberlin, M. T. (2005). Teachers' discussion of students' thinking: Meeting the challenge of attending to students' thinking. *Journal of Mathematics Teacher Education, 8*(2), 141–170.

Chapman, A. (1995). Intertextuality in school mathematics: The case of functions. *Linguistics and Education, 7,* 243–262.

Chapman, A. (1997). Towards a model of language shifts in mathematics learning. *Mathematics Education Research Journal, 9*(2), 152–173.

Civil, M., & Planas, N. (2004). Participation in the mathematics classroom: Does every student have a voice? *For the Learning of Mathematics, 24*(1), 7–12.

Cobb, P. (1999). Individual and collective mathematical development: The case of statistical data analysis. *Mathematical Thinking and Learning, 1*(1), 5–43.

Cobb, P., Wood, T., Yackel, E., & McNeal, B. (1992). Characteristics of classroom mathematics traditions: An interactional analysis. *American Educational Research Journal, 29*(3), 573–604.

Cobb, P., Wood, T., Yackel, E., Nicholls, J., Wheatley, G., Trigatti, B., & Perlwitz, M. (1991). Assessment of a problem-centered second-grade mathematics project. *Journal for Research in Mathematics Education, 22*(1), 3–29.

Cobb, P., & Yackel, E. (1996). Constructivist, emergent, and sociocultural perspectives in the context of developmental research. *Educational Psychologist, 31*(3–4), 175–190.

Cobb, P., Yackel, E., & Wood, T. (1992). Interaction and learning in mathematics classroom situations. *Educational Studies in Mathematics, 23,* 99–122.

Conner, A., Singletary, L. M., Smith, R. C., Wagner, P. A., & Francisco, R. T. (2014). Teacher support for collective argumentation: A framework for examining how teachers support students' engagement in mathematical activities. *Educational Studies in Mathematics, 86*(3), 401–429.

Crawford, K. (1990). Language and technology in classroom settings for students from non-technological cultures. *For the Learning of Mathematics, 10*(1), 2–6.

Crespo, S. (2006). Elementary teacher talk in mathematics study groups. *Journal of Mathematics Teacher Education, 63*(1), 91–102. doi:10.1007/s10857-006-9006-8

Cuevas, G. J. (1984). Mathematics learning in English as a second language. *Journal for Research in Mathematics Education, 15*(2), 134–144.

Davies, B., & Harré, R. (1990). Positioning: The discursive production of selves. *Journal for the Theory of Social Behaviour, 20*(1), 43–63.

Dawe, L. (1983). Bilingualism and mathematical reasoning in English as a second language. *Educational Studies in Mathematics, 14,* 325–353.

de Freitas, E., Wagner, D., Esmonde, I., Knipping, C., Lunney Borden, L., & Reid, D. (2012). Discursive authority and sociocultural positioning in the mathematics classroom: New directions for teacher professional development. *Canadian Journal of Science, Mathematics and Technology Education, 12*(2), 137–159. doi:10.1080/14926156.2012.679994

Dowling, P. (1996). A sociological analysis of school mathematics texts. *Educational Studies in Mathematics, 31,* 389–415.

Edwards, D. (1993). But what do children really think?: Discourse analysis and conceptual content in children's talk. *Cognition and Instruction, 11*(3&4) 207–225.

Edwards, D. (1997). *Discourse and cognition.* London, United Kingdom: Sage.

Edwards, D., & Mercer, N. (1987). *Common knowledge.* London, United Kingdom: Methuen.

Edwards, D., & Potter, J. (1992). *Discursive psychology.* London, United Kingdom: Sage.

Ellerton, N. F., & Clarkson, P. C. (1996). Language factors in mathematics teaching and learning. In A. J. Bishop, M. K. Clements, C. Keitel, J. Kilpatrick, & C. Laborde (Eds.),

International handbook of mathematics education (pp. 987–1033). Dordrecht: Kluwer Academic.

Empson, S. (2003). Low-performing students and teaching fractions for understanding: An interactional analysis. *Journal for Research in Mathematics Education, 34*(4), 305–343.

Enyedy, N. (2003). Knowledge construction and collective practice: At the intersection of learning, talk, and social configurations in a computer-mediated mathematics classroom. *Journal of the Learning Sciences, 12*(3), 361–407.

Erickson, F. (1982). Classroom discourse as improvisation. In L. C. Wilkinson (Ed.), *Communicating in the classroom* (pp. 153–181). New York: Academic Press.

Esmonde, I. (2009a). Explanations in mathematics classrooms: A discourse analysis. *Canadian Journal of Science, Mathematics and Technology Education, 9*(2), 86–99.

Esmonde, I. (2009b). Mathematics learning in groups: Analyzing equity in two cooperative activity structures. *Journal of the Learning Sciences, 18*, 247–284.

Esmonde, I., Brodie, K., Dookie, L., & Takeuchi, M. (2009). Social identities and opportunities to learn: Student perspectives on group work in an urban mathematics classroom. *Journal of Urban Mathematics Education, 2*(2), 18–45.

Esmonde, I., & Langer-Osuna, J. M. (2013). Power in numbers: Student participation in mathematical discussions in heterogeneous spaces. *Journal for Research in Mathematics Education, 44*(1), 288–315.

Evans, J. (1999). Building bridges: Reflections on the problem of transfer of learning in mathematics. *Educational Studies in Mathematics, 39*, 23–44.

Evans, J., Morgan, C., & Tsatsaroni, A. (2006). Discursive positioning and emotion in school mathematics practices. *Educational Studies in Mathematics, 63*, 209–226.

Fairclough, N. (1992). *Discourse and social change* (Vol. 73). Cambridge, United Kingdom: Polity Press.

Font, V., Bolite, J., & Acevedo, J. (2010). Metaphors in mathematics classrooms: Analyzing the dynamic process of teaching and learning of graph functions. *Educational Studies in Mathematics, 75*(2), 131–152.

Font, V., Godino, J. D., Planas, N., & Acevedo, J. I. (2010). The object metaphor and synecdoche in mathematics classroom discourse. *For the Learning of Mathematics, 30*(1), 15–19.

Foote, M. Q., & Bartell, T. G. (2011). Pathways to equity in mathematics education: How life experiences impact researcher positionality. *Educational Studies in Mathematics, 78*, 45–68.

Forman, E. (2003). A sociocultural approach to mathematics reform: Speaking, inscribing and doing mathematics within communities of practice. In J. Kilpatrick, G. Martin, & D. Schifter (Eds.), *A research companion to Principles and Standards for School Mathematics* (pp. 333–352). Reston, VA: National Council of Teachers of Mathematics.

Forman, E., & Ansell, E. (2002). Orchestrating the multiple voices and inscriptions of a mathematics classroom. *Journal of the Learning Sciences, 11*(2), 251–274.

Forman, E., Larreamendy-Joerns, J., Stein, M. K., & Brown, C. A. (1998). "You're going to want to find out which and prove it": Collective argumentation in a mathematics classroom. *Learning and Instruction 8*(6), 527–548.

Forrester, M., & Pike, C. (1998). Learning to estimate in the mathematics classroom: A conversation-analytic approach. *Journal for Research in Mathematics Education, 29*(3), 334–356.

Freeman, J. B. (2005). Systematizing Toulmin's warrants: An epistemic approach. *Argumentation, 19*(3), 331–346.

Garfinkel, H. (1967). *Studies in ethnomethodology*. Englewood Cliffs, NJ: Prentice Hall.

Gee, J. P. (1996). *Social linguistics and literacies: Ideology in discourses*. London, United Kingdom: Falmer Press.

Gee, J. P. (2011a). *An introduction to discourse analysis: Theory and method* (3rd ed.). London, United Kingdom: Routledge.

Gee, J. P. (2011b). *How to do discourse analysis: A toolkit*. London, United Kingdom: Routledge.

Gerofsky, S. (1996). A linguistic and narrative view of word problems in mathematics education. *For the Learning of Mathematics, 16*(2), 36–45.

Giannakoulias, E., Mastorides, E., Potari, D., & Zachariades, T. (2010). Studying teachers' mathematical argumentation in the context of refuting students' invalid claims. *The Journal of Mathematical Behavior, 29*(3), 160–168.

González, G., & DeJarnette, A. F. (2012). Agency in a geometry review lesson: A linguistic view on teacher and student division of labor. *Linguistics and Education, 23*, 182–199.

Goodwin, C. (1994). Professional vision. *American Anthropologist, 96*, 606–633.

Goos, M. (2004). Learning mathematics in a classroom community of inquiry. *Journal for Research in Mathematics Education, 35*(4), 258–291.

Goos, M., Galbraith, P., & Renshaw, P. (2002). Socially mediated metacognition: Creating collaborative zones of proximal development in small group problem solving. *Educational Studies in Mathematics, 49*, 193–223.

Gray, E. M., & Tall, D. O. (1994). Duality, ambiguity, and flexibility: A "proceptual" view of simple arithmetic. *Journal for Research in Mathematics Education, 25*(2), 116–140.

Gresalfi, M. S. (2009). Taking up opportunities to learn: Constructing dispositions in mathematics classrooms. *Journal of the Learning Sciences, 18*, 327–369.

Gresalfi, M. S., & Cobb, P. (2006). Cultivating students' discipline-specific dispositions as a critical goal for pedagogy and equity. *Pedagogies: An International Journal, 1*(1), 49–57.

Grice, H. P. (1975). Logic and conversation. In P. Cole & J. L. Morgan (Eds.), *Syntax and semantics: Speech acts* (Vol. 3, pp. 41–58). New York, NY: Academic Press.

Grumet, M. R. (1989). Generations: Reconceptualist curriculum theory and teacher education. *Journal of Teacher Education, 40*(1), 13–17.

Güçler, B. (2013). Examining the discourse on the limit concept in a beginning-level calculus classroom. *Educational Studies in Mathematics, 82*(3), 439–453.

Halliday, M. A. K. (1978). *Language as social semiotic*. London, United Kingdom: Edward Arnold.

Halliday, M. A. K., & Hasan, R. (1989). *Language, context, and text: Aspects of language in a social-semiotic perspective*. Geelong, Australia: Deakin University Press.

Han, Y., & Ginsburg, H. P. (2001). Chinese and English mathematics language: The relation between clarity and mathematics performance. *Mathematical Thinking and Learning, 3*(2&3), 201–220.

Hand, V. (2012). Seeing culture and power in mathematical learning: Toward a model of equitable instruction. *Educational Studies in Mathematics, 80*, 233–247.

Harré, R., & Gillett, G. (1994). *The discursive mind.* Thousand Oaks, CA: Sage.

Harré R., & Moghaddam, F. M. (Eds.). (2003). *The self and others: Positioning individuals and groups in personal, political, and cultural contexts.* Westport, CT: Praeger.

Harré, R., & Slocum, N. (2003). Disputes as complex social events. *Common Knowledge, 9*(1), 100–118.

Harré, R., & van Langenhove, L. (Eds.). (1999). *Positioning theory: Moral contexts of intentional action.* Oxford, United Kingdom: Blackwell.

Hartshorne, C., & Weiss, P. (1931). *Collected papers of Charles Sanders Peirce.* Cambridge, MA: Harvard University Press.

Herbel-Eisenmann, B. (2007). From intended curriculum to written curriculum: Examining the "voice" of a mathematics textbook. *Journal for Research in Mathematics Education, 38*(4), 344–369.

Herbel-Eisenmann, B., Drake, C., & Cirillo, M. (2009). "Muddying the clear waters": Teachers' take-up of the linguistic idea of revoicing. *Teaching and Teacher Education, 25*(2), 268–277.

Herbel-Eisenmann, B., & Otten, S. (2011). Mapping mathematics in classroom discourse. *Journal for Research in Mathematics Education, 42*(5), 451–485.

Herbel-Eisenmann, B., & Wagner, D. (2007). A framework for understanding the way a textbook may position the mathematics learner. *For the Learning of Mathematics, 27*(2), 8–14.

Herbel-Eisenmann, B., & Wagner, D. (2010). Appraising lexical bundles in mathematics classroom discourse: Obligation and choice. *Educational Studies in Mathematics, 75*, 43–63.

Herbel-Eisenmann, B., Wagner, D., & Cortes, V. (2010). Lexical bundle analysis in mathematics classroom discourse: The significance of stance. *Educational Studies in Mathematics, 75*, 23–42.

Herbel-Eisenmann, B., Wagner, D., Johnson, K. R., Suh, H., & Figueras, H. (2015). Positioning in mathematics education: Revelations on an imported theory. *Educational Studies in Mathematics, 89*(2), 185–204.

Heyd-Metzuyanim, E. (2013). The co-construction of learning difficulties in mathematics teacher–student interactions and their role in the development of a disabled mathematical identity. *Educational Studies in Mathematics, 83*(3), 341–368.

Heyd-Metzuyanim, E., & Sfard, A. (2012). Identity struggles in the mathematics classroom: On learning mathematics as an interplay of mathematizing and identifying. *International Journal of Educational Research, 51–52*, 128–145.

Hoadley, U. (2007). The reproduction of social class inequalities through mathematics pedagogies in South African primary schools. *Journal of Curriculum Studies, 39*(6), 679–706.

Hoadley, U., & Ensor, P. (2009). Teachers' social class, professional dispositions and pedagogic practice. *Teaching and Teacher Education, 25*, 876–886.

Holland, D. C., & Eisenhart, M. A. (1990). *Educated in romance: Women, achievement, and college culture.* Chicago, IL: University of Chicago Press.

Holland, D., Lachiotte, W., Jr., Skinner, D., & Cain, C. (1998). *Identity and agency in cultural worlds.* Cambridge, MA: Harvard University Press.

Hoyles, C. (1985). What is the point of group discussion in mathematics? *Educational Studies in Mathematics, 16*, 205–214.

Hufferd-Ackles, K., Fuson, K., & Sherin, M. G. (2004). Describing levels and components of a math-talk learning community. *Journal for Research in Mathematics Education, 35*(2), 81–116.

Hunter, R., & Anthony, G. (2011). Forging mathematical relationships in inquiry-based classrooms with Pasifika students. *Journal of Urban Mathematics Education, 4*, 98–119.

Hussain, M. A., Monaghan, J., & Threlfall, J. (2013). Teacher-student development in mathematics classrooms: Interrelated zones of free movement and promoted actions. *Educational Studies of Mathematics, 82*, 285–302.

Hutchins, E. (1995). *Cognition in the wild.* Cambridge, MA: MIT Press.

Inglis, M., Mejia-Ramos, J. P., & Simpson, A. (2007). Modelling mathematical argumentation: The importance of qualification. *Educational Studies in Mathematics, 66*(1), 3–21.

Jacobsen, E. (Ed.). (1975). Interactions between linguistics and mathematical education: Final report of the symposium sponsored by UNESCO, CEDO and ICMI, Nairobi, Kenya, September 1–11, 1974 (UNESCO Report No. ED-74/CONF.808). Paris, France: UNESCO.

Jankvist, U. (2011). Anchoring students' metaperspective discussions of history in mathematics. *Journal for Research in Mathematics Education, 42*(4), 346–385.

Jefferson, G. (Ed.). (1992). *Harvey Sacks: Lectures on conversation.* Oxford, United Kingdom: Blackwell.

Ju, M. K., & Kwon, O. N. (2007). Ways of talking and ways of positioning: Students' beliefs in an inquiry-oriented differential equations class. *The Journal of Mathematical Behavior, 26*(3), 267–280.

Jurow, A. (2005). Shifting engagements in figured worlds: Middle school mathematics students' participation in an architectural design project. *The Journal of the Learning Sciences, 14*(1), 35–67.

Kazemi, E., & Stipek, D. (2001). Promoting conceptual thinking in four upper-elementary mathematics classrooms. *The Elementary School Journal, 102*(1), 59–80.

Kieran, C. (2001). The mathematical discourse of 13-year-old partnered problem solving and its relation to the mathematics that emerges. *Educational Studies in Mathematics, 46*, 187–228.

Kim, D. J., Ferrini-Mundy, J., & Sfard, A. (2012). How does language impact the learning of mathematics? Comparison of English and Korean speaking university students' discourses on infinity. *International Journal of Educational Research, 51–52*, 86–108.

Kjeldsen, T. H., & Blomhøj, M. (2012). Beyond motivation: History as a method for learning meta-discursive rules in mathematics. *Educational Studies in Mathematics, 80*(3), 327–349.

Kotsopoulos, D. (2014). The case of Mitchell's cube: Interactive and reflexive positioning during collaborative learning in mathematics. *Mind, Culture, and Activity, 21*(1), 34–52.

Krummheuer, G. (1995). The ethnography of argumentation. In P. Cobb & H. Bauersfeld (Eds.), *The emergence of mathematical meaning: Interaction in classroom cultures* (pp. 229–269). Hillsdale, NJ: Lawrence Erlbaum.

Krummheuer, G. (2000). Studies of argumentation in primary mathematics education. *ZDM—The International Journal on Mathematics Education, 5*, 155–161.

Krummheuer, G. (2007). Argumentation and participation in the primary mathematics classroom: Two episodes and related theoretical abductions. *Journal of Mathematical Behavior, 26,* 60–82.

Lakoff, G., & Núñez, R. (1997). The metaphorical structure of mathematics: Sketching out cognitive foundations for a mind-based mathematics. In L. English (Ed.), *Mathematical reasoning: Analogies, metaphors, and images* (pp. 21–89). Mahwah, NJ: Lawrence Erlbaum Associates.

Lakoff, G., & Núñez, R. (2000). *Where mathematics comes from: How the embodied mind brings mathematics into being.* New York, NY: Basic Books.

Lampert, M. (1990). When the problem is not the question and the solution is not the answer: Mathematical knowing and teaching. *American Educational Research Journal, 27,* 29–63.

Lampert, M., & Cobb, P. (2003). Communication and language. *A research companion to Principles and Standards for School Mathematics* (pp. 237–249). Reston, VA: National Council of Teachers of Mathematics.

Langer-Osuna, J. M. (2011). How Brianna became bossy and Kofi came out smart: Understanding the trajectories of identity and engagement for two group leaders in a project-based mathematics classroom. *Canadian Journal of Science, Mathematics, and Technology Education, 11,* 207–225.

Langer-Osuna, J. M. (2014). From getting "fired" to becoming a collaborator: A case on the co-construction of identity and engagement in a project-based mathematics classroom. *Journal of the Learning Sciences, 24*(1), 53–92.

Langer-Osuna, J. M., & Esmonde, I. (2017). Identity in research on mathematics education. In J. Cai (Ed.), *Compendium for research in mathematics education* (pp. 637–648). Reston, VA: National Council of Teachers of Mathematics.

Lappan, G., Fey, J. T., Fitzgerald, W. M., Friel, S. N., & Phillips, E. D. (2006). *The Connected Mathematics Project.* Pearson Prentice Hall.

Lau, P., Singh, P., & Hwa, T. (2009). Constructing mathematics in an interactive classroom context. *Educational Studies in Mathematics, 72,* 307–324.

Lave, J., & Wenger, E. (1991). *Situated learning: Legitimate peripheral participation.* New York, NY: Cambridge University Press.

Lean, G. A., Clements, M. A., & Del Campo, G. (1990). Linguistic and pedagogical factors affecting children's understanding of arithmetic word problems: A comparative study. *Educational Studies in Mathematics, 21*(2), 165–191.

Lerman, S. (2001). Cultural, discursive psychology: A sociocultural approach to studying the teaching and learning of mathematics. *Educational Studies in Mathematics, 46,* 87–113.

LeRoux, K. (2008). A critical discourse analysis of a real-world problem in mathematics: Looking for signs of change. *Language and Education, 22*(5), 307–326.

Levenson, E., Tirosh, D., & Tsamir, P. (2009). Students' perceived sociomathematical norms: The missing paradigm. *The Journal of Mathematical Behavior, 28*(2), 171–187.

Lobato, J. (2012). "Noticing" as an alternative transfer of learning process. *Journal of the Learning Sciences, 21*(3), 433–482.

Lobato, J., Ellis, A. B., & Munoz, R. (2003). How "focusing phenomena" in the instructional environment support individual students' generalizations. *Mathematical Thinking and Learning, 5*(1), 1–36.

Males, L. M., Otten, S., & Herbel-Eisenmann, B. A. (2010). Challenges of critical colleagueship: Examining and reflecting on mathematics teacher study group interactions. *Journal of Mathematics Teacher Education, 13*(6), 459–471.

Marks, G., & Mousley, J. (1990). Mathematics education and genre: Dare we make the process writing mistake again? *Language and Education, 4*(2), 117–135.

Martin, D. B., Anderson, C. R., & Shah, N. (2017). Race and mathematics education. In J. Cai (Ed.), *Compendium for research in mathematics education* (pp. 607–636). Reston, VA: National Council of Teachers of Mathematics.

McClain, K. (2002). Teacher's and students' understanding: The role of tools and inscriptions in supporting effective communication. *Journal of the Learning Sciences, 11*(2–3), 217–249.

McCrone, S. S. (2005). The development of mathematical discussions: An investigation in a fifth-grade classroom. *Mathematical Thinking and Learning, 7*(2), 111–133.

Meaney, T. (2005). Mathematics as text. In A. Chronaki & I. M. Christiansen (Eds.), *Challenging perspectives in mathematics classroom communication* (pp. 109–141). Westport, CT: Information Age.

Meaney, T. (2006). Really that's probably about roughly what does down hesitancies and uncertainties in mathematics assessment interactions. *Language and Education, 20*(5), 374–390.

Meaney, T. (2007). Weighing up the influence of context on judgements of mathematical literacy. *International Journal of Science and Mathematics Education, 5*(4), 681–704.

Mejía-Ramos, J. P., & Inglis, M. (2011). Semantic contamination and mathematical proof: Can a non-proof prove? *The Journal of Mathematical Behavior, 30,* 19–29.

Mercer, N. (2008). The seeds of time: Why classroom dialogue needs a temporal analysis. *Journal of the Learning Sciences, 17*(1), 33–59.

Mercer, N., & Sams, C. (2006). Teaching children how to use language to solve maths problems. *Language and Education, 20*(6), 507–528.

Mesa, V., & Chang, P. (2010). The language of engagement in two highly interactive undergraduate mathematics classrooms. *Linguistics and Education, 21,* 83–100.

Middleton, J., Jansen, A., & Goldin, G. A. (2017). The complexities of mathematical engagement: Motivation, affect, and social interactions. In J. Cai (Ed.), *Compendium for research in mathematics education* (pp. 667–699). Reston, VA: National Council of Teachers of Mathematics.

Milner, H. R. (2007). Race, culture, and researcher positionality: Working through dangers seen, unseen, and unforeseen. *Educational Researcher, 36*(7), 388–400.

Monaghan, F. (1999). Judging a word by the company it keeps: The use of concordance software to explore aspects of the mathematics register. *Language and Education, 13*(1), 59–70.

Mordant, I. (1993). Psychodynamics of mathematics texts. *For the Learning of Mathematics, 13*(1), 20–23.

Morgan, C. (1996). "The language of mathematics": Towards a critical analysis of mathematics texts. *For the Learning of Mathematics, 16*(3), 2–10.

Morgan, C. (1998). *Writing mathematically: The discourse of investigation.* London, United Kingdom: Falmer.

Morgan, C. (2005). Word, definitions and concepts in discourses of mathematics, teaching and learning. *Language and Education, 19*(2), 102–116.

Morgan, C. (2006). What does social semiotics have to offer mathematics education research? *Educational Studies in Mathematics, 61*(1–2), 219–245.

Moschkovich, J. (2007). Examining mathematical discourse practices. *For the Learning of Mathematics, 27*(1), 24–30.

Moschkovich, J. N. (2008). "I went by twos, he went by one": Multiple interpretations of inscriptions as resources for mathematical discussions. *Journal of the Learning Sciences, 17*(4), 551–587.

Nachlieli, T., & Tabach, M. (2012). Growing mathematical objects in the classroom—The case of function. *International Journal of Educational Research, 51–52*, 10–27.

Nardi, E., Biza, I., & Zachariades, T. (2011). "Warrant" revisited: Integrating mathematics teachers' pedagogical and epistemological considerations into Toulmin's model for argumentation. *Educational Studies in Mathematics, 79*(2), 157–173.

Nathan, M. J., Eilam, B., & Kim, S. (2007). To disagree, we must also agree: How intersubjectivity structures and perpetuates discourse in a mathematics classroom. *Journal of the Learning Sciences, 16*(4), 523–563.

Nathan, M. J., Long, S. D., & Alibali, M. W. (2002). The symbol precedence view of mathematical development: A corpus analysis of the rhetorical structure of textbooks. *Discourse Processes, 33*(1), 1–21.

National Council of Teachers of Mathematics. (2000). *Principles and standards for school mathematics*. Reston, VA: Author.

Newton, J. A. (2012). Investigating the mathematical equivalence of written and enacted middle school standards-based curricula: Focus on rational numbers. *International Journal of Educational Research, 51–52*, 66–85.

Nystrand, M., Wu, L. L., Gamoran, A., Zeiser, S., & Long, D. A. (2003). Questions in time: Investigating the structure and dynamics of unfolding classroom discourse. *Discourse Processes, 35*(2), 135–198.

O'Connor, M. C. (2001). "Can any fraction be turned into a decimal?" A case study of mathematical group discussions. *Educational Studies in Mathematics, 46*, 143–185.

O'Connor, M. C., & Michaels, S. (1993). Aligning academic task and participation status through revoicing: Analysis of a classroom discourse strategy. *Anthropology & Education Quarterly, 24*(4), 318–335.

Ogden, C. K., & Richards, I. A. (1923). *The meaning of meaning*. New York, NY: Harcourt, Brace, and World.

O'Grady, J. (2010, January 10). Stephen Toulmin obituary. *The Guardian*. Retrieved from http://www.theguardian.com/theguardian/2010/jan/10/stephen-toulmin-obituary

O'Halloran, K. O. (2000). Classroom discourse in mathematics: A multisemiotic analysis. *Linguistics and Education, 10*(3), 359–388.

Olson, J. C., & Knott, L. (2013). When a problem is more than a teacher's question. *Educational Studies in Mathematics, 83*, 27–36.

Pedemonte, B. (2007). How can the relationship between argumentation and proof be analysed? *Educational Studies in Mathematics, 66*(1), 23–41.

Pedemonte, B. (2008). Argumentation and algebraic proof. *ZDM—The International Journal on Mathematics Education, 40*(3), 385–400.

Peirce, C. S. (1978). *Écrits sur le signe [Writings on the sign]* (G. Deledalle, Ed.). Paris, France: Éditions du Seuil.

Peshkin, A. (1988). In search of subjectivity. One's own. *Educational Researcher, 17*, 17–21.

Peshkin, A. (2001). Angles of vision: Enhancing perception in qualitative research. *Qualitative Inquiry, 7*, 238–253.

Pickering, A. (1995). *The mangle of practice: Time, agency, and science*. Chicago, IL: University of Chicago Press.

Pimm, D. (1987). *Speaking mathematically: Communication in mathematics classrooms*. London, United Kingdom: Routledge & Kegan Paul.

Pinar, W. F., & Grumet, M. R. (1982). Socratic caesura and the theory-practice relationship. *Theory Into Practice, 21*(1), 50–54.

Planas, N., & Gorgorió, N. (2004). Are different students expected to learn norms differently in the mathematics classroom? *Mathematics Education Research Journal, 16*(1), 19–40.

Pressley, M., & Afflerbach, P. (1995). *Verbal protocols of reading: The nature of constructively responsive reading*. Hillsdale, NJ: Lawrence Erlbaum.

Price, A. (2000). The role of real world scripts in the teaching and learning of addition. *ZDM—The International Journal on Mathematics Education, 32*(5) 131–137.

Radford, L., Arzarello, F., Edwards, E., & Sabena, C. (2017). The multimodal material mind: Embodiment in mathematics education. In J. Cai (Ed.), *Compendium for research in mathematics education* (pp. 700–721). Reston, VA: National Council of Teachers of Mathematics.

Rasmussen, C., & Kwon, O. N. (2007). An inquiry-oriented approach to undergraduate mathematics. *The Journal of Mathematical Behavior, 26*(3), 189–194.

Rasmussen, C., & Marrongelle, K. (2006). Pedagogical content tools: Integrating student reasoning and mathematics in instruction. *Journal for Research in Mathematics Education, 37*(5), 388–420.

Rezat, S. (2006). The structures of German mathematics textbooks. *ZDM—The International Journal on Mathematics Education, 38*(6), 482–487.

Richards, J. (1991). Mathematical discussions. In E. von Glasersfeld (Ed.), *Radical constructivism in mathematics education* (pp. 13–52), Dordrecht, The Netherlands: Kluwer.

Roberts, T. (1998). Mathematical registers in Aboriginal languages. *For the Learning of Mathematics, 18*(1), 10–16.

Robotti, E. (2012). Natural language as a tool for analyzing the proving process: The case of plane geometry proof. *Educational Studies in Mathematics, 80*(3), 433–450.

Rowland, T. (1992). Pointing with pronouns. *For the Learning of Mathematics, 12*(2), 44–48.

Rowland, T. (1995). Hedges in mathematical talk: Linguistic pointers to uncertainty. *Educational Studies in Mathematics, 29*(4), 327–353.

Rowland, T. (1999). Pronouns in mathematics talk: Power, vagueness and generalization. *For the Learning of Mathematics, 19*(2), 19–26.

Ryve, A. (2004). Can collaborative concept mapping create mathematically productive discourses? *Educational Studies in Mathematics, 56*(3), 157–177.

Ryve, A. (2006). Making explicit the analysis of students' mathematical discourses—Revisiting a newly developed methodological framework. *Educational Studies in Mathematics, 62*, 191–209.

Ryve, A. (2011). Discourse research in mathematics education: A critical evaluation of 108 journal articles. *Journal for Research in Mathematics Education, 42*(2), 167–199.

Ryve, A., Nilsson, P., & Pettersson, K. (2013). Analyzing effective communication in mathematics group work: The role of visual mediators and technical terms. *Educational Studies in Mathematics, 82*, 497–514.

Schleppegrell, M. J. (2007). The linguistic challenges of mathematics teaching and learning: A research review. *Reading and Writing Quarterly, 23*, 139–159.

Searle, J. R. (1979). *Expression and meaning: Studies in the theory of speech acts.* Cambridge, UK: Cambridge University Press.

Setati, M. (2005). Teaching mathematics in a primary multilingual classroom. *Journal for Research in Mathematics Education, 36*(5), 447–466.

Setati, M. (2008). Access to mathematics versus access to the language of power: The struggle in multilingual mathematics classrooms. *South African Journal of Education, 28*, 103–116.

Setati, M., & Adler, J. (2000). Between languages and discourses: Language practices in primary multilingual mathematics classrooms in South Africa. *Educational Studies in Mathematics, 43*(3), 243–269.

Sfard, A. (2000a). On reform movement and the limits of mathematical discourse. *Mathematical Thinking and Learning, 2*, 157–189.

Sfard, A. (2000b). Steering (dis)course between metaphors and rigor: Using focal analysis to investigate an emergence of mathematical objects. *Journal for Research in Mathematics Education, 31*(3), 296–327.

Sfard, A. (2001). There is more to discourse than meets the ears: Looking at thinking as communicating to learn more about mathematical learning. *Educational Studies in Mathematics, 46*, 13–57.

Sfard, A. (2002). The interplay of intimations and implementations: Generating new discourse with new symbolic tools. *Journal of the Learning Sciences, 11*(2–3), 319–357.

Sfard, A. (2007). When the rules of discourse change, but nobody tells you: Making sense of mathematics learning from a commogitive standpoint. *The Journal of the Learning Sciences, 16*, 565–613.

Sfard, A., & Kieran, C. (2001). Cognition as communication: Rethinking learning-by-talking through multi-faceted analysis of students' mathematical interactions. *Mind, Culture, and Activity, 8*(1), 42–76.

Sfard, A., & Lavie, I. (2005). Why cannot children see as the same what grown-ups cannot see as different?—Early numerical thinking revisited. *Cognition and Instruction, 23*(2), 237–309.

Shepard, R. G. (1993). Writing for conceptual development in mathematics. *The Journal of Mathematical Behavior, 12*, 287–293.

Shepherd, M. D., Selden, A., & Selden, J. (2012). University students' reading of their first-year mathematics textbooks. *Mathematical Thinking and Learning, 14*(3), 226–256.

Shield, M., & Galbraith, P. (1998). The analysis of student expository writing in mathematics. *Educational Studies in Mathematics, 36*(1), 29–52.

Shreyar, S., Zolkower, B., & Pérez, S. (2010). Thinking aloud together: A teacher's semiotic mediation of a whole-class conversation about percents. *Educational Studies in Mathematics, 73*(1), 21–53.

Shulman, L. (1987). Knowledge and teaching: Foundations of the new reform. *Harvard Educational Review, 57*(1), 1–23.

Siegel, M., Borasi, R., & Fonzi, J. (1998). Supporting students' mathematical inquiries through reading. *Journal of Research in Mathematics Education, 29*(4), 378–413.

Sinclair, N., & Moss, J. (2012). The more it changes, the more it becomes the same: The development of the routine of shape identification in dynamic geometry environment. *International Journal of Educational Research, 51–52*, 28–44.

Skovsmose, O. (2012). Symbolic power, robotting, and surveilling. *Educational Studies in Mathematics, 80*, 119–132.

Solomon, Y., & O'Neill, J. (1998). Mathematics and narrative. *Language and Education, 12*(3), 210–221.

Staats, S. (2008). Poetic lines in mathematics discourse: A method from linguistic anthropology. *For the Learning of Mathematics, 28*(2), 26–32.

Staples, M., & Truxaw, M. (2010). The mathematics learning discourse project: Fostering higher order thinking and academic language in urban mathematics classrooms. *Journal of Urban Mathematics Education, 3*(1), 27–56.

Staples, M. & Truxaw, M. (2012). An initial framework for the language of higher-order thinking mathematics practices. *Mathematics Education Research Journal, 24*, 257–281.

Stephan, M., & Rasmussen, C. (2002). Classroom mathematical practices in differential equations. *Journal of Mathematical Behavior, 21*, 459–490.

Straehler-Pohl, H., & Gellert, U. (2013). Towards a Bernsteinian language of description for mathematics classroom discourse. *British Journal of Sociology of Education, 34*(3), 313–332.

Stylianides, G. J., Stylianides, A. J., & Weber, K. (2017). Research on the teaching and learning of proof: Taking stock and moving forward. In J. Cai (Ed.), *Compendium for research in mathematics education* (pp. 237–266). Reston, VA: National Council of Teachers of Mathematics.

Toulmin, S. (2003). *The uses of argument.* Cambridge, United Kingdom: Cambridge University Press. (Original work published 1958)

Truxaw, M. P., & DeFranco, T. C. (2007). Mathematics in the making: Mapping verbal discourse in Polya's "Let us teach guessing" lesson. *Journal of Mathematical Behavior, 26*, 96–114.

Truxaw, M. P., & DeFranco, T. C. (2008). Mapping mathematics classroom discourse and its implications for models of teaching. *Journal for Research in Mathematics Education, 39*(5), 489–525.

Turner, E., Gutiérrez, R. J., & Sutton, T. (2011). Student participation in collective problem solving in an after-school mathematics club: Connections to learning and identity. *Canadian Journal of Science, Mathematics, and Technology Education, 11*, 226–246.

van Oers, B. (2001). Educational forms of initiation in mathematical culture. *Educational Studies in Mathematics, 46*, 59–85.

Veel, R. (1999). Language, knowledge, and authority in school mathematics. In F. Christie (Ed.), *Pedagogy and the shaping of consciousness* (pp. 185–216). London, United Kingdom: Cassell.

Vygotsky, L. S. (1978). *Mind in society: The development of higher psychological processes*. Cambridge, MA: Harvard University Press.

Vygotsky, L. (1986). *Thought and language*. Cambridge, MA: The MIT Press.

Wagner, D. (2007). Students' critical awareness of voice and agency in mathematics classroom discourse. *Mathematical Thinking and Learning, 9*(1), 31–50.

Wagner, D. (2008). "Just go": Mathematics students' critical awareness of routine procedure. *Canadian Journal of Science, Mathematics and Technology Education, 8*(1), 35–48.

Wagner, D., & Herbel-Eisenmann, B. (2008). "Just don't": The suppression and invitation of dialogue in mathematics classrooms. *Educational Studies in Mathematics, 67*(2), 143–157.

Wagner, D., & Herbel-Eisenmann, B. (2009). Remythologizing mathematics through attention to classroom positioning. *Educational Studies in Mathematics, 72*, 1–15.

Walkerdine, V. (1988). *The mastery of reason: Cognitive development and the production of rationality*. London, United Kingdom: Taylor & Frances/Routledge.

Walshaw, M., & Anthony, G. (2008). The teacher's role in classroom discourse: A review of recent research into mathematics classrooms. *Review of Educational Research, 78*(3), 516–551.

Walter, J. G., & Barros, T. (2011). Students build mathematical theory: Semantic warrants in argumentation. *Educational Studies in Mathematics, 78*(3), 323–342.

Walter, J. G., & Johnson, C. (2007). Linguistic invention and semantic warrant production: Elementary teachers' interpretation of graphs. *International Journal of Science and Mathematics Education, 5*, 705–727.

Wartofsky, M. W. (Ed.). (1979). *Models: Representation and the scientific understanding* (Vol. 129). Dordrecht, The Netherlands: D. Reidel.

Waywood, A. (1994). Informal writing-to-learn as a dimension of a student profile. *Educational Studies in Mathematics, 27*, 321–340.

Webb, N. M. (1991). Task-related verbal interaction and mathematics learning in small groups. *Journal for Research in Mathematics Education, 22*(5), 366–389.

Weber, K., Maher, C., Powell, A., & Lee, H. S. (2008). Learning opportunities from group discussions: Warrants become the objects of debate. *Educational Studies in Mathematics, 68*(3), 247–261.

Webster, J. (Ed.). (2003). *On language and linguistics. Collected works of MAK Halliday* (Vol. 3). London, United Kingdom: Continuum.

Wertsch, J. V. (1991). A sociocultural approach to socially shared cognition. In L. B. Resnick, J. M. Levine, & S. D. Teasley (Eds.), *Perspectives on socially shared cognition* (pp. 85–100). Washington, DC: American Psychological Association.

Williams, J., & Wake, G. (2007). Metaphors and models in translation between college and workplace mathematics. *Educational Studies in Mathematics, 64*, 345–371.

Wittgenstein, L. (1953). *Philosophical investigations* (G. E. M. Anscombe, Trans.). Oxford, United Kingdom: Blackwell.

Wood, M. B., & Kalinec, C. A. (2012). Student talk and opportunities for mathematical learning in small group interactions. *International Journal of Educational Research, 51–52*, 109–127.

Xu, L., & Clarke, D. (2013). Meta-rules of discursive practice in mathematics classrooms from Seoul, Shanghai and Tokyo. *ZDM—The International Journal on Mathematics Education, 45*(1), 61–72.

Yackel, E. (2002). What we can learn from analyzing the teacher's role in collective argumentation. *The Journal of Mathematical Behavior, 21*(4), 423–440.

Yackel, E., & Cobb, P. (1996). Sociomathematical norms, argumentation, and autonomy in mathematics. *Journal for Research in Mathematics Education, 27*(4), 458–477.

Yackel, E., Rasmussen, C., & King, K. (2000). Social and sociomathematical norms in an advanced undergraduate mathematics course. *The Journal of Mathematical Behavior, 19*(3), 275–287.

Zack, V., & Graves, B. (2001). Making mathematical meaning through dialogue: "Once you think of it, the Z minus three seems pretty weird." *Educational Studies in Mathematics, 46*, 229–271.

Zolkower, B., & deFreitas, E. (2012). Mathematical meaning-making in whole-class conversation: Functional-grammatical analysis of a paradigmatic text. *Language and Dialogue, 2*(1), 60–79.

Zolkower, B., & Shreyar, S. (2007). A teacher's mediation of a thinking-aloud discussion in a 6th grade mathematics classroom. *Educational Studies in Mathematics, 65*(2), 177–202.

Appendix A

Set of Journal Titles Initially Searched for This Review

American Education Research Journal
Canadian Journal of Science
Cognition & Instruction
Culture and Activity
Discourse Processes
Educational Studies in Mathematics
Elementary School Journal
For the Learning of Mathematics
International Journal of Science and Mathematics Education
Journal for Research in Mathematics Education
Journal of Learning Sciences
Journal of Mathematical Behavior
Journal of Mathematics Teacher Education
Journal of Teacher Education
Journal of Urban Mathematics Education
Language and Education
Linguistics and Education
Mathematics and Technology Education
Mathematics Education Research Journal

Mathematics Thinking and Learning
Mind
Reading and Writing Quarterly
Teachers' College Record
Urban Education
ZDM—the International Journal on Mathematics Education

Appendix B

Books Excluded From This Review

Adler, J. B. (2001). *Teaching mathematics in multilingual classrooms* (Vol. 26). New York, NY: Springer.

Alrø, H., & Skovsmose, O. (2004). *Dialogue and learning in mathematics education: Intention, reflection, critique* (Vol. 29). Dordrecht, The Netherlands: Kluwer Academic Publishers.

Barton, B. (2008). *The language of mathematics: Telling mathematical tales.* New York, NY: Springer.

Chapin, S. H., O'Connor, M. C., & Anderson, N. C. (2009). *Classroom discussions: Using math talk to help students learn* (2nd ed.). Sausalito, CA: Math Solutions Publications.

Chapman, A. P. (2003). *Language practices in school mathematics: A social semiotic approach* (Vol. 75). Lewiston, NY: Edwin Mellen Press.

Chronaki, A., & Christiansen, I. M. (Eds.). (2005). *Challenging perspectives on mathematics classroom communication.* Charlotte, NC: IAP.

Cobb, P., & Bauersfeld, H. (Eds.). (1995). *The emergence of mathematical meaning: Interaction in classroom cultures.* Hillsdale, NJ: Lawrence Erlbaum.

Cobb, P., Yackel, E., & McClain, K. (Eds.). (2000). *Symbolizing and communicating in mathematics classrooms: Perspectives on discourse, tools, and instructional design.* Mahwah, NJ: Lawrence Erlbaum.

Forman, E. A., Minick, N., & Stone, C. A. (Eds.). (1993). *Contexts for learning: Sociocultural dynamics in children's development.* New York, NY: Oxford University Press.

Gerofsky, S. (2004). *A man left Albuquerque heading east: Word problems as genre in mathematics education* (Vol. 5). New York, NY: Peter Lang.

Herbel-Eisenmann, B., Choppin, J., Wagner, D., & Pimm, D. (Eds.). (2011). *Equity in discourse for mathematics education: Theories, practices, and policies.* In A. Bishop (Series Ed.). Mathematics Education Library (Vol. 55). New York, NY: Springer.

Herbel-Eisenmann, B., & Cirillo, M. (Eds.). (2009). *Promoting purposeful discourse: Teacher research in mathematics classrooms.* Reston, VA: National Council of Teachers of Mathematics.

Meaney, T., Trinick, T., & Fairhall, U. (2011). *Collaborating to meet language challenges in Indigenous mathematics classrooms* (Vol. 52). London, United Kingdom: Springer.

Morgan, C. (1998). *Writing mathematically: The discourse of "investigation"* (Vol. 9). London, United Kingdom: Routledge.

Mousley, J., & Marks, G. D. (1991). *Discourses in mathematics.* Geelong, Victoria, Australia: Deakin University.

O'Halloran, K. (2005). *Mathematical discourse: Language, symbolism and visual images.* London, United Kingdom: Bloomsbury.

Pimm, D. (1987). *Speaking mathematically: Communication in mathematics classrooms.* London, United Kingdom: Routledge & Kegan Paul.

Pimm, D. (1995). *Symbols and meanings in school mathematics.* London, United Kingdom: Routledge.

Rowland, T. (2000). *Pragmatics of mathematics education* (Vol. 14). London, United Kingdom: Falmer Press.

Sfard, A. (2008). *Thinking as communicating: Human development, the growth of discourses, and mathematizing.* Cambridge, United Kingdom: Cambridge University Press.

Stein, M. K., & Smith, M. (2011). *5 practices for orchestrating productive mathematics discussions.* Reston, VA: National Council of Teachers of Mathematics.

28 K-12数学教学核心实践的研究

维多利亚·R.雅各布斯
美国北卡罗来纳大学格林斯伯勒分校
丹尼斯·A.斯潘格勒
美国乔治亚大学
译者：吴颖康
华东师范大学数学科学学院

"高质量的教学，无论采用什么形式，均聚焦于重要的数学内容，以完整的方式呈现和展开。它基于对学生现有知识和思维方式以及这些知识和思维方式如何发展的细致考虑。这样的教学对于大部分学生来说都是有效的，而且这样的教学有助于发展我们所说的数学精熟水平，即关于数学的知识、技能、能力和倾向（美国国家研究协会[NRC]，2001，第315页）。

本章的开始，我们提出，高效的教学必须从五个相互关联的方面促进数学精熟水平的发展，即概念理解、过程流畅、策略能力、适合的推理和积极的情感（NRC，2001）。尽管我们知道有关高效教学的构成已经是并仍将是有争议的讨论对象（参见Franke，Kazemi，& Battey，2007），但是数学精熟水平自2001年被提出以来一直广泛出现在关于高效教学的讨论和研究中。必须注意，发展数学精熟水平的教学并不和某种特殊的教学方法相关，而被认为比只关注上述五方面中的某一方面的教学更为复杂。

在本章中，我们将基于经常被引用的D. K.科恩、劳登布什和鲍尔（2003）的模型，探究高效教学的某一个侧面。在他们的模型中，教学被描述为"在教师和学生之间围绕内容进行的交互活动"（第122页）。根据这个模型，教师并不是单向地作用于学生，相反地，教师和学生之间是相互依赖、相互作用的（Hiebert & Grouws，2007）。对于师生之间相互依赖的交互作用的重视也与格罗斯曼、康普顿和同事们（2009）关于教学是一种关联性的专业的认识相一致。他们认为教学实践是复杂的，不可预测的，而且依赖于人际关系的质量。在他们重大的、跨专业的关联性专业研究中，他们发现将一门专业分解为一些核心的实践对促进该专业知识的发展是有必要的。我们采纳了核心实践的说法，并用它来组织本章关于K-12数学课堂教学的研究，[1]但是我们仅局限在与以学生数学思维为中心的师生交互直接相关的核心实践上。接下来，我们先解释本章为什么只聚焦在师生交互中学生的数学思维上，然后给出本章的概览。

聚焦学生的数学思维

数学教育研究给出了许多具体的有关高效教学的

我们感谢苏珊·恩普森、梅甘·弗兰克、兰迪·菲利普、米里亚姆·谢林、梅甘·斯特普尔斯、埃琳·特纳、霍尔特·威尔逊和四位匿名评审，感谢他们在撰写本章期间提供的深思熟虑的反馈。感谢娜奥米·杰瑟普在收集和汇总文献中提供的帮助，感谢邦尼·沙佩尔出色的编辑工作。本章的写作得到了美国国家科学基金会（DRL-1316653）的部分支持，但所表达的观点不一定反映该支持机构的立场、政策或认可。

28 Research on Core Practices in K–12 Mathematics Teaching

VICTORIA R. JACOBS
University of North Carolina at Greensboro, United States

DENISE A. SPANGLER
University of Georgia, United States

High-quality instruction, in whatever form it comes, focuses on important mathematical content, represented and developed with integrity. It takes sensitive account of students' current knowledge and ways of thinking as well as ways in which those develop. Such instruction is effective with a range of students and over time develops the knowledge, skills, abilities, and inclinations that we term mathematical proficiency. (National Research Council [NRC], 2001, p. 315)

We take as a starting point for this chapter the idea that high-quality instruction should promote the development of the five interrelated strands that have been identified as *mathematical proficiency*: conceptual understanding, procedural fluency, strategic competence, adaptive reasoning, and productive disposition (NRC, 2001). Although we recognize that what constitutes high-quality instruction has been and will likely remain controversial (see also Franke, Kazemi, & Battey, 2007), mathematical proficiency has been widely included in visions of high-quality instruction and research on teaching since the construct was put forth in 2001. Note that teaching for mathematical proficiency is not linked to a particular teaching method but has generally been considered more complex than teaching that is focused on only a subset of the five strands.

In this chapter, we explore research on a particular slice of high-quality instruction, based on the oft-cited model of D. K. Cohen, Raudenbush, and Ball (2003), in which instruction is depicted as "interactions among teachers and students around content" (p. 122). According to this model, teachers do not act on students unidirectionally but instead act in interdependent ways (Hiebert & Grouws, 2007). Emphasis on these interdependent interactions is also consistent with Grossman, Compton, and colleagues' (2009) idea that teaching is a *relational* profession—a profession in which practice is complex, unpredictable, and dependent on the quality of human relationships. In their seminal, cross-profession study of relational professions, they found decomposition of a profession into *core practices* was required to support the development of expertise. We use this idea of core practices to frame our chapter about research on mathematics teaching in K–12 classrooms,[1] but we limit ourselves to core practices directly involved in teacher–student interactions in which students' mathematical thinking is central. Below we expand on our choice to focus on students' mathematical thinking in these interactions before providing an overview of the chapter.

Focus on Students' Mathematical Thinking

Research in mathematics education has provided multiple detailed accounts of high-quality instruction that have illustrated how productive teaching moves are often con-

---

We thank Susan Empson, Megan Franke, Randy Philipp, Miriam Sherin, Megan Staples, Erin Turner, Holt Wilson, and four anonymous reviewers for their thoughtful feedback during the writing of this chapter. Thanks also to Naomi Jessup for her assistance in gathering and summarizing literature and to Bonnie Schappelle for her exceptional editing. The writing of this chapter was supported in part by the National Science Foundation (DRL–1316653), but the opinions expressed do not necessarily reflect the position, policy, or endorsement of the supporting agency.

研究实例，解释了有成效的教学活动如何取决于课堂教学过程中学生的意见、问题和策略（例如Ball, 1993；Lampert, 2001；Schoenfeld, 2011）。这样的教学能使教师和学生都受益。当教师试图理解学生推理方式的时候，他们可以通过了解学生的那些不同的但在数学上同样有效的推理策略而获益；学生可以通过对自己推理过程的表述和反思以及对自己或同学的推理的评价而获益。

在过去几十年间数学教育领域都非常重视学生的数学思维。这可以从一系列的研究和专业发展活动所做的努力中看出（有关评论参见Goldsmith, Doerr, & Lewis, 2014），包含了诸如雄心教学（Forzani, 2014；Kazemi, Franke, & Lampert, 2009；Lampert等, 2013）、适应性教学（Cooney, 1999；Daro, Mosher, & Corcoran, 2011；Sherin, 2002）和响应性教学（Edwards, 2003；Jacobs & Empson, 2016；Robertson, Scherr, & Hammer, 2016）等工作。"估计学生的想法并给出回应"（Ball, Lubienski, & Mewborn, 2001, 第453页）在诸如《教学研究手册》（Richardson, 2001）等高知名度出版物中被认为是教学的核心活动。类似地，学生的数学思维在政策性文件中也发挥了显著作用。例如，全美数学教师理事会（NCTM, 2014）在《行动原则》（参见Association of Mathematics Teacher Educators & National Council of Supervisors of Mathematics, 2014；Daro等, 2011；NRC, 2001, 2005）中指出"引发并利用学生思维"（第10页）是推荐的八大数学教学实践之一。

最近几十年有关学生数学思维的研究发展迅速。该研究领域是这套书（Cai, 2017）许多章节的主题，也是许多其他数学教育领域知名度较高的综合性出版物（Grouws, 1992；Lester, 2007）的核心内容，它也反映在迅猛发展的基于学习轨迹（和学习进阶）的研究中。学习轨迹（和学习进阶）描述了学生知识和技能可能的成长和发展路径中的关键性阶段（Daro等, 2011, 第12页；另见这套书Clements & Sarama, 2004；Lobato & Walters, 2017那一章；Maloney, Confrey, & Nguyen, 2014；Simon, 1995；Sztajn, Confrey, Wilson, & Edgington, 2012）。

聚焦学生思维的教学还得到了众多研究成果的支持。这些研究（例如Bobis等, 2005；Carpenter, Fennema, Peterson, Chiang, & Loef, 1989；Clements, Sarama, Spitler, Lange, & Wolfe, 2011；Fennema等, 1996；Jacobs, Franke, Carpenter, Levi, & Battey, 2007；Simon & Schifter, 1993；Sowder, 2007；Wilson & Berne, 1999）指出它对于教师学习和学生学习都有积极效果。其中最为重要的研究发现是学生的数学思维可以成为贯穿教师职业生涯的专业发展活动的生成源，这是因为教师可以从学生的思维中学会如何学习，而且他们每天都有机会进行这样的活动（Franke, Carpenter, Levi, & Fennema, 2001）。

总之，我们关于学生数学思维在高效教学中能起到重要作用的观点来源于多年的研究发现。我们进一步发现，随着学习轨迹研究的激增，近来很多关于教学的研究都关注学生的数学思维。然而，我们同时注意到学生的数学思维只是解决问题的一个方面。弗兰克和他的同事们（2007）认为有必要考虑但不能局限于学生的数学思维："为学生创造学习机会的教学必须了解学生的身份、历史、文化背景和学校经历，所有这些都和数学学习有关"（第243页）。我们同意这一观点。教师必须要了解他们的学生是谁，只有这样他们才能利用学生语言的、文化的以及社会背景的资源和其数学思维进行教学。更进一步地，对于这一拓展的学生资源的更为直接的关注，不仅有助于提高针对某些学生群体的教学成效，而且对课堂、学校和社会水平的现状提出了挑战。因此，我们尽可能地在这一拓展的视野下讨论研究进展。

本章概览

在本章中，我们聚焦于以学生思维为中心的、与K-12数学课堂教学交互部分有关的核心实践。这一取向的建立归结于弗兰克和他的同事们（2007）关于数学教学和课堂实践的研究综述："将学生思维研究和课堂实践结合起来，从而确定教师可以遵循的核心实践"（第250页）。

在本章开始，我们探索了教学中与核心实践有关的研究。为此，我们突出了核心实践团队的研究工作，这

tingent on comments, questions, and strategies offered by students in the midst of instruction (e.g., Ball, 1993; Lampert, 2001; Schoenfeld, 2011). This type of teaching has benefits for both teachers and students. When teachers explore students' ways of reasoning, they benefit by gaining a window into students' reasoning, which can be mathematically powerful but often differs from teachers' reasoning. Students benefit because they not only have opportunities to articulate and reflect on their reasoning but also learn to value their own and peers' sense making.

The educational landscape over the past several decades has included an emphasis on students' mathematical thinking. This emphasis can be seen in a variety of research and professional development efforts (see Goldsmith, Doerr, & Lewis, 2014, for a review), including work targeting *ambitious teaching* (Forzani, 2014; Kazemi, Franke, & Lampert, 2009; Lampert et al., 2013), *adaptive teaching* (Cooney, 1999; Daro, Mosher, & Corcoran, 2011; Sherin, 2002), and *responsive teaching* (Edwards, 2003; Jacobs & Empson, 2016; Robertson, Scherr, & Hammer, 2016). "Sizing up students' ideas and responding" (Ball, Lubienski, & Mewborn, 2001, p. 453) has also been identified as a core activity of teaching in high-profile publications such as the *Handbook of Research on Teaching* (Richardson, 2001). Similarly, students' mathematical thinking has played a prominent role in policy documents. For example, the National Council of Teachers of Mathematics (NCTM, 2014) identified "elicit and use evidence of student thinking" (p. 10) as one of the eight recommended mathematics teaching practices in *Principles to Actions* (see also Association of Mathematics Teacher Educators & National Council of Supervisors of Mathematics, 2014; Daro et al., 2011; NRC, 2001, 2005).

The past several decades have also included a growing body of research detailing students' mathematical thinking. This research is the focus of numerous chapters in the current volume (Cai, 2017) as well as in other high-profile syntheses of research in mathematics education (Grouws, 1992; Lester, 2007). It is also reflected in the fast-growing research base on *learning trajectories* (and *learning progressions*)—descriptions of students' thinking that are structured to "identify key waypoints along the path in which students' knowledge and skills are likely to grow and develop" (Daro et al., 2011, p. 12; see also Clements & Sarama, 2004; Lobato & Walters, 2017, this volume; Maloney, Confrey, & Nguyen, 2014; Simon, 1995; Sztajn, Confrey, Wilson, & Edgington, 2012).

Teaching focused on students' mathematical thinking is further supported by research that has shown positive outcomes for both student learning and teacher learning (e.g., Bobis et al., 2005; Carpenter, Fennema, Peterson, Chiang, & Loef, 1989; Clements, Sarama, Spitler, Lange, & Wolfe, 2011; Fennema et al., 1996; Jacobs, Franke, Carpenter, Levi, & Battey, 2007; Simon & Schifter, 1993; Sowder, 2007; Wilson & Berne, 1999). Of particular importance is the finding that students' mathematical thinking can become a generative source of professional development throughout teachers' careers because teachers learn how to learn from their students' thinking, and they have opportunities to do so on a daily basis (Franke, Carpenter, Levi, & Fennema, 2001).

In summary, our emphasis on the important role that students' mathematical thinking plays in high-quality instruction stems from years of research evidence. Further, with the surge in research on learning trajectories, we found that much of the recent research on teaching has included attention to students' mathematical thinking. We note, however, that students' mathematical thinking is only one piece of the puzzle. We agree with Franke and colleagues (2007), who argued for the need to consider, but also to go beyond, students' mathematical thinking: "Teaching to open opportunities for participation requires getting to know students' identities, histories, and cultural and school experiences, all in relation to the mathematical work" (p. 243). Teachers need to get to know who students are as individuals so that they can draw on students' linguistic, cultural, and community-based resources as well as their mathematical ideas. Further, more direct attention to this broadened view of students' resources has the potential not only to improve instruction for particular students but also to provide opportunities for challenging the status quo at the classroom, institutional, and societal levels. Thus, when available, we include research that is grounded in this broader perspective.

Chapter Overview

In this chapter, we focus on research on core practices related to the interactive part of K–12 mathematics teaching in which students' thinking is central. The approach builds on the charge to researchers from Franke and colleagues (2007) in their synthesis of research on mathematics teaching and classroom practice: "Marry the research on student thinking with classroom practice by beginning to identify core practices that teachers could follow" (p. 250).

We begin the chapter by exploring what is involved in research on core practices in teaching. To do so, we highlight the work of the Core Practice Consortium, a large-scale collaboration among educators who take a core-practices approach to working with novice teachers.

是以核心实践为途径、与新手教师合作的教育者之间的大型合作项目。我们利用该研究团队大量的合作性工作确定教学中核心实践研究需考虑的方面。随后，我们确定了两个具体的教学实践，即教师关注和引领讨论。我们认为这两个教学实践是以学生思维为中心的高效数学课堂教学中课堂交互部分的核心。教师关注的研究捕获了教师注意和理解学生所说与所做的等方面内容，但未回应之前的一些即时的做法。引领讨论的研究捕获了教师支持学生解释自己的想法、理解并批判他人推理时的即时工作。我们选择这两个实践的原因是因为这两个实践不仅在过去十年中得以广泛研究，而且还提供了可供对照的具体案例。教师关注反映了数学教育中一个相对比较新的研究热点，而引领讨论长期以来一直是研究热点，但随着对学生数学思维的重视，在这些讨论中有关教师角色的概念化持续得到关注。在本章结束的时候概述了我们关于这两个核心实践研究的收获，并给出了未来的研究建议。

在选择本章的焦点内容时，我们对应该排除哪些内容作出了艰难的抉择。在本节内容结束时，我们将明确解释3个被我们排除的研究领域。尽管当这些领域与我们选择的焦点内容有交叉时，我们有时会稍有论及，但是对于他们的陈述必定是不完整的。

第一，教师的工作是复杂的、多方面的，在聚焦围绕师生互动的教师工作时，我们选择不直接讨论诸如长期或短期的教学计划、建立与课堂管理有关的行为规范、与同事合作、与家长互动等教师工作的其他方面。然而，我们认为围绕师生互动的教师专长为教师在其他方面的工作提供了信息，而教师其他方面的工作也为师生互动的进行提供了信息，因而有时这两方面的研究是有重叠的。尤其是，我们认为有价值的任务选择（NCTM，1991）是教师在交互环节能参与学生数学思维的基础。比如说，当选择的任务有多个切入点而且有多种得到结果的方法时，那么更多的学生就有可能投入到任务解决的活动中，而且有机会使他们的思维过程显性化。因为任务选择不包含在我们课堂交互的目标指向中，因此在本章中我们没有对其进行评介，但是可以参见斯坦、雷米拉德和斯密斯（2007）就这方面的总结。

第二，我们选择不脱离教学实践单独讨论教师的个人特征（例如，知识、信念、从教年限等）。考虑到有许多例子表明，有着类似个人特征的教师在教学中采用了截然不同的教学方法，我们同意希伯特和格劳斯（2007）的观点，即尽管这些个人特征在教学中可以发挥作用，但是不能决定教师如何教学。进一步地，我们认识到每个教师都有潜力以一种富有成效的方式成长这一观点具有的可贵之处。在选择教学实践（而不是教师特征）的焦点时，我们认同那些指出将改善教学作为提高课堂教学质量的有效途径的学者的看法（Hiebert & Morris, 2012; Lewis, Perry, Friedkin, & Roth, 2012）。

第三，尽管教学工作和教学专长发展研究经常交织在一起，我们选择只关注教学工作。当有些研究涉及专业发展或大学课程设置时，我们只关注这些研究为我们理解教学带来哪些益处，而不是它们在支持教师发展、情境对教师学习的影响或大规模能力培养需要什么等方面的贡献（见这套书中Sztajn, Borko, & Smith, 2017中关于专业发展的讨论）。

理解有关教学中核心实践的研究

为按照格罗斯曼、康普顿和他们的同事们（2009）关于核心实践的想法架构本章，我们还参考了有关核心实践对于发展和促进高效教学有积极作用的其他研究者的工作。教学具有内在的复杂性，当试图通过实践以获得教学专长时，从整体上接触教学是相当困难的。核心实践的工作有助于减少这种复杂性，但是确定一些特殊的核心实践必然具有争议性，因为这一专业的某些内容被认为处于最为突出的位置。就像格罗斯曼和麦克唐纳（2008）所警告我们的，把一个复杂系统在错误的部位进行分解有摧毁整体完整性的危险。为了探索将数学教学进行有效分解需要考虑的领域，我们参考了核心实践团队的工作。他们的大量工作不仅加深了我们对核心实践研究的理解，而且还为该领域的研究者提供了一个具有潜在开创性的模型，有助于合作，产生新知识并改进教学。

We use the Consortium's extensive collaborative efforts to identify considerations in research on core practices in teaching. We then address two specific teaching practices that we consider to be core to the interactive part of high-quality mathematics instruction in which students' mathematical thinking is central: teacher noticing and leading discussions. Research on teacher noticing captures the in-the-moment work of teachers when they are focusing their attention on and making sense of what students say and do but before they actually respond. Research on leading discussions captures the in-the-moment work of teachers when they are supporting students in explaining their thinking and making sense of and critiquing the reasoning of others. We chose these two practices because they not only have been extensively studied in the past decade but also provide contrasting cases. Teacher noticing reflects a relatively new emphasis in mathematics education whereas leading discussions has long been a focus of study, but with the increasing emphasis on students' mathematical thinking, the field's conceptualization of teachers' roles in these discussions has continued to grow. We conclude the chapter with a summary of what we have learned from the research on these two core practices and provide suggestions for future research.

In choosing our focus for this chapter, we made difficult decisions about what to exclude, and we close this introduction by explicitly acknowledging three of these areas. Although we sometimes address these areas when they intersect with our chosen emphasis, their treatment is not comprehensive.

First, teachers' work is complex and multifaceted, and by focusing on teachers' work during teacher–student interactions around content, we chose not to address directly other features of teachers' work, such as long- and short-term planning, establishing norms related to classroom management, collaborating with colleagues, working with families, and so on. However, we argue that expertise in teacher–student interactions around content informs, and is informed by, teachers' work in these other areas, and thus at times the research is overlapping. In particular, we acknowledge that the selection of worthwhile tasks (NCTM, 1991) is foundational for teachers' engagement with students' mathematical thinking during the interactive part of instruction. For example, when selected tasks have multiple points of entry and multiple routes to a solution, more students will be likely to engage with the tasks and have an opportunity to make their thinking visible. Because task selection is not part of our targeted focus on the interactive part of teaching, we do not review that research in this chapter, but see Stein, Remillard, and Smith (2007) for a summary of this work.

Second, we chose not to address teachers' individual characteristics (e.g., knowledge, beliefs, or years of experience) separately from the practice of teaching. Given the many examples of teachers with similar characteristics teaching in dramatically different ways, we agree with Hiebert and Grouws (2007), who argued that these characteristics, although they can play a role in teaching quality, do not determine how a teacher teaches. Further, we see benefits to viewing all teachers as having the potential to grow in productive ways. In choosing to emphasize a slice of the practice of teaching (instead of teachers' characteristics), we join others who have advocated for a focus on teaching as the more productive path for improving classroom instruction (Hiebert & Morris, 2012; Lewis, Perry, Friedkin, & Roth, 2012).

Third, although research on the work of teaching and the development of teaching expertise are often intertwined, we chose to focus only on the work of teaching. When studies did involve professional development or university coursework, we focused our discussion on what these studies contributed to our understanding of teaching rather than their contributions in support of, for example, teachers' growth, the effect of context on teachers' learning, or what is needed for large-scale capacity building. (See Sztajn, Borko, & Smith, 2017, this volume, for a discussion of research on professional development.)

Understanding Research on Core Practices in Teaching

In framing the chapter around Grossman, Compton, and colleagues' (2009) idea of core practices, we join others who have found research on core practices to be productive for understanding and promoting high-quality instruction. Teaching is inherently complex, and when one is working to gain expertise, approaching teaching holistically can be overwhelming. Core-practices work has the benefit of reducing this complexity, but identifying specific core practices is necessarily controversial because some parts of the profession are chosen to be foregrounded over others. As Grossman and McDonald (2008) warned, breaking apart a complex system in the wrong places has the possibility of destroying the integrity of the whole. To explore what the field needs to consider to productively decompose mathematics teaching, we draw on the work of the Core Practice Consortium. Their extensive collaboration not only furthers our understanding of research on core practices in teaching but also pro-

核心实践团队

来自8个不同院校和不同专业的研究者[2]组成了核心实践团队,并开展了历史上较大规模的合作。该团队致力于将数学教学分解为一些核心实践的集合,从而新手教师可以讨论并了解这些核心实践,除此之外,该团队还探索有助于新手教师在这些核心实践上获得专业发展的教师教育教学法。开始于2012年的这个合作项目仍旧处于初级阶段,而且是在早期的、更小范围的工作基础上发展起来的。参与项目院校的教育者正在重构他们的教师教育计划。尽管他们采用了不同方式且处于重构的不同阶段,但他们一致承诺重构的教师教育计划是基于实践的,而且他们愿意一起面对教师教育过程中经常出现的挑战。

该团队的研究和发展工作建立在现有研究基础上,而且他们认为教学是建立在学生关于某个学科的认识基础上的、具有交互性和生成性的工作(Forzani,2014)。更进一步地,该团队的成员在教师教育中都采纳了这样一种教学法(Grossman,Compton 等,2009),即新手教师不仅要讨论这些核心实践,而且有机会在不同的情境下,在有人提供即时辅导的情况下,反复演练这些核心实践(Ball & Forzani,2009;Ghousseini,Beasley,& Lord,2015;Kazemi,Ghousseini,Cunard,& Turrou,2016;Lampert 等,2013)。考虑到这项工作涉及不同院校和不同学科(包括科学和人文科学等领域),不同组别的研究者进行了针对情境和学科因素的调整,因此该团队也在探索这些有目的的调整。

这种基于核心实践重构数学教师教育的取向在国际上已经被认同和采纳。比如说,新西兰两所大学的研究者在此基础上建立了一个横跨多年的名为学习雄心数学教学工作的研究项目,试图提高最初的数学教师教育(Anthony & Hunter,2012,2013;Hunter,Hunter,& Anthony,2013)。初步的反馈结果是积极的,包括职前教师对核心实践的演练和即时辅导(Averill,Drake,& Harvey,2013)。

在本节中,我们从核心实践团队项目中有选择地指出了以下4项核心实践教学研究需要考虑的方面:(1)判断核心实践的标准,(2)根据目标确定核心实践,(3)用共同的且精确的语言表述核心实践,(4)关注核心实践的关联性本质。

判断核心实践的标准

为研究教学中的核心实践,研究者必须考虑究竟什么是核心实践,并且需要持续讨论如何才能更好地确认这些实践。格罗斯曼、哈默尼斯和麦克唐纳(2009)给出了判断教师教育中核心实践标准的初步清单:

- 在教学中经常发生的实践;
- 新手教师在不同课程或不同教学方法的课堂中都能实施的实践;
- 新手教师可以实际掌握的实践;
- 能够让新手教师更多了解学生和教学的实践;
- 能保持教学的完整性和复杂性的实践;
- 基于研究的且有提高学生成就潜力的实践。(第277页)

简单地说,教学实践行为要想被认为是核心实践,必须处于教学的中心,而且新手教师可以在不同的情境下接触和学习。此外,它们还必须是基于研究的,而且能够支持学生和教师教学的潜力发展(参见Ball,Sleep,Boerst,& Bass,2009)。

确定恰当的标准是确认核心实践最根本的挑战。每一个核心实践行为必须保持足够的复杂性来抓住教学的关联性本质,这样才能感觉到是教学而不是被简单提取出来用于研究的技能。然而这种复杂性又不能太过,以至于新手教师无法学习(Ball & Forzani,2011)。明确判断究竟什么是核心实践的问题还留待解决,目前正在研究的核心实践在范围上覆盖很广而且有交叉,也就是说有些实践行为可以被看作其他实践行为的子集。例如,与一个教学片段相联系的核心实践是"将学生指向参考他人的数学想法"(Lampert 等,2013,第232页),然而,一个更大范围下的、不限制于情境的核心实践是"为学生参考外部基准设立长期和短期目标"(O'Flahavan & Boté,2014,第22页)。

vides a potentially groundbreaking model for how those in the field can work together to produce new knowledge and improve teaching.

Core Practice Consortium

In a historically large-scale collaboration—across institutions and disciplines—researchers from eight institutions[2] have formed the Core Practice Consortium to decompose mathematics teaching into sets of core practices that can be discussed and accessed by novice teachers as well as to explore teacher education pedagogies that can help novices gain expertise in these practices. This collaborative venture, which began in 2012, is still in the early stages and builds on prior, often more localized, work. Educators at the participating institutions are redesigning their teacher education programs, and although they are using different approaches and are in various stages of redesign, they share a commitment to practice-based teacher education and a desire to confront perennial challenges of teacher education collectively.

The research and development work of the Consortium draws on existing research and is connected to a vision of teaching in which the interactive, contingent work of building on students' ideas about particular subject matter is central (Forzani, 2014). Further, members of the Consortium are working toward a common pedagogy of enactment in teacher education (Grossman, Compton, et al., 2009) in which novice teachers not only discuss these core practices but also have repeated opportunities to rehearse them in multiple settings with in-the-moment coaching (Ball & Forzani, 2009; Ghousseini, Beasley, & Lord, 2015; Kazemi, Ghousseini, Cunard, & Turrou, 2016; Lampert et al., 2013). Because this work is multi-institutional and multidisciplinary (including such domains as science and literacy), the researchers are also exploring the purposeful adaptations various groups have made in response to contextual or disciplinary factors.

This core-practices approach to redesigning mathematics teacher education has already been recognized and adopted internationally. For example, researchers in two New Zealand universities are building on this work in their multiyear project, *Learning the Work of Ambitious Mathematics Teaching*, to improve initial mathematics teacher education (Anthony & Hunter, 2012, 2013; Hunter, Hunter, & Anthony, 2013). Preliminary feedback has been positive, including uniformly strong endorsement from prospective teachers for the rehearsals and in-the-moment coaching of core practices (Averill, Drake, & Harvey, 2013).

In this section, we draw selectively from the Consortium projects to highlight four considerations in research on core practices in teaching: (1) determine criteria for core practices, (2) identify sets of core practices to target, (3) use common and precise language for core practices, and (4) attend to the relational nature of core practices.

Determine Criteria for Core Practices

To study core practices in teaching, researchers must consider what counts as a core practice, and debate continues about how best to identify these practices. Grossman, Hammerness, and McDonald (2009) provided an initial list of criteria for identifying core practices for teacher education:

- Practices that occur with high frequency in teaching;
- Practices that novices can enact in classrooms across different curricula or instructional approaches;
- Practices that novices can actually begin to master;
- Practices that allow novices to learn more about students and about teaching;
- Practices that preserve the integrity and complexity of teaching; and
- Practices that are research-based and have the potential to improve student achievement. (p. 277)

In short, for instructional practices to be considered core, they need to be central to teaching and accessible to and learnable by novice teachers in a variety of settings. They also need to be research-based and have potential to support both student and teacher learning (see also Ball, Sleep, Boerst, & Bass, 2009).

Determining the appropriate grain size is a fundamental challenge to identifying core practices. Each core practice needs to maintain enough complexity to capture the relational nature of teaching so that it feels like teaching rather than a skill that is simply extracted for study. However, this complexity must not be too overwhelming or it will not be learnable (Ball & Forzani, 2011). Determining exactly what counts as a core practice remains unresolved, and core practices currently under investigation are both wide ranging in scope and overlapping in that some practices could be viewed as subsets of other practices. For example, a core practice that is tied to a single instructional episode is "orienting students toward each other's mathematical ideas" (Lampert et al., 2013, p. 232), whereas a core practice of a larger grain size—not situation-bound—is "setting long- and short-term

根据目标确定核心实践

为了研究教学中的核心实践，研究者必须考虑他们要调查哪些核心实践。核心实践团队中的研究者已经确定了许多核心实践，许多相关信息可以在相关院校（或研究项目）的网站上查找到。这些核心实践的集合并不是一般教学技能的检查表，就像19世纪60年代和70年代基于教学技能的教师教育那样。相反地，这些核心实践与学科内容以及该学科下学生的思维有很强的关联（Forzani，2014）。尽管如此，团队中有些教师教育工作强调跨学科专业的实践，这样做基于的假设是这些核心实践可以在不同的学科中用不同的方式被具体化。虽然这种简化教师教育计划的方式很有吸引力，但是奥夫拉哈文和博特（2014）反对重点关注跨学科内容的实践，这是因为，将核心实践的呈现与基于学科内容的情境和问题紧密相连，有助于新手教师更轻松地学习。

尽管根据目标确定核心实践有现实必要性，但是团队研究者最终的目的并不是像我们所想的那样确立一些核心实践。相反地，他们的目的是发展核心实践概念的共同理解，从而使这个概念本身就能成为组织和实施基于实践的教师教育创新举措的工具（McDonald, Kazemi, & Kavanagh，2013，第380页）。因此，首要目标并不是达成对核心实践的共识，而是核心实践的想法能够成为改进该领域的工具。

用共同的且精确的语言表述核心实践

为共同研究教学中的核心实践，研究者所使用的专业术语必须统一，这样才能在不同研究院校和不同学科之间进行足够精确的对话（Ball & Forzani, 2011；Grossman & McDonald, 2008）。这种共同的且精确的语言，也是研究团队的一个重大目标，它有助于在更广泛的团体中进行有关核心实践的交流，而且可以避免一些常用术语由于其多种含义引起歧义而变得不可用。

为实现共同的且精确的语言这一目标，团队研究者正在把来自不同项目具有交叉含义的概念进行合并（和精炼），从而使术语可以变得更简单合理。比如说，在目前的文献中，核心实践经常和高影响力实践交换使用，但团队的研究者同意在今后的写作中唯一使用核心实践（M.L.Franke，2015年2月6日的个人交流）。类似地，团队也在努力使对所选择术语的含义有共同的和精确的理解。比如说，他们目前的焦点之一是更好地理解核心实践引领讨论，作为其中的一项工作，他们在不同的学科内容下对课堂讨论的含义进行了统一。在这一共同的界定下，研究者正在探索来自不同院校参与研究的新手教师是如何投入到引领讨论的实践中去的。

这项为达成术语和含义共识的努力并不寻常，而且与各种为获得认同而致力于独特性研究的学术研究小组的典型工作不同。核心实践团队的研究者并没有假设各种个人努力会最终汇集起来推动整个领域的发展，而是采取了早期合作这一并不常见的方式。

关注核心实践的关联性本质

为研究教学中的核心实践，研究者必须关注核心实践的关联性本质——个体的核心实践并不孤立发生，而是在其他实践的情境中发生，而且这种关联是非常关键的。关于核心实践关联性本质的证据来自一个具有先驱性的核心实践研究小组的工作，即在教学实践中学习、从教学实践中学习、为教学实践而学习。从意大利语的教学方式（Lampert & Graziani，2009）中受到启发，该研究小组利用了策略性选择的教学活动，即针对某个教学目标的较短的教学片段，为学生和新手教师提供学习机会。这些教学活动通过提供一个"教学实践的各个方面都关联着存在的有限空间"（Ghousseini, Franke, & Turrou，2014，第5页），捕捉到了核心实践的关联性本质。

比如说，考虑他们关于大声计数的教学活动：

> 在本活动中，教师带领全班同学大声进行跳跃式的计数，通过决定从哪个数开始、每次跳跃多少（例如，每次跳跃10，每次跳跃19，每次跳跃3/4），是向前数还是向后数以及什么时候停止，来教不同的概念和技能。教师把计数的结果记录在黑板上，然后停下来，引出学生得到下一个数字的策略，让他们观

learning goals for students referenced to external benchmarks" (O'Flahavan & Boté, 2014, p. 22).

Identify Sets of Core Practices to Target

To study core practices in teaching, researchers must consider which sets of core practices they will investigate. Multiple sets have already been identified by researchers in the Consortium, and information about many of them can be found at institutional (or program-level) websites. These sets of core practices are not checklists of general teaching competencies, typical of the competency-based teacher education efforts of the 1960s and 1970s. Instead, the practices are strongly linked to disciplinary content and students' thinking in a discipline (Forzani, 2014). Nonetheless, some teacher education work in the Consortium has highlighted practices that do cut across disciplines, with the assumption that they may be instantiated in different ways in different disciplines. Although recognizing the appeal of this approach for streamlining teacher education programs, O'Flahavan and Boté (2014) have argued against an emphasis on cross-cutting practices because they believe that presenting core practices as tightly linked to discipline-based situations and problems will enable novices to learn more easily.

Despite the practical necessity of identifying sets of core practices to target, the ultimate goal for researchers in the Consortium is *not* a single universal set of core practices as one might expect. Instead, their goal is to develop "a common understanding of the concept of core practice so that the concept itself might become a field-wide tool for the organization and implementation of practice-based teacher education initiatives" (McDonald, Kazemi, & Kavanagh, 2013, p. 380). Thus, the overarching goal is not consensus on practices but rather that the idea of core practices could become the field's vehicle for improvement.

Use Common and Precise Language for Core Practices

To work collectively to study core practices in teaching, researchers must agree on a shared technical language that is sufficiently precise for conversations across institutions and disciplines (Ball & Forzani, 2011; Grossman & McDonald, 2008). This common and precise language—a major goal of the Consortium—also facilitates communication about core practices in the broader community and prevents often-used terms from becoming almost unusable because their multiple meanings create confusion about what is being discussed.

To accomplish the goal of common and precise language, the Consortium researchers are working to merge (and refine) overlapping ideas from multiple projects so that terminology can be streamlined. For example, in the current literature, *core practices* is often used interchangeably with *high-leverage practices,* but researchers in the Consortium have agreed to use the term *core practices* exclusively in future writing (M. L. Franke, personal communication, February 6, 2015). Similarly, they are working toward a common and precise understanding of what selected terminology signifies. For example, one of their current foci is to better understand the core practice of *leading discussions,* and as part of that work, they have developed a common definition of *classroom discussion* across subject areas. Building on this common definition, researchers are currently exploring how novice teachers engage in the practice of leading discussions at each of the participating institutions.

This consensus-building effort around terminology and meaning is unusual and contrary to typical academic work for diverse research groups, who often gain recognition for doing something unique. Researchers in the Consortium, rather than assuming that diverse individual efforts will eventually converge to move the field forward, are taking a less common approach by collaborating initially.

Attend to the Relational Nature of Core Practices

To study core practices in teaching, researchers must attend to the relational nature of core practices—individual core practices do not occur in isolation but rather in the context of other practices, and this interconnectedness is critical. Evidence for the relational nature of core practices comes from one of the pioneering groups in the core-practices work, the *Learning In, From, and For Teaching Practice* project. Inspired by a similar approach for teaching Italian (Lampert & Graziani, 2009), this group has made use of strategically selected instructional activities—short episodes of teaching addressing an instructional goal—that are meant to be learning opportunities for not only students but also novice teachers. These instructional activities capture the relational nature of core practices by providing a "bounded space in which all of the aspects of practice exist relationally" (Ghousseini, Franke, & Turrou, 2014, p. 5).

For example, consider their instructional activity of choral counting:

> In this activity, the teacher leads the class in skip counting aloud, teaching different concepts and skills by deciding what number to start with, what to count by (e.g., by 10s,

察在计数过程中出现的模式，以及共同建构对模式中出现的数学的解释（第5页）。

大声计数教学活动的目标是超越死记硬背的计数，让学生有机会在计数时讨论数学模式。要实施这个教学活动需要同时介入和联系多个核心实践，例如引出和回应学生的数学想法，将学生指向参考他人的数学想法，通过书面形式表示数学想法，以及在不同的对话和表示中建立联系。来自该研究小组的量化数据进一步支撑了核心实践的关联性本质。在他们关于大学课堂中辅导新手教师排练教学活动的分析中，在超过1200个教师教育者/新手教师的互动中有几乎三分之二的互动涉及多于一个的核心实践（Lampert等，2013）。

教学中核心实践研究的结束语

在本节中，我们从核心实践团队项目中有选择地指出了以下四项核心实践教学研究需要考虑的方面：（1）判断核心实践的标准；（2）根据目标确定一组核心实践；（3）用共同的且精确的语言表述核心实践；（4）关注核心实践的关联性本质。核心实践团队的研究者也提出核心实践方式的部分力量在于超越了这些特殊的实践活动本身。核心实践的研究可以支持帮助教师发展在自身的课堂实践中，向课堂实践学习的技能和认同感，从而使他们的学习具有生成性（Chan，2010；Kazemi等，2009；McDonald等，2013）。

目前核心实践团队的工作聚焦在新手教师上，但是利用核心实践的方式关注经验教师的研究和专业发展已经在进行中了（例如Gibbons, Hintz, Kazemi, & Hartmann, 2016；J. Webb, Wilson, Reid, & Duggan, 2015）。在下面的内容中，我们将转向两个具体的核心实践，即教师关注和引领讨论，在过去十年，关于新手教师和经验教师的这两个核心实践都已经被广泛研究。

在结束本节时，我们重申教学中核心实践研究的潜力。这项工作抓住了教学的复杂性，从而可以帮助研究者在以往棘手的问题上取得进步。然而，任何聚焦都会带来过于聚焦的危险，或者是仅以一些特殊的方式进行的话，可能会限制发展，只会强化现状。因此，我们必须警惕，这项很有发展前景的工作必须与本领域中其他教与学的问题联合进行考虑。

教师关注：数学教学中一个隐藏的核心实践

数学教育者在数学教师关注这一领域正表现出日益浓厚的兴趣。关注指的是将注意力集中到视觉上较为复杂的世界中的某一情境的特征上并对其进行意义建构的行为。这个术语通常可以用来描述个体的日常行为（如，关注到天气或某个朋友的反应），或者更为特殊地，可以用来辨别某一类专业人士共同具有的独特的关注行为模式。在专业领域内关于关注的研究有不同的名称，包括专业愿景（Goodwin，1994）和学科感知（Stevens & Hall，1998），而且受到心理学家关于日常关注的长期研究的启发（例如，容量限制[Schneider & Shiffrin，1977]，个人感知对所见的影响[Bartlett，1932]，未被看见的对象的特征[Most等，2001；Simons & Chabris，1999]）。在本节中，我们将特别聚焦在数学教师的关注上。

数学教师需要关注专长，从而在复杂的课堂环境中发现并理解具有教学重要性的特征。然而在历史上，这个核心实践通常会被忽略掉。教师关注通常被界定为在实际回应学生之前集中注意力并弄清楚学生说了什么和做了什么，它在瞬间迅速发生，不能被直接观察到。由于其内隐的本质，研究关注并获得关注这一专长是非常复杂的。在本节中，我们将描述本领域中对于教师关注的逐步理解和在数学教师关注研究领域中所采用的方法（参见Sherin, Jacobs, & Philipp，2011a和Schack, Fisher, & Wilhelm，待出版，以全面了解关于数学教师关注的研究）。本节的其余内容被分成三个组成部分：（1）教师关注值得研究的理由，（2）作为核心实践的教师关注的概念化，（3）研究教师关注的方法。

by 19s, by ¾ s), whether to count forward or backward, and when to stop. The teacher records the count on the board, stopping to elicit children's strategies for figuring out the next number, their observations about patterns that emerge in the count and to co-construct explanations of the mathematics that arise in the patterns. (p. 5)

The goal of choral counting is to go beyond rote counting to provide students opportunities to discuss mathematical patterns while the count proceeds. Enacting this instructional activity requires simultaneously engaging with and connecting multiple core practices, such as eliciting and responding to students' mathematical talk, orienting students toward one another's mathematical ideas, representing mathematical ideas in writing, and making connections between the conversation and representations. Quantitative data from this group further support the relational nature of core practices. In their analysis of the coaching of novice teachers during rehearsals of instructional activities in university classes, almost two thirds of more than 1,200 teacher-educator/novice-teacher exchanges involved attention to more than one core practice (Lampert et al., 2013).

Closing Thoughts on Research on Core Practices in Teaching

In this section, we drew selectively from the Consortium's projects to highlight four considerations in research on core practices in teaching: determine criteria for core practices, identify sets of core practices to target, use common and precise language for core practices, and attend to the relational nature of core practices. Researchers in the Consortium have also argued that part of the power in a core-practices approach is that the benefits go beyond the specific practices themselves. Research on core practices can support efforts to help teachers develop the skills and identities to learn in and from their own classroom practices, thus making their learning generative (Chan, 2010; Kazemi et al., 2009; McDonald et al., 2013).

Current work of the Consortium has focused on novice teachers, but already underway are research and professional development efforts that use a core-practices approach with practicing teachers (e.g., Gibbons, Hintz, Kazemi, & Hartmann, 2016; J. Webb, Wilson, Reid, & Duggan, 2015). In the next sections, we turn to two specific core practices—teacher noticing and leading discussions—that have been extensively studied with both novice and experienced teachers in the past decade.

In closing this section, we reiterate the potential of research on core practices in teaching. This work holds great promise for containing the complexity of teaching so that researchers can make progress on previously intractable issues. However, any narrowing brings with it a danger of narrowing too much or in particular ways that might constrain progress and simply reinforce the status quo. Thus, we caution that this promising work needs to be considered in connection with other teaching and learning issues in the field.

Teacher Noticing: A Hidden Core Practice of Mathematics Teaching

Mathematics educators have shown growing interest in the noticing of mathematics teachers. *Noticing* refers to the act of focusing attention on and making sense of situational features in a visually complex world. This term can be used generally to describe what individuals do on a daily basis (e.g., noticing the weather or a friend's reaction) or, more specifically, to identify distinct patterns of noticing that are common to particular groups of professionals. Research on noticing within professions has been investigated under a variety of names, including *professional vision* (Goodwin, 1994) and *disciplined perception* (Stevens & Hall, 1998), and has been informed by the long history of psychologists' investigations of everyday noticing (e.g., capacity limitations [Schneider & Shiffrin, 1977], the influence of personal perspectives on what is seen [Bartlett, 1932], and the characteristics of what is not seen [Most et al., 2001; Simons & Chabris, 1999]). In this section, we focus specifically on the noticing of mathematics teachers.

Mathematics teachers need noticing expertise to detect and make sense of instructionally important features in complex classroom environments. Historically, however, the field has often overlooked this core practice. *Teacher noticing*, which is often conceptualized as focusing attention on and making sense of what students say or do before actually responding to them, happens quickly, in the moment, and is not directly observable. Because of this hidden nature, studying noticing and gaining noticing expertise are complicated. In this section, we describe the field's growing understanding of teacher noticing and the methodologies used to study it (see also Sherin, Jacobs, & Philipp, 2011a, and Schack, Fisher, & Wilhelm, in press, for compilations of research on mathematics teacher noticing). The remainder of this section is organized in three parts: (1) reasons teacher noticing is worthy of study, (2) conceptualizations of the

为什么教师关注值得研究?

我们把教师关注看作是高效数学教学的核心实践,这是因为关注对教师根据教学中学生生成的各种想法作出即时性决策有奠基性的作用。因而,从概念层面来说,教师关注是值得研究的,因为教师只会对关注到的事物作出回应。现有研究已给出了将教师关注视作核心实践的其他证据。

第一,关注专长与其他感兴趣的结果有积极的联系。比如说,肖邦(2011)发现能够关注到学生思维细节(而不是将学生思维评估为正确或错误的)的教师更能用有成效的方式改造具有挑战性的数学任务,比如说提高任务的复杂程度和增加学生投入数学概念学习的机会。类似地,科斯廷、吉文、索特罗和施蒂格勒(2010)研究了教师对课堂教学实录中师生互动的分析后发现,那些聚焦在改进教学的建议上的教师分析和那些教师的学生学习得分增量之间有正相关。

第二,关注专长有助于教师的学习。课堂观察——包括现场观察和通过录像观察——被广泛用来帮助职前和在职教师学习教学技艺,而且从观察中学到什么取决于关注到了什么(Star & Strickland, 2008)。此外,教师关注专长还有助于发展从经验中学习的能力。梅森是该领域的奠基者,他认为当教师用某些方式学习即时关注时,他们在今后能更好地利用他们的经验更有目的性地实施教学(Mason, 2002, 2011)。

第三,关注专长并不是教师常规拥有的,但是已有研究一致地发现它是可以学习的实践。教师通常并不仅仅通过教学经验就能发展关注专长(Dreher & Kuntze, 2015;Jacobs, Lamb, & Philipp, 2010),研究给出了很多证据表明,在一定的支持下,职前教师(例如 Fernández, Llinares, & Valls, 2012;Roth McDuffie 等, 2014;Schack 等, 2013)和在职教师(例如 Goldsmith & Seago, 2011;Jacobs 等, 2010;van Es & Sherin, 2008)的关注专长都能得到提高。谢林和范·伊斯(2009)也给出了证据说明当教师在专业发展活动中提高了他们对课堂外学生数学思维的关注时,他们在课堂内的关注专长也会得到提高。

第四,教育之外的专业领域为理解关注实践的重要性提供了存在性证据。例如,米勒(2011)强调了深刻理解体育和航空领域中专家关注的益处。关注球的运动路径或仪器面板并理解其中所涉及的要素,对于开发模拟和视频以帮助个体获得这些领域的专长很有帮助。数学教育正努力向这个方向前进,但是相关研究项目和出版物的数量已经有了飞速的增长。在下一节,我们将探索研究者是如何对数学教师关注进行概念化的。

作为核心实践的教师关注是如何概念化的?

研究者对数学教师关注的概念化,不论是其构成要素还是感兴趣的关注的类型,有不同的理解。这种多样性反映了这是一个新兴的领域。接下来我们将展示其中的一些主要工作,以说明这一概念化的范围。

教师关注的要素。关注发生于某一瞬间,而且是循环往复出现的。特别地,教师在课堂里时,先集中注意力于某一情境的特征,并解释其意义,之后再作出回应。他们的回应会影响之后的教学事件,并导致产生了一组需要注意的新的情境特征,就这样不断循环下去(Sherin, Jacobs, & Philipp, 2011b)。然而,不同的研究者对这一循环中的教师关注有不同的考虑。我们将描述教师关注的三种概念化(参见图28.1),每一种反映了不同的范围,取决于包括了以下三要素中的哪些要素。这三种要素是(a)注意,(b)解释,和(c)下一步行为的决策。我们将描述每一种教师关注的概念化形成,并用一个常常被引用的研究进行说明。

概念化1:注意。在第一种概念化中,教师关注本质上等同于注意,即什么吸引了或者没有吸引教师的注意力。例如,斯塔尔和他的同事们(Star, Lynch, & Perova, 2011;Star & Strickland, 2008)让职前教师在观

core practice of teacher noticing, and (3) methods for studying teacher noticing.

Why Is Teacher Noticing Worthy of Study?

We consider teacher noticing to be a core practice of high-quality mathematics instruction because it is foundational for teachers' in-the-moment decision making that takes into account the variety of students' ideas that emerge during instruction. Thus, on a conceptual level, teacher noticing is worthy of study because teachers can be responsive only to what has been noticed. Research has provided additional evidence for considering teacher noticing a core practice.

First, noticing expertise has been positively linked to other outcomes of interest. For example, Choppin (2011) found that teachers who noticed the details in students' thinking (vs. evaluating their thinking as correct or incorrect) were able to adapt challenging mathematical tasks in more productive ways, such as enhancing task complexity and increasing students' opportunities to engage with mathematical concepts. Similarly, Kersting, Givvin, Sotelo, and Stigler (2010) examined teachers' analyses of teacher–student interactions in a collection of classroom videos, and they found that teachers' analyses that were focused on suggestions for instructional improvement were positively linked to the learning gains of the teachers' own students.

Second, noticing expertise can enable teacher learning. Classroom observation—both live and via video—has been used extensively to help prospective and practicing teachers learn their craft, and what can be learned from observation depends directly on what is noticed (Star & Strickland, 2008). In addition, teachers' noticing expertise can play a role in their abilities to learn from experience. Mason, one of the founders of this field of study, has argued that when teachers learn to notice in particular ways in the moment, they are better able to use their experiences to act more purposefully in the future (Mason, 2002, 2011).

Third, noticing expertise is not something that teachers routinely possess, but research has consistently shown that it is a learnable practice. Teachers generally do not develop noticing expertise through teaching experience alone (Dreher & Kuntze, 2015; Jacobs, Lamb, & Philipp, 2010), but research has provided ample evidence that, with support, noticing expertise can improve for both prospective teachers (e.g., Fernández, Llinares, & Valls, 2012; Roth McDuffie et al., 2014; Schack et al., 2013) and practicing teachers (e.g., Goldsmith & Seago, 2011; Jacobs et al., 2010; van Es & Sherin, 2008). Sherin and van Es (2009) have also provided evidence that when teachers improve their noticing of students' mathematical thinking *outside* the classroom in professional development, noticing *inside* their classrooms also improves.

Fourth, professions outside of education provide an existence proof of the importance of understanding the practice of noticing. For example, Miller (2011) highlighted the benefits of a deep understanding of expert noticing in the fields of athletics and aviation. Understanding what is involved with noticing the path of a ball or a panel of instruments has been useful for developing simulations and videos to help individuals gain expertise in these fields. Mathematics education is still working toward this understanding, but the number of related research projects and publications has been rapidly increasing, and in the next section, we explore how researchers are currently conceptualizing mathematics teacher noticing.

How Is the Core Practice of Teacher Noticing Conceptualized?

Researchers differ in how they conceptualize mathematics teacher noticing, both in terms of the components and the types of noticing that are of interest. This variety is reflective of an emerging field of study, and below we present some of the major efforts to illustrate the range of conceptualizations.

Components of teacher noticing. Noticing is enacted in the moment and is cyclical. Specifically, when teachers are in a classroom, they notice by focusing their attention on and making sense of situational features prior to responding. Their responses shape subsequent instructional events, resulting in a new set of situational features to notice, and the cycle continues (Sherin, Jacobs, & Philipp, 2011b). However, what is considered teacher noticing in this cycle varies across researchers. We describe three conceptualizations of teacher noticing (see Figure 28.1), each reflecting a different scope depending on how many of these three components are included: (*a*) attention, (*b*) interpretation, and (*c*) decisions about next steps. We describe each conceptualization and share an oft-cited research program to illustrate each.

Conceptualization 1: Attention. In the first conceptualization, *teacher noticing* is essentially synonymous with attention—what does and does not attract teachers' attention. For example, Star and colleagues (Star, Lynch, & Perova, 2011; Star & Strickland, 2008) asked prospective teachers to recall classroom features and events after watching a lesson on video. They tracked both how

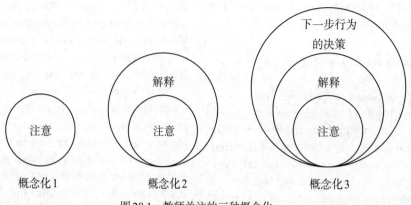

图28.1 教师关注的三种概念化

看一个课堂教学视频后回忆课堂特征和事件，同时追踪了这些职前教师可以回忆多少课堂特征和事件以及哪些类型的课堂特征和事件对于他们来说是突出或不突出的。研究者用这种方式捕捉了职前教师的注意的范围，这些信息将有助于后续课程中运用观察。斯塔尔和他的同事们认为注意是有时包括在教师关注中的其他实践行为的基础，值得单独拿出来研究。

概念化2：注意和解释。在第二种概念化中，教师关注除了包括什么吸引或者没有吸引教师的注意力，还延展到对事件的解释。例如，谢林和范·伊斯（2005，2009；另见van Es & Sherin，2002，2008）通过一系列与视频俱乐部相关的研究探究了教师关注，视频俱乐部是一种教师专业发展形式，在职教师合作讨论来自他们自己课堂的教学视频。研究者记录了教师关注什么（例如，关注授课教师或学生）和教师如何解释他们所看到的，最令人感兴趣的是教师如何利用他们所知道的背景对课堂情境进行推理的能力，和如何在具体的课堂事件与更为宽泛的教与学原则之间建立联系的能力。研究者还追踪了教师的立场，看他们的立场是描述性的、评估性的还是解释性的。因而，对于谢林和范·伊斯来说，教师如何解释教学情境与他们把注意力放在何处一样重要。

概念化3：注意，解释和下一步行为的决策。在第三种概念化中，教师关注不仅包括上述所提及的注意和对注意到的内容的解释，还包括对教师下一步教学决策的考虑。例如，雅各布和他的同事们（2010）提出了对儿童数学思维专业关注的构念，用以突出包含以下三个相互关联的实践的关注类型：（a）注意儿童策略的细节，（b）解释这些策略所反映的儿童的理解，（c）决定如何根据这些理解进行回应。因为教师具有考虑其下一步教学行动的自然倾向（Erickson，2011；Schoenfeld，1998），他们选择在教师关注的概念化中包括决定如何应对这个成分。更进一步地，决定如何应对在概念上与注意和解释有暂时的联系——这三种实践在课堂里融合在一起，而且几乎同时发生（Jacobs, Lamb, Philipp, & Schappelle，2011）。在一项典型的研究中，这些研究者利用视频和书面作业，评估职前教师和三组在参与专注于学生思维的专业发展活动年限上有差异的在职教师的关注专长。研究揭示了一系列的关注技能，因而为在何时和以何种方式发展对儿童数学思维的专业关注这一专长提供了初步的启示（Jacobs等，2010）。

教师关注的类型。除了在教师关注的概念化中所包括的要素不同之外，研究者选择调查的教师关注类型也不同。大多数研究聚焦在教师如何或者在多大程度上关注研究者感兴趣的内容——德雷赫和孔泽（2015）将其称为特定主题的关注。感兴趣的主题通常是已有研究表明对于高效数学教学非常重要的一般性教学特征，我们给出两个已经被广泛研究的主题：关于学生数学思维的教师关注和关于公平学习环境指标的教师关注。与这两个主题有关的研究并不是互相排斥的，而是相互重叠的，只是在哪个处于更显著的位置上有差异（Turner & Drake，

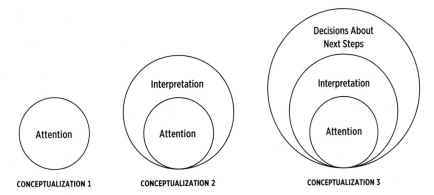

FIGURE 28.1. *Three conceptualizations of teacher noticing.*

much the prospective teachers were able to recall and what types of classroom features and events were and were not salient to them. In this way, the researchers captured the range of where prospective teachers placed their attention, which could inform the use of observation in future courses. Star and colleagues have argued that because attention is foundational to other practices that are sometimes included in teacher noticing, it is worthy of separate study.

Conceptualization 2: Attention and interpretation. In the second conceptualization, teacher noticing extends beyond what does and does not attract teachers' attention to include teachers' interpretations of events. For example, Sherin and van Es (2005, 2009; see also van Es & Sherin, 2002, 2008) explored teacher noticing in a series of studies involving video clubs—a form of professional development in which practicing teachers collaboratively discuss videos from their own classrooms. The researchers documented where teachers focused their attention (e.g., on the teacher or students) and how teachers interpreted what they saw. Of particular interest were teachers' abilities to use what they knew about the context to reason about the situation and to make connections between specific classroom events and broader principles of teaching and learning. The researchers also tracked whether the teachers' stance was descriptive, evaluative, or interpretive. Thus, for Sherin and van Es, how teachers interpreted instructional situations was as important as where they placed their attention.

Conceptualization 3: Attention, interpretation, and decisions about next steps. In the third conceptualization, teacher noticing includes not only the attending and interpreting components of noticing discussed above but also consideration of teachers' next instructional steps. For example, Jacobs and colleagues (2010) proposed the construct of *professional noticing of children's mathematical thinking* to highlight a particular type of noticing that includes three interrelated practices: (*a*) attending to the details of children's strategies, (*b*) interpreting children's understandings reflected in those strategies, and (*c*) deciding how to respond on the basis of those understandings. They chose to include *deciding how to respond* in their conceptualization of teacher noticing because of teachers' natural orientations toward considering their next moves (Erickson, 2011; Schoenfeld, 1998). Further, deciding how to respond is linked conceptually and temporally to attending and interpreting—all three practices are integrated and occur almost simultaneously, in the moment, in the classroom (Jacobs, Lamb, Philipp, & Schappelle, 2011). In a cross-sectional study, these researchers used video and written work to assess the noticing expertise of prospective teachers and three groups of practicing teachers who differed according to their number of years of professional development focused on children's thinking. Findings revealed a range of noticing expertise, thus providing initial insights into when and in what ways expertise in professional noticing of children's mathematical thinking can develop (Jacobs et al., 2010).

Types of teacher noticing. In addition to the number of components included in their conceptualizations of teacher noticing, researchers also vary in terms of the types of teacher noticing they choose to investigate. Most studies have focused on how, and the extent to which, teachers notice something of interest to researchers—what Dreher and Kuntze (2015) called *theme-specific noticing*. Themes of interest are generally instructional features that research has shown to be important to high-quality mathematics instruction, and we highlight two themes that have been widely studied: teacher noticing of students' mathematical thinking and teacher noticing of indicators of equitable learning environments. Studies related to these two themes are not mutually exclusive but instead overlap and differ in what is foregrounded (Turner & Drake, 2016). Other researchers have

2016）。其他研究者选择研究教师关注的多样性，因此不会采用特定主题的方式。下面我们将结合实例分享这三种方式。

特定主题的关注：关于学生数学思维的教师关注。最常见的特定主题的教师关注，是教师如何和在多大程度上关注学生的数学思维。对于该主题的普遍关注很可能是对教师的教学反馈要在引发和理解学生想法的基础上进行的这一观点日益得到认可的反映。然而，对于应当指向哪些学生的思维还存在很大差异，我们把相关的讨论分成特定内容领域和一般内容领域两种方式。

在特定内容领域方式下，研究者探究关于学生在某一数学内容领域下的思维的教师关注。有些聚焦在教师关注与基于研究的学生思维的知识之间的一致性，比如说源于长期研究项目的学习轨迹或框架。例如，沙克和他的同事们（2013）用斯特菲（1992）早期算术学习阶段来研究对于儿童算术的思维的教师关注，雅各布和他的同事们（2010）用源于认知指导教学（CGI）项目（Carpenter, Fennema, Franke, Levi, & Empson, 2015）的学生思维框架来研究关于儿童整数运算的思维的教师关注。其他采用特定内容领域方式的研究者聚焦在一些重要的数学内容上以及研究教师如何通过学生的发言或书面作业关注到这些重要的数学内容。这里所说的重要的数学内容包括比例问题中的乘法推理（Fernández, Llinares, & Valls, 2013），代数思维（Walkoe, 2015），运算性质的推广（Schifter, 2011）和加法式增长模式的推广（Zapatera & Callejo, 2013）。

在一般内容领域方式下，研究者用更大的视角调查关于学生思维的教师关注，强调关注可以运用到不同的数学内容领域。例如，利瑟姆、皮尔逊、斯托克罗和范泽斯特（2015；另见Stockero, 2014）曾讨论关于MOSTs——基于学生思维的数学上有重大意义的教学机会的教师关注的重要性。MOSTs是同时融合了学生的数学思维、具有重要意义的数学和教学机会的可以在任何数学内容领域下的课堂实例。研究者的假设是追踪这些课堂实例在促进学生学习上比其他实例更有潜力。类似地，桑塔伽塔（2011）也观察了跨数学内容领域的关于学生思维的教师关注，但是是在分析整堂课的背景下进行的；因此，她的研究包括了但同时也超越了即时的交互活动。她探索了教师关注在与课例分析框架（Santagata, Zannoni, & Stigler, 2007）相关方面的作用，其中，教师被敦促确定该堂课的学习目标，分析学生的思维从而决定学生是否已经在达成学习目标上有所进步，以及什么样的教学策略支持了这种进步，同时基于对学生思维的分析提出其他的教学策略并确定这些策略对学生学习有潜在的作用。

特定主题的关注：关于公平学习环境指标的教师关注。在特定主题关注方面，一个正在发展的研究领域是教师如何以及在多大程度上关注公平学习环境的指标。有些研究已经聚焦在教师所关注的特定的指标上。例如，汉德（2012）研究了已经在实施公平教学的教师，并记录了他们所关注的内容。她发现这些教师有规律地关注到课堂中权力和地位的作用，包括关注到学生如何参与课堂活动和他们如何在与数学、课堂任务等相关的活动中看待自己、以及别人是如何看待他们的。韦杰（2014）也聚焦在关于学生参与的教师关注上，她发现教师关注与他们如何将自己作为公平的数学教育者的自我身份的认识有关系。

其他研究者已经调查了与特定学生群体相联系的有关公平学习环境指标的教师关注。例如，费尔南德斯（2012）研究了教师所关注的英语学习者在建构意义和交流他们的数学思维时面临的语言挑战和可用资源的问题，而多明格斯和亚当斯（2013）探索了教师对学生双语能力在学生注意和参与数学任务中的作用的关注。

在这个领域中被引用得最多的研究项目之一是教数学，该项目是多个大学的合作项目，目的是对学前K-8年级的数学教师培养进行改革，以促进公平教学的实践。这项工作可以被看作是联系两种特定主题关注的桥梁，这是因为尽管研究者讨论了学生的数学思维，他们仍然认为对于学生思维的教师关注不足以应对针对各种学习者的有效数学教学。相反地，为了创设公平的学习

chosen to study the variety in what teachers notice, thereby not taking a theme-specific approach at all. We share examples of each of these three approaches below.

Theme-specific noticing: Teacher noticing of students' mathematical thinking. The most common focus of theme-specific noticing has been how and the extent to which teachers notice students' mathematical thinking. The prevalence of this focus is likely reflective of the growing endorsement for mathematics teaching in which students' ideas are elicited and understood before teachers respond. However, great variability is still found in terms of what student thinking is being targeted, and we have organized the discussion into domain-specific and domain-general approaches.

In the domain-specific approach, researchers explore teacher noticing of students' thinking in particular mathematical domains. Some have focused on the consistency of teacher noticing with research-based knowledge of students' thinking, such as learning trajectories or frameworks that derive from long-standing research programs. For example, Schack and colleagues (2013) used Steffe's (1992) stages of early arithmetic learning to study teacher noticing of children's early-numeracy thinking, and Jacobs and colleagues (2010) used children's-thinking frameworks from the Cognitively Guided Instruction (CGI) project (Carpenter, Fennema, Franke, Levi, & Empson, 2015) to study teacher noticing of children's thinking about whole-number operations. Other researchers taking a domain-specific approach have focused on key mathematical ideas and how teachers notice these ideas in students' comments or written work. Sample target mathematical ideas have included multiplicative reasoning in proportional problems (Fernández, Llinares, & Valls, 2013), algebraic thinking (Walkoe, 2015), generalizations about the behavior of operations (Schifter, 2011), and generalizations about additive-growth patterns (Zapatera & Callejo, 2013).

In the domain-general approach, researchers examine teacher noticing of students' thinking with a broader lens, highlighting noticing ideas that apply across mathematical domains. For example, Leatham, Peterson, Stockero, and Van Zoest (2015; see also Stockero, 2014) have argued for the importance of teacher noticing of MOSTs—mathematically significant pedagogical opportunities to build on student thinking. MOSTs are classroom instances, in any mathematical domain, that simultaneously involve students' mathematical thinking, significant mathematics, and pedagogical opportunities. The researchers posit that following up on these classroom instances has more potential for promoting student learning than following up on other instances. Santagata (2011) similarly looked at teacher noticing of students' thinking across mathematical domains, but in the context of analyzing entire classroom lessons; thus, her investigation included, but also moved beyond, in-the-moment interactions. She explored the role of teacher noticing in relation to a lesson analysis framework (Santagata, Zannoni, & Stigler, 2007) in which teachers were prompted to identify the learning goals for the lesson, analyze students' thinking to determine whether students had made progress toward achieving those goals and what instructional strategies supported that progress, and build on the analysis of students' thinking to propose alternative instructional strategies and identify their potential effects on student learning.

Theme-specific noticing: Teacher noticing of indicators of equitable learning environments. A growing area of research in theme-specific noticing has been how and the extent to which teachers notice indicators of equitable learning environments. Some of this research has focused on identifying specific indicators that teachers notice. For example, Hand (2012) studied teachers who were already teaching equitably and documented what they noticed. She found that these teachers regularly noticed the roles of power and status in the classroom, including noticing how students participated in classroom activities and how they positioned themselves and were positioned by others in relation to the mathematics, classroom tasks, and one another. Wager (2014) also focused on teacher noticing of students' participation, and she found that teachers' noticing was linked to their own identities in terms of how they positioned themselves as equitable mathematics educators.

Other researchers have investigated teacher noticing of indicators of equitable learning environments in relation to specific populations of students. For example, Fernandes (2012) studied teacher noticing of the linguistic challenges and resources available to English learners when they were building meaning and communicating their mathematical thinking, and Dominguez and Adams (2013) explored teacher noticing of the role that students' bilingualism played in students' noticing of and engagement with mathematical tasks.

One of the most cited projects in this strand of research is TEACH MATH—a multi-university collaboration to transform pre-K–8 mathematics teacher preparation in ways that promote equitable instructional practices. This work can be viewed as a bridge between the two types of theme-specific noticing because although these researchers addressed students' mathematical thinking, they also argued that teacher noticing of this thinking is insufficient to teach mathematics effectively to diverse learners. Instead, to create equitable learning environments, they

环境，他们强调教师需要关注更广范围的学生资源，即学生的多种数学知识基础，具体包括学生的数学思维，基于家庭、文化、社区的与数学有关的知识（Turner等，2012）。在与职前教师展开教学工作时，教数学项目的研究者鼓励职前教师与这些广泛的学生资源建立实质性的联系，并利用这些资源作为教师关注的基础（Roth McDuffie等，2014；Turner等，2014）。

超越特定主题的关注：教师关注的多样性。有些研究者并不聚焦在教师对某个特定内容的关注能力，而是研究教师关注的多样性。例如，在黄和李（2009）关于中国教师对视频课中重要课堂事件关注的新手-专家研究中，他们通过提出以下问题引发了这里所提到的多样性，"这些课的特征是什么？"（第424页）。其他研究者提出了更有针对性的问题，但是仍旧以捕捉跨课堂特征的教师关注为目标。例如，斯塔和他的同事（Star等，2011；Star & Strickland，2008）在五个类别（课堂环境、课堂管理、任务、数学内容和交流）下提出了问题，这些类别以及类别下的问题是研究者有意选择的，范围从研究者认为相对不重要的方面（例如，教室是否有窗户）到涉及数学本质的内容（例如，数学内容是如何被解释和表示的）。

我们没有发现教师关注多样性的研究中提到教师关注的内容是聚焦的，相反地，这些研究结果表明教师关注包括什么，超出了我们这一章所关注的教学交互。国际比较研究的结果因为揭示隐含的文化价值和假设，因而在理解这种多样性上特别有意义。例如，米勒和周（2007）报告了中国和美国小学教师的关注，这些教师被要求观看课堂教学视频并写下他们对所发现的有价值的内容的描述和评估。相比中国教师，美国教师更倾向于对与一般教学问题有关（如课堂结构，呈现方式和动机策略）的视频片段进行评论。相反地，中国教师更倾向于对其中的数学内容和与教学内容知识有关（例如，学生的理解，学生对于内容的困难和如何克服这些困难）的视频片段进行评论。研究者总结说这些在教师关注上的差异反映了两国在教师培养和学校组织上的差异，中国的小学教师更有可能是数学专业的，而且在内容领域受过很强的训练。

教师关注是如何被研究的？

研究教师关注很有挑战性，这是因为教师关注这一实践与特定的情境相联系而且在这些情境下并不是可以被直接观察到的。为了捕捉瞬时的教师关注，研究者需要中断课堂、提问教师或者让教师用出声想的方式说出他们的关注（Ericsson & Simon，1993），但是这些类型的中断很难在不打扰课堂教学的情况下实现（见Fredenberg，2015提到的一个例外情况）。

谢林、拉斯和科尔斯托克（2011）提到了三种研究教师关注的主要方式：（a）教师投入到研究者选择的展示其他教师课堂实践的人工制品，并分享他们对其的关注；（b）教师用课后回顾的方式分享他们对自己课堂的关注；（c）研究者审视教师在课堂中可见的行为，并在这些行为的基础上推断教师关注到了什么。每一种方式都有其优势和不足，需要更多的研究来理解可以用每一种方式捕捉的教师关注的类型及其之间的关系（见图28.2）。

第一，不同的情境为教师关注提供了不同的机会，只有第一种方式涉及使用共同的人工制品，有助于比较不同教师的教师关注（反映了早期专家-新手研究中使用的研究方法；见Berliner，1994，以获得概述）。第二，由于教师关注并不可直接观察，因而教师自己的看法有助于理解教师关注。在前两种方式中教师的声音尤为重要，但第三种方式需要基于教师的行为进行实质性的推断。研究者可能会基于教师的行为作出错误的推断，也可能不能辨别教师可能关注到的但没有转化成教学行为的事件和交互。第三，我们不知道教师对自己学生的关注和教师对缺乏历史和情境信息的其他教师的学生的关注有何质的区别。然而，只有第二种和第三种方式让教师们在关注时可以使用他们作为局内人的知识。最后，我们不知道教师的即时关注与时间不再是问题的专业发展活动（或其他离开课堂的情境）中的关注有何质量上的差异。然而，只有第三种方式保持了教师在课堂教学中进行关注活动时所感受到的即时性压力。

emphasized the need for teachers to notice a broader set of student resources—students' *multiple mathematical knowledge bases*, which include not only students' mathematical thinking but also their home, cultural, and community-based funds of knowledge relevant to mathematics (Turner et al., 2012). In their work with prospective teachers, the TEACH MATH researchers have encouraged substantive connections to this broader set of student resources, including using them as the basis for teacher noticing (Roth McDuffie et al., 2014; Turner et al., 2014).

Beyond theme-specific noticing: Variety in teacher noticing. Rather than focus on teachers' abilities to notice something in particular, some researchers have explored the variety in what teachers notice. For example, in Huang and Li's (2009) expert–novice study of Chinese teachers' noticing of important classroom events in video lessons, they elicited this variety by posing such general questions as "What are the characteristics of these lessons?" (p. 424). Other researchers have posed more targeted questions but still with the goal of capturing teacher noticing across a range of classroom features. For instance, Star and colleagues (Star et al., 2011; Star & Strickland, 2008) posed questions in five broad categories (classroom environment, classroom management, tasks, mathematical content, and communication), and the categories, as well as the questions within them, were purposefully selected to range from what the researchers considered to be relatively unimportant (e.g., whether the classroom had windows) to the mathematically substantive (e.g., how mathematical content was explained and represented).

We did not find the studies about the variety in what teachers noticed to be convergent, but results did indicate that what teachers notice includes, but also goes beyond, our chapter focus on the interactive part of teaching. Results from international comparisons are of particular interest in making sense of this variety because they can be useful for uncovering hidden cultural values and assumptions. For example, Miller and Zhou (2007) reported on the noticing of Chinese and U.S. elementary school teachers who were asked to view classroom videos and then write a description and evaluation of what they found noteworthy. Compared to Chinese teachers, U.S. teachers were more likely to comment on parts of the videos related to general pedagogical issues (e.g., classroom structure, presentation style, and motivational strategies). In contrast, Chinese teachers were more likely to comment on the mathematical content of the lessons and parts of the videos related to pedagogical content knowledge issues (e.g., students' understandings as well as students' difficulties with the content and how to overcome those difficulties). The researchers concluded that these differences in the teachers' noticing were reflective of the teacher training and school organization in the two countries, given that Chinese elementary school teachers are more likely to be specialized in mathematics and to have had training focused heavily on content-related instruction.

How Is Teacher Noticing Studied?

Studying teacher noticing is challenging because the practice is tied to specific situations but is not directly observable in those situations. To capture teacher noticing in the moment, researchers would need to stop a lesson midstream to question teachers or have teachers think aloud about their noticing (Ericsson & Simon, 1993), but these types of interruptions are difficult to enact without disrupting the lesson itself (see Fredenberg, 2015, for an exception).

Sherin, Russ, and Colestock (2011) have identified three major approaches to studying teacher noticing: (*a*) teachers engage with researcher-selected artifacts of practice from other teachers' classrooms and share their noticing about those artifacts; (*b*) teachers retrospectively share their noticing about their own lessons after they have finished teaching those lessons; and (*c*) researchers examine teachers' visible actions during instruction and infer, on the basis of those actions, what teachers noticed. Each approach has advantages and disadvantages, and more research is needed to understand the types of teacher noticing that can be captured by each approach and their relationships (see Figure 28.2).

First, different situations provide different opportunities for teachers to notice so only the first approach, which involves the use of common artifacts, facilitates comparing teacher noticing across teachers (reflective of the methodology used in earlier expert–novice studies; see Berliner, 1994, for a summary). Second, because teacher noticing is not directly observable, teachers' perspectives can be helpful in understanding their noticing. Teachers' voices are prominent in the first two approaches, but the third approach requires substantial inference based on teachers' actions. Researchers not only might infer incorrectly from teachers' actions but will also fail to identify events and interactions that teachers may have noticed but not (yet) converted into instructional actions. Third, we do not know if teacher noticing of their own students is qualitatively different from their noticing of other teachers' students for whom they lack historical and contextual information. However, only the second and third approaches provide access to teacher noticing in which teachers can use their insider knowledge. Finally, we do not know if teacher noticing in the moment is qualitatively different from their noticing in professional development (or other situations

		研究教师关注的方式		
		方式1 教师分享他们对研究者选择的人工制品的关注	方式2 教师课后回顾并分享他们对自己的课堂的关注	方式3 研究者从教师行为推断他们的关注
优势	研究者能轻易比较不同教师的教师关注	X		
	研究者能从教师的角度触及教师关注	X	X	
	研究者能从局内者关于学生知识的角度触及教师关注		X	X
	研究者能触及涉及即时压力的教师关注			X

图28.2　用于研究教师关注的三种方式的优势和不足

最为常见的方法。目前研究者一般采用前两种方式来研究教师关注——让教师（或不同类别的教师）参与到课堂实践的人工制品（来自于他们自己的课堂或其他教师的课堂）并用口头或书面方式分享他们的关注。技术上的发展促进了从教师与人工制品的在线交互过程中收集数据（例如 Chao & Murray, 2013；Fernández 等, 2012；Walkoe, 2015）。

在教师关注研究中使用的课堂实践的人工制品通常是教学视频（全班，小组，或一对一的交互）和学生的书面作业。除此以外，研究者还使用照片（Oslund & Crespo, 2014）、转录记录（Dreher & Kuntze, 2015；Scherrer & Stein, 2013）和课堂故事的动画（Chieu, Herbst, & Weiss, 2011）等方式提供实践的影像。

人工制品的选择是复杂和重要的。需要考虑的问题包括了人工制品是由研究者还是教师提供，视频片段的长度或书面作业的量，视频是否需要编辑，被描绘的学生数学思维的可视性和清晰度，等等。各研究项目选择人工制品的标准各不相同，并且往往会成为研究目标之一（例如 Goldsmith & Seago, 2011；Sherin, Linsenmeier, & van Es, 2009）。我们还注意到许多研究使用同样的方法和人工制品来捕获教师关注，以帮助教师发展这一专长。人工制品的选择在教师学习的情境中同样重要，但根据主要目的究竟是教师评价还是教师发展会有不同的决定。

有前景的方法。技术的发展为捕捉教师关注带来了激动人心的新选择。一些研究者使用教师可以穿戴的（例如可以附在帽子或眼镜上）可移动的视频摄像机来生成源于教师视角的视频（区别于更为传统的观察者视角）。有些摄像机具有选择性归档功能，因而教师在教学中可以选择某个瞬间进行迅速摄影，如果教师认为那个瞬间是非常重要的。这些教师选择的视频片段可以被用来进行后继关于关注的讨论，从而帮助教师回忆一些特定的片段并解释当时他们为什么认为这些片段是重要的（Colestock, 2009；Sherin, Russ, Sherin, & Colestock, 2008；Stockero, 2013）。这类技术的使用能帮助研究者在真实的情境下更近距离地研究瞬时发生的教师关注（Sherin 等, 2011）。

关于教师关注的结束语

在本节中，我们讨论了教师关注的概念化和使用的研究方法。教师关注是本章中强调的两个核心实践中的

		Approaches to Studying Teacher Noticing		
		Approach 1 Teachers share their noticing about researcher-selected artifacts of practice.	**Approach 2** Teachers share their noticing about their own lessons retrospectively.	**Approach 3** Researchers infer teachers' noticing from teachers' actions.
Benefits	Researchers can easily compare teacher noticing across teachers.	X		
	Researchers can access teachers' perspectives on their noticing.	X	X	
	Researchers can access teacher noticing that allows for insider knowledge about students.		X	X
	Researchers can access teacher noticing that involves in-the-moment pressures.			X

FIGURE 28.2. *Advantages and disadvantages of three approaches to studying teacher noticing.*

outside the classroom) when time is less of an issue. However, only the third approach preserves the in-the-moment pressures of competing demands that teachers feel when noticing in the midst of instruction.

Most common methodologies. Researchers currently studying teacher noticing have generally adopted one of the first two approaches—asking teachers (or groups of teachers) to engage with artifacts of practice (from their own or other teachers' classrooms) and then to share their noticing verbally or in writing. Technological advances have also facilitated data collection from teachers' online interactions with artifacts (e.g., Chao & Murray, 2013; Fernández et al., 2012; Walkoe, 2015).

The artifacts of practice used in teacher-noticing research are most often video (of whole-class, small-group, or one-on-one interactions) and student written work. In addition, researchers have explored the use of photographs (Oslund & Crespo, 2014), transcripts (Dreher & Kuntze, 2015; Scherrer & Stein, 2013), and animations of classroom stories (Chieu, Herbst, & Weiss, 2011) to provide images of practice.

Selection of artifacts is complex and significant. Issues to consider include whether the artifacts will be provided by researchers or teachers, the length of the video-recorded episode or the amount of written work to be shared, whether video will be edited or unedited, and visibility and clarity of the student mathematical thinking portrayed, to name a few. Selection criteria vary across projects and are often an object of study themselves (e.g., Goldsmith & Seago, 2011; Sherin, Linsenmeier, & van Es, 2009). We also note that many of the same methodologies and artifacts used to capture teacher noticing have been used to help teachers develop this expertise. Selection of artifacts is equally important in these teacher-learning situations, but different decisions might be made depending on whether teacher assessment or development is the major goal.

Promising methodologies. Advances in technology have provided an exciting new option for capturing teacher noticing. Several researchers have used portable video cameras that teachers can wear (e.g., attached to a hat or glasses) to generate video from the teacher perspective (vs. the more traditional observer perspective). Some of the cameras are equipped with selective archiving capabilities so that teachers—while teaching—can choose moments of instruction to capture on video immediately after they occur if teachers deem them important. These teacher-selected video clips can then be used in follow-up conversations about noticing to help teachers recall specific episodes and explain why they deemed these episodes important in the moment (Colestock, 2009; Sherin, Russ, Sherin, & Colestock, 2008; Stockero, 2013). The use of this technology brings researchers closer to being able to study teacher noticing in the moment in authentic contexts (Sherin et al., 2011).

Closing Thoughts on Teacher Noticing

In this section, we explored the conceptualization of and methodologies used to study teacher noticing, the

第一个。我们只聚焦在K-12年级职前和在职数学教师关注的研究，但是数学教师关注之外的关于关注的广泛的（同时也是快速发展的）研究领域同样也有助于本研究工作的开展。例如，我们能从其他内容领域的关注研究和专业发展活动中受益，比如语文（Garner & Rosaen，2009；Rosaen 等，2010；Ross & Gibson，2010）和科学（Russ & Luna，2013；Seidel，Stürmer，Blomberg，Kobarg，& Schwindt，2011；Talanquer，Tomanek，& Novodvorsky，2013）。类似地，数学教师关注的研究工作也受到非教师类群体关注研究的启发，比如专业发展领导者（Kazemi，Elliott，Mumme，Lesseig，& Kelley-Peterson，2011）、校长（Johnson，Uline，& Perez，2011）、大学数学教师（Breen，McCluskey，Meehan，O'Donovan，& O'Shea，2014）、研究者（Wager，2014）和学生（Lobato，Hohensee，& Rhodehamel，2013）。最后，还可以从旨在联系我们知道的教师关注和其他感兴趣的教育构念的研究中受益，这些教育构念包括教师的态度（Fisher 等，2014）、倾向（Hand，2012）、专业知识（Dreher & Kuntze，2015）、学科特定的知识（Blomberg，Stürmer，& Seidel，2011；Stürmer，Könings，& Seidel，2013）和职位认同（Wager，2014）。

尽管关注的重要性已经逐渐得到认同，但是我们用一条警示结束本节。我们担心教师关注作为描述教学核心实践的一个概念可能处于失去其力量的边缘，因为术语关注如何被使用往往缺乏准确性。在核心实践团队项目研究者通过合作研究发展精确的语言的基础上，我们鼓励研究者在调查教师关注时要清晰地表达他们是如何定义教师关注并对其进行可操作化界定的，这样的话其他人能够理解他们的概念化，而且教师关注这个构念本身不会落入针对所有教学的通用的同义词。这个警示并不意味着将本领域聚焦到一个单一的教师关注观点上。数学教师关注还是一个正在发展的领域，因为根据情境的不同概念化可以提供不同的好处，我们相信这些差异性值得探究。然而，实现教师关注研究的潜力依赖于对核心实践是什么（和不是什么）的清晰认识。

引领讨论：数学教学中的一个核心实践

数学教育者长期以来对课堂交流都有兴趣，近年来这种兴趣得到了增长，这是因为目前对高效数学课堂的憧憬包括了给予学生解释他们的思维和理解并评论其他学生推理的机会。为确保所有学生都有这样的机会需要教师在引领课堂讨论时有意识地进行实施。[3]

我们在本章中使用讨论这个术语，但是同时认识到该领域研究者使用了各种各样的术语来描述课堂交流，包括探究性的发言（Mercer & Sams，2006），教学对话（Moschkovich，2008）和集体辩论（Conner，Singletary，Smith，Wagner，& Francisco，2014；Forman，Larreamendy-Joerns，Stein，& Brown，1998）。我们选择使用术语讨论，从而将本节与这套书中关于话语的那一章区分开来。本节与赫贝尔-艾森曼、米尼、毕肖普和海德-梅朱亚尼姆（2017，这套书）话语那一章的区别在于，我们只关注教师在引领讨论时的作用，而他们则采纳了一种与话语有关的不同理论下的更为宽泛的视角。

我们选择术语讨论也是为了与核心实践文献保持一致性。正如上文所述，核心实践团队项目中的研究者正在将引领讨论作为一个教学的核心实践进行探索，而且，为了与他们使用共同的且精确的术语所付出的努力保持一致，他们集体采用了下面关于课堂讨论的工作定义，我们也将采纳这个定义：

> 在全班讨论中，教师和所有学生共同致力于某一特定内容，使用相互的想法作为资源。讨论的目的是建立与特定教学目标相关的共同的知识和能力，以及让学生能进行听、说和解释的练习。在富有教学成效的讨论中，教师和大部分学生口头发表自己看法、积极聆听，并且对其他人的发言进行反馈并从中学习（Grossman 等，2014，引言，10:07—11:10）。

本节的其他内容分为三个部分：（1）引领讨论的实践值得研究的理由，（2）核心实践引领讨论的概念化，（3）研究教师如何引领讨论的方法。

first of two core practices of teaching highlighted in this chapter. We focused exclusively on research on noticing by K–12 prospective and practicing mathematics teachers, but the extensive (and rapidly growing) field of noticing outside of mathematics teacher noticing could also inform this work. For example, we could learn from research and professional development efforts focused on noticing in other content areas, such as literacy (Garner & Rosaen, 2009; Rosaen et al., 2010; Ross & Gibson, 2010) and science (Russ & Luna, 2013; Seidel, Stürmer, Blomberg, Kobarg, & Schwindt, 2011; Talanquer, Tomanek, & Novodvorsky, 2013). Similarly, work on mathematics teacher noticing could be informed by research on the noticing of nonteacher groups, such as leaders of professional development (Kazemi, Elliott, Mumme, Lesseig, & Kelley-Peterson, 2011), principals (Johnson, Uline, & Perez, 2011), university mathematics faculty (Breen, McCluskey, Meehan, O'Donovan, & O'Shea, 2014), researchers (Wager, 2014), and students (Lobato, Hohensee, & Rhodehamel, 2013). Finally, lessons can be learned from research designed to connect what we know about teacher noticing with other educational constructs of interest, such as teachers' attitudes (Fisher et al., 2014), dispositions (Hand, 2012), professional knowledge (Dreher & Kuntze, 2015), subject-specific knowledge (Blomberg, Stürmer, & Seidel, 2011; Stürmer, Könings, & Seidel, 2013), and positional identities (Wager, 2014).

Despite this growing endorsement for the importance of noticing, we end this section with a cautionary note. We worry that teacher noticing may be on the verge of losing its power as a construct that describes a core practice of teaching because of the lack of precision in how the term *noticing* is sometimes used. Building on the efforts of researchers in the Core Practice Consortium to develop precise language in their collaborative work, we encourage researchers investigating teacher noticing to clearly describe how they are defining and operationalizing *teacher noticing* so that others will understand their conceptualizations and the construct itself will not devolve into a generic synonym for all of teaching. This caution is not meant to suggest that the field converge on a singular view of teacher noticing. Mathematics teacher noticing is still an emerging field, and because different conceptualizations can offer different benefits depending on the situation, we believe that those differences are worthy of investigation. However, realizing the potential of research on teacher noticing is dependent on maintaining a clear focus on what this core practice is (and is not).

Leading Discussions: A Core Practice of Mathematics Teaching

Mathematics educators have long been interested in classroom communication, and this interest has increased in recent years because contemporary visions of high-quality mathematics instruction include opportunities for students to explain their thinking and make sense of and critique the reasoning of others. Ensuring that these opportunities arise for all students requires teachers to make intentional moves when leading discussions.[3]

We use the term *discussion* in this chapter, but we recognize that researchers have used a variety of terms to describe classroom communication, including *exploratory talk* (Mercer & Sams, 2006), *instructional conversations* (Moschkovich, 2008), and *collective argumentation* (Conner, Singletary, Smith, Wagner, & Francisco, 2014; Forman, Larreamendy-Joerns, Stein, & Brown, 1998). We chose the term *discussion* to distinguish this section from a chapter on discourse in this volume. This section differs from the discourse chapter by Herbel-Eisenmann, Meaney, Bishop, and Heyd-Metzuyanim (2017, this volume) in that we focus exclusively on teachers' roles in leading discussions, whereas they take a broader view of various theoretical orientations to discourse.

We also chose the term *discussion* for consistency with the core-practices literature. As noted in an earlier section, researchers in the Core Practice Consortium are currently exploring leading discussions as a core practice of teaching, and in keeping with their efforts to use common and precise terminology, they have collectively adopted the following working definition of *classroom discussion*, which we also adopt:

> In a whole-class discussion, the teacher and all of the students work on specific content together, using one another's ideas as resources. The purposes of a discussion are to build collective knowledge and capability in relation to specific instructional goals and to allow students to practice listening, speaking, and interpreting. In instructionally productive discussions, the teacher and a wide range of students contribute orally, listen actively, and respond to and learn from others' contributions. (Grossman et al., 2014, Introduction, 10:07–11:10)

The remainder of this section is organized in three parts: (1) reasons the practice of leading discussions is worthy of study, (2) conceptualizations of the core practice of leading discussions, and (3) methods for studying how teachers lead discussions.

为什么引领讨论值得关注？

我们认为引领讨论是高效数学课堂教学的一个核心实践，这是因为教师组织安排数学发言的方式直接影响学生学习数学的机会。为引发学生的思维和使学生参与到其他人的想法中，教师通过学习学生如何看待数学、学生能理解的表征方式和学生所看到的联系来形成他们的教学决策。引领讨论值得研究的想法与当前呼吁学生分享他们的思维并对他人的推理进行批判的政策文件（例如NCTM, 2014；National Governors Association Center for Best Practices & Council of Chief State School Officers, 2010）相符，而且研究给出了额外的证据说明引领讨论是一个核心实践。

研究表明学生参与数学讨论是高效教学的基本要素（见Walshaw & Anthony, 2008，以获得完整了解）。更进一步地，引领讨论的专长很重要，这是因为学生在讨论中的参与和认知及情感获得都有正向关系。这样的正向结果可以发生在所有学生身上，有一些研究表明对于长期以来缺少关注的学生群体还能获得特殊的益处。

针对认知获得，博尔勒和斯特普尔斯（2008）发现，与学校中不使用讨论的对照组相比，小组和整班情境下讨论数学对于所有学生来说能导致更好的成绩，而且该学校中充分使用讨论的小组之间的成绩差距非常接近。[4] 类似地，N.M.韦伯和他的同事（2014）发现，让学生将自己的思维与同伴提出来的想法联系起来，能促进学生在讨论中的参与程度，而学生在讨论中的参与程度与学习的提高相联系，这种学习的提高可以用标准化的测试和研究者开发的评估测量出来，不过这两者之间的因果关系还没有建立起来（见N.M. Webb等, 2014，以获得学生在讨论中的参与与认知获得之间关系的研究工作的汇总）。

针对情感获得，研究表明在有丰富课堂讨论活动的教室中学习数学的学生跟其他学生相比更喜欢数学（Boaler & Staples, 2008；Goos, 2004），对数学有更积极的倾向（Gresalfi, 2009）。此外，学生还发展了与数学相关的权威性、能动性、认同感和拥有感（Bishop, 2012；Boaler & Staples, 2008；Goos, 2004；Gresalfi, 2009；Hunter, 2010；Turner, Dominguez, Maldonado, & Empson, 2013）。

最后，和教师关注的核心实践一样，引领讨论的专长也不是教师常规拥有的（例如Boston & Smith, 2009；Clarke, 2013；Khisty & Chval, 2002），但是研究持续表明教师可以学会有效地引领讨论（例如Boston & Smith, 2009；Hunter, 2010；Nathan & Knuth, 2003；Silver, Ghousseini, Gosen, Charalambous, & Strawhun, 2005；Staples & Truxaw, 2010）。现在我们转向对引领讨论核心实践的概念化的具体描述。

引领讨论的核心实践是如何概念化的？

绝大多数调查教师引领课堂讨论方式的研究者，通过质性研究描述当讨论发生时课堂上的情况。这些描述通常包括记录教师做了什么，学生如何回应和根据学生的回应教师做了什么，还常常包括对于所讨论的数学内容的丰富性和准确性的评价。考虑到本章的目的，我们聚焦在教师做了什么，我们把教师做了什么称为教学行动，这些行动想要的效果称为目标。

我们把教学行动定义为教师采取的、观察者可以看到或听到的行为，比如说提问、给出表征形式或者修改任务。教学行动有各种不同的尺度，从诸如复述（教师重复或解释学生已经说过的内容）这样特定的行为到诸如提问（可以被进一步分解为很多精确的类别）这样更大的、更一般的行为。我们需要警惕，教学活动是相当微妙的，因为它们可以以对学生产生积极或消极影响的方式实施。换句话说，一个行动是好的还是坏的，并不由行动本身所固有，而是取决于它是出于何种目的、在何种情境下、用什么样的方式实施（参见Moschkovich, 1999和Turner等, 2013，以了解复述的使用方式如何积极地或消极地影响英语学习者）。

我们把目标定义为教师对一堂课中的特定时刻、一堂课的片段、整堂课、一系列课或一学年课的意图。与教学行动可以被观察到不同，教师的目标一定是研究者推断的，因为它们通常不是明确给出的。然而，与教学行动相类似，目标也有各种不同的尺度。例如，教师可能有许多很具体的目标，有的与身边的数学有关（比如说保证某种特定的表征方式得以分

Why Is Leading Discussions Worthy of Study?

We consider leading discussions to be a core practice of high-quality mathematics instruction because the ways that teachers orchestrate mathematical talk directly influence students' opportunities to learn mathematics. In eliciting students' thinking and engaging students with one another's ideas, teachers shape their instructional decisions by learning how students are viewing the mathematics, the representations that make sense to students, and the connections students see. The idea that leading discussions is worthy of study is consistent with contemporary policy documents that call for students to share their thinking and critique the reasoning of others (e.g., NCTM, 2014; National Governors Association Center for Best Practices & Council of Chief State School Officers, 2010), and research has provided additional evidence for considering leading discussions a core practice.

Research has shown that student engagement in mathematical discussions is an essential component of high-quality instruction (see Walshaw & Anthony, 2008, for a comprehensive review). Further, expertise in leading discussions is important because student engagement in discussions has been positively linked to both cognitive and affective outcomes. These positive outcomes occur for all students, and some studies have shown particular benefits for students from historically underserved groups.

Related to cognitive outcomes, Boaler and Staples (2008) found that discussing mathematics in small-group and whole-class settings led to higher achievement for all students, when compared with a matched group in a school that did not use discussion, and closed achievement gaps between some subgroups in the school that used discussion extensively.[4] Similarly, N.M. Webb and colleagues (2014) found that asking students to connect their thinking to ideas advanced by their peers enhanced student engagement in discussions, which correlated with increased student learning as measured by both standardized tests and researcher-developed assessments, although clear causality was not established. (See N.M. Webb et al., 2014, for a summary of work on the link between students' participation in discussion and cognitive outcomes.)

Related to affective outcomes, research has shown that students who learn mathematics in discussion-rich classrooms are more likely than others to enjoy mathematics (Boaler & Staples, 2008; Goos, 2004) and have productive dispositions toward mathematics (Gresalfi, 2009). In addition, students develop senses of authority, agency, identity, and ownership related to mathematics (Bishop, 2012; Boaler & Staples, 2008; Goos, 2004; Gresalfi, 2009; Hunter, 2010; Turner, Dominguez, Maldonado, & Empson, 2013).

Finally, as with the core practice of teacher noticing, expertise in leading discussions is not something teachers routinely possess (e.g., Boston & Smith, 2009; Clarke, 2013; Khisty & Chval, 2002), but research has consistently shown that teachers can learn to lead discussions effectively (e.g., Boston & Smith, 2009; Hunter, 2010; Nathan & Knuth, 2003; Silver, Ghousseini, Gosen, Charalambous, & Strawhun, 2005; Staples & Truxaw, 2010). We now turn to an elaboration of the ways that researchers are conceptualizing the core practice of leading discussions.

How Is the Core Practice of Leading Discussions Conceptualized?

Most researchers who investigate the ways teachers lead discussions conduct qualitative studies in which they describe what happens in classrooms when discussions take place. These descriptions usually include transcripts that show what teachers do, how students respond, and what teachers do in light of students' responses, and they often include assessments of the richness and accuracy of the mathematics discussed. For purposes of this chapter, we focus on what teachers do, and we call what teachers do *teaching moves* and the desired outcomes of these moves *goals*.

We define *teaching moves* as actions that teachers take that observers can see or hear, such as asking a question, providing a representation, or modifying a task. Teaching moves come in a variety of grain sizes, ranging from specific moves, such as *revoicing* (in which teachers repeat or paraphrase something students have said), to larger, more general moves, such as questioning (which can be further decomposed into more precise categories). We caution that teaching moves are considerably nuanced in that they can be enacted in ways that have either positive or negative effects on students. In other words, whether a move is good or bad is not inherent in the move but is determined by how it is enacted, for what purpose(s), and in what context. (See Moschkovich, 1999, and Turner et al., 2013, for examples of ways in which revoicing can be used to position English learners both positively and negatively.)

We define *goals* as the intentions teachers have for particular moments in a lesson, larger segments of a lesson, an entire lesson, a series of lessons, or a school year. Unlike teaching moves, which can be observed, teachers' goals typically must be inferred by researchers because they are not usually stated explicitly. However, similar

享），有的与社会规范有关（比如说保证某位学生在课堂中能够发言）。较大的目标包括培养评定学生的能力，鼓励他们在面对困难时的坚韧性，以及帮助学生遵守纪律规范。

教学行动和目标之间的关系在三个方面体现出复杂性。第一，使用同样的教学行动可以达成不同的目标，不同的教学行动可以达成相同的目标。例如，复述的教学行动可以被用来实现评定学生的能力，帮助确保每个人都遵循讨论的要点，或者为讨论提供支持等目的。类似地，复述、提供表征、修改任务和策略性地使用黑板等行动都可以达成保证每个人都遵循讨论的要点这一教学目标。

第二，教师可以使用一种或多种行动同时实现多个目标，目标和行动之间的确切关系对于观察者来说可能是不明确的，其中的部分原因是目标必定是被推断的。例如，教师可以使用追问（提出追根究底的问题）的行动实现以下一个或多个目标：评价发言的学生实际上是否理解，评定学生为有能力的，或者确保班上的其他同学能跟上被呈现的想法。更进一步地，基于学生的回应或其他学生的反应，在一堂课上教师的目标可能会多次发生变化，提高了推断改变的目标的挑战性。

第三，与教学行动和目标有关的语言在文献中经常被替换使用，可能会使用同样的术语表示行动和目标。例如，追问学生以获得解释，既可以是一系列行动，也可以是一系列目标，每一个都有很多层级。作为行动，追问包括了一系列的追根究底的问题，每一个新问题都是学生前一个回答的回应。作为目标，追问是从学生那里获得更多信息，而这个目标常常被分解为许多精确的类别——例如，教师可能会因为不知道学生在想什么而追问学生去理解其意义，或者教师可能会追使学生去获得正当理由，让学生发言解释程序、计算或论断背后的原因（Brodie, 2010）。为意义和为理由的追问行动很有可能是不同的。文献当中使用的其他术语表明，行动和目标包括了脚手架、支持和拓展。

考虑到行动和目标之间关系的复杂本质，找到一个统一的架构来组织对教师在引领讨论时所做工作的研究是非常具有挑战性的。尽管文献中有一些用来理解教师在引领讨论中的作用的组织框架（例如Cengiz, Kline, & Grant, 2011；Fraivillig, Murphy, & Fuson, 1999；Herbel-Eisenmann, Steele, & Cirillo, 2013；Hufferd-Ackles, Fuson, & Sherin, 2004；Jacobs & Empson, 2016；Ozgur, Reiten, & Ellis, 2015），但是这些框架一般被用在一些个人的研究中，经常是与特定的数学内容相关的，因此不能推广到整个相关文献。我们将从四个方面组织本节，这四个方面与教师在引领讨论时的4个目标有关：（1）使学生参与到其他学生的数学思维中，（2）追问，（3）脚手架，（4）评定学生为有能力的。我们选择这些目标的原因是，它们已经被广泛研究，而且我们的选择是为了说明问题，而不是详尽无遗。针对每个目标，我们首先对其进行详细阐述，然后描述教师用以实现这些目标的行动。

使学生参与到其他学生的数学思维中。使学生参与到其他学生的数学思维中，是一个必要的教学目标。如果没有它，课堂的谈话并不一定会是讨论。在讨论时，学生解释他们的思维，理解和批判其他同学的推理。使学生参与到其他学生的数学思维中这一目标与一系列定义明确的行动相联系。

例如，弗兰克和他的同事（2015）将六个教学行动定义为他们所谓的邀请性行动：

> 解释别人的解法，讨论这些解法的差异，对另一学生的想法提出建议，将一个学生的想法与另一个学生的想法相联系，和其他学生一起创造一个解法，或者使用一个由其他学生提供的解法。（第133页）

他们发现，这六个邀请性行动在不同的教师，不同的年级，不同的数学内容和不同的课堂结构（例如，整班、小组和同桌）上的使用具有一致性。最常见的邀请性行动是让学生将他们的想法与另一些学生的想法联系起来，而最不常见到的行动是让学生使用一个由其他学生提出的策略。然而，这些研究者指出这些教学行动的类型和学生在同龄人的数学思维中的参与程度之间没有相关性。这表明教师按照稿子照搬的教学行动并不总是导致高水平的参与，相反地，教师必须用依赖于情境的

to teaching moves, goals also come in a variety of grain sizes. For example, teachers may have fine-grained goals related to the mathematics at hand (such as ensuring that a specific representation is shared) or related to social norms (such as ensuring that a particular student speaks during the lesson). Goals of a larger grain size include positioning all students as competent, encouraging persistence in the face of difficulty, and introducing students to the norms of the discipline.

The relationship between teaching moves and goals is complicated in three ways. First, the same move can be used to meet different goals, and multiple moves can meet the same goal. For example, the move of revoicing can be used to accomplish the goals of positioning students as competent, helping to ensure that everyone is following the main points of the discussion, or scaffolding the discussion. Similarly, the moves of revoicing, providing a representation, modifying a task, and making strategic use of the board can all be used to meet the goal of ensuring that everyone is following the main points of the discussion.

Second, teachers may use one or more moves to work toward multiple goals simultaneously, and the precise relationship between goals and moves may be unclear to observers, in part because goals must be inferred. For instance, teachers may use the move of *pressing* (probing questions) to accomplish one or more of the following goals: assessing what the student speaking actually understands, positioning that student as competent, or ensuring that other students in the class are able to follow the ideas being presented. Further, on the basis of a student's response or the reactions of other students, teachers' goals may change multiple times during a lesson, increasing the challenge of inferring shifting goals.

Third, the language surrounding teaching moves and goals is often used interchangeably in the literature, with the same term being used to describe both a move and a goal. For example, pressing students for explanations is both a set of moves and a set of goals, each with multiple layers. As a move, pressing is enacted as a series of probing questions with each new question being specifically responsive to what students have just said. Stated as a goal, pressing is the acquisition of more information from students, and this goal is often decomposed into more precise categories—for example, teachers may press students for *meaning* because the teachers do not know what students are thinking, or teachers may press students for *justification* to get students to articulate the reasons behind a procedure, computation, or claim (Brodie, 2010). The specific pressing moves (questions) used to meet the goals of pressing for meaning versus pressing for justification are likely to be different. Other terms used in the literature that simultaneously indicate moves and goals include *scaffolding, supporting,* and *extending.*

Given the complicated nature of the relationship between moves and goals, finding a unifying structure to organize the research on the work that teachers do when leading discussions is challenging. Although some organizational schemes for understanding the teachers' role in leading discussions have been advanced in the literature (e.g., Cengiz, Kline, & Grant, 2011; Fraivillig, Murphy, & Fuson, 1999; Herbel-Eisenmann, Steele, & Cirillo, 2013; Hufferd-Ackles, Fuson, & Sherin, 2004; Jacobs & Empson, 2016; Ozgur, Reiten, & Ellis, 2015), these schemes have typically been applied to individual studies, often in particular mathematical content areas, and thus do not scale up to a review of the literature. We have chosen to organize this section by illustrating the work around four goals that teachers can have when leading discussions: (1) engaging students with their peers' mathematical thinking, (2) pressing, (3) scaffolding, and (4) positioning all students as competent. We selected these goals because they have been widely studied, and our selection is meant to be illustrative, not exhaustive. For each goal, we first elaborate on the goal and then describe the moves teachers use to meet the goals.

Engaging students with their peers' mathematical thinking. Engaging students with the ideas of their peers is an essential goal of instruction because without this element, the *talk* in the classroom is not necessarily *discussion,* in which students are explaining their thinking and making sense of and critiquing the reasoning of others. Engaging students with their peers' mathematical thinking is a goal that is often associated with a set of well-defined moves.

For instance, Franke and colleagues (2015) identified six teaching moves that they called *invitation moves*:

> Explain someone else's solution, discuss differences in solutions, make a suggestion about another student's idea, connect one student's idea to another student's idea, create a solution together with other students, or use a solution that was offered by another student. (p. 133)

They found that the six invitation moves were used consistently across teachers, grade levels, mathematical content, and classroom structures (e.g., whole-class, small-group, and pair work). The most common invitation move was asking students to connect their ideas to another student's ideas, while the least frequently used move was asking students to use a strategy shared by a classmate. However, the researchers found no correlation between the type of teaching move and the level of

方式从丰富的行动中抽取想要的教学行动。这些研究者还表示，学生参与的水平并不仅仅是由教师最初的邀请性行动决定的，后续的行动同样很重要。

文献中还给出了其他的一些邀请性行动和一个非邀请性行动。邀请其他学生向一位同学的解法提问（Turner 等，2013），让学生比较他们与另一学生的思维（N.M. Webb 等，2014），以及让学生就另一学生的思维进行细节的补充（N.M. Webb 等，2014）已经被认为在使学生参与到其他学生的数学思维方面是有效的。相比之下，使用"只是"来表示简单的意思（如交叉相乘）会让学生不参与，因为它表明这个任务太简单了，他们应该会做（Wagner & Herbel-Eisenmann，2008）。

追问。正像前面所说的，追问在文献中既是目标，也是行动。教师追问学生的目的是为了帮助他们澄清、解释、证明其思维是合理的，这对教师、分享的学生和其他学生都有好处。研究表明追问学生可以指向更为完整和正确的解释（Franke 等，2009；N.M. Webb 等，2009），而这些又和讨论中高水平的参与相联系（Kazemi & Stipek，2001；Smith & Stein，1998；N.M. Webb 等，2014）。

如前面所述，追问有两个较为宽泛的目标，分别是为了意义和为了理由（Brodie，2010）。有关追问的其他的具体目标包括保持任务的认知要求（Stein, Engle, Smith, & Hughes, 2008）和揭示学生中间存在理解不到位的情形从而在深入下一步的讨论之前能通过对话和协商加以解决（Anghileri，2006；Staples，2007）。沃夏尔（2015）也指出追问学生可以使他们保持在有成效的挣扎状态，即"为理解数学，弄懂一些不显然的问题所做的努力"（Hiebert & Grouws，2007，第387页）。

作为行动，追问是一系列持续的师生活动，期间教师通过以下方式不断探寻以获得额外信息：让学生为自己的论断提供理由，通过比较不同的解法核实自己的答案，通过文字、图像和数值解法的三角互证说明自己的解法正确（Kazemi & Stipek，2001）。为了有效性，追问的行动必须是明确的（而不是一般的，例如"为什么？"或"你是怎么知道的？"）。此外，由于学生并不习惯于被追根究底，教师应该在追问的时候给予情感支持（Staples，2007）。教师应该对正确的和错误的答案都进行追问，但是文献显示了教师们不同的用法。例如，N.M.韦伯和他的同事（2008）发现有些教师并不倾向于追问学生正确的回答，而施莱芬巴赫、弗莱瓦瑞斯、西莫斯、和佩里（2007）指出跟中国教师相比，美国教师不大可能追问学生错误的回答，而是倾向于为学生纠正他们的错误。与此相反，斯特普尔斯（2007）给出了一名教师的例子，该教师既对正确的回答也对错误的回答进行追问，强调对想法而不是对答案的追根究底。

脚手架。脚手架在文献中也被描述为既是目标也是行动。脚手架的目的是当学生不能独立完成任务时提供支持，使之能有所进步（Wood, Bruner, & Ross, 1976）。脚手架原来是从个体学生数学思维的角度进行研究的，但是最近，研究者考察教师在利用脚手架帮助学生学习聆听、理解其他学生的想法，并在讨论中基于其他学生的想法进行建构的作用（Baxter & Williams, 2010；Chapin, O'Connor, & Anderson, 2009；Franke 等，2015；Mendez, Sherin, & Louis, 2007）。巴克斯特和威廉姆斯（2010）将这个大目标分解为具体的类别，即分析性的脚手架（帮助学生发展数学理解）和社会性的脚手架（帮助学生学习合作），他们认为这两个类别互为补充，且经常同时使用。针对数学讨论中的脚手架，其他研究者给出了许多具体的目标，具体包括——

- 使得讨论的结构对于所有参与者来说都清晰可见（Ghousseini, 2009）；
- 保持班级学生在认知上处于一个状态（维持一个共同的基础；Staples, 2007）；
- 使学生保持在有成效的挣扎状态（Warshauer, 2015）；
- 使得对话向某个特定的数学方向行进（Sleep, 2012）；
- 使某个观点更为明显（Land, Drake, Sweeney, Franke, & Johnson, 2014）；
- 帮助学生看到一般化（Land 等，2014）；
- 帮助学生采用数学的思维（Walshaw & Anthony, 2008）。

student engagement with peers' mathematical thinking, suggesting that teachers cannot invoke a scripted set of moves that will always lead to high levels of engagement but must instead draw from a repertoire of moves in ways that are context dependent. The researchers also noted that the level of student engagement was not determined merely by teachers' first invitation moves but that follow-up moves played an important role.

Additional invitation moves and, in one case, a disinvitation move have been identified in the literature. Inviting other students to ask a peer questions about his or her solution (Turner et al., 2013), asking students to compare their thinking with another student's thinking (N. M. Webb et al., 2014), and asking students to add details to another student's thinking (N. M. Webb et al., 2014) have been shown to be effective in engaging students with one another's ideas. In contrast, the use of the word *just* to mean *simply* (as in "just multiply across") serves to *disinvite* students to participate because it communicates that the task is easy and they should *just* be able to do it (Wagner & Herbel-Eisenmann, 2008).

Pressing. As noted above, *pressing* is described in the literature as both a goal and a move. Teachers press students with a goal of helping them clarify, elaborate, and justify their thinking—for the benefit of the teacher, the student sharing, and other students. Research has shown that pressing students leads to more complete and correct explanations (Franke et al., 2009; N. M. Webb et al., 2009), which correlates with higher levels of student engagement in discussion (Kazemi & Stipek, 2001; Smith & Stein, 1998; N. M. Webb et al., 2014).

As described above, two broad goals of pressing are to press for meaning or for justification (Brodie, 2010). Other specific goals for pressing include maintaining the cognitive demand of a task (Stein, Engle, Smith, & Hughes, 2008) and uncovering instances of misalignment of understandings among students that require further conversation and negotiation before moving forward with the discussion (Anghileri, 2006; Staples, 2007). Warshauer (2015) also found that pressing students can meet the goal of keeping them engaged in productive struggle—an "effort to make sense of mathematics, to figure something out that is not immediately apparent" (Hiebert & Grouws, 2007, p. 387).

As a move, pressing is a series of sustained teacher–student interactions in which teachers probe for additional information by asking students to provide reasons for their claims; verify their answers by comparing solutions with peers; and justify their solutions by triangulating verbal, graphical, and numerical solutions (Kazemi & Stipek, 2001). To be effective, pressing moves must be specific (not generic, such as "why?" or "how do you know?"), and because students are often unaccustomed to having their thinking pressed, teachers should provide emotional support when pressing (Staples, 2007). Teachers should also press on both correct and incorrect answers, but the literature shows mixed use. For instance, N. M. Webb and colleagues (2008) showed that some teachers tended not to press on students' correct answers, and Schleppenbach, Flevares, Sims, and Perry (2007) found that U.S. teachers were less likely to press students on their incorrect answers than were Chinese teachers, with U.S. teachers instead tending to correct students' errors for them. Staples (2007), in contrast, provided an example of a teacher who pressed on both correct and incorrect answers, emphasizing pressing on ideas rather than answers.

Scaffolding. *Scaffolding* is also described in the literature as both a goal and a move. The goal of scaffolding is to provide support that enables students to make progress on tasks that they are not yet fully capable of completing on their own (Wood, Bruner, & Ross, 1976). Scaffolding has traditionally been studied in relation to the mathematical thinking of an individual student, but, more recently, researchers have studied the role of the teacher in using scaffolding to help students learn to listen to, make sense of, and build on the ideas of others during a discussion (Baxter & Williams, 2010; Chapin, O'Connor, & Anderson, 2009; Franke et al., 2015; Mendez, Sherin, & Louis, 2007). Baxter and Williams (2010) have decomposed this broad goal into the more precise categories of *analytic scaffolding* (helping students develop mathematical understanding) and *social scaffolding* (helping students learn to work together), which they suggest are complementary and often used simultaneously. Other researchers have identified a range of more precise goals for scaffolding during a mathematical discussion that can include—

- making the structure of the discussion visible for all participants (Ghousseini, 2009);
- keeping the class together intellectually (maintaining common ground; Staples, 2007);
- keeping students engaged in productive struggle (Warshauer, 2015);
- steering the conversation in a particular mathematical direction (Sleep, 2012);
- making a particular point more obvious (Land, Drake, Sweeney, Franke, & Johnson, 2014);
- helping students see a generalization (Land et al., 2014); and
- helping students adopt the register of mathematics (Walshaw & Anthony, 2008).

教师可以通过实施一系列的行动达到脚手架的目标。一般来说，脚手架行动为数学或讨论本身提供了结构，而不是默写解法（Staples，2007）。教师用以支撑数学讨论的行动可以包括——

- 让学生复述、澄清他们的想法或解释他们的想法成立的理由（Anghileri，2006；Goos，2004；Staples，2007）；
- 当众记录学生的想法（Cengiz等，2011；Staples，2007）；
- 有考虑地使用黑板（Staples，2007）；
- 重新表述任务的目的和子任务（Cengiz等，2011；Staples，2007）；
- 总结较长的讨论（Baxter & Williams，2010）；
- 做一件与数学有关的事情（Franke等，2015）；
- 在学生解法、表示方式、游戏中的数学想法和任务的情景之间建立显性的联系（Anghileri，2006；Baxter & Williams，2010；Staples，2007；Stein等，2008）；
- 对某一特定的想法引起注意（Staples，2007）；
- 控制讨论的节奏或速度（Staples，2007）。

为课堂讨论搭建脚手架是满足个体学生的需求和推动小组集体思维前进之间，何时在数学上取得进步和何时建立社会数学规范之间的一种美妙的跳跃（见Baxter & Williams，2010和Nathan & Knuth，2003，以获得关于在课堂上这种平衡的详细描述）。

评定学生为有能力的。 讨论有加强或破坏现有权利结构、学生认同、关于谁擅长或不擅长数学的思维定势的潜力。教师可以围绕将课堂上每一名学生评定为有能力的意义建构者这样一个目标来实施教学行动，从而达成增强以往被较少关注的学生的自我效能感，并向同班同学表明这些学生拥有有价值的数学想法这样更为具体的目的。

在一个非常有名的评定行动中，教师注意到学习表现较靠后的学生有内容要分享，并有意和有策略性地对此引起注意，这个多侧面的行动在复杂教学（E. Cohen，1994）中被称为分配能力。分配能力具体包括以下几个方面：（a）邀请那些学生分享、说明和澄清他们的想法；（b）明确地指出那些学生分享的内容的独特性、价值，或者那些学生说了什么，所说的又能如何推动小组前进；（c）邀请其他学生向分享想法的学生提问（E. Cohen，1994，2000；Featherstone等，2011；Franke等，2015；Turner等，2013）。为了使这些教学行动能有效解决学生学习表现的地位问题，它们必须是公开的、有智慧的、具体的、和小组任务相关的（E. Cohen，1994）。所有这些举动都将学习表现靠后的学生评定为有能力的，因为它们增强了发言者的信心，并向其他学生发出信号，表明某个想法值得坚持去理解，尽管它看上去有些混乱，一时难以跟上。特纳和他的同事（2013）发现当教师使用这些将英格兰学习者评定为有能力且擅长数学的行动时，他们的同伴也开始这么做。

引领讨论是如何被研究的？

在本节中，我们将描述用以研究教师引领讨论方式的方法。我们首先叙述最常见的、目前正在使用的数据收集方法，然后我们的注意力将转向具有推动该领域向前发展希望的、用在更为一般的数学教学研究中的、近期开发的研究工具。

最为常见的方法。 绝大多数关于教师如何引领讨论的研究都是归纳的，意思是，研究者收集课例信息，然后寻找教学目标和行动的类别来分析数据，经常也会在某种程度上将这两者联系起来（如上文所述，这种联系是复杂的）。在有些情况下，研究者一开始就根据文献中所描述的教学行动的类别，然后对已有类别进行补充，有时候把教学行动细分为更为具体的类别，有时候也加入一些新的类别。在另外一些情形中，研究者最初有目标，然后寻找、辨别和记录教师用以支撑目标的行动。在有些情况下，教师在课后接受访谈，以确认他们特殊教学行动背后的目标，但这种考虑教师的做法并不常见。

捕捉教师和学生在讨论中的交互，一般被局限在描述可观察到的教师教学行为和在这些行动之前或之后的学生活动。这一局限性的产生是由于研究者通常并不了

Teachers can enact a variety of moves to reach scaffolding goals. In general, scaffolding moves provide structure to the mathematics or the discussion without dictating solutions (Staples, 2007). Moves that teachers use to scaffold mathematical discussions can include—

- asking students to repeat, clarify, or justify their ideas (Anghileri, 2006; Goos, 2004; Staples, 2007);
- recording students' ideas publicly (Cengiz et al., 2011; Staples, 2007);
- making thoughtful use of the board (Staples, 2007);
- restating the purpose of a task or subtask (Cengiz et al., 2011; Staples, 2007);
- summarizing a lengthy discussion (Baxter & Williams, 2010);
- doing a piece of the mathematics (Franke et al., 2015);
- making explicit connections—between student solutions, representations, or the mathematical ideas in play and the context of the task (Anghileri, 2006; Baxter & Williams, 2010; Staples, 2007; Stein et al., 2008);
- drawing attention to particular ideas (Staples, 2007); and
- controlling the flow or pace of the discussion (Staples, 2007).

Scaffolding a discussion is a delicate dance between meeting the needs of individual students and moving the collective thinking of the group forward, as well as between when to privilege making progress on the mathematics versus building sociomathematical norms. (See Baxter & Williams, 2010, and Nathan & Knuth, 2003, for detailed descriptions of how this balancing looks in classrooms.)

Positioning all students as competent. Discussions have the potential to reinforce or disrupt existing power structures, student identities, and stereotypes about who is and is not proficient in mathematics. Teachers can enact teaching moves with a deliberate goal of positioning all students in the classroom as competent sense makers and thereby accomplish the more specific goals of bolstering the self-efficacies of students from historically underserved populations and signaling classmates that these students have worthwhile mathematical ideas.

In one well-known positioning move, teachers notice when students who have low academic status have something to contribute and draw attention to it deliberately and strategically, a multifaceted move called *assigning competence* in complex instruction (E. Cohen, 1994). Specific aspects of assigning competence include (*a*) inviting those students to share, justify, or clarify their ideas; (*b*) explicitly noting the uniqueness, value, or both of what those students are sharing and how it can help move the group forward; and (*c*) inviting other students to ask questions of a student who has shared ideas (E. Cohen, 1994, 2000; Featherstone et al., 2011; Franke et al., 2015; Turner et al., 2013). For these teaching moves to be effective in addressing status issues, they must be public, intellectual, specific, and relevant to the group's task (E. Cohen, 1994). All these moves position students with low academic status as competent because they bolster the speaker's confidence and signal other students that an idea is worth persisting to understand, even if it seemed a bit confusing and hard to follow in the moment. Turner and colleagues (2013) found that when teachers used these types of moves to position English learners as competent and capable mathematicians, their peers began to do so as well.

How Is Leading Discussions Studied?

In this section, we describe the methods that have been used to study the ways in which teachers lead discussions. We first describe the most common data-collection methods currently in use and then turn our attention to recently developed instruments used in more general studies of mathematics teaching that show promise for moving the field forward.

Most common methodologies. Most studies of how teachers lead discussions are inductive, meaning that the researchers collect data on classroom lessons and then analyze the data by looking for categories of teaching goals and moves, often linking the two to some extent (although as noted above, this linking is complicated). In some cases, researchers begin with categories of teaching moves described in the literature and then seek to add nuances to the information already available, sometimes by subdividing moves into more precise categories and other times by adding new categories. In other cases, researchers start with a goal and seek to identify and document moves that teachers make in support of the goal. In a few cases teachers are interviewed after the lesson to ascertain their goals behind particular moves, but including teachers' voices is not common.

Capturing the interaction between students and teachers during discussion is generally limited to describing observable moves made by teachers and the student activity that preceded or followed the moves. This limitation occurs because researchers typically are not privy to the reasons teachers made particular moves, the ways teachers saw the moves as being responsive to student thinking, or how the moves might have been influenced by competing goals. In addition, teachers may not always be conscious of the reasons they are making particular

解教师采用某些教学行动的理由，教师将教学行动视为对学生思维回应的方式，或者这些行动可能如何受到其他竞争目标的影响。此外，教师可能并不是一直清楚他们在某个瞬间采取某些教学行动背后的理由。近来有些研究者已经开始记录与学生学习和投入有关的教学行动的效果（例如Clarke, 2013；N.M. Webb 等, 2009），但是因为课堂交互的复杂性，这些结果只是表明了相关性，并不能做出因果关系的论断（N. M. Webb 等, 2008）。

具有发展前景的方法。近年来，研究者已经开发了用于描述被广泛解释的数学课堂教学质量的观察量表，有些量表把数学讨论包含在内。这些量表的不同方面有助于教师引领讨论的研究领域厘清引领讨论的要素、这些要素的定义以及判断这些要素出现或不出现的证据。尽管使用这样的量表一定会导致将复杂的教学简单化，但是研究者大量使用共同量表有两个主要的优势。第一，研究者可以在不同的关于核心实践的研究项目之间进行更为清晰的交流。第二，在大型教学研究中使用的这些量表有利于辨别重要的模式，例如最为常见的行动，在某些教学行动之前和之后的情况，以及系列行动，等等。

在最近十年中开发的，包含了教师在引领数学讨论中产生作用的量表有教学质量评估（IQA；Boston, 2012；Matsumura, Garnier, Slater, & Boston, 2008），教学的数学质量（MQI；National Center for Teacher Effectiveness, 2014），数学扫描（M-Scan；Berry 等, 2013；Walkowiak, Berry, Meyer, Rimm-Kaufman, & Ottmar, 2014）。由于这三个量表是直接与数学相关的，而且已经建立了信度和效度数据，所以我们将具体的分析限制在这些量表上。这些量表是在大型的、多层面的项目中开发的，它们不仅仅包含讨论的质量这一个方面。[5]

这三个量表都包括一系列有不同细节和复杂程度的评估准则。这些评估准则的要素表明量表的开发者对于什么是高效讨论的标志的认识，这些标志与上述文献相一致。我们的概述将限制在量表中与教师引领数学讨论的作用直接相关的方面，但是我们需要提醒读者的是，这些量表是用在整体性的研究中，这里的子量表并没有经过独立的论证。

教学质量评估量表测量的是，在多大程度上教师不断地追问学生提供概念性的解释；解释他们的推理；并且在他们和其他学生的想法、表示或策略之间建立联系。该量表包括了一个检查清单，在这个清单上评估者会看到各种不同的教学行动，包括复述、标记（唤起一个重要的想法）、为准确性和推理而追问，以及让学生与他人的想法联系起来或者基于之前的知识基础。

教学的数学质量量表测量的是，在多大程度上教师使用学生的数学贡献来推动课堂教学向前。教师使用学生数学贡献的指标如下：聆听学生说了什么并给出恰当的回应，评论学生的数学想法，引发学生对想法的澄清，让其他学生对同伴的贡献进行评论或扩充，强调学生语句或问题中的重要想法，或者用学生的名字标记他们的想法。

数学扫描量表测量的是，与引领讨论有关的两个要素：教师在话语中的作用和教师的问题。教师在话语中的作用从以下方面进行测量：在多大程度上学生的想法被引发，在多大程度上课堂发言是基于学生的想法而不是教师导向的，以及在多大程度上学生有机会和同伴分享他们的数学想法。教师的问题从在多大程度上教师鼓励数学思维和在多大程度上教师要求数学解释和说明两方面进行测量。

除了测量教师在引领讨论中的作用以外，这三个量表还包括了与学生在讨论中做了什么（例如，解释、联系、说明和提问）相关的评估准则。然而，教师和学生的活动目前是分开测量的，因而不能捕捉讨论的交互性内涵，而这正是本节的焦点。尽管如此，现有的量表确实为开发更为显性化地捕捉这种交互性的工具提供了起点，从而将教师行动和学生在教师行动之前和之后的行为联系起来。例如，教学质量评估量表既包括了评价在多大程度上教师追问学生提供证据和说明的评估准则，也包括了评价在多大程度上学生提供了证据和说明的评估准则。有了这些评估准则，研究者可以捕获教师和学生关于某个教学行动（追问）的行为，但是目前，这里的联系是在一节课层面上的整体分析。新的研究量表可以在类似这些捕捉教师学生一系列行为关系的评估准则基础上进行，但是需要在更细微的层面上进行，从而能够保持讨论的交

moves in the moment. Recently some researchers have begun to document the outcomes of teaching moves related to student learning and engagement (e.g., Clarke, 2013; N. M. Webb et al., 2009), but because of the complex nature of classroom interactions, these results lead only to correlations from which one cannot conclude causation (N. M. Webb et al., 2008).

Promising methodologies. In recent years, scholars have developed observation instruments for describing the quality of classroom mathematics instruction broadly construed, and some include dimensions of mathematical discussions as part of the instrument. These aspects of the instruments show promise for helping the field articulate particular elements of leading discussions, definitions of these elements, and evidence to determine the presence or absence of these elements. Although using such instruments necessarily leads to streamlining and simplifying complex parts of instruction, there are two main advantages to researchers making widescale use of common instruments. First, researchers can more clearly communicate across projects about the core practice under study. Second, instruments that also lend themselves to use in large-scale studies of instruction can facilitate the identification of important patterns, such as the most commonly used moves, the antecedents and successors of particular moves, and sequences of moves, to name a few.

Examples of instruments developed within the last decade that include a dimension related to teachers' roles in leading mathematics discussions are the Instructional Quality Assessment, (IQA; Boston, 2012; Matsumura, Garnier, Slater, & Boston, 2008), Mathematical Quality of Instruction (MQI; National Center for Teacher Effectiveness, 2014), and Mathematics Scan (M-Scan; Berry et al., 2013; Walkowiak, Berry, Meyer, Rimm-Kaufman, & Ottmar, 2014). We limit our analysis to these three instruments because they are mathematics-specific and have established reliability and validity data. Because these instruments were developed in large, multifaceted projects, they assess more elements than just the quality of the discussion.[5]

All three instruments consist of a series of rubrics of varying levels of detail and complexity. The elements of the rubrics provide a sense of what the developers consider to be the markers of high-quality discussion, and these markers are consistent with the research literature reviewed above. We confine our summary here to the aspects of the instruments directly related to teachers' roles in leading mathematical discussions, but we caution readers that the instruments are intended to be used in their entirety and that subscales have not been validated for stand-alone use.

The IQA measures the extent to which teachers consistently press students to provide conceptual explanations; explain their reasoning; and make connections between their own and other students' ideas, representations, or strategies. The instrument includes a checklist on which raters note a variety of teaching moves including revoicing, marking (calling out an important idea), pressing for accuracy or reasoning, and asking students to relate to one another's ideas or build on prior knowledge.

The MQI measures the extent to which teachers use students' mathematical contributions in moving the instruction forward. The following are offered as indicators of teachers' use of students' mathematical contributions: hear what students are saying and respond appropriately, comment on students' mathematical ideas, elicit students' clarification of ideas, ask other students to comment or expand on contributions of peers, highlight key ideas in students' statements or questions, or label students' ideas with their names.

The M-Scan measures two elements related to leading discussions: the teachers' role in discourse and the teachers' questions. The teachers' role in discourse is measured by the extent to which students' ideas are solicited, the classroom talk is built on students' ideas rather than being teacher-directed, and students have opportunities to share mathematical ideas with their peers. The teachers' questions are measured by the extent to which they encourage mathematical thinking and require mathematical explanation and justification.

In addition to measuring teachers' roles in leading discussions, these three instruments also include rubrics related to what students do during discussions (e.g., explain, connect, justify, and question). However, the activities of teachers and students are currently measured separately, thereby missing the opportunity to capture the *interactive* nature of discussions, the focus of this chapter. Nonetheless, the existing instruments do provide potential starting places for developing new instruments that could capture this interaction by more explicitly linking teachers' moves to students' actions that precede and follow the teachers' moves. For example, the IQA includes both a rubric to assess the extent to which teachers press students to provide evidence and justification and a rubric to assess the extent to which students provide evidence and justification. With these rubrics, researchers can capture the actions of both teachers and students around a particular type of teaching move (pressing), but currently, this connection is done holistically, at the lesson level. New instruments could build on rubrics like these to capture connected sets of teacher–student actions, but at a more fine-grained level that preserves the interactive nature of discussions.

互性本质。

关于引领讨论的结束语

在本节中，我们通过阐明教师目标和他们用以实现目标的教学活动讨论了教师引领讨论的工作。我们试图解释与教师具有的目标、实施的教学行动和这些教学行动对学生所起的认知和情感作用有关的研究所固有的复杂性。我们简要地强调了这些复杂性的两个要点。第一，将某个教学行动与某个教学目标（反之亦然）对应起来是具有情境性的，这是因为，依赖于情境，一个教学行动可以对应许多目标，而一个目标可能可以通过使用不同的教学行动达到。第二，教学行动和学生学习成效之间没有直接的联系。教学行动是在其他课堂活动中实施的，所有这些都会影响学生的学习效果。

在结束本节时，我们承认引领讨论是一个多方面的、交互性的活动，在话语付诸文字的时候，它的一些细微差别必然会消失。因为讨论发生于课堂上的多个学生之间，因而当教师决定探究谁的想法、什么时候提供信息、对话要持续多久以及在推动全班学生向前的时候如何满足个别学生的需求时，必然会有许多不可避免的困境（见O'Connor，2001和Brodie，2010，以获得这些困境的具体例子）。为达到某个课程目标，教师也会面对很多实时的限制和期望。然而，研究发现参与丰富的课堂讨论对学生具有巨大的益处，所以进一步的研究对于理解教师在引领讨论时的工作细节和支持教师学习如何实施这一核心教学实践均非常重要。

结论和进一步研究

我们用核心实践的想法来组织本章关于数学教学的研究，我们强调了两个实践，即教师关注和引领讨论，这两个实践是教学交互部分中不可或缺的，而且在最近十年有了广泛的研究。总的来说，目前研究的突出发现是教学的复杂性和研究者深入理解这种复杂性的方式。进一步地，与包含对教师（和教育）负面看法的全国背景形成鲜明对比的是，该领域的研究表明关于高效教学的案例正不断增长。

在最后这一节，根据已经知道的，我们确立了一些值得进一步研究的领域。我们首先给出了一些有关于教师关注或引领讨论研究的具体建议，然后我们用针对核心教学实践研究的更为一般的建议作为结束。

关于教师关注的进一步研究

尽管在近几年数学教师关注已经获得了许多研究者的关注，但它仍旧是一个刚刚起步的研究领域。它描绘并赋予了一种语言来刻画以前被隐藏的，但目前已经被表明是重要且可习得的教学实践。然而关于如何最有成效地对教师关注进行概念化和研究的讨论还在持续中。我们指出两个可进一步研究的方面：（1）探索在研究教师关注时使用的方法论的内在得失，（2）追求教师关注的新兴概念化。

探索方法论上的得失。由于交互的情境性和动态化的本质，研究教学中的交互具有挑战性。因为教师关注不是可以直接观察到的，因此捕捉教师关注就更具挑战性。我们认为科技的进步，例如教师可穿戴的、可由教师控制生成视频的可携带摄像机，可以推动我们更近距离地研究教师自己课堂中某个瞬间的教师关注。因而，我们鼓励研究者继续使用这些可获得的进步科技。

然而，研究者不可能完整地捕获某个瞬间的教师关注，这是因为，由定义可知，隐藏的教师关注在教师可观察到的回应学生的行为之前就已发生。因而，我们建议进一步的研究可以探究用以研究教师关注的不同方法（见图28.2）之间的方法论得失问题。例如，感兴趣的研究问题可能包括如下这些：教师关注其他人的学生怎样与教师关注自己的、有额外历史和情境信息的学生相联系？教师在教学中的关注怎样与几乎没有时间限制的情境（例如，在专业发展活动中）下的教师关注相联系？这项工作不仅能增加对这一构念的理论理解，而且具有支持教师发展关注专长的实践意义。

继续教师关注已有的概念化。考虑到教学的复杂性，教师不可能回应（甚至是意识到）课堂上发生的所有情况，教师关注研究的主要贡献之一是思考教师如何投入

Closing Thoughts on Leading Discussions

In this section, we provided a glimpse into teachers' work in leading discussions through the explication of teachers' goals and the teaching moves they make to meet those goals. We have attempted to illustrate the complexities inherent in research on the goals teachers have, the teaching moves they enact, and teaching moves' cognitive and affective effects on students. We briefly highlight two points about these complexities. First, mapping a particular teaching move onto a particular goal (or vice versa) is situation-specific because a single teaching move can meet a variety of goals, and a single goal may be met through the use of multiple teaching moves, depending on the context. Second, no direct link exists between a teaching move and student outcomes. Teaching moves are made in the midst of a host of other classroom activities, all of which affect student outcomes.

In closing this section, we acknowledge that leading discussions is a multifaceted, interactive activity, and some of its nuances are necessarily lost in words committed to paper. Because discussions take place in classrooms with multiple students, dilemmas inherently arise when teachers decide whose ideas to pursue, when to provide information, how long to let conversations go, and how to meet the needs of individual students while moving the collective forward. (See O'Connor, 2001, and Brodie, 2010, for examples of these dilemmas.) Teachers also face very real time constraints and expectations to meet particular curricular goals. However, the research shows clearly the tremendous benefits to students from participating in rich mathematical discussions, so further research is important both to understand in more detail the work that teachers do in leading discussions and to support teachers in learning to enact this core practice of teaching.

Conclusions and Future Research

We used the idea of core practices to frame this chapter on research on mathematics teaching, and we highlighted two practices—teacher noticing and leading discussions—that are integral to the interactive part of teaching and have been extensively researched in the last decade. Overall, what stands out about this research is the complexity of teaching and the ways in which researchers have made substantial progress in understanding this complexity. Further, in contrast to a national backdrop that often includes a negative view of teachers (and education in general), this research has showcased a growing number of examples of high-quality instruction.

In this final section, we build on what we have learned to identify some areas in which more research would be helpful. We begin with suggestions specific to either research on teacher noticing or research on leading discussions, and we conclude with suggestions for the broader idea of researching core practices in teaching.

Future Research on Teacher Noticing

Mathematics teacher noticing is still an emerging field of study that has captured the attention of many researchers in recent years. This work depicts and gives language to a previously hidden part of teaching that has been shown to be important and learnable, yet conversation continues about the most productive ways to conceptualize and study teacher noticing. We suggest two areas for future research: (1) explore the tradeoffs inherent in the methodology used to study teacher noticing and (2) pursue emerging conceptualizations of teacher noticing.

Explore methodological tradeoffs. Studying the interactive part of teaching is challenging because of its situational and dynamic nature. Capturing teacher noticing is even more challenging because teacher noticing is not directly observable. We recognize that technological advances—such as portable video cameras that teachers wear to generate video from the teacher perspective—can move us closer to studying teacher noticing in the moment in teachers' own classrooms. Thus, we encourage researchers to continue to capitalize on these technological advances when they become available.

However, researchers will never be able to completely capture teacher noticing in the moment because it is, by definition, hidden—teacher noticing occurs before teachers' visible responses to students. Thus, we suggest that future research explore the methodological tradeoffs of various approaches to studying teacher noticing (see Figure 28.2). For example, research questions of interest might include the following: How does teachers' noticing of others' students relate to noticing of their own students for whom they have additional historical and contextual information? How does teachers' noticing in the midst of teaching relate to teachers' noticing in situations with fewer time constraints (e.g., in professional development)? This work would not only inform our theoretical understanding of the construct but also have practical implications for supporting teachers in developing noticing expertise.

Pursue emerging conceptualizations of teacher noticing. Given the complexity of teaching, teachers cannot respond to (or even be aware of) everything that occurs in classrooms, and a major contribution of research on

到具有教学重要性的细节中的方式。然而，什么细节最具关键性还在争议中。我们将这种争议视为有成效的，且强烈要求研究者对教师关注的新类别持开放态度，这些新类别会随着不同类别的细节对于本领域变得重要而出现。

我们也鼓励研究者考虑将两种或更多种的关注类别进行整合的潜在益处，从而可以利用每一个类别的长处。本章提供一种可能性，即将本章中强调的两个与具体主题相关的关注整合会带来好处。这两个关注分别是：教师关注学生的思维和教师关注公平学习环境的指标。关于教师关注学生的数学思维的研究已经有很长历史，且富有成效，它对教师和学生都有正向作用，且它的主要优势之一是对学生思维的数学细节的明确聚焦。最近，关于教师关注公平学习环境指标的研究也强调关注这些学生是谁和除了数学思维之外他们还带来了什么资源的重要性，其重点是支持历史上被忽略的那部分学生群体。尽管我们承认考虑更多的因素给教师和研究者带来的实际挑战，但是我们认为诸如教数学（Turner 等，2012）这样的项目已经为探索卓有成效地增加被考虑因素的数量的创新性方式打开了思路。侧重细节的教师关注的研究是这种对话发生的一个自然场所。

关于引领讨论的进一步研究

很久以来引领讨论就被认为是一个重要的教学实践，但是最近研究者在揭示讨论中教学行动和它们与特定目标之间的联系方面取得了进展。尽管如此，为获得高质量的、聚焦在学生数学思维上的讨论，关于对所需的目标和行动进行最有成效的分类和研究的对话仍在继续。我们建议两个未来研究的领域：（1）开发可以捕捉引领讨论交互性工作细节的研究工具，（2）在引领讨论的研究中更为明确地包括教师的声音。

开发捕捉引领讨论交互性工作细节的研究工具。最近各种不同的捕捉数学教学的量表（例如教学质量评估量表，教学的数学质量量表和数学扫描量表）已经被开发，但是为了确定其优势和不足，还需要更广泛地使用这些工具。然而，这些工具并不是为捕捉引领讨论的交互性本质的细节而特意设计的，并不总是考虑到教学行动的时机，而且教师行为和学生行为通常是分开编码的，所以教学活动并不总是和特定的学生回应相联系的。进一步地，用这些工具记录教师和学生之间长时间的交互序列并不总是可能的，而这种遗漏是有问题的，因为研究表明教师丰富他们最初行动的方式通常比最初的行动本身更重要（N.M. Webb 等，2014）。我们承认设计一个可以注意到交互细节的、而且可以被不同研究者在包括大规模研究中所使用的工具，是非常具有挑战性的。尽管如此，我们鼓励研究者朝这个目标努力，在现有工具优势的基础上设计一个新的研究工具。此外，在给出主张时，研究者需要认识到他们的工具在多大程度上，以何种方式可以（或不可以）捕捉到交互细节。

在研究中包括教师的声音。我们对教师关注和引领讨论这两个核心实践中教师的不同作用印象深刻。教师的声音在教师关注的研究中处于核心地位。这种重要性的存在可能是必要的，因为教师关注是一个隐藏的实践，因而需要教师自己来揭示。然而，与教师引领讨论实践相平行的另一个要素是在教师作出某种行动时他们的目标也是隐藏的，所以在引领讨论的研究中缺少教师的声音着实让人震惊。

未来的研究可以在从教师的角度考虑教学目标、教学行动和它们之间的联系这些方面受益。教师对他们的教学目标和所实施的教学行动所作出的决策通常基于大量的信息，而这些信息对于研究者而言并不了解，具体包括关于他们所教授的学生的知识，关于课程资料的知识，当地的要求，昨天教了什么以及明天、下周或下个月准备教什么，等等。更进一步地，教师会决定把某些目标在特定时间点加以突出（例如，在一节课的某个特定部分中，给予发展社交数学规范以特权，凌驾于追求数学深度之上），但是这些做法以及背后的原因研究者并不知道。

最后，在目前高效教学图景中的假设是，教师在课

teacher noticing is a way to think about how teachers engage with instructionally important *details*. However, which details are most critical to notice remains controversial. We view this controversy as productive and urge researchers to be open to new types of teacher noticing that could emerge as different categories of details become important to the field.

We also encourage researchers to consider the potential benefits of integrating two or more types of noticing, thereby drawing on the strengths of each. We offer one possibility for consideration. We see potential benefits to integrating the two strands of theme-specific noticing highlighted in this chapter: teacher noticing of students' mathematical thinking and teacher noticing of indicators of equitable learning environments. Research on teacher noticing of students' mathematical thinking has a long and productive history that includes positive effects for both students and teachers, and one of the major strengths of this strand is the explicit focus on the mathematical details of students' thinking. More recently, research on teacher noticing of indicators of equitable learning environments has highlighted the importance of also focusing on who students are and what resources they bring beyond their mathematical thinking—emphases that have been shown to be particularly important for supporting students from historically underserved populations. Although we acknowledge the practical challenges for teachers and researchers when asked to foreground an increasing number of factors, we suggest that projects such as TEACH MATH (Turner et al., 2012) have opened a space for the field to explore creative ways of productively broadening the number of factors to be considered. Research on teacher noticing, with its emphasis on details, is a natural place for this conversation to occur.

Future Research on Leading Discussions

Leading discussions has long been identified as an important practice in teaching, but recently researchers have made progress in illuminating teaching moves and their connections to particular goals during discussions. Nonetheless, conversation continues about the most productive ways to categorize and study the goals and moves needed for achieving high-quality discussions focused on students' mathematical thinking. We suggest two areas for future research: (1) develop instruments that can capture details of the interactive work of leading discussions and (2) include teacher voice more explicitly in the research on leading discussions.

Develop instruments that capture the interactive work of leading discussions. A variety of mathematics-specific instruments for capturing teaching (e.g., IQA, MQI, and M-Scan) have recently been developed, but more widespread use is needed to identify the strengths and limitations of each. These instruments are not, however, specifically designed to capture in detail the interactive nature of leading discussions—the timing of teaching moves is not always considered, and teacher actions and student actions are often coded separately so that teaching moves are not always linked with specific student responses. Further, documenting a protracted sequence of interactions between teachers and students is not usually possible with these instruments, and this omission is problematic because research has shown that the ways in which teachers enrich their initial moves are often more important than the initial moves themselves (N. M. Webb et al., 2014). We acknowledge that designing instruments that attend to interactive details and have the potential to be used by multiple researchers, including possibly in large-scale studies, is extremely challenging. Nonetheless, we encourage researchers to work toward this goal by designing new instruments that build on the strengths of the existing instruments. In addition, when making claims, researchers need to acknowledge the ways in which and the extent to which their instruments are (and are not) able to capture interactive details.

Include teacher voice in research. We were struck by the differing roles of teacher voice in the two core practices of teacher noticing and leading discussions. Teachers' voice is central to the research on teacher noticing. Perhaps this emphasis exists by necessity given that teacher noticing is a hidden practice and thus is most often revealed via teachers' voices. However, a parallel in the practice of leading discussions is that the goals teachers have when making particular moves are also hidden, so the lack of teacher voice in the research on leading discussions was striking.

Future research could benefit from consideration of teachers' perspectives on instructional goals, teaching moves, and their connections. Teachers base their decisions about goals and how to enact moves on a host of information to which researchers are generally not privy—knowledge of the particular students they are teaching, knowledge of curriculum materials, knowledge of local expectations, and knowledge of what happened yesterday and what is planned for tomorrow or next week or next month, to name a few. Further, teachers make decisions to foreground certain goals at particular times (e.g., privileging the development of sociomathematical norms over the pursuit of mathematical depth during a specific part of a lesson), but these decisions and the reasons for them usually are not transparent to researchers.

堂上的行为是基于他们对学生数学思维的解释的。然而，关于引领讨论的大部分研究都没有捕捉到教师如何解释学生的回答或者这些解释又是如何影响教师的目标和行动。因而，在未来研究中加入教师的声音可以促进该领域理解教师如何解释学生对于某个教学行动的回应，如何思考为追求某个目标所做出的一系列教学行动，以及当学生的回应不在教师预料中的时候教师如何调整他们的目标和行动。

关于教学中核心实践的进一步研究

将教学分解为核心实践的工作已经有了很大的发展，而且对教学有了更为具体和细腻的理解。因此，我们觉得将数学教学研究的讨论建立在核心实践工作的基础上是富有成效的。这项工作发展很快，但是仍处于它的婴儿期，我们建议两个未来研究的领域：（1）调查公平和社会公正的目标如何与核心实践的工作建立起显性联系，（2）将那些对于改进教与学有高影响力的核心实践有策略地与特定的对象联系起来。

将公平和社会公正与核心实践的工作联结起来。数学教学长期以来并不关注某些特定人群，而有策略地选择核心实践可以被视作一种能够打破长久以来低效学习环境的方式（Ghousseini 等，2014；McDonald 等，2013）。然而，还需要更多的研究来明确公平和社会公正的目标可以如何与核心实践的工作建立联系。

有些初见成效的工作已经在进行中了。比如说，古西尼和他的同事（2014）已经关注到塑造教师身份的重要性，其中包括对数学教学的承诺，具体涉及"学生是谁和他们是如何被看待的，以及形成学生的经验、知识和机会的学校内外制度体系"（第3页）。这种方式表明数学教学需要的不仅仅是某些特定实践的熟练度，还需要考虑这些实践使用的情境——这些学生是谁以及影响学生作为学习者的身份和经验的学校制度。通过用更为宽泛的视角来考虑课堂中实施的核心实践，研究者还可以开始致力于改进制度和社会结构，这些制度和社会结构在历史上忽视了某些学生群体。为落实学校有效地服务于更多学生人群的承诺，就需要更多的课堂研究，包括针对历史上被忽略的学习者的课堂，从而研究者可以明确支持这些学生学习的核心实践和保证对所有学生带入课堂的知识和资源的了解。

有策略地将核心实践定位为高影响力。我们承认将教学分解为核心实践的困难，这些实践要足以促进学习又不能过于束缚从而丢失了与真实教学实践的联系。我们建议要有更多的研究，从而更好地理解那些已经在研究中的实践，以及明确更多并没有被视作为核心的实践。如果一些核心实践，例如教师关注，不能被直接观察到，那么确认的过程就会变得更具挑战性。就算一些核心实践被确认了，还是必须要对哪些核心实践更为重要作出判断。我们建议用一种实用主义的视角看待高影响力的实践：哪些实践——在什么样的尺度下——有可能给予我们更大的动力促进针对特定受众的教与学？

核心实践的研究对于职前和在职教师的改进教与学都有潜力，但是哪些核心实践更容易获得和更为重要以及个体如何能更好地投入这些实践，可能要依赖于受众。核心实践团队的工作为作出这样的决策提供了一个模型。核心实践团队的研究者最初把他们的受众限制在职前教师，已经可以有策略地将他们的关注点聚焦在那些对于新手教师来说可以获得而且特别重要的实践上。对于其他的人群或特定的情境哪些焦点更为有效？例如，对于小学和中学教师，职初和经验教师，研究者是否应该有不同的聚焦点？类似地，核心实践的目标是否会随着内容领域（例如，教有理数和教几何）而变化？我们承认许多具有不同尺度的实践可能是富有成效的，但是我们强调，明确哪些实践可能对改进某个目标受众的教与学更有影响力具有潜在好处。有策略地进行选择更有可能支持教师，教师不仅可

Finally, in the current vision of high-quality instruction, the assumption is that what teachers do in the classroom should be informed by their interpretations of students' mathematical thinking. However, most research on leading discussions does not capture how teachers interpret students' responses or how those interpretations inform teachers' goals and moves. Thus, including teachers' voices in future research could assist the field in understanding how teachers interpret students' responses to particular moves, how they think about clustering teaching moves in pursuit of particular goals, and how they adjust their goals and moves when students' responses are not among those anticipated.

Future Research on Core Practices in Teaching

Efforts to decompose teaching into core practices have proliferated and are resulting in a more detailed and nuanced understanding of teaching. As such, we found grounding our discussion of research on mathematics teaching in the work on core practices to be fruitful. This work is progressing rapidly but is still in its infancy, and we suggest two areas for future research: (1) investigate how the goals of equity and social justice can be explicitly connected to the work on core practices and (2) strategically target core practices that are likely to be leverage points for improving teaching and learning with particular audiences.

Connect equity and social justice to work on core practices. Mathematics teaching has a long history of underserving certain populations, and strategically selecting core practices can be viewed as a way to disrupt long-standing practices that may have contributed to unproductive learning environments (Ghousseini et al., 2014; McDonald et al., 2013). However, more research is needed to make explicit how the goals of equity and social justice can be connected to the work on core practices.

Some promising efforts are already underway. For example, Ghousseini and colleagues (2014) have drawn attention to the importance of crafting teachers' identities to include a commitment to mathematics teaching in which teachers attend to "who the students are and how they are positioned, as well as the structures in and outside of school that shape students' experiences, knowledge, and opportunities" (p. 3). This approach illuminates how mathematics teaching requires not only proficiency with particular practices but also consideration of the context in which practices are used—who the students are and the structure of schooling that affects students' identities and experiences as learners. By using this broader lens for considering the core practices enacted in classrooms, researchers can also begin to work on improving the institutional and societal structures that have historically underserved certain groups of students. In moving toward the promise of schools that effectively serve a wider population of students, more research is also needed in classrooms that include learners from these historically underserved populations so that researchers can identify core practices that support their learning and ensure recognition of the knowledge and resources each and every student brings to the classroom.

Strategically target core practices as leverage points. We acknowledge the difficulty in decomposing teaching into core practices that are sufficiently constrained to facilitate learning without being so constrained that the connection to authentic practice is lost. We suggest that more research is needed to better understand practices already under study as well as to identify other practices that may not yet be recognized as core. This identification is particularly challenging when some core practices—like teacher noticing—may not be directly observable. Even when a range of core practices has been identified, decisions must be made about which core practices receive the most attention. We suggest a pragmatic lens of leverage points: Which practices—and of what grain size—are likely to give us the most traction for improving teaching and learning with specific audiences?

Research on core practices has the potential for informing teaching and learning with both prospective and practicing teachers, yet which core practices are most accessible and significant, and how individuals are likely to engage with those practices, may vary depending on the audience. The work of the Core Practice Consortium provides a model for making these decisions. By initially restricting their audience to prospective teachers, the Consortium researchers have been able to strategically narrow their focus to practices that are especially accessible and significant to novice teachers. What foci might be productive for other populations or particular contexts? For example, should researchers focus on different core practices for elementary versus secondary mathematics teachers or early-career teachers versus experienced teachers? Similarly, should targeted core practices vary by content areas (e.g., teaching rational numbers vs. geometry)? We acknowledge that many practices, of multiple grain sizes, could be productive, but we underscore the potential benefits of identifying which practices are likely to give the most leverage for improving teaching and learning with a target audience. Strategic choices are more likely to support teachers in not only gaining expertise with specific core practices but also in

以获得某些核心实践的专长，还可以学习如何从他们的课堂实践中进行学习，从而使他们的学习成为生成性的（Kazemi 等，2009）。

在结束本节时，我们重申进行教学中的核心实践研究四个方面的重要性：判断核心实践的标准，根据目标确定核心实践，用共同的且精确的语言表述核心实践，关注核心实践的关联性本质。研究者只有通过关注这四个方面才能开始到达麦克唐纳和他的同事们（2013）提出的愿景——围绕核心实践的工作能够成为推动该领域进步的工具这一想法。

结语

在本章中，我们展示了教学的复杂性，现在我们又回到这个想法。教学工作是艰辛的，改变个人教学实践所需的工作是艰辛的，研究教学的工作也是艰辛的（参见Berliner，2002）。这项复杂的工作不可能孤立地完成。我们必须找到高效协同工作的方式，而且我们可以在核心实践团队的合作中找到这种有前景的方式。我们还要强调将教师作为研究工作合作者的重要性——教师的声音是必须的。最后，我们提出一个长远目标，即教学中核心实践的研究是一种相对较新又很有前途的，帮助我们理解和改进教学的方式。为确定分解教学中交互部分最有成效的方法，开发和提炼可用于捕获和理解这些实践的研究方法，以及找到支持教师在追求高质量教学时使用这些实践的方式，还有许多工作要做。

注释

1. 参见拉斯马森和瓦洛（2017，这套书）关于本科数学课堂教学的讨论。

2. 核心实践团队的参与者包括波士顿教师驻校学院；斯坦福大学；加州大学；洛杉矶分校；科罗拉多大学；密西根大学；圣母大学；华盛顿大学；威斯康星大学。

3. 我们承认，对于公开讨论（由教师协调）和只在学生之间进行的私下讨论的相应价值，可能存在文化差异（Clark，2013）。本章中，我们聚焦在全班讨论上，这是因为它与我们关于教学交互性的兴趣点相一致，在互动教学中，教师会对学生的数学思维作出回应。

4. 应该要注意，讨论只是本研究中教学的一个部分；异质分组和丰富的课程资源在教学中也起到了突出的作用。

5. 为了获得对教学的数学质量、教学质量评估和改革的教学观察方案（Sawada 等，2000）的深入分析及其奠基性的研究基础，参见波士顿、博斯蒂克、莱塞格和舍曼（2015）。

learning how to learn in and from their own classroom practices so that their learning is generative (Kazemi et al., 2009).

In closing this section, we reiterate the importance of the four considerations in research on core practices in teaching: determine criteria for core practices, identify sets of core practices to target, use common and precise language for core practices, and attend to the relational nature of core practices. Only by attention to these four considerations can researchers begin to reach the vision put forth by McDonald and colleagues (2013)—the idea that work around core practices could become the field's vehicle for improvement.

Final Thoughts

Throughout this chapter, we have showcased the complexity of teaching, and we return to that idea here. The work of teaching is hard, the work needed to change one's teaching practice is hard, and the work of studying teaching is hard (see also Berliner, 2002). None of this complex work can be done in isolation. We must find ways to work together productively, and a promising approach can be seen in the collaborative efforts of the Core Practice Consortium. We also emphasize the importance of involving teachers as partners in this work—teachers' voices are essential. Finally, we suggest a long-term perspective. Research on core practices in teaching is a relatively new, yet promising, approach to understanding and improving teaching. Additional work is needed to determine the most productive ways to decompose the interactive part of teaching, to develop and refine research methods for capturing and understanding these practices, and to find ways to support teachers in using these practices in their pursuit of high-quality instruction.

Notes

1. See Rasmussen and Wawro (2017, this volume) for a discussion of teaching in undergraduate mathematics classrooms.

2. Participants in the Core Practice Consortium include the Boston Teacher Residency; Stanford University; University of California, Los Angeles; University of Colorado; University of Michigan; University of Notre Dame; University of Washington; and University of Wisconsin.

3. We acknowledge that there may be cultural variations in views about the relative value of public discussions (mediated by the teacher) versus private discussions exclusively among students (Clarke, 2013). In this chapter, we focus on public discussion because it aligns with our interest in the interactive work of teaching, in which teachers are responsive to students' mathematical thinking.

4. It should be noted, however, that discussions were only one part of the instruction in this study; heterogeneous grouping and rich curriculum materials also played prominent roles in instruction.

5. For a thorough analysis of and the foundational research base for the MQI and IQA as well as the Reformed Teaching Observation Protocol (Sawada et al., 2000), see Boston, Bostic, Lesseig, and Sherman (2015).

References

Anghileri, J. (2006). Scaffolding practices that enhance mathematics learning. *Journal of Mathematics Teacher Education, 9*(1), 33–52.

Anthony, G., & Hunter, R. (2012). (Re)thinking and (re)forming initial mathematics teacher education. *New Zealand Journal of Educational Studies, 47*(1), 145–151.

Anthony, G., & Hunter, R. (2013). Learning the work of ambitious mathematics teaching. In V. Steinle, L. Ball, & C. Bardini (Eds.), *Mathematics education: Yesterday, today and tomorrow. Proceedings of the 36th annual conference of the Mathematics Education Research Group of Australasia* (pp. 699–702). Melbourne, Australia: MERGA.

Association of Mathematics Teacher Educators & National Council of Supervisors of Mathematics. (2014). *Improving student achievement in mathematics through formative assessment in instruction* (Joint Position Paper). Retrieved from http://amte.net/group/resources/05-14/position-improving-student-achievement-mathematics-through-formative-assessmen

Averill, R., Drake, M., & Harvey, R. (2013). Coaching pre-service teachers for teaching mathematics: The views of students. In V. Steinle, L. Ball, & C. Bardini (Eds.), *Mathematics education: Yesterday, today and tomorrow. Proceedings of the 36th annual conference of the Mathematics Education Research Group of Australasia* (pp. 707–710). Melbourne, Australia: MERGA.

Ball, D. L. (1993). With an eye on the mathematical horizon: Dilemmas of teaching elementary school mathematics. *The Elementary School Journal, 93*(4), 373–397.

Ball, D. L., & Forzani, F. M. (2009). The work of teaching and the challenge for teacher education. *Journal of Teacher Education, 60*(5), 497–511.

Ball, D. L., & Forzani, F. M. (2011). Building a common core for learning to teach and connecting professional learning to practice. *American Educator, 35*(2) 17–21, 38–39.

Ball, D. L., Lubienski, S. T., & Mewborn, D. S. (2001). Research on teaching mathematics: The unsolved problem of teachers' mathematical knowledge. In V. Richardson (Ed.), *Handbook of research on teaching* (4th ed., pp. 433–456). Washington, DC: American Educational Research Association.

Ball, D. L., Sleep, L., Boerst, T. A., & Bass, H. (2009). Combining the development of practice and the practice of development in teacher education. *The Elementary School Journal, 109*(5), 458–474.

Bartlett, F. (1932). *Remembering: A study in experimental and social psychology.* Cambridge, United Kingdom: Cambridge University Press.

Baxter, J. A., & Williams, S. (2010). Social and analytic scaffolding in middle school mathematics: Managing the dilemma of telling. *Journal of Mathematics Teacher Education, 13*(1), 7–26.

Berliner, D. C. (1994). Expertise: The wonder of exemplary performances. In J. N. Mangieri & C. C. Block (Eds.), *Creating* powerful thinking in teachers and students: Diverse perspectives (pp. 161–186). Fort Worth, TX: Harcourt Brace.

Berliner, D. C. (2002). Educational research: The hardest science of all. *Educational Researcher, 31*(8), 18–20.

Berry, R. Q., III, Rimm-Kaufman, S. E., Ottmar, E. M., Walkowiak, T. A., Merritt, E., & Pinter, H. H. (2013). *The Mathematics Scan (M-Scan): A measure of Standards-based mathematics teaching practices* (Unpublished measure). University of Virginia, Charlottesville, VA.

Bishop, J. P. (2012). "She's always been the smart one. I've always been the dumb one.": Identities in the mathematics classroom. *Journal for Research in Mathematics Education, 43*(1), 34–74.

Blomberg, G., Stürmer, K., & Seidel, T. (2011). How pre-service teachers observe teaching on video: Effects of viewers' teaching subjects and the subject of the video. *Teaching and Teacher Education, 27*(7), 1131–1140.

Boaler, J., & Staples, M. (2008). Creating mathematical futures through an equitable teaching approach: The case of Railside School. *Teachers College Record, 110*(3), 608–645.

Bobis, J., Clarke, B., Clarke, D., Thomas, G., Wright, R., Young-Loveridge, J., & Gould, P. (2005). Supporting teachers in the development of young children's mathematical thinking: Three large-scale cases. *Mathematics Education Research Journal, 16*(3), 27–57.

Boston, M. D. (2012). Assessing the quality of mathematics instruction. *Elementary School Journal, 113*(1), 76–104.

Boston, M. D., Bostic, J., Lesseig, K., & Sherman, M. (2015). A comparison of mathematics classroom observation protocols. *Mathematics Teacher Educator, 3*(2), 154–175.

Boston, M. D., & Smith, M. S. (2009). Transforming secondary mathematics teaching: Increasing the cognitive demands of instructional tasks used in teachers' classrooms. *Journal for Research in Mathematics Education, 40*(2), 119–156.

Breen, S., McCluskey, A., Meehan, M., O'Donovan, J., & O'Shea, A. (2014). A year of engaging with the discipline of noticing: Five mathematics lecturers' reflections. *Teaching in Higher Education, 19*(3), 289–300.

Brodie, K. (2010). Pressing dilemmas: Meaning making and justification in mathematics teaching. *Journal of Curriculum Studies, 42*(1), 27–50.

Cai, J. (Ed.). (2017). *Compendium for research in mathematics education*. Reston, VA: National Council of Teachers of Mathematics.

Carpenter, T. P., Fennema, E., Franke, M., Levi, L., & Empson, S. B. (2015). *Children's mathematics: Cognitively Guided Instruction* (2nd ed.). Portsmouth, NH: Heinemann.

Carpenter, T. P., Fennema, E., Peterson, P. L., Chiang, C. P., & Loef, M. (1989). Using knowledge of children's mathematics thinking in classroom teaching: An experimental study. *American Educational Research Journal, 26*(4), 499–531.

Cengiz, N., Kline, K., & Grant, T. J. (2011). Extending students' mathematical thinking during whole-group discussions. *Journal of Mathematics Teacher Education, 14*(5), 355–374.

Chan, A. G. (2010). *Identity and practice: Preservice teacher learning within a mathematics methods course* (Unpublished doctoral dissertation). University of California at Los Angeles.

Chao, T., & Murray, E. (2013). Teacher asynchronous noticing to foster students' mathematical thinking. In M. V. Martinez & A. Castro Superfine (Eds.), *Proceedings of the 35th annual meeting of the North American Chapter of the International Group for the Psychology of Mathematics Education* (p. 1199). Chicago: University of Illinois at Chicago.

Chapin, S., O'Connor, C., & Anderson, N. (2009). *Classroom discussions: Using math talk to help students learn, Grades K–6* (2nd ed.). Sausalito, CA: Math Solutions.

Chieu, V. M., Herbst, P., & Weiss, M. (2011). Effect of an animated classroom story embedded in online discussion on helping mathematics teachers learn to notice. *Journal of the Learning Sciences, 20*(4), 589–624.

Choppin, J. (2011). The impact of professional noticing on teachers' adaptations of challenging tasks. *Mathematical Thinking and Learning, 13*(3), 175–197.

Clarke, D. J. (2013). Contingent conceptions of accomplished practice: The cultural specificity of discourse in and about the mathematics classroom. *ZDM—The International Journal on Mathematics Education, 45*(1), 21–33.

Clements, D. H., & Sarama, J. (2004). Learning trajectories in mathematics education. *Mathematical Thinking and Learning, 6*(2), 81–89.

Clements, D. H., Sarama, J., Spitler, M. E., Lange, A. A., & Wolfe, C. B. (2011). Mathematics learning by young children in an intervention based on learning trajectories: A large-scale cluster randomized trial. *Journal for Research in Mathematics Education, 42*(2), 127–166.

Cohen, D. K., Raudenbush, S. W., & Ball, D. L. (2003). Resources, instruction, and research. *Educational Evaluation and Policy Analysis, 25*(2), 119–142.

Cohen, E. (1994). *Designing groupwork*. New York, NY: Teachers College Press.

Cohen, E. G. (2000). Equitable classrooms in a changing society. In M. Hallinan (Ed.), *Handbook of the sociology of education* (pp. 265–283). New York, NY: Kluwer.

Colestock, A. (2009). A case study of one secondary mathematics teacher's in-the-moment noticing of student thinking while teaching. In S. L. Swars, D. W. Stinson, & S. Lemons-Smith (Eds.), *Proceedings of the 31st annual meeting of the North American Chapter of the International Group for the Psychology of Mathematics Education* (Vol. 5, pp. 1459–1466). Atlanta: Georgia State University.

Conner, A., Singletary, L. M., Smith, R. C., Wagner, P. A., & Francisco, R. T. (2014). Teacher support for collective argumentation: A framework for examining how teachers support students' engagement in mathematical activities. *Educational Studies in Mathematics, 86*(3), 401–429.

Cooney, T. J. (1999). Conceptualizing teachers' ways of knowing. *Educational Studies in Mathematics, 38*(1–3), 163–187.

Daro, P., Mosher, F. A., & Corcoran, T. (2011). *Learning trajectories in mathematics: A foundation for standards, curriculum, assessment, and instruction* (Consortium for Policy Research in Education, Research Report #RR-68). Philadelphia, PA: Consortium for Policy Research in Education.

Dominguez, H., & Adams, M. (2013). Más o menos: Exploring estimation in a bilingual classroom. *Teaching Children Mathematics, 20*(1), 36–41.

Dreher, A., & Kuntze, S. (2015). Teachers' professional knowledge and noticing: The case of multiple representations in the mathematics classroom. *Educational Studies in Mathematics, 88*(1), 89–114.

Edwards, A. (2003). Learning to see in classrooms: What are student teachers learning about teaching and learning while learning to teach in schools? *British Educational Research Journal, 29*(2), 227–242.

Erickson, F. (2011). On noticing teacher noticing. In M. G. Sherin, V. R. Jacobs, & R. A. Philipp (Eds.), *Mathematics teacher noticing: Seeing through teachers' eyes* (pp. 17–34). New York, NY: Routledge.

Ericsson, K. A., & Simon, H. A. (1993). *Protocol analysis: Verbal reports as data* (Rev. ed.). Cambridge: Massachusetts Institute of Technology Press.

Featherstone, H., Crespo, S., Jilk, L., Oslund, J., Parks, A., & Wood, M. (2011). *Smarter together! Collaboration and equity in the elementary mathematics classroom*. Reston, VA: National Council of Teachers of Mathematics.

Fennema, E., Carpenter, T., Franke, M. L., Levi, L., Jacobs, V. R., & Empson, S. B. (1996). Mathematics instruction and teachers' beliefs: A longitudinal study of using children's thinking. *Journal for Research in Mathematics Education, 27*(4), 403–434.

Fernandes, A. (2012). Mathematics preservice teachers learning about English language learners through task-based interviews and noticing. *Mathematics Teacher Educator, 1*(1), 10–22.

Fernández, C., Llinares, S., & Valls, J. (2012). Learning to notice students' mathematical thinking through on-line discussions. *ZDM—The International Journal on Mathematics Education, 44*(6), 747–759.

Fernández, C., Llinares, S., & Valls, J. (2013). Primary school teacher's noticing of students' mathematical thinking in problem solving. *The Mathematics Enthusiast, 10*(1–2), 441–468.

Fisher, M. H., Schack, E. O., Thomas, J., Jong, C., Eisenhardt, S., Tassell, J., & Yoder, M. (2014). Examining the relationship between preservice elementary teachers' attitudes toward mathematics and professional noticing capacities. In J.-J. Lo, K. R. Leatham, & L. R. Van Zoest (Eds.), *Research trends in mathematics teacher education* (pp. 219–237). New York, NY: Springer International.

Forman, E. A., Larreamendy-Joerns, J., Stein, M. K., & Brown, C. (1998). "You're going to want to find out which and prove it": Collective argumentation in a mathematics classroom. *Learning and Instruction, 8*(6), 527–548.

Forzani, F. M. (2014). Understanding "core practices" and "practice-based" teacher education: Learning from the past. *Journal of Teacher Education, 65*(4), 357–368.

Fraivillig, J. T., Murphy, L. A., & Fuson, K. A. (1999). Advancing children's mathematical thinking in *Everyday Mathematics* classrooms. *Journal for Research in Mathematics Education, 30*(2), 148–170.

Franke, M. L., Carpenter, T. P., Levi, L., & Fennema, E. (2001). Capturing teachers' generative change: A follow-up study of professional development in mathematics. *American Educational Research Journal, 38*(3), 653–689.

Franke, M. L., Kazemi, E., & Battey, D. (2007). Understanding teaching and classroom practice in mathematics. In F. K. Lester Jr. (Ed.), *Second handbook of research on mathematics teaching and learning* (pp. 225–256). Charlotte, NC: Information Age; Reston, VA: National Council of Teachers of Mathematics.

Franke, M. L., Turrou, A. C., Webb, N. M., Ing, M., Wong, J., Shim, N., & Fernandez, C. (2015). Student engagement with others' mathematical ideas: The role of teacher invitation and support moves. *Elementary School Journal, 116*(1), 126–148.

Franke, M. L., Webb, N. M., Chan, A. G., Ing, M., Freund, D., & Battey, D. (2009). Teacher questioning to elicit students' mathematical thinking in elementary school classrooms. *Journal of Teacher Education, 60*(4), 380–392.

Fredenberg, M. D. (2015). *Factors considered by elementary teachers when developing and modifying mathematical tasks to support children's mathematical thinking* (Unpublished doctoral dissertation). University of California, San Diego, and San Diego State University.

Garner, G., & Rosaen, C. (2009). Strengthening partnerships and boosting conceptual connections in preservice field experience. *Teaching Education, 20*(4), 329–342.

Ghousseini, H. (2009). Designing opportunities to learn to lead classroom mathematics discussions in pre-service teacher education: Focusing on enactment. In D. S. Mewborn & H. S. Lee (Eds.), *Scholarly practices and inquiry in the preparation of mathematics teachers* (pp. 203–218). Association of Mathematics Teacher Educators (AMTE) Monograph No. 6. San Diego, CA: AMTE.

Ghousseini, H., Beasley, H., & Lord, S. (2015). Investigating the potential of guided practice with an enactment tool for supporting adaptive performance. *Journal of the Learning Sciences, 24*(3), 461–497.

Ghousseini, H., Franke, M. L., & Turrou, A. G. (2014, April). *A practice-based design for teacher preparation to support ongoing examination and practice in relation to equity*. Paper presented at the annual meeting of the American Educational Research Association, Philadelphia, PA.

Gibbons, L., Hintz, A., Kazemi, A., & Hartmann, E. (2016). *Supporting innovation in teaching mathematics: Examining a routine that fosters professional learning across the system*. Manuscript submitted for publication.

Goldsmith, L. T., Doerr, H. M., & Lewis, C. A. (2014). Mathematics teachers' learning: A conceptual framework and synthesis of research. *Journal of Mathematics Teacher Education, 17*(1), 5–36.

Goldsmith, L. T., & Seago, N. (2011). Using classroom artifacts to focus teachers' noticing. In M. G. Sherin, V. R. Jacobs, & R. A. Philipp (Eds.), *Mathematics teacher noticing: Seeing through teachers' eyes* (pp. 169–187). New York, NY: Routledge.

Goodwin, C. (1994). Professional vision. *American Anthropologist, 96*(3), 606–663.

Goos, M. (2004). Learning mathematics in a classroom community of inquiry. *Journal for Research in Mathematics Education, 35*(4), 258–291.

Gresalfi, M. S. (2009). Taking up opportunities to learn: Constructing dispositions in mathematics classrooms. *Journal of the Learning Sciences, 18*(3), 327–369.

Grossman, P., Compton, C., Igra, D., Ronfeldt, M., Shahan, E., & Williamson, P. W. (2009). Teaching practice: A cross-professional perspective. *Teachers College Record, 111*(9), 2055–2100.

Grossman, P., Franke, M. L., Kavanagh, S. S., Windschitl, M. A., Dobson, J., Ball, D. L., & Bryk, A. S. (2014, April). *Enriching research and innovation through the specification of professional practice: The Core Practice Consortium*. Symposium presented at the annual meeting of the American Educational Research Association, Philadelphia, PA. Webcast retrieved from http://www.aera.net/EventsMeetings/PreviousAnnualMeetings/2014AnnualMeeting/2014AnnualMeetingWebcasts/EnrichingResearchandInnovation/tabid/15496/Default.aspx

Grossman, P., Hammerness, K., & McDonald, M. (2009). Redefining teaching, re-imagining teacher education. *Teachers and Teaching, 15*(2), 273–289.

Grossman, P., & McDonald, M. (2008). Back to the future: Directions for research in teaching and teacher education. *American Educational Research Journal, 45*(1), 184–205.

Grouws, D. A. (1992). *Handbook of research on mathematics teaching and learning.* New York, NY: Macmillan.

Hand, V. (2012). Seeing culture and power in mathematical learning: Toward a model of equitable instruction. *Educational Studies in Mathematics, 80*(1–2), 233–247.

Herbel-Eisenmann, B., Meaney, T., Bishop, J. P., & Heyd-Metzuyanim, E. (2017). Highlighting heritages and building tasks: A critical analysis of mathematics classroom discourse literature. In J. Cai (Ed.), *Compendium for research in mathematics education* (pp. 722–765). Reston, VA: National Council of Teachers of Mathematics.

Herbel-Eisenmann, B. A., Steele, M. D., & Cirillo, M. (2013). (Developing) teacher discourse moves: A framework for professional development. *Mathematics Teacher Educator, 1*(2), 181–196.

Hiebert, J., & Grouws, D. A. (2007). The effects of classroom mathematics teaching on students' learning. In F. K. Lester Jr. (Ed.), *Second handbook of research on mathematics teaching and learning* (pp. 371–404). Charlotte, NC: Information Age; Reston, VA: National Council of Teachers of Mathematics.

Hiebert, J., & Morris, A. K. (2012). Teaching, rather than teachers, as a path toward improving classroom instruction. *Journal of Teacher Education, 63*(2), 92–102.

Huang, R., & Li, Y. (2009). What matters most: A comparison of expert and novice teachers' noticing of mathematics classroom events. *School Science and Mathematics, 112*(7), 420–432.

Hufferd-Ackles, K., Fuson, K. C., & Sherin, M. G. (2004). Describing levels and components of a math-talk learning community. *Journal for Research in Mathematics Education, 35*(2), 81–116.

Hunter, R. (2010). Changing roles and identities in the construction of a community of mathematical inquiry. *Journal of Mathematics Teacher Education, 13*(5), 397–409.

Hunter, R., Hunter, J., & Anthony, G. (2013). Using instructional activities to learn the work of ambitious mathematics in pre-service teacher educator settings. In V. Steinle, L. Ball, & C. Bardini (Eds.), *Mathematics education: Yesterday, today and tomorrow. Proceedings of the 36th annual conference of the Mathematics Education Research Group of Australasia* (pp. 703–706). Melbourne, Australia: MERGA.

Jacobs, V. R., & Empson, S. B. (2016). Responding to children's mathematical thinking in the moment: An emerging framework of teaching moves. *ZDM—The International Journal on Mathematics Education, 48*(1–2), 185–197.

Jacobs, V. R., Franke, M. L., Carpenter, T. P., Levi, L., & Battey, D. (2007). Professional development focused on children's algebraic reasoning in elementary school. *Journal for Research in Mathematics Education, 38*(3), 258–288.

Jacobs, V. R., Lamb, L. L. C., & Philipp, R. A. (2010). Professional noticing of children's mathematical thinking. *Journal for Research in Mathematics Education, 41*(2), 169–202.

Jacobs, V. R., Lamb, L. L. C., Philipp, R. A., & Schappelle, B. P. (2011). Deciding how to respond on the basis of children's understandings. In M. G. Sherin, V. R. Jacobs, & R. A. Philipp (Eds.), *Mathematics teacher noticing: Seeing through teachers' eyes* (pp. 97–116). New York, NY: Routledge.

Johnson, J., Uline, C., & Perez, L. (2011). Expert noticing and principals of high-performing urban schools. *Journal of Education for Students Placed at Risk (JESPAR), 16*(2), 122–136.

Kazemi, E., Elliott, R., Mumme, J., Lesseig, K., & Kelley-Petersen, M. (2011). Noticing leaders' thinking about videocases of teachers engaged in mathematics tasks in professional development. In M. G. Sherin, V. R. Jacobs, & R. A. Philipp (Eds.), *Mathematics teacher noticing: Seeing through teachers' eyes* (pp. 188–203). New York, NY: Routledge.

Kazemi, E., Franke, M., & Lampert, M. (2009). Developing pedagogies in teacher education to support novice teachers' ability to enact ambitious instruction. In B. Hunter, B. Bicknell, & T. Burgess (Eds.), *Crossing divides: Proceedings of the 32nd annual conference of the Mathematics Education Research Group of Australasia* (Vol. 1, pp. 11–29). Palmerston North, New Zealand: MERGA.

Kazemi, E., Ghousseini, H., Cunard, A., & Turrou, A. (2016). Getting inside rehearsals: Insights from teacher educators to support work on complex practice. *Journal of Teacher Education, 67*(1), 18–31.

Kazemi, E., & Stipek, D. (2001). Promoting conceptual thinking in four upper-elementary mathematics classrooms. *Elementary School Journal, 102*(1), 59–80.

Kersting, N. B., Givvin, K. B., Sotelo, F. L., & Stigler, J. W. (2010). Teachers' analyses of classroom video predict student learning of mathematics: Further explorations of a novel measure of teacher knowledge. *Journal of Teacher Education, 61*(1–2), 172–181.

Khisty, L. L., & Chval, K. B. (2002). Pedagogic discourse and equity in mathematics: When teachers' talk matters. *Mathematics Education Research Journal, 14*(3), 154–168.

Lampert, M. (2001). *Teaching problems and the problems of teaching.* New Haven, CT: Yale University Press.

Lampert, M., Franke, M. L., Kazemi, E., Ghousseini, H., Turrou, A. C., Beasley, H., . . . Crowe, K. (2013). Keeping it complex: Using rehearsals to support novice teacher learning of ambitious teaching. *Journal of Teacher Education, 64*(3), 226–243.

Lampert, M., & Graziani, F. (2009). Instructional activities as a tool for teachers' and teacher educators' learning. *The Elementary School Journal, 109*(5), 491–509.

Land, T. J., Drake, C., Sweeney, M., Franke, N., & Johnson, J. M. (2014). *Transforming the task with number choice: Kindergarten through Grade 3.* Reston, VA: National Council of Teachers of Mathematics.

Leatham, K. R., Peterson, B. E., Stockero, S. L., & Van Zoest, L. R. (2015). Conceptualizing mathematically significant pedagogical opportunities to build on student thinking. *Journal for Research in Mathematics Education, 46*(1), 88–124.

Lester, F. K., Jr. (Ed.). (2007). *Second handbook of research on mathematics teaching and learning.* Charlotte, NC: Information Age; Reston, VA: National Council of Teachers of Mathematics.

Lewis, C. C., Perry, R. R., Friedkin, S., & Roth, J. R. (2012). Improving teaching does improve teachers: Evidence from lesson study. *Journal of Teacher Education, 63*(5), 368–375.

Lobato, J., Hohensee, C., & Rhodehamel, B. (2013). Students' mathematical noticing. *Journal for Research in Mathematics Education, 44*(5), 809–850.

Lobato, J., & Walters, C. D. (2017). A taxonomy of approaches to learning trajectories and progressions. In J. Cai (Ed.), *Compendium for research in mathematics education* (pp. 74–101). Reston, VA: National Council of Teachers of Mathematics.

Maloney, A. P., Confrey, J., & Nguyen, K. H. (2014). *Learning over time: Learning trajectories in mathematics education*. Charlotte, NC: Information Age.

Mason, J. (2002). *Researching your own practice: The discipline of noticing*. London, United Kingdom: RoutledgeFalmer.

Mason, J. (2011). Noticing: Roots and branches. In M. G. Sherin, V. R. Jacobs, & R. A. Philipp (Eds.), *Mathematics teacher noticing: Seeing through teachers' eyes* (pp. 35–49). New York, NY: Routledge.

Matsumura, L. C., Garnier, H. E., Slater, S. C., & Boston, M. D. (2008). Toward measuring instructional interactions "at-scale." *Educational Assessment, 13*(4), 267–300.

McDonald, M., Kazemi, E., & Kavanagh, S. S. (2013). Core practices and pedagogies of teacher education: A call for a common language and collective activity. *Journal of Teacher Education, 64*(5), 378–386.

Mendez, E. P., Sherin, M. G., & Louis, D. A. (2007). Multiple perspectives on the development of an eighth-grade mathematical discourse community. *The Elementary School Journal, 108*(1), 41–61.

Mercer, N., & Sams, C. (2006). Teaching children how to use language to solve maths problems. *Language and Education, 20*(6), 507–528.

Miller, K. F. (2011). Situation awareness in teaching. In M. G. Sherin, V. R. Jacobs, & R. A. Philipp (Eds.), *Mathematics teacher noticing: Seeing through teachers' eyes* (pp. 51–65). New York, NY: Routledge.

Miller, K., & Zhou, X. (2007). Learning from classroom video: What makes it compelling and what makes it hard. In R. Goldman, R. Pea, B. Barron, & S. J. Derry (Eds.), *Video research in the learning sciences* (pp. 321–334). Mahwah, NJ: Erlbaum.

Moschkovich, J. (1999). Supporting the participation of English language learners in mathematical discussions. *For the Learning of Mathematics, 19*(1), 11–19.

Moschkovich, J. N. (2008). "I went by twos, he went by one": Multiple interpretations of inscriptions as resources for mathematical discussions. *Journal of the Learning Sciences, 17*(4), 551–587.

Most, S. B., Simons, D. J., Scholl, B. J., Jimenez, R., Clifford, E., & Chabris, C. F. (2001). How not to be seen: The contribution of similarity and selective ignoring to sustained inattentional blindness. *Psychological Science, 12*(1), 9–17.

Nathan, M. J., & Knuth, E. J. (2003). A study of whole classroom mathematical discourse and teacher change. *Cognition and Instruction, 21*(2), 175–207.

National Center for Teacher Effectiveness. (2014). *Mathematical quality of instruction*. Retrieved from http://isites.harvard.edu/icb/icb.do?keyword=mqi_training

National Council of Teachers of Mathematics. (1991). *Professional standards for teaching mathematics*. Reston, VA: Author.

National Council of Teachers of Mathematics. (2014). *Principles to actions: Ensuring mathematical success for all*. Reston, VA: Author.

National Governors Association Center for Best Practices & Council of Chief State School Officers. (2010). *Common Core State Standards for Mathematics*. Washington, DC: Authors.

National Research Council. (2001). *Adding it up: Helping children learn mathematics*. Washington, DC: National Academy Press.

National Research Council. (2005). *How students learn: Mathematics in the classroom*. Washington, DC: National Academy Press.

O'Connor, M. C. (2001). "Can any fraction be turned into a decimal?" A case study of a mathematical group discussion. *Educational Studies in Mathematics, 46*(1), 143–185.

O'Flahavan, J. F., & Boté, L. (2014, April). *Chickens and eggs, carts and horses, and a holy grail or two: In search of the nexus between core teaching practices and student learning*. Paper presented at the annual meeting of the American Educational Research Association, Philadelphia, PA.

Oslund, J. A., & Crespo, S. (2014). Classroom photographs: Reframing what and how we notice. *Teaching Children Mathematics, 20*(9), 564–572.

Ozgur, Z., Reiten, L., & Ellis, A. B. (2015). On framing teacher moves for supporting student reasoning. In T. G. Bartell, K. N. Bieda, R. T. Putnam, K. Bradfield, & H. Dominguez (Eds.), *Proceedings of the thirty-seventh annual meeting of the North American Chapter of the International Group for the Psychology of Mathematics Education* (pp. 1062–1069). East Lansing: Michigan State University.

Rasmussen, C., & Wawro, M. (2017). Post-calculus research in undergraduate mathematics education. In J. Cai (Ed.), *Compendium for research in mathematics education* (pp. 551–579). Reston, VA: National Council of Teachers of Mathematics.

Richardson, V. (Ed.). (2001). *Handbook of research on teaching* (4th ed.). Washington, DC: American Educational Research Association.

Robertson, A. D., Scherr, R. E., & Hammer, D. (Eds.). (2016). *Responsive teaching in science and mathematics*. New York, NY: Routledge.

Rosaen, C. L., Lundeberg, M., Terpstra, M., Cooper, M., Fu, J., & Rui, N. (2010). Seeing through a different lens: What do interns learn when they make video cases of their own teaching? *The Teacher Educator, 45*(1), 1–22.

Ross, P., & Gibson, S. A. (2010). Exploring a conceptual framework for expert noticing during literacy instruction. *Literacy Research and Instruction, 49*(2), 175–193.

Roth McDuffie, A., Foote, M. Q., Bolson, C., Turner, E. E., Aguirre, J. M., Bartell, T. G., . . . Land, T. (2014). Using video analysis to support prospective K–8 teachers' noticing of students' multiple mathematical knowledge bases. *Journal of Mathematics Teacher Education, 17*(3), 245–270.

Russ, R. S., & Luna, M. J. (2013). Inferring teacher epistemological framing from local patterns in teacher noticing. *Journal of Research in Science Teaching, 50*(3), 284–314.

Santagata, R. (2011). From teacher noticing to a framework for analyzing and improving classroom lessons. In M. G. Sherin, V. R. Jacobs, & R. A. Philipp (Eds.), *Mathematics teacher noticing: Seeing through teachers' eyes* (pp. 152–168). New York, NY: Routledge.

Santagata, R., Zannoni, C., & Stigler, J. W. (2007). The role of lesson analysis in pre-service teacher education: An empirical investigation of teacher learning from a virtual video-based field experience. *Journal of Mathematics Teacher Education, 10*(2), 123–140.

Sawada, D., Piburn, M., Turley, J., Falconer, K., Benford, R., & Bloom, I. (2000). Measuring reform practices in science and mathematics classrooms: The reformed teaching observation protocol. *School Science and Mathematics, 102*(6), 245–253.

Schack, E. O., Fisher, M. H., Thomas, J. N., Eisenhardt, S., Tassell, J., & Yoder, M. (2013). Prospective elementary school teachers' professional noticing of children's early numeracy. *Journal of Mathematics Teacher Education, 16*(5), 379–397.

Schack, E. O., Fisher, M. H., & Wilhelm, J. A. (Eds.). (2017). *Teacher noticing: Bridging and broadening perspectives, contexts, and frameworks.* New York, NY: Springer International.

Scherrer, J., & Stein, M. K. (2013). Effects of a coding intervention on what teachers learn to notice during whole-group discussion. *Journal of Mathematics Teacher Education, 16*(2), 105–124.

Schifter, D. (2011). Examining the behavior of operations: Noticing early algebraic ideas. In M. G. Sherin, V. R. Jacobs, & R. A. Philipp (Eds.), *Mathematics teacher noticing: Seeing through teachers' eyes* (pp. 204–220). New York, NY: Routledge.

Schleppenbach, M., Flevares, L. M., Sims, L. M., & Perry, M. (2007). Teachers' responses to student mistakes in Chinese and U.S. mathematics classrooms. *The Elementary School Journal, 108*(2), 131–147.

Schneider, W., & Shiffrin, R. M. (1977). Controlled and automatic human information processing: I. Detection, search, and attention. *Psychological Review, 84*(1), 1–66.

Schoenfeld, A. H. (1998). Toward a theory of teaching-in-context. *Issues in Education, 4*(1), 1–94.

Schoenfeld, A. H. (2011). *How we think: A theory of goal-oriented decision making and its educational applications.* New York, NY: Routledge.

Seidel, T., Stürmer, K., Blomberg, G., Kobarg, M., & Schwindt, K. (2011). Teacher learning from analysis of videotaped classroom situations: Does it make a difference whether teachers observe their own teaching or that of others? *Teaching and Teacher Education, 27*(2), 259–267.

Sherin, M. G. (2002). When teaching becomes learning. *Cognition and Instruction, 20*(2), 119–150.

Sherin, M. G., Jacobs, V. R., & Philipp, R. A. (Eds.). (2011a). *Mathematics teacher noticing: Seeing through teachers' eyes.* New York: Routledge.

Sherin, M. G., Jacobs, V. R., & Philipp, R. A. (2011b). Situating the study of teacher noticing. In M. G. Sherin, V. R., Jacobs, & R. A. Philipp (Eds.), *Mathematics teacher noticing: Seeing through teachers' eyes* (pp. 3–13). New York, NY: Routledge.

Sherin, M. G., Linsenmeier, K. A., & van Es, E. A. (2009). Selecting video clips to promote mathematics teachers' discussion of student thinking. *Journal of Teacher Education, 60*(3), 213–230.

Sherin, M. G., Russ, R. S., & Colestock, A. A. (2011). Accessing mathematics teachers' in-the-moment noticing. In M. G. Sherin, V. R. Jacobs, & R. A. Philipp (Eds.), *Mathematics teacher noticing: Seeing through teachers' eyes* (pp. 79–94). New York, NY: Routledge.

Sherin, M. G., Russ, R. S., Sherin, B., & Colestock, A. (2008). Professional vision in action: An exploratory study. *Issues in Teacher Education, 17*(2), 27–46.

Sherin, M. G., & van Es, E. A. (2005). Using video to support teachers' ability to notice classroom interactions. *Journal of Technology and Teacher Education, 13*(3), 475–491.

Sherin, M. G., & van Es, E. A. (2009). Effects of video club participation on teachers' professional vision. *Journal of Teacher Education, 60*(1), 20–37.

Silver, E. A., Ghousseini, H., Gosen, D., Charalambous, C., & Strawhun, B. T. F. (2005). Moving from rhetoric to praxis: Issues faced by teachers in having students consider multiple solutions for problems in the mathematics classroom. *Journal of Mathematical Behavior, 24*(3–4), 287–301.

Simon, M. A. (1995). Reconstructing mathematics pedagogy from a constructivist perspective. *Journal for Research in Mathematics Education, 26*(2), 114–145.

Simon, M., & Schifter, D. (1993). Toward a constructivist perspective: The impact of a mathematics teacher in-service program on students. *Educational Studies in Mathematics, 25*(4), 331–340.

Simons, D. J., & Chabris, C. F. (1999). Gorillas in our midst: Sustained inattentional blindness for dynamic events. *Perception, 28*(9), 1059–1074.

Sleep, L. (2012). The work of steering instruction toward the mathematical point: A decomposition of teaching practice. *American Educational Research Journal, 49*(5), 935–970.

Smith, M. S., & Stein, M. K. (1998). Selecting and creating mathematical tasks: From research to practice. *Mathematics Teaching in the Middle School, 3*(5), 344–350.

Sowder, J. T. (2007). The mathematical education and development of teachers. In F. K. Lester Jr. (Ed.), *Second handbook of research on mathematics teaching and learning* (pp. 157–223). Charlotte, NC: Information Age; Reston, VA: National Council of Teachers of Mathematics.

Staples, M. (2007). Supporting whole-class collaborative inquiry in a secondary mathematics classroom. *Cognition and Instruction, 25*(2–3), 161–217.

Staples, M. E., & Truxaw, M. P. (2010). The Mathematics Learning Discourse Project: Fostering higher order thinking and academic language in urban mathematics classrooms. *Journal of Urban Mathematics Education, 3*(1), 27–56.

Star, J. R., Lynch, K., & Perova, N. (2011). Using video to improve preservice mathematics teachers' abilities to attend to classroom features: A replication study. In M. G. Sherin, V. R. Jacobs, & R. A. Philipp (Eds.), *Mathematics teacher noticing: Seeing through teachers' eyes* (pp. 117–133). New York, NY: Routledge.

Star, J. R., & Strickland, S. K. (2008). Learning to observe: Using video to improve preservice mathematics teachers' ability to notice. *Journal of Mathematics Teacher Education, 11*(2), 107–125.

Steffe, L. (1992). Learning stages in the construction of the number sequence. In J. Bideaud, C. Meljac, & J. Fischer (Eds.), *Pathways to number: Children's developing numerical abilities* (pp. 83–88). Hillsdale, NJ: Erlbaum.

Stein, M. K., Engle, R. A., Smith, M. S., & Hughes, E. K. (2008). Orchestrating productive mathematical discussions: Five practices for helping teachers move beyond show and tell. *Mathematical Thinking and Learning, 10*(4), 313–340.

Stein, M. K., Remillard, J., & Smith, M. S. (2007). How curriculum influences student learning. In F. K. Lester Jr. (Ed.), *Second handbook of research on mathematics teaching and learning* (pp. 319–369). Charlotte, NC: Information Age; Reston, VA: National Council of Teachers of Mathematics.

Stevens, R., & Hall, R. (1998). Disciplined perception: Learning to see in technoscience. In M. Lampert & M. L. Blunk (Eds.),

Talking mathematics in school: Studies of teaching and learning (pp. 107–150). New York, NY: Cambridge University Press.

Stockero, S. L. (2013). Student teacher noticing during mathematics instruction. In A. M. Lindmeier & A. Heinze (Eds.), *Proceedings of the 37th Conference of the International Group for the Psychology of Mathematics Education* (Vol. 4, pp. 249–256). Kiel, Germany: PME.

Stockero, S. L. (2014). Transitions in prospective mathematics teacher noticing. In J.-J. Lo, K. R. Leatham, & L. R. Van Zoest (Eds.), *Research trends in mathematics teacher education* (pp. 239–259). New York, NY: Springer International.

Stürmer, K., Könings, K. D., & Seidel, T. (2013). Declarative knowledge and professional vision in teacher education: Effect of courses in teaching and learning. *The British Journal of Educational Psychology, 83*(3), 467–483.

Sztajn, P., Borko, H., & Smith, T. (2017). Research on mathematics professional development. In J. Cai (Ed.), *Compendium for research in mathematics education* (pp. 793–823). Reston, VA: National Council of Teachers of Mathematics.

Sztajn, P., Confrey, J., Wilson, P. H., & Edgington, C. (2012). Learning trajectory based instruction: Toward a theory of teaching. *Educational Researcher, 41*(5), 147–156.

Talanquer, V., Tomanek, D., & Novodvorsky, I. (2013). Assessing students' understanding of inquiry: What do prospective science teachers notice? *Journal of Research in Science Teaching, 50*(2), 189–208.

Turner, E. E., Aguirre, J. M., Bartell, T. G., Drake, C., Foote, M. Q., & Roth McDuffie, A. (2014). Making meaningful connections with mathematics and the community: Lessons from prospective teachers. In T. G. Bartell & A. Flores (Eds.), *Embracing resources of children, families, communities and cultures in mathematics learning* (pp. 30–49). TODOS Research Monograph 3. San Bernardino, CA: TODOS: Mathematics for All.

Turner, E., Dominguez, H., Maldonado, L., & Empson, S. (2013). English learners' participation in mathematicas discussion: Shifting positions and dynamic identities. *Journal for Research in Mathematics Education, 44*(1), 199–234.

Turner, E. E., & Drake, C. (2016). A review of research on prospective teachers' learning about children's mathematical thinking and cultural funds of knowledge. *Journal of Teacher Education, 67*(1), 32–46.

Turner, E. E., Drake, C., Roth McDuffie, A., Aguirre, J., Bartell, T. G., & Foote, M. Q. (2012). Promoting equity in mathematics teacher preparation: A framework for advancing teacher learning of children's multiple mathematics knowledge bases. *Journal of Mathematics Teacher Education, 15*(1), 67–82.

van Es, E. A., & Sherin, M. G. (2002). Learning to notice: Scaffolding new teachers' interpretations of classroom interactions. *Journal of Technology and Teacher Education, 10*(4), 571–596.

van Es, E. A., & Sherin, M. G. (2008). Mathematics teachers' "learning to notice" in the context of a video club. *Teaching and Teacher Education, 24*(2), 244–276.

Wager, A. A. (2014). Noticing children's participation: Insights into teacher positionality toward equitable mathematics pedagogy. *Journal for Research in Mathematics Education, 45*(3), 312–350.

Wagner, D., & Herbel-Eisenmann, B. (2008). "Just don't": The suppression and invitation of dialogue in the mathematics classroom. *Educational Studies in Mathematics, 67*(2), 143–157.

Walkoe, J. (2015). Exploring teacher noticing of student algebraic thinking in a video club. *Journal of Mathematics Teacher Education, 18*(6), 523–550.

Walkowiak, T. A., Berry, R. Q., Meyer, J. P., Rimm-Kaufman, S. E., & Ottmar, E. R. (2014). Introducing a standards-based observational measure of mathematics teaching practices: Evidence of validity and score reliability. *Educational Studies in Mathematics, 85*(1), 109–129.

Walshaw, M., & Anthony, G. (2008). The teacher's role in classroom discourse: A review of recent research into mathematics classrooms. *Review of Educational Research, 78*(3), 516–551.

Warshauer, H. K. (2015). Productive struggle in middle school mathematics classrooms. *Journal of Mathematics Teacher Education, 18*(4), 375–400.

Webb, J., Wilson, P. H., Reid, M., & Duggan, A. (2015). Learning instructional practices in professional development. In T. G. Bartell, K. N. Bieda, R. T. Putnam, K. Bradfield, & H. Dominguez (Eds.), *Proceedings of the thirty-seventh annual meeting of the North American Chapter of the International Group for the Psychology of Mathematics Education* (pp. 836–843). East Lansing: Michigan State University.

Webb, N. M., Franke, M. L., De, T., Chan, A. G., Freund, D., Shein, P., & Melkonian, D. K. (2009). "Explain to your partner": Teachers' instructional practices and students' dialogue in small groups. *Cambridge Journal of Education, 39*(1), 49–70.

Webb, N. M., Franke, M. L., Ing, M., Chan, A., De, T., Freund, D., & Battey, D. (2008). The role of teacher instructional practices in student collaboration. *Contemporary Educational Psychology, 33*(3), 360–381.

Webb, N. M., Franke, M. L., Ing, M., Wong, J., Fernandez, C. H., Shin, N., & Turrou, A. C. (2014). Engaging with others' mathematical ideas: Interrelationships among student participation, teachers' instructional practices, and learning. *International Journal of Educational Research, 63*, 79–93.

Wilson, S. M., & Berne, J. (1999). Teacher learning and the acquisition of professional knowledge: An examination of research on contemporary professional development. *Review of Research in Education, 24*, 173–209.

Wood, D., Bruner, J. S., & Ross, G. (1976). The role of tutoring in problem solving. *Journal of Child Psychology and Psychiatry, 17*(2), 89–100.

Zapatera, A., & Callejo, M. L. (2013). Preservice primary teachers's noticing of students' generalization process. In A. M. Lindmeier & A. Heinze (Eds.), *Proceedings of the 37th conference of the International Group for the Psychology of Mathematics Education* (Vol. 4, pp. 425–432). Kiel, Germany: PME.

29 数学专业发展研究

葆拉·斯坦
美国北卡罗莱纳州立大学
希尔达·博尔科
美国斯坦福大学
托马斯·M.史密斯
美国加州大学河滨分校
译者：韩继伟
　　　东北师范大学数学与统计学院

近期的研究表明了教师对学生学习的贡献（Hargreaves，2004），以及教师的学科内容知识和教学知识的重要性（Hill，Rowan，& Ball，2005）。因为教师和他们的教学事关重大，所以对于学区、学校，当然也包括教师们自己来说，通过专业发展（PD）来提高教师的知识和他们的教学已经变得越来越重要。目前，美国公立学校的大约350万教师中，99%的人每年都会参加某种形式的专业发展活动（National Center for Education Statistics，2013）。人们对教师专业发展越来越多的关注，以及教师专业发展与政策议程（例如提高学生的成绩）越来越多的联系推动了这一问题的研究。在数学领域已经开展了很多这方面的研究。因此，数学专业发展近来已经成为数学教育这个大研究领域中一个独立的研究领域（Sztajn，2011），这方面的学术出版物也越来越多。在这一章里，我们将评述这个新的研究领域里的最新研究结果和研究方法走向。

我们强调研究一词是因为我们认识到：虽然关于教师数学专业发展的研究显著增长，但是提供给教师的大多数专业发展机会从来没有被系统地研究过，它们也不一定是建立在研究结果的基础上的。因此，在人们所知道的专业发展与教师实际经历的专业发展项目二者之间存在差距。本章侧重于前者，也就是说，我们试图从数学专业发展的研究中考查目前该领域中已有的结论，而且我们也承认这些结论并不总是能够告诉教师专业发展项目该如何进行。而我们对这些研究的关注，对我们组织本章的方式，评述论文的选择，以及结论中我们所强调和讨论的内容等，都具有重要的意义。

博尔科（2004）提出用三个阶段来描绘专业发展研究领域的图景：第一个阶段是，考察在一个地方实施的个别专业发展项目的设计，由此为这个项目的潜在影响力提供存在性证明。第二个阶段是，研究在多个地方推行的独立专业发展项目，把理解专业发展项目由一个地方推广到多个地方的必要条件作为主要关注点。第三个阶段是，通过项目的资源要求和项目效果的异同来比较各种专业发展模式。索德（2007）在其评述性文章的结论中也提到了前两个阶段。关于专业发展的设计，她指出一个新的范式出现了，那就是由提供工作坊、关注技术（第215页）转换到连接教师学习和教学实践的专业发展活动。她也同时指出扩大专业发展仍然是"教育者面临的最大的挑战"（第213页）。这一章是以索德的工作为基础，根据博尔科的三个研究阶段来组织的。本文聚焦于这些研究中的研究问题，并据此提出我们自己的三个研究问题：

1. 我们从讨论个别数学专业发展项目影响的研究中知道了什么？
2. 我们从考察数学专业发展项目规模化的必要条件

29 | Research on Mathematics Professional Development

PAOLA SZTAJN
North Carolina State University, United States

HILDA BORKO
Stanford University, United States

THOMAS M. SMITH
University of California Riverside, United States

Recent research has demonstrated the contributions teachers make to student learning (Hargreaves, 2014) and shown the importance of teachers' content and pedagogical content knowledge (Hill, Rowan, & Ball, 2005). Because teachers and their teaching matter, improving their knowledge and instructional practice through professional development (PD) has become increasingly important for districts, schools, and, of course, teachers themselves. Currently, 99% of the approximately 3.5 million U.S. public school teachers attend some form of PD every year (National Center for Education Statistics, 2013). The growing attention to teacher PD and its increasing connection to policy agendas, such as increasing student achievement, have propelled research on this topic. Much of this research has been conducted in the area of mathematics. As a result, PD in mathematics has recently become its own field of research within the larger field of mathematics education research (Sztajn, 2011), with a growing body of scholarly publications. In this chapter, we review recent emerging findings and methodological trends in this new *research* field.

We highlight the word *research* to recognize that, despite the significant growth of research on teachers' mathematics PD, most of the PD opportunities offered to teachers are never systematically investigated. Nor are they necessarily based on research findings. Thus, a gap exists between what is known about PD and what teachers actually experience in PD programs. Our chapter focuses on the former; that is, we seek to examine what the field knows from research on mathematics PD, acknowledging that such knowledge does not always inform the array of PD programs offered to teachers. Our focus on research has implications for the way we organized this chapter, the selection of papers we included in this review, and what we highlighted and discussed in the concluding remarks.

Mapping the terrain of PD research, Borko (2004) suggested three phases. The first examines the design of individual PD programs offered at one site as an existence proof for the potential of the program. The second includes research on a single PD program enacted by more than one facilitator at more than one site and focuses on understanding what it takes to scale PD from one site to many. The third is the comparison across PD models to examine similarities and differences in their resource requirements and impact on outcomes. Sowder (2007), in the conclusion of her review, addressed the first two phases. Regarding the design of PD, she noted the emergence of a new paradigm that moved "from providing workshops and focusing on techniques" (p. 215) to connecting teacher learning and the practice of teaching. She also suggested that scaling up PD remained "the greatest challenge we educators face" (p. 213). This chapter builds on the work of Sowder and is organized around Borko's three research phases. We focus on the research questions addressed in the studies reviewed and ask ourselves three questions:

1. What do we know from research that addresses the impact of a single mathematics PD program?
2. What do we know from research that examines what it takes to bring mathematics PD programs to scale?
3. What do we know from research that compares different mathematics PD programs?

In what follows, we first present how PD and research are defined in this chapter. Next, we explain our review process and its limitations. We then address these three

3. 我们从比较不同数学专业发展项目的研究中知道了什么？

在下文中，我们首先介绍了在本章中专业发展和研究是如何定义的。然后，解释我们的评述过程和局限性。接着，我们回答这三个问题，从每个研究阶段中标示出中心主题和研究发现的模式。最后在结论部分，我们将讨论数学专业发展研究领域的现况，以及推动这个研究领域向前发展所能够采取的步骤的一些建议。

专业发展和专业发展研究

30多年前，反思性实践者和行动中的知识（Schön，1983）这些概念使人们关注到教师的实践和他们的知识之间的联系。从那时起，由于理解了教师学习是积极的、情境化的和社会性的（Putnam & Borko，2000），人们对实践和知识之间联系的关注也加强了。如今，众所周知实践是教师学习的重要途径。教师既通过根植在他们日常专业生活中的真实学习情境学习（Webster-Wright，2009），也通过他们参与教学的各个方面或因素来获得学习机会（Ball & Cohen，1999）。在这一章里，我们将仔细考查那些提供给在职的幼儿园到12年级数学教师的专业发展项目，这些项目为教师提供了有组织的学习经验。这些项目有具体的与数学教学有关的学习目标，有以完成这些目标为目的的计划，包括一些预期的培养学生学习的促进措施，允许检查、提炼和重复数学中的专业发展的模型。

本章的展开由两个重要问题来引导。第一个是研究方法。学校和地区面临提高学生数学成绩的巨大压力，而目前对问责的关注已经导致人们对学生成绩测量关注的提高。这个问责风潮影响专业发展的研究。由于专业发展已成为实现更大计划（如2002年的《不让一个孩子掉队法案》或"力争上游"教改计划）所设立的目标的政策机制，因此越来越强调把实验设计作为研究专业发展的首要方法，而且最好是将教师学习和学生成绩联系起来。这就导致人们对那些精细的描述性研究和其他类型的教育研究的价值的认识降低了，而这些又是基础性的、探索性的、发展性的、设计性的和改进性的研究（Earle等，2013；National Science Foundation，2015）。

在教育政策的讨论中，过分强调实验研究方法可能也意味着我们对教师专业发展研究知之甚少。随着对实验设计的关注，适用于总结已有专业发展研究的元分析或综述性研究方法也受到了关注。由于在新兴的数学专业发展研究领域里只有一小部分研究做了随机试验，所以这些分析通常会报告说在专业发展文献中缺少"严谨的"研究，这些分析所得结论对这个领域并未有很多了解（例如Gersten, Taylor, Keys, Rolfhus, & Newman-Gonchar, 2014；Scher & O'Reilly, 2009；Yoon, Duncan, Lee, Scarloss, & Shapley, 2007）。在这一章中，我们将区分适合元分析的研究（例如建立起参加专业发展项目与教师、学生预期效果之间联系的研究）和采用系列恰当教育研究方法的严谨研究，以回答其他一些重要的问题（例如显示特定的专业发展途径或工具的有关观念的证明，描述各种组织或政治背景如何影响专业发展中所实施的内容）。我们认为数学专业发展研究已经开始积累了一些研究结果，已经有了不断增长的知识基础。

引导这一章深入下去的第二个问题涉及专业发展项目的设计和使它们有效的因素。当前专业发展论述中关于有效专业发展特征的共识——关注内容和学生的内容学习、教师主动学习的机会、连贯性、持续时间和集体参与（Desimone, 2009；Garet, Porter, Desimone, Birman, & Yoon, 2001）——来自于一系列相关性和影响的研究。在这一章中，我们以这些取得广泛共识的特征为有效专业发展的必要条件，但这些远远不是让我们理解专业发展有效的充分条件。为了帮助该领域理解高质量的数学教师专业发展，本综述力图基于并超越目前专业发展有效的共识去考察和描述专业发展设计的特征。

综述过程

在数学专业发展研究的文献综述中，我们寻求包容性，并且考虑包括那些超出了我们这个领域早期认识的研究项目和在美国期刊发表之外出版的项目。我们首先

questions, highlighting central themes and patterns of findings from each of the research phases. We conclude with a discussion about the current state of the field of research on mathematics PD and suggestions for next steps in moving research in this field forward.

Professional Development and Professional Development Research

Over 30 years ago, the concepts of reflective practitioner and knowledge in action (Schön, 1983) brought attention to the connection between teachers' practices and their knowledge. Since then, the focus on the link between practice and knowledge has been strengthened by the understanding that teacher learning is active, situated, and social (Putnam & Borko, 2000). Today, we know that practice is the premier space for teacher learning. Teachers learn both through authentic learning situations embedded in their daily professional lives (Webster-Wright, 2009) and through learning opportunities that engage them with various facets of or artifacts from their teaching (Ball & Cohen, 1999). In this chapter we examine PD programs that offer organized learning experiences to in-service K–12 mathematics teachers. These programs have specified learning goals related to mathematics teaching and learning; are purposefully planned to achieve such goals; include some form of facilitation to foster the learning envisioned; and allow for the examination, refinement, and replication of a model for PD in mathematics.

Two important issues guided the development of this chapter. The first was research methods. Schools and districts are under intense pressure to improve student performance in mathematics, and the current attention to accountability has led to an increased focus on student achievement measures. This accountability wave affects research on PD. As PD becomes a policy mechanism for accomplishing goals established in larger initiatives, such as the No Child Left Behind Act of 2002 or Race to the Top, there is greater emphasis on experimental designs as the premier approach for studying PD, preferably connecting teacher learning to student achievement. Such discourse has led to reduced recognition for the value of careful descriptive studies and other types of educational research that fit under the foundational, exploratory, development, design, and improvement research categories (Earle et al., 2013; National Science Foundation, 2015).

An implication of the overprivileging of experimental research methodologies in the policy debate is the rising rhetoric that we know very little about PD. Accompanying the focus on experimental designs is attention to meta-analyses or synthesis research as the appropriate way to summarize what is known about PD. Because only a small number of randomized trials have been conducted in the emerging field of mathematics PD research, these analyses often report on the lack of "rigorous" studies in the PD literature and conclude that little is known (e.g., Gersten, Taylor, Keys, Rolfhus, & Newman-Gonchar, 2014; Scher & O'Reilly, 2009; Yoon, Duncan, Lee, Scarloss, & Shapley, 2007). In this chapter, we differentiate between appropriate studies for meta-analysis (i.e., studies that causally establish a link between participation in PD and desired teacher and student outcomes) and rigorous studies that employ an array of appropriate educational research methods to address other important questions (e.g., demonstrating proof of concept for a particular PD approach or tool, describing how various organizational or political contexts can influence what is implemented in a PD). We hold the perspective that research on mathematics PD has begun to accumulate findings and the field has a growing knowledge base.

The second issue that guided the development of this chapter relates to the design of PD programs and what makes them effective. The consensus in the current PD discourse about features of effective PD—namely, a focus on content and students' learning of content, active learning opportunities for teachers, coherence, duration, and collective participation (Desimone, 2009; Garet, Porter, Desimone, Birman, & Yoon, 2001)—comes from a series of correlational and impact studies. In this chapter, we take these broad consensus features to be necessary for effective PD but far from sufficient in developing our understanding of what makes PD effective. To help the field understand high-quality mathematics PD for teachers, this review seeks to examine and describe features of PD design that build on and go beyond the current consensus features of PD effectiveness.

Review Process

In our review of the mathematics PD research literature, we sought to be inclusive and consider programs both beyond our initial knowledge of the field and beyond what has been published in U.S. journals. We began with a search of the ERIC database, using *professional development* and *mathematics* as our search terms. We looked for peer-reviewed papers published after 2005, considering that earlier papers had been examined by Sowder (2007). Given the space constraints for the chapter, we narrowed the initial search by selecting 20 peer-reviewed journals from the ERIC search that we considered appropriate

用专业发展和数学作为检索词在ERIC数据库里检索。考虑到早期论文已经被索德（2007）分析过，本文只查找2005年以后发表的那些同行评审的论文。考虑到本章的篇幅限制，我们从ERIC检索中选择了20个我们认为合适的同行评议期刊，作为对数学专业发展实证研究的综述的起点，从而缩小了初始检索范围。附录提供了我们所挑选的初始期刊列表。当对这个领域有了进一步的理解，我们的检索就扩大到了其他期刊和出版物。

最初的检索确定了681篇论文，其中237篇是来自被挑选的20个期刊。通过摘要的分析我们预先筛选了这237篇论文，剔除了那些不是针对在职K–12数学教师发展的论文。通过阅读筛选出来的作为结果的202篇论文，又除掉58篇不是专业发展研究的论文。根据专业发展研究的三个阶段给剩下的144篇论文编码：根据他们是否回答了个别的专业发展所产生的影响这一问题（阶段1），将项目推广到与规模化有关的问题（阶段2），或者是多个项目的比较（阶段3），我们将其中的113篇论文分别归类为阶段1，2或3。剩下的31篇论文不属于这三种阶段，被归入到更大的一个类别中，该类别中包括了诸如理论框架的讨论或者大型教师专业发展经历的调查，但几乎没有任何关于特定专业发展项目的信息。

我们把这一最初的处理结果作为考察数学专业发展研究的起点。虽然我们的文献检索并不彻底，但在这个过程中有数学专业发展方面的广泛研究，使得我们能够致力于评述新的研究发现和研究方法的趋势。我们用那31篇"其他"类别中的论文来帮助我们形成对专业发展研究相关的更广泛的问题的理解，并根据所要思考的问题分析了那113篇论文。在每个阶段，本文首先分析所有的论文来理解该类别研究的内容、形式、背景和工具。然后确定出在每个阶段有大量论文公开发表的研究项目。对于这些项目，我们扩大检索范围到那些不在我们最初所列名单的期刊上发表的论文、书中的一些章节和调查报告。这样一个从广泛检索发表的期刊论文开始，然后扩大到检索包括所关注的研究项目有关的所有文字的过程使我们收集了更大范围的专业发展项目，并对几个有大量公开发表研究成果的专业发展项目有了深入的理解。

需要指出的是：尽管我们的组织框架聚焦在博尔科（2004）三个阶段的研究上，但单篇论文的分类依据是我们所要回答的研究问题，而不是论文所写的时间是在哪个专业发展项目阶段。比如说，博尔科的阶段1专注于"在一个场地的个人专业发展项目"（第4页），在本综述中，我们运用阶段1的意义是聚焦在那些关于个人专业发展项目影响的论文。因此，当在多个地方推行的单一专业发展项目的研究没有关注或解决跨地方有效实施专业发展的问题时，这些研究被归为阶段1的研究。并且如果关于这同一个项目的其它论文是回答和项目推广有关的问题的，那么这些论文就被归为阶段2的研究。因此在下文中，同一个项目或者相似的项目可以根据特定论文中的研究问题而在不同的研究阶段中被讨论，对于有很多出版文献的较大的项目尤其如此。

考察单个专业发展项目影响的研究

迄今为止，针对某个特定数学专业发展项目的设计或效果问题的研究是文献中最普遍的，在我们所划分的阶段1，2，3的论文中有73%是这类论文。这些研究主要探索特定专业发展项目的执行及其影响（广义上的）。他们探索了专业发展项目的本质、教师作为学习者、教师在专业发展中的参与和他们学习之间的关系"（Borko，2004，第5页）。研究这类问题的研究人员以多元视角研究教师学习，他们所使用的方法包括质性的专业发展的个案研究方法（例如Gal，2011；Muir & Beswick，2007；Muñoz-Catalán, Yañez, & Rodríguez, 2010；Ross & Bruce, 2007；Witterholt, Goedhart, Suhre, & van Streun, 2012），也包括使用控制组和比较组的量化比较方法（Antoniou & Kyriakides, 2013；Jacobs, Franke, Carpenter, Levi, & Battey, 2007；McMeeking, Orsi, & Cobb, 2012；Stone, Alfeld, & Pearson, 2008）。大多数的研究关注教师知识或者教学的改变，也有少量的一些研究考察了这些方面对学生成绩的影响。

本文中有关阶段1的一些研究强调了数学专业发展项目的多面性本质。正如这部分评述的项目所示范的一样，数学专业发展项目通常会提出多个学习目标（比如提高教师的学科知识和教师的学科教学知识），使用多种工具

as a starting point for our review of empirical studies about mathematics PD. The appendix presents the list of selected initial journals. Our search was augmented to include other journals and publications once we developed further understanding of the field.

The initial search identified 681 articles, with 237 from the selected 20 journals. We prescreened the 237 articles through an analysis of their abstracts and excluded the ones that were not about PD for in-service K–12 mathematics teachers. We read and screened the resulting 202 articles, eliminating 58 more papers that were not about PD research. We coded the remaining 144 papers according to the three phases of PD research: 113 of them were classified as Phase 1, 2, or 3 on the basis of whether they addressed questions about the impact of a single PD (Phase 1), issues related to bringing programs to scale (Phase 2), or comparisons across multiple programs (Phase 3). The 31 remaining papers did not fit these phases and were placed into a broad group that included, for example, discussions about theoretical frameworks or large surveys about teachers' PD experiences with little information on any specific PD program.

We took the results of this initial process as our starting point for examining mathematics PD research. We acknowledged that our literature search was far from exhaustive but considered that the wide array of research on mathematics PD that surfaced from this process enabled us to start working toward our goal of reviewing emerging findings and methodological trends. We used the 31 papers in the category "other" to frame our understanding of broader issues related to PD research and examined the 113 papers according to the questions considered. Within each phase, we first examined all papers to understand the range of content, formats, settings, and tools examined within that category. We then identified PD programs in each phase that had generated a substantial body of published research. For these programs, we broadened our search to include papers published in journals not on our initial list, book chapters, and other research reports. This process of starting with a broad search of published journal articles and then expanding the search to include what has been written on focused programs of research resulted in both a collection of studies about a wide range of PD programs and an in-depth understanding of several PD programs with an extensive record of published research.

It is important to note that although our organizing framework focused on Borko's (2004) three phases of research, our categorizing of individual articles was based on the research questions addressed rather than the phase of the PD program at the time that the article was written. For example, whereas Borko's Phase 1 attended to "an individual professional development program at a single site" (p. 4), our application of Phase 1 in this review focused on papers that asked questions about the impact of an individual PD program. As a consequence, studies about a single PD program offered at multiple sites were classified as Phase 1 research when they did not attend to or problematize issues of effectively implementing the PD across sites. Further, other articles about this same program were classified as Phase 2 if they addressed questions related to scalability. Therefore, in what follows, the same PD program or similar PD programs can be discussed in different research phases on the basis of the research questions asked in a particular article; this was particularly true for larger PD programs with larger numbers of publications.

Research That Examines the Impact of Individual Professional Development Programs

Studies that address questions about the design or effect of one specific mathematics PD program are, by far, the most prevalent in the literature and represented 73% of the papers we classified as Phase 1, 2, or 3. These studies attend to the implementation of a particular PD program and its impact, defined broadly. They "explore the nature of the professional development program, teachers as learners, and the relationship between teachers' participation in the professional development and their learning" (Borko, 2004, p. 5). The researchers for studies in this category embraced a variety of perspectives on teacher learning and used methodologies that ranged from qualitative cases of one PD participant (e.g., Gal, 2011; Muir & Beswick, 2007; Muñoz-Catalán, Yañez, & Rodríguez, 2010; Ross & Bruce, 2007; Witterholt, Goedhart, Suhre, & van Streun, 2012) to quantitative comparisons using control or comparison groups (Antoniou & Kyriakides, 2013; Jacobs, Franke, Carpenter, Levi, & Battey, 2007; McMeeking, Orsi, & Cobb, 2012; Stone, Alfeld, & Pearson, 2008). Most studies attended to changes in teachers' knowledge or instruction, with a few also examining impact on student achievement.

Our review of Phase 1 studies highlighted the multifaceted nature of mathematics PD programs. As the programs reviewed in this section exemplify, mathematics PD programs often address more than one learning goal (e.g., enhancing teachers' subject matter and pedagogical content knowledge), use multiple tools (e.g., working with tasks to understand the underlying mathematics combined with watching videos of classroom practice), and include more than one format (e.g., a summer institute

（比如使用理解基础数学并观看课堂实践视频相结合的方式），包括多种形式（例如暑期学校和接下来的和教师的每月一次的见面和教练的支持）。而且，在这一类型的数学专业发展项目中有两个重要的相似之处非常明显。首先，阶段1中的大部分研究都考察了专业发展项目，这些专业发展项目都采用了相似的视角来考察从幼儿园到12年级的数学教学——这可能也表明了数学教学研究者们对于促进学生学习的广泛特征有了一个更大的共识。这些项目设计出来是用来支持教师实施以下形式的教学，即这种教学建立在丰富的数学任务基础上、关心学生思维、主张把互动作为学习机制，将提高所有学生的数学理解作为其重要的目标（Lampert, Beasley, Ghousseini, Kazemi, & Franke, 2010）。第二，阶段1中被检验的专业发展项目与有效专业发展项目有很多一致性特征（例如Desimone, 2009；Garet等, 2001）。这些项目关注数学教学和数学学习、倡导促进教师主动学习的教学方法、实施很多个小时并坚持很多个月，这意味着专业发展研究人员对值得设计和研究的专业发展类型有了共识。

尽管有这些相似之处，但我们所讨论的阶段1研究中所考察的专业发展项目在项目设计者选择什么工具，为了达到想要取得的教师学习目标在什么时候以及如何使用这些工具等方面还是不同的。因此，我们将围绕工具来组织讨论阶段1中的研究。这为我们考察专业发展项目里面发生了什么以及关注共性特征之外的设计方面提供了一个视角。从设计的角度来看，为了一个特定的目标而选择合适的物理性的、概念性的或符号性的工具用在专业发展项目中，这是设计者所要作的主要决策之一（Tirosh, 2008）。

我们挑选了三个特定的工具，它们在我们所分析的论文中较为突出，因此有必要对这些新的重要的知识做以下总结：(1)学生数学思维的框架，(2)数学教学的短片，(3)数学任务。这些工具应该添加到以往索德（2007）所给出的其他工具中，包括学生工作、课程材料、案例和教案。对于所选择的每个工具，我们首先要做一个关于它在不同专业发展项目中应用的一般性的讨论。然后重点强调了两个专业发展项目，该工具是这两个专业发展设计的核心。对于每个工具，在选择重点强调的两个专业发展项目时，我们的目标是对专业发展设计和在研究中所使用的研究方法进行较为详细的描述。

学生数学思维的框架

对学生思维的关注是戈德史密斯、多尔和路易斯（2013）的数学教师专业学习的研究综述中所提及的新兴的类别之一，他们所编码的论文中有40%都包括学生思维。这些研究探查了教师对学生数学思维的关注和理解所产生的变化、教师在启发学生思维能力上的变化，以及在教学决策中利用数学思维的变化。在本节中，我们将考察一个明确设计用来将教师注意力集中在学生思维上的工具：总结那些已有研究中有关学生学习特定数学内容的研究框架。在数学中，学生思维的框架已经引导专业发展项目的设计近三十年（Carpenter, Fennema, Peterson, Chiang, & Loef, 1989）。最近，由于这些框架（以学习进展或学习轨迹的形式）和《州际共同核心标准》（National Governors Association Center for Best Practices & Council of Chief State School Officers, 2010）之间明确的联系，这些框架再次得到关注。共同核心标准的倡议者呼吁将以这些基于研究的框架转化为"给教师的实用工具"（Daro, Mosher, & Corcoran, 2011, 第57页）。

使用学生思维框架的早期专业发展项目关注数的概念和算术。最近，使用这些框架的专业发展项目兼顾到了在理解学生在其它内容领域的数学知识发展的研究上取得的新进展（例如Clements & Sarama, 2011；Jacobs等, 2007；Wickstrom, Baek, Barrett, Cullen, & Tobias, 2012，P.H.Wilson, Sztajn, Edgington & Confrey, 2014）。然而，发展这些框架的大部分工作仍集中在小学数学。

那些围绕学生数学思维框架而组织起来的专业发展项目，是最早将教师在专业发展环境下的学习与学生在专业发展参与者的教室中的学习联系起来的项目之一。在前期工作中，认知指导教学（CGI）项目的设计者（Carpenter等, 1989；Fennema等, 1996）建立了这样的联系。到目前为止，他们的研究被专业发展项目的综合研究者认定为（例如Gersten等, 2014；Yoon等, 2007）少数几个将数学教师专业发展和学生成绩紧密联系起来

followed by monthly meetings with teachers and the support of a coach). Still, two important similarities were evident among the mathematics PD programs studied within this category. First, the majority of the Phase 1 studies examined PD that embraced a similar vision of K–12 mathematics instruction—which is probably an indication of a larger consensus among researchers about broad features of the mathematics teaching that leads to learning. These programs were designed to support teachers in implementing instruction that builds on rich mathematical tasks, attends to student thinking, values interactions as a learning mechanism, and is considered "ambitious" (Lampert, Beasley, Ghousseini, Kazemi, & Franke, 2010) in its goal of advancing mathematical understanding for all students. Second, the PD programs examined in Phase 1 were consistently in line with many of the consensus features of effective PD (e.g., Desimone, 2009; Garet et al., 2001). These programs had a content focus on mathematics teaching and learning, embraced pedagogies that promote teachers' active learning, and were implemented over a significant number of hours and work over many months—which suggests an agreement among PD researchers about the types of PD worth designing and investigating.

Despite these similarities, the PD programs examined in our discussion of Phase 1 research differed in program designers' choices of *what* tools to use as well as *how* and *when* to use them to achieve the desired learning goals for teachers. Thus, we chose to organize our discussion of Phase 1 research around tools, which provided a lens for us to examine what is happening inside PD programs and focus on design aspects that go beyond the consensus features. From a design perspective, choosing the appropriate physical, conceptual, or symbolic tool to use in a PD program for a particular purpose is one of the main decisions designers make (Tirosh, 2008).

We selected three specific tools that were prominent in the papers we examined and for which we considered there was important new knowledge to be summarized: (1) frameworks of student mathematical thinking, (2) video clips of mathematics instruction, and (3) mathematical tasks. These tools add to other tools previously considered by Sowder (2007), including student work, curriculum materials, cases, and lesson plans. For each of the selected tools, we first offer a general discussion about its use in different PD programs. We then highlight two programs of PD research in which the tool is central to the PD design. In choosing to highlight only two PD programs for each tool, we aim to offer, within the space constraints for this chapter, more detailed descriptions about both PD designs and the research methods used in studying them.

Frameworks of Student Mathematical Thinking

Attention to student thinking was one of the emergent categories in Goldsmith, Doerr, and Lewis's (2013) review of research on mathematics teachers' professional learning and was present in 40% of the papers they coded. These studies examined changes in teachers' attention to and understanding of students' mathematical thinking, capacity to elicit student thinking, and use of student thinking in instructional decisions. In this section we examine one tool explicitly designed to focus teachers' attention on student thinking: frameworks that summarize what we know from research about how students learn specific mathematics content topics. In mathematics, frameworks of student thinking have guided the design of PD programs for almost three decades (Carpenter, Fennema, Peterson, Chiang, & Loef, 1989). Recently, these frameworks have received renewed attention given the explicit connection between such frameworks (in the form of learning progressions or learning trajectories) and the Common Core State Standards (National Governors Association Center for Best Practices & Council of Chief State School Officers, 2010). Proponents of the Common Core called for translation of these research-based frameworks into "usable tools for teachers" (Daro, Mosher, & Corcoran, 2011, p. 57).

Early PD using frameworks of student thinking focused on number concepts and arithmetic. More recently, PD programs using such frameworks have taken into account new research advances in understanding the development of students' mathematical knowledge in other content areas (e.g., Clements & Sarama, 2011; Jacobs et al., 2007; Wickstrom, Baek, Barrett, Cullen, & Tobias, 2012; P. H. Wilson, Sztajn, Edgington, & Confrey, 2014). Most of the work in developing such frameworks, however, continues to focus on elementary school mathematics.

PD programs organized around frameworks of student mathematical thinking were among the first to connect teacher learning in PD settings to student learning in the classrooms of PD participants. In their pioneering work, the designers of the Cognitively Guided Instruction (CGI) program (Carpenter et al., 1989; Fennema et al., 1996) established such connections and, to date, their work is identified in syntheses of PD research (e.g., Gersten et al., 2014; Yoon et al., 2007) as one of a handful of programs causally connecting mathematics PD to student achievement. More recent studies of PD organized around frameworks of student thinking use methodologies such as cluster randomized trials (e.g., Clements, Sarama, Spitler, Lange, & Wolfe, 2011) and other experimental designs (e.g., Jacobs et al., 2007) to show the

的项目之一。最近的围绕学生思维框架而组织的专业发展研究使用了诸如集群随机试验的方法（例如Clements, Sarama, Spitler, Lange & Wolfe, 2011）和其他的一些实验性设计的方法（例如Jacobs等, 2007），以此来展现围绕学生思维框架而设计的专业发展对学生成绩的因果影响。

作为如何使用学生思维这个工具的例子有两个专业发展项目需要被重点提及，即认知指导教学（CGI）项目和澳大利亚的把我也算在内（CMIT）项目。这些项目被挑选出来是因为它们在研究专业发展的设计和影响上有很长的历史，这些长期的项目也允许研究者去考虑"当老师参与进专业发展项目很多年以后会发生什么"这样一些新出现的问题。

认知指导教学项目。20多年来，CGI 专业发展项目和教师共享了学生数学思维的研究成果，参加该项目的教师也把基于研究的学生数学思维框架应用到了课堂教学中。CGI项目最初的工作致力于一个框架，这个框架用来描述各种各样的加法和减法问题，并给教师提供不同数学难度水平下学生解决这些问题所用的不同策略的理解。CGI项目的设计者已经将专业发展拓展到更广泛地理解与整数相关的学生思维框架（Carpenter, Fennema, Franke, Levi, & Empson, 1999），早期代数（Carpenter, Franke, & Levi, 2003），和分数（Empson & Levi, 2011）相关的内容。

早期的CGI研究出现了几个现代的延伸和分支，其中许多都包含了学习的情境视角。（Jacobs等, 2007；Jacobs, Lamb, & Philipp, 2010）。这些专业发展项目关注于教师对专业发展研究框架的理解和掌握的发展。这些项目是为支持教师集体参与这些框架而设计的。在专业发展研究中，该框架指导教师讨论问题类型和对展示学生解决这些问题的策略的录像的分析。教师给自己的学生提出问题，并在专业发展活动上讨论学生在课堂上产生的作业。专业发展的讨论往往集中在确定什么是学生知道的，可以做的，以此来反驳那些占主导地位的某些特定学生（或很多学生）不能学习复杂数学的言论。（Battey & Franke, 2013；Jacobs等, 2007）。

在最近的一项CGI专业发展研究中，即教师的演变视角研究（STEP）项目（Jacobs等, 2010），研究者们考察了三组教师，他们在过去参与CGI专业发展项目的时间是不同的（刚刚参与CGI专业发展项目、参与CGI专业发展项目2年，参与CGI 专业发展项目4年或4年以上）。这些教师和同一个专业发展促进者一起工作。每组教师每年都会碰几次面（Lamb, Philipp, Jacobs, & Schappelle, 2009）。STEP项目的研究者感兴趣的是教师在CGI专业发展项目中的参与是如何支持他们课堂教学中对学生思维的关注的。他们将专业实践中的注意概念化成三个相互联系的技能：（1）关注孩子们的思维，（2）解释孩子们的理解，（3）根据孩子的理解来决定如何回应。

研究小组使用了两种教师注意的书面测量，一个围绕课堂录像来设计，另一个围绕学生样本的行为（Jacobs, Lamb, Philipp, & Schappelle, 2011）。在对拥有不同CGI经验的教师进行横向分析时，雅各布斯等人（2010）发现，虽然几乎没有CGI专业发展经验的教师也能学着去关注孩子的思维，并且形成早期的解释学生思维的能力，但能够决定如何在学生思维的基础上立刻回应则是一个更难一些的注意技能，这需要大量的CGI专业发展的经验才能形成。STEP项目还显示，随着参与专业发展项目的不断进行，教师可以进一步发展他们对于自己学生和教学方面的探究性立场（Lamb等, 2009）。

把我也算在内项目。自20世纪90年代以来澳大利亚和新西兰开展了大量的以学生数学思维为框架的专业发展项目，最初的专业发展项目之一是澳大利亚的把我也算在内（CMIT），这个设计是为了促进教师对学生的数学发展的理解（Bobis等, 2005）。CMIT专业发展项目是为校本实施而设计的。对这个项目的研究既涉及专业发展项目对教师的影响，也涉及专业发展的校本实施和促进。由于它对我们理解以学生数学思维为框架进行专业发展设计有主要贡献，因此本文在这里考查了他们在阶段1中所思考的一些问题。

在参与CMIT的学校中，由3到5名教师组成的小组与一名数学顾问一起工作了10到20周，并得到了基于课堂的支持，因为这些老师学会了用两种工具来计划和实施教学：一种是与学生思考有关的框架——数学学习框架；另一个是符合框架的诊断评估。学习框架描述了学

causal impact on student achievement of PD designed around frameworks of student thinking.

The two PD programs of research highlighted as examples of how this tool is used are CGI and the Australian Count Me In Too (CMIT). These programs were selected because of their sustained histories of research on the design and impact of the PD. The longevity of these programs has also allowed researchers to consider emerging questions about what happens when teachers participate in a PD program over many years.

Cognitively Guided Instruction program. For more than 20 years, CGI PD has shared with teachers research findings about student thinking in mathematics, and teachers in CGI PD have applied these research-based frameworks of student mathematical thinking in their classrooms. The initial work in CGI focused on a framework describing various types of addition and subtraction problems and providing teachers with an understanding of different strategies, at different levels of mathematical sophistication, that students use to solve these problems. CGI designers have expanded the PD to focus on frameworks for understanding student thinking related to whole numbers more broadly (Carpenter, Fennema, Franke, Levi, & Empson, 1999), early algebra (Carpenter, Franke, & Levi, 2003), and fractions (Empson & Levi, 2011).

Several current continuations and branching of the initial CGI research exist, with many of them embracing a situated perspective on learning (Jacobs et al., 2007; Jacobs, Lamb, & Philipp, 2010). These PD programs focus on the development of teachers' understanding and ownership of the research-based framework examined in the PD. The programs are designed to support teachers' collective engagement with these frameworks. During the PD, the framework guides teachers' discussions about types of problems and analysis of videos that showcase students' strategies for solving these problems. Teachers pose problems to their own students and discuss in PD meetings the student work generated in their classrooms. PD discussions often focus on identifying what students know and can do to counteract predominant narratives that particular students (or groups of students) cannot learn sophisticated mathematics (Battey & Franke, 2013; Jacobs et al., 2007).

In a recent CGI PD study, the Studying Teachers' Evolving Perspectives (STEP) project (Jacobs et al., 2010), researchers examined three cohorts of teachers who differed in the amount of previous participation in CGI PD (teachers just starting in CGI PD, teachers who had 2 years of CGI PD experience, and teachers with 4 or more years of CGI PD). These cohorts worked with the same PD facilitator, and each cohort met several times during the year (Lamb, Philipp, Jacobs, & Schappelle, 2009). Researchers in the STEP project were interested in how teachers' participation in the CGI PD supported the development of noticing children's mathematical thinking during instruction. They conceptualized the professional practice of noticing as three connected skills: (1) attending to children's thinking, (2) interpreting children's understandings, and (3) deciding how to respond on the basis of children's understandings.

The research team used two written measures of teacher noticing, one designed around a classroom video and the other around sample student work (Jacobs, Lamb, Philipp, & Schappelle, 2011). In a cross-sectional analysis of data from the teachers with different amounts of experience with CGI, Jacobs et al. (2010) found that although teachers with little CGI PD experience could learn to attend to children's thinking and develop early expertise in interpreting student thinking, being able to decide how to respond in-the-moment and on the basis of student thinking was a much harder noticing skill that developed through sustained experience with CGI PD. The STEP project also showed that with continued participation in PD, teachers can further develop their inquiry stance toward their own students and instruction (Lamb et al., 2009).

Count Me In Too program. A substantial amount of PD using frameworks of student mathematical thinking has been conducted in Australia and New Zealand since the 1990s. One of the initial PD projects was the Australian Count Me In Too (CMIT), which was designed to promote teachers' understanding of students' mathematical development (Bobis et al., 2005). The CMIT PD program was designed for school-based implementation, and its research attended to both the impact of the PD on teachers and the PD school-based implementation and facilitation. Because of its key contributions to our understanding of the design of PD using frameworks of student mathematical thinking, we examine here some of the Phase 1 questions they considered.

In CMIT-participating schools, groups of three to five teachers worked with a mathematics consultant for 10–20 weeks and received classroom-based support as they learned to plan and implement instruction using two tools: a framework of student thinking—the *Learning Framework in Number*—and a diagnostic assessment aligned with the framework. The learning framework provides descriptions of major stages of student mathematical development. The assessment is a task-based interview that generates profiles of students' knowledge. These profiles form a "basis for instruction that is

生数学发展的主要阶段。评估是一个基于了解学生知识生成的相关任务的面试。这些了解构成了"针对每个学生当前知识和策略水平的教学基础"（Bobis等，2005，第30页）。

对95名教师的调查和访谈数据支持CMIT项目有效的结论（Bobis，2003）。95名教师中，72%的教师报告其在关于学生是怎么学习数学的理解上发生了改变，并且78%的教师表明他们在教数学的方法上发生了改变。在教师实践发生改变的自我报告中，最普遍的改变是增加了对学生的数学思维的关注度，同时增加了在班级中使用能力分组的频率——这与之前的实施结果相类似（例如Bobis，1996，1999）。

另外三个个案学校（Bobis，2009）的教师调查和访谈数据显示：教师关于CMIT项目的经验的多少、对理解框架的自我评价的信心水平与他们使用这种框架去指导评估和教学的能力之间有关系。教师们认为参与专业发展项目最重要的直接结果之一是他们提高了针对不同能力水平的学生而布置不同任务的能力。伯比斯（2009）认为仅在项目中获得的经验的多少并不是支持他们自信心水平的决定性因素。相反，专业支持的时间和质量的结合是重要的影响因素。这种对于不同地区教师经验质量的关注将CMIT项目阶段1的分析与该项目阶段2的研究联系起来。

以学生数学思维为框架的专业发展。CGI和CMIT的研究都表明这些框架能让教师弄清楚他们的学生的数学思维。这些框架能充当激发教师在教学中关注学生思维，在课堂上理解学生思维，以及在教学中针对不同学生使用不同的策略的工具。这些结果与其他项目的结论相似，那些项目表明围绕着学生数学思维框架的教师专业发展可以帮助教师利用即时的机会去构建学生的想法（Clements等，2011）、在课堂讨论中实施具体的教学实践（P. H. Wilson, Sztajn, Edgington, & Myers, 2015）、质疑并分析学生的数学推理（Norton & McCloskey, 2008）、发展他们自己的数学知识（P. H. Wilson等，2014）。

CGI和CMIT的研究表明，教师对学生数学思维框架的使用和他们围绕学生评估所组织的教学方式之间有着有趣的联系。而CGI项目强调有明确目标地进行专业发展记录的必要性，以打消某些学生只能在框架的某些层次上学习的想法；CMIT更注重帮助教师明确学生的具体水平，并制定有针对性的个性化教学计划。还需要进一步的研究来调查学生数学思维框架在专业发展环境中的使用，以及教师之后在组织全体、小组和个性化课堂教学中使用这些框架的情况。

从CGI STEP和CMT研究中突显出的另外一个重要的问题就是教师长期坚持参与到专业发展项目中的重要性。这两个项目都表明，当教师能够多年坚持参与到一个持久的专业发展项目中时，他们基于学生如何学习的知识而开展教学的能力会持续发展。更重要的是，教师在持久的专业发展项目中所达到的关于学生数学思维的理解水平，几乎不可能通过一学年的专业发展项目所实现。需要更多的纵向与横向研究来检验教师在长期的专业发展项目中的学习。

基于多年对长期数学专业发展的关注，引发出另外一个重要的问题：教师专业发展的差异化。这一概念来自CGI和CMIT项目的研究结果，这一概念提出为对学生数学思维框架有着不同专业经验水平的教师提供不同的专业发展。还需要更进一步的研究来理解教师随时间变化的专业学习轨迹，以及如何在他们学习学生数学思维的不同阶段为他们提供最好的支持。总之，差异化教师专业发展这一概念尤为重要，并且需要进一步研究。

数学教学录像片段

回顾相关文献，在大量的专业发展研究设计中，录像有着广泛的应用。从我们最初检索中查找出的研究论文大约有三分之一都是研究以录像作为工具的专业发展项目。这些研究中有少数几个运用教师在专业发展背景下讨论的录像，以此达到让参与者了解其他教师的专业发展经验，从而促进对专业发展讨论的目的（例如Clark, Moore, & Carlson, 2008；Santagata, 2009）。而大多数使用录像作为工具的专业发展项目都让教师参与到数学教学录像片段的讨论中——这也是这部分的重点。在这些专业发展项目中，教学录像的呈现可以提供一个经验分享的机会，通过这些经验，教师可以共同探索关

targeted to each student's current levels of knowledge and strategies" (Bobis et al., 2005, p. 30).

Survey and interview data from 95 teachers support claims of effectiveness of CMIT (Bobis, 2003). Of the 95 teachers, 72% reported changes in their understanding of how students learn mathematics, and 78% of the teachers indicated they had changed the ways they taught mathematics. Among the most common self-reported changes to teachers' practice were increased attention to students' mathematical thinking and the increased use of ability grouping in the classroom—similar to findings from previous implementations (e.g., Bobis, 1996, 1999).

Another set of survey and interview data from teachers in three case schools (Bobis, 2009) showed a relation between the amount of teachers' experience with CMIT, the levels of teachers' self-rated confidence in understanding the framework, and their ability to use the framework to guide assessment and instruction. Teachers reported that their perceived increased capacity to differentiate tasks to suit their students' various ability levels was one of the most important direct results from participation in the PD program. Bobis (2009) suggested that teachers' amount of experience with the project alone was not the determining factor in supporting their confidence level. Rather, it was the combination of time and the quality of the professional support that was influential; this attention to the quality of teachers' experiences across sites connects CMIT Phase 1 analysis to Phase 2 research also conducted in this program.

PD using frameworks of student mathematical thinking. Research on CGI and CMIT suggests that these frameworks allow teachers to make sense of their students' mathematical thinking. They can serve as tools to empower teachers to attend to student thinking in their teaching, interpret student thinking in their classrooms, and make use of different students' strategies in their instruction. These findings are similar to those of other programs showing that PD organized around frameworks of student mathematical thinking help teachers capitalize on in-the-moment opportunities to build on students' ideas (Clements et al., 2011), implement specific instructional practices during classroom discussions (P. H. Wilson, Sztajn, Edgington, & Myers, 2015), question and analyze student mathematical reasoning (Norton & McCloskey, 2008), and develop their own knowledge of mathematics (P. H. Wilson et al., 2014).

CGI and CMIT research suggest interesting connections between teachers' use of frameworks of student mathematical thinking and the ways in which they organize instruction around their assessment of where students are in relation to the framework. Whereas CGI programs highlighted the need to purposefully create a PD narrative that counteracted the idea that certain students can only work at certain levels of the framework, CMIT focused on helping teachers identify students' specific levels and plan for targeted, individualized instruction. Additional research is needed to examine the uses of frameworks of student mathematical thinking in PD settings and teachers' subsequent uses of these frameworks in organizing whole-group, small-group, and individualized classroom instruction.

Another important issue emerging from the CGI STEP study and CMIT is the importance of teachers' participation in a PD program over multiple years. Both programs showed that when teachers participate in a sustained PD program over many years, their capacity to teach based on knowledge of how students learn continues to develop. More important, the levels of understandings teachers develop about students' mathematical thinking through sustained PD does not seem to be achievable with a PD program that spans a single school year. More longitudinal and cross-sectional studies are needed to examine teacher learning in sustained PD programs.

Attention to sustained mathematics PD over many years brings forth another important issue: that of differentiated PD for teachers. The concept of providing different PD for teachers with different levels of PD experience with frameworks of student mathematical thinking follows from findings in both CGI and CMIT. Further research is needed to understand teachers' professional learning trajectories over time and how to best support them at different points in their learning about students' mathematical thinking. In general, the concept of differentiated PD for teachers is important and requires further examination.

Video Clips of Mathematics Instruction

Our review of the literature indicated widespread use of video across a large range of PD designs. About one third of the research papers that emerged from our initial search studied a PD program that used video as a tool. A handful of these programs used videos of discussions among teachers in PD settings with the goal of engaging participants with the PD experiences of other teachers to foster participation in PD discussions (e.g., Clark, Moore, & Carlson, 2008; Santagata, 2009). However, the majority of PD programs using video as a tool engaged teachers in discussions about video clips of mathematics instruction—which is the focus of this section. In these PD programs, video representations of teaching are used to provide a shared experience through which teachers

于数学内容、教学手段或者学生思维的问题（Borko, Koellner, Jacobs, & Seago, 2011；Sherin, 2004）。

我们发现在专业发展项目中使用数学教学录像片段的通常包括两种常见的设计系列。在第一种设计系列中，参与者首先关注数学内容，然后关注学生思维或教学实践。通常，教师首先作为一个学习者参与到一个具体的数学内容中，然后再对录像中关于该内容的教学进行讨论。这个系列已经被用在面对面的专业发展项目中（例如 Jacobs 等，2007；Koellner 等，2007），同时也出现在运用在线论坛讨论多媒体案例的专业发展项目中（例如 Koc, Peker, & Osmanoglu, 2009；McGraw, Lynch, Kos, Budak, & Brown, 2007）。一些使用该设计系列的专业发展项目使用从大量可利用的录像资源中预先挑选好的录像片段（例如 Schifter, Bastable, & Russell, 1999a, 1999b；Seago, Mumme, & Branca, 2004），或者从之前参加专业发展项目的教师教学录像中挑选（例如 Santagata, 2009；Swan, 2007）。我们称这两个例子为教师观看"其他人的录像"。其他的专业发展项目使用的录像来自于参与专业发展项目的教师在课堂上教授某一个特定概念并拍摄下的录像（例如 Jacobs 等，2007；Koellner 等，2007；Taylor, 2011）。我们称这种为教师观看"他们自己的录像"。

我们发现第二种设计系列在专业发展中使用数学教学录像片段来引发讨论，包括对数学内容的讨论。因此，关注教师们的数学理解可能会紧跟数学教学录像片段的探讨。在这些案例中，专业发展项目让教师参与观察和反思数学课堂教学背景下学生思维的录像片段（Sherin & van Es, 2009；van Es & Sherin, 2008）、教学实验（Norton & McCloskey, 2008），或实验性的教学（Ticha & Hospesova, 2006），然后以教师关于学生的数学学习的讨论为切入点讨论数学内容。一些项目还通过使用教学录像片段作为行动研究项目的工具（例如 Males, Otten, & Herbel-Eisenmann, 2010；Muir & Beswick, 2007；Scott, Clarkson, & McDonough, 2012）或专业学习共同体的工具（例如 Brodie & Shalem, 2011；Nickerson & Moriarty, 2005），让教师参与他们自己的教学实践。在这些项目中，专业发展关注的是教学实践，这些内容来自于对学生数学推理的讨论或教师们所观察到的实践。

强调录像工具的两个专业发展项目分别是录像俱乐部和问题解决圈。在这两个项目中，教师观看他们自己课堂的录像。然而，录像俱乐部更关注录像中呈现的有关学生思维的教学问题的讨论，问题解决圈则以讨论数学内容为起始点，然后再关注学生的思维和教学。这些项目已经产生了大量关于专业发展模式的研究，并且提升了我们在专业发展背景下使用数学教学录像片段的意义的理解。

录像俱乐部项目。录像俱乐部是为了帮助教师学习观察和理解学生的数学思维（Jacobs & Spangler, 2017, 这套书；Sherin & van Es, 2009；van Es & Sherin, 2008）。参与的教师定期会面以观看和分析研究者挑选的来自某位参与教师的课堂录像片段，这些片段描述了近期讲授课程中学生的数学思维。教师在录像中通常扮演的是提供课堂背景信息的角色，而主持会议的研究者则以一个教师关注到的一般性的问题为起点推动这场讨论。在这个问题之后，会给出与学生思考相关的更具体的提示（Sherin, Linsenmeier, & van Es, 2009）。在专业发展会议中，组织者努力建立一个支持性的学习环境，使教师的观点被重视并且被用来引导讨论（van Es, 2009）。

在一个关于录像俱乐部的研究中，薛琳和万·埃斯对一个为期一年的专业发展项目进行研究，该俱乐部由七个小学的教师组成，他们共见面10次来讨论各自的教学录像片段（Sherin & van Es, 2009；van Es & Sherin, 2008, 2010）。研究人员定性地对参与教师之间的相互影响以及对每位参与者被干预前后所进行采访的转录本进行分析。研究者从四个维度对录像进行编码：（1）参与者（教师讨论的对象），（2）主题（教师讨论的内容），（3）态度（教师是描述、评价或解释），（4）策略（教师所讨论的学生的思维方式）。薛琳和万·埃斯寻找了教师在录像俱乐部中推理能力的改变。他们发现通过一年的课程，教师会更频繁地讨论学生的数学思维并且提出关于学生思维的更复杂的推理。薛琳和万·埃斯（2009）在一个关于初中教师俱乐部的研究中也报告了类似的结果。

根据万·埃斯和薛琳（2008）的研究，教师在学习关

can collectively explore issues concerning mathematical content, instructional practices, or students' thinking (Borko, Koellner, Jacobs, & Seago, 2011; Sherin, 2004).

We found two frequently reported design sequences in the use of video clips of mathematics instruction in PD programs. In the first one, participants focus on mathematics content first and then on student thinking or instruction. Typically, teachers first engage as learners with a specific mathematics topic, followed by a discussion of videos showcasing instruction of the targeted content. This sequence has been used in face-to-face PD programs (e.g., Jacobs et al., 2007; Koellner et al., 2007) and also in emerging PD using online discussion forums with multimedia cases (e.g., Koc, Peker, & Osmanoglu, 2009; McGraw, Lynch, Koc, Budak, & Brown, 2007). Some PD programs that implement this sequence used preselected video clips from a growing set of available video resources (e.g., Schifter, Bastable. & Russell, 1999a, 1999b; Seago, Mumme, & Branca, 2004) or from the classrooms of teachers who participated in previous implementations of the PD program (e.g., Santagata, 2009; Swan, 2007). We refer to these two instances as teachers watching "videos of others." Other PD programs used videos collected in the classrooms of PD participants who implemented particular ideas in their classroom and had their teaching videotaped (e.g., Jacobs et al., 2007; Koellner et al., 2007; Taylor, 2011). We refer to this as teachers watching "videos of their own teaching."

The second design sequence we identified uses video clips of mathematics instruction to generate discussions in the PD, including discussions of mathematics content. Thus, attention to teachers' mathematical understandings may follow the examination of video clips of mathematics instruction. In these cases, the PD engaged teachers in noticing and reflecting on video clips that showed student thinking in the context of mathematics classrooms (Sherin & van Es, 2009; van Es & Sherin, 2008), teaching experiments (Norton & McCloskey, 2008), or experimental teaching (Ticha & Hospesova, 2006) and then used teachers' discussions of students' mathematics as an entry point into discussions about mathematics content. Several programs also engaged teachers with their own teaching practices through the use of video clips of instruction as tools for action research projects (e.g., Males, Otten, & Herbel-Eisenmann, 2010; Muir & Beswick, 2007; Scott, Clarkson, & McDonough, 2012) or professional learning communities (e.g., Brodie & Shalem, 2011; Nickerson & Moriarty, 2005). In these programs, the PD typically focused on instruction, with content addressed as it emerged from discussions of students' mathematical reasoning or teachers' observed practices.

The two PD programs of research highlighted for this tool are video clubs and the Problem-Solving Cycle. In both of these programs, teachers watch videos from their own classrooms. However, whereas the video clubs focus discussions on instructional issues related to student thinking showcased in the videos, the Problem-Solving Cycle starts with a discussion of the mathematics content and then attends to student thinking and instruction. These programs have generated a significant body of research about their PD models and enhanced our understanding of what it means to use video clips of mathematics instruction in PD settings.

Video Club program. Video clubs are designed to help teachers learn to notice and interpret students' mathematical thinking (Jacobs & Spangler, 2017, this volume; Sherin & van Es, 2009; van Es & Sherin, 2008). Participating teachers meet regularly to watch and analyze excerpts the researchers select to portray students' mathematical thinking in a recently taught lesson from a participating teacher's classroom. The teacher portrayed in the video typically provides background information about the classroom, and the researcher facilitating the meeting starts the discussion with a general question about what the teachers notice. This question is then followed with more specific prompts related to student thinking (Sherin, Linsenmeier, & van Es, 2009). In the PD meetings, the facilitator works to develop a supportive learning environment in which teachers' ideas are valued and used to guide the discussions (van Es, 2009).

In one study about video clubs, Sherin and van Es investigated a yearlong PD in which a group of seven elementary school teachers met 10 times to discuss excerpts from their own teaching (Sherin & van Es, 2009; van Es & Sherin, 2008, 2010). The researchers qualitatively analyzed transcripts from interactions among participating teachers as well as preintervention and postintervention interviews with each participant. The researchers coded their data across four dimensions: (1) actor (whom the teachers talked about), (2) topic (what the teachers talked about), (3) stance (whether teachers were describing, evaluating, or interpreting), and (4) strategy (the ways in which teachers talked about students' thinking). Sherin and van Es looked for changes in teachers' reasoning during the video club. They reported that over the course of the year, teachers talked more frequently about students' mathematical thinking and exhibited more sophisticated reasoning about student thinking. Sherin and van Es (2009) also reported similar findings in a video club with middle grade teachers.

According to van Es and Sherin (2008), there were three different paths teachers followed as they learned

注和解释学生的数学时有三条不同的途径：（1）一条是直接的途径，教师关注的内容和讨论的方式有了明显变化；（2）一条是环状的途径，教师在他所关注的内容上来回反复；（3）一条是渐进的途径，教师在专业发展课程中逐步地改变。万·埃斯（2009）的一个研究表明当录像俱乐部发展起来后，教师在专业发展讨论中对自己有不同的定位并在讨论中担任新的角色。教师们开始推动团体关注学生思维的细节、采取尝试的态度、使用证据、提出替代性的解释并且质疑其他每个人的解释。

通过分析专业发展项目中参与教师早期和后期的课堂录像，录像俱乐部的研究者发现了这些教师在课堂教学中的改变（Sherin & van Es, 2009; van Es & Sherin, 2010）。研究发现，在项目后期，教师给学生更多的机会拓展他们的思维。这些教师在教学过程中加强了对学生策略的探索。在后续的访谈中，教师们反映倾听学生想法成为了他们教学中重要的一部分，它能帮助教师认识到学生们有很多有趣的数学想法。

问题解决循环项目。问题解决循环（PSC）围绕着教师自己的教学录像展开，并且它非常关注丰富的数学任务以及使用任务的教学。PSC是一个循环往复的专业发展方法，它旨在提高教师数学教学的知识和改善他们的教学。每个循环周期都是由三个相互关联的系列专业发展工作坊组成，每个工作坊都聚焦于一个任务。每个循环周期的第一个工作坊旨在通过分析挑选出的PSC任务，帮助教师发展专业知识内容。第一个工作坊专业知识内容之后，教师和自己的学生一起完成教学任务并且进行录像。专业发展促进者挑选授课的录像片段以供第二个和第三个工作坊使用，第二个和第三个工作坊分别聚焦于学生的数学推理和教师的教学行为的分析。这些工作坊旨在帮助教师学习如何引导、关注和建立学生的思维，而录像片段能够帮助教师在课堂实践中开展研究工作（Borko, Jacobs, Koellner, & Swackhamer, 2015; Jacobs等，2007; Koellner等，2007）。

PSC项目最初在几个不同学区的中学数学教师中实施。博尔科、雅各布斯、埃伊特豪尔赫和皮特曼（2008）分析了专业发展项目两年来所有的基于录像的讨论。研究发现，随着时间的推移这些对话变得越来越富有成效：教师以更集中、更深入和更分析的方式讨论了与数学思维和教学策略有关的具体问题。在项目后期的访谈中，教师们对于观看和分析他们自己在专业发展项目中的录像持积极态度。许多教师都认为这些讨论是这个项目中最有价值的部分。

PSC的第二次分析研究了一个教师参与PSC后对其教学实践影响的个案（Koellner, Jacobs, & Borko, 2011）。研究者分析这位教师参与专业发展项目两年来教授的14节课的录像、在每一节录像课后的访谈，以及每次PSC工作坊后的书面反思。这一深入的个案研究显示出教师教学中大量的变化，而这些改变正反映了项目的关注焦点。例如，专业发展的一个中心思想就是通过鼓励学生解释和证明他们的想法，来促进课堂中的数学思维。在第一次分析的课堂录像中，教师主要关注学生回答的准确性。而在随后的课程中，他提问的目的在于鼓励学生解释和证明他们解决问题的策略。该个案中的教师逐渐在小组上分配更多的时间，利用小组讨论的机会促进更深入的、以学生为导向的对话。

运用数学教学录像片段的专业发展。录像俱乐部和PSC项目的研究表明，参加专业发展项目的教师运用数学教学的录像片段，能够改变他们与学生的数学专业话语。录像支持的分析心态（Sherin, 2004），再加上对专业发展有关视频的讨论的精心促进，有助于教师学着将他们的专业对话集中于学生与他们所学数学两方面。教师变得越来越具有分析性和解释性，更频繁地使用证据来支持他们在学生思维和数学学习与教学上的看法。教师也学着用尊重性的和批判性的方式讨论教学，由此产生有效的专业对话。

录像俱乐部和PSC项目还展示了专业发展利用使用数学教学录像片段来改变教师教学实践的前景。参与项目的教师会在自己的课堂上用更多的时间倾听学生的解答，并且开始重视、引导和探索学生的数学思维。虽然选取的教师样本不多，但是两个项目的研究均显示了，在专业发展中教师分析数学教学录像片段深度的变化与教师在课堂实践中的改变之间的联系。在阶段2的后期

to attend to and interpret students' mathematics: (1) a direct path in which there was a marked shift in what teachers noticed and how they talked about it, (2) a cyclical path in which teachers went back and forth in what they noticed, and (3) an incremental path in which teachers changed progressively over the course of the PD. A study by van Es (2009) also showed that as the video club unfolded, teachers positioned themselves differently and took new roles during PD discussions. Teachers began to prompt the group to attend to the particulars of student thinking, adopt tentative stances, use evidence, propose alternative explanations, and question each other's interpretations.

Video club researchers also looked for changes in participating teachers' classroom practices through the analysis of videos from the classrooms of participating teachers early and late in the PD program (Sherin & van Es, 2009; van Es & Sherin, 2010). They showed that, later in the program, teachers allowed students more opportunities to explore their thinking. These teachers increased their probing of students' strategies during instruction. In follow-up interviews, teachers reported that listening to students became important in their teaching, helping them recognize that students have interesting mathematical ideas.

Problem-Solving Cycle program. The Problem-Solving Cycle (PSC) is organized around videos of teachers' own instruction and has a strong focus on rich mathematical tasks and instruction that uses such tasks. PSC is an iterative PD approach designed to increase teachers' knowledge of mathematics for teaching and improve their instruction. Each cycle consists of a series of three interconnected PD workshops focused on a single task. The first workshop of each cycle is designed to support the development of teachers' specialized content knowledge through the analysis of the selected PSC task. After the first workshop, teachers implement the task with their own students, and their lessons are videotaped. PD facilitators select clips from these lessons for use in Workshops 2 and 3, which focus on the analysis of students' mathematical reasoning and teachers' pedagogical moves, respectively. These workshops are designed to help teachers learn how to elicit, attend to, and build on student thinking, and the video clips help situate teachers' explorations in their classroom practice (Borko, Jacobs, Koellner, & Swackhamer, 2015; Jacobs et al., 2007; Koellner et al., 2007).

The PSC was initially implemented with middle school mathematics teachers across several school districts. Borko, Jacobs, Eiteljorg, and Pittman (2008) analyzed full-group video-based discussions over 2 years of PD. They showed that, over time, these conversations became more productive: Teachers talked in a more focused, indepth, and analytic manner about specific issues related to mathematical thinking and instructional strategies. In interviews at the end of the project, teachers were positive about the impact of watching and analyzing their own videos in the PD. Many teachers identified these discussions as the most valuable aspect of the project.

A second analysis of the PSC examined the impact of participation in the PSC on the instructional practice of one case teacher (Koellner, Jacobs, & Borko, 2011). The researchers analyzed videos from 14 lessons taught by the teacher over 2 years of participation in the PD, interviews after each videotaped lesson, and the teacher's written reflections after each PSC workshop. This indepth case showed a number of changes in instruction that were reflective of the project's foci. For example, one central focus of the PD was promoting mathematical thinking in the classroom by encouraging students to explain and justify their ideas. In the first videotaped lessons analyzed, the teacher primarily attended to the accuracy of students' answers. In later lessons, his questions were aimed at encouraging students to explain and justify their problem-solving strategies. The case teacher gradually increased the amount of time spent in small groups, using group time to promote in-depth, student-led conversations.

PD using video clips of mathematics instruction. Research on video clubs and PSC programs shows that teachers in PD programs that use video clips of mathematics instruction can change their professional discourse about students' mathematics. The analytic mindset the videos support (Sherin, 2004), coupled with careful facilitation of PD discussions about the videos, assist teachers in learning to focus their professional conversations on both students and the mathematics they are learning. Teachers become more analytical and interpretive and use evidence more frequently to support their claims about students' thinking and mathematics learning and teaching. Teachers also learn to talk about instruction in ways that are both respectful and critical, which allows for productive professional conversations.

The video club and PSC also show the promise of PD using videos of mathematics instruction to change teachers' instructional practices. Teachers in these programs spent more time listening to students' solutions and began to value, elicit, and probe students' mathematical thinking in their own classrooms. Although working with small samples of teachers, research conducted on both programs pointed to connections between changes in the depth of teachers' analysis of videos clips of mathematics

讨论分析中，有证据显示PSC对学生的学习也有影响，具体来说参加专业发展的教师的学生在州数学评估中的成绩高于未参加的老师的学生成绩（Koellner & Jacobs, 2015）。

为了了解使用教师自己的数学教学录像或者他人录像的局限性和代价，仍需要做进一步的研究。克莱因奈施和施耐德（2013）的一个研究中应用了配对的实验设计，实验设计中的五对教师被分配到独立的基于电脑的专业发展实验中，让他们观看自己或者他人的录像。观看他人录像的教师组比观看自己录像的教师组表现出更深入的思考过程、更愿意思考其他可能的教学方法，并且表现出和录像中教师有相同的或者更多情感和动机方面的投入。这些结果与针对科学学科的一个类似研究的结论相反（Seidel, Stürmer, Blomberg, Kobarg, & Schwindt, 2011）。克莱因奈施和施耐德（2013）承认样本数量偏小和简单的个人专业发展经验限制了他们研究结论的普遍性。

许多使用录像片段的专业发展项目都同时结合其他工具一起使用（例如：数学任务、学生的作业等）。在这些案例中，考虑工具之间的一致性很重要。例如，在PSC基础上，泰勒（2011）设计了一个反思连结循环来帮助教师构建学生的课外知识。专业发展项目的教师开会讨论他们自己教学的录像片段，然后在两次专业发展会议的间隙设计一个新的教学实践活动。同时也会给教师发几篇文献来阅读，以帮助他们将自己的教学和学生的课外知识联系起来。通过分析教师在专业发展过程的体验，泰勒指出专业发展工具（例如：录像、课的设计和文献阅读等）的结合并不能支持那些预想的专业发展的主要目标的实现。他指出专业发展项目的设计者需要进一步探索如何使多种专业发展活动联合起来帮助专业发展目标的实现。泰勒关于专业发展工具结合的相关结论可以概括为：需要更多的研究去探索在专业发展设计中不同工具是如何组合在一起，以促进（或阻碍）教师在数学专业发展项目中的学习的。

数学任务

课堂教学的研究已经系统地表明数学任务对学生学习的重要性。由于我们对高认知要求的任务在数学教学和学习中作用的理解不断增加，所以不难发现：在我们所检索的专业发展文献中很大一部分是围绕着数学任务而组织的。华生和梅森（2007）指出，在专业发展中使用数学任务的方式上，"全世界"达成了强烈的一致（第207页）。他们指出，围绕数学任务而设计的专业学习活动应包括经验的结合，例如解决数学任务的经验、学习者对解决任务过程反思的经验、为所给任务发展框架的经验、与学生一起解决任务而提出策略的经验、与学生一起尝试解决任务的经验、与学生一起尝试解决任务的经验、倾听学生在解决问题时想法的经验，以及分析那些围绕任务使用而组织起来的教学经验。

许多强烈关注数学任务的专业发展项目已经使教师将解答这些任务视为一项"开放性活动"（Silver, Clark, Ghousseini, Charalambous, & Sealy, 2007, 第264页）。在一些项目中，设计这些任务的主要目的是为了发展教师的数学专业知识，而不一定是为了在K-12背景下使用它们（例如Chamberlin, 2009；Chamberlin, Farmer, & Novak, 2008；Thompson, Carlson, & Silverman, 2007）。这些项目的研究更多关注的是教师的数学工作以及他们数学知识的发展。例如，奥尔和布朗（2012）关注教师如何用比例推理弄清楚双数轴的表征。他们研究了六位使用这种表征的专业发展参与者，结果发现教师知识中这些独立的要素会促进或是阻碍他们的数学理解。他们的结论是，即使教师完成问题的方法是错误的但也不失逻辑性，教师知识的多少与组织方式对他们在任务中的理性认识都是关键的。

在其他专业发展项目中，数学任务与教师关于学生、教学和课程的知识紧密联系。在这些案例中，尽管最初的任务也是要挑战教师的数学理解，但是它在K-12的教学中经常作为连接教师专业实践发展的一种方式。在这些项目中，教师有时被要求用多种方法解决同一个问题（例如Ferrini-Mundy, Burrill, & Schmidt, 2007），或者推测学生可能会用的解决问题的不同方法（例如Kabasakalian, 2007），或者分析一组假设的学生解答（例如Jacobs等, 2007），或者研究强调促进或限制学生

instruction in the PD and changes in teachers' classroom practices. There is emerging evidence, from studies later discussed in Phase 2, of the impact of PSC on student learning; specifically, students of participating teachers scored higher on the state mathematics assessment than students of nonparticipating teachers (Koellner & Jacobs, 2015).

Further research is needed to understand the constraints and affordances of using videos from teachers' own mathematics instruction or videos from others. In one study, Kleinknecht and Schneider (2013) used a matched experimental design with five pairs of teachers who were assigned to an individual, computer-based PD experience using either their own videos or videos from others. The teachers watching videos of others displayed deeper reflection processes, were more likely to consider alternative instructional approaches, and showed similar or higher emotional and motivational involvement than teachers in the own-video group. These results contradict findings from a similar study in science (Seidel, Stürmer, Blomberg, Kobarg, & Schwindt, 2011). Kleinknecht and Schneider (2013) acknowledged that the small sample size and the brief individualized PD experience limited the generalizability of their findings.

Many PD programs that use video clips do so in combination with other tools (e.g., mathematics tasks, student work). In these cases, considering the alignment between tools is important. For example, building on the PSC, Taylor (2011) designed the reflection connection cycle to help teachers build on students' out-of-school knowledge. Teachers in the PD met to discuss video clips from their own instruction and then plan a new activity for implementation between PD sessions. Teachers were also assigned several papers to read to help them connect their instruction to students' out-of-school knowledge. Analyzing teachers' experiences in the PD, Taylor concluded that the combination of PD tools (e.g., videos, lesson design, and readings) did not support the main PD goals as envisioned. He concluded that PD designers needed to further examine how various PD activities came together in support of the PD goals. Generalizing Taylor's conclusion about the use of PD tools in combination, more research is needed to examine how different tools come together in the design of PD experiences to foster (or hinder) teacher learning in mathematics PD programs.

Mathematical Tasks

Research on classroom instruction has systematically shown the importance of mathematical tasks for student learning. Given our increasing understanding of the role cognitively demanding tasks play in mathematics teaching and learning, it is not surprising to find, in our search of the PD literature, a large number of PD programs organized around tasks. Watson and Mason (2007) noted a strong agreement "throughout the world" (p. 207) in the way mathematical tasks have been used in PD. They suggested that, often, professional learning activities designed around mathematical tasks include combinations of experiences such as solving mathematical tasks, reflecting on the experiences of solving tasks as learners, developing frameworks for selecting tasks, creating strategies for working with students on tasks, trying out tasks with students, listening to students as they solve tasks, and analyzing teaching that is organized around the use of tasks.

Many PD programs with a strong focus on mathematical tasks have engaged teachers in solving these tasks as an "opening activity" (Silver, Clark, Ghousseini, Charalambous, & Sealy, 2007, p. 264). In some of these programs, these tasks have been designed with the main goal of developing teachers' subject matter knowledge of mathematics and are not necessarily designed for use in K–12 settings (e.g., Chamberlin, 2009; Chamberlin, Farmer, & Novak, 2008; Thompson, Carlson, & Silverman, 2007). Research in these PD programs has often attended to teachers' mathematical work and the development of their knowledge of mathematics. For example, Orrill and Brown (2012) focused on how teachers use their knowledge of proportional reasoning to make sense of the double number line representation. They examined the work of six PD participants on tasks using this representation and showed that separate components of teachers' knowledge supported or hindered their mathematical understanding. They concluded that the ways in which teachers approached the tasks were logical even when incorrect and suggested that both the quantity and the organization of teachers' knowledge were critical to their sense making during the tasks.

In other PD programs, the mathematical tasks have been strongly connected to the development of teachers' knowledge of students, teaching, and curriculum. In these cases, although the initial task was also meant to challenge teachers' understanding of mathematics, it could often be adapted for use in K–12 instruction as a way to connect the professional development to teachers' practices. In these programs, teachers were sometimes asked to solve a task in multiple ways (e.g., Ferrini-Mundy, Burrill, & Schmidt, 2007), suggest different ways in which their students might solve it (e.g., Kabasakalian, 2007), examine a set of hypothetical students' solutions to the task (e.g., Jacobs et al., 2007), or work on variations of the

学习的问题变式（例如Horoks & Robert，2007）。

我们认为这个研究工具有两个需要特别指出的专业发展项目是："加强中学教师准备项目"和（我们称之为）"设计专业发展资源项目"。选择这些项目是因为他们是建立在大型的、多年关于数学教学研究的基础上，并且这些项目使用了之前在K-12背景下的研究结论，并以此来指导围绕任务而组织的专业发展设计。

加强中学数学教师准备项目。加强中学数学教师准备（ESP）项目（Boston，2013；Boston & Smith，2009，2011）旨在帮助7—12年级的教师改善他们的教学实践。在专业发展的第一年里，参与者在这一年里开会6天。并运用数学任务分析框架（Stein，Grover，& Henningsen，1996）指导专业发展的设计。这一框架表明课程中的数学任务与学生学习之间的联系都取决于教师如何设计并在课堂中实施任务。专业发展中的教师们处理并评估数学任务的认知需要，在自己的课堂上挑选和实施任务，并分析斯坦、史密斯、汉宁森和希尔弗（2000）的案例叙述中所描述的任务实施情况。项目中的所有教师都是未来数学教师的导师，并且专业发展会持续到第二年，来继续关注参与项目教师的实习指导工作。

为了考察教师在参加ESP项目的第一年里在任务的选择和实施的实践方面是否发生了改变，波士顿和史密斯（2009）分析了18位教师在参加项目的第一个学年早期、中期与后期的数据。在每一个时间节点，教师都要提交连续五天的教学任务和参与这些教学任务的三组学生的学习材料，并且在这期间他们有一整节课被参与者观察。在对照组，同样收集了10位教师的教学任务与课堂观察。通过使用教育质量评估的测量工具（Boston & Wolf，2006；Matsumura，Garnier，Slater，& Boston，2008），研究者发现实验组教师对认知需求任务的使用显著提高，并且在教学中能够坚持任务的认知需求。在学年末这些教师与对照组教师在设置和实施任务方式上有显著差异。教师所在的学校所使用的不同的课程材料不再是教师任务选择和实施的主要影响因素了。

为了寻找教师对数学任务的认知需求知识的改变和参与专业发展项目提供的学习机会之间的联系，波士顿（2013）进行了一个混合方法的研究，主要分析18位教师在专业发展项目早期和后期对任务中认知需求知识进行分类评估的反应，同时也分析了所有专业发展会议中的教师参与的录像。在专业发展项目结束后，教师关于数学任务认知需求的知识明显增加，并且分析这些任务时所使用的语言也和早期不一样。教师能够识别出任务的哪些方面能够为那些以不同数学理解水平来解决任务的学生提供学习机会。研究者也指出，教师关于解决问题的经验会影响他们对任务认知需求水平的判断。

在接下来的研究中，波士顿和史密斯（2011）研究了7位参与数学任务专业发展项目两年的实验组教师的教学任务、学生活动和课堂观察。研究发现这些教师保持了对高认知需求任务的选择与实施的技能。通过对其中四位教师的数据进行进一步分析，研究者们发现在实践中坚持不断提高的教师是自我反思型的，并且与专业发展项目提供的想法和工具更易产生共鸣，同时对于专业发展的讨论也经常有所贡献。

设计专业发展资源项目。基于对数学课堂的研究，斯旺（2000）调查了教师对能引起学生讨论的任务的使用，比如在不同的表征之间建立联系、概括性的检验、诊断常见错误或者产生认知冲突的任务。以此为基础，斯旺多年来已经通过好几个课题来研究如何促进教师在教学中使用这样的任务。这些课题形成了一个数学专业发展的研究项目，为此在本章我们将其命名为设计专业发展资源（DPDR）项目。通过这些工作，斯旺（2011）提出一个四阶段的专业发展模式，包括：（1）了解参与教师的实践与信念；（2）让教师解决任务并且观看在课堂上实施类似任务的录像；（3）使教师相信学生能解决这样认知水平的任务并且促进任务的实施；（4）让教师对使用那些认知需求任务的教学经验进行反思。

task that highlight particular affordances or constraints for student learning (e.g., Horoks & Robert, 2007).

The two PD programs of research we highlight for this tool are the Enhancing Secondary Mathematics Teachers Preparation program and (what we are calling) the Designing Professional-Development Resources program. We selected these programs because they build on larger, multiple-year research studies about teaching mathematics with tasks and use findings from previous research in K–12 settings to inform the design of PD organized around tasks.

Enhancing Secondary Mathematics Teachers Preparation program. The Enhancing Secondary Mathematics Teachers Preparation (ESP) program (Boston, 2013; Boston & Smith, 2009, 2011) was designed to help grades 7–12 teachers improve their instructional practices. In the first year of the PD, participants met for 6 days throughout the year. The mathematical tasks analysis framework (Stein, Grover, & Henningsen, 1996) guided the design of the PD. This framework shows that the link between a mathematical task in the curriculum and student learning depends on both how the task is set up by the instructor and how the task is implemented in the classroom. Teachers in the PD solved and assessed the cognitive demand of mathematical tasks, selected and implemented tasks in their own classrooms, and analyzed the implementation of tasks depicted in case narratives from Stein, Smith, Henningsen, and Silver (2000). All teachers in this program served as mentor teachers to future mathematics teachers, and the PD continued for a second year, with a focus on participants' mentoring practices.

To examine whether teachers in the ESP changed their practices of task selection and implementation during the first year of the program, Boston and Smith (2009) analyzed a set of data collected from 18 participating teachers early, midway, and late in the school year. At each point, teachers submitted 5 consecutive days of instructional tasks and three sets of student work from these tasks and were observed for one full lesson during the 5-day period. Instructional tasks and lesson observations were also collected for 10 teachers in a comparison group. Using measures from the Instructional Quality Assessment (Boston & Wolf, 2006; Matsumura, Garnier, Slater, & Boston, 2008), the researchers found that treatment teachers significantly increased their use of cognitively demanding tasks and could maintain the demand of the task during instruction. At the end of the school year, these teachers differed significantly from teachers in the comparison group in the ways they set up and enacted the tasks. Different curriculum materials adopted in the teachers' schools were not a significant predictor of teachers' task selection or implementation.

Seeking to connect changes in teachers' knowledge about the cognitive demand of mathematical tasks and their participation in the learning opportunities offered during the PD program, Boston (2013) conducted a mixed method study that analyzed responses from the 18 teachers to a task-sort assessment of their knowledge about tasks' cognitive demand, conducted early and late in the PD, as well as videotapes of teachers' participation in all PD meetings. At the end of the PD program, teachers had significantly increased their knowledge about the cognitive demand of tasks and were using language for analyzing tasks that differed from the language used earlier in the PD. Teachers could identify aspects of tasks that provided learning opportunities for students who approached the task with different levels of mathematical understanding. The researcher suggested that teachers' experiences in solving tasks influenced their learning about task levels of cognitive demand.

In a follow-up study, Boston and Smith (2011) collected and analyzed a new set of instructional tasks, student work, and lesson observations from seven teachers in the treatment group 2 years after they had participated in the initial PD about tasks. They found that teachers maintained the gains they had made in the selection and implementation of high cognitive-demand tasks. Further examining data from four of these teachers, the researchers concluded that teachers who sustained improvements to their practice were self-reflective, resonated with the ideas and tools presented in the PD, and made frequent contributions to PD discussions.

Designing Professional Development Resources program. Working in mathematics classrooms, Swan (2000) examined teachers' use of tasks designed to generate discussion among students, such as tasks that connect representations, examine generalizations, diagnose common mistakes, or generate cognitive conflict. On the basis of this work, Swan conducted several projects over the course of many years that investigated how to support teachers' use of such tasks in instruction. These projects come together into a program of research on mathematics PD that, for the purpose of this chapter, we are labeling Designing Professional Development Resources (DPDR). Through this work, Swan (2011) developed a four-stage PD model that encompassed (1) recognizing participating teachers' existing practices and beliefs, (2) having the teachers solve tasks and watch videos of similar tasks implemented in classrooms, (3) engaging teachers in suspending their potential disbelief that their students could solve such cognitively demanding

与44位代数教师合作，斯旺（2007）提出了一个专业发展项目，它包括一个为期2天的初始工作坊（了解信念并分析任务），接下来是在6个月的周期中有两个为期1天的会议，教师可以分享他们使用数学任务的课堂经验。共有36位老师完成了关于他们在项目前后的信仰和实践的调查问卷（Swan, 2006a）。斯旺发现，教师报告了他们的信念朝着把数学看成是"教师和学生通过合作讨论一起构建的思想的网络"的方向改变（Swan, 2007, 第226页）。教师还报告了他们的教学朝着以学生为中心的实践转变。在参与的教师中，有28位提供了来自他们学生的数据（Swan, 2006b）。没有参与专业发展项目的14位对照组教师也提供了学生的数据。学生在代数学习中的收获与教师在课堂上实施的专业发展中讨论的任务数量有关。更加关注以学生为中心，使用专业发展任务更多的教师所教授的学生会有进一步收获。作为对照组，使用标准代数课程的教师所教的学生没有进步（Swan, 2007）。

在之后的专业发展模式实施过程中，斯旺和斯温（2010）和24位算术教师一起工作，这些老师所教学生都是16岁或者16岁以上。为了改变教师的信念和实践，在9个月里组织教师参加了4到6天的专业发展项目。研究者通过问卷、访谈和课堂观察收集数据。参与这个研究的教师也表示获得了更多的以学生为中心的教学实践，同时获得了新的数学信念，即数学是通过教师和学习者之间的讨论而形成的思想体系。通过分析几个研究的结果，斯旺、比德、多尔曼和穆迪克（2013）指出许多教师最开始认为当实施以学生中心的教学实践时，他们需要撤掉对学生的支持；之后他们才意识到，他们需要重新定义教师的角色和他们提供给学生的支持。斯旺等（2013）认为经历了这一变化过程的专业发展的参与者们在以学生为中心的教学中使用任务时可能"起初变得不那么有效，然后才变得更有效"（第951页）。

使用数学任务的专业发展项目。让教师解决和分析数学任务的专业发展可以促进教师教学的改变。许多参加了ESP项目和DPDR项目的教师都反映他们学会了在更加关注学生的教学方法中选择和使用更多高认知需求的任务。此外，对于参加ESP项目的一些教师，教学的改变在专业发展项目结束以后仍在持续。数学专业发展的研究通常会在专业发展结束时显示出某种程度的结果变化。然而在考虑规模问题之前，理解这种变化是如何随着时间而持续的，这也是证明专业发展项目阶段1有效性的一个重要组成部分。

ESP项目和DPDR项目都是建立在之前研究的重要结论之上的，并且使用研究的基本框架来支持教师在参与专业发展项目期间的工作。其他使用数学任务的专业发展项目也是围绕这一研究框架进行设计的。例如，鲁贝尔和褚（2011）设计了一个使用与文化相关的数学教学法框架的专业发展项目。他们研究了来自城市低收入社区教师的教学，尽管他们中的大多数仍拒绝改变并继续使用低认知需求的任务，但已经有部分教师开始使用一些高认知需求的任务并强调概念性的理解，来为教育资源匮乏的社区的高中生提供一些学习数学的机会。

ESP项目和DPDR项目的不同在于如何界定专业发展和教学实践变化之间的关系。ESP项目的设计基于的假设是专业发展首先导致教师知识的改变，从而导致教学实践的变化；而DPDR项目的设计原则是实践的改变促进信念的改变。这些项目的结果表明，这两种方法都可以支持教师知识或信念的改变，以及教学实践的变化。专业发展项目、教师改变和教学实践之间如此复杂的关系已经促使部分研究者改变其研究方法，从专业发展背景下的教师学习概念化的单向模型转向理解教师变化的相互关联的方法（例如Clarke & Hollingsworth, 2002; Richardson & Anders, 1995）。未来还需要更多的研究去考察参与专业发展项目的教师在教师知识、信念和教学实践改变之间的关系。

阶段1研究的讨论

在本节中，我们围绕在当前许多专业发展项目使用的三个工具来组织阶段1研究的分析：（1）学生数学思维的框架，（2）数学教学的录像片段，（3）数学任务。我们分享了两个应用这些工具的专业发展项目的设计以及相关的研究发现。我们还建议在不同的领域进行阶段1的额外研究，比如研究教师多年在一个专业发展项目中的

tasks and supporting their implementation of tasks, and (4) having teachers reflect on the experience of teaching using demanding tasks.

Working with 44 algebra teachers, Swan (2007) offered a PD program that included a 2-day initial workshop (recognize beliefs and analyze tasks), followed by two 1-day meetings during a 6-month period for teachers to share their classroom experiences using tasks. A total of 36 teachers in the PD completed questionnaires about their beliefs and practice before and after the program (Swan, 2006a). Swan found that teachers reported changes in their beliefs toward a view of mathematics as "a network of ideas that the teacher and student construct together through collaborative discussion" (Swan, 2007, p. 226). Teachers also reported changes in their instruction toward more student-centered practices. Among participating teachers, 28 provided data from their students (Swan 2006b). A group of 14 comparison teachers, who did not attend the PD, also provided student data. Gains in students' algebra learning were associated with the number of tasks discussed in the PD that teachers implemented in their classrooms. Greater gains were made by students of teachers who used more of the PD tasks in more student-centered ways. No gains were made in the comparison group that used the standard algebra curriculum (Swan, 2007).

In a later implementation of the PD model, Swan and Swain (2010) worked with 24 numeracy teachers who teach students who are 16 years of age or older. Teachers attended from 4 to 6 days of PD organized over 9 months with the goal of changing their beliefs and practice. The researchers collected data through questionnaires, interviews, and classroom observations. Teachers in this study also reported embracing more student-centered practices as well as beliefs that mathematics is a connected body of ideas developed through discussions among teachers and learners. Analyzing findings across several studies, Swan, Pead, Doorman, and Mooldijk (2013) suggested that many teachers first think that they need to remove their support of students when implementing student-centered instruction and only later realize they need to redefine their role and the supports they offer students. Swan et al. (2013) suggested that PD participants who experience this change process might "at first become less effective, before they become more effective" (p. 951) in the use of tasks during student-centered instruction.

PD using mathematical tasks. PD that engages teachers in solving and analyzing mathematical tasks can support changes in teachers' instruction. Many of the teachers who participated in the ESP and DPDR programs learned or reported that they learned to select and use more cognitively demanding tasks in more student-centered approaches to instruction. Further, for some teachers in ESP, changes in instruction were sustained after the PD was over. Research on mathematics PD often shows some level of change in outcomes at the conclusion of the PD. Understanding how this change is sustained over time is, again, an important component of demonstrating the efficacy of Phase 1 PD programs prior to considering issues of scale.

Both ESP and DPDR build on significant bodies of previous research and use research-based frameworks to support teachers' work during the PD. Other PD programs using tasks are also designed around such research frameworks. For example, Rubel and Chu (2011) designed a PD using a framework for culturally relevant mathematics pedagogy. They examined the instruction of participating teachers from schools in urban, low-income communities and showed that whereas many of the teachers remained resistant to change and continued to use low-demand tasks, others came to use higher demand tasks and emphasize conceptual understanding, offering students from high schools in underserved neighborhoods several opportunities to learn mathematics.

ESP and DPDR differ in how they conceptualize the relationship between PD and changes in practice. Whereas ESP was designed around the conjecture that PD leads to changes in teachers' knowledge, which then leads to changes in practice, DPDR design was conceptualized around the principle that changes in practice foster changes in beliefs. Results from these programs seem to indicate that both approaches can support changes in teachers' knowledge or beliefs, as well as changes in practices. Such complex relations between PD programs, teacher change, and teaching practice have supported several researchers in shifting away from unidirectional models for conceptualizing teacher learning in PD settings and toward interconnected approaches to understanding teacher change (e.g., Clarke & Hollingsworth, 2002; Richardson & Anders, 1995). More research is needed to examine the relations between changes in teacher knowledge, beliefs, and practice as they participate in PD.

Discussion of Phase 1 Research

In this section, we organized our analysis of Phase 1 studies around three tools currently used in many PD programs: (1) frameworks of student mathematical thinking, (2) video clips of mathematics instruction, and (3) mathematical tasks. We shared the design of two different PD programs using each of these tools and research

参与情况、对有不同专业发展经验水平的教师实施区别化的专业发展、专业发展结果的长期可持续性以及在专业发展项目中工具的综合使用。阶段1研究仍然有许多待完成的工作。

在我们的综述中，研究计划包括利用多种研究方法来论证一个特定专业发展项目的效果，包括个案研究、干预前和后的调查、准实验和实验设计。然而，大多数阶段1研究采取的仍然是质化研究或者对小样本教师的相关性分析。尽管这些研究的发现提供了许多有效专业发展项目设计的深刻见解，但是扩大阶段1使其包括更多的准实验、实验和纵向研究设计，可以促进对专业发展的一些重要方面有更好的理解，比如如何将参与专业发展的教师与不参与的教师作比较，多年参与专业发展的影响，在专业发展项目中的收获是否可以持续，以及专业发展对学生学习的影响。

通过关注6个专业发展项目的设计，可以明确贯穿数学专业发展研究的一个层面，即专业发展设计与教学实践之间的关联。正如我们前面所论述的，实践是教师学习的首要途径，将这些专业发展项目建立在教师实践的基础上，是我们深入探讨的那些项目都有的重要特征。这些专业发展项目使用研究教学法让教师参与分析和批判实践的表现形式和使用执行教学法让教师参与计划和实践教学的某些特定方面——都是在专业发展背景下和自己的课堂中（Grossman, Hammerness, & McDonald, 2009）。通过特定的工具，特定的教法，将专业发展和教学实践联系起来，并且在实际课堂环境中真实地实施明确的实践，这是我们所认为的专业发展设计有效的典型特征。

我们所讨论的6个项目还强调研究者对于参与专业发展项目教师的教学实践变化的关注。探索教学实践的变化是连接教师专业发展和学生学习的一个基本联系，将研究转向不仅关注教师的变化还要关注学生的变化是很重要的。另外，我们所讨论的专业发展项目还显示出研究者研究的焦点也转向对于教师如何参与专业发展项目这一视角，包括教师在专业发展会议中讨论话语的变化。对于专业发展参与性变化的关注与该领域研究的转变相一致，即进一步采用情境视角来看待学习，并表明对教师学习过程的理解更加集中。

将这些专业发展项目与数学教育研究更广泛的联系起来，对我们所探讨的项目也尤为重要。学生如何学习、任务的认知需求与实施的严格性、以及学生为中心的教学实践等这些研究结果引领着阶段1中所研究的很多专业发展项目的设计。这些发现经常被编入专业发展项目设计者用来促进教师学习的工具和协议中。或许对研究结果的关注是我们对专业发展定义和用于搜索的参数的产物。进一步思考专业发展的设计以及研究和实践之间的联系，对于数学专业发展研究领域是很重要的。在继续重视和加强教师在教学时所产生的知识的同时，寻找联结教师和基于研究的知识的方法也是一个值得重视的方向。

专业发展项目规模化

为了满足中小学教师日益增长的专业发展机会的需求，我们要求教育社区创建有效的专业发展项目，并且是规模化和可持续的。一个规模化并且可持续的专业发展项目应该可以在保证其核心原则一致的条件下，由不同的促进者在多个地点有效地实施。博尔科（2004）将研究专业发展项目的规模化和可持续性作为阶段2的研究。正如她所阐述的，"在阶段2研究者们研究由不同的促进者在不同的背景下实施的单独的专业发展项目，由此来探索促进者、专业发展项目、以及作为学习者的教师之间的关系"（第4页）。

在我们最初的专业发展文献的综述中，我们将113篇期刊文献中的24%的文献定义为阶段2的研究。这些研究解决了一系列广泛的问题，当我们从阶段1研究一个专业发展项目到阶段3比较不同项目时，学术界已经开始着手解决这些问题。这些问题关注专业发展项目的三个特征，这些特征对确定项目是否可以规模化和可持续以及如何实现项目的扩展与持续是至关重要的：（1）促进和准备专业发展项目的引导者或专业发展项目领导人的特点；（2）规模化的方法；（3）背景。因此，将本章最开始提出的问题，即我们从探索专业发展项目规模化的研究中学习到了什么，分成三个独立的小问题：

findings related to the use of such tools in these programs. We also suggested various areas for additional Phase 1 research, such as examining teachers' participation in one PD program over multiple years, implementation of differentiated PD for teachers with different levels of PD experience, long-term sustainability of PD results, and use of combinations of tools in PD. Much work remains to be done in Phase 1 research.

The research programs included in our review utilized a wide array of research methods to demonstrate the promise of a particular PD program, including case studies, pre- and post-intervention surveys, quasi-experiments, and experimental designs. However, most Phase 1 studies have used qualitative or correlational analysis of small samples of teachers. Whereas findings from these studies provide many insights into the design of effective PD, expanding Phase 1 research to include more quasi-experimental, experimental, and longitudinal designs can help the field better understand important aspects of PD, such as how participants compare to teachers who do not participate in the PD, the impact of participation over multiple years, whether the gains achieved in the PD are sustainable, and the impact of PD on student learning.

By focusing on the design of six PD programs, we make clear an aspect of mathematics PD that cuts across the programs we reviewed: the connection between PD design and teaching practice. As we suggested earlier, practice is the premier space for teacher learning, and grounding the PD in aspects of teachers' practice was an important design feature for all the programs we examined in depth. These PD programs used pedagogies of investigation to engage teachers in analyzing and critiquing representations of practice and pedagogies of enactment to engage teachers in planning and practicing specific aspects of teaching—both in the PD setting and in their own classrooms (Grossman, Hammerness, & McDonald, 2009). Connecting PD to practice through specific tools, particular pedagogies, and actual implementation of well-defined aspects of practice in classrooms are among what we consider effective features of PD design.

The six programs we discussed also highlight the attention PD researchers have paid to examining changes in participating teachers' instructional practice. Detecting changes in instruction is a fundamental link connecting teacher PD to student learning, and shifting research to attend not only to changes in teachers but also to changes in teaching is important. In addition, the PD programs discussed show the increasing attention researchers have paid to how teachers participate in the PD itself, including changes in teachers' discourse during PD meetings. This attention to changes in PD participation aligns with a shift in the field to further embrace a situative perspective on learning and signals an increased focus in understanding the processes of teacher learning.

Connections between these PD programs and mathematics education research more broadly were also important in the programs we examined. Findings from research on how students learn, tasks' cognitive demand and rigor of implementation, and student-centered instructional practices guided the design of many PD programs examined in Phase 1. These findings were often codified in the tools and protocols PD designers used to facilitate teacher learning. Perhaps such attention to research results is an artifact of our definition of PD and the parameters used for our search. Still, further thinking about the role of research in the design of PD and the connections between research and practice is important for the field of mathematics PD research. Finding ways to connect teachers to research-based knowledge as we continue to value and strengthen the knowledge teachers generate when teaching is an important charge for the field.

Bringing Professional Development Programs to Scale

To meet the growing demand for PD opportunities for K–12 teachers, the education community has been asked to create effective PD programs that are scalable and sustainable. A PD program that is scalable and sustainable can be effectively enacted by multiple facilitators and in multiple contexts while maintaining consistency with its core principles. Borko (2004) labeled research investigating the scalability and sustainability of PD programs as Phase 2 research. As she explained, "In Phase 2, researchers study a single professional development program enacted by more than one facilitator at more than one site, exploring the relationships among facilitators, the professional development program, and teachers as learners" (p. 4).

In our initial review of the PD literature, we categorized 24% of the 113 journal articles we examined as Phase 2 research. These studies addressed a wide array of issues that the research community has begun to grapple with as they move from examining one PD program in Phase 1 to comparing across different programs in Phase 3. These issues focus on three features of PD programs that are central to determining both whether programs are scalable and sustainable and what it takes to implement scalable, sustainable programs: (1) the nature of facilitation and preparation of PD facilitators or PD leaders, (2) approaches to scale, and (3) context.

- 专业发展项目的促进者需要了解将要做和可以做的工作，以及在准备过程中什么是必要的？
- 用什么方法来将专业发展项目规模化？
- 更大的背景特征对规模化专业发展项目的影响是什么？

我们将探讨以上每一个问题，讨论几个对于我们知识增长有益的研究与项目。正如前面所提到的，我们选择在本节中强调的研究是基于它们所涉及的问题。尽管建立在比阶段1少得多的研究基础上，但我们对阶段2的分析却表明这个阶段包含了丰富的新兴研究。

专业发展项目促进和促进者的准备

识别对教师和学生有影响的专业发展项目尽管很重要，但这只是将专业发展规模化的一个影响因素。熟练的促进者对于确保专业发展项目有效性起了核心作用（例如Bell，Wilson，Higgins，& Mccoach，2010；Bobis，2011）。因此，了解促进者为了有效领导专业发展所必须知道和能够做什么，对于数学教育界准备和支持促进者至关重要。关注促进者是如何准备的同时，要更好地理解促进者的工作可以确保专业发展项目实施是由促进者而不是项目的设计者完成的。

专业发展促进研究。尽管专业发展的实践和专业发展领导者的角色还没有被定义和研究，但是对于数学专业发展领导者的知识和实践的关注近年来在持续增加（例如Borko，Koellner，& Jacobs，2011；Elliott等，2009）。在我们的检索中，我们发现了一些文献回答了一个专业发展促进者应该知道什么和能够做什么的问题——一些是基于研究者自己实施或研究专业发展的经验（例如Coles，2012），另一些是基于对专业发展领导者或教师的调查（例如Rogers等，2007）。我们发现极少有研究系统分析专业发展领导者的促进实践，研究者没有使这样的实践成为研究的核心。我们本节从讨论一个经验丰富的专业发展领导者的促进实践开始。

研究者万·埃斯、腾尼、戈德史密斯和西戈（2014）在两个基于录像的专业发展项目中，研究了有经验的专业发展领导者的促进实践：线性函数的教与学（LTLF；Seago等，2004）和录像俱乐部（Sherin & Han，2002）。尽管这两个项目之前从教师学习的视角被研究过，但这部分我们主要关注"促进者塑造、关注和支持教师使用录像来探讨数学学习和教学的方法"（van Es等，2014，第340页）。作者分析了两个项目录像中记录的所有专业发展的对话，其中有教师持续讨论学生数学思维细节。他们将讨论中的推动措施分为实践中的四个方面。首先，将小组引向录像分析任务，为参与者制定了活动框架；其次，保持探究的立场及持续关注录像和数学的行动，使得讨论始终建立在录像中捕捉到的事件和互动的基础上，并始终侧重于提出基于证据的关于教与学的主张；然后通过鼓励小组合作的方式，促进者鼓励所有参与者加入讨论并确保不同的观点和视角可以被表达和重视。分析表明了，这些实践活动的组织协调，而不是使用特定的措施，能支持教师学习中录像使用的有效性。

其他的研究使用了多种不同的研究设计来解决推进实践的本质或促进者的作用问题。谢林等（2009）分析了和万·埃斯等人（2014）相同的录像俱乐部会议，以调查思维促进者对录像片段的选择以及这些片段的特征，这些特征促进了教师对学生数学思维的实质性研究。同样关注基于录像的讨论，科尔斯（2012）分析了他与系里数学教师一起主持的专业发展会议的录像，并找出了促进者角色做决策的那些点。林德（2011）对参与不同专业发展项目的小学教师进行了访谈，让他们描述一个既能激发他们又可以支持他们学习的促进者的特点。萨克（2008）分析了一个经验丰富的专业发展领导者和数学领导机构中的一群高中教师产生冲突的原因。

当我们分析这些研究的结论时，发现几个明显的模式。成功的促进者可以和与他们合作的教师建立一种协作的、信任的关系，并促进富有挑战性但有支持性的对话。在专业发展会议期间，他们精心地安排了对话，在鼓励更广泛的参与，并回应参与者贡献的同时，保持对专业发展项目目标的关注。正如万·埃斯和同事们（2014）指出，专业发展领导者用来发展共同体和协调讨论的具体举措，会因不同的专业发展项目和不同推进者而有所不同。在多项研究中，常见的行动和策略包括制

Thus, we separated the question posed at the beginning of the chapter—What do we know from research that seeks to understand what it takes to bring PD programs to scale?—into three subquestions that the field is beginning to examine:

- What do PD facilitators need to know and be able to do, and what is entailed in their preparation?
- What approaches have been used to bring PD programs to scale?
- What is the impact of larger contextual features on scaling PD programs?

We examine each of these questions, discussing several research studies and programs that have contributed to our growing knowledge base. As we noted previously, our choice of research studies to highlight in this section is based on the questions they addressed. Although building on a much smaller number of studies than Phase 1, our analysis of Phase 2 shows the richness of the emerging research subsumed within this phase.

Professional Development Facilitation and Facilitator Preparation

Identifying PD programs that can affect teachers and their students, although critical, is only one factor in bringing effective PD to scale. Skillful facilitators play a central role in ensuring the effectiveness of a PD program (e.g., Bell, Wilson, Higgins, & Mccoach, 2010; Bobis, 2011). Thus, an understanding of what facilitators must know and be able to do to effectively lead PD is crucial to the mathematics education community's ability to prepare and support facilitators. Together with attention to how facilitators are prepared, a better understanding of the work of facilitators ensures that PD programs can be enacted by facilitators other than the program designers.

Studies of PD facilitation. Although the practice of PD and the role of PD leaders are underdefined and understudied, attention to the knowledge and practices of mathematics PD leaders has increased in recent years (e.g., Borko, Koellner, & Jacobs, 2011; Elliott et al., 2009). In our search, we found several articles that addressed what PD facilitators must know and be able to do—some based on the researchers' own experiences conducting or studying PD (e.g., Coles, 2012) and others based on surveys of PD leaders or teachers (e.g., Rogers et al., 2007). We found a much smaller number of studies that systematically analyzed the facilitation practices of PD leaders, making such practice the focus of their research. We begin this section with a discussion of one such study of experienced PD leaders' facilitation practices.

Researchers van Es, Tunney, Goldsmith, and Seago (2014) studied the facilitation practices of experienced PD leaders in two video-based PD programs: *Learning and Teaching Linear Functions* (LTLF; Seago et al., 2004) and video clubs (Sherin & Han, 2002). Although these two programs had been previously studied from the perspective of teacher learning, this research attended to the "ways that facilitators shape, focus, and support teachers' use of video to inquire into mathematics learning and teaching" (van Es et al., 2014, p. 340). The authors analyzed video recordings of all PD conversations, from both programs, in which teachers engaged in sustained discussions of the details of students' mathematical thinking. They categorized facilitation moves during these discussions into four practices. *Orienting the group to the video analysis task* framed the activity for participants. Moves to *sustain an inquiry stance* and *maintain a focus on the video and the mathematics* served to keep the discussion grounded in events and interactions captured in the video and to maintain a focus on making evidence-based claims about learning and teaching. By *supporting group collaboration*, the facilitators encouraged all participants to engage in the discussion and ensured that their different perspectives and expertise were represented and valued. The analysis suggested that the coordination of these practices, rather than the use of particular moves, was key to supporting the productive use of video for teacher learning.

Additional studies used a variety of research designs to address the nature of facilitation practices or the role of the facilitator. Sherin et al. (2009) analyzed the same video club meetings as van Es et al. (2014) to investigate facilitators' selection of video clips and the features of the clips that promoted teachers' substantive exploration of students' mathematical thinking. Also focusing on video-based discussions, Coles (2012) analyzed video of PD sessions that he led with mathematics teachers in his department and identified decision points within the role of the facilitator. Linder (2011) conducted interviews with elementary teachers who had participated in different PD programs, asking them to describe characteristics of facilitators who were able to both motivate and support their learning. Sack (2008) analyzed reasons for conflicts that arose between an experienced PD leader and a group of high school teachers in a mathematics leadership institute.

Several patterns are apparent as we examine findings across these studies. Successful facilitators create a collegial, trusting relationship with the teachers they are

定参与的规范、质疑、澄清、形成联系、重新制定框架和确认参与者的想法。

不同专业发展方式中促进者的角色是不同的。一个例子是，在一个适应性的共同体中，位于不同地点的专业发展项目促进者的特点（Borko, Koellner, & Jacobs, 2011）。在这个共同体的一端是高度适应性的方法，比如录像俱乐部。这些项目的特点是关注当地的背景，并根据一般准则，而不是预设的活动和材料来推进。适应性专业发展项目的促进措施包括，对于每一个专业发展会议，确定教师学习的目标，选择工具来主持专业发展项目以及准备那些引导讨论的问题。在这个共同体的另一端是高度详细的项目，比如线性函数的教与学，其中的目标、资源和促进材料是对于一个特定的、预先确定的专业发展经验而言的。对于一个高度详细的专业发展项目促进者的一个主要任务是熟悉材料并且确保理解设计者的核心目标。

促进研究的新发现提供了一些例子，说明促进者做了什么，以及促使他领导专业发展项目的一般特征。这些研究所确定的系列举措只是一个开始。进一步的研究需要确定另外的实践，明确构成这些实践的促进举措的特征，以及确定它们所需要的知识。进一步研究如何促进专业发展成为阶段2研究的核心。

关于专业发展促进者准备的研究。关于如何准备并支持专业发展领导者的研究非常少。聚焦于领导者准备的研究项目通常强调以下两个方面，即专业发展领导者的准备以及研究领导者的学习过程。在本节中，我们重点介绍以下两个项目——面向数学专业发展规模化模型：促进者实施问题解决循环的准备工作的田野研究（iPSC；Koellner 等，2011），另一个是中学数学和教学的机构设定（MIST；Cobb, Jackson, Smith, Sorum, & Henrick, 2013；另见 Cobb, Jackson, & Dunlap, 2017，这套书）。接下来，我们将分析这两个项目和聚焦专业发展领导者准备的其它项目的共同特点。

iPSC。在开发和研究了问题解决循环的专业发展模型后，博尔科、克尔纳和雅各布斯（2014）实施了 iPSC 项目以研究 PSC 的可推广性与可持续性。他们开发了一个数学领导力准备模型（MLP），包括一个暑期领导力学校和三次领导人支持会议。在暑期领导力学校中，专业发展的领导者观看和讨论之前实施 PSC 的录像片段，这些片段展示了促进者高效或低效的实践。他们也参加 PSC 模拟活动，这些模拟活动使用的是为下学年挑选的数学任务。在这一年里，PSC 的领导者在工作坊之前会参加领导者支持会议。这些会议包括推进者使用多种表征共同解决 PSC 数学问题，讨论不同解决策略间的关系，为基于录像的讨论挑选录像片段，单独和合作制定计划，演练他们计划的专业发展活动，接收他人和 MLP 领导者的反馈的时间。

博尔科、克尔纳、雅各布斯和同事们曾与一个大郊区的学区一起开发和测试 MLP 模型（Borko 等，2015）。他们做了一系列分析考察了参与 MLP 模型的学校领导者组成的 PSC 的促进作用，该研究使用了一个观察协议来对他们 PSC 工作坊的录像进行评定（Borko 等，2014）。他们发现专业发展领导者在做工作坊时与 PSC 模型的核心原则是一致的，这为 PSC 是一个可推广的专业发展模型和 MLP 模型在 PSC 领导者的准备工作中的有效性提供了初始证据。专业发展领导者在创造一个尊重和信任的工作坊氛围，以及建立教师间的合作关系方面尤为成功。通常，他们也能够选择恰当的录像片段作为讨论的跳板，同时使团队成员讨论的 PSC 问题聚焦于多种数学表征和多种问题解决策略。相比之下，专业发展领导者在将讨论引向深入分析教学实践或学生思维方面有更大的困难。

MIST。科布和同事们的 MIST 项目，通过与两个大型城市学区的数学教师、数学教练、学校领导以及学区领导合作，实施数学综合教学提升（Cobb & Jackson, 2011；Cobb & Smith, 2008）。作为他们与其中一个学区合作的一个内容，杰克逊和同事们（2015）进行了一个实验设计，他们支持三位数学专业发展领导者的学习，并研究了他们所提供的支持和领导者专业发展课程之间的关系。这些专业发展领导者都是优秀的数学教师，但是在支持教师学习上缺乏经验。

MIST 研究小组基于先前专业课程的分析，为领导者的学习确定了三个目标：（1）把教师学习看作一个发展轨迹，（2）通过对教师教学实践的持续评估为教师的

working with and promote conversations that are challenging yet supportive. During the PD sessions, they orchestrate discussions that encourage broad participation and are responsive to participants' contributions while maintaining a focus on the goals for the PD activities. As van Es and colleagues (2014) noted, the specific moves that PD leaders use to develop community and orchestrate discussions vary across PD programs and facilitators. Moves and strategies that were common across multiple studies include setting norms for participation, questioning, clarifying, making connections, reframing, and validating participants' ideas.

The facilitator's role differs for different approaches to professional development. One example is the nature of facilitation for PD programs located at different places on a continuum of adaptability (Borko, Koellner, & Jacobs, 2011). On one end of this continuum are highly adaptive approaches such as video clubs. These programs are characterized by attention to local context and facilitation based on general guidelines rather than preset activities and materials. Facilitation of adaptive PD involves, for each PD session, identifying goals for teacher learning, selecting tools to anchor PD activities, and preparing questions to guide the discussions. On the other end of the continuum are highly specified programs such as *Learning and Teaching Linear Functions*, where goals, resources, and facilitation materials are specified for a particular, predetermined PD experience. A major planning task for the facilitators of highly specified PD is to familiarize themselves with the materials and make sure they understand the designers' core goals.

Emerging findings from studies of facilitation provide a few examples of what facilitators do and general features of what it takes to lead PD. The set of practices identified in these studies is only a beginning. Research is needed to identify additional practices, characterize the facilitation moves they are composed of, and determine the knowledge required to enact them. Further examining what it takes to facilitate PD effectively is central to Phase 2 research.

Studies about the preparation of PD facilitators. Research on how to prepare and support PD leaders is sparse. Research projects focused on leader preparation typically have a dual emphasis on preparing PD leaders and studying the process of leader learning. In this section, we highlight two such projects—Toward a Scalable Model of Mathematics Professional Development: A Field Study of Preparing Facilitators to Implement the Problem-Solving Cycle (iPSC; Koellner et al., 2011) and Middle-School Mathematics and the Institutional Setting of Teaching (MIST; Cobb, Jackson, Smith, Sorum, & Henrick, 2013; see also Cobb, Jackson, & Dunlap, 2017, this volume). We then consider features that the two projects share with other programs focused on preparing PD leaders.

iPSC. After developing and studying the Problem-Solving Cycle PD model, Borko, Koellner, and Jacobs (2014) conducted the iPSC project to investigate the scalability and sustainability of the PSC. They developed the mathematics leadership preparation model (MLP), consisting of a summer leadership academy and cycles of three leader support meetings. During the summer leadership academy, the PD leaders typically viewed and discussed video clips from prior implementations of the PSC that illustrated more- and less-effective facilitation practices. They also participated in PSC simulations using mathematical tasks selected for the upcoming school year. During the year, PSC leaders attended leader support meetings prior to workshops. These meetings included time for the facilitators to collectively solve the PSC mathematics problems using multiple representations, discuss relationships among solution strategies, select video clips to use for video-based discussions, plan individually and collaboratively, rehearse the PD activities they were planning, and receive feedback from one another and the MLP leaders.

Borko, Koellner, Jacobs, and colleagues worked with a large suburban school district to develop and test the MLP model (Borko et al., 2015). One set of analyses they conducted examined facilitation of the PSC by the school-based PD leaders who participated in the MLP model, using an observation protocol to rate videos of their PSC workshops (Borko et al., 2014). They found that the PD leaders conducted their workshops with integrity to the core principles of the PSC model, thus providing initial evidence that the PSC is a scalable PD model and the MLP model is effective in preparing PSC leaders. The PD leaders were particularly successful in creating a climate of respect and trust in their workshops and establishing collaborative working relationships among the teachers. They also were generally able to select video clips that were appropriate for use as springboards for discussion and to engage their groups in discussions of the PSC problems focused on multiple mathematical representations and multiple solution strategies. In contrast, the PD leaders had more difficulty facilitating discussions to deeply analyze instructional practices or student thinking.

MIST. Cobb and colleagues on the MIST project collaborated with mathematics teachers, mathematics coaches, school leaders, and district leaders in two large urban school districts implementing comprehensive instructional improvement programs in mathematics (Cobb &

学习提供支持，（3）通过改进教师想法来促进专业发展。为实现这些目标，研究者和学区的中学数学领导者（领导者）设计了一个四阶段的学习周期，来支持专业发展领导者的学习，并且在一学年中进行四次。每个周期包括以下几个阶段：（a）研究团队的成员和领导者共同计划的数学领导者专业发展会议，（b）数学领导者专业发展讲习会，（c）由数学领导者组织的教师专业发展讲习会，（d）研究团队观看教师专业发展讲习会录像的会议，目的是告知以后数学领导者专业发展讲习会的计划。数学领导者专业发展讲习会的活动通常包括：观看和讨论之前教师专业发展讲习会的录像片段、为接下来的专业发展讲习会一起做计划。在计划的过程中数学领导者处理并分析了教师在专业发展过程中将要面对的数学任务，并确定重点解决方案。

与iPSC项目一样，在MIST项目中，专业发展领导者在引导的某些方面比其他方面改善得更多。他们开始将教师专业发展看成是对教学实践发展轨迹越来越复杂的支持。他们设计的专业发展课程反映了随着时间的推移，对教学改进核心问题的更实质性的关注。然而，他们要求的教学改进的质量会因在一个周期与跨周期中的活动而有所不同（Jackson等，2015）。

尽管在iPSC和MIST项目指导下的专业发展项目有所区别，研究者提供给专业发展促进者的支持也有所不同，但在专业发展促进者准备方式上却有明显的共性。两个研究团队都领导了一些活动，在这些活动中，专业发展领导者分析了他们将在接下来的专业发展课程中讨论的数学任务和学生解题策略，并观看和分析了他们之前的专业发展课程的录像。和阶段1中所描述的一些专业发展项目如何让教师研究和执行教学法类似（Grossman等，2009），这些项目在专业发展项目促进者的准备过程中，也使用了研究和执行教学法。

研究和执行教学法也是其它数学专业发展领导者准备工作的典型组成部分。例如，艾略特和同事们在研究数学领导者学习（RMLL）项目中，支持专业发展领导者学习，从而为教师创造丰富的专业发展机会（Elliott等，2009）。他们的三个为期两天的RMLL系列研讨会包括以下活动：解决一个数学任务、分析一个与教师解决同样任务的数学专业发展领导者的录像案例、考虑如何将这些想法应用于他们自己的促进实践中，这些都与教师和学生在专业发展环境中一起围绕一个数学任务而做准备有很多类似之处。尽管研究领导者学习不是一个核心问题，但"项目通道"（Clark等，2008）还是关注到了专业发展领导者为其专业学习共同体（PLC）所做的准备。专业学习共同体促进者参加了暑期工作坊、后续会议和每周的训练课程。在这期间，他们收到了先前专业学习共同体课程在推进方面的反馈。

在这些项目中，专业发展领导者的学习机会类似于教师学习活动，但是这些项目更关注于领导专业发展和促进教师间对话的实践。杰克逊和同事们（2015）解释道："当他们为数学领导者学习设计支持活动时，可以根据高质量的教师专业发展和职前教师教育的文献进行推断"（第94页）。鉴于关于如何帮助专业发展领导者设计和推进高质量专业发展活动的研究基础薄弱，很有可能其他的专业发展领导者准备项目的设计者也采用了类似的方法。需要更多的关注专业发展领导者准备的方法以及它们的影响的研究，特别是在专业发展领导者的准备工作中使用研究和执行教学法。

实现专业发展规模化的途径

在分析大规模实施的专业发展研究项目时，我们发现了把这些项目推广到更多地区的两种途径。一些项目为促进者开发了专业发展课程和相关材料；其他的项目与学区、州、国家一起合作，建立当地开展大规模专业发展项目的能力。我们将讨论这两种途径，并提供每种途径的实例。

专业发展课程。因为这类课程一开始就被设计成能够被许多促进者在多个地区使用，所以专业发展课程

Jackson, 2011; Cobb & Smith, 2008). As one component of their collaboration with one of the districts, Jackson and colleagues (2015) conducted a design experiment in which they supported the learning of three mathematics PD leaders and studied the relationship between the support they provided and the leaders' PD sessions. These PD leaders were accomplished mathematics teachers but were inexperienced in supporting teacher learning.

The MIST research team identified three goals for the leaders' learning based on an analysis of prior PD sessions: the PD leaders should (1) come to view teacher learning as a developmental trajectory, (2) design supports for teacher learning that are informed by ongoing assessments of their instructional practices, and (3) facilitate PD by pressing on teachers' ideas. To address these goals, the researchers and the district's director of secondary mathematics (director) designed a four-phase cycle for supporting PD leaders' learning, which they enacted four times during the school year. Each cycle of support included the following phases: (*a*) a meeting in which members of the research team and the director coplanned a math leader PD session, (*b*) a math leader PD session, (*c*) teacher PD sessions led by the math leaders, and (*d*) a meeting in which the research team viewed videos of the teacher PD sessions to inform planning of future math leader PD sessions. Activities in the math leader PD sessions typically included viewing and discussing clips from the previous teacher PD sessions and then coplanning the upcoming PD sessions. During planning, the math leaders solved and analyzed the mathematical tasks that the teachers would work with during the teacher PD and identified key solutions to highlight.

As was true in iPSC, the PD leaders in the MIST project improved in some aspects of facilitation more than others. They began to develop a vision of teacher PD as supporting a developmental trajectory of increasingly sophisticated forms of instructional practice, and the PD sessions they designed reflected a more substantial focus on issues central to instructional improvement over time. However, the quality of their press for instructional improvement varied across activities within a cycle and across cycles (Jackson et al., 2015).

Although there were differences in the PD programs underlying the iPSC and MIST projects, as well as in the support that the researchers provided PD facilitators, there are several notable commonalities in the ways in which both programs approached the preparation of PD facilitators. Both research teams led activities in which the PD leaders analyzed the mathematical tasks and student solution strategies they would address in their upcoming PD sessions and viewed and analyzed video from their previous PD sessions. Similarly to how some PD programs described in Phase 1 engaged teachers in pedagogies of investigation and enactment (Grossman et al., 2009), these programs used pedagogies of investigation and enactment in the preparation of PD facilitators.

Pedagogies of investigation and enactment were typical components of other programs for preparing mathematics PD leaders, as well. For example, Elliott and colleagues in the Researching Mathematics Leader Learning (RMLL) project supported PD leaders learning to cultivate mathematically rich PD opportunities for teachers (Elliott et al., 2009). Their series of three 2-day RMLL seminars included activities such as working on a mathematical task, analyzing video cases of mathematics PD leaders engaging in the same task with teachers, and considering how to apply the ideas in the cases to their own facilitation practice—a sequence that in many ways parallels sequences used in PD settings to prepare teachers to work with students around mathematical tasks. Although studying leader learning was not a central focus, Project Pathways (Clark et al., 2008) also attended to preparation of PD leaders for their professional learning communities (PLC). PLC facilitators participated in a summer workshop, follow-up meetings, and weekly coaching sessions during which they received specific feedback on the facilitation of previous PLC sessions.

Many of the learning opportunities for PD leaders in these programs were similar to learning activities for teachers but focused on practices of leading PD and facilitating conversations among teachers. Jackson and colleagues (2015) explained that they "extrapolated from the literature on high-quality teacher PD and pre-service teacher education when designing supports for math leader learning" (p. 94). Given the thin research base on how to support PD leaders in designing and facilitating high-quality PD, it is likely that other designers of programs for preparing PD leaders used a similar approach. More research is needed on approaches to preparing PD leaders, as well as their impact, and specifically on the use of pedagogies of investigation and enactment in the preparation of PD leaders.

Approaches to Bringing Professional Development to Scale

In examining projects that implemented PD at scale, we found two different approaches used to bring these programs to multiple sites. Several projects developed a PD curriculum with accompanying materials for facilitators; others worked with school districts, states, or countries to build local capacity to conduct PD at scale.

的开发从本质上是有推广性的。一些专业发展课程关注提高教师的数学教学知识，比如发展数学思想（例如Schifter等，1999a，1999b），线性函数的教与学（Seago等，2004），以及几何的教与学（Seago等，待出版）。这些课程主要包括以下内容：供教师分析和解决的数学任务、处理相同数学任务的课堂教学案例或学生作业范例，促进者为专业发展领导者提供的支持材料如建议的日程、讲义、幻灯片。我们将主要描述两个已经被实证研究证实有效的专业发展课程。

发展数学思想（DMI）是一个为K-8年级教师设计的专业发展课程，它致力于帮助教师发展教学中会用到的专业数学知识。课程的七个模块强调了不同的数学思想，还包括一个促进者的指导手册、教师的个案记录本以及中小学数学课堂教学的录像片段。在典型的专业发展研习会中，教师需要解决数学问题，讨论文字的或基于录像的课堂片段，以及检查学生作业的范例。

贝尔、威尔逊、希金斯和麦蔻驰（2010）进行了一个准实验研究，比较了那些参加DMI和在同一地区的一切照常的教师的数学教学知识。所有参与研究的DMI的促进者都有促进DMI的经验。然而他们在学习DMI上的机会的广度是不同的，研究者通过总结对评估促进者关于DMI的经验广度的9个调查问题的回答创造了一个变量，这9个问题包括他们是否参加过DMI领导力协会，是否给另外一位推进者当过学徒或者是否写过关于他们的推进工作的案例。推进者和教师的调查结果表明DMI专业发展课程在实施中高度忠于专业发展开发者所重视的特征。尽管在专业发展项目的结构性上有一些变化，比如进行会议的频率和学校的课程内容，但是项目的核心方面，比如涵盖的所有内容、使用的家庭作业、提供给教师的书面反馈都没有改动。实验组教师比对照组教师在评估教学知识的多项选择问题上的前后测中稍微强一些，在一个结果开放的评估上明显地要好于对照组教师。另外，尽管促进者的DMI的经验广度对于多项选择题测试没有显著的调节作用，但是对于开放性测试而言它却有显著的调节作用。具体而言，在促进者有更多的DMI经验的地区，实验组和对照组的教师分数有更大的差异。这表明如果促进者有更多的机会学习DMI，他也能让教师学得更多。

几何的教与学（LTG）课程的专业发展材料是为5—8年级的教师而设计的，它由五个模块组成，旨在培养教师使用几何变换的观点来教相似形。每个模块都包含以下活动：教师探索数学；观看、分析和讨论录像案例；比较和对比不同案例的主要问题；与他们自身教学实践相联系。为促进者提供的资料包括详细的讲习会日程、幻灯片、课程图表、录像案例中所描述的以时间编码的课堂片段文本以及关键数学术语的现场指导。

在两个周期的测试和修正之后，LTG的基础模块在两个学年后由九个有经验的促进者在八个地区进行现场测试。在所有测试地区实验组的教师在几何内容知识的多项选择测试中都比对照组教师取得了显著性的更大进步。同样，在三分之二的基于课程内容的数学推理问题的前测和后测中，实验组教师表现明显更好。在三个嵌入式评价教师描述一个学生解决相似形问题的方法以及应用这个方法解决相关联问题的能力的录像分析上，实验组的教师也明显表现的更好。对于嵌入式评价的任何一个问题对照组教师在分数上没有显著的变化。由研究者开发的多项选择几何测试中，实验组教师的学生比对照组教师的学生分数更高。在写作本章时，基础模块和四个扩展模块的最终版本还没有被测试（Seago等，待出版）。

DMI和LTG都是高度具体化的专业发展项目，它们为教师描绘了特定的学习目标并提供专业发展材料（例如数学任务、带有学生成绩单的录像案例、学生作业的样本）及促进资源（比如，日程、幻灯片）以实施一整套预定的，旨在实现这些目标的专业发展经验。关于这些项目的有限研究为以下结论提供了初始证据，即这种高度具体化、资源密集型的专业发展课程，在实施过程中可以保持开发者设定的关键特征。此外，当由有经验的推进者可以严格实施时他们能使教师和学生的学习都提高。

开发局部专业发展能力。提高教与学的一些工作集

We discuss these two approaches and provide examples of each.

Professional development curricula. The creation of PD curricula is inherently about scalability, in that the curricula are designed to be used by multiple facilitators and in multiple sites. Several PD curricula focus on improving teachers' knowledge of mathematics for teaching, such as *Developing Mathematical Ideas* (e.g., Schifter et al., 1999a, 1999b), *Learning and Teaching Linear Functions* (Seago et al., 2004), and *Learning and Teaching Geometry* (Seago et al., in press). These curricula typically include mathematical tasks for teachers to solve and analyze; cases of classroom teaching or samples of student work that address the same mathematical tasks; and a facilitator's guide with support materials for the PD leaders such as suggested agendas, handouts, and PowerPoint slides. We highlight two PD curricula whose effectiveness has been investigated empirically.

Developing Mathematical Ideas (DMI) is a PD curriculum for K–8 teachers that focuses on helping teachers develop the specialized knowledge of mathematics used in teaching. The seven modules in the curriculum address different mathematical ideas and include a facilitator's guide, a casebook for teachers, and video clips of elementary and middle school mathematics classrooms. During a typical PD session, teachers solve mathematical problems, discuss written or video-based classroom episodes, and examine examples of student work.

Bell, Wilson, Higgins, and Mccoach (2010) conducted a quasi-experimental study comparing the mathematical knowledge for teaching of teachers who participated in DMI with that of a business-as-usual group of teachers at the same sites. All DMI facilitators who participated in the study had prior experience facilitating DMI. They varied, however, in their *breadth of opportunities to learn* DMI, a variable the researchers created by summing responses to nine survey questions assessing facilitators' breadth of DMI experiences, such as whether they had attended a DMI leadership institute, apprenticed to another facilitator, or written cases about their facilitation. Facilitators' and teachers' survey responses indicated that the DMI PD curriculum was implemented with a high degree of fidelity to the features valued by the PD developers. Although there were variations in some structural features of the PD program, such as frequency of meetings and the schools' curricular contexts, core aspects of the program, such as covering all the content, using homework, and providing teachers with written feedback, remained intact. The treatment teachers moderately outgained the comparison group from pretest to posttest on a multiple-choice assessment of knowledge for teaching and substantially outgained them on an open-ended assessment. In addition, although the facilitator's breadth of DMI experiences was not a significant moderator for the multiple-choice test, it was for the open-ended test. Specifically, the difference between gain scores of the treatment and comparison teachers was greater at the sites with facilitators who had more DMI experiences, suggesting that the more opportunities a facilitator had to learn about DMI, the more learning that facilitator was able to generate among teachers.

The *Learning and Teaching Geometry* (LTG) PD materials are designed for mathematics teachers in grades 5–8. They consist of five modules aimed at fostering mathematical knowledge for teaching similarity using a transformational view of geometry. Each module includes activities in which teachers explore the mathematics; view, analyze, and discuss video cases; compare and contrast issues across cases; and make links to their own instructional practice. Resources for facilitators include detailed session agendas, PowerPoint slides, lesson graphs, time-coded transcripts of the classroom episodes depicted in the video cases, and a field guide of key mathematical terms.

After two cycles of pilot testing and revisions, a field test of the LTG Foundation Module was conducted at eight sites with nine experienced facilitators, over 2 academic years. Across all the field test sites, treatment teachers made significantly greater gains than comparison teachers on a multiple-choice test of geometry content knowledge. They also made significant pretest to posttest gains on two of three questions on an assessment of mathematical reasoning embedded in the curriculum and on all three video analysis questions on an embedded assessment of their ability to describe a student's approach to solving a similarity problem and apply that approach to solving two associated problems. Comparison group teachers did not demonstrate significant changes in scores for any of the questions on either of the embedded assessments. The students of treatment teachers also made greater gains on a multiple-choice geometry test developed by the researchers than students of comparison teachers. At the time this chapter was written, the final version of the Foundation Module and the four Extension Modules had not been tested (Seago et al., in press).

DMI and LTG are both highly specified PD programs that delineate particular learning goals for teachers and provide professional development materials (e.g., math tasks, video cases with transcripts, samples of student work) and facilitation resources (e.g., agendas, PowerPoint slides) for the implementation of a predetermined

中在开发能够为教师提供持续的专业学习机会的学校系统的能力上。大多数情况下，专业发展是一个更大的教学改进工作的组成部分。我们发现了在地区、州或省和国家水平实施这些工作的例子。有些工作由系统本身启动和运行，例如，新西兰基础数学知识发展项目（NDP；Higgins & Parsons, 2011a, 2011b）和由加拿大安大略省教育部发起的大规模专业学习项目（Bruce, Esmonde, Ross, Dookie, & Beatty, 2010）。其他的项目，比如上一节讨论的MIST和iPSC，由学校、学区和系统外的研究伙伴合作进行。我们将NDP发展项目及其影响描述为将专业发展规模化的系统启动方法的一个例子。

NDP发展项目是从2000到2009年由政府资助的一个大规模的项目，旨在通过提高教师的专业知识、技能和信心来提高小学数学的教学和学习。新西兰97%的小学参加了这个项目，包括25 000多名教师和690 000名学生。专业发展是这些项目的核心。NDP专业发展项目关注三个目标：（1）改进教师的数学知识，（2）提高教师对于学生如何学习数学的理解，（3）提高教师关于如何表征一个数学概念的理解。为了实现这些目标，NDP发展项目的设计包括三个嵌套的领导层：（1）国家或者区域协调员，（2）学校外的专业发展促进者，（3）校本的数学教师领导者。

NDP专业发展项目旨在帮助教师使用一套三个互相联系的教学工具：（1）数字框架；（2）诊断访谈；（3）策略教学模型。数字框架是日益复杂的数学思维阶段的图解表征，它所起的作用是指导教学决策的学习进程。由教师来做的诊断访谈可以帮助他们确定学生的知识和策略、在数字框架中为学生定位并制定教学顺序。策略教学模型采用从数学理念的具体表征到意象，然后再到抽象的数学原理的过程，旨在促进适合不同数学思维阶段的学生的问题解决策略的显式教学。这个项目的开展是基于之前提到的澳大利亚把我也算在内（CMIT）的专业发展项目的经验。

NDP专业发展项目的一个重要部分是一个为期15周的集中的专业发展项目，这个项目由80到90名外部的促进者领导，由全国区域性团队组织。所有促进者都参加一年一度的为期三天的基础数学会议，为他们提供开发促进技能、发展专业知识网络的机会，并帮助确保在整个系统内举办工作坊的一致性。另外，NDP发展项目网站上也会为促进者提供所需的资源。由推进者组织的专业发展活动主要包括向教师介绍教学工具的工作坊和参观学校，在参观期间，他们展示诊断访谈、模拟教学策略、观察教师并为教师的教学实践提供反馈。此外，校本的数学领导教师可负责支持NDP发展项目提高教师知识，并且保持全校范围内对学生成绩的关注这一目标的达成。例如，在教师专业发展讲习会上发挥领导作用，收集、分析和报告学生成绩。

在项目中进行了大量的研究和评估。研究者进行了内容分析，使用了1000多份由教师、校长和推进者完成的调查问卷，并对21位教师进行了访谈以确定来自8个独立研究的数据集的主题。这些自我报告的结论是：这个项目对教师的知识和课堂实践有显著的影响。例如，一些教师表示参加专业发展项目改变了他们关于学生对数学概念理解的信念，并且改变了教师讲授这些概念的方式。学校内部的紧密联系和与外部推进者的持续接触是支持教师的极为有效的方式。形成性评价数据的使用，如数学能力诊断访谈，有助于维持和加强学校对提高学生成绩的关注。对使用诊断性访谈所得到的数据的量化分析表明，学生的成绩有所提升（Higgins & Parsons, 2009, 2011a, 2011b；New Zealand Numeracy Development Projects, 2010）。例如，永-洛弗里奇（2008）报告说，1—8年级学生的教师参与了项目，无论其性别、种族、民族或社会经济地位如何，他们在项目前到项目后的面试中都取得了实质性进展。

set of PD experiences that address those goals. The limited research on these programs provides initial evidence that such highly specified, resource-intensive PD curricula can be implemented with fidelity to the key features intended by the developers. Further, when implemented with fidelity by experienced facilitators, they can result in gains in teacher and student learning.

Building local professional development capacity. Several efforts to improve teaching and learning focus on building the capacity of school systems to provide sustainable professional learning opportunities. In most cases, the PD is one component of larger instructional improvement efforts. We found examples of such efforts implemented at the district, state or provincial, and national levels. Some of these efforts were initiated and run by the system itself, for example the New Zealand Numeracy Development Project (NDP; Higgins & Parsons, 2011a, 2011b) and the large-scale professional learning program launched by the Ontario, Canada, Ministry of Education (Bruce, Esmonde, Ross, Dookie, & Beatty, 2010). Other projects, such as the MIST and iPSC projects discussed in the previous section, were conducted collaboratively between the school and district and research partners external to the systems. We describe NDP and its impact as an example of a system-initiated approach to bringing PD to scale.

NDP was a large-scale, government-funded program implemented from 2000 to 2009 to improve the teaching and learning of mathematics in primary grades by improving the professional knowledge, skills, and confidence of teachers. Of the primary schools in New Zealand, 97% participated, including more than 25,000 teachers and 690,000 students. PD was the centerpiece of the project. The NDP PD focused on three objectives: (1) improving teachers' knowledge of mathematics, (2) enhancing their understanding of how students learn mathematics, and (3) enhancing their understanding of how to represent mathematical concepts. To accomplish these objectives, the NDP design included three nested levels of leadership: (1) national or regional coordinators, (2) PD facilitators external to the schools, and (3) school-based numeracy teacher leaders.

The NDP PD was designed to help teachers use a set of three interconnected pedagogical tools: (1) the *number framework*, (2) *diagnostic interviews*, and (3) the *strategy-teaching model*. The number framework, a diagrammatic representation of increasingly sophisticated stages of mathematical thinking, served as a learning progression to guide pedagogical decisions. Diagnostic interviews conducted by the teachers helped them identify students' knowledge and strategies, locate students on the number framework, and plan instructional sequences. The strategy-teaching model was designed to promote explicit teaching of problem-solving strategies appropriate for students at different stages of mathematical thinking, using a progression from concrete representations of mathematical ideas to mental imaging and then to abstract mathematical principles. The project built on previous experiences of the Australian Count Me In Too (CMIT) PD program, highlighted earlier in the chapter.

One major component of the NDP PD was an intensive 15-week PD program led by 80 to 90 external facilitators, organized in regional teams throughout the country. All facilitators attended an annual 3-day numeracy conference, which provided an opportunity for them to build their facilitation skills, develop professional networks, and help ensure consistency in the delivery of workshops across the system. In addition, resources for facilitators were provided on the NDP website. PD activities conducted by the facilitators consisted primarily of workshops to introduce teachers to the pedagogical tools and visits to schools during which they demonstrated diagnostic interviewing, modeled teaching strategies, observed the teachers, and provided feedback on their teaching practices. In addition, school-based numeracy lead teachers were responsible for supporting the NDP goals of enhancing teacher knowledge and maintaining a school-wide focus on student achievement, for example, by taking a lead role in teacher development sessions and collecting, analyzing, and reporting student achievement data.

Numerous research and evaluation studies were conducted on the project. Researchers conducted a content analysis, using over 1,000 surveys completed by teachers, principals, and facilitators and interviews with 21 teachers to identify themes across data sets from eight separate studies. They concluded, based on these self-reports, that the project significantly influenced teachers' knowledge and classroom practices. For example, teachers reported that participation in the PD led to changes in their beliefs about students' understanding of number concepts and in the ways that they taught these concepts. Strong within-school networks and continued access to external facilitators were particularly effective ways of supporting teachers. Use of formative assessment data, such as the numeracy diagnostic interviews, contributed to sustaining and strengthening a school's focus on improving students' achievement. Quantitative analyses of data gathered using the diagnostic interview demonstrated improvements in student achievement (Higgins & Parsons, 2009, 2011a, 2011b; New Zealand

几个其他系统的教学改进项目与NDP有相同的特点。例如，中学数学人才发展项目（Balfanz, Mac Iver, & Byrnes, 2006）就把全校范围内使用基于研究的教学材料和一个包括暑期及月度的专业发展工作坊，以及课堂辅导支持系统结合起来。这个项目达到了一个中等程度的实施水平，该项目中教师的学生在多项成就测试中的表现均优于控制组学校的学生。本章前面讨论过的CMIT项目是一个专业发展项目的例子，它提出了阶段1的研究问题，也进行了阶段2的研究，比如鲍里斯（2011）开展的一项持续性的纵向研究。他识别出了对于项目很重要的几个因素，其中影响最大的是驻扎在学校的促进者的存在，他们被认为是对项目所需的教学工具知识和学生学习知识有很深了解的人。基于对MIST项目（在阶段2的关于专业发展领导者准备的部分进行了讨论）中区域教学改进工作的初步分析，科布及其同事（Cobb & Jackson, 2011; Cobb 等, 2013）推测大规模的教学改进需要5个相关因素的协调，其中两个因素为：（1）由学校外部的促进者领导的专业发展工作坊；（2）由数学教练提供的工作嵌入式支持。科布及其同事发现，教师接触具有教学专长的教练是教学改进的最强的预测因素之一，这一发现表明支持系统的重要性是显而易见的。

在我们所讨论的系统级的项目中，也有一些有趣的新兴模式。作为更为宽泛的教学改进工作的一部分，专业发展包括多层次的教师支持：暑期学院、学年中的工作坊和课堂辅导的结合。这些层面的支持是由外部专业发展促进者和学校内的领导者提供的。为了使项目持续发展，他们使用了"培训者的培训者"模式，来为让教师成为校本领导者而做准备。在所有的案例中，研究者都认识到有效的专业发展对于项目取得成功的重要性。

关注数学专业发展的背景

为了使专业发展项目可以成功地大规模实施，最新的研究表明在保证核心原则一致性的前提下，必须要考虑当地背景下关键的社会和政策特征（Cobb, McClain, Lamberg, & Dean, 2003）。因此，考虑与专业发展规模化相关的问题时，专业发展设计者和政策制定者都需要明确并计划学校的社会和政策背景是如何影响专业发展的实施与影响的。自从索德（2007）的评述发表以后，研究者开始关注这些问题，在这个领域里有一些少量的但却是不断增长的研究。这些研究可以分为两类，第一类是关注背景在影响教师参与专业发展中的作用，第二类关注背景如何影响专业发展项目的实施和有效性。我们用这些问题来组织数学专业发展项目的背景研究。

背景和专业发展中的教师参与。之前的研究表明，教师愿意自己选择他们的专业发展（Garet 等, 2001）。因此，了解那些能够预测数学教师参与专业发展的因素对于设计出教师会选择参与的专业发展项目是至关重要的。德西蒙、史密斯和菲利普斯（2007）在分析教师参与专业发展项目中，研究了政策背景。他们使用政策归因理论来描述专业发展项目中政策背景的特性（Porter, Archbald, & Tyree, 1990; Porter, Floden, Freeman, Schmidt, & Schwille, 1988）。这个理论表明，政策背景归因可以促进政策实施，比如权威（政策对于那些实施它的人是否是可以接受且有说服力的）、问责制（奖励和制裁）和一致性（某项政策与同一学校、地区和州的其他政策保持一致的程度，以及与其实施者的观念和信仰保持一致的程度）。

德西蒙等（2007）对国家教育统计中心的学校和教师调查（SASS）中的高中数学和科学教师的全国样本数据进行了二次分析，他们使用三级分层线性模型（HLM）来预测在不同类型的专业发展活动中教师的参与程度。和教师的稳定性（由低流失率来测量）一样，教师在学校政策、课堂实践和计划与呈现专业发展方面的影响预测了聚焦于内容的专业发展项目的参与程度。与此相反，与专业发展相关的正式问责制（例如关于专业发展如何

Numeracy Development Projects, 2010). For example, Young-Loveridge (2008) reported that students in their schooling years 1–8 whose teachers participated in the project made substantial progress across levels of the framework from pre-project to post-project interview, regardless of their gender, race, ethnicity, or socioeconomic status.

Several other system-level instructional improvement efforts have features in common with the NDP. The Talent Development Middle School Mathematics Program (Balfanz, Mac Iver, & Byrnes, 2006), for example, combined schoolwide use of research-based instructional materials with a support system that included summer and monthly PD workshops and in-class coaching. The program achieved a moderate level of implementation, and students of teachers in the program outperformed students from control schools on multiple measures of achievement. The Count Me In Too project, discussed earlier in the chapter as an example of a PD program asking Phase 1 research questions, also conducted Phase 2 studies such as that by Bobis (2011), which focused on sustainability. Bobis identified several factors that were important to the program, the most influential of which was the presence of facilitators who were based in the schools and were perceived as having deep knowledge of the program's pedagogical tools and student learning. On the basis of their initial analyses of districts' instructional improvement efforts in the MIST project, discussed in the Phase 2 section on the preparation of PD leaders, Cobb and colleagues (Cobb & Jackson, 2011; Cobb et al., 2013) conjectured that instructional improvement at scale requires coordination of five interrelated components, two of which are (1) PD workshops led by facilitators external to the schools and (2) job-embedded support provided by math coaches. The importance of a system of support is evident in Cobb and colleagues' finding that teachers' access to a coach with instructional expertise was one of the strongest predictors of instructional improvement.

There are interesting emerging patterns across the system-level programs we discussed. The PD, which was one component of a broader instructional improvement effort, consisted of multiple tiers of support for teachers, including combinations of summer institutes, workshops during the school year, and in-classroom coaching. These levels of support were provided by both external PD facilitators and school-based leaders. To make the programs sustainable over time, a "trainer-of-trainers" model was used to prepare teachers to become school-based leaders. In all cases, the researchers noted the importance of effective PD to the success of the program.

Attending to the Context of Mathematics Professional Development

For PD programs to be successfully implemented at scale, recent research has suggested that they must take into account key social and policy features of local settings while maintaining integrity to their core principles (Cobb, McClain, Lamberg, & Dean, 2003). Therefore, to consider issues related to bringing PD to scale, both PD designers and policymakers must understand, and plan for, how the social and policy contexts of schools influence PD implementation and impact. Since the publication of the Sowder (2007) review, researchers have begun to focus on these issues, and there is a small but growing set of studies in this area. These studies can be grouped into two types, with the first focusing on the role of context in influencing teacher participation in PD and the second on how context can influence the implementation and effectiveness of PD programs. We use these issues to organize the studies that examine the context of mathematics PD programs.

Context and teacher participation in PD. Prior research has shown that teachers tend to choose their own professional development (Garet et al., 2001). Therefore, understanding factors that predict mathematics teachers' participation in PD is critical for designing PD that teachers will choose to attend. Desimone, Smith, and Phillips (2007) examined the policy context in their analysis of teacher participation in PD. They used *policy attributes theory* to characterize the policy environment of PD programs (Porter, Archbald, & Tyree, 1990; Porter, Floden, Freeman, Schmidt, & Schwille, 1988). This theory suggests that attributes of the policy environment such as authority (whether policy is accepted and persuasive to those who have to implement it), accountability (rewards and sanctions), and consistency (the extent to which a policy is aligned with other policies in the same school, district, and state, and with the perceptions and beliefs of its implementers) can increase policy implementation.

Desimone et al. (2007) conducted a secondary analysis of the National Center for Education Statistics' Schools and Staffing Survey (SASS) data from a national sample of high school mathematics and science teachers, using a three-level hierarchical linear model (HLM) to predict teachers' level of participation in different types of PD activities. Teachers' influence over school policy, classroom practices, and planning and presenting PD predicted participation in content-focused PD, as did teacher stability, measured by low turnover rate. In contrast, formal accountability linked to PD participation

影响教育实践或者学生成绩的正式评估）与参与的增长没有关联。德西蒙等（2007）认为情境特征与专业发展的参与度呈正相关，这些背景特征如教学的自主性和积极的合作环境。

通过将SASS数据与州教育政策属性的项目数据库合并，菲利普斯、德西蒙和史密斯（2011）研究发现：学校层面和州层面的政策背景——例如州标准和州评估的一致性——对于教师参加一个高利害学科（数学）的预测性比参加一个低利害学科（科学）的预测性要更好。这一结果表明，专业发展参与的政策背景受问责系统力度的影响，教师对正向和负向刺激均会有所回应。

背景与专业发展实施和有效性。在最初的文献综述中，我们只确定了几个关于社会和政策背景如何影响数学专业发展项目实施和有效性的研究。这些研究使用不同的研究方法来回答背景问题，比如质性社会网络分析、案例比较研究和设计实验。接下来我们将讨论这些研究。

科伯恩和拉塞尔（2008）使用质性社会网络分析，来研究在采用一套新课本的两个学区中，学区政策如何影响小学数学教师获得教学支持的，其中包括专业发展。为了支持课程采用，每个学区设计一个包括地区范围因素和学校层面辅导的专业发展项目。研究关注背景如何影响地区专业发展在不同学校的跟进方式。这个研究包括每个学区的四所小学，是根据学校所在学区数学指导者对其不同水平的专业团体和教师专业知识的建议而有目的地进行抽样的。

作者认为学校和学区政策可以影响教师所获得的教学支持。例如，学校领导者通过在学校分配辅导资源的方式影响教师辅导知识的获得（例如有多少教师可以被辅导）。在教练观察课堂并定期提供反馈的地方，教练和教师之间有着更深入的互动。另外，被积极辅导的教师和同事也有更深入的交流，比如即使在教练不在的情况下，他们也和同事讨论数学的本质或学生是如何学习的问题。作者认为，互动深度的不同，是因为有更多机会接触教练的教师更有可能吸取不同类型的专业发展的经验套路，并且通过和教练的交流使其具体化。

研究者还发现，校长通过谈论对新采纳的、关注探究的课程实施，影响着专业发展的内容对教学实践影响的程度。通过分析每个学区的两所学校，研究者们发现：在校长关于课程的言论与信息（例如，强调考试准备，或如何促使学生获得更有效率的问题解决策略）与该学区志向更高的学生的学习目标不一致的地方，教师们关于课程的言论也与这些目标不太一致。不同地区间数学领导者和学校领导者之间的对话频率和内容，也影响教师们是否进行数学更深层次的讨论（Stein & Coburn, 2008）。

通过将不同学校教师案例比较研究嵌入到一个设计研究课题中，科布和同事们研究学校和学区背景（或称作教学的制度背景（Cobb, McClain 等, 2003））是如何影响教师参与专业发展项目的方式，以及专业发展项目对他们课堂实践的影响程度。根据科布、赵和迪恩（2009）的研究，制度设定包括地区和学校为教学所实施的政策，比如课程资料的采纳和使用指南；教师需要对哪些事情和人负责；专业发展、辅导和非正式的专业网络所提供的社会支持；来自教学领导者的协助。案例研究的数据来自于一个五年的教师发展实验，实验设计的目的是为了支持来自同一个市区的五所学校的初中数学教师改进统计数据分析的课堂教学实践。作者运用设计实验的方法（Cobb, Confrey, diSessa, Lehrer, & Schauble, 2003）来验证和修正关于中学统计学教师学习所需支持的猜想。

基于设计实验的每一个环节的分析，科布和同事们认为设计实验的方法可以处理专业发展设计过程中不同阶段背景的复杂性，包括初始设计阶段和在每个周期的设计修改过程中，假设将根据所学知识进行测试和修订的阶段。例如，政策和领导力的文献显示：教师间促进教学实践提高的有效合作需要一个共享的高质量教学的愿景。对于背景这一方面的预期，使得研究者在专业发展项目的设计阶段就将该学区教学愿景的评估写进它们的专业发展项目中（Cobb 等, 2009）。一旦设计实验启动，研究者就开始理解背景的另一个方面——缺少数学

(e.g., formal evaluation of how PD influences instructional practices or student achievement) was not associated with increased participation. Desimone et al. (2007) concluded that contextual features such as instructional autonomy and a positive environment for collaboration are positively associated with PD participation.

Building on these results by merging the SASS data with a project-created database of state education policy attributes, Phillips, Desimone, and Smith (2011) found that the policy context at both the school level and the state level—for example, alignment between state standards and state assessments—is more predictive of teacher participation in PD in a high-stakes subject (mathematics) than a low-stakes subject (science). This finding suggests that the policy context of PD participation is influenced by the strength of the accountability system; teachers respond to both positive and negative incentives.

Context and PD implementation and effectiveness. Our initial review of the literature identified a few research projects that examined how social and political contexts influence the implementation and effectiveness of mathematics PD programs. These studies used a variety of research methods to address issues of context, such as qualitative social network analyses, comparative case studies, and design experiments. We discuss these studies in what follows.

Coburn and Russell (2008) used qualitative social network analysis to study how district policies influence elementary mathematics teachers' access to instructional supports, including PD, in two districts implementing a new textbook adoption. To support the curriculum adoption, each district designed a PD program that included a district-wide component and coaching at the school level. The research focused on how context influenced the ways in which the district PD was followed up in different schools. The study included four elementary schools per district, purposively sampled on the basis of recommendations from the district directors of mathematics for their varying levels of professional community and teachers' expertise.

The authors found that school and district policy can influence teachers' access to instructional supports. For example, school leaders influenced teachers' access to coaching expertise by the ways in which they allocated coaching resources in their schools (e.g., how many teachers would be coached). Where coaches observed classrooms and regularly provided feedback, there was a greater depth of interaction between teachers and coaches. In addition, teachers who were actively coached tended to have greater depth of interaction with their colleagues, such as discussing the nature of mathematics or how students learn, even when the coach was not around. The authors suggested that the difference in depth of interactions was because teachers with greater access to coaches were more likely to draw on the types of routines experienced in the PD and reified through interactions with the coaches.

The researchers also found that principals influenced the degree to which the content of PD affected instruction by the way in which they talked about the implementation of the newly adopted, inquiry-focused curriculum. Analyzing two schools in each district, they found that where principals' talk and messages about the curriculum (e.g., emphasizing test preparation or how to move the students on to more efficient problem-solving strategies) were incongruent with the districts' more ambitious student learning goals, teachers' talk about the curriculum was also less congruent with these goals. The frequency and content of conversations between district mathematics leaders and school leaders also influenced whether teachers engaged in deeper discussions about the mathematics (Stein & Coburn, 2008).

Embedding comparative case studies of teachers from different schools into a design research study, Cobb and colleagues examined how school and district context, or what they called the *institutional setting of teaching* (Cobb, McClain et al., 2003), influenced the ways that teachers participated in PD and the degree to which the PD influenced their classroom practice. According to Cobb, Zhao, and Dean (2009), the institutional setting includes policies implemented by districts and schools for instruction, such as the adoption of curriculum materials and guidelines for their use; what teachers are held accountable for and to whom they are held accountable; the social supports provided through access to PD, coaching, and informal professional networks; and assistance from instructional leaders. Data for the case studies came from a 5-year teacher development experiment designed to support a group of middle school mathematics teachers from five schools in the same urban district in improving their classroom practices for teaching statistical data analysis. The authors applied design experiment methodology (Cobb, Confrey, diSessa, Lehrer, & Schauble, 2003) to test and revise conjectures about supports for teacher learning of middle school statistics.

On the basis of analysis conducted during each iteration of the design experiment, Cobb and colleagues concluded that the design experiment approach can address the complexity of context at multiple stages of the PD design process, including the initial design and the design revisions undertaken in each cycle as conjectures are tested and revised based on what is learned.

教学领导能力是如何阻碍教师在专业发展项目中实施所学习到的内容。作为修正专业发展设计的一部分，研究者和参与项目的教师一起合作，制定支持措施，帮助他们的学校领导了解有效数学教学所涉及的挑战。这个设计修正对于专业发展影响教师教学实践至关重要。

意识到当地环境对于一个专业发展项目会产生一定的挑战，斯特格达、克斯廷、吉文和斯蒂格勒（2011）研究了政策和组织性背景是如何影响专业发展项目的有效性的。设计专业发展项目主要是为了解决三个障碍，作者假定这些障碍在阻止美国教师有效实施"建立关联"问题上至关重要：（1）缺少学科知识和学科教学知识，（2）缺少模型（即可供选择的教学策略知识以及如何在课堂中实施这些策略的知识），（3）缺少背景支持（即支持有效使用"建立关联"问题的教学材料）。这些障碍在一系列的三个专业发展模块中被解决了，每一个模块的内容（即分式、比和比例、代数式和方程）都在将要讲授给学生之前进行。每个模块包含三个阶段：（1）内容探究，（2）课程分析，（3）联系实践。对于三个模块中的每一个，内容探究和课程分析两阶段每个用一整天讨论，通常相隔一周。接下来是一个教学窗口，其中教师讲授他们分析的课程，并收集和分析学生的活动，然后在促进者领导的学校会议上分享学生的活动。

专业发展项目的有效性在一个田野试验中得以检验，试验中的被试来自五个低表现水平的城区内的学校里的64位六年级的教师，他们被随机地分到实验组和对照组。研究测量的内容有实施的忠实度、教师的知识和实践（包括用录像记录一个建立关联的数学问题），以及学生数学学习（包括学区范围内的季度测评和年终的州测评）。忠实度测验显示全天的、跨场地的专业发展项目出席率更高，但是在学校范围内的共享会议会出现更多的问题。教师来开会的时候经常还没有教过目标课程，并且很少带来学生的作业。实验组和对照组教师数据的比较显示：专业发展项目对教师知识或教学实践没有显著影响。

研究者猜想教师感受到学校和地区管理者将项目强加到他们身上，这样会伤害他们的积极性，同时缺乏来自学校管理者和教练的支持，也是导致项目无效的一个原因。例如，一些教师"认为这个项目和它的研究同学校与地区管理者强加在他们身上的很多其他事情一样，没有给教师机会表达他们的专业需要"（Santagata等，2011，第11页）。其他教师担心自己的测试成绩会在整个学区共享，因而缺乏信任。研究者还提到，对于某些校长和学区管理者而言，对项目的支持"不是一直都在"（第11页）。研究者认为教师教学实践或学生成绩没有得到提高的部分原因在于未能将专业发展与教师的感知需求和管理者的优先事项适当地结合起来，而这是学区背景的两个重要方面。

阶段2研究的讨论

博尔科（2004）报告：一小部分项目虽然实现了大范围的实施，但是"并没有产生任何能够被充分证明可以被多个推进者或在多个背景下完美实施的专业发展课程。"（第10页）。在她的评述之后，数学研究界在这一领域持续取得了一些进步。关于DMI和LTG项目的研究证明，课程为本的专业发展项目可以成功地被不同的促进者在不同的地区实施。NDP和iPSC的研究显示：有可能建立学校系统的能力，为教师提供可持续的专业学习机会。

人们对于支持和阻碍专业发展规模化的政策和组织背景方面的理解也取得了进步。对专业发展背景的综述研究显示：教育者和政策制定者不应该期待对不同背景的教师都产生同样影响的专业发展模型——当讨论科学专业发展的质量时，S.M.威尔逊（2013）也提出过这个问题。教师教学的自主权、与那些了解专业发展在努力帮助教师学习什么的专家接触、合作的机会、组织的氛

For example, the policy and leadership literature suggests that a shared vision of high-quality instruction is needed for collaboration between teachers to be effective in leveraging instructional improvement. Anticipating this aspect of the context allowed the researchers in this study to build an assessment of the district's instructional vision into their PD during the design phase (Cobb et al., 2009). Once the design experiment was underway, the researchers came to understand how an additional aspect of context—lack of instructional leadership capacity in mathematics—inhibited teachers' implementation of what they learned in the PD. As part of revising the PD design, the researchers collaborated with the participating teachers in formulating supports for developing their school leaders' understanding of the challenges involved in teaching mathematics effectively. This design revision was seen as critical for the PD to affect teachers' instructional practice.

Signaling the challenges that local context can have on the implementation of a PD program, Santagata, Kersting, Givvin, and Stigler (2011) examined how policy and organizational context can influence the effectiveness of a PD program. The PD in focus was designed to address three barriers the authors hypothesized to be most critical in preventing U.S. teachers from implementing "making connection" problems effectively: (1) lack of content and pedagogical content knowledge, (2) lack of models (i.e., knowledge of alternative instructional strategies and how to implement them in the classroom), and (3) lack of contextual support (e.g., instructional materials that support effective use of "making connection" problems). These barriers were addressed in a series of three PD modules, each timed to be covered immediately before the content (i.e., fractions, ratio and proportion, and expressions and equations) would be taught to students. Each module included three phases: (1) content exploration, (2) lesson analysis, and (3) link to practice. For each of the three modules, content exploration and lesson analyses were each focused on for a full pullout day, typically a week apart. This was followed by a teaching window, during which teachers taught the lesson they had analyzed, collected and analyzed their students' work, and shared student work in facilitator-led meetings at their school.

The effectiveness of the PD program was tested in a field trial with 64 sixth-grade teachers from five low-performing inner-city schools randomly assigned to treatment and control groups. The study measures included fidelity of implementation, teacher knowledge and practice (including video recording a making-connection mathematical problem), and student mathematics learning (including district-wide quarterly assessments and the end-of-year state assessment). The fidelity measures showed that attendance was high at full-day, across-site PD but more problematic at school-based sharing meetings. Teachers often came without having taught target lessons and rarely brought student work. A comparison of outcome data for teachers in the treatment and control groups showed that the PD program did not significantly affect teacher knowledge or practices.

The authors conjectured that teachers' perceptions that school and district administrators imposed the program on them hurt teachers' buy in and that a lack of support from some school administrators and coaches contributed to null effects. For example, some teachers "perceived the program and its research as one of the many things being imposed on them by school and district administrators, who did not provide opportunities for teachers to express their professional needs" (Santagata et al., 2011, p. 11). Other teachers expressed concerns that their individual test scores would be shared with the district, contributing to a lack of trust. The authors also noted that for some principals and district administrators, support for the program "was not always present" (p. 11). The authors suggested that the lack of improvement in teachers' instructional practices or their students' achievement may be partially explained by the failure to adequately align the PD with teachers' perceived needs and administrators' priorities—two important aspects of the district context.

Discussion of Phase 2 Research

Borko (2004) reported that her literature search identified a small number of projects that had achieved widespread enactment but "did not yield any professional development programs for which there is adequate evidence that they can be enacted with integrity by multiple facilitators or in multiple settings" (p. 10). The mathematics research community has made consistent progress in this area since her review. Research on the DMI and LTG projects has provided evidence that curriculum-based PD programs can be successfully implemented by multiple facilitators at multiple sites. Investigations of NDP and iPSC have demonstrated that it is possible to build the capacity of school systems to provide sustainable professional learning opportunities for teachers.

Progress has also been made in understanding the political and organizational contexts that allow or inhibit the scaling of PD. An implication of the reviewed studies of PD context, however, is that educators and policy mak-

围以及来自地区和学校领导的支持都可以影响教师对专业发展项目的参与度,以及专业发展对于他们教学实践的效果。尽管关注背景可以看作研究阶段2正在发展的一个领域,但很明显,专业发展项目的设计应该把背景因素考虑进去。研究者和开发者开始认识到,当外部开发的专业发展项目没有把当地的教师工作背景考虑进去时,就更有可能在实施中遇到问题。与此同时,那些设计中从不同方面考虑了当地背景的项目,则更可能被成功地实施。

解决阶段2问题的不同类型的项目,显示了成功实施大规模专业发展项目中的几个重要因素:对于专业发展目标知识和更大规模的教学改进工作有深入了解的促进者、专业发展领导者的准备、持续不断的支持、以及在专业发展设计中将当地的社会和政策背景考虑进去的能力。然而,进一步的研究需要明确具体且重要的促进行动、背景特征、专业发展领导者的支持、以及确定这些是如何影响专业发展项目实施的。

跨专业发展项目的比较

博尔科(2004)的研究把旨在提供关于"定义明确的专业发展项目的实施、效果和资源需求"的比较信息的专业发展研究归类为阶段3的研究(第11页)。基于这个定义,如果他们检查了来自多个数学专业发展项目的数据,我们就把这些研究纳入到这节。这个广义的定义既包括使用随机对照试验(RCTs)的两个或多个专业发展项目的直接比较,也包括几个单独专业发展项目的综合分析。我们只评述了三个研究。这也反映了阶段3研究在数学专业发展研究领域的匮乏。进一步地,由于三个研究在目的、测量结果和比较方法方面有所不同,所以很难将它们的研究发现综合起来。接下来,我们将单独评述每一个研究,并检查他们开始数学阶段3研究时所描述的框架。

专业发展研究的综述

第一个比较专业发展项目的研究是一个科学、技术、工程和数学(STEM)领域中的170个关于性别平等项目研究结果的综述,这些项目在1993到2001年间受到资助(Battey, Kafai, Nixon, & Kao, 2007)。作者采用个案调查的方法分析项目报告,并回答了三个研究问题:STEM项目中基于性别平等的专业发展是什么样的?怎样才能成功地实施?如何成功地持续?专业发展项目与以下这些项目进行比较:与探究有关的、长期可持续的以及将专业发展与课堂整合的项目。从这个比较中所获得的主要结论是,包含以下四个部分的项目对探究实践的课堂实施的影响始终大于那些没有包含全部的四个部分的项目,其中的四个部分分别是:对STEM中性别平等的意识、对STEM中性别平等的最佳实践、主题培训以及探究重点。在综合研究中,只有12%的基于性别平等的STEM的专业发展项目包含了这四个组成部分的组合。

我们研究的第二个综述将专业发展的导向框架应用到了14个科学和数学的专业发展项目中(Marra 等, 2011)。为了和州的要求一致,所有这些专业发展项目都是为了提高参与者对数学或科学内容,教学和学习的理解。同时,项目的设计还要满足学校的需要,包含工作嵌入的成分。其中的13个项目包括暑期学校,所有的项目都会举办不同类型的学年跟进会议。总之,这些项目服务了4—8年级的369名教师。作者研究了不同导向的项目(活动驱动、内容驱动、教法驱动、课程材料驱动和需要驱动)和教师报告结果的差异(对于教学实践的影响、专业发展对于教学实践的贡献以及教师在学科知识和教学知识中的信心)之间的关系。

14个项目结果的数据主要来源是参与者对由研究者开发的一项调查的回答,该调查在夏天结束时对所有参与者进行在线管理,并在每个专业发展项目结束时再次进行管理。教师们对他们的教学实践在未来一年如何因专业发展而改善的看法进行了评估。他们需要回答12个有关教学实践的问题,比如,内容知识的潜在改进或提高学生的学习动机。问题采用从没有到非常的4分等级评价。作者使用他们的导向框架来分析这些结果,发现来自平衡的专业发展导向项目(即支持教师学习科学或数学内容的同时也支持他们学习恰当的数学策略的专业发展项目)的教师报告说,他们打算在教学实践上比其他

ers should not expect a PD model to have the same influence on teachers across settings—an issue S. M. Wilson (2013) suggested when discussing the quality of science PD. Teachers' instructional autonomy, access to someone who is an expert in what the PD is trying to help teachers learn, collaborative opportunities, organizational climate, and support from district and school leaders can affect teachers' participation in PD and the impact of the PD on their instructional practice. Although the focus on context can be considered a developing area of Phase 2 research, clearly the design of PD programs needs to take context into account. Researchers and developers are beginning to understand that externally developed PD programs are more likely to face implementation problems when they do not take into account the local context of teachers' work, and, at the same time, that programs designed to consider the local context in various ways can be successfully implemented.

The various types of projects addressing Phase 2 questions suggest several factors that are important to successful large-scale implementation of PD programs: facilitators with deep knowledge of the PD goals and the larger instructional improvement efforts, preparation and ongoing support for PD leaders, and the ability to take local social and policy contexts into account in the design of the PD. Additional research is needed, however, to identify specific important facilitation practices, contextual features, and support for PD leaders and to determine how they impact PD implementation.

Comparing Across Professional Development Programs

Borko (2004) categorized as Phase 3 PD research the studies that aim to provide comparative information about the "implementation, effects, and resource requirements of well-defined professional development programs" (p. 11). On the basis of this definition, we included studies in this section if they examined data from multiple mathematics PD programs. This broad definition allowed for direct comparisons of two or more PD programs using randomized control trials (RCTs), as well as syntheses across studies of individual PD. Only three studies were reviewed. This small number reflects the paucity of Phase 3 studies within the field of mathematics PD research. Further, differences in the three studies regarding objectives, outcomes measured, and methods of comparison make it difficult to aggregate findings across them. In what follows, we review each study separately and examine the frameworks they have begun to delineate for conducting Phase 3 research in mathematics PD.

Syntheses of PD Studies

The first study comparing PD programs is a synthesis of findings from 170 PD programs on gender equity in the areas of science, technology, engineering, and mathematics (STEM) that were funded between 1993 and 2001 (Battey, Kafai, Nixon, & Kao, 2007). The authors used case survey methods to analyze project reports and answer three research questions: What does PD on gender equity in STEM look like? How can it be successfully implemented? How is it successfully sustained? The PD programs were compared in relation to student engagement in inquiry, sustainability over the long term, and integration of the PD and the classroom. The key finding from this comparison is that programs that included awareness of gender equity in STEM, best practices in gender equity in STEM, subject-matter training, and an inquiry focus consistently affected classroom implementation of inquiry practices more than programs that did not include all four of these components. Only 12% of the PD programs on gender equity in STEM examined in the synthesis included this combination of all four components.

The second synthesis we examined applied a PD orientation framework to 14 science and mathematics PD projects (Marra et al., 2011). In accord with state requirements, all these PD programs addressed the need to improve participants' understanding of mathematics or science content, teaching, and learning; were designed to meet school needs; and included job-embedded components. Thirteen of the projects included a summer institute, and all projects held some type of school-year follow-up sessions. Together, they served 369 teachers in grades 4–8. The authors examined the relationship between projects of differing orientations (activity driven, content driven, pedagogy driven, curriculum-materials driven, and needs driven) and with differences in teacher-reported outcomes (impact on teaching practices, contributions of PD aspects to teaching practice, and teacher confidence in subject matter and teaching knowledge).

The primary sources of outcomes data for the 14 projects were participants' responses on a survey developed by the researchers, administered online to all participants at the end of the summer and again near the end of each PD project. Teachers rated their perceptions of how their teaching practices would improve in the coming year due to PD. They responded to 12 items regarding teaching practices, such as potential improvements in content knowledge or increasing student motivation, on a 4-point scale from *none* to *very much*. The authors analyzed the results using their orientations framework and

项目的老师有更大的改进。另外，来自平衡的专业发展导向项目的教师还报告说，相比于来自活动导向或教法导向的专业发展项目的教师，专业发展对他们的专业实践的贡献更大。

随机对照试验

第三组论文介绍了初中数学专业发展影响研究（Garet等，2010，2011）的结果。在这个研究中，两个提供者"美国的选择"和"皮尔森成就解决方案"使用他们自己的专业发展材料，根据教育科学机构（IES）提供的旨在提高教师对有理数教学能力的框架设计了专业发展项目。为了满足教育科学机构的规范，两个提供者独立设计了他们的干预措施。两个项目的设计都是为了发展教师的数学学科知识和学科教学知识。这些项目由68小时的接触时间组成，包括一个为期3天的暑期学校和一系列在学年中进行的为期一天的跟进研讨班。校内辅导紧随着每个研讨日，辅导提供五次为期二天的学校参观，共持续十天的时间。每个专业发展提供者被分配到参与研究的12个学区中的六个学区去工作。这种分配是为了平衡不同提供者之间使用格伦科/PH数学和关联的数学项目课程的地区安排的均衡。

由两个提供者提供的专业发展活动包括：教师单独或小组合作解决数学问题的机会、用简短的口头报告解释他们是如何解决问题的、接收关于他们如何解决问题和提出解决方案的反馈、参与学生对于有理数内容常见的误解的讨论以及设计在随后的辅导参观期间他们将要讲授的课程。辅导部分是为了帮助教师将在学会和研讨班中所学的材料应用到他们的课堂实践而设计的。

尽管两个专业发展提供者都包含这些设计特征，但他们设计的专业发展在一些具体细节上还是有所不同。例如，在"美国的选择"专业发展环节中，教师被要求独立或以小组形式解决一组数学问题。这组问题的设计是为了明确或巩固重要的有理数概念的定义，以及说明学生在有理数中常见的理解误区。促进者被指示在随后的结构化讨论中强调这些相同的想法。相反地，"皮尔森成就解决方案"使用一个单独的、更开放的问题或任务来建构每一个专业发展环节。每项任务都是为了引出多种方法，并支持对核心思想、常见的学生方法以及与任务相关的潜在误解的扩展讨论而设计的。在这些讨论之后，教师将合作设计他们如何讲授与这个任务有关的课程。

在分析了来自两个提供者的联合数据后发现，研究者报告实施的忠实度在研究的第一年很高。即暑期学校、研讨班和辅导的时间平均为67.6小时，教师报告平均参加了83%的暑期学校和研讨班。然而，参与专业发展项目对于教师数学学科知识或学科教学知识的影响没有显著的统计性效果。研究者还开发了一个基于数学教学各个方面的和专业发展目标有关的教学实践的测量量表。在暑期学校和研讨班的八个预计日程的五个进行完了以后，他们观察对照组和实验组教师。他们发现专业发展项目对教师参与引导学生思考的活动的频率有统计上的显著积极影响，但教师在使用不同的表征方面（比如数轴、比率表格、面积模型）没有统计性的显著差异。在教师参与关注数学推理的活动频次上（比如，教师证明一个步骤或解答或让学生去证明或解释的次数）也没有统计性的显著差异。同样地，在由"西北评价协会"实施的一个有理数定制化的评价中，学生得分的表现也没有显著差异。

尽管把专业发展项目的影响与不同设计元素作比较，使得阶段3的研究是可取的，但是作者建议"直接比较两个提供者亚组之间的影响结果是不合适的，因为研究区域不是随机分配给提供者的"（Garet等，2010，第51页）。然而，在此我们要提及，两个提供者影响的大小是有不同的。与"皮尔森成就解决方案"的样本相比，在"美国的选择"所使用的样本中，对照组和实验组的教师在研究者开发的教师教学实践的量表中的表现差异更大。具体而言，由"美国的选择"管理的专业发展项目对教学实践的三分之二的量表都有统计上的显著影响："教师引导学生思维"量表（效应值=0.63），以及"教师使用表征"量表（效应值=0.60）。但是没有一个项目在第三

found that teachers from projects with a balanced orientation to PD, that is, PD supporting teacher learning of science or mathematics content in tandem with learning-appropriate pedagogical strategies, reported an intention to make greater improvements in their teaching practices than teachers from the other projects. Teachers from balanced PD also reported that the PD would make more of a contribution to their professional practice than did teachers in the activity- or pedagogy-driven PD projects.

Randomized Control Trial

The third set of papers presented findings from the Middle School Mathematics Professional Development Impact Study (Garet et al., 2010, 2011). In this study, two providers, America's Choice and Pearson Achievement Solutions, using their own set of PD materials, designed PD programs within a common framework provided by the Institute for Education Sciences (IES) to increase teachers' capacity to teach rational numbers. Whereas the providers independently designed their interventions to meet IES specifications, both programs were designed to develop teachers' mathematics subject matter knowledge and pedagogical content knowledge. The programs consisted of 68 contact hours, including a 3-day summer institute and a series of 1-day follow-up seminars held during the school year. In-school coaching followed each seminar day for a total of 10 days of coaching provided through five 2-day visits to each school. Each PD provider was assigned to work with 6 of the 12 districts participating in the study. These assignments were made to balance the allocation of districts using Glencoe/PH Mathematics and Connected Mathematics Project curricula across providers.

PD activities offered by both providers included opportunities for teachers to solve mathematics problems individually and in groups, make short oral presentations to explain how they solved problems, receive feedback on how they solved the problems and presented their solutions, engage in discussions about the most common student misconceptions associated with topics in rational numbers, and plan lessons they would teach during the follow-up coaching visits. The coaching component was designed to help teachers apply material covered in the institutes and seminars to their classroom instruction.

Although both PD providers incorporated these overall design features, their planned PD differed in several specific elements. For example, in America's Choice PD segments, teachers were asked to solve sets of mathematics problems, working individually or in small groups. The problem sets were designed to define or reinforce definitions of key rational number concepts and to illustrate common student misconceptions. Facilitators were directed to emphasize these same ideas in subsequent structured discussions. In contrast, Pearson Achievement Solutions used a single, more open-ended problem or task to structure each PD segment. Each task was designed to elicit multiple approaches and to support extended discussions about the core ideas, common student approaches, and potential misconceptions associated with the task. Following these discussions, teachers collaboratively planned how they would teach a lesson related to the task.

Analyzing the combined data from both providers, the authors reported that fidelity of implementation was high during the first year of the study; that is, the average number of hours of institutes, seminars, and coaching delivered was 67.6 hours, and teachers reported participating in an average of 83% of the institutes and seminars. There were no statistically significant effects of participation in the PD program on teachers' mathematics content knowledge or pedagogical content knowledge, however. The researchers also developed a measure of instructional practice based on aspects of mathematics instruction that were related to the goals of the PD, and they observed teachers in treatment and control schools once after 5 of the 8 scheduled days of institutes and seminars had occurred. They found a statistically significant, positive impact of the PD program on the frequency with which teachers engaged in activities that elicited student thinking, although there were no statistically significant differences in teachers' use of representations (e.g., number lines, ratio tables, area models) or in the frequency with which teachers engaged in activities that focused on mathematical reasoning (e.g., the number of times the teacher justified a procedure or solution or asked a student to justify or explain). There were also no statistically significant differences in student achievement on a customized assessment of rational number constructed by the Northwest Evaluation Association.

Although comparing the impact of PD programs with different design elements is what makes a Phase 3 study desirable, the authors suggest that "it is not appropriate to directly compare the impact findings between the two provider subgroups, because the study districts were not randomly assigned to the providers" (Garet et al., 2010, p. 51). We note here, however, that there were reported differences in the size of the effects across the two providers. Differences between the treatment and control schools on the researcher-developed measures of teachers' instructional practices were larger in the America's Choice sample than in the Pearson Achievement Solu-

个测验即"教师对数学推理的关注"上有统计意义上的显著影响。

在研究的第二年，预算只能允许原有12学区中的6个地区作为样本。这样的选择是为了保持专业发展提供者间的平衡，并包括最多数量的学校，以提高统计的精度，同时也考虑到要去掉那些可能会因重组计划而改变研究学校结构的学区。这些学区是在知道第一年的研究结果之前就被选定的。这些被选学区的教师第一年是在实验组的，他们额外参加了的一个为期2天的暑期学校、为期3天的研讨班以及为期8天的校内集中辅导。专业发展提供者为第二年的课程选择学习材料，这些材料是他们认为可以加强和加深教师理解的，尤其是在那些第一年中教师表现最薄弱的领域（Garet等，2011）。另外，在第二年研究中新加入实验学校的七年级的数学教师接受了暑期学校形式的两天的"弥补"式专业发展课程。实施忠实度的量表显示：专业发展项目的供给和出席率都接近预期。然而，教师的流动却限制了实验组学校教师可以接受的最大可能的专业发展。与第一年的结果相反，参加两年的专业发展项目对教师的学科教学知识有积极影响。根据"西北评价协会有理数测验"的总分，该课程无论是对教师的数学学科知识还是对学生的平均学业水平都没有统计意义上的显著影响（Garet等，2011）。且他们在第二年的研究中没有进行课堂观察。

阶段3研究的讨论

本节评述的三个研究都使用了不同的方法来比较专业发展项目，为阶段3的研究提供了不同路径的例证。巴蒂等（2007）的研究阐明了整合不同专业发展研究报告的数据的可能性和挑战性。专业发展项目描述方式与结果分析方式的不同，限制了我们从这些形式的比较中获取信息。相反地，马拉等（2011）的研究采用一个通用的工具（即教师调查）来收集不同研究结果中可比较的数据，这提升了结果的可比性。然而，主要结果的测量是教师在教学改进意图方面的自我报告。研究界更多地运用普通的测量方法和工具，这可以对专业发展研究进行有价值的比较，即使这些比较在研究开展前并没有被事先计划好。

加雷特和同事们（2010，2011）的研究是唯一的利用两个不同的专业发展设计者或提供者来进行大规模有效性的研究，但是由于这个研究的设计限制，我们得不到那种通过对不同提供者进行直接比较所能得到的结论，也限制了我们去理解两个提供者在材料组织和内容上的不同是怎样影响教师学习或学生成绩的。因为只有三个研究，所以关于数学教师专业发展项目阶段3研究的模式可以讨论的内容很少。尽管在开展科学教师专业发展项目的阶段3研究方面已有了一些进展（例如Banilower, Heck, & Weiss, 2007；Borman, Gamoran, & Bowdon, 2008；Heller, Daehler, Wong, Shinohara, & Miratrix, 2012；Penuel, Gallagher, & Moorthy, 2011），但对于数学教师专业发展的研究，仍需进行不同专业发展模式的比较。

未来数学专业发展研究的方向

无论从政策的角度还是研究的角度来看，专业发展作为数学教师的学习机会其重要性都不能被低估。本章所评述的这些研究，展现出了数学教师专业发展作为一个研究领域的丰富性，以及该领域自索德2007年的评述后所经历的发展。我们的评述也揭示了专业发展研究领域的知识基础不断增长的强劲势头，反击了对数学教师专业发展所知甚少的说法。虽然还有很多需要我们弄清楚的问题，但我们对新兴的发现和方法论趋势的总结，突显了被研究的数学教师专业发展项目的若干重要设计特征，以及研究问题的类型和研究时所使用的方法。当许多利益相关者均对数学教师专业发展有效性的构成感兴趣时，研究者需要提供关于以下内容的详细信息来引导讨论，比如：有关教师在专业发展背景中所接受的学习机会的详细信息，呈现关于这些学习机会是如何帮助教师和学生学习的研究结果，将专业发展扩展至不同场地或背景、以及与其他项目的比较。

博尔科（2004）将专业发展研究划分为三个阶段时，她指出这些阶段将有助于整个领域"向着为所有教师提

tion sample. Specifically, the PD program administered by America's Choice had statistically significant impacts on two of the three measures of instructional practice: the "teacher elicits student thinking" scale (effect size = 0.63) and the "teacher uses representations" scale (effect size = 0.60). Neither program had a statistically significant impact on the third measure, "teacher focuses on mathematical reasoning."

In the second year of the study, budget issues allowed for a sample that included 6 of the original 12 districts, chosen to maintain the balance among PD providers, to include the largest number of schools to increase statistical precision, and to eliminate districts where restructuring initiatives might change the composition of study schools. These districts were chosen before first-year results were known, and teachers in these districts who were part of the first-year treatment group participated in an additional 2-day summer institute, 3 days of seminars during the school year, and 8 days of intensive in-school coaching. The PD providers selected material for the second year that they believed would reinforce and deepen the teachers' understanding, particularly in areas where the teachers had seemed weakest during the first year (Garet et al., 2011). In addition, seventh-grade math teachers who were new to the treatment schools in the second year of the study received "make up" PD in the form of a 2-day special summer institute. Fidelity of implementation measures suggested that provision of PD and attendance rates were close to those expected. Teacher turnover, however, limited the maximum possible PD that teachers in the treatment schools could receive. In contrast to findings from Year 1, there was a positive impact of participating in 2 years of the PD on teachers' pedagogical content knowledge. There was no statistically significant impact on either teachers' mathematics content knowledge or average student achievement, as measured by the Northwest Evaluation Association Rational Number Test total score (Garet et al., 2011). No classroom observation was conducted in the second year of the study.

Discussion of Phase 3 Research

The three studies reviewed in this section used different approaches in their comparisons across PD programs, exemplifying different paths for conducting Phase 3 research. Battey et al. (2007) illustrated the possibilities and challenges of trying to pool reported data across PD studies. Differences in the way that PD is described and in the outcomes analyzed limit what we can learn from these forms of comparisons. In contrast, Marra et al. (2011) were able to employ a common instrument (i.e., a teacher survey) to collect comparable data on outcomes across studies, which increased the comparability of the outcomes. The major outcomes measures, however, were teacher self-reports of their intention to make more improvements in their teaching. Increased use of common measures and instrumentation by the research community would allow for valuable comparisons across PD studies, even if the comparisons were not planned before the studies were conducted.

The study by Garet and colleagues (2010, 2011) is the sole, large-scale effectiveness study that utilized two different PD designers/providers, but the study design limited conclusions that could be drawn from a direct comparison across providers, restricting our ability to understand how differences in the organization of materials and content by the two providers influenced teacher learning or student achievement. With only three studies to examine, there is little to discuss regarding patterns in Phase 3 research on mathematics PD programs. Although there has been some progress in conducting Phase 3 research in science PD (e.g., Banilower, Heck, & Weiss, 2007; Borman, Gamoran, & Bowdon, 2008; Heller, Daehler, Wong, Shinohara, & Miratrix, 2012; Penuel, Gallagher, & Moorthy, 2011), those researching mathematics PD have yet to fully engage in comparative studies of different PD models.

Future Directions for Research on Mathematics PD

The importance of PD as a learning opportunity for mathematics teachers cannot be underestimated—both from a policy and from a research perspective. The studies reviewed in this chapter show the richness of mathematics PD as a research field and the growth the field has experienced since Sowder's review in 2007. Our review also shows the strength of the growing knowledge base in PD research, counteracting claims that little is known about mathematics PD. Although there is still much more to be understood, our summary of emerging findings and methodological trends highlights several important design features of the mathematics PD programs being researched, as well as the types of research questions and methods used to study them. At a time when so many stakeholders are interested in what constitutes "effectiveness" in mathematics PD, researchers must lead the conversation by offering detailed information about the specifics of the learning opportunities offered to teachers in PD settings and presenting findings about how these learning opportunities support teacher and student

供高质量的专业发展这个目标"迈进（第4页），并强调"在全部三个阶段中都有重要的工作需要完成"（第12页）。围绕第一、二、三阶段来组织本章能够让我们展示出数学教育研究在这些阶段的进展。仅仅在十年前，关于阶段2的研究还很少，也没有阶段3的研究。我们的评述表明，尽管这一领域仍主要关注阶段1的研究，但是阶段2的知识积累已经不断涌现，特别是在专业发展的推进及使大规模专业发展取得成功所需要考虑的实施背景方面的知识等方面。正如我们在结尾部分所强调的那样：加强阶段1的研究，并关注从阶段1到阶段2研究的过渡将继续成为数学教师专业发展研究中的首要任务。阶段3研究的数量仍是最少的，并且因为相关研究甚少，目前尚不清楚这一领域的新发现所揭示的"定义明确的专业发展计划的实施、效果和资源需求"的内容（Borko，2004，第11页）。

在将注意力转回专业发展研究的三个阶段前，需要提醒读者的是，我们将论文划分不同阶段的定义与最初由博尔科（2004）所提出的定义不同，部分原因在于所研究的专业发展项目的性质与我们最初检索出的文章所提出的研究问题不同。有几个研究被归类为阶段1，不仅仅是在一个场地进行专业发展；对于那些在多场地进行的专业发展，如果在研究中对专业发展和参与教师之间的关系进行了考察，而没有将多地实施中的促进或变化等主题提出疑问，则仍被归类为阶段1。因此，对这些研究进行分类时，我们更关注的是其提出的研究问题而不是专业发展项目的设计。我们讨论一系列阶段2的研究时，提出了将专业发展扩展至一定规模的重要因素是什么的问题，比如，促进者需要知道什么及如何帮助他们做好领导准备。在阶段2的研究中也包含了背景问题，因为设计者和研究者在他们进行专业发展项目规模化时不能忽略当地的实际情况。最后，因为我们希望在这个评述中包括专业发展项目比较的最新工作，所以我们将那些使用多种方法对多个专业发展项目进行研究的归类为阶段3。

阶段1是一个多产的研究领域，并且在该阶段中我们已经分享了许多重要发现。我们所讨论的研究大多数使用的是质化研究方法，他们在证明专业发展设计可有效帮助教师学习的同时，还可以提供关于专业发展模式的详细记述，以及对专业发展中教师经验的充分描述。这些研究证明了基础性的、探索性的、发展性的、基于数学专业发展的知识库生成设计研究具有持续的价值。但是，阶段1的研究极少研究专业发展参与和教师或学生参与后的效果之间的因果关系。因此，阶段1需要更多的有效性研究，特别是对于那些想参与到为数学教师提供专业发展的政策讨论中的研究者来说，这一需要则更为迫切。

为了理解专业发展的有效性，研究者提供更好的关于专业发展项目的描述同样重要（Sztajn，2011）。我们在评述中突出了几个发表成果比较多的专业发展项目，为了描述这些项目，我们浏览了论文集。然而，在初次文献检索出的论文中，有许多没有提供充足的关于专业发展设计和实施的信息，这使得研究的内容有时难以理解。对专业发展项目的详细描述以及对专业发展有效性的细致分析，能够有助于该领域进一步明确有效的数学教师专业发展的构成要素，而不仅仅是识别出几个选择性设计特征。

我们对于工具的关注为以下研究内容提供了清晰的例子，比如：关于学生思维、课堂教学录像以及数学任务的框架，如何运用到专业发展项目中以促进教师的学习。研究者对教师学习过程越来越多的理解，以及在专业发展和课堂背景中更频繁的对学习的测量，使得该领域能够开始将专业发展设计的具体特征（比如，工具和围绕这些工具所设计的活动）和教师专业学习轨迹联系起来。尽管如此，测量教师在任何背景中的学习仍是一个费时费力的过程。这个领域可以从更好的测量方法中获益。在阶段1研究中，使用更好的、可共享的教师学习的测量可以增强我们的知识基础，并为研究者开展下一阶段的研究提供坚实基础。

当研究进入阶段2，与阶段1有关的问题并没有消失，尤其是关于如何将内容、工具、形式及排列顺序进行整合，以最大程度促进教师学习的问题仍旧至关重要。不同的是要弄清专业发展的需要和机会，不但要理解内容、工具、形式及排列顺序如何与教师和学生的学习效果产生关联，而且要理解这种关联如何随着实施地点和背景的变化而改变。从阶段2研究中，可以知道促进和环境对于扩大专业发展规模尤为重要。在众多阶段2研究项

learning, scale to several PD sites or contexts, and compare to others.

When Borko (2004) organized PD research into three phases, she suggested these phases could support the field in making progress "toward the goal of providing high-quality professional development for all teachers" (p. 4) and stressed that there was "important work to be done in all three phases" (p. 12). Organizing this chapter around Phases 1, 2, and 3 allowed us to show the progress mathematics education research has made across these phases. Just over 10 years ago, there were very few studies in Phase 2 and none in Phase 3. Our review shows that although the field continues to focus mostly on Phase 1 research, there is an emerging accumulation of knowledge in Phase 2, particularly about facilitation and aspects of contexts that need to be attended to for PD to be successful at scale. As we highlight in this concluding section, strengthening Phase 1 studies and attending to the transition from Phase 1 to Phase 2 research continue to be important priorities in mathematics PD research. Phase 3 research remains minimal, and with so few studies, it is less clear what the emerging findings in this area reveal about the "implementation, effects, and resource requirements of well-defined professional development programs" (Borko, 2004, p. 11).

Before returning our attention to the three phases of PD research, we remind readers that the definitions we used to classify articles across these phases differ from what was originally proposed by Borko (2004)—in part due to the nature of the PD programs examined and the research questions asked in the articles that emerged from our initial search of the literature. Several of the studies we classified as Phase 1 were not solely about PD offered at a single site: PD offered at several sites were considered under Phase 1 when the reported research examined the relation between the PD and participating teachers without problematizing issues of facilitation or variability in implementation across sites. Thus, we focused on the research questions asked, rather than the design of the PD program, to classify these studies. We examined under Phase 2 a wide array of studies that asked questions about what is important in bringing PD to scale, such as what facilitators need to know and how to prepare them to lead. We also included issues of context under Phase 2, as designers and researchers cannot ignore the local settings into which they bring a PD program as they scale up. Finally, as we sought to include in our review emerging efforts toward comparing PD programs, we classified as Phase 3 studies that used quite different research methodologies in their attempts to examine more than one PD program.

Phase 1 continues to be a productive area for research, and we have shared many important findings from this phase. Most studies we examined primarily utilized qualitative methods, which can offer detailed accounts of PD models and rich descriptions of teachers' experiences in the PD while providing proof of concept that the PD design can effectively support teacher learning. These studies demonstrated the continued value of foundational, exploratory, development, and design research for growing the knowledge base on mathematics PD. Still, very few Phase 1 studies investigated causal links between participation in PD and teacher or student outcomes. Thus, there is a need for more effectiveness studies in Phase 1, particularly if mathematics PD researchers want to participate in policy discussions about the PD offered to mathematics teachers.

To understand PD effectiveness, it is also important that researchers provide better descriptions of PD programs (Sztajn, 2011). We highlighted in our review several PD programs for which there are many publications, and we scanned the collection of publications to provide descriptions of these programs. However, many of the papers that emerged in our initial search of the literature offered insufficient information on PD design or implementation, making it sometimes difficult to understand what was being studied. Detailed descriptions of PD programs, together with careful analysis of PD effectiveness, can support the field in further defining what constitutes effective mathematics PD beyond the identification of a few selective design features.

Our focus on tools provides clear examples of how frameworks of student thinking, videos of classroom instruction, and mathematical tasks can be used in PD to promote teacher learning. Researchers' increasing understanding of processes of teacher learning, together with more frequent measurement of learning in both the PD and classroom settings, are enabling the field to begin to connect specific features of PD design (such as tools and the activities designed around such tools) to teacher professional learning trajectories. Measuring teacher learning in any setting, nonetheless, continues to be a resource- and time-intensive process, and the field can benefit from better measures. The use of better, shared measures of teacher learning across Phase 1 studies can enhance the field's knowledge base and provide a solid foundation for researchers in the next phases.

As research moves into Phase 2, issues that were of concern in Phase 1 do not go away; in particular, questions about how content, tools, formats, and sequencing are integrated to best support teacher learning continue to be important. What is different is the need, and opportu-

目中，典型的关键因素是专业发展中，对专业发展方案和其它大规模教学改进措施方面有深厚知识储备的推进者、对推进者系统的准备和支持、对社会和政策背景的适应。一些阶段2的项目强调了对参与教师多层支持的重要性，比如：暑期学校、学年中的工作坊和课堂内的辅导的联合。这些层次的支持由外部推进者和学校内的领导提供。这个综述研究强调了专业发展对专业发展领导者的重要性。促进者和教师都能获得在期望的实践中取得成就的人的专业知识，这是支持专业发展模式规模化的一个重要因素。

对专业发展背景作出解释的研究都得到了共同的结论：专业发展设计者在开发支持教师学习的各项措施时，必须注意教师的工作背景。在设计教学支持时还要考虑规模化的复杂性，包括专业发展，这种复杂性表明：不仅要理解背景，还要影响背景。当前的研究并没有给那些想提高数学教学水平的学区和学校的领导们提供一个发展路线图。教学上的提高需要的是一个整合的支持系统。争论的焦点不仅在于跨多个角色小组和资源的努力，而且在于支持教师、教练、专业发展提供者、学校和地区领导人重新组织他们的实践的组织学习框架。这样的一个学习框架，对于成功地扩大专业发展的规模尤为重要。

由于阶段2研究数量有限，而比较研究又费时费力，因此在我们的研究中几乎找不到阶段3的研究也并不奇怪。考虑到这些研究中不同的研究问题及回答问题的不同方法，我们无法确定阶段3研究的共同主题。但是，阶段3研究强调：在有着高教师流失率的学校背景下，使教师持续地参与专业发展具有挑战性。正如许多阶段1研究中证明了的结论：鉴于多年参与一个专业发展模式的价值，教师的流失成为专业发展设计者和研究者必须考虑的一项重要的背景特征。设计和预算限制也阻碍了不同专业发展设计间的比较，以及这些设计与教师和学生在随机对照试验中的结果之间关系。这些限制制约了我们从阶段3的研究中获得更多信息的能力，也进一步证明了在这一水平上开展阶段3研究的难度。只能解决"什么是有效的"，但是不能解决"为什么有效以及在什么情境下有效"的大规模的研究，对提升未来专业发展设计的贡献是有限的。

通过对阶段1，阶段2和阶段3研究的讨论，我们指出未来数学专业发展的几个重要研究领域。总结本章，我们将提出数学专业发展中几个额外的、没有出现在本文中的，但对我们知识基础的发展很重要的领域。第一个是将技术作为专业发展中的一种交互方式来使用。第二个问题涉及我们研究中没有考虑到的一些新兴研究方法。

在我们的文献的检索中，几乎没有出现数学专业发展项目为教师提供全程在线的或者线上线下混合的学习机会，而这些专业发展方式又变得越来越普遍，并且被宣传成是扩大专业发展的途径。在线讨论社区、学习共同体、同步的和非同步的工作组、线上监督、虚拟案例、大规模网上开放课程（慕课），以及许多其他工具和资源可供教师参与并非完全（或根本不是）以面对面的形式提供的专业发展。在我们考虑如何让每个教师都参与专业发展以及为教师提供高质量的数学专业发展时，进一步将这些形式纳入专业发展研究对数学专业发展共同体非常重要。

至于新兴的研究专业发展的方法，研究者们开始使用合作的、以改进为中心的方法，这种方法在考虑扩大规模的同时能直接关注设计上的问题。长期以来，以教师发展为中心的改革举措未能产生足够的教师参与度，资源保证或是领导层支持等问题，至少在一定程度上促使了越来越多的基于设计的实施研究（DBIR；Fishman, Penuel, Allen, Cheng, & Sabelli, 2013；Penuel, Fishman, Cheng, & Sabelli, 2011）。基于设计的实施研究是一种新兴的方法，以合作的、反复的和基于系统调查的方式将研究和实践结合起来（见 http://learndbir.org/）。基于设计的实施研究方法的特点是，通过系统的探究来发展与课堂教学和学习以及实施相关的理论，并发展维持系统变化的能力。当在K-12教育背景下实施时，它通常要涉及研究者、学区和学校领导以及校内教练和专业发展推进者的多年协作（例如Borko & Klingner, 2013；Cobb 等, 2013）。研究者及实践者在设计与实施中的紧密合作，有助于提高教师参与及成功扩大规模的可能性。

另一个教育领域内新兴的关于设计、测试、改正、测量创新的方法是改进科学（Bryk, Gomez, Grunow, &

nity, to understand not only how content, tools, formats, and sequencing relate to teacher and student learning outcomes, but also how this relation varies across sites and contexts. From Phase 2 research, we are learning about the importance of facilitation and context to scaling up PD. Key factors that were typical of multiple Phase 2 projects were PD facilitators with deep knowledge of the PD initiative and other larger instructional-improvement initiatives, systematic preparation and support for the facilitators, and adaptation to social and policy contexts. Several Phase 2 projects also highlighted the importance of multiple tiers of support for participating teachers, such as combinations of summer institutes, workshops during the school year, and in-classroom coaching. These tiers of support were provided by external facilitators and school-based leaders. The research reviewed emphasizes the importance of PD for PD leaders. Access by both facilitators and teachers to the expertise of someone who is accomplished in the desired practices emerged as an important factor supporting the scaling of PD models.

The common implication of studies that account for the context of PD is that professional development designers must attend to the contexts in which teachers work when developing supports for teachers' learning. The complexity of designing instructional supports with scale-up in mind, including professional development, implicates not just understanding context, but influencing it. Current research does not provide a roadmap for district and school leaders who aim to improve mathematics teaching, and integrated systems of supports are needed for instructional improvement. The argument is not just for alignment of efforts across multiple role groups and resources, but for organizational learning frameworks that support teachers, coaches, PD providers, and school and district leaders in reorganizing their practices. Such an organizational learning framework is key for successful implementation of PD at scale.

With a limited number of studies in Phase 2 and the costs and complexity of mounting comparative research, it is not really surprising that we found few Phase 3 studies in our search. Given the studies' different research questions and methods for answering them, we identified no common themes in Phase 3 research. Still, Phase 3 research highlighted the challenge of engaging teachers in sustained professional development in the context of schools with high teacher turnover rates. Given the value of participation in a single PD model over multiple years, as demonstrated in a number of Phase 1 studies, teacher turnover becomes an important context feature that PD designers and researchers must take into account. Design

and budget constraints also prevent comparisons of different PD designs and their relationship to teacher and student outcomes in RCTs. These constraints limit our ability to learn from Phase 3 research, further highlighting how difficult it is to conduct research at this level. Large-scale studies that can only address "what works" but not "why and under what conditions" are limited in their contribution to improving the design of future PD.

Throughout our discussions of Phase 1, 2, and 3 research studies, we pointed to several important areas for future research in mathematics PD. To conclude this chapter, we suggest several additional important areas for mathematics PD research that did not emerge in our review yet are important for the development of our knowledge base. The first one is the use of technology as a mode of interaction in PD. The second one relates to several emerging research methodologies that were not considered in the studies we examined.

Very few of the mathematics PD programs emerging from our search of the research literature engaged teachers in fully online or blended PD opportunities. These approaches to PD are becoming more prevalent and are being advertised as a way to provide PD at scale. Online discussion boards and learning communities, synchronous and asynchronous working groups, online mentoring, virtual cases, massive online open courses (MOOCS), and many other tools and resources exist for teachers to participate in PD that is not completely (or not at all) offered in a face-to-face format. Further including these formats for PD research is important for the mathematics PD community as we consider how to reach every teacher and offer high-quality mathematics PD.

Regarding emerging methodologies for examining PD, researchers have begun to employ collaborative, improvement-focused methodologies that attend directly to design issues while keeping scale-up in mind. The long history of reform initiatives focused on teacher PD that have failed to produce sufficient teacher buy in, resource commitments, or leadership support has motivated, at least in part, what is increasingly becoming known as *design-based implementation research* (DBIR; Fishman, Penuel, Allen, Cheng, & Sabelli, 2013; Penuel, Fishman, Cheng, & Sabelli, 2011). DBIR is an emerging approach to connecting research and practice in a way that is collaborative, iterative, and grounded in systematic inquiry (see http://learndbir.org/). The DBIR approach is characterized by a dual focus on developing theory related both to classroom teaching and learning and implementation through systematic inquiry and to developing capacity for sustaining change in systems. When conducted in K–12 educational settings, it typically entails a multiyear

LeMahieu，2015；Langley 等，2009）。与基于设计的实施研究相似，它使研究者与实践者可以合作进行反复设计、测试并改进一个教育系统的组成部分，这使得该设计在扩大规模时能适应当地背景。虽然据我们所知，改进科学方法尚未应用于K-12阶段数学教师专业发展的研究中，但与设计实验及基于设计的实施研究的相似之处是显而易见的。这些方法为研究者研究背景、建立学区进行不断改进的能力增添了途径，而这些改进是可持续的、并在条件允许和特定背景限制下是可达到的。随着越来越多的来自不同背景和环境的教师不断增多，专业发展中的变化都需要被检测、修正和实施。基于设计的实施研究和"改进科学"均模糊了阶段1和阶段2研究间的界限，这为适应背景变化提供了一种方法。

最后一个重要的新兴专业发展研究方法是用"大数据"的方法，理解专业发展的设计如何与参与者的特点和背景相互作用，当数学专业发展研究共同体提供和研究在线专业发展的机会时，这个方法将会引起大家的兴趣。我们预计这些领域中的研究将会在未来的十年内趋于成熟，因此将会在下一次专业发展文献的述评中讨论。

collaborative effort among researchers, district and school leaders, and school-based coaches or PD facilitators (e.g., Borko & Klingner, 2013; Cobb et al., 2013). The close collaboration between researchers and practitioners in design and implementation can help increase the likelihood of participants' buy in and successful scale-up.

Another emerging methodology for designing, testing, revising, and scaling innovations in education is *improvement science* (Bryk, Gomez, Grunow, & LeMahieu, 2015; Langley et al., 2009). Similar to DBIR, it allows researchers to collaborate with practitioners to iteratively design, test, and improve components of an education system—adapting the design to local contexts as it is scaled. Although, to our knowledge, improvement science methods have yet to be utilized in K–12 mathematics PD research, the parallels with design experiments and DBIR are tangible. These methods add to the ways in which researchers can study context and build the capacity of school districts to engage in continuous improvement that is sustainable and can be achieved within the affordances and constraints of a particular environment. Both DBIR and improvement science blur the lines between Phase 1 and Phase 2 research, providing a means to adapt to context as the changes to the PD are tested, revised, and implemented with increasing numbers of teachers across diverse settings and contexts.

One final important emerging methodology for PD research, which will gain interest as the mathematics PD research community offers and studies online PD opportunities, is the use of "big data" methodologies to understand how the design of PD interacts with the characteristics and contexts of the participants. We anticipate that research in these areas will mature over the next decade so that it can be covered in the next review of PD literature.

References

Antoniou, P., & Kyriakides, L. (2013). A dynamic integrated approach to teacher professional development: Impact and sustainability of the effects on improving teacher behaviour and student outcomes. *Teaching and Teacher Education, 29,* 1–12.

Balfanz, R., Mac Iver, D., & Byrnes, V. (2006). The implementation and impact of evidence-based mathematics reforms in high-poverty middle schools: A multi-site, multi-year study. *Journal for Research in Mathematics Education, 37*(1), 33–64.

Ball, D. L., & Cohen, D. K. (1999). Developing practices, developing practitioners: Toward a practice-based theory of professional development. In G. Sykes & L. Darling-Hammond (Eds.), *Teaching as the learning profession: Handbook of policy and practice* (pp. 30–32). San Francisco, CA: Jossey-Bass.

Banilower, E. R., Heck, D. J., & Weiss, I. R. (2007). Can professional development make the vision of the standards a reality? The impact of the National Science Foundation's local systemic change through teacher enhancement initiative. *Journal of Research in Science Teaching, 44,* 375–395.

Battey, D., & Franke, M. (2013). Integrating professional development on mathematics and equity: Countering deficit views of students of color. *Education and Urban Society, 47,* 433–462.

Battey, D., Kafai, Y., Nixon, A., & Kao, L. (2007). Professional development for teachers on gender equity in the sciences: Initiating the conversation. *The Teachers College Record, 109,* 221–243.

Bell, C. A., Wilson, S. M., Higgins, T., & Mccoach, D. B. (2010). Measuring the effects of professional development on teacher knowledge: The case of developing mathematical ideas. *Journal for Research in Mathematics Education, 41*(5), 479–512.

Bobis, J. (1996). *Report of the evaluation of the Count Me In Too Project*. Sydney, Australia: NSW Department of Education and Training.

Bobis, J. (1999). *Count Me In Too: The impact of the Count Me In Too Project on the professional knowledge of teachers*. Sydney, Australia: NSW Department of Education and Training.

Bobis, J. (2003). *Count Me In Too: Evaluation of stage 2*. Sydney, Australia: NSW Department of Education and Training.

Bobis, J. (2009). *Count Me In Too. The learning framework in number and its impact on teacher knowledge and pedagogy*. Sydney, Australia: NSW Department of Education and Training.

Bobis, J. (2011). Mechanisms affecting the sustainability and scale-up of a system-wide numeracy reform. *Mathematics Teacher Education and Development, 13*(1), 34–53.

Bobis, J., Clarke, D., Clarke, B., Thomas, G., Young-Loveridge, J., & Wright, R. (2005). Supporting teachers in the development of young children's mathematical thinking: Three large scale cases. *Mathematics Education Research Journal, 16*(3), 27–57.

Borko, H. (2004). Professional development and teacher learning: Mapping the terrain. *Educational Researcher, 33*(8), 3–15.

Borko, H., Jacobs, J., Eiteljorg, E., & Pittman, M. E. (2008). Video as a tool for fostering productive discussions in mathematics professional development. *Teaching and Teacher Education, 24,* 417–436.

Borko, H., Jacobs, J., Koellner, K., & Swackhamer, L. (2015). *Mathematics professional development: Improving teaching using the problem-solving cycle and leadership preparation models*. New York, NY: Teachers College Press.

Borko, H., & Klingner, J. K. (2013). Supporting teachers in schools to improve their instructional practice. *National Society for the Study of Education Yearbook, 112,* 274–297.

Borko, H., Koellner, K., & Jacobs, J. (2011). Meeting the challenges of scale: The importance of preparing professional development leaders. *Teachers College Record.* Published March 4, 2011. http://www.tcrecprd.org, ID Number: 16358.

Borko, H., Koellner, K., & Jacobs, J. (2014). Examining novice teacher leaders' facilitation of mathematics professional development. *The Journal of Mathematical Behavior, 33,* 149–167.

Borko, H., Koellner, K., Jacobs, J., & Seago, N. (2011). Using video representations of teaching in practice-based professional development programs. *ZDM—The International Journal on Mathematics Education, 43,* 175–187.

Borman, G. D., Gamoran, A., & Bowdon, J. (2008). A randomized trial of teacher development in elementary science: First-year achievement effects. *Journal of Research on Educational Effectiveness, 1*, 237–264.

Boston, M. D. (2013). Connecting changes in secondary mathematics teachers' knowledge to their experiences in a professional development workshop. *Journal of Mathematics Teacher Education, 16*, 7–31.

Boston, M. D., & Smith, M. S. (2009). Transforming secondary mathematics teaching: Increasing the cognitive demands of instructional tasks used in teachers' classrooms. *Journal for Research in Mathematics Education, 40*(2), 119–156.

Boston, M. D., & Smith, M. S. (2011). A "task-centric approach" to professional development: Enhancing and sustaining mathematics teachers' ability to implement cognitively challenging mathematical tasks. *ZDM—The International Journal on Mathematics Education, 43*, 965–977.

Boston, M. D., & Wolf, M. K. (2006). *Assessing academic rigor in mathematics instruction: The development of the instructional quality assessment toolkit: CSE technical report 672*. Los Angeles, CA: University of California, National Center for Research on Evaluation, Standards, and Student Testing (CRESST).

Brodie, K., & Shalem, Y. (2011). Accountability conversations: Mathematics teachers' learning through challenge and solidarity. *Journal of Mathematics Teacher Education, 14*, 419–439.

Bruce, C. D., Esmonde, I., Ross, J., Dookie, L., & Beatty, R. (2010). The effects of sustained classroom-embedded teacher professional learning on teacher efficacy and related student achievement. *Teaching and Teacher Education, 26*, 1598–1608.

Bryk, A. S., Gomez, L. M., Grunow, A., & LeMahieu, P. G. (2015). *Learning to improve: How America's schools can get better at getting better*. Cambridge, MA: Harvard Education Press.

Carpenter, T. P., Fennema, E., Franke, M. L., Levi, L., & Empson, S. B. (1999). *Children's mathematics: Cognitively guided instruction*. Portsmouth, NH: Heinemann.

Carpenter, T. P., Fennema, E., Peterson, P. L., Chiang, C., & Loef, M. (1989). Using knowledge of children's mathematics thinking in classroom teaching: An experimental study. *American Educational Research Journal, 26*, 499–531.

Carpenter, T. P., Franke, M. L., & Levi, L. (2003). *Thinking mathematically: Integrating arithmetic and algebra in elementary schools*. Portsmouth, NH: Heinemann.

Chamberlin, M. (2009). Teachers' reflections on their mathematical learning experiences in a professional development course. *Mathematics Education Research Journal, 11*, 22–35.

Chamberlin, M. T., Farmer, J. D., & Novak, J. D. (2008). Teachers' perceptions of assessments of their mathematical knowledge in a professional development course. *Journal of Mathematics Teacher Education, 11*, 435–457.

Clark, P. G., Moore, K. C., & Carlson, M. P. (2008). Documenting the emergence of "speaking with meaning" as a sociomathematical norm in professional learning community discourse. *The Journal of Mathematical Behavior, 27*, 297–310.

Clarke, D., & Hollingsworth, H. (2002). Elaborating a model of teacher professional growth. *Teaching and Teacher Education, 18*, 947–967.

Clements, D. H., & Sarama, J. (2011). Early childhood teacher education: The case of geometry. *Journal of Mathematics Teacher Education, 14*, 133–148.

Clements, D. H., Sarama, J., Spitler, M. E., Lange, A. A., & Wolfe, C. B. (2011). Mathematics learned by young children in an intervention based on learning trajectories: A large-scale cluster randomized trial. *Journal for Research in Mathematics Education, 42*, 127–166.

Cobb, P., Confrey, J., diSessa, A. A., Lehrer, R., & Schauble, L. (2003). Design experiments in educational research. *Educational Researcher, 32*(1), 9–13.

Cobb, P., & Jackson, K. (2011). Towards an empirically grounded theory of action for improving the quality of mathematics teaching at scale. *Mathematics Teacher Education and Development, 13*(1), 6–33.

Cobb, P., Jackson, K., & Dunlap, C. (2017). Conducting design studies to investigate and support mathematics students' and teachers' learning. In J. Cai (Ed.), *Compendium for research in mathematics education* (pp. 208–233). Reston, VA: National Council of Teachers of Mathematics.

Cobb, P., Jackson, K., Smith, T., Sorum, M., & Henrick, E. (2013). Design research with educational systems: Investigating and supporting improvements in quality of mathematics teaching and learning at scale. In B. J. Fishman, W. R. Penuel, A.-R. Allen, & B. H. Cheng (Eds.), *Design based implementation research: Theories, methods, and exemplars*. National Society for the Study of Education Yearbook (Vol. 112, pp. 320–349). New York, NY: Teachers College Record.

Cobb, P., McClain, K., Lamberg, T., & Dean, C. (2003). Situating teachers' instructional practices in the institutional setting of the school and district. *Educational Researcher, 32*(6), 13–24.

Cobb, P., & Smith, T. (2008). The challenge of scale: Designing schools and districts as learning organizations for instructional improvement in mathematics. In T. Wood, B. Jaworski, K. Krainer, P. Sullivan, & D. Tirosh (Eds.), *International handbook of mathematics teacher education*. (Vol. 3, pp. 231–254). Rotterdam, The Netherlands: Sense.

Cobb, P., Zhao, Q., & Dean, C. (2009). Conducting design experiments to support teachers' learning. *The Journal of the Learning Sciences, 18*, 165–199.

Coburn, C. E., & Russell, J. L. (2008). District policy and teachers' social networks. *Educational Evaluation and Policy Analysis, 30*, 203–235.

Coles, A. (2012). Using video for professional development: The role of the discussion facilitator. *Journal of Mathematics Teacher Education, 16*, 165–184.

Daro, P., Mosher, F., & Corcoran, T. (2011). *Learning trajectories in mathematics. Research report 68*. Madison, WI: Consortium for Policy Research in Education.

Desimone, L. M. (2009). Improving impact studies of teachers' professional development: Toward better conceptualizations and measures. *Educational Researcher, 38*(3), 181–199.

Desimone, L. M., Smith, T. M., & Phillips, K. J. (2007). Does policy influence mathematics and science teachers' participation in professional development? *The Teachers College Record, 109*, 1086–1122.

Earle, J., Maynard, R., Curran Neild, R., Easton, J. Q., Ferrini-Mundy, J., Albro, E., . . . Winter, S. (2013). *Common guidelines for education research and development*. Washington, DC: Institute of Education Sciences.

Elliott, R., Kazemi, E., Lesseig, K., Mumme, J., Carroll, C., & Kelley-Petersen, M. (2009). Conceptualizing the work of leading mathematical tasks in professional development. *Journal of Teacher Education, 60*, 364–379.

Empson, S. B., & Levi, L. (2011). *Extending children's mathematics: Fractions and decimals.* Portsmouth, NH: Heinemann.

Fennema, E., Carpenter, T. P., Franke, M. L., Levi, L., Jacobs, V. R., & Empson, S. B. (1996). A longitudinal study of learning to use children's thinking in mathematics instruction. *Journal for Research in Mathematics Education, 27*, 403–434.

Ferrini-Mundy, J., Burrill, G., & Schmidt, W. H. (2007). Building teacher capacity for implementing curricular coherence: Mathematics teacher professional development tasks. *Journal of Mathematics Teacher Education, 10*, 311–324.

Fishman, B. J., Penuel, W. R., Allen, A.-R., Cheng, B. H., & Sabelli, N. (2013). Design-based implementation research: An emerging model for transforming the relationship of research and practice. In B. Fishman, W. R. Penuel, A. Allen, & B. H. Cheng (Eds.), *Design-based implementation research: Theories, methods, and exemplars.* National Society for the Study of Education Yearbook (Vol. 112, pp. 136–156). New York, NY: Teachers College Record.

Gal, H. (2011). From another perspective—Training teachers to cope with problematic learning situations in geometry. *Educational Studies in Mathematics, 78*, 183–203.

Garet, M., Porter, A. C., Desimone, L., Birman, B. F., & Yoon, K. S. (2001). What makes professional development effective? Results from a national sample of teachers. *American Educational Research Journal, 38*, 915–945.

Garet, M., Wayne, A., Stancavage, F., Taylor, J., Eaton, M., Walters, K., . . . Doolittle, F. (2011). *Middle school mathematics professional development impact study: Findings after the second year of implementation* (NCEE 2011–4024). Washington, DC: U.S. Department of Education, Institute of Education Sciences, National Center for Education Evaluation and Regional Assistance.

Garet, M., Wayne, A., Stancavage, F., Taylor, J., Walters, K., Song, M., . . . Doolittle, F. (2010). *Middle school mathematics professional development impact study: Findings after the first year of implementation* (NCEE 2010–4009). Washington, DC: U.S. Department of Education, Institute of Education Sciences, National Center for Education Evaluation and Regional Assistance.

Gersten, R., Taylor, M. J., Keys, T. D., Rolfhus, E., & Newman-Gonchar, R. (2014). *Summary of research on the effectiveness of math professional development approaches* (REL 2014–010). Washington, DC: U.S. Department of Education, Institute of Education Sciences, National Center for Education Evaluation and Regional Assistance, Regional Educational Laboratory Southeast.

Goldsmith, L. T., Doerr, H. M., & Lewis, C. C. (2013). Mathematics teachers' learning: A conceptual framework and synthesis of research. *Journal of Mathematics Teacher Education, 17*, 5–36.

Grossman, P., Hammerness, K., & McDonald, M. (2009). Redefining teacher: Re-imagining teacher education. *Teachers and Teaching: Theory and Practice, 15*, 273–290.

Hargreaves A. (2014). Foreword: Six sources of change in professional development. In L. E. Martin, S. Kragler, D. J. Quatroche, & K. L. Bauserman (Eds.), *Handbook of professional development in education* (pp. x–xix). New York, NY: Guilford Press.

Heller, J. I., Daehler, K. R., Wong, N., Shinohara, M., & Miratrix, L. W. (2012). Differential effects of three professional development models on teacher knowledge and student achievement in elementary science. *Journal of Research in Science Teaching, 49*, 333–362.

Higgins, J., & Parsons, R. (2009). A successful professional development model in mathematics: A system-wide New Zealand case. *Journal of Teacher Education, 60*, 231–242.

Higgins, J., & Parsons, R. (2011a). Improving outcomes in mathematics in New Zealand: A dynamic approach to the policy process. *International Journal of Science and Mathematics Education, 9*, 503–522.

Higgins, J., & Parsons, R. (2011b). Professional learning opportunities in the classroom: Implications for scaling up system-level professional development in mathematics. *Mathematics Teacher Education and Development, 13*(1), 54–76.

Hill, H., Rowan B., & Ball, D. (2005). Effects of teachers' mathematical knowledge for teaching on student achievement. *American Educational Research Journal, 42*, 371–406.

Horoks, J., & Robert, A. (2007). Tasks designed to highlight task-activity relationships. *Journal of Mathematics Teacher Education, 10*, 279–287.

Jackson, K., Cobb, P., Wilson, J., Webster, M., Dunlap, C., & Applegate, M. (2015). Investigating the development of mathematics leaders' capacity to support teachers' learning on a large scale. *ZDM—The International Journal on Mathematics Education, 47*, 93–104.

Jacobs, V. R., Franke, M. L., Carpenter, T. P., Levi, L., & Battey, D. (2007). Professional development focused on children's algebraic reasoning in elementary school. *Journal for Research in Mathematics Education, 38*, 258–288.

Jacobs, V. R., Lamb, L. L., & Philipp, R. A. (2010). Professional noticing of children's mathematical thinking. *Journal for Research in Mathematics Education, 41*, 169–202.

Jacobs, V. R., Lamb, L. L. C., Philipp, R. A., & Schappelle, B. P. (2011). Deciding how to respond on the basis of children's understandings. In M. G. Sherin, V. R. Jacobs, & R. A. Philipp (Eds.), *Mathematics teacher noticing: Seeing through teachers' eyes* (pp. 97–116). New York, NY: Routledge.

Jacobs, V. R., & Spangler, D. A. (2017). Research on core practices in K–12 mathematics teaching. In J. Cai (Ed.), *Compendium for research in mathematics education* (pp. 766–792). Reston, VA: National Council of Teachers of Mathematics.

Kabasakalian, R. (2007). Language and thought in mathematics staff development: A problem probing protocol. *The Teachers College Record, 109*, 837–876.

Kleinknecht, M., & Schneider, J. (2013). What do teachers think and feel when analyzing videos of themselves and other teachers teaching? *Teaching and Teacher Education, 33*, 13–23.

Koc, Y., Peker, D., & Osmanoglu, A. (2009). Supporting teacher professional development through online video case study discussions: An assemblage of preservice and inservice teachers and the case teacher. *Teaching and Teacher Education, 25*, 1158–1168.

Koellner, K., & Jacobs, J. (2015). Distinguishing models of professional development: The case of an adaptive model's impact on mathematics teachers' knowledge, instruction, and student achievement. *Journal of Teacher Education, 66*, 51–67.

Koellner, K., Jacobs, J., & Borko, H. (2011). Mathematics professional development: Critical features for developing leadership skills and building teachers' capacity. *Mathematics Teacher Education and Development, 13*, 115–136.

Koellner, K., Jacobs, J., Borko, H., Schneider, C., Pittman, M., Eiterljorg, E., ... Frykholm, J. (2007). The problem-solving cycle—A model to support the development of teachers' professional knowledge. *Mathematical Thinking and Learning, 9*, 273–303.

Lamb, L. C., Philipp, R. A., Jacobs, V. R., & Schappelle, B. P. (2009). Developing teachers' stance of inquiry. In D. Slavit, T. Holmlund Nelson, & A. Kennedy (Eds.), *Perspectives on supported collaborative teacher inquiry* (pp. 16–45). New York, NY: Routledge.

Lampert, M., Beasley, H., Ghousseini, H., Kazemi, E., & Franke, M. L. (2010). Using designed instructional activities to enable novices to manage ambitious mathematics teaching. In M. K. Stein & L. Kucan (Eds.), *Instructional explanations in the disciplines* (pp. 129–141). New York, NY: Springer.

Langley, G. L., Moen, R., Nolan, K. M., Nolan, T. W., Norman, C. L., & Provost, L. P. (2009). *The improvement guide: A practical approach to enhancing organizational performance* (2nd ed.). San Francisco, CA: Jossey-Bass.

Linder, S. (2011). The facilitator's role in elementary mathematics professional development. *Mathematics Teacher Education and Development, 13*(2), 44–66.

Males, L. M., Otten, S., & Herbel-Eisenmann, B. A. (2010). Challenges of critical colleagueship: Examining and reflecting on mathematics teacher study group interactions. *Journal of Mathematics Teacher Education, 13*, 459–471.

Marra, R., Arbaugh, F., Lannin, J., Abell, S., Ehlert, M., Smith, R., ... Rogers, M. P. (2011). Orientations to professional development design and implementation: Understanding their relationship to PD outcomes across multiple projects. *International Journal of Science and Mathematics Education, 9*, 793–816.

Matsumura, L. C., Garnier, H. E., Slater, S. C., & Boston, M. D. (2008). Toward measuring instructional interactions "at-scale." *Educational Assessment, 13*, 267–300.

McGraw, R., Lynch, K., Koc, Y., Budak, A., & Brown, C. A. (2007). The multimedia case as a tool for professional development: An analysis of online and face-to-face interaction among mathematics pre-service teachers, in-service teachers, mathematicians, and mathematics teacher educators. *Journal of Mathematics Teacher Education, 10*, 95–121.

McMeeking, L. B. S., Orsi, R., & Cobb, R. B. (2012). Effects of a teacher professional development program on the mathematics achievement of middle school students. *Journal for Research in Mathematics Education, 43*, 159–181.

Muir, T., & Beswick, K. (2007). Stimulating reflection on practice: Using the supportive classroom reflection process. *Mathematics Teacher Education and Development, 8*, 74–93.

Muñoz-Catalán, M. C., Yañez, J. C., & Rodríguez, N. C. (2010). Mathematics teacher change in a collaborative environment: To what extent and how. *Journal of Mathematics Teacher Education, 13*, 425–439.

National Center for Educational Statistics. (2013). *Characteristics of public and private elementary and secondary school teachers in the United States: Results from the 2011–12 schools and staffing survey* (Report 2013-314). Washington, DC: National Center for Education Statistics.

National Governors Association Center for Best Practices & Council of Chief State School Officers. (2010). *Common Core State Standards for Mathematics*. Washington, DC: Author. Retrieved from http://www.corestandards.org/read-the-standards/

National Science Foundation. (2015). *Discovery research preK-12: Program solicitation* (NSF 15-592). Washington, DC: Author. Retrieved from http://www.nsf.gov/publications/pub_summ.jsp?ods_key=nsf15592

New Zealand Numeracy Development Projects. (2010). *Findings from the New Zealand Numeracy Development Projects, 2009*. Wellington, New Zealand: Learning Media.

Nickerson, S. D., & Moriarty, G. (2005). Professional communities in the context of teachers' professional lives: A case of mathematics specialists. *Journal of Mathematics Teacher Education, 8*, 113–140.

Norton, A. H., & McCloskey, A. (2008). Teaching experiments and professional development. *Journal of Mathematics Teacher Education, 11*, 285–305.

Orrill, C. H., & Brown, R. E. (2012). Making sense of double number lines in professional development: Exploring teachers' understandings of proportional relationships. *Journal of Mathematics Teacher Education, 15*, 381–403.

Penuel, W. R., Fishman, B., Cheng, B. H., & Sabelli, N. (2011). Organizing research and development at the intersection of learning, implementation, and design. *Educational Researcher, 40*(7), 331–337.

Penuel, W. R., Gallagher, L. P., & Moorthy, S. (2011). Preparing teachers to design sequences of instruction in earth science: A comparison of three professional development programs. *American Educational Research Journal, 48*, 996–1025.

Phillips, K. J., Desimone, L., & Smith, T. M. (2011). Teacher participation in content-focused professional development & the role of state policy. *Teachers College Record, 113*, 2586–2630.

Porter, A. C., Archbald, D., & Tyree, A., Jr. (1990). Reforming the curriculum: Will empowerment policies replace control? In S. H. Fuhrman & B. Malen (Eds.), *The politics of curriculum and testing: The 1990 yearbook of the Politics of Education Association* (pp. 11–36). New York, NY: Falmer.

Porter, A. C., Floden, R., Freeman, D., Schmidt, W., & Schwille, J. (1988). Content determinants in elementary school mathematics. In D. Grouws & T. Cooney (Eds.), *Perspectives on research on effective mathematics teaching* (pp. 96–113). Reston, VA: National Council of Teachers of Mathematics.

Putnam, R. T., & Borko, H. (2000). What do new views of knowledge and thinking have to say about research on teacher learning? *Educational Researcher, 29*(1), 4–15.

Richardson, V., & Anders, P. (1995). *A theory of teacher change and the practice of staff development*. New York, NY: Teachers College Press.

Rogers, M., Abell, S., Lannin, J., Wang, C., Musikul, K., Barker, D., & Dingman, S. (2007). Effective professional development in science and mathematics education: Teachers' and facilitators' views. *International Journal of Science and Mathematics Education, 5*, 507–532.

Ross, J. A., & Bruce, C. D. (2007). Teacher self-assessment: A mechanism for facilitating professional growth. *Teaching and Teacher Education, 23*, 146–159.

Rubel, L. H., & Chu, H. (2011). Reinscribing urban: Teaching high school mathematics in low income, urban communities of color. *Journal of Mathematics Teacher Education, 15,* 39–52.

Sack, J. J. (2008). Commonplace intersections within a high school mathematics leadership institute. *Journal of Teacher Education, 59,* 189–199.

Santagata, R. (2009). Designing video-based professional development for mathematics teachers in low-performing schools. *Journal of Teacher Education, 60,* 38–51.

Santagata, R., Kersting, N., Givvin, K. B., & Stigler, J. W. (2011). Problem implementation as a lever for change: An experimental study of the effects of a professional development program on students' mathematics learning. *Journal of Research on Educational Effectiveness, 4,* 1–24.

Scher, L., & O'Reilly, F. (2009). Professional development for K–12 math and science teachers: What do we really know? *Journal of Research on Educational Effectiveness, 2,* 209–249.

Schifter, D., Bastable, V, & Russell, S. J. (with Cohen, S., Lester, J. B., & Yaffee, L.). (1999a). *Developing mathematical ideas: Number and operations, part 1. Building a system of tens: Casebook.* Parsippany, NJ: Dale Seymour.

Schifter, D., Bastable, V, & Russell, S. J. (with Yaffee, L., Lester, J. B., & Cohen, S.). (1999b). *Developing mathematical ideas: Number and operations, part 2. Making meaning for operations: Casebook.* Parsippany, NJ: Dale Seymour.

Schön, D. A. (1983). *The reflective practitioner.* New York, NY: Basic Books.

Scott, A., Clarkson, P., & McDonough, A. (2012). Professional learning and action research: Early career teachers reflect on their practice. *Mathematics Education Research Journal, 24,* 129–151.

Seago, N., Jacobs, J., Driscoll, M., Callahan, P., Matassa, M., & Nikula, J. (in press). *Learning and teaching geometry: Videocase materials for mathematics professional development.* San Francisco, CA: WestEd; Reston, VA: National Council of Teachers of Mathematics.

Seago, N., Mumme, J., & Branca, N. (2004). *Learning and teaching linear functions: Video cases for mathematics professional development, 6–10.* Portsmouth, NH: Heinemann.

Seidel, T., Stürmer, K., Blomberg, G., Kobarg, M., & Schwindt, K. (2011). Teacher learning from analysis of videotaped classroom situations. Does it make a difference whether teachers observe their own teaching or that of others? *Teaching and Teacher Education, 27,* 259–267.

Sherin, M. G. (2004). New perspectives on the role of video in teacher education. In J. Brophy (Ed.), *Using video in teacher education* (pp. 1–27). New York, NY: Elsevier.

Sherin, M. G., & Han, S. Y. (2002). Teacher learning in the context of a video club. *Teaching and Teacher Education, 20,* 163–183.

Sherin, M. G., Linsenmeier, K., & van Es, E. A. (2009). Selecting video clips to promote mathematics teachers' discussion of student thinking. *Journal of Teacher Education, 60,* 213–230.

Sherin, M. G., & van Es, E. A. (2009). Effects of video club participation on teachers' professional vision. *Journal of Teacher Education, 60,* 20–37.

Silver, E. A., Clark, L. M., Ghousseini, H. N., Charalambous, C. Y., & Sealy, J. T. (2007). Where is the mathematics? Examining teachers' mathematical learning opportunities in practice-based professional learning tasks. *Journal of Mathematics Teacher Education, 10,* 261–277.

Sowder, J. T. (2007). The mathematical education and development of teachers. In F. K. Lester Jr. (Ed.), *Second handbook of research on mathematics teaching and learning* (pp. 157–223). Charlotte, NC: Information Age; Reston, VA: National Council of Teachers of Mathematics.

Stein, M. K., & Coburn, C. E. (2008). Architectures for learning: A comparative analysis of two urban school districts. *American Journal of Education, 114,* 583–626.

Stein, M. K., Grover, B., & Henningsen, M. (1996). Building student capacity for mathematical thinking and reasoning: An analysis of mathematical tasks used in reform classrooms. *American Educational Research Journal, 33,* 455–488.

Stein, M. K., Smith, M. S., Henningsen, M., & Silver, E. A. (2000). *Implementing standards-based mathematics instruction: A casebook for professional development.* New York, NY: Teachers College Press.

Stone, J. R., III, Alfeld, C., & Pearson, D. (2008). Rigor and relevance: Enhancing high school students' math skills through career and technical education. *American Educational Research Journal, 45,* 767–795.

Swan, M. (2000). GCSE mathematics in further education: Challenging beliefs and practice. *The Curriculum Journal, 11,* 199–223.

Swan, M. (2006a). Designing and using research instruments to describe the beliefs and practices of mathematics teachers. *Research in Education, 75,* 58–70.

Swan, M. (2006b). Learning GCSE mathematics through discussion: What are the effects on students? *Journal of Further and Higher Education, 30,* 229–241.

Swan, M. (2007). The impact of task-based professional development on teachers' practices and beliefs: A design research study. *Journal of Mathematics Teacher Education, 10,* 217–237.

Swan, M. (2011). Designing tasks that challenge values, beliefs and practices: A model for the professional development of practicing teachers. In P. Sullivan & O. Zaslavski (Eds.), *Constructing knowledge for teaching secondary mathematics: Tasks to enhance prospective and practicing teacher learning* (pp. 57–71). Dordrecht, The Netherlands: Springer.

Swan, M., Pead, D., Doorman, M., & Mooldijk, A. (2013). Designing and using professional development resources for inquiry-based learning. *ZDM—The International Journal on Mathematics Education, 45,* 945–957.

Swan, M., & Swain, J. (2010). The impact of a professional development programme on the practices and beliefs of numeracy teachers. *Journal of Further and Higher Education, 34,* 165–177.

Sztajn, P. (2011). Standards for reporting mathematics professional development in research studies. *Journal for Research in Mathematics Education, 42,* 220–236.

Taylor, E. V. (2011). Supporting children's mathematical understanding: Professional development focused on out-of-school practices. *Journal of Mathematics Teacher Education, 15,* 271–291.

Thompson, P. W., Carlson, M. P., & Silverman, J. (2007). The design of tasks in support of teachers' development of coherent mathematical meanings. *Journal of Mathematics Teacher Education, 10,* 415–432.

Ticha, M., & Hospesova, A. (2006). Qualified pedagogical reflection as a way to improve mathematics education. *Journal of Mathematics Teacher Education, 9*, 129–156.

Tirosh, D., (2008). Tools and processes in mathematics teacher education: An introduction. In D. Tirosh & T. Wood (Eds.), *International handbook of mathematics teacher education: Vol. 2. Tools and processes in mathematics teacher education* (pp. 1–11). Amsterdam, The Netherlands: Sense.

van Es, E. A. (2009). Participants' roles in the context of a video club. *Journal of the Learning Sciences, 18*, 100–137.

van Es, E. A., & Sherin, M. G. (2008). Mathematics teachers' "learning to notice" in the context of a video club. *Teaching and Teacher Education, 24*, 244–276.

van Es, E. A., & Sherin, M. G. (2010). The influence of video clubs on teachers' thinking and practice. *Journal of Mathematics Teacher Education, 13*, 155–176.

van Es, E. A., Tunney, J., Goldsmith, L. T., & Seago, N. (2014). A framework for the facilitation of teachers' analysis of videos. *Journal of Teacher Education, 65*, 340–356.

Watson A., & Mason, J. (2007). Taken-as-shared: A review of common assumptions about mathematical tasks in teacher education. *Journal of Mathematics Teacher Education, 10*, 205–215.

Webster-Wright, A. (2009). Reframing professional development through understanding authentic professional learning. *Review of Educational Research, 72*, 702–739.

Wickstrom, M. H., Baek, J., Barrett, J. E., Cullen, C. J., & Tobias, J. M. (2012). Teachers' noticing of children's understanding of linear measurement. In *Proceedings of the 34th annual meeting of the North American Chapter of the International Group for the Psychology of Mathematics Education* (pp. 488–494). Kalamazoo, MI: Western Michigan University.

Wilson, P. H., Sztajn, P., Edgington, C., & Confrey, J. (2014). Teachers' use of their mathematical knowledge for teaching in learning a mathematics learning trajectory. *Journal of Mathematics Teacher Education, 17*, 149–175.

Wilson, P. H., Sztajn, P., Edgington, C., & Myers, M. (2015). Teachers' uses of a learning trajectory in student-centered instructional practices. *Journal of Teacher Education, 66*, 227–244.

Wilson, S. M. (2013). Professional development for science teachers. *Science, 340*(6130), 310–313.

Witterholt, M., Goedhart, M., Suhre, C., & van Streun, A. (2012). The interconnected model of professional growth as a means to assess the development of a mathematics teacher. *Teaching and Teacher Education, 28*, 661–674.

Yoon, K. S., Duncan, T., Lee, S. W.-Y., Scarloss, B., & Shapley, K. (2007). *Reviewing the evidence on how teacher professional development affects student achievement* (REL 2007–No. 033). Washington, DC: U.S. Department of Education, Institute of Education Sciences, National Center for Education Evaluation and Regional Assistance, Regional Educational Laboratory Southwest.

Young-Loveridge, J. (2008). *Patterns of performance and progress of NDP students in 2008*. Retrieved from http://nzmaths.co.nz/findings-nz-numeracy-development-projects-2008

Appendix

Initial List of Journals

American Educational Research Journal
American Journal of Education
Cognition and Instruction
Educational Evaluation and Policy Analysis
Educational Researcher
Educational Studies in Mathematics
For the Learning of Mathematics
International Journal of Science and Mathematics Education
Journal for Research in Mathematics Education
Journal of Mathematical Behavior
Journal of Mathematics Teacher Education
Journal of Research on Educational Effectiveness
Journal of Teacher Education
Journal of the Learning Sciences
Mathematical Thinking and Learning: An International Journal
Mathematics Education Research Journal
Mathematics Teacher Education and Development
Review of Educational Research
Teachers College Record
Teaching and Teacher Education

30 课程研究的若干重要问题：基于证据的发现和今后的研究方向

格温德林·M.劳埃德
美国宾夕法尼亚州立大学
蔡金法
美国特拉华大学
詹姆斯·E.塔尔
美国密苏里大学
译者：聂必凯
美国特拉华大学

一个世纪以来，数学课程一直被认为是教育改革的主要杠杆。为了应对学生成绩落后的问题，教育改革方案寻求通过加宽或加深学校数学课程的方式，以提高学生的成绩，为他们的大学学习和工作做准备，以及增强一个国家在日益全球化的社会中的竞争优势。从历史上看，学校数学课程一直是改进全世界学生学习的核心机制（Cai & Howson, 2013；Howson, Keitel, & Kilpatrick, 1981；Reys & Reys, 2010；Senk & Thompson, 2003；Stein, Remillard, & Smith, 2007；van den Heuvel-Panhuizen, 2000）。事实上，数学教育改革的提倡者常常试图通过修改课程来改变课堂教学实践，进而影响学生的学习。鉴于此，课程一直是数学教育学术探究的焦点问题。近年来，随着政策环境的变化，包括世界各国增加的教师问责制，关于课程在数学教育中的作用的研究变得更加重要。归根结底，课程研究的价值在于它提供了理解数学在课堂中是如何教以及如何学的机会。

本章将给读者提供当前数学课程研究的一个历史背景。我们关注的是2005年以来的有关研究，缘由有二：其一，在过去十年中，有关数学课程的研究急剧增长，研究的方法也更为复杂。其二，斯坦等人（2007）综述了2004年以及之前的研究，本章是对他们工作的拓展。尽管本章旨在综述过去十年的数学课程研究，只要合适，我们也会追溯课程研究的历史。在本章的中心部分，我们确定了自2005年以来发表在重要研究期刊上的课程研究论文所呈现的数学课程研究的进展和趋势。在此过程中，我们通过对美国和世界其他国家的重要期刊论文和学术专著进行回顾，致力于从国际视野来思考数学课程的研究。再者，我们讨论了数学课程研究中方法论上的挑战和重大问题，并对数学课程未来研究的方向提出了建议。

总的来说，在数学教育研究文献中，课程这一术语的用法很多。本章中的课程特指那些书面的课程材料和教科书。如斯坦等人（2007）那样，我们将关注那些学生和教师在数学课堂里使用最频繁的课程资源。尽管我们承认教育政策和标准与数学课程有紧密的联系，但在我们的回顾中将不涉及对政策和标准的研究。再者，由于多数研究文献关注的是纸质的课程材料，本章也不对数字材料进行综述。（有关现代技术与课程的讨论，请参见这套书 Roschelle, Noss, Blikstein, & Jackiw, 2017那一章。）

本章中的一些研究是受美国国家科学基金资助的（ESI-0454739，DRL-1008536，EHR-9983393，REC-0532214，DUE-0536678）。本文所表达的任何观点均为作者的观点，并不代表美国国家科学基金会。

30 | Issues in Curriculum Studies: Evidence-Based Insights and Future Directions

GWENDOLYN M. LLOYD
Pennsylvania State University, United States

JINFA CAI
University of Delaware, United States

JAMES E. TARR
University of Missouri, United States

For nearly a century, mathematics curriculum has been considered a primary lever for educational reform. In response to lagging student performance, reform initiatives have sought to increase the breadth or depth of school mathematics to boost student achievement, prepare students for college and the workplace, and increase a nation's competitive edge in an increasingly global society. Historically, the school mathematics curriculum has been a central mechanism for improving student learning around the world (Cai & Howson, 2013; Howson, Keitel, & Kilpatrick, 1981; Reys & Reys, 2010; Senk & Thompson, 2003; Stein, Remillard, & Smith, 2007; van den Heuvel-Panhuizen, 2000). In fact, advocates of mathematics education reform often attempt to modify the curriculum to change instructional practice and thereby influence student learning. For these reasons, curriculum has long been the focus of scholarly inquiry in mathematics education. With changing policy contexts that include increased teacher accountability in countries around the world, this scholarship about the role of curriculum in mathematics education has become even more important in recent years. Ultimately, however, the value of curriculum research lies in the opportunities it provides to understand how mathematics is taught and learned in classrooms.

In this chapter, we provide a historical context for current research about mathematics curriculum. We focus our review on research since 2005 for two reasons. First, in the past decade, the body of scholarship about mathematics curriculum has grown considerably and research methods have become more sophisticated. Second, we build our chapter on Stein et al.'s (2007) review in which research through about 2004 is considered. While our chapter aims to synthesize the research literature on mathematics curriculum conducted in the past 10 years, whenever appropriate, we provide a historical trace of curriculum studies. In the central sections of this chapter, we identify advances and trends in mathematics curriculum studies published in major research journals since 2005. In so doing, we make a deliberate effort to consider mathematics curriculum scholarship internationally by reviewing major research journals and books from the United States and countries around the world. Furthermore, we discuss methodological challenges and issues in this area and suggest future directions for research about mathematics curriculum.

Within the body of mathematics education research literature, the term *curriculum* is used in a variety of ways. In this chapter, *curriculum* refers to written curricular materials and textbooks. Like Stein et al. (2007), we have chosen to focus on the resources with which students and teachers work most closely in mathematics classrooms. We do not include research about policies and standards in our review, although we acknowledge their relationship with mathematics curriculum. Furthermore, because the majority of the research literature to date is based on print curricular materials, we do not address digital materials in this review. (For a discussion of technology and curriculum, please see the chapter by Roschelle, Noss, Blikstein, & Jackiw, 2017, this volume.)

Some of the research included in this chapter was supported by grants from the National Science Foundation (ESI-0454739, DRL-1008536, EHR-9983393, REC-0532214, and DUE-0536678). Any opinions expressed herein are those of the authors and do not necessarily represent the views of the National Science Foundation.

我们将从以下三个层面来分析数学课程——预期课程、执行课程，以及获得课程（Gehrke, Knapp, & Sirotnik, 1992；Husén, 1967；Schmidt, McKnight, & Raizen, 1997；Valverde, Bianchi, Wolfe, Schmidt, & Houang, 2002）。图30.1（改编自Cai, 2014）是对课程这一概念的解读，预期课程是教育系统层面（例如，国家、地区或学校水平）对数学学习的期望，包含教科书和标准。由于本章关注的是书面课程材料和教科书，因此，预期课程被用来指教科书的作者对数学应该教什么和如何教的期望。在数学课堂里，执行课程指数学的教与学的过程，表现为师生与课程材料和教学任务之间的互动（Snyder, Bolin, & Zumwalt, 1992）。最后，获得课程指学生实际所学到的。对数学课程概念三个层面的解读描述了教科书作者的预期、课堂教学的实际发生，以及学生的实际所学。

这一框架曾用于第一次国际数学研究（FIMS；Husén, 1967）和第二次国际数学研究（SIMS；Travers, 1992），已被数学教育研究界广泛接受和使用，并用于指导数学教育的研究（例如，美国国家研究协会[NRC]，2004）。由于课程分析框架的每一层面（预期课程、执行课程、获得课程）支持着其他两个层面，因此理解每一层面及其与其他层面之间的关系是课程研究的一个重要目标。事实上，美国国家研究协会（2004）指出高质量的课程评价研究往往始于阐明课程的重要构成部分，这些部分可以用于区分该课程材料与其他的课程（预期课程）。课程研究也要透彻地调查课程实施的特点，包括是否所有学生体验了该课程、课程材料被实施的情况以及教学材料中所包括的评估所起的作用（执行课程）。最后，美国国家研究协会指出课程研究必须仔细测评课程对学生数学学习的影响（获得课程）。与之相应地，本章的展开也将基于预期、执行和获得课程这样一个概念框架。

本章一开始，就将近来数学课程的研究置于一定的历史背景之下。其中三个主要的部分将关注与预期课程、执行课程和获得课程相关的数学教育研究的重要发现和走向。同时我们还考虑了每个课程层面在概念上和方法论上的发展和挑战。最后，本章总结了过去十年来课程研究的主要走向，并提出了未来几年课程研究中所需要关注的领域。

历史背景下的数学课程研究

预期课程

课程影响学生的学习已成为广泛的共识，因此，分析预期课程已成为近一百年来的一个学术研究方向（例如 Clapp, 1924；O.L. Davis, 1962；Husén, 1967）。早在1924年，克拉普就分析了学校数学教科书中基本的数字问题出现的频次和顺序。他对分析的过程给出了详细的描述，并对教科书中基本的数字问题的出现频次和顺序与学生数学学习的困难之间的关系给出了详细的研究结果。大约60年后，哈曼和阿什克拉夫特（1986）报告了与克拉普相似的关于教科书中基本的加法问题的相对难度和频次的发现。对预期课程的分析作为学术研究的一个方向得到广泛认同还应归功于1959年的第一次国际数学研究。尽管第一次国际数学研究的目的是关注不同国家的学生在不同数学内容上的成绩，但是预期课程的构成仍被作为是导致学生成绩显著差异的其中一个潜在因素而被加以探究（Husén, 1967）。

课程分析被接纳为数学教育学术研究的一个方向的

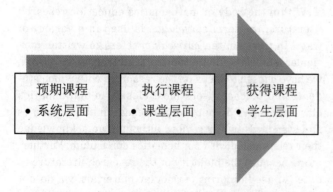

图30.1 课程三个层面框架图。改编自 "Searching for Evidence of Curricular Effect on the Teaching and Learning of Mathematics: Some Insights From the LieCal Project" by J.Cai 2014, Mathematics Education Research Journal, 26(4), p.812。

We examine mathematics curriculum at three levels—the intended curriculum, the enacted curriculum, and the attained curriculum (Gehrke, Knapp, & Sirotnik, 1992; Husén, 1967; Schmidt, McKnight, & Raizen, 1997; Valverde, Bianchi, Wolfe, Schmidt, & Houang, 2002). In this conceptualization of curriculum, illustrated in Figure 30.1 (adapted from Cai, 2014), the *intended curriculum* refers to educational-system level (e.g., national, district, or school level) expectations for mathematics learning, including textbooks and standards. Because we focus on written curricular materials and textbooks, the intended curriculum is used to refer to textbook authors' intentions about what and how mathematics should be taught. At the classroom level, the *enacted curriculum* refers to the processes of teaching and learning mathematics that emerge as teachers and students interact with curricular materials and tasks in classrooms (Snyder, Bolin, & Zumwalt, 1992). Finally, the *attained curriculum* refers to what is learned by students. The three-level conceptualization of mathematics curriculum delineates authors' intentions, what happens in classrooms, and what students learn.

This framework, used in the First International Mathematics Study (FIMS; Husén, 1967) and the Second International Mathematics Study (SIMS; Travers, 1992), has been widely accepted and used to guide research in mathematics education (e.g., National Research Council [NRC], 2004). Because each level of curriculum (intended, enacted, attained) supports the other two in the curriculum process, understanding the three levels—individually and in relation to one another—is an important goal of curriculum research. In fact, the NRC (2004) recommended that high-quality curriculum evaluation studies begin with an articulation of the program components that differentiate curricular materials from one another (*intended curriculum*). Studies must also thoughtfully examine implementation characteristics, including whether all students experience the program, how well curricular materials are implemented, and the role of assessments embedded in instructional materials (*enacted curriculum*). Finally, the NRC recommended that research must carefully measure the impact of curriculum on students' mathematics learning (*attained curriculum*). Accordingly, our chapter is organized based on this conceptualization of intended, enacted, and attained curriculum.

We begin the chapter by situating recent research about mathematics curriculum in a historical context. Three main sections then focus on major findings and trends in mathematics education research with respect to intended curriculum, enacted curriculum, and attained curriculum. We also consider conceptual and methodological advances and challenges at each level of curriculum. Finally, the chapter summarizes major trends in curriculum research of the past decade and suggests areas in need of work in the years to come.

Situating Mathematics Curriculum Research in a Historical Context

Intended Curriculum

Because of the common view that curriculum can influence student learning, analysis of the intended curriculum has been a line of scholarly inquiry for nearly a century (e.g., Clapp, 1924; O. L. Davis, 1962; Husén, 1967). In 1924, Clapp analyzed the frequency and order of presentation of basic number problems in school mathematics textbooks. He provided detailed descriptions of the analysis procedures and findings for his investigation about the relationships between the frequency and order of the basic number problems and related student difficulties. About 60 years later, Hamann and Ashcraft (1986) reported findings, similar to those of Clapp, about the relative difficulty and frequency of basic addition problems in textbooks. Wider acceptance of the analysis of intended curriculum as a line of scholarly inquiry can be credited to FIMS in 1959. Although the purpose of FIMS was to examine students' achievement in various mathematics topics in different countries, the composition of the intended curriculum was explored as one of the underlying factors contributing to notable differences in student achievement (Husén, 1967).

The acceptance of curriculum analysis as a line of scholarly inquiry in mathematics education is further

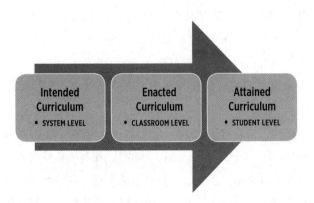

FIGURE 30.1. *Framework of three levels of curriculum. Adapted from "Searching for Evidence of Curricular Effect on the Teaching and Learning of Mathematics: Some Insights From the LieCal Project" by J. Cai, 2014, Mathematics Education Research Journal, 26(4), p. 812.*

进一步实证是重要的数学教育研究杂志都已发表过有关预期课程分析的文章。对预期数学课程的分析报告也已发表在一些非特定学科的教育研究期刊上。《数学教育研究》早期发表的文章所关注的重要主题之一就是对课程的设计和内容的讨论（例如B.H. Griffiths, 1971; Leung, 1987; Turnau, 1980）。例如，图尔纳曾关注波兰数学教科书的设计原则，并指出随版次变化而产生的教科书特点的变化。梁（1987）也详细描述了中国预期课程的目标和内容。然而，《认知和教学》是最早发表基于预期数学课程分析的实证研究的杂志之一（Hamann & Ashcraft, 1986; Stigler, Fuson, Ham, & Kim, 1986）。《数学教育研究学报》在1988年出版了第一篇有关预期数学课程分析的实证研究。在该研究中，富森、施蒂格勒和巴奇（1988）分析了小学阶段引入加减法内容的年级水平。这些有关预期课程的早期研究为我们考查数学教科书的内容和设计提供了新的视野，同时也引发了关于书面的课程材料与学生学习机会之间关系的重要问题的讨论。

执行课程

无论课程设计得多好，如果没能在课堂中得以实施，它们也不会有多大的价值（Cai, 2014）。一个多世纪以来，有关教学方法和课程材料对教与学的影响一直是数学教育课程开发者和研究者所关注的问题。20世纪一些最广为人知的课程探索发生在20世纪50年代和60年代的"新数学"时期，不少新颖的课程材料得以在这一时期开发和研究（Fey & Graeber, 2003）。其中一个著名的研究叫全国数学能力的纵向研究，该研究对学校数学研究小组（SMSG）所编写的教科书进行了大规模的有效性评价（Osborne, 1975）。尽管所用的研究方法在20世纪50年代和60年代算得上是成熟的，但评价者在探寻执行课程的性质这方面的能力有限。长期以来，关联执行课程与获得课程的研究往往有相当的难度，特别是研究方法上的重大挑战（Fey, 1980; Fullan & Pomfret, 1977）。其实，在1968年，比格即沮丧地认为课程执行中涉及的变量太多，找到一种能够对不同课程处理方法进行有意义的比较的途径来解释这些变量是随之而来的挑战。

直到最近，对执行课程的系统研究仍然很少见。富兰和庞弗雷特（1977）对美国、加拿大和英国有关课程执行的定量研究进行了综述，他们发现这些研究关注数学课程的很少。有一个例外，即埃文斯和谢弗（1974，被富兰和庞弗雷特所引用）所做的"执行的程度——初步近似"的研究，他们使用了一个11道题目（还含小题）的量表来测量一个强调个性化的数学课程的执行程度。即使在更为宽泛的课程研究领域中，出于对一些改革项目的评价的需要，对执行课程的研究直到20世纪60年代和70年代才开始成为主流研究方向之一（Snyder等，1992）。要决定教育革新项目为何成功或不成功，对课程执行有较深入的了解是非常必要的。

芝加哥大学学校数学项目（UCSMP）作为持续时间最长（一直还在进行）的数学课程项目之一，自20世纪80年代以来，一直进行着教科书评价方面的研究（例如 Hirschhorn, 1993; Mathison, Hedges, Stodolsky, Flores, & Sarther, 1989; Thompson & Senk, 2001）。正如汤普森和申克（2001）所描述的那样，在早期的评估研究中，UCSMP试图记录所有覆盖的内容，但多年以来，他们不断地扩展研究量表的种类，来考查课程的执行。这些研究量表包括教师问卷调查表、教师访谈方案、课堂观察指导和学生表格。UCSMP对课程执行忠实度的考虑早于之后颁布的指令（例如NRC, 2004），这些指令要求研究人员在研究课程有效性时，必须分析执行课程。

在20世纪80年代末和20世纪90年代初，对教师使用数学教科书的不同调查研究产生了多样化的结果，有些研究表明教师的教科书使用方式是照本宣科（例如Barr, 1988），而有些研究则认为教师在使用数学教科书时进行了精心考量（Freeman & Porter, 1989; Sosniak & Stodolsky, 1993; Stodolsky, 1989）。加利福尼亚州的州级改革政策引导了新数学教科书的出现，在该州进行的有关政策研究中，一系列案例研究阐明了教师有权对教科书进行解读的观点（例如Ball & Cohen, 1996; Cohen, evidenced by the fact that all the major research journals in mathematics education have published articles reporting analyses of intended curriculum. Reports of analyses of intended mathematics curriculum have also been published in non-discipline-specific educational research journals. Discussion of the design and content of curricula was one of the major themes among early articles published in *Educational Studies in Mathematics* (e.g., B. H. Griffiths, 1971; Leung, 1987; Turnau, 1980). For example, Turnau focused on the design principles of mathematics textbooks in Poland and identified characteristic changes from one edition to another. Leung (1987) described in detail the objectives and content of the intended curriculum in China. However, *Cognition and Instruction* was one of the earliest journals to publish empirical research based on analyses of intended mathematics curriculum (Hamann & Ashcraft, 1986; Stigler, Fuson, Ham, & Kim, 1986). In 1988, the *Journal for Research in Mathematics Education* published its first empirical study on the analysis of intended mathematics curriculum. In that study, Fuson, Stigler, and Bartsch (1988) analyzed the grade levels at which the topics of addition and subtraction were introduced in the primary grades. These early studies of the intended curriculum provided new insights into the content and design of mathematics textbooks and generated important questions about relationships between written curricular materials and students' opportunities to learn.

Enacted Curriculum

Regardless of how well curricular materials are designed, they have little value outside of their implementation in classrooms (Cai, 2014). Questions about the impact of instructional methods and curricular materials on teaching and learning have been a focus of mathematics education curriculum developers and researchers for over a century. Some of the most widely publicized curriculum efforts of the 20th century took place during the "new math" era of the 1950s and 1960s, a period in which novel curricular materials were developed and studied (Fey & Graeber, 2003). One well-known study was the large-scale evaluation of the effectiveness of the textbooks of the School Mathematics Study Group (SMSG) in the National Longitudinal Study of Mathematical Abilities (Osborne, 1975). Although the research methods of the 1950s and 1960s were sophisticated for their time, the evaluators' capacity to capture the nature of the enacted curriculum was limited. Research linking the enacted and attained curriculum has long been seen as difficult, with significant methodological challenges (Fey, 1980; Fullan & Pomfret, 1977). Indeed, in 1968, Begle lamented the large number of variables involved with curriculum enactment and the associated challenge of accounting for them in a way that allows meaningful comparison of different curricular treatments.

Systematic research about the enacted curriculum was quite sparse until fairly recently. Fullan and Pomfret's (1977) review of quantitative studies that measured implementation in the United States, Canada, and the United Kingdom revealed few studies focused on mathematics curriculum. One exception was "Degree of Implementation: A First Approximation," a study conducted by Evans and Scheffler (1974, as cited in Fullan & Promfret), who used an 11-item scale (with subitems) to measure the degree of implementation of an individualized mathematics curriculum. Even in the broader field of curriculum studies, research about curriculum enactment did not become a major line of inquiry until the 1960s and 1970s when demand for evaluations of innovative programs increased (Snyder et al., 1992). Richer understandings of curriculum enactment became viewed as essential in determining why innovations were or were not successful.

As one of the longest running (and ongoing) mathematics curriculum projects, the University of Chicago School Mathematics Project (UCSMP) has engaged in research to evaluate its textbooks since the 1980s (e.g., Hirschhorn, 1993; Mathison, Hedges, Stodolsky, Flores, & Sarther, 1989; Thompson & Senk, 2001). As Thompson and Senk (2001) describe, beginning with early evaluation studies, UCSMP has attempted to document content coverage and, over the years, has expanded its corpus of instruments to examine curriculum enactment. Among these instruments are teacher questionnaires, teacher interview protocols, classroom observation guidelines, and student forms. UCSMP's consideration of the fidelity of implementation predated imperatives (e.g., NRC, 2004) to attend to the enacted curriculum in studies of curricular effectiveness.

Investigations of teachers' use of mathematics textbooks in the late 1980s and early 1990s produced mixed findings, with some studies suggesting that teachers progress through textbook lessons in a lockstep manner (e.g., Barr, 1988), and others finding that teachers are deliberate in their use of mathematics textbooks (Freeman & Porter, 1989; Sosniak & Stodolsky, 1993; Stodolsky, 1989). The notion that textbooks are open to interpretation by teachers was illustrated in a series of case studies that emerged from policy research

1990；Wilson，1990）。类似地，佩平和哈格蒂的研究记录了英国、法国和德国在课程材料以及教师和学生在课堂教学中使用课程材料方式的差异（Pepin，2014；Pepin & Haggarty，2001）。这些研究通过不同的方式预示了在一般的课程实施方面，尤其是教师使用课程材料方面，在未来二十年可能会显现的那些发现和问题。

正如斯坦等人（2007）的综述所指出的，数学教育领域自20世纪90年代以来，对课程的研究活动出现了显著的增长。对数学课程研究的日益关注可能得益于多方面的影响力。在20世纪90年代和21世纪头十年，世界各地涌现了新课程改革的浪潮（例如Australian Education Council，1990；Ministry of Education，1992；全美数学教师理事会[NCTM]，1989）。接着，大量的新课程材料得以开发，以影响课堂教学和提高学生成绩。随着这些课程材料在课堂上得以采用和实施，研究者对这些材料中的数学和教学法的新表征形式对学生、教师和学校意味着什么产生了浓厚的兴趣。在美国，由于没有国家课程，加上美国学生的数学表现令人失望，因此，一些专业组织制定了一系列课程标准文件（例如NCTM，1989，1991），这些标准为学校数学开辟了一个新的视野，还推动了美国国家科学基金会（NSF）投资近1亿美元用于开发基于标准的课程材料（也称为美国国家科学基金会资助的课程材料）。这些课程材料与传统的出版商开发的美国数学教科书在根本上有所不同，即它们更加强调培养数学思考、数学推理和解决问题的能力，而较少强调常规技能和符号操作（Nathan，Long，& Alibali，2002；Schoenfeld，2006；Senk & Thompson，2003）。

虽然一个主要的问题是，并将继续是，新课程材料的执行会如何影响学生的数学学习，20世纪90年代和21世纪头十年的研究者也开始对教师使用新课程材料的经历感兴趣。关于学生和教师与新课程材料互动的研究将有可能为专业发展和教师教育、未来的课程开发和各地选取教科书的决策提供指导。

获得课程

在美国和全球教育政策转变的背景下，提供课程项目影响学生数学学习的证据变得尤为重要。以新的联邦教育政策为标志的美国问责运动（No Child Left Behind Act，2002），也对新课程方案有效性研究提出了很高的要求（Schoenfeld，2002）。由于基于标准的课程材料开始占据更多美国市场份额，这些联邦政策要求各州开发严格的标准，并要求学校，至少是那些接受联邦资助的学校，使用科学的、基于研究的干预措施来改善学生的学习和缩小学习成绩之间的差距。在其他国家，国家授权或推动的课程改革也旨在影响数学教学的质量，并提高学生的学习成绩（例如，Boesen等，2014；Charalambous & Philippou，2010；Gooya，2007；Ponte，Matos，Guimarães，Leal，& Canavarro，1994）。另外，20世纪90年代的改革时代以这样一种观点为标志，即认为教育研究缺乏质量和科学的严谨性，这个观点也同样适用于描述数学课程方面的研究。一个重要讯息就是美国教育科学研究院院长在美国教育研究协会2003年年会的特邀报告中指出的，有关课程有效性的科学研究十分稀缺：

> 最近，一位学区总监问我，最好的小学数学课程是什么。我说现有的研究还没能回答这个问题；我所能提供的一切都是我的观点[斜体为另加]。他说他已经有足够的理论观念了，一线的人们……希望我们的研究界能够帮助他们理解这些观念并为这些观念提供证据。他们感到他们还没有这些。（Whitehurst，2003，第9—10页）

虽然对高质量研究的需求是显而易见的，但致力于判定一个课程方案有效性的研究是非常复杂的，部分原因在于书面材料可以通过多种方式转化为课堂教学。

要对数学课程对学生学习的影响做出结论，就要求研究者明确地解释在课堂中执行课程材料这一过程的性质。正如美国国家研究协会（2004）研究综述所发现的，许多数学教科书和课程材料有效性的研究有一个共同的弱点，就是没能系统地关注课程的执行。在综述了对19

conducted in California, where new mathematics textbooks emerged in response to state-level reform policies (e.g., Ball & Cohen, 1996; Cohen, 1990; Wilson, 1990). Similarly, research by Pepin and Haggarty documented differences in both curricular materials and their uses by teachers and students in classrooms in England, France, and Germany (Pepin, 2014; Pepin & Haggarty, 2001). These studies, in many ways, foreshadowed findings and questions about curriculum enactment in general and teachers' use of curricular materials in particular that would emerge over the next two decades.

The field of mathematics education has experienced a significant surge in research activity about curriculum since the 1990s, as reviewed by Stein et al. (2007). There have likely been multiple forces contributing to this increased attention to the study of mathematics curriculum. During the 1990s and 2000s, new curriculum reform efforts emerged around the world (e.g., Australian Education Council, 1990; Ministry of Education, 1992; National Council of Teachers of Mathematics [NCTM], 1989). In turn, numerous new curricular materials were developed to influence classroom instruction and improve student achievement. As these curricular materials were adopted and implemented in classrooms, researchers became interested in what the new representations of mathematics and pedagogy in these materials would entail for students, teachers, and schools. In the United States, in the absence of a national curriculum and in the wake of disappointing mathematics performance by American students, professional organizations produced a series of standards documents (e.g., NCTM, 1989, 1991) that provided a new vision for school mathematics. The Standards were an impetus for the National Science Foundation (NSF) to invest nearly $100 million in the development of Standards-based curricular materials (also called NSF-funded materials) that differed from traditional publisher-developed U.S. mathematics textbooks in fundamental ways; namely, by placing increased emphasis on the development of mathematical thinking, reasoning, and problem solving, and less emphasis on routine skills and symbol manipulation (Nathan, Long, & Alibali, 2002; Schoenfeld, 2006; Senk & Thompson, 2003).

Although a primary question was, and continues to be, how the enactment of new curricular materials might affect students' learning of mathematics, researchers of the 1990s and 2000s also became interested in teachers' experiences using new curricular materials. Research on students' and teachers' interactions with new curricular materials would have the potential to provide implications for professional development and teacher education, future curriculum development, and local decision making about textbook adoption.

Attained Curriculum

Providing evidence about the impact of curriculum programs on student learning in mathematics became decidedly more important given the shifting educational policy context in the United States and around the world. The accountability movement in the United States, marked by new federal educational policies (No Child Left Behind Act, 2002), placed research about the effectiveness of new curriculum programs in high demand (Schoenfeld, 2002). As Standards-based curricular materials began to comprise more of the U.S. market share, these policies required states to develop rigorous standards and schools—at least those receiving federal funding—to use scientific, research-based interventions to improve student learning and close the achievement gap. In other countries, nationally mandated or promoted curricular changes were also intended to affect the quality of mathematics instruction and increase student achievement (e.g., Boesen et al., 2014; Charalambous & Philippou, 2010; Gooya, 2007; Ponte, Matos, Guimarães, Leal, & Canavarro, 1994). In addition, the reform era in the 1990s was marked by a view, broadly applied to curriculum research in mathematics, that educational research lacks quality and scientific rigor. The dearth of scientific research on curricular effectiveness was a key message in the 2003 invited address of the director of the U.S. Institute of Educational Sciences at the annual meeting of the American Educational Research Association:

> Recently, a district superintendent asked me what was the best mathematics curriculum for elementary school students. *I said there was no research that provided an answer; that all I could offer was my opinion* [italics added]. He said he had enough opinions already. The people on the front lines... want us, the research community, to provide them with a way to cut through the opinion and advocacy with evidence. They feel they aren't getting that. (Whitehurst, 2003, pp. 9–10)

Although the demand for high-quality research was clear, research that aims to determine the effectiveness of a curriculum program is enormously complex, in part due to the multitude of ways that written materials can be transformed into classroom instruction.

Drawing conclusions about the effects of mathematics programs on student learning demands that

个美国数学课程项目（13个得到美国国家科学基金的支持，另外6个是商业性的）的实证研究之后，美国国家研究协会提出如下建议：

> 课程评估应该对课程材料实施的范围、质量和类型给出可信和有效的指标作为评估的证据。至少，课程评估应该论及课程材料的覆盖范围（某些研究者称之为"学习的机会"）以及教师专业发展所需达到的程度和类型。（第194页）

此外，该综述委员会还指出，利用有针对性的案例研究来记录和理解课程的执行是有意义的。他们写道："案例研究常常能揭示课程方案构成与课程执行要素的各个方面，以及这两者之间的相互作用，特别是与课程方案设计者的意图相左的相互作用，因此，案例研究能对课程的有效性提供重要的见解"（第201页）。

在国际上，关于数学课程的问题一直是国际教育协会（IEA）大规模跨国研究的重点，这些研究包括1959年的第一次国际数学研究，1981年的第二次国际数学研究，以及1995年的第三次国际数学与科学研究。值得一提的是第二次国际数学研究收集了教师和学生的课堂活动问卷，并通过这些数据来研究执行课程（Travers, 1992）。为了对教学实践和学生学习提供重要的见解，参与这些研究的国家在国际数学与科学趋势研究项目中进行了更为频繁的合作研究（每四年进行一次）。他们的研究结果在全球受到了高度的期待和监测，并经常由媒体宣布"获胜者"（de Lange, 2007）。政策制定者利用国际研究成果，努力地改进各自国家的学校数学，认真审查学生表现优异的那些国家的课程材料和教学方法，并倡导更严格的标准。例如，美国《州共同核心数学标准》的第一句话："十多年来，针对高分国家的数学教育的研究表明，美国数学课程必须变得更加集中和连贯，以提高这个国家学生的数学成绩"（National Governors Association Center for Best Practices & Council of Chief State School Officers, 2010, 第1页）。在这一意义上，数学课程研究的发现可以对本地和全球的数学教育都产生重大的影响。

在本节中，我们为本章的其余部分提供了一个宽广的历史背景。对数学课程及其对教学和学习的潜在影响的注重，可以追溯到教育研究的早期阶段。随着教育研究领域的成熟，课程研究出现了清晰的研究路线和进展方向。为了突显课程研究的主要发现和走向，我们将从预期课程开始。

关于预期课程的主要发现和走向

一些研究者对国内外不同类型的课程进行了分析和比较，甚至还对课程进行了历史性的分析和比较（例如Baker 等, 2010; Grevholm, 2011; R. Griffiths, 1987; Pepin, Gueudet, & Trouche, 2013）。这样的研究有助于我们理解数学思想的引入、呈现和定义的不同方法，并帮助我们认识特定课程的独特性和文化背景。对预期课程分析最有吸引力的价值也许是其对数学教学和学习的潜在影响（Cai & Cirillo, 2014）。

已有充足的研究证实，许多教师完全依靠教科书进行教学（Thomson & Fleming, 2004）。魏斯（1987）的调查显示，约90%的美国小学和中学教师依靠单一的数学教科书进行数学教学。此外，教师报告说教科书以外的内容很少会在课堂教学里出现（Ball & Feiman-Nemser, 1988; Tarr, Chávez, Reys, & Reys, 2006; Weiss, 1987）。二十五年后的后续调查发现，情况基本未变，近85%的小学和中学教师使用单一的数学教科书（Banilower等, 2013）。

鉴于教科书是教师的主要资源，研究者试图通过研究预期课程来了解学生的学习机会是不足为奇的（例如Porter, 1995; Schmidt, McKnight, Valverde, Houang, & Wiley, 1997）。虽然课程不应该被视为能够直接管理和安排学生的学习经历，但预期课程确实为探查学生学习机会提供了见解。

researchers account, in some explicit way, for the nature of the enactment of curricular materials in the classroom. As an NRC (2004) review found, systematic attention to curriculum enactment has been a weakness in many studies of the effectiveness of mathematics textbooks and curricular materials. Based on their review of empirical studies of 19 American mathematics curriculum programs (13 developed with NSF support and 6 generated commercially), the NRC's recommendations included the following:

> Evaluations should present evidence that provides reliable and valid indicators of the extent, quality, and type of the implementation of the materials. At a minimum, there should be documentation of the extent of coverage of curricular material (what some investigators refer to as "opportunity to learn") and of the extent and type of professional development required. (p. 194)

Furthermore, the review committee pointed to the value of utilizing targeted case studies to document and understand curriculum enactment, writing, "case studies often reveal aspects of program components, implementation components, and interactions among these two that behave differently than intended by program designers, and therefore provide essential insights into curricular effectiveness" (p. 201).

Internationally, questions about mathematics curriculum have long been the focus of the large-scale cross-national studies of the International Education Association (IEA), including FIMS in 1959, SIMS in 1981, and the Third International Mathematics and Science Study in 1995. Notably, the enacted curriculum was studied in SIMS via teacher and student questionnaires about classroom activities (Travers, 1992). To provide key insights into instructional practices and student learning, these cooperative ventures among participating nations are being conducted more frequently (every 4 years) in the Trends in International Mathematics and Science Study. Results are highly anticipated and monitored across the globe, and the media often declare "winners" (de Lange, 2007). Policy makers use results of international studies in efforts to improve school mathematics in their respective countries, carefully scrutinizing the curricular materials and instructional approaches of high-performing nations and lobbying for more rigorous standards. Consider, for instance, the very first sentence of the Common Core State Standards for Mathematics in the United States: "For over a decade research studies of mathematics education in high-performing countries have pointed to the conclusion that the mathematics curriculum in the United States must become substantially more focused and coherent in order to improve mathematics achievement in this country" (National Governors Association Center for Best Practices & Council of Chief State School Officers, 2010, p. 1). In this sense, findings of mathematics curriculum studies can have significant impacts, both locally and globally.

In this section, we offered a broad historical backdrop for the remainder of the chapter. Attention to mathematics curriculum and its potential to influence teaching and learning can be traced to the early days of educational research. As the field has matured, clear lines of inquiry and advances in research on curriculum have emerged. To highlight the major findings and trends about curriculum studies, we begin with the intended curriculum.

Major Findings and Trends About the Intended Curriculum

Researchers have conducted analyses and comparisons of different types of curricula within and across nations, as well as historical analyses and comparisons of curricula (e.g., Baker et al., 2010; Grevholm, 2011; R. Griffiths, 1987; Pepin, Gueudet, & Trouche, 2013). Such research helps us understand ways to introduce, present, and define mathematical ideas and recognizes the unique features and cultural contexts of particular curricula. Perhaps the most compelling value of analysis of intended curriculum is its potential impact on mathematics teaching and learning (Cai & Cirillo, 2014).

It has been well documented that many teachers rely exclusively on textbooks to teach (Thomson & Fleming, 2004). A survey by Weiss (1987) showed that about 90% of elementary and secondary teachers in the United States relied on a single textbook for mathematics instruction. In addition, teachers reported that content not found in textbooks was rarely taught (Ball & Feiman-Nemser, 1988; Tarr, Chávez, Reys, & Reys, 2006; Weiss, 1987). A follow-up survey 25 years later found the situation largely unchanged; nearly 85% of elementary and secondary teachers used a single mathematics textbook (Banilower et al., 2013).

Given the role of textbooks as a primary resource for teachers, it is no wonder that researchers have tried to examine the intended curriculum to understand students' opportunity to learn (e.g., Porter, 1995; Schmidt, McKnight, Valverde, Houang, & Wiley, 1997). Although curriculum should not be viewed as directly managing and ordering students' learning experiences, the

过去，研究者已经研究了书面课程材料中涵盖的数学主题、主题在各年级的安排、从一个年级到另一个年级对新内容的介绍、跨年级数学内容的引入和发展，以及某一特定数学内容的学习目标。自2005年以来，在预期课程的研究中出现了以下三个走向：（1）对书面课程中问题的分析；（2）对具体内容领域或数学专题的分析；（3）对数学教科书历史发展的分析。在本节中，我们将讨论与这三个走向相关的主要研究发现。

对课程中数学问题的分析

分析预期课程中的问题，其动机来源于多伊尔（1983）关于学术任务的研究。多伊尔认为不同认知需求的问题可能会引起不同类型的学习。问题不仅能控制学生对某些内容特定方面的关注，而且还主导学生处理信息的方式。无论情境如何，有价值的任务应该是有趣的，并具有一定挑战性，从而能激发问题解决者的探索、猜想和努力（NCTM，2000）。具有真实疑难并且涉及重要数学的数学任务具有为学生数学学习提供智力情境的可能性。多伊尔（1983）定义了四种一般类别的学术任务：记忆性任务、程序性或常规性任务、理解性任务和意见性任务（Doyle，1983）。斯坦、格罗弗和亨宁森（1996）扩展了多伊尔（1983）的工作，并提出了一个数学任务框架（MTF），用来检查预期课程的问题及其在数学课堂实施中的问题的认知需求。在这个框架下，每个任务都可被划归到以下四个逐渐递增的认知需求水平之一：记忆、无关联的程序、有关联的程序和做数学。

琼斯和塔尔（2007）利用这一框架分析了在数学教育改革的四个不同时代（"新数学"运动、回到基础、问题解决和课程标准）出版的教科书中概率任务所需的认知需求水平。他们从每个时代选出了两个教科书系列：一个是在当时拥有最大市场份额的流行系列，另一个是与当时创新的改革理念相适应的替代系列。琼斯和塔尔发现课程标准时代的教科书系列比前三个时代的系列更加重视概率。事实上，他们分析的所有概率任务中有一半以上都出自课程标准时代的教科书中，或流行系列或替代系列。这一发现与覆盖统计和概率内容的国际趋势是一致的（Cai & Howson，2013）。在任务的认知需求方面，课程标准时代的两个系列含有需要高认知需求的（做数学和有关联的程序）任务的比例要高得多。

蔡、聂和莫耶（2010）在他们的课程对代数学习影响的纵向研究（LieCal）项目中也使用了数学任务框架。研究者分析了五套初中数学课程中的代数任务，而不是概念任务，其中包括了琼斯和塔尔（2007）所选择的两个课程标准时代的教科书系列。蔡等人发现关联数学项目（CMP）课程，也是琼斯和塔尔研究中代表课程标准时代的替代系列教科书，比流行的格伦科课程具有更高的较高认知需求的任务的百分比。这些研究结果表明，CMP和格伦科的课程设计者在撰写各自教科书系列时似乎有不同的意图。这些例子还表明了分析预期课程中的任务的价值，因为更高认知需求的任务为学生提供潜在的学习机会，同时，教师非常依赖课程材料进行教学是不争的事实。

还有一些研究者也分析了预期课程中的任务（Bieda, Ji, Drwencke, & Picard, 2014；Fan & Zhu, 2007；Lithner, 2004；Thompson, Senk, & Johnson, 2012；Vincent & Stacey, 2008）。例如，汤普森等人（2012）分析了所选的美国高中教科书中的练习和问题，以了解高中代数1，代数2和微积分预备教科书中以下代数三大主题中与证明相关的推理的学习机会：（1）指数，（2）对数，（3）多项式、方程和函数。他们发现，在大约6%的练习中，学生有机会参与与证明相关的推理，而从代数1到微积分预备，这一百分比翻倍了。范和朱（2007）的研究分析了教科书中的数学问题，以考查中国、新加坡和美国的中学数学教科书中所要求使用的问题解决的策略有哪些。他们发现，与中国和新加坡教科书相比，美

intended curriculum does provide insights into opportunities for students' learning.

In the past, researchers have examined the mathematical topics included in written curricular materials, the grade level placement of topics, the introduction of new topics from one grade to another, the introduction and development of mathematical topics across grades, and the learning goals for a particular mathematical topic. Since 2005, three trends have emerged in research on the intended curriculum: (1) analysis of the problems in written curricula, (2) analysis of specific content areas or mathematical topics, and (3) analysis of the historical development of mathematics textbooks. In this section, we discuss key findings related to these three trends.

Analyses of the Mathematics Problems in Curricula

The motivation for analyzing problems from the intended curriculum is rooted in Doyle's (1983) research on academic tasks. Doyle argued that problems with different cognitive demands are likely to induce different kinds of learning. Problems govern not only students' attention to particular aspects of content but also their ways of processing information. Regardless of the context, worthwhile tasks should be intriguing, and with a level of challenge that invites exploration, speculation, and hard work (NCTM, 2000). Mathematical tasks that are genuinely problematic and involve significant mathematics have the potential to provide the intellectual contexts for students' mathematics learning. Doyle (1983) defined four general categories of academic tasks: memory tasks, procedural or routine tasks, comprehension or understanding tasks, and opinion tasks (Doyle, 1983). Stein, Grover, and Henningsen (1996) extended the work of Doyle (1983) and developed a mathematical task framework (MTF) to examine the cognitive demand of the problems both in the intended curriculum and as it is implemented in mathematics classrooms. In this framework, each task is classified into four increasingly demanding categories of cognition: memorization, procedures without connections, procedures with connections, and doing mathematics.

Jones and Tarr (2007) used this framework to analyze the levels of cognitive demand required by probability tasks in textbooks published across four eras of mathematics education reform (new math, back to basics, problem solving, and standards). They selected two textbook series from each era: a popular series that had the largest market share during that era and an alternative series that was developed to align with the innovative ideas of the respective era. Jones and Tarr found that standards-era textbook series placed a much greater emphasis on probability than series from the previous three eras. In fact, more than half of all probability tasks analyzed were included in standards-era textbooks, popular or alternative. This finding is consistent with international trends of content coverage in statistics and probability (Cai & Howson, 2013). In terms of the cognitive demand of the tasks, the two series in the standards era included much larger percentages of tasks requiring high cognitive demand (doing mathematics and procedures with connections).

Cai, Nie, and Moyer (2010) also used the MTF in their Longitudinal Investigation of the Effect of Curriculum on Algebra Learning (LieCal) project. Instead of probability tasks, the researchers analyzed the algebra tasks in five middle school mathematics curricula, including the same two standards-era textbook series selected by Jones and Tarr (2007). Cai et al. found that the *Connected Mathematics Project* (*CMP*) curriculum, the alternative series representing the standards era in Jones and Tarr's study, had a greater percentage of tasks with higher cognitive demand than those in the popular Glencoe curriculum. These findings suggest that the curriculum designers of *CMP* and Glencoe appear to have had different intentions when writing their textbook series. These examples also demonstrate the value of analyzing tasks in intended curriculum because of the potential learning opportunities that higher cognitive demand tasks afford students and the fact that teachers rely heavily on curricular materials to teach.

Other researchers have also analyzed the tasks in intended curricula (Bieda, Ji, Drwencke, & Picard, 2014; Fan & Zhu, 2007; Lithner, 2004; Thompson, Senk, & Johnson, 2012; Vincent & Stacey, 2008). For example, Thompson et al. (2012) analyzed exercises and problems in selected U.S. high school textbooks to understand the learning opportunities for proof-related reasoning on three broad topics in algebra from high school algebra 1, algebra 2, and precalculus textbooks: (1) exponents; (2) logarithms; and (3) polynomial expressions, equations, and functions. They found that in about 6% of the exercises, students have opportunities to engage with proof-related reasoning, with the percent doubling from algebra 1 to precalculus. Fan and Zhu (2007) analyzed mathematical problems to understand which problem-solving strategies selected middle school mathematics textbooks in China, Singapore, and the United States required. They found that the U.S. textbooks were much more likely to adopt "draw a diagram" and "make a table"

国教科书采用"画一个示意图"和"做一个表格"这些问题解决策略的可能性更大。

利瑟内（2004）分析了瑞典微积分教科书中的练习题，以检查这些习题中所涉及的推理类型。他发现，近70%的练习问题可以通过搜寻类似的情况来解决，甚至只需要模仿教科书中的类似例题来解决。蔡、罗和渡边（2002）将所选的日本、中国等亚洲国家和美国课程中的例题和练习题分为三类，以分析每一课程如何促进学生对"平均数"这一概念的理解。A类问题要求学生直接应用平均数的公式。B类问题要求学生使用多个步骤并灵活地应用算法。C类问题要求学生能恰当地使用并在统计情境中解释平均数概念。他们发现美国的两套改革系列的教科书（《关联数学》和《情境中的数学》）更侧重于平均数概念的统计意义（即平均数是数据集的一个代表），而三个亚洲教科书系列和美国的传统教科书系列更侧重于把这一概念当作求平均数的一个算法。这些课程材料之间的差异是非常重要的，因为它们可以反映不同课程中学生学习机会的潜在差异。

对预期课程中特定内容的处理的分析

研究预期课程的第二个走向是通过精细的分析关注特定的数学内容。这些数学内容包括比例推理（Shield & Dole, 2013）、方程和解方程（Cai 等, 2010）、推理与证明（Otten, Gilbertson, Males, & Clark, 2014）、分配律（Ding & Li, 2010）、分数（Alajmi, 2012; Charalambous, Delaney, Hsu, & Mesa, 2010; Son & Senk, 2010）、文字题（Xin, 2007）、二次方程（Hong & Choi, 2014），等号（McNeil 等, 2006）和变量（Nie, Cai, & Moyer, 2009）。虽然这些分析包括不同年级的不同内容，并且包括来自不同国家的课程，但这些分析的重点是：课程如何为学生学习这些数学内容创造潜在的机会。

在回顾了大量文献之后，希尔德和多尔（2013）提出了一个框架用来考察教科书是否为学生提供了深入学习比例推理的机会。该框架由五个课程目标组成：（1）使用真实的生活情境来对比加法式比较和乘法式比较，（2）识别乘法结构和比例推理，（3）使用有意义的符号表征，（4）明确地联系相关的分数思想，（5）有效地使用一系列表征。对于每个目标，他们还确定了三个具体指标。他们对澳大利亚的5套教科书系列按照每个指标（高，中，低）进行编码，以确定课程促进学生深度学习的机会。他们发现这些教科书系列"不大具备支持教师促进学生在比例推理上的健全发展的能力"（第196页）。他们进一步指出，教科书系列特别缺乏多种表征的使用以及对乘法结构和比例推理的辨别。根据分析，他们得出的结论是：不常使用表格表征可能会导致学生没有充足的机会来体验比例推理情境里的乘法结构。

丁和李（2010）选取了两套在美国广泛使用的小学教科书系列和一套主要的中国教科书系列，并从以下三个维度分析了有关分配律的引入和呈现：（1）问题的情境，（2）每一问题情境下的典型问题类型，（3）使用分配律的变式。虽然他们的研究被置于比较的背景下，但他们分析的主要目的是理解教科书中分配律的引入和呈现是如何为促进学生学习提供机会的。丁和李首先回顾了认知心理学中对使用例子、自我解释、具体和抽象表征以及深层次问题的研究结果。他们发现，美国的这两套教科书系列都偏向计算，但中国系列则偏向于概念。在美国教科书系列中，分配律的例子主要涉及整数，而中国教科书系列的例子则涉及整数、小数、分数和百分数。类似地，查拉兰布斯等人（2010）对塞浦路斯和爱尔兰等使用的小学数学教科书中的分数加法和减法的内容进行了比较分析。他们发现，这些教科书对分数加法和减法的处理在内容覆盖、安排顺序、分数的建构、样例、任务的认知需求以及希望学生给出的回答类型等方面有很多相似和差异之处。

strategies than those of the Chinese and Singaporean textbooks.

Lithner (2004) analyzed exercise problems from a calculus textbook used in Sweden to examine the kinds of reasoning involved. He found that nearly 70% of the exercise problems could be solved by searching for similar situations and even by searching only the worked-out examples in the textbook. Cai, Lo, and Watanabe (2002) from selected Asian (Japan, China) and U.S. curricula into three categories to capture the and U.S. curricula into three categories to capture the kinds of understanding each curriculum promotes about the concept of average. Category A includes problems that require students to directly apply the average algorithm to solve them. Category B includes problems that require students to apply the algorithm flexibly in several steps to solve them. Category C includes problems that focus on the appropriate use and interpretation of the concept of average in the statistical context. They found that the two U.S. reform series (*CMP* and *Mathematics in Context*) focused more on the statistical aspect of the concept of average (as a representative of a data set) whereas the three Asian series and the U.S. traditional series focused more on the concept as an algorithm to find the average. These differences in curricular materials are significant because they may reflect potential differences in learning opportunities for students.

Analyses of the Treatment of Specific Topics in the Intended Curriculum

A second trend in research on intended curriculum is a focus on particular mathematical topics through fine-grained analyses. These mathematical topics include proportional reasoning (Shield & Dole, 2013), equations and equation solving (Cai et al., 2010), reasoning and proving (Otten, Gilbertson, Males, & Clark, 2014), the distributive property (Ding & Li, 2010), fractions (Alajmi, 2012; Charalambous, Delaney, Hsu, & Mesa, 2010; Son & Senk, 2010), word problems (Xin, 2007), quadratic equations (Hong & Choi, 2014), the equals sign (McNeil et al., 2006), and variables (Nie, Cai, & Moyer, 2009). Although the analyses cut across different topics at different grade levels and include curricula from different countries, the focus of these analyses is on the potential opportunities created for students to learn these mathematical topics.

Based on an extensive review of the literature, Shield and Dole (2013) developed a framework to examine whether textbooks provide opportunities for students' deep learning of proportional reasoning. The framework consists of five curriculum goals: (1) contrasting additive and multiplicative comparison through the use of authentic life-related situations, (2) identifying multiplicative structure and proportional reasoning, (3) using meaningful symbolic representation, (4) explicitly connecting related fraction ideas, and (5) effectively using a range of representations. For each goal, they also identified three specific indicators. They coded five textbook series from Australia for each of the indicators (in terms of high, medium, or low) to identify opportunities for promoting students' deep learning. They found that these textbook series were "limited in their capacity to promote teaching that supports a robust development of students' proportional reasoning" (p. 196). They further indicated that the textbook series were particularly lacking in the use of multiple representations and the identification of multiplicative structures and proportional reasoning. Based on their analysis, they concluded that the sparse use of tabular representations may provide limited opportunity for students to see the multiplicative structure of proportional reasoning situations.

Ding and Li (2010) analyzed the introduction and presentation of the distributive property in two widely used U.S. elementary textbook series and one main Chinese textbook series along three dimensions: (1) problem contexts, (2) typical problem types within each problem context, and (3) variability in using the distributive property. Although they situated their study in a comparative context, the main goal of their analysis was to understand how the presentation and introduction of the distributive property in the textbooks provide opportunities to promote students' learning. Ding and Li first reviewed the findings from cognitive psychology about using worked-out examples, self-explanation, concrete and abstract representations, and deep questions. They found that, whereas the two U.S. textbook series had a computational orientation, the Chinese series had a conceptual orientation. In the U.S. series, the instances of the distributive property mainly involved whole numbers, whereas the Chinese textbook series' instances involved whole numbers, decimals, fractions, and percentages. Similarly, Charalambous et al. (2010) conducted a comparative analysis of the addition and subtraction of fractions in elementary mathematics textbooks used in Cyprus and Ireland. They found a number of similarities and differences among the textbooks regarding the topics of addition and subtraction of fractions covered and their sequencing, the constructs of fractions, the worked examples, the cognitive demands of the tasks, and the types of responses required of students.

洪和崔（2014）用类似的方法分析了美国和韩国高中教科书，重点是二次方程。研究者发现，一套美国高中改革教科书采用了函数取向的方法，其中变量被定义为一个变化的、描述二次关系的量，而韩国教科书采取结构取向的方法，强调符号和代数表征。虽然韩国教科书早于美国教科书引入了与二次方程相关的概念，但是，与韩国教科书相比，美国教科书中包括了更高比例的高认知需求的二次方程问题。对这些研究的综述结果揭示了课程向学生提供具体内容的潜在学习机会的类型。

对教科书的历史分析

预期课程研究的第三个走向是从历史的角度进行分析（例如 Baker 等，2010；Cai, Jiang, Hwang, Nie, & Hu，2016；Donoghue，2003；Jones & Tarr，2007；Nicely，1991；Sinclair，2008）。一些研究者追溯了一个世纪以来预期课程的历史发展，而另一些研究者则研究了一个国家改革运动之间或之中的课程变化。最著名的两项研究是贝克等人（2010）的研究和辛克莱（2008）的研究。

贝克等人（2010）研究分析了1900年至2000年间美国出版的141种小学数学教科书。在他们的分析中，从教科书的抽样到确定要分析的课程要素，研究者克服了许多来自方法论的困难。他们的研究发现揭示了美国课程发展的许多里程碑，例如，他们发现，20世纪头二十年的教科书页数（平均87页）比后几十年的教科书（平均300多页）页数明显要少。在上世纪中叶之前，整数的基本运算是主要内容，然而，从20世纪60年代中期开始，几何和测量的内容明显增多，算术也更多地涉及小数、分数、百分数以及比和比例。另外，从20世纪60年代中期到90年代，教科书中问题解决策略的内容也在增加。20世纪最后35年的教科书减少了对死记硬背、机械地应用算法和背记算术事实的要求，相反，在20世纪后期，小学数学教科书中问题认知需求水平不断提升是有目共睹的。20世纪70年代中期，统计、概率和数据分析的内容已出现在幼儿园和一年级的教科书中，而这些内容在上世纪中叶时直到四年级才出现。历史上，越来越宽泛的美国数学课程被认为是"一英里宽、一英寸深"（Schmidt, McKnight, & Raizen, 1997, 第62页）。然而，贝克等人（2010）则提出了一些不同的观点：自20世纪60年代中期以来，美国课程在广度和深度方面似乎都有所增加。

辛克莱（2008）采取了不同的方法来研究过去150年来美国几何课程的演进。与贝克等人（2010）在研究中使用精细的内容分析方法不同，她按时间顺序分析了导致几何课程实质性变化的"重大事件"。这些重大事件是经过考查了重要的研究报告或委员会报告、有影响力的教科书和学习理论的新发展后才确定的。这些事件影响了几何学习的目标和几何课程的组成。她发现在过去150年中几何课程的发展有好几个具有里程碑意义的事件。例如，她发现在20世纪初期，几何是教科书中测量章节中的一部分。在"新数学"时期（20世纪60年代至20世纪70年代），小学低年级阶段开始通过使用学具和模型来教授立体几何。直到20世纪80年代，学校的数学教科书才将几何大量地从测量内容中分离出来。

贝克等人（2010）和辛克莱（2008）的研究分析表明，课程发展趋势与美国主要的数学教育改革运动是一致的。虽然目前有关预期课程的历史分析大都是关于美国课程的，但是蔡等人（2016）的分析是一个例外，他们分析了中国小学教科书两个版本中（相应于国家课程标准发布前后的版本）问题提出的任务。让学生提出问题是中国国家课程标准的建议之一。蔡等人发现即使问题提出的任务在两个版本教科书中均只占任务总数很小的一部分，但课程标准发布后的版本却包括了更多明确的问题提出的活动和例题。

预期课程的研究方法问题

在数学教育中，迄今为止，对预期课程的研究方法的讨论几乎没有。然而，随着这条学术研究线路的发展，

Hong and Choi (2014) used similar approaches to analyze U.S. and Korean high school textbooks with a focus on quadratic equations. These researchers found that a U.S. high school reform-based textbook took a functional approach in which *variable* is defined as a quantity that varies and describes quadratic relations, whereas the Korean textbooks took a structural approach, emphasizing symbolic and algebraic representations. Although the Korean textbooks introduced the concepts related to quadratic equations earlier than the U.S. textbook series, the U.S. textbooks included larger percentages of problems involving quadratic equations of high cognitive demand than did the Korean textbooks. The findings from the studies reviewed reveal the kinds of potential topic-specific learning opportunities offered to students.

Historical Analyses of Textbooks

A third trend in research on intended curriculum is analysis from a historical perspective (e.g., Baker et al., 2010; Cai, Jiang, Hwang, Nie, & Hu, 2016; Donoghue, 2003; Jones & Tarr, 2007; Nicely, 1991; Sinclair, 2008). Some researchers have traced the historical development of the intended curriculum over a century while others have studied the changes of curriculum between or among the reform movements in a nation. Two of the most notable studies are those by Baker et al. (2010) and Sinclair (2008).

Baker et al. (2010) analyzed 141 elementary school mathematics textbooks in the United States published between 1900 and 2000. In their analysis, the researchers overcame a number of methodological challenges, from sampling the textbooks to identifying the curricular elements to be analyzed. Their findings revealed a number of important milestones in curricular development in the United States. For example, they found that textbooks had fewer pages in the first two decades of the 20th century (87 pages, on average) than the later decades (more than 300 pages, on average). Before midcentury, basic arithmetic with whole numbers was the dominant content. However, beginning in the mid-1960s, substantially more geometry and measurement appeared, as did more arithmetic involving decimals, fractions, percentages, and ratios and proportions. In addition, the coverage for problem-solving strategies increased from the mid-1960s to 1990s. Textbooks of the last 35 years of the 20th century decreasingly required rote drill and mechanical application of algorithms and memorization of arithmetic facts. Instead, an increased cognitive demand was evident in the problems of elementary mathematics textbooks throughout the rest of the 20th century. In the mid-1970s, the content topics of statistics, probability, and data analysis started to be introduced in kindergarten and first-grade textbooks instead of in fourth grade, as was the practice at midcentury. Historically, by becoming increasingly broad, the U.S. mathematics curriculum has been perceived as "a mile wide and an inch deep" (Schmidt, McKnight, & Raizen, 1997, p. 62). However, the findings of Baker et al. (2010) suggest something different: since the mid-1960s, the U.S. curriculum appears to have increased both in breadth *and* depth.

Sinclair (2008) took a different approach to investigating the evolution of the geometry curriculum in the United States in the past 150 years. Instead of the fine-grained content analysis in the study by Baker et al. (2010), Sinclair chronologically analyzed the "notable events" leading to substantial changes in geometry curriculum. Notable events were identified by locating important reports or commissions, influential textbooks, and new developments in learning theories. These events affected the goals of learning geometry and the composition of the curriculum. She discovered several milestones of the geometry curriculum in the past 150 years. For example, she found that in the early 1900s, geometry was included in textbooks in the measurement sections. In the new-math period (1960s–1970s), solid geometry was to be taught in lower elementary grades using manipulatives and models. It was not until the 1980s that geometry was largely separated from measurement in school mathematics textbooks.

The analyses by Baker et al. (2010) and Sinclair (2008) demonstrate that trends in curriculum are aligned with the major mathematics education reform movements in the United States. Although most historical analyses of the intended curriculum have been about U.S. curricula, an exception is the analysis by Cai et al. (2016) of problem-posing tasks in two editions of Chinese elementary textbooks (the editions corresponding to before and after the release of the national curriculum standards). Engaging students in problem posing is one of the recommendations in the Chinese national curriculum standards. Cai et al. found that even though problem-posing tasks occupy a small proportion of the total number of tasks in both editions of the textbooks, the later edition included more purposeful problem-posing activities and worked out examples.

Methodological Issues in Research on Intended Curriculum

In mathematics education there has been, to date, little discussion about methodological issues in research

一些研究者最近探讨了该研究领域的方法论问题（Cai & Cirillo, 2014；Stylianides, 2014）。这些研究者指出，相对于学生学习和教师教学的研究，预期课程的研究是很不成熟的。这是因为"制定明确的通用准则来指导文本分析，这方面的工作是非常稀缺的"（Nicholls, 2003，第11页）。

蔡和西里洛（2014）确定了研究预期课程时需要考虑的几个基本问题：

- 我们应该分析多少种教科书？
- 我们应该分析哪种教科书？
- 我们应该分析教科书中的什么文本？阐述部分？练习部分？
- 我们应该分析多少文本？
- 我们应该如何分析（即，我们用什么研究框架进行分析）？
- 哪些研究问题可以引导我们的分析？

类似地，斯蒂联尼德斯（2014）讨论了分析预期课程中"推理和证明"时在方法论上的三个挑战：（1）如何确定去哪里寻找推理和证明的活动，（2）如何在分析中确定教师的指导作用，（3）如何确定分析应取的理论视角。虽然他的讨论是关于"推理和证明"的，但其他数学内容领域的预期课程研究也面临着类似的挑战。

除了需要仔细说明为什么课程分析是一种适当的方法选择之外，选择适当的理论框架来分析课程也是至关重要的。框架不仅将决定要分析什么，还将决定如何进行分析。在上述三个预期课程研究走向中，研究者利用不同的资源来开发各自的课程分析框架。为了分析预期课程中的任务，数学任务框架（Stein等，1996）是使用较频繁的一个框架。分析某一特定的数学内容是预期课程研究的第二个走向，研究者通常会针对特定数学内容来开发相应内容的分析框架。比例推理的框架（Shield & Dole, 2013）和推理与证明（Stylianides, 2009）的框架就是两个很好的典范。在预期课程的历史分析中，研究者将重点放在从不同的维度来分析预期课程。美国国家研究协会（2004）已经确定了近20个可能的预期课程分析维度，其中包括内容的罗列、内容呈现的顺序、内容呈现的方式、内容的重点、初次引入的年级水平、所需的阅读水平、解释的类型和使用、形式化的途径、信息技术或学具的使用、评价类型及其与课堂实践的关系。

到目前为止，对预期课程所做的研究对课程材料进行了不同的抽样和分析，有的分析许多书籍，有的则分析一个年级的整本书，有的分析一些章节或模块，有的仅分析几节课或几个数学内容。课程材料的抽样应由研究问题决定。但是，需要抽样和分析多少课程材料才能回答这些研究问题呢？我们建议将来的预期课程研究应提供有力的论证来阐述是如何对所分析的特定材料进行抽样的。

类似地，预期课程的研究还要使用不同的分析单位，包括页面、任务、句子和关键词。恰当的分析单位应该是什么？使用不同的分析单位会对推广研究发现带来挑战。因此，未来的预期课程研究应提供有力的论证来阐述选择的分析单位是合适的。

执行课程研究的主要发现和走向

教师和学生在课堂上执行课程时，势必会改变书面课程材料中的预期要求（Snyder等，1992）。虽然预期课程与执行课程之间会存在差异是一直存在的看法，但近年来，研究者对这些差异的界定、测量和理解取得了重大进展（Heck, Chval, Weiss, & Ziebarth, 2012；Thompson & Usiskin, 2014）。虽然一些大型研究项目的主要目的是建立课程材料与学生成绩之间的关系，而研究课程执行只是其中的一部分，但也有研究把课程执行作为研究重点，这些研究考查了由教师计划的，并在特定的学校和社区环境中由教师和学生在课堂上实际执行的课程。这些从不同角度进行的研究，采用了不同的研究问题、概念框架和研究方法，从而形成了越来越成熟的关于数学课程执行的知识体系。

about intended curriculum. However, as this line of scholarly inquiry advances, some researchers have recently begun to discuss methodological issues (Cai & Cirillo, 2014; Stylianides, 2014). These researchers have pointed out that, relative to research about students' learning and teachers' instruction, research about the intended curriculum is fundamentally underdeveloped. This is because "surprisingly little work has been done in terms of setting out clear generic guidelines for analyzing texts" (Nicholls, 2003, p. 11).

Cai and Cirillo (2014) identified several fundamental questions to consider when engaging in research on intended curriculum:

- How many textbooks should we analyze?
- Which textbook(s) should we analyze?
- What text in the textbook(s) should we analyze? The exposition? The exercises?
- How much of that text should we analyze?
- How should we analyze it (i.e., what framework will we use to conduct our analysis)?
- What research questions should guide our analysis?

Similarly, Stylianides (2014) discussed three methodological challenges related to the analysis of the intended curricula on the topic of reasoning and proving: (1) deciding where to look for reasoning and proving activities, (2) deciding the roles of teachers' guides in the analysis, and (3) deciding the perspectives from which the analysis should be conducted. While his discussion is situated in the context of reasoning and proving, analogous challenges apply to other mathematical content areas as well.

In addition to the need for careful justification of why curriculum analysis is an appropriate methodological choice, the selection of an appropriate framework to analyze the curriculum is also essential. The framework will determine not only *what* to analyze, but also *how* to analyze it. In the aforementioned three trends of research on intended curriculum, researchers have drawn upon different resources to develop their framework. For the analysis of tasks in intended curriculum, the MTF (Stein et al., 1996) has been used frequently. For the second trend of research on intended curriculum, analyzing specific topics, researchers generally develop topic-specific frameworks. The frameworks for proportional reasoning (Shield & Dole, 2013) and reasoning and proving (Stylianides, 2009) are two good examples. In the historical analysis of intended curriculum, researchers have focused on different dimensions to analyze the intended curriculum. The NRC (2004) has identified nearly 20 possible dimensions along which to analyze intended curriculum, including the listing of topics, sequence of topics, topic presentation, emphasis of topics, grade level of introduction, reading level required, type and use of explanation, approach to formalization, use of technology or manipulatives, and types of assessment and their relation to classroom practice.

To date, studies of the intended curriculum have sampled curricular materials for analysis differently, from analyzing a number of books, to analyzing an entire book in one grade, to analyzing some chapters or modules, to analyzing only a few lessons or topics. The sampling of curricular materials should be determined by the research questions. But, how much material must be sampled and analyzed to answer those questions? We recommend that future studies on the intended curriculum have strong justifications for sampling particular materials for analysis.

Similarly, studies on the intended curriculum have used different units of analysis, including pages, tasks, sentences, and keywords. What should be the appropriate unit of analysis? It can be challenging to generalize the findings when different units of analysis are used. Thus, future studies on intended curriculum should have strong justifications for the appropriateness of the chosen unit of analysis.

Major Findings and Trends About the Enacted Curriculum

The expectations represented in written curricular materials become transformed as teachers and students enact the curriculum in the classroom (Snyder et al., 1992). Although the idea that there is a difference between the intended curriculum and the enacted curriculum is not new, the field has made significant progress in recent years in conceptualizing, measuring, and understanding these differences (Heck, Chval, Weiss, & Ziebarth, 2012; Thompson & Usiskin, 2014). Although some studies examine curriculum enactment as one part of a larger research program that aims to establish relationships between curricular materials and student outcomes, other studies examine curriculum enactment as a primary focus—investigating the curriculum as it is planned by teachers and enacted in the classroom by teachers and students within particular school and community contexts. Studies from these different perspectives have employed diverse research questions, conceptual frames, and research methods to contribute to an increasingly sophisticated body of knowledge about the enactment of mathematics curriculum.

教师课程材料的使用和课程执行的视角

教师如何使用书面课程材料为他们自己在某个特定的课堂和学校的学生设计数学教学,这是过去二十年来具有显著发展的一个研究领域(Gueudet, Pepin, & Trouche, 2012;Remillard, Herbel-Eisenmann, & Lloyd, 2009;Stein 等,2007)。教师使用课程材料的研究,即"教师如何利用或与这些课程资源互动,以及教师如何,且在多大程度上,依靠或利用这些资源来制定教学计划和执行教学"(Lloyd, Remillard, & Herbel-Eisenmann, 2009,第7页),为帮助我们了解教师利用书面课程材料这项工作的复杂性提供了必要的见解。例如,M.W.布朗(2009)描述了教师在进行教学设计时使用课程材料的三种不同方式(完全照搬、部分改编和即兴发挥)。这些构念为分析和比较教师的课程使用提供了得力的工具。例如,劳埃德(2008)利用这些构念刻画了一位师范生如何对两种不同课程资源采取不同的使用方法。

在一项颇有影响的文献分析工作中,雷米拉德(2005)发现课程研究者对教师使用课程材料的框架设计方式存在差异。在一定程度上,雷米拉德总结出教师课程材料使用方式的四个概念性类别:(1)遵循或颠覆文本,(2)利用文本,(3)解释文本,(4)参与文本,它们反映了执行课程研究中的概念性演变。虽然早期的研究倾向于将教师的课程使用方式仅视为遵循或颠覆课程材料(例如Freeman & Porter, 1989;Manouchehri & Goodman, 1998)或利用文本(例如M.S. Smith, 2000;Sosniak & Stodolsky, 1993),但最近更多的研究已转向将教师的课程使用视为解释或参与课程材料(例如M.W. Brown, 2009;Lloyd, 2008;Sherin & Drake, 2009;Stein & Kaufman, 2010)。运用参与视角的研究旨在诠释教师自身和课程材料这两个角色在从书面课程材料到教师教学计划,以及从教学计划到课堂实际执行的转变中所起的作用(Remillard, 2005;Remillard等, 2009;Stein等, 2007)。

尽管认识到教师参与课程材料的主动性是重要的,但描述教师使用课程材料的特征并不能为课程的执行提供一幅完整的画面。教师使用课程材料(或教师对课程的使用)和课程的执行这两个短语通常被人们互换使用,实际上,它们是不同的。在特定背景下,当教师、学生和课程材料之间进行互动时,执行课程便出现了。而教师使用课程材料只是课程实施过程的一部分。教师通过对课程材料进行解读和转化从而制定教学计划,还要与学生一起在一定的学校环境中将这些预期课程进一步转化为课堂活动。这些学生和教师的实际课堂体验构成了执行课程。由于不同的研究者在实证研究中会关注执行课程的不同方面,雷米拉德和赫克(2014)最近提出了四个维度来涵盖执行课程,为今后执行课程的研究设计提供了基础。这四个维度是:(1)数学,(2)教学互动和规范,(3)教师的教学行动,(4)资源和工具的使用。

执行课程的各种变异

20世纪90年代和21世纪头十年涌现出许多研究报告,证实了数学课堂活动的变异。早期的报告描述了教师和学生的数学活动与课程作者的预期之间的差异,以及相同的课或方案在不同课堂执行所出现的差异(参见Stein等人,2007的综述)。与课程执行相关的最重要的问题之一也许是:为什么变异会发生在从书面课程转变到执行课程的过程中?事实上,过去十年来具有重大进展的研究领域之一是确定执行课程的因素及其潜在影响。研究者确定的主要影响因素包括(a)教师的特征,(b)学生的特征,(c)书面课程材料的性质,(d)课程执行的背景。

教师的特征。对执行课程有潜在影响的各因素中,教师的特征被研究得最多。研究者研究了许多不同的教师特征,包括知识、信念、取向、课程观和教学设计能力。

一组最近的研究探讨了教师的数学教学知识(MKT)、所使用的课程材料的特质和数学教学质量之间的关系(Charalambous & Hill, 2012;Charalambous,

Perspectives on Teachers' Use of Curricular Materials and Curriculum Enactment

How teachers use written curricular materials to design mathematics instruction for their students in particular classroom and school contexts is an area of research that has grown significantly in the past two decades (Gueudet, Pepin, & Trouche, 2012; Remillard, Herbel-Eisenmann, & Lloyd, 2009; Stein et al., 2007). Research about *teachers' use of curricular materials*, that is, "how teachers engage or interact with these resources as well as how and the extent to which they rely on them in planning and enacting instruction" (Lloyd, Remillard, & Herbel-Eisenmann, 2009, p. 7), has provided needed insights into the complexity of teachers' work as they engage with written curricular materials. For example, M. W. Brown (2009) characterized different ways that teachers might use curricular materials (*offloading, adapting,* and *improvising*) as they design instruction. These constructs provide helpful tools for characterizing and comparing teachers' curriculum use. Lloyd (2008), for example, drew on these constructs to characterize a student teacher's contrasting uses of two different curriculum resources.

In a seminal analysis of literature, Remillard (2005) identified differences in the ways that curriculum researchers have framed teachers' use of curricular materials. To some extent, the four conceptual categories developed by Remillard—teachers' use as (1) *following or subverting the text*, (2) *drawing on the text*, (3) *interpreting the text*, and (4) *participating with the text*—reflect a conceptual evolution in research about the enacted curriculum. Whereas early research tended to view teachers' curriculum use as subverting or following curricular materials (e.g., Freeman & Porter, 1989; Manouchehri & Goodman, 1998) or drawing on the text (e.g., M. S. Smith, 2000; Sosniak & Stodolsky, 1993), more recent studies have moved toward viewing teachers' curriculum use as interpreting or participating with curricular materials (e.g., M. W. Brown, 2009; Lloyd, 2008; Sherin & Drake, 2009; Stein & Kaufman 2010). Studies assuming the participation perspective aim to take into account the roles of both teachers and curricular materials in transformations from written curricular materials to teachers' instructional plans and from plans to classroom enactment (Remillard, 2005; Remillard et al., 2009; Stein et al., 2007).

Although recognizing the active role of teachers as they engage with curricular materials is important, characterizations of teachers' use of curricular materials do not provide a complete picture of curriculum enactment. The phrases *teachers' use of curricular materials* (or *teachers' curriculum use*) and *curriculum enactment* are often applied interchangeably; however, they are different. The enacted curriculum emerges through interactions among teachers, students, and curricular materials in a given context. Teachers' use of curricular materials is only one part of this process. Whereas teachers interpret and transform curricular materials into instructional plans, the intended curriculum is even further transformed into classroom activities by teachers and students within particular school contexts. These actual classroom experiences of students and teachers make up the enacted curriculum. Because different researchers have attended to different aspects of the enacted curriculum in empirical studies, Remillard and Heck (2014) recently proposed four dimensions—(1) the mathematics, (2) instructional interactions and norms, (3) pedagogical moves by the teacher, and (4) the use of resources and tools—that encompass the enacted curriculum and around which future studies can be designed.

Variations in the Enacted Curriculum

During the 1990s and 2000s, numerous reports emerged documenting variations in mathematics classroom activities. Early reports described differences between teachers' and students' mathematical activities and those intended by curriculum writers as well as differences in the enactment of the same lesson or program in different classrooms (see the review by Stein et al., 2007). Perhaps one of the most important questions related to curriculum enactment is, why does variation occur in the transformation from the written to enacted curriculum? Indeed, one of the areas of significant research advancement over the past decade has been the identification of factors and potential influences on the enacted curriculum. Key influencing factors identified by researchers include (*a*) teacher characteristics, (*b*) student characteristics, (*c*) the nature of written curricular materials, and (*d*) the context in which the curriculum is enacted.

Teacher characteristics. Of the potential influences on the enacted curriculum, teacher characteristics is the one that has been studied most extensively. Researchers have examined many different teacher characteristics, including knowledge, beliefs, orientations, curriculum vision, and pedagogical design capacity.

A set of recent studies explored relationships among teachers' mathematics knowledge for teaching (MKT), the nature of the curricular materials in use, and the mathematical quality of instruction (Charalambous &

Hill, & Mitchell, 2012; Hill & Charalambous, 2012; Sleep & Eskelson, 2012）。这些研究表明, 面向教学的数学知识既可以限制（Charalambous等, 2012）也可以支持（Hill & Charalambous, 2012）教师为学生创造从数学课程任务中学习数学的机会。从这些研究中得到的一个有价值的发现是, 当教师的数学教学知识水平低时, 使用具有教育性的课程材料（这些课程材料专门为教师学习提供书面材料, 参见E.A. Davis & Krajcik, 2005）可以提高数学教学质量。

其中的一个数学教学知识研究还提出了对教师信念的潜在作用的新见解。斯利普和埃斯科尔森（2012）比较了两名初中老师执行同样课程任务的方式, 其中一名老师被认为具有高水平的数学教学知识, 另一名教师的数学教学知识处于中等水平。研究者发现, 虽然教师的数学教学知识和课程材料两者似乎都对教学质量有影响, 但教师对数学教与学的信念也影响了教学质量。这位具有高水平数学教学知识的老师把数学程序和技能看作学生学习目标的至关重要的部分, 这也可以从他程序性极强的数学课中看得出来。对于另一位老师, 虽然概念理解是她的核心信念, 但她中等水平的数学教学知识制约了她可以为学生提供的数学机会。两位教师的教学都比课程设计者的预期显得更为程序化, 这表明教师的数学教学知识和课程材料可能不是课程执行的仅有影响因素。

教师对所用课程的哲学观和设计的信念与理解也对执行课程有影响。戈雅（2007）在研究伊朗课程改革时, 发现了教师对新几何教科书的一系列信念, 并描述了"传统学派"对课程改革的愿望不大, 而"渐进主义者"和"创新者"则对部分或全部变革持开放的态度。类似地, 博森等人（2014）指出, 瑞典的教师没有弄清他们被要求在课堂上实施的课程改革的意义, 从而导致了他们固守以程序为导向的数学教学方法。

查拉兰布斯和菲利普欧（2010）研究了塞浦路斯国家数学课程改革的影响, 他们分析了教师关切（参见Christou, Menon, & Philippou, 2009）和自我效能信念——即教师对自己达到改革目标感知的能力, 之间的相互作用。研究者发现, 那些在改革前的实践中感觉更成功的教师, 他们更担心自己改变教学的能力和学生应对教学改革的学习能力。在一项基于旨在帮助小学教师将代数思维纳入其数学教学的项目的研究中, 唐克斯和韦勒（2009）发现, 参加专注于理解预期变化的持续专业发展活动, 教师的担忧发生了变化, 对创新课程的执行也得到了改进。

教师对课程创新的理解涉及德雷克和谢林（2009）所指的教师课程观, 即教师如何看待其使用的特定课程材料给学生提供的学习机会。通过对参与教师的观察和访谈分析, 研究者刻画了参与教师两年间在教学前、教学中和教学后的课程策略（阅读、评价和改编课程材料）。教师在教学之前阅读和评估课程材料的方式随时间而发生的变化, 被用来作为教师课程观成长的证据, 因为教师获得了课程材料中数学和教学方法的知识。在这项研究工作的基础上, 肖邦（2011）考查了三位教师的课程观以及他们在几个系列的数学教学活动中对课程的使用。他发现对课程使用最大的影响发生在这样的教师身上, 该教师的课程观最多地整合了学生在课堂活动中学习的详细证据。

从社会文化的视角出发, M.W.布朗（2009）认为, 教师的教学设计能力——他们感知课程材料、做出精心而明智的决策, 并全面遵循计划的能力——可以解释具有类似知识、信念和责任心的教师在课程执行上的差异。尽管德雷克和谢林（2009）对课程观的观念描述了教师对课程材料的特定方法或哲学观的认识, 但布朗所指的这个教学设计能力构念, 为考虑教师的知识, 更重要的是考虑教师用这些知识进行创新的能力提供了一个视角。正如布朗所指的那样, 当教师与学生一起执行新的课程材料时, 他们以有效的方式感知和利用现有资源的能力对执行课程的质量有很大的影响。

研究发现, 教师如何看待学生使用新课程材料的能力, 也会影响课程的执行（Arbaugh, Lannin, Jones, &

Hill, 2012; Charalambous, Hill, & Mitchell, 2012; Hill & Charalambous, 2012; Sleep & Eskelson, 2012). These studies suggest that this knowledge can serve to both limit (Charalambous et al., 2012) and support (Hill & Charalambous, 2012) teachers' efforts to create opportunities for students' learning of mathematics with curricular tasks. A valuable finding from these studies is that when teachers have low levels of MKT, use of *educative* curricular materials (those offering material written specifically for teachers' learning; see also E. A. Davis & Krajcik, 2005) can increase the mathematical quality of instruction.

One of these studies of MKT offered new insights about the potential role of teachers' beliefs as well. Sleep and Eskelson (2012) compared the ways that two middle grades teachers, one considered to have high-level MKT and the other mid-level MKT, enacted the same curricular task. The researchers found that although the teachers' MKT and curricular materials both seemed to play a role in instructional quality, the teachers' beliefs about mathematics teaching and learning also influenced instructional quality. Mathematical procedures and skills were central to the goals for student learning of the teacher with high-level MKT, as reflected by a highly procedural lesson. For the other teacher, although conceptual understanding played a central role in her beliefs, her mid-level MKT limited the mathematical opportunities she was able to offer her students. Both teachers' instruction was considered more procedural than what was intended by the curriculum designers, suggesting that teachers' MKT and curricular materials may not be the only influences on curriculum enactment.

Teachers' beliefs and understandings about the philosophy and design of the curriculum they are using have also been shown to play a role in the enacted curriculum. In her study of curriculum reform in Iran, Gooya (2007) found a range of teachers' beliefs about new geometry textbooks and described how "the traditionalists" felt little need for curricular change, whereas the "incrementalists" and "innovators" were open to some or all of the changes. Similarly, Boesen et al. (2014) suggested that teachers in Sweden did not identify the meaning of a curricular change they were asked to implement in their classrooms, resulting in the persistence of procedure-oriented mathematics instruction.

In a study of the impact of changes to the national mathematics curriculum in Cyprus, Charalambous and Philippou (2010) identified interactions between teachers' concerns (see also Christou, Menon, & Philippou, 2009) and their efficacy beliefs—that is, teachers' perceived ability to meet reform goals. The researchers found that teachers who felt more successful with prereform practices expressed strong concerns about their own abilities to change instruction and their students' abilities to learn in response to reformed instruction. A study by Tunks and Weller (2009), based on a program intended to help elementary teachers incorporate algebraic thinking into their mathematics instruction, found that as teachers' concerns changed through sustained professional development focused on understanding intended changes, enactment of the innovation improved as well.

Teachers' understandings of curricular innovations relate to what Drake and Sherin (2009) have termed teachers' *curriculum vision*, namely teachers' views of opportunities for student learning offered by the particular curricular materials in use. Through analysis of observations and interviews with teachers, these researchers characterized teachers' curriculum strategies (reading, evaluating, and adapting curricular materials) before, during, and after instruction over a 2-year period. Changes over time in the ways teachers read and evaluated curricular materials prior to instruction were used as evidence of teachers' growing curriculum vision, as teachers gained knowledge of the mathematical and pedagogical approaches of the curricular materials. Building on this work, Choppin (2011) examined three teachers' curriculum vision and their curriculum use over several enactments of instructional sequences. He found that the greatest impact on curriculum use occurred in the case of the teacher whose curriculum vision most incorporated specific evidence of student learning from classroom activities.

Assuming a sociocultural perspective, M. W. Brown (2009) suggested that teachers' *pedagogical design capacity*—their capacity to perceive affordances of curricular materials, make deliberate and informed decisions, and follow through on their plans—can explain differences in curriculum enactment by teachers with similar knowledge, beliefs, and commitments. Although Drake and Sherin's (2009) notion of curriculum vision describes teachers' knowledge of the particular approach or philosophy of curricular materials, Brown's construct of pedagogical design capacity provides a lens for considering not only teachers' knowledge but more so teachers' ability to do new things with their knowledge. As Brown suggested, when teachers enact novel curricular materials with students, their ability to perceive and mobilize available resources in productive ways influences the quality of the enacted lesson.

Teachers' perceptions of students' ability to be successful with new curricular materials has been found to affect curriculum enactment (Arbaugh, Lannin, Jones,

Park-Rogers, 2006; Charalambous & Philippou, 2010; Eisenmann & Even, 2009; Lloyd, 2008）。在艾森曼和埃文（2009）的案例研究中，一位教师在两所不同的学校教授同一本教科书，但作出了不同的课程决策。研究者发现，在面向全班的授课过程中，两所学校的学生学到了不同版本的代数，因为在其中一所学校的全班授课中，老师经历了学生缺乏合作和全班活动频繁地被打断的情况。虽然这个例子和其他例子可能表明教师对课程材料需求的看法会限制课程的执行，但也有一些证据表明，教师关注学生的经历和学习也会对课程的执行有积极的影响。在肖邦（2011）的研究中，一位老师根据学生的经历对数学任务做出了调整，使得任务的复杂性增强了，并且增加了学生有效学习特定数学概念的机会。

学生的特征。虽然学生的知识、信念和态度很有可能影响他们在课堂上参与课程任务的方式，但是关于学生在课程执行中的作用的研究几乎很少。因为理解学生的贡献是理解在课堂上实施预期课程这个过程的一个重要因素，所以这方面需要进一步的研究。

通过一系列关于学生对美国新课程项目看法的报告，已经形成了一条与学生特征相关的明确的调查线。（Jansen, Herbel-Eisenmann, & Smith, 2012; J. P. Smith & Star, 2007; Star, Smith, & Jansen, 2008）。学生对新课程材料的看法可能会影响他们在课堂活动中的参与度，但研究者也可能将学生的看法视为课程整体影响的一个重要方面。施塔尔等（2008）描述了当一批中学生和大学生从传统课程过渡到改革课程后，他们是如何看待两种数学课程之间的差异的。有趣的是，他们发现学生的感知差异有很大的差别（参见 J. P. Smith & Star, 2007），但与课程开发者和研究者的观点却比较一致。

课程材料的性质。长期以来，研究者们对在数学和教学法方面有创新的课程材料进行了研究，认为这些材料的特点影响了它们的执行（例如 Stake & Easley, 1978）。然而，直到最近，数学教育研究者尝试解释了课程特征是如何影响书面课程到执行课程的转变的。在过去十年中，一些研究考虑了课程特征对课程执行的影响，如：设计特征（M.W. Brown, 2009; E. A. Davis & Krajcik, 2005; Hirsch, 2007）、数学表征及数学表征的习得（Johansson, 2007）和教学法取向（Chval, Chávez, Reys, & Tarr, 2009; Grouws 等, 2013; Tarr 等, 2008; Tarr, Grouws, Chávez, & Soria, 2013）。他们还注意到了课程材料的其他特点，包括设计者没有预料到的特征，例如教科书的"声音"及教科书如何定位学生（Herbel-Eisenmann, 2007; Herbel-Eisenmann & Wagner, 2007）。虽然这项研究引起了人们对以下观点的关注：课程材料的差异会使得课程在执行时产生重大差异，但是这一领域的研究还没能从预期课程的研究中找到关于课程特征的重要发现。

在过去十年中，有关课程特征对课程执行的作用的研究呈显著增长的态势。这类研究探讨了课程材料（即教师用书）中提供给教师的信息，以及教师获取此类信息后对课程执行的影响（例如 Cengiz, Kline, & Grant, 2011; Charalambous 等, 2012; E. A. Davis & Krajcik, 2005; Doerr & Chandler-Olcott, 2009; Grant, Kline, Crumbaugh, Kim, & Cengiz, 2009; Stein & Kaufman, 2010; Stein & Kim, 2009; Superfine, 2009）。查拉兰布斯等人（2012）描述了一个数学教学知识水平较低的老师在使用对教师支持度较低的课程材料时是如何执行一堂课的。实际上，这堂课的质量是不高的。相比之下，另一名数学教学知识水平较低的老师使用了支持度较高的课程材料，其执行课程被认为质量更高。

斯坦和金（2009）分析了两套基于标准的具有不同设计特征的课程材料，这些设计特征的区别在于：（1）每次课的主要任务的认知需求水平；（2）教科书的透明度，即教科书作者的设计意图的显性程度，这些设计包括：选择任务、给任务排序以及解读与任务相关的数学思想；（3）教师用书是否要求教师预先设想学生完成任务的方法，是否给教师提供学生回答样例和学生可能遇到的困难。研究发现，与另一个课程相比，其中一个课程的任务认知需求水平较低，透明度较低，教师预测学生体验的机会也较少。斯坦和金指出，当教师不了解课程活

& Park-Rogers, 2006; Charalambous & Philippou, 2010; Eisenmann & Even, 2009; Lloyd, 2008). In a case study by Eisenmann and Even (2009), one teacher made different curricular decisions when teaching with the same textbook at two different schools. The researchers found that, during whole-class work, students at the two schools learned different versions of algebra because the teacher experienced a lack of student cooperation and frequent interruptions in whole-class activities at one of the schools. Whereas this example and others may suggest that teachers' perceptions of the demands of curricular materials can constrain classroom enactments, there is also some evidence that teachers' attention to students' experiences and learning can also affect enactment in positive ways. In Choppin's (2011) study, the adaptations that one teacher made in response to students' experiences enhanced the task complexity and opportunities for students to engage productively with particular mathematical concepts.

Student characteristics. Although students' knowledge, beliefs, and attitudes are likely to affect the way they engage with curricular tasks in the classroom, there has been little research about the role of students in curriculum enactment. Because understanding students' contributions is an important element of understanding what happens as the intended curriculum becomes enacted in the classroom, further research is needed in this area.

One clear line of inquiry related to student characteristics has developed across a set of reports about students' perceptions of new curriculum programs in the United States (Jansen, Herbel-Eisenmann, & Smith, 2012; J. P. Smith & Star, 2007; Star, Smith, & Jansen, 2008). Students' perceptions of novel curricular materials may influence their engagement with classroom activities, but researchers may also view students' perceptions as one important aspect of the overall impact of the curriculum. Star et al. (2008) described how secondary and university students perceived differences between traditional and reform mathematics curriculum programs as the students made transitions from one program to another. Interestingly, students' perceived differences were found to vary widely (see also J. P. Smith & Star, 2007) but aligned moderately well with curriculum developers' and researchers' views.

Nature of curricular materials. Researchers studying curricular materials containing novel approaches to mathematics and pedagogy have long considered that characteristics of the materials influence their enactment (e.g., Stake & Easley, 1978). Yet it is only recently that mathematics education researchers have attempted to account for *how* curricular features might influence the transformation from written to enacted curriculum. In the past decade, studies have considered the impact on enactment of curricular characteristics such as design features (M. W. Brown, 2009; E. A. Davis & Krajcik, 2005; Hirsch, 2007), representations of mathematics and how it is learned (Johansson, 2007), and pedagogical orientations (Chval, Chávez, Reys, & Tarr, 2009; Grouws et al., 2013; Tarr et al., 2008; Tarr, Grouws, Chávez, & Soria, 2013). They have also attended to other features of curricular materials, including those not anticipated by designers, such as the "voice" of the textbook and how it positions students (Herbel-Eisenmann, 2007; Herbel-Eisenmann & Wagner, 2007). Although this work has drawn attention to the important notion that differences in curricular materials may lead to important differences in enactment, research in this area has not yet drawn on key findings about curricular features from studies of the intended curriculum.

One line of inquiry related to the role of curricular features in enactment has grown significantly in the past decade. This research explores the information provided for teachers in curricular materials, namely in teachers' editions, and the influence of teachers' access to this information on curriculum enactment (e.g., Cengiz, Kline, & Grant, 2011; Charalambous et al., 2012; E. A. Davis & Krajcik, 2005; Doerr & Chandler-Olcott, 2009; Grant, Kline, Crumbaugh, Kim, & Cengiz, 2009; Stein & Kaufman, 2010; Stein & Kim, 2009; Superfine, 2009). Charalambous et al. (2012) illustrated how a teacher with a low level of MKT and using curricular materials that were considered less supportive for teachers enacted a lesson that was characterized as having low instructional quality. In contrast, another teacher who had low MKT but used a more supportive edition of the materials enacted a lesson that was characterized as having higher instructional quality.

Stein and Kim (2009) analyzed two sets of standards-based curricular materials with distinct design features according to (1) the level of cognitive demand of each lesson's main task; (2) the *transparency* of the teacher materials—that is, the visibility of textbook authors' rationales for the selection of tasks and sequences of tasks as well as explanations of mathematical ideas related to tasks; and (3) whether the teachers' editions invited teachers to anticipate students' approaches to tasks and offered examples of student work and possible difficulties students might have. One curriculum program was found to offer less demanding tasks, less transparency, and fewer opportunities for teachers to anticipate students' experiences than the other

动的根本目的——当教师缺乏发展课程愿景（Drake & Sherin, 2009）的机会时，在课堂执行出现与课程所设置的路径相异的情况下，他们可能会遇到各种困难。

在一项相关研究中，斯坦和考夫曼（2010）考查了48个小学教师执行斯坦和金（2009）研究的那两个课程方案时的质量。研究者发现，执行课程质量更高的那个课程方案正是高认知需求任务比例较高的课程方案，并且它也比另一个课程在帮助教师落实重要的数学思想方面提供了更多的支持。虽然课程的特点可能会以不同的方式影响课程的执行，但是，显然这一因素，像其他因素一样，也不能被孤立起来看，例如，在斯坦和考夫曼（2010）的研究中，为教师提供的专业发展以及不同学校和地区的环境也可能影响教学质量。

情境。过去十年的研究文献越来越多地关注情境影响课程执行的理念。研究表明，课程的执行会受到各种情境因素的影响，例如利益攸关的考试（Au, 2007; Boesen, Lithner, & Palm, 2010; Lloyd, 2007）和家长的压力（Gadanidis & Kotsopoulos, 2009; Herbel-Eisenmann, Lubienski, & Id-Deen, 2006; Sherin & Drake, 2009）。虽然在过去十年的多个课程研究中都讨论了情境因素，但情境因素通常被描述为解释课程执行差异的次要因素，只有少数研究将情境作为课程执行研究的重点。

赫贝尔-艾森曼等人的研究（2006）提醒大家注意课程情境在课程执行中的作用，这些情境包括学生和父母的期望以及学校数学教学的历史。研究者阐述了当一名教师处于两个课程情境时，他所用的两个课程方案的差异（一个是改革取向的课程，另一个是更为传统的课程）并不能完全解释这位教师课堂教学方式的差异。这些作者提请注意"由教师工作环境条件（包括学生的变化，行政或家长的压力，或课程材料的变化）导致的班级之间和日常发生的波动"的影响（第315页）。一位教师可能会接受某种特定的教学原则，这些或许是符合课程开发者的意图的，然而，这位教师在一天中不同时间段，与不同的学生一起，使用不同的材料，或在不同的学校教学时，他的教学可能会很不一样（参见Eisenmann & Even, 2009）。

最近几项关于专业学习的研究也关注了课程执行中情境的作用。当课程的执行发生在为教师提供连贯的学习机会的背景下，尤其是当这些学习机会是系统的，并且与正在使用的课程材料有关联时，课程执行似乎会受到积极的影响（Boston & Smith, 2009; McDuffie & Mather, 2009; Ponte, 2012）。迈克达菲和马瑟（2009）在有关教师合作专业发展活动的研究中，记录了教师参与课程推理的过程。值得注意的是，课程材料处于专业发展团队活动的核心，教师们从学习者的角度分析课程材料，作为学习者一起完成任务，详细绘制学习轨迹，并且在教学过程中根据与学生合作的情况来修订教学计划。庞特（2012）的研究描述了葡萄牙的一个规模较大的国家专业发展项目，这一项目为在课堂上使用新数学课程的教师提供了结构化的支持。

执行课程研究方法的问题

过去十年，关于执行课程的研究形成了如下几个议题：需要对正在研究的关键构念更好地进行概念化，使概念化与研究设计之间保持一致，现有文献中有关执行课程的薄弱之处，以及在将来的研究中需要在预期课程、执行课程和获得课程之间建立更强的联系。

概念化和研究设计。在过去十年中，该领域对课程执行的复杂性的认识有了长足的进步。这种进步的证据不仅在先前的研究发现中可以找到，而且体现在课程执行研究中使用的问题和理论构念的范围中。例如，研究者对教师的知识持有不同的观点，从而导致他们对课程执行中知识的作用的衡量标准也不同（Son & Senk, 2014）。另一个例子涉及课程实施的忠实度，尽

program. Stein and Kim suggest that when teachers do not understand the underlying purpose of curricular activities—that is, when teachers lack opportunities to develop curriculum vision (Drake & Sherin, 2009)—they may experience difficulties when classroom enactments differ from the pathways laid out in the curriculum.

In a related study, Stein and Kaufman (2010) examined the quality of 48 elementary teachers' implementations of the same two curriculum programs that Stein and Kim (2009) studied. The researchers found that the curriculum program that was implemented at a high level contained a higher percentage of tasks of high cognitive demand and seemed to provide more support to help teachers locate the big mathematical ideas than did the other program. Although particular characteristics of curriculum programs may affect enactment in different ways, it seems clear that this factor, like others, cannot be isolated. For example, in the Stein and Kaufman (2010) study, the professional development offered to teachers and the environments in the different schools and districts likely influenced instructional quality as well.

Context. The notion that context affects curriculum enactment has garnered increased attention in the research literature of the past decade. Studies have suggested that curriculum enactment may be affected by a variety of contextual factors, such as high-stakes tests (Au, 2007; Boesen, Lithner, & Palm, 2010; Lloyd, 2007) and parental pressures (Gadanidis & Kotsopoulos, 2009; Herbel-Eisenmann, Lubienski, & Id-Deen, 2006; Sherin & Drake, 2009). Although contextual factors are discussed in multiple curriculum studies of the past decade, they are typically portrayed as possible secondary explanations for variations in curriculum enactment. In only a few studies is context a primary focus of research related to curriculum enactment.

A study by Herbel-Eisenmann et al. (2006) drew attention to the role of *curricular context*, including student and parent expectations and the history of mathematics instruction within a school, in curriculum enactment. These researchers illustrated how one teacher's different classroom practices in two curricular contexts were not entirely explained by differences between the two curriculum programs in use (one reform-oriented and one more conventional). These authors draw attention to the influence of "fluctuations that occur from class-to-class and from day-to-day based on the contextual conditions in which a teacher works, including changes in students, administrative or parental pressure, or curriculum materials" (p. 315). A teacher might embrace particular pedagogical principles, perhaps consistent with curriculum developers' intentions; however, that teacher's instruction may look quite different at different times of day or when working with different students, using different materials, or teaching at different schools (see also Eisenmann & Even, 2009).

Several recent studies of professional learning have also attended to the role of context in curriculum enactment. Curriculum enactment appears to be positively influenced when it occurs in a context that offers coherent opportunities for teachers' learning, particularly when those opportunities are systemic and connected to curricular materials in use (Boston & Smith, 2009; McDuffie & Mather, 2009; Ponte, 2012). In a study of collaborative professional development activities, McDuffie and Mather (2009) documented teachers' engagement in *curricular reasoning*. Notably, curricular materials were the centerpiece of the professional development team's activities as teachers analyzed their curricular materials from learners' perspectives, completed tasks together as learners, mapped learning trajectories, and revised plans based on work with students during instruction. Ponte (2012) described a larger scale, national professional development program in Portugal that also provided structured support to teachers as they used a new mathematics curriculum in their classrooms.

Methodological Issues in Research on the Enacted Curriculum

Several issues emerge when considering the past decade of research about the enacted curriculum: the need for better conceptualizations of key constructs under study and alignment between conceptualizations and research design, weaknesses in the existing literature about the enacted curriculum, and the need for future research that establishes stronger links among the intended, enacted, and attained curriculum.

Conceptualizations and research design. Over the past decade, the field has advanced considerably in its awareness of the complexity of curriculum enactment. Evidence of this advancement appears not only in the research findings discussed previously but also in the range of questions and theoretical constructs used in studies of curriculum enactment. For example, researchers have assumed different perspectives on teachers' knowledge, resulting in different measures of the role of knowledge in curriculum enactment (Son & Senk, 2014). Another example involves fidelity of implementation, a construct about which researchers hold varying perspectives (e.g., S. A. Brown, Pitvorec, Ditto, &

管研究者普遍认为这是一个研究学生学习结果的重要变量，但是研究者们对这个构念持有不同的观点。（例如 S. A. Brown, Pitvorec, Ditto, & Kelso, 2009；Chval 等，2009；Heck 等，2012）

尽管视角和概念的多样性是一个促使研究领域活跃和不断增长的指标，但在比较、推广一项研究的结论到另一项研究，或者要整合多个研究的结果时也可能给研究者带来挑战。最近出版的几卷文献探讨了数学执行课程的研究现状，并呼吁创建更有力的概念和方法工具（见 Heck 等，2012；Remillard 等，2009；Thompson & Usiskin, 2014）。为了构造一个共同的语言和一组构念，雷米拉德和赫克（2014）提出了一个包含四个维度的执行课程的模型：（1）数学，（2）教学互动和规范，（3）教师的教学行为，（4）资源和工具的使用。诸如此类的概念模型为再次检视现有研究，以及帮助研究者设计关于学生学习的未来研究提供了有用的视角，从而明确地探讨课程执行的各个方面。

在雷米拉德和赫克（2014）的框架范围内，研究者以许多不同的方式来把理论构念操作化[①]。研究者开发并调整了一系列广泛的访谈方案、观察方案、调查问卷和其他数据收集工具（例如教师日志），以收集有关课程执行的信息（Heck 等，2012；Ziebarth, Fonger, & Kratky, 2014）。这些工具有很大的差异，这些差异在于研究者所关注的是哪些课程执行特征，以及他们如何测量这些特征（S. A. Brown 等，2009；Chval 等，2009；Heck 等，2012）。例如，在"比较高中数学课程的选取：对不同课程的探索（COSMIC）"的研究中，研究者运用了"教师的教科书使用日记"来确定课堂教学中教科书内容的重点领域（Tarr, McNaught, & Grouws, 2012）。另外，莫耶、蔡、聂和王（2011），舍恩、凯布拉、芬恩和法（2003），以及塔尔等人（2013）所使用的观察方案对课堂活动进行了评级，以描述教学中贯彻改革理念的程度。显然，研究者用了这些工具中的每一种，并与其他工具结合以测量课程的执行情况，然而，他们的目的是完全不同的。有关各种研究课程执行的测量工具的详细信息，请参阅以下文献：汤普森和斯金（2014）及齐巴思等人（2014）。

执行课程研究文献的薄弱之处。 从教师使用课程材料的研究中，我们获得了大量有关数学执行课程的知识。这些研究考查了教师课程使用的许多方面，详细记录了不同课堂里课程执行的方方面面，探索了有助于解释不同课程执行方式差异的因素。然而，在这方面的研究中，对于某几个教师群体课堂上的课程执行的研究要比其他教师群体的更多。例如，索恩和申克（2014）综述了教师知识在课程执行中的作用，并注意到，关于经验丰富的老师课程使用的研究发现比关于新手教师的研究发现更多。类似地，有许多研究发现针对的是小学教师，小学教师的数学背景与中学数学教师的数学背景是完全不同的。

另外，比起教师和学生长期使用课程材料的研究，教师初次使用课程材料的研究更为常见。肖邦（2011）提出"在教师真正理解学生如何通过这些材料进行学习之前，他们需要多次反复使用课程材料"（第332页）。然而，很少有研究收集有关课程执行的纵向数据。此外，近几年来，研究者更清楚地意识到课程执行可以为教师提供学习机会，但过去十年对这一领域的研究仅有几例（例如 Choppin, 2011；J. D. Davis, 2009）。缺乏纵向研究和对教师学习的关注表明，我们在课程执行方面的知识基础薄弱。

也许是由于对教师使用课程材料的研究日益突出，相比其他影响课程执行的因素，教师的特征得到了更广泛的研究。同时，需要研究学生的特征、预期课程的特征以及情境因素是如何影响课程执行的。到目前为止，对执行课程的研究似乎没有充分利用预期课程研究的具体成果。例如，研究人员可以考虑如何通过对书面课程材料的微观分析来确定将一个数学主题安排在一个特定年级的方案，从而影响课堂中课程执行的质量。另外，尽管教科书一直是数学教学的主要资源，但新的信息技术，包括数字课程，给教师提供了其它的选择，对电子书的研究才刚刚在法国、日本、新加坡和美国开始（Howson, 2013）。最近召开的第三届国际课程会议

[①] 也就是把抽象的理论构念转化为可操作或可测量的具体指标。——译者注。

Kelso, 2009; Chval et al., 2009; Heck et al., 2012), despite general agreement that it is an important variable in studies of student outcomes.

Although this diversity in perspectives and constructs is an indicator of a vibrant and growing research area, it can also create challenges for researchers to compare or extend findings from one study to another and aggregate results. Several recent edited volumes explore the state of the field of research on the enacted mathematics curriculum and call for more robust conceptual and methodological tools (see Heck et al., 2012; Remillard et al., 2009; Thompson & Usiskin, 2014). With the purpose of suggesting a common language and set of constructs, Remillard and Heck (2014) propose a model for the enacted curriculum that includes four dimensions: (1) the mathematics, (2) instructional interactions and norms, (3) pedagogical moves by the teacher, and (4) the use of resources and tools. Conceptual models such as this one provide useful lenses for revisiting extant studies and helping researchers design future studies of student learning that make explicit the aspects of curriculum enactment under investigation.

Within the domains of Remillard and Heck's (2014) framework, researchers have operationalized constructs in many different ways. Researchers have developed and adapted a wide range of interview protocols, observation protocols, surveys, and other data collection instruments (e.g., teacher logs) to gather information about curriculum enactment (Heck et al., 2012; Ziebarth, Fonger, & Kratky, 2014). These tools vary considerably in terms of what aspects of curriculum enactment they attend to as well as how those aspects are measured (S. A. Brown et al., 2009; Chval et al. 2009; Heck et al., 2012). For instance, whereas the "textbook-use diary" was used in the Comparing Options in Secondary Mathematics: Investigating Curriculum (COSMIC) study to identify areas of emphasis of the textbook content in classroom instruction (Tarr, McNaught, & Grouws, 2012), the observation protocols used by Moyer, Cai, Nie, and Wang (2011), Schoen, Cebulla, Finn, and Fi (2003), and Tarr et al. (2013) rated classroom activities to describe the extent to which instruction involved reform-oriented practices. Clearly, researchers used each of these tools—in conjunction with other tools—to measure curriculum enactment; however, their purposes were quite different. For detailed information about different instruments for studying curriculum enactment, see Thompson and Usiskin (2014) and Ziebarth et al. (2014).

Weaknesses in the enacted curriculum literature. Much of our current knowledge about the enacted mathematics curriculum has emerged from studies of teachers' use of curricular materials. These studies have investigated numerous aspects of teachers' curriculum use, documented in detailed ways aspects of curriculum enactment in different classrooms, and explored factors that can help explain variation across curriculum enactment. In this line of research, however, curriculum enactment in the classrooms of certain groups of teachers has been studied more often than in the classrooms of others. For example, in reviewing studies of the role of teacher knowledge in curriculum enactment, Son and Senk (2014) observed that there are more research findings about experienced teachers' curriculum use than there are about novice teachers. Similarly, many results have emerged from studies of elementary teachers, whose mathematics background is typically quite different than that of secondary mathematics teachers.

In addition, studies of teachers' initial use of curricular materials are more common than studies of teachers and students using materials over several years. Choppin (2011) suggested "it would take multiple iterations of the use of curricular materials before teachers develop an understanding of how students engage with those materials" (p. 332). Yet, few studies have collected longitudinal data about curriculum enactment. Moreover, although awareness of opportunities for teacher learning through curriculum enactment has grown in recent years, the past decade has seen little research activity in this area, with a few exceptions (e.g., Choppin, 2011; J. D. Davis, 2009). The lack of longitudinal studies and scarce recent attention to teacher learning suggest weaknesses in our knowledge base about curriculum enactment.

Perhaps due to the increased prominence of research about teachers' use of curricular materials, teacher characteristics have been studied more extensively than other factors influencing curriculum enactment. Research about how student characteristics, features of the intended curriculum, and contextual factors influence the enacted curriculum is needed. To date, studies of the enacted curriculum do not appear to capitalize on specific research findings about the intended curriculum. For instance, researchers might consider how unique grade-level placements of a mathematical topic, identified through microanalysis of written curricular materials, influence qualities of curriculum enactment in classrooms. In addition, although textbooks have been a staple of mathematics instruction, new technologies, including digital curricula, present teachers with important alternatives, and research on e-books has only recently begun to be carried out in France, Japan, Singapore, and the United States (Howson, 2013). The importance of digital curriculum as an emerging line of

进一步证实了数字课程作为一个新兴的研究课题的重要性。这次课程大会汇集了来自澳大利亚、加拿大、丹麦、法国、以色列、日本、韩国、荷兰、瑞典、英国和美国的研究者，他们探讨了数学课堂中数字课程材料的设计、传递和执行，并提出了一些重要的见解，这些已在这个致力于指明课程未来研究方向的国际会议的论文集中作了报道（Bates & Usiskin，2016）。

预期课程、执行课程和获得课程之间的联系。在过去的十年中，通过利用在教师使用课程材料的定性研究中所获得的构念和发现等途径，研究人员对学生的学习结果进行了大规模的研究，在开发更成熟的方法用来解释课程执行的本质方面取得了进展（Heck等，2012；Ziebarth等，2014）。然而，将书面课程材料、执行课程和学生成绩联系起来的研究十分少（Fan，Zhu，& Miao，2013；Hunsader & Thompson，2014；NRC，2004）。当研究者继续探索这一领域时，他们需要建立一个更为全面的知识库来解释教师是如何解读和执行课程材料的，以及影响课程解读和执行这些过程的因素有哪些。此外，还需要进一步的研究来增进我们对该领域的理解，即执行课程的变化是如何与学生学习的变化相联系的，如果有联系，什么时候它们之间有联系。

获得课程研究的主要发现和走向

如前面所述，执行一个数学课程是一个复杂现象，教师和学生的课堂活动会受到许多因素的影响。因此，要记录学生所学到的内容并将其归因于某一特定的课程方案在方法论方面充满了挑战性。2002年，美国国家研究协会呼吁要在教育领域进行更多的教育科学研究（NRC，2002），之后的十年，关于课程有效性方面的研究已经有了许多进展，还出现了几个备受瞩目的研究。

科学地建立课程的有效性

课程研究的一个主要走向是越来越重视科学地评价课程（Clements，2007；NRC，2002，2004；Slavin，Lake，& Groff，2009）。以下几个框架为课程研究的研究设计工作提供了潜力，每一框架对设计课程研究都具有可用性和局限性。

有效教育策略资料中心。有效教育策略资料中心（WWC）是美国教育部教育科学研究所于2002年成立的一个机构，有效教育策略资料中心制定标准并按标准审阅教育研究，从而评价干预措施有效性研究的严谨性，这些干预也包括数学课程方案。有效教育策略资料中心提供的证据标准旨在直接影响联邦政府的基金资助，从而最终产生更高质量的教育研究。有效教育策略资料中心审查协议本质上是筛选干预研究，步骤基本上是从以下这个问题开始："团体成员是否是通过随机过程来确定的？"（U.S. Department of Education Institute for Educational Sciences，2014，第9页）。按有效教育策略资料中心的指示，"只有精心设计和实施良好的随机对照试验（RCTs）才被认为是有力的证据，而与之大致相当的准实验设计（QEDs）可能只有在一定件下才算符合标准"（第11页）。通过使用严格的证据标准，有效教育策略资料中心在2007年发布了初中数学干预的报告，随后提供了高中数学干预、幼儿和小学数学干预的报告。有效教育策略资料中心的研究报告得到了一些不一样的回应，有人为他们欢呼，因为他们为"哪一套课程是最好的"这个问题提供了答案；也有人批评他们的研究是"失败"的，因为他们将课程评估的复杂性过于简单化了（见Schoenfeld，2006）。

无论如何，有效教育策略资料中心的证据标准导致绝大多数数学教育的课程干预研究被排除在外，例如，虽然CMP课程对学生学习的影响一直是20多年以来的研究重点，但79项调查中只有1项符合有效教育策略资料中心的证据标准（而且有保留），基于这仅有的一项符合标准的研究结果（Schneider，2000），有效教育策略资料中心（2010，第1页）给出了以下结论："CMP课程对数学成绩没有明显的影响"。正如美国国家研究协会（2004）认为的，"一个精心设计的研究就能确定课程的有效性，这种想法看似简单，实则带有欺骗性"（第96页），但这恰恰是有效教育策略资料中心确定CMP课程有效性所使用的方法。

国家研究委员会。可能是受有效教育策略资料中心的目标的影响，美国国家研究协会（2004）评价了一些

research is further evidenced by the recent Third International Curriculum Conference that brought together researchers from Australia, Canada, Denmark, France, Israel, Japan, Korea, the Netherlands, Sweden, the United Kingdom, and the United States. Key insights into the design, delivery, and enactment of digital curriculum materials in mathematics classrooms are reported in proceedings of an international conference dedicated to future directions in curriculum research (Bates & Usiskin, 2016).

Links among the intended, enacted, and attained curriculum. In part by drawing upon constructs and findings from qualitative studies about teachers' use of curricular materials, researchers conducting large-scale studies about student outcomes have made progress over the past decade in developing more sophisticated methods of accounting for the nature of curriculum enactment (Heck et al., 2012; Ziebarth et al., 2014). However, we continue to have a limited body of studies that link written curricular materials, enacted curriculum, and student outcomes (Fan, Zhu, & Miao, 2013; Hunsader & Thompson, 2014; NRC, 2004). As researchers continue to explore this area, they will need to build a more comprehensive knowledge base about how written curricular materials are interpreted by teachers and enacted in classrooms, as well as what factors influence this process. Moreover, additional research is needed to advance the field's understanding of how variations in the enacted curriculum relate—if and when they do—to variations in students' learning.

Major Findings and Trends About the Attained Curriculum

As reviewed in the previous section, enactment of a mathematics curriculum is a complex phenomenon and numerous factors contribute to shaping teachers' and students' classroom activities. Accordingly, documenting what students learn and attributing it to a particular curricular program is replete with methodological challenges. In the decade following calls for more scientific research in education (NRC, 2002), there have been numerous advances and several high-profile studies of curricular effectiveness.

Scientifically Establishing Curricular Effectiveness

A key trend in curriculum research is the greater emphasis placed on evaluating curricula scientifically (Clements, 2007; NRC, 2002, 2004; Slavin, Lake, & Groff, 2009). Several frameworks offer the potential to inform the research design of curriculum studies, and each provides affordances and constraints.

What Works Clearinghouse. In 2002, the What Works Clearinghouse (WWC) was established as an initiative of the Institute for Educational Sciences of the U.S. Department of Education to assess the rigor of research on the effectiveness of interventions, including mathematics programs, by providing standards and reviewing educational research. The evidence standards were intended to directly influence federal funding and ultimately yield higher quality educational research. The WWC review protocol essentially filters intervention studies by beginning with the question, "Is group membership determined through a random process?" (U.S. Department of Education Institute for Educational Sciences, 2014, p. 9). According to the WWC, "only well-designed and well-implemented randomized controlled trials (RCTs) are considered strong evidence, while quasi-experimental designs (QEDs) with equating may only meet standards with reservations" (p. 11). Applying their strict evidence standards, the WWC issued reports of interventions in middle school in 2007 with subsequent reports for high school mathematics and early-childhood and elementary mathematics interventions. The WWC reports faced a mixed reception; they were heralded for purportedly answering, "which curriculum is best?" but also proclaimed a "failure" for their gross oversimplification of the complexities in conducting curriculum evaluation (see Schoenfeld, 2006).

Regardless of perception, WWC's evidence standards resulted in the exclusion of the vast majority of studies of curricular interventions in mathematics education. For example, although the impact of *CMP* on student learning has been the focus of research for more than 20 years, only 1 of 79 investigations satisfied WWC evidence standards (albeit *with reservations*). Results of the single study (Schneider, 2000) led the WWC (2010, p. 1) to declare, "*CMP* was found to have no discernible effects on math achievement." As the NRC (2004) argues, "it is deceptively simple to imagine that a curriculum's effectiveness could be easily determined by a single well-designed study" (p. 96), and yet this is precisely what WWC yielded in its determination regarding the curricular effects of *CMP*.

National Research Council. Somewhat related to the goals of WWC, the NRC (2004) assessed the quality of evaluation studies of 13 NSF-funded curricula and 6 commercially developed programs, including *UCSMP* and *Saxon Math*. In contrast to WWC's highly exclusionary filter, the NRC identified a set of minimum conditions

课程评估研究的质量，这些课程有 13 个美国国家科学基金会资助的课程和 6 个商业开发课程，包括 UCSMP 和萨克森数学。与有效教育策略资料中心高淘汰的筛选机制相反，美国国家研究协会制定了一组最低条件，当研究满足这些条件时，它将被归类为"最低限度地满足了研究方法上的要求"（第101页）。为了达到这个合格水平，研究需要包括可量化可测量的结果，提供足够的细节来判断样本的可比性，并至少包括以下某个要素："执行忠实度或专业发展活动的报告，按内容领域或分组后学生成绩来分列的研究结果，以及/或者用多个结果测量，或对可测量的构念进行精准的理论分析，如数感、证明或比例推理等"（第102页）。在所综述的 95 项比较研究中，67 项涉及美国国家科学基金会资助的课程，11 项涉及 UCSMP 课程和 14 项涉及萨克森数学。令人惊讶的是，95 项研究中只有 3 项涉及其他商业机构开发的课程，包括美国最受欢迎的数学教科书系列。

总的来说美国国家研究协会小组评定 95 项比较研究中的 63 项研究为"最低限度地满足了研究方法上的要求"，这比有效教育策略资料中心的筛选尺度要低得多。该小组的主要结论之一是"对于美国国家科学基金会支持的课程和 UCSMP 课程，它们的评估数据库在数量和质量上都大大超过了商业开发课程的评估数据库"（第202页）。广义而言，美国国家科学基金会资助的以标准为基础的课程的研究基础比商业开发的课程的研究基础明显更强，主要是因为（1）资助机构要求对所支持的课程进行评估研究，（2）课程开发人员中包括了有做研究要求的大学教师，（3）出版商资助的针对商业性（传统）教科书的研究更多地关注市场因素，而不是学生学习。美国国家研究协会小组指出，尽管美国国家科学基金会资助的课程的证据基础较强，但 100% 的课程评估研究使用了准实验研究设计，报告最后建议联邦政府大量投资，招募数学教育家、数学家、测量专家和一线教师参与，以"提高国家在数学课程评价方面的能力"（第201页）。

教育研究和发展共同指导原则。在美国教育部教育科学研究院和美国国家科学基金会的一项跨机构项目中，《教育研究与发展共同指导原则》（2013）确定了以下六种类型的研究，这六种类型的研究是为教育干预提供证据的途径：（1）基础研究，（2）早期或探索性研究，（3）设计和开发研究，（4）功效研究，（5）有效性研究，（6）推广研究。应该指出的是，功效研究、有效性研究和推广研究在很大程度上类似于克莱门茨（2007）的课程研究框架：首先在"理想"情境下小规模地对策略或干预进行测试（功效研究），然后在"典型"或"常规"情境下以有限的规模进行检测（有效性研究），最终在广泛的学生和教师群体中，在课堂和学校环境以及各种情况下进行大规模的测试，将研究结论加以一般化（推广研究）。"共同指导原则"遵循了《教育科学研究》（NRC，2002）的原则，旨在为撰写 STEM 教育领域的基金申请书提供建议，但后来被更广泛地运用于其他领域的申请。共同指导原则的一个重要规定是运用逻辑模型——投入、活动、产出、短期和长期成果以及情境因素——这一模型将计划中的研究置于"方案理论"或"实践理论"框架下。这个规定也回应了对数学教育研究的一个共同批评，即数学教育研究明显缺乏理论来指导研究的各个阶段（Lester，2005；Schoenfeld，2007）。

数学课程评价

数学课程对制衡教育改革起着关键的作用，这些关键作用体现在：近期在编制和综述课程有效性研究的共同努力（例如 NRC，2004；Slavin 等，2009），用于对课程进行科学评估的联邦基金不断增加，数学教育以外的研究群体（如经济政策的、教育政策的与社会政策的研究者）的更多参与（Stephan 等，2015）。以下三个部分将讨论近期评估小学、初中和高中数学课程计划的研究。

小学数学课程。在美国，研究者已经对诸如以下的综合性小学数学课程进行了大量的研究：数、数据和空间探索（Agodini, Harris, Thomas, Murphy, & Gallagher, 2010；Gatti & Giordano, 2010）、数学愿景（Resendez & Azin, 2008）、数学进展（Beck Evaluation & Testing Associates, 2005）、萨克森数学（Agodini 等，2010；Resendez & Azin, 2006）和斯科特·福尔斯曼-艾迪生·韦斯利小学数学（Agodini 等，2010；Resendez & Manley, 2005）。

在教育科学研究院提供的 2100 万美元的资助下，社

that, when satisfied, classified studies as "at least minimally methodologically adequate" (p. 101). To earn this qualifying status, studies needed to include quantifiable measurable outcomes, provide sufficient details to judge the comparability of samples, and include at least one of the following components: "a report of implementation fidelity or professional development activity, results disaggregated by content strands or by performance of student subgroups, and/or multiple outcome measures or precise theoretical analysis of a measured construct, such as number sense, proof, or proportional reasoning" (p. 102). Among the 95 comparative studies reviewed, 67 involved NSF-funded curricula, 11 *UCSMP*, and 14 *Saxon Math*; astonishingly, only 3 of the 95 involved other commercially developed curricula, including the most popular mathematics textbook series in the United States.

In total, the NRC panel judged 63 of 95 comparative studies "minimally methodologically adequate," considerably less restrictive than WWC filters. Among its key findings, the panel concluded, "the database of evaluations for the NSF-supported curricula and for UCSMP greatly exceeded the database for the commercially generated materials in quantity and quality" (p. 202). Broadly speaking, the research base is markedly stronger for NSF-funded, standards-based curricula than commercially developed curricula primarily because (1) evaluation studies were required by the funding agency, (2) curriculum developers included university faculty who are expected to conduct research, and (3) publisher-sponsored research on commercial (traditional) textbooks focused more on market factors than on student learning. Despite a stronger evidentiary base for NSF-supported curricula, the NRC panel noted that 100% of the studies used a quasi-experimental design. The report concludes by recommending substantial federal investment "to improve the nation's capacity in mathematics curriculum evaluation" (p. 201) by enlisting mathematics educators, mathematicians, measurement experts, and classroom teachers.

Common Guidelines for Education Research and Development. In a cross-agency project between the U.S. Department of Education Institute of Education Sciences and the National Science Foundation, *Common Guidelines for Education Research and Development* (2013) identifies six types of research that constitute a "pipeline" of evidence on educational interventions: (1) foundational research, (2) early-stage or exploratory research, (3) design and development research, (4) efficacy research, (5) effectiveness research, and (6) scale-up research. It should be noted that efficacy, effectiveness, and scale-up research are largely analogous to Clements's (2007) curriculum research framework in which strategies or interventions are first tested on a small scale under "ideal" circumstances (efficacy research); then examined on a limited scale under "typical" or "routine" circumstances (effectiveness research); and finally tested on a large scale in a wide spectrum of student and teacher populations, classroom and school contexts, and circumstances to generalize research findings (scale-up research). The *Common Guidelines* document adheres to the principles of *Scientific Research in Education* (NRC, 2002) and is intended to inform the development of grant proposals, initially in STEM education but later expanded to apply more broadly. A key stipulation is the inclusion of a *logic model*—comprising inputs, activities, outputs, short-term and long-term outcomes, and contextual factors—that grounds the proposed research in a "program theory" or "theory of practice." Such a provision addresses a common criticism of research in mathematics education, namely the explicit lack of theory to guide all phases of inquiry (Lester, 2005; Schoenfeld, 2007).

Evaluations of Mathematics Curricula

The critical role of mathematics curriculum in leveraging reform is evidenced by recent concerted efforts to compile and synthesize research on curricular effectiveness (e.g., NRC, 2004; Slavin et al., 2009), increased federal funding to conduct scientific curricular evaluations, and greater involvement of research communities outside of mathematics education such as economics and educational and social policy (Stephan et al., 2015). The three sections below discuss recent studies evaluating curricular programs for elementary, middle, and high school mathematics.

Elementary mathematics programs. In the United States, researchers have conducted numerous studies of comprehensive elementary mathematics programs such as *Investigations in Number, Data, and Space* (Agodini, Harris, Thomas, Murphy, & Gallagher, 2010; Gatti & Giordano, 2010), *enVisionMATH* (Resendez & Azin, 2008), *Progress in Mathematics* (Beck Evaluation & Testing Associates, 2005), *Saxon Math* (Agodini et al., 2010; Resendez & Azin, 2006), and *Scott Foresman-Addison Wesley Elementary Mathematics* (Agodini et al., 2010; Resendez & Manley, 2005).

Funded by a $21-million grant from the Institute for Educational Sciences, social policy researchers Agodini et al. (2010) conducted a large-scale comparative study

会政策研究者阿戈蒂尼等人（2010）对四套小学数学课程进行了大规模的比较研究：（1）数、数据和空间探索（简称"探索"），（2）数学表达，（3）萨克森数学和（4）斯科特·福尔斯曼-艾迪生·韦斯利数学（SFAW）。选取这些课程的原因之一是这些教科书在以学生为中心还是以教师为中心的取向上有差异。探索更倾向于以学生为中心的"建构主义"方式，萨克森更强调教师的直接教学，而另外两套课程则提供了更多的混合教学方式。该研究提出了对课程研究的诸多建议，比如实验设计，对10个州的110所小学进行了随机对照试验。

为了建立学生的学习结果与数学课程之间的联系，阿戈蒂尼等发现至少98%的教师使用了分配给他们的课程。通过课堂观察，他们发现使用萨克森的教师明显比使用其他课程的教师花更多时间教授数学。他们还考虑了可能解释学生表现差异的许多教师特征，包括经验、专业发展（即针对特定课程的培训）和教学支持（例如数学辅导）。研究者使用了国家标准化测试来评估小学生在年级内和年级间的成绩增长，在相应的分层模型中，研究者发现，使用数学表达课程的一年级学生的平均成绩显著高于同年级使用探索和SFAW课程的学生的平均成绩；使用数学表达和萨克森课程的二年级学生的平均成绩显著高于同年级使用SFAW课程的学生；而其他课程两两比较没有发现显著差异。此外，为了回答对何人有效这个问题（Clements, 2007；NRC, 2004），研究者把学生的学习结果按成绩（最低、中等和最高三等分）和贫困状况分组（在参与研究的学校中，不超过40%的学生有资格获得免费和低价午餐[FRL]，超过40%的学生具有获得免费和低价午餐的资格），以了解每个课程方案的最大受益群体。

对于那些有国家课程的国家，教科书的选择余地很小，课程之间的差别也较小（Ni, Li, Cai, & Hau, 2015）。在这些国家中，当国家课程发生变化时，进行相应的课程评估是必需的，中国就是一个例子，2001年，中国提出了针对（小学）义务教育的新的数学课程标准。到2006年年底，随着基于新标准的教科书的推出，对新标准的执行也成为了强制性的。新的内容标准更加强调建模、推理和交流；新的教学方法倾向于多用问题导向教学，少用直接讲解；新的学生评价方式包括具有多个问题解决路径和多重表征的开放式任务。

研究者通过一系列研究（Li & Ni, 2011；Ni, Li, Li, & Zhang, 2011；Ni等，2015），测查了新课程对课堂教学和学习结果的影响（按学生的社会和经济背景分组），并考查了课程实施和课堂话语模式对课程影响的调节作用。与阿戈蒂尼等人（2010）对学生学习结果的单一的学生成绩测量标准相反，这一评估采用以下三种成绩测量方式：（1）数学计算（多项选择题），（2）常规问题解决（多项选择题），（3）复杂问题解决（开放型问题）。每种评估方式分三个时间点进行，为期18个月，从五年级开始，到六年级结束。尽管学生使用不同的课程：改革的和传统的，但是随着时间的推移，所有学生的总体表现在三项测量上都有所进步，但是，这两组学生的成绩增长模式是不同的。最初的评估结果显示，使用改革课程的学生在计算和复杂问题解决方面的表现优于传统组，而两组学生在常规问题解决上表现类似。在常规问题解决和复杂问题解决上，两组学生从第一次评估到第三次评估的增长率也相似。在数学计算上，使用传统课程的学生从第一次评估到第三次评估的增长率显著地高于使用改革课程的学生，而且在最终评估上的表现大大超越了使用改革课程的学生。为了体现美国国家研究协会（2004）的关键原则，倪及其同事为大规模研究课程改革的影响提供了一个概念框架和研究设计。

初中数学课程。为了提高数学成绩，美国和其他国家的一些学区采用了数学高成就国家的课程材料。新加坡数学是一套利用问题来发展概念和技能的课程，被当做是能够出国际学生评估项目（PISA）最高成就的课程。虽然有精明的市场推销和新加坡学生在国际比较中引人注目的表现，但缺乏新加坡数学在美国学校实施情况的科学研究，并且没有一项研究能满足有效教育策略资料中心的证据标准。基于现实数学的理论的综合中学数学课程，包括情境中的数学，是由弗洛登塔尔学院（荷兰）与威斯康星大学合作开发的数学课程。尽管研究者在十年间对这套课程进行了广泛的横向和纵向研究（例如

of four elementary school mathematics curricula: (1) *Investigations in Number, Data, and Space* (*Investigations*), (2) *Math Expressions*, (3) *Saxon Math*, and (4) *Scott Foresman-Addison Wesley Mathematics* (*SFAW*). The curricular programs were selected, in part, because of differences in their relative emphasis on student-centered or teacher-centered approaches—*Investigations* using a more student-centered, "constructivist" approach, *Saxon* emphasizing direct instruction from the teacher, and the other two programs offering more of a blended approach. The study addresses numerous recommendations for curriculum research including an experimental design, namely a randomized controlled trial involving 110 elementary schools in 10 states.

To establish linkages between student outcomes and mathematics curricula, Agodini et al. found at least 98% of teachers used their assigned curriculum. Through classroom observations, they found that teachers of *Saxon* spent significantly more time teaching mathematics than teachers of other programs. They also took into account numerous teacher characteristics, including experience, professional development (i.e., curriculum-specific training), and instructional supports (e.g., math coaching), any one of which might offer alternative explanations for differential student performance. Using a nationally normed test designed to assess learning gains within and across elementary grades, in hierarchical models researchers found that at grade 1, average achievement of *Math Expressions* students was significantly higher than that of students using *Investigations* and *SFAW*; at grade 2, average achievement of *Math Expressions* and *Saxon* students was significantly higher than that of *SFAW* students; and none of the other curriculum-pairs were significantly different. Furthermore, addressing the question of efficacy *for whom* (Clements, 2007; NRC, 2004), student outcomes were disaggregated by subgroups based on achievement (lowest, middle, and highest third) and poverty status (up to 40% of students in a participating school eligible for free and reduced-price lunches [FRL], greater than 40% eligible for FRL) to understand who benefitted most from each curriculum program.

In countries with a national curriculum, there are fewer textbook options and relatively modest differences between them (Ni, Li, Cai, & Hau, 2015). In such counties, evaluations are primarily necessitated when changes in the national curriculum occur. Such was the case in China when, in 2001, new mathematics curriculum standards were put forward for compulsory (elementary) education. By late 2006, with the introduction of textbooks embodying the new standards, implementation became mandatory. New content standards placed greater emphasis on modeling, reasoning, and communication; new instructional methods favored problem-oriented teaching over direct explication; and new student assessments included open-ended tasks with multiple solution paths and representations.

In a series of studies (Li & Ni, 2011; Ni, Li, Li, & Zhang, 2011; Ni et al., 2015), researchers gauged the impact of the new curriculum on classroom instruction and learning outcomes (disaggregated by students' social and economic backgrounds) and examined how curriculum implementation and classroom discourse moderated curricular influence. In contrast to Agodini et al. (2010), who used a single measure of student outcomes, this evaluation employed three achievement measures: (1) mathematical computation (multiple-choice format), (2) routine problem solving (multiple-choice format), and (3) complex problem solving (open-ended task format). Each assessment was administered at three time-points over 18 months, beginning in grade 5 and ending in grade 6. Although students in both curricular treatments, reform and traditional, demonstrated overall performance increases over time on all three outcome measures, there were differentiated achievement patterns across the two groups. At the initial assessment, the reform group outperformed the traditional group on calculation and complex problem solving, but the groups scored similarly on routine problem solving. Growth rates from the first assessment to the third assessment were similar for the two groups on the routine problem solving and complex problem solving, but the growth rate of the traditional group was significantly higher on the computation test with performance eclipsing students in the reform group on the final assessment. By addressing key tenets of NRC (2004), Ni and colleagues provide a conceptual framework and research design for studying the impact of curriculum reform on a large scale.

Middle school mathematics programs. In an effort to improve mathematics achievement, curricular materials of high-performing nations have been adopted by school districts in the United States and in other countries. *Singapore Math*, utilizing problem-based development of concepts and skills, is marketed as producing among the highest achievement on the Program for International Student Assessment (PISA). Shrewd marketing and the compelling performance of Singapore in international studies notwithstanding, there is a dearth of scientific studies of *Singapore Math* in U.S. schools; no studies satisfied the WWC evidence standards. Comprehensive middle grades mathematics

Romberg, Webb, Folgert, & Shafer, 2005），但是没有一项研究符合有效教育策略资料中心的证据标准。类似地，关于数学专题（另一个美国国家科学基金会资助的改革课程）的研究也没能满足该证据标准，而只有一项针对CMP的研究是令人满意的。塔尔等人（2008）研究了这三套美国国家科学基金会资助的初中数学课程对学生学习和课堂学习环境的影响。研究者通过分层线性模型发现，课程类型（课程由美国国家科学基金会资助还是由出版社开发）并不是预测学生成绩的重要因素，而基于标准的学习环境（SBLE）则会缓和课程效果。当改革的课程与中等或更高水平的基于标准的学习环境相结合时，学生在数学推理、问题解决和交流方面的成绩表现会显著更好。塔尔等人（2008）把三个改革的课程归作一类来研究并得到了以上的结果，但这项研究仍然未能满足有效教育策略资料中心的证据标准，因为"有效性测量不能仅仅归因于干预——这项干预是与另一项干预连结在一起的"（WWC, 2010，第10页）。

课程对代数学习影响的纵向研究（LieCal）项目研究了美国国家科学基金会资助的一套课程（CMP）和另外几套由出版商开发的课程对一个大型的城市学区的效果（Cai, Wang, Moyer, Wang, & Nie, 2011）。该研究仔细分析了数学课程的性质，探究了课程执行的范围和性质，观察了初中年级的数学课堂教学，并采用了受课程目标影响的多项结果测量。在使用两级线性模型增长曲线和重复测量方差的相关分析中，研究人员发现，使用美国国家科学基金会资助课程的学生在问题解决技能方面的增长明显更高，但使用不同课程类型的学生在符号操作技能方面的增长没有差异。此外，他们还发现，至少根据一些测量结果，课堂教学的性质和质量是预测学生初中三年成绩增长的一个显著因素。

高中数学课程。针对高中阶段的课程有效性研究较少，但也有一些值得注意的例外。卡内基学习公司出版的《卡内基学习课程和认知向导》，结合教科书和互动软件，根据学生的需求提供个性化、自定步调的教学。针对这一课程的三个研究使用了不同方式的随机分配方法。卡巴罗、贾思武和吴（2007）将卡内基代数1课程或基于标准的课程分配给22个教室使用。坎普萨诺、迪纳尔斯基、阿戈蒂尼和拉尔（2009）随机分配了18名教师使用卡内基代数1课程或传统的教学方法。潘恩、麦卡弗里、斯劳特、斯蒂尔和池本（2010）随机分配了8所高中的学生使用卡内基几何课程或标准的（传统）几何课程。另外三项研究（Shneyderman, 2001; J. E. Smith, 2001; Wolfson, Koedinger, Ritter, & McGuire, 2008）使用了准实验设计或随机对照试验，但由于一些学生被排除在分析之外，这些试验受到了影响。这六项研究的结果非常不一致：一项研究认为这套课程具有统计学上的积极影响，四项研究不能确定是否有影响，一项研究认为与传统课程相比，它具有统计学上的负面影响。这些相互不一致的研究结果突显了课程研究中的一个关键问题，即提问带有的确定性：什么课程有效？此外，这些看似矛盾的结果印证了美国国家研究协会（2004）的主张，即判断课程的有效性需要做一系列精心设计的研究——包括课程分析、比较研究和案例研究。

由美国国家科学基金会资助的COSMIC项目，对两套高中数学课程材料进行了三年纵向比较研究：从学习内容的组织方式来看，一套课程采用了综合的方式，另一套课程采用了按特定主题方法来组织内容。在按数学主题来组织学习内容的课程中，学生先学习代数1，接着是几何，然后是代数2；在采用综合方式的课程中，学生使用的是核心-加的课程，该数学课程每一年都融合了高中的代数、几何、统计和离散数学。COSMIC项目采用准实验设计，并且考虑了学生的人口统计特征和教师的特征，研究者通过课堂观察、关于学习机会的数据和教师调查来关注课程执行，还使用了多种测量学生成绩的方法来评估中学生的数学学习。

在对第一年的数据作横向分析中，课程类型对学生在标准化测量以及共同目标和数学推理评估测试的成绩有显著的预测作用，参加综合课程的学生表现优于参加专题课程的学生（Grouws等，2013）。第二年，综合课程的学生只在标准化测试中得分显著较高；课程类型不再是预测学生在共同目标或数学推理测试中的成绩的重要因素了（Tarr等，2013）。第三年，综合课程的学生

programs built on a program theory of realistic mathematics include *Mathematics in Context*, developed by the Freudenthal Institute (the Netherlands) in partnership with the University of Wisconsin. Despite extensive cross-sectional and longitudinal studies (e.g., Romberg, Webb, Folgert, & Shafer, 2005) spanning a decade, no studies of *Mathematics in Context* satisfied WWC evidence standards. Similarly, no studies of *MATH Thematics* (another NSF-funded reform program) satisfied evidence standards, while only one investigation of *CMP* was deemed satisfactory. Tarr et al. (2008) examined the impact of all three NSF-funded middle grades mathematics curricula on student learning and the classroom learning environment. Using hierarchical linear modeling, researchers found that curriculum type (NSF-funded vs. publisher developed) was not a significant predictor of student outcomes. However, the standards-based learning environment (SBLE) moderated curricular effects; when reform curricula were coupled with either moderate or high SBLE, students performed significantly higher on a test of mathematical reasoning, problem solving, and communication. Despite these findings, by combining three reform curricula into one category, Tarr et al. (2008) failed to satisfy WWC evidence standards because "the measures of effectiveness cannot be attributed solely to the intervention—the intervention was combined with another intervention" (WWC, 2010, p. 10).

The Longitudinal Investigation of the Effect of Curriculum on Algebra Learning (LieCal) project studied the effects of an NSF-funded curriculum, *CMP*, and multiple publisher-developed curricula in one large, urban school district (Cai, Wang, Moyer, Wang, & Nie, 2011). The study carefully examined the nature of the mathematics curricula, investigated the extent and nature of curriculum enactment, made observations in middle grade classrooms, and utilized multiple outcome measures sensitive to the goals of the curriculum programs. In a related analysis using a two-level hierarchical linear model growth curve and repeated measures ANOVA, the researchers found significantly greater growth in problem-solving skills for students using the NSF-funded curriculum but detected no differences of gains in symbolic manipulation skills across curriculum types. Additionally, they found the nature and quality of classroom instruction to be significant predictors of students' achievement gains over the 3 middle school years on at least some outcome measures.

High school mathematics programs. Fewer studies of curricular effectiveness are available at the high-school level, with notable exceptions. The *Carnegie Learning Curricula and Cognitive Tutor*, published by Carnegie Learning, combines textbooks and interactive software to provide individualized, self-paced instruction based on student needs. Three studies of this program used random assignment but in different ways. Cabalo, Jaciw, and Vu (2007) assigned 22 *classrooms* to the Carnegie Algebra 1 program or the standard curriculum, Campuzano, Dynarski, Agodini, and Rall (2009) randomly assigned 18 *teachers* to the Carnegie Algebra 1 program or traditional instructional methods, and Pane, McCaffrey, Slaughter, Steele, and Ikemoto (2010) randomly assigned *students* in eight high schools to Carnegie Geometry or the standard (traditional) geometry curriculum. Three additional studies (Shneyderman, 2001; J.E. Smith, 2001; Wolfson, Koedinger, Ritter, & McGuire, 2008) used quasi-experimental designs or randomized controlled trials that were compromised by the exclusion of some students from analyses. Results of the six studies were largely inconsistent: One study with a statistically positive effect, four with indeterminate effects, and one with a statistically negative effect relative to the traditional program. Such inconsistent findings highlight a key problem in curriculum research, namely the deterministic nature of asking, *what works?* Moreover, these seemingly conflicting results underscore NRC's (2004) assertion that a collection of well-designed studies—including curriculum analyses, comparative studies, and case studies—are needed to judge the effectiveness of curricular programs.

The COSMIC project was a 3-year longitudinal comparative study, funded by the NSF, of two types of high school mathematics curricular materials: an integrated approach and a subject-specific approach to content organization. In the subject-specific path, students progressed through algebra 1, geometry, and algebra 2; in the integrated pathway, students studied from *Core-Plus*, a program that integrates elements of algebra, geometry, statistics, and discrete mathematics in each year of high school mathematics. Employing a quasi-experimental design and taking into account student demographics and teacher characteristics, the COSMIC project evaluated secondary-school-students' mathematics learning using multiple measures of student achievement while carefully attending to curriculum implementation via classroom observations, opportunity to learn data, and teacher surveys.

In a cross-sectional analysis of year 1 data, curriculum type was a significant predictor of student achievement on a standardized measure as well as on a test of common objectives and assessment of mathematical reasoning, with students in the integrated program

在共同目标测试中表现更好，但两组学生在标准化测试上却没有什么差别（Chávez, Tarr, Grouws, & Soria, 2013）。

COSMIC项目确定了几个可以解释学生成绩显著变化的变量，这些变量包括学生层面的以往成绩和教师层面的为学生提供的学习机会。值得注意的是，在格劳斯等人（2013）和塔尔等人（2013）的研究中，课程类型与学生以往成绩之间存在着明显的跨层次互动关系，有的学习结果测量表明，先前成绩较高的学生相比成绩较低的学生从综合课程中获益更多。有人一直持有一种热情洋溢的信念，即传统的（按学科内容）编排的高中数学课程组织会更好地服务于成绩好的学生，但这些作者认为，他们的结果削弱了这一令人兴奋的信念。

超越年级水平的课程的影响

课程对学生数学学习的影响的研究已不是什么新鲜事物，但一种新的趋势是关注学生在课程年级之外的表现。例如，蔡、莫耶、王等（2013年）研究了作为初中生曾参加过LieCal项目并在同一学区就读高中的学生的长期课程效应。研究者的报告表明，那些在初中阶段使用CMP课程的11年级学生在问题解决和问题提出方面显著地比非CMP课程的学生表现得更好。在一项相关的研究中，蔡、莫耶和王（2013）发现使用CMP课程的学生在9年级的开放性问题测试、10年级全州组织的测试及11年级的问题提出任务上都表现得更好。

哈维尔、波斯特、麦汉妮、迪普伊和勒博（2013）在对高中数学课程有效性的延伸研究中，考察了大学数学成绩和选课方式。研究者发现学生的高中数学课程与大学数学成绩没有显著的关系，此外，高中数学课程与那些进入大学即修"大学代数"或"微积分预备"或更高级的课程的学生的选课模式无关。然而，在高中使用核心-加数学课程的人更有可能在进入大学时选一门发展性数学课程①。在一项相关研究中，哈维尔、麦汉妮、

波斯特、诺曼和迪普伊（2011）分析了一个大型研究型大学的1588名学生8个学期的选课和成绩等数据。这些研究人员发现，对于那些至少修了两个学期微积分预备或更高水平数学课的学生，无论他们使用的是哪套高中课程，每一个高中课程都一样为之后更严格的大学数学课程做好了准备。在随后的几年中，他们追踪了使用某个数学课程的学生，收集的学生数据给课程效果提供了另外的补充测量工具，并对课程材料的改进提供了可能性。

获得课程研究的方法问题

尽管对获得课程的研究非常感兴趣，但在研究课程与学生成绩之间的因果关联时，仍然存在许多挑战。虽然美国新一代的评估手段提供了检测新课程标准影响的机会（Heck等，2012；Tarr等，2013），但提高学生成绩的压力可能会阻碍他们参与此类研究，因为学校已经承担着繁重的强制性测试，所以不愿意参与无论大小的进一步评估（Chval, Reys, Reys, Tarr, & Chávez, 2006）。第二，尽管该领域已经对课程有效性进行了额外的、更严格的研究，但随机对照试验——研究的"黄金标准"——在大多数教育环境中仍然是不切实际的，在大规模研究中尤其如此。即使准实验设计更为实际，但仍然可能受到外部有效性威胁的困扰（NRC, 2004）。此外，如果没有外部基金资助，昂贵的研究成本会使得大规模的"科学"研究很难进行。例如，阿戈蒂尼等人（2010）的研究费用是2100万美元，超出了其所评价的课程的开发成本（Confrey & Maloney, 2015）。虽然纵向研究对于衡量课程多年的影响至关重要，但这些研究的成本往往更高，并且面临更多额外的挑战，如学生和教师退出，数据的缺失以及对更深入、复杂的分析的要求，并且一些结果往往是难以解读的。

另一个重要的挑战是精心设计区分教学和课程各自影响的实证研究——实施、测量和理解课程与教学的独

---

①相当于补旧性质的数学课。——译者注

outperforming those in the subject-specific curriculum (Grouws et al., 2013). In year 2, students in the integrated program scored significantly higher on the standardized measure only; curriculum type was not a significant predictor on the common objectives or mathematical reasoning tests (Tarr et al., 2013). In year 3, there was a differential effect in favor of the integrated program on the common objectives test but not on the standardized measure (Chávez, Tarr, Grouws, & Soria, 2013).

The COSMIC project identified several variables that accounted for significant variation in student outcomes, including prior achievement at the student level and opportunity to learn at the teacher level. Notably, in Grouws et al. (2013) and Tarr et al. (2013), there was a significant cross-level interaction between curriculum type and students' prior achievement; on some outcome measures, students with higher prior achievement benefitted more from the integrated program than students with lower prior achievement. The authors assert that this result undermines an impassioned belief, held by some, that higher-achieving students are better served by a more traditional (subject-specific) organization of the high school mathematics curriculum.

Effects Beyond Grade Levels of the Curriculum

Although investigations of the effects of a curriculum program on students' mathematical learning are far from new, an emerging trend is a focus on students' performance *beyond* the grade levels of the program. For example, Cai, Moyer, Wang, et al. (2013) studied the long-term curricular effects of students who participated in the LieCal Project as middle-school students and were attending high schools in the same school district. They reported that grade 11 students who studied from *CMP* in middle school performed significantly higher on problem-solving and problem-posing tasks than students of the non-*CMP* program. In a related study, Cai, Moyer, and Wang (2013) detected differential curricular effects in favor of *CMP* on open-ended problem tasks in grade 9, on a state-mandated test in grade 10, and on problem-posing tasks in grade 11.

In an extension of research into the effects of various high school mathematics programs, Harwell, Post, Medhanie, Dupuis, and LeBeau (2013) examined college mathematics achievement and course-taking patterns. Researchers found no significant relationship between students' high school mathematics curricula and college mathematics achievement. Moreover, high school mathematics curricula were unrelated to course-taking patterns for students who began college in college algebra/precalculus or a more advanced course. However, those who studied *Core-Plus* in high school were more likely to begin college in a developmental mathematics course. In a related study, Harwell, Medhanie, Post, Norman, and Dupuis (2011) analyzed course-taking and achievement data of 1,588 students across 8 semesters at a large research university. For students who took at least two semesters of mathematics at or beyond precalculus, researchers found each high school curricular program equally prepared students for subsequent, more rigorous college mathematics coursework. Data collected on students in subsequent years following a mathematics program provide alternative and supplemental measures of curricular effects and offer the potential to inform refinements to the curricular materials.

Methodological Issues in Research on the Attained Curriculum

Despite great interest in studies of the attained curriculum, numerous challenges remain in conducting research that provides causal links between curriculum and student outcomes. Although a new generation of assessments in the United States affords opportunities to gauge the impact of new curriculum standards (Heck et al., 2012; Tarr et al., 2013), pressures to increase student achievement may inhibit participation in such studies because schools are already overburdened by mandatory testing programs and therefore are reluctant to commit to further assessments, however large or small (Chval, Reys, Reys, Tarr, & Chávez, 2006). Second, although the field has responded to the call for additional, more rigorous studies of curricular effectiveness, randomized controlled trials—considered the "gold standard" of research—remain impractical in most educational settings, and this is particularly true for large-scale studies. Even quasi-experimental designs, although more practical, are nonetheless potentially plagued by threats to external validity (NRC, 2004). Moreover, the exorbitant cost makes conducting large-scale "scientific" research prohibitive without external funding. For example, at $21 million, the cost of the Agodini et al. (2010) study exceeded the developmental costs of the curricular programs it evaluated (Confrey & Maloney, 2015). And although longitudinal studies are critical to measuring the impact of curricula over years, their costs are often even greater and present additional challenges such as student and teacher attrition, missing data, and requirements for more sophisticated, complex analyses, the results of which are often difficult to interpret.

特贡献，以及二者如何一起协同工作来促进学生的学习。LieCal和COSMIC项目的最新方法为如何通过实证研究实现这一目标提供了见解，但这方面的研究仍需要进一步扩展和深入。没有适当的研究设计，研究结果可能会将数学课程的影响与教学效果混为一谈。

有效性研究必须使用课程测量效度，即"全面抽样课程目标，有效测量与这些目标有关的内容，确保为测试而教（而非为课程而教）不可行或不大可能混淆研究结果，并关注课程变化带来的影响"（NRC，2004，第6页）。如雷米拉德、哈里斯和阿戈蒂尼（2014）在考察了当时可用的州课程标准和课程材料后确定了阿戈蒂尼等人（2010）使用的单一的结果测量是适当的，也就是说，研究者认为所选择的测试没有偏向任何一个数学课程。然而，鉴于数感、数的性质和数的运算占一年级测试问题的75%并且占二年级的50%，很难确定学生成绩增长的显著差异是否归因于每套课程方案中内容模块某些特定的优势（或短处）。具体来说，虽然使用数学表达课程的学生表现优于使用其他课程的学生，但无法确定课程的一些特定要素（例如几何课）是否产生了优异的表现，因为单一的结果测量不够敏感。

最终，一些关键的和复杂的问题，如，学生是否、为何和怎样从在课堂执行的数学课程的方方面面中获益？可能无法通过实验研究得到明确的答案。因此，美国国家研究协会（2004）认为，"没有任何一种单一的研究方法足以确认课程的有效性，使用多种评价方法加强了有效性的确定"（第191页）。克莱门茨（2007）同样倡导在数学课程的发展中进行循环研究，包括用形成性评价来帮助了解课程材料某个具体要素的有效性，即可以让一位教师对个别学生、小组或全班执教课程材料，接着让一组有多样知识背景和经验的教师执教，采用小规模的随机临床试验法进行终结性评价，以了解学生在现实的课堂环境中向典型教师学到了什么，并用大规模的随机临床试验来确定不同责任方所执行课程的忠实度。

小结和未来方向

长期以来，理解课程如何影响教与学，一直是教育研究的核心目标。本章中，我们采用了三个层面的课程概念化模型来研究近期数学课程研究的走向和进展。我们讨论了十年来关于预期课程、执行课程和获得课程的研究，以强调最近的关于以下这些方面的实证研究结果的丰富贡献：课程特点和设计、教师和学生的相关课堂活动，以及课程对学生数学学习的影响。尽管评估这些领域的研究现状是很重要的，但是同样很重要的是，研究者从预期课程、执行课程和获得课程之间的相互作用中学到了什么——每一层面如何影响其他课程。事实上，研究者越来越重视课程过程的多层面问题，这是过去十年数学课程研究中最引人注目的主题之一。

在过去十年中，对预期数学课程的研究取得了十足的进展，自2005年以来发表了大量的研究报告。在此期间，研究者分析了各种课程材料的问题和任务以及数学内容和重点。任务和内容分析为研究学生的学习机会提供了新的思路，包括分析预期课程如何反映不同文化或国家对数学应如何教与学的期望。最近的研究也从历史的视角考察了预期课程，分析了数学教科书的内容与主要的改革运动之间的关系，从而明确了过去100—150年来数学教育的课程走向。

20世纪90年代和21世纪初，随着美国和国际课程开发活动及相关改革的激增，相应地，我们看到针对执行数学课程的研究也在稳步增长。虽然关于课程执行的早期研究，包括教师使用课程材料的研究，提供了许多从书面课程转变到执行课程时发生变异的例证，但过去十年的研究使我们对变异发生的过程和原因有了更多的理解。已经研究了影响课程执行的各种因素，包括教师和

Another critical challenge is to carefully design empirical studies that *disentangle* the effects of teaching and curriculum—to operationalize, measure, and understand the unique contributions of curriculum and instruction as well as how they work in tandem to foster students' learning. Recent approaches of the LieCal and COSMIC projects offer insights into how this can be achieved empirically, but further advances are required. Without an adequate research design, results may conflate the effects of a mathematics curriculum with the effects of instruction.

It is imperative that studies of effectiveness use *curricular validity of measures* that "comprehensively sample the curricular objectives in the course, validly measure the content within those objectives, ensure that teaching to the test (rather than the curriculum) is not feasible or likely to confound the results, and be sensitive to curricular changes" (NRC, 2004, p. 6). As reported in Remillard, Harris, and Agodini (2014), the singular outcome measure in Agodini et al. (2010) was declared appropriate after an examination of state curriculum standards and curricular materials available at the time; that is, the researchers argued the selected test was not biased toward any one mathematics program. However, given that number sense, properties, and operations comprised about 75% of the test items at grade 1 and 50% at grade 2, it is difficult to know whether significant differences in performance gains were attributable to particular strengths (or limitations) of content modules that comprised each program. Specifically, although students using *Math Expressions* outperformed those using other programs, it cannot be determined whether particular elements of the curriculum (e.g., geometry lessons) yielded superior performance because the single outcome measure was not sensitive enough.

Ultimately, experimental studies may not provide definitive answers to critical and complex research questions regarding whether, why, and how students benefit from elements of a mathematics curriculum as enacted in the classroom. For that reason, the NRC (2004) argues, "No single methodology by itself is sufficient to establish a curricular program's effectiveness. The use of multiple methodologies of evaluation strengthens the determination of effectiveness" (p. 191). Clements (2007) likewise advocates for cyclic research in the development of mathematics curricula, including formative evaluations that seek to understand the effectiveness of specific elements of the curricular materials as enacted by a teacher with individuals or small groups, whole classes, and subsequently by a diverse group of teachers, as well as summative evaluations using randomized clinical trials on both a small scale, to understand what students attain with typical teachers in realistic classroom settings, and a large scale, to ascertain the fidelity of implementation by various stakeholders.

Summary and Future Directions

Understanding how curriculum affects teaching and learning has long been a central goal of educational research. In this chapter, we employed a three-level conceptualization of curriculum to examine trends and advances in recent scholarship about mathematics curriculum. We discussed a decade of research about the intended, enacted, and attained curriculum to highlight the rich contributions of recent empirical findings about curricular features and designs, related classroom activities of teachers and students, and impacts on students' mathematical learning. Although it is important to appraise the current state of research in each of these areas, it is also essential to consider what researchers have learned about interactions among the intended, enacted, and attained curriculum—namely, how each level can influence the others. In fact, researchers' increased attention to multiple levels of the curriculum process is one of the most notable themes in studies of mathematics curriculum of the past decade.

Research about the intended mathematics curriculum has advanced considerably over the past decade, with the majority of research reports published since 2005. During this period, researchers have analyzed the problems and tasks as well as the mathematical content and emphases of a variety of curricular materials. Both task and content analyses have yielded new insights into students' opportunities to learn, including emerging understandings about how the intended curriculum represents different cultural or national expectations of how mathematics should be taught and learned. Recent studies have also examined the intended curriculum from a historical perspective, analyzing relationships between the content of mathematics textbooks and the major reform movements, thereby identifying curricular trends in mathematics education over the past 100–150 years.

In response to a surge in curriculum development activity and related reforms in the United States and internationally during the 1990s and 2000s, we have seen a steady growth in research about the enacted mathematics curriculum. Whereas earlier studies of curriculum enactment, including studies of teachers' use of curricular materials, provided numerous illustrations of variations in the transformation from written to enacted curriculum, research of the past decade has

学生的特性、课程特征和情境。在这些因素中，教师的特性在课程执行中的作用得到了最为广泛的研究；这方面的研究说明了教师的知识、信念和价值取向对课堂中的课程执行的影响。

最近关于课程执行复杂性的研究都支持系统地检测学生使用课程材料的学习。在过去十年中，该领域对获得课程的研究在质量上和严谨性方面取得了显著的进步，而这一研究领域历史上曾遭遇过许多方法论上的挑战。近期少数有关课程有效性的研究回应了高质量研究的诉求，仔细研究了书面课程材料的设计、课堂执行的特点和学生数学学习结果之间的关系。这些研究项目（例如，COSMIC，LieCal）不仅报告了与某些特定的课程材料和教学特征有关的学生学习的新发现，而且还提供了新的方法论工具和研究方法，未来研究可以继续改进这些工具和研究方法。

回顾过去十年的数学课程研究，不仅要涉及现有研究的走向和进展，还要考虑需要做更多研究工作的领域。展望未来的研究方向可以促进我们对数学课程的现有认识，我们建议需要特别关注以下五个领域：（1）跨国研究，（2）纵向研究，（3）概念性探讨，（4）研究方法的改进，（5）刻意构建知识库。

需要更多的跨国研究

虽然一些针对预期课程的研究都是跨国的，但迄今为止，大多数对执行课程和获得课程的研究还都仅限于一个国家，而且研究通常由来自该国的研究人员进行。严谨的跨国研究可提供一些新的视角和发现，这是仅在一个国家进行的数学课程研究无法获得的。预期课程的跨国比较研究揭示了数学教科书以前未被认识的一些特征，反过来，这些特点还有可能会增加对执行课程和获得课程各方面的理解。

尽管研究者已对数学教科书进行了跨国的内容分析和历史分析，但我们仍然缺乏对预期课程某些特定数学方面的跨国的历史的分析。例如，对美国和中国教科书在过去一个世纪中对统计等主题的处理进行历史分析，可能会得出有价值的见解。从书面课程材料所反映的情况来看，中国和美国学生学习统计的机会是如何随着时间的推移而改变的？另一个可能有用的跨国研究是分析书面课程材料中向教师提供的信息的内容和呈现方式，这些材料描述了不同国家是如何通过课程材料提供教学支持的。

学生数学表现的变异与预期课程或执行课程的差异有什么关系？跨国研究也许能回答这一类重要问题。因此，我们鼓励国际研究者合作研究课程对学生学习的影响。我们注意到，在中国、日本、韩国、新加坡等成绩高的国家，很少做过大规模的对获得课程的研究。尽管国际教育协会的研究（例如，第二次国际数学研究，TIMSS）比较了各国学生的数学成就，但我们对于高成绩国家的实施课程和获得课程之间的关系还知之甚少。高成绩国家学生的成绩究竟在多大程度上归功于课程或其他因素，如学生的学习动机和校外教育项目呢？

需要纵向研究

缺乏对执行课程和获得课程的纵向调查有损于我们的文献基础。大部分关于课程执行的研究，包括教师对数学课程材料的使用，大多分析的是在课程使用初期收集的数据，很少有研究检查一段时间以来课程的执行，包括教师和学生的角色以及情境因素。教师在数年中使用同一课程材料，其课程实施的本质是什么？另一个可以考虑的是，比如对于四年高中段使用同一课程的学生，他们的角色在课程执行过程中有变化吗？有怎样的变化？这些问题之所以是重要的，不仅因为它们提出了教

extended our understanding of how and why these variations occur. A variety of factors affecting curriculum enactment have been investigated, including teacher and student characteristics, curricular features, and context. Of these factors, the role of teacher characteristics in curriculum enactment has been investigated most extensively; and this research illustrates the influence of teachers' knowledge, beliefs, and orientations on classroom enactments of curriculum.

Recent research about the complexity of curriculum enactment has supported systematic efforts to examine students' learning with curricular materials. Over the past decade, the field has seen significant advancement in the quality and rigor of research about the attained curriculum, an area of research historically fraught with methodological challenges. A small number of recent studies of curricular effectiveness have responded to calls for high-quality research that carefully examines relationships among the design of written curricular materials, characteristics of classroom enactments, and students' learning of mathematics. These research projects (e.g., COSMIC, LieCal) have provided not only new findings about student outcomes as they relate to particular curricular materials and instructional features, but also new methodological tools and research approaches that future studies may continue to refine.

Reviewing the past decade of research about mathematics curriculum involves not only identifying trends and advances in existing studies but also considering areas in which additional work is needed. Looking ahead to ways in which future studies can advance our existing understanding of mathematics curriculum, we suggest five areas of particular need: (1) cross-national studies, (2) longitudinal studies, (3) conceptual questioning, (4) methodological refinement, and (5) deliberate building of the knowledge base.

The Need for Additional Cross-National Studies

Whereas many studies of the intended curriculum have been cross-national, most studies of the enacted and attained curricula to date have been conducted in one nation, typically by researchers from that country. Rigorous cross-national studies of mathematics curriculum can provide new perspectives and findings that simply cannot be gained by research conducted in one country alone. Cross-national comparisons of the intended curriculum have revealed previously unrecognized features of the mathematics textbooks that, in turn, have the potential to increase understanding of aspects of both the enacted and attained curricula.

Although researchers have conducted cross-national content analyses of mathematics textbooks as well as historical analyses of textbooks, we continue to lack cross-national, historical analyses of the intended curriculum in particular mathematical strands. For example, valuable insights are likely to emerge from a historical analysis of U.S. and Chinese textbooks' treatment of a topic such as statistics over the past century. How have Chinese and American students' opportunities to learn statistics, as represented in written curricular materials, changed over time? Another potentially useful cross-national investigation would be an analysis of the content and presentation of information provided to teachers in written curricular materials that characterizes how different nations provide instructional support through curricular materials.

Cross-national studies could potentially address key questions about how variations in the mathematical performance of students might relate to differences in the intended or enacted curriculum. For this reason, we encourage international researchers to collaborate on studies of the impact of curriculum on student learning. We note that few large-scale studies of the attained curriculum have been conducted in high-performing nations such as China, Japan, Korea, and Singapore. Although the IEA's studies (e.g., SIMS, TIMSS) compare students' mathematics achievement across countries, we continue to know little about the relationship between the enacted and attained curriculum in high-performing nations. To what extent can student achievement in high-performing nations be attributed to curriculum or to other factors such as student motivation and out-of-school educational programs?

The Need for Longitudinal Studies

Our literature base suffers from a lack of longitudinal investigations of the enacted and attained curricula. The majority of research about curriculum enactment, including teachers' use of mathematics curricular materials, is based on analysis of data collected during early stages of curriculum use. Few studies have examined curriculum enactment, including teachers' and students' roles as well as contextual factors, over time. What is the nature of curriculum enactment over multiple years of teachers' use of the same curricular materials? Another consideration is whether and how students' roles in curriculum enactment might vary over years of engagement with a 4-year high school curriculum program, for example. These questions are important not only because they raise issues about teachers' and students' classroom

师和学生课堂体验随着时间的推移而变化的问题，而且还因为在分析获得课程时，考虑课程执行在一段时间上的差异也越来越重要。

与之相关的是，直到近几年来，对学生学习结果的研究才开始考查课程的跨年级影响，我们仍需要进一步系统研究课程的跨年级影响，来了解课程对学生数学学习的长期影响。

概念性质疑的需要。虽然在不同研究中可以看到一些共同的元素，但是对数学课程的一些研究都采用了不同的概念和分析框架。例如较早前我们注意到，在研究预期课程时，研究者通常会制定具体的研究框架来支持他们计划进行的分析（例如支持内容分析的数学框架）。我们还注意到研究者对某些构念（例如教师知识、课程执行的忠实度、教师课程材料的使用和课程的执行）的概念化具有多样性。虽然随着研究者努力解读课程研究的各个方面的发现时，众多的概念和框架也会带来挑战，但是这些不同的观点为个别研究或相关的系列研究提供了重要的指导。

我们认为，该领域将受益于更多关注不同观点之间的关系，以及更多关注使用特定概念和框架的影响。不同的观点可能会影响研究者从数学课程研究中所获得的信息。雷米拉德（2005）提供了这种分析的一个范例，她从教师使用课程材料的研究中提炼出了几种不同的观点。她的分析使得研究者在设计和报告执行课程与获得课程的时候更多地关注他们对教师课程使用的假设。

另一个例子就是本章图30.1所示的课程分析的三层面模型。这个模型反映了长期以来常见的课程观，但并不是所有的研究者都使用它。事实上，许多学者在这个课程分析模型中加入了另外的或替代的层面（例如 Clements, 2002; Remillard & Heck, 2014; Stein 等；2007; Tarr 等, 2006）。例如，斯坦等人（2007）使用了一个四层面模型，在那个模型中，他们区分了书面课程和教师的预期课程，即教师在上课之前制定的授课计划。不同模型的存在引发了需要进一步思考的问题：对于某个特定的课程概念，它会产生什么独特见解？这些不同的概念如何协同促进数学课程的研究，为数学课程的研究提供信息？当这些问题被用于不同的概念和分析框架时，可以帮助研究人员开始评估为研究设计提供信息的概念选择，为研究设计提供信息，并且认识到对知识库的影响。

需要改进研究方法。过去十年出现了针对数学课程的众多研究活动，一个结果是收获了大批日益成熟的方法论工具和技术。研究方法的进展反映了研究者对正在研究的理论构念的理解的增长（即特定的课程方面、过程或正在调查的影响）以及对如何实施和测量这些构念的理解的增长。例如，获得课程研究的明显走向是使用多种工具（例如调查、教科书使用日志、访谈和课堂观察）来建立预期课程、执行课程与学生的学习结果之间的联系。这与早期的研究形成了鲜明的对比，早期的研究往往依赖于单一的测量方法（例如，只有调查）来描述执行课程，从而限制了将学生成绩归因于数学课程的推断。

研究者更多致力于改进研究工具可以加强未来的课程研究。尽管研究者倾向于为每项新研究开发新的工具（和理论构念，如上所述），但是也有许多研究者立足于现有的工具，并且不断改进和测试它们而取得了进展。系统地改进研究工具可以提高我们应对已知方法挑战的能力，例如，总结执行课程的特征或使用对课程目标敏感的结果测量（即课程测量效度）。通过分析使用相同工具或相关工具的多项研究，我们可以积累一些特定研究工具优缺点的信息，并持续改进我们测量某些理论构念的能力（Ziebarth, Wilson, Chval, Heck, & Weiss, 2012）。

需要有计划地建设知识库。正如我们在本章中探讨的那样，数学课程的研究文献在过去十年中有了显著的增长。然而，为了这一领域的持续进步，我们建议研究人员应有意识地寻求一定的方法来推进现有的知识。在我们展望下一个十年的研究时，我们希望新的文献能提

experiences over time, but also because it is important to take differences in curriculum enactment over time into consideration when examining the attained curriculum.

Relatedly, it is only in recent years that studies of student outcomes have begun to examine the impact of a curriculum program beyond the program's grade levels. Additional systematic studies of effects beyond the grade levels of the curriculum are needed to develop an understanding of long-term implications of curriculum on students' mathematics learning.

The need for conceptual questioning. Although common elements can be seen across studies, many studies of mathematics curriculum employ distinct conceptualizations and analytic frameworks. For example, earlier we noted that in studies of the intended curriculum, researchers often must develop study-specific frameworks to support the kind of analysis they plan to conduct (e.g., a mathematical framework to support a content analysis). We have also noted diversity in researchers' conceptualizations of particular constructs such as teacher knowledge, fidelity of implementation, teachers' use of curriculum materials, and curriculum enactment. These different perspectives provide important guides for individual studies or sets of related studies, although the multitude of conceptualizations and frameworks can also create challenges as researchers strive to interpret findings across bodies of curriculum research.

Our view is that the field would benefit from more attention to questions about relationships among different perspectives and the impact of using particular conceptualizations and frameworks. Different perspectives might influence what is learned from particular research studies about mathematics curriculum. Remillard (2005) provided an exemplar of this kind of analysis when she identified several different perspectives from research studies about teachers' use of curricular materials. Her analysis helped researchers be more aware of their assumptions about teachers' use when they design and report research on enacted and attained curriculum.

As another illustration, we return to our choice in this chapter to conceptualize curriculum using the three-level model shown in Figure 30.1. This model reflects a long-standing, commonly used perspective on curriculum, but one that not all researchers share. In fact, numerous scholars have proffered additional or alternative levels in the curriculum model (e.g., Clements, 2002; Remillard & Heck, 2014; Stein et al., 2007; Tarr et al., 2006). For example, Stein et al. (2007) used a four-level model in which they distinguished between the written curriculum and the *teachers' intended curriculum,* namely the plans that teachers make prior to enactment in the classroom. The existence of different models raises some questions that warrant further consideration: What unique insights might emerge from particular conceptualizations of curriculum? How might these different conceptualizations work together to inform research about mathematics curriculum? Questions such as these, applied to different conceptualizations and analytic frameworks, could help researchers begin to assess the conceptual choices that inform research design and recognize the consequences for the knowledge base.

The need for methodological refinement. One result of the past decade of intensive research activity about mathematics curriculum is an expansive collection of increasingly sophisticated methodological tools and techniques. Advances in research methods reflect growth in researchers' understanding of what constructs they are studying (i.e., the particular curricular aspects, processes, or effects under investigation) and how those constructs can be operationalized and measured. For instance, a clear trend in research about the attained curriculum is the use of multiple tools (e.g., surveys, textbook use logs, interviews, classroom observations) to help establish links between the intended and enacted curriculum and students' learning. This contrasts with earlier research that often relied upon single measures (e.g., surveys alone) to develop descriptions of the enacted curriculum, thus limiting the inferences that could be made about student outcomes attributable to a mathematics curriculum.

Future curriculum research would be enhanced by greater commitment by researchers to the refinement of instruments. Although there is a tendency for researchers to develop new tools (and constructs, as discussed above) for each new study, many researchers have also made progress by working with existing tools, refining, and further testing them. Systematic refinement of instruments can improve our ability to address known methodological challenges such as characterizing the enacted curriculum or using outcome measures that are sensitive to curricular goals (i.e., curricular validity of measures). By looking across multiple studies that have used the same instrument or related instruments, we can accumulate information about the strengths and weaknesses of particular instruments and seek continued improvements in our ability to measure particular constructs (Ziebarth, Wilson, Chval, Heck, & Weiss, 2012).

The need for deliberate building of the knowledge base. As we have explored in this chapter, the body of research

供有计划地建设与数学课程有关的知识库的证据。

以上,我们鼓励研究者质疑和阐明自己和他人对数学课程研究的观点和理论基础,并继续改进用于分析的工具和技术。我们还阐述了对数学课程作进一步跨国研究和纵向研究的必要性。除了之前提到的其他几个领域(例如,针对中学水平的研究、学生在课程执行中的作用研究、系统地调查学校环境对执行课程和获得课程的影响以及关于数字课程材料的可行性和制约因素的研究),跨国研究和纵向研究也是需要进一步探究的领域。关于这些要素(理论构想和分析框架、方法论工具和技术以及一些新兴结果的调查领域)中的每一个进步都有可能帮助研究者扩展数学课程的知识库。我们面临的挑战是在不同研究之间建立更紧密的联系,以就数学课程在改善教和学方面的作用建立一个更为连贯的知识库。

literature about mathematics curriculum has grown significantly over the past decade. Yet, for continued progress in this area, we recommend that researchers intentionally seek ways to advance existing knowledge. As we look to the next decade of research, we hope that emergent literature shows evidence of deliberate building of the knowledge base related to mathematics curriculum.

Above, we encouraged researchers to question and make explicit the perspectives and underpinnings of their own and others' studies of the mathematics curriculum and to continue to refine the tools and techniques used for analysis. We also described the need for further cross-national studies and longitudinal studies of mathematics curriculum. These areas in need of further investigation are in addition to several others mentioned previously (e.g., studies at the secondary level, studies of students' roles in curriculum enactment, systematic investigation of the influence of school context on the enacted and attained curricula, and research about the affordances and constraints of digital curriculum materials). Advancements with respect to each of these elements—constructs and analytic frameworks, methodological tools and techniques, and areas of investigation with emergent results—have the potential to help researchers extend the knowledge base about mathematics curriculum. The challenge before us is to establish stronger connections across individual research studies so that we build a more coherent knowledge base about the role of mathematics curriculum in improving teaching and learning.

References

Agodini, R., Harris, B., Thomas, M., Murphy, R., & Gallagher, L. (2010). *Achievement effects of four early elementary school math curricula: Findings for first and second graders—Executive summary* (NCEE 2011–4002). U.S. Department of Education, Institute of Education Sciences, National Center for Education Evaluation and Regional Assistance.

Alajmi, A. H. (2012). How do elementary textbooks address fractions? A review of mathematics textbooks in the USA, Japan, and Kuwait. *Educational Studies in Mathematics, 79*(2), 239–261.

Arbaugh, F., Lannin, J., Jones, D. L., & Park-Rogers, M. (2006). Examining instructional practices in Core-Plus lessons: Implications for professional development. *Journal of Mathematics Teacher Education, 9*, 517–550.

Au, W. (2007). High-stakes testing and curricular control: A qualitative metasynthesis. *Educational Researcher, 36*(5), 258–267.

Australian Education Council. (1990). *A national statement on mathematics for Australian schools*. Canberra, Australia: Curriculum Corporation.

Baker, D., Knipe, H., Collins, J., Leon, J., Cummings, E., Blair, C., & Gamson, D. (2010). One hundred years of elementary school mathematics in the United States: A content analysis and cognitive assessment of textbooks from 1900 to 2000. *Journal for Research in Mathematics Education, 41*(4), 383–423.

Ball, D. L., & Cohen, D. K. (1996). Reform by the book: What is—or might be—the role of curriculum materials in teacher learning and instructional reform? *Educational Researcher, 25*(9), 6–8, 14.

Ball, D. L., & Feiman-Nemser, S. (1988). Using textbooks and teacher's guides: A dilemma for beginning teachers and teacher educators. *Curriculum Inquiry, 18*, 401–423.

Banilower, E. R., Smith, P. S., Weiss, I. R., Malzahn, K. A., Campbell, K. M., & Weis, A. M. (2013). *Report of the 2012 National Survey of Science and Mathematics Education*. Chapel Hill, NC: Horizon Research.

Barr, R. (1988). Conditions influencing content taught in nine fourth-grade mathematics classrooms. *Elementary School Journal, 88*(4), 387–411.

Bates, M., & Usiskin, Z. (Eds.). (2016). *Digital curricula in school mathematics*. Charlotte, NC: Information Age.

Beck Evaluation & Testing Associates. (2005). *Progress in Mathematics © 2006: Grade 1 pre-post field test evaluation study*. New York, NY: Sadlier-Oxford Division, William H. Sadlier.

Begle, E. G. (1968). Curriculum research in mathematics. *Journal of Experimental Education, 37*, 44–48.

Bieda, K., Ji, X., Drwencke, J., & Picard, A. (2014). Reasoning-and-proving opportunities in elementary mathematics textbooks. *International Journal for Education Research, 64*, 71–80.

Boesen, J., Helenius, O., Bergqvist, E., Bergqvist, T., Lithner, J., Palm, T., & Palmberg, B. (2014). Developing mathematical competence: From the intended to the enacted curriculum. *The Journal of Mathematical Behavior, 33*, 72–87.

Boesen, J., Lithner, J., & Palm, T. (2010). The relation between types of assessment tasks and the mathematical reasoning students use. *Educational Studies in Mathematics, 75*(1), 89–105.

Boston, M. D., & Smith, M. S. (2009). Transforming secondary mathematics teaching: Increasing the cognitive demands of instructional tasks used in teachers' classrooms. *Journal for Research in Mathematics Education, 40*(2), 119–156.

Brown, M. W. (2009). The teacher-tool relationship. In J. T. Remillard, B. A. Herbel-Eisenmann, & G. M. Lloyd (Eds.), *Mathematics teachers at work: Connecting curriculum materials and classroom instruction* (pp. 17–36). New York, NY: Routledge.

Brown, S. A., Pitvorec, K., Ditto, C., & Kelso, C. R. (2009). Reconceiving fidelity of implementation: An investigation of elementary whole-number lessons. *Journal for Research in Mathematics Education, 40*(4), 363–395.

Cabalo, J. V., Jaciw, A., & Vu, M.-T. (2007). *Comparative effectiveness of Carnegie Learning's Cognitive Tutor Algebra I curriculum: A report of a randomized experiment in the Maui School District*. Palo Alto, CA: Empirical Education.

Cai, J. (2014). Searching for evidence of curricular effect on the teaching and learning of mathematics: Some insights from the LieCal project. *Mathematics Education Research Journal, 26*(4), 811–831.

Cai, J., & Cirillo, M. (2014). What do we know about reasoning and proving? Opportunities and missing opportunities from curriculum analysis. *International Journal of Educational Research, 64*, 132–140.

Cai, J., & Howson, A. G. (2013). Toward an international mathematics curriculum. In M. A. Clements, A. Bishop, C. Keitel, J. Kilpatrick, & K. S. F. Leung (Eds.), *Third international handbook of mathematics education research* (pp. 949–978). New York, NY: Springer.

Cai, J., Jiang, C., Hwang, S., Nie, B., & Hu, D. (2016). Does textbook support the implementation of mathematical problem posing in classrooms? An international comparative perspective. In P. Felmer, J. Kilpatrick, & E. Pehkonen (Eds.), *Problem solving in mathematics education: New advances and perspectives* (pp. 3–22). New York, NY: Springer.

Cai, J., Lo, J. J., & Watanabe, T. (2002). Intended treatments of arithmetic average in U.S. and Asian school mathematics. *School Science and Mathematics, 102*(8), 391–404.

Cai, J., Moyer, J. C., & Wang, N. (2013). Longitudinal investigation of the effect of middle school curriculum on learning in high school. In A. M. Lindmeier & A. Heinze (Eds.), *Proceedings of the 37th Conference of the International Group for the Psychology in Mathematics Education* (pp. 137–144). Kiel, Germany: PME.

Cai, J., Moyer, J. C., Wang, N., Hwang, S., Nie, B., & Garber, T. (2013). Mathematical problem posing as a measure of curricular effect on students' learning. *Educational Studies in Mathematics, 83*(1), 57–69.

Cai, J., Nie, B., & Moyer, J. (2010). The teaching of equation solving: Approaches in Standards-based and traditional curricula in the United States. *Pedagogies: An International Journal, 5*(3), 170–186.

Cai, J., Wang, N., Moyer, J. C., Wang, C., & Nie, B. (2011). Longitudinal investigation of the curriculum effect: An analysis of student learning outcomes from the LieCal Project. *International Journal of Educational Research, 50*(2), 117–136.

Campuzano, L., Dynarski, M., Agodini, R., & Rall, K. (2009). *Effectiveness of Reading and Mathematics Software Products: Findings From Two Student Cohorts—Executive Summary* (NCEE 2009-4042). Washington, DC: U.S. Department of Education, Institute of Education Sciences, National Center for Education Evaluation and Regional Assistance.

Cengiz, N., Kline, K., & Grant, T. J. (2011). Extending students' mathematical thinking during whole-group discussions. *Journal of Mathematics Teacher Education, 14*, 355–374.

Charalambous, C. Y., Delaney, S., Hsu, H.-Y., & Mesa, V. (2010). A comparative analysis of the addition and subtraction of fractions in textbooks from three countries. *Mathematical Thinking and Learning, 12*, 117–151.

Charalambous, C. Y., & Hill, H. C. (2012). Teacher knowledge, curriculum materials, and quality of instruction: Unpacking a complex relationship. *Journal of Curriculum Studies, 44*(4), 443–466.

Charalambous, C. Y., Hill, H. C., & Mitchell, R. N. (2012). Two negatives don't always make a positive: Exploring how limitations in teacher knowledge and the curriculum contribute to instructional quality. *Journal of Curriculum Studies, 44*(4), 489–513.

Charalambous, C. Y., & Philippou, G. N. (2010). Teachers' concern and efficacy beliefs about implementing a mathematics curriculum reform: Integrating two lines of inquiry. *Educational Studies in Mathematics, 75*, 1–21.

Chávez, Ó., Tarr, J., Grouws, D., & Soria, V. (2013). Third-year high school mathematics curriculum: Effects of content organization and curriculum implementation. *International Journal of Science and Mathematics Education, 13*(1), 1–24.

Choppin, J. (2011). Learned adaptations: Teachers' understanding and use of curriculum resources. *Journal of Mathematics Teacher Education, 14*(5), 331–353.

Christou, C., Menon, M. E., & Philippou, G. (2009). Beginning teachers' concerns regarding the adoption of new mathematics curriculum materials. In J. T. Remillard, B. A. Herbel-Eisenmann, & G. M. Lloyd (Eds.), *Mathematics teachers at work: Connecting curriculum materials and classroom instruction* (pp. 223–244). New York, NY: Routledge.

Chval, K. B., Chávez, O., Reys, B. J., & Tarr, J. (2009). Considerations and limitations related to conceptualizing and measuring textbook integrity. In J. T. Remillard, B. A. Herbel-Eisenmann, & G. M. Lloyd (Eds.), *Mathematics teachers at work: Connecting curriculum materials and classroom instruction* (pp. 70–84). New York, NY: Routledge.

Chval, K. B., Reys, R., Reys, B. J., Tarr, J. E., & Chávez, Ó. (2006). Pressures to improve student performance: A context that both urges and impedes school-based research. *Journal for Research in Mathematics Education, 37*(3), 158–166.

Clapp, F. L. (1924). The number combinations: Their relative difficulty and the frequency of their appearance in textbooks. University of Wisconsin Bureau of Educational Research Bulletin No. 2.

Clements, D. H. (2002). Linking research and curriculum development. In L. D. English (Ed.), *Handbook of international research in mathematics education* (pp. 599–636). Mahwah, NJ: Lawrence Erlbaum.

Clements, D. H. (2007). Curriculum research: Toward a framework for "research-based curricula." *Journal for Research in Mathematics Education, 38*(1), 35–70.

Cohen, D. K. (1990). A revolution in one classroom: The case of Mrs. Oublier. *Educational Evaluations and Policy Analysis, 23*(2), 145–170.

Confrey, J., & Maloney, A. (2015). Engineering [for] effectiveness in mathematics education: Intervention at the instructional core in an era of Common Core Standards. In J. Middleton, J. Cai, & S. Hwang (Eds.), *Large-scale studies in mathematics* (pp. 373–403). New York, NY: Springer.

Davis, E. A., & Krajcik, J. S. (2005). Designing educative curriculum materials to promote teacher learning. *Educational Researcher, 34*(3), 3–14.

Davis, J. D. (2009). Understanding the influence of two mathematics textbooks on prospective secondary teachers' knowledge. *Journal of Mathematics Teacher Education, 12*(5), 365–389.

Davis, O. L. (1962). Textbooks and other printed materials. *Review of Educational Research, 32*(2), 127–140.

de Lange, J. (2007). Large-scale assessment and mathematics education. In F. K. Lester Jr. (Ed.), *Second handbook of research on mathematics teaching and learning* (pp. 1111–1142). Charlotte, NC: Information Age; Reston, VA: National Council of Teachers of Mathematics.

Ding, M., & Li, X. (2010). A comparative analysis of the distributive property in the US and Chinese elementary mathematics textbooks. *Cognition and Instruction, 28*, 146–180.

Doerr, H. M., & Chandler-Olcott. (2009). Negotiating the literacy demands of standards-based curriculum materials: A site for teachers' learning. In J. T. Remillard, B. A. Herbel-Eisenmann, & G. M. Lloyd (Eds.), *Mathematics teachers at work: Connecting curriculum materials and classroom instruction* (pp. 283–301). New York, NY: Routledge.

Donoghue, E. F (2003). Algebra and geometry textbooks in twentieth-century America. In G. Stanic & J. Kilpatrick (Eds.), *A history of school mathematics* (Vol. 1, pp. 329–398). Reston, VA: National Council of Teachers of Mathematics.

Doyle, W. (1983). Academic work. *Review of Educational Research, 53*(2), 159–199.

Drake, C., & Sherin, M. G. (2009). Developing curriculum vision and trust: Changes in teachers' curriculum strategies. In J. T. Remillard, B. A. Herbel-Eisenmann, & G. M. Lloyd (Eds.), *Mathematics teachers at work: Connecting curriculum materials and classroom instruction* (pp. 321–337). New York, NY: Routledge.

Eisenmann, T., & Even, R. (2009). Similarities and differences in the types of algebraic activities in two classes taught by the same teacher. In J. T. Remillard, B. A. Herbel-Eisenmann, & G. M. Lloyd (Eds.), *Mathematics teachers at work: Connecting curriculum materials and classroom instruction.* (pp. 152–170). New York, NY: Routledge.

Fan, L., & Zhu, Y. (2007). Representation of problem-solving procedures: A comparative look at China, Singapore, and US mathematics textbooks. *Educational Studies in Mathematics, 66*(1), 61–75.

Fan, L., Zhu, Y., & Miao, Z. (2013). Textbook research in mathematics education, development status and directions. *ZDM—International Journal on Mathematics Education, 45*(5), 633–646.

Fey, J. T. (1980). Mathematics education research on curriculum and instruction. In R. Shumway (Ed.), *Research in mathematics education* (pp. 388–432). Reston VA: National Council of Teachers of Mathematics.

Fey, J. T., & Graeber, A. O. (2003). From the new math to the agenda for action. In G. Stanic and J. Kilpatrick (Eds.), *A history of school mathematics* (Vol. 1, pp. 521–558). Reston, VA: National Council of Teachers of Mathematics.

Freeman, D. J., & Porter, A. C. (1989). Do textbooks dictate the content of mathematics instruction in elementary school? *American Educational Research Journal, 26*, 403–421.

Fullan, M., & Pomfret, A. (1977). Research on curriculum and instruction implementation. *Review of Educational Research, 47*(2), 335–397.

Fuson, K. C., Stigler, J. W., & Bartsch, K. (1988). Grade placement of addition and subtraction topics in Japan, China, the Soviet Union, and the United States. *Journal for Research in Mathematics Education, 19*, 449–456.

Gadanidis, G., & Kotsopoulos, D. (2009). "This is how we do this and this is the way it is." Teachers' choice of mathematical compass. *For the Learning of Mathematics, 29*, 29–34.

Gatti, G., & Giordano, K. (2010). *Pearson Investigations in Number, Data, & Space efficacy study: Final report.* Pittsburgh, PA: Gatti Evaluation.

Gehrke, N. J., Knapp, M. S., & Sirotnik, K. A. (1992). In search of the school curriculum. *Review of Research in Education, 18*, 51–110.

Gooya, Z. (2007). Mathematics teachers' beliefs about a new reform in high school geometry in Iran. *Educational Studies in Mathematics, 65*, 331–347.

Grant, T. J., Kline, K., Crumbaugh, C., Kim, O., & Cengiz, N. (2009). How can curriculum materials support teachers in pursuing student thinking during whole-group discussions? In J. T. Remillard, B. A. Herbel-Eisenmann, & G. M. Lloyd (Eds.), *Mathematics teachers at work: Connecting curriculum materials and classroom instruction* (pp. 103–117). New York, NY: Routledge.

Grevholm, B. (2011). Network for research on mathematics textbooks in the Nordic countries. *Nordic Studies in Mathematics Education, 16*(4), 91–102.

Griffiths, B. H. (1971). Mathematical insight and mathematical curricula. *Educational Studies in Mathematics, 4*(2), 153–165.

Griffiths, R. (1987). A tale of horses: Arithmetic teaching in Victoria 1860–1914. *Educational Studies in Mathematics, 18*(2), 191–207.

Grouws, D. A., Tarr, J. E., Chávez, Ó., Sears, R., Soria, V., & Taylan, R. D. (2013). Curriculum and implementation effects on high school students' mathematics learning from curricula representing subject-specific and integrated content organizations. *Journal for Research in Mathematics Education, 44*(2), 416–463.

Gueudet, G., Pepin, B., & Trouche, L. (Eds.). (2012). *From textbooks to "lived" resources: Mathematics curriculum materials and teacher development.* New York, NY: Springer.

Hamann, M. S., & Ashcraft, M. H. (1986). Textbook presentations of the basic addition facts. *Cognition and Instruction, 3*, 173–192.

Harwell, M. R., Medhanie, A., Post, T. R., Norman, K. & Dupuis, D. (2011). The preparation of students completing a Core-Plus or commercially developed high school mathematics curriculum for intense college mathematics coursework. *Journal of Experimental Education, 80*(1), 96–112.

Harwell, M. R., Post, T. R., Medhanie, A., Dupuis, D. N., & LeBeau, B. (2013). A multi-institutional study of high school mathematics curricula and college mathematics achievement and course taking. *Journal of Research in Mathematics Education, 44*, 742–774.

Heck, D. J., Chval, K. B., Weiss, I. R., & Ziebarth, S. W. (Eds.). (2012). *Approaches to studying the enacted mathematics curriculum.* Charlotte, NC: Information Age.

Herbel-Eisenmann, B. A. (2007). From intended curriculum to written curriculum: Examining the "voice" of a mathematics textbook. *Journal for Research in Mathematics Education, 38*(4), 344–369.

Herbel-Eisenmann, B. A., Lubienski, S. T., & Id-Deen, L. (2006). Reconsidering the study of mathematics instructional practices: The importance of curricular context in understanding local and global teacher change. *Journal of Mathematics Teacher Education, 9*, 313–345.

Herbel-Eisenmann, B., & Wagner, D. (2007). A framework for uncovering the way a textbook may position the mathematics learner. *For the Learning of Mathematics, 27*(2), 8–14.

Hill, H. C., & Charalambous, C. Y. (2012). Teacher knowledge, curriculum materials, and quality of instruction: Lessons learned and open issues. *Journal of Curriculum Studies, 44*(4), 559–576.

Hirsch, C. (Ed.). (2007). *Perspectives on the design and development of school mathematics curricula*. Reston, VA: National Council of Teachers of Mathematics.

Hirschhorn, D. B. (1993). A longitudinal study of students completing four years of UCSMP mathematics. *Journal for Research in Mathematics Education, 24*, 136–158.

Hong, D. S., & Choi, K. M. (2014). A comparison of Korean and American secondary school textbooks: The case of quadratic equations. *Educational Studies in Mathematics, 85*(2), 241–263.

Howson, A. G., Keitel, C., & Kilpatrick, J. (1981). *Curriculum development in mathematics*. Cambridge, United Kingdom: Cambridge University Press.

Howson, G. (2013). The development of mathematics textbooks: Historical reflections from a personal perspective. *ZDM—International Journal on Mathematics Education, 45*(5), 647–658.

Hunsader, P. D., & Thompson, D. R. (2014). Influence of mathematics curriculum enactment on student achievement. In D. R. Thompson & Z. Usiskin (Eds.), *Enacted Mathematics Curriculum* (pp. 47–74). Charlotte, NC: Information Age.

Husén, T. (1967). *International study of achievement in mathematics: A comparison of twelve countries, Volumes I & II*. New York, NY: Wiley.

Jansen, A., Herbel-Eisenmann, B., & Smith, J. P., III. (2012). Detecting students' experiences of discontinuities between middle school and high school mathematics programs: Learning during boundary crossing. *Mathematical Thinking and Learning, 14*(4), 285–309.

Johansson, M. (2007). Mathematical meaning making and textbook tasks. *For the Learning of Mathematics, 27*(1), 45–51.

Jones, D. L., & Tarr, J. E. (2007). An examination of the levels of cognitive demand required by probability tasks in middle grades mathematics textbooks. *Statistics Education Research Journal, 6*(2), 4–27.

Lester, F. (2005). The place of theory in mathematics education research. In H. Chick et al. (Eds.), *Proceedings of the 29th Annual PME*. Melbourne, Australia (Vol. 1, pp. 172–178).

Leung, F. K. S (1987). The secondary school mathematics curriculum in China. *Educational Studies in Mathematics Education, 18*(1), 35–57.

Li, Q., & Ni, Y. (2011). Impact of curriculum reform: Evidence of change in classroom practice in China. *International Journal of Educational Research, 50*, 71–86.

Lithner, J. (2004). Mathematical reasoning in calculus textbook exercises. *The Journal of Mathematical Behavior, 23*(4), 405–427.

Lloyd, G. M. (2007). Strategic compromise: A student teacher's design of kindergarten mathematics instruction in a high-stakes testing climate. *Journal of Teacher Education, 58*(4), 328–347.

Lloyd, G. M. (2008). Curriculum use while learning to teach: One student teacher's appropriation of mathematics curriculum materials. *Journal for Research in Mathematics Education, 39*(1), 63–94.

Lloyd, G. M., Remillard, J. T., Herbel-Eisenmann, B. A. (2009). Teachers' use of curriculum materials. In J. T. Remillard, B. A. Herbel-Eisenmann, & G. M. Lloyd (Eds.), *Mathematics teachers at work: Connecting curriculum materials and classroom instruction* (pp. 3–14). New York, NY: Routledge.

Manouchehri, A., & Goodman, T. (1998). Mathematics curriculum reform and teacher: Understanding the connections. *Journal of Educational Research, 92*, 27–41.

Mathison, S., Hedges, L. V., Stodolosky, S. S., Flores, P., & Sarther, C. (1989). *Teaching and learning algebra: An evaluation of UCSMP Algebra*. Chicago, IL: University of Chicago School Mathematics Project.

McDuffie, A. R., & Mather, M. (2009). Middle school mathematics teachers' use of curricular reasoning in a collaborative professional development project. In J. T. Remillard, B. A. Herbel-Eisenmann, & G. M. Lloyd (Eds.), *Mathematics teachers at work: Connecting curriculum materials and classroom instruction* (pp. 302–320). New York, NY: Routledge.

McNeil, N. M., Grandau, L., Knuth, E. J., Alibali, M. W., Stephens, A. C., Hattikudur, S., & Krill, D. E. (2006). Middle-school students' understanding of the equal sign: The books they read can't help. *Cognition and Instruction, 24*, 367–385.

Ministry of Education. (1992). *Mathematics in the New Zealand Curriculum*. Wellington, New Zealand: Learning Media.

Moyer, J. C., Cai, J., Nie, B., & Wang, N. (2011). Impact of curriculum reform: Evidence of change in classroom instruction in the United States. *International Journal of Educational Research, 50*(2), 87–99.

Nathan, M. J., Long, S. D., & Alibali, M. W. (2002). The symbol precedence view of mathematical development: A corpus analysis of the rhetorical structure of textbooks. *Discourse Processes, 33*(1), 1–21.

National Council of Teachers of Mathematics. (1989). *Curriculum and evaluation standards for school mathematics*. Reston, VA: Author.

National Council of Teachers of Mathematics. (1991). *Professional standards for teaching mathematics*. Reston, VA: Author.

National Council of Teachers of Mathematics. (2000). *Principles and standards for school mathematics*. Reston, VA: Author.

National Governors Association Center for Best Practices & Council of Chief State School Officers. (2010). *Common core state standards for mathematics*. Washington, DC: Author. Retrieved from http://www.corestandards.org

National Research Council. (2002). *Scientific research in education*. Washington, DC: National Academy Press.

National Research Council. (2004). *On evaluating curricular effectiveness: Judging the quality of K–12 mathematics evaluations*. Washington, DC: National Academy Press.

Ni, Y., Li, Q., Cai, J., & Hau, K. (2015). Has curriculum reform made a difference in the classroom? An evaluation of the new mathematics curriculum in China. In B. Sriraman, J. Cai, K. Lee, L. Fan, Y. Shimizu, C. S. Lim, & K. Subramaniam (Eds.), *The first sourcebook on Asian research in mathematics education: China, Korea, Singapore, Japan, Malaysia, and India* (pp. 141–168). Charlotte, NC: Information Age.

Ni, Y. J., Li, Q., Li, X., Zhang, Z. H. (2011). Influence of curriculum reform: Evidence of student learning outcomes in China. *International Journal of Educational Research, 50*, 100–116.

Nicely, R. F. (1991). Higher-order thinking skills in mathematics textbooks: A research summary. *Education, 111*(4), 456–460.

Nicholls, J. (2003). Methods in school textbook research. *International Journal of Historical Learning, Teaching and Research, 3*(2), 11–26.

Nie, B., Cai, J., & Moyer, J. C. (2009). How a standards-based mathematics curriculum differs from a traditional curriculum: With a focus on intended treatments of the ideas of variable. *ZDM—International Journal on Mathematics Education, 41,* 777–792.

No Child Left Behind Act of 2002 Pub. L. No. 107–110.

Osborne, A. R. (Ed.). (1975). Critical analyses of the NLSMA reports [Special issue]. *Investigations in Mathematics Education, 8*(3).

Otten, S., Gilbertson, N. J., Males, L. M., & Clark, D. L. (2014). The mathematical nature of reasoning-and-proving opportunities in geometry textbooks. *Mathematical Thinking and Learning, 16,* 51–79.

Pane, J. F., McCaffrey, D. F., Slaughter, M. E., Steele, J. L., & Ikemoto, G. S. (2010). An experiment to evaluate the efficacy of cognitive tutor geometry. *Journal of Research on Educational Effectiveness, 3*(3), 254–281.

Pepin, B. (2014). Re-sourcing curriculum materials: In search of appropriate frameworks for researching the enacted mathematics curriculum. *ZDM—International Journal on Mathematics Education, 46*(5), 837–842.

Pepin, B., Gueudet, G., & Trouche, L. (2013). Investigating textbooks as crucial interfaces between culture, policy and teacher curricular practice: Two contrasted case studies in France and Norway. *ZDM—International Journal on Mathematics Education, 45,* 685–698.

Pepin, B., & Haggarty, L. (2001). Mathematics textbooks and their use in English, French, and German classrooms: A way to understand teaching and learning cultures. *ZDM—International Journal on Mathematics Education, 33*(5), 158–175.

Ponte, J. P. (2012). A practice-oriented professional development programme to support the introduction of a new mathematics curriculum in Portugal. *Journal of Mathematics Teacher Education, 15,* 317–327.

Ponte, J. P., Matos, J. F., Guimarães, H. M., Leal, L. C., & Canavarro, A. P. (1994). Teachers' and students' views and attitudes towards a new mathematics curriculum: A case study. *Educational Studies in Mathematics, 26*(4), 347–365.

Porter, A. C. (1995). The uses and misuses of opportunity-to-learn standards. *Educational Researcher, 24*(1), 21–27.

Remillard, J. T. (2005). Examining key concepts in research on teachers' use of mathematics curricula. *Review of Educational Research, 75,* 211–246.

Remillard, J. T., Harris, B., & Agodini, R. (2014). The influence of curriculum material design on opportunities for student learning. *ZDM—International Journal on Mathematics Education, 46*(5), 735–749.

Remillard, J. T., & Heck, D. J. (2014). Conceptualizing the curriculum enactment process in mathematics education. *ZDM—International Journal on Mathematics Education, 46,* 705–718.

Remillard, J. T., Herbel-Eisenmann, B. A., & Lloyd, G. M. (Eds.). (2009). *Mathematics teachers at work: Connecting curriculum materials and classroom instruction.* New York, NY: Routledge.

Resendez, M., & Azin, M. (2006). *Saxon Math randomized control trial: Final report.* Jackson, WY: PRES Associates.

Resendez, M., & Azin, M. (2008). *A study on the effects of Pearson's 2009 enVisionMATH program. 2007–2008: First year report.* Jackson, WY: PRES Associates.

Resendez, M., & Manley, M. A. (2005). *Final report: A study on the effectiveness of the 2004 Scott Foresman-Addison Wesley Elementary Math program.* Jackson, WY: PRES Associates.

Reys, B. J., & Reys, R. E. (Eds.). (2010). *Mathematics curriculum: Issues, trends, and future directions,* 72nd Yearbook of the National Council of Teachers of Mathematics (NCTM). Reston, VA: NCTM.

Romberg, T. A., Webb, D. C., Folgert, L., & Shafer, M. C. (2005). *The longitudinal/cross-sectional study of the impact of teaching mathematics using Mathematics in Context on student achievement: Differences in performance between Mathematics in Context and conventional students.* Monograph No. 6. Madison, WI: Wisconsin Center for Education Research.

Roschelle, J., Noss, R, Blikstein, P., & Jackiw, N. (2017). Technology for learning mathematics. In J. Cai (Ed.), *Compendium for research in mathematics education* (pp. 853–876). Reston, VA: National Council of Teachers of Mathematics.

Schmidt, W. H., McKnight, C. C., & Raizen, S. A. (1997). *A splintered vision: An investigation of U.S. science and mathematics education.* Boston, MA: Kluwer.

Schmidt, W. H., McKnight, C. C., Valverde, G., Houang, R. T., & Wiley, D. E. (1997). *Many visions, many aims: A cross-national investigation of curricular intentions in school mathematics.* Dordrecht, The Netherlands: Kluwer.

Schneider, C. L. (2000). Connected Mathematics and the Texas Assessment of Academic Skills. *Dissertation Abstracts International, 62*(02), 503A. (UMI No. 3004373)

Schoen, H., Cebulla, K. J., Finn, K. F., & Fi, C. (2003). Teacher variables that relate to student achievement when using a standards-based curriculum. *Journal for Research in Mathematics Education, 34*(3), 228–259.

Schoenfeld, A. H. (2002). Making mathematics work for all children: Issues of standards, testing, and equity. *Educational Researcher, 31,* 13–25.

Schoenfeld, A. H. (2006). What doesn't work: The challenge and failure of the What Works Clearinghouse to conduct meaningful reviews of studies of mathematics curricula. *Educational Researcher, 35*(2), 13–21.

Schoenfeld, A. H. (2007). Method. In F. Lester Jr. (Ed.), *Handbook of research on mathematics teaching and learning* (2nd ed., pp. 69–107). Charlotte, NC: Information Age; Reston, VA: National Council of Teachers of Mathematics.

Senk, S. L., & Thompson, D. R. (Eds.). (2003). *Standards-based school mathematics curricula: What are they? What do students learn?* Mahwah, NJ: Lawrence Erlbaum Associates.

Sherin, M. G., & Drake, C. (2009). Curriculum strategy framework: Investigating patterns in teachers' use of a reform-based elementary mathematics curriculum. *Journal of Curriculum Studies, 41*(4), 467–500.

Shield, M., & Dole, S. (2013). Assessing the potential of mathematics textbooks to promote deep learning. *Educational Studies in Mathematics, 82*(2), 183–199.

Shneyderman, A. (2001). *Evaluation of the Cognitive Tutor Algebra I program.* Miami, FL: Miami-Dade County Public Schools, Office of Evaluation and Research.

Sinclair, N. (2008). *The history of the geometry curriculum in the United States.* Charlotte, NC: Information Age.

Slavin, R. E., Lake, C., & Groff, C. (2009). Effective programs in middle and high school mathematics: A best-evidence synthesis. *Review of Educational Research, 79*(2), 839–911.

Sleep, L., & Eskelson, S. L. (2012). MKT and curriculum materials are only part of the story: Insights from a lesson on fractions. *Journal of Curriculum Studies, 44*(4), 537–558.

Smith, J. E. (2001). *The effect of the Carnegie Algebra Tutor on student achievement and attitude in introductory high school algebra* (Unpublished dissertation). Virginia Polytechnic Institute and State University, Blacksburg.

Smith, J. P., III, & Star, J. R. (2007). Expanding the notion of impact of K–12 Standards-based mathematics and reform calculus programs. *Journal for Research in Mathematics Education, 38*(1), 3–34.

Smith, M. S. (2000). Balancing old and new: An experienced middle school teacher's learning in the context of mathematics instructional reform. *Elementary School Journal, 100*(4), 351–375.

Snyder, J., Bolin, F., & Zumwalt, K. (1992). Curriculum implementation. In P. W. Jackson (Ed.), *Handbook of research on curriculum: A project of the American Educational Research Association* (pp. 402–435). New York, NY: Macmillan.

Son, J., & Senk, S. (2010). How reform curricula in the USA and Korea present multiplication and division of fractions. *Educational Studies in Mathematics, 74*(2), 117–142.

Son, J., & Senk, S. L. (2014). Teachers' knowledge and the enacted mathematics curriculum. In D. R. Thompson & Z. Usiskin (Eds.), *Enacted Mathematics Curriculum* (pp. 75–96). Charlotte, NC: Information Age.

Sosniak, L. A., & Stodolsky, S. S. (1993). Teachers and textbooks: Materials use in four fourth-grade classrooms. *Elementary School Journal, 93*(3), 249–275.

Stake, R. E., & Easley, J. (1978). *Case studies in science education*. Urbana-Champaign: University of Illinois.

Star, J. R., Smith, J. P., III, & Jansen, A. (2008). What students notice as different between reform and traditional mathematics programs. *Journal for Research in Mathematics Education, 39*(1), 9–32.

Stein, M. K., Grover, B. W., & Henningsen, M. (1996). Building student capacity for mathematical thinking and reasoning: An analysis of mathematical tasks used in reform classrooms. *American Educational Research Journal, 33*(2), 455–488.

Stein, M. K., & Kaufman, J. H. (2010). Selecting and supporting the use of mathematics curricula at scale. *American Educational Research Journal, 47*(3), 663–693.

Stein, M. K., & Kim, G. (2009). The role of mathematics curriculum materials in large-scale urban reform: An analysis of demands and opportunities for teacher learning. In J. T. Remillard, B. A. Herbel-Eisenmann, & G. M. Lloyd (Eds.), *Mathematics teachers at work: Connecting curriculum materials and classroom instruction* (pp. 37–55). New York, NY: Routledge.

Stein, M. K., Remillard, J. T., & Smith, M. S. (2007). How curriculum influences student learning. In F. K. Lester Jr. (Ed.), *Second handbook of research on mathematics teaching and learning* (pp. 319–369). Charlotte, NC: Information Age; Reston, VA: National Council of Teachers of Mathematics.

Stephan, M. L., Chval, K. B., Wanko, J. J., Civil, M., Fish, M. C., Herbel-Eisenmann, B., . . . Wilkerson, T. L. (2015). Grand challenges and opportunities in mathematics education research. *Journal for Research in Mathematics Education, 46*(2), 134–146.

Stigler, J. W., Fuson, K. C., Ham, M., & Kim, M. S. (1986). An analysis of addition and subtraction word problems in American and Soviet elementary mathematics textbooks. *Cognition and Instruction, 3*, 153–171.

Stodolsky, S. S. (1989). Is teaching really by the book? In P. W. Jackson & S. Haroutunian-Gordon (Eds.), *From Socrates to software* (88th Yearbook, Pt. 1, pp. 159–184). Chicago, IL: National Society for the Study of Education.

Stylianides, G. J. (2009). Reasoning-and-proving in school mathematics textbooks. *Mathematical Thinking and Learning, 11*, 258–288.

Stylianides, G. J. (2014). Textbook analyses on reasoning-and-proving: Significance and methodological challenges. *International Journal of Educational Research, 64*, 63–70.

Superfine, A. C. (2009). The "problem" of experience in mathematics teaching. *School Science and Mathematics, 109*(1), 7–19.

Tarr, J. E., Chávez, Ó., Reys, R. E., & Reys, B. J. (2006). From the written to the enacted curricula: The intermediary role of middle school mathematics teachers in shaping students' opportunity to learn. *School Science and Mathematics, 106*(4), 191–201.

Tarr, J. E., Grouws, D. A., Chávez, Ó., & Soria, V. M. (2013). The effects of content organization and curriculum implementation on students' mathematics learning in second-year high school courses. *Journal for Research in Mathematics Education, 44*(4), 683–729.

Tarr, J. E., McNaught, M. D., & Grouws, D. A. (2012). The development of multiple measures of curriculum implementation in secondary mathematics classrooms: Insights from a 3-year curriculum evaluation study. In D. J. Heck, K. B. Chval, I. R. Weiss, & S. W. Ziebarth (Eds.), *Approaches to studying the enacted mathematics curriculum* (pp. 89–116). Charlotte, NC: Information Age.

Tarr, J., Reys, R., Reys, B., Chávez, Ó., Shih, J., & Osterlind, S. (2008). The impact of middle grades mathematics curricula and the classroom learning environment on student achievement. *Journal for Research in Mathematics Education, 39*, 247–280.

Thompson, D. R., & Senk, S. L. (2001). The effects of curriculum on achievement in second-year algebra: The example of the University of Chicago School Mathematics Project. *Journal for Research in Mathematics Education, 32*(1), 58–84.

Thompson, D. R., Senk, S. L., & Johnson, G. J. (2012). Opportunities to learn reasoning and proof in high school mathematics textbooks. *Journal for Research in Mathematics Education, 43*(3), 253–295.

Thompson, D. R., & Usiskin, Z. (2014). (Eds.). *Enacted Mathematics Curriculum*. Charlotte, NC: Information Age.

Thomson, S., & Fleming, N. (2004). *Summing it up: Mathematics achievement in Australian schools in TIMSS 2002*. Melbourne: Australian Council for Educational Research.

Travers, K. J. (1992). Overview of the longitudinal version of the Second International Mathematics Study. In L. Burstein (Ed.), *The IEA study of mathematics III: Student growth and classroom processes* (pp. 1–14). Oxford, England: Pergamon.

Tunks, J., & Weller, K. (2009). Changing practice, changing minds, from arithmetical to algebraic thinking: An application of the concerns-based adoption model (CBAM). *Educational Studies in Mathematics, 72,* 161–183.

Turnau, S. (1980). The mathematics textbook for young students. *Educational Studies in Mathematics, 11*(4), 393–410.

U.S. Department of Education, Institute of Education Sciences. (2014). *What Works Clearinghouse: Procedures and standards handbook version 3.0.* Retrieved from http://ies.ed.gov/ncee/wwc/DocumentSum.aspx?sid=19

U.S. Department of Education, Institute of Education Sciences & National Science Foundation. (2013). *Common Guidelines for Education Research and Development.* (Publication No. NSF13126). Retrieved from http://www.nsf.gov/pubs/2013/nsf13126/nsf13126.pdf

Valverde, G. A., Bianchi, L. J., Wolfe, R. G., Schmidt, W. H., & Houang, R. T. (2002). *According to the book: Using TIMSS to investigate the translation of policy into practice in the world of textbooks.* Dordrecht, The Netherlands: Kluwer.

van den Heuvel-Panhuizen, M. (2000). *Mathematics education in the Netherlands: A guided tour.* Freudenthal Institute CD-ROM for ICME9. Utrecht, The Netherlands: Utrecht University.

Vincent, J., & Stacey, K. (2008). Do mathematics textbooks cultivate shallow teaching? Applying the TIMSS Video Study criteria to Australian eighth-grade mathematics textbooks. *Mathematics Education Research Journal, 20*(1), 82–107.

Weiss, I. R. (1987). *Report of the 1985–86 National Survey of Science and Mathematics Education.* Research Triangle Park, NC: Research Triangle Institute.

What Works Clearinghouse. (2010). *WWC intervention report: Connected Mathematics Project (CMP).* Retrieved from http://ies.ed.gov/ncee/wwc/pdf/intervention_reports/wwc_cmp_012610.pdf

Whitehurst, G. J. (2003, April). *The Institute of Education Sciences: New wine, new bottles.* Paper presented at the annual meeting of the American Research Association, Chicago, IL.

Wilson, S. M. (1990). A conflict of interests: The case of Mark Black. *Educational Evaluation and Policy Analysis, 12,* 293–310.

Wolfson, M., Koedinger, K., Ritter, S., & McGuire, C. (2008). *Cognitive Tutor Algebra I: Evaluation of results (1993–1994).* Pittsburgh, PA: Carnegie Learning.

Xin, Y. P. (2007). Word problem solving tasks in textbooks and their relation to student performance. *Journal of Educational Research, 100,* 347–359.

Ziebarth, S. W., Fonger, N. L., & Kratky, J. L. (2014). Instruments for studying the enacted mathematics curriculum. In D. Thompson & Z. Usiskin (Eds.), *Enacted Mathematics Curriculum* (pp. 97–120). Charlotte, NC: Information Age.

Ziebarth, S. W., Wilson, L. D., Chval, K. B., Heck, D. J., & Weiss, I. R. (2012). Issues to consider in measuring enactment of curriculum materials. In D. J. Heck, K. B. Chval, I. R. Weiss, & S. W. Ziebarth (Eds.), *Approaches to studying the enacted mathematics curriculum* (pp. 195–203). Charlotte, NC: Information Age.

31 学习数学的技术

杰里米·罗谢尔
美国SRI国际
理查德·诺斯
英国伦敦大学学院
保罗·布利克斯坦
美国斯坦福大学
尼古拉斯·杰基
美国SRI国际
译者：顾非石
　　　上海徐汇区教育学院
　　　贺真真
　　　上海师范大学教育学院

本章是卡普特（1992）里程碑式综述"技术与数学教育"的续篇。在他的开创性评论发表近25年之后，技术在数学教育中的应用以不同的方式蓬勃发展，其中许多应用是人们所预期的，但是也有许多应用都超出了卡普特的想象。卡普特的综述之所以经久不衰，不仅是因为他能够审视当代研究和发展趋势，还因为他能够从这些进步中预测将来。我们写作本文的目标是遵循这样的精神。

二十年前，卡普特（1992）指出："任何试图描述技术在数学教育中的作用的人，都会面临类似于描述一个被新激活的火山一样的挑战"（第515页）。他通过审视"潜在的原则和过程"以及确定保持稳定的问题来应对不断变化的形势所带来的挑战。卡普特于1992年发表的综述比Mosaic浏览器早了一年，如果说在他实际写作时万维网已经存在，那也是在大众意识之外的。然而，尽管出现了像互联网这样的大爆发，卡普特的理论依然经受了时间的考验。虽然他讨论的所有具体技术现在已经被超越，但其中的原则、过程和问题今天仍然是适用的。

像卡普特一样，我们的目标通过在数学学习中强调那些支持有意义地选择和使用技术的，长期而潜在的因素来为实践者服务，而不是全面调查现有的或新兴的技术。作为手册中的一章综述，我们无法论述每一个数学工具或应用程序的优点，也不希望为教师提供一个实用的指南，帮助他们决定使用或购买当今的哪些工具。实用性层面的评价最好留给那些可以经常发表新评论的出版物，以适应技术更新的速度。我们还旨在通过提供一个框架，为整理技术在数学学习中的各种作用、将各种可能性与学习理论和学习研究联系起来、为阐明尚未解决的问题和挑战作出贡献。

建立在卡普特理论的基础上

在卡普特的理论基础上，借鉴那些今天仍然适用的观点，我们阐明了数学教育技术应用变革的关键力量。这种力量是从静态媒体（例如纸、黑板）向动态媒体（例如触摸屏、电子白板）发生了转变。纸页上的标记仍然是固有不变的，但利用技术的数学交互使得数学的表示形式随时间的推移而变化。动态媒体的使用正在改变数学实践和数学教育：例如，我们如何去体验连续性和变异性等核心数学思想。因此，新兴技术改变了数学认

本材料部分基于受美国国家科学基金会在拨款编号为REC-1055130和IIS-1441631支持项目的工作。本材料中表达的任何观点、发现、结论或建议都是作者的观点，不一定反映美国国家科学基金会的观点。我们还要感谢巴西莱曼创业与教育创新中心对这项工作的支持。我们还要感谢来自Castilleja学校（加利福尼亚州帕洛阿尔托）的希瑟·班和安吉·周为"历史纪念碑"项目所做的设计，以及茱莉亚·梅斯基塔。

31 | Technology for Learning Mathematics

JEREMY ROSCHELLE
SRI International, United States

RICHARD NOSS
University College London, United Kingdom

PAULO BLIKSTEIN
Stanford University, United States

NICHOLAS JACKIW
SRI International, United States

This chapter follows up on Kaput's (1992) landmark review, "Technology and Mathematics Education." In the almost 25 years since the publication of his seminal review, the application of technology in mathematics education has blossomed in diverse ways, many of which were anticipated by—but many of which go beyond—Kaput's fertile imagination. An enduring strength of Kaput's work derives from his ability not only to review contemporary research and development advances but also to project forward from those advances. We aim to follow in this spirit.

Two decades ago, Kaput (1992) stated "anyone who presumes to describe the roles of technology in mathematics education faces challenges akin to describing a newly active volcano" (p. 515). He responded to the challenges of a changing landscape by looking at "underlying principles and processes" and by identifying issues that remain stable. Published in 1992, Kaput's piece predates the Mosaic browser by a year; if the World Wide Web existed when he actually wrote his chapter, it was outside the public's consciousness. Yet, Kaput's approach has stood the test of time despite such major eruptions as the Internet. Although all the specific technologies Kaput discussed have now been surpassed, the principles, processes, and issues remain relevant.

Like Kaput, we aim to serve practitioners by highlighting long-term, underlying factors that support meaningful selection and use of technology in mathematics learning rather than comprehensively surveying available or emerging technologies. In the scope of a compendium chapter, we cannot review each and every mathematics tool or app on its merits, nor can we hope to offer teachers a practical guide for making decisions about which of today's tools to use or purchase. Practical ratings are best left to publications that can provide frequently updated reviews, matching the pace of change. We also aim to contribute by providing a framework for sorting out the varied roles for technology in mathematics learning and connecting the possibilities to learning theory and learning research and by articulating unresolved issues and challenges.

Building on Kaput's Foundations

Building on the foundations that Kaput laid and that are still relevant today, we elucidate key forces for change in the application of technology in mathematics education. One such force is the shift from static media (e.g., paper, chalkboards) to dynamic media (e.g., touchscreens, electronic whiteboards). Whereas marks on a paper page remain fixed, mathematical interactions with technology result in representations that change over time. The availability of dynamic media is changing both the practice of mathematics and mathematics education: for example, how we experience core mathematical ideas such as continuity and variation. Thus, emerging technology changes

---

This material is based in part on work supported by the National Science Foundation under Grant Number REC-1055130 and IIS-1441631. Any opinions, findings, conclusions, or recommendations expressed in this material are those of the authors and do not necessarily reflect the views of the National Science Foundation. We also thank the Lemann Center for Entrepreneurship and Educational Innovation in Brazil for their support of this work. We further thank Heather Bang and Angi Chau from the Castilleja School (Palo Alto, CA) for the design of the "Historical Monuments" project, as well as Juliana Mesquita.

识论，即对数学思想价值的认识、理解和审视。

接下来，基于卡普特的理论，我们还将研究专门用于数学，而不是在学校科目中通用的技术方法。通常，形成性评价技术在数学测验和反馈上的结构和特征与在其他主题上的测验和反馈大致相同，但卡普特是从技术可以实现的具体数学实践和方法的角度来看待技术的，例如，他将"动态联系"视为帮助学生将符号与模型连接起来，支持建模的数学实践（Kaput & Roschelle, 1998）。同样，他也重视技术如何突显数学结构，隐藏不太重要的细节（Kaput, 1997）。

卡普特（1992）并没有把技术看作是对教育的孤立投入。在他看来，当技术与基于科学的学习理论相联系时，它就变得有意义了。虽然技术与学习之间最早的联系是基于诸如"强化"之类的行为主义概念，但卡普特（1992）转向了以认知科学为基础的概念。在其中一个例子中，卡普特在讨论计算器时介绍了减负思想：如果计算器能让学生减轻常规计算的负担，那么学生就可以更多地关注更深层次的学习目标，这与已知的认知负荷原理相吻合（Paas, Renkl, & Sweller, 2004）。然而，在讨论动态表征时，卡普特引入了"心理模型"（例如Johnson-Laird, 1983）的观点，如果技术可以使数学现象的组成部分之间的动态关系更明显、更具体，那么学生就可以更容易地学习这些关系，从而创建自己的数学模型。

后来，卡普特和同事（例如Stroup等，2002）将动态表征作为社会参与的一种形式进行了探讨，在这种形式中，课堂中的学生可以为动态演变的数学表征作出贡献。这种方法鼓励学生通过数学论证，在共同展示中实现数学共识（Bishop, 2013; Dalton & Hegedus, 2013）。这种方法的一个新奇之处在于它将一时的变化重新定位在了社会维度的变化中（Stroup等，2002）：学生个体的工作与班级集体工作之间的关系可以模拟成一个特定数学函数与它所属的函数族之间的关系。此外，当技术允许课堂里的学生可以看到通过技术手段汇总的每个学生的具体工作时，梅森（1992）从特殊情形中看到普遍规律的概念也就可以实现了。

本章继续从这个方向努力，会提请读者注意更广泛的基于科学的学习理论，用于指导数学学习中的技术应用。例如，社会文化理论塑造了我们对工具如何增强人类认知的理解，就像我们稍后讨论的工具化概念一样。认知学习理论的其他方面有助于理解如何最好地组织数学实践和反馈。此外，视觉和动态表征的作用继续受到学习科学的影响，而学习科学将认知与感知、具体的交互、话语、论证和协作以及多重表征在学习中的作用联系在一起。自1992年以来，关于兴趣、参与和发展的学习理论也发生了变化，并贯穿于我们对学习数学的技术的讨论之中。

最后，我们注意到卡普特对"设计"的思考。与当今许多数学教育的参与者将新技术工具视为"产品""应用程序"或"解决方案"不同，卡普特认为先进技术的主要贡献是基础设施（Kaput, Hegedus, & Lesh, 2007）。技术为我们提供了新媒体、新功能、新支持，允许我们设计新的学习活动和新的课程途径（Roschelle, Knudsen, & Hegedus, 2010）以及人们可以参与做数学和学数学的新方式，但技术本身不能解决数学学习中的长期困难，需要额外的课程、活动和教学设计。

按目的组织技术

本章的结构灵感来自于德林维斯（2012）基于三大主要教学目的来组织数学学习中的技术（见图31.1）：（1）做数学，（2）实践技能，（3）发展概念。显然，技术和研究的分类有很多不同的方法，没有一种方案是完美的，我们发现这个方案简单易懂、简明扼要，并且有助于将相关的研究文献分组在一起。

关于"做数学"，我们将把技术视为数字和符号运算的工具。该类别与德林维斯框架中的其他两个类别不同，因为工具（例如计算器或电子表格）同时适用于工作场

mathematical epistemology, that is, how people come to know, understand, and see the value of mathematical ideas.

Continuing to build on Kaput, we also look at the ways technology can be specifically mathematical and not generic across school subjects. Typically, formative assessment technologies have roughly the same structure and features for mathematics quizzes and feedback as for quizzes and feedback in other topics. Kaput, however, looks at technology from the viewpoint of the specific mathematical practices and approaches technology can enable. For example, he sees "dynamic linking" as helping students connect notation to models, supporting the mathematical practice of modelling (Kaput & Roschelle, 1998). Likewise, he values how technology can highlight mathematical structures and hide less important detail (Kaput, 1997).

Kaput (1992) did not see technology as an isolated input to education. In his view, technology becomes meaningful when it relates to scientifically based learning theories. Although the earliest connections between technology and learning were based on behaviorist concepts such as "reinforcement," Kaput (1992) shifted to cognitive science-based concepts. In one example, Kaput introduced *offloading* in a discussion of calculators: If the calculator allows students to offload routine calculations, students may be able to focus more on deeper learning goals. This connects with the established principle of cognitive load (Paas, Renkl, & Sweller, 2004). In contrast, when discussing dynamic representations, Kaput introduced a "mental models" (e.g., Johnson-Laird, 1983) point of view. If the technology can render the dynamic relationships among components of a mathematical phenomenon more visible and tangible, it can become easier for students to learn those relationships and thus create their own model of the mathematics.

Later, Kaput and colleagues (e.g., Stroup et al., 2002) explored dynamic representations as a form of social participation, where students in a classroom could contribute to a dynamically evolving mathematical representation. This approach encourages students' mathematical argumentation as a means of achieving a mathematical consensus in a common display (Bishop, 2013; Dalton & Hegedus, 2013). A novelty of this approach is that it relocates variation from a temporal to a social dimension (Stroup et al., 2002): The relationship between a student's own work and the work of the classroom can model the relationship between a particular mathematical function and the family of functions to which it belongs. Further, Mason's (1992) concept of seeing the general in the specific can be instantiated when technology allows a classroom of students to see the general across each students' specific work, which has been aggregated by technological means.

The chapter continues in this direction by drawing attention to the now broader range of scientifically based learning theories that inform uses of technology in mathematics learning. For example, sociocultural theories shape our understanding of how tools come to augment human cognition, as in the concept of *instrumentalization* that we discuss later. Additional facets of cognitive learning theory inform understanding of how to best organize mathematical practice and feedback. Further, the role of visual and dynamic representations continues to be influenced by learning sciences that connect cognition to perception, embodied interaction, discourse, argumentation and collaboration, and the role of multiple representations in learning. Learning theories regarding interest, engagement, and development have also evolved since 1992 and figure throughout our discussion of technologies for learning mathematics.

Finally, we note how Kaput thought about *design*. Whereas many involved in mathematics education today see new technology tools as "products," "apps," or "solutions," Kaput saw advancing technology's primary contribution as *infrastructural* (Kaput, Hegedus, & Lesh, 2007). Technology gives us new media, new capabilities, new affordances that allow the design of new learning activities and new curricular approaches (Roschelle, Knudsen, & Hegedus, 2010), and new ways that people can participate in doing and learning mathematics—but the technology itself cannot address the enduring difficulties of mathematics learning without the additional layers of curricular, activity, and pedagogical design.

Organizing Technologies by Purpose

Our structure for this chapter is inspired by Drijvers's (2012) organization of technologies for mathematics learning by three dominant pedagogical purposes (see Figure 31.1): (1) do mathematics, (2) practice skills, and (3) develop concepts. Clearly, there are many different ways to categorize technologies and research, and no scheme is perfect. We have found this scheme to be simple to understand, parsimonious, and useful in grouping related research literature together.

With respect to "do mathematics," we will consider technology a tool for numeric and symbolic computation. This category differs from the other two categories in Drijvers's framework because the tools, such as a calculator or spreadsheet, fit into both workplace and edu-

图31.1 数学教育技术的三个目的。源自 Digital Technology in Mathematics Education: Why it Works (or Doesn't) by Paul Drijvers, Juty 2012, a paper presented at the 12th International Congress on Mathematics Education, Seoul, Korea. 下载自 http://www.icme12.org/upload/submission/2017_F.pdf。保罗·德林维斯同意使用。

所和教育环境。相比之下,"学习数学"中的两个类别涉及专门用于教育情境的技术。在"实践技能"中,我们考虑旨在更好地组织学生实践的工具,例如支持学生在线做作业的工具。在"发展概念"中,我们考虑侧重于学生意义形成和理解的方法,如动态表征。

请注意,德林维斯对"做数学"的使用与QUASAR(量化理解:提高学生成绩和推理)项目的"做数学"概念(Henningsen & Stein, 1997)不同。QUASAR项目的"做数学"的概念是指学生模仿专业数学家推理的理想参与水平。这种理想水平可以适用于德林维斯框架的所有方面,因为数学实践的最高水平可能被运用在具有不同目的的数学经验中。

在本章中,我们也扩展了德林维斯的框架,认识到在课堂中使用技术的第四个目的。数学教育现在可以被扩展到教学化组织(正式)和兴趣驱动(非正式)的学习环境中。在第四个目的中,我们将技术视为参与数学的背景。例如,机器人、制造工具、程序设计和其他环境,这些现在在许多年轻人的生活中是常见的,而且数学推理的重要机会通常可以通过技术媒体而出现在这些环境中。当从科学、技术、工程和数学(STEM)角度考虑学习活动时,这些活动可能会强调其他学科,但也为学习数学提供了充足而重要的机会。现在这是一个重要的方向,因为学校和其他机构正在通过数千个物理空间来建立新型的STEM学习,例如机器人实验室、创客空间

和FabLabs。这些混合空间的出现正在模糊我们今天所说的"正式"和"非正式"教育之间的界限,为数学学习创造了新的机会。因此,我们不应将用于学习数学的技术局限于数学课堂或工作场所应用的技术。

在每个类别中,我们分析了设计的两个维度:一个侧重于生产力,另一个侧重于变革。生产力维度强调技术在使数学学习更有效率或富有成果方面的作用,变革维度则强调技术如何让我们重新认识数学的教与学,并问一问什么样的数学会受到威胁。生产力维度倾向于改变达到给定目标的手段,变革维度往往旨在通过明确分析数学知识的本质来改变游戏。我们将看到,关于生产力和变革的论述出现在技术的每一个目标上,只是方式有些不同。

我们在讨论四个目的时会尽可能地寻求平行的结构,例如,对于每个目的,我们都会讨论历史、重大研究、学习理论、政策观点和未来机会。

做数学的技术

无论在工作场所(Hoyles, Wolf, Molyneux-Hodgson, & Kent, 2002),还是对数学家,技术在做数学方面发挥着越来越大的作用。通常在工作场所使用的技术也可以具有强大的教学用途,一个显而易见的历史例子是算盘,这是一个工具,对于进行计算和学习数字系统的组织是

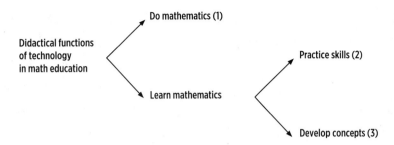

FIGURE 31.1. *Three purposes for technology in mathematics education. From* Digital Technology in Mathematics Education: Why It Works (or Doesn't) *by Paul Drijvers, July 2012, a paper presented at the 12th International Congress on Mathematics Education, Seoul, Korea. Retrieved from http://www.icme12.org/upload/submission/2017_F.pdf. Used with permission of Paul Drijvers.*

cational settings. In contrast, the two categories within "learn mathematics" involve technology designed specifically for use in educational situations. Within "practice skills," we consider tools aimed at better organizing student practice, such as tools that support students online as they do homework. Within "develop concepts," we consider approaches focused on student sense making and understanding, such as dynamic representations.

Please note that Drijvers's use of "do mathematics" is not the same as the QUASAR (Quantitative Understanding: Amplifying Student Achievement and Reasoning) project's concept of "doing mathematics" (Henningsen & Stein, 1997). The QUASAR concept of "doing mathematics" refers to an aspirational level of engagement in which students emulate the reasoning of professional mathematicians. This aspiration can apply across all aspects of Drijvers's framework, because the highest levels of mathematical practices are possible in mathematical experiences with varied purposes.

In this chapter, we also go beyond Drijvers's framework in recognizing a fourth purpose for the use of technology in the classroom. Mathematics education can now extend across both didactically organized (formal) and interest-driven (informal) learning contexts. In the fourth purpose, we consider technology as a context for engaging with mathematics. For example, robotics, fabrication tools, programming, and other contexts, are now common in the lives of many youth, and important opportunities for mathematical reasoning can emerge in these contexts, often through the mediation of technology. When learning activities are considered from a science, technology, engineering, and mathematics (STEM) perspective, the activities may emphasize other disciplines yet provide ample and important opportunities to learn mathematics. This is an important dimension now because schools and other institutions are building—by the thousands—physical spaces for new types of STEM learning, such as robotics labs, makerspaces, and FabLabs. The emergence of these hybrid spaces is blurring the boundaries between what we today term "formal" and "informal" education and creating new opportunities for mathematics learning. Consequently, we should not restrict our view of technologies for learning mathematics to only those technologies that have their home in mathematics classrooms or workplaces.

Within each of these categories, we analyze two dimensions of design: one focused on *productivity* and another focused on *transformation*. The productivity dimension emphasizes the role of technology in making mathematics learning more efficient or fruitful. The transformation dimension emphasizes how technology can allow reconceptualization of mathematics teaching and learning and calls into question what mathematics is at stake. Whereas the productivity dimension tends to vary the means to address given ends, the transformative dimension often aims to change the game by explicitly analyzing the nature of mathematics knowledge. We will see that discourses about productivity and transformation occur in each of the purposes of technology, but in somewhat different ways.

Where possible, we seek parallelism across our discussion of the four purposes. For example, we discuss history, key research, learning theories, policy perspectives, and opportunities for the future with respect to each purpose.

Technology for Doing Mathematics

Technology plays an increasing role in doing mathematics, both in the workplace (Hoyles, Wolf, Molyneux-Hodgson, & Kent, 2002) and for mathematicians. Technologies that are commonly used in the workplace can also have

有用的（Hatano，1997；Miller & Stigler，1991）。其他相关的例子有计算器和图形计算器（Kaput & Thompson，1994）、电子表格（J. E. Baker & Sugden，2003）、计算机代数系统（Ruthven，2002），包括Methematcia和Maple等工具，甚至诸如尺、量角器和圆规等。

继卡普特之后，我们也在思考这样使用技术的核心基础作用是什么？从做数学的工具来看，工具不仅提供了进行数字和符号运算的基础，而且在数学概念化方面发挥着关键作用（例如，给学生一把有匀称刻度的尺，那么相等单位概念的必要性可能就变得不太明显了）。从政策的角度来看，在教育环境中配置普通数学工具具有天然的吸引力。由于教师是为学生工作或继续教育做准备的，所以使他们不仅具备知识技能，还具备使用常用工具的实用技能是有意义的。

自《学校数学的原则和标准》（NCTM，2000）颁布起，计算器和其他工具一直是全美数学教师理事会（NCTM）建议的一个明确组成部分。最近《州共同核心数学标准》（CCSSM；全国州长联合会最佳实践中心和全美州首席教育官员理事会［NGA & CCSSO］，2010）在实践标准中包括了这些工具：

> 精通数学的学生在解决数学问题时会考虑可用的工具，这些工具可能包括铅笔和纸张、具体模型、尺子、量角器、计算器、电子表格、计算机代数系统、统计软件包或动态几何软件。（第7页）

从广泛的政策角度来看，同样重要的因素是，这类工具越来越多地在考试评价中被接受或授权，在测试中接受技术可以成为更广泛采用技术的一个强大因素。

而且，技术正在改变数学家的工作方式。在20世纪下半叶，技术工具从根本上加强了符号计算、可视化和逻辑证明，而这些技术工具又在许多数学学科中变得至关重要，数据科学、图论和计算几何学等学科等，仅列几例而已，它们已经出现了或已经被计算的可能性和应用所改变了，虽然这些变革在很大程度上还没有影响到学校课程的数学教学内容，但是课程已经越来越明显有接受技术所蕴含机会的趋势。

下面，我们将强调关于做数学的技术的两个研究视角，一个强调生产力，另一个强调变革。在实践中，可以将它们组合起来，但为了使思想更加清晰，我们把它们区分开，并分别予以讨论。

生产力：劳动分工

劳动分工是生产力视角下有用的组织隐喻。数学工作可以在人和工具之间进行分配，工具不仅可以更快、更精确地进行计算，而且可以解放人类思维去更多地思考做数学中策略方面的内容。

相关的学习原则建立在"认知负荷"的概念之上。认知科学研究早就已经确立了人类在工作记忆、信息处理和相关特征方面都存在着局限，当任务要求人们超过这些限制时，人类的表现就会受到影响（Sweller, van Merrienboer, & Paas, 1998）。从学习的角度来看，技术可以潜在地减轻那些不重要的学习任务，并将学生的认知努力集中在要学习的数学方面（Paas, Renkl, & Sweller, 2004）。例如，《州共同核心数学标准》（NGA & CCSSO, 2010）使用短语"策略性地使用适当的工具"，这对于数学学习来说可能就是如何战略性地解决数学问题，同时将常规计算委托给设备。不过，我们稍后会再来讨论这个可能过于简单化的"委托"比喻。

另一个重要的学习观点是"多媒体原理"（Fletcher & Tobias, 2005），其重点是为学习者提供相关图形和语言表征的价值。在图形计算器的情况下，语言表征可以表示为符号表达式的函数，图形表征可以是该函数的图像。在电子表格中，单元格中的公式偏向语言，但是数字数组或图形图表中的模式通常更加直观，统计工具提供了类似的表征配对。

关于计算器和图形计算器在学习中的作用已经进行了大量的研究，实际上，已经进行了相当多的研究，有几个元分析（即对独立发表的研究的聚合分析）。埃林顿（2003）的一项包含54项高质量实验研究的元分析发现，

a powerful pedagogical use. One obvious historical example is the abacus, a tool that is useful both for doing calculations and for learning about the organization of the number system (Hatano, 1997; Miller & Stigler, 1991). Other relevant examples are the calculator and graphing calculator (Kaput & Thompson, 1994); spreadsheets (J. E. Baker & Sugden, 2003); computer algebra systems (Ruthven, 2002), including tools such as Mathematica, Maple, and the like; and even supports such as rulers, protractors, and compasses.

Following Kaput, we consider what is the core infrastructural role of this use of technology? In the case of tools for doing mathematics, tools provide an infrastructure for doing numeric and symbolic computation but also play a key role in how mathematics is conceptualized (for example, if students are given a ruler with even marks, the conceptual necessity for equal units may become less obvious). From a policy perspective, including commonplace mathematical tools in educational settings has a natural appeal. As teachers prepare students for work or further education, it makes sense to equip them not only with intellectual skills but also with practical skills in using commonplace tools.

Calculators and other tools have been an explicit component of the National Council of Teachers of Mathematics (NCTM) recommendations since *Principles and Standards for School Mathematics* (NCTM, 2000). The more recent Common Core State Standards for Mathematics (CCSSM; National Governors Association Center for Best Practices and Council of Chief State School Officers [NGA & CCSSO], 2010) include tools such as these in a practice standard:

> Mathematically proficient students consider the available tools when solving a mathematical problem. These tools might include pencil and paper, concrete models, a ruler, a protractor, a calculator, a spreadsheet, a computer algebra system, a statistical package, or dynamic geometry software. (p. 7)

An equally important factor from a broad policy perspective is that tools of this sort are increasingly accepted or mandated on assessments. The acceptance of technology on tests can be a powerful factor in wider adoption.

Further, technology is changing how mathematicians do their work. In the second half of the 20th century, symbolic computation, visualization, and logical proof have each been radically enhanced by technological tools, which have in turn become essential in many mathematical disciplines. The disciplines of data science, graph theory, and computational geometry—to name only a few—have emerged or been transformed by computational possibilities and applications. While these transformations have largely yet to affect the mathematical content of school curricula, curricula have become progressively more receptive to the opportunities they imply.

Below, we highlight two research perspectives on technologies for doing mathematics, one emphasizing productivity and the other emphasizing transformation. In practice, these can be combined. We distinguish and separate the views for intellectual clarity.

Productivity: Division of Labor

Division of labor is a useful organizing metaphor for the productivity perspective. Mathematical effort can be divided between people and tools. Tools not only can do calculations faster and more accurately but also can free up human minds for strategic aspects of doing mathematics.

A related learning principle builds on the concept of "cognitive load." Cognitive science research has long established human limits in working memory, processing, and related characteristics; human performance suffers when tasks require people to exceed these limits (Sweller, van Merrienboer, & Paas, 1998). From a learning perspective, technology can potentially offload tasks that are unimportant to learning in that moment and keep students' cognitive effort focused on the aspect of mathematics to be learned (Paas, Renkl, & Sweller, 2004). For example, CCSSM (NGA & CCSSO, 2010) uses the phrase "use appropriate tools strategically," and one aspect of this for mathematics learning could be how to approach mathematical problems strategically while delegating routine calculations to a device. However, we will later revisit this perhaps overly simplistic "delegation" metaphor.

Another important learning perspective is the "multimedia principle" (Fletcher & Tobias, 2005), which focuses on the value of providing learners with related graphical and linguistic representations. In the case of graphing calculators, the linguistic representation can be a function represented as a symbolic expression and the graphical representation can be a graph of the function. In spreadsheets, the formulas in cells are more linguistic, but the patterns, either in arrays of numbers or in graphical charts, are often more visual. Statistical tools provide similar pairings of representations.

Considerable research has been conducted on the role of calculators and graphing calculators in learning; indeed, so much research has been conducted that there are several meta-analyses (i.e., analytic aggregations of indepen-

基于图形计算器的干预对学生成就有可靠的积极的影响。此外，研究表明，当允许在测试中应用图形计算器时，从计算和运算到概念理解和问题解决，被试的表现都得到了提高。第二个元分析特别关注了图形计算器在代数教学中的应用（Ellington，2007）。科竹、加西和米勒（2005）使用美国教育部有效教育策略资料中心（美国教育部，n.d.）发布的严格的质量控制标准筛选了现有研究，他们选择了8项高质量研究来考察图形计算器对K-12数学学习的影响，其中的4项研究专门评估了图形计算器对代数学习的影响。在各种学生群体和教学条件下，使用配套于教学材料的图形计算器对代数成就具有强大的积极影响。

教育工作者和政策制定者往往想要知道的不仅仅是什么有作用，他们希望知道的是对如何有效实施的具体指导。幸运的是，强大的图形计算器研究可以解决这些问题。海勒、柯蒂斯、加菲和维博库尔（2005）的一项研究描述了一个实施的模型，其中包括新的教科书、教师专业发展和评估，所有这些都与图形技术和动态代数主题相一致，这项研究发现，使用图形计算器最频繁的老师和学生学到的最多。国家教育统计中心（2001）关于美国教育进展评估的报告（NAEP；"国家报告卡"）让人们相信，在八年级（但不是四年级）频繁使用图形技术与获得更高数学成绩相关，它说：

> 八年级学生的老师报告说，几乎每天都使用计算器的学生成绩最好，每周使用计算器的学生也比不频繁使用计算器的学生的平均得分更高。此外，允许不受限制地使用计算器的教师和允许在考试中使用计算器的教师，其八年级学生的平均分数要高于不允许在课堂上使用计算器的教师的学生。频繁使用图形计算器和高成就之间的关系适用于富裕的学生和贫穷的学生，男学生和女学生，不同种族和民族的学生，以及具有不同政策和课程的不同州。（第144页）

数学工具的流行已经同时为职前教师（例如Stohl Drier, Harper, Timmerman, Garofalo, Shockey 2000）和在职教师（Lawless & Pelligrino, 2007）的专业发展和问题处理提供了充足的研究成果。例如，技术教学内容知识框架（TPACK；Koehler & Mishra，2005）表明，教师不仅需要关于技术的知识，而且需要关于技术如何与内容知识和教学知识相结合的知识。PURIA框架[1]提出了数学教师加强技术采用的发展过程（PURIA是个人游戏、个人使用、推荐、实施和评估的首字母缩写，Zbiek & Hollebrands, 2008）。同样，鲁斯文（2014）提出，数学技术的实施框架必须关注问题的多个层面。教室的工作环境经常需要调整以利于技术的充分运用，例如，有投影的界面以及学生和教师展示他们如何使用技术的方式，包括教科书、习题集和其他教学材料在内的资源系统必须与技术的可用性协调一致，必须设计和测试利用技术学习的活动和课程脚本。最后，鲁斯文指出"时间经济"是永远存在的挑战，因为课堂上的时间是宝贵的，技术不能消耗太多的时间，也不能造成太多的中断（例如技术工具的崩溃）。

电子白板提供了一个案例研究，其中技术被定位为数学的工具，但没有什么积极的影响。在早期的浪潮中，英国广泛安装了交互式白板，之后美国也掀起了广泛采用白板的浪潮，交互式白板通常包括用于做数学的特定工具，例如屏幕上的计算和测量。研究发现，白板的使用产生了积极的效果，但这种效果主要体现在教师和学生对技术的看法上（Higgins, 2010），而白板对于学生学习成绩的影响却并不显著。在很多情况下，教学实践发生的变化并不大（H. J. Smith, Higgins, Wall, & Miller, 2005）。在鲁斯文（2014）的框架中，工作环境发生了变化，时间经济也得到了管理，但在资源系统、活动或课程脚本中只发生了轻微的变化。特别是，学生不会因为在课堂上添加了一块交互式白板，就以新的强度或新的方式积极地做数学，学生未能积极地与数学工具进行互动，因而他们的学习不太可能得到改善。

dently published studies). A meta-analysis by Ellington (2003) of an inclusive set of 54 high-quality experimental studies found a reliable positive effect of graphing-calculator-based interventions on student achievement. In addition, the studies suggested that when graphing calculators are allowed on tests, gains extend from calculations and operations to conceptual understanding and problem-solving performance. A second meta-analysis looked specifically at graphic calculator use in teaching algebra (Ellington, 2007). Khoju, Jaciw, and Miller (2005) screened available research using stringent quality-control criteria published by the U.S. Department of Education's What Works Clearinghouse (U.S. Department of Education, n.d.). They selected eight high-quality studies examining the impact of graphing calculators on K–12 mathematics achievement. Four of these studies specifically assessed the effect on algebra learning. Across a wide variety of student populations and teaching conditions, use of graphing calculators with aligned instructional materials had a strong, positive effect on algebra achievement.

Educators and policymakers often want to know more than what works—they want concrete guidance on how to achieve an effective implementation. Fortunately, strong graphing calculator research is available to address these concerns. A study by Heller, Curtis, Jaffee, and Verboncoeur (2005) described and studied a model implementation, which included a new textbook, teacher professional development, and assessment—all aligned with the graphing technology and the theme of dynamic algebra. This study established that the teachers and students who used graphing calculators most frequently learned the most. The National Center for Educational Statistics (2001) report on the National Assessment of Educational Progress (NAEP; "the Nation's Report Card") provides confidence that frequent use in the eighth grade (but not in the fourth grade) is associated with greater mathematics achievement, stating,

> Eighth-graders whose teachers reported that calculators were used almost every day scored highest. Weekly use was also associated with higher average scores than less frequent use. In addition, teachers who permitted unrestricted use of calculators and those who permitted calculator use on tests had eighth-graders with higher average scores than did teachers who did not indicate such use of calculators in their classrooms. The association between frequent graphing calculator use and high achievement holds for both richer and poorer students, for both girls and boys, for varied students with varied race and ethnicity, and across states with varied policies and curricula. (p. 144)

The prevalence of tools for doing mathematics has generated a considerable body of research on related teacher professional development and implementation issues for both preservice teachers (e.g., Stohl Drier, Harper, Timmerman, Garofalo, & Shockey 2000) and in-service teachers (Lawless & Pelligrino, 2007). For example, the technological pedagogical content knowledge framework (TPACK; Koehler & Mishra, 2005) suggests that teachers need not only knowledge about technology but also knowledge about how technology relates to content knowledge and pedagogical knowledge. The PURIA framework[1] suggests a developmental process of intensified technology adoption by mathematics teachers (PURIA stands for personal Play, personal Use, Recommendation, Implementation, and Assessment; Zbiek & Hollebrands, 2008). Likewise, Ruthven (2014) suggests that an implementation framework for technology in mathematics must address multiple layers of concerns. The working environment of the classroom often must be adjusted to take advantage of technology: for example, including surfaces for projection and ways for students and a teacher to show how they are using technology. The resource system, which includes textbooks, problem sets, and other instructional materials, must be coordinated with the availability of technology. Activities and curricular scripts, which leverage technology for learning, must be designed and tested. Finally, Ruthven notes the ever-present challenge of "time economy"—time in classrooms is precious, and technology cannot consume too much time or cause too much disruption (e.g., by breaking down).

Electronic whiteboards offer a case study in which technology has been positioned as a tool for doing mathematics but with little positive impact. In an early wave, interactive boards were widely installed in the United Kingdom, followed by a later wave of broad adoption in the United States. The interactive boards often included specific tools for doing mathematics, such as on-screen calculation and measurement. Positive effects were found, but primarily in terms of teachers' and students' perceptions of the technology (Higgins, 2010). Results on student achievement were hard to detect, and in many cases, little change in teaching practices occurred (H. J. Smith, Higgins, Wall, & Miller, 2005). In Ruthven's (2014) framework, the working environment is changed and time economy managed, but only slight changes occur in the resource system, activities, or curricular scripts. In particular, students do not tend to be active in doing mathematics with new intensity or in new ways just because of the addition of an interactive board to the classroom, and without active student engagement with mathematical tools, it is unlikely that learning will be enhanced.

变革：工具化

在关注人与做数学的工具之间的"伙伴关系"的"共同演化"方面，工具化的观点与生产力的观点形成了鲜明的对照。工具化观点强调技术如何增强和改变人类智力的本质（Engelbart，1995），而不仅仅是使工作更有效率，特别是劳动分工的隐喻需要小心对待，在采用计算器或计算机代数系统（CAS）等新技术的情况下，将人的处理能力外包给机器会导致工作的重新分配问题。但实际上，情况并不简单，随着学生把技术深入地融入他们的数学策略和实践中，这些策略和实践就会以不同于以往的纸笔媒介的形式出现。

阿蒂格（2002）强调了这一点：

> 技术不仅仅具有可以产生结果的**实用价值**，而且还具有**认识论价值**，因为它们部分地由对对象的理解所构成，这也是新问题的根源。（第248页，重点为原文所加）

以计算机代数系统为例，阿蒂格指出实用价值（获得答案）和认识论价值（发展和扩展概念）是相互交织的，这样的交织方式在技术的存在下变得更加突出。数学家们意识到越来越强大的技术不会自动地、毫不费力地增进数学洞察力，而是需要改变他们的工作方式。这些变化并不仅仅是简单的过程，就像你已经知道如何用铅笔书写了，却要学习用钢笔写字那样；这些变化也可能是复杂和深远的，就像一个已经知道如何骑自行车的人学习驾驶汽车时会发生什么。当人们可以驾驶汽车而不是只会骑自行车时，他们组织日常活动的方式就会改变。在威利龙和拉巴德尔（1995）之后，阿蒂格称之为"工具起源"的过程：对学习者（包括专业用户）来说学习过程是必要的，以适应数学活动和数学本身因技术而产生的变革——技术如何塑造数学知识，技术又如何塑造学习（见 Noss & Hoyles，1996）。

研究工具性起源（Noss，Hoyles，& Kent，2004；Trouche，2003）的一个最有趣的副产品是让人们更好地理解，技巧随着时间的推移变得越来越寻常，但同时它们也失去了数学本质。尽管像解方程这样的技巧经常随着时间的推移逐渐发展，但求解过程的常规化（回忆用于求解二次方程的公式、分部积分，等等）导致了技巧的去数学化，这其中很容易忽视产生该技巧的理论思想。正如阿蒂格（2002）优雅地指出的："最终被认为是数学的，只是我们实际数学活动那座冰山被简化成的一角，而这种急剧性的简化，强烈地影响着我们对数学和数学学习的看法以及与之相关的价值观"（第249页）。因此，虽然人与强大技术之间的伙伴关系可能带来好处，但是我们也必须警惕这一伙伴关系会变得日常化，从而可能会丧失意义。

特罗切（2000）详细分析了每个使用计算机代数系统或图形计算器的数学老师都会熟悉的一种情况，这个例子涉及一组学生，他们被要求使用图形计算器来计算当 x 趋近无穷大时一个4次多项式的极限，其中 x^4 的系数为0.03。由于系数这么小，函数的图形表示与学生的期望不一致。正如阿蒂格（2002）指出的：

> 这些差异（在预期的和实际发生的事情之间）对图形计算器的工具起源有明显的影响，图形计算器是年初可用的唯一计算器，然后对符号计算器的工具起源产生了明显的影响。（第6页）

因此，技术不仅在于能更有效地产生答案，而且是培养观察数学、将数学概念化的新途径。

我们认为，对人与数学工具之间的长期合作伙伴关系的共同进化缺乏关注，是教育实践（与专业数学实践形成鲜明对比）在适应新技术带来的变革如此缓慢的原因所在。在学校，数学工具通常以替代的观点引入（例如以电子表格替代纸质表格），将技术概念化为可以改变人类思维与工具之间关系的本质，这为学习创造了新的目标，这种目标可以是学生在数学方面所能做的发展进步，因为他们可以将工具深深地融入到他们的概念库中。

Transformation: Instrumentalization

The instrumentalization view contrasts with the productivity view by focusing on the "coevolution" of the "partnership" between a person and a tool in doing mathematics. An instrumentalization view emphasizes how technology can augment and change the nature of human intelligence (Engelbart, 1995), not just make the work more efficient. In particular, the division of labor metaphor needs to be treated with care. In adopting a new technology, such as the calculator or a computer algebra system (CAS), outsourcing processing power from the human to the machine results in a redistribution of who does what. But in reality, the picture is not so simple. As students integrate technologies deeply into their mathematical strategies and practices, the strategies and practices can come to take a different form than they ever would with paper-and-pencil media.

Artigue (2002) makes the point forcibly:

> Techniques do not have only a *pragmatic value* which permits them to produce results, they have also an *epistemological value*, as they are partly made up of the understanding of the objects, and are also a source of new questions. (p. 248, emphasis in original)

Using the example of computer algebra systems, Artigue points to the ways in which the pragmatic (getting the answer) and the epistemological (developing and extending concepts) are intertwined—ways which become more salient in the presence of technology. Mathematicians appreciate how increasingly powerful technologies do not automatically and effortlessly enhance mathematical insight but require changes to the way they work. The changes are not just relatively simple, like learning to write with pen if you already know how to write with a pencil. The changes can also be complex and far-reaching, like what happens when a person who already knows how to ride a bicycle learns to drive a car. How people organize their daily activities changes when they can drive a car and not just ride a bicycle. Following Verillon and Rabardel (1995), Artigue calls the process "instrumental genesis": The learning process that is necessary for the learner (including the professional user) to accommodate how mathematical activity and mathematics itself can be transformed by technology—how mathematical knowledge is shaped by technology and how it shapes learning (see Noss & Hoyles, 1996).

One of the most interesting by-products of the study of instrumental genesis (Noss, Hoyles, & Kent, 2004; Trouche, 2003) has been a greater understanding of how techniques lose their mathematical essence as they become more and more routine over time. Even though a technique such as solving an equation has often evolved incrementally over time, the routinization of the solution process (recall the formula for solving a quadratic, integrate by parts, and so on), leads to a de-mathematization of the technique, in which it is easy to ignore the theoretical ideas that gave rise to the technique. As Artigue (2002) elegantly puts it, "what is finally considered as mathematical is reduced to being the tip of the iceberg of our actual mathematical activity, and this dramatic reduction strongly influences our vision of mathematics and mathematics learning and the values attached to these" (p. 249). Hence, while there can be benefits to a partnership between person and powerful technology, we must also be alert to the possible loss of meaning as the partnership becomes routine.

Trouche (2000) analyzes, in detail, a situation that will be familiar to every mathematics teacher who has used a CAS or graphic calculator with students. The example concerns a group of students required to use their graphic calculators to find the limit as x approaches $+\infty$ of a polynomial function of degree 4 whose term in x^4 has a coefficient of 0.03. With such a small coefficient, the graphical representation of the function does not align with what the student might expect. As Artigue (2002) notes:

> These differences [between what is expected and what happens] have evident effects on the instrumental genesis of graphic calculators, which are the sole calculators available at the beginning of the year, and then on the instrumental genesis of symbolic calculators. (p. 6)

Thus technology is not only about more efficiently producing answers, but also about cultivating new ways of seeing and conceptualizing mathematics.

We argue that lack of attention to the coevolution of the long-term partnership between people and mathematical tools is why educational practices—in stark contrast to the practices of professional mathematics—have been so slow to accommodate transformative change in the wake of new technologies. In school, mathematical tools are often introduced with a substitution point of view (e.g., a digital worksheet substitutes for a paper worksheet). Conceptualizing technology as allowing a change in the nature of the relationship between the human mind and tools creates new goals for learning as a developmental progression in what students can do in mathematics as they deeply accommodate tools into their conceptual repertoire.

扩展和问题

在过去十年中,工具已经被整合到课堂网络中,可以在课堂上快速共享数学信息。例如,借助德州仪器的导航系统(TI公司的Navigator system),教师可以看到并分享学生的工作页面、分发和收集学生的作业,并进行快速的形成性评价。快速的形成性评价功能包括多项选择调查,收集学生的答案并绘制为直方图。教师还可以收集数学表达式、数据点、图形等,并对这些数学对象进行比较。这有助于学生分享他们的作业。因为有可能匿名,所以他们可以专注于数学思维而不是个人表现(Davis,2003)。显示学生在课堂上处理数学任务的相似之处和差异,可以促进对问题解决的多种策略和误解的讨论(J. P. Smith, diSessa & Roschelle, 1994),并可为数学论证提供有用的刺激(Stroup, Ares and Hurford, 2005)。此外,当网络更容易呈现学生的思维时,教师可以采用适应性教学法,调整教学的步调或内容以适应学生的需要(Noss & Hoyles, 1996)。一个大而成功的研究项目(Pape等,2012)通过对全国教师抽样并随机分配实验来调查联网计算器在代数教学中的形成性评价能力,总体影响是积极的,研究人员还发现,教师专业发展与技术的融合很重要。

从工具角度来看,进一步研究的两个关键问题是将技术整合到专业发展和评价中。首先,正如我们上面讨论的那样,尽管现在有了更强大的框架,但提供足够的专业发展机会使教师有效地使用技术仍然是一个"棘手的问题"(Borko, Whitcomb, & Liston, 2009)。无论是职前培训还是在职的专业发展,学习如何使用技术来推进学习,往往与学习如何有效地教授数学这一核心问题脱节。第二,随着技术在高风险评估中成为一个更为突出的因素,关于计算机数学表达的约束如何影响学生的工作存在一些严肃而重要的问题(Heiten, 2014),例如,计算机通常要求学生以计算机常用的符号输入数学表达(例如,总是强制使用"*"作为显式乘法符号),并且可能不支持学生在纸上使用的相同的数学表征。对技术评估的效度以及如何减少人们的顾虑我们知道得还不够。

实践数学的技术

网络无处不在的一个显著影响就是开发了具体的教学技术来支持学生的数学技能实践,实现某些数学技能的快速、准确和自动化显然是数学发展的关键部分。技术可以为学生提供快速的反馈,并且当他们需要帮助时,可以提供针对当前问题的教学资源,这导致长期以来人们认为,技术可以扮演个人的、适应性强的导师角色(Brown & Sleeman, 1982)。实践数学的技术可以追溯到程序教学时代(Zoll, 1969),它基于学生表现,提供不同的技能训练,尽管是通过复杂的指令(例如,"如果你的答案是x,下一题请做问题y")。这种方法体现在萨佩斯(1971)计算机辅助教学早期的一个典型例子中,最终它带来了一系列颇为成功的商业化产品。

今天实践数学的典型技术包括广泛使用的测试题库,在学生完成一系列练习后通常会向他们提供反馈意见,并向老师提供报告。智能辅导系统,如认知导师(Ritter, Anderson, Koedinger, & Corbett, 2007)也属于这一类。这些系统,包括专家对数学任务进行表征的方式,通过比较学生与专家的表现,可以诊断学生的干预需求(Corbett & Anderson, 1994)。该类别中的其他系统重点是根据学生给出的答案,提供适当的教程、提示和指导(Koedinger, McLaughlin, & Heffernan, 2010)。

从政策的角度来看,在数学实践过程中提供反馈的技术被广泛接受并采纳,在某种程度上,是因为这些技术作为帮助学生准备标准化测试的工具,很容易融入到强调问责制的政策背景下。随着高风险评估越来越多地从纸质转向数字化、网络媒体,测试准备工作也可能会在网上进行。此外,教师和管理人员重视向学生提供反馈意见,并提供报告以支持决策,这些系统很容易适应一种流行的技术叙述,即提供"个性化"的学习资源。据称,通过在学生练习期间收集关于学生表现的数据,未来给学生提供的学习机会能更好地符合他们的能力。

技术用于实践,其作用不同于前一节所提到的技术作为做数学的工具的作用。显然,实践可以包括做的

Extensions and Issues

In the past decade, tools have been integrated into classroom networks that allow rapid sharing of mathematical information in a classroom. For example, with Texas Instrument's Navigator system, a teacher can see and share screenshots of students' work, distribute and collect work from students, and engage in quick formative assessments. The quick formative assessment capabilities include multiple-choice polls, where students' answers are collected and plotted as a histogram. Teachers can also collect mathematical expressions, data points, graphs, and so forth, and aggregate these mathematical objects for comparison. This can enable students to share their work, potentially anonymously, so they can focus on mathematical thinking and not personal performance (Davis, 2003). Revealing similarities and differences in how students in a classroom approached a mathematical task can foster discussion of multiple solution strategies as well as misconceptions (J. P. Smith, diSessa & Roschelle, 1994) and can provide useful stimulus for mathematical argumentation (Stroup, Ares, & Hurford, 2005). Further, when the network more readily makes student thinking visible, teachers can engage in adaptive pedagogies, adjusting the pace or content of instruction to suit what students need (Noss & Hoyles, 1996). A large and successful research project (Pape et al., 2012) investigated the formative assessment capabilities of networked calculators for teaching algebra via a random assignment experiment with a national sample of teachers. The overall impacts were positive; researchers also found that the integration of teacher professional development with the technology was important.

Two key issues for further research from the point of view of the tool perspective relate to integration of technology into professional development and assessment. First, although, as we discussed above, stronger frameworks are now available, providing adequate teacher professional development so teachers use technology effectively remains a "wicked problem" (Borko, Whitcomb, & Liston, 2009). Both in preservice and in-service professional development, learning how to use technology to advance learning is too often separated from the core of learning to teach mathematics effectively. Second, as technology becomes a more prominent factor in high-stakes assessment, there are serious and important questions about how constraints of computer math representations may affect students' work (Heiten, 2014). For example, computers often require students to enter mathematics in usual notations (for example, always forcing use of "*" as an explicit multiplication symbol) and may not support the same range of mathematical representations that a student might use on paper. Not enough is known about the validity of assessments with technology and what might be done to mitigate concerns.

Technology for Practicing Mathematics

A clear impact of the ubiquity of the Web has been the development of specific pedagogical technologies to support students' practice of mathematical skills. Achieving speed, accuracy, and automaticity of certain mathematical skills is obviously a key part of mathematical development. Technology can provide students with rapid feedback and when they need help, can provide access to tutorial resources focused on the problem at hand, leading to long-standing visions of technology as taking the role of personal, adaptive tutor (Brown & Sleeman, 1982). Technology for practicing mathematics goes back to the era of programmed instruction (Zoll, 1969), which provided differential skill practice based on student performance, albeit by complex instructions to the student (e.g., "if your answer was x, go to problem y next."). This approach was embodied in an early classic example of computer-aided instruction by Suppes (1971), which eventually led to a series of commercially successful products.

Today's prototypical technologies for practicing mathematics include widely used test item banks, which typically give feedback to students as they practice a series of problems and provide reports to their teachers. Intelligent tutoring systems, such as the Cognitive Tutor (Ritter, Anderson, Koedinger, & Corbett, 2007), also naturally fit in this category. These systems include a built-in representation of how an expert would do a mathematical task and, by comparing student performance to the expert performance, can diagnose students' need for intervention (Corbett & Anderson, 1994). Other systems in this category focus on providing appropriate tutorials, hints, and guidance based on the answers students give (Koedinger, McLaughlin, & Heffernan, 2010).

From a policy perspective, technologies for practicing mathematics with feedback are widely accepted and adopted. In part, this is because these technologies fit easily into a policy context that emphasizes accountability, as standardized tools for preparing students for the test. As high-stakes assessment increasingly shifts from paper-based to digital, online media, preparation for tests will also likely take place online. Further, teachers and administrators value providing feedback to students and generating reports to support decision making. These systems easily fit a prevailing narrative of technology

工具，然而，从这个角度来看，关键且基础的功能是用计算机可以存储、处理和分析的形式语言描述数学技能的层次结构。通过在一个根据数学技能之间的关系组织起来的数据库中跟踪数学技能，可以捕获学生的详细进度，并可加以补救。这些表现形式已经在智能辅导系统（Corbett & Anderson, 1994）中被形式化了，又更一般地通过知识空间理论（Falmagne, Cosyn, Doignon 和 Thiéry, 2006）提供了一种自动确定学生学到了什么、遇到了什么困难以及下一步应该学习什么的方法。

下面，与上一节讨论工具相类似，我们再次强调了在使用技术过程中的生产力观点和变革观点。

生产力：技术作为适应工具

从生产力的角度来看，技术在实践技能方面的作用可以通过嵌套循环来理解（Koedinger, Brunskill, de Baker, McLaughlin & Stamper, 2013）。持续时间较短的循环包括当学生在线表现数学技能时为他们提供快速的形成性的反馈，这个循环可以包括就某一数学技能的每一步给予学生反馈。事实上，研究发现，仅仅作为特定的步骤，而不是对总体答案提供反馈时，反馈会特别有效（Van Lehn, 2011）。持续时间较长的循环，会涉及挑选要完成的下一个数学任务或要提供的信息。教师可以根据学生的表现来调整课堂时间的使用，参与这个循环。一个更全面的循环可以根据系统所显示的大量学生的数据来调整系统本身，从而更好地为学生和教师服务。

大量且不断增长的认知科学文献为组织有效的技能实践提供了原则（Koedinger, Booth, & Klahr, 2013，见图31.2），这些原则可用于调整这些循环所提供的学习机会。例如，认知科学家发现，随着时间的推移，"间隔"练习可以提高学习和记忆力（S.K. Carpenter, Cepeda, Rohrer, Kang和Pashler, 2012），技术可以比教师或学生自己更容易地实现间隔机制（例如定期重返较早练习过的技能部分），从这种技术中获得更大生产力的希望也来自更好的反馈，例如，一个在线系统不是等待老师批改作业，而是可以在学生做作业时立即给他们反馈，提供分步的实例也是可行的，这些都符合现有文献提出的建议。

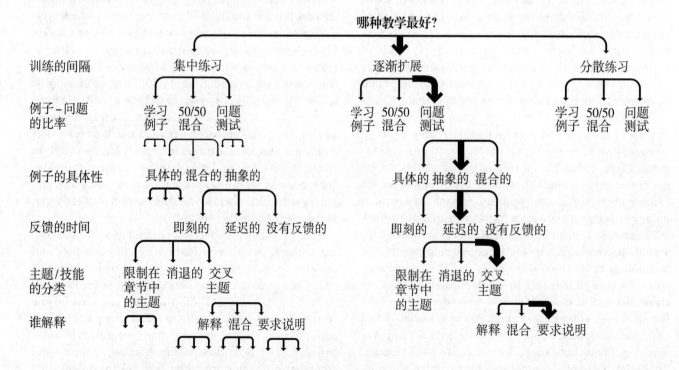

图31.2 组织有效技能实践的科学原则。摘自 "Instructional Complexity and the Science to Constrain It" by K. R. Koedinger, J.L.Booth, and D.Krahr, 2013, Science,342,P..935。转载已经许可。

as providing "personalization" of learning resources: by collecting data on students' performance during practice sessions, future learning opportunities can, it is claimed, be adapted to fit students better.

The role of technology for practice is different from its role in our prior section on technology as a tool for doing mathematics. Obviously, practice can include tools for doing. However, the key infrastructural capability in this perspective is one of describing hierarchies of mathematical skills in formal language that a computer can store, process, and analyze. By tracking mathematical skills in a database organized in terms of the relationships among mathematical skills, students' detailed progress can be captured and potentially remediated. These representations have been formalized in intelligent tutoring systems (Corbett & Anderson, 1994) and more generically via knowledge space theory (Falmagne, Cosyn, Doignon, & Thiéry, 2006), thus providing a way to automatically determine what a student has learned, is having trouble with, and should study next.

Below, we again highlight the perspectives of productivity and transformation in this use of technology, parallel to the prior section on tools.

Productivity: Technology as Adaptive

From a productivity perspective, the role of technology in practicing skills can be understood in terms of nested loops (Koedinger, Brunskill, de Baker, McLaughlin, & Stamper, 2013). The shorter duration loop involves providing students with rapid, formative feedback as they perform mathematical skills online. This loop can include giving students feedback on each step of a mathematical skill; indeed, research has found that feedback is particularly effective when provided for a particular step and not just for the overall answer (Van Lehn, 2011). The longer duration loop involves selecting the next mathematical tasks to be done or information to be provided. Teachers may be involved in this loop by adapting their use of classroom time based upon how students are performing. A yet more encompassing loop can adapt the system itself to better serve both students and teachers on the basis of data showing how the system performs with very large numbers of students.

A large and growing body of cognitive science literature provides principles for organizing effective practice of skills (Koedinger, Booth, & Klahr, 2013; see Figure 31.2), and these principles can be used to adjust the opportunities to learn provided in these loops. For example, cognitive scientists have found that "spacing" practice over time can increase learning and retention (S. K. Carpenter, Cepeda, Rohrer, Kang, & Pashler, 2012), and technology can implement spacing regimes (such as returning to skills practiced in an earlier session on a regular schedule) more easily than teachers or students can on their own. The promise of greater productivity from this type of technology also arises from better feedback. For example, instead of waiting for a teacher to grade homework, an online system can give students feedback instantly as they do the homework. Providing step-by-step worked examples is also possible and fits with the recommendations of the established literature.

A large body of research exists on the effectiveness of systems supporting the practice of mathematical skills.

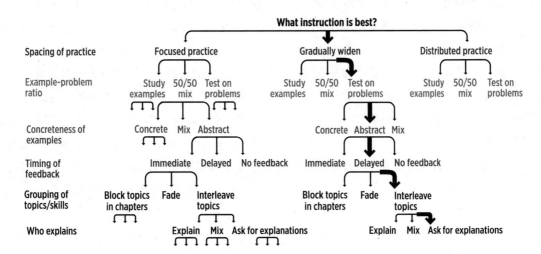

FIGURE 31.2. *Scientific principles for organizing effective practice of skills. From "Instructional Complexity and the Science to Constrain It" by K. R. Koedinger, J. L. Booth, and D. Klahr, 2013,* Science, *342, p. 935. Reprinted with permission.*

对支持数学技能实践系统的有效性，目前已经有了大量研究。例如，张和斯莱文（2013）对数学教育技术的使用做过一个元分析，他们只考虑了符合严格高标准的一些研究。他们发现在对计算机辅助教学系统的研究中，最大的效果其效应量为+0.19（这在教育研究中是一个非常值得注意的效应值）。然而，最近对这类商业产品的一项大规模而且严格的研究发现，这类产品没有显著效果（Campuzano, Dynarski, Agodini, & Rall, 2009）。至于认知导师这一在数学中最广泛使用的智能辅导系统，其中有效教育策略资料中心（2013）对6项不同严谨程度的研究报告了积极的、不确定的和消极的影响这样褒贬不一的结论，造成这些混合结果的一个因素似乎是教师们得到的准备和支持，这些能够让他们按照期望来使用该系统（Pane, McCaffrey, Slaughter, Steele, & Ikemoto, 2010）。有可能通过统一高质量的课堂实施，效果会更好。一些较新的方法，例如ASSISTments，包括广泛的专业发展，以使教师将该技术整合到有效的课堂常规中，并有希望取得初步成果（Koedinger等，2010）。

变革：运用时间和空间

虽然伴随适应系统而来的言词通常侧重于生产力，但是围绕一些较新的数学实践工具的愿景引入了变革的言词，这样的言词关注的是在不同环境下（通常是以技术为中心的环境和以教师为中心的环境）重新安排学习的机会，以便以不同的方式使用时间或空间。这种言词的一个广泛的主题是，教师和技术在学习中各自具有互补的、积极的作用，而继续在课堂中按传统方式使用空间和时间，这样并没有最大限度地利用教师主导和基于技术对学习作出的贡献。因此，需要使用教师主导和基于技术的"混合"学习方式。

新兴的"混合式学习"方法的研究已经确定了四种模式（Staker & Horn, 2012）：

1. 轮换模式，其中教师（或系统）在不同时间将学生分配给不同的模式（包括"翻转"模式，即学生在家里阅读入门材料并在老师的支持下在学校练习）。
2. 灵活的模型，学生主要在线学习，按照自己的速度学习，但需要课堂老师的支持和监督。
3. 自混合模式，学生选择在线资源（例如可汗学院）来补充教师主导的课堂教学。
4. 虚拟学校模式，学生主要在线学习，但也具有特定的有计划的课堂内体验机会。

混合学习研究目前处于早期阶段，虽然在以教师与技术主导的教学和不同学习空间的最佳利用之间取得适当平衡方面有明确的希望，但是开发可以由不同背景的教师来实施的混合学习模式是困难的。例如，虽然对目前流行的可汗学院资源的课堂使用情况进行的仔细研究发现，学习结果是乐观的，但同时也发现，寻找连续一致的学习影响结论还为时尚早（Murphy, Gallagher, Krumm, Mislevy, & Hafter, 2014）。在一定程度上，很难找到对学生的一致影响，因为教师正在以许多不同的方式实施混合式学习，广泛接受的教学方法尚未建立。例如，关于比较广泛的混合式学习技术的一项研究确定了一系列实施细节，这些实施细节对于所产生的学习机会的质量很重要，但取得的成就还不一致，例如将可用资源整合到学生连贯一致的学习体验中去（Murphy, Snow, Mislevy, Gallagher, Krumm, & Wei, 2014）。

混合式学习的另一个研究领域涉及完全数字化的课程材料，即用完全的在线解决方案取代纸质教科书。例如，"推理的大脑"课程提供了一门完整的五年级数学课程，学生在课堂学习中花费大量时间在线学习数学，然而，"推理的大脑"也为课堂上的教师规定了一个明确而重要的角色，并将学生各自独立完成的作业报告给教师，指导教师的行动。该系统的早期结果令人鼓舞（Ocumpaugh, Baker, Gaudino, Labrum, & Dezenhorf, 2013），但尚未在严格的大规模研究中进行测试。

还有一种研究集中在远离教学来源的在线数学学习。对51项研究的元分析发现，在线学习可以比面对面学习产生更好的学习效果（Means, Toyama, Murphy, Bakia, & Jones, 2009），不过，仅从在线的角度不能解释这一优势，课程必须重新设计，以充分利用在线功能。一项研

For example, Cheung and Slavin (2013) conducted a meta-analysis of the use of technology in mathematics education, considering only studies that met high standards for rigor. They found the largest effects in studies of computer-aided instruction systems, which had an effect size of +0.19 (considered a noteworthy effect in educational research). However, a very large, recent, rigorous study of commercial products of this type found no significant effect (Campuzano, Dynarski, Agodini, & Rall, 2009). With regard to the Cognitive Tutor, which is the most widely deployed intelligent tutoring system in mathematics, the What Works Clearinghouse (2013) reported a mix of positive, indeterminate, and negative effects across six different rigorous studies. One contributing factor in these mixed results seems to be the preparation and support teachers received so they could use the system as intended (Pane, McCaffrey, Slaughter, Steele, & Ikemoto, 2010). It is possible that with uniformly high classroom implementation, results would be better. Some newer approaches, such as ASSISTments, include extensive professional development to prepare teachers to integrate this technology into effective classroom routines and have promising initial results (Koedinger et al., 2010).

Transformation: Use of Time and Space

Whereas the rhetoric accompanying adaptive systems has typically focused on productivity, the visions surrounding some of the newer tools for mathematical practice introduce a rhetoric of transformation. This rhetoric focuses on the opportunity to rearrange learning across different contexts (typically a technology-centered context and a teacher-centered context) so as to use time or space differently. A broad theme in this rhetoric is the idea that teachers and technology each have a complementary, positive role in learning—and that continuing with the conventional uses of space and time in classrooms does not take maximal advantage of both teacher-led and technology-based contributions to learning. Thus, "blended" ways of using both teacher-led and technology-based learning are needed.

Research on emerging "blended learning" approaches has identified four models (Staker & Horn, 2012):

1. A rotation model, in which the teacher (or system) assigns students to different modalities at different times (including the "flipped" model, where students review introductory material at home and practice in school with the support of a teacher).
2. A flex model, where students learn primarily online and at their own pace but with support and oversight from a classroom teacher.
3. A self-blend model, where students choose online resources (such as the Khan Academy) to supplement teacher-led classroom instruction.
4. A virtual school model, where students learn primarily online but with specific scheduled opportunities for within-classroom experiences.

Research on blended learning is currently in the early stages. While there is clearly promise in finding the right balance between teacher- and technology-led instruction and the best use of different learning spaces, developing a blended learning model that can be implemented by a range of teachers across diverse settings is difficult. For example, careful research on classroom use of the popular Khan Academy resources found cause for optimism but also found that it was too early to look for consistent effects on learning (Murphy, Gallagher, Krumm, Mislevy, & Hafter, 2014). In part, it is hard to find consistent effects on students because teachers are implementing blended learning in so many different ways that widely accepted pedagogical routines have not yet been established. For example, a study on a broader range of blended learning technologies identified a set of implementation details that are inconsistently achieved yet important to the quality of the resulting learning opportunities, such as the integration of available resources into a cohesive and coherent learning experience for students (Murphy, Snow, Mislevy, Gallagher, Krumm, & Wei, 2014).

A further area of research on blended learning relates to fully digital curriculum material, which replaces paper textbooks with completely online solutions. For example, Reasoning Mind offers a comprehensive full course in fifth-grade mathematics, where students spend substantial time during class learning and doing mathematics online. Reasoning Mind, however, also specifies a well-defined and important role for the classroom teacher and provides reports to the teacher from students' independent work to guide what the teacher does. Early results from this system are promising (Ocumpaugh, Baker, Gaudino, Labrum, & Dezenhorf, 2013), but not yet tested in a rigorous, large-scale study.

Alternative approaches have focused on learning mathematics online at a distance from the source of instruction. A meta-analysis of 51 studies found that online learning can produce better learning than can face-to-face learning (Means, Toyama, Murphy, Bakia, & Jones, 2009); however, merely being online does not explain the advantage—the course must be redesigned to take substantial advantage of online capabilities. One study found that, for middle schools that do not offer algebra already, an online course can be an effective option (Heppen et al., 2011). However, learning in a fully online

究发现，对于那些还没有开设代数课程的初中来说，网络课程可以是一个有效的选择（Heppen等，2011）。然而，对于在学校上学感到困难的学生来说，在完全在线的环境中学习可能是一个挑战，在线环境可能不如教师主导的课堂那么有支持性，使用起来也较为困难。总体而言，根据数学教学和实践技术的可用性来设计教学模式的机遇仍处于早期的探索阶段，现在下结论还为时过早。

扩展和问题

在支持学生数学技能实践的同时，技术也可用于收集有关特定学生和学生之间的丰富数据，当问题的格式使任务标准化时，解释数据可能比开放式工具或探索性动态表征方式更容易。目前，这正在引起人们对大数据方法的关注，这些方法能支持学生数学技能的掌握，例如基于数据，检测学生挫败的时间（R. S. Baker, D'Mello, Rodrigo, & Graesser, 2010）。更普遍的是，实践数学技能的技术正在迅速成为教育数据挖掘和学习分析研究的重要测试平台。相对于反馈和任务选择循环的讨论，教育数据挖掘（R. S. Baker & Yacef, 2009）可以被看作是一个长期循环，在这个循环中，研究人员分析学生的学习进度记录，以寻求更优的学习进度，这些进度可能会根据学生的初始表现进行区分或个性化。

学习分析是与数据挖掘相关的一个概念，但它强调"检测器"的定义，可以分析学生的击键、鼠标点击和其他输入，并归纳出与学生学习相关的更高级别的特性。在这项研究中，通过统计和机器学习技术对自动检测器进行训练（Calvo & D'Mello, 2010）。研究人员已经能够建立非认知行为的探测器，如甄别表现出犹豫、困惑或开小差的学生。在学生练习数学技能的同时自动检测情绪状态，可能会引发更好地支持学习的干预措施。最近，在新兴的多模态学习分析领域，研究人员已经开始采用新型传感器和技术，如生物传感器、眼球跟踪器、手势跟踪器、语音识别和自动视频分析，以建立多模态检测器来考察学生屏幕外的行为（Blikstein, 2013b；Worsley & Blikstein, 2013）。

另一个扩展的领域是在社会或协作环境中练习数学技能。例如，在TechPALS项目中，学生们以三人小组进行工作，小组每个成员都有一个手持设备，这些设备用于交流，在三个学生之间分配解决一个有凝聚力的任务各自要担当的角色，要求他们讨论这一数学问题以取得进展。在涉及三所学校的一项实验中，使用TechPALS的学生比在类似的个人实践环境中学习的学生学得多（Roschelle, Rafanan等，2010）。计算机支持的合作学习研究还探讨了如何通过在线讨论来支持团队学习数学（Stahl, 2009）。广义上讲，合作学习具有良好的理论和经验基础，技术可以在协调学生之间的有效合作学习活动方面以及与教师的作用相结合方面发挥宝贵的作用（Kirschner & Erkens, 2013）。

虽然优化实践的认知科学很强大，但这一领域长期存在的挑战（实际上在Kaput, 1992提出）可能是对数学是什么的狭隘观点。数学不仅仅是顺畅运用技能的能力，它还包括概念理解、策略能力、适合的推理和丰富的情感倾向（Kilpatrick, Swafford, & Findell, 2001）。将这些额外的维度简化到认知技能获取框架，则限制了学生在数学发展的这些附加维度上取得进步的机会。混合方法可以通过嵌入式技术在较大的数学学习课程中增强实践机会来潜在地克服这种制约，无论是在混合模型还是在非混合模型中，关于影响类研究所得到的不一致的结果以及在混合学习评估中观察到的一些困难均指向了实施方面的挑战。根据鲁斯文（2014）的框架，教师需要额外的支持来将实践系统整合到更大的活动系统和课程脚本中。通常，关于混合学习系统的讨论会将成本降低和质量改进的可能性相结合，既然它们是两个正交的维度，我们认为研究人员和政策制定者应该把它们作为研发的独立方向，同时考虑到这两方面，创新者可以寻求使程序更流畅更有效率的方法，以及采用技术来提高数学学习质量和丰富程度的方式。

技术促进数学概念化

现在我们转而专门讨论为教学目的（而不是作为工作场所的工具）而设计的技术，而且专注于学生在日常

environment can be challenging for students who struggle in school; online environments can be less supportive than a teacher-led classroom and hard to use. Overall, the opportunity to design instructional models based on the availability of tutorial and practice technologies for mathematics is still in an early, exploratory phase, and it is too early to draw firm conclusions.

Extensions and Issues

While supporting students' practice of mathematical skills, technology can also be used to collect copious data about particular students and across students. When the problem format standardizes the task, interpreting data can be somewhat easier than in the case of open-ended tools or exploratory dynamic representations. This is presently drawing attention to big data approaches that support students' learning of mathematical skills, based on, for example, detecting when a student is frustrated (R. S. Baker, D'Mello, Rodrigo, & Graesser, 2010). More generally, technologies for practicing mathematical skills are rapidly becoming important test beds for research on educational data mining and learning analytics. Relative to the discussion of the feedback and task-selection loops, educational data mining (R. S. Baker & Yacef, 2009) can be seen as a long-term loop. In this loop, researchers analyze records of student progress to seek more optimal learning progressions, which may be potentially differentiated or personalized based on students' initial performance.

Learning analytics is a concept related to data mining, but it emphasizes definition of "detectors" that can analyze student keystrokes, mouse-clicks, and other inputs and induce higher level features relevant to student learning. In this research, automated detectors are trained via statistical, machine learning techniques (Calvo & D'Mello, 2010). Researchers have been able to build detectors for noncognitive behaviors, such as students who exhibit uncertainty, confusion, or disengagement. Being able to automatically detect emotional states while students practice mathematical skills may lead to interventions that better support learning. More recently, in the nascent field of multimodal learning analytics, researchers have started to employ new types of sensors and technologies, such as biosensors, eye-trackers, gesture trackers, speech recognition, and automated video analysis, to build multimodal detectors that take into consideration students' off-screen behaviors (Blikstein, 2013b; Worsley & Blikstein, 2013).

Another area of extension has been to practice math skills in social or collaborative environments. For example, in the TechPALS project, students worked in groups of three, and each member of a group had a handheld device. The devices communicated to distribute roles in a cohesive mathematical task among the three students, requiring them discuss the mathematics to make progress. In an experiment across three schools, students using TechPALS learned more than students in a comparable individual practice environment (Roschelle, Rafanan, et al., 2010). Research in computer-supported collaborative learning has also explored how to support teams in learning mathematics through online discussions (Stahl, 2009). Broadly speaking, there is a sound theoretical and empirical basis for collaborative learning, and technology can play a valuable role in orchestrating productive collaborative learning activities among students and in conjunction with the role of a teacher (Kirschner & Erkens, 2013).

While the cognitive science of optimizing practice is strong, a long standing challenge in this area (indeed, noted in Kaput, 1992) can be a narrow view of what mathematics is. Mathematics is not just the ability to perform skills fluently; it also includes conceptual understanding, strategic competency, adaptive reasoning, and productive dispositions (Kilpatrick, Swafford, & Findell, 2001). Reducing these additional dimensions to a cognitive skill acquisition framework limits the opportunities for students to progress along these additional dimensions of mathematical development. Potentially, blended approaches may overcome this objection by embedded technology-enhanced practice opportunities in a larger mathematics learning curriculum. Both in blended and nonblended models, the findings of mixed outcomes in impact studies as well as some of the difficulties observed in evaluations of blended learning point to challenges of implementation. Per Ruthven's (2014) framework, teachers need additional support to integrate practice systems into larger activity systems and curricular scripts. Often, the discussion about blended learning systems conflates the potentials for cost reduction and for quality improvement. Since those are two orthogonal dimensions, we believe that researchers and policy makers should treat them as independent directions for research and development. With both in mind, innovators can look for ways to make acquiring procedural fluency more efficient and ways to use technologies to raise the bar in the quality and richness of mathematical learning.

Technology for Conceptualizing Mathematics

Now we turn to technologies that are designed specifically for a pedagogical purpose—rather than as workplace tools—and that focus on students' conceptual

技能之上或之外的概念理解。在设计运用技术促进数学概念化的时候，往往会提升对认识论的关注（Hoyles, Noss, & Kent, 2004），从广义上讲，研究学习的科学家和数学教育研究人员将概念理解视为观念之间的关系或联系（T. P. Carpenter 等, 1997）。技术提供了显示、运作和观察这些关系的媒介（Heid & Blume, 2008）。

在学习理论方面，关注概念的数学技术往往与心智模型理论有关（Gentner & Stevens, 1983; Johnson-Laird, 1983），在这个过程中，人们在脑海中形成一个画面，在这个画面上，人们对一个表象进行操作，并预期其结果。在数学中，对表象的操作会被形式地定义（例如，控制代数变换和几何变换的法则）。在正确地应用形式化的数学规则时，学生经常会遇到困难，更困难的是难以形成关于法则如何起作用的直觉，但电脑可以准确地执行数学变换，同时计算机可以允许学生与表征的可变部分进行交互作用，并且可以立即显示变换的视觉结果。在数学学习中使用动态表征的基本思想是，通过让学生参与数学活动，使用直观的、可操作的、基于计算机的模型，这些模型能够正确地遵循形式规则，学生将能够更好地理解数学表征如何运作（Heid & Blume, 2008）。

使用这种技术来加深对概念的理解也是许多国家长期以来政策建议的一个重要特征。在美国，《州共同核心数学标准》（NGA & CCSSO, 2010）鼓励使用动态几何软件，就像早期的 NCTM 标准（2000）一样。在美国以外，地方和国家的决策者已经超越了文本建议的层面，开始明确将其付诸实施，在加拿大、新加坡、韩国、马来西亚、泰国等地已经有大规模的实践。这些建议遵循了许许多多评估教学实验和"动态几何环境"或"动态几何学软件"对学习影响的研究（King & Schattschneider, 1997; Laborde, Kynigos, Hollebrands & Strässer, 2006; Sinclair & Robuti, 2012）。这些研究既包括记录教师积极性的大型国家技术评估（例如 Becker, Ravitz, & Wong, 1999），也包括证明对学生表现影响的严格的随机对照试验（Jiang, White, & Rosenwasser, 2011）。

在之前的主题中，我们先考虑生产力，然后是变革，在这里，我们改变一下顺序。在大多数情况下，用于概念理解的技术设计者优先考虑变革目的，例如，让学生能够在更小的年纪，或更深入地，或在与其他数学思想建立更强的联系中学习数学概念。关于如何在典型的学校条件下有效地使用所产生的动态表征的研究通常会在之后进行。

变革：动态可以作为概念理解的一条新途径

在讨论动态数学时，动态是什么意思？卡普特（1992）将新出现的数学的"动态表征"与书面和印刷的、静态的"基于字符串"的数学符号形式进行对比，所谓"字符串"表征，他意指传统的被写成符号序列的代数表达式，如 $f(x)= 2x + 4$。他所指的动态，类似于在基于计算机的表征中，线性函数图像的斜率可以被视为动画中某一移动物体的速度。

对学生来说，从"动态的"运动角度看斜率，其简单的意义在于，它相较于在纸上写出同一函数，并有条不紊地通过取点、绘制图形会更加令人激动和吸引人。在这个意义上，"动态"与"数字""互动"和"多媒体"等其他一些通用的描述互相结合，突出了技术媒体的新可能性。在这种比较弱的动态意义下，观看关于数学函数的视频比阅读打印出来的有关数学函数的文本更具有动态感，虽然这听起来似乎有道理，也有吸引力，但这并不是驱使数学教育研究去探索动态表征的那种动态意义。

相比之下，我们需要一个关于动态的更为精准的意义，其中要将时间维度纳入学生具有明确数学意义的数学表征经验中，要求在为学生理解数学关系的新机会方面也有更为准确的构想。在这个基于研究的观点中，一个关键的问题就是，数学表征如何与学生的时间经验相联系，以提高对数学关系的理解？

动态几何。第一个探索如何将数学意义注入学生关于形式化的数学系统的体验中的并广泛使用的技术是动态几何，它是在20世纪80年代后期出现的，（关

understanding above or apart from routine skills. When designing technology for conceptualizing mathematics, attention to epistemology is often heightened (Hoyles, Noss, & Kent, 2004). Broadly, learning scientists and mathematics education researchers see conceptual understanding as the relationships or connections among ideas (T. P. Carpenter et al., 1997). Technology provides a medium for displaying, acting upon, and observing these relationships (Heid & Blume, 2008).

In connection to learning theory, conceptually focused mathematics technologies often relate to the theory of mental models (Gentner & Stevens, 1983; Johnson-Laird, 1983), in which people formulate a mental tableau on which to play out operations on a representation and anticipate the results. In mathematics, operations on representations are defined formally (e.g., rules govern algebraic and geometric transformations). Students often encounter difficulty correctly applying formal mathematical rules and even greater difficulty developing intuitions about how the rules work. Computers, however, can faithfully enact mathematical transformations. Further, a computer can allow students to interact with the changeable parts of the representation and can display the visual results of transformations immediately. The idea underlying the use of dynamic representations for mathematics learning is that by engaging students in mathematical activities with visual, manipulable computer-based models that correctly follow formal rules, the students will be able to develop stronger understandings of how mathematical representations work (Heid & Blume, 2008).

Use of such technology to enhance conceptual understanding has also been an important feature of policy recommendations in many countries for a long time. In the United States, the Common Core State Standards for Mathematics (NGA & CCSSO, 2010) encourage dynamic geometry software use, as did the early NCTM Standards (2000). Outside the United States, provincial and national policy makers have moved beyond recommendation to explicit implementation, with large-scale adoptions in Canada, Singapore, Korea, Malaysia, Thailand, and elsewhere. These recommendations follow numerous research studies evaluating teaching experiments and the learning impact of "dynamic geometry environments" or "dynamic geometry software" (King & Schattschneider, 1997; Laborde, Kynigos, Hollebrands, & Strässer, 2006; Sinclair & Robuti, 2012). These studies are as varied as large-scale national technology assessments documenting teacher enthusiasm (e.g., Becker, Ravitz, & Wong, 1999) and rigorous, randomized controlled trials demonstrating impact on student performance (Jiang, White, & Rosenwasser, 2011).

Whereas in the prior topics we considered productivity followed by transformation, we reverse the order here. For the most part, designers of technology for conceptual understanding have prioritized a transformation purpose—for example, to enable students to learn a mathematical concept at a younger age, or more deeply, or with greater connection to other mathematical ideas. Research on how to productively use the resulting dynamic representations in typical school conditions has usually come later.

Transformation: Dynamism as a New Path to Conceptual Understanding

What is meant by dynamism in a discussion of dynamic mathematics? Kaput (1992) juxtaposed newly emerging "dynamic representations" of mathematics and the written and printed, static, "character-string" based forms of mathematical notation. By "character-string" representations, he meant traditional algebraic expressions that are written as a sequence of symbols, such as $f(x) = 2x + 4$. By dynamic, he meant, for example, that in a computer-based representation, the slope of the graph of a linear function could be visualized as the speed of a moving object in an animation.

To a student, seeing the slope as movement can be "dynamic" in the simple sense of being more exciting and engaging than it is to write the same functions on paper and methodically plot their graphs, point by point. In this sense, "dynamic" joins other somewhat generic descriptors like "digital," "interactive," and "multimedia" in highlighting the new possibilities of technological media. In this weak sense of dynamic, watching a video about mathematical functions is more dynamic than reading a printed page about mathematical functions. While plausible and attractive, this is not the sense of dynamic that has driven mathematics education research to study dynamic representations.

In contrast, we argue for a more refined sense of dynamic where the incorporation of a time dimension into a student's experience of a mathematical representation has a clear mathematical meaning and a more precise formulation in terms of new opportunities for students to make sense of mathematical relationships. In this research-based view, a key question becomes, how is a mathematical representation being connected to a student's experience of time to advance understanding of mathematical relationships?

Dynamic Geometry. The first broadly used technology to explore how time could infuse mathematical meaning into a student experience of a formal mathematical

于动态几何教学上而不是技术上的叙述,参见这套书Sinclair, Cirillo, & de Villiers, 2017)。动态几何这一术语是由杰基和拉斯马森在1990年创造设计的(King & Schattschneider, 1997,第ix页),用以描述几何画板(Jackiw, 1989),但是很快,动态几何成为文献中对以Sketchpad和Cabri(Baulac, Bellemain & Laborde, 1988)为先驱的一类软件的总的描述。在这样的软件中,学生通过指定元素之间的数学关系,构建数学图像,如几何图形,但也构造图像、图表、平面图和其他不太规范的直观化表达,通常以定义的形式将新元素与现有元素联系起来。一旦构造了这样的图像,动态就进入画面,于是学生可以操纵视觉配置(例如,通过用鼠标拖动几何作图中的点),整个配置动态地、连续地变换到共享一组相同数学关系所定义的新配置。因此,视觉图像不再是对一组定义的一个可能实例的说明,而是可以动态转换成任何可能的实例,也就是所有可能的实例。这是一种具体的细节,通过它,学习者可以亲身参与到数学的抽象和一般化之中。

回到什么是动态这一宽泛问题上,动态几何环境的设计师将学生的时间经验和他们与几何图形的交互融合在一起,使学生能够从相对容易理解的单一实例过渡到更困难的理解一般性的挑战。将时间注入几何可以帮助学生理解比具体和一般(或具体和抽象的相关维度)关系更大的其他重要思想。当学生在一个几何作图过程中持续地变动一个点(通常是通过鼠标)并感受其他相关几何元素的变化时,变化和连续性的思想将会精密地融合。让学生轻松地参与探索变化反过来会凸显不变性——数学上相同的模式,而连续性通常立足于某种概念上的或数学上的兴趣的渐变,并且随着这种渐变而变化。

看一下三角形高的例子,它们交于一点(图31.3)。通过观察许多例子,例如,通过动态改变顶点,一个学生迅速意识到三条高交于垂心这一点,在某些构造中垂心落在三角形的内部,在某些构造中垂心落在三角形的外部。但是,在动态连续的方法中(与检查许多但不相关的单个三角形不同),垂心从三角形内部到外部之间交替时,一定有一些特定的过渡时刻,也就是说,有一扇"门",垂心在这里交替地进入或离开三角形。任何在动态几何学习中探索构造的学生都可以在几秒钟之内就在动态变化中找到这样的门,并且发现它们每次正好出现在当顶点动态地移动到三角形从锐角三角形变化到钝角三角形时。在这样的时刻出现的直角三角形及其对毕达哥拉斯定理(该定理包含了对偏向演绎分析的真知灼见)的引用,用一种动态的方法显化了自己,直角三角形不是在概念上独立于锐角三角形与钝角三角形的第三种情况,而是作为两个不同数学范畴之间的一个重要边界或平衡点。

以这种方式,动态几何引进了代数(变化)和实分析(连续性)这些强大的数学思想到几何的传统研究中,同时在教学方面,又将几何直观解释的概念价值,传播到广泛的数学课程,包括早期的数字(例如Sinclair & Crespo, 2006)、代数(Olive, 1998)、三角学(Shaffer, 1995)、微积分(Gorini, 1997)、线性代数(Meel & Hern, 2005)、群论(Schattschneider, 1997)、拓扑(Hawkins & Sinclair, 2007)和复分析(Jackiw, 2003)。

在过去30年中,动态几何已经获得了广泛的应用和影响力,Sketchpad和Cabri都开发了多个主要软件版本,其他软件包也已经将自己的动态功能应用于创新和强大的新的数学领域。班维尔和拉博尔德(2004)将Cabri扩展到三维坐标,将动态操作引入到Cabri 3D立体几何中。

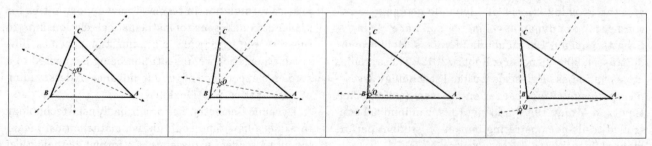

图31.3 当C被拖动到左边时,垂心O从B处离开△ABC。

system was dynamic geometry, which emerged in the late 1980s. (For a pedagogically—rather than technologically—focused account of dynamic geometry, see Sinclair, Cirillo, & de Villiers, 2017, this volume.) The term *dynamic geometry* was coined by Jackiw and Rasmussen in 1990 (King & Schattschneider, 1997, p. ix) to describe The Geometer's Sketchpad (Jackiw, 1989), but quickly entered the literature as a generic description of a class of software pioneered by Sketchpad and Cabri (Baulac, Bellemain & Laborde, 1988). In such software, students construct mathematical images—geometric figures, but also graphs, charts, plots, and other less canonical visualizations—by specifying mathematical relationships between elements, usually in the form of definitions relating new elements to existing ones. Dynamism enters the picture once such an image has been constructed: The student can then manipulate the visual configuration (e.g., by dragging a point in the geometric construction with a mouse), and the entire configuration dynamically and continuously transforms into new configurations that share the same defining set of mathematical relationships. The visual image is thus no longer an illustration of one possible instance of a set of definitions but rather, since dynamically transformable into *any possible* instance, equivalent to *all possible* instances. It is a concrete particular through which a learner can physically engage with the abstract and the mathematically general.

Returning to the broad question about what dynamism is, the designers of dynamic geometry environments integrate a student's experience of time and their interaction with geometric figures to enable the student to bridge from relative ease of understanding single instances to the more difficult challenge of understanding generalities. Infusing time into geometry can help students understand other big ideas beyond the relationships between the specific and general (or the related dimension of concrete and abstract). The ideas of variation and continuity critically combine as a student continuously varies a point (usually with the mouse) in a geometric construction and experiences the consequences for other related geometric elements. The ease of engaging students in exploring variation in turn makes invariance—the mathematical pattern of sameness—stand out, while continuity often establishes some sort of gradient of conceptual or mathematical interest along which to vary.

Consider the example of a triangle's altitudes, which coincide in a point (Figure 31.3). By looking at numerous examples—for instance, by dynamically varying a vertex—a student rapidly realizes that this point—the orthocenter—falls interior to the triangle in some configurations and exterior to it in others. But in a dynamically *continuous* approach (unlike, say, in the inspection of numerous but unrelated case-by-case triangles), an orthocenter that alternates between interior and exterior must have some specific moment of transition in between—that is, a "door" that leads the orthocenter alternately into or out of the triangle. Any student exploring the configuration in dynamic geometry can find such doors within seconds with dynamic variation—and discover that they occur at each vertex, exactly as that vertex passes dynamically from acute to obtuse. The right angle triangle appearing at such moments—and its invocation of Pythagoras, which contains the critical insight to a more deductive analysis—reveals itself in a dynamic approach not as a third case, conceptually separate from acute and obtuse, but instead as a crucial border or balance point between two different mathematical regimes.

In this way, dynamic geometry imports into the traditional study of geometry powerful mathematical ideas from algebra (variation) and real analysis (continuity), while at the same time it exports, pedagogically, the conceptual value of geometric visualization and explanation across a broad swath of mathematics curriculum, including early number (e.g., Sinclair & Crespo, 2006), algebra (Olive, 1998), trigonometry (Shaffer, 1995), calculus (Gorini, 1997), linear algebra (Meel & Hern, 2005), group theory (Schattschneider, 1997), topology (Hawkins & Sinclair, 2007), and complex analysis (Jackiw, 2003).

Over the past 30 years, dynamic geometry has enjoyed wide adoption and impact. Both Sketchpad and Cabri have evolved multiple major software versions, and other

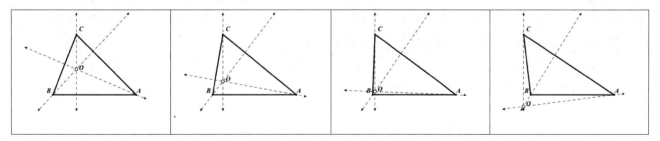

FIGURE 31.3. *The orthocenter O exits △ABC at B as C is dragged to the left.*

里希特-格伯特和科滕坎普（1999）在Cinderella中使用复齐次坐标来统一欧几里得几何和非欧几里得几何的动态方法。像GeoGebra这样的开源软件和维基百科上列出的50多个其他动态几何环境中的一些软件已经从这些系统中大量复制了动态操作，如果不是清楚地复制的话。

几何之外的动态表征。虽然Sketchpad和Cabri建立了流派，但动态表征并不限于几何内容。卡普特自己与SimCalc的工作重点是描述变化和变异的数学，以及传统的高中阶段的微积分课程如何通过对其基础表征作适当的改变，从而能够在小学阶段加以实施（Kaput & Roschelle，1998）。现在已经开发了可以用于探索数据的动态应用程序，例如Fathom和TinkerPlots（Konold, Harradine, & Kazak, 2007），以支持学生在数据分析中的推理。此外，诸如TI-Nspire的商业产品已经将动态表征纳入到比计算机更便宜的手持产品中，并且可能更容易在数学课堂中采用（Clark-Wilson, 2010）。此外，动态表征已经被开发出来，将通常被认为是高级的主题，复杂性纳入到更年少的学生的学习领域中（Jacobson & Wilensky, 2006）。与本章讨论的其他技术一样，我们不考察所有的可能性，而是关注这种类型的技术对数学教与学的核心贡献。因此，我们研究了MiGen项目（Noss等，2009），将其作为具有动态表征的较新研究和设计的例子，特别地，我们用这个例子来强调，研究学生在尝试学习一个困难的数学主题时所遇到的挑战和动态表征的设计之间的相互作用，以使新的通向该主题的方式成为可能。因此，我们的目的是表明动态表征不仅是技术的和数学的，而且在认识论上也是有战略意义的：旨在创造出新的方法来认识以前难以学习的数学概念。

将MiGen作为认识论的设计。MiGen在引入数学一般化的过程中利用了技术的动态潜力（Mavrikis, Noss, Hoyles, & Geraniou, 2013）。众所周知，一般化对于新手来说很难，他们在掌握未知数和变量的作用方面有相当大的困难，可怕的"x"，往往（但不总是！）代表无穷多的数值。（Noss等，2009给出了基本原理和设计的细节）。目前，我们专注于动态的三个方面，它们形成了设计成果的基础，每个方面都针对一个学习困难。

首先，我们考虑动态的问题表征的想法。MiGen团队选择了一类数学问题作为概括的目标，这些问题涉及计数平铺图案中的瓷砖数量。选择平铺问题是因为这些是容易理解且具有吸引力的工作，并且因为一般化的问题容易在这种情况下进行表述，以一种直接涉及学生知识中有问题的方面的方式表达出来。

MiGen团队发现，当学生被要求概括出平铺图案有多少个瓷砖时（当图案被扩大或缩小以填补更多或更少的空间时），有一种压倒性的趋势就是只基于数字的计数和"概括"，没有考虑到图案的结构。通过将原始问题呈现为动态或动画序列，计数的可能性大大降低，学习者必须通过识别结构来应对挑战，毕竟对一个移动的目标进行计数很难！因此，动态的方法被用来防止学生陷入糟糕的策略，也是为了呈现有规律的变化，但这次是跨越离散量而不是连续量（图31.4）。

动态的第二个用途涉及MiGen的一般作图工具的想法。虽然有很多方法来构建平铺图案的一个实例，但只有使用适当的作图工具创建的实例才能避免在出现"混乱"的问题（Healy, Hoelzl, Hoyles, & Noss, 1994）。混乱一词抓住了这样一个思想，即除非结构之间的关系是明确和正确的，否则只要改变变量的值就可以破坏它们。通过区分可能被混淆和不能被混淆的图案，设计旨在鼓励学生从总体上思考，考虑在作图阶段和之后一个变量（或特定情况下表征一个变量）可以取值的无限范围。

动态的第三个用途是同时建立一般情况与具体情况。当学生在处理具体情况时，计算机会构造出一个一般情况，其中所选参数的值是随机分配的。这个与My Model一起发展的计算机模型允许学习者在处理具体情况的同时留意一般情况，从而参与一个有效地产生解决方案并与数学家的实际工作产生共鸣的策略。

最终，这三种动态的使用结合在一起，使学生们努

software packages have applied their own dynamic behaviors to innovative and powerful new mathematical domains. Bainville and Laborde (2004) extended Cabri to three-dimensional coordinates, which brought dynamic manipulation to solid geometry in Cabri 3D; Richter-Gebert and Kortenkamp (1999) used complex homogeneous coordinates in Cinderella to unify dynamic approaches to Euclidean and non-Euclidean geometries. Open source software like GeoGebra and some of the more than 50 other dynamic geometry environments listed on Wikipedia have copied liberally, if not always coherently, from these systems.

Dynamic representations beyond geometry. While Sketchpad and Cabri established the genre, dynamic representation is not limited to geometric content. Kaput's own work with SimCalc focused on the mathematics of change and variation and how material from the traditional upper-secondary calculus curriculum became accessible at the primary level with a suitable change in representation infrastructure (Kaput & Roschelle, 1998). Dynamic applications for exploring data, such as Fathom and TinkerPlots (Konold, Harradine, & Kazak, 2007) have been developed to support student reasoning in data analysis. Further, commercial products such as TI-Nspire have incorporated dynamic representations into handheld products that are more affordable than a computer and potentially easier to adopt in mathematics classrooms (Clark-Wilson, 2010). In addition, dynamic representations have been developed to bring topics that are normally considered advanced, such as complexity, into the reach of younger students (Jacobson & Wilensky, 2006). As with other technologies discussed in this chapter, we do not survey all the possibilities, but rather focus on the core contribution of this genre of technology to teaching and learning mathematics. In that light, we examine the MiGen project (Noss et al., 2009) as an example of newer research and design with dynamic representations. In particular, we use the example to highlight the interplay between research on the challenges that students experience while trying to learn a difficult mathematical topic and the design of dynamic representations to enable new means of access to the topic. Thus, we aim to demonstrate that dynamic representations are not only technological and mathematical, but also strategically epistemological: Aimed at enabling new ways of coming to know mathematical concepts that were formerly difficult to learn.

MiGen as an epistemological design. MiGen exploits the dynamic potential of technology in the process of introducing mathematical generalization (Mavrikis, Noss, Hoyles, & Geraniou, 2013). Generalization is notoriously difficult for novices, who have considerable trouble grasping the role of unknowns and variables—the dreaded "x"—which often (but not always!) represent an infinity of values. (Details of the rationale and design are given in Noss et al., 2009.) For the moment, we focus on three aspects of dynamism that form the bedrock of the design effort, each of which addresses a learning difficulty.

First, we consider the idea of dynamic problem representation. The MiGen team chose a class of mathematics problems as the target for generalization. These problems involve counting the number of tiles in tiling patterns. Tiling problems were chosen because these are easily understood and attractive to work with and because the problem of generalization can be expressed naturally in this context—and in a way that directly engages the aspects of students' knowledge that are problematic.

The MiGen team found that when students are asked to generalize how many tiles are involved in a tiling pattern (as the pattern is expanded or contracted to fill more or less space), there is an overwhelming tendency simply to count and to "generalize" on a numerical basis without taking account of the structure of the pattern. By presenting the original problem as a dynamic or animated sequence, the possibility of counting is much less attractive, and the learner has to confront the challenge by recognizing structure—counting a moving target is hard! Hence dynamism was used to prevent students from getting stuck with a poor strategy—again for principled variation, but this time across discrete quantities rather than continuous ones (Figure 31.4).

A second use of dynamism involves MiGen's idea of general construction tools. While there are many ways to construct one instance of a tiling pattern, only those created using the appropriate construction tools avoid the problem of "messing up" (Healy, Hoelzl, Hoyles, & Noss, 1994) when the pattern is animated. The term *messing up* captures the idea that unless the relationships between structures are explicit and correct, they can be disrupted just by changing a variable's value. By distinguishing between patterns that can and cannot be messed up, the design aims to encourage students to think generally: to consider the infinite range of values that a variable (or a specific case representing a variable) can take during and after the construction phases.

A third exploitation of dynamism is the simultaneous construction of the general case alongside the specific. As the student works on the specific case, the computer constructs a general case where the values of the chosen parameters are randomly assigned. This Computer's Model, which evolves alongside My Model, allows learners to keep an eye on the general as they works on the

力了解变量概念在代数抽象中的作用。令人遗憾的是，目前的课堂教学视变量和代数为形式化运算中的练习，没有过多提及代数的其中一个目的是表达模式的一般化结构的思想。在MiGen中，动态表征的变革作用是提供一种工具，将一个基本的，但被忽略的代数教学目的引入教学。

专业发展中的动态表征。动态表征也可用于改变教师专业发展（PD）。Bridging PD项目的目标是帮助教师去鼓励低收入学校的学生参与数学论证（Knudsen, Vahey, Lara-Meloy & Shechtman, 2014）。虽然数学论证现在被认为是对所有学生的期望（见《州共同核心数学标准》，NGA & CCSSO，2010，第6—7页），但数学论证在为低收入学生提供的课堂中是相当罕见的。

Bridging PD项目主要是教师专业发展的计划，但我们在此讨论它，是因为它出于两个原因包含了动态表征（Stevens & Lara-Meloy, 2014）。首先，研究人员了解到，为了支持论证，教师必须深化他们自己的概念理解，用动态表征来探究数学可以是一个很好的教师活动，正如动态表征对于学生对概念理解的发展而言是丰富有效的，所以他们也可以为教师发展数学理解提供全

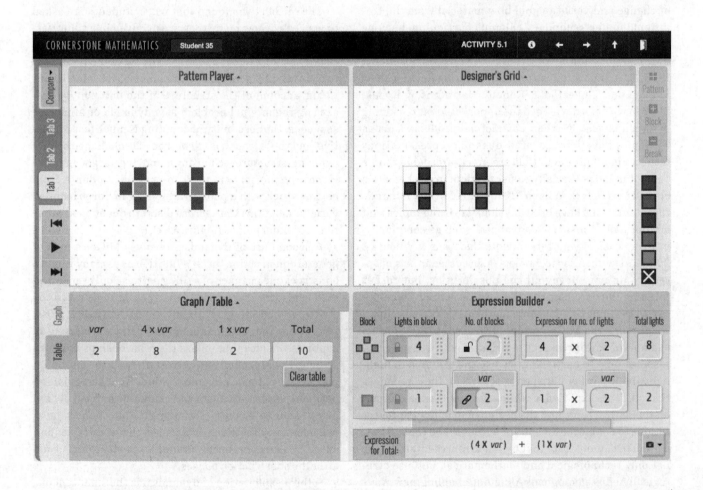

图31.4　截屏于基础数学项目。右上方区域是设计师的网格，学生在这里建立一个灯光模式（类似于MiGen的"瓷砖"）。右下方区域显示每种颜色灯光的总数和整体总数的表达式。左上方区域是模式播放器（替代MiGen的计算机模型），其中计算机选择要显示的构建块数量，从而"概括"设计师对变量值的选择，如果错误的话，模式会被弄乱。左下方区域显示了对所选的灯数，可变的方块数目的值。

particular, thus participating in a strategy that is both effective in producing a solution and resonant with what mathematicians actually do.

Ultimately, these three uses of dynamism work together to engage students in striving to understand the role that a concept of variable plays in algebraic generalization. Sadly, much current classroom instruction treats variables and algebra as exercises in formal manipulation, without much reference to the notion that one purpose of algebra is to express the general structure of patterns. In MiGen, the transformative role of dynamic representations is to provide a vehicle to bring a fundamental but neglected purpose of algebra into instruction.

Dynamic representations in professional development. Dynamic representation can also be used to transform teacher professional development (PD). The goal of the Bridging PD program has been to prepare teachers to engage students at lower-income schools in mathematical argumentation (Knudsen, Vahey, Lara-Meloy, & Shechtman, 2014). Although mathematical argumentation is now considered an expectation for all students (i.e., in the Common Core State Standards, NGA & CCSSO, 2010, pp. 6–7), mathematical argumentation is quite rare in classrooms serving lower income students.

Bridging PD is primarily a program of teacher professional development, but we discuss it here because it incorporates dynamic representations for two reasons (Stevens & Lara-Meloy, 2014). First, the researchers understand that for teachers to support argumentation, they have to deepen their own conceptual understanding, and investigating mathematics with dynamic representations can be a sound teacher activity. Just as dynamic representations are fertile for students' development of conceptual understanding, so can they be fertile for

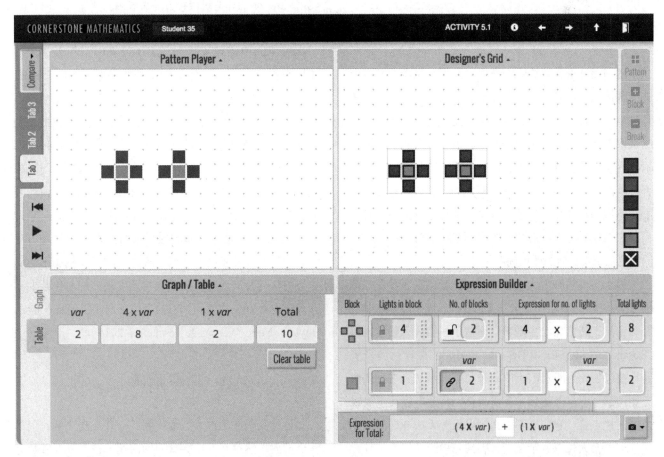

FIGURE 31.4. *Screen shot from the Cornerstone Mathematics project. The top-right quadrant is the Designer's Grid where the student builds a pattern of lights (similar to MiGen's "tiles"). The bottom-right quadrant shows the total number of lights for each color and an expression for the total overall. The top-left quadrant is the Pattern Player (replacing MiGen's Computer Model), in which the computer chooses how many building blocks to display and thus "generalizes" the designer's choice of variable value—if she is wrong, the pattern will be messed up. The bottom-left quadrant shows the number of lights for chosen values of the variable number of blocks.*

面丰富的基础。第二，研究人员意识到，现有的许多课堂任务并不能激发学生阐明理由、提出看法、明白对方的想法，例如，很多课堂练习单中几乎没有什么可争论的。使用动态表征进行探究可以给学生一个数学结构或活动进行调查，以便他们提出自己的猜想和解释，从而有一些东西可以争论（Knudsen, Lara-Meloy, Stevens, & Shechtman, 2014）。"Bridging PD"项目研究发现，参加这个专业发展计划的教师的课堂上，论证的质量和数量都有所增加（Shechtman, Knudsen, Michalchik, Kim, & Stevens, 2015），这完成了一件非常罕见的事情：通过专业发展来改变低收入学生的课堂，把重点放在论证上。因此，Bridging PD项目提出动态表征可以为教师职业发展提供一种具有变革意图的有效的资源。

生产力：课程活动系统

动态表征通常首先在有意愿的早期采用者的教室中进行测试，志愿者老师们往往坚定地致力于概念性理解和对技术的冒险，即使这种方法可能不会导致在标准化测试中取得最高的考试成绩，他们也会投入测试这种技术。然而，随着时间的推移，一些研究团队决定调查是否可以由更广泛的教师群体来实施动态表征，包括那些不是典型的早期采用者的教师，以及那些更关心现有课程期望的教师。这种调查需要平衡动态表征的变革意图，同时更加重视目前的课程进度和问责制度。

为了实现这一更大的平衡，创新者侧重于开发"课程活动系统"，集软件、工作簿、教师指南、教师专业发展和专注于特定概念主题教学评估于一体的完整集成包（Roschelle, Knudsen & Hegedus, 2010）。课程活动系统通常被包装成为期1至3周的课堂教学模块，旨在强调概念性理解目标以及进度和问责目标。将动态表征包装成模块的优点是，它们既不那么短以至于无法开展深刻学习，也不会太长以致于让老师感到是中断他们教学日程的一大障碍。此外，通过将技术、教学资源、教师专业发展和评估的紧密配合，可以减轻教师在技术和其他教学资源之间寻找匹配的负担。

在Scaling Up SimCalc项目（Roschelle, Shechtman等，2010）的一系列实验中，动态表征和这种课程活动系统方法进行了大规模的测试。在这个项目中，他们从得克萨斯州学校招聘了大量教师，并随机分配他们使用标准教学材料或使用SimCalc课程活动系统方法教授某一特定课题，为期2—3周。学生们参加了前测和后测，测试旨在衡量得克萨斯州评估的内容，以及更符合附加概念理解目标的国际测试的内容。在与得克萨斯州州级考试相对应的评估量表中，得分略高（表示生产力没有下降），在与概念性理解和数学推理相对应的量表上，教师使用SimCalc的那些班级得分显著较高。

使用SimCalc以及其他动态表征的类似结果在进一步的研究中得到了重现，上述报告中七年级学生的结果与得克萨斯州八年级学生的结果相同（Roschelle, Shechtman等，2010）。佛罗里达的研究人员（Vahey, Roy, & Fueyo, 2013）将这种方法扩展到几何领域，也得到了类似的结果。英国的基础数学项目（Hoyles等，2013）也发现了类似的结果，他们还将计划作了扩展，包括与上面讨论的MiGen项目类似的Simcalc相关的表征、几何表征和与一般化相关的表征。在英国，超过125所学校正在使用由基础数学项目开发的用这种方法编制的材料。除了结果表明这种方法在大规模实施时对学生的学习是有效的，研究人员还发现，精心包装的动态表征也可以成为教师学习的有用资源（Clarke-Wilson, Hoyles, Noss, Vahey, & Roschelle, 2014）。

扩展和问题

虽然我们在生产力方面提到了一些成功之例，但是整个世界仍然存在一个更普遍的问题，在不改变教学方

teachers' development of mathematical understanding. Second, researchers realized that many existing classroom tasks do not stimulate students to articulate reasons, take positions, and make sense of each other's understandings. With many classroom worksheets, for example, there is very little to argue about. Investigations with dynamic representations can give students a mathematical structure or activity to investigate so that they come up with their own conjectures and explanations and thus have something to argue about (Knudsen, Lara-Meloy, Stevens, & Shechtman, 2014). Research with Bridging PD has found an increase in the quality and quantity of argumentation in the classrooms of teachers who have participated in this PD program (Shechtman, Knudsen, Michalchik, Kim, & Stevens, 2015) This accomplishes something that is quite rare: transforming lower-income classrooms through professional development to include a focus on argumentation. Hence, Bridging PD suggests that dynamic representations can provide an effective resource for teacher professional development with a transformative intent.

Productivity: Curricular Activity Systems

Dynamic representations are often first tested in classrooms with willing early adopters. Teachers who volunteer are often strongly committed to conceptual understanding and adventuresome with regards to technology and would commit to testing this technology even if this approach might not result in the highest test scores on standardized tests. Over time, however, some research teams decided to investigate whether dynamic representations could be implemented by a broader set of teachers, including teachers who are not classic early adopters and who are more concerned with existing curricular expectations. Such investigations have required balancing the transformational intent of dynamic representations with greater attention to current curriculum pacing and accountability regimes.

To achieve this greater balance, innovators have focused on developing "curricular activity systems"—complete, integrated packages of software, workbooks, teacher guides, teacher professional development, and assessments focused on teaching a particular conceptual topic (Roschelle, Knudsen, & Hegedus, 2010). Curricular activity systems have often been packaged as modules of classroom instruction lasting 1 to 3 weeks, intended to address both conceptual understanding goals as well as pacing and accountability objectives. An advantage of packaging dynamic representations into modules is that these are neither so short as to have little chance of deep learning impact nor so long as to be experienced by teachers as a major disruption to their teaching schedule. Further, by aligning technology, instructional resources, teacher professional development, and assessment in a tight package, teachers are relieved of the burden of figuring out the fit among technologies and other instructional resources.

Dynamic representations and this curricular activity system approach were tested at scale in a series of experiments under the Scaling Up SimCalc project (Roschelle, Shechtman, et al., 2010). In this project, a large number of teachers from schools in Texas were recruited and randomly assigned to use either their standard instructional materials or a SimCalc-based curricular activity system approach for a 2–3 week period while teaching a particular topic. Students took a pretest and a posttest, and the tests were designed to measure both the content on the Texas state assessment as well as content from international tests that was more consistent with additional conceptual understanding goals. On the assessment scale corresponding to the Texas state test, scores were marginally higher (indicating that no productivity was lost). On scales corresponding to conceptual understanding and mathematical reasoning, classrooms whose teachers used SimCalc scored significantly higher.

Similar results using SimCalc as well as other dynamic representations have been replicated in further research. The results reported above for seventh grade students were replicated with Texas students in eighth grade (Roschelle, Shechtman, et al., 2010). Researchers in Florida (Vahey, Roy, & Fueyo, 2013) extended the approach to geometry with similar results. The Cornerstone Mathematics program (Hoyles et al., 2013) in the United Kingdom found similar results, and the program there expanded to include SimCalc-related representations, geometry representations, and generalization-related representations comparable to the MiGen project discussed above. Over 125 schools in the United Kingdom are now using materials developed by the Cornerstone Mathematics program using this approach. In addition to results showing this approach can be productive for student learning when implemented at scale, researchers have also found that well-packaged dynamic representations can be a useful resource for teacher learning (Clarke-Wilson, Hoyles, Noss, Vahey, & Roschelle, 2014).

Extensions and Issues

Although we cited some successes with regard to productivity, a more general issue remains in the world at large. It is increasingly common to sprinkle "interactives"

法以支持学生发展概念性理解的情况下，将"互动"带到其他传统教学材料中的做法越来越普遍。例如，教师有时会使用动态表征来"说明"数学思想，但这种方式对于让学生参与到理解数学的过程中并没有什么帮助，在交互式对象中仅使用时间维度可以提高学生的兴趣，而不会增加他们对数学关系的概念性理解。

我们还提请注意三个扩展方向，它们似乎是未来研究的沃土。首先，卡普特（1992）认为动态性表现在时间上，后来与其他同事的研究则认为变异要通过社会安排来表达。例如，在一些NetLogo（Stroup & Wilensky, 2014）活动中，每个学生都控制一个点，所有的点一起参与了一个更大的数学结构。同样，在一些相互联系的SimCalc活动中，每个学生控制一个线性函数的斜率，所有的函数一起形成了一个函数族（所有函数截距相同，但是斜率不同）。动态表征研究的核心与建构主义和心理模型理论相结合，而动态表征使用中的社会转向则与强调合作在学习中的作用的学习理论以及更一般的社会建构主义相联系（Goodyear, Jones, & Thompson, 2014）。

第二，用于学习的游戏是一个受到很多关注的领域（例如Squire, 2011），游戏和动态表征之间的可能连接值得进一步调查研究。通常的游戏化例子（Hamari, Koivisto, & Sarsa, 2014），引入了诸如积分、故事情节和排行榜等游戏元素，以吸引学生对那些不感兴趣的数学问题的注意。有一个名为"数据游戏"的项目采取了不同的方向，"数据游戏"（Finzer, 2013）开始于容易识别但很简单的游戏，因为太简单，它们将很快变成无聊的独立游戏。例如，一个数据游戏使用了简单的"寻找隐藏的宝藏"故事背景，这在许多基本游戏中很常见。这些游戏提供了一种立即可识别的但人为约束的情境，在这个情境下，能让学生们很好地认识到自己该做什么。

在每个游戏中，"数据游戏"团队都强调挑战的数学，因此，理解数学成为吸引和维持学习者兴趣的主要问题。在上述"寻找隐藏的宝藏"游戏中，具有挑战性的数学是用于感知宝藏可能位置的测量仪器存在不确定的误差，在一组不确定的数据点之间发现中心趋势是一个重大的挑战。尽管许多产品试图将常规数学练习"游戏化"，但"数据游戏"展示了一个引人注目的替代方案，将常规游戏"数学化"的过程会驱使学生学习。

第三，很明显，数学表征中的动态部分反映了技术投入和产出的进步。早期Macintosh计算机的二维位图图形屏幕输出设备和二维基于鼠标的输入设备为直接和动态地访问二维欧几里得平面创造了机会。动态几何软件（此后不久就出现）可以被看作是对下面这个问题的创造性回应："当鼠标出现时，传统的三角形会发生什么变化？"现在，用户界面的范式正在转向关注多点触控（即，设备可以同时识别用户在两个或多个点触摸屏幕，并且可以识别类似于使用多个手指的捏合等之类的手势）。多点触控改变了互动的本质，为不强调连续性的动态新形式创造了可能性，一个例子是一个名为TouchCounts的iPad应用程序，它是针对正在学习计数的儿童的，TouchCounts（Sinclair, 2011）能够进行互动，学生可以用一个或多个手指触摸屏幕，并听到所呼叫的数的名称。因此，用三根手指触摸会听到"三"，这种交互作用是不可能用鼠标实现的，鼠标不能识别手指的数量，这种交互确实在三个手指中"三"的化身和触摸屏幕听到一个数字的行为之间建立了重要的联系。此外，TouchCounts使学生可以根据多点触控手势继续进行计数。组成数字10的一种方法是用三根手指，然后三根手指，然后两根手指，然后两根手指触摸，另一种方法是同时触摸所有10个手指。我们期望，随着超越键盘和鼠标的用户体验变得越来越普遍，动态数学的可能性将继续发展。

技术作为兴趣驱动数学学习的背景

数学教育一个越来越重要的领域是我们所说的"兴趣驱动的数学学习"。兴趣驱动活动的特点是自愿参加课外活动，或博物馆，或在K-12学校中开展相对开放的科

into otherwise traditional instructional materials without changing the pedagogy to support students' development of conceptual understanding. For example, dynamic representations are sometimes used by teachers to "illustrate" mathematical ideas but in a way that shows little promise for engaging students in the process of making sense of mathematics. The mere use of a time dimension in an interactive object may increase students' interest without enhancing conceptual understanding of mathematical relationships.

We also call attention to three extensions that seem to be fertile grounds for future research. First, whereas Kaput (1992) considers dynamism as expressed in time, later work with additional colleagues considered variation as expressed through social arrangements. For example, in some NetLogo (Stroup & Wilensky, 2014) activities, each student controls one point, and all the points together participate in a larger mathematical structure. Likewise, in some connected SimCalc activities, each student controls the slope of one linear function, and all the functions together form a family of functions (all with the same intercept but different slopes). Whereas the core of dynamic representations research connects with theories of constructivism and mental models, a social turn in the use of dynamic representations connects with learning theories that emphasize the role of collaboration in learning, and social constructivism more generally (Goodyear, Jones, & Thompson, 2014).

Second, games for learning is an area receiving much attention (e.g., Squire, 2011), and the possible connections between games and dynamic representations is worthy of further investigation. In commonplace examples of gamification (Hamari, Koivisto, & Sarsa, 2014), gaming elements such as points, narratives, and leader boards are introduced to motivate student attention to otherwise uninteresting mathematics problems. A project called Data Games takes a different direction. Data Games (Finzer, 2013) begins with games that are recognizable but simple—so simple that they would rapidly become boring as stand-alone games. One data game, for example, uses a simple "search for hidden treasure" metaphor, common to many basic games. These games are useful in providing an immediately recognizable, yet artificially constrained, context in which students have a good sense of what to do.

Within each game, the Data Games team created an emphasis on challenging mathematics so that coming to understand the mathematics becomes the dominant issue that grabs and sustains the learner's interest. Thus, in the "search for hidden treasure" game mentioned above, the challenging mathematics becomes the uncertain variance in the measurement instruments used to sense the possible locations of treasure. Finding central tendencies among a cluster of uncertain data points becomes the focal challenge. Whereas many products attempt to "gamify" routine mathematical exercises, Data Games illustrates a compelling alternative in which the process of "mathifying" a routine game drives student learning.

Third, it is clear that dynamism in mathematical representations is in part a reflection of advances in technological input and output. The 2-dimensional bitmapped graphics screen output device and the 2-dimensional mouse-based input device of early Macintosh computers created the opportunities for direct and dynamic mathematical access to the 2-dimensional Euclidean plane. Dynamic geometry software (which emerged shortly thereafter) can be seen as a creative response to the question, "What happens to the traditional triangle in the presence of a mouse?" Now the user interface paradigm is shifting to focus on multitouch (i.e., the device can recognize the user touching the screen in two or more points at once and can recognize gestures like pinching, that use multiple fingers). Multitouch shifts the nature of interaction, enabling possibilities for new forms of dynamism that do not emphasize continuity. One example is an iPad app called TouchCounts, which is aimed at young children who are learning to count. TouchCounts (Sinclair, 2011) enables an interaction whereby students touch one or more fingers on the screen and hear the name of the number called out. Thus, touching with three fingers results in hearing "three." This interaction is not possible with a mouse, which cannot recognize the number of fingers down—but it does make an important connection between the embodiment of "three" in three fingers and the act of touching a screen to hear a number. Further, TouchCounts enables students to count up according to multitouch gestures. One way to make the number 10 is to touch with three fingers, then three fingers, then two fingers, and then two fingers. Another way is to touch simultaneously with all 10 fingers. We expect that the possibilities of dynamic mathematics will continue to evolve as user experiences beyond keyboard and mouse become increasingly common.

Technology as a Context for Interest-Driven Mathematics

An increasingly significant area of mathematics education is what we will refer to as "interest-driven mathematical learning." Interest-driven activities are characterized by either voluntary participation in after-

学和工程学方面的活动。这些活动很少被明确地确定为数学问题，但可能包含丰富的数学学习机会（Blikstein，2013a）。我们预计，这种类型的活动对于数学教育将变得越来越重要，由于（a）科学博物馆、课后以及暑期课程的激增；（b）"创客运动"和FabLabs在学校的迅速发展；和（c）数字制造设备（3D打印机、激光切割机、乙烯切割机、铣床）和诸如Arduino（www.arduino.cc）、Raspberry Pi（www.raspberrypi.org）、Lego Mindstorms/NXT、GoGo Board（gogoboard.org）和LilyPad（lilypadarduino.org）的物理计算设备的成本快速下降。这个趋势是重要的，因为它也突出了数学教育在过去已经发挥了重要作用（并将继续发挥重要作用）的两个要素，它们是：（1）计算机编程，（2）有形物体的构建。

在创客空间和FabLabs中学习数学

"创客"空间中的学生有什么类型的数学学习机会？设想一个在数字制造实验室的学生建造了一个自动化的花盆，当土壤干燥时，花盆能自动灌溉，并考虑在项目的整个开发过程中，学生可能会遇到几个需要数学的步骤。为了设计和3D打印花盆，学生可以测量现有的花盆并进行调整以弥补材料的改变（例如从粘土变到3D打印聚合物）。然后，她可以利用对几何体（球、棱柱及其衍生物）的理解，并执行一系列复杂的3D布尔加法/减法运算，以在3D设计软件中创建对象。同时也可能会需要其他的考虑：材料的变化是否会移动重心？悬挂时花盆会变形吗？她如何可以把它做得尽可能的轻，但也足够结实？

如果花盆的建造是以传统的手工方式（即手工成型粘土）进行的，那么将这个过程数学化的机会根本就不存在，这个过程将会更具有试探性，更易于制定解决方案，而测量、设计、优化和预测的机会就更小了。

这种"无脚本"学习（Collins & Halverson，2009）的一个额外的好处是激励性的，因为当学生真正需要数学思想和方法来完成一个世纪任务时，才向他们介绍这些经验（Papert，1980），于是引起高度的参与、主动性和目的感。但它也需要新的，可能更为复杂的课堂（或学习环境）动态、编排（Dillenbourg，2013）和课程设计。一个常见的误解是，在这种学习环境中"什么都可以"，然而，仅仅让学生在实验室里随意地使用新奇的工具，不一定会促进学习和参与性，甚至可能会吓跑那些不太认同这些工具和实践的缺乏代表性的少数人群。

深入研究一个例子可以说明个中的机遇和挑战。布利克斯泰英（2013a）描述了一个案例研究，其中历史和数学教师与制造实验室协调员合作，创建了一个关于美国历史上女性人物的单元，其主要任务是为一位重要的女性历史人物建立一个公共广场或纪念碑的模型，这些模型可以有雕像、长凳、树木、楼梯和草坪，学生可以使用手工材料、激光切割机和3D打印机。然而，教师不是创建一个完全开放的项目，而是引进了结构：历史纪念碑的木质底座是标准化的（15×15平方规定尺度的方格，见图31.5左），该项目的一个关键规则是所有东西都必须按比例建造。尽管这个活动最初是关于历史的，但它最终变得更多的是关于数学的。当教学团队把底座标准化并强化了比例（明确的设计决策）时，通过设计，他们强调了测量、比例以及二维和三维几何。学生们意识到，所有元素的相对尺寸必须是精确的，它们的尺寸必须是一致的。因为学生深深地投入到自己的项目中，强制比例的想法也是合理的，不是人为的约束，所以，学习和使用数学不仅是自然的而且最终被大多数学生认为是完成项目最快、最优的方式。

从这个例子，我们可以看到其他兴趣驱动的数学项目也可能具有的属性。一方面，创客空间和FabLabs为学习和使用数学提供了非常丰富的环境，因为与传统的"模拟"动手活动相反，数字设计和制作凸显了数学概念的有用性。另一方面，把这些益处变为现实将需要大量的工作和研究，例如，这些学习环境的无剧本、基于项目的性质可能需要对数学课程进行彻底的反思，我们

school programs or museums or relatively open-ended experiences in science and engineering in K–12 schools. These activities are rarely explicitly identified as mathematical but may contain rich mathematical learning opportunities (Blikstein, 2013a). We foresee that this type of activity will be increasingly important for mathematics education given (a) the proliferation of science museums and after school and summer programs; (b) the rapid growth of the "maker movement" and FabLabs in schools; and (c) the fast decrease in the cost of digital fabrication machines (3D printers, laser cutters, vinyl cutters, milling machines) and physical computing devices such as Arduino (www.arduino.cc), Raspberry Pi (www.raspberrypi.org), Lego Mindstorms/NXT, GoGo Board (gogoboard.org), and the LilyPad (lilypadarduino.org). This trend is significant because it also highlights two elements of mathematics education that have had important roles in the past (and continue to be relevant): (1) computer programming and (2) the construction of tangible objects.

Learning Mathematics in Makerspaces and FabLabs

What types of mathematical learning opportunities are there for students in "maker" spaces? Consider a student in a digital fabrication lab building an automated flowerpot that waters itself when the dirt is dry, and consider how throughout the development of the project the student might come across several steps that will require mathematics. To design and 3D print the flowerpot, the student may measure an existing flowerpot and make adaptations to compensate for the change in material (e.g., clay to 3D printing polymer). Then, she may apply an understanding of geometrical solids (spheres, prisms, and their derivatives) and perform a series of complex 3D Boolean addition/subtraction operations to create the object in 3D design software. Other considerations might arise: Did the change in material move the center of gravity? Will the flowerpot deform when hung? How can she make it as light as possible but also strong enough?

If the construction of the flowerpot was to happen in a traditional hands-on fashion (i.e., manually shaping clay), opportunities to mathematize the process would simply not be there—the process would be much more tentative and amenable to make-do solutions, with fewer opportunities to measure, design, optimize, and predict.

One additional benefit of this type of "unscripted" learning (Collins & Halverson, 2009) is motivational, since these experiences introduce students to mathematical ideas and procedures when they actually need them to accomplish a real task (Papert, 1980), thus promoting high engagement, agency, and sense of purpose. But it also requires new—and possibly more complex—forms of classroom (or learning environment) dynamics, orchestration (Dillenbourg, 2013), and curricular design. One common misconception is that in these kind of learning environments, "anything goes." Merely letting students loose in a lab with fancy tools will not necessarily promote learning or engagement. It might even scare away underrepresented populations that identify less with these tools and practices.

Looking at one example in depth can illuminate the opportunities and challenges. Blikstein (2013a) describes a case study in which history and math teachers teamed up with the fabrication lab coordinator to create a unit on female figures in American history. The main task was to build a prototype for a public square or monument for one important historic female character. These prototypes could have statues, benches, trees, stairs, and grass, and students could use craft materials, laser cutters, and 3D printers. However, instead of creating a completely open-ended project, the teachers introduced structure: The wooden base of the historical monuments was standardized (a 15-by-15-square grid with defined scale, see Figure 31.5, left). One key rule of the project was that everything had to be built to scale. Even though the activity was initially about history, it eventually became much more about mathematics. When the teaching team standardized the base and enforced scale (an explicit design decision), they emphasized, by design, measurement, proportion, and geometry, both in two and three dimensions. Students realized that the relative sizes of all elements had to be exact, and their dimensions consistent. Because students were deeply invested in their projects, and the idea of enforcing scale was reasonable and not an artificial constraint, learning and using mathematics was not only natural but ended up being regarded by most students as the fastest and most optimal way to get the project done.

From this example, we can see attributes that other interest-driven mathematics projects are also likely to have. On the one hand, makerspaces and FabLabs offer an extremely rich context for learning and using mathematics, since digital design and fabrication—as opposed to traditional, "analog" hands-on activities—foreground the usefulness of mathematical concepts. On the other hand, bringing those benefits to life will require considerable work and research. For example, the unscripted, project-based nature of those learning environments might require a radical rethinking of what a mathematics curriculum looks like—we would not be able to predict when or in which order students would come across

【第870页】

图31.5 公共广场项目的模型。一个正在进行中的项目,网格清晰可见(左上);以及学生完成的两个项目(右上和底部)。

无法预测学生何时或以什么顺序会遇到特定的内容主题,学生不是坐在教室里而是在嘈杂的实验室里走动时才会暴露问题,在这种环境中进行的数学活动和教学方案的种类可能与传统的有极大差异。在接下来的几十年中,我们预计数学教育者将会面对在这些技术丰富的环境中实现接受强大数学学习的挑战。

在编程环境中学习数学

兴趣驱动的数学活动领域的另一个中心主题是计算机编程。Logo编程语言在20世纪70年代和80年代爆炸式发展之后,学校在教学编码方面几乎没有兴趣。幸运的是,过去的十年中,几个国家对在学校推广编程重新

FIGURE 31.5. *Prototype public-square projects. One in-progress project with the grid clearly visible (top left) and two of the projects done by students (top right and bottom).*

a particular content topic, and their exposure will happen not while sitting in a classroom but while roaming around a noisy lab. The kinds of mathematical activities and teaching protocols that would work in such environments could be quite different from the traditional ones. Over the coming decades, we anticipate mathematics educators rising to the challenge of enabling strong mathematics learning in these technology-rich settings.

Learning Mathematics in a Programming Context

Another central theme in the realm of interest-driven mathematical activities is computer programming. The explosion of the Logo programming language in the 1970s and 1980s was followed by a period of little interest in teaching coding in schools. Fortunately, the last decade has seen a renewed interest in promoting coding in schools

产生了兴趣,政府和基金会投入大量资金和热情。

同样值得注意的是一个常见的问题,"为什么Logo没有兑现它的承诺?"现在听起来既空洞又过时,大约有1000万上传的Scratch项目和Scratch程序(是Logo的直系后代),目前在许多国家或多或少获得授权,特别是在英国。我们对早期Logo和其他编程语言所面临的问题的解释是,学习基于文本的语言及其语法对于大多数学校来说都非常费时,而没有在线或同伴支持技术来减轻对教师时间的需求。Scratch(实际上它只是试图解决编程语法问题的系统中最新的一个)的发展及其庞大的在线社区,已被证明是一个转折点。考虑到这些学科的相关概念,计算机编程的第二次到来为数学教育者提供了机会(见Noss & Hoyles, 1996)。然而,目前对职业技能的重视可能使教育工作者和政策制定者走上软件工程的道路,并且与聚焦于建模的编程应用相比,软件工程与数学的联系较少。在计算机科学初期,教师常常来自数学系,现在我们猜想,数学教育工作者很可能有新的理由参与到那些能够帮助学生利用数学学习机会的活动中来,而这些学习机会可能来自于他们对编程的兴趣。

结论和注意事项

过去20年来,无论在生产力方面还是在变革方面,关于如何利用技术来促进数学学习方面的知识得到了迅速发展。这种知识已经从基于课堂的教学环境扩展到兴趣驱动的非正式学习环境。对于教育工作者来说,技术本质上可能是令人兴奋的,但是如何实现对学生学习数学有长期且有意义的大规模变革仍然很困难。我们认为,数学教育工作者如果懂得那些基于研究得到的原则是如何将技术特征与学习机会联系起来的,那么他们就最有可能促进学生的学习。

扩展德林维斯(2012)的框架,我们将未来的工作方法分为四大类,每一类都汇集了一系列历史上的工作:(1)学习理论,(2)设计研究,(3)实施研究,(4)其他方面的必要知识。我们已经从各种各样的设计视角探索了这些类别的工作,将技术的目的定位在教育环境中,从关注生产力(技术会支持或在某种程度上优化现有实践、内容和课程)到关注变革(技术用新的方法取代或革新这些实践、数学内容和课程)。

第一类,做数学的技术,强调在学习和日常生活中都有用的工具。这个类别的生产力观点建议将劳动转移给计算器或其他工具,以便学习者更好地关注学习任务。然而,变革观点强调人与工具交织在一起发展以及增强认知的出现。第二类,实践技能的技术,解决组织数学任务的有效排序、提供有用的反馈和适合的教学法等挑战。生产力视角强调优化实践效率的可能性,而变革视角则利用实践技术来重组学生和教师如何在课堂上使用时间和空间。第三类是概念性理解技术,它研究技术支持下的数学学习材料设计如何帮助学生认识到数学概念是相关的、有联系的和有意义的。这项工作的生产力方面集中在如何将这些技术打包成为实际而高效的学校实践,而变革观点则看重技术如何创造更深入地学习数学的机会,使许多学生以前不能接近的数学变得清楚。最后,我们研究了不包括在德林维斯框架中的第四个类别,在这个类别中,动手操作、学生对丰富技术的体验,如制造材料、设计机器人、编程等,通过将数学学习定位在学生的项目和兴趣中,为数学学习提供真实的动机背景。

我们将这四个类别中的每一个都视为是有力而且可能是重要的学习途径,它们不仅支持开展更好的数学学习,而且极为重要地,支持扩展更好的数学学习机会。尽管现有研究有其优势,但基于研究的技术是否将被用来提高生产力和实现数学学习的变革,这一点尚不能确定。一个庞大的数学技术消费市场的出现,意味着在以研究为基础的知识之外的因素可能会被大力推动接受和使用,对领导层的一大挑战就是将基于研究的知识置于

in several countries, with significant funding and enthusiasm coming both from governments and foundations.

It is also worth noting that the common question, "Why didn't Logo deliver what it promised?" now sounds both hollow and dated, with some 10 million uploaded Scratch projects and Scratch—a direct descendant of Logo—now more or less mandated in many countries, most recently in the United Kingdom. Our interpretation of the problems faced by Logo and other programming languages in the early days is that learning the text-based languages, along with their syntax, was very time consuming for most schools, and there were no technologies for online or peer support to alleviate the demand for teacher time. The development of Scratch—actually only one of the latest in a trajectory of systems that try to address the problem of syntax in programming—together with its huge online community, has proved to be a turning point. This second coming of computer programming offers opportunities for mathematics educators, given the strongly related concepts in these disciplines (see Noss & Hoyles, 1996). However, the current emphasis on job skills can lead educators and policy makers down a software engineering path, and software engineering can be less related to mathematics than applications of programming that focus on modeling. In the early days of computer science, instructors often came from mathematics faculty. Now we suspect there may be new reasons for mathematics educators to become engaged in helping students harness the mathematical learning opportunities that can arise from their interests in coding.

Conclusions and Cautions

The past 20 years have produced a rapid expansion in knowledge about how technology *can* be used to enable mathematics learning, in terms of both productivity and transformation. This knowledge has expanded from classroom-based didactical contexts to interest-driven, informal learning contexts. Technology can be intrinsically exciting to educators, but achieving lasting, meaningful large-scale change in how students learn mathematics remains hard. We have argued that mathematics educators who understand how research-based principles connect technology features to learning opportunities are in the strongest position to advance the learning of their students.

Extending Drijvers' (2012) framework, we have organized approaches into four broad categories, each of which draws together a constellation of historical precedents: (1) learning theory, (2) design research, (3) implementation research, and (4) other aspects of necessary knowledge.

We have explored work in each of these categories across a spectrum of design perspectives that locate technology's purpose in educational settings and that range from a focus on *productivity* (where technology abets or in some sense optimizes existing practice, content, and curricula) to a focus on *transformation* (where technology replaces or revolutionizes these practices, mathematical content, and curricula with new approaches).

The first category, technology for doing mathematics, emphasizes tools that are useful both while learning and in everyday life. A productivity perspective in this category suggests offloading labor to calculators or other tools so the learner can better focus on learning tasks. However, a transformational perspective highlights the developmental intertwining of person and tool and the emergence of augmented cognition. The second category, technology for practicing skills, tackles the challenge of organizing effective sequencing of mathematics tasks, provision of useful feedback, and adaptive pedagogies. A productivity perspective emphasizes the possibility of optimizing the efficiency of practice, whereas a transformational perspective leverages practice technologies to reorganize how students and teachers use time and space in classrooms. A third category, technology for conceptual understanding, examines how technology-enabled designs of materials for mathematics learning can help students understand mathematical concepts as related, connected, and meaningful. A productivity aspect of this work has focused on how these technologies can be packaged for effective and efficient school implementation, whereas transformational perspectives look at how technology creates opportunities to learn mathematics more deeply and to bring into focus mathematics that many students previously could not access. Finally, we examined a fourth category not included in Drijvers' framework, in which hands-on, technology-rich student experiences—fabricating materials, designing robots, writing programs—provide authentic motivational context for mathematics learning by locating mathematics learning within students' projects and interests.

We see each of these four categories as having cogent, and potentially important, trajectories supporting better mathematics learning as well as—critically—the expansion of opportunities to learn better mathematics. Despite the strengths of the existing research, whether research-based technologies *will* be used to increase productivity and enable transformation of mathematics learning has seldom been more uncertain. The emergence of a vast consumer marketplace for mathematics technology means that factors beyond research-based knowledge may powerfully drive adoption and use. A challenge for leadership

数学教育中使用哪些技术以及如何使用它们的决策前沿。

事实上，一旦技术的规模超越早期的成功案例，那么在数学教育中实现对技术的合理的教学使用已被证明是一项挑战。当技术能够使课程、教学法、教师专业发展和评价方面协调一致地发生变化时，它就发挥了重大的影响。因此，学习数学时使用技术，提高学生学习成果的因果关系并不直接，而是遵循一条路径，该路径明显地由设计和实施更广泛的数学教与学系统来决定，由推动学校与地区从现有体系向新体系转变的领导层来决定。在这种观点下，不管技术的生产力和变革的影响如何，技术都必须被视为基础设施，设计和实施运用技术的新的教与学系统仍然很困难，因此，从基于研究的方法中获得在数学教育中使用技术的大规模影响仍然是一个重大的挑战。

is to keep research-based knowledge at the forefront of decisions about what technologies to use in mathematics education and how to use them.

Indeed, implementing sound pedagogical use of technologies in mathematics education has proven to be challenging once technology scales beyond early cases of success. Technology achieves its greatest impact when it enables coordinated changes in curriculum, pedagogy, teacher professional development, and assessment. Thus, a causal arrow from using technology to learn mathematics to increased student learning outcomes is not straight but rather follows a path significantly determined by design and implementation of broader systems for teaching and learning mathematics and by the leadership required to move schools and districts from existing systems to new ones. In this view, technology—regardless of its productive and transformative impact—must be seen as infrastructural. Designing and implementing new systems of teaching and learning that leverage technology remains hard, and, thus, achieving large-scale impact from research-based approaches to the use of technology in mathematics education remains an important challenge.

References

Artigue, M. (2002). Learning mathematics in a CAS environment: The genesis of a reflection about instrumentation and the dialectics between technical and conceptual work. *International Journal of Computers for Mathematical Learning, 7*(3), 245–274.

Bainville, E., & Laborde, J. M. (2004). Cabri 3D [computer software]. Grenoble, France: Cabrilog.

Baker, J. E., & Sugden, S. J. (2003). Spreadsheets in education: The first 25 years. *Spreadsheets in Education (eJSiE), 1*(1), 18–43.

Baker, R. S., D'Mello, S. K., Rodrigo, M. M. T., & Graesser, A. C. (2010). Better to be frustrated than bored: The incidence, persistence, and impact of learners' cognitive–affective states during interactions with three different computer-based learning environments. *International Journal of Human-Computer Studies, 68*(4), 223–241.

Baker, R. S., & Yacef, K. (2009). The state of educational data mining in 2009: A review and future visions. *Journal of Educational Data Mining, 1*(1), 3–17.

Baulac, Y., Bellemain, F., & Laborde, J. M. (1988). *Cabri-géomètre, un logiciel d'aide à l'enseignement de la géométrie, logiciel et manuel d'utilisation*. Paris, France: Cedic-Nathan.

Becker, H. J., Ravitz, J. L., & Wong, Y. (1999). *Teacher and teacher-directed student use of computers and software* (No. 3). Irvine: Center for Research on Information Technology and Organizations, University of California.

Bishop, J. P. (2013). Mathematical discourse as a process that mediates learning in SimCalc classrooms. In S. J. Hegedus & J. Roschelle (Eds.), *The SimCalc vision and contributions* (pp. 233–249). Dordrecht, The Netherlands: Springer.

Blikstein, P. (2013a). Digital fabrication and "making" in education: The democratization of invention. In J. Walter-Herrmann & C. Büching (Eds.), *FabLabs: Of machines, makers and inventors*. Bielefeld, Germany: Transcript Publishers.

Blikstein, P. (2013b, April). *Multimodal learning analytics*. Paper presented at the Proceedings of the Third International Conference on Learning Analytics and Knowledge, Leuven, Belgium.

Borko, H., Whitcomb, J., & Liston, D. (2009). Wicked problems and other thoughts on issues of technology and teacher learning. *Journal of Teacher Education, 60*(1), 3–7.

Brown, J. S., & Sleeman, D. (Eds.). (1982). *Intelligent tutoring systems*. New York, NY: Academic Press.

Calvo, R. A., & D'Mello, S. (2010). Affect detection: An interdisciplinary review of models, methods, and their applications. *IEEE Transactions on Affective Computing, 1*(1), 18–37.

Campuzano, L., Dynarski, M., Agodini, R., & Rall, K. (2009). *Effectiveness of reading and mathematics software products: Findings from two student cohorts*. Washington, DC: Institute of Education Sciences.

Carpenter, S. K., Cepeda, N. J., Rohrer, D., Kang, S. H., & Pashler, H. (2012). Using spacing to enhance diverse forms of learning: Review of recent research and implications for instruction. *Educational Psychology Review, 24*(3), 369–378.

Carpenter, T. P., Hiebert, J., Fennema, E., Fuson, K. C., Wearne, D., & Murray, H. (1997). *Making sense: Teaching and learning mathematics with understanding*. Portsmouth, NH: Heinemann.

Cheung, A. C., & Slavin, R. E. (2013). The effectiveness of educational technology applications for enhancing mathematics achievement in K–12 classrooms: A meta-analysis. *Educational Research Review, 9*, 88–113.

Clark-Wilson, A. (2010). Emergent pedagogies and the changing role of the teacher in the TI-Nspire Navigator-networked mathematics classroom. *ZDM—The International Journal on Mathematics Education, 42*(7), 747–761.

Clark-Wilson, A., Hoyles, C., Noss, R., Vahey, P., & Roschelle, J. (2014). Scaling a technology-based innovation: Windows on the evolution of mathematics' teachers practices. *ZDM—The International Journal on Mathematics Education*. doi:10.1007/s11858-014-0635-6

Collins, A., & Halverson, R. (2009). The second education revolution: From apprenticeship to schooling to lifelong learning. *Journal of Computer Assisted Learning, 26*(1), 18–27.

Corbett, A. T., & Anderson, J. R. (1994). Knowledge tracing: Modeling the acquisition of procedural knowledge. *User Modeling and User-Adapted Interaction, 4*(4), 253–278.

Dalton, S., & Hegedus, S. (2013). Learning and participation in high school classrooms. In S. J. Hegedus & J. Roschelle (Eds.), *SimCalc vision and contributions* (pp. 145–166). Dordrecht, The Netherlands: Springer.

Davis, S. (2003). Observations in classrooms using a network of handheld devices. *Journal of Computer Assisted Learning, 19*(3), 298–307.

Dillenbourg, P. (2013). Design for classroom orchestration. *Computers & Education, 69*, 485–492.

Drijvers, P. (2012, July). *Digital technology in mathematics education: Why it works (or doesn't)*. Paper presented at the 12th International Congress on Mathematics Education, Seoul, Korea. Retrieved from http://www.icme12.org/upload/submission/2017_F.pdf

Ellington, A. J. (2003). A meta-analysis of the effects of calculators on students' achievement and attitude levels in precollege mathematics classes. *Journal for Research in Mathematics Education, 34*(5), 433–463.

Ellington, A. J. (2007). The effects of non-CAS graphing calculators on student achievement and attitude levels in mathematics: A meta-analysis. *School Science and Mathematics, 106*(1), 16–26.

Engelbart, D. C. (1995). Toward augmenting the human intellect and boosting our collective IQ. *Communications of the ACM, 38*(8), 30–32.

Falmagne, J. C., Cosyn, E., Doignon, J. P., & Thiéry, N. (2006). The assessment of knowledge, in theory and in practice. In B. Ganter & R. Willie (Eds.), *Formal concept analysis* (pp. 61–79). Berlin, Germany: Springer.

Finzer, W. (2013). The data science education dilemma. *Technology Innovations in Statistics Education, 7*(2). Retrieved from http://escholarship.org/uc/item/7gv0q9dc on March 2, 2016.

Fletcher, J. D., & Tobias, S. (2005). The multimedia principle. In R. E. Mayer (Ed.), *The Cambridge handbook of multimedia learning* (pp. 117–133). Cambridge, United Kingdom: Cambridge University Press.

Gentner, D., & Stevens, A. L. (Eds.). (1983). *Mental models.* Hillsdale, NJ: Erlbaum.

Goodyear, P., Jones, C., & Thompson, K. (2014). Computer-supported collaborative learning: Instructional approaches, group processes and educational designs. In J. M. Spector, M. D. Merrill, J. Elen, & M. J. Bishop (Eds.), *Handbook of research on educational communications and technology* (pp. 439–451). New York, NY: Springer.

Gorini, C. (1997). Dynamic visualization in calculus. In J. King & D. Schattschneider (Eds.), *Geometry turned on! Dynamic software in learning, teaching, and research* (pp. 89–94). Washington, DC: The Mathematical Association of America.

Hamari, J., Koivisto, J., & Sarsa, H. (2014). Does gamification work?—A literature review of empirical studies on gamification. In R. H. Sprague Jr. (Ed.), *2014 47th Hawaii International Conference on System Sciences* (pp. 3025–3034). Piscataway, NJ: The Institute of Electrical and Electronics Engineers.

Hatano, G. (1997). Learning arithmetic with an abacus. In P. Bryant & N. Terezinha (Eds.), *Learning and teaching mathematics: An international perspective* (pp. 209–231). Hove, United Kingdom: Psychology Press.

Hawkins, A., & Sinclair, N. (2007). Explorations in topogeometry using Sketchpad. *International Journal of Computers for Mathematics Learning, 13*(1), 71–82.

Healy, L., Hoelzl, R., Hoyles, C., & Noss, R. (1994). Messing up: Reflections on introducing Cabri Géomètre. *Micromath, 10*(1), 14–16.

Heid, M. K., & Blume, G. W. (Eds.). (2008). *Research on technology and the teaching and learning of mathematics: Research syntheses. Volume 1.* Charlotte, NC: Information Age; Reston, VA: National Council of Teachers of Mathematics.

Heiten, L. (2014, September 23). Will Common Core testing platforms impede math tasks? *Education Week.* Retrieved from http://www.edweek.org/ew/articles/2014/09/24/05math.h34.html on March 2, 2016.

Heller, J. L., Curtis, D. A., Jaffe, R., & Verboncoeur, C. J. (2005). *Impact of handheld graphing calculator use on student achievement in algebra 1.* Oakland CA: Heller Research Associates.

Henningsen, M., & Stein, M. K. (1997). Mathematical tasks and student cognition: Classroom-based factors that support and inhibit high-level mathematical thinking and reasoning. *Journal for Research in Mathematics Education, 28*(5), 524–549.

Heppen, J. B., Walters, K., Clements, M., Faria, A. M., Tobey, C., Sorensen, N., & Culp, K. (2011). *Access to algebra I: The effects of online mathematics for grade 8 students. NCEE 2012-4021.* Washington, DC: National Center for Education Evaluation and Regional Assistance.

Higgins, S. E. (2010). The impact of interactive whiteboards on classroom instruction and learning in primary schools in the UK. In M. Thomas & E. C. Schmid (Eds.), *Interactive whiteboards for education: Theory, research and practice* (pp. 86–101). Hershey, PA: IGI Global.

Hoyles, C., Noss, R., & Kent, P. (2004). On the integration of digital technologies into mathematics classrooms. *International Journal of Computers for Mathematical Learning, 9*(3), 309–326.

Hoyles, C., Noss, R., Vahey, P., & Roschelle, J. (2013). Cornerstone mathematics: Designing digital technology for teacher adaptation and scaling. *ZDM—The International Journal on Mathematics Education, 45*(7), 1057–1070.

Hoyles, C., Wolf, A., Molyneux-Hodgson, S., & Kent, P. (2002). *Mathematical skills in the workplace: Final report to the Science, Technology and Mathematics Council.* London, United Kingdom: University of London, Institute of Education.

Jackiw, N. (1989). The Geometer's Sketchpad [computer software]. Berkeley CA: Key Curriculum Press.

Jackiw, N. (2003). Visualizing complex functions with the Geometer's Sketchpad. In T. Triandafillidis & K. Hatzikiriakou (Eds.), *Proceedings of the 6th International Conference on Technology in Mathematics Teaching* (pp. 291–299). Volos: University of Thessaly, Greece.

Jacobson, M. J., & Wilensky, U. (2006). Complex systems in education: Scientific and educational importance and implications for the learning sciences. *The Journal of the Learning Sciences, 15*(1), 11–34.

Jiang, Z., White, A., & Rosenwasser, A. (2011, Fall–Winter). Randomized control trials on the dynamic geometry approach. *Journal of Mathematics Education at Teachers College, 2*, 8–17.

Johnson-Laird, P. N. (1983). *Mental models: Towards a cognitive science of language, inference, and consciousness.* Cambridge, MA: Harvard University Press.

Kaput, J. (1992). Technology and mathematics education. In D. Grouws (Ed.), *A handbook of research on mathematics teaching and learning* (pp. 515–556). New York, NY: Macmillan.

Kaput, J. (1997). Rethinking calculus: Learning and thinking. *The American Mathematical Monthly, 104*(8), 731–737.

Kaput, J., Hegedus, S., & Lesh, R. (2007). Technology becoming infrastructural in mathematics education. In R. Lesh, E. Hamilton, & J. Kaput (Eds.), *Foundations for the future in mathematics education* (pp. 173–192). Mahwah, NJ: Lawrence Erlbaum Associates.

Kaput, J., & Roschelle, J. (1998). The mathematics of change and variation from a millennial perspective: New content, new context. In C. Hoyles, C. Morgan, & G. Woodhouse (Eds.), *Rethinking the mathematics curriculum* (pp. 13–26). London, United Kingdom: Falmer Press.

Kaput, J. J., & Thompson, P. W. (1994). Technology in mathematics education research: The first 25 years in the JRME. *Journal for Research in Mathematics Education, 25*(6), 676–684.

Khoju, M., Jaciw, A., & Miller, G. I. (2005). *Effectiveness of graphing calculators in K–12 mathematics achievement: A systematic review.* Palo Alto, CA: Empirical Education.

Kilpatrick, J., Swafford, J., & Findell, B. (Eds.). (2001). *Adding it up: Helping children learn mathematics.* Washington, DC: National Academy Press.

King, J., & Schattschneider, D. (1997). *Geometry turned on: Dynamic software in learning, teaching, and research.* Washington, DC: Mathematical Association of America.

Kirschner, P. A., & Erkens, G. (2013). Toward a framework for CSCL research. *Educational Psychologist, 48*(1), 1–8.

Knudsen, J., Lara-Meloy, T., Stevens, H. S., & Shechtman, N. (2014). Advice for mathematical argumentation. *Mathematics Teaching in the Middle School, 19*(8), 494–500.

Knudsen, J., Vahey, P., Lara-Meloy, T., & Shechtman, N. (2014, April). *Teacher support of mathematical argumentation in the middle school classroom: A case study.* Paper presented at the annual meeting of the American Educational Research Association, Philadelphia, PA.

Koedinger, K. R., Booth, J. L., & Klahr, D. (2013). Instructional complexity and the science to constrain it. *Science, 342*, 935–937.

Koedinger, K. R., Brunskill, E., de Baker, R. S. J., McLaughlin, E. A., & Stamper, J. C. (2013). New potentials for data-driven intelligent tutoring system development and optimization. *AI Magazine, 34*(3), 27–41.

Koedinger, K. R., McLaughlin, E., & Heffernan, N. (2010). A quasi-experimental evaluation of an on-line formative assessment and tutoring system. *Journal of Educational Computing Research, 43*(1), 489–510.

Koehler, M. J., & Mishra, P. (2005). What happens when teachers design educational technology? The development of technological pedagogical content knowledge. *Journal of Educational Computing Research, 32*(2), 131–152.

Konold, C., Harradine, A., & Kazak, S. (2007). Understanding distributions by modeling them. *International Journal of Computers for Mathematical Learning, 12*(3), 217–230.

Laborde, C., Kynigos, C., Hollebrands, K., & Strässer, R. (2006). Teaching and learning geometry with technology. In A. Gutiérrez & P. Boero (Eds.), *Handbook of research on the psychology of mathematics education: Past, present and future* (pp. 275–304). Rotterdam, The Netherlands: Sense.

Lawless, K. A., & Pellegrino, J. W. (2007). Professional development in integrating technology into teaching and learning: Knowns, unknowns, and ways to pursue better questions and answers. *Review of Educational Research, 77*(4), 575–614.

Mason, J. (1992). Geometric tools. *Micromath, 8*(3), 24–27.

Mavrikis, M., Noss, R., Hoyles, C., & Geraniou, E. (2013). Sowing the seeds of algebraic generalization: Designing epistemic affordances for an intelligent microworld. *Journal of Computer Assisted Learning, 29*(1), 68–84.

Means, B., Toyama, Y., Murphy, R., Bakia, M., & Jones, K. (2009). *Evaluation of evidence-based practices in online learning: A meta-analysis and review of online learning studies.* U.S. Department of Education.

Meel, D., & Hern, T. (2005). Tool building: Web-based linear algebra modules. *Journal of Online Mathematics and Its Applications.*

Miller, K. F., & Stigler, J. W. (1991). Meaning of skill: Effects of abacus expertise on number representation. *Cognition and Instruction, 8*(1), 26–67.

Murphy, R., Gallagher, L., Krumm, A., Mislevy, J., & Hafter, A. (2014). *Research on the use of Khan Academy in schools: Research brief.* Menlo Park, CA: SRI International.

Murphy, R., Snow, E., Mislevy, J., Gallagher, L., Krumm, A., & Wei, X. (2014). *Blended learning report.* Dallas, TX: Michael and Susan Dell Foundation.

National Center for Education Statistics. (2001). *The nation's report card: Mathematics 2000.* (*No. NCES 2001-571*). Washington, DC: U.S. Department of Education.

National Council of Teachers of Mathematics. (2000). *Principles and standards for school mathematics.* Reston, VA: Author.

National Governors Association Center for Best Practices and Council of Chief State School Officers. (2010). *Common Core State Standards for Mathematics.* Washington, DC: Author. Retrieved from http://www.corestandards.org/wp-content/uploads/Math_Standards1.pdf

Noss R., & Hoyles, C. (1996). *Windows on mathematical meanings: Learning cultures and computers.* Dordrecht, The Netherlands: Kluwer.

Noss, R., Hoyles, C., & Kent, P. (2004). On the integration of digital technologies into mathematics classrooms. *International Journal of Computers for Mathematical Learning, 9*(3), 309–326.

Noss, R., Hoyles, C., Mavrikis, M., Geraniou, E., Gutierrez-Santos, S., & Pearce, D. (2009). Broadening the sense of "dynamic": A microworld to support students' mathematical generalization. *ZDM—The International Journal on Mathematics Education, 41*(4), 493–503.

Ocumpaugh, J., Baker, R. S. J. D., Gaudino, S., Labrum, M., & Dezendorf, T. (2013, July). *Field observation of engagement in Reasoning Mind.* Paper presented at the 16th International Conference on Artificial Intelligence in Education, Memphis, TN.

Olive, J. (1998). Opportunities to explore and integrate mathematics with the Geometer's Sketchpad. In R. Lehrer & D. Chazan (Eds.), *Designing learning environments for developing understanding of geometry and space* (pp. 395–418). Mahwah, NJ: Lawrence Erlbaum Associates.

Paas, F., Renkl, A., Sweller, J. (2004). Cognitive load theory: Instructional implications of the interaction between information structures and cognitive architecture. *Instructional Science, 32*, 1–8.

Pane, J. F., McCaffrey, D. F., Slaughter, M. E., Steele, J. L., & Ikemoto, G. S. (2010). An experiment to evaluate the efficacy of Cognitive Tutor geometry. *Journal of Research on Educational Effectiveness, 3*(3), 254–281.

Pape, S. J., Irving, K. E., Owens, D. T., Boscardin, C. K., Sanalan, V. A., Abrahamson, A. L., … Silver, D. (2012). Classroom connectivity in algebra I classrooms: Results of a randomized control trial. *Effective Education, 4*(2), 169–189.

Papert, S. (1980). *Mindstorms: Children, computers, and powerful ideas.* New York, NY: Basic Books.

Richter-Gebert, J., & Kortenkamp, U. H. (1999). *Cinderella* [computer software]. New York, NY: Springer.

Ritter, S., Anderson, J. R., Koedinger, K. R., & Corbett, A. (2007). Cognitive Tutor: Applied research in mathematics education. *Psychonomic Bulletin & Review, 14*(2), 249–255.

Roschelle, J., Knudsen, J., & Hegedus, S. (2010). From new technological infrastructures to curricular activity systems: Advanced designs for teaching and learning. In M. J. Jacobson & P. Reimann (Eds.), *Designs for learning environments of the future: International perspectives from the learning sciences* (pp. 233–262). New York, NY: Springer.

Roschelle, J., Rafanan, K., Bhanot, R., Estrella, G., Penuel, W. R., Nussbaum, M., & Claro, S. (2010). Scaffolding group explanation and feedback with handheld technology: Impact on students' mathematics learning. *Educational Technology Research and Development, 58*(4), 399–419. doi:10.1007/s11423-009-9142-9

Roschelle, J., Shechtman, N., Tatar, D., Hegedus, S., Hopkins, B., Empson, S., & Gallagher, L. P. (2010). Integration of technology, curriculum, and professional development for advancing middle school mathematics: Three large-scale studies. *American Educational Research Journal, 47*(4), 833–878.

Ruthven, K. (2002). Instrumenting mathematical activity: Reflections on key studies of the educational use of computer algebra systems. *International Journal of Computers for Mathematical Learning, 7*, 275–291.

Ruthven, K. (2014). Frameworks for analysing the expertise that underpins successful integration of digital technologies into everyday teaching practice. In A. Clark-Wilson, O. Robutti, & N. Sinclair (Eds.), *The mathematics teacher in the digital era* (pp. 373–393). Dordrecht, The Netherlands: Springer.

Schattschneider, D. (1997). Visualization of group theory concepts with dynamic geometry software. In J. King & D. Schattschneider (Eds.), *Geometry turned on: Dynamic software in learning, teaching, and research* (pp. 121–128). Washington, DC: Mathematical Association of America.

Shaffer, D. (1995). *Exploring trigonometry with the Geometer's Sketchpad*. Berkeley, CA: Key Curriculum Press.

Shechtman, N., Knudsen, J., Michalchik, V., Kim, H., & Stevens, H. (2015). *Teacher professional development to support classroom mathematical argumentation*. Manuscript submitted for review.

Sinclair, N. (2011). Touchcounts: An embodied, digital approach to learning number. In E. Faggiano & A. Montone (Eds.), *Proceedings of the 11th International Conference on Technology in Mathematics Teaching* (pp. 262–267). Bari, Italy: Department of Mathematics, University of Bari.

Sinclair, N., Cirillo, M., & de Villiers, M. (2017). The learning and teaching of geometry. In J. Cai (Ed.), *Compendium for research in mathematics education* (pp. 457–489). Reston, VA: National Council of Teachers of Mathematics.

Sinclair, N., & Crespo, S. (2006). Learning mathematics in dynamic computer environments. *Teaching Children Mathematics, 12*(9), 436–444.

Sinclair, N., & Robuti, O. (2012). Technology and the role of proof: The case of dynamic geometry. In A. J. Bishop, M. A. Clements, C. Keitel, & F. Leung (Eds.), *Third international handbook of mathematics education* (pp. 571–596). Dordrecht, The Netherlands: Kluwer Academic.

Smith, H. J., Higgins, S., Wall, K., & Miller, J. (2005). Interactive whiteboards: Boon or bandwagon? A critical review of the literature. *Journal of Computer Assisted Learning, 21*(2), 91–101.

Smith, J. P., III, Disessa, A. A., & Roschelle, J. (1994). Misconceptions reconceived: A constructivist analysis of knowledge in transition. *The Journal of the Learning Sciences, 3*(2), 115–163.

Squire, K. (2011). *Video games and learning: Teaching and participatory culture in the digital age*. (Technology, Education—Connections). New York, NY: Teachers College Press.

Stahl, G. (2009). *Studying virtual math teams*. Berlin, Germany: Springer.

Staker, H., & Horn, M. (2012). *Classifying K–12 blended learning*. Palo Alto, CA: Innosight Institute.

Stevens, H., & Lara-Meloy, T. (2014, April). Middle schoolers engaged in argumentation. In *Reasoning and sense-making with technology in middle school*. Invited panel at the meeting of the National Council of Teachers of Mathematics, New Orleans, LA.

Stohl Drier, H., Harper, S., Timmerman, M. A., Garofalo, J., & Shockey, T. (2000). Promoting appropriate uses of technology in mathematics teacher preparation. *Contemporary Issues in Technology and Teacher Education, 1*(1), 66–88.

Stroup, W. M., Ares, N. M., & Hurford, A. C. (2005). A dialectic analysis of generativity: Issues of network supported design in mathematics and science. *Mathematical Thinking and Learning, 7*(3), 181–206.

Stroup, W. M., Kaput, J., Ares, N., Wilensky, U., Hegedus, S. J., Roschelle, J., . . . Hurford, A. (2002, October). *The nature and future of classroom connectivity: The dialectics of mathematics in the social space*. Paper presented at the Psychology and Mathematics Education North America conference, Athens, GA.

Stroup, W. M., & Wilensky, U. (2014). On the embedded complementarity of agent-based and aggregate reasoning in students' developing understanding of dynamic systems. *Technology, Knowledge and Learning, 19*(1), 19–52.

Suppes, P. (1971). *Computer-assisted instruction at Stanford. Technical Report No. 174*, Psychology and Education series. Stanford, CA: Institute for Mathematical Studies in the Social Sciences, Stanford University.

Sweller, J., van Merrienboer, J. J. G., & Paas, F. G. W. C. (1998). Cognitive architecture and instructional design. *Educational Psychology Review, 10*(3), 251–296.

Trouche L. (2000). La parabole du gaucher et de la casserole à bec verseur: étude des processus d'apprentissage dans un environnement de calculatrices symboliques. *Educational Studies in Mathematics, 41*, 239–264.

Trouche, L. (2003, June). *Managing the complexity of human/machine interaction in a computer based learning environment: Guiding student's process command through instrumental orchestrations*. Plenary presentation at the Third Computer Algebra in Mathematics Education Symposium, Reims, France.

U.S. Department of Education. (n.d.). *What Works Clearinghouse procedures and standards handbook, version 3.0*. Retrieved March 2, 2015, from http://ies.ed.gov/ncee/wwc/pdf/reference_resources/wwc_procedures_v3_0_standards_handbook.pdf

Vahey, P., Roy, G. J., & Fueyo, V. (2013). Sustainable use of dynamic representational environments: Toward a district-wide adoption of SimCalc-based materials. In S. J. Hegedus & J. Roschelle (Eds.), *The SimCalc vision and contributions* (pp. 183–202). Dordrecht, The Netherlands: Springer.

Van Lehn, K. (2011). The relative effectiveness of human tutoring, intelligent tutoring systems, and other tutoring systems. *Educational Psychologist, 46*(4), 197–221.

Verillon, P., & Rabardel, P. (1995). Cognition and artifacts: A contribution to the study of thought in relation to instrumented activity. *European Journal of Psychology of Education, 10*(1), 77–101.

What Works Clearinghouse. (2013). *Carnegie Learning Curricula and Cognitive Tutor: WWC intervention report.* Washington, DC: U.S. Department of Education, Institute of Education Sciences.

Worsley, M., & Blikstein, P. (2013, April). *Towards the development of multimodal action based assessment.* Paper presented at the Third International Conference on Learning Analytics and Knowledge, Leuven, Belgium.

Zbiek, R. M., & Hollebrands, K. (2008). A research-informed view of the process of incorporating mathematics technology into classroom practice by in-service and prospective teachers. In M. K. Heid & G. W. Blume (Eds.), *Research on technology and the teaching and learning of mathematics: Research syntheses* (Vol. 1, pp. 287–344). Charlotte, NC: Information Age; Reston, VA: National Council of Teachers of Mathematics.

Zoll, E. J. (1969). Research in programmed instruction in mathematics. *The Mathematics Teachers, 62*(2), 103–110.